Cometography
A Catalog of Comets
Volume 2: 1800–1899

Cometography is a four-volume catalog of every comet observed throughout history. It is the most complete and comprehensive collection of data on comets available. Volume 2 provides a complete discussion of the comets seen during the nineteenth century. *Cometography* uses the most reliable orbits known to determine the distances from the Earth and Sun at the time a comet was discovered and last observed, as well as the largest and smallest angular distance to the Sun, most northerly and southerly declination, closest distance from the Earth, and other details to enable the reader to understand the physical appearance of each well-observed comet. The book also provides non-technical details to help the reader appreciate better how the comet may have influenced various cultures at the time of its appearance. *Cometography* will be valuable to historians of science as well as providing amateur and professional astronomers with a definitive reference on comets through the ages.

Gary Kronk has held a life-long passion for astronomy, and has been researching historical information on comets ever since sighting Comet Kohoutek in 1973/74. His work has been published in numerous magazines and journals, and in two previous books – *Comets: A Descriptive Catalog* (1984), and *Meteor Showers: A Descriptive Catalog* (1988). Kronk holds positions in various astronomical societies, including those of Coordinator of the Comet Section of the Association of Lunar and Planetary Observers, and Consultant for the American Meteor Society.

Cometography

A Catalog of Comets

VOLUME 2:   1800–1899

Gary W. Kronk

CAMBRIDGE
UNIVERSITY PRESS

PUBLISHED BY THE PRESS SYNDICATE OF THE UNIVERSITY OF CAMBRIDGE
The Pitt Building, Trumpington Street, Cambridge, United Kingdom

CAMBRIDGE UNIVERSITY PRESS
The Edinburgh Building, Cambridge CB2 2RU, UK
40 West 20th Street, New York, NY 10011–4211, USA
477 Williamstown Road, Port Melbourne, VIC 3207, Australia
Ruiz de Alarcón 13, 28014 Madrid, Spain
Dock House, The Waterfront, Cape Town 8001, South Africa

http://www.cambridge.org

First published 2003

Printed in the United Kingdom at the University Press, Cambridge

Typeface Palatino 9.5/13 pt.   *System* LATEX 2 [TB]

A catalog record for this book is available from the British Library

ISBN 0 521 58505 8 hardback

Contents

Introduction

As would be expected, the nineteenth century was a period of change with respect to the study of comets. Although visual observations were still the primary method of acquiring data, bigger telescopes and new methods of observing greatly added to our understanding of these objects. The field of celestial mechanics also saw improvements, as astronomers were able to provide more accurate ephemerides to allow comets to be followed longer, and new techniques were applied to the orbits of periodic comets which enabled their perihelion dates to be predicted to within just a few hours.

Comet discoveries

Continuing with the tradition of the late eighteenth century, the two French rivals, C. Messier and P. F. A. Méchain, independently found the first comet of the nineteenth century, but they were beaten by a new kid on the block. A Frenchman named J. L. Pons found the comet before anyone. The discovery marked a changing of the guard among comet hunters. Messier and Méchain would never get credit for a comet discovery again, while Pons would go on to surpass everyone and become the greatest visual comet discoverer of all time.

Marseille Observatory ranked as the dominant observatory for comet discoveries during the century. A total of 48 comets held the names of Marseille observers at one time, although comets later identified as periodic by J. F. Encke, W. von Biela, A. C. D. Crommelin, and others would drop the current official total to 42. In addition, another 13 comets were independently discovered at Marseille before word arrived of their previous discovery elsewhere. Marseille astronomers are credited with at least one comet discovery in every decade of the nineteenth century except the 1840s.

Despite the dominance of the Marseille observers, the French were not the most prolific comet discoverers during this century. That honor went to astronomers within the USA, whose names are on more than 70 comets found during the nineteenth century. The earliest US discovery of the century was by M. Mitchell, who found a comet in 1847. The period of 1862–99 was quite incredible as 59 comets were found by L. Swift, E. E. Barnard, W. R. Brooks, and C. D. Perrine in the USA.

Comet observations

The visual observations of comets changed little from the eighteenth century through the first half of the nineteenth century. Observers continued to provide measurements of the coma diameter and the tail length, while estimates were made of the brightness of the nucleus. The only estimates of the comet's total magnitude were made at times when naked-eye observations were possible. The astronomers providing the longest series of such data during this century were J. F. J. Schmidt, J. Holetschek, and T. W. Backhouse. The latter two astronomers were among the first to provide extensive series of total magnitude estimates of comets fainter than naked-eye visibility. No special technique was used to obtain these estimates, as they basically made observations with the smallest instrument possible. Beginning in the early 1880s, many other astronomers began using the same technique.

There were many attempts to try to obtain a better indication of the comet's brightness and/or appearance throughout the century. One of the earliest techniques was to compare a comet to the objects in William Herschel's catalog of deep sky objects. Herschel's catalog categorized 2514 objects into eight classes. These were as follows:

Class I: Bright nebulae
Class II: Faint nebulae
Class III: Very faint nebulae
Class IV: Planetary nebulae
Class V: Very large nebulae
Class VI: Very compressed and rich clusters of stars
Class VII: Compressed clusters of small and large stars
Class VIII: Coarsely scattered clusters of stars

This method of classification was not widely used, but does occasionally come up in the literature throughout the century. It has since been shown that comparing the appearance of a comet with a deep sky object is generally not accurate because of the potentially varying intensity of the central condensations.

A more interesting technique was the attempt to accurately determine a comet's brightness by comparing it to stars. The technique, now known as the Bobrovnikoff method, appears to have first been utilized during the last decades of the nineteenth century. The English amateur astronomer G. Knott specifically noted that while he was observing comet C/1882 F1 (Wells) on 1882 May 11, he threw a telescope out of focus and compared the head of the comet to a nearby star. Before the end of the century, the same technique was also being utilized by Backhouse, as well as the American astronomers E. F. Sawyer and L. Boss. This technique increased in popularity early in the twentieth century and modifications were later made which improved the accuracy.

Astronomical periodicals

The dissemination of comet discovery announcements, observations, and research underwent a notable change during the first couple of decades of the nineteenth century. The French journal *Memoirs of the Academie des Sciences* was no longer a major monthly source of comet information as the century began, but the *Connaissance des Temps* and *Berliner Astronomische*

Jahrbuch continued to carry such information. The better of these two journals was the *Berliner Astronomische Jahrbuch*, but, as it was an astronomical almanac, it was only published once a year, so that the included astronomical papers were not published in a timely fashion. This was remedied for a short time by monthly periodicals such as the *Monatliche Correspondence*, the *Correspondance Astronomique, Geographique, Hydrographique et Statistique du Baron de Zach* and others, but it was not until the 1820s that significant long-term changes took place.

The 1820s saw the beginning of two of astronomy's oldest periodicals: the German periodical *Astronomische Nachrichten* and the British periodical *Monthly Notices of the Royal Astronomical Society*. These would be joined by the French periodical *Comptes Rendus Hebdomadaires des Séances de l'Académie des Sciences* during the 1830s. Although all were fine journals, the *Astronomische Nachrichten* grew in popularity as the years passed. Most important was its conversion from a monthly periodical to more of a weekly, with a publication date that floated to enable a rather quick distribution of information. For most of the nineteenth century, the *Astronomische Nachrichten* was the clearing house of comet news and research.

Another important journal came on the scene in 1849. Named the *Astronomical Journal*, it became an important periodical for publishing observations made by observatories in the USA. Financial problems forced a temporary end to the journal early in 1861, but it resumed late in 1886 and continues to the present time.

Observations

Visual observers still dominated the field, with the greatest advances being the building of larger telescopes. Refractors dominated the century, with those between 15 cm and 20 cm being most prominent; however, during the last couple of decades, refractors quickly increased in size and culminated with the 102-cm refractor of Yerkes Observatory around the middle of the 1890s. The result was that early in the century comets were usually lost after having faded to magnitude 11 or 12, while at the end of the century comets were being followed until near magnitude 16.

Two new tools came into use during the nineteenth century that greatly increased our knowledge of comets. The first was the spectroscope. The Italian astronomer G. B. Donati first applied a spectroscope to a comet on 1864 August 5 and 6. The comet was C/1864 N1 (Tempel) and Donati reported the appearance of three bright bands, which are now known as the Swan bands of carbon. Although many astronomers began observing bright comets with this new instrument, it was the English astronomer W. Huggins who quickly advanced this field and identified some of the molecules astronomers were detecting within the comets. Spectroscopic observations of C/1882 R1 (Great September Comet) ultimately revealed the presence of carbon, sodium, iron, and nickel.

The first photograph was a 7-second exposure of C/1858 L1 (Donati) made by the English photographer W. Usherwood on 1858 September 27. Being made with a portrait lens, the photograph succeeded in revealing part of the comet's bright tail. The next night, Harvard astronomer G. P. Bond became the first astronomer to photograph a comet through a telescope. His 6-minute exposure revealed only the bright region surrounding the nucleus of comet Donati. Interestingly, photographs would not again be taken of a comet until the appearance of C/1881 K1 (Great Comet). At that time, W. Huggins became the first person to photograph the spectrum of a comet, while the French astronomer P. J. C. Janssen revealed how valuable photography could be when his long exposures revealed details in the comet's tail which could not be seen with the naked eye or with a telescope. The US astronomer E. E. Barnard made the first photographic discovery of a comet in 1892 and the Germany astronomer M. F. J. C. Wolf was conducting a regular photographic survey of the sky before the century ended.

The most interesting comets of the nineteenth century

The nineteenth century was noted for some very interesting comets. The Great Comet of 1811 was visible to the naked eye for nine months, setting a record that would not be broken until comet Hale–Bopp moved through our skies in 1996 and 1997. The Great Comet of 1843 was the first well-studied sungrazer and exhibited one of the longest tails in history. This comet and its brighter, sungrazer cousin, the Great September Comet of 1882, both became easy objects to see with the naked eye in broad daylight.

One of the most interesting of the nineteenth century comets was not among the brightest. Back in January of 1846 the well-known periodic comet Biela was suddenly found to have split into two pieces. Both pieces returned in 1852, but neither was seen in the years that followed. Interestingly, huge meteor storms occurred at the end of November of 1872, 1885, and 1892. Astronomers realized the storms were caused by the debris of comet Biela, thus indicating the comet had probably broken up entirely

Celestial mechanics

The art of calculating the motions of comets underwent some improvements during the nineteenth century. Progress during the eighteenth century had enabled astronomers to calculate orbits based on three positions, where the middle position was perfectly represented and the first and third were represented by residuals. During the first year of the nineteenth century, the Germany astronomer C. F. Gauss developed a technique to calculate an orbit from three positions, which precisely fits the first and third position, while determining residuals for only the middle position.

The next improvement came during 1819, when J. F. Encke investigated the orbit of a comet discovered by Pons on 1818 November 26. He noted the similarity to the orbit of another comet found by Pons on 1805 October 20. Curiosity in a possible identity between these two comets prompted Encke to calculate an elliptical orbit for the 1818 comet, followed by an elliptical orbit for the 1805 comet. He was convinced the two were identical. During the next few months Encke looked further back in time and found comets seen in 1795 and then 1786, which he successfully linked to the two nineteenth century comets. Encke's ultimate paper on the subject applied planetary perturbations to his calculations, which enabled the four apparitions to be very accurately linked.

Encke predicted the comet would return on 1822 May 24. The comet was recovered in Australia on 1822 June 2, just a short distance from his predicted position. Encke suggested in 1823 that the apparent decrease in the comet's calculated orbital period was due to a resisting medium. Because of Encke's success in predicting the return of this comet, the comet that had been seen in 1786, 1795, 1805, 1818, and 1822 was later named after him.

The resisting medium theory was first applied to the investigation of the motion of comet Encke by F. E. von Asten in 1877. He used positions obtained between 1818 and 1868, determined the perturbations by six planets, and added rough nongravitational terms based on Encke's theory. Although the early nongravitational terms determined by Encke and von Asten were an interesting first attempt to explain the variation in the motion of a comet that cannot be explained by the gravitational influences of the planets, the cause was not a resisting medium. F. W. Bessel (1836) introduced a new idea that came about as a result of his observations of 1P/Halley in 1835. Bessel had noticed material apparently exiting the nucleus on the sunward side and wondered whether it could be responsible for slowing the comet. Unfortunately, von Asten favored Encke's resisting medium and it was not until the mid-twentieth century that Bessel's idea was refined by F. L. Whipple (1950) and B. G. Marsden (1968).

Cometography

Overall, this volume of *Cometography* follows the same basic format as did the first volume, but there are two significant changes. First, I have altered the style in which the "Sources" are given for each comet. As is normally the case with books and papers, the author's name is given at the beginning of each reference, but, to keep the main text readable, I have put the names of astronomers whose observations were discussed at the beginning of each reference, so if you wish to look for a particular astronomer's observations you will generally only have to check a couple of sources. The second change followed a suggestion by Marsden, I have reduced the precision of the positions, as there was no real benefit to keeping them as precise as given for the last few comets discussed in volume 1.

Acknowledgments

I would like to express my gratitude to those individuals who played important roles in helping me finish this second volume of *Cometography*.

Thanks go to the librarians who assisted me at Linda Hall Library (Kansas City, Missouri), Northwestern University (Evanston, Illinois), St. Louis University (Missouri), and Washington University (St. Louis, Missouri). At Washington University's Bernard Becker Medical Library, I especially wish to thank Nancy Bennett, of interlibrary loan, for working to acquire everything I requested.

Special thanks go to all of those people who helped me acquire sources, especially Maik Meyer (Kelkheim, Germany), Jonathan Shanklin (Cambridge, England), Wayne Orchiston (Anglo-Australian Observatory, Australia), Robert Burnham (Hales Corners, Wisconsin, USA), Hartmut Frommert (Germany), Antonio Giambersio (Potenza, Italy), Gernot Burkhardt (Heidelberg, Germany), James Caplan (Marseille Observatory, France), Josette Alexandre (Paris Observatory, France), and Jerry Grover (Cheshire, England).

Special thanks go to John Bortle of New York for looking over segments of the manuscript with a critical eye during the last couple of years. The completeness of his own comet observations, now extending over four decades, was an initial inspiration for this project.

Special thanks go to Don Yeomans (Jet Propulsion Laboratory, California), and Brian G. Marsden and Daniel W. E. Green (Harvard-Smithsonian Center for Astrophysics, Massachusetts) who so graciously answered questions and sent me hard-to-find articles over the years. Brian's own work of writing the annual comet summaries for the *Quarterly Journal of the Royal Astronomical Society* was another early inspiration for this project.

Special thanks go to my great friends Kurt Sleeter, Eric Young, and Ed Cunnius. Over the years, their sense of humor did wonders to get my mind off the book for a while, and our observing the night sky together always had a calming effect that reminded me of why I love astronomy so much.

Most of all, I wish to thank my wife, Karen, as well as my two boys, David and Michael. We are a very close family, which made it difficult for me to devote extensive periods of time to writing this book. Nevertheless, they were patient and always knew when to drag me out of the house for a while to see a movie or do anything that would get my mind off of things for a while.

Catalog of Comets

C/1801 N1 *Discovered:* 1801 July 11.90 ($\Delta = 0.37$ AU, $r = 0.84$ AU, Elong. $= 52°$)
(Pons) *Last seen:* 1801 July 23.91 ($\Delta = 0.60$ AU, $r = 0.56$ AU, Elong. $= 28°$)
  Closest to the Earth: 1801 July 12 (0.3738 AU)
1801 *Calculated path:* CAM (Disc), UMa (Jul. 13), LMi (Jul. 19), LEO (Jul. 23)

Not long after the nineteenth century began, J. J. de Lalande offered a prize of 600 francs to the first person to discover a comet in the new century. On July 11 and 12 of 1801, four of the greatest names in early cometary astronomy responded with that first comet, with the winner being a man who would become the greatest visual comet discoverer of all time.

J. L. Pons (Marseille, France) discovered this comet in Camelopardalis on 1801 July 11.90 and, on the next evening, three independent discoveries were made from Paris, France: P. F. A. Méchain on July 12.92, C. Messier on July 12.93, and A. Bouvard on July 12.93. Messier and Bouvard found the comet almost simultaneously, and this was about 15 minutes after Méchain. Messier described the comet as very faint, while Bouvard described it as small and round, without a tail. Méchain measured the comet's position on July 12.94 as $\alpha = 7^h\ 22.0^m$, $\delta = +69°\ 40'$.

The comet was not kept under observation for very long. Méchain saw the comet on July 16, 18, 19, 21, and 23. The last date marked the final time the comet was seen, with Méchain giving a position of $\alpha = 10^h\ 06.0^m$, $\delta = +28°\ 29'$ on July 23.91.

Three very similar parabolic orbits have been calculated over the years. Méchain (1802) determined the perihelion date as 1801 August 9.04. J. K. Burckhardt (1806) determined it as August 9.06. A. W. Doberck (1873) determined it as August 9.06. The last orbit is given below and indicates the comet passed only 5° from the sun on August 12.

Interestingly a fifth independent discovery was reported. In a letter written to J. E. Bode, Reissig (Kassel, Germany) said he had seen a small nebulous comet through a break in the clouds on June 30 "between the head of the Great Bear and the Giraffe." He said it remained visible for only 5 or 6 minutes before clouds covered it. Bode wrote that there was too little information to conclude with any certainty that Reissig saw the same comet discovered by Pons and others, and that Lalande's prize should go to Pons. The story did not end there. Over a century and a half later, J. Ashbrook computed the position of C/1801 N1 for 1801 June 30 and found it would have actually been situated in Andromeda. He suggested Reissig's comet was a fabrication. Interestingly, Reissig reported another comet in 1803, which remained visible for eight days at magnitude 5–6, but was seen by no one else.

T	ω	Ω (2000.0)	i	q	e
1801 Aug. 9.0565 (UT)	219.840	45.313	159.267	0.25641	1.0

ABSOLUTE MAGNITUDE: $H_{10} = 9$ (V1964)

FULL MOON: Jun. 26, Jul. 25

SOURCES: C. Messier, P. F. A. Méchain, and A. Bouvard, *MC*, **4** (1801 Aug.), pp. 179–80; P. F. A. Méchain and Reissig, *BAJ for 1805*, **30** (1802), pp. 128–30; *MINS*, **6** (1806), p. 62; J. K. Burckhardt, *BAJ for 1809*, **34** (1806), p. 272; J. L. Pons, *MC*, **18** (1808 Sep.), p. 250; *American Almanac*. Boston: James Munroe & Co. (1847), p. 88; A. W. Doberck, *AN*, **81** (1873 May 27), pp. 321–4; A. W. Doberck, *MNRAS*, **34** (1874 Jun.), p. 426; V1964, p. 53; Reissig, *ST*, **37** (1969 Apr.), p. 230.

C/1802 Q1

(Pons)

1802

Discovered: 1802 August 26.90 ($\Delta = 0.39$ AU, $r = 1.12$ AU, Elong. $= 96°$)

Last seen: 1802 October 5.84 ($\Delta = 0.89$ AU, $r = 1.17$ AU, Elong. $= 76°$)

Closest to the Earth: 1802 August 18 (0.3626 AU)

Calculated path: OPH (Disc), HER (Sep. 8)

J. L. Pons (Marseille, France) discovered this comet on 1802 August 26.90, at a position of $\alpha = 16^h 35.3^m$, $\delta = -10° 35'$. He described it as small. Pons confirmed his find on August 27.92. P. F. A. Méchain (Paris, France) independently discovered this comet on August 28.88. He said it was much fainter than the nearby globular clusters M10 and M12, and exhibited no distinct nucleus or tail. H. W. M. Olbers (Bremen, Germany) independently discovered this comet in the evening sky on September 2.85 and described it as faint, diffuse, with a brighter central region 2–3′ across.

Olbers said the comet was easily observed on September 4, but noted it appeared fainter on the 5th. On the last date, K. L. Harding (Lilienthal, Germany) observed with a refractor of 3-foot focal length and described the comet as a whitish nebulosity with a stellar nucleus. From the eastern side of the coma extended a faint trace of a tail. Moonlight interfered with observations during the period of September 7–13, with Olbers simply describing the comet as very faint on the first date and hardly visible on the last.

Once moonlight ceased to be a problem, observations picked up once again. Olbers said the comet was rather well observed on September 19, while J. E. Bode (Berlin, Germany) described the comet as an extremely faint nebulosity with indistinct edges on the 20th. By September 23, Olbers remarked on the appreciable decrease in both brightness and size since September 2, although he noted that there was a fairly bright, nearly stellar, central condensation.

The comet had become a difficult object to see by September 25, with Olbers remarking on the poorly-defined center. Bode could no longer see the comet on the 29th, and Olbers caught his final glimpses of the comet on September 30 and October 2, when he noted it was very faint.

Part of the reason for the ending of observations was the reappearance of the moon in the evening sky, with the comet being seen for the final time on October 5.84, when Messier found it at a position of $\alpha = 17^h$ 19.6^m, $\delta = +36° 23'$.

Olbers had calculated two parabolic orbits while the comet was still being followed. Around mid-September he determined the perihelion date as September 14.39, and near the end of the month he revised this to September 10.29. Very similar orbits were computed by Olbers (1802) and Méchain (1802) later in the year. K. Lundmark (1916, 1917) noted a strong similarity to the orbit of C/1909 L1, but added that a positive identification was not possible. Lundmark's orbit is given below.

T	ω	Ω (2000.0)	i	q	e
1802 Sep. 10.3554 (UT)	21.8478	313.0142	57.0108	1.094249	1.0

ABSOLUTE MAGNITUDE: $H_{10} = 8.3$ (V1964)

FULL MOON: Aug. 13, Sep. 11, Oct. 11

SOURCES: P. F. A. Méchain, *BAJ for 1805*, **30** (1802), pp. 229–30; H. W. M. Olbers, *BAJ for 1805*, **30** (1802), pp. 232–3; H. W. M. Olbers, *BAJ for 1805*, **30** (1802), pp. 247–8; K. L. Harding, *BAJ for 1805*, **30** (1802), p. 257; J. E. Bode, *BAJ for 1805*, **30** (1802), p. 266; J. L. Pons, P. F. A. Méchain, and H. W. M. Olbers, *MC*, **6** (1802 Oct.), pp. 376–81; H. W. M. Olbers, *MC*, **6** (1802 Nov.), pp. 506–7; P. F. A. Méchain, *MC*, **6** (1802 Dec.), pp. 584–7; P. F. A. Méchain, *BAJ for 1806*, **31** (1803), pp. 129–31; J. L. Pons, *MC*, **18** (1808 Sep.), p. 250; O1899, pp. 74–6; K. Lundmark, *AN*, **202** (1916 Feb. 3), pp. 65–70; K. Lundmark, *AMAF*, **12** (1917 Sep. 12), pp. 1–53; V1964, p. 53.

C/1804 E1   *Discovered:* 1804 March 7.11 ($\Delta = 0.23$ AU, $r = 1.13$ AU, Elong. $= 122°$)

(Pons)   *Last seen:* 1804 April 1.85 ($\Delta = 0.51$ AU, $r = 1.32$ AU, Elong. $= 119°$)

  Closest to the Earth: 1804 March 9 (0.2229 AU)

1804   *Calculated path:* LIB (Disc), VIR (Mar. 11), BOO (Mar. 13)

J. L. Pons (Marseille, France) discovered this comet on 1804 March 7.11. On March 8.20, Pons' colleague J. J. C. Thulis (Marseille, France) measured the position as $\alpha = 14^h$ 35.1^m, $\delta = -15° 56'$. A. Bouvard (Paris, France) independently discovered this comet on March 11.16, while H. W. M. Olbers (Bremen, Germany) independently discovered it on March 12.98. Olbers described it as larger and brighter than M5 (a globular cluster in Serpens of about magnitude 6.2 and about 13′ across), but without a distinct boundary.

Messier and Olbers consistently managed to acquire observations throughout the remainder of March. On the 13th Olbers said it was not visible to the naked eye, although his cometseeker revealed it as a bright object. Upon observing it with his refractor, the comet was seen as a "pale and diffuse light." Olbers said poor seeing caused the comet to appear faint on the 14th, while bright moonlight made it very difficult to find on the 20th. Messier wrote that his observations during the period of March 11–17 always revealed the comet as very faint, with a round coma 5–6' across. He added that the nucleus was hardly apparent. Once the moon had left the sky, Olbers reported the comet was easy to see on March 27 and 28, and he even noted a nucleus was occasionally seen on the last date. But his observation on the 29th revealed a much fainter comet than on the previous nights. Messier saw the comet for the final time on the 31st, when he found it after much difficulty, because it was "almost completely invisible."

The comet was last detected on April 1.85, when Olbers determined the position as $\alpha = 14^h\ 33.0^m$, $\delta = +51° 52'$. He said he expected the comet to be similar in brightness to that observed on March 29, but was puzzled when he failed to find it with the cometseeker. Another sweep with the same telescope revealed "a small insignificant nebulosity near a star of 8 or 9 magnitude." Examination with a large refractor separated the two objects and Olbers noted the comet's center was a little northeast of the star. Olbers searched for the comet on April 8, but was unable to find it.

C. F. Gauss (1804) used ten positions obtained between March 13 and April 1 and computed a parabolic orbit with a perihelion date of 1804 February 14.09. This orbit is given below. Very similar orbits were also determined by D. von Wahl (1804) and A. Bouvard (1808).

T	ω	Ω (2000.0)	i	q	e
1804 Feb. 14.0881 (UT)	331.9459	179.5344	56.4522	1.071168	1.0

ABSOLUTE MAGNITUDE: $H_{10} = 8.0$ (V1964)

FULL MOON: Feb. 25, Mar. 26, Apr. 24

SOURCES: J.-J. L. de Lalande, J. L. Pons, J. J. C. Thulis, *BAJ for 1807*, **32** (1804), p. 225; H. W. M. Olbers, C. F. Gauss, and D. von Wahl, *BAJ for 1807*, **32** (1804), pp. 229–33; H. W. M. Olbers, *MC*, **9** (1804 Apr.), p. 344; H. W. M. Olbers, J. L. Pons, A. Bouvard, C. Messier, and C. F. Gauss, *MC*, **9** (1804 May), pp. 432–5; H. W. M. Olbers and A. Bouvard, *MC*, **9** (1804 Jun.), pp. 503–7; C. F. Gauss, *CDT* (1807), pp. 374–9; A. Bouvard, *CDT* (1808), p. 338; J. L. Pons, A. Bouvard, and H. W. M. Olbers, *MC*, **18** (1808 Sep.), p. 250; *American Almanac*. Boston: James Munroe & Co. (1847), p. 88; O1899, pp. 76–7; V1964, p. 54.

2P/1805 U1 *Discovered:* 1805 October 20.1 ($\Delta = 0.44$ AU, $r = 0.84$ AU, Elong. $= 56°$)
(Encke) *Last seen:* 1805 November 20.25 ($\Delta = 1.03$ AU, $r = 0.34$ AU, Elong. $= 19°$)
Closest to the Earth: 1805 October 16 (0.4352 AU)
1805 *Calculated path:* UMa (Disc), LEO (Oct. 22), COM (Oct. 24), VIR (Oct. 30)

J. L. Pons (Marseille, France) discovered this comet near ν Ursae Majoris on 1805 October 20.1. His colleague, J. J. C. Thulis, measured the position as $\alpha =$ $11^h\ 06.6^m$, $\delta = +33°\ 29'$ on October 20.20. An independent discovery was made by J. S. Huth (Frankfurt an der Oder, Germany) on October 20.08. He found the object with a cometseeker, describing it as a very bright nebulosity, but immediately identified it as a comet when seen in a refractor of 3.5-foot focal length. Huth noted that it appeared "as large and bright as the famous nebula in Andromeda, is nearly round, very bright in the center . . . and more sharply defined to the north." He estimated the diameter as 4' or 5' across and said no nucleus was seen. Another independent discovery was made by A. Bouvard (Paris, France) on October 20.17.

Huth said the comet appeared larger and brighter on the 22nd than when seen on the 20th, and was visible to the naked eye. On October 23, J. E. Bode (Berlin, Germany) described the comet as a rather bright nebulosity in his refractor of 3.5-foot focal length. On October 25, Huth said he saw a bright flickering within the coma, but no distinct nucleus. The coma gradually faded from the center toward the edges, although it seemed to extend toward the south-southwest. On October 29, Huth said the comet's motion had decreased, but the size and brightness had not noticeably changed since the 25th. A stellar nucleus was seen near the northern edge, which seemed to occasionally flicker. On October 30, H. W. M. Olbers (Bremen, Germany) said the comet was very bright, about 4' across, with a diffuse central condensation. It also displayed a faint tail about 1.5° long.

On November 1, Huth said the cometseeker revealed a vapor-like tail over 3° long and about 1.7' wide. Its edges were sharply defined near the comet, but showed no clear boundary near the end. The coma was about 3' across. Olbers said the tail appeared about 3° long in a cometseeker. Bode saw a small, thin tail in the finder. On November 3, Bode said the comet was seen "quite clearly, but the tail was less visible." On November 6, Huth said the nucleus occasionally flickered, while the tail was short and considerably fainter than the coma. On November 16, Huth saw the comet for the final time. On November 13 and 14, Olbers said moonlight and morning twilight interfered with his observations. Nevertheless, he estimated the comet shone with a brightness equal to stars of magnitude 4.

The comet was last detected on November 20.25, when Thulis determined the position as $\alpha = 14^h\ 23.3^m$, $\delta = -13°\ 30'$. Bode searched for this comet in the evening sky on December 5, 9, 10, and 11, but no trace was found. The comet passed 7.0° from the sun on December 22.

Very similar parabolic orbits were computed by F. W. Bessel (1806), C. F. Gauss (1806), and A. M. Legendre (1806). These indicated a perihelion date of 1805 November 18. J. F. Encke (1819a) suspected this comet was identical to one found by Pons in 1818. He was confident that the 1818 comet moved in a short-period elliptical orbit, but he was interested to see if the 1805 comet did as well. He ultimately found that an orbit with a perihelion date of November 22.00 and a period of 3.36 years fitted the positions for the 1805 comet. Encke (1819b) announced that comets seen in 1786 and 1795 were previous returns of this comet. The most recent investigation into this orbit was conducted by B. G. Marsden and Z. Sekanina (1974). It gave nongravitational terms of $A_1 = -0.29$ and $A_2 = -0.03847$ and is given below.

T	ω	Ω (2000.0)	i	q	e
1805 Nov. 21.9864 (TT)	182.3076	337.1517	13.5936	0.340645	0.846528

ABSOLUTE MAGNITUDE: $H_{10} = 8.0$ (V1964)

FULL MOON: Oct. 8, Nov. 7, Dec. 6

SOURCES: J. S. Huth and J. J. C. Thulis, *MC*, **12** (1805 Nov.), pp. 499–502; J. S. Huth and A. Bouvard, *BAJ for 1809*, **34** (1806), pp. 127–31; H. W. M. Olbers and F. W. Bessel, *BAJ for 1809*, **34** (1806), pp. 134–6; J. E. Bode, *BAJ for 1809*, **34** (1806), pp. 261–2; H. W. M. Olbers, F. W. Bessel, and C. F. Gauss, *MC*, **13** (1806 Jan.), pp. 79–83; J. J. C. Thulis, *MC*, **13** (1806 Feb.), p. 194; F. W. Bessel, J. J. C. Thulis, and A. M. Legendre, *MC*, **14** (1806 Jul.), pp. 68–71; J. L. Pons, *MC*, **18** (1808 Sep.), p. 250; J. J. C. Thulis, *CDT* (1809), pp. 325–6; J. F. Encke, *CA*, **2** (1819a), pp. 305–8, 316; J. F. Encke, *CA*, **2** (1819), p. 415; J. F. Encke, *BAJ for 1822*, **47** (1819b), pp. 188–91; O1899, pp. 77–8; V1964, p. 54; B. G. Marsden and Z. Sekanina, *AJ*, **79** (1974 Mar.), pp. 415–16.

3D/1805 V1 *Discovered:* 1805 November 10.9 ($\Delta = 0.25$ AU, $r = 1.20$ AU, Elong. $= 145°$)
(Biela) *Last seen:* 1805 December 14.70 ($\Delta = 0.05$ AU, $r = 0.95$ AU, Elong. $= 53°$)
Closest to the Earth: 1805 December 9 (0.0366 AU)
1806 I *Calculated path:* AND (Disc), PSC (Dec. 2), CET–AQR (Dec. 7), SCL (Dec. 8), GRU (Dec. 9)

J. L. Pons (Marseille, France) discovered this comet on 1805 November 10.9. His colleague, J. J. C. Thulis confirmed the comet on November 10.91 and estimated the position as $\alpha = 1^h\ 06.6^m$, $\delta = +40°\ 43'$. He described the comet as very small, with a fairly strong nucleus, and a weak coma. Independent discoveries were made by A. Bouvard (Paris, France) on November 16.95 and by J. S. Huth (Frankfurt an der Oder, Germany) on November 22.75. Huth noted the comet was visible to the naked eye as a star of magnitude 5 or 6. Huth described the comet as large, only a little fainter than M31, with a nucleus. When looked at with the refractor the nucleus showed a faint, planetary disk.

The comet was heading toward both the sun and Earth, and observers noted that it steadily brightened and grew larger in the days following discovery. Huth said the comet was not much fainter than M31 on November 23, while on the 30th he noted it was found at the same time the galaxy was. Interestingly, while both of these observations were naked eye, the observation on the last date was made in moonlight. Huth consistently remarked on the planetary appearance of the nucleus in his refractor, and he watched the coma grow from 6–7′ on the 23rd, to 20′ on the 30th. Although no tail was seen on the 23rd, Huth noted a slight extension of the coma on the northern edge on the 26th. By the 30th, he said the coma continued to spread to the east-northeast into a possible tail. J. E. Bode (Berlin, Germany) saw the comet on the 28th and said the coma was thinner and the nucleus easier to see than for comet 2P/1805 U1 (Encke).

As December began, the comet's increasingly rapid, southward motion was quite apparent. Moonlight still deterred observations, although Huth commented on the 1st, "If I hold my hand between my eye and the moon, I can clearly see the comet as much larger and brighter than [M31]." The comet was described as easily visible to the naked eye on December 8, even after the rising of the moon. C. F. Gauss (Brunswick, Germany) said the brightness of the comet was similar to a star of magnitude 3 or 4. H. W. M. Olbers (Bremen, Germany) described the comet as very beautiful, and added that a telescope revealed a very small, but distinct, planet-like nucleus situated within a large coma. J. Schroeter (Lilienthal, Germany) observed with a reflector of 13-foot focal length and said the coma was 5.5′ across. He measured the "whole nucleus" as 6.419″ across, while the "central brightest portion of the nucleus" was 4.052″ across. No tail was reported by any observer.

The comet was last detected in Europe on December 9.74, when Thulis determined the position as $\alpha = 23^h\ 16.2^m$, $\delta = -35°\ 22'$. Bode looked for the comet on December 9 and 10, but failed to find it. He suggested this was because it was then too close to the southern horizon.

Observers in more southerly latitudes began seeing the comet on December 10. Around December 10.6, W. Morison (Madras, India) saw an object easily visible to the naked eye and similar in brightness to a star of magnitude 4. He wrote that, although it looked like a comet, he was skeptical of its nature because of its very rapid motion. That evening, Dabadie and Laprie (Royal College of Port Louis, Mauritius) saw a "beautiful nebulous star . . . traversing the space between the constellations of Grus and Pavo."

The final observations of this comet were made by observers on the Isle de France (now Mauritius). Dupeloux roughly measured the comet's distances from Achernar on December 13.70 and 14.70, and from Canopus on December 13.72 and 14.69. Without knowledge of Dupeloux's observations, but aware of those of Dabadie and Laprie a few days earlier, Malavois precisely measured the comet's distances from Achernar on December 14.68 and Canopus on December 14.69. Using those distances, he roughly plotted the comet on a celestial globe and concluded its position was about $\alpha = 19^h$, $\delta = -74.5°$. Malavois said his first observation occurred when weak twilight was still present and noted the coma was about 20–25′ across. During the second observation, with dark skies, the coma was closer to 45′ across. Malavois was traveling on the 15th and was not able to look for the comet, and skies were cloudy on the 16th and 17th. He was not able to see the comet thereafter.

The first parabolic orbit was calculated by F. W. Bessel, and it revealed a perihelion date of 1806 January 1.49. Olbers immediately pointed out its similarity to the orbit of a comet seen in 1772. Gauss then took positions obtained during the period of November 16 to December 8, and determined the perihelion date as 1805 December 31.78. He also noted the similarity to the orbit of the comet of 1772, but added, "one cannot bring the elements substantially closer without disfiguring the agreement with the observations. The comet of 1772 came close to no planet that could have caused a large modification of its elements." Similar parabolic orbits were calculated by Bessel, Gauss, A. M. Legendre, and A. Bouvard during the next few years, but attempts were underway to establish a link between this comet and that of 1772.

Gauss redetermined the orbit of the comet of 1772 during February 1806 and became convinced that it was identical to Pons' current comet. Although the first elliptical orbit was calculated by Bessel during April 1806, the period had been *assumed* to be 33.86 years. The first true calculation of an elliptical orbit must be credited to Gauss, who determined the perihelion date as 1806 January 2.93 and the period as 4.74 years during May. Further investigation into the orbit of this comet basically ended because of the short period of time the comet was under observation.

Work on this orbit finally resumed in 1826 after W. von Biela had discovered a comet on February 27. Shortly after mid-March, several astronomers were pointing out the similarity between the orbits of their 1826 comet, and those seen in 1772 and 1805. That comet was followed until May 9, which provided enough observations for a fairly accurate determination of the orbit. Based on the assumption that the 1805 and 1826 comets were the same, J. F. A. Gambart (1826) calculated orbits for both comets. For the 1805 comet, Gambart determined a perihelion date of January 2.47 and a period of 6.74 years.

Further observed returns of this comet allowed additional refinements to the orbit of the 1805 apparition. Some of the astronomers who published later investigations include J. S. Hubbard (1860), J. von Hepperger (1900), B. G. Marsden and Z. Sekanina (1971), and W. Landgraf (1986). Landgraf's orbit is given below.

T	ω	Ω (2000.0)	i	q	e
1806 Jan. 2.4028 (TT)	218.0800	254.0782	13.5913	0.907144	0.745847

ABSOLUTE MAGNITUDE: $H_{10} = 7–8$ (V1964)

FULL MOON: Nov. 7, Dec. 6, Jan. 5

SOURCES: J. L. Pons, A. Bouvard, J. S. Huth, J. J. C. Thulis, H. W. M. Olbers, F. W. Bessel, and C. F. Gauss, *MC*, **13** (1806 Jan.), pp. 83–91; J. J. C. Thulis, *MC*, **13** (1806 Feb.), p. 195; C. F. Gauss and F. W. Bessel, *MC*, **13** (1806 Mar.), pp. 310–13; F. W. Bessel, A. M. Legendre, and C. F. Gauss, *MC*, **14** (1806 Jul.), pp. 71–86; C. F. Gauss, *MC*, **14** (1806 Aug.), pp. 181–6; J. J. C. Thulis, *MC*, **14** (1806 Oct.), pp. 382–3; J. S. Huth, *BAJ for*

1809, **34** (1806), pp. 131–4; H. W. M. Olbers and F. W. Bessel, *BAJ for 1809*, **34** (1806), pp. 135–6; C. F. Gauss, J. J. C. Thulis, and N. Maskelyne, *BAJ for 1809*, **34** (1806), pp. 137–40; J. Schroeter, *BAJ for 1809*, **34** (1806), pp. 140–2; J. E. Bode, *BAJ for 1809*, **34** (1806), pp. 262–3; A. Bouvard, *CDT* (1808), p. 340; J. L. Pons, *MC*, **18** (1808 Sep.), p. 251; J. J. C. Thulis, *CDT* (1809), pp. 326–7; A. Bouvard, *CDT* (1824), pp. 316–20; J. F. A. Gambart, *MAS*, **2** (1826), pp. 504–6; *Briefwechsel zwischen W. Olbers und F. W. Bessel*, **1**, edited by A. Erman. Leipzig: Avenarius & Mendelssohn (1852), p. 22; J. S. Hubbard, *AJ*, **6** (1860 Jun. 26), pp. 115–17; F. A. T. Winnecke, Dabadie, Laprie, Malavois, and Dupeloux, *VJS*, **15** (1880), pp. 372–8; W. T. Lynn and W. Morison, *The Observatory*, **14** (1891 Oct.), pp. 345–7; O1899, pp. 78–9; J. von Hepperger, *SAWW*, **109** Abt. IIa (1900), pp. 623–55; V1964, p. 54; B. G. Marsden and Z. Sekanina, *AJ*, **76** (1971 Dec.), pp. 1138–41; W. Landgraf, *QJRAS*, **27** (1986), p. 604.

C/1806 V1 *Discovered:* 1806 November 10.24 ($\Delta = 1.81$ AU, $r = 1.34$ AU, Elong. $= 47°$)
(Pons) *Last seen:* 1807 February 12.78 ($\Delta = 1.71$ AU, $r = 1.31$ AU, Elong. $= 50°$)
Closest to the Earth: 1806 December 29 (0.4520 AU)
1806 II *Calculated path:* VIR (Disc), CRV (Dec. 4), CRT (Dec. 9), HYA (Dec. 15), CEN (Dec. 19), ANT-VEL (Dec. 21), CAR (Dec. 26), VOL (Dec. 28), DOR (Dec. 31), RET (Jan. 2), HOR (Jan. 5), ERI (Jan. 8), PHE (Jan. 10), FOR (Jan. 15), SCL (Jan. 17), CET (Jan. 31)

J. L. Pons (Marseille, France) discovered this comet in the morning sky on 1806 November 10.24. His colleague, J. J. C. Thulis, confirmed the comet on November 11.17 and measured the position as $\alpha = 12^h 09.1^m$, $\delta = +2° 15'$.

The comet was heading southward at discovery and, although several astronomers in Germany, France, and England measured positions, few descriptive observations were made. F. W. Bessel (Lilienthal, Germany) saw the comet in unfavorable skies on December 8 and noted a faint nucleus and a tail that was hardly visible. On the same night, J. E. Bode (Berlin, Germany) said the comet was "obvious in the finder, while the telescope shows a small nucleus and weak traces of a tail." H. W. M. Olbers (Bremen, Germany) observed the comet on the morning of the 9th and said it displayed a central nucleus and faint traces of a tail. Although the comet was approaching both the sun and Earth, its increasingly southern declination brought observations to an end shortly after mid-December. The comet was last detected in Paris on December 18 and in Marseille on the 20th.

The comet was closest to both the sun and Earth on December 29, and it attained a maximum southern declination of $-67°$ on December 31. So, as the new year began, the comet was moving northward and fading. Pons accidentally found the comet in the evening sky on 1807 January 17, less than $10°$ above the horizon. He considered it his seventh comet discovery, even after finding out it was the same comet he had discovered the previous November. Before the end of the month, Bessel had resumed his observations and the comet was seen for the first time by C. Herschel (Slough, England).

During the first days of February, the comet was still under observation at Marseille, Paris, Lilienthal, and Slough. W. Herschel saw the comet with the reflector of 10-foot focal length on the 1st and wrote, "There was no visible nucleus, nor did the light which is called the coma increase suddenly towards the centre, but was of an irregular round form, and with this low power extended to about 5, 6, or 7 minutes in diameter. When I magnified 169 times it was greatly reduced in size, which plainly indicated that a farther increase of magnifying power would be of no service for discovering a nucleus." The comet was last seen on February 12.78, when Thulis determined the position as $\alpha = 1^h\ 08.9^m$, $\delta = -20°\ 57'$.

The first parabolic orbit was calculated by Bessel before the comet was recovered in January of 1807. The perihelion date was determined as 1806 December 29.80, and Bessel provided an ephemeris for the period of December 21 to February 19. Almost identical parabolic orbits that took advantage of the complete observational arc were later calculated by Bessel (1807) and J. K. Burckhardt (1819). Burckhardt's orbit is given below.

Another orbit was calculated over 40 years later. F. Hensel (1862) took 61 positions obtained during the period of November 11 to February 12, and determined seven Normal positions. He then calculated a hyperbolic orbit with a perihelion date of December 29.42 and an eccentricity of 1.0101820. No planetary perturbations were applied. Interestingly, the orbit does not fit the positions as well as the parabolas.

T	ω	Ω (2000.0)	i	q	e
1806 Dec. 29.4315 (UT)	225.22645	324.9969	144.9746	1.081571	1.0

ABSOLUTE MAGNITUDE: $H_{10} = 6$ (V1964)

FULL MOON: Oct. 27, Nov. 26, Dec. 25, Jan. 24, Feb. 22

SOURCES: J. E. Bode, *BAJ for 1810*, **35** (1807), p. 238; W. Herschel, *PTRSL*, **97** (1807), pp. 264–6; F. W. Bessel, *MC*, **15** (1807 Jan.), pp. 85–8; F. W. Bessel, *MC*, **15** (1807 Apr.), pp. 373–6; F. W. Bessel and J. J. C. Thulis, *MC*, **16** (1807 Aug.), pp. 176–82; J. L. Pons, *MC*, **18** (1808 Sep.), p. 251; *CDT* (1810), pp. 298–9; J. K. Burckhardt, *CDT* (1819), p. 378; *American Almanac*. Boston: James Munroe & Co. (1847), p. 88; F. Hensel, *AN*, **58** (1862 Aug. 8), pp. 89–92; O1899, pp. 79–80; V1964, p. 54.

C/1807 R1
(Great Comet)

1807

Discovered: 1807 September 9.7 ($\Delta = 1.20$ AU, $r = 0.68$ AU, Elong. $= 34°$)
Last seen: 1808 March 27.87 ($\Delta = 3.68$ AU, $r = 3.12$ AU, Elong. $= 49°$)
Closest to the Earth: 1807 September 26 (1.1533 AU)
Calculated path: VIR (Disc), LIB (Sep. 28), VIR (Sep. 29), SER (Oct. 3), HER (Oct. 17), LYR (Nov. 18), CYG (Dec. 4), LAC (Jan. 8), AND (Jan. 22), CAS (Feb. 4), AND (Mar. 9), PER (Mar. 24)

This comet should probably have been seen in the Southern Hemisphere weeks before its actual discovery, but no records of any such observations exist. For observers in Australia, for example, the comet would have been poised 10° or more above the western horizon at the beginning of evening twilight throughout the month of August, with a magnitude that should have brightened steadily from 3 to 1. The comet's altitude for such observers would have started to decrease noticeably early in September as it headed for its close approaches to both the sun and Earth later in the month.

The actual discovery of this comet is credited to C. Giovanni (Sicily, Italy), who saw the comet in evening twilight, very low over the west-southwest horizon on 1807 September 9.7. Giovanni benefited by observing at a latitude 5–15° farther south than most astronomers in Europe and elsewhere. Moonlight interfered during the next 11 days, and then several independent discoveries were made, the order of which closely paralleled the latitudes of observers. S. Pease was exploring the Mississippi River in the USA, when he saw the comet in the evening sky on September 21.05. J. L. Pons (Marseille, France) saw the comet in evening twilight on September 21.79, and his colleague, J. J. C. Thulis, obtained the first measurement of the comet's position on September 21.80, when he determined it as $\alpha = 14^{\mathrm{h}} 05.4^{\mathrm{m}}$, $\delta = -6° 35'$. Further independent discoveries were made by the following: a professor of mathematics at Vesoul, France, on September 26.8, J. Vidal (Mirepoix, France) and H. Flaugergues (Viviers, France) on September 27.8, E. Pigott (Bath, England) on the 28th, J. S. Huth (Frankfurt an der Oder, Germany) on the 29th, and Eule (Dresden, Germany) and Gonzalez (Madrid, Spain) on the 30th. Vidal was preparing for an observation of one of Jupiter's moons when he noted a train of light partially blocked by the bell tower of a nearby cathedral. Leaving the observatory for a better look, he saw it was a bright, naked-eye comet with a tail 7–8° long and a nucleus equal in brightness to a star of magnitude 1 or 2. Flaugergues indicated the nucleus was about magnitude 2, while the tail was about 1° long. Huth said the comet was visible to the naked eye and noted, "The tail was situated almost in the direction of the equator, but turned slightly northward. In the reflector the body of the comet appeared like a perfect ball the size of Jupiter with only a slight coma." He added that the tail was sharply defined near the coma, but diffused quickly and was only 5° long, with a width of about 0.5° at the end. Eule said the comet had a tail and exhibited a nucleus that looked like a nebulous star of the first magnitude. By the end of September, the comet was steadily moving away from both the sun and Earth. Gonzalez said the tail extended 1.5–2°.

The comet remained a naked-eye object throughout the month of October and the two most distinct features discussed by observers were the tail and nucleus. On the 1st, Huth said the tail was long and curved slightly southward, while J. E. Bode estimated the length as 5°. Huth noted the tail was more strongly curved on the 2nd. By the 4th, Huth reported the tail had split into two components, with a straight tail extending over 6°, and the curved tail not quite as long. Both tails were still visible on October 20, when H. W. M. Olbers noted they were separated by about 1.5°. He said the northern tail was very slender, faint, and straight, with a length of about 10°, while the southern tail was short, wide, and brighter, with a length of about 4.5°. The southern tail was also more intense on the southern side, while the concave side of the tail was very poorly defined. The tails apparently became indistinguishable from one another a few days later. W. Dunbar (near Natchez, Mississippi) gave the tail length as about 2.7° on the 24th, while W. Herschel said the tail was "considerably longer on the south-preceding, than on the north-following side" on the 26th. On October 28 and 31, Herschel noted that the tail's south-preceding side was better defined than its north-following side.

The nucleus was described by Huth as large, light yellow, round, and sharply limited on October 1, while Dunbar said it was similar in bright-ness to a star of magnitude 2 or 3 on the 3rd. On the last date, Dunbar observed the comet with a reflector of 8-foot focal length (128×) and wrote that the nucleus and coma were shown "with tolerable distinctness; the idea produced in the mind of the observer was that of a round body in com-bustion, which had produced so much smoke as to obscure the nucleus; the smoke seemed to be emitted in every direction; but, as if it met on one side with a gentle current of air, the smoke seemed to be repelled and bent around the nucleus, escaping on the opposite side, in the direction of the tail." On the 4th Herschel estimated the diameter of the nucleus as 3″ in his 14-cm reflector, while it was slightly smaller than 2.8″ on the 5th in his 23-cm reflector. Estimates of 2.8″ and less than 2.5″ were obtained by Herschel on the 18th and 19th, respectively, while using his 61-cm reflector. Smaller tele-scopes with the lesser resolving capability revealed a larger nucleus. Such was the case on October 20, when Olbers saw a distinct nucleus of 8″ or 9″ diameter, which he said indicated a diameter of 900 miles. No observer, at any time during the month, described the nucleus as stellar. In fact, J. H. Fritsch (Quedlinburg, Germany) best summarized the descriptions of other astronomers when he noted that the nucleus "more resembled a condensed ball of light, rather than a stellar body" on the 9th.

Although Herschel said the coma was 6′ across on October 19, J. H. Schröter (Göttingen, Germany) produced a fine series of measurements spanning October 20–31. These indicated the coma steadily increased in diameter from 3.5′ on the 20th to 4.6′ on the 31st. Schröter's last measurement came on November 3, when the diameter was given as 5.6′.

The comet was still a naked-eye object throughout November, although observers did remark that it was fading and the tail was becoming less conspicuous. Bode described the tail as short and distorted on the 1st, while Herschel estimated the length as 2.5° on the 20th. On the last date, C. Lofft (Troston, England) said the comet could only be distinguished from the faint stars by its hazy appearance, as the tail could no longer be seen with the naked eye. Bode noted the nucleus was still rather bright on the 4th, while Herschel reported the nucleus was distinctly visible in his 14-cm reflector as "a mere point" on the 20th.

The comet was still a naked-eye object into December, although it was near the limit of visibility. Bode said on the 4th that the comet was smaller and fainter than on previous evenings, and Dunbar wrote on the 7th that "the comet has become so dim, as to be seen with the naked eye only in a pure atmosphere, with favourable circumstances." Observers were reporting that the tail was no longer visible, except for Herschel, who noted the tail was 23′ long in his 61-cm reflector on the 6th and that he saw faint nebulosity on the south-preceding side on the 16th. The coma was centrally condensed throughout the month, with Herschel noting it was about 4.75′ across on the 6th. In addition, he described the coma as round on the 6th and somewhat irregular in shape on the 16th.

It is uncertain whether any naked-eye observations were made as 1808 began. Several observers were still describing the comet as a very bright telescopic object early in the month, but Dunbar specifically noted on January 6 that the comet was no longer visible to the naked eye. The comet was also frequently described as very large and centrally condensed, but no tail was reported. Dunbar's observations with a reflector revealed a nucleus of magnitude 7 on the 6th, and magnitude 8 on the 16th. Herschel demonstrated the advantage of using a large telescope on January 1. He initially looked at the comet with a 14-cm reflector and saw that the center consisted of very small stars. Upon checking with the 61-cm reflector he noted that there were "several small stars shining through the nebulosity of the coma."

Although Herschel again reported the comet as "very bright" in his 61-cm reflector as February began, other astronomers were beginning to have difficulty in finding it. Olbers last saw the comet on the 19th, F. W. Bessel reported the comet as very faint on the 24th, and Dunbar located a very faint object in the expected position on the 26th after a long search. Herschel seems to have detected the tail during the month. On the 2nd he saw a faint, diffused nebulosity on the north-preceding side, and surmised it might have been "the vanishing remains of the comet's tail." On February 19 and 20, he noted, "The faint nebulosity in the place where the tail used to be, still projects a little farther from the center than in other directions." Herschel also reported the coma was 3.4′ across on the 19th and 20th.

The comet was last detected on March 27.87, when V. Wisniewski (St. Petersburg, Russia) estimated the position as $\alpha = 1^h 43.3^m$, $\delta = +48° 54'$.

The earliest parabolic orbits were calculated using positions obtained during the first half of October. Bode determined the perihelion date as 1807 September 21.32, J. K. Burckhardt determined it as September 26.12, and F. v. P. Triesnecker determined it as September 16.86. Triesnecker's revision near the end of October included positions obtained until October 24 and gave the perihelion date as September 18.98. Very similar orbits were later calculated by Bessel, M. C. T. Damoiseau, B. Oriani, C. F. Gauss, Burckhardt, F. Lemaur, J. J. de Ferrer, N. Bowditch, and N. Cacciatore.

After the comet had been under observation for over five months, Bessel noted that an elliptical orbit fitted the positions best. His initial computations used positions up to February 24 and revealed a perihelion date of September 19.24 and a period of 1953 years. Further revisions were made by Bessel over the next two years. Bessel (1810) used positions spanning the period of September 22 to March 27 to determine a perihelion date of September 19.24 and a period of 1713 years. This orbit is given below.

T	ω	Ω (2000.0)	i	q	e
1807 Sep. 19.2389 (UT)	4.0970	269.4837	63.1762	0.646124	0.995488

ABSOLUTE MAGNITUDE: $H_{10} = 1.6$ (V1964)

FULL MOON: Aug. 18, Sep. 16, Oct. 16, Nov. 15, Dec. 15, Jan. 13, Feb. 12, Mar. 12, Apr. 10

SOURCES: *CDT* (1807), pp. 494–6; *Gentleman's Magazine*, **77** (1807 Oct.), p. 972; C. Lofft, *Gentleman's Magazine*, **77** (1807 Nov.), p. 1072; J. Oltmanns, J. E. Bode, and J. K. Burckhardt, *MC*, **16** (1807 Nov.), pp. 484–94; A. Bouvard, H. W. M. Olbers, J. H. Schröter, F. W. Bessel, C. F. Gauss, and B. Oriani, *MC*, **16** (1807 Dec.), pp. 562–7; J. S. Huth, *BAJ for 1811*, **36** (1808), pp. 116–19; H. W. M. Olbers, *BAJ for 1811*, **36** (1808), pp. 119–24; F. v. P. Triesnecker, *BAJ for 1811*, **36** (1808), pp. 125–7; M. A. David and A. Bittner, *BAJ for 1811*, **36** (1808), pp. 128–30; T. Bugge, *BAJ for 1811*, **36** (1808), pp. 130–4; C. F. Gauss, *BAJ for 1811*, **36** (1808), pp. 135–6; T. Derfflinger, *BAJ for 1811*, **36** (1808), pp. 140, 146; J. H. Fritsch, *BAJ for 1811*, **36** (1808), pp. 149–51; F. W. Bessel, *BAJ for 1811*, **36** (1808), pp. 153–63; J. E. Bode, *BAJ for 1811*, **36** (1808), pp. 163–70; Eule, *BAJ for 1811*, **36** (1808), pp. 254–5; W. Herschel, *PTRSL*, **98** (1808), pp. 145–59; F. W. Bessel, *MC*, **17** (1808 Jan.), pp. 80–7; C. F. Gauss, *MC*, **17** (1808 Feb.), pp. 182–3; H. W. M. Olbers, *MC*, **17** (1808 Feb.), pp. 189–90; F. W. Bessel, *MC*, **17** (1808 May), pp. 471–3; F. W. Bessel, *MC*, **17** (1808 Jun.), pp. 551–7; T. Bugge, *MC*, **18** (1808 Jul.), pp. 87–9; V. Wisniewski, *MC*, **18** (1808 Aug.), p. 171; F. W. Bessel, *MC*, **18** (1808 Sep.), pp. 237–44; J. L. Pons, *MC*, **18** (1808 Sep.), pp. 251–2; G. Santini, *MC*, **18** (1808 Oct.), pp. 360–1; F. T. Schubert and V. Wisniewski, *BAJ for 1812*, **37** (1809), pp. 95–103; F. W. Bessel and V. Wisniewski, *BAJ for 1812*, **37** (1809), pp. 125–6; W. Herschel, *BAJ for 1812*, **37** (1809), pp. 230–1; J. L. Pons, J. K. Burckhardt, J. Vidal, H. Flaugergues, and Gonzalez, *CDT* (1809), pp. 494–6; *TAPS*, **6** (1809), pp. 345–7, 368–74; V. Wisniewski, *MC*, **19** (1809 May), p. 521; W. Herschel, *MC*, **20** (1809 Dec.), pp. 512–14; F. W. Bessel, *BAJ for 1813*, **38** (1810), pp. 187–9; W. Herschel, *BAJ for 1813*, **38** (1810), p. 218; J. J. de Ferrer, *BAJ for 1813*, **38** (1810), pp. 245–8; J. Vidal, P. Ciera, M. C. T. Damoiseau, and H. W. M. Olbers, *CDT* (1810), pp. 376–82; F. W. Bessel and V. Wisniewski, *MC*, **21** (1810 Feb.), pp. 189–90; F. W. Bessel, *MC*, **22** (1810 Sep.), pp. 205–12; N. Bowditch, *BAJ for 1814*, **39** (1811), pp. 148–9; N. Cacciatore, *BAJ for 1815*, **40** (1812), pp. 122–4; J. H. Schröter, *MC*, **25** (1812 Apr.), pp. 362–7; J. J. de Ferrer, *MC*, **25** (1812 Jun.), p. 525; W. Dunbar and S. Pease, *MC*, **25** (1812 Jun.), pp. 529–30; H. Flaugergues, *CA*, **2** (1819), pp. 345–52; J. J. de Ferrer and F. Lemaur, *MAS*, **3** (1829), pp. 1–4; *American Almanac*. Boston: James

Munroe & Co. (1847), pp. 88–9; O1899, pp. 80–1; W. T. Lynn, C. Giovanni, J. L. Pons, and E. Pigott, *Observatory*, **25** (1902 Sep.), pp. 336–7; V1964, p. 54.

26P/1808 C1   *Discovered:* 1808 February 6.15 ($\Delta = 0.12$ AU, $r = 0.99$ AU, Elong. $= 88°$)
(Grigg–   *Last seen:* 1808 February 9.19 ($\Delta = 0.13$ AU, $r = 0.96$ AU, Elong. $= 73°$)
Skjellerup)   *Closest to the Earth:* 1808 February 5 (0.1165 AU)
1808 III   *Calculated path:* SER (Disc), OPH (Feb. 7)

J. L. Pons (Marseille, France) discovered this comet in the morning sky between "the neck of Serpens and the tongue of Libra" on 1808 February 6.15. He simply described it as "very small." Pons only managed to observe the comet until February 9.19, by which time it was situated in Ophiuchus and was seen with difficulty because of moonlight. The moon prevented additional observations.

Several years later, after Pons had moved to Florence, Italy, H. C. Schumacher, editor of the *Astronomische Nachrichten*, requested that Pons go through his notes and provide any additional information that he had about this comet. Pons' colleague, G. Inghirami, finally extracted a few paragraphs and a drawing from Pons' papers and the details were included in the January 1829 issue of the *Astronomische Nachrichten*. Inghirami's letter stated that the comet was one of the few to escape without the calculation of an orbit, "because one only had some very doubtful positions by comparison with other nebulae. It was very weak and difficult to see. Its nebulosity was round; it extended about one degree and one suspected at certain intervals a very weak nucleus in two parts. Its movement was rather fast towards the south and one could see it only 3 days because the moonlight was very strong, so that in spite of very persistent searches, one could not even suspect it on the 10th." The drawing was said to have been made on the 9th at 5 a.m. in the morning using a large seeker with a field of view of $3°$. It showed two nebulae, which were said to have been "on the belly of Ophiuchus a little south of the equator." The comet was situated very near the lower of the two nebulae.

T. R. von Oppölzer (1869) suggested the two nebulosities in Pons' drawing were M10 and M12, two globular clusters included by C. Messier in his nineteenth century catalog of nebulae. He also suggested the drawing should be flipped because of the reversal of the field of view by the seeker, so that M12 was at the top with the comet close by. Oppölzer went on to suggest the comet might have been a previous appearance of P/Pons–Winnecke, which was discovered by Pons on 1819 June 12, and rediscovered by F. A. T. Winnecke on 1858 March 9. Oppölzer noted that if the perihelion date of P/Pons–Winnecke was 1808 April 12, it would have been very close to the rough positions Pons gave for the 1808 comet.

During 1880, an anonymous author (possibly J. R. Hind) writing in *Nature* examined Oppölzer's suggested link. He surmised that Pons' statement of a 1° diameter and a rapid southward motion would indicate the comet was near Earth. In order to place P/Pons–Winnecke in the positions observed by Pons, it was necessary to assume a perihelion date of 1808 April 12, but the resulting distance from Earth amounted to 1.04–1.01 AU for the interval of the observations, with the motion being towards the east-southeast.

The story of this comet resumed in 1986, when L. Kresák (Astronomical Institute, Slovak Academy of Sciences) suggested the two objects in Pons' drawing were M10 and M12, and that the comet was at a position of $\alpha = 16^h 45.0^m$, $\delta = -1° 26'$ on February 9.22. He said a backward integration of P/Grigg–Skjellerup's orbit was carried out by N. A. Belyaev and coworkers on the assumption of constant nongravitational effects, and a subsequent correction of less than two days in the resulting perihelion date indicated the comet passed perihelion on 1808 March 17.10, and passed 0.12 AU from Earth at the time of Pons' observations. B. G. Marsden said, "In view of the encounters with Jupiter in 1881, 1845, and 1809, the positional residuals seem more than satisfactory, and the identification appears to be certain."

Belyaev, Kresák, and A. Carusi (1986) used positions obtained during the apparitions between 1808 and 1922, as well as perturbations by all nine planets, and computed an elliptical orbit with a perihelion date of 1808 March 17.10, and a period of 4.82 years.

T	ω	Ω (2000.0)	i	q	e
1808 Mar. 17.10 (UT)	193.59	18.50	3.50	0.7315	0.7437

ABSOLUTE MAGNITUDE: $H_0 = 12.5$, $n = 13.5$ (Kronk)
FULL MOON: Jan. 13, Feb. 12
SOURCES: J. L. Pons, MC, 18 (1808 Sep.), p. 252; J. L. Pons, CA, 12 (1825), p. 509; H. C. Schumacher, J. L. Pons, and G. Inghirami, AN, 7 (1829 Jan.), p. 113; T. R. von Oppölzer, AN, 75 (1869 Dec. 10), pp. 107–10; Nature, 21 (1880 Jan. 15), p. 264; L. Kresák, N. A. Belyaev, and A. Carusi, IAUC, No. 4234 (1986 Jul. 22); CCO, 6th ed. (1989), pp. 11, 50.

C/1808 F1 *Discovered:* 1808 March 25.86 ($\Delta = 0.58$ AU, $r = 1.20$ AU, Elong. $= 94°$)
(Pons) *Last seen:* 1808 April 3.03 ($\Delta = 0.61$ AU, $r = 1.04$ AU, Elong. $= 77°$)
Closest to the Earth: 1808 March 27 (0.5773 AU)
1808 I *Calculated path:* DRA (Disc), CAM (Mar. 26)

J. L. Pons (Marseille, France) discovered this comet on 1808 March 25.86. He described it as round and very faint, and determined a position of $\alpha = 9^h 58.6^m$, $\delta = +80° 54'$ on March 26.04.

The September 1822 issue of the *Astronomische Nachrichten* printed a letter from H. W. M. Olbers which gave positions obtained on the evenings of March 26, 28, 29, and 31. He attributed these positions to both J. J. C. Thulis

(Marseille, France) and F. X. von Zach. Olbers also noted that from the time of discovery until April 1 the brightness and size of the comet did not display any significant variation.

The comet was independently discovered by V. Wisniewski (St. Petersburg, Russia) on March 29.83. He described it as round, diffuse, and 3' across, but no trace of a tail was detected. Wisniewski measured a position on April 1, and described the comet as faint on April 2. The comet was last detected on April 3.03, when Wisniewski said it was very faint in moonlight. He obtained measurements with nearby stars which later allowed J. F. Encke to establish the position as $\alpha = 5^h\ 31.9^m$, $\delta = +62°\ 42'$.

Encke (1826) used ten positions obtained between March 26 and April 3, and computed a parabolic orbit with a perihelion date of 1808 May 13.45. This orbit is given below.

T	ω	Ω (2000.0)	i	q	e
1808 May 13.453 (UT)	253.743	325.642	134.303	0.38986	1.0

ABSOLUTE MAGNITUDE: $H_{10} = 6.8$ (V1964)

FULL MOON: Mar. 12, Apr. 10

SOURCES: H. W. M. Olbers and V. Wisniewski, *BAJ for 1811*, **36** (1808), pp. 215–17; J. L. Pons and H. W. M. Olbers, *MC*, **17** (1808 Jun.), pp. 557–8; V. Wisniewski, *MC*, **18** (1808 Aug.), p. 172; J. L. Pons, *MC*, **18** (1808 Sep.), p. 252; H. W. M. Olbers and J. L. Pons, *AN*, **1** (1822 Sep.), pp. 307–10; J. L. Pons and V. Wisniewski, *CA*, **12** (1825), pp. 509–11; J. F. Encke, *AN*, **5** (1826 Jun.), pp. 1–8; *American Almanac*. Boston: James Munroe & Co. (1847), p. 89; V1964, p. 54.

C/1808 M1   *Discovered:* 1808 June 24.96 ($\Delta = 0.74$ AU, $r = 0.72$ AU, Elong. $= 45°$)
(Pons)   *Last seen:* 1808 July 4.04 ($\Delta = 0.65$ AU, $r = 0.64$ AU, Elong. $= 37°$)
   Closest to the Earth: 1808 July 2 (0.6469 AU)
1808 II   *Calculated path:* CAM (Disc), LYN-CAM (Jul. 1), LYN (Jul. 2), UMa (Jul. 4)

J. L. Pons (Marseille, France) discovered this comet in the evening sky on 1808 June 24.96. He originally thought it was a nebula and described it as very small and faint. On June 25.9, Pons reobserved the object and noted it had moved almost 3° due north. On June 26.90, Pons estimated the position as $\alpha = 4^h\ 21.2^m$, $\delta = +61°\ 26'$.

Although J. J. C. Thulis (Marseille, France) had usually acted to confirm and measure precise positions for all of Pons' comets, he was sick during the time of this comet's appearance and Pons remained the only observer. The comet remained a faint object during its short period of visibility, but Pons noted its rapid eastward movement across the sky, which amounted to 8–9° per day.

The comet was last detected near the horizon and in moonlight on July 4.04, when Pons estimated the position as $\alpha = 8^h\ 07.8^m$, $\delta = +57°\ 38'$.

Only one orbit has ever been determined for this comet. F. W. Bessel (1808) used three positions obtained between June 26 and July 4, and computed a parabolic orbit with a perihelion date of 1808 July 12.67.

T	ω	Ω (2000.0)	i	q	e
1808 Jul. 12.6677 (UT)	131.5607	26.8770	140.7050	0.607953	1.0

ABSOLUTE MAGNITUDE: $H_{10} = 9.0$ (V1964)

FULL MOON: Jun. 8, Jul. 7

SOURCES: J. L. Pons, *MC*, **18** (1808 Sep.), pp. 245–9; F. W. Bessel, *MC*, **18** (1808 Oct.), pp. 358–9; F. W. Bessel and J. L. Pons, *BAJ for 1812*, **37** (1809), pp. 128–9; J. L. Pons, *CA*, **12** (1825), pp. 511–12; V1964, p. 54.

X/1808 N1 After having made a very difficult observation of comet C/1808 M1 near the horizon on 1808 July 4, J. L. Pons (Marseille, France) turned his telescope to other objects of interest and on July 4.11 he discovered another comet. He estimated the position as $\alpha = 3^h\ 10.2^m$, $\delta = +56°\ 36'$, and noted it was whiter and more diffuse than comet C/1808 M1, but still faint. No nucleus was detected. The only other observation of this comet was also the final observation, which Pons obtained on July 6.12. He estimated the position as $\alpha = 3^h\ 31.8^m$, $\delta = +58°\ 19'$. As with comet C/1808 M1, J. J. C. Thulis (Marseille, France) was sick and unable to backup Pons' observations with precise positions. No further observations were made. The moon was full on July 8.

SOURCES: J. L. Pons, *MC*, **18** (1808 Sep.), p. 249; J. L. Pons, *CA*, **12** (1825), p. 512.

C/1810 Q1 *Discovered:* 1810 August 23.1 ($\Delta = 1.01$ AU, $r = 1.23$ AU, Elong. $= 75°$)
(Pons) *Last seen:* 1810 October 8.2 ($\Delta = 1.26$ AU, $r = 0.97$ AU, Elong. $= 49°$)
Closest to the Earth: 1810 August 12 (0.9904 AU)
1810 *Calculated path:* CAM (Disc), DRA (Aug. 28), UMa (Sep. 10), CVn (Sep. 23), UMa (Sep. 28), CVn (Oct. 3), UMa (Oct. 10)

J. L. Pons (Marseille, France) discovered this comet in the morning sky on 1810 August 23.1. He described it as small, round, and faint. The first position was obtained by Pons on August 30.08, and was estimated as $\alpha = 12^h\ 47.3^m$, $\delta = +75°\ 46'$. The comet had passed just 4° from the north celestial pole on August 18.

Pons remained the only observer of this comet. He determined further positions on August 31, as well as September 1–10. The full moon then interrupted his observations until September 16. He measured the comet's position for the final time on September 21.99, with the location given as $\alpha = 12^h\ 05.4^m$, $\delta = +54°\ 23'$.

Pons saw the comet for the final time on October 8.2. It seems apparent that further observations during the following days were interrupted by moonlight.

F. W. Bessel (1811) used positions obtained between August 30 and September 21 and computed a parabolic orbit with a perihelion date of 1810 October 6.32. F. v. P. Triesnecker (1812) used positions obtained on August 30, September 4, and 9 and calculated a parabolic orbit with a perihelion date of September 30.67.

A. Thraen (1881) reevaluated Pons' ten meridian observations obtained between August 30 and September 21 and computed a parabolic orbit with a perihelion date of October 6.74. This orbit is given below.

T	ω	Ω (2000.0)	i	q	e
1810 Oct. 6.7379 (UT)	114.9175	311.5035	62.9454	0.969623	1.0

ABSOLUTE MAGNITUDE: $H_{10} = 6.3$ (V1964)

FULL MOON: Aug. 14, Sep. 13, Oct. 12

SOURCES: F. W. Bessel, *BAJ for 1814*, **39** (1811), pp. 178–80; J. L. Pons, *MC*, **23** (1811 Mar.), p. 302; F. W. Bessel, *MC*, **24** (1811 Jul.), pp. 71–3; F. v. P. Triesnecker, *BAJ for 1815*, **40** (1812), pp. 127–8; A. Thraen, *AN*, **99** (1881 May 7), pp. 343–8; V1964, p. 54.

C/1811 F1 (Great Comet, also Flaugergues)

1811 I

Discovered: 1811 March 25.9 ($\Delta = 2.16$ AU, $r = 2.72$ AU, Elong. $= 114°$)

Last seen: 1812 August 17.97 ($\Delta = 3.55$ AU, $r = 4.54$ AU, Elong. $= 167°$)

Closest to the Earth: 1811 October 16 (1.2213 AU)

Calculated path: PUP (Disc), MON (Apr. 28), CMi (May 21), HYA (Jun. 5), CNC (Jun. 8), LEO (Aug. 2), LMi (Aug. 21), UMa (Sep. 7), CVn (Sep. 21), UMa (Oct. 2), BOO (Oct. 5), HER (Oct. 15), SGE (Nov. 14), AQL (Nov. 15), SGE (Nov. 16), AQL (Nov. 19), DEL (Dec. 18), AQL (Dec. 25), AQR (Dec. 26), CAP (1812 Jul. 30), PsA (Aug. 16)

Numerous astronomers have declared this comet among the most impressive in history, with its naked-eye visibility beginning in mid-April 1811 and lasting until the first week of January 1812. In addition, observers with telescopes followed it for 511 days, which more than doubled the record of 201 days set by the Great Comet of 1807. But the comet's impact outside of the astronomical community was also noteworthy.

Napoleon I (Emperor of the French) considered the comet's spectacular appearance as an omen indicating his success in his planned invasion of Eastern Europe and Russia in 1812 (Brown, 1974). According to the *General Evening Post* for 1811 September 21–24 (Lynn, 1893), "The present comet must be deemed ominous to Bonaparte from the length of time that it will be visible; no comet ever continued longer except that which appeared in the reign of the monster Nero." Even more interesting is the appearance of "comet wine" on the lists of wine merchants for several years following the appearance of the Great Comet of 1811. It seems the year 1811 saw the appearance of several particularly good vintages of wine. According to *The Great Vintage Wine Book* by Michael Broadbent (1981), the red and white wines from Bordeaux, France, were considered five-star vintages (on a five-star scale), with the 1811 Chateau Lafite "considered the finest red Bordeaux ever made." In addition, the burgundy from the Cote d'Or region near Beaune, France, and the port from the Douro region of Portugal were also rated as five-star wines in 1811.

H. Flaugergues (Viviers, France) discovered this comet on 1811 March 25.9. He said it was situated in Argo Navis, a huge constellation that had actually been broken into several smaller constellations during the previous century. On March 26.9, he estimated the position as $\alpha = 8^h 01.1^m$, $\delta = -29° 15'$. Flaugergues further observed the comet on the evenings of March 28–31, as well as on April 1.

Observations temporarily ceased after April 1, as the moon began interfering on its way to its full phase on April 8, but they resumed on April 11. On that date, J. L. Pons (Marseille, France), not having received word of the discovery, accidentally found the comet on April 11.82. Meanwhile, F. X. von Zach (St. Peyre, France) was able to confirm Flaugergues' discovery on April 11.83.

The comet was a naked-eye object during the remainder of April and was fairly easy to see without optical aid during May. During this time its solar elongation was steadily decreasing as it moved slowly northward.

William J. Burchell was in Cape Town (South Africa) from late 1810 until mid-1811. On the evening of 1811 June 2, an earthquake hit the region and Burchell wrote in his journal that many of the people "coupled the comet, which had been seen every night since the 12th of the foregoing month, and the earthquake together, and drew from this two-fold portentous sign, the certain prognostics of the annihilation of the Cape."

By the end of May, observers were already finding this naked-eye object difficult to see because of its low altitude and entrance into evening twilight. Flaugergues last detected the comet on May 29, when it was 54° from the sun. Zach last detected it on June 2, at an elongation of 52°. Don J. J. de Ferrer (Havana, Cuba) last determined the comet's position on June 11 and last saw the comet on June 15, by which time the elongation had decreased to 41°.

The comet's final observer before conjunction with the sun was Alexander von Humboldt (Paris, France). He last caught a glimpse of the comet in strong twilight on June 16.9, at which time the elongation was 40°.

De Ferrer determined positions of the comet on six evenings during the period of May 19–27. Zach determined the comet's position on 14 nights during the period of May 3 to 28.

The Earth's steady motion away from the comet ended on June 25 when their distances had increased to a maximum of 2.4142 AU. Thereafter, the distance between our planet and the comet decreased. Meanwhile, the comet's angular distance from the sun continued to decrease and reached a minimum of just under 10° during the last days of July and first days of August.

By mid-August the comet was situated almost due north of the sun. It was a little less than 19° from the sun on the evening of August 18, and Flaugergues and Olbers were independently searching for the comet shortly after sunset. Olbers was unsuccessful, but Flaugergues was able to spot it very close to the horizon. The comet was then 2.03 AU from Earth and 1.12 AU from the sun. On August 21, Olbers made another attempt to see it in the evening, but was again unsuccessful, adding that his "horizon was not widely free enough"; however, just a few hours later on the morning of the 22nd he found the comet very near the horizon, situated near 20 Leo Minoris and 21° from the sun. Olbers said the comet was visible before 20 Leo Minoris (magnitude 5.36 – SC2000.0) and was visible at about the same time as 46 Leo Minoris (magnitude 3.83 – SC2000.0). He added that the nebulosity "brightened toward the middle, but haze and twilight prevented me from distinguishing if it exhibited a nucleus and also something of a tail."

Several independent recoveries were made. J. E. Bode (Berlin, Germany) found the comet with a telescope on the evening of August 22. It was then in the north-northwest and was bright enough to be seen for a short time before it went below the horizon. A few hours later, on the morning of the 23rd, he saw the comet after it had risen above the horizon. It then appeared brighter to the naked eye. Bode also became the first person to detect the comet's tail on this morning and he simply described it as short. F. W. Bessel (Königsberg, now Kaliningrad, Russia) recovered the comet on the evenings of August 22 and 23. He gave some interesting details about the comet in letters to the *Berliner Astronomisches Jahrbuch* (dated 1811 August 26) and the *Monatliche Correspondenz* (dated 1811 August 29). He said a Dollond telescope of 7-foot focal length failed to show a nucleus on the 23rd, but did reveal a very compact coma that allowed the comet to be seen with the naked eye without much trouble despite an altitude of under 4°. An independent recovery was also made on the evening of August 23 by J. S. Huth (Frankfurt an der Oder, Germany).

Olbers obtained a good look at the tail with a cometseeker during the last days of August. On the evening of the 28th he saw two rays which he said "formed a parabola, or even a hyperbola." They were separated by an angle of 80–85° and each extended 30–40'. On the 29th he saw a more distinct tail that was broad and 3° long. He added that he still could not distinguish a nucleus.

As the comet began clearing evening twilight, its full splendor was seen by many for the first time. Alexander Ross (a member of the John Jacob Astor expedition traveling down the Columbia River in Oregon, USA) saw the comet on September 1. He "observed, for the first time, about 20 degrees above the horizon, and almost due west, a very brilliant comet, with a tail about 10° long. The Indians at once said it was placed there by the Good Spirit – which they called Skom-malt-squisses – to announce to them the glad tidings of our arrival; and the omen impressed them with a reverential awe for us, implying that we had been sent to them by the Good Spirit, or Great Mother of Life."

The moon was full on September 2, but W. Herschel (Glasgow, Scotland) was still able to observe the comet with a 14-foot focal length reflector. Although the tail was not visible because of moonlight, hazy skies, and the comet's low altitude, Herschel still noted the comet was "like a very brilliant nebula, gradually brighter in a large place about the middle." On September 6, Matthieu (Paris, France) said the nucleus was bright, while the tail was 5–6° long and divided into two close branches. On the 8th S. Perkins (Liverpool, Nova Scotia, Canada) wrote, "at Evening I observe a Comet or Some New appearance of a Star that has an appearance of a Light tail or Blaze it was Nearly in the N.N.W. about one Hour high at 8 o'clock and Set further Northward about – . there was a thin Cloud or haize about it So that I could Not discern the Body of the Star by the Naked Eye but I looked with a Glass and Saw it and an appearance of Light but could not discern any tail or Blaze. It has been observed by Several people for two or three Evenings past." On September 9, Herschel (Alnwick, England) saw the comet with a refractor at a magnification of 65× and noted, "the planetary disk-like appearance seen with the naked eye was transformed into a bright cometic nebula, in which, with this power, no nucleus could be perceived." He estimated the conspicuous tail as 9° or 10° long and noted a "very considerable" curvature. On September 13, Perkins and his daughters "were up at 3 o'Clock to observe a remarkable Star which they had been told rose towards morning they say it had the same appearance as that which is Seen in the Evening and as the motion of that when we See it in the Evening as it is Setting is to the Eastward. I conclude it is the Same it Sets by Nine So it is 5 or 6 Hours under the Horison." On September 17 Bode estimated the tail to be 10° long in his Dollond telescope of 3.5-foot focal length, and by September 20, he said it was over 10° long.

Herschel was back in Glasgow on September 18 and obtained several detailed observations of the comet up to the end of the month. With a reflector of 10-foot focal length, he noted on the evening of the 18th that the stellar head took on the appearance of a globular nebula when viewed at 110×. He estimated that its diameter was about 5′ or 6′, "of which one or two minutes about the centre were nearly of equal brightness." He added that the tail was 11° or 12° long and remarked "that towards the end of the tail its curvature had the appearance as if, with respect to the motion of the comet, that part of the tail were left a little behind the head." In addition, "The appearance of the nebulosity ... perfectly resembled the milky nebulosity of the nebula in the constellation of Orion, in places where the brightness of the one was equal to that of the other." With a night glass with a field of view of 4° 41′, Herschel noted the tail was accompanied by a stream on each side. He noted "that the two streams or branches arising from the sides of the head scattered a considerable portion of their light as they proceeded towards the end of the tail, and were at last so much diluted that the whole of the farthest part of the tail, contained only scattered light." During the first half of the nineteenth century, Juan Pío Pérez (Yucatan, Mexico) included a note in the *Codex Pérez* which stated a comet was seen in the northeast on September 18. It was referred to as "God's sign." On September 20 and 23, F. T. Schubert (St Petersburg, Russia) saw the comet with the naked eye through breaks in the clouds. On September 29, Herschel observed with a reflector of 10-foot focal length and noted the head was 3′ 00″ across.

The comet reached its most northerly apparent declination of +49.5° on October 3, and G. Piazzi (Palermo, Italy) said the "nucleus" was 2′ 30″ across, although it seems more likely that this was a bright inner portion of the coma. On October 6, Herschel observed with a reflector of 20-foot focal length and noted the head was 3′ 45″ across, while a fainter outer coma was estimated as 15′ across. He added that the tail was about 25° long. On October 8, Bode found the tail was 12° long in his telescope. On October 11, Olbers said the tail was 12° 51′ long. On October 12, Herschel estimated that the tail was 17° long. He added, "its breadth in the broadest part was 6 3/4 degrees, and about 5 or 6 degrees from the head it began to be a little contracted." Herschel observed with his night glass and remarked "that the two streams remained sufficiently condensed in their diverging course to be distinguished for a length of about six degrees, after which their scattered light began to be pretty equally spread over the tail." On October 13, Olbers measured the tail as 12° 28′ long. On October 14, Herschel estimated the tail length as 17.5° and Bode said the tail extended to η Draconis, which is about 17°. On October 15, Herschel commented, "in a very clear atmosphere, I found the tail to cover a space of 23 1/2 degrees in length." He added that his night glass showed the preceding branch of the tail was 7° 01′ long, while the following was only 4° 41′ long.

On October 16 Herschel noted a well-defined luminous point in the center of the coma and measured its diameter as 0.79″. He added, "that part of the head which was towards the sun was a little brighter and broader than that towards the tail, so that the planetary disk or point was a little eccentric." On October 17, Herschel found the bright point within the coma to have been "a little beyond the centre." He added that "the tail appeared to be more curved than it had been at any time before." On October 18, Piazzi measured the "nucleus" as 2′ 15″ in diameter. On October 19, Herschel examined the comet with a reflector of 10-foot focal length. At a magnification of 169×, he noted the bright point within the coma was 1.39″ across. At 600×, he estimated it was between 0.68″ and 1.06″ across. Bode said the tail extended to μ Draconis, which amounted to about 14°. On the 30th, Piazzi measured the "nucleus" as 2′ 08″ across. The comet reached a maximum solar elongation of 67° on October 31, and Schubert said the comet was distinctly seen with a tail 5° long.

On November 3, Herschel observed with his night glass and noted, "The two branches were nearly of an equal length." On November 4, W. J. Burchell, near the Vaal River about 50 miles west of present day Kimberley, South Africa, wrote, "as I lay waiting for sleep, and amusing myself in observing the constellations above my head, I noticed a faint nebulous star of the third magnitude, which I had not been used to see in that part of the heavens. Looking at it more attentively, it appeared plainly to be a comet." Herschel found the nucleus "more eccentric than I had ever seen it before" and showed a slight disk in the reflector of 10-foot focal length with a magnification of 289×. On November 5, Herschel estimated the tail was not longer than 12.5°. He added that the preceding stream was 5° 16′ long, while the following was 4° 41′ long. Bode estimated the tail as 10° in length. On November 9, Herschel noted, "The two branches might still be seen to extend full 4°, but their light was much scattered." He added, "The tail of the comet being very near the milky-way, the appearance of the one compared to that of the other, in places where no stars can be seen in the milky-way, was perfectly alike." He estimated the tail's length as 10°. In the reflector of 10-foot focal length, Herschel saw the nucleus "imperfectly" with a magnification of 169×, but "it was more visible" with a magnification of 240×; however, "the nebulosity of the envelope overpowered its light already so much that no good observations could be made of it." On November 10, Herschel obtained only a glimpse of the nucleus in a reflector of 10-foot focal length and noted it was as eccentrically placed as on the 4th. He added that the preceding branch was 5° 16′ long, while the following one was 3° 31′ long. On November 13, Herschel could no longer see the nucleus. He did notice that the following stream was now longer and 4° 06′ in length, while the preceding stream was 3 degrees 31 minutes long. Piazzi measured the "nucleus" as 2′ 15″ across. On November 14, Herschel found both streams equal in length and 3° 31′ long. On November 15, the comet entered Aquila. Herschel noted the following stream was 4° 06′ long, while the preceding one was 3° 31′ long. On the 16th Herschel noted the tail was about 7.5° long to the naked eye and found the following stream 3° 48′ long, while the preceding one was 3° 13′ long. On November 19, Herschel found the two streams to be of equal length and 4° 23′ long. The tail was estimated as 6° 10′ long.

On December 2, Herschel noted the tail was "hardly 5 degrees long and of a very feeble light." He said the streams were both 3° 12′ long. He added, "they joined more to the sides than the vertex, and had lost their former vivid appearance; their colour being changed into that of scattered light." The comet passed less than one-half degree from Altair on December 3. On December 4, Bode observed the comet with his Dollond telescope of 3.5-foot focal length and said the comet was "noticeably smaller with the coma seeming more diffuse." He estimated the tail as 5° long. On December 9, Herschel wrote that the tail length had changed little since the 2nd. He noted, "The branches were already so much scattered that observations of them could no longer be made with any accuracy." Piazzi again said he saw the "nucleus," and gave its diameter as 2'. As in October and November, this was probably an inner coma. On December 14, Herschel wrote that the tail "still remained as before, but the end of it was much fainter."

As 1812 began the comet was moving slowly southeastward through Aquarius some 37° from the sun. On January 2, Herschel commented that the comet "could only be distinguished from a bright globular nebula by the scattered light of its tail, which was still 2° 20′ long." Ferrer determined positions of the comet on six evenings during the period of January 5–10, and noted, "the sky was very clear, but the light of the comet was so weak that it could scarcely be distinguished with the naked eye." He also pointed out that on January 8, the comet was first seen when its altitude was 16° or 17°, and was last seen when its altitude was only 5°. B. Oriani (Milan, Italy) determined positions on January 7 and 10. On January 8, J. H. Fritsch (Quedlinburg, Germany) described the comet as very bright, with a tail 3° long. Zach's last sighting came on January 11.76, when he was able to make only a semi-precise determination of the comet's position. The comet was then 29° from the sun.

The comet's solar elongation decreased as January continued, dropping to 25° by the 17th and 20° by the 24th. The elongation had decreased to 15° as February began and had dropped to 10° by the 12th. On February 17 the comet passed only 9.5° from the sun, and the elongation began to increase thereafter.

Ferrer began looking for the comet again in early July. He used the refractor of 4.5-foot focal length, "but I could not discover it on account of the little light it had at that time." However, while using a 10-cm refractor on July 11.31, Ferrer spotted the comet with a magnification of only 5×. The subsequent field of view was given as 5°. Ferrer wrote "some stars of the 10th and 12th magnitude surrounded" the comet. He added, "the extremity of its nucleus was in contact with one of these stars, and its centre 2′ towards the south, and in the same right ascension." He continued, "The comet appeared as a very slight vapour, its tail opposed to the sun scarcely looked 10 minutes in length." The comet was again observed by Ferrer on July 13 and July 14, but he was not able to determine an accurate position. He even tried using a 30-cm "repeating-circle," but whenever the threads were illuminated, the comet would disappear. Ferrer last saw the comet on July 15.31, and noted it was "in contact with a star of 10th magnitude."

V. Wisniewski (Novocherkassk, Russia) found the comet with a 3.5-foot focal length refractor and described it as faint and blurred, with a coma scarcely 1.5′ across. No tail was seen. He added that it appeared yellowish. On August 11 Wisniewski observed under not so clear skies with his telescope of 3.5-foot focal length and described the comet as extremely faint. On August 12 Wisniewski said the sky was clearer than on the previous night and noted the comet was consequently more distinctly seen. It was about 1′ across. He added, "The comet had scarcely the brightness of an

11th-magnitude star." On August 15 Wisniewski said the sky was not very clear and the comet was subsequently extremely faint.

The comet was last detected on 1812 August 17.97, when Wisniewski said a strong wind was shaking the telescope and the comet could hardly be seen.

J. K. Burckhardt computed the first orbit for this comet. Using three positions obtained between March 26 and April 19, he determined a parabolic orbit with a perihelion date of 1811 September 22.26.

Burckhardt computed a new orbit during June. Although still parabolic, it indicated the comet would pass perihelion on September 15.91. From this orbit, H. W. M. Olbers (Bremen, Germany) noted the comet would become a very bright object during October.

The first elliptical orbit was computed by Flaugergues during October. He determined a period of 510 years and suggested this was a return of the comet seen by the Chinese in September of 1301. Meanwhile, an excerpt of a letter written by Bessel on October 20 gave details of Bessel's determination of an elliptical orbit with a period of 3383 years.

Several astronomers took this time to refine the orbit and provide more reliable ephemerides to help in the comet's recovery later in the year. Bouvard determined a parabolic orbit during March which had a perihelion date of September 12.81. His ephemeris covered the period of April 12 to December 19. Piazzi took some of the positions he had measured during the period of September 9 to January 8 and calculated a parabolic orbit with a perihelion date of September 12.88. Noting that some positions deviated from positions predicted by this orbit, he worked to adapt an elliptical orbit and found that one with a period of 2620 years satisfied the measured positions better than a parabola did.

During March 1812, Ferrer took positions he had determined during the period of May 21 to January 8, and computed an elliptical orbit with a period of 3757 years. He wrote that the comet would arrive at opposition at the beginning of August when the distance from Earth would decrease to 3.14 AU. Ferrer pointed out that on January 8 the comet had been situated 2.86 AU from Earth, so that "it can be scarcely doubted therefore, that it will be visible in its opposition, and in the meridian." He computed an ephemeris for the period of June 1 to August 25. Francisco Lemaur also used Ferrer's observations to compute an elliptical orbit, which Ferrer later remarked gave "a closer approximation between the observed and calculated places than mine." The perihelion date was determined as September 12.

A. Conti (1813) took the available positions and calculated an elliptical orbit with a perihelion date of September 12.74 and a period of about 3056 years.

During 1825, F. W. A. Argelander evaluated the observations at hand and computed an elliptical orbit with a period of 3065 years, but even he did not have the benefit of Ferrer's observations, which were not published in their final form until 1829.

The comet's orbit was finally reexamined when Norbert Herz (1893) used 984 positions obtained between 1811 March 31 and 1812 August 17, as well as perturbations by two planets, and computed an elliptical orbit with a perihelion date of September 12.76, and a period of about 3094 years.

In the years that followed, numerous people looked back on this comet.

Ferrer wrote, "I used all attention to discover the nucleus of this comet" with a refractor of 4.5-foot focal length, "yet never could perceive more than a luminous point from time to time, which can no how be supposed to arise from defect of clearness of sky in the Isle of Cuba." He concluded "it is beyond a doubt that the diameters of these bodies [referring to comets C/1807 R1, C/1811 F1, and C/1813 G1] are exceedingly small, and we much fear therefore that the greater part of those who have observed them have confounded the nucleus with the nebula...." Ferrer specifically noted Herschel's observation of October 16 and wrote "not to mention the difficulty of measuring such small quantities, radiation must augment considerably the luminous disc."

J. R. Hind (1857) wrote, "The finest comets which have been observed during the present century are those of 1811 and 1843. The former one was more remarkable for its brilliancy and the length of time it continued visible, than for the apparent extent of the tail; indeed, we have frequently met with eye-witnesses of that comet who have no recollection of any vestige of a tail."

W. H. Smyth (1858) compared this comet to comet C/1858 L1 (Donati) and said of this comet, "As a mere *sight*-object, the branched tail was of greater interest, the nucleus with its 'head-veil' was more distinct, and its circumpolarity was a fortunate incident for gazers."

Even the French writer Jules Verne (1878) knew of this comet as he wrote, "The great comet of 1811...has caused the year of its appearance to be familiarly recognized as 'the comet-year'...."

T	ω	Ω (2000.0)	i	q	e
1811 Sep. 12.7562 (UT)	65.4097	143.0497	106.9342	1.035412	0.995125

ABSOLUTE MAGNITUDE: $H_{10} = 0.0$ (V1964)

FULL MOON: Mar. 10, Apr. 8, May 8, Jun. 6, Jul. 6, Aug. 4, Sep. 2, Oct. 2, Oct. 31, Nov. 30, Dec. 29, Jan. 28, Feb. 27, Mar. 27, Apr. 26, May 26, Jun. 24, Jul. 24, Aug. 22

SOURCES: H. W. M. Olbers, *BAJ for 1814*, **39** (1811), pp. 242–6; C. F. Gauss, *BAJ for 1814*, **39** (1811), pp. 254–7; F. W. Bessel, *BAJ for 1814*, **39** (1811), pp. 257–9; J. E. Bode, *BAJ for 1814*, **39** (1811), pp. 262–4; J. L. Pons, *MC*, **23** (1811 Apr.), p. 422; H. Flaugergues and J. K. Burckhardt, *MC*, **23** (1811 Jun.), pp. 598–602; J. K. Burckhardt, *MC*, **24** (1811 Jul.), pp. 93–4; H. W. M. Olbers and J. K. Burckhardt, *MC*, **24** (1811 Jul.), pp. 95–7; C. F. Gauss, *MC*, **24** (1811 Aug.), pp. 180–2; F. X. von Zach, *MC*, **24** (1811 Aug.), p. 191; *Gentleman's Magazine*, **81** (1811 Sep.), p. 280; H. W. M. Olbers, F. W. Bessel, J. K. Burckhardt, F. X. von Zach, Harding, and B. Oriani, *MC*, **24** (1811 Sep.), pp. 289–318; C. F. Gauss, J. K. Burckhardt, H. W. M. Olbers, F. W. Bessel, and F. T. Schubert, *MC*,

24 (1811 Oct.), pp. 406–23; H. Flaugergues, F. W. Bessel, and B. Oriani, *MC*, **24** (1811 Nov.), pp. 507–22; *Louisiana Gazette*, St. Louis ed. (1811 Nov. 2); J. L. Pons, F. X. von Zach, and C. F. Werner, *MC*, **24** (1811 Dec.), pp. 525–56; J. S. Huth, *BAJ for 1815*, **40** (1812), pp. 104–8; F. W. Bessel, *BAJ for 1815*, **40** (1812), pp. 111–15; J. E. Bode, *BAJ for 1815*, **40** (1812), pp. 115–17; H. W. M. Olbers, *BAJ for 1815*, **40** (1812), pp. 119–20; F. v. P. Triesnecker and J. T. Bürg, *BAJ for 1815*, **40** (1812), pp. 129–32, 136–7; J. B. Sniadecki, *BAJ for 1815*, **40** (1812), p. 142; J. E. Bode, *BAJ for 1815*, **40** (1812), pp. 167–71; M. A. David and A. Bittner, *BAJ for 1815*, **40** (1812), pp. 186–90; J. E. Bode, *BAJ for 1815*, **40** (1812), pp. 205–8; F. W. Bessel, *BAJ for 1815*, **40** (1812), pp. 254–5; F. T. Schubert, *BAJ for 1815*, **40** (1812), p. 260; J. H. Fritsch, *BAJ for 1815*, **40** (1812), p. 261; Eule, *BAJ for 1815*, **40** (1812), p. 273; W. Herschel, *PTRSL*, **102** (1812), pp. 115–43; W. Herschel, *PTRSL*, **102** (1812), p. 232; H. W. M. Olbers, *MC*, **25** (1812 Jan.), pp. 3–22; F. X. von Zach, *MC*, **25** (1812 Jan.), pp. 85–8; B. Oriani, *MC*, **25** (1812 Jan.), p. 93; H. W. M. Olbers, *MC*, **25** (1812 Jan.), pp. 98–9; F. X. von Zach, *MC*, **25** (1812 Feb.), pp. 183–92; F. T. Schubert, *MC*, **25** (1812 Feb.), pp. 201–5; F. W. Bessel, F. T. Schubert, and H. W. M. Olbers, *MC*, **25** (1812 Mar.), pp. 283–90; A. Bouvard, *MC*, **25** (1812 Apr.), pp. 380–1; M. A. David and A. Bittner, *MC*, **25** (1812 Apr.), pp. 382–8; T. Bugge, *BAJ for 1816*, **41** (1813), p. 126; T. Derfflinger, *BAJ for 1816*, **41** (1813), p. 177; W. Herschel, *BAJ for 1816*, **41** (1813), pp. 185–202; G. Piazzi, *BAJ for 1816*, **41** (1813), pp. 214–16; F. W. Bessel, *BAJ for 1816*, **41** (1813), pp. 234–8; V. Wisniewski, *BAJ for 1816*, **41** (1813), pp. 261–5; Matthieu, *MC*, **27** (1813 Mar.), pp. 299–301; G. Piazzi, *MC*, **27** (1813 Apr.), pp. 356–65; F. X. von Zach and A. Conti, *MC*, **28** (1813 Jul.), pp. 24–32; F. W. Bessel, *MC*, **28** (1813 Jul.), pp. 92–6; W. Herschel, *MC*, **28** (1813 Nov.), pp. 455–69; W. Herschel, *MC*, **28** (1813 Dec.), pp. 558–68; T. Derfflinger, *BAJ for 1817*, **42** (1814), p. 149; F. T. Schubert, *BAJ for 1818*, **43** (1815), pp. 159–63; F. Pond, *BAJ for 1820*, **45** (1817), pp. 197–8; *ZA*, **3** (1817), pp. 221–3; W. J. Burchell, *Travels in the Interior of Southern Africa*. London: Longman, Hurst, Rees, Orme, and Brown (1822), pp. 159–60, 431; J. J. de Ferrer and F. Lemaur, *MAS*, **3** (1829), pp. 9–38; J. R. Hind, *The Comet of 1556*. London: John W. Parker and Son (1857), p. 55; W. H. Smyth, *MNRAS*, **19** (1858 Nov.), p. 27; J. Verne, *Hector Servadac*. Translated by Ellen E. Frewer. London: Sampson Lowe, Marston, Searle & Rivington (1878), p. 216; N. Herz, *The Observatory*, **16** (1893 Feb.), pp. 98–9; W. T. Lynn, *The Observatory*, **16** (1893 Sep.), p. 327; G1894, pp. 34–5, 186–7; O1899, pp. 81–5; A. Ross, "Adventures of the First Settlers on the Oregon or Columbia River," *Early Western Travels: 1748–1846*, Volume 7. Edited by Reuben Gold Thwaites. Ohio: The Arthur H. Clark Company (1904), p. 151; V1964, p. 54; *CCO*, 1st ed. (1972), pp. 15, 41; Peter Lancaster Brown, *Comets, Meteorites, and Men*. New York: Taplinger Publishing Co., Inc. (1974), p. 145; *The Diary of Simeon Perkins*. Edited by Charles Bruce Ferguson. Great Britain: Robert MacLehose and Company Limited (1978), pp. 335–6; *CCO*, 3rd ed. (1979), pp. 15, 44; *The Codex Pérez and The Book of Chilam Balam of Maní*. Translated and edited by Eugene R. Craine and Reginald C. Reindorp. Norman: University of Oklahoma Press (1979), p. 58; SC2000.0, pp. 258, 278; Michael Broadbent, *The Great Vintage Wine Book*. New York: Alfred A. Knopf (1981), pp. 35, 165, 193, 303, 308, 402.

C/1811 W1   *Discovered:* 1811 November 16.99 ($\Delta = 0.74$ AU, $r = 1.58$ AU, Elong. $= 132°$)
(Pons)   *Last seen:* 1812 February 16.99 ($\Delta = 1.51$ AU, $r = 2.04$ AU, Elong. $= 107°$)
  Closest to the Earth: 1811 December 2 (0.7130 AU)
1811 II   *Calculated path:* ERI (Disc), TAU (Dec. 21), AUR (Feb. 9)

J. L. Pons (Marseille, France) discovered this comet in Eridanus on 1811 November 16.96. He described it as a small and very faint spot that vanished when the reticle was illuminated to measure the position. F. X. von Zach (Capellete, France) confirmed Pons' find on November 17.92 and estimated the position as $\alpha = 4^h\ 29.7^m$, $\delta = -25°\ 52'$. J. J. Blanpain (Marseille, France) confirmed Pons' find on November 17.98.

Although the comet was well observed during the remainder of November and throughout December by Zach, B. A. von Lindenau (Seeberg, Germany), H. W. M. Olbers (Bremen, Germany), C. F. Gauss (Göttingen, Germany), and J. K. Burckhardt (Paris, France), actual descriptions are almost non-existent. In fact, Olbers' simple statement that the comet was very faint in hazy skies on December 9 appears to be it. It is probable that the comet was bright enough to be visible in binoculars during this time, as Burckhardt's observations were made with a small meridian circle telescope.

The comet was moving away from both the sun and Earth as the new year began and astronomers were providing more information on the comet's appearance – especially W. Herschel (Slough, England). He observed the comet with several telescopes on January 1 and said it had "a considerable nucleus" surrounded by a faint coma which looked like a "faint haziness, which although of some extent, was not much brighter near the nucleus than at a distance from it." Herschel estimated the diameter of the nucleus as 5″ or 6″ on the 2nd and noted the transition of the light from the nucleus to the coma was so abrupt that it "plainly pointed out that the nucleus and its [coma] were two distinct objects." He added that this comet "seemed to be all nucleus" and that "had it not been for an extremely faint light in a direction opposite to the sun, it would hardly have been entitled to the name of a comet." On the 3rd, Eule (Dresden, Germany) described the telescopic appearance of the comet as a round nebulosity, without a tail, and added that the nucleus was about magnitude 6 or 7. Gauss saw the comet on January 3 and 4 and simply noted it was still rather bright. Herschel's observation on the 8th revealed a "pretty well defined nucleus" with a very faint coma. He examined the nucleus with a magnification of 170× and found it less bright, but "rather better defined." The same magnification caused the coma to be "nearly lost." Herschel saw little change when he observed the comet on the 18th, although he reported the tail was 9′ 40″ long. On January 20, Herschel observed the comet in "uncommonly clear" skies. He remarked, "I saw the body of the comet well defined." He noted that low power revealed a "bright image of the nucleus," while a magnification of 170× "showed the nucleus of a larger diameter, but much less bright, and not so well defined." The tail was not visible because of the moonlight.

Gauss saw the comet on February 2 and described it as very faint. Zach found the comet on the 4th and described it as extremely faint. B. Oriani (Milan, Italy) saw the comet on February 12 and said it appeared very faint.

This comet was last detected on February 16.99, when Olbers measured the position as $\alpha = 4^h\ 51.3^m$, $\delta = +30°\ 30'$.

The first computation of a parabolic orbit should be attributed to C. F. Werner. He took positions obtained on November 21, December 5, and December 19, and determined the perihelion date as 1811 November 9.74. This proved a very good orbit, with later orbits by Werner, F. B. G. Nicolai, and Oriani managing to adjust the perihelion date upward to around November 11–12 during January and February.

The first elliptical orbit was calculated by Nicolai. He used positions obtained from November 18 though February 16, and determined the perihelion date as November 11.49 and the period as about 875 years. This orbit stood for over 90 years, until A. Nekrassow (1909) redetermined it. He took 69 positions, determined nine Normal positions, applied perturbations by Jupiter, and calculated a perihelion date of November 11.54 and a period of about 755 years. This orbit is given below.

T	ω	Ω (2000.0)	i	q	e
1811 Nov. 11.5427 (UT)	314.5022	95.6302	31.2553	1.581701	0.980916

ABSOLUTE MAGNITUDE: $H_{10} = 5.2$ (V1964)

FULL MOON: Oct. 31, Nov. 30, Dec. 29, Jan. 28, Feb. 27

SOURCES: J. L. Pons and F. X. von Zach, *MC*, **24** (1811 Dec.), pp. 551–6; H. W. M. Olbers, J. L. Pons, and J. J. Blanpain, *BAJ for 1815*, **40** (1812), pp. 118–19, 120–1; J. E. Bode, *BAJ for 1815*, **40** (1812), p. 171; C. F. Gauss and F. B. G. Nicolai, *BAJ for 1815*, **40** (1812), pp. 192–3; Eule, *BAJ for 1815*, **40** (1812), p. 273; W. Herschel, *PTRSL*, **102** (1812), pp. 229–37; F. X. von Zach and C. F. Werner, *MC*, **25** (1812 Jan.), pp. 89–92; C. F. Gauss, F. B. G. Nicolai, and B. A. von Lindenau, *MC*, **25** (1812 Jan.), pp. 94–7; H. W. M. Olbers and J. K. Burckhardt, *MC*, **25** (1812 Jan.), pp. 99–100; F. X. von Zach and C. F. Werner, *MC*, **25** (1812 Feb.), pp. 193–200; C. F. Gauss, *MC*, **25** (1812 Feb.), pp. 206–7; F. X. von Zach and C. F. Werner, *MC*, **25** (1812 Mar.), pp. 291–3; B. Oriani, *MC*, **26** (1812 Nov.), pp. 530–1; W. Herschel, *BAJ for 1816*, **41** (1813), pp. 203–8; F. B. G. Nicolai, *MC*, **27** (1813 Mar.), pp. 201–21; O1899, p. 85; A. Nekrassow, *AN*, **182** (1909 Aug. 19), pp. 65–70; V1964, p. 54.

12P/1812 O1 *Discovered:* 1812 July 21.07 ($\Delta = 1.78$ AU, $r = 1.28$ AU, Elong. $= 45°$)
(Pons–Brooks) *Last seen:* 1812 September 28.18 ($\Delta = 1.23$ AU, $r = 0.81$ AU, Elong. $= 41°$)
Closest to the Earth: 1812 September 21 (1.2157 AU)
1812 *Calculated path:* CAM (Disc), LYN (Jul. 21), CNC (Aug. 24), HYA (Sep. 14)

J. L. Pons (Marseille, France) discovered this comet during a casual sweep through the morning sky on 1812 July 21.07. He described it as a small, ill-defined nebulosity, without a tail, and not visible to the naked eye. Pons confirmed this was a comet when he saw it on the morning of the 22nd, and sent word of the discovery to F. X. von Zach (Capellete, France). Zach first observed the comet in the head of Lynx on July 24.10, and measured the position as $\alpha = 6^h\ 12.4^m$, $\delta = +58°\ 34'$. He added that the comet was very difficult to see because of the full moon. Independent discoveries were made by V. Wisniewski on either July 31 or August 1, and by A. Bouvard (Paris, France) on August 2.01.

The comet steadily brightened after the discovery as it approached both the sun and Earth. Bouvard first saw the comet with the naked eye in the morning sky on August 19. He said the nucleus was then bright and enveloped in a coma, while the comet exhibited a tail 1.5–2° long. In discussing the positions he obtained during the period of August 14 to September 1, Zach added, "The comet at present is visible to the naked eye." By the end of the month, however, twilight was beginning to become a factor, as noted by F. v. P. Triesnecker (Vienna, Austria) on the 26th, who said the comet was rather difficult to find, as it had just become visible shortly before morning twilight began. On September 4, Bouvard noted the tail was 3° long, while on the 14th he found it had maintained the same length, but consisted of two parallel branches. Also on the 14th, J. E. Bode said the coma was "considerable" and the faint tail was 1° long.

Bode had received word of this comet's discovery toward the end of August, but the notice did not contain a position or direction of motion. Working only with the knowledge that the comet was faint and in the morning sky, he began searching in the northeast during the final days of August and the beginning of September, but nothing was found. He read in a local newspaper that the comet had been seen by A. Stark (Augsburg, Germany) on August 20 in an asterism known as "Herschel's Telescope." On the morning of September 9, Bode found the comet with a refractor, just 3° from Venus. He said there was a bright coma and a weak trace of tail. Bode also noted the comet was visible to the naked eye.

The comet's brightening trend would have halted by the middle of September as it passed perihelion, but while the comet was still inching closer to Earth for a few days thereafter, it was also moving into morning twilight. Zach measured the tail as 2° 17′ long on September 15 and indicated the nucleus was 1.3′ across. M. A. David and A. Bittner (Prague, Czech Republic) last detected the comet on September 17, with bad weather occurring the next few mornings. When the skies cleared on the 22nd they could find no trace of the comet. Final observations were made by Bode on the 19th, Bouvard on the 22nd, and Triesnecker on the 27th. Zach described the comet as very faint on the 22nd.

This comet was last seen on September 28.18, when Zach measured the position as $\alpha = 9^h\ 38.0^m$, $\delta = -13°\ 21'$. He said the faint comet was seen "regardless of the moon, strong morning dawn, and the haze of the horizon."

The first parabolic orbit was calculated by C. F. Werner, using positions obtained during the period of July 24–29. The resulting perihelion date was 1812 September 14.16. Near the end of September, both Werner and J. N. Nicollet provided orbits using positions spanning the period of July 24 to early September. Werner's orbit gave a perihelion date of September

15.35, while Nicollet's gave it as September 15.49. As further positions were measured, additional orbits were calculated by Bouvard, Triesnecker, and Werner. The resulting perihelion date had then been determined as about September 15.7.

The first elliptical orbit was calculated by J. F. Encke (1816). He determined the perihelion date as September 15.81 and the period as 70.69 years. No further investigation into the orbit of this comet occurred for several decades. L. Schulhof and J. F. Bossert (1882) took the available positions, considered planetary perturbations, and determined the period as 73.18 years. Even though they provided an ephemeris, the two-month observation arc did not allow a precise determination of the period and that of Schulhof and Bossert was nearly six months too long. Nevertheless, the comet was accidentally rediscovered in 1884 and was recovered in 1954. Thereafter, several astronomers were able to use multiple apparitions to firmly determine the orbit for the 1812 apparition. The most recent orbits were calculated by P. Herget and H. J. Carr (1972), D. K. Yeomans (1985, 1986), and K. Kinoshita (2000). These all indicated an orbital period closer to 72.6 years. Kinoshita determined nongravitational terms of $A_1 = -0.20$ and $A_2 = -0.0270$. This orbit is given below.

T	ω	Ω (2000.0)	i	q	e
1812 Sep. 15.8283 (TT)	199.2865	255.6387	73.9561	0.777102	0.955345

ABSOLUTE MAGNITUDE: $H_{10} = 4.2$ (V1964)

FULL MOON: Jun. 24, Jul. 24, Aug. 22, Sep. 20, Oct. 20

SOURCES: J. E. Bode, *BAJ for 1815*, **40** (1812), pp. 259–60; J. L. Pons, *Gentleman's Magazine*, **82** (1812 Aug.), p. 176; J. L. Pons, F. X. von Zach, and C. F. Werner, *MC*, **26** (1812 Sep.), pp. 270–2, 283–4; F. X. von Zach and C. F. Werner, *MC*, **26** (1812 Oct.), pp. 408–9; F. X. von Zach, H. W. M. Olbers, and A. Bouvard, *MC*, **26** (1812 Oct.), pp. 410–11; C. F. Werner and J. N. Nicollet, *MC*, **26** (1812 Nov.), pp. 486–7; F. X. von Zach and C. F. Werner, *MC*, **26** (1812 Dec.), pp. 582–3; F. v. P. Triesnecker, *BAJ for 1816*, **41** (1813), pp. 154–5; M. A. David and A. Bittner, *BAJ for 1816*, **41** (1813), p. 170; A. Bouvard, *BAJ for 1816*, **41** (1813), pp. 238–9; V. Wisniewski, *BAJ for 1816*, **41** (1813), pp. 254–5; A. Bouvard, *MC*, **27** (1813 Mar.), pp. 290–1; M. A. David and A. Bittner, *MC*, **27** (1813 May), p. 488; J. F. Encke, *ZA*, **2** (1816 Nov.–Dec.), pp. 377–403; H. Flaugergues, *CA*, **5** (1820), pp. 549–51; J. N. Nicollet, *CDT* (1820), 418; A. Bouvard, *OAP*, **1** (1825), p. 128; L. Schulhof and J. F. Bossert, *AN*, **103** (1882 Nov. 1), pp. 289–98; O1899, p. 85; V1964, p. 54; P. Herget and H. J. Carr, *QJRAS*, **13** (1972), pp. 428–9, 434; D. K. Yeomans, *QJRAS*, **26** (1985), p. 91; D. K. Yeomans, *QJRAS*, **27** (1986), p. 604; personal correspondence from K. Kinoshita (2000).

C/1813 C1   *Discovered:* 1813 February 4.8 ($\Delta = 0.37$ AU, $r = 0.90$ AU, Elong. $= 66°$)
(Pons)   *Last seen:* 1813 March 11.79 ($\Delta = 1.46$ AU, $r = 0.71$ AU, Elong. $= 26°$)
   Closest to the Earth: 1813 February 1 (0.3376 AU)
1813 I   *Calculated path:* LAC (Disc), AND (Feb. 6), PSC (Feb. 17)

J. L. Pons (Marseille, France) discovered this comet in the evening sky on 1813 February 4.8. He described it as small, without a tail, and noted it was uncondensed and could not withstand lighting. Pons informed F. X. von Zach (Capellete, France) of his new find on the morning of February 5, and Zach confirmed the discovery on February 5.80. He measured the position as $\alpha = 22^h \, 42.8^m$, $\delta = +45°\, 23'$.

Pons, Zach, and A. Bouvard (Paris, France) remained the only observers of this comet and no further physical descriptions were made. The comet was last detected on March 11.79 by Zach. He determined the position as $\alpha = 1^h \, 03.8^m$, $\delta = +5°\, 35'$. The comet was then fading as it moved away from both the sun and Earth. It was also dropping into evening twilight.

This comet was not observed long enough to calculate anything but a parabolic orbit. The first was determined by C. F. Werner using positions obtained at Capellete on February 5, 6, and 7. The resulting perihelion date was 1813 March 6.19. Early in March, Werner took positions obtained on February 8, 12, and 28, and determined a parabolic orbit with a perihelion date of March 5.08. After the final observation by Zach, he took all of that astronomer's positions and determined the perihelion date as March 5.02. This last orbit was basically confirmed by later calculations by J. N. Nicollet (1820), J. Holetschek (1907), and H. A. Peck (1908). Peck said his orbit best represented the observations, but added, "the elements are very uncertain, the probable error of the time of perihelion passage being ±0.03760." No planetary perturbations were applied. This orbit is given below.

T	ω	Ω (2000.0)	i	q	e
1813 Mar. 5.0220 (UT)	170.5629	243.1457	158.8486	0.699314	1.0

ABSOLUTE MAGNITUDE: $H_{10} = 9.0$ (V1964)

FULL MOON: Jan. 16, Feb. 15, Mar. 17

SOURCES: J. E. Bode, J. L. Pons, and A. Bouvard, *BAJ for 1816*, **41** (1813), p. 230; J. L. Pons, F. X. von Zach, and C. F. Werner, *MC*, **27** (1813 Feb.), pp. 194–5; F. X. von Zach and C. F. Werner, *MC*, **27** (1813 Mar.), pp. 282–6; A. Bouvard, *MC*, **27** (1813 Mar.), p. 290; H. W. M. Olbers, *MC*, **27** (1813 Apr.), p. 393; F. X. von Zach and C. F. Werner, *MC*, **27** (1813 Jun.), pp. 568–70; A. Stark, *BAJ for 1818*, **43** (1815), p. 280; J. N. Nicollet, *CDT* (1820), p. 419; *American Almanac*. Boston: James Munroe & Co. (1847), p. 89; J. Holetschek, *AN*, **176** (1907 Oct. 10), pp. 67–70; H. A. Peck, *AJ*, **26** (1908 Apr. 17), pp. 7–10; V1964, p. 54.

X/1813 D1 A. Stark (Augsburg, Germany) discovered this object with his 3.5-foot focal length Dollond refractor on 1813 February 19.78. He described it as a "very small, extremely faint comet, without a tail." It was very near the star Mira, and Stark used that star to give the object's position as $\alpha = 2^h \, 05.2^m$, $\delta = -1°\, 52'$. He reobserved the object on February 20.78 and gave the position as

$\alpha = 2^h\ 15.1^m,\ \delta = -5°\ 49'$. Stark said clouds prevented observations after the 20th. He added that he was surprised that no other astronomer had noticed the comet.

SOURCES: A. Stark, *BAJ for 1818*, **43** (1815), p. 280.

C/1813 G1   *Discovered:* 1813 April 3.1 ($\Delta = 0.87$ AU, $r = 1.41$ AU, Elong. $= 98°$)
(Pons)   *Last seen:* 1813 May 18.06 ($\Delta = 0.58$ AU, $r = 1.22$ AU, Elong. $= 95°$)
   Closest to the Earth: 1813 April 30 (0.2672 AU)
1813 II   *Calculated path:* OPH (Disc), SER (Apr. 14), OPH (Apr. 15), SER (Apr. 20), OPH (Apr. 22), SCO (Apr. 25), LUP (Apr. 27), CEN–LUP (Apr. 30), CEN (May 1), VEL (May 7)

J. L. Pons (Marseille, France) had been hampered by a long stretch of bad weather during March of 1813, but once the skies cleared on March 28 he resumed his comet sweeps. He concentrated on the northern sky, but was fooled by nebulosities on the 28th and 29th. He switched to searching the southern part of the sky in April, and finally discovered a real comet on April 3.1. He noted it was visible to the naked eye, and said a telescope revealed a diffuse coma and a well-condensed nucleus. An independent discovery was made by K. L. Harding (Göttingen, Germany) on April 4.07, and he measured the first position of $\alpha = 18^h\ 09.4^m,\ \delta = +7°\ 35'$. Harding described the comet as small and round, with a bright nucleus, but with no tail.

The comet was heading toward both the sun and Earth, but it was also heading quickly southward. F. X. von Zach (Capellete, France) saw the comet with a telescope on April 8 and noted a star-like nucleus. He added that it was not visible to the naked eye. H. W. M. Olbers (Bremen, Germany) battled moonlight to see the comet around mid-month. He said the comet's light was weakened by moonlight on the 15th and was still faint in moonlight on the 19th. Olbers managed to see the comet before moonrise on the 24th and found it was easily visible to the naked eye. He described its light as similar to a 3rd-magnitude star. The comet's rapid southern motion caused European observations to end before April came to a close, with the final observations being made by Olbers on April 25.98 and Harding on April 26.06. Harding noted the comet was still visible to the naked eye despite the low altitude.

At the lower latitude of Havana, Cuba, J. J. de Ferrer observed the comet on April 30.11. He said the tail was "scarcely 8' long." Ferrer obtained precise measurements of the comet's location on seven evenings during May using a refractor of 4.5-foot focal length. He added that magnifications of $90\times, 200\times$, and $280\times$ revealed "a luminous point in the middle of the nebulosity, which sparkled now and then. If this point could have been seen without interruption, it might, perhaps, have been supposed to have a diameter of a second, but as that was not the case, we must infer that it was much smaller." Ferrer obtained the final observation of this comet on May 18.06 and indicated a position of $\alpha = 8^h\ 59.8^m$, $\delta = -39°\ 52'$. The comet had attained its most southerly declination of $-50°$ on May 5.

The first parabolic orbits were calculated by C. F. Gauss and J. F. Encke using positions obtained on April 8, 10, and 12. Gauss gave the perihelion date as 1813 May 20.83, while Encke gave it as May 20.26. Both orbits quite closely predicted the comet's future motion, with Encke's proving the better of the two. Later orbits by Gauss, Olbers, J. N. Nicollet, Encke, P. Daussy, Ferrer, C. L. Gerling, and C. F. Werner, pushed the perihelion date closer to May 20. Werner's orbit is given below.

T	ω	Ω (2000.0)	i	q	e
1813 May 20.0872 (UT)	205.1622	45.2750	99.0577	1.214703	1.0

ABSOLUTE MAGNITUDE: $H_{10} = 4.4$ (V1964)
FULL MOON: Apr. 15, May 15, Jun. 14
SOURCES: J. E. Bode and K. L. Harding, *BAJ for 1816*, **41** (1813), p. 231; F. X. von Zach, K. L. Harding, J. L. Pons, C. F. Gauss, and J. F. Encke, *MC*, **27** (1813 Apr.), pp. 386–91; F. X. von Zach and C. F. Werner, *MC*, **27** (1813 May), pp. 489–91; C. F. Gauss and J. F. Encke, *MC*, **28** (1813 Jul.), pp. 97–9; K. Burckhardt and P. Daussy, *MC*, **28** (1813 Jul.), pp. 100–2; C. F. Gauss, K. L. Harding, C. L. Gerling, and H. W. M. Olbers, *MC*, **28** (1813 Dec.), pp. 501–13; H. W. M. Olbers, *BAJ for 1817*, **42** (1814), pp. 97–100; J. N. Nicollet, *CDT* (1820), p. 420; J. J. de Ferrer, *MAS*, **3** (1829), pp. 6–9; G1894, pp. 36–7, 188; O1899, pp. 85–6; V1964, p. 54.

13P/1815 E1
(Olbers)

1815

Discovered: 1815 March 6.89 ($\Delta = 1.41$ AU, $r = 1.43$ AU, Elong. $= 70°$)
Last seen: 1815 August 25.85 ($\Delta = 2.36$ AU, $r = 2.10$ AU, Elong. $= 63°$)
Closest to the Earth: 1815 January 7 (1.2656 AU), 1815 May 25 (1.3683 AU)
Calculated path: PER (Disc), AUR (Apr. 13), CAM (Apr. 16), AUR (Apr. 18), CAM (Apr. 24), LYN (May 1), CAM (May 9), UMa (May 17), CVn (Jun. 25), COM (Jul. 10), BOO (Jul. 25), VIR (Aug. 19)

H. W. M. Olbers (Bremen, Germany) discovered this comet on 1815 March 6.89. He simply described it as a small comet and measured the position as $\alpha = 3^h\ 16.5^m$, $\delta = +32°\ 07'$ on March 6.93. He confirmed the discovery and motion on March 7.80 and commented, "The comet goes thus slowly to the north and the east to the body of Perseus. It is small, has a badly defined nucleus, and a very pale transparent coma, and was visible in the cometseeker."

C. F. Gauss (Göttingen, Germany) received word of the discovery on March 13 and briefly observed the comet that evening, but no position was determined. A stretch of bad weather followed and his next observation was not made until March 20, at which time he determined a precise position.

On March 29 and 30, Olbers noted a faint tail-like extension in the direction away from the sun.

On April 2, J. E. Bode (Berlin, Germany) said the comet appeared faint, like a confused nebula. On April 6, Olbers said the tail was 8–10′ long. During the last half of April and the first part of May, Olbers noted that the preceding edge of the nucleus was quite well defined, while a very bright nebulosity followed the nucleus and ran into a short, pale tail. He added that at the beginning of May the nucleus was about 8″ in diameter, while the tail was "very pale, and not over 25 to 30 minutes long." J. H. Fritsch (Quedlinburg, Germany) said the tail was 20′ long on May 3. He also noted a bright nucleus that appeared diffuse in the dense nebulosity. The comet attained its most northerly declination of +62° on May 17. Bode noted the comet "seemed to have increased in light" on May 27.

Having made its closest approaches to both the sun and Earth, the comet faded after May. Bode described it as faint on June 10 and said it was no longer visible in his telescope's finder on July 3. F. v. P. Triesnecker (Vienna, Austria) described the comet as very faint on July 1 and 2. Olbers, F. W. Bessel (Königsberg, now Kaliningrad, Russia), and M. A. David (Prague, Czech Republic) independently made their final observations on July 13, with Olbers noting the comet was "uncommonly faint in a telescope," but suggested a hazy sky and moonlight degraded observing conditions.

Gauss continued to observe the comet after the July full moon, as he measured positions on July 27, 29, and August 4. David tried to find the comet on August 5, but could find no trace with his refractor of 7-foot focal length. The August full moon again interrupted observations, but Gauss obtained one final view on August 25.85, when the position was measured as $\alpha = 14^h\,28.1^m$, $\delta = +5°\,34'$.

Olbers briefly summarized the comet's appearance as exhibiting a nucleus that was nearly always diffuse, but distinct, which allowed good measures of the position. He added that at the end of April, the comet was seen with the naked eye by observers in St Petersburg (Russia) and Dorpat (now Tartu, Estonia).

Gauss used positions he had acquired during the period of March 20–30, and calculated a parabolic orbit on March 31 with a perihelion date of 1815 April 25.17. He added in a letter dated 1815 April 24 and published in the 1815 issue of the *Berliner Astronomisches Jahrbuch*, "I hope the comet will continue to be observed up to July, so that a reliable elliptical orbit can be calculated." Using positions he had obtained during the period of March 6 to April 2, inclusive, Olbers calculated a parabolic orbit with a perihelion date of 1815 April 27.34. Bessel took Olbers' positions from March 6 and 16, as well as his own position from April 1, and calculated a parabolic orbit with a perihelion date of April 25.25. In a letter dated 1815 April 10 (1815a), he provided an ephemeris for the period of April 8 to July 5, indicating the comet would be brightest around the beginning of May. During the next few weeks, very similar parabolic orbits were published by Triesnecker and B. A. von Lindenau.

The first elliptical orbit was calculated by Bessel and he put much effort into trying to pin down the comet's period. His work appeared as a series of letters published in the first volume of the *Zeitschrift für Astronomie* early in 1816 (Bessel, 1816a). Bessel began with positions obtained during the period of March 6 to April 1 and determined a perihelion date of April 26.50 and a period of 73.00 years. He then added a position obtained on June 12 and calculated a perihelion date of April 26.50 and a period of 73.90 years. Bessel's last attempt was discussed in a letter dated 1816 January 10. It used positions spanning the entire period of visibility and determined a perihelion date of April 26.49 and a period of 74.05 years.

Other astronomers worked on the comet's orbit, both while it was being observed and in the months following its final observation. Gauss determined the period as 77.34 years and F. B. G. Nicolai (1816) determined it as 72.56 and 74.79 years. After J. N. Nicollet (1817) published a revised orbit with a period of 72.99 years, several decades passed before astronomers were again interested in this comet. F. K. Ginzel (1882) finally re-examined the orbit of this comet and, using 12 Normal positions, determined a perihelion date of April 26.49 and a period of 73.93 years. He also provided a prediction that the comet would return to perihelion in December of 1886, which would ultimately prove to be ten months early. Later calculations using positions from the comet's 1815, 1887, and 1956 apparitions would reveal that while Ginzel's orbit was generally correct, the period was nearly a year too short. The most recent investigations of this comet's orbit were conducted by H. Q. Rasmusen (1967), D. K. Yeomans (1978, 1986), and K. Kinoshita (2000). Kinoshita's orbit is given below.

T	ω	Ω (2000.0)	i	q	e
1815 Apr. 26.4955 (TT)	65.5907	86.0329	44.4979	1.212 910	0.931710

ABSOLUTE MAGNITUDE: $H_{10} = 4.3$ (V1964), $H_0 = 5.0$, $n = 6$ (Kronk)

FULL MOON: Feb. 23, Mar. 25, Apr. 23, May 23, Jun. 21, Jul. 21, Aug. 19, Sep. 18

SOURCES: H. W. M. Olbers, *BAJ for 1818*, **43** (1815), pp. 152–6; C. F. Gauss, *BAJ for 1818*, **43** (1815), pp. 172–3; F. W. Bessel, *BAJ for 1818*, **43** (1815a), pp. 182–3; F. W. Bessel, *BAJ for 1818*, **43** (1815b), pp. 204–10; F. v. P. Triesnecker, *BAJ for 1818*, **43** (1815), pp. 215–18; H. W. M. Olbers, *BAJ for 1818*, **43** (1815), pp. 218–29; C. F. Gauss, *BAJ for 1818*, **43** (1815), pp. 229–32; B. A. v. Lindenau, *BAJ for 1818*, **43** (1815), pp. 244–6; F. B. G. Nicolai, *BAJ for 1818*, **43** (1815), pp. 264–9; J. H. Fritsch, *BAJ for 1818*, **43** (1815), p. 281; J. E. Bode, H. W. M. Olbers, and F. Pond, *BAJ for 1819*, **44** (1816), pp. 104–9; M. A. David, *BAJ for 1819*, **44** (1816), pp. 157–8; F. G. W. von Struve, *BAJ for 1819*, **44** (1816), pp. 255–6; F. B. G. Nicolai, H. W. M. Olbers, C. F. Gauss, F. W. Bessel, G. Santini, and B. A. von Lindenau, *ZA*, **1** (1816), pp. 283–305; M. A. David, *ZA*, **1** (1816), pp. 339–41; F. W. Bessel, *ZA*, **1** (1816a), pp. 342–50; F. v. P. Triesnecker, *ZA*, **2**

(1816), pp. 320–5; J. N. Nicollet, *ZA*, **4** (1817), pp. 62–3; H. W. M. Olbers and J. N. Nicollet, *CDT* (1820), pp. 420–1; F. K. Ginzel, *VJS*, **17** (1882), pp. 109–14; V1964, p. 54; H. Q. Rasmusen, *The Definitive Orbit of Comet Olbers for the Periods 1815–1887–1956*. Copenhagen: Arnold Busck (1967), p. 52; D. K. Yeomans, *QJRAS*, **19** (1978), pp. 80–1, 88; D. K. Yeomans, *QJRAS*, **27** (1986), p. 604; personal correspondence from K. Kinoshita (2000).

C/1816 B1   *Discovered:* 1816 January 22.8 ($\Delta = 0.48$ AU, $r = 1.22$ AU, Elong. $= 108°$)

(Pons)   *Last seen:* 1816 February 1.82 ($\Delta = 0.40$ AU, $r = 0.99$ AU, Elong. $= 79°$)

   Closest to the Earth: 1816 February 3 (0.3944 AU)

1816   *Calculated path:* UMi (Disc), DRA (Jan. 23), CEP (Jan. 25), LAC (Feb. 2)

This was a very poorly observed comet and the only orbit ever calculated was not published until 36 years after the comet was seen. It was discovered by J. L. Pons (Marseille, France) on 1816 January 22, near the border separating Ursa Minor and Camelopardalis. Pons said he was only able to see the comet on one other occasion. Although the discovery position seems to have not been published in any of the major journals of the time, J. Holetschek (1913) calculated it as $\alpha = 16^h\ 04.0^m$, $\delta = +85°\ 54'$.

The comet was last detected on February 1.82, when it was seen at Paris Observatory at a position of $\alpha = 22^h\ 38.9^m$, $\delta = +59°\ 31'$.

J. E. Bode did not receive word of the discovery until February 17. The notice had said the comet was originally found between "the small bear and the giraffe," but there was no indication of the movement. That evening Bode searched the northern sky with his refractor, but found nothing. Further unsuccessful searches were made on February 21 and 22. He later learned the comet had moved quickly and had left that region by the time he began his searches.

The only orbit ever calculated was by J. K. Burckhardt. He took the positions obtained during the period spanning January 22 to February 1 and determined a parabolic orbit with a perihelion date of 1816 March 1.85. Interestingly, the orbit went unpublished until A. Erman published the correspondence of H. W. M. Olbers and F. W. Bessel in 1852. H. L. d'Arrest (1852) published the details in the 1852 July 6 issue of the *Astronomische Nachrichten*. This orbit is given below.

T	ω	Ω (2000.0)	i	q	e
1816 Mar. 1.8456 (UT)	304.326	325.829	43.111	0.048503	1.0

ABSOLUTE MAGNITUDE: $H_{10} = 8$ (V1964)

FULL MOON: Jan. 15, Feb. 13

SOURCES: J. E. Bode and J. L. Pons, *BAJ for 1820*, **45** (1817), p. 100; H. L. d'Arrest and J. K. Burckhardt, *AJ*, **2** (1852 Jun. 4), p. 131; *Briefwechsel zwischen W. Olbers und F. W. Bessel*, **2**. Edited by A. Erman. Leipzig: Avenarius & Mendelssohn (1852), pp. 32, 434;

H. L. d'Arrest and J. K. Burckhardt, *AN*, **34** (1852 Jul. 6), p. 377; J. Holetschek, *DAWW*, **88** (1913), p. 783; V1964, p. 54.

X/1817 V1 J. E. Bode received a letter from C. F. Scheithauer (Chemnitz, now Karl Marx Stadt, Germany) on November 3 giving details of a comet he found on November 1.73. He simply gave the location as to the left and above κ Ophiuchi.

H. W. M. Olbers (Bremen, Germany) discovered this comet with a comet-seeker on 1817 November 1.75. He noted it was in the west shoulder of Ophiuchus near the star κ. On November 1.77, he determined the position as α = 16^h 52.4^m, δ = +9° 16′, while on November 1.79 he determined it as α = 16^h 52.5^m, δ = +9° 13′. Olbers described the comet as small and brilliant, especially towards the middle, although he was uncertain as to whether there was a nucleus. No tail was visible. Olbers said the night of the 2nd was fairly clear except for the sky around Ophiuchus. He said that although κ Ophiuchi was visible, fainter stars seen the night before were not and the comet was not seen. Olbers said the sky was cloudy until the 5th, and he searched for a few hours on that evening, as well as on the evenings of the 6th and 7th, but could not find the comet. He suggested the comet might have moved into the haze of the southern horizon by the 5th.

Bode received word of Olbers' discovery on November 7. He searched with his refractor that evening but found nothing. Further unsuccessful searches were conducted between 6 and 7 p.m. local time on November 10 and 12.

FULL MOON: Oct. 25, Nov. 23
SOURCES: H. W. M. Olbers, J. E. Bode, and C. F. Scheithauer, *BAJ for 1821*, **46** (1818), pp. 143–4; J. E. Bode, *BAJ for 1821*, **46** (1818), p. 179; H. W. M. Olbers, *Gentleman's Magazine*, **88** (1818 Jan.), p. 63; O1899, pp. 86–7.

C/1817 Y1 *Discovered:* 1817 December 26.9 (Δ = 1.34 AU, *r* = 1.53 AU, Elong. = 80°)
(Pons) *Last seen:* 1818 May 2.01 (Δ = 0.86 AU, *r* = 1.55 AU, Elong. = 112°)
Closest to the Earth: 1817 December 3 (1.2460 AU), 1818 May 18 (0.7691 AU)
1818 II *Calculated path:* CYG (Disc), VUL (Feb. 17), SGE (Mar. 15), AQL (Mar. 31)

J. L. Pons (Marseille, France) began his usual routine of sweeping for comets near the end of evening twilight when he found a "very faint nebula between Cepheus and the left wing of Cygnus" on 1817 December 26.9. He said it did not seem to be a comet, but he decided to check on it later. There was no prominent star nearby, so he noted the object was about 1.5° northwest of a triangle of 8th-magnitude stars. When Pons went to check on the object a few hours later, a hazy sky and its proximity to the horizon prevented him from seeing it. Observing conditions were poor on the evening of the 27th, but the 28th was very clear. Pons found the small triangle but the nebula was no longer visible. He began sweeping for the object and soon found two small, faint nebulae quite similar to the one he had seen on the 26th. Although he believed one of them was his object, a change in the weather put an end to his observing session before motion was detected. The night of the 29th was again clear and Pons visited the position of the two nebulae and found everything as it was on the previous evening; therefore, neither of these objects was the object seen on the 26th. He tracked back to the triangle of stars and again began a sweep. He very quickly found the object, and noted it had not moved as far as he had expected. The date was then December 29.82 and the position was roughly measured as $\alpha = 19^h\ 50.9^m$, $\delta = +52°\ 28'$.

As the new year began, the comet was approaching the sun, but it was also moving away from Earth. Consequently, there was little change in brightness. J. J. Blanpain (Marseille, France) saw the comet on January 4 and described it as very small and very faint, without a nucleus or tail. Pons observed the comet on the evenings of January 7 and 10 and noted, "This nebulous star does not have anything which distinguishes it as a comet. No beard, no tail, no nucleus. It is a small, faint, light spot, which in size and light, at least so for the time I observed it, neither decreased nor increased. It is extremely difficult to find and recognize, and I see it only in my largest cometseeker. It is the faintest of all my comets discovered so far." Blanpain saw the comet on January 19 and noted it was brighter and larger than during previous observations, with traces of a tail.

The comet steadily moved away from Earth until February 13. At that time it was situated 1.59 AU away. Thereafter, the distance between the two bodies decreased. In addition, the comet passed perihelion near the end of the month. As February began Pons said the comet seemed to have brightened in the middle and he added that it appeared to exhibit a trace of a tail on the 6th. Throughout the month Pons said the comet could not be seen in his cometseeker of 2-foot focal length, but was visible, with some difficulty, in his seeker of 4-foot focal length. Blanpain saw the comet on the 14th and noted a strong nucleus, but no tail.

Although the comet continued to approach Earth throughout March and April, its increasing distance from the sun was taking a toll on its brightness. H. W. M. Olbers (Bremen, Germany) said the centrally condensed comet was "extraordinarily pale and weak" on March 4, while it was difficult to see on the 10th. Meanwhile, A. Stark (Augsburg, Germany) independently discovered the comet on March 5 near 22 Vulpeculae. Olbers noted the comet was easier to see on the 14th than on the 10th. Moonlight blocked the comet from view for a few days during the last half of March, but once the moonlight had left the sky B. A. von Lindenau and J. F. Encke (Seeberg, Germany) saw it on the 31st and described it as very weak, with indistinct borders. They observed no nucleus or tail. J. N. Nicollet described the comet as very weak on April 3, while F. B. G. Nicolai (Mannheim, Germany) simply noted it was faint on the 3rd and 5th. Lindenau and Encke saw the comet on April 10 and described it as weak. Olbers continued to follow the comet throughout April. He noted it was easy to see on the 4th, and difficult to see on the 16th and 28th. He noted that moonlight interfered with seeing on the 16th.

The comet was last detected on May 2.01, when Olbers described it as "extraordinarily weak." He determined the position as $\alpha = 19^h 07.0^m$, $\delta = -3° 19'$.

The first parabolic orbit calculated for this comet came from J. N. Nicollet. He took positions obtained at Marseille from late December through February and determined the perihelion date as 1818 March 3.96. C. F. Gauss (1818) took positions from January 19 to March 14 and determined the perihelion date as February 27.37. Olbers (1818a) used similar positions and determined the perihelion date as February 27.90. Encke (1818b) took positions obtained to April 10 and calculated a perihelion date of February 26.47. Encke (1818c) revised this to February 26.46 after Olbers' final observation.

J. Mrazek (1928) determined an orbit that was a better fit to the available positions. He ignored the approximate positions obtained between 1817 December 29 and 1818 February 25, and instead used 26 positions obtained between 1818 March 4 and May 2, as well as perturbations by five planets. The result was a perihelion date of February 26.47. This orbit is given below.

T	ω	Ω (2000.0)	i	q	e
1818 Feb. 26.4690 (UT)	112.3619	72.9950	89.7518	1.197821	1.0

ABSOLUTE MAGNITUDE: $H_{10} = 4.5$–6.8 (V1964), $H_{10} = 5.5$ (Kronk)

FULL MOON: Dec. 23, Jan. 22, Feb. 21, Mar. 22, Apr. 21, May 20

SOURCES: F. X. von Zach and J. L. Pons, *ZA*, **4** (1817 Nov.–Dec.), pp. 483–6; H. W. M. Olbers and J. F. Encke, *BAJ for 1821*, **46** (1818), pp. 145–7; J. F. Encke, J. J. Blanpain, J. N. Nicollet, and H. W. M. Olbers, *BAJ for 1821*, **46** (1818), pp. 158–66; F. B. G. Nicolai, *BAJ for 1821*, **46** (1818), p. 205; J. L. Pons, *BAJ for 1821*, **46** (1818), p. 233; J. L. Pons, H. W. M. Olbers, J. F. Encke, and B. A. von Lindenau, *ZA*, **5** (1818a), pp. 148–50, 152–4; B. A. von Lindenau and J. F. Encke, *ZA*, **5** (1818b), pp. 181–7; J. F. Encke, H. W. M. Olbers, *ZA*, **5** (1818c), pp. 253–5; C. F. Gauss, *ZA*, **5** (1818), pp. 276–7; J. L. Pons, *Gentleman's Magazine*, **88** (1818 Feb.), p. 166; J. L. Pons and A. Stark, *BAJ for 1822*, **47** (1819), pp. 169–70; *PMJ*, **53** (1819 Apr.), pp. 300–1; J. N. Nicollet and J. J. Blanpain, *CDT* (1821), pp. 337–8; O1899, pp. 87–90; J. Mrazek, *AN*, **232** (1928 Mar. 27), pp. 177–90; V1964, p. 54.

27P/1818 D1   *Discovered:* 1818 February 23.78 ($\Delta = 0.69$ AU, $r = 0.82$ AU, Elong. $= 54°$)

(Crommelin)   *Last seen:* 1818 February 27.78 ($\Delta = 0.67$ AU, $r = 0.85$ AU, Elong. $= 58°$)

Closest to the Earth: 1818 March 8 (0.6577 AU)

1818 I   *Calculated path:* CET (Disc), ERI (Feb. 27)

J. L. Pons (Marseille, France) discovered this comet during a routine search for comets on 1818 February 23.78, at a position of $\alpha = 2^h 05.0^m$, $\delta = -15° 15'$. The comet was described as smaller than C/1817 Y1, and invisible to the naked eye. In fact, Pons commented that the comet could not be seen when the measuring device was illuminated. He added that the coma was not very extensive, but was brighter towards the middle. There was no tail.

Pons remained the only observer of this comet and saw it on only three other occasions. On February 24, Pons only caught a glimpse of the comet because of bad weather, but it was enough to establish that it had moved toward the southeast. Pons also saw the comet on February 26, not far from σ Ceti. His final observation of the comet came on February 27.78, when he estimated the position as $\alpha = 2^h 25.2^m$, $\delta = -18° 15'$. Pons remarked that a period of bad weather followed, which lasted nearly two weeks, and the comet was not seen again. He surmised that it had probably dropped below the horizon during that time.

J. F. Encke (1818) tried to determine a parabolic orbit, but every attempt left large errors. He added, "one must remain completely uncertain as to the course of this body." Years later, both N. R. Pogson (1850) and J. R. Hind (1872) were able to produce rough parabolic orbits with perihelion dates of 1818 February 7.90 and February 3.72, respectively. Hind noted that the ascending node (Ω) was close to that of the orbit of Biela's comet. He then took the orbits of Biela's comet for 1772 and 1826 and tried to apply them to the path of the 1818 comet. In each case he noted that the comet was too far from perihelion and too far removed in latitude. Hind finally concluded, "These large differences appear conclusive against the idea of a possible connexion of the first Comet of 1818 and the Comet of Biela." Despite Hind's conclusion, E. Weiss (1873) became intrigued by the orbit of this comet and, in particular, the orbit determined by Pogson. He also mentioned the possibility that this comet might have been part of comet 3D/Biela at one time.

Weiss' involvement with Pons' 1818 comet played a very important role when a comet discovered on 1873 November 10 only remained visible until November 16 (see 27P/1873 V1). Weiss was intrigued by the short period of visibility and decided to see how its parabolic orbit would fit the 1818 positions. He found a very good agreement if the perihelion date was assumed to be February 7.10.

The two comets would eventually be proven identical by A. C. D. Crommelin only after the comet was accidentally discovered for a third time in 1928. For this apparition, Crommelin (1932) determined the perihelion date as February 6.55 and the period as 27.73 years. Later orbits were determined by B. G. Marsden (1973) and K. Kinoshita (2000). Kinoshita's orbit is given below.

T	ω	Ω (2000.0)	i	q	e
1818 Feb. 6.753 (UT)	195.597	252.058	29.171	0.74627	0.91848

ABSOLUTE MAGNITUDE: $H_{10} = 9$ (V1964)

FULL MOON: Jan. 22, Feb. 21, Mar. 22

SOURCES: J. F. Encke and J. L. Pons, *BAJ for 1821*, **46** (1818), p. 166; F. X. von Zach and J. L. Pons, *ZA*, **5** (1818 Jan.–Feb.), pp. 150–1; N. R. Pogson, *MNRAS*, **10** (1850 Apr.), p. 135; J. R. Hind, *MNRAS*, **33** (1872 Nov.), pp. 50–1; E. Weiss, *AN*, **83** (1874 Jan. 9), pp. 5–8; V1964, p. 54; B. G. Marsden, *CCO*, 1st ed. (1972), pp. 15, 41; B. G. Marsden, *QJRAS*, **14** (1973), pp. 404–5; *CCO*, 2nd ed. (1975), pp. 15, 43; personal correspondence from K. Kinoshita (2000).

2P/1818 W1   *Discovered:* 1818 November 26.76 ($\Delta = 0.80$ AU, $r = 1.30$ AU, Elong. $= 93°$)

(Encke)   *Last seen:* 1819 January 12.73 ($\Delta = 0.62$ AU, $r = 0.51$ AU, Elong. $= 27°$)

Closest to the Earth: 1819 January 17 (0.6022 AU)

1819 I   *Calculated path:* PEG (Disc), AQR (Dec. 28)

This cometary apparition proved important in the study of comets. Since the discovery by E. Halley in 1705 that a comet was returning at intervals of roughly 76 years, numerous astronomers had been trying to identify other comets that were returning at regular intervals, but none were successful. After J. F. Encke began investigating the orbit of this comet, he soon realized it was moving in a short-period orbit and had previously been seen in 1786, 1795, and 1805.

J. L. Pons (Marseille, France) discovered this comet in Pegasus on 1818 November 26.76. On November 27.9, he estimated the position as $\alpha = 22^h\ 09.5^m$, $\delta = +8°\ 02'$. He described it as a very faint, shapeless nebulosity, with an ill-defined border.

The comet steadily brightened and increased in size during the following weeks. Observations were made by observers in Marseille, Kremsmünster, Vienna, Mannheim, Göttingen, and Seeberg. Pons reported on December 28 that the comet had increased in brightness, was of an oval shape, and may have been visible to the naked eye. On January 5, J. F. Encke (Seeberg, Germany) saw the comet very near the globular cluster M2 and noted that it "resembled its exterior almost perfectly." The diameter of this cluster is typically given as about 13′.

The comet was last detected on 1819 January 12, with K. L. Harding (Göttingen, Germany) seeing it on January 12.727 and Encke seeing it on January 12.732. The position on the final date was given as $\alpha = 21^h\ 02.4^m$, $\delta = -5°\ 35'$. Bode received word of the comet's discovery on January 16. His skies remained cloudy until January 22, at which time his diligent searches for this comet were in vain.

The first parabolic orbit was published by Encke in the February 1819 issue of the *Correspondance Astronomique*. He took positions obtained on December 22, January 1, and 6, and determined a perihelion date of 1819 January 25.40. The article added that the orbit "resembles a little that of the first comet of 1805; perhaps later calculations will teach us something on their identity." Encke published another article in the same issue of the *Correspondance Astronomique* which contained the first elliptical orbit. He took positions from November 30 to January 12 and determined a perihelion date of January 27.60 and a period of 4.15 years. He compared his orbit for the 1819 comet with the parabolic orbit calculated by Bessel for the 1805 comet and said the differences between the two were within the limits of what planetary perturbations can produce. In the March 1819 issue of the *Correspondance Astronomique*, Encke showed the results of his calculations of an elliptical orbit for the 1805 comet. A comparison of the orbits of the comets of 1805 and 1819 now showed very little difference, and Encke was sure the two were identical. He continued working on the problem and, in May of 1819, Encke published details of his attempts to link the comets of 1805 and 1819 to a comet observed in 1795. During June of 1819, Encke had successfully linked this comet to one seen on only two nights in 1786. Later orbits calculated by F. E. von Asten (1877) and B. G. Marsden and Z. Sekanina (1974) gave a perihelion date of January 27.75 and a period of 3.29 years. Marsden and Sekanina also gave nongravitational terms of $A_1 = +0.69$ and $A_2 = -0.03962$ and their orbit is given below.

T	ω	Ω (2000.0)	i	q	e
1819 Jan. 27.7540 (TT)	182.4134	337.1036	13.6430	0.335087	0.848631

ABSOLUTE MAGNITUDE: $H_{10} = 8$ (V1964)

FULL MOON: Nov. 12, Dec. 12, Jan. 11, Feb. 10

SOURCES: J. L. Pons, *CA*, **1** (1818), pp. 518–19; J. L. Pons, *CA*, **1** (1818), pp. 602–3; J. L. Pons, *Proces-Verbaux Academie des Sciences*, **6** (1818 Dec. 7), p. 388; J. L. Pons, J. T. Bürg, and B. A. von Lindenau, *BAJ for 1822*, **47** (1819), p. 135; Derfflinger, *BAJ for 1822*, **47** (1819), p. 157; J. F. Encke, *BAJ for 1822*, **47** (1819), pp. 180–202; F. Carlini, *CA*, **2** (1819), p. 110; F. W. Bessel, F. B. G. Nicolai, K. L. Harding, and J. F. Encke, *CA*, **2** (1819), pp. 187–90, 206–8; J. F. Encke, *CA*, **2** (1819), pp. 305–7; *PMJ*, **53** (1819 Apr.), pp. 300–1; J. F. Encke, *CA*, **2** (1819), pp. 496–506; J. F. Encke, *CA*, **2** (1819), pp. 600–10; J. E. Bode, *BAJ for 1823*, **48** (1820), p. 154; J. N. Nicollet, *CDT* (1822), p. 349; J. F. Encke, *CDT* (1823), p. 317; F. E. von Asten, *BASP* (Series 4), **22** (1877), p. 559; F. L. Whipple, *Astrophysical Journal*, **111** (1950), pp. 375–94; V1964, p. 54; B. G. Marsden, *AJ*, **73** (1968 Jun.), pp. 367–79; B. G. Marsden and Z. Sekanina, *AJ*, **79** (1974 Mar.), pp. 415–16.

C/1818 W2 *Discovered:* 1818 November 28.19 ($\Delta = 0.67$ AU, $r = 0.87$ AU, Elong. $= 59°$)
(Pons) *Last seen:* 1819 January 30.76 ($\Delta = 1.55$ AU, $r = 1.32$ AU, Elong. $= 57°$)
  Closest to the Earth: 1818 December 15 (0.1525 AU)
1818 III *Calculated path:* HYA (Disc), VIR (Dec. 8), LIB–VIR (Dec. 12), LIB (Dec. 13), VIR–SER (Dec. 14), HER (Dec. 15), LYR (Dec. 18), CYG (Dec. 20), LAC (Jan. 5)

J. L. Pons (Marseille, France) discovered this comet near β Hydra on 1818 November 28.19. He confirmed his find on November 30.23, and estimated the comet's position as $\alpha = 11^h 55.0^m$, $\delta = -29° 45'$. Pons described the comet as a round, pale nebulosity, with a diameter of 5' or 6'. Pons obtained additional positions on December 1 and 2. This was the third comet Pons had found in 1818 and he received a medal from Lalande as a reward.

Bad weather followed, but Pons apparently caught sight of the comet again on the mornings of December 14 and 19. No positions were obtained on either occasion, although he did note it was in Lyra on the 19th and moving quickly. With the comet now at a high northern declination, Pons was also able to see the comet on the evening of the 19th and described it as round and barely visible to the naked eye.

Other observatories were not immediately successful at finding the comet. One reason was poor communication, as only a handful of observatories even received word of Pons' discovery and several of those reported long periods of bad weather. Another reason was because the comet had unexpectedly quickened its pace across the sky, because, as was later discovered, it passed very close to Earth on December 15. This rapid motion resulted in the comet moving far from the earlier announced discovery positions.

News of a new comet began circulating after mid-December. C. F. Scheithauer (Chemnitz, now Karl Marx Stadt, Germany) found a comet on December 20.76 and F. W. Bessel (Königsberg, now Kaliningrad, Russia) found one on December 22.74. Scheithauer described the comet as "a large nebula, without a nucleus." Bessel simply described it as small. Bessel continued providing positions through the remainder of December and well into January.

J. E. Bode initially believed Scheithauer and Bessel had discovered a comet other than that seen by Pons because it was situated more than $120°$ from the location Pons' comet had been at near the beginning of December. He received news of Scheithauer's discovery on December 27, and searched the area of Cygnus and Lyra that evening, but nothing was found. He received news of Bessel's find on December 30 and assumed it was the same object as that seen by Scheithauer, but again believed it different than that seen by Pons. He said bad weather prevented his finding the comet on the evenings of December 30 and 31.

As the new year began, the comet continued being observed, but it was steadily fading. F. Carlini (Milan, Italy) tried to find it on January 16, but was unsuccessful. Bessel described the comet as faint when he last saw it on the 27th. The comet was last detected on January 30.76 by K. L. Harding (Göttingen, Germany) while using the 22.9-cm Herschel reflector. He described it as "quite faint and very difficult." The position was determined as $\alpha = 22^h\ 25.1^m$, $\delta = +35°\ 23'$.

The first orbit was calculated by Carlini (1818). He said Milan astronomers had been clouded out since the announcement of this comet's discovery was received and their first attempt to find the comet came almost one month after its discovery. To search for the comet he took Pons' positions of November 30, December 1, and 2, and calculated a parabolic orbit with a perihelion date of 1818 December 29. A much more accurate orbit was calculated by Carlini at the beginning of January, when he used positions from November 30 to December 27 and determined the perihelion date as December 5.72. This was quite close to the later calculations of Bessel (1819), J. N. Nicollet (1822), and himself. O. A. Rosenberger and H. F. Scherk (1821) used positions obtained between 1818 November 30 and 1819 January 30, and computed a parabolic orbit with a perihelion date of 1818 December 5.43. No planetary perturbations were applied. The same positions were also used to determine a hyperbolic orbit with a perihelion date of December 5.53, and an eccentricity of 1.011617. Because of the short period of visibility, the parabolic orbit of Rosenberger and Scherk is given below.

T	ω	Ω (2000.0)	i	q	e
1818 Dec. 5.4347 (UT)	348.1071	92.5372	116.9058	0.855096	1.0

ABSOLUTE MAGNITUDE: $H_{10} = 8$ (V1964)

FULL MOON: Nov. 12, Dec. 12, Jan. 11, Feb. 10

SOURCES: J. L. Pons, *CA*, **1** (1818), pp. 518–19; F. Carlini, *CA*, **1** (1818), pp. 592–3; J. L. Pons and F. Carlini, *CA*, **1** (1818), pp. 601–3; J. E. Bode, C. F. Scheithauer, and F. W. Bessel, *BAJ for 1822*, **47** (1819), p. 170; F. W. Bessel, *BAJ for 1822*, **47** (1819), pp. 171–2; C. F. Scheithauer, *BAJ for 1822*, **47** (1819), p. 248; F. Carlini, *CA*, **2** (1819), pp. 106–10; F. Carlini and F. W. Bessel, *CA*, **2** (1819), pp. 186–7; *PMJ*, **53** (1819 Jan.), p. 74; *PMJ*, **53** (1819 Apr.), pp. 300–1; F. W. Bessel, *BAJ for 1823*, **48** (1820), p. 154; F. W. Bessel, O. A. Rosenberg, H. F. Scherk, and K. L. Harding, *BAJ for 1824*, **49** (1821), pp. 141–5; J. N. Nicollet, *CDT* (1822), p. 349; V1964, p. 54.

7P/1819 L1
(Pons–Winnecke)

1819 III

Discovered: 1819 June 12.92 ($\Delta = 0.79$ AU, $r = 0.98$ AU, Elong. $= 64°$)
Last seen: 1819 July 19.84 ($\Delta = 0.40$ AU, $r = 0.77$ AU, Elong. $= 43°$)
Closest to the Earth: 1819 August 21 (0.1317 AU)
Calculated path: LEO (Disc)

J. L. Pons (Marseille, France) discovered this comet on 1819 June 12.92, in the western sky. He measured the position as $\alpha = 10^h 08.8^m$, $\delta = +25° 23'$ on June 13.95 and said the comet was small and tailless, with only a slight condensation within the faint coma.

Pons said the comet was not a naked-eye object, but it did brighten in the days following discovery. He kept it under observation until June 29, after which time moonlight blocked it from view. After the moon had left the sky, F. Carlini (Brera Observatory, Milan, Italy) picked up the comet on July 14.87. Unfortunately, the comet was moving toward evening twilight and was last seen by Carlini on July 19.84 at a position of $\alpha = 10^h 51.8^m$, $\delta = +14° 01'$.

G. Santini (Padova, Italy) received word of this comet at the beginning of July, but met with cloudy skies on the evenings of July 1 and 2. With clear skies on July 3 he began looking for the comet in Leo, but his colleague F. Bertirossi-Busatta noticed a naked-eye comet toward the area of sunset and they moved the instrument to begin securing positions of it instead (see C/1819 N1).

The first parabolic orbit was calculated by J. F. Encke (1819a). He took positions obtained at Marseille from June 13 to June 29 and determined a perihelion date of 1819 July 19.72. Carlini took all of the positions spanning the period of June 13 to July 19 and determined the perihelion date as July 21.17.

The first elliptical orbit was calculated by Encke (1819b) after he noted residuals amounted to several arc minutes for some positions. His initial orbit utilized the same June positions used for his parabolic orbit and resulted in a perihelion date of July 31.64 and a period of 2.35 years. Encke (1819c) revised his orbit using all of the positions from Marseille and Milan. The resulting perihelion date was July 19.40 and the period was 5.62 years. Unfortunately, with no additional observations at hand, a more precise orbit was not possible and the comet was lost for several apparitions. Upon its accidental rediscovery in 1858, astronomers realized the accepted 1819 orbit had an orbital period that was nearly one month too long. Additional orbits were calculated for this apparition following its rediscovery in 1858 and subsequent recoveries that indicated a perihelion date of July 19.71 and a period of about 5.55 years. K. Kinoshita (2000) determined the comet's non-gravitational terms as $A_1 = +0.70$ and $A_2 = +0.1462$ and his orbit is given below.

T	ω	Ω (2000.0)	i	q	e
1819 Jul. 19.713 (UT)	162.015	115.481	10.748	0.77164	0.75410

ABSOLUTE MAGNITUDE: $H_{10} = 8.8$ (V1964)

FULL MOON: Jun. 8, Jul. 7, Aug. 5

SOURCES: J. F. Encke, *BAJ for 1822*, **47** (1819), p. 207; J. F. Encke, *BAJ for 1822*, **47** (1819a), pp. 243–4; J. L. Pons, *CA*, **2** (1819), p. 519; G. Santini and F. Bertirossi-Busatta, *CA*, **2** (1819), p. 584; J. L. Pons, *CA*, **2** (1819), pp. 611–14; J. F. Encke and F. Carlini, *CA*, **3** (1819b), pp. 195–200; J. F. Encke, *CA*, **3** (1819c), pp. 292–4; J. F. Encke and F. Carlini, *BAJ for 1823*, **48** (1820), pp. 221–2; A. Bouvard, *CDT* (1822), pp. 350–1; T. R. von Oppölzer, *SAWW*, **62** Abt. II (1870), pp. 655–75; J. F. Encke, *AN*, **77** (1871 May 30), p. 313; V1964, p. 54; B. G. Marsden, *AJ*, **75** (1970 Feb.), pp. 80–1; B. G. Marsden, *QJRAS*, **26** (1985 Sep.), p. 309; personal correspondence from K. Kinoshita (2000).

C/1819 N1 (Great Comet or Tralles)

1819 II

Discovered: 1819 July 1.89 ($\Delta = 0.77$ AU, $r = 0.36$ AU, Elong. $= 17°$)

Last seen: 1819 October 25.8 ($\Delta = 2.10$ AU, $r = 2.38$ AU, Elong. $= 94°$)

Closest to the Earth: 1819 June 25 (0.6679 AU)

Calculated path: AUR (Disc), LYN (Jul. 4), UMa (Aug. 9)

This very bright comet holds an important place in the history of the study of comets in that equipment used by D. F. J. Arago on 1819 July 3 represented a significant first step in the understanding of what comets were made of.

The initial discovery was apparently made by J. G. Tralles (Berlin, Germany) in evening twilight, around July 1.89, but it was not until July 1.96 that any measurements were made. At that time he said the comet was 2° 20′ above the northern horizon and described it as very brilliant. Shortly thereafter, independent discoveries were reported by people in Edinburgh, York, and Leeds in the British Isles. The first official confirmation of the comet came on July 2.96 when J. E. Bode (Berlin, Germany) determined its position as $\alpha = 6^h 47.0^m$, $\delta = +41° 56′$. Tralles saw the comet again on the evening of the 2nd and said a telescope revealed a coma 40″ across.

The comet came under intensive observation on July 3. F. G. W. Struve (Dorpat, now Tartu, Estonia) independently discovered it late in the evening. He said the coma was 40″ across and the nucleus was about 8″ across. The tail extended a few degrees. F. Bertirossi-Busatta (Padova, Italy) spotted the comet with the naked eye near the area of sunset, while G. Santini and himself were attempting to observe Pons' comet of June 12 (see 7P/1819 L1). William Burney (Gosport, England) said the comet was first detected with the naked eye in the evening "with a lucid train projecting upwards or from the sun, and nearly in a perpendicular direction." Burney said a refractor revealed, "its body appeared more confused, or had a greater nebulosity, than when seen with the naked eye, perhaps from the dewy haze then descending. Though the brilliancy of moonlight was not favourable to observations, yet the nucleus of the comet appeared of a pale white light, and was sometimes brighter than at others, as was also the tail, which expanded upwards at intervals from 6 deg. to 10 deg. in length by the sextant." He added that the comet was similar in appearance to the great comet of 1811, "but the train is much longer and wider." H. W. M. Olbers (Bremen, Germany) said the nucleus was about magnitude 1–2, and the tail extended about 7–8°. A. Bouvard (Paris, France) gave the tail length as 6° or 7°.

Other observations were given on July 3, but the most historical was that by Arago. Using his recently developed polarimeter, he looked at the comet's tail region. The doubly refracting prism was attached to a small refractor and revealed two tail images of differing intensities. This difference indicated the comet's tail was polarized, which indicated that at least some of its light came from reflected sunlight. To test his idea, he also looked at Capella and noted the two images did not vary in intensity at all.

Additional descriptions of the comet were given during July. From observations obtained up to the 7th, Burney noted, "its head is globular" and "sometimes it appears as small as a star of the second or third magnitude, at others equal to Saturn in apparent diameter, but of a lighter colour than that planet. The breadth of the head, including diffused coma, is nearly half the moon's apparent diameter. The tail is well connected with the head, without any perceptible aperture; but has not appeared on any evening to be so long by several degrees as on Saturday night, when it measured from six to ten degrees in length, and upwards of two degrees in breadth at its extremity." Stöpel (Tangermünde, Germany) wrote to the *Berliner Astronomisches*

Jahrbuch on July 10 and said the comet was observed there and exhibited a tail 3° long, with a strong, well-defined nucleus.

The comet had faded enough by August that the moon interfered with observations early in the month. When astronomers next turned their telescopes toward the comet during the second half of August, they found the comet had faded considerably. J. Soldner (Munich, Germany) saw the comet on August 20 and noted it was faint. On August 26, J. E. Bode (Berlin, Germany) described the comet as a pale, somewhat oblong nebulosity, with a faint trace of tail extending northward. J. Leski (Krakow, Poland) last saw the comet on the 28th, when he described it as very faint. T. Derfflinger (Kremsmünster, Austria) last saw the comet on August 23, 26, and 30, and said it was very difficult to observe, appearing very faint in a telescope. On the 30th, Bode noted the comet was extremely faint and hard to detect. He attributed this appearance to moonlight and the comet's nearness to the horizon.

The only descriptive details published for the month of September came from A. Bittner (Prague, Czech Republic). He saw the comet on the 12th, 14th, 15th, 16th, and 18th, and simply described it as faint and ill-defined.

The comet's position was last measured by Olbers on October 12.80 and by Struve on October 12.84. Struve measured it as $\alpha = 8^h 53.4^m$, $\delta = +53° 00'$. Olbers saw the comet for the final time on October 20. Although he noted the coma was still 2' in diameter, the comet was so faint as to not allow a position measurement. The final observation of the comet came on October 25.8, when Struve noted it was too faint for a position to be determined.

The comet was impressive enough to have warranted its entry into diaries and journals of people who were not astronomers. John Woods (an English farmer sailing to Baltimore, Maryland) wrote for July 5, "A little before day-break the mate discovered a comet." Members of the Stephen Harriman Long expedition traveling across the mid-west toward the Rocky Mountains spotted this comet low in the northwestern sky on two evenings in early July. Edwin James (botanist and geologist for the expedition) wrote that on the evening of July 7 (July 8, universal time) "dense cumulostratus and cirrostratus clouds skirted the horizon: above these we observed a comet bearing north-west by north." A meteorological journal kept for the expedition specifically said the comet was discovered at 9 p.m. and its distance from the North Star was measured as 49° 38'. The next evening at 8:57 p.m. (July 9, universal time) the comet was measured at 48° 46' from Polaris and at an altitude of 7°.

This comet transited the face of the sun on June 26. It took four hours to cross from the southern limb to the northern, and on June 26.3 it was situated just over 2' from the sun's center. This transit was first mentioned by Olbers in a letter he wrote to Bode on July 27, which was published in the 1822 edition of the *Berliner Astronomisches Jahrbuch*. After the publication of the letter, several solar observers went back to their records to check if they might have seen the comet and/or its nucleus. J. R. Hind (1876) investigated the reports and said a negative observation came from General von Lindener (Glatz, now Klodzko, Poland), while reports of something unusual on the face of the sun that morning came from F. von P. Gruithuisen (Munich, Germany), Wildt (Hanover, Germany), C. J. Pastorff (Buchholtz, Germany), and Augustin Stark (Augsburg, Germany). Gruithuisen reported that he had seen five unidentified dark spots on the sun during 1819, one of which was seen on June 26. Hind was most impressed by the reports of Pastorff and Stark. Pastorff was particularly concerned how his observation would be treated and commented, "it is not an observation made à la D'Angos," referring to uncertain objects reported by Jean Auguste d'Angos in 1784 and 1798, but never substantiated by precise records. Pastorff said, "On the 26th of June, 1819, on looking at the Sun, I saw very distinctly upon its disk a nebulous spot and three black ones. The nebulous spot was perfectly round and slightly luminous. The roundness, the nebulosity, the luminous point in the centre appeared to me so remarkable, that I made a sketch as correctly as possible." Stark said he first observed a spot which "was neither like an opening nor shallow" on June 26 at 7:15 a.m. (local time). He noted that when he again looked at the sun at noon this spot was no longer visible. Hind commented that Stark's image was much closer to the calculated position at that time than was Pastorff's. In addition, he believed, "It appears very improbable that a comet seen in projection upon the Sun's disk would present the cometary aspect which Pastorff describes; it is far more likely that the nucleus only would be discerned."

Several parabolic orbits were determined using the July observations. A. Bouvard (1819) calculated one of the first and indicated a perihelion date of 1819 August 3. He only used positions obtained during the first half of July and noted, "these elements are as yet but approximations." Using positions obtained on July 3, 7, and 11, C. K. L. Rümker determined that the comet passed perihelion on June 28.87. J. B. Sniadecki took positions from July 6, 10, and 14 and determined the perihelion date as June 28.21. This last orbit was very close to later parabolic solutions.

The first intimation that the comet was not moving in a parabolic orbit came from J. R. Hind (1876). Although he calculated a parabolic orbit, he commented, "I found reason to infer that the comet's motion was not closely parabolic...." The first elliptical orbit was determined by H. A. Peck (1906). He took about 400 positions obtained between July 2 and October 15, and established a perihelion date of June 28.21, and a period of 66 509 years. Peck (1907) revised his calculations and determined both an elliptical and a parabolic orbit. The elliptical orbit had a period of 151 810 years. The parabolic orbit included perturbations by six planets, and Peck noted it "represents the entire series of observations nearly as well as either one of the ellipses given." This parabolic orbit is given below.

T	ω	Ω (2000.0)	i	q	e
1819 Jun. 28.2181 (UT)	13.4158	276.2349	80.7517	0.341514	1.0

ABSOLUTE MAGNITUDE: $H_{10} = 4.0$ (V1964)

FULL MOON: Jun. 8, Jul. 7, Aug. 5, Sep. 4, Oct. 3, Nov. 2

SOURCES: H. W. M. Olbers, *BAJ for 1822*, **47** (1819), pp. 177–80; J. F. Encke, *BAJ for 1822*, **47** (1819), pp. 202–4; J. E. Bode, *BAJ for 1822*, **47** (1819), pp. 214–16; F. B. G. Nicolai, *BAJ for 1822*, **47** (1819), pp. 223–6; H. W. M. Olbers, J. E. Bode, and A. Bouvard, *BAJ for 1822*, **47** (1819), pp. 228–31; C. F. Gauss, *BAJ for 1822*, **47** (1819), pp. 235–6; J. Leski, *BAJ for 1822*, **47** (1819), p. 241; C. K. L. Rümker, *BAJ for 1822*, **47** (1819), pp. 244–5; F. G. W. Struve, *BAJ for 1822*, **47** (1819), p. 250; J. G. Tralles, *BAJ for 1822*, **47** (1819), pp. 252–3; Stöpel, *BAJ for 1822*, **47** (1819), p. 256; G. Santini and F. Bertirossi-Busatta, *CA*, **2** (1819), pp. 584–5; J. G. Tralles, J. L. Pons, G. Santini, F. Carlini, G. Inghirami, F. B. G. Nicolai, C. F. Gauss, J. Soldner, B. A. von Lindenau, J. F. Encke, C. K. L. Rümker, and A. Bouvard, *CA*, **2** (1819), pp. 614–31; J. L. Pons, F. Carlini, and A. Bouvard, *CA*, **3** (1819), pp. 201–7; *PMJ*, **54** (1819 Jul.), pp. 75–7; *PMJ*, **54** (1819 Sep.), p. 238; *Annales de chimie et de physique*, **13** (1820), pp. 104–10; J. Soldner, *BAJ for 1823*, **48** (1820), pp. 99–100; J. B. Sniadecki and N. Cacciatore, *BAJ for 1823*, **48** (1820), pp. 120–4; A. Bittner, *BAJ for 1823*, **48** (1820), pp. 129–31; H. W. M. Olbers and F. G. W. Struve, *BAJ for 1823*, **48** (1820), pp. 133–9; J. T. Bürg, *BAJ for 1823*, **48** (1820), pp. 144–6; F. I. C. Hallaschka, *BAJ for 1823*, **48** (1820), p. 148; F. G. W. Struve, *BAJ for 1823*, **48** (1820), pp. 169–70; Derfflinger, *BAJ for 1823*, **48** (1820), pp. 173–6; F. Carlini, *CA*, **5** (1820), pp. 551–6; A. M. Fisher, *AJS* (Series 1), **2** (1820 Nov.), p. 374; H. W. M. Olbers, *PMJ*, **57** (1821 Jun.), pp. 444–6; A. Bouvard, *CDT* (1822), p. 351; J. R. Hind, *MNRAS*, **36** (1876 May), pp. 309–13; O1899, pp. 91–3; John Woods, "Two Years' Residence in the Settlement on the English Prairie in the Illinois Country," *Early Western Travels: 1748–1846*, Volume 10. Edited by Reuben Gold Thwaites. Cleveland: The Arthur H. Clark Company (1904), p. 186; Edwin James, "Account of an Expedition from Pittsburgh to the Rocky Mountains, Performed in the Years 1819, 1820," *Early Western Travels: 1748–1846*, Volume 14. Edited by Reuben Gold Thwaites. Cleveland: The Arthur H. Clark Company (1905), p. 145; Edwin James, "Part IV of James' Account of S. H. Long's Expedition," *Early Western Travels: 1748–1846*, Volume 17. Edited by Reuben Gold Thwaites. Cleveland: The Arthur H. Clark Company (1905), p. 266; *AJ*, **25** (1906 Apr. 10), pp. 61–72; *AJ*, **25** (1907 Apr. 29), pp. 137–8; V1964, p. 54; *CCO*, 1st ed. (1972), pp. 15, 41.

D/1819 W1 *Discovered:* 1819 November 28.19 ($\Delta = 0.24$ AU, $r = 0.90$ AU, Elong. $= 62°$)

(Blanpain) *Last seen:* 1820 January 25.14 ($\Delta = 0.51$ AU, $r = 1.30$ AU, Elong. $= 116°$)

Closest to the Earth: 1819 October 31 (0.1101 AU)

1819 IV *Calculated path:* VIR (Disc), COM (Jan. 16)

J. J. Blanpain (Marseille, France) discovered this comet in Virgo on 1819 November 28.19. The comet was then situated in the morning sky and Blanpain estimated the position as $\alpha = 12^h\ 02.0^m$, $\delta = -0°\ 20'$. He estimated the diameter as 6' or 7', and said a "very small and confused nucleus" was present. No tail was observed. Blanpain confirmed his find on November 29.24.

Blanpain only followed the comet until December 2, but an independent discovery was made on December 5.14, by J. L. Pons (Marlia, Italy). He described the comet as small and faint, with no tail or condensation. He continued making observations and other astronomers began to observe the comet around mid-December, with A. Bouvard (Paris, France) first detecting the comet on the 14th and P. Caturegli (Bologna, Italy) making his first observation on the 22nd. Bouvard described the comet as very faint in a 6.5-cm refractor. Bouvard again saw the comet in strong moonlight on the 30th and noted it was very faint. Pons obtained his final observation of the comet on the 31st, the night of a full moon. He described it as very faint, but the sky clouded before he could measure its position.

As 1820 began, Bouvard and F. Carlini (Milan, Italy) were continuing observations. Bouvard's last came on the morning of January 15, when he noted the comet was extraordinarily faint in the 6.5-cm refractor. Carlini obtained the last observation of the comet on January 25.14, when he measured the position as $\alpha = 12^h 55.9^m$, $\delta = +17° 20'$.

Some astronomers could not find the comet during January. Pons battled clouds during January, but did have clear mornings on the 3rd and 15th. On both occasions he searched for the comet in the predicted area, as well as outside that area, but failed to locate it. He concluded that it had become too faint for him to observe. On the last date, Pons noted the comet had entered a region containing several known nebulae. C. K. L. Rümker (Hamburg, Germany) also experienced considerable bad weather, but did look for the comet on January 10, 11, and 12. He found nothing. Rümker concluded, "either the comet is no longer visible, or it can easily be confused with a nebula, of which this part of the sky abounds."

Early parabolic orbits were calculated by F. Carlini (1819) and J. F. Encke (1820). Their resulting perihelion dates were 1819 November 17.38 and November 21.54, respectively. Encke published the first elliptical orbit in the 1824 volume of the *Berliner Astronomisches Jahrbuch*. Using seven positions obtained between December 14 and January 15, he determined a perihelion date of November 20.75 and a period of 4.81 years. Several other astronomers have calculated elliptical orbits over the years, with the most recent being I. Lagarde (1907). He began with the orbit computed by Encke in 1824 and revised it to better fit the seven observations obtained between December 14 and January 15. The result was an elliptical orbit with a perihelion date of November 20.85, and a period of 5.10 years. This orbit is given below.

T	ω	Ω (2000.0)	i	q	e
1819 Nov. 20.8474 (UT)	350.2613	79.8119	9.1080	0.892318	0.698752

ABSOLUTE MAGNITUDE: $H_{10} = 8.5$ (V1964)

FULL MOON: Nov. 2, Dec. 1, Dec. 31, Jan. 30

SOURCES: J. L. Pons and C. K. L. Rümker, *CA*, **3** (1819), pp. 193–5; F. Carlini, *CA*, **3** (1819), pp. 297–8; J. J. Blanpain, *PMJ*, **54** (1819 Dec.), p. 469; F. Carlini and J. F. Encke, *CA*, **4** (1820), pp. 519–20; J. L. Pons, *BAJ for 1824*, **49** (1821), p. 202; J. F. Encke, *BAJ for 1824*, **49** (1821), pp. 216–21; A. Bouvard, *OAP*, **1** (1825), p. 143; T. Clausen and J. F. Encke, *AN*, **10** (1833 Jan. 29), p. 345; F. de Vico, *AN*, **14** (1836 Aug. 24), pp. 61–4; *AJS* (Series 1), **28** (1835 Jul.), p. 398; I. Lagarde, *CR*, **144** (1907 Jan. 28), pp. 181–3; V1964, p. 54.

C/1821 B1 *Discovered:* 1821 January 21.76 ($\Delta = 1.69$ AU, $r = 1.59$ AU, Elong. $= 67°$)
(Nicollet–Pons) *Last seen:* 1821 May 3.97 ($\Delta = 2.11$ AU, $r = 1.26$ AU, Elong. $= 24°$)
Closest to the Earth: 1821 March 23 (0.9271 AU)
1821 *Calculated path:* PEG (Disc), PSC (Mar. 19), CET (Mar. 23), ERI (Apr. 9)

This comet was discovered almost simultaneously by J. N. Nicollet (Royal Observatory, Paris, France) and J. L. Pons (Marlia, Italy). The observations of both astronomers indicate they found the comet on 1821 January 21.76 near γ Pegasi. Nicollet described the comet as very small and faint, without an apparent nucleus. A tail extended about one-half degree. Pons said the comet looked like a white spot, with a very small tail. Nicollet measured the position as $\alpha = 0^h \, 02.4^m$, $\delta = +17° \, 00'$ on January 21.84 and was the first to announce the discovery.

The comet was heading toward both the sun and Earth when discovered, but even though astronomers were not yet aware of this fact, they noted very quickly that the comet was brightening. Pons said the comet seemed slightly brighter on January 22, with a tail extending about 2°. Nicollet said the tail was 2° long on the 23rd. J. J. Blanpain (Marseille, France) independently discovered this comet with a small telescope on January 25.8. He said the comet was 4' in diameter and exhibited a "very marked" nucleus of magnitude 7 or 8, as well as a tail 1.5° long. By the 30th, Nicollet noted the comet was brighter than when previously seen and exhibited the beginnings of a nucleus, as well as a tail about 3.5° long.

Unaware of this comet's discovery, H. W. M. Olbers (Bremen, Germany) independently discovered it on January 30.77. Back on 1820 September 27, he had noticed that star charts by K. L. Harding lacked stars of magnitude 6 and 7 in this region, so he periodically studied the area. On this occasion he immediately noticed a comet-like object with his cometseeker. His refractor revealed a small, faint, and diffuse comet, with a tail estimated as between 0.75° and 1° long. Olbers added that within the nebulosity of the head he occasionally noticed a very small, diffuse nucleus.

The comet continued to approach the sun and Earth during February and brightened. G. Santini (Padova, Italy) reported it was visible to the naked eye on the evening of February 19 and the Chinese texts *Ch'ing shih lu ching chi tzu liao chi yao* and *Ch'ing Ch'ao Hsü Wên Hsien Thung Khao* (HA1970) reported the discovery of a naked eye "broom star" in the western sky on the evening of February 20. J. H. Fritsch (Quedlinburg, Germany) reported a tail about 1.5° long on the 7th.

The comet's slow motion during January and February is best exemplified by J. F. W. Herschel (Slough, England) who spotted the comet with a "night-glass" near γ Pegasi on the evening of February 27. This star is the same star Pons and Nicollet had noticed the comet near when discovered on January 21. Herschel said the comet was observed near the end of twilight, and was not then visible to the naked eye, while γ Pegasi was. In the telescope, Herschel noted the comet was no more than 10° above the horizon, and appeared as a "mere misty mass" without a "central star-like point." He estimated the tail as 2.5° long and added, "I imagined the tail to be somewhat less bright along its axis, but the dark space certainly was not very decided. The head seemed rather obtuse, and appeared, I thought, to have lateral portions of light, which seemed to go off at a greater angle than the tail."

The comet was becoming a more difficult object to observe as March began, because of its steadily decreasing altitude following sunset. C. F. Gauss (Göttingen, Germany) saw the comet in twilight on the evening of March 2 and noted it had the appearance of a star of magnitude 3 or 4. A few days later, on March 6, Olbers, F. W. Bessel and F. W. A. Argelander (Königsberg, now Kaliningrad, Russia), and J. F. Encke (Seeberg, Germany) all made their final observations of the comet. Santini saw the comet for the last time on March 9, and F. Carlini (Milan, Italy) made his final observations on March 9 and 10. Carlini's last observation was also the final time the comet was seen in the Northern Hemisphere. The comet passed only 4° from the sun on March 22. F. B. G. Nicolai (Mannheim, Germany) tried to find the comet with a telescope in daylight on March 17 and 25, but, despite very clear skies, "not the slightest trace of the comet" was found.

The comet exited the sun's glare at the beginning of April and was first spotted by crewmembers of an English ship near Valparaiso, Chile, on April 1.9. The crewmembers reporting observations included Captain B. Hall, Lieutenant W. Robertson, and Midshipman H. Foster. They reported the tail extended about 7° on April 2, with the northern part being longest. They also said the comet was visible to the naked eye in moonlight on April 17. Hall and his crew reported 11 positions between April 8 and 30, and noted that the comet faded and the tail grew shorter with each passing day. The comet was also observed from Sydney, Australia on April 7 and from the island of St Helena.

The final observations of this comet were obtained by Hall on May 2 and 3. He measured the position as $\alpha = 3^h 44.3^m$, $\delta = -3° 26'$ on May 3.97.

The earliest parabolic orbit was calculated by F. Carlini using positions covering the months of January, February, and March. He revealed a perihelion date of 1821 March 22.15. During the next year, similar orbits would be calculated by Encke, J. E. B. Valz, H. von Staudt, Nicollet, Nicolai, Olbers, J. Brinkley, and Bessel. Not only were these orbits very similar, but the later ones didn't use positions obtained after March. Brinkley (1822) and O. A. Rosenberger (1822) independently calculated orbits which utilized the positions of April and May. Brinkley gave the perihelion date as March 21.97, while Rosenberger determined it as March 22.04. Rosenberger's orbit is given below.

T	ω	Ω (2000.0)	i	q	e
1821 Mar. 22.0366 (UT)	169.2121	51.1843	106.4613	0.091823	1.0

ABSOLUTE MAGNITUDE: $H_{10} = 3.4$ (V1964)

FULL MOON: Jan. 18, Feb. 17, Mar. 18, Apr. 17, May 17

SOURCES: J. F. Pons, J. N. Nicollet, and J. J. Blanpain, *CA*, **4** (1820), pp. 413–15; G. Santini, *CA*, **4** (1820), p. 508; F. Carlini, *CA*, **4** (1820), pp. 619, 622; F. Carlini, J. F. Encke, H. W. M. Olbers, F. B. G. Nicolai, and F. W. Bessel, *CA*, **5** (1820), pp. 80–8; H. W. M. Olbers and J. N. Nicollet, *BAJ for 1824*, **49** (1821), pp. 99–100; F. B. G. Nicolai, F. I. C. Hallaschka, and H. W. M. Olbers, *BAJ for 1824*, **49** (1821), pp. 168–76; J. F. Encke, *BAJ for 1824*, **49** (1821), pp. 221–2; F. W. Bessel, *BAJ for 1824*, **49** (1821), pp. 241–2; J. H. Fritsch, *BAJ for 1824*, **49** (1821), p. 252; J. L. Pons, *PMJ*, **57** (1821 Feb.), p. 151; J. N. Nicollet and H. W. M. Olbers, *MAS*, **1** (1821), pp. 154–7; C. F. Gauss and H. von Staudt, *BAJ for 1825*, **50** (1822), pp. 104–5; B. Hall and J. Brinkley, *BAJ for 1825*, **50** (1822), pp. 254–5; J. Brinkley, B. Hall, W. Robertson, and H. Foster, *PTRSL*, **112** (1822), pp. 46–63; O. A. Rosenberger, *AN*, **1** (1822 Nov.), pp. 425–30; J. E. B. Valz, *CA*, **10** (1824), pp. 274–9; J. N. Nicollet, *CDT* (1824), p. 355–7; *Briefwechsel zwischen W. Olbers und F. W. Bessel*, **2**. Edited by A. Erman. Leipzig: Avenarius & Mendelssohn (1852), pp. 217–19; O1899, p. 93; V1964, p. 54; HA1970, p. 86.

C/1822 J1 *Discovered:* 1822 May 12.9 ($\Delta = 0.88$ AU, $r = 0.53$ AU, Elong. $= 32°$)
(Gambart) *Last seen:* 1822 June 22.91 ($\Delta = 1.94$ AU, $r = 1.16$ AU, Elong. $= 29°$)
   Closest to the Earth: 1822 April 30 (0.6416 AU)
1822 I *Calculated path:* TAU (Disc), AUR (May 13), LYN (Jun. 19)

J. F. A. Gambart (Marseille, France) discovered this comet on 1822 May 12.9. Independent discoveries were made by J. L. Pons (Marlia Observatory, Italy) on May 14.9, W. von Biela (Prague, Czech Republic) on May 16.9, and C. F. Scheithauer (Chemnitz, now Karl Marx Stadt, Germany) on May 21.9. Scheithauer noted it had a short tail.

At discovery the comet had already passed closest to the sun and Earth and should have been fading. It was also flirting with the edge of evening twilight. After attaining a maximum solar elongation of 33° on May 19, the comet began very slowly sinking into twilight again. F. I. C. Hallaschka (Prague, Czech Republic) observed a bright, distinct nucleus on May 18 and then reported the comet was visible to the naked eye on May 21. On the last date he also said the comet appeared as a faint, white nebulosity, with a center slightly brighter than the surrounding coma. Later in May, F. X. von Zach (Capellete, France) reported the comet was visible to the naked eye in moonlight, although Gambart said the moonlight weakened the light of the comet on May 28.

J. E. Bode received word on May 28 that Scheithauer had discovered a comet in Auriga, and then read in the newspaper on May 30 that discoveries were made in Marseille and Prague as well. He realized that the comet was then near the northwest horizon on the evening of the 30th, but twilight and moonlight prevented his finding it.

Gambart and Hallaschka remained the most prolific observers of this comet and obtained several observations into June. Gambart said the comet appeared "very bright" when seen in excellent skies on the 3rd, and he noted it was easily seen on the 4th. Despite cloudy skies on the 5th, Gambart said the comet was still quite apparent. Gambart last saw the comet on June 17, and said clouds were present thereafter. The comet was last detected on June 22.92, when Hallaschka gave the position as $\alpha = 6^h\ 28.0^m$, $\delta = +51°\ 40'$. He said the comet had become very difficult to see. Hallascka tried to observe the comet on June 25, but nothing was found. His colleague, M. A. David looked for the comet on the 28th, but noted the comet could no longer be seen.

Nearly identical parabolic orbits were computed by P. A. Hansen (1822), G. F. K. Ursin (1822), J. N. Nicollet (1822, 1826), J. F. Encke (1822), and Gambart (1826). Encke used 18 positions obtained between May 17 and June 14, and computed a parabolic orbit with a perihelion date of 1822 May 6.08. No planetary perturbations were applied. This orbit is given below.

Interestingly, this comet had actually attained a maximum solar elongation of 43° on April 9. Assuming a normal brightness trend, it could have been at its discovery brightness as early as April 20, at which time the solar elongation would have been about 6° greater than at discovery. On the other hand, the comet was then at a declination of −26°, or about 37° directly south of the sun, and would not have been above the horizon before sunrise or after sunset. Any observers in the southern hemisphere could have seen the comet at this time, since it rose about 1.5 hours before the sun in the morning sky, but no observations have ever come to light.

T	ω	Ω (2000.0)	i	q	e
1822 May 6.0793 (UT)	344.6905	179.9345	126.3969	0.504429	1.0

ABSOLUTE MAGNITUDE: $H_{10} = 6.7$ (V1964)

FULL MOON: May 6, Jun. 4, Jul. 4

SOURCES: J. F. A. Gambart, J. L. Pons, W. von Biela, and F. X. von Zach, *CA*, **6** (1821), pp. 381–4; F. B. G. Nicolai, *BAJ for 1825*, **50** (1822), p. 150; J. F. Encke and F. I. C. Hallaschka, *BAJ for 1825*, **50** (1822), pp. 152–8; M. A. David, *BAJ for 1825*, **50** (1822), p. 171; F. I. C. Hallaschka, *AN*, **1** (1822 Aug.), p. 297; P. A. Hansen, G. F. K. Ursin, J. N. Nicollet, and J. F. Encke, *AN*, **1** (1822 Sep.), pp. 309–12; C. F. Scheithauer, *BAJ for*

1826, **51** (1823), p. 175; J. F. A. Gambart, *CDT* (1826), pp. 222, 236–8, 246; J. N. Nicollet, *CDT* (1826), p. 278; V1964, p. 54.

C/1822 K1 *Discovered:* 1822 May 31.1 ($\Delta = 0.68$ AU, $r = 1.19$ AU, Elong. $= 87°$)
(Pons) *Last seen:* 1822 June 25.9 ($\Delta = 0.35$ AU, $r = 0.93$ AU, Elong. $= 65°$)
   Closest to the Earth: 1822 June 17 (0.1308 AU)
1822 III *Calculated path:* PSC (Disc), AQR (Jun. 5), CET (Jun. 12), SCL (Jun. 13), PHE (Jun. 15), ERI (Jun. 16), HOR–DOR–PIC (Jun. 17), PUP (Jun. 18), VEL–PYX (Jun. 21), ANT (Jun. 25)

J. L. Pons (Royal Park La Marlia Observatory, Italy) was on the lookout for P/Encke when he found a slightly condensed nebulous object on 1822 May 31.1 that exhibited neither a tail nor a nucleus. Pons said the comet might have been visible to the naked eye, if not for moonlight. Since Marlia Observatory had just opened, Pons then had no instruments to determine a position. In fact, the first position was not measured until June 9.10, when P. Caturegli (Bologna, Italy) gave it as $\alpha = 23^h 10.6^m$, $\delta = -8° 49'$.

The comet was not well observed, as it quickly moved southward and below the horizon for observers in the Northern Hemisphere. J. F. A. Gambart (Marseille, France) described the comet as extremely faint on June 11, when it was at a low altitude with some morning twilight. Caturegli acquired additional rough positions on June 11, 12, and 13, with the last being the final observation from the Northern Hemisphere.

The story of this comet would have ended here if not for W. Robertson and C. Drinkwater (HMS *Creole*, in harbor at Rio de Janeiro, Brazil). They were the only observers of this comet once it entered the skies of the Southern Hemisphere and obtained measurements of its distance from various stars with the aid of a sextant. On the evening of June 18, they first noticed the "bright orbicular nebula" in the sky near Canopus. Upon directing a telescope to it, they recognized it as a comet. On June 19, they said the comet appeared fainter than on the previous night, but added that a thin haze covered the sky. The comet was still seen with the naked eye on June 22, despite moonlight, and a telescope revealed it was round, without a nucleus or tail.

The final measurements of the comet's position were made by Robertson and Drinkwater on June 24.89. They indicate a position of $\alpha = 9^h 11.5^m$, $\delta = -26° 47'$. They said the comet exhibited no nucleus or tail. The comet was last detected on June 25.9 by Robertson and Drinkwater. They said moonlight prevented a determination of the comet's position. They searched for the comet again on June 27 and 29, in moonlight, but nothing was observed.

Only parabolic orbits were calculated for this comet because of the short period of visibility and the lack of a long series of precise positions. The first orbits were determined by A. von Heiligenstein (1826). Using only the positions obtained by Caturegli and Gambart during the period of June 9–13, he calculated two orbits with perihelion dates of 1822 July 17.03 and July 16.52. A few years later, T. Henderson (1831) calculated an orbit. He was sent the observations of Robertson and Drinkwater for comment and ultimately converted the sextant observations into ecliptic latitudes and longitudes. Henderson then calculated an orbit using the positions of June 19, 22, and 24, and found a perihelion date of July 16.15.

The first calculation to use positions spanning the period of June 9–24 was by J. R. Hind (1880), in which he determined a perihelion date of July 16.34. A very similar orbit was determined by H. A. Peck (1907), in which all 11 positions were utilized. The resulting perihelion date was July 16.33. Peck's orbit is given below.

T	ω	Ω (2000.0)	i	q	e
1822 Jul. 16.3337 (UT)	237.7758	100.2415	143.6913	0.847130	1.0

ABSOLUTE MAGNITUDE: $H_{10} = 7.0$ (V1964)
FULL MOON: May 6, Jun. 4, Jul. 4
SOURCES: J. L. Pons, *CA*, **6** (1821), p. 385; J. F. A. Gambart, *CDT* (1826), pp. 238, 246; A. von Heiligenstein, *AN*, **4** (1826 May), p. 533; W. Robertson, C. Drinkwater, T. Henderson, and P. Caturegli, *PTRSL*, **121** (1831), pp. 1–7; J. R. Hind, *Nature*, **22** (1880 Jul. 1), p. 205; H. A. Peck, *AJ*, **25** (1907 Jul. 31), pp. 165–8; V1964, p. 54.

2P/1822 L1
(Encke)

1822 II

Recovered: 1822 June 2.33 ($\Delta = 0.77$ AU, $r = 0.42$ AU, Elong. $= 22°$)
Last seen: 1822 June 29.4 ($\Delta = 0.29$ AU, $r = 0.89$ AU, Elong. $= 56°$)
Closest to the Earth: 1822 July 4 (0.2692 AU)
Calculated path: GEM (Rec), ORI (Jun. 2), GEM (Jun. 3), MON (Jun. 8), CMi (Jun. 14), MON (Jun. 18), PUP (Jun. 24), PYX (Jun. 27)

This is the first predicted appearance of this comet. J. F. Encke (1819) suggested the comet would likely return during May of 1822. Encke (1820) took the positions from the 1818–19 apparition and determined a perihelion date of 1822 May 24.97 and a period of 3.32 years. There was an ephemeris for the period of 1822 February 25 to July 27.

One of the earliest documented searches was made by J. E. Bode (Berlin, Germany) on the evening of 1822 February 15. Using a 3.5-foot focal length refractor he swept the area in which the comet was predicted to be located, which was the region south of ω Piscium. No trace of the comet was found. Calculations reveal the comet was in the predicted region, but was then too faint for recovery. It was moving quite slowly because of its distance from the Earth and sun, and that area of the sky was carried into twilight during March and then daylight by the end of April. It passed 10° from the sun on April 24, and then slowly moved away.

C. K. L. Rümker (Paramatta, New South Wales, Australia) recovered this comet with the aid of Encke's ephemeris on 1822 June 2.33. The comet was then situated very low in the evening sky, and Rümker determined its position as $\alpha = 6^h$ 10.9^m, $\delta = +17°$ 40′. Rümker confirmed the recovery and therefore the identity with the expected comet on June 3.33. The comet was about a half day earlier than Encke's prediction.

Physical descriptions were few during this apparition, with the bulk coming from W. Robertson and C. Drinkwater (HMS *Creole*, in harbor at Rio de Janeiro, Brazil), who managed to observe the comet after sunset for nearly two weeks during June. They described the comet as "like a faint nebula of a round form" on June 7. They said it appeared faint in a telescope on the 12th, while on the 13th they remarked that the comet had not increased in brightness since it was seen on the 7th. On June 17, Robertson and Drinkwater noted the comet had "the same nebulous, orbicular appearance as when first seen." They again described it as "very faint" in a telescope on the 18th. Further attempts to see the comet on the 19th and 22nd proved fruitless, first because of hazy skies, and then because of moonlight.

The comet's position was measured for the final time on June 23.37, when Rümker determined it as $\alpha = 7^h$ 43.8^m, $\delta = -9°$ 10′. He noted its faintness and the fact that increasing moonlight would hamper further observations. The last apparent observation came on June 29.4, when Rümker detected a faint glow near the expected position, but no precise measurements were possible. Rümker's searches in early July revealed no trace of the comet.

Later investigations into the orbit of this comet indicated a perihelion date of May 24.46 and a period of 3.32 years. Although Encke continued doing work on the orbit of this comet for years, F. E. von Asten (1877) began his own elaborate work. Asten's initial paper determined the orbit of every apparition from 1818 to 1868. The next major advance in the understanding of this comet's motion was by B. G. Marsden and Z. Sekanina (1974). Although they determined an orbit very similar to that of Asten, they also determined the comet's nongravitational terms of $A_1 = +1.14$ and $A_2 = -0.03996$. Their orbit is given below.

T	ω	Ω (2000.0)	i	q	e
1822 May 24.4577 (TT)	182.7625	336.9209	13.3611	0.345886	0.844439

ABSOLUTE MAGNITUDE: $H_{10} = 7.0$ (V1964)

FULL MOON: May 6, Jun. 4, Jul. 4

SOURCES: J. F. Encke *BAJ for 1822*, **47** (1819), p. 201; J. F. Encke, *BAJ for 1823*, **48** (1820), pp. 211–23; J. F. Encke, *BAJ for 1824*, **49** (1821), pp. 225–6; C. K. L. Rümker, *BAJ for 1826*, **51** (1823), pp. 106–7; J. E. Bode, *BAJ for 1826*, **51** (1823), p. 175; C. K. L. Rümker, *PTRSL*, **119** (1829), pp. 54–5; W. Robertson and C. Drinkwater, *PTRSL*, **121** (1831), pp. 7–8; F. E. von Asten, *BASP* (Series 4), **22** (1877), p. 559; A. Berberich, *AN*, **119**

(1888 Apr. 24), p. 51; V1964, p. 54; B. G. Marsden and Z. Sekanina, *AJ*, **79** (1974 Mar.), pp. 415–16.

C/1822 N1   *Discovered:* 1822 July 13.9 ($\Delta = 1.80$ AU, $r = 1.93$ AU, Elong. = 81°)
(Pons)   *Last seen:* 1822 November 11.40 ($\Delta = 2.07$ AU, $r = 1.18$ AU, Elong. = 19°)
   Closest to the Earth: 1822 August 29 (1.0206 AU)
1822 IV   *Calculated path:* CAS (Disc), CEP (Jul. 17), DRA (Aug. 5), HER (Aug. 23), OPH (Sep. 22), SCO (Oct. 4)

J. L. Pons (Royal Park La Marlia Observatory, Italy) discovered this comet on 1822 July 13.9. Independent discoveries were made by J. F. A. Gambart (Marseille, France) on July 16 and A. Bouvard (Paris, France) on July 20. Gambart confirmed his find in very difficult observing conditions on July 17.93, and gave the comet's position as $\alpha = 23^{\mathrm{h}}\ 32.0^{\mathrm{m}}$, $\delta = +65°\ 21'$. He noted the comet was then very faint and only occasionally appeared for a few moments.

Although three astronomers had independently discovered this comet in July, word traveled slowly and no one but these three made observations for the remainder of July and well into August. Gambart was the only observer to provide descriptions of the comet during this time. The comet initially remained a faint object, with Gambart's observations revealing it as very faint on the 19th and unable to withstand much illumination within the eyepiece. Although Gambart noticed little improvement on the 20th, he reported the comet had increased in brightness by the 21st, and showed a condensation that seemed slightly off center. The comet was "rather visible" on the 22nd. On July 26 he occasionally noted a stellar object near the middle of the coma that was similar to a star of magnitude 9 or 10. Gambart again noted the comet was excessively faint under good skies on August 1, but it was again well seen on the 3rd, with a bright point-like central condensation. On August 8 Gambart said the comet was at least as bright as the globular cluster M13. The comet had attained its greatest northern declination of +70° on August 2.

Observations by other astronomers began on the evening of August 19, when W. von Biela (Prague, Czech Republic) independently discovered this object with the naked eye while searching for comets. He confirmed motion the following evening and announced his discovery the next day. Interestingly, on August 20, H. W. M. Olbers (Bremen, Germany) finally received notice of this comet's discovery by Gambart and Bouvard. That evening he quickly found the comet and surmised, "It must have increased very much in light, because it was brighter" than M13. He added, "if one knew its place it was easily detected with the naked eye." Gambart saw the comet on the same night and saw a trace of a narrow tail extending about 40′ away from the sun. Observations were subsequently begun by K. L. Harding (Göttingen, Germany) on August 21 and F. W. Bessel (Königsberg, now Kaliningrad, Russia) on August 29. Olbers noted on the 27th that the comet was visible to the naked eye in spite of unfavorable weather. A tail was also visible.

The comet continued to approach perihelion during September and most of October, but having passed closest to Earth at the end of August, the brightness should have slowly faded. Olbers noted the comet was still brighter than M13 on September 1, and he said it was still a naked-eye object on September 11 and 14. J. E. Bode (Berlin, Germany) noted it "had become fainter" by the 14th. On the 15th Harding remarked that the nucleus appeared brighter and more distinct than on previous evenings. This observation was followed by announcements from astronomers in Milan on September 16 and Olbers on the 19th and 20th that the comet had brightened and become smaller. The tail was well observed with telescopes during September, with its length estimated as 2° by Gambart on the 8th, 1.5° by Olbers on the 11th, 2.5° by Olbers on the 13th, 1.5° by G. K. F. Kunowsky (Germany) on the 14th, and 1° by Bode on the 14th. Olbers said it was 4° long on the 20th. Olbers also noted the coma was 2.5′ across around mid-September. N. Cacciatore (Palermo, Italy) estimated the coma diameter as 2.3′ on the 15th and added that the nucleus was 7″ across. On the 23rd, he said the coma was 2′ across, while the nucleus was less than 7″ across. The comet was also moving southward during this same period and was spotted by C. K. L. Rümker (Paramatta, New South Wales, Australia) on September 21. Olbers said it was still a naked-eye object on this same date, but was not during the last days of the month as moonlight began interfering.

As October began, some physical descriptions were still being made. Cacciatore said the coma was 1′ across on the 5th and exhibited a nucleus 4″ across. As the month progressed observers in the Northern Hemisphere began losing the comet as it moved closer to the horizon. Olbers last spotted it on the 14th, an astronomer in Milan last saw it on the 21st, and an astronomer in Florence last saw it on the 22nd. The last observation was the final one in the Northern Hemisphere. Although Rümker continued measuring the comet's position through the remainder of October and into November, he did not provided details as to the comet's appearance.

The comet was last detected on November 11.40, when Rümker determined the position as $\alpha = 16^h\ 04.5^m$, $\delta = -30°\ 27′$. The comet was then situated very low over the western horizon in the evening sky.

The first parabolic orbit was calculated by P. A. Hansen. He used positions obtained on September 2–6 and determined the perihelion date as 1822 October 24.07. In the coming weeks, additional orbits were calculated by Olbers, F. W. A. Argelander, F. B. G. Nicolai, Gambart, and Bouvard in which the derived perihelion date was ultimately increased by a few hours.

The first elliptical orbit was published by J. F. Encke during October of 1822. Using positions spanning the period of July–September, he determined the perihelion date as October 25.46 and the period as 194 years. The comet was still under observation for several weeks after Encke's orbit was published. After it was no longer visible, astronomers began publishing revised elliptical solutions, with much longer periods. The period was determined as 1817 years by Rümker (1823), 1554 years by Encke (1822), and 5449 years by Encke (1824). A definitive orbit by A. Stichtenoth (1898) used 456 positions obtained between July 18 and November 11, and determined a perihelion date of October 24.27 and a period of about 5449 years. It is given below.

T	ω	Ω (2000.0)	i	q	e
1822 Oct. 24.2663 (UT)	181.1061	95.2422	127.3426	1.145098	0.996302

ABSOLUTE MAGNITUDE: $H_{10} = 3.0$ (V1964)

FULL MOON: Jul. 4, Aug. 3, Sep. 1, Sep. 30, Oct. 30, Nov. 28

SOURCES: W. von Biela and G. K. F. Kunowsky, *BAJ for 1825*, **50** (1822), pp. 259–60; *CA*, **7** (1822), pp. 375–88; J. F. A. Gambart, A. Bouvard, and H. W. M. Olbers, *AN*, **1** (1822 Sep.), p. 307; P. A. Hansen, H. W. M. Olbers, and K. L. Harding, *AN*, **1** (1822 Sep.), pp. 337–52; J. F. Encke, *AN*, **1** (1822 Oct.), pp. 371–6; F. W. A. Argelander, F. B. G. Nicolai, and H. W. M. Olbers, *AN*, **1** (1822 Oct.), pp. 393–8; K. L. Harding and H. W. M. Olbers, *AN*, **1** (1822 Nov.), pp. 419–22; F. W. A. Argelander, *AN*, **1** (1822 Nov.), p. 431; F. I. C. Hallaschka and H. W. M. Olbers, *BAJ for 1826*, **51** (1823), pp. 155–61; J. E. Bode, *BAJ for 1826*, **51** (1823), p. 176; C. K. L. Rümker, *BAJ for 1826*, **51** (1823), pp. 180–1; C. K. L. Rümker, *AN*, **2** (1823 Jul.), pp. 207–12; J. F. Encke, *AN*, **3** (1824 Apr.), pp. 107–10; N. Cacciatore and G. Piazzi, *Del Reale Osservatorio di Palermo*. Palermo: Dalla Tipografia di Filippo Solli (1826), p. 212; J. F. A. Gambart, *CDT* (1826), pp. 224, 238–45, 247; A. Bouvard, *CDT* (1826), p. 279; C. K. L. Rümker, *PTRSL*, **119** (1829), pp. 55–7; A. Stichtenoth, *AN*, **145** (1898 Mar. 5), p. 383; O1899, pp. 93–103; V1964, p. 54.

C/1823 Y1 (Great Comet)

1823

Discovered: 1823 December 29.2 ($\Delta = 0.82$ AU, $r = 0.64$ AU, Elong. $= 40°$)

Last seen: 1824 April 1.9 ($\Delta = 2.20$ AU, $r = 2.38$ AU, Elong. $= 87°$)

Closest to the Earth: 1824 January 23 (0.4861 AU)

Calculated path: OPH (Disc), HER (Dec. 31), CrB (Jan. 13), HER (Jan. 15), BOO (Jan. 19), DRA (Jan. 20), UMi–DRA (Jan. 25), UMi (Jan. 26), DRA (Jan. 27), UMa (Jan. 30), CAM (Feb. 7), LYN (Feb. 10), AUR (Feb. 21), GEM (Mar. 14)

Nell de Breautè (Dieppe, France) discovered this comet on 1823 December 29.2, but little additional information is available. J. L. Pons (Marlia, Italy) independently discovered it with his naked eye on December 30.17. He was out looking at the morning sky and noted what appeared to be chimney smoke rising over a small hill to the northeast. He became suspicious because the "smoke" did not change in appearance. Eventually a "beautiful nucleus" cleared the hill and Pons knew this was a new comet. He estimated the tail length as 3–4° and said the comet was situated about a degree and a half from κ and ι Serpentis. Pons also remarked that a northward motion was suspected, but he could not measure a position. Other independent discoveries were made at about the same time as Pons' observation. W. von Biela

(Prague, Czech Republic) found this comet on 1823 December 30.2, and, with the help of his colleague M. A. David, he measured the position as $\alpha = 16^h 51.9^m$, $\delta = +12° 28'$ on December 31.19. Biela noted that the comet was brighter than comet C/1819 N1 had been, with a tail about 3° long. Another discovery was also reported by Schulz (Düren, Germany) who also saw the comet in the morning sky on December 30.23. Other independent discoveries were made during the following days, including that by G. Santini (Padova, Italy) with the naked eye on January 4.25. Calculations reveal the comet passed 8° from the sun on December 9.

Two earlier observations have been reported over the years, but these were probably not observations of this comet. First, shortly after the comet's discovery, reports surfaced about an object seen on December 1.8 by Swiss hunters. The few details available are given in Appendix 1, but H. W. M. Olbers and others have pointed out that C/1823 Y1 would then have been south of the sun and not visible in Europe. M. F. Baldet (1950) reported the comet was seen in the morning sky on December 23, but provided no references. The Author has found no trace of such an observation in contemporary journals. In addition, no such observation was mentioned in the catalogs of J. G. Galle (1894) and J. Holetschek (1913). It must be concluded that the December 23 date was an error.

Numerous observations were made as the new year began. Following a period of bad weather, F. B. G. Nicolai (Mannheim, Germany) was finally able to observe the comet on January 4 and 5. He noted a scintillating nucleus and a tail about 1.5° long. H. W. M. Olbers (Bremen, Germany) saw the comet with the naked eye in somewhat hazy skies on the 5th. He said it looked like a star of magnitude 3 and exhibited a tail at least 5° long. On the same date, K. L. Harding (Göttingen, Germany) said the nuclear condensation was "very bright and sharply defined" in the finder, with a pale coma surrounding it. The tail was more than 5° long. Harding saw the comet on the 6th and estimated the tail as 4.75° long, while his 4-foot focal length refractor (126×) revealed a nuclear condensation about as bright as a 6th-magnitude star and nearly 3″ across. Nicolai saw the comet on 10th and noted the comet had not increased in brightness. N. Cacciatore (Palermo, Italy) viewed the comet with a telescope on the same date and said the nucleus was 8″ across, while the coma measured 1′ 40″. The tail appeared about 6° long. Harding saw the comet on January 11 and said the tail was hardly 3° long, and the nucleus was less pronounced than when previously seen. A less distinct nucleus was also reported by Olbers on the 12th and Cacciatore on the 13th. Also on the 12th, Olbers said the comet was still visible to the naked eye, despite moonlight, while J. F. A. Gambart (Marseille, France) described the comet as extremely beautiful. Gambart found the tail was 5° long on the 13th. Moonlight made observations difficult during the next several days.

The comet was better seen after the moon had left the sky. Gambart found the tail was 4–5° long on January 20. Cacciatore estimated the coma as 1.3′

across on the 21st, and said it exhibited a nucleus 10″ across, and a tail about 5° long. The very next night Biela reported a bright, sunward tail, in addition to the main tail. On January 23, the sunward tail was observed by several astronomers. Harding said it was 2.5° long. Westphal (Egypt) said it was visible to the naked eye and expressed his puzzlement at missing it on the 22nd. He gave the length as about 4.25°, while the main tail extended about 3.5°. Westphal added that the comet's overall shape was similar to that of M31. Biela and his colleague F. I. C. Hallaschka noted the sunward tail was equally as bright as the main tail. Meanwhile, Cacciatore said the comet had slightly faded since the 21st, while Harding said the comet was brighter than M31 (magnitude 3.5 – NGC2000.0). On January 24, the sunward tail still dominated the descriptions given by astronomers. Harding said it was 2.5° long, while the main tail extended about 4.5°. Interestingly, Westphal agreed that the main tail was 4.5° long, but gave the length of the sunward tail as no less than 7°. He added that the sunward tail was half as wide as the main tail. Gambart said the comet's naked-eye appearance was like a nebulous band directed parallel to the celestial equator. He said the whole comet seemed somewhat curved, although he suspected it was because the main tail and sunward tail were not directly opposite to one another. P. A. Hansen (Altona, Germany) saw the sunward tail on the evenings of January 24 and 25. He noted that it seemed longer than on previous nights and appeared to end in a point. Biela said the sunward tail was still visible on the 25th, but Harding only noted that the main tail was 7° long. Biela and Hallaschka were the only astronomers to report the sunward tail on January 27, but they noted it was more difficult to see than on previous nights. Cacciatore said the comet was faint, hardly 1′ across, with a diffuse nucleus. He estimated the tail length as about 2.5°. Although Westphal saw the comet on the nights of the 27th and 28th, he only saw the sunward tail on the 28th. He said it was then barely visible and formed an angle of about 170° to the main tail. Meanwhile, on both nights, he noted the nucleus was like a star of magnitude 5 or 6, while the main tail was broad and about 3° long. Olbers saw the comet with the naked eye on January 29, despite hazy skies. His cometseeker revealed a tail extending 2.5–3° in length. The comet attained its most northerly declination of +73° on January 30 and Westphal noted the comet had faded since the 28th and exhibited no trace of the sunward tail. Nevertheless, Westphal reported one final observation of the sunward tail on the 31st, saying it was relatively bright and formed an angle of about 138.5° with the main tail.

Other observations were made during January that were not assigned to specific dates. Chinese astronomers reported a "broom star" was visible in the lunar month spanning January 1–30. Pons wrote a letter to the *Correspondance Astronomique* on the last day of January, which was published in the 1824 volume. He said, "I am quite surprised that this beautiful comet, which is so apparent to the naked eye, could only be seen with great difficulty as it passed through the meridian telescope. For with the slightest lighting of the wire, the comet disappeared, or was only seen with great difficulty." He added, "what surprises me even more, is that this comet has two tails of which one is almost opposed to the other."

There was no longer any trace of the sunward tail as February began and, as Cacciatore reported on the 5th, the nucleus was no longer visible. The main tail was also rapidly fading, with Cacciatore reporting it as hardly 1° long on the 5th and Gambart noting that he only suspected a fan-shaped tail on the 17th. The brightness steadily faded, with Gambart reporting it was quite faint, despite clear skies on the 7th and remarking that it was very faint on the 10th. Gambart did note on the 17th that the comet was as well seen in his telescope's finder as the globular cluster M4 (magnitude 5.9 – NGC2000.0). On February 18, Hansen noted the comet was so faint that it disappeared in the meridian scope as soon as the measuring thread was illuminated. Olbers simply described the comet as faint on the 22nd, while on the 28th, F. G. W. von Struve (Dorpat, now Tartu, Estonia) said it was very faint and Gambart noted it was "fainter than ever." Struve added that the comet was about 1.5′ across.

There were still no reports of a tail or nucleus as March began. Struve saw the comet on the 4th and said it was faint, but of a large diameter. Santini saw the comet on the 6th and described it as very faint, but exhibiting "a sensible diameter." He remarked that he could have continued observations beyond this date if not for moonlight. On March 7, Gambart said the comet was "fainter than ever," while David described it as a "hardly noticeable, shapeless nebulosity." Physical descriptions were absent during the next week and a half because of moonlight.

The comet was quite faint during the last half of March. Pons said it was "very faint" on the 18th, and Olbers said it looked like a weak, insignificant nebula on the 20th. Harding last detected the comet on the 23rd and 24th, using the 13-foot focal length Schröter reflector. Harding said it was still bright, with a distinct central condensation. He said he believed it would easily have been followed into April, if not for a long stretch of bad weather. V. Wisniewski (St Petersburg, Russia) saw the comet for the final time on the 28th. The position was measured for the final time on March 31.90, when V. Knorre (Nicolajew, now Mykolayiv, Ukraine) measured it as $\alpha = 6^h\ 44.8^m$, $\delta = +31°\ 05'$. The comet was last detected on April 1.9, when Pons noted that he only suspected its existence.

The first parabolic orbit was calculated by Nicolai. He took his positions from January 4 to January 10 and determined the perihelion date as 1823 December 9.80. Slightly closer representations were obtained by J. F. Encke and Hansen within the next few days. Encke took positions from January 2 to January 11, and calculated a perihelion date of December 9.98, while Hansen took positions obtained during the period of January 2–12, and found a perihelion date of December 9.97. These two orbits very accurately represented the comet's motion, with further computations by Carlini, Gambart, J. N. Nicollet, Nicolai, Encke, Hansen, and J. C. E. Schmidt offering only slight improvements during the next couple of months. The most complete analysis to date was worked out by A. Hnatek (1912). He took about 200 positions spanning the period of December 31 to March 7, applied perturbations by Mercury to Jupiter, and determined the orbit below.

T	ω	Ω (2000.0)	i	q	e
1823 Dec. 9.9340 (UT)	28.4867	305.5054	103.8194	0.226742	1.0

ABSOLUTE MAGNITUDE: $H_{10} = 4.2$ (V1964)

FULL MOON: Dec. 17, Jan. 16, Feb. 14, Mar. 15, Apr. 13

SOURCES: J. L. Pons, G. Santini, F. Carlini, B. A. von Lindenau, J. F. Encke, Schulz, *CA*, **9** (1823), pp. 595–601; J. E. Bode, Schulz, W. von Biela, M. A. David, F. I. C. Hallaschka, J. C. E. Schmidt, Westphal, and K. L. Harding, *BAJ for 1827*, **52** (1824), pp. 122–35; H. W. M. Olbers, *BAJ for 1827*, **52** (1824), pp. 184–5; F. B. G. Nicolai, *BAJ for 1827*, **52** (1824), pp. 218–19; J. L. Pons, *CA* (1824), pp. 89–90; Nell de Breautè, *CA*, **10** (1824), p. 186; G. Santini, J. F. Encke, and J. L. Pons, *CA*, **10** (1824), pp. 289–94; W. von Biela, M. A. David, and F. B. G. Nicolai, *AN*, **2** (1824 Jan.), p. 455; P. A. Hansen, W. von Biela, H. W. M. Olbers, F. B. G. Nicolai, and K. L. Harding, *AN*, **2** (1824 Jan.), pp. 465–80; J. F. Encke, P. A. Hansen, and F. B. G. Nicolai, *AN*, **2** (1824 Jan.), pp. 491–6; H. W. M. Olbers and Schulz, *AN*, **2** (1824 Jan.), circular No. 48; H. W. M. Olbers and K. L. Harding, *AN*, **3** (1824 Feb.), pp. 5–10; W. von Biela, F. I. C. Hallaschka, P. A. Hansen, and J. Soldner, *AN*, **3** (1824 Feb.), pp. 27–32; J. N. Nicollet and H. W. M. Olbers, *AN*, **3** (1824 Mar.), pp. 45–8; H. W. M. Olbers, *AN*, **3** (1824 Apr.), p. 89; F. B. G. Nicolai and J. F. Encke, *AN*, **3** (1824 Apr.), pp. 109–14; J. F. Encke, K. L. von Littrow, and M. A. David, *AN*, **3** (1824 Apr.), pp. 113–18; M. A. David, *AN*, **3** (1824 Apr.), pp. 117–20; F. G. W. von Struve, *AN*, **3** (1824 Jun.), pp. 183–6; K. L. Harding, *AN*, **3** (1824 Jul.), p. 193; V. Wisniewski, *AN*, **3** (1824 Oct.), p. 285; J. Taylor, F. B. G. Nicolai, H. C. Schumacher, P. A. Hansen, F. Carlini, J. Brinkley, and W. Colburn, *AJS* (Series 1), **8** (1824 Aug.), pp. 315–16; N. Cacciatore, *Del Reale Osservatorio di Palermo* libri VII, VIII e IX (1826), p. 28; V. Knorre, *AN*, **5** (1826 Aug.), pp. 101–10; J. F. A. Gambart, *CDT* (1827), pp. 313–15; J. F. A. Gambart, *CDT* (1828), pp. 273–7; H. W. M. Olbers, *AN*, **8** (1831 Jan.), pp. 469–72; *American Almanac*. Boston: James Munroe & Co. (1847), p. 90; G1894, p. 193; O1899, p. 103; A. Hnatek, *DAWW*, **87** (1912), pp. 1–91; J. Holetschek, *DAWW*, **88** (1913), p. 805; M. F. Baldet, *Annuaire Pour l'an 1950*. Paris: Gauthier-Villars (1950), p. B.56; V1964, p. 55; HA1970, p. 86; NGC2000.0, pp. 7, 191.

C/1824 N1   *Discovered:* 1824 July 14.4 ($\Delta = 0.76$ AU, $r = 0.59$ AU, Elong. = 35°)

(Rümker)   *Last seen:* 1824 August 11.37 ($\Delta = 1.54$ AU, $r = 0.88$ AU, Elong. = 32°)

Closest to the Earth: 1824 July 6 (0.6684 AU)

1824 I   *Calculated path:* SEX (Disc), LEO (Jul. 18)

Just three days after full moon, C. K. L. Rümker (Picton, New South Wales, Australia) discovered this comet from his farm (named Stargard) on 1824 July 14.4. He said it exhibited a nucleus and a very faint tail. Rümker confirmed his discovery on July 15.38 and estimated the position as $\alpha = 9^h\ 45.3^m$, $\delta = +3°\ 27'$.

Rümker soon informed T. M. Brisbane (Paramatta, New South Wales, Australia), who made his first observation on the evening of July 28. These two astronomers remained the only observers of this comet, which was then steadily moving away from both the sun and Earth. Its altitude also decreased in the days that followed discovery. Rümker obtained his final observation on August 6. Brisbane saw the comet for the final time on August 11.37 and measured the position as $\alpha = 11^h 38.7^m$, $\delta = +22° 54'$. He said it was hardly visible.

The short period of visibility was only sufficient for a parabolic orbit. Rümker (1826) calculated the first orbit using his positions obtained during the period of July 15 to August 6. The resulting perihelion date was 1824 July 12.01. A. W. Doberck (1874, 1895) has calculated the only other orbits, which used most of the positions determined by Rümker and Brisbane. These differed little from Rümker's orbit. Doberck's 1895 orbit is given below.

T	ω	Ω (2000.0)	i	q	e
1824 Jul. 12.0188 (UT)	334.0800	236.7952	125.4324	0.591671	1.0

ABSOLUTE MAGNITUDE: $H_{10} = 7$ (V1964)

FULL MOON: Jul. 11, Aug. 9, Sep. 8

SOURCES: C. K. L. Rümker, *AN*, **3** (1825 Jan.), p. 455; T. M. Brisbane, *MAS*, **2** (1826), pp. 281–2; C. K. L. Rümker, *MAS*, **2** (1826), p. 284; C. K. L. Rümker, *PTRSL*, **119** (1829), pp. 58–9; A. W. Doberck, *MNRAS*, **34** (1874 Jun.), p. 426; A. W. Doberck, *AN*, **84** (1874 Jun. 16), pp. 75–80; A. W. Doberck, *AN*, **138** (1895 Aug. 31), pp. 321–31; V1964, p. 55.

C/1824 O1
(Scheithauer)

1824 II

Discovered: 1824 July 23.9 ($\Delta = 0.64$ AU, $r = 1.52$ AU, Elong. $= 132°$)
Last seen: 1824 December 25.92 ($\Delta = 0.82$ AU, $r = 1.74$ AU, Elong. $= 148°$)
Closest to the Earth: 1824 July 23 (0.6351 AU)
Calculated path: OPH (Disc), HER (Jul. 24), CrB (Aug. 22), HER (Aug. 27), BOO (Sep. 10), DRA (Oct. 10), UMa (Oct. 26), DRA (Oct. 30), UMi (Nov. 13), DRA (Nov. 15), CAM (Dec. 1), AUR (Dec. 23)

C. F. Scheithauer (Chemnitz, now Karl Marx Stadt, Germany) discovered this comet on 1824 July 23.9 and estimated the position as $\alpha = 17^h 56^m$, $\delta = +14°$. Independent discoveries were made by J. L. Pons (Royal Park La Marlia Observatory, Italy) on July 24.89, J. F. A. Gambart (Royal Observatory, Marseille, France) on July 27.90, and K. L. Harding (Göttingen, Germany) on August 2.91. Pons simply described the comet as very small, while Gambart said the comet was extremely faint, without a nucleus or a tail. Harding said it was not visible to the naked eye, but a seeker revealed an excellent nucleus, but no tail.

The comet remained a very faint object for most of August, but showed signs of brightening late in the month. Harding reported the nucleus he saw on the 2nd was no longer visible on the 3rd and 5th. H. W. M. Olbers (Bremen, Germany) said the moon hampered his early observations on the morning of August 7 and the comet was extremely difficult to see. Olbers said the comet was still very faint because of moonlight on the 12th, but seemed slightly brighter on the 15th. W. von Biela (Prague, Czech Republic) saw the comet on August 13 and described it as "a very faint nebulosity, without a nucleus or tail." After the moon was no longer an issue, Olbers saw the comet on August 17 and described it as very faint and without a certain nucleus. Under excellent skies on the 23rd, Olbers said the comet was easy to see and exhibited a well-defined nucleus. Olbers said the nucleus seemed to vanish as the comet passed very close to a star of magnitude 11 or 12 on the 26th. On August 28 Olbers noted the comet had "obviously increased in light." The only hints of a tail for the entire apparition came on August 20 and 25. On the first date, Pons suspected a very faint extension, while, on the second date, Olbers said the very faint coma seemed to extend away the sun. Some brightness fluctuations were reported during the later half of the month by E. Capocci (Naples, Italy) and F. Carlini (Milan, Italy). Carlini also noted the comet was continually near the limit of visibility in his telescope, so it is uncertain whether the fluctuations were a real phenomenon involving the comet, or simply changing conditions in the atmosphere.

The comet was obviously brighter in September, with Olbers reporting on the 2nd that it was easy to see despite moonlight. With stronger moonlight present on the 11th, Olbers said the comet was faintly seen, but later noted it was still brighter and easier to see than on the 14th. K. L. von Littrow (Vienna, Austria) reported a distinct nucleus on September 15 and 16, while Olbers reported the nucleus was bright, but diffuse on the 18th. Capocci noted the comet was at its brightest around September 20. Olbers described the comet as very bright on the 26th.

The comet was still an apparently bright object at the beginning of October, as Olbers noted on the 3rd that the comet was easy to see despite bright moonlight. Capocci described the comet as faint at the end of October and even fainter as November began. Olbers said the comet was small and hard to detect in the cometseeker on November 15, although a central condensation was still present. Gambart found the comet faint and difficult to observe on the 27th. The comet attained its most northerly declination of +77° on November 28. On November 29, Gambart said the comet was faint and difficult to observe.

Pons saw the comet for the final time on December 24. The comet was last detected on December 25.92, when Capocci determined the position as $\alpha = 5^h 19.0^m$, $\delta = +52° 52'$.

One of the earliest orbits was calculated by O. A. Rosenberger using the European positions gathered from late July up to mid-August. His parabolic orbit revealed a perihelion date of 1824 October 2.10. A few weeks later, A. Bouvard took positions obtained during the period of August 4 to September 15 and determined the perihelion date as September 29.74. This orbit was a much closer representation of the comet's true motion. Later orbits were calculated by P. A. Hansen, J. F. Encke, and F. W. A. Argelander.

Encke had also calculated hyperbolic orbits early on. Even though he had determined a parabolic orbit from positions obtained during the period of July 27 to September 21, he also found that a hyperbolic orbit fitted as well. The result was a perihelion date of September 29.49 and an eccentricity of 1.006046. A month later he added positions through October 26 and found a perihelion date of September 29.56 and an eccentricity of 1.0017345. After the final observations of the comet had been obtained, Encke (1825) took 150 positions obtained during the period of July 26 to December 25. He found that a parabolic orbit fitted the positions best and determined a perihelion date of September 29.57. This orbit is given below.

T	ω	Ω (2000.0)	i	q	e
1824 Sep. 29.5664 (UT)	85.2292	281.7415	54.5985	1.049835	1.0

ABSOLUTE MAGNITUDE: $H_{10} = 6.5$ (V1964)

FULL MOON: Jul. 11, Aug. 9, Sep. 8, Oct. 8, Nov. 6, Dec. 6, Jan. 4

SOURCES: C. F. Scheithauer, K. L. Harding, W. von Biela, and Rosenberger, *BAJ for 1827*, **52** (1824), pp. 200–1; J. L. Pons, F. Carlini, E. Capocci, *CA*, **11** (1824), pp. 98, 192–3, 384, 588–90; K. L. Harding, *AN*, **3** (1824 Aug.), p. 241; J. F. A. Gambart and A. Bouvard, *AN*, **3** (1824 Sep.), p. 257; A. Bouvard, *AN*, **3** (1824 Oct.), p. 313; P. A. Hansen and J. F. Encke, *AN*, **3** (1824 Oct.), pp. 321–6; F. W. A. Argelander, *AN*, **3** (1824 Nov.), p. 353; K. L. von Littrow, *AN*, **3** (1824 Dec.), p. 367; *BAJ for 1828*, **53** (1825), p. 125; J. F. Encke, *CA*, **12** (1825), pp. 503–4; J. F. A. Gambart, *AN*, **3** (1825 Jan.), p. 455; J. F. Encke and E. Capocci, *AN*, **4** (1825 Jun.), p. 123; F. W. A. Argelander, *AN*, **4** (1825 Nov.), pp. 283–6; T. M. Brisbane, *MAS*, **2** (1826), pp. 283–4; O1899, pp. 104–12; V1964, p. 55.

C/1825 K1
(Gambart)

1825 I

Discovered: 1825 May 19.11 ($\Delta = 1.15$ AU, $r = 0.91$ AU, Elong. $= 49°$)
Last seen: 1825 July 15.38 ($\Delta = 1.51$ AU, $r = 1.20$ AU, Elong. $= 52°$)
Closest to the Earth: 1825 June 10 (0.7808 AU)
Calculated path: CAS (Disc), CEP (Jun. 3), CAM (Jun. 4), CEP (Jun. 5), CAM (Jun. 6), DRA (Jun. 10), UMa (Jun. 11), LEO (Jul. 1)

J. F. A. Gambart (Marseille, France) discovered this comet in the morning sky on 1825 May 19.11, at a position of $\alpha = 0^h\ 20.0^m$, $\delta = +48°\ 22'$. He described it as round and 2′ across, with a condensation near the center. Gambart confirmed his discovery on May 19.90, when it was then a very difficult object to observe at low altitude in the evening sky.

Only a handful of descriptions came during the remainder of May. Gambart described the comet as round and about 2′ across on the 20th, while it was about 5′ across on the 28th. Although he did not see a tail on the first date, he said a possible trace was seen in a refractor on the second date. H. C. Schumacher (Altona, Germany) observed the comet near 50 Cassiopeiae on May 31. He described the comet as fairly distinct, with a tail.

The comet attained its maximum northern declination of +81° on June 6. J. L. Pons (Marlia, Italy) saw the comet on the 7th and said it was centrally condensed, but did not display a nucleus or tail. He did suspect the coma was slightly elongated away from the sun. The tail was very well seen on the 8th by both Gambart and Pons. Gambart said the tail was very faint and about 40′ long. Pons said it was faint, slightly "frayed", and about 1.5° long. Pons found the tail unchanged on the 9th. The coma was estimated as 7′ across by F. M. Schwerd (Speyer, Germany) on June 10. The comet was very close to M81 and M82 on June 12. H. W. M. Olbers (Bremen, Germany) said the comet "far exceeded these objects in size, brightness, and luminosity" (M81 is magnitude 6.9 and 25.7′ across; M82 is magnitude 8.4 and 11.2′ long – NGC2000.0). On both the 12th and 14th Gambart noted the comet was diminishing in brightness. On the 18th Gambart said the central condensation was becoming less distinct. The comet attained a maximum elongation of 61° on the 20th. On June 25.01, Olbers noted that the comet passed over a star and completely disappeared except for the slightest trace of nebulosity that probably would have gone unnoticed if he had not known the comet was in that position. The Author identified this star as SAO 62382, which has a magnitude of 7.86. Gambart described the comet as faint on June 26 and Olbers last saw the comet on the 27th. Moonlight blocked the comet from view for F. I. C. Hallaschka (Prague, Czech Republic) on June 30.

Hallaschka saw only a faint trace of the comet on July 2. C. K. L. Rümker (Stargard, New South Wales, Australia) independently discovered this comet on July 9.40. He reobserved the comet every night thereafter, and obtained the final observation of this comet on July 15.38, when the position was $\alpha = 11^h\ 16.0^m$, $\delta = +13°\ 54'$.

The first parabolic orbit was calculated by Schumacher using positions obtained on May 20, 21, and 22. He determined the perihelion date as 1825 May 31.44. This was just 10 hours from what the perihelion date would ultimately prove to be. Very similar orbits were determined during the following weeks and months by Schwerd (1825), Gambart (1825), F. Carlini (1825), K. L. Harding (1825), F. B. G. Nicolai (1825), T. Clausen (1825), and Rümker (1826). Olbers and Gambart remarked on the similarity between the orbit of this comet and that of the third comet of 1790, but a link was ruled out when it became obvious that C/1825 K1 was not moving in a short-period orbit.

An elliptical orbit was calculated by H. Boegehold (1908). Using 132 positions obtained during the period of May 19 to July 15, he applied perturbations by three planets and determined the perihelion date as May 31.02 and the period as 3873 years. Boegehold said the eccentricity was uncertain because of the two-month arc.

T	ω	Ω (2000.0)	i	q	e
1825 May 31.0226 (UT)	106.1910	22.5831	123.3414	0.889011	0.996395

ABSOLUTE MAGNITUDE: $H_{10} = 5.6$ (V1964)

FULL MOON: May 2, Jun. 1, Jun. 30, Jul. 29

SOURCES: H. W. M. Olbers, F. B. G. Nicolai, and T. Clausen, *BAJ for 1828*, **53** (1825), pp. 150–2; K. L. Harding and F. B. G. Nicolai, *BAJ for 1828*, **53** (1825), pp. 192–5; J. L. Pons, *CA*, **12** (1825), pp. 513–14, 609–11; F. Carlini and J. F. A. Gambart, *CA*, **13** (1825), p. 84–7; J. F. A. Gambart and H. C. Schumacher, *AN*, **4** (1825 Jul.), (circular); F. M. Schwerd, *AN*, **4** (1825 Aug.), pp. 177–82; J. F. A. Gambart, *AN*, **4** (1825 Oct.), pp. 225–8; F. I. C. Hallaschka and W. von Biela, *BAJ for 1829*, **54** (1826), pp. 107–9; C. K. L. Rümker, *AN*, **4** (1826 Jun.), p. 511; J. F. A. Gambart, *CDT* (1829), pp. 322–8; C. K. L. Rümker, *MAS*, **3** (1829), p. 101; C. K. L. Rümker, *PTRSL*, **119** (1829), p. 60; J. F. A. Gambart, *CDT* (1830), pp. 121–5; O1899, p. 112; H. Boegehold, *EAN*, **2** (1908), p. 25; V1964, p. 55; NGC2000.0, p. 86.

C/1825 N1 *Discovered:* 1825 July 15.1 ($\Delta = 2.74$ AU, $r = 2.34$ AU, Elong. $= 56°$)
(Pons) *Last seen:* 1826 July 8.87 ($\Delta = 2.90$ AU, $r = 3.11$ AU, Elong. $= 92°$)
Closest to the Earth: 1825 October 12 (0.6177 AU)
1825 IV *Calculated path:* TAU (Disc), CET (Sep. 27), ERI (Sep. 28), CET (Oct. 4), FOR (Oct. 9), SCL (Oct. 11), PHE (Oct. 18), GRU (Oct. 22), IND (Nov. 5), MIC (Nov. 16), SGR (Nov. 21), CrA (Feb. 21), SCO (Mar. 31), LUP (Apr. 23), CEN (May 6), HYA (May 8), VIR (May 21)

J. L. Pons (Marlia, Italy) discovered this comet on 1825 July 15.1, while searching for 2P/Encke. The comet was independently discovered by W. von Biela (Josephstadt, Austria) on July 20.00 at a position of $\alpha = 4$h 06.9m, $\delta = +26°$ 06'. He described the comet as small and pale. Another independent discovery was made by J. Dunlop (Paramatta, New South Wales, Australia) on July 21.81. He noted a faint tail about 2' long.

Only a few physical descriptions were provided during the remainder of July. Biela saw the comet on the 22nd and said the coma contained a small nucleus and exhibited a trace of a tail. Pons described the comet as very faint and slightly elongated on July 26. Dunlop said the tail extended 15' on the 27th.

As August began, the comet was still located 2.54 AU from Earth and 2.27 AU from the sun. Although the comet and its features were still faint that changed as the month progressed. K. L. Harding (Göttingen, Germany) said the comet appeared as an extended nebulosity, with a very faint tail on August 10. H. W. M. Olbers (Bremen, Germany) saw the comet on the same night and described it as a shapeless, faint, nebulous mass, with a central condensation. Harding found the tail was 1.5° long on the 11th, while Olbers described it as "short" on the same night. Both men noted the tail was curved. Olbers observed that the comet seemed a little more obvious on the 14th, and brighter still on the 15th. On the last date, Olbers said the comet exhibited a small, diffuse nucleus and very weak traces of a tail. The moon began causing problems a few days later. Olbers said the faint tail was clearly seen after the moon had set on August 23, and the comet was faint in bright moonlight on the 24th. On the 25th, Harding said the tail was completely straight and about 1.75° long. During the period of August 26–29, E. Capocci (Naples, Italy) said the comet was surrounded by an extensive nebulosity and exhibited a broad tail over 1° long. He added that it was very faint in moonlight. In a letter written to the *Correspondance Astronomique* on August 27, Pons said the comet had become visible to the naked eye before the moon began interfering. Olbers said the comet could only be seen with difficulty in the bright moonlight on August 30.

The comet was still beyond the orbit of Mars as September began, but additional naked-eye observations were being reported. Olbers said the comet was visible to the naked eye on the 8th, if you knew exactly where to look. He added that a telescope revealed a diffuse nucleus and a faint tail extending about 3°. N. Cacciatore (Palermo, Italy) saw the comet on September 9 and reported it was about 2' across with a tail 2° long. He noted that the comet vanished in the telescope when the wires were illuminated to measure its position. Interestingly, Pons would report the same thing on the 17th. Göbel (Coburg, Germany) described the comet as an ill-defined, centrally condensed, round nebulosity on September 11, and noted an occasionally visible stellar nucleus. The tail steadily grew as the month progressed, with Dunlop reporting the tail as 2.5° long on the 12th, and Olbers reporting its length as 5° on the 15th and 6° on the 19th. Olbers added that the comet was an easy naked-eye object on the last date. Pons reported several changes in the tail before moonlight began interfering. On the 18th, he found the tail had become narrower and sharper since previously seen, and it fanned out "like the tail of a peacock" at the end. On the 20th, Pons found the tail's center more luminous and noted that the point mid-way between the coma and the tip of the tail was "inflated." The tail had decreased in size by the 21st, although the coma had grown larger and was displaying a stronger nucleus. Pons found the tail very faint on the 22nd, but noted it was divided into several branches. Meanwhile, the comet appeared brighter and seemed surrounded by a large, faint nebulosity. This large, faint nebulosity was gone by the next night, but the tail had lengthened and sharpened. Pons said the tail had widened toward the end when seen on September 24. Olbers saw the comet in strong moonlight on the 29th and found an occasionally visible stellar nucleus.

The comet was brightest and best observed during October. Pons said the comet was visible to the naked eye, despite moonlight, on the 1st. On October 3, Olbers said the comet was just as bright to the naked eye as light clouds illuminated by the moon. That same night, J. F. W. Herschel observed the comet with a 15-cm reflector when it was very close to the moon and wrote, "Head large, and pretty dense, – not however condensing very much in the approach towards the centre, but suddenly coming to a distinct stellar nucleus, which is nearly as bright, only not quite so sharp, as a star of the 9th or 10th magnitude in the field with it – The nucleus is not quite in the centre of the head, but rather toward the tail side." Herschel saw the comet again on the 4th. He said the nucleus was stellar in the 15-cm reflector, but took on the appearance of a planetary disk without a sharp outline in the 46-cm reflector. He said the brightest part of the "nucleus" was 10–15″ across. Herschel summarized saying that the comet was "a magnificent sight" and added that the tail was 7° or 8° long prior to moonrise. That same night, Dunlop found the tail was clearly 10–11° long. With the moon no longer an issue, observers were reporting the comet was easily seen with the naked eye on October 5. In addition, the tail also appeared brighter, and was estimated as 5° long by F. I. C. Hallaschka (Prague, Czech Republic), over 7° long by Olbers, and 10° long by Göbel and Dunlop. The next night, Cacciatore reported a coma 3.5′ across and a central stellar nucleus. Pons also saw a very small, round nucleus, but G. Santini (Padova, Italy) saw three bright points at the position of the nucleus when the comet was viewed in the 4-foot focal length refractor. The tail length was reported as 7° long by Santini, 7–8° by Cacciatore, 8° by Pons, and nearly 10° by A. Lang (St. Croix, Virgin Islands). A change was noted in the tail by October 7, when Herschel reported it was divided into two distinct branches and Pons noted three. Dunlop gave the tail length as 9°. Herschel added that the 46-cm reflector revealed "no sharp, star-like centre, but a much brighter yet quite milky, round kernel of about 15″ to 20″ in diameter, shading insensibly but almost suddenly away." The comet remained a bright naked-eye object for the next several days, with telescopes revealing a stellar nucleus. Naked-eye tail lengths were given as 11° by Olbers on the 8th, 13° by Lang and 13.5° by C. J. Pastorff (Buchholz, Germany) on the 9th, and 9° by Santini and 11–12° by Dunlop on the 10th.

The comet was probably at its best around mid-October, thanks, in part, to it passing 0.6 AU from Earth on the 12th. On the 11th, Pons wrote, "The comet was in grand costume, never before so elegant; – with trailing robe, it only needs arms to resemble a lady of high rank of the past century." The coma was given as 3′ across by Cacciatore, while the tail length was given as 7–8° by Dunlop and 11° by Cacciatore. When closest to Earth on the 12th, Pons reported the tail was curved, appearing thin near the coma and broad toward the end. Harding said the tail was then 14° long, while the nucleus was about magnitude 10. Many of the more northerly observers in the Northern Hemisphere did not see the comet after this date, but those in more southerly latitudes and in the Southern Hemisphere continued to watch the comet closely. Pons was the only observer to report a physical description on the 13th and it was an interesting one. He wrote that "its adornings appeared to have fallen to rags. . . ." He added that the tail was very formless and strongly curved near the middle. The next night, Pons wrote that the comet "seemed to have a new robe without tear; it was modestly veiled like a nun." Hallaschka saw only part of the tail extending above the horizon, while Dunlop reported the length was 10–11°. During the next few nights, descriptions were only provided by Pons and Dunlop. On the

15th, Pons said the tail was little changed, except for the appearance of two rays. Dunlop gave the length as 11–12°. On the 16th, Pons said the comet's nearness to the horizon caused the two rays to appear very weak. On the 17th, Pons said the tail appeared faint and slightly curved. This marked Pons' final observation of 1825. The comet was last seen in the Northern Hemisphere on October 20.86, when Cacciatore found it about 8° above the horizon.

Southern Hemisphere observers continued their observations, though physical descriptions were few. For the remainder of October, Dunlop gave tail lengths of 12° on the 18th, 8–9° on the 19th, 7° on the 20th, and 3.5° on the 24th. C. K. L. Rümker (Stargard, New South Wales, Australia) began observing the comet around this time and measured precise positions at every possible opportunity. This included measurements, when clear, during both morning and evening hours from October 19 to the end of the month. J. Reeves was on a ship sailing just north of Indonesia on October 30 when he saw the comet so close to α Gruis that the star "appeared to form the nucleus of the comet." On October 31, G. Peard (HMS *Blossom*, sailing with Frederick William Beechey off the coast of Chile) described the comet as "extremely brilliant...and rather curved towards the east."

The comet attained its most southerly declination of −47° on November 1, at which time Dunlop gave the tail length as 7°. Dunlop reported the tail was unchanged the next night, while Troughton (Buenos Aires, Argentina) reported the tail was curved toward the north on November 3. There are indications that the tail began lengthening again as November progressed, but this is based on only a handful of observations. Dunlop reported a tail 9° long on the 7th and 10° long on the 8th, while Troughton gave it as 20° long on the 13th. Moonlight and increasingly poor positioning in the sky reduced descriptive observations for the rest of the year. The remaining observers began losing sight of the comet from mid-December onward. Lang last saw the comet on December 14.95, when it was about 5° above the horizon. Troughton last saw the comet on December 19.05 and noted it had greatly decreased in brightness since his earlier observations. Rümker last detected the comet low over the horizon on December 20.43 and Dunlop last detected it low over the horizon on December 24.44.

The comet passed 18° from the sun on 1826 January 9 and was lost in the sun's glare for some time. Previous comet catalogs do not report a recovery prior to the first days of April, but the Author has uncovered a much earlier observation. Aboard HMS *Blossom*, Peard was sailing among the islands of French Polynesia in the South Pacific, when on February 7.5 he saw a comet "in the East & close to the Constellation of the Southern Crown." Although the editor of the 1973 edition of Peard's journal included a footnote correcting the constellation of the "Southern Crown" to the Southern Cross, the fact is that this comet was very near the "Southern Crown," better known as Corona Australis, on this date and actually entered that constellation on February 21.

One of the earliest documented attempts to find the comet in European skies came from Cacciatore. He searched for the comet in the morning sky on March 20. The comet would then have been at a low altitude, and Cacciatore said he found it after the beginning of twilight. Unfortunately, a period of bad weather followed and it was not until the morning of April 3 that he could again look. Cacciatore then realized he had not seen the comet, but one of the globular clusters in that region of the sky.

The comet was finally recovered on April 2.09, when Pons remarked, "It has been totally ruined . . . it has neither tail nor beard, neither envelope nor nucleus,—it is only a ghost, a white smoke!" J. E. B. Valz (Nîmes, France) failed to see the comet on the 2nd, but was able to see a weak nebulosity at low altitude on the morning of the 3rd. The comet was finally seen by Cacciatore on April 4 and 5. He said it was about 3′ across and exhibited a tail extending about 40′ southwestward.

Few descriptive observations are available after the comet's recovery. Pons noted the tail was about 1° long on April 17, and even suspected a central condensation. A. Stark (Augsburg, Germany) said he "discovered" the comet on the evening of May 3. He said it was "uncommonly faint," with a condensation, but no noticeable nucleus. G. Inghirami was the only observer of the comet after June 8. His final observation came on July 8.87, when he was able to determine that $\alpha = 13^h\ 04.2^m$, but did not obtain a value for δ.

Early computations of the orbit of this comet came slowly primarily because of poor communications. When orbits began being published during November of 1825, they did not include positions obtained prior to September. One of the first parabolic orbits was calculated by P. A. Hansen using positions obtained on September 2, 9, and 18. The resulting perihelion date was 1825 December 10.89. Additional parabolic orbits were determined over the next couple of months by Hallaschka, C. A. F. Peters, Tallquist, E. Capocci, F. M. Schwerd, and J. J. Morstadt, which revised the perihelion date to December 11.02.

The first elliptical orbit was calculated by P. A. Hansen. He used positions obtained on September 2, 17, and 30, and determined a perihelion date of December 11.96 and a period of 382 years. Hansen revised the orbit a couple of months later and determined the perihelion date as December 11.77 and the period as 556 years. Another revision by Hansen in the middle of 1826, took positions into 1826 and gave the perihelion date as December 11.18 and the period as about 4386 years. The next revision came from Rümker (1826), who used positions from October 2, 30, and December 20, to determine a perihelion date of December 11.28 and a period of 146 years.

The most comprehensive determination of this comet's orbit came from J. S. Hubbard (1859). He used about 275 positions obtained between 1825 July 21 and 1826 July 8, as well as perturbations by five planets, and computed an elliptical orbit with a perihelion date of December 11.18 and a period of about 4472 years. This orbit is given below.

T	ω	Ω (2000.0)	i	q	e
1825 Dec. 11.1849 (UT)	256.9173	218.1285	146.4354	1.240846	0.995429

ABSOLUTE MAGNITUDE: H_{10} − 2.2 (V1964)

FULL MOON: Jun. 30, Jul. 29, Aug. 28, Sep. 27, Oct. 26, Nov. 25, Dec. 25, Jan. 23, Feb. 22, Mar. 23, Apr. 22, May 21, Jun. 19, Jul. 19

SOURCES: K. L. Harding, *BAJ for 1828*, **53** (1825), p. 192; H. E. E., *BAJ for 1828*, **53** (1825), p. 221; J. L. Pons, G. Plana, E. Capocci, P. A. Hansen, G. Santini, and N. Cacciatore, *CA*, **13** (1825), pp. 183–5, 279–83, 386–96, 487–98, 589–96; W. von Biela, *AN*, **4** (1825 Aug.), (circular); P. A. Hansen, *AN*, **4** (1825 Nov.), pp. 257–62; F. W. A. Argelander, F. M. Schwerd, Tallquist, and Göbel, *AN*, **4** (1825 Nov.), pp. 281–4, 291–5; F. I. C. Hallaschka, *BAJ for 1829*, **54** (1826), pp. 100–1; F. I. C. Hallaschka and J. Schwarzenbrunner, *BAJ for 1829*, **54** (1826), pp. 109–14; C. K. L. Rümker and H. W. M. Olbers, *BAJ for 1829*, **54** (1826), pp. 142–4; C. J. Pastorff, *BAJ for 1829*, **54** (1826), pp. 148–9; F. B. G. Nicolai, *BAJ for 1829*, **54** (1826), p. 171; A. Stark, *BAJ for 1829*, **54** (1826), pp. 188–9; G. Plana, J. L. Pons, J. E. B. Valz, and N. Cacciatore, *CA*, **14** (1826), pp. 94–5, 402–11; J. F. W. Herschel, *MAS*, **2** (1826), pp. 486–7; F. M. Schwerd, *AN*, **4** (1826 Jan.), pp. 343–6; P. A. Hansen, F. I. C. Hallaschka, and C. A. F. Peters, *AN*, **4** (1826 Feb.), pp. 359–62, 379; Morstadt, *AN*, **4** (1826 Mar.), p. 395; A. Lang, *AN*, **4** (1826 Apr.), pp. 449–52; P. A. Hansen, *AN*, **5** (1826 Jun.), p. 32; G. Inghirami and K. L. Harding, *AN*, **5** (1826 Sep.), pp. 145–54; J. Schwarzenbrunner, *AN*, **5** (1826 Oct.), pp. 171–4; K. L. Harding, *AN*, **5** (1826 Dec.), p. 269; J. Dunlop, *Edinburgh Journal of Science*, **6** (1827), p. 93; Reeves, *MNRAS*, **1** (1827 May 11), p. 27; C. K. L. Rümker, *MAS*, **3** (1829), pp. 100–1, 379–84; C. K. L. Rümker, *PTRSL*, **119** (1829), pp. 60–5; H. C. Dwerhagen and E. Troughton, *AN*, **10** (1832 Oct. 9), pp. 253–6; J. S. Hubbard, J. L. Pons, W. von Biela, J. F. W. Herschel, G. Santini, F. W. A. Argelander, H. W. M. Olbers, Göbel, and G. Inghirami, *AJ*, **6** (1859 May 10), pp. 17–22; J. S. Hubbard, J. L. Pons, K. L. Harding, W. von Biela, F. B. G. Nicolai, J. Soldner, E. Capocci, J. E. B. Valz, G. Santini, N. Cacciatore, J. Dunlop, David, F. I. C. Hallaschka, and P. A. Hansen, *AJ*, **6** (1859 May 23), pp. 26–31; J. S. Hubbard, F. M. Schwerd, C. K. L. Rümker, A. Lang, and G. Plana, *AJ*, **6** (1859 Jun. 14), pp. 33–7; O1899, pp. 113–23; V1964, p. 55; *To the Pacific and Arctic with Beechey: The Journal of Lieutenant George Peard of H.M.S. 'Blossom.'* Edited by Barry M. Gough. Cambridge: Cambridge University Press (1973), pp. 69–70, 108.

C/1825 P1   *Discovered:* 1825 August 9.05 (Δ = 0.95 AU, r = 0.90 AU, Elong. = 54°)

(Pons)   *Last seen:* 1825 August 27.11 (Δ = 0.67 AU, r = 0.90 AU, Elong. = 60°)

Closest to the Earth: 1825 September 2 (0.6431 AU)

1825 II   *Calculated path:* AUR (Disc), TAU–AUR–GEM (Aug. 18), ORI (Aug. 21), MON (Aug. 26)

J. L. Pons (Florence, Italy) discovered this comet on 1825 August 9.05. He said it was not visible to the naked eye and was "so faint, it was almost not worth speaking of." He observed it for one hour, but noted no sensible movement. G. Inghirami (Florence, Italy) confirmed Pons' find on August 11.09, and gave the position as $\alpha = 5^h\ 32.1^m$, $\delta = +39°\ 22'$.

The comet remained a morning sky object, which steadily moved south-ward throughout the period of visibility. Inghirami obtained additional precise positions on August 12 and 13. Following a break of one week, he resumed observations on August 21, and obtained his final observations on August 23, 24, 25, and 26.

K. L. Harding (Göttingen, Germany) independently discovered this comet near γ Geminorum on August 24.05. He described it as a round nebulosity with a distinct nucleus, but no tail. He obtained additional observations on August 25 and 26, before seeing the comet for the final time on August 27.11, at which time he measured the position as $\alpha = 6^h$ 15.9^m, $\delta = +9°\ 27'$. The comet was initially lost because of moonlight, but when dark skies returned it was too far south for observations.

The first orbit was computed by H. W. M. Olbers. His parabolic solution was based on positions obtained by Inghirami on August 11, 21, and 24, and indicated a perihelion date of 1825 August 18.86. Although later orbits by C. A. F. Peters and T. Clausen indicated a perihelion date during the last days of August, Clausen ultimately revised his calculations using positions spanning the period of August 12–26. This included a revision of Inghirami's positions. The result was a perihelion date of August 19.21. This orbit is given below.

T	ω	Ω (2000.0)	i	q	e
1825 Aug. 19.2111 (UT)	177.2965	195.3795	89.6749	0.883471	1.0

ABSOLUTE MAGNITUDE: $H_{10} = 6.5$ (V1964)

FULL MOON: Jul. 29, Aug. 28

SOURCES: K. L. Harding, *BAJ for 1828*, **53** (1825), p. 191; J. L. Pons, *CA*, **13** (1825), pp. 185–7, 284–5; K. L. Harding, *AN*, **4** (1825 Sep.), (circular); H. W. M. Olbers, *BAJ for 1829*, **54** (1826), pp. 120–1; J. L. Pons, *BAJ for 1829*, **54** (1826), p. 222; G. Inghirami, C. A. F. Peters, and T. Clausen, *AN*, **4** (1826 Jan.), pp. 321–8; V1964, p. 55.

2P/Encke *Recovered:* 1825 July 13.07 ($\Delta = 1.79$ AU, $r = 1.35$ AU, Elong. $= 48°$)

Last seen: 1825 September 7.15 ($\Delta = 1.27$ AU, $r = 0.43$ AU, Elong. $= 17°$)

1825 III *Closest to the Earth:* 1825 August 28 (1.2347 AU)

Calculated path: TAU (Rec), AUR (Jul. 21), GEM (Aug. 11), CNC (Aug. 22), LEO (Aug. 31)

During June 1825, J. F. Encke applied planetary perturbations for the period of 1819–25, and predicted this comet would next arrive at perihelion on 1825 September 16.78. He supplied a daily ephemeris for July and August, but noted circumstances would be unfavorable. J. E. B. Valz (Nîmes, France) "suspected" the comet in the vicinity of 42 Tauri on 1825 July 13.07. Although this was very close to the predicted position, no formal measurement was possible and no announcement was made. Unfortunately, a long period of bad weather followed and Valz was not able to reobserve the comet until July 25.07. It was then about 1.5° south of ι Aurigae, which was once again close to the predicted position, but, although the comet had brightened slightly, it was still too faint to secure a precise position, and Valz still refrained from announcing the recovery. The comet was independently recovered by K. L. Harding (Göttingen, Germany) on July 27.05, and the position was given as $\alpha = 4^h 56.5^m$, $\delta = +31° 18'$. Valz finally measured the comet's precise position on July 27.10.

The comet attained its most northerly declination of +32° on August 6. Additional independent recoveries were made by G. Plana (Turin, Italy) on August 10 and J. L. Pons (Florence, Italy) on August 14. Plana's follow-up observations on the 11th and 12th revealed the comet as a small, circular nebulosity, with a small, but bright nucleus. H. W. M. Olbers (Bremen, Germany) said the comet was very easy to see on August 15, and was much brighter than 3D/Biela had been. N. Cacciatore (Palermo, Italy) described the comet as a faint nebulosity, not visible to the naked eye, and about 1.5' across on the 17th. Olbers simply noted the comet was very bright on the 23rd. Encke saw the comet on August 16, 24, 25, and 26, and noted it appeared very round and centrally condensed. F. W. A. Argelander (Åbo, now Turku, Finland) saw the comet on August 23 and 25. He said it was so bright it could be seen regardless of moonlight. F. M. Schwerd (Speyer, Germany) said the comet was 2' across on the 23rd and 24th. He added that although it was centrally condensed, no nucleus was seen. Argelander saw the comet in strong morning twilight on the 31st and said it appeared small and planet-like with no trace of nebulosity.

The comet's position was measured for the final time on September 7.15, when E. Capocci (Naples, Italy) gave it as $\alpha = 10^h 03.0^m$, $\delta = +15° 44'$. The comet was then at a very low altitude in morning twilight. He said it was impossible to find it in the following days, despite his efforts.

Minor refinements of this comet's orbit using positions from multiple apparitions as well as planetary perturbations have been published by numerous astronomers. Of most significance were the studies of Encke, F. E. von Asten (1877), and B. G. Marsden and Z. Sekanina (1974). These have typically revealed a perihelion date of September 16.77 and a period of 3.31 years. Marsden and Sekanina determined the nongravitational terms as $A_1 = +1.14$ and $A_2 = -0.03996$. The orbit of Marsden and Sekanina is given below.

T	ω	Ω (2000.0)	i	q	e
1825 Sep. 16.7717 (TT)	182.7699	336.9130	13.3794	0.344765	0.844869

ABSOLUTE MAGNITUDE: $H_{10} = 7$–8 (V1964)

FULL MOON: Jun. 30, Jul. 29, Aug. 28, Sep. 27

SOURCES: J. E. Bode and K. L. Harding, *BAJ for 1828*, **53** (1825), pp. 200–2; J. F. Encke, *CA*, **12** (1825), pp. 505–8; J. L. Pons, G. Santini, G. Plana, F. Carlini, J. E. B. Valz, and

E. Capocci, *CA*, **13** (1825), p. 187–92, 285–92, 379–86, 498–501, 596–7; J. F. Encke, *AN*, **4** (1825 Jun.), pp. 123–8; J. F. Encke, K. L. Harding, and F. W. A. Argelander, *AN*, **4** (1825 Sep.), p. 227; F. W. A. Argelander and F. M. Schwerd, *AN*, **4** (1825 Nov.), pp. 281, 285–92; F. I. C. Hallaschka, *BAJ for 1829*, **54** (1826), p. 109; F. B. G. Nicolai, *BAJ for 1829*, **54** (1826), pp. 170–1; N. Cacciatore, *Del Reale Osservatorio di Palermo libri VII, VIII e IX* (1826), p. 221; F. E. von Asten, *BASP* (Series 4), **22** (1877), p. 559; O1899, pp. 112–13; V1964, p. 55; B. G. Marsden and Z. Sekanina, *AJ*, **79** (1974 Mar.), pp. 415–16.

C/1825 V1 *Discovered:* 1825 November 7.01 ($\Delta = 1.92$ AU, $r = 2.80$ AU, Elong. $= 147°$)
(Pons)   *Last seen:* 1826 April 10.82 ($\Delta = 2.22$ AU, $r = 2.01$ AU, Elong. $= 65°$)
     Closest to the Earth: 1825 December 3 (1.8343 AU)
1826 II *Calculated path:* ERI (Disc), LEP (Mar. 26), ERI (Mar. 27), LEP (Mar. 28)

Although more than 200 positions were obtained for this comet during its 5 months of observation, actual published physical descriptions are sparse.

J. L. Pons (Florence, Italy) discovered this comet on 1825 November 7.01 about 4° west of γ Eridani. He described it as small and round, with a central condensation and a possible nucleus. G. Inghirami (Florence, Italy) confirmed the comet on November 16.91 and determined the position as $\alpha = 3^{h} 28.1^{m}$, $\delta = -17° 23'$. Inghirami noted the comet was not visible to the naked eye, but did exhibit a tail and a nucleus.

For the remainder of November and throughout December only a handful of descriptions were published. Pons saw the comet on November 19 and noted it was still faint, but seemed to have slightly brightened since the discovery. Another observation by Pons on December 9 revealed the comet was still faint, but seemed to have brightened further. Interestingly, when Pons next saw the comet on December 16, he described it as "extremely faint" and noted it seemed to have faded since the 9th. Meanwhile, early in December, F. B. G. Nicolai (Mannheim, Germany) tried to find the comet, but was unsuccessful.

Two descriptions were given during early January, before moonlight began interfering. Pons saw the comet on January 1 and noted it appeared larger, with a very apparent nucleus. Nicolai also saw the comet that night and described it as extremely faint. The comet attained its most southerly declination of $-23°$ on January 3.

After the January moon had left the evening sky, observations quickly continued. F. M. Schwerd (Speyer, Germany) saw the comet on January 26 and estimated the coma as 2–3' across. Nicolai said his final observations on January 26 and 27 indicated the comet had brightened. H. W. M. Olbers (Bremen, Germany) saw the comet on the 27th and described it as very faint and ill-defined, with an occasionally visible condensation. His next observations on January 28, February 2 and 7 revealed the comet was easy to see in a cometseeker. Schwerd noted that his observations on January 27, February 2, 3, and 10 revealed a pale nebulosity with a small bright point. J. F. Encke (Berlin, Germany) said the comet was well seen in the cometseeker on January 31, but very faint in a telescope. During the period of January 28 to February 3, T. Clausen and Nehus (Altona, Germany) noted a faint, minute nucleus. Olbers said the comet was very easy to see on March 5. A. Stark (Augsburg, Germany) observed this comet on March 6 and 13, but moonlight and bad weather prevented further observations.

The comet was last detected on April 10.82, when Inghirami indicated a position of $\alpha = 5^h\ 22.9^m$, $\delta = -12°\ 24'$. Although the comet's brightness should have remained above its discovery magnitude until the end of June, assuming no abnormal brightness changes, it was not seen after the April full moon.

The first orbit calculated for this comet was actually an elliptical one by Clausen. He used positions spanning the period of November 16 to December 17 and determined a perihelion date of 1826 April 22.70 and a period of about 265 years. A short time later, he published a parabolic orbit with a perihelion date of April 22.46. Interestingly, this orbit would hold its own against the larger more definitive investigations published during the twentieth century. Nicolai published the results of his investigation in March of 1826. Using positions from November 16, January 1, and February 11, he actually determined a hyperbolic orbit with a perihelion date of April 22.48 and an eccentricity of 1.0089597. Two months later, utilizing positions obtained into April, Nicolai published a parabolic orbit with a perihelion date of April 22.42.

Two elaborate analyses of this comet's orbit were conducted during the twentieth century. E. B. Cowley and J. Whiteside (1907) took 228 positions from the period of November 16 to April 10 and determined a perihelion date of April 22.41. R. J. Buckley (1976) took 59 positions from the period of November 16 to April 8 and calculated a perihelion date of April 22.46. Buckley's orbit is given below.

T	ω	Ω (2000.0)	i	q	e
1826 Apr. 22.4581 (TT)	279.3841	200.0590	39.9903	2.007 734	1.0

ABSOLUTE MAGNITUDE: $H_{10} = 2.4$ (V1964)

FULL MOON: Oct. 26, Nov. 25, Dec. 25, Jan. 23, Feb. 22, Mar. 23, Apr. 22

SOURCES: J. L. Pons, CA, **13** (1825), pp. 597–606; F. B. G. Nicolai, BAJ for 1829, **54** (1826), pp. 171–3; J. L. Pons, G. Inghirami, F. M. Schwerd, and A. Stark, BAJ for 1829, **54** (1826), pp. 213–16; J. L. Pons, N. Cacciatore, F. Carlini, and T. Clausen, CA, **14** (1826), pp. 84–93, 389–92; J. L. Pons and G. Inghirami, AN, **4** (1826 Jan.), p. 321; G. Inghirami and T. Clausen, AN, **4** (1826 Feb.), pp. 361–6; T. Clausen, Nehus, H. W. M. Olbers, P. A. Hansen, and F. B. G. Nicolai, AN, **4** (1826 Feb.), pp. 369–72; J. F. Encke, F. M. Schwerd, and T. Clausen, AN, **4** (1826 Feb.), pp. 379–82; F. B. G. Nicolai, AN, **4** (1826 Mar.), p. 415; F. B. G. Nicolai, AN, **4** (1826 May), p. 531; G. Inghirami, AN, **5** (1826 Jun.), pp. 21–6; F. M. Schwerd, Astronomische Beobachtungen angestellt auf des Sternwarte des königl. Lyzeums in Speyer, Abtheilung I: Beobachtungen des Jahres 1826 (1829), p. 105; O1899, pp. 125–8; E. B. Cowley and J. Whiteside, EAN, **2** (1907), p. 18; V1964, p. 55; R. J. Buckley, QJRAS, **26** (1985), p. 91.

3D/1826 D1   *Discovered:* 1826 February 27.79 ($\Delta = 1.21$ AU, $r = 0.95$ AU, Elong. $= 50°$)
(Biela)   *Last seen:* 1826 May 9.85 ($\Delta = 1.03$ AU, $r = 1.19$ AU, Elong. $= 71°$)
   Closest to the Earth: 1826 April 19 (0.9599 AU)
1826 I   *Calculated path:* PSC (Disc), CET (Mar. 1), ARI (Mar. 4), TAU (Mar. 19), ORI
   (Apr. 3), MON (Apr. 21), CMi (Apr. 29)

W. von Biela (Josephstadt, Austria) discovered this comet on the evening of 1826 February 27.79 at a position of $\alpha = 1^h 47.3^m$, $\delta = +9° 28'$. He described it as a small, round nebulosity, with a very faint, central point of light. After reobserving the comet on February 28.79, he noted it had moved about 1° eastward and announced his discovery. Despite this announcement, Biela remained the only observer of this comet until an independent discovery was made by J. F. A. Gambart (Marseille, France) on March 9.83, who described it as "very faint" and about 1.5′ across. The night of the 10th was cloudy, but Gambart confirmed his find on March 11.81. An independent discovery was also apparently made in China around this time. According to the Chinese historical document *Ching-chao Hsü Wen-hsien tung-kao*, a "broom-star" comet was observed sometime during the lunar month of 1826 February 7 to March 8. No further details are available.

K. L. Harding (Göttingen, Germany) saw the comet on March 12 and noted it was an easy object in a cometseeker. The 6-day old moon was above the horizon on the 14th, when Harding noted a short tail. J. L. Pons (Florence, Italy) found the comet without trouble in strong moonlight on March 19 and described it as a nebulosity without a tail or nucleus. The comet attained its most northerly declination of +11° on March 26. Observations by G. Santini (Padova, Italy) on March 25, 29, and 31 revealed a strong nucleus and indicated the comet was steadily brightening. Interestingly, C. Brioschi (Naples, Italy) said the comet was very difficult to see in his telescope on March 31, even though the threads of the micrometer were barely illuminated.

The comet was described as faint by Santini on April 7 and Brioschi on the 8th. Positions continued to be gathered until the 15th, at which time moonlight began to interfere. Observations resumed on the 27th, with Harding noting the comet was relatively bright and distinct.

G. Inghirami (Florence, Italy), Brioschi, Gambart, and Santini were still observing the comet as May began, but it was fading from view. The final observation was made on May 9.85, when Brioschi gave the position as $\alpha = 7^h 54.3^m$, $\delta = +5° 12'$. He noted the comet had become so faint as to prevent the use of the micrometer during his observations.

The first parabolic orbit was calculated by T. Clausen using positions measured by Harding on March 12, 13, and 14. The resulting perihelion date was 1826 March 6.67. In a letter written on 1826 March 22 and published in the *Berliner Astronomische Jahrbuch for 1829*, Biela gave the results of his calculations. Using positions he measured on February 28, March 7, and 12, he determined the perihelion date as March 15.91. More importantly,

Biela noted the strong similarity between the orbit of this comet and those of comets seen in 1772 and 1805, and suggested a period of about 6.75 years. Interestingly, Gambart wrote a letter to the same periodical on March 22. He used positions obtained on March 9–21 and determined a perihelion date of March 18.92. He commented, "The orbit . . . so nearly resembles the orbits of the Comets of 1772 and 1805, particularly that of the last year, as to merit the attention of Astronomers." Shortly thereafter, H. W. M. Olbers came to a similar conclusion.

The first elliptical orbit was calculated by Clausen using positions obtained on February 28, March 9, and 20. The result was a perihelion date of March 18.62 and a period of 3.46 years. Using positions obtained up to the end of March, Clausen and Gambart determined the period as 6.67 years and 6.57 years, respectively. Gambart noted that the orbit "perfectly represents the actual observations" and "also represent those of the Comet of 1805." Gambart published another orbit in May that indicated a period of 6.74 years.

Several astronomers later calculated an orbit for this comet after it had been observed at more than one apparition, including G. Santini (1835), J. S. Hubbard (1860), J. von Hepperger (1898), and B. G. Marsden and Z. Sekanina (1971). Marsden included nongravitational terms of $A_1 = +2.78$ and $A_2 = -2.5037$ in his calculation and his orbit is given below.

T	ω	Ω (2000.0)	i	q	e
1826 Mar. 18.9498 (TT)	218.2621	253.9823	13.5627	0.902430	0.746575

ABSOLUTE MAGNITUDE: $H_{10} = 7.5$ (V1964)

FULL MOON: Feb. 22, Mar. 23, Apr. 22, May 21

SOURCES: W. von Biela and J. F. A. Gambart, *BAJ for 1829*, **54** (1826), pp. 114–20; H. W. M. Olbers, *BAJ for 1829*, **54** (1826), pp. 122–4; J. E. Bode, *BAJ for 1829*, **54** (1826), pp. 139–42; C. Brioschi, *BAJ for 1829*, **54** (1826), pp. 150–4; F. B. G. Nicolai, *BAJ for 1829*, **54** (1826), p. 172; J. F. A. Gambart, W. von Biela, J. L. Pons, and G. Santini, *CA*, **14** (1826), pp. 393–401; J. F. A. Gambart, *MAS*, **2** (1826), pp. 503–6; K. L. Harding, *AN*, **4** (1826 Mar.), p. 435; T. Clausen, K. L. Harding, and J. F. A. Gambart, *AN*, **4** (1826 Apr.), pp. 465–70; J. F. A. Gambart, *AN*, **4** (1826 May), pp. 501–8; J. F. A. Gambart, *AN*, **5** (1826 Aug.), p. 125; K. L. Harding, *AN*, **5** (1826 Sep.), p. 151; T. Clausen and J. F. A. Gambart, *CDT* (1830), pp. 52–5; G. Santini, *AN*, **12** (1835 Jan. 14), pp. 113–18; G. Santini, *AJS* (Series 1), **28** (1835 Jul.), p. 398; J. S. Hubbard, *AJ*, **6** (1860 Jun. 26), pp. 117–18; J. S. Hubbard, *AJ*, **6** (1860 Jul. 17), pp. 121–4; J. von Hepperger, *SAWW*, **107** Abt. IIa (1898), pp. 377–489; O1899, pp. 123–5; V1964, p. 55; HA1970, p. 86; B. G. Marsden and Z. Sekanina, *AJ*, **76** (1971 Dec.), pp. 1138–41.

C/1826 P1 (Pons) *Discovered:* 1826 August 7.1 ($\Delta = 0.84$ AU, $r = 1.40$ AU, Elong. $= 98°$)

Last seen: 1826 December 11.18 ($\Delta = 1.25$ AU, $r = 1.40$ AU, Elong. $= 77°$)

1826 IV *Closest to the Earth:* 1826 September 14 (0.5212 AU)

Calculated path: FOR (Disc), ERI (Aug. 12), LEP (Aug. 27), ORI (Sep. 2), MON (Sep. 6), ORI-MON (Sep. 9), CMi (Sep. 16), CNC (Sep. 23), LEO (Oct. 6), LMi (Oct. 19), LEO (Oct. 25), COM (Nov. 8)

J. L. Pons (Florence, Italy) discovered this comet in the morning sky near 11 Eridani on 1826 August 7.1. Pons' colleague, G. Inghirami, measured the comet's position as $\alpha = 3^h 10.1^m$, $\delta = -25° 34'$ on August 9.12. J. F. A. Gambart (Marseille, France) independently discovered the comet in the morning sky near 27 Eridani on August 15.1. He described the comet as inconspicuous, round, small, and without a nucleus.

Moonlight interrupted observations for most of the last half of August, but details of the comet's appearance were recorded at the end of the month. G. Santini (Padova, Italy) said the comet's faintness and moonlight prevented him from finding it until August 30. J. Schwarzenbrunner (Kremsmünster, Austria) saw the comet from August 31 to September 2 and said it was very faint and round, with a diffuse nucleus. K. L. Harding (Göttingen, Germany) saw the comet on September 1, 2, and 4, and described it as quite bright, with a recognizable nucleus. On the last date he also suspected a trace of tail. F. M. Schwerd (Speyer, Germany) saw the comet on several occasions during the period of September 1–13. He noted it remained a round nebulosity about 4' across, without a nucleus or tail. News was slow reaching observers in the Southern Hemisphere and an independent discovery was made on September 4 by C. K. L. Rümker (Paramatta, New South Wales, Australia). The comet was "very bright" when seen by H. W. M. Olbers (Bremen, Germany) on the 12th. Although he also noted a central condensation, he was uncertain about a nucleus. The comet crossed the celestial equator on September 13. The comet was rather easy to see in strong moonlight on the 14th, according to Olbers, and it was still easy to see in the refractor on the 18th.

Following the departure of the moon from the sky, Olbers observed the comet on September 30 and described it as very bright and visible to the naked eye. He added that a telescope revealed a diffuse nucleus and weak traces of a short tail. The appearance was unchanged on October 1. Olbers saw only weak traces of a tail on the 7th. The comet was difficult for Olbers to see in the large refractor on the 14th because of hazy skies, but it was much easier to see the next morning. Curiously, Olbers then noted that the comet had "obviously lost the brightness it showed during the full moon in September." The comet reached a minimum solar elongation of 57° on the 15th. Olbers said the comet was easy to see in both the refractor and a cometseeker on October 21, despite moonlight, while it was a difficult object to see on the next morning because the moon had moved closer. The comet attained its most northerly declination of +25° on the 31st. Olbers saw the comet on November 7 and said it was easy to see, with a distinct, but diffuse nucleus. No tail was present.

The comet's position was measured for the final time on November 27.03, when Olbers gave it as $\alpha = 12^h\ 39.8^m$, $\delta = +25°\ 18'$. The final observations of this comet were obtained by L. Del Re (Naples, Italy) on December 7.19, 10.20, and 11.18. He noted the comet was seen with great difficulty on the 7th, but was only suspected on the 10th and 11th. The comet had attained a minimum declination of $+24°$ on December 8, before turning to a more northerly motion.

Several astronomers calculated parabolic orbits during September. Among the first were F. B. G. Nicolai and Schwerd, who independently determined perihelion dates of October 9.53 and October 9.43, respectively. Additional orbits came from W. von Biela and F. W. A. Argelander during the next few weeks. Despite observations continuing into December, revised orbits using positions later than September were a rarity until Rümker (1829) used only his positions obtained between September 4 and October 5, and computed a parabolic orbit with a perihelion date of October 9.29.

The first complete analysis of all of the available positions was not conducted until over 80 years later, when R. Klug (1907) collected about 300 positions obtained during the period spanning August 9 to November 27, reduced them to eight Normal positions, and applied perturbations by three planets. The result was an elliptical orbit with a perihelion date of October 9.48 and a period of about 6279 years.

T	ω	Ω (2000.0)	i	q	e
1826 Oct. 9.4761 (UT)	13.8101	46.4025	25.9496	0.852878	0.997492

ABSOLUTE MAGNITUDE: $H_{10} = 6.5$ (V1964)

FULL MOON: Jul. 19, Aug. 17, Sep. 16, Oct. 15, Nov. 14, Dec. 14

SOURCES: J. F. A. Gambart, *BAJ for 1829*, **54** (1826), p. 194; F. B. G. Nicolai, *BAJ for 1829*, **54** (1826), pp. 224–6; G. Inghirami, J. L. Pons, J. F. A. Gambart, and K. L. Harding, *AN*, **5** (1826 Sep.), pp. 145, 151–4; F. M. Schwerd and J. Schwarzenbrunner, *AN*, **5** (1826 Oct.), pp. 169–74; F. B. G. Nicolai, *AN*, **5** (1826 Oct.), pp. 179–82; W. von Biela and L. Del Re, *AN*, **5** (1827 Jan.), pp. 297–300; F. W. A. Argelander, *AN*, **5** (1827 Mar.), pp. 357–60; L. Del Re, *AN*, **5** (1827 Sep.), pp. 425–9; F. M. Schwerd, *Astronomische Beobachtungen angestellt auf des Sternwarte des königl. Lyzeums in Speyer, Abtheilung I: Beobachtungen des Jahres 1826* (1829), p. 108; G. Santini, *MAS*, **3** (1829), pp. 104–5; C. K. L. Rümker, *PTRSL*, **119** (1829), pp. 65–7; O1899, pp. 128–33; R. Klug, *DAWW*, **80** (1907), p. 314; V1964, p. 55.

C/1826 U1   *Discovered:* 1826 October 22.8 ($\Delta = 1.05$ AU, $r = 0.97$ AU, Elong. $= 56°$)
(Pons)   *Last seen:* 1827 January 6.22 ($\Delta = 1.79$ AU, $r = 1.43$ AU, Elong. $= 53°$)
            Closest to the Earth: 1826 November 11 (0.7987 AU)

1826 V   *Calculated path:* BOO (Disc), SER (Nov. 5), LIB (Nov. 13), SCO (Nov. 21), OPH (Nov. 23), HER (Dec. 1), OPH (Dec. 3), HER (Dec. 10), OPH (Dec. 11), HER (Dec. 12), LYR (Jan. 4)

J. L. Pons (Florence, Italy) discovered this comet on 1826 October 22.8. G. Inghirami and P. Tanzini (Florence, Italy) gave the comet's position as $\alpha = 14^h\ 21.7^m$, $\delta = +43°\ 36'$ on October 23.79. The comet was independently discovered by T. Clausen (Hamburg, Germany) on October 26.80 and J. F. A. Gambart (Marseille, France) on October 28.8. Gambart said it was rather apparent, with an extension in the shape of a tail.

The comet was brightening as it moved toward both the sun and Earth, and was first seen with the naked eye on November 1 by J. N. Nicollet (Paris, France). A further sign that the comet was becoming more impressive came from G. Santini (Padova, Italy), who saw the comet during the period of November 6–12 and wrote, "It was very bright, with a round and distinct nucleus; also accompanied by a sensible tail." The comet was moving almost directly southward during this time and Santini's observation on November 12.70 was the last time it was seen until late November.

As astronomers worked to determine the orbit of this comet, Gambart discovered early in November that the comet was going to transit the face of the sun on November 18. Gambart worked out the particulars of the event just a couple of days before it was to occur. He sent out a few letters, but some did not arrive until the evening of the 17th and, consequently, few astronomers were prepared. Gambart wrote that the transit could be expected in the morning hours and suggested searching not long after sunrise. Unfortunately, as Gambart later wrote, "On the 18th November the whole of Europe was enveloped in the same clouds. . . . " Only two observers ever reported details of their observations – Gambart and H. Flaugergues (Viviers, France). Gambart's observation was continuous from 8:35 a.m. until 8:56 a.m., but he noted, "The comet was not visible." Likewise, Flaugergues' observations, which began about 10 minutes later than Gambart's, also revealed no trace of the comet. Gambart personally worked to refine the comet's orbit in the following days and weeks. By the beginning of 1827, he computed a new orbit, using observations extending from October 26 to December 11, which revealed the comet would have left the sun's disk at 9:06 a.m. Thus, Gambart was confident that the comet was invisible and concluded that the comet was either "too small or too rare to be visible in that situation." Its minimum distance from the sun's center was 0.1°. By the end of November 18 the comet reached a maximum solar elongation of 1.6°. The comet was also at its most southerly declination of −21°. The comet passed behind the sun on November 19, and once again came within 0.1° of the sun's center.

The comet was recovered coming out of the sun's glare on November 28.6, when F. W. A. Argelander (Åbo, now Turku, Finland) saw it at a very low altitude in the evening sky. He said the tail was visible, but the coma and nucleus were hidden by clouds. Argelander saw the comet again on November 29 and noted that in addition to the main tail, a short second tail might have been present, although he was uncertain because of the comet's low altitude. Santini noted the nucleus was somewhat diffuse during the period of December 1–5, while the tail was well defined. H. W. M. Olbers (Bremen, Germany) saw the comet with his naked eye on December 2. He also noted a faint tail 6–8° long. Inghirami also reported a naked-eye observation on the 6th and added that the comet had a very beautiful tail. The comet was seen in the refractor by Olbers on the 10th with great difficulty because of strong moonlight. He noted the comet appeared as a small, faint, hardly noticeable cloud. Santini saw the comet for the final time on December 25 and noted it was very small, but still surrounded by a bright nebulosity.

The comet was last detected on 1827 January 6.22, when Argelander gave the position as $\alpha = 18^{\mathrm{h}}\ 23.1^{\mathrm{m}}$, $\delta = +29°\ 36'$. He noted the comet was then exceedingly faint and exhibited a tail 15–20′ long. Moonlight interfered thereafter.

The first orbit was calculated by W. von Biela using positions obtained during the period spanning October 23 to November 5. The resulting perihelion date was 1826 November 19.42. T. Clausen followed up by calculating the perihelion date as November 18.95. Almost identical orbits were independently determined by Gambart, Santini, and Clüver.

A definitive orbit was calculated over 80 years later. A. Hnatek (1908) took 28 positions from the period of October 23 to January 6, determined seven Normal positions, and calculated both a parabolic and an elliptical orbit. Both orbits had perihelion dates of November 18.91, while the period of the elliptical orbit was about 27 843 years. Although the elliptical orbit fitted the positions marginally better, Hnatek preferred the parabolic orbit.

T	ω	Ω (2000.0)	i	q	e
1826 Nov. 18.9074 (UT)	279.5835	237.5503	90.6235	0.026904	1.0

ABSOLUTE MAGNITUDE: $H_{10} = 7.0$ (V1964)

FULL MOON: Oct. 15, Nov. 14, Dec. 14, Jan. 13

SOURCES: T. Clausen, J. F. A. Gambart, J. L. Pons, G. Inghirami, H. W. M. Olbers, and J. N. Nicollet, *AN*, **5** (1826 Nov.), pp. 241–4; T. Clausen, *AN*, **5** (1826 Nov.), p. 251; G. Santini, J. L. Pons, and P. Tanzini, *AN*, **5** (1826 Dec.), p. 257; G. Inghirami, *AN*, **5** (1827 Jan.), p. 289; W. von Biela, *AN*, **5** (1827 Jan.), p. 297; G. Santini and F. W. A. Argelander, *AN*, **5** (1827 Mar.), pp. 354–60; J. F. A. Gambart and H. Flaugergues, *MNRAS*, **1** (1827 May 11), p. 27; Clüver, *AN*, **5** (1827 Sep.), p. 433; J. F. A. Gambart, *MAS*, **3** (1829), pp. 85–7; O1899, pp. 133–4; A. Hnatek, *AN*, **178** (1908 Aug. 14), pp. 337–50; V1964, p. 55.

C/1826 Y1 *Discovered:* 1826 December 26.2 ($\Delta = 1.30$ AU, $r = 1.04$ AU, Elong. $= 52°$)

(Pons) *Last seen:* 1827 January 26.72 ($\Delta = 1.16$ AU, $r = 0.56$ AU, Elong. $= 29°$)

1827 I *Closest to the Earth:* 1827 January 14 (1.0765 AU)

Calculated path: HER (Disc), SGE (Jan. 12), AQL (Jan. 19), DEL (Jan. 20), EQU (Jan. 25)

J. L. Pons (Florence, Italy) discovered this comet in the morning sky on 1826 December 26.2, following nearly two months of cloudy weather. G. Inghirami (Florence, Italy) measured the position as $\alpha = 16^h \, 41.6^m$, $\delta = +21° \, 21'$ on December 27.2. An independent discovery was made by J. F. A. Gambart (Marseille, France) on December 27. The comet had attained its maximum solar elongation of 53° just one week before discovery.

The comet was not followed for long by the initial discoverers because of bad weather and moonlight, with Gambart last detecting the comet on December 30 and the Florence observers last seeing it on January 1.

After announcements of the discovery had circulated to several European observatories, the comet again came under observation in the evening sky shortly after mid-January, when K. L. Harding (Göttingen, Germany) found it on January 17. He described it as a small, very bright nebulosity, without a point-like nucleus, but with a short tail. H. W. M. Olbers (Bremen, Germany) saw the comet on January 18, 19, and 22. He said a nucleus developed during this period, while traces of a tail were recognizable.

J. Schwarzenbrunner (Kremsmünster, Austria) observed the comet on 1827 January 20, 21, 24, and 26 and described it as very small, with a strong nucleus, but little coma. There was also no tail. The last date also marked the final observation of the comet and on January 26.72, Schwarzenbrunner gave the position as $\alpha = 21^h \, 00.5^m$, $\delta = +9° \, 03'$. The comet passed 13° from the sun on February 18 and should have faded below the capabilities of available telescopes by the time it exited the sun's glare.

The first orbit was calculated by Gambart using positions he had obtained on December 27, 28, and 29. The resulting perihelion date was 1827 February 3.97. During August of 1827, A. von Heiligenstein took positions from December 28 to January 26 and determined the perihelion date as February 5.42.

Elis Strömgren (1902) used 13 positions obtained between December 28 and January 26, and computed a parabolic orbit with a perihelion date of February 5.41. No planetary perturbations were applied.

T	ω	Ω (2000.0)	i	q	e
1827 Feb. 5.4110 (UT)	151.0340	186.9924	102.3898	0.506166	1.0

ABSOLUTE MAGNITUDE: $H_{10} = 6.3$ (V1964)

FULL MOON: Dec. 14, Jan. 13, Feb. 11

SOURCES: J. L. Pons and G. Inghirami, *AN*, **5** (1827 Jan.), p. 301; J. Schwarzenbrunner, *AN*, **5** (1827 Feb.), p. 343; J. F. A. Gambart, *MNRAS*, **1** (1827 Mar. 9), pp. 10–11; A. von Heiligenstein, *AN*, **5** (1827 Sep.), pp. 435–8; O1899, pp. 134–5; E. Strömgren, K. L. Harding, and H. W. M. Olbers, *AN*, **160** (1902 Nov. 22), pp. 241–50; V1964, p. 55.

D/1827 M1
(Pons–Gambart)

1827 II

Discovered: 1827 June 21.04 ($\Delta = 0.53$ AU, $r = 0.84$ AU, Elong. $= 56°$)
Last seen: 1827 July 21.87 ($\Delta = 1.26$ AU, $r = 1.13$ AU, Elong. $= 58°$)
Closest to the Earth: 1827 June 22 (0.5301 AU)
Calculated path: CAS (Disc), CEP (Jun. 23), CAM–CEP (Jun. 24), CAM (Jun. 25), DRA (Jun. 27), UMa–DRA (Jun. 30), UMa (Jul. 2), CVn (Jul. 9)

J. L. Pons (Florence, Italy) discovered this comet on 1827 June 21.04 at a position of $\alpha = 2^h 01.7^m$, $\delta = +66° 14'$. An independent discovery was made by J. F. A. Gambart (Marseille, France) just minutes later.

Few physical descriptions were published for this comet. Pons was the most prolific observer, measuring three additional positions in June and a total of 17 in July. The comet attained its most northerly declination of $+83°$ on June 26. J. E. B. Valz (Nîmes, France) saw the comet on the morning of July 5 and 6, and on the evening of July 6, with a nearly full moon in the sky. The comet attained its maximum observed elongation of $63°$ on July 7.

The comet was last detected on July 21.87, when Pons determined the position as $\alpha = 12^h 24.3^m$, $\delta = +39° 45'$. Pons said the comet was extremely faint. He added that the comet could not be observed thereafter because of its faintness and the fact that it had entered a region of the sky containing numerous nebulosities.

As with comet C/1826 Y1, a month of observations did not generate great interest in examining the orbit of this comet. It was not until 1828 that the first parabolic orbits appeared in the *Astronomische Nachrichten*. During March, Valz took positions spanning the entire period of visibility and determined the perihelion date as 1827 June 8.83. One month later, A. von Heiligenstein essentially used the same positions and determined a perihelion date of June 8.37.

There was no further interest until the twentieth century. S. Ogura (1917) took 74 positions spanning the period of visibility, and found the orbit was distinctly elliptical. His best orbit had a perihelion date of June 7.69 and a period of 63.83 years, with a likely uncertainty in the period of ± 10 years. A second orbit with a period of 46.0 years was calculated, but it did not fit the positions as well. S. Nakano (1979) took 68 positions, but applied the perturbations of five planets to his calculations. The result was a perihelion date of June 7.64 and a period of 57.46 years. He said an uncertainty of ± 10 years was still present in the calculations.

I. Hasegawa (1979) introduced an interesting twist to the investigation of this orbit. He pointed out that S. Kanda had noted in 1972 that there was a "probable" link between comet Pons–Gambart and a comet seen in 1110. Hasegawa began examining the observations published in ancient Chinese and Korean texts and was able to derive three rough positions, with which he then calculated a parabolic orbit. He confirmed the resemblance to comet Pons–Gambart.

T	ω	Ω (2000.0)	i	q	e
1827 Jun. 7.6376 (TT)	19.1895	320.0292	136.4601	0.806508	0.945838

ABSOLUTE MAGNITUDE: $H_{10} = 7.0$ (V1964)

FULL MOON: Jun. 9, Jul. 8, Aug. 7

SOURCES: J. L. Pons, *AN*, **6** (1828 Jan.), pp. 159–64; J. E. B. Valz, *AN*, **6** (1828 Mar.), p. 251; A. von Heiligenstein, *AN*, **6** (1828 Apr.), pp. 305–8; J. L. Pons and J. E. B. Valz, *AN*, **7** (1828 Dec.), pp. 55–8; S. Ogura, *AOAT*, **5** No. 3 (1917), pp. 1–12; V1964, p. 55; I. Hasegawa, *PASJ*, **31** (1979), pp. 260–1, 263–4; S. Nakano, *QJRAS*, **26** (1985), p. 104.

C/1827 P1   *Discovered:* 1827 August 3.03 ($\Delta = 1.29$ AU, $r = 1.15$ AU, Elong. $= 58°$)
(Pons)   *Last seen:* 1827 October 17.14 ($\Delta = 1.59$ AU, $r = 1.06$ AU, Elong. $= 40°$)
   Closest to the Earth: 1827 August 31 (0.6582 AU)
1827 III   *Calculated path:* CAM (Disc), LYN (Aug. 12), CAM (Aug. 14), LYN (Aug. 17), UMa (Aug. 20), LMi (Aug. 27), LEO (Sep. 2)

J. L. Pons (Florence, Italy) discovered this comet in the morning sky on 1827 August 3.03. He described it as small, but fairly bright.

Observations outside of Italy did not begin until the evening of August 17, when the comet was observed by F. M. Schwerd (Speyer, Germany). He said the comet was small and exhibited no tail. Schwerd saw the comet again on the 20th and noted a coma 4′ across and the beginning of a tail. F. B. G. Nicolai (Mannheim, Germany) saw the comet on August 22 and 23. He described it as rather bright, with a diameter of about 4′ and a faint tail. Schwerd noted a very thin tail that extended 10′ on the 22nd and 15′ on the 23rd. He added that the nucleus was brighter on the 23rd and the coma was 6–7′ across. The comet was last detected prior to perihelion in morning twilight on August 30 by Schwerd.

After passing less than 4° from the sun on September 17, the comet slowly moved into the morning sky. Only one observation was made after it had cleared morning twilight and that was by Nicolai on October 17.14. He gave the position as $\alpha = 11^h 40.4^m$, $\delta = +22° 25'$. Using the 4.5-foot focal length refractor, he described the comet as a small, very faint nebulosity.

The earliest parabolic orbits were independently calculated by C. A. F. Peters and Clüver. Both astronomers used positions from August 20, 21, and 22. Peters determined the perihelion date as 1827 September 12.74 and Clüver determined it as September 12.34. Similar orbits were calculated during the next few months by Schwerd, J. E. B. Valz, Nicolai, and Clüver. Before the comet had vanished, the perihelion date had been pinpointed to September 12.17.

The only published orbit that used positions spanning the entire period of visibility was calculated by Clüver (1828). His resulting parabolic orbit had a perihelion date of September 12.17. He also calculated an elliptical orbit with a perihelion date of September 12.19 and a period of about 2611 years. Clüver's parabolic orbit is given below.

T	ω	Ω (2000.0)	i	q	e
1827 Sep. 12.1723 (UT)	258.7037	152.1008	125.8787	0.137758	1.0

ABSOLUTE MAGNITUDE: $H_{10} = 7.3$ (V1964)
FULL MOON: Jul. 8, Aug. 7, Sep. 5, Oct. 5, Nov. 3
SOURCES: F. M. Schwerd, *AN*, **5** (1827 Sep.), pp. 469–72; C. A. F. Peters, F. B. G. Nicolai, Clüver, and F. M. Schwerd, *AN*, **6** (1827 Sep.), pp. 43–50; F. B. G. Nicolai, *AN*, **6** (1828 Feb.), p. 211; J. E. B. Valz, *AN*, **6** (1828 Mar.), p. 251; H. W. M. Olbers, F. B. G. Nicolai, and Clüver, *AN*, **7** (1828 Dec.), pp. 61–4; F. M. Schwerd, *Astronomische Beobachtungen angestellt auf des Sternwarte des königl. Lyzeums in Speyer, Abtheilung II: Beobachtungen des Jahres 1827* (1829), pp. 115–16; F. B. G. Nicolai, *AN*, **7** (1829 Feb.), p. 148; J. L. Pons, *AN*, **7** (1829 May), pp. 291–4; V1964, p. 55.

2P/Encke

1829

Recovered: 1828 September 16.91 ($\Delta = 1.09$ AU, $r = 1.97$ AU, Elong. = 139°)
Last seen: 1828 December 27.64 ($\Delta = 0.51$ AU, $r = 0.49$ AU, Elong. = 14°)
Closest to the Earth: 1828 December 12 (0.4723 AU)
Calculated path: PSC (Rec), AND (Oct. 10), PEG (Oct. 21), EQU (Nov. 26), DEL (Dec. 3), AQL (Dec. 11)

Two predictions were supplied for this apparition. J. F. Encke calculated the comet's return using planetary perturbations and applying the hypothesis that it encountered a resistance as it moved through space. He gave the perihelion date as 1829 January 10.21. M. C. T. Damoiseau (1827) calculated the comet's return using only planetary perturbations and predicted a perihelion date of January 10.86. F. G. W. von Struve (Dorpat, now Tartu, Estonia) recovered this comet on 1828 September 16.91, and measured the position as $\alpha = 1^h 34.8^m$, $\delta = +26° 23'$. He was using a 24-cm refractor and described the comet as a faint nebulosity. Encke's prediction proved the more accurate.

The comet must have been a very faint object upon discovery and, despite heading toward close approaches with the sun and Earth, it remained quite faint through most of October. Evidence for this comes from K. L. Harding (Göttingen, Germany) and H. W. M. Olbers (Bremen, Germany). Harding said he began searching for this comet on August 19, but failed to see it until October 27. Olbers wrote, "I sought in vain for Encke's comet in September and October."

The comet was nevertheless followed by some astronomers during October. Struve still described the comet as faint on October 2, 6, and 13, while J. F. Encke also acknowledged the difficulty in seeing this comet during a visit to Dorpat Observatory (now Tartu, Estonia) on the 13th. The comet had attained its most northerly declination of +29° on October 11 and was lost in moonlight after the 13th. Struve picked up the comet again on October 25 and, with the moon above the horizon, he noted it was extremely faint; however, the next evening, Struve saw the comet in a moonless sky and said it was bright enough to withstand illuminating the threads of the filar micrometer. The comet was still described as extremely faint by F. B. G. Nicolai (Mannheim, Germany) on October 28 and J. South's (Kensington, England) observation on the 30th revealed it as a luminous patch that was so extremely faint as to be visible only by averted vision. Meanwhile, Struve estimated the coma diameter as 3' on both the 28th and 29th.

The comet continued closing in on Earth and the sun during November and was widely observed. Struve said the comet was an easy object in the cometseeker on November 1, while J. F. A. Gambart (Marseille, France) still described the comet as a hardly visible spot. Olbers found the comet very faint on the 2nd, which was also the first night the comet was seen in the Southern Hemisphere by C. K. L. Rümker (Paramatta, New South Wales, Australia). On the 3rd, Olbers referred to it as "a very diffuse and feeble nebula," while South said it was a distinctly seen "nebulous spot." W. Richardson (Greenwich, England) independently recovered the comet on November 4, while South described the comet as a "faint nebulous spot." Olbers found the comet more obvious in both his cometseeker and refractor on November 5, but then found it less noticeable the next night. On November 7, Struve saw the comet in a cometseeker and noted the coma was 9' across, while the bright condensation was 4' across. A look with the 24-cm refractor revealed a tail-like extension about 18' long. South and two friends saw the comet when it was only a few degrees from the 7-day-old moon on the 12th. After full moon had passed, Olbers saw the comet before moonrise on the 24th and noted it had increased much in size and brightness. Olbers described the comet as centrally condensed and 4–5' across on the 25th, but noted no nucleus. Struve said the comet looked like a star of magnitude 6 to the naked eye on the 30th, while a cometseeker revealed a coma 9' across.

The comet passed closest to Earth around mid-December and was brightening. Olbers described it as very bright and elongated on the 1st, while Struve said it was visible to the naked eye as a star of magnitude 5 on the 7th. Struve also noted the coma was 6' across in the refractor on both the 7th and 8th. Olbers described the comet as brighter and larger than the globular cluster M13 on the 9th. He added that no tail was present and the refractor revealed a diffuse nucleus. Moonlight was again interfering when Struve saw the comet on the 14th. He said the coma was just 3.5' across. Olbers simply described the comet as "very diffuse" on the 15th. Struve saw the comet through a telescope in bright evening twilight on the 25th and described it as "very beautiful." Nicolai also saw the comet low over the western horizon on the same night, but clouds covered the comet before a position could be measured.

As the comet continued to move toward perihelion, it entered evening twilight. Struve last measured the position on December 26.64, and found it to be $\alpha = 18^h 58.4^m$, $\delta = -10° 17'$. He was the last person to see the comet when, on December 27.64, he found it close to the western horizon and watched it until it disappeared low over the horizon.

Olbers (1830) deduced the following points from his personal observations in 1828. First, he suggested the comet "has no light of its own, and shines only by the reflection of that of the sun." Second, he surmised that "this comet actually does suffer a resistance in that part of space wherein it moves." He said he was very impressed by the accuracy of Encke's prediction, which was "calculated on the hypothesis of such resistance," and noted that Damoiseau's prediction, assuming no resistance, suffered errors which greatly increased after mid-November.

Minor refinements of this comet's orbit using positions from multiple apparitions, as well as planetary perturbations, have been published by numerous astronomers. Of most significance were the studies of Encke, F. E. von Asten (1877), and B. G. Marsden and Z. Sekanina (1974). These have typically revealed a perihelion date of January 10.24 and a period of 3.32 years. Marsden and Sekanina determined the nongravitational terms as $A_1 = +1.14$ and $A_2 = -0.03996$. The orbit of Marsden and Sekanina is given below.

T	ω	Ω (2000.0)	i	q	e
1829 Jan. 10.2434 (TT)	182.7893	336.9003	13.3652	0.345447	0.844640

ABSOLUTE MAGNITUDE: $H_{10} = 8.5$ (V1964)

FULL MOON: Aug. 25, Sep. 23, Oct. 23, Nov. 21, Dec. 21, Jan. 20

SOURCES: M. C. T. Damoiseau, *CDT* (1827), pp. 219–24; K. L. Harding and J. F. A. Gambart, *AN*, **7** (1828 Nov.), pp. 49–54; J. F. W. Herschel, and J. South, *MNRAS*, **1** (1828 Nov. 14), pp. 87–8; H. W. M. Olbers, *AN*, **7** (1828 Dec.), p. 61; J. F. Encke, *MNRAS*, **1** (1829 Jan. 9), p. 95; C. K. L. Rümker, *PTRSL*, **119** (1829), pp. 67–9; J. F. Encke, *AN*, **7** (1829 Jan.), pp. 115–18; F. B. G. Nicolai, *AN*, **7** (1829 Feb.), pp. 143–8; F. G. W. von Struve and J. F. Encke, *AN*, **7** (1829 Feb.), pp. 153–84; J. F. Encke and M. C. T. Damoiseau, *CDT* (1830), pp. 272–4; J. South, J. Dunlop, W. Richardson, H. W. M. Olbers, J. F. Encke, and M. C. T. Damoiseau, *MAS*, **4** (1830), pp. 185–9; F. E. von Asten, *BASP* (Series 4), **22** (1877), p. 560; O1899, pp. 135–9; V1964, p. 55; B. G. Marsden and Z. Sekanina, *AJ*, **79** (1974 Mar.), pp. 415–16.

C/1830 F1
(Great Comet)

1830 I

Discovered: 1830 March 16.7 ($\Delta = 0.19$ AU, $r = 1.02$ AU, Elong. $= 91°$)

Last seen: 1830 August 17.85 ($\Delta = 1.38$ AU, $r = 2.26$ AU, Elong. $= 141°$)

Closest to the Earth: 1830 March 26 (0.1471 AU)

Calculated path: MEN (Disc), OCT (Mar. 18), IND (Mar. 22), GRU (Mar. 28), MIC (Mar. 30), CAP (Apr. 2), AQR (Apr. 6), EQU (Apr. 15), PEG (Apr. 25), VUL (May 19), SGE (Jul. 27), AQL (Aug. 14)

Faraguet (Port Louis College, Mauritius) discovered this comet "between the Chameleon and the Larger Magellan's Nebula", about 20° from the south celestial pole, on 1830 March 16.7. He confirmed the find on March

17, when he found the comet had moved about 5° toward the north. It is interesting to note that Faraguet failed to get credit for this observation for over a century and a half. His observations were communicated to the Royal Society by Dabadie (Professor of the Royal College of Port Louis, Mauritius), and, through the years, previous compilers of comet catalogs have attributed the comet's discovery to Dabadie.

Several independent discoveries were made during the next few nights. R. T. Paine (Boston, Massachusetts, USA) passed on a letter from an unnamed person to the September 1830 issue of the *Astronomische Nachrichten*. This person had sailed from Calcutta to Boston and first saw the comet when the ship was off the southern tip of Africa. The exact date was given as March 17.73, and the observer said the comet was about 3rd magnitude, with a tail 7° or 8° long. He described it as "a mass of luminous matter, brilliant in the centre and becoming fainter towards the edges." J. C. Wickham (HMS *Adventure*) first saw this comet near the south celestial pole on March 18. He described it as very bright and large. Reid (master of the merchant vessel *William Brown*) discovered the comet on March 18.9. He noted the comet was at an altitude of 35° toward the south-southwest, with an ecliptic latitude of −30°.

At discovery the comet was only visible from the Southern Hemisphere and was brightening rapidly as it neared Earth. It attained its maximum southern declination of –85° on March 19 and then moved northward at a rate of over 5° per day. It also began to fade after this close approach to Earth. Another independent discovery was made on March 29.23, when Captain H. Foster (Ascension Island) found the comet with the naked eye in the morning sky. He noted the comet was "quite visible" on April 1.

The first observation in the Northern Hemisphere was an independent discovery on April 21.1 by J. F. A. Gambart (Marseille, France). He simply described the comet as "very apparent." News spread rapidly from Marseille. J. N. Nicollet (Paris, France) saw the comet on April 26 and noted a tail about 0.75° long and a brilliant nucleus. He said the appearance to the naked eye was of a "beautiful lengthened nebula." K. L. Harding (Göttingen, Germany) said the tail was about 2° long on April 27, while a very bright, well-defined nucleus was seen on the 27th, 28th, and 29th. H. W. M. Olbers (Bremen, Germany) saw the comet on the 29th and said the comet exhibited a "brilliant head, a small diffuse nucleus, and, after the setting of the moon, a tail 2.5° long." On the 30th, Harding said the nucleus still appeared sharp and bright at a magnification of 180×. Olbers said the comet was still well seen despite the increasing moonlight.

The comet was fading as May began, since its distances from the sun and Earth were increasing. J. South (Kensington, England) easily saw it in bright moonlight on May 2 while using his 5-foot focal length telescope. He described it as circular, about 2′ across, with a well-defined nucleus. With his 20-foot focal length refractor he noted a faint diffuse tail. Olbers reported that the comet steadily faded through the 12th, and continued to be an easy object to see until moonlight began interfering after the 23rd. Olbers added that the tail was still over 1° long in his cometseeker on the 16th. South continued to scrutinize the nucleus with the 20-foot focal length refractor. On the 4th he noted that a magnification of 1126× and higher revealed the nucleus as "an apparently uniform luminous cloud." When seen on the 15th, South said a magnification of 773× revealed the nucleus as a disk, while magnifications of 1126× and 2440× revealed the nucleus as "a mere luminous cloud."

The comet was no longer an easy object in moonlight as June began, as Olbers found it very difficult to see on the 4th. Once the moon had left the skies, Olbers reported the nucleus was seen in the cometseeker on the 10th, and then noted the comet was better seen on the 14th than on previous evenings. On the last date he also noted the coma seemed larger, but the nucleus was more difficult to see. Interestingly, Olbers reported on the 16th that he obtained one of his best views of the comet. The comet attained its greatest northern declination of +28° on June 18. Olbers found the comet very faint under clear skies on the 21st and he saw the comet for the final time on the 24th, noting that his measured positions were probably not very accurate because of the faintness of the comet.

Only a handful of astronomers continued their observations into July and August, but no physical descriptions were apparently made. Some of these observations were made by F. B. G. Nicolai (Mannheim, Germany) and P. Tanzini (Florence, Italy). The comet was last detected on August 17.85, when Tanzini indicated a position of $\alpha = 19^h 55.9^m$, $\delta = +14° 37'$.

The first published parabolic orbit came from Olbers. Using positions from April 22 to April 30, he determined the perihelion date of April 9.82, which ended up being an excellent representation of the comet's true orbit. During the remainder of 1830, minor revisions were published by F. M. Schwerd, F. B. G. Nicolai, J. E. B. Valz, G. Santini, and C. Conti. During December, Faraguet began notifying others of a parabolic orbit he had calculated using rough positions from March 19 to April 15. It indicated a perihelion date of April 12.22. Haedenkampf and Mayer (1831) not only published a parabolic orbit, but an elliptical one as well. The elliptical orbit gave the perihelion date as April 9.80 and the period as 58 466 years. The definitive orbit was determined by L. R. Schulze (1873). He took about 300 positions from the period spanning March 23 to August 17, added the perturbations of four planets, and calculated a parabolic orbit with a perihelion date of April 9.80.

T	ω	Ω (2000.0)	i	q	e
1830 Apr. 9.7951 (UT)	5.7962	208.7642	21.2570	0.921424	1.0

ABSOLUTE MAGNITUDE: $H_{10} = 5.2$ (V1964)

FULL MOON: Mar. 9, Apr. 8, May 7, Jun. 6, Jul. 6, Aug. 4, Sep. 2

SOURCES: J. F. A. Gambart and J. N. Nicollet, *AN*, **8** (1830 May), p. 251; J. F. A. Gambart and J. South, *MNRAS*, **1** (1830 May 14), pp. 180–1; K. L. Harding and H. W. M. Olbers, *AN*, **8** (1830 Jun.), pp. 253–6; F. M. Schwerd, *AN*, **8** (1830 Jul.), p. 299; F. B. G. Nicolai, *AN*, **8** (1830 Aug.), pp. 317–20; J. E. B. Valz, F. B. G. Nicolai, and R. T. Paine, *AN*, **8** (1830 Sep.), pp. 339–42, 349–52; J. C. Wickham, P. King, and Dabadie, *MNRAS*, **1** (1830 Dec. 10), pp. 195–6; G. Santini and C. Conti, *MNRAS*, **2** (1831 Jan. 14), pp. 2–3; F. B. G. Nicolai, G. Santini, and J. South, *MAS*, **4** (1831), pp. 623–6; P. Tanzini, *AN*, **9** (1831 Apr. 26), pp. 149–60; Haedenkampf and Mayer, *AN*, **9** (1831 May 4), pp. 165–74; H. Foster, *MRAS*, **8** (1835), pp. 191–6; F. Fallows, *MRAS*, **19** (1849–50), p. 102; L. R. Schulze, *AN*, **82** (1873 Aug. 6), pp. 97–102; O1899, pp. 139–45; McIntyre, Donald, *Comets in Old Cape Records*. Cape Town: Cape Times Limited (1949), pp. 13–15; V1964, p. 55.

C/1831 A1
(Great Comet)

1830 II

Discovered: 1831 January 7.25 ($\Delta = 1.02$ AU, $r = 0.42$ AU, Elong. $= 24°$)

Last seen: 1831 March 19.8 ($\Delta = 1.06$ AU, $r = 1.95$ AU, Elong. $= 143°$)

Closest to the Earth: 1830 December 9 (0.6856 AU), 1831 February 16 (0.5335 AU)

Calculated path: SER (Disc), OPH (Jan. 14), LIB (Jan. 30), SER (Feb. 2), LIB (Feb. 6), VIR (Feb. 7), LEO (Feb. 25)

This comet was discovered by J. Herapath (Hounslow Heath, England) on 1831 January 7.25. He said, "The tail was then nearly perpendicular to the horizon, inclining towards the south, and of a white colour, apparently between 1° and 2° long." He added, "The head was of the same colour as the tail, but, in proportion, far more splendid. To me, it appeared to equal in light stars of the second magnitude, while it exceeded them in size." On January 7.27, Herapath estimated the position as $\alpha = 17^h 36.7^m$, $\delta = -12° 33'$. The comet had passed only 4° from the sun on December 29. Independent discoveries were made by R. T. Paine (Boston, Massachusetts, USA) on January 7.42, J. South (Kensington, England) on January 9.25, and by W. von Biela (Botzen, Austria) on January 14.20. Biela described it as visible to the naked eye and exhibiting a tail about 2.5° long.

Even though the comet was widely observed, only a few physical descriptions were published. G. Santini (Padova, Italy) noted the comet was visible to the naked eye through most of January. N. Cacciatore (Palermo, Italy) saw the comet in the morning sky on January 23. He said the bright nuclear condensation was about 20″ across and was situated with a nebulosity about 3′ long. The tail was about 3° long. F. B. G. Nicolai (Mannheim, Germany) saw the comet on February 11, 12, 14, and 15. He described it as nearly a round, homogeneously lit nebulosity about 3–4′ across, without noticeable condensation. H. W. M. Olbers (Bremen, Germany) described the comet as faint and diffuse, with a diameter of more than 20′ on the 17th.

The comet's position was measured for the final time by C. K. L. Rümker (Hamburg, Germany) on March 8.91, and by Santini on March 9.00. Santini gave the location as $\alpha = 10^h 25.8^m$, $\delta = +16° 44'$. The comet was last detected on March 19.8, when V. Knorre (Nicolajew, now Mykolayiv, Ukraine) described it as very faint. No position could be measured. Nicolai said a long period of bad weather followed his February observations and when he next looked for the comet "in the middle of March" he could no longer see it.

Parabolic orbits were calculated by several astronomers. During March, C. A. F. Peters determined the perihelion date as 1830 December 28.17. At the beginning of May, Knorre determined the perihelion date as December 28.19. Around mid-July, Santini determined it as December 28.16.

J. P. Wolfers (1832) used 61 positions obtained between January 21 and March 8, and computed a parabolic orbit with a perihelion date of December 28.16. No planetary perturbations were considered. This orbit is given below.

T	ω	Ω (2000.0)	i	q	e
1830 Dec. 28.1604 (UT)	26.8884	340.2392	135.2630	0.125887	1.0

ABSOLUTE MAGNITUDE: $H_{10} = 6.2$ (V1964)

FULL MOON: Dec. 29, Jan. 28, Feb. 26, Mar. 28

SOURCES: J. Herapath, *MAS*, **4** (1831), p. 626; W. von Biela and J. Herapath, *AN*, **8** (1831 Jan.), p. 475; J. Herapath and J. South, *MNRAS*, **2** (1831 Jan. 14), pp. 6–7; C. A. F. Peters, *AN*, **9** (1831 Mar. 18), p. 83; C. K. L. Rümker, C. A. F. Peters, and R. T. Paine, *AN*, **9** (1831 Apr. 26), p. 147; V. Knorre and F. B. G. Nicolai, *AN*, **9** (1831 May 4), pp. 173–80, 187; N. Cacciatore, *AN*, **9** (1831 Jul. 7), p. 281; G. Santini, *AN*, **9** (1831 Jul. 18), pp. 287–90; J. P. Wolfers, *AN*, **10** (1832 Mar. 20), pp. 67–72; J. Herapath and W. T. Lynn, *The Observatory*, **16** (1893 Jan.), pp. 70–1; O1899, p. 145; V1964, p. 55.

2P/Encke

1832 I

Recovered: 1832 June 2.39 ($\Delta = 0.34$ AU, $r = 0.77$ AU, Elong. $= 36°$)

Last seen: 1832 June 30.1 ($\Delta = 0.30$ AU, $r = 1.22$ AU, Elong. $= 127°$)

Closest to the Earth: 1832 June 18 (0.2568 AU)

Calculated path: ERI (Rec), FOR (Jun. 8), ERI (Jun. 15), PHE (Jun. 16), ERI-PHE (Jun. 18), TUC (Jun. 23)

J. F. Encke (1831) tried to improve upon the orbit of this comet by attempting to explain an interesting decrease of about 2.5 hours in the perihelion date at every return. Encke believed the decrease was due to the comet moving through a resisting medium and he created a mathematical model to account for this. Encke (1832) applied his new model, with the addition of planetary perturbations, and predicted the comet would next arrive at perihelion on 1832 May 4.47. This orbit indicated the comet would remain lost in the sun's glare from late February until late May or early June, and would pass only 14° from the sun on May 18.

The comet was recovered in the morning sky by O. F. Mossotti (Buenos Aires, Argentina) on 1832 June 2.39. He estimated the position as $\alpha = 3^h 46.6^m$, $\delta = -11° 20'$ and described the comet as "a nebulosity of very small diameter, without nucleus, and very faint...." The next three mornings were cloudy, but on June 6.39, Mossotti again observed it, thus confirming this was comet Encke. He described it as faint. An independent recovery was made by T. Henderson (Cape of Good Hope, South Africa) on June 3.17. Using a 9-cm refractor (32×), he simply described it as faint. Henderson had tried to recover the comet on May 31, but nothing was found and he surmised the comet "was not sufficiently disengaged of the twilight." Encke's prediction was only 0.01 day off.

The comet moved across the sky at a rate of over 2° per day during the remainder of June. Henderson simply described the comet as very faint on June 4, 5, 6, and 9. Mossotti glimpsed the comet on the 10th, but noted "it disappeared at intervals, and it was impossible to observe it well." Moonlight effectively blocked the comet for nearly two weeks thereafter, with Henderson reporting he could not see the comet during the period of June 10–23.

Henderson was the only person to see the comet after the moon had left the sky. His observations on June 24, 27, 28, and 29 revealed a very faint comet that was barely visible in a 9-cm refractor. The final measurement of the comet's position was made on June 29.12, but although the faintness prevented the measurement of δ, he did give $\alpha = 22^h\ 32.7^m$. Henderson saw the comet for the last time on June 30.1, when he noted "occasional glimpses of the comet were suspected" in a 13-cm refractor.

Interestingly, some sources in the late nineteenth and early twentieth centuries have reported that the comet was last seen on August 21 by K. L. Harding (Göttingen, Germany). H. W. M. Olbers originally gave Harding's positions in a letter that was published in the 1832 October 9 issue of the *Astronomische Nachrichten*. He began the letter by noting how the faintness of comet Encke had prevented it from being seen before perihelion. In the next paragraph he wrote, "Professor Harding was lucky to observe the comet still on August 21. In case he has not yet reported his observations to you I will give them." What followed were four positions measured on August 21. But the Author notes that the positions almost exactly match those expected for comet C/1832 O1 (Gambart), while comet 2P/Encke was about 60° to the southeast. J. G. Galle (1894) was one of the first astronomers to assign Harding's positions to 2P/Encke, but J. Holetschek (1908) caught the mistake and gave the details in a letter that was published in the 1908 April 29 issue of the *Astronomische Nachrichten*. Nevertheless, the August 21 date still occasionally appears in some catalogs as belonging to 2P/Encke.

J. J. Littrow (1833) compared the orbit of this comet with that of P/Biela and noted that, at a point 1.5087 AU from the sun, the paths of these comets pass only 12 350 miles of one another. He said, "A slight change in the elements, which, especially those of Biela's comet, are subject to great disturbances, may greatly diminish, and even totally annihilate, this distance, in which case the collision of the two comets would take place in the direction of the line of the centers."

Minor refinements of this comet's orbit using positions from multiple apparitions, as well as planetary perturbations, have been published by numerous astronomers. Of most significance were the studies of Encke, F. E. von Asten (1877), and B. G. Marsden and Z. Sekanina (1974). These have typically revealed a perihelion date of May 4.48 and a period of 3.31 years. Marsden and Sekanina determined the nongravitational terms as $A_1 = -0.63$ and $A_2 = -0.03759$. Their orbit is given below.

T	ω	Ω (2000.0)	i	q	e
1832 May 4.4790 (TT)	182.7813	336.9147	13.3910	0.343404	0.845466

ABSOLUTE MAGNITUDE: $H_{10} = 9$–10 (V1964)

FULL MOON: May 14, Jun. 13, Jul. 12

SOURCES: J. F. Encke, AN, **9** (1831 Aug. 29), pp. 317–48; J. F. Encke, MNRAS, **2** (1832 May 11), p. 108; H. W. M. Olbers, K. L. Harding, and O. F. Mossotti, AN, **10** (1832 Oct. 9), pp. 253–8; O. F. Mossotti, MNRAS, **2** (1833 Jan. 11), p. 142; J. J. Littrow, AJS (Series 1), **24** (1833 Jul.), pp. 346–8; T. Henderson, AN, **11** (1833 Jul. 9), pp. 25–8; O. F. Mossotti and T. Henderson, AN, **11** (1833 Aug. 6), p. 39; T. Henderson, AN, **11** (1834 Feb. 12), pp. 293–5; T. Henderson and O. F. Mossotti, MRAS, **8** (1835), pp. 243–50; F. W. Bessel, AN, **13** (1836), pp. 345–50; F. E. von Asten, BASP (Series 4), **22** (1877), p. 560; G1894, p. 197; J. Holetschek, AN, **177** (1908 Apr. 29), p. 343; V1964, p. 55; B. G. Marsden and Z. Sekanina, AJ, **79** (1974 Mar.), pp. 415–16.

C/1832 O1 *Discovered:* 1832 July 19.94 ($\Delta = 0.85$ AU, $r = 1.58$ AU, Elong. $= 116°$)
(Gambart) *Last seen:* 1832 August 27.83 ($\Delta = 1.45$ AU, $r = 1.27$ AU, Elong. $= 59°$)
Closest to the Earth: 1832 July 19 (0.8463 AU)
1832 II *Calculated path:* HER (Disc), SER (Jul. 28), VIR (Aug. 7), LIB (Aug. 12), VIR (Aug. 14), LIB (Aug. 24), VIR (Aug. 26)

J. F. A. Gambart (Royal Observatory, Marseille, France) discovered this comet on 1832 July 19.94 at a position of $\alpha = 16^h 54.3^m$, $\delta = +25° 55'$. He said the observation was difficult due to the comet's low brightness. Gambart added that it exhibited no tail or nucleus, although a magnification of 50× revealed a pronounced point of central condensation. Additional observations by Gambart on the 22nd and 25th revealed the comet might have been slightly brighter, but it was still a difficult object to see. The comet was independently discovered by K. L. Harding (Göttingen, Germany) on July 29.89. He described it as a round, washed out, and pale nebulosity, with a faint, star-like central nucleus. There was no tail.

The comet remained a difficult object throughout August. Gambart's observation on the 1st was made under exceptional skies and was considered by Gambart as the most reliable of the entire apparition. Gambart had to use an ephemeris to find the comet on the 8th. He reported it was at the limit of visibility in moonlight. F. B. G. Nicolai (Mannheim, Germany) received news of the comet's discovery on this day, but attempts to detect the comet on that evening as well as during the next few nights were not successful. Following the full moon, Gambart found the comet on the 13th and reported it was hardly visible. Nicolai finally detected the comet on the 16th and noted it was fairly close to the southwestern horizon. Gambart observed it under "very fine skies" on August 20, but still had a difficult time seeing the comet. On August 21, Harding described the comet as small and faint. Nicolai saw the comet very close to the horizon on the 25th, but was unable to measure its position. On August 26, Gambart said the coma was near the limit of visibility, while its central condensation was still "rather bright" and presented "a kind of scintillation." He expressed doubt that the scintillation was caused by a faint star.

The comet was last detected on August 27 by three different astronomers. They were Gambart on August 27.76, G. Santini (Padova, Italy) on August 27.80, and C. A. F. Peters (Hamm, Germany) on August 27.83. Peters measured the position as $\alpha = 14^h\ 09.5^m$, $\delta = -9°\ 30'$. Gambart said the comet was too feeble for an accurate position to be measured. Although the comet was still a month from perihelion, it became lost in the sun's glare and remained hidden by twilight until it was too faint to be detected.

The first parabolic orbits were independently calculated by Peters and H. W. M. Olbers, with both astronomers using positions from July 20, 29, and August 4. Peters determined a perihelion date of 1832 September 25.36, while Olbers determined it as September 25.78. Olbers' orbit would later prove to be only about 7 hours early. Continued refinements of the orbit were made in the following months by Gambart, Peters, A. von Heiligenstein, E. Bouvard, G. Santini, and C. Conti. L. R. Schulze (1873) calculated an orbit using 32 positions from July 19 to August 27. His parabolic orbit included the perturbations by three planets and gave the perihelion date as September 26.07. This orbit is given below.

T	ω	Ω (2000.0)	i	q	e
1832 Sep. 26.0735 (UT)	204.6342	74.8269	136.6727	1.183005	1.0

ABSOLUTE MAGNITUDE: $H_{10} = 6.0$ (V1964)

FULL MOON: Jul. 12, Aug. 11, Sep. 10

SOURCES: J. F. A. Gambart, K. L. Harding, C. A. F. Peters, and H. W. M. Olbers, *AN*, **10** (1832 Sep. 4), pp. 217, 227; H. W. M. Olbers and K. L. Harding, *AN*, **10** (1832 Oct. 9), p. 253; J. F. A. Gambart and F. B. G. Nicolai, *AN*, **10** (1832 Oct. 9), pp. 259–62; A. von Heiligenstein, *AN*, **10** (1832 Oct. 9), p. 267; J. F. A. Gambart and C. A. F. Peters, *AN*, **10** (1832 Oct. 23), pp. 269–72; E. Bouvard, *AN*, **10** (1832 Dec. 25), pp. 305–8; G. Santini and C. Conti, *AN*, **10** (1833 Jan. 8), pp. 319–22; AJS (Series 1), **24** (1833 Jul.), p. 348; G. Santini, *MRAS*, **6** (1833), pp. 228–9; E. Bouvard, *CDT* (1835), pp. 30–6; *American Almanac*. Boston: James Munroe & Co. (1847), p. 91; L. R. Schulze, *AN*, **82** (1873 Aug. 6), pp. 101–10; V1964, p. 55.

3D/1832 S1   *Recovered:* 1832 September 24.15 ($\Delta = 0.74$ AU, $r = 1.30$ AU, Elong. $= 95°$)
(Biela)   *Last seen:* 1833 January 4.09 ($\Delta = 1.04$ AU, $r = 1.06$ AU, Elong. $= 63°$)
   Closest to the Earth: 1832 October 24 (0.5514 AU)
1832 III   *Calculated path:* AUR (Rec), GEM (Oct. 1), CNC (Oct. 14), LEO (Oct. 24), SEX (Nov. 4), LEO (Nov. 9), CRT (Nov. 19), VIR (Nov. 22), CRV (Nov. 24), VIR (Dec. 8), HYA (Dec. 15)

Predictions for the return of this comet were independently provided by M. C. T. Damoiseau and G. Santini (Padova, Italy). Santini determined a likely perihelion date of 1832 November 24.70, while Damoiseau determined it as November 27.47. Although Santini, K. L. Harding, and F. B. G. Nicolai (Mannheim, Germany) searched for the comet early in September of 1832, it was J. F. W. Herschel (Slough, England) who first spotted the comet.

Herschel found the comet by means of the predictions of both Santini and Damoiseau. He determined positions for several dates using both orbits and then averaged each date's positions. His first search was made on the morning of September 23 using a 48-cm reflector, but increasing haze and, finally, clouds prevented any possible success. The next morning skies were perfectly clear and after about five minutes of sweeping about the predicted position, Herschel recovered the comet on 1832 September 24.15, at a position of $\alpha = 5^h 40.2^m$, $\delta = +36° 18'$. He described it as a bright, conspicuous nebula about 2.5–3' across, but with no tail. He added, "though its rate of increase of density towards the centre was rather more rapid and decided than about the circumference, . . . there was nothing in the least entitled to the name of a nucleus." He knew of no bright nebula in the area, so his belief that this was P/Biela was fairly certain. Herschel confirmed the object was moving on September 24.19. Another observation was made by Herschel on September 25.07 when he wrote, "It was very little, if at all, perceptibly brighter or larger than the preceding night."

There had been an announcement that the comet had been recovered by observers at Roman College on August 23.13 and that it was observed from there until September 20, but no confirmation of this appeared. Herschel commented that if the Roman observations were verified it would "be a great proof of the advantage of an Italian sky," however, the faintness in his own telescope prompted Herschel to wonder "with what instrument this observation can have been made." Santini later wrote that the comet had been "uselessly searched for in August and September."

Herschel's telescope was apparently the only telescope in the world at that time that could see this comet, but he did not follow up with additional observations during October and he did not immediately report his observations. Subsequently, several astronomers continued to try to recover the comet and a flurry of independent recoveries were made shortly after mid-October. The first came on the morning of October 20, when the comet was found by both J. F. A. Gambart (Marseille, France) and J. E. B. Valz (Nîmes, France). Both astronomers described the comet as very faint. F. W. Bessel

(Königsberg, now Kaliningrad, Russia) and F. G. W. Struve (Dorpat, now Tartu, Estonia) found the comet on October 21. Bessel said moonlight and clouds had prevented previous searches, and added that the comet could hardly be seen because of hazy skies. Struve first saw the comet through the finder and said the refractor revealed a circular coma about 3' across, while there was a condensation near, but not in, the center. No tail was present. Another independent recovery was made by F. B. G. Nicolai (Mannheim, Germany) on October 22, after several previous attempts had failed. He described the comet as a small, extremely faint nebulous mass. Struve reported the comet was somewhat brighter on the 24th.

Although the comet had passed Earth, it was still approaching perihelion and continued to brighten for most of November. Santini saw the comet on the 1st and noted it was difficult to observe because of a turbulent atmosphere which sometimes rendered it invisible. On November 4, Herschel saw the comet for the first time since September 25. He described it as a "large and very bright nebula" in the 48-cm reflector and estimated the coma diameter as 4'. Herschel added, "The condensation towards the middle was considerable, and the centre itself was occupied by a bright point about equal to a star of the 13th magnitude. It had no tail, but only a feeble trace of some extension of its nebulosity. . . . " Herschel saw the comet at a higher altitude on the 5th and said the comet "was proportionately brighter, and was, indeed, a very fine and brilliant object. The trace of a tail or branch in the same direction as last night, though extremely feeble, was now unequivocal, and the central point not to be overlooked." He estimated the coma diameter as 5' and he "suspected some degree of nebulosity even beyond that limit." Struve saw the comet on November 7 and estimated the coma was 3' across. Following the full moon, T. Henderson (Cape of Good Hope, South Africa) first observed the comet on November 19 using a 9-cm refractor. He described it as very faint. Santini described the comet as "very faint and very difficult to observe" on November 20. On the 29th, Struve said the comet had increased in light, but was not substantially different in size from the 7th. The condensation was not centralized, but was shifted toward the sun.

Most of the European astronomers were no longer able to see the comet as December began, primarily because of its increasingly southward placement, but the Italian astronomers continued to track it through most of December. The comet was last detected on the 26th by K. Kreil (Milan, Italy), on the 27th by Santini, and on the 28th by G. Inghirami (Florence, Italy). The comet had reached a minimum elongation of 59° during the first week of December and then began increasing its apparent distance from the sun. Henderson resumed his observations on December 27 and continued into the first days of January.

The comet was last detected on 1833 January 4.09, by Henderson with the help of a 9-cm refractor (32×). He gave the position as $\alpha = 14^h 15.3^m$, $\delta = -27° 55'$. Henderson was not able to see the comet again "owing to the increasing moonlight." The comet was also very near the southern horizon.

J. J. Littrow (1833) compared the orbits calculated for this comet with that of P/Encke and noted that, at a point 1.5087 AU from the sun, the paths of these comets pass only 12 350 miles from one another. He said a slight change in the orbit of Biela's comet "may greatly diminish" this distance so that a "collision of the two comets would take place in the direction of the line of the centers."

Despite the success of the predicted orbits of Damoiseau and Santini, several astronomers still performed analyses during the months and years following recovery. Some of these included Nicolai, Santini (1833, 1835), and J. Baranowski (1837). Santini (1833) re-examined the differences in the orbital computations by Damoiseau and himself and noted "an error... in the constants of the variation of the daily motion" used for his computation. He pointed out that this brought the orbital elements closer together, but "a difference still remains between the observed times of perihelion passage in 1826 and 1832, such as would arise from a resisting medium."

Several astronomers published later investigations of the orbit of this comet, including G. Santini (1835), J. S. Hubbard (1860), J. von Hepperger (1898, 1900), B. G. Marsden and Z. Sekanina (1971), and W. Landgraf (1986). Marsden included nongravitational terms of $A_1 = +3.86$ and $A_2 = -2.5370$ in his calculation and his orbit is given below.

T	ω	Ω (2000.0)	i	q	e
1832 Nov. 26.6152 (TT)	221.6588	250.6690	13.2164	0.879073	0.751299

ABSOLUTE MAGNITUDE: $H_{10} = 8.2$ (V1964)

FULL MOON: Sep. 10, Oct. 9, Nov. 8, Dec. 7, Jan. 6

SOURCES: M. C. T. Damoiseau, AN, 6 (1828 Jan.), pp. 155–60; M. C. T. Damoiseau, CDT (1830), p. 55; G. Santini and M. C. T. Damoiseau, AN, 10 (1832 Sep. 4), p. 219; J. F. W. Herschel, MNRAS, 2 (1832 Nov. 9), pp. 117–24; G. Santini, F. B. G. Nicolai, F. W. Bessel, J. F. A. Gambart, and J. E. B. Valz, AN, 10 (1832 Dec. 7), pp. 293–6, 299; G. Santini, MNRAS, 2 (1832 Dec. 14), p. 129; F. B. G. Nicolai, AN, 10 (1832 Dec. 25), pp. 303–6; J. F. W. Herschel, MRAS, 6 (1833), pp. 99–109; T. Henderson, MRAS, 6 (1833), pp. 159–68; F. W. Bessel, F. M. Schwerd, F. B. G. Nicolai, and G. Santini, MRAS, 6 (1833), pp. 230; J. F. W. Herschel and G. Santini, AN, 10 (1833 Jan. 8), pp. 317–22; F. B. G. Nicolai, MNRAS, 2 (1833 Jan. 11), pp. 141–2; PMJS (Series 3), 2 (1833 Mar.), pp. 222–9; T. Henderson, MNRAS, 2 (1833 Jun. 14), p. 189; J. J. Littrow, AJS (Series 1), 24 (1833 Jul.), pp. 346–8; T. Henderson, AN, 11 (1833 Jul. 9), p. 27; G. Inghirami, AN, 11 (1833 Aug. 6), pp. 39–42; G. Santini, AN, 11 (1833 Dec. 24), p. 195; K. Kreil, Effemeridi Astronomiche di Milano (1834), p. 68; J. F. W. Herschel and T. Henderson, AN, 11 (1834 Feb. 12), pp. 293–6; F. G. W. Struve, AN, 12 (1834 Sep. 3), pp. 17–32; PMJS (Series 3), 5 (1834 Oct.), pp. 301, 310–11; F. G. W. Struve, AN, 12 (1834 Oct. 31), pp. 33–44; T. Henderson, MRAS, 8 (1835), pp. 240–2; G. Santini, AN, 12 (1835 Jan. 14), pp. 113–18; G. Santini, AJS (Series 1), 28 (1835 Jul.), p. 398; F. W. Bessel, AN, 13 (1835 Dec. 5), pp. 81–90; F. W. Bessel, F. G. W. Struve, and J. Baranowski, AN, 14 (1837 Feb. 9), pp. 177–80;

American Almanac. Boston: James Munroe & Co. (1847), p. 91; J. S. Hubbard, *AJ*, **6** (1860 Jul. 17), pp. 124–6; J. von Hepperger, *SAWW*, **107** Abt. IIa (1898), pp. 377–489; J. von Hepperger, *SAWW*, **109** Abt. IIa (1900), pp. 647–53; V1964, p. 55; W. Landgraf, *QJRAS*, **27** (1986), p. 604.

C/1833 S1 *Discovered:* 1833 September 30.4 ($\Delta = 1.06$ AU, $r = 0.67$ AU, Elong. $= 37°$)
(Dunlop) *Last seen:* 1833 October 16.42 ($\Delta = 1.09$ AU, $r = 0.95$ AU, Elong. $= 54°$)
  Closest to the Earth: 1833 October 6 (1.0434 AU)
1833 *Calculated path:* LIB (Disc), SCO (Oct. 6), OPH (Oct. 10)

Shortly after sunset on 1833 September 30.4, and just two days after full moon, J. Dunlop (Paramatta, New South Wales, Australia) "found a very small body resembling a comet...." A brief comparison with nearby stars revealed a position of $\alpha = 14^h 44^m$, $\delta = -18° 32'$, but measuring a more precise position was impossible due to approaching clouds. Dunlop confirmed the comet on October 1.42, and described it as "very small and faint." He went on to say, "I can perceive that the nebulous matter at the head, or preceding extremity, is considerably denser than the following extremity or end of the tail. The nebulous matter appears to be about two minutes in length and one minute in breadth. The head rounded off and the tail terminating in a blunt point."

Dunlop remained the only observer of this comet and battled clouds and hazy conditions during the next couple of weeks to acquire observations. He noted it was extremely faint on October 2 in hazy skies, and under similar conditions on the 3rd, he described the comet as very faint, about 1' across, and with no tail. Dunlop said it was so small and faint on the 4th, that "it would not bear the least illumination of the wire." On the 6th, he said it was nearly as bright as when discovered, but added that the condensation had shifted from the "preceding extremity" to near the center of the coma. Dunlop again described the comet as faint on the 7th and noted it was "more confused and scattered on the preceding extremity, and the tail following is shorter and fainter, the whole length does not exceed 2 minutes." On October 12, Dunlop found the comet "pretty bright" through an opening in the clouds. On the 14th, he found it "extremely faint" because of haze and clouds. Moonlight was interfering when he saw the comet on the 15th, but he nevertheless described it as "pretty bright."

The comet was last detected on October 16.42, when Dunlop measured the position as $\alpha = 17^h 01.6^m$, $\delta = -23° 43'$. He then described it as "very faint, on account of the moonlight." The moon prevented additional observations in the days that followed, and, after the full moon, Dunlop "searched diligently about the head of Sagittarius," but was unable to locate the comet.

These observations were first presented in the 1835 January 14 issue of the *Astronomische Nachrichten* by T. Henderson, who also included a parabolic orbit using positions spanning the period of October 1–16, and derived a perihelion date of 1833 September 11.55. In the 1835 January 21 issue of the same periodical, C. A. F. Peters used positions from the same period and computed a parabolic orbit with a perihelion date of September 10.69.

E. Hartwig (1857) computed three parabolic orbits using different combinations of positions. The first orbit used 15 positions obtained between October 1 and 16, and produced a perihelion date of September 10.70. The second orbit used nine positions obtained between October 1 and 14, and revealed a perihelion date of September 10.70. The third orbit used six positions obtained between October 1 and 8, and revealed a perihelion date of September 10.90. The final orbit, which used only the most reliable positions, fitted the used positions best.

L. Schulhof (1888) used 15 positions obtained between October 1 and 16, and computed a parabolic orbit, with a perihelion date of September 10.69. Schulhof also considered a short-period orbit and said the eccentricity could be as small as 0.8, indicating a period as small as 3.5 years. No planetary perturbations were applied. This orbit is given below.

T	ω	Ω (2000.0)	i	q	e
1833 Sep. 10.6912 (UT)	259.5795	325.5873	7.3488	0.458122	1.0

ABSOLUTE MAGNITUDE: $H_{10} = 7.0$ (V1964)

FULL MOON: Sep. 28, Oct. 28

SOURCES: J. Dunlop and T. Henderson, *AN*, **12** (1835 Jan. 14), pp. 117–20; C. A. F. Peters, *AN*, **12** (1835 Jan. 21), p. 127; J. Dunlop and T. Henderson, *MRAS*, **8** (1835), pp. 251–9, 264–5; *American Almanac*. Boston: James Munroe & Co. (1847), p. 91; T. Henderson and J. Dunlop, *AN*, **42** (1855 Sep. 27), p. 61; T. Henderson and J. Dunlop, *AN*, **42** (1855 Oct. 12), pp. 75–80; T. Henderson and J. Dunlop, *AN*, **42** (1855 Oct. 18), pp. 93–6; T. Henderson and J. Dunlop, *AN*, **42** (1855 Oct. 23), pp. 105–8; E. Hartwig, *AN*, **47** (1857 Aug. 29), pp. 37–44; L. Schulhof, *BA*, **5** (1888), p. 537; V1964, p. 55.

C/1834 E1   *Discovered:* 1834 March 8.2 ($\Delta = 0.60$ AU, $r = 0.79$ AU, Elong. $= 53°$)

(Gambart)   *Last seen:* 1834 April 14.80 ($\Delta = 1.34$ AU, $r = 0.58$ AU, Elong. $= 24°$)

Closest to the Earth: 1834 February 28 (0.5688 AU)

1834   *Calculated path:* SGR (Disc), CAP (Mar. 9), AQR (Mar. 22), PSC (Apr. 22)

This comet was discovered by J. F. A. Gambart (Marseilles, France) on 1834 March 8.2. It was then in the morning sky near M75 and was described as nebulous, round, pale, and 4–5′ across. No position could be determined because of its nearness to the horizon. Gambart reobserved the comet on March 10.20, and estimated the position as $\alpha = 20^h\ 09.7^m$, $\delta = -22°\ 33'$. Unfortunately, the comet was moving toward evening twilight and Gambart could not find it on the following mornings. The announcement brought no further Northern Hemisphere observations. Fortunately, an independent discovery was made by J. Dunlop (Paramatta, New South Wales, Australia) on March 19.8. Dunlop said the comet was visible in the east before sunrise, but twilight prevented the determination of a position. He did note it was in the same field as 42 Capricorni. Dunlop confirmed the discovery on March 21.79 and described it as resembling "a small bright nebula," about 1.5′ across, "with a very faint stream of light proceeding from the head, at intervals, exceedingly rare, and of a very pale bluish colour, remarkably different from that of the head."

Dunlop described the comet as "faint" on March 22, but his observations during the remainder of March and up to April 6 were only for positions and did not include physical descriptions.

The comet was last detected on April 14.80, when Dunlop measured the position as $\alpha = 23^h 58.4^m$, $\delta = +2° 40'$. He noted that the comet's "brightness was about equal to a star of the 7th magnitude," while its round, well-defined coma was no more than 1′ across. He said it resembled a pretty bright, small, round nebula."

Dunlop's observations were first presented in the 1835 January 14 issue of the *Astronomische Nachrichten* by Thomas Henderson. In that issue, Henderson determined a parabolic orbit using the ten positions obtained by Dunlop between March 20 and April 13, and derived a perihelion date of 1834 April 11.76. He noted this comet might be the same as Gambart's. The same issue also contained orbits by A. C. Petersen and C. A. F. Peters, which combined the positions given by Gambart and Dunlop. Petersen determined a perihelion date of April 3.16, while Peters determined it as April 3.32.

L. Schulhof (1889) used 11 positions obtained between March 10 and April 14, and computed a parabolic orbit with a perihelion date of April 3.29. He added that the eccentricity is probably not less than 0.996, indicating a minimum period of 1400 years. No planetary perturbations were applied. This orbit is given below.

T	ω	Ω (2000.0)	i	q	e
1834 Apr. 3.2926 (UT)	49.9876	229.0377	5.9759	0.513106	1.0

ABSOLUTE MAGNITUDE: $H_{10} = 7.0$ (V1964)

FULL MOON: Feb. 23, Mar. 25, Apr. 23

SOURCES: J. F. A. Gambart, *AN*, **11** (1834 May 7), p. 373; J. F. A. Gambart, *AJS* (Series 1), **26** (1834 Jul.), p. 402; J. Dunlop, T. Henderson, A. C. Petersen, and C. A. F. Peters, *AN*, **12** (1835 Jan. 14), pp. 117–20; J. Dunlop and T. Henderson, *MRAS*, **8** (1835), pp. 259–64, 266–7; *American Almanac*. Boston: James Munroe & Co. (1847), p. 91; L. Schulhof, *BA*, **6** (1889), pp. 109–15; V1964, p. 55.

C/1835 H1	*Discovered:* 1835 April 20.90 ($\Delta = 1.12$ AU, $r = 2.06$ AU, Elong. $= 152°$)
(Boguslawski)	*Last seen:* 1835 May 27.87 ($\Delta = 1.98$ AU, $r = 2.17$ AU, Elong. $= 86°$)
	Closest to the Earth: 1835 April 11 (1.0540 AU)
1835 I	*Calculated path:* CRV (Disc), CRT (Apr. 22), LEO (May 1), SEX (May 7)

This comet was discovered in the evening sky by P. H. L. von Boguslawski (Wroclaw, Poland) on 1835 April 20.90. He gave the position as $\alpha = 11^h$ 58.2^m, $\delta = -12° \ 07'$ on April 20.98. Boguslawski described it as a small, round, and very washed out nebulosity. He also noted a star-like nucleus and a broad tail extending eastward. During moments of very transparent air, the tail seemed much longer. Boguslawski confirmed his discovery on April 21.84 and said the comet was 3–4' across and diffuse, with a somewhat bright, nearly star-like condensation.

Boguslawski had difficulty measuring the comet on April 22, because of clouds and the indistinct shape of the comet, which now lacked a nucleus. M. Weisse (Krakow, Poland) said the comet was barely seen in the meridian telescope on April 30 and could not withstand the illumination of the micrometer wire to measure a precise position.

With the comet moving away from both the sun and Earth as May began, the number of observers quickly dwindled. Following the full moon, K. Kreil (Milan, Italy) described the comet as faint on the 18th, while Boguslawski said it was faint in a heliometer on the 20th. The comet was last detected on May 27.87, when Kreil gave a position of $\alpha = 10^h \ 04.4^m$, $\delta = +2° \ 49'$.

The comet had a perihelion distance of over 2 AU and this initially made it very difficult for astronomers to pinpoint its perihelion date. The comet was moving away from both the sun and Earth when discovered and could only be observed for just over a month. The first orbit was calculated by J. F. Encke using positions obtained during the period of April 20–28. The resulting perihelion date was 1835 April 4.89. Using the slightly longer arc of April 20–30, C. A. F. Peters determined the perihelion date as March 27.93. During the next couple of months additional orbits were calculated by C. K. L. Rümker, F. B. G. Nicolai, and Boguslawski, with perihelion dates ranging from March 25 to April 3. About a year later, F. W. Bessel (1836) took positions he had received which covered the period of April 20 to May 20 and computed a parabolic orbit with a perihelion date of March 28.08. The orbit of this comet was again investigated by G. Rechenberg (1897). He used 38 positions obtained between April 20 and May 27, and computed a parabolic orbit with a perihelion date of March 27.71. This orbit is given below.

T	ω	Ω (2000.0)	i	q	e
1835 Mar. 27.7052 (UT)	210.5630	60.7681	170.8846	2.040153	1.0

ABSOLUTE MAGNITUDE: $H_{10} = 4.2$ (V1964)

FULL MOON: Apr. 13, May 12, Jun. 10

SOURCES: P. H. L. von Boguslawski, J. F. Encke, and C. A. F. Peters, *AN*, **12** (1835 May 16), pp. 253–6; F. B. G. Nicolai, *AN*, **12** (1835 Jul. 1), pp. 281–4; C. K. L. Rümker, *CR*, **1** (1835 Aug. 5), p. 10; P. H. L. von Boguslawski, M. Weisse, and C. K. L. Rümker, *AN*, **12** (1835 Aug. 26), pp. 409–16; P. H. L. von Boguslawski, *CR*, **1** (1835 Sep. 14), p. 111; F. W. Bessel, *AN*, **13** (1836 Jun. 1), pp. 339–42; K. Kreil, *AN*, **13** (1836 Jul. 16), pp. 383–6; G. Rechenberg, *AN*, **143** (1897 Mar. 10), pp. 11–14; V1964, p. 55.

2P/Encke *Recovered:* 1835 July 23.10 ($\Delta = 1.53$ AU, $r = 0.87$ AU, Elong. $= 33°$)
Last seen: 1835 September 24.8 ($\Delta = 1.38$ AU, $r = 0.77$ AU, Elong. $= 33°$)

1835 II *Closest to the Earth:* 1835 September 10 (1.3137 AU)
Calculated path: AUR (Rec), GEM (Jul. 29), CNC (Aug. 8), LEO (Aug. 18),
SEX-LEO (Aug. 29), VIR (Sep. 4), CRV-VIR (Sep. 14)

J. F. Encke (1833) examined the observed apparitions of this comet for the
period of 1818–32 and noted that it would next arrive at perihelion during
the last half of August in 1835. He added that it would remain hidden in
the sun's rays. During March of 1835, Encke published a formal prediction. Taking the orbit for the 1832 apparition, he applied perturbations by
Jupiter and determined that the next perihelion would occur on 1835 August
26.88. The comet was recovered by K. Kreil (Milan, Italy) on 1835 July 23.10.
The position was only partially given as $\alpha = 5^h$ 46.5^m, but δ was then near
$+30.7°$. Encke's prediction ultimately proved only about 42 minutes late.

The comet was then brightening rapidly as its distances from the sun and
Earth decreased, but its elongation from the sun was also decreasing, so it
did not remain visible for long in the Northern Hemisphere. P. H. L. von
Boguslawski (Wroclaw, Poland) found the comet on July 31 after morning
twilight had begun. He described it as a diffuse, oval nebulosity, which
was about one-half to one-third the diameter of comet Biela at the end of
November 1832. In the 4.5-foot focal length Fraunhofer telescope, he noted
the brightness of the comet resembled stars of magnitude 8. On August 2,
Boguslawski said the sky was not very transparent near the horizon, but
during some moments he did see the comet southeast of the comparison
star, appearing very pale. He added it was visible when two 9th-magnitude
stars were barely seen.

Encke (1835) wrote about the possibility that the comet could be picked
up at the Cape of Good Hope sometime around the middle of September
and he provided ephemerides to help in the searches. The comet had passed
$0.2°$ from the sun on August 26, the day it passed perihelion, and the solar
elongation increased thereafter. J. F. W. Herschel was at the Cape at that time
and looked for the comet on the evening of the 18th with the 48-cm reflector.
He actually had several oak trees cut away in anticipation of the comet's low
altitude, but was still unsuccessful. Herschel later received word from his
friend T. Maclear (Royal Observatory, Cape of Good Hope, South Africa),
who reported that the comet had been found on the evening of September
14.8 with a 14-foot focal length reflector. The solar elongation was then
$24°$. Maclear also saw the comet on September 19.8 and 24.8. According
to Herschel, Maclear said, "It was *in* or *near* the calculated place, but no
measure could be got. It looked *'as he saw it in England'*."

Minor refinements of this comet's orbit using positions from multiple
apparitions, as well as planetary perturbations, have been published by
numerous astronomers. Of most significance were the studies of Encke,

F. E. von Asten (1877), and B. G. Marsden and Z. Sekanina (1974). These have typically revealed a perihelion date of August 26.86 and a period of 3.31 years. Marsden and Sekanina determined the nongravitational terms as $A_1 = -0.63$ and $A_2 = -0.03759$. Their orbit is given below.

T	ω	Ω (2000.0)	i	q	e
1835 Aug. 26.8553 (TT)	182.7775	336.9149	13.3758	0.344362	0.845103

ABSOLUTE MAGNITUDE: $H_{10} = 7.7$ (V1964)

FULL MOON: Jul. 10, Aug. 8, Sep. 7, Oct. 6

SOURCES: J. F. Encke, *AN*, **11** (1833 Dec. 24), p. 193; J. F. Encke, *AN*, **12** (1835 Mar. 7), pp. 179–82; P. H. L. von Boguslawski, *AN*, **12** (1835 Aug. 26), p. 407; K. Kreil, *AN*, **13** (1836 Jul. 16), pp. 383–6; K. Kreil, *Effemeridi Astronomiche di Milano* (1837), p. 64; F. E. von Asten, *BASP* (Series 4), **22** (1877), p. 560; V1964, p. 55; J. F. W. Herschel and T. Maclear, *Herschel at the Cape*. Austin: University of Texas Press (1969), pp. 186–92; B. G. Marsden and Z. Sekanina, *AJ*, **79** (1974 Mar.), pp. 415–16.

1P/1835 P1
(Halley)

1835 III

Recovered: 1835 August 5.12 ($\Delta = 2.46$ AU, $r = 1.96$ AU, Elong. $= 49°$)

Last seen: 1836 May 19.9 ($\Delta = 2.73$ AU, $r = 3.02$ AU, Elong. $= 96°$)

Closest to the Earth: 1835 October 12 (0.1865 AU)

Calculated path: TAU (Rec), GEM (Sep. 1), AUR (Sep. 15), GEM-AUR (Sep. 28), LYN (Oct. 4), UMa (Oct. 8), BOO (Oct. 12), CrB (Oct. 14), HER (Oct. 16), OPH (Oct. 19), SER (Nov. 2), OPH (Nov. 17), SCO (Dec. 30), LUP (Jan. 27), CEN (Feb. 13), HYA (Mar. 12), CRT (Mar. 23), HYA (Apr. 11), SEX (Apr. 26)

The number of predictions for this comet's 1835 apparition is unparalleled in cometary astronomy, with only the 1986 apparition of this same comet coming close, but still not surpassing it in number. Five astronomers produced ten predictions that ranged from 1835 October 28 to November 26.

M. C. T. Damoiseau (1820) meticulously calculated the perturbations of Jupiter, Saturn, and Uranus upon the comet's orbit for the period of 1682–1835 and predicted the comet would next pass perihelion on 1835 November 17.15. Although this later proved to be about 17 hours late, discoveries over a century later would reveal the comet's orbit was affected by nongravitational forces, which were defined as a jet-like outgassing from the nucleus that had the capability of slightly deflecting, slowing, or speeding up the comet's motion. This could change the perihelion date of a purely gravitational prediction of 1P/Halley by several days. Without knowledge of this effect, the accuracy of the purely gravitational solutions was not high. Damoiseau (1829) published another prediction that added the perturbations by Earth for the same period and found a perihelion date of November 4.81.

During the next few years similar predictions were made. Philippe Gustave Le Doulcet, comte de Pontécoulant (1830, 1833, 1834, and 1835) determined perihelion dates of November 7.5, November 13.1, November 10.8, and November 12.9. J. W. Lubbock (1831) determined the perihelion date as October 31.19. O. A. Rosenberger (1834, 1835) determined it as November 4.34, and November 12.07. J. W. H. Lehmann (1835) determined it as November 26.73.

The first attempts to find the comet began with an ephemeris published by C. K. L. Rümker (1834). Covering the period of 1834 December 16 to 1835 April 3, it indicated the comet would become lost in twilight after the last date. Rümker suggested searches made with the largest telescopes might prove successful during December 1834. The only well-documented search was that by J. F. W. Herschel (Cape of Good Hope, South Africa). He began searching Rümker's positions on 1835 January 29 with an 48-cm reflector shortly after evening twilight had ended. He said he "thoroughly examined the region where it [the comet] is. Could find no trace of it...." He was also unsuccessful on February 17 and March 21.

The faint comet became hopelessly lost in the sun's glare early in April and passed only 4° from the sun on June 2. Although it entered the morning sky thereafter, its trek was slow, and a combination of the comet's faintness and the uncertainty as to its exact position made finding it a difficult task. It was finally recovered on August 5.12 by E. Dumouchel (Rome, Italy). He was joined by F. de Vico a few minutes later and the comet's position was then given as $\alpha = 5^h 26^m$, $\delta = +22° 17'$. They described it as extremely faint. They confirmed the recovery on August 6, and noted a definite movement toward the east. The moon interfered with additional observations for several days after. An independent recovery was made by F. G. W. von Struve (Dorpat, now Tartu, Estonia) on August 21. He described it as a centrally condensed nebulosity about 1.5' across in his 23-cm refractor. He added that a bright point-like nuclear condensation was eccentrically situated in the sunward half of the coma.

The comet seemed to brighten fairly quickly during the remainder of August. A. Bouvard, P. A. E. Laugier, D. F. J. Arago, and E. Plantamour (Paris, France) saw the comet on August 21 and described it as very faint and 2' across. They said it was centrally condensed, but saw no trace of a tail. T. J. Hussey (Hages, England) independently recovered the comet on August 23 and said it was very large and at the limit of his 17-cm refractor. He noted it was like the "finest smoke." J. South (London, England) saw the comet on the same morning and said it was extremely faint and about 2' across. On the 24th, E. Loomis (Yale College, Connecticut, USA) was still unable to find the comet. W. H. Smyth (Bedford, England) found the comet on the 25th after sweeping around the predicted position for a few minutes. He was using a refractor of 8.5-foot focal length (66×) and said he was initially hesitant in believing this was 1P/Halley because it was easier to find than expected. He described it as a "nebulous blot of indistinct form and misty appearance." E. J. Cooper (Sligo, Ireland) found the comet in a 12-cm refractor on the 26th and said his 34-cm telescope revealed a distinct nucleus. Struve saw the comet in a cometseeker on the 28th. Smyth described the comet as a well-defined "mass of pale light" when next seen on the 29th. He added that the coma was 2' across and contained a distinct nucleus of about magnitude 12 "in the following portion" of the coma. On August 30 C. J. Pastorff (Buchholz, Germany) described the comet as very faint and only about 0.75' across, while Smyth noted the nucleus was similar in brightness to a star of magnitude 10 and that it was "nearly in the centre of its gaseous envelope, which had its most concentrated part in the direction of the approaching sun." E. Loomis and D. Olmsted saw the comet on the morning of the 31st and said it was a round, faint, nebulous-looking object about 2' across, with a central condensation.

The comet changed little in appearance during the first days of September, until just before the moon began interfering. Smyth saw the comet with his refractor of 8.5-foot focal length on the 1st and noted the nucleus was very distinct and around magnitude 9. That same morning, T. Maclear (Royal Observatory, Cape of Good Hope) swept for the comet with his reflector of 14-foot focal length and, with his fingers numb from the cold, he finally caught a glimpse of a faint nebulous body before morning twilight began. Herschel tried to find the comet on the morning of the 2nd, after having been alerted by Maclear, but was unsuccessful. That same morning, Müller (Geneva, Switzerland) described the comet as excessively faint, while Struve found the comet 3' across, with a nucleus situated within the sunward half of the coma. Müller said the centrally condensed coma was elongated and about 4' across on the 3rd. Pastorff found the coma 2–2.5' across on the 4th, without a nucleus. Smyth found the comet on the 6th and noted it had "increased both in size and brightness, so that I could distinguish it in the finder." He added that the nucleus was near the center of the coma.

Moonlight interfered for several days, but Struve found the comet bright in moonlight on September 14. He noted the nucleus was situated towards the sunward side of the coma, and was at first thought to be a star of magnitude 8 shining through the coma. On the 18th, Smyth noted the comet "had now increased so as to be easily and clearly bisectable by the cross-wires of the finder" despite moonlight. He added that the nucleus was "in the following portion of the envelope." The comet was finally seen without moonlight on the 19th and H. J. Anderson (Columbia College New York, USA) said it was faintly visible to the naked eye. Müller added that the coma was then 6' across. The nucleus was estimated as magnitude 4 and 6 by Smyth and Loomis, respectively, on the 21st. Both observers said the nucleus was in the south-following portion of the coma and that more coma extended toward the sun than away from it. Loomis added that the coma was 6' across. Struve and his son O. W. von Struve saw the comet with the naked eye on September 23 and noted it was fainter than 48 Aurigae (magnitude 5.55 – SC2000.0). By the 24th, Struve, senior, said the comet was brighter than 48 Aurigae, but a little fainter than χ Aurigae (magnitude 4.76

– SC2000.0). He added that the nucleus was starlike and 1″ to 2″ across. Struve, senior, also said a trace of a tail was noted on the preceding side of the coma. On the 25th, Loomis noted the comet was "distinctly visible to the naked eye." Smyth said the nucleus was about magnitude 3 on the 26th and "decidedly in the south following portion of the gaseous envelope." He also noted "decided indications of a tail." Struve, senior, said the tail was 34′ long on the 29th.

The comet was at its greatest solar elongation on October 2 (89°). On that same night, F. W. Bessel (Königsberg, now Kaliningrad, Russia) observed an emanation from the nuclear region which extended towards a point opposite to the direction of the tail. The sides of this emanation were separated by an angle of 90°, and the emanation extended toward PA 87° 50′ on October 2.97. Several observers reported the comet was an easy naked-eye object on the 3rd. Smyth added that the nucleus was elliptical and measured about 2″ by 4″. Struve, senior, said the tail was then 76′ long and curved. B. F. Joslin (Union College, Schenectady, New York, USA) said the comet's naked-eye brightness on the 4th was about equal to a star of magnitude 1. He added that the comet remained visible to the naked eye until about a half hour before sunrise. Joslin also said the tail was only visible to the naked eye by averted vision and was invisible in his 10-cm refractor. Müller found the tail to be 0.3° long on the same night and Smyth said the nucleus seemed oval and occasionally showed "a minute point of light on the central disc, but so flittingly, as to render the fact almost uncertain." Struve, senior, observed the comet with a 23-cm refractor on the 5th and said the nucleus was bright and about 2″ across at 320×. He also noted a fan-shaped "flame" near the nucleus. Bessel said the comet was more pronounced on the 8th than on the 2nd, and extended 15–20″ toward PA 135.3°. The same night, Struve, senior, said, "The nucleus appeared like a glowing coal of oblong shape," with a dimension of 2.25″ by 0.9″. He added, "The fan-shaped flame is once more seen and very conspicuous." Despite moonlight, Struve, senior, faintly detected the tail with the naked eye on October 9 and added that the "flame" was double and more extensive than on previous days. Smyth wrote, "A very singular gleamy light was perceptible on the following portion of the nucleus, stretching into the nebulosity, and of a different intensity from that which I considered to indicate a tail. To this I attributed the apparent oval form of the body." On the 10th, Loomis said the comet was as conspicuous in the moonlight as α Ursae Majoris (magnitude 1.79 – SC2000.0). Olmsted said the comet appeared slightly elongated to the naked eye, while Joslin said the tail was distinctly visible to the naked eye. Struve, senior, observed with a cometseeker and noted the tail was 1° 57′ long, while the coma was 18′ across. He added that, although the nucleus was more vague than on previous nights, the "flame" was no less distinctive. Smyth wrote, "The comet . . . presented an extraordinary phenomenon. The brush, fan, or gleam of light before mentioned [on the 9th] was clearly perceptible issuing from the nucleus, which was now about 17″ in diameter, and shooting into the coma; the glances at times being very strong, and of a different aspect from the other parts of the luminosity."

The moon left the sky just as the comet was passing closest to Earth and the comet quickly became quite spectacular. On the night of October 11, Joslin and Smyth both noted the comet had increased in size and brightness during the preceding 24 hours. Olmsted and Loomis apparently agreed as they gave tail lengths in the range of 8–10°, while Loomis said the coma was nearly 38′ across. Smyth reported that within the coma, on the side opposite to the tail, "there proceeded from the nucleus across the coma a luminous band, or lucid sector, more than 60 or 70 seconds in length, and about 25 broad, with two obtuse-angled rays, the nucleus being its central point. The light of this singular object was more brilliant than the other parts of the nebulosity, and considerably more so than the tail; it was, therefore, amazingly distinct." Smyth increased the magnification and noted, "the nucleus appeared to be rather gibbous than perfectly round; but with the strange sector impinging, it was a question of difficulty." The comet was closest to Earth on the 12th and G. B. Amici (Florence, Italy) observed it with the naked eye and said it was brighter than the stars of the "Big Dipper", the brightest of which are magnitude 1.79. Several observers remarked on the brightness of the nucleus and their estimates of the tail length generally ranged from 6° to 9°. Joslin noted that the tail was 24–27° when viewed with averted vision. Bessel observed the jet from the nuclear region which extended 30″ towards a point opposite to the direction of the tail. He said the jet's rotation about the nucleus was noted as it extended toward PA 208° 06′ on October 12.70, PA 222° 20′ on October 12.88, and PA 233° 58′ on October 12.97. Struve, senior, said the "flame" was well seen and appeared "like a flash of fire from the nucleus." He also noted a second weak "flame". Olmsted said the "peculiar emanation of light" resembled "the brush of electric light from a pointed wire when highly charged and seen in a dark room." Observers reported the tail as 9–12° long on the 13th, but Joslin said it was only slightly longer when seen with averted vision. Although Olmsted said the jet had vanished, it was still well seen by others. Joslin said his 10-cm refractor showed a "very distinct and regular conical brush of light, the axis of which was directed downwards, and a little to the right, making an angle of about 120° with the long tail. . . . " He said it was the same width as the coma's condensation at the beginning, but over twice as wide at its extremity. Bessel said the jet extended 45″ toward a point opposite to the direction of the tail. He said the jet's rotation about the nucleus was noted as it extended toward PA 250° 23′ on October 13.04, and PA 280° 00′ on October 13.73. Although most observers said the tail was 6–7° long on the 14th, Struve estimated the length as 20°. Bessel observed the jet and said it extended 45″ toward PA 222° 25′ on October 14.74. Observations on the 15th revealed the tail was generally 6–10° long, with averted vision revealing a length of 12–20°. Bessel said the comet appeared brighter than on the 14th and indicated the tail extended nearly 28° toward PA 50°. F. Kaiser (Leiden,

Netherlands) estimated the brightness as magnitude 1. Müller said the coma was 15' across, with a brilliant nuclear condensation about 1' across. The jet was spotted for the first time by D. F. J. Arago (Paris, France), and he said the borders were distinctly defined and it was the brightest feature within the coma. Bessel said the jet extended toward PA 176° 55' on October 15.72.

Although the comet showed some signs of fading during the remainder of October, it still maintained an impressive appearance. Joslin said the head was "rather larger, though fainter, than a star of the first magnitude" on the 18th. The casual observer still noted a tail length of between 6° and 12° from the 16th to the 21st, but experienced observers noted a much longer tail when seen with averted vision. P. H. L. von Boguslawski (Breslau, now Wroclaw, Poland) estimated the tail as 24° long on the 16th. Joslin estimated it as 45° long on the 17th, 35° long on the 18th and 18° long on both the 19th and 20th. Smyth estimated it as 15–20° long on the 19th. Arago obtained several observations of the "sector" during this period. Observing the sector with a telescope on the 16th, he noted it "was divisible into three others, viz., two equal lateral ones and a more brilliant central one embraced by the two former, as though a brilliant cone was surrounded by a conical brush of less brilliancy but still greatly exceeding in brightness the rest of the nebulosity." The three sectors remained visible on the 17th and 21st. On the last date, he noted the most faint and narrow of the three was situated upon the axis of the tail. Arago noted the nucleus was bright and well defined. He added, "Some divergent gleams of light broke from the nucleus; and at times it seemed as if a small wedge-shaped luminosity was perceptible opposite the larger one, but both broke down under high powers." Smyth said the nucleus had a gibbous aspect on the 19th and 21st. Müller gave the coma diameter as 8' on the 18th and 7' on the 21st.

The comet continued to decrease in size during the remaining days of October. It continued to be a bright, naked-eye object throughout this period, although moonlight diminished its appearance at the end of the month. Joslin said the head was "about as large as a star of the second magnitude" on the 26th. Bessel indicated the tail extended about 3° toward PA 71° on the 22nd. Other observers reported similar tail lengths, until moonlight became an issue, although averted vision was still revealing lengths of 7–12° during most of this period. Bessel said the nucleus was brighter on the 22nd than on the 20th and noted that the "emanation" extended 35" toward the west. Arago found the nucleus large and diffuse on the 23rd and said the "sectors" noted on previous days were no longer visible. Bessel found the "emanation" very faint on the 25th and Struve said the "flame" was no longer visible on the 27th, while no distinct nucleus was present. Interestingly, Struve found the nucleus larger on the 29th than on previous days and said the "flame" was again observed.

The comet was steadily moving toward evening twilight during November. It was still described at "very bright" with a tail 2–3° long by several astronomers, until its decreasing altitude became a hindrance. Maclear reported that the tail seemed to broaden as the month progressed, while E. J. Cooper (Sligo, Ireland) noted the tail was bifurcated a short distance from the head on the 10th. Maclear also generally reported the coma as "nearly circular." Müller reported the coma diameter as 1' on the 2nd. Struve said the nucleus was 2.3" across on the 5th and added that two "flames" were visible near the nucleus. On November 11, Maclear noted that stars of magnitude 7 and 8 were "nearly invisible" at the comet's altitude. Maclear said observations were difficult on the 14th, as the comet was only visible in twilight. Maclear saw the comet for the final time on November 17 and simply described the comet as faint in twilight. M. Koller (Kremsmünster, Austria) was the last person to see the comet before conjunction with the sun, when he found it 3.5° above the horizon on November 22.68.

The comet passed only 2° from the sun on December 6 and remained lost in the sun's glare for nearly all of December. Herschel looked for the comet before sunrise on December 22, but failed to see it. The comet was independently picked up in the morning sky on December 31 by K. Kreil (Milan, Italy) and Olmsted. The solar elongation was then 32°. Olmsted said it was situated about 10° above the horizon, but it was almost immediately lost it in morning twilight.

The comet was again under widespread observation as 1836 began, but its appearance was quite different from what it had been two months earlier, as the comet was steadily moving away from both the sun and Earth. Moonlight and low altitude initially affected the comet's appearance: it was not visible to the naked eye and no tail was present. After the moon had left the sky and the comet's altitude increased, its appearance changed for the better. W. R. Dawes (Ormskirk, England) said the comet was "exceedingly faint" on the 16th and 19th. Boguslawski said the "comet" was no brighter than a star of magnitude 6 on the 22nd, while Maclear said it equaled a star of magnitude 2 on the 25th. Loomis reported the comet was irregular in outline on the 15th, while many astronomers reported a strongly parabolic shape from the 26th to the end of the month. The coma was estimated as 2–3' across by Loomis and Müller on the 15th, 4' across by Müller on the 21st, 4' across by Herschel on the 26th, and 5' across by Herschel on the 27th. Interestingly, during the last days of the month, a circular "halo" was noted by several observers as surrounding the coma. Diameter estimates of 8' and 12' came from Maclear on the 25th and 31st, respectively, while Loomis said it was 15' across on the 26th. Although the tail was distinctly absent from the reports of most observers during the month, Maclear noted on the 29th that his use of a large aperture "sweeper" revealed "the outline of a faint tail."

The comet's fading light and moonlight at both the beginning and end of the month took a toll on the observations made during February. Maclear found the general figure of the comet still parabolic on the 2nd. Although he said the comet was faint in moonlight on the 6th, he noted that the nucleus was still "bright." Maclear said the comet continued to be faint in moonlight on the 8th, but was visible to the naked eye without moonlight on the 18th. Interestingly, the comet was frequently described as tenuous or, according to Maclear on the 18th, like a "veil of gossamer." This led to some difficulty in observing it with telescopes. Maclear said the comet was extinguished in the telescope when the wire micrometer was illuminated on the 15th and 16th. The tail was still visible to Maclear on the 13th, when he noted a length of nearly 41′ and added, "The termination of the tail is irregular." Müller estimated the coma as 3.5–4′ across on the 14th, while Loomis said it was 6′ across on the 22nd. Although moonlight nearly hid the comet from view during the last days of the month, Maclear remarked on the 29th that, after the moon had set, "the whole figure of the comet could be distinctly traced through a sweeper."

Although March began with the comet again being almost completely invisible in moonlight, some naked-eye observations were still made. Loomis saw it with the naked eye on the 21st, while Dumouchel reported that it was still a naked-eye object under dark, clear skies on the 18th, 19th, 21st, and 24th. Müller reported the coma was 7′ long and 3′ wide on March 16, and 3–4′ across on the 21st. On the 25th, Herschel remarked that the comet was rapidly fading, while the nucleus was "still pretty bright."

Although Loomis was still able to see the comet in the finder of his 13-cm refractor on April 5, the comet obviously faded rapidly as April progressed. K. L. von Littrow (Vienna, Austria) said a telescope was needed to see the comet on the 7th, but while the coma was hardly perceptible, the nucleus still seemed rather bright. Maclear described the comet as "very faint" on the 14th. Müller found it "extremely faint and indistinct" on the 17th, with a diameter of 2′. Müller reported it as "excessively faint in moonlight" on the 21st.

Few observers attempted to see the comet in May, but those who did either failed or grabbed one last glimpse of this great comet. Maclear and Herschel independently succeeded in finding the comet on the 5th, but it was the last time for both astronomers. Herschel thought he had found the comet one last time on the 11th, but eventually realized the "comet" was probably a nebula.

The comet's final observations came on May 17 and 19. J. von Lamont (Munich, Germany) obtained the final measurement of the comet's position on May 17.92, when he gave it as $\alpha = 10^h\ 06.1^m$, $\delta = -6° 46′$. The comet was seen for the final time on May 19.9, when Boguslawski saw it very low over the western horizon at the end of twilight.

Using positions from August and September, Pontécoulant calculated an orbit with a perihelion date of November 16.4. Additional revisions using positions from this apparition were published by J. E. B. Valz, H. C. Schumacher, Pontécoulant, Laugier and Plantamour (1837), Rosenberger (1837), and H. Westphalen (1847) in the coming months and years. During the twentieth century, elaborate studies involving several apparitions of this comet were published by P. H. Cowell and A. C. D. Crommelin (1910), J. L. Brady and E. Carpenter (1971), T. Kiang (1972), D. K. Yeomans (1984), W. Landgraf (1986), G. Sitarski (1988), and Yeomans (1990). All of these revealed a perihelion date of November 16.44 for this apparition. Yeomans' last orbit is given below.

T	ω	Ω (2000.0)	i	q	e
1835 Nov. 16.4397 (TT)	110.7042	57.5185	162.2587	0.586564	0.967395

ABSOLUTE MAGNITUDE: $H_{10} = 4.4$ (V1964)

FULL MOON: Jul. 10, Aug. 8, Sep. 7, Oct. 6, Nov. 5, Dec. 5, Jan. 4, Feb. 2, Mar. 3, Apr. 1, May 1, May 30

SOURCES: M. C. T. Damoiseau, *Memorie della Reale Accademia delle Scienze di Torino*, **24** (1820), pp. 1–76; M. C. T. Damoiseau, *CDT* (1829), pp. 25–34; J. W. Lubbock and P. G. Le Doulcet, comte de Pontécoulant, *MNRAS*, **2** (1831 Jan. 14), p. 5; J. W. Lubbock, *MAS*, **4** (1831), pp. 509–16; M. C. T. Damoiseau, *CDT* (1832), p. 34; P. G. Le Doulcet, comte de Pontécoulant, *CDT* (1833), p. 112; *PMJS* (Series 3), **5** (1834 Oct.), p. 284; *PMJS* (Series 3), **6** (1835 Jan.), pp. 45–52; O. A. Rosenberger, *AN*, **12** (1835 Mar. 21), pp. 187–94; J. W. H. Lehmann, *AN*, **12** (1835 Jul. 25), pp. 369–400; E. Dumouchel, A. Bouvard, and F. de Vico, *CR*, **1** (1835 Aug. 17), pp. 40–1; D. F. J. Arago, *CR*, **1** (1835 Aug. 24), p. 66; O. A. Rosenberger, *AN*, **12** (1835 Aug. 26), pp. 401–8; E. Dumouchel, *AN*, **12** (1835 Aug. 26), p. 415; D. F. J. Arago, *CR*, **1** (1835 Aug. 31), p. 87; T. J. Hussey and J. South, *PMJS* (Series 3), **7** (1835 Sep.), p. 236; P. H. L. von Boguslawski and D. F. J. Arago, *CR*, **1** (1835 Sep. 7), pp. 96–7; P. G. Le Doulcet, comte de Pontécoulant, *CR*, **1** (1835 Sep. 14), pp. 103, 112–13; P. G. Le Doulcet, comte de Pontécoulant, D. F. J. Arago, and J. E. B. Valz, *CR*, **1** (1835 Sep. 21), pp. 129–31; P. G. Le Doulcet, comte de Pontécoulant, *CR*, **1** (1835 Oct. 12), pp. 205–8; G. Inghirami, K. L. von Littrow, F. B. G. Nicolai, F. W. Bessel, C. J. Pastorff, and H. C. Schumacher, *AN*, **13** (1835 Oct. 17), pp. 1–10; P. G. Le Doulcet comte de Pontécoulant and D. F. J. Arago, *CR*, **1** (1835 Oct. 19), pp. 234–6, 241–2; D. F. J. Arago, *CR*, **1** (1835 Oct. 26), pp. 255–8; J. E. B. Valz, *CR*, **1** (1835 Nov. 9), pp. 321–2; P. G. Le Doulcet, comte de Pontécoulant, *CR*, **1** (1835 Nov. 23), pp. 361–5; F. W. Bessel, P. H. L. von Boguslawski, F. W. A. Argelander, O. A. Rosenberger, and E. Dumouchel, *AN*, **13** (1835 Nov. 28), pp. 65–74; G. B. Amici, *CR*, **1** (1835 Dec. 21), p. 503; D. Olmsted and E. Loomis, *AJS* (Series 1), **29** (1836 Jan.), pp. 155–6; F. W. Bessel and D. F. J. Arago, *CR*, **2** (1836 Jan. 18), pp. 67–8; M. Koller, *AN*, **13** (1836 Jan. 28), p. 143; *PMJS* (Series 3), **8** (1836 Feb.), p. 173; F. Kaiser, *AN*, **13** (1836 Feb. 3), pp. 177–80; K. L. von Littrow and E. J. Cooper, *CR*, **2** (1836 Feb. 15), pp. 155, 160; F. W. Bessel, *AN*, **13** (1836 Feb. 20), pp. 185–232; F. G. W. von Struve and O. W. von Struve, *AN*, **13** (1836 Mar. 2), pp. 233–40; C. Darlu and K. L. von Littrow, *CR*, **2** (1836 May 9), pp. 474–5; E. Dumouchel, *AN*, **13** (1836 Jun. 1), pp. 341–4; E. Loomis, *AJS* (Series 1), **30** (1836 Jul.), pp. 209–21; J. von Lamont, *AN*, **14** (1836 Aug. 24), p. 57; J. F. W. Herschel, *CR*, **3** (1836 Oct. 31), pp. 505–6; W. H. Smyth, *MRAS*, **9** (1836), pp. 229–46; R. Owen, *MRAS*, **9** (1836), pp. 269–71; F. W. Bessel and P. G. Le Doulcet, comte de Pontécoulant, *CDT* (1837), p. 104; B. F. Joslin, *AJS* (Series 1), **31** (1837 Jan.), pp. 142–55, 324–32; P. A. E. Laugier, E. Plantamour, O. A. Rosenberger, and G. Santini, *CR*, **5** (1837 Oct. 16), pp. 553–4; P. G. Le Doulcet, comte de Pontécoulant, *CDT* (1838), p. 115; T. Maclear, *MRAS*, **10** (1838), pp. 91–151; J. F. W. Herschel and T. G. Taylor, *MRAS*, **10** (1838), pp. 325–35; Müller, *MRAS*, **12** (1842), pp. 385–96; G. B. Airy, *MRAS*, **16** (1847), pp. 337–47; H. Westphalen, *AN*, **25** (1847 Feb. 11), pp. 187–92; E. Loomis, J. Herschel, and P. H. L.

von Boguslawski, *AJS* (Series 2), **5** (1848 May), pp. 370–2; P. G. Le Doulcet, comte de Pontécoulant, *CR*, **58** (1864), pp. 706–9; P. H. Cowell and A. C. D. Crommelin, *AN*, **185** (1910 Aug. 11), pp. 265–8; *PA*, **42** (1934 Apr.), p. 201; V1964, p. 55; *Herschel at the Cape*. Austin: University of Texas Press (1969), pp. 134–240; J. L. Brady and E. Carpenter, *AJ*, **76** (1971 Oct.), p. 733; T. Kiang, *MRAS*, **76** (1972), p. 35; T. Maclear and J. F. W. Herschel, *Maclear & Herschel: Letters & Diaries at the Cape of Good Hope 1834–1838*. Edited by B. Warner and N. Warner. Cape Town: A. A. Balkema (1984), pp. 99–100; W. Landgraf, *AAP*, **163** (1986), p. 258; *JRASC*, **80** (1986 Apr.), p. 63; G. Sitarski, *AA*, **38** (1988), p. 263; Donald K. Yeomans, "Investigating the motion of Comet Halley over several millennia," *Comet Halley: Investigations, Results, Interpretations, Volume 2*. Edited by J. W. Mason. England: Ellis Horwood Limited (1990), p. 230; SC2000.0, pp. 119, 149, 282.

2P/Encke

1838

Recovered: 1838 August 15.0 ($\Delta = 1.64$ AU, $r = 2.09$ AU, Elong. $= 102°$)

Last seen: 1838 December 17.25 ($\Delta = 0.85$ AU, $r = 0.35$ AU, Elong. $= 20°$)

Closest to the Earth: 1838 November 7 (0.2192 AU)

Calculated path: ARI (Rec), TRI (Sep. 8), PER (Sep. 18), AND (Sep. 30), PER (Oct. 14), CAS (Oct. 18), CEP (Oct. 27), DRA (Nov. 3), HER (Nov. 8), OPH (Nov. 25), SER (Nov. 28), OPH (Nov. 30), SCO (Dec. 4), OPH (Dec. 17)

The recovery of this comet began with the prediction of J. F. Encke (1838). He examined the motion of the comet during the apparitions of 1829, 1832, and 1835, and then applied perturbations by Jupiter. About the middle of 1838 he published his prediction that the comet would pass perihelion on 1838 December 19.46. Using Encke's ephemeris, P. H. L. von Boguslawski (Breslau, now Wroclaw, Poland) recovered this comet on 1838 August 15.0. He said it looked like an extremely faint, shapeless nebulosity, which was very difficult to observe. The position was determined as $\alpha = 2^h\ 15.3^m$, $\delta = +24°\ 38'$ on August 15.04. An independent discovery was made by J. G. Galle (Berlin, Germany) on September 17.06. He said the refractor of 14-foot focal length revealed the comet as an extremely faint object about 2.5' in diameter. The comet was unchanged when Galle confirmed his recovery on the 20th.

The comet was not well observed before the early October full moon began to interfere. Galle saw the comet on several occasions during the period of September 22 to October 2 and always found the comet "very faint." He also estimated the coma diameter as 1–2' on the 24th and 25th, and 2'–3' on the 26th. S. H. Schwabe (Dessau, Germany) saw the comet with a refractor of 6-foot focal length on September 29 and October 2. He also described the comet as very faint, noting it was invisible at a magnification of 45×, but visible at 54× and 64×.

After the moon had left the sky, astronomers quickly resumed observations, but initially found the comet a faint object. F. B. G. Nicolai (Mannheim, Germany) described it as faint on October 9 and 10. Müller (Geneva, Switzerland) described it as an extremely weak, very indistinct nebulosity, about

4' across when seen in a 10-cm telescope on the 10th. Schwabe said the comet was invisible in a refractor of 2.5-foot focal length on the 11th, but was visible in the same telescope on the 12th. Also on the 11th, M. Koller (Kremsmünster, Austria) said that the comet looked like a collection of light spots near the center of the telescope's field of view and that it was impossible to determine its diameter. The coma diameter was given as 6' by both Galle on the 12th and Müller on the 13th. Müller noted a slight brightening toward the center. The comet seemed to undergo a change around mid-month, as Müller found it slightly brighter on the 16th and about 7' across, and then he found it brighter still on the 22nd and about 9' across. Müller added that there was an appreciable nucleus on the 22nd. Schwabe found the comet in a small telescope with a magnification of only 8× on the 25th. The coma diameter was estimated as 10' by Galle on the 25th and 11' by Müller on the 27th. Moonlight was interfering at the end of the month and Müller said it reduced the coma diameter to 2'. Some interesting observations were made by J. E. B. Valz (Marseille, France) during this month, which were somewhat contrary to other observations, especially with respect to the coma diameter. He described the comet as 20' across on October 9 and 10, with a trace of a nucleus visible on the 9th. He noted the coma was elongated toward the sun on the 15th. On the 25th, Valz said the coma was 15' across, with the length nearly doubling its width. The most luminous area of the coma was not central, but in the half of the coma directed away from the sun. Valz said the coma was 25' across on the 28th.

Moonlight did not put a halt to observations as November began. On the 2nd, with the moon barely passed full, Müller said the comet was 7–8' across in a 10-cm telescope. Schwabe said sharp-sighted people could see the comet with the naked eye as a faint nebula on the 5th. Koller reported it was visible to the naked eye on the 7th, and Encke said it was visible to the naked eye as a nebulous star of magnitude 5 on the 8th. On the 6th, the coma was estimated as about 13' across by Valz and 10–11' across by Müller. Müller also noted the nucleus was not in the center of the coma. Koller reported the eastern edge of the coma was well defined on the 7th, while the western edge was almost imperceptible. He noted that the brightest section of the coma seemed to show signs of scintillation on the 9th. Nicolai saw the comet with the naked eye using averted vision on the 10th. Schwabe noted a fine nucleus was visible on the 10th and 12th. Müller said the comet was visible to the naked eye on the 14th, while the 10-cm telescope revealed the nucleus was a little eccentric. He said it was only slightly visible to the naked eye on the 20th. Müller added that the comet appeared round in the finder, but was elongated in the 10-cm telescope, with a nucleus situated left of center. On the 23rd, Müller saw the comet at low altitude and reported it was not visible to the naked eye. The coma diameter was apparently at its greatest around mid-November and seemed to steadily shrink as the month progressed, both because of its increasing distance from Earth and its decreasing altitude. Valz reported it was 11' across on the 13th and Müller estimated it as 9′ across on the 14th. Valz next reported it was 8–9′ across on the 16th. On the 20th, Müller said it appeared 8′ across, while Valz estimated it as 6–7′ across. The final estimates of the month came from Valz, as he estimated the coma diameter as 4′ on the 23rd and 3′ on the 24th. Koller saw the comet for the final time on November 27 and remarked that it could not be measured because of low altitude and the beginning of twilight. Valz said the comet was no longer visible in evening twilight on November 29.

Valz was the only observer to continue providing physical descriptions as the comet reemerged in the morning sky during December. He first picked up the comet on the 7th and said the brightness was similar to a star of 4th magnitude. On the 12th, Valz said the comet's brightness was similar to a 5th-magnitude star, while the coma was less than 20″ across. The comet's appearance was like that of a 6th-magnitude star when Valz saw it on the 14th. He then estimated the coma diameter as 15″. Valz described the comet as similar to a 7th-magnitude star on the 16th, and gave the coma diameter as 10″ to 12″.

The comet was last detected on December 17.25 by Valz. He said the comet's brightness was similar to an 8th-magnitude star, while the coma was 7–8″ across. He gave the position as $\alpha = 16^h\ 11.3^m$, $\delta = -19° 58'$, with δ determined 6.55 minutes after α. Valz tried to observe the comet on December 18, but was unsuccessful, despite stars of magnitude 7 and 8 being visible near the comet's expected position. The comet was lost in the sun's glare thereafter and remained within 20° of the sun until February 3; however, it was then too faint to be seen.

Minor refinements of this comet's orbit using positions from multiple apparitions, as well as planetary perturbations, have been published by numerous astronomers. Of most significance were the studies of Encke, F. E. von Asten (1877), and B. G. Marsden and Z. Sekanina (1974). These have typically revealed a perihelion date of December 19.51 and a period of 3.31 years. Marsden and Sekanina determined the nongravitational terms as $A_1 = -0.63$ and $A_2 = -0.03759$. Their orbit is given below.

T	ω	Ω (2000.0)	i	q	e
1838 Dec. 19.5091 (TT)	182.8106	336.8958	13.3790	0.343962	0.845246

ABSOLUTE MAGNITUDE: $H_{10} = 9.4$ (V1964)

FULL MOON: Aug. 5, Sep. 4, Oct. 3, Nov. 1, Dec. 1, Dec. 31

SOURCES: J. F. Encke, *AN*, **15** (1838 Jun. 7), pp. 281–6; P. H. L. von Boguslawski, *CR*, **7** (1838 Sep. 3), pp. 536–7; P. H. L. von Boguslawski, *AN*, **15** (1838 Sep. 20), p. 367; J. G. Galle, *CR*, **7** (1838 Oct. 8), pp. 687–8; J. G. Galle, *AN*, **16** (1838 Oct. 18), pp. 5–10; J. E. B. Valz, *CR*, **7** (1838 Oct. 22), p. 745; J. G. Galle and Müller, *CR*, **7** (1838 Nov. 5), pp. 795–7; Müller, *CR*, **7** (1838 Nov. 19), pp. 898–9; J. E. B. Valz, *CR*, **7** (1838 Dec. 3), pp. 974–6; J. E. B. Valz, *CR*, **7** (1838 Dec. 24), p. 1124; P. H. L. von Boguslawski and F. B. G. Nicolai, *AN*, **16** (1839 Jan. 11), p. 167; S. H. Schwabe, *AN*, **16** (1839 Feb. 14), pp. 181–6; M. Koller, *AN*, **16** (1839 Oct. 3), pp. 387–92; J. G. Galle, *ABSB*, **1** (1840), pp. 152–8; J. F. Encke, J. E. B. Valz, and Müller, *MRAS*, **11** (1840), pp. 205–17; *PMJS* (Series 3), **16**

(1840 Feb.), p. 151; F. E. von Asten, *BASP* (Series 4), **22** (1877), p. 560; A. Berberich, *AN*, **119** (1888 Apr. 24), p. 54; V1964, p. 55; B. G. Marsden and Z. Sekanina, *AJ*, **79** (1974 Mar.), pp. 415–16.

C/1839 X1   *Discovered:* 1839 December 3.20 ($\Delta = 0.82$ AU, $r = 0.92$ AU, Elong. $= 60°$)
(Galle)   *Last seen:* 1840 February 10.21 ($\Delta = 1.79$ AU, $r = 0.98$ AU, Elong. $= 24°$)
          Closest to the Earth: 1839 December 11 (0.7838 AU)
1840 I   *Calculated path:* VIR (Disc), SER (Dec. 18), OPH (Dec. 26), SER (Jan. 13), AQL (Jan. 17)

J. G. Galle (Berlin, Germany) discovered this comet in the morning sky near 8 Virginis on 1839 December 3.20. He determined the position as $\alpha = 12^h$ 38.5^m, $\delta = -2°$ 10′ on December 3.22. Galle said the coma contained a well-defined point or nucleus, and the tail extended away from the sun.

Although this comet was widely observed, physical descriptions were not plentiful. Following the discovery observation, the very next descriptions came on December 29 and 30 by R. Snow. Snow was observing from Ashurst (England) and found the comet without difficulty on the first morning. He said the comet was 58″ across, while the tail "extended beyond the field of view, and with a large telescope would have been a magnificent object." Snow found the comet brighter on the second night and noted a nearly stellar nucleus. He said the tail extended "a great way beyond the field" and added that the comet was "a very fine object, perfectly visible when there was daylight enough to allow its being observed over the wire without a lamp."

Snow made his next observations from Dulwich (England) on January 6 and 7. He said the comet was very bright on both mornings. Snow remarked that the nucleus was "very large but not stellar" on the 6th, while the tail was nearly 7° long on the 7th. M. Koller (Kremsmünster, Austria) saw the comet on the 8th and said it appeared to the naked eye like a nebulous star of magnitude 4–5. He noted a slight tail. He reported the comet changed little in appearance on January 10, 11, and 12.

Although the comet was best observed during the first half of January, observations continued until the end of the month. The comet was even seen during the time around full moon. By the end of January, the comet's increasing distances from the sun and Earth, as well as its steadily decreasing elongation from the sun, were beginning to make it more difficult to follow and fewer observatories were providing positions. The last observations came from F. W. A. Argelander (Bonn, Germany) on February 9.22 and P. H. L. von Boguslawski (Breslau, now Wroclaw, Poland) on February 10.21. Argelander simply described the comet as very faint. Boguslawski found the comet at a very low altitude in the morning sky and gave the position as $\alpha = 20^h$ 00.3^m, $\delta = -6°$ 29′.

One of the earliest parabolic orbits was calculated by A. von Humboldt using positions from December 3 to December 11. The resulting perihelion date was 1840 January 4.74. A. C. Petersen followed a little later with a slightly longer arc of December 3–15 and determined the perihelion date as January 5.13. Interestingly, after C. K. L. Rümker calculated an almost identical orbit to Petersen's using the same positions, Petersen revised his orbit using the same observational arc as before and found a perihelion date of January 4.97. This orbit would prove very close to the actual orbit, and further calculations by Koller, J. F. Encke, Rümker, T. Henderson, R. Kysaeus, G. Lundahl, Petersen, and J. P. Wolfers offered little improvement. Wolfers noted the orbit was similar to that of comet C/1764 A1, except for a large difference in the inclination.

Astronomers later noted a slight deviation from a parabola. C. A. F. Peters and O. von Struve (1843) were the first to investigate this. They exclusively used the positions obtained at Pulkovo Observatory (St. Petersburg, Russia) and noted a slight hyperbolic motion. The resulting perihelion date was January 4.97 and the eccentricity was 1.0002050. Nearly 50 years later, G. Rechenberg (1892) investigated the comet's motion and noted an elliptical motion. He used 103 positions obtained during the period of December 3 to February 10 and determined a perihelion date of January 4.97 and a period of about 597 000 years. No planetary perturbations were applied.

The matter of whether the comet was moving in a hyperbolic or elliptical orbit has apparently been settled by R. J. Buckley (1979). He used 59 positions obtained between December 3 and February 10, and applied perturbations by Venus to Neptune. The result was a perihelion date of January 4.98 and a period near 2.8 million years. This orbit is given below.

T	ω	Ω (2000.0)	i	q	e
1840 Jan. 4.9752 (TT)	72.2670	122.1768	53.0715	0.618430	0.999969

ABSOLUTE MAGNITUDE: $H_{10} = 6.3$ (V1964)

FULL MOON: Nov. 21, Dec. 20, Jan. 19, Feb. 17

SOURCES: J. G. Galle, *MNRAS*, **5** (1839 Dec.), p. 8; J. G. Galle, *AN*, **17** (1839 Dec. 19), p. 47; A. von Humboldt, *CR*, **9** (1839 Dec. 23), pp. 822–4; H. C. Schumacher, C. K. L. Rümker, J. F. Encke, and A. C. Petersen, *AN*, **17** (1839 Dec. 28), supplement; H. Lawson and R. Snow, *MNRAS*, **5** (1840 Jan.), p. 9; A. C. Petersen, *CR*, **10** (1840 Jan. 6), pp. 17–18; J. F. Encke and J. G. Galle, *AN*, **17** (1840 Jan. 9), p. 95; H. C. Schumacher, P. H. L. von Boguslawski, C. K. L. Rümker, and J. P. Wolfers, *AN*, **17** (1840 Jan. 16), pp. 109–12; A. C. Petersen, *CR*, **10** (1840 Jan. 20), p. 115; *PMJS* (Series 3), **16** (1840 Feb.), pp. 151–2; A. C. Petersen, R. Kysaeus, and G. Lundahl, *AN*, **17** (1840 Feb. 13), pp. 113–18; M. Koller, *AN*, **17** (1840 Mar. 5), pp. 157–60; F. W. A. Argelander and G. Lundahl, *AN*, **17** (1840 Mar. 12), pp. 171–6; J. F. Encke, H. C. Schumacher, and C. K. L. Rümker, *AJS* (Series 1), **38** (1840 Apr.), p. 378; F. W. A. Argelander, *AN*, **17** (1840 May 7), pp. 233–8; R. Snow, *MRAS*, **11** (1840), pp. 291–4; C. K. L. Rümker, *MRAS*,

12 (1842), pp. 422–3; C. A. F. Peters and O. von Struve, *MASP* (Series 6), **1** (1843), pp. 327–78; G. Rechenberg, *AN*, **131** (1892 Dec. 15), pp. 249–60; G1894, pp. 52–3; V1964, p. 56; R. J. Buckley, *JBAA*, **89** (1979 Apr.), pp. 260, 262.

C/1840 B1 *Discovered:* 1840 January 25.91 ($\Delta = 1.17$ AU, $r = 1.42$ AU, Elong. $= 82°$)
(Galle) *Last seen:* 1840 April 3.06 ($\Delta = 2.09$ AU, $r = 1.26$ AU, Elong. $= 25°$)
Closest to the Earth: 1840 January 30 (1.1587 AU)
1840 II *Calculated path:* DRA (Disc), CEP (Jan. 26), CAS (Feb. 5), AND (Feb. 17), TRI (Mar. 3), ARI (Mar. 15)

J. G. Galle (Berlin, Germany) discovered this comet in Draco on 1840 January 25.91. He gave the position as $\alpha = 20^h\ 17.6^m$, $\delta = +63°\ 07'$ on January 25.95. Galle said the comet was round and faint, without an appreciable tail.

Although the comet was widely observed, E. Plantamour (Geneva, Switzerland) provided most of the details about the comet's physical appearance during February and March. His first observation came on February 23, when he described the comet as more than 1′ across, with a condensation slightly northeast of the center. From that night until March 8, he said the comet changed little in appearance, except for an appreciable increase in brightness. Moonlight caused the comet to appear faint to Plantamour during the period of March 8–16, but his observations on the night of March 19 revealed the comet was brighter than when last seen in dark skies on March 7. Plantamour did not notice a difference in the brightness of the comet on March 19, 20, and 21, and suggested this was when the comet reached its greatest luminosity.

The last descriptions of the comet came from Plantamour and Elias Loomis (Hudson Observatory, Ohio, USA). Loomis obtained his first observation of the comet on March 19. He said it was "faint, but brightest in the central parts, resembling a small nebula, nearly circular, and about one minute in diameter; but its margin was exceedingly ill-defined." On March 20, Loomis remarked that the nucleus was "somewhat eccentric, and on the lower side of the comet, as seen in an inverting telescope." Plantamour said the comet rapidly faded after March 23 and suggested this was because of its steadily decreasing altitude and entrance into evening twilight.

The comet was last detected on April 3.06, by Loomis. He noted that "no remarkable change in the comet's appearance was subsequently observed [since his first observation], except that its brightness diminished somewhat more rapidly than had been anticipated." Loomis tried to find the comet on April 8, but "searched for it in vain."

The first orbits were determined by A. C. Petersen and J. F. Encke. Both astronomers used positions obtained during the period of January 25–30 and calculated parabolic orbits, with Petersen giving the perihelion date as 1840 March 12.11 and Encke giving it as March 11.73. By mid-March, P. A. E. Laugier used positions from January and February, and determined the perihelion date as March 13.43. This orbit was a good representation of the comet's motion and later calculations by Petersen, C. K. L. Rümker, Encke, F. V. Mauvais, E. Bouvard, R. Kysaeus, M. Koller, and Plantamour, offered only slight revisions.

A couple of years later, astronomers realized the comet was moving in a long-period orbit. Loomis (1843) used 78 positions obtained between January 25 and April 3, as well as perturbations by seven planets, and determined the perihelion date as March 13.62 and the period was about 2423 years. Using a similar set of positions, Plantamour (1843) determined the perihelion date as March 13.49 and the period was about 13 864 years. J. Kowalczyk (1876) took 182 positions spanning the period of January 25 to April 1 and determined a perihelion date of March 13.58 and a period of about 3789 years. Kowalczyk's orbit is given below.

T	ω	Ω (2000.0)	i	q	e
1840 Mar. 13.5756 (UT)	156.5777	239.0574	120.7744	1.220789	0.994977

ABSOLUTE MAGNITUDE: $H_{10} = 6.1$ (V1964)

FULL MOON: Jan. 19, Feb. 17, Mar. 18, Apr. 16

SOURCES: J. G. Galle and A. C. Petersen, *AN*, **17** (1840 Feb. 13), supplement; J. G. Galle, *CR*, **10** (1840 Mar. 2), pp. 376–7; F. W. A. Argelander, *AN*, **17** (1840 Mar. 12), p. 171; P. A. E. Laugier and J. F. Encke, *CR*, **10** (1840 Mar. 16), pp. 466–7; A. C. Petersen, C. K. L. Rümker, and J. F. Encke, *AN*, **17** (1840 Mar. 26), pp. 189–92; F. V. Mauvais, *CR*, **10** (1840 Apr. 13), pp. 625–6; E. Bouvard, *CR*, **10** (1840 Apr. 27), pp. 711–12; F. W. A. Argelander and R. Kysaeus, *AN*, **17** (1840 May 7), pp. 233–40; M. Koller, *AN*, **18** (1841 Feb. 4), pp. 85–8; C. K. L. Rümker, *MRAS*, **12** (1842), p. 424; E. Loomis, *TAPS* (NS), **8** (1843), pp. 146–54; E. Plantamour, *AN*, **20** (1843 Apr. 22), pp. 305–23; E. Plantamour, *AN*, **20** (1843 Apr. 29), pp. 329–32; *American Almanac*. Boston: James Munroe & Co. (1847), p. 91; J. Kowalczyk, *AN*, **87** (1876 Mar. 6), pp. 225–32; V1964, p. 56.

C/1840 E1 (Galle)   *Discovered:* 1840 March 7.17 ($\Delta = 1.36$ AU, $r = 0.91$ AU, Elong. $= 42°$)

Last seen: 1840 March 28.1 ($\Delta = 1.56$ AU, $r = 0.76$ AU, Elong. $= 23°$)

1840 III   *Closest to the Earth:* 1840 February 24 (1.3223 AU)

Calculated path: CYG (Disc), PEG (Mar. 8)

J. G. Galle (Berlin, Germany) discovered this telescopic comet near μ Cygni on 1840 March 7.17, and said it exhibited a bright, straight tail 3° long, which was pointing away from the sun. He used a refractor on March 7.19 to determine the position as $\alpha = 21^h 31.9^m$, $\delta = +29° 19'$. Galle confirmed his discovery on March 8.10.

Observatories in Hamburg, Breslau, Altona, Bonn, Kremsmünster, and St. Petersburg obtained numerous observations of this comet during the period spanning March 11 to 26, but no physical descriptions were published.

The comet's position was obtained for the final time on March 27.10, when O. von Struve (Pulkovo Observatory, Russia) measured it as

$\alpha = 23^h\ 27.0^m$, $\delta = +22°\ 23'$. The comet was last detected on March 28.1, when Struve said he was unable to find a suitable comparison star to measure its position because of the comet's low altitude in morning twilight.

The first parabolic orbits were published by A. C. Petersen and J. F. Encke. Petersen used positions from March 7 to March 11 and found a perihelion date of 1840 April 3.58, while Encke used positions spanning the period of March 7–12 and found a perihelion date of April 2.82. Petersen noted a similarity between the orbit of this comet and that of comet C/1097 T1, as determined by J. K. Burckhardt. After seeing Encke's orbit, Galle independently saw the similarity to the orbit of comet C/1097 T1, as well as comet C/1468 S1. He suggested a period of 370 years was likely. Later orbits by A. Bouvard, V. Mauvais, P. A. E. Laugier, C. K. L. Rümker, and D. F. J. Arago confirmed the general correctness of the initial orbits of Petersen and Encke. Arago also noted a similarity to Burckhardt's orbit of C/1097 T1, and a similarity to the orbit of C/1774 P1, except for the perihelion distance.

Definitive parabolic orbits were independently determined by A. W. Doberck and J. Kowalczyk during 1873. Doberck took 27 positions spanning the period between March 7 and 27, and computed an orbit with a perihelion date of April 2.91. Kowalczyk used 24 positions obtained between March 7 and 27, and computed an orbit with a perihelion date of April 2.94. No planetary perturbations were applied in either case. Kowalczyk's orbit fit the positions best and is given below.

T	ω	Ω (2000.0)	i	q	e
1840 Apr. 2.9378 (UT)	138.0439	188.2712	79.8512	0.748504	1.0

ABSOLUTE MAGNITUDE: $H_{10} = 6.0$ (V1964)
FULL MOON: Feb. 17, Mar. 18, Apr. 16
SOURCES: J. G. Galle, *CR*, **10** (1840 Mar. 16), pp. 467–8; J. G. Galle, A. C. Petersen, and J. F. Encke, *AN*, **17** (1840 Mar. 26), pp. 185–8; J. F. Encke, A. Bouvard, P. A. E. Laugier, and V. Mauvais, *CR*, **10** (1840 Mar. 30), pp. 534–6; C. K. L. Rümker and A. C. Petersen, *AN*, **17** (1840 May 7), pp. 229–32; A. W. Doberck, *AN*, **80** (1873 Jan. 31), p. 377; J. Kowalczyk and O. von Struve, *AN*, **81** (1873 Mar. 22), pp. 129–34; A. W. Doberck, *MNRAS*, **34** (1874 Jun.), p. 426; V1964, p. 56.

C/1840 U1
(Bremiker)

1840 IV

Discovered: 1840 October 26.81 ($\Delta = 1.11$ AU, $r = 1.50$ AU, Elong. = 90°)
Last seen: 1841 February 16.78 ($\Delta = 1.90$ AU, $r = 1.96$ AU, Elong. = 78°)
Closest to the Earth: 1840 December 4 (0.8659 AU)
Calculated path: DRA (Disc), CEP (Nov. 10), CYG (Nov. 14), CEP (Nov. 15), LAC (Nov. 27), AND (Dec. 3), PSC (Dec. 23), ARI (Jan. 9), CET (Jan. 30)

K. Bremiker (Berlin, Germany) discovered this comet in the evening sky on 1840 October 26.81, and described it as a faint nebulosity. He gave the position as $\alpha = 18^h\ 41.1^m$, $\delta = +60°\ 36'$ on October 27.89.

The comet attained its most northerly declination of $+61°$ on October 30 and remained circumpolar for most observatories in the Northern Hemisphere throughout November and well into December. The comet remained a faint object throughout its apparition and was only observable at observatories possessing large telescopes. W. R. Dawes (Bishop's Observatory, London, England) was the most prolific provider of descriptions for this comet. Observing with an 18-cm refractor, he first saw the comet on November 14 and said the nucleus resembled a star of magnitude 10 or 11, which "abruptly [diffused] itself into the nebulosity around it." Dawes said magnifications of $63\times$ and $105\times$ best revealed the nucleus. The comet itself only endured a slight illumination of the micrometer wires. Dawes noted the nucleus was distinctly visible as a small star on November 16, 19, and 24. He said the coma was extinguished with only a little illumination of the micrometer wire on the 16th and noted the comet "was certainly fainter than before" on the 19th. Also, on the 19th, Dawes said the nucleus was in the "north following side" of the coma. Dawes also noted that F. Baily and T. Galloway both noted the nucleus "looked like a small star visible through the middle of the comet." Another astronomer providing a physical description during November was G. Santini (Padova, Italy). He described it as extremely faint on the 22nd, "without sensible trace of a nucleus."

Dawes saw the comet on a couple of occasions during early December, but it appeared very faint each time, because of hazy conditions. Dawes' next good observation came on December 22. Even though stars of magnitude 4 were barely visible to the naked eye, he believed the comet must have been brighter than when previously seen. He said the stellar nucleus was still apparent, while the overall size of the comet had also increased. He saw a "decided stellar nucleus" on the 29th and noted the nebulosity "appeared more extended and dense." He indicated that the coma was about 1.3' across.

Cloudy skies prevailed for Dawes until 1841 January 11. On that evening the sky partially cleared and he looked for the comet in its expected position. He noted a thick haze across that region of the sky and no trace of the comet was found. The next evening was clearer, but there was still no trace of the comet. M. Koller (Kremsmünster, Austria) saw only traces of the comet in his telescope on January 22 and 23.

The comet was last detected on February 16.78, when Bremiker gave the position as $\alpha = 3^h \, 06.9^m$, $\delta = +5° \, 34'$.

The first orbit came from A. C. Petersen during the first days of November. Using positions from October 27 to October 31, he calculated a parabolic orbit with a perihelion date of 1840 November 16.39. Later parabolic orbits by C. K. L. Rümker, J. F. Encke, P. A. E. Laugier, F. V. Mauvais, Santini, and Koller steadily narrowed down the perihelion date to about November 14.4. Once orbital calculations began using positions spanning the comet's entire apparition, it became obvious that the comet was moving in an elliptical orbit. The first person to discover this was W. C. Götze (1844) when he determined the perihelion date as November 14.16 and the period as 359 years. Götze later revised this orbit and found a period of 344 years, while an analysis using 252 total positions was published by C. A. Schultz-Steinheil (1890) and revealed a period of 367 years. R. J. Buckley (1979) tried to eliminate the poor observations for his analysis and used only 81 positions from this apparition. After considering perturbations by Venus to Neptune, he determined the perihelion date as November 14.07 and the period as 286 years. Buckley added that because of the large number of poor observations, the period could range from 250 to 400 years. This orbit is given below.

T	ω	Ω (2000.0)	i	q	e
1840 Nov. 14.0652 (TT)	133.5127	251.2145	57.9043	1.479951	0.965941

ABSOLUTE MAGNITUDE: $H_{10} = 6.5$ (V1964)
FULL MOON: Oct. 11, Nov. 9, Dec. 9, Jan. 7, Feb. 6, Mar. 7
SOURCES: K. Bremiker, J. G. Galle, A. Bouvard, and P. A. E. Laugier, *CR*, **11** (1840 Nov. 9), p. 768; P. A. E. Laugier and F. V. Mauvais, *CR*, **11** (1840 Nov. 16), pp. 821–2; K. Bremiker and A. C. Petersen, *AN*, **18** (1840 Nov. 19), p. 63; F. V. Mauvais, *CR*, **11** (1840 Dec. 14), pp. 986–7; C. K. L. Rümker and J. F. Encke, *AN*, **18** (1841 Jan. 16), pp. 67–70; G. Santini and M. Koller, *AN*, **18** (1841 Feb. 4), pp. 83–7; J. F. Encke and F. W. A. Argelander, *AN*, **18** (1841 Mar. 4), pp. 139–42; M. Koller, *AN*, **18** (1841 Apr. 1), p. 183; G. Santini, *MNRAS*, **5** (1841 Nov.), p. 123; W. R. Dawes, *MRAS*, **12** (1842), pp. 225–30; *PMJS* (Series 3), **19** (1842 Jan.), pp. 577–8; W. C. Götze, *AN*, **21** (1844 May 9), pp. 353–8; W. C. Götze, *MNRAS*, **6** (1844 May 10), p. 78; W. C. Götze and K. Bremiker, *AN*, **22** (1844 Nov. 9), pp. 241–8; C. A. Schultz-Steinheil, *Kungliga Svenska Vetenskapsakademiens Handlingar*, **23** No. 14 (1890), 28 pp.; V1964, p. 56; R. J. Buckley, *JBAA*, **89** (1979 Apr.), pp. 260, 262.

2P/Encke *Recovered:* 1842 February 8.78 ($\Delta = 1.89$ AU, $r = 1.31$ AU, Elong. $= 40°$)
Last seen: 1842 May 23.2 ($\Delta = 0.56$ AU, $r = 0.97$ AU, Elong. $= 69°$)
1842 I *Closest to the Earth:* 1842 May 4 (0.5329 AU)
Calculated path: PSC (Rec), ARI (Mar. 25), CET (Apr. 17)

J. F. Encke published his prediction for the upcoming return of this comet during January of 1842, with the perihelion date given as 1842 April 12.38. The comet was recovered by J. G. Galle (Berlin, Germany) on 1842 February 8.78. Galle said the comet was so faint it could not withstand any illumination within the field of view. Nevertheless, he gave the position as $\alpha = 23^h 46.7^m$, $\delta = +6° 47'$.

Although positions were measured at several observatories, physical descriptions were few. Following the recovery, the next description was not made until March 12, when P. A. E. Laugier and F. V. Mauvais (Paris, France) said the comet appeared faint. The coma was 2–3′ across, and contained a central condensation. The condensation was all that could be observed during the period of March 19–26, according to astronomers at Greenwich Observatory (England), because of bright moonlight. S. C. Walker and E. O.

Kendall (High School Observatory, Philadelphia, Pennsylvania, USA) observed with a 9-foot focal length refractor on March 29 and noted a faint tail extending about 3.3'. Walker and Kendall said the tail was 7' long on April 2. The Greenwich astronomers estimated the coma diameter as about 1' on April 5 and 6. The Greenwich astronomers saw the comet for the final time on April 9. They said it was then in bright twilight and noted that atmospheric conditions were not favorable. Nevertheless, they noted that the comet could occasionally be seen better than nearby 27 Arietis (magnitude 6.23 – SC2000.0). E. Loomis (Hudson Observatory, Ravenna, Ohio, USA) last detected the comet on April 12.07, in twilight and about 3° above the horizon. The observation was so short he was only able to measure the comet's right ascension.

The comet passed less than 8° from the sun on April 21 and finally emerged into the morning sky on May 3, when W. Mann (Royal Observatory, Cape of Good Hope, South Africa) first detected it with a 9.5-cm refractor. He described it as "like a faint nebulous blotch." Mann found the comet very faint on May 12 and his colleague, T. Maclear, described it on the 14th as resembling "a faint nebula, somewhat oval, and rather condensed towards the centre. . . . " Maclear found the comet very faint on the 18th and even fainter on the 19th.

The last complete position was obtained on May 21.14, when Maclear determined it as $\alpha = 0^h\ 17.6^m$, $\delta = -20°\ 42'$. A partial position of $\alpha = 0^h\ 14.7^m$ was measured by Mann on May 22.18. He described the comet as very faint. The comet was seen for the last time on May 23.2, when Maclear saw the comet after the moon had set. Bright twilight was then present and Maclear said the comet was extremely faint and only visible by gently bumping the telescope.

Minor refinements of this comet's orbit using positions from multiple apparitions, as well as planetary perturbations, have been published by numerous astronomers. Of most significance were the studies of Encke (1845), F. E. von Asten (1877), and B. G. Marsden and Z. Sekanina (1974). These have typically revealed a perihelion date of April 12.52 and a period of 3.31 years. Marsden and Sekanina determined the nongravitational terms as $A_1 = -0.63$ and $A_2 = -0.03759$. The orbit of Marsden and Sekanina is given below.

T	ω	Ω (2000.0)	i	q	e
1842 Apr. 12.5200 (TT)	182.8118	336.8905	13.3613	0.344933	0.844850

ABSOLUTE MAGNITUDE: $H_{10} = 9$–10 (V1964)

FULL MOON: Jan. 26, Feb. 25, Mar. 26, Apr. 24, May 24

SOURCES: *Astronomical Observations made at the Royal Observatory at Greenwich* (1842), pp. 64–78; J. F. Encke, *CR*, **14** (1842 Jan. 24), p. 172; J. G. Galle, *CR*, **14** (1842 Feb. 21), pp. 314–15; J. G. Galle, *AN*, **19** (1842 Mar. 10), pp. 185–8; P. A. E. Laugier and F. V. Mauvais, *CR*, **14** (1842 Mar. 14), pp. 406–8; J. F. Encke and J. G. Galle, *AN*, **19** (1842

Jun. 16), p. 305; *TAPS* (NS), **8** (1843), pp. 311–14; T. Maclear and W. Mann, *MNRAS*, **6** (1844 Apr. 12), pp. 68–9; E. Loomis, *AN*, **22** (1844 Oct. 12), p. 203; J. F. Encke, *AN*, **23** (1845 Apr. 19), pp. 81–8; T. Maclear and W. Mann, *MRAS*, **15** (1846), pp. 211–28; *TAPS* (NS), **10** (1853), pp. 10–11; F. E. von Asten, *BASP* (Series 4), **22** (1877), p. 560; V1964, p. 56; B. G. Marsden and Z. Sekanina, *AJ*, **79** (1974 Mar.), pp. 415–16; SC2000.0, p. 50.

C/1842 U1	*Discovered:* 1842 October 28.79 ($\Delta = 0.69$ AU, $r = 1.17$ AU, Elong. $= 86°$)
(Laugier)	*Last seen:* 1842 November 27.72 ($\Delta = 0.61$ AU, $r = 0.67$ AU, Elong. $= 42°$)
	Closest to the Earth: 1842 November 14 (0.4447 AU)
1842 II	*Calculated path:* DRA (Disc), HER (Nov. 6), LYR (Nov. 8), HER (Nov. 13), AQL (Nov. 16), SGR (Nov. 23)

P. A. E. Laugier (Paris, France) discovered this comet on 1842 October 28.79. He described it as extremely faint and without a tail and determined the position as $\alpha = 16^h 41^m$, $\delta = +68° 44'$ on October 28.92. Laugier confirmed the discovery on October 30.80 and said the nucleus seemed slightly brighter in the direction opposite to the sun. He suggested this was the beginnings of a tail.

Laugier reported a tail about 10′ long on November 2, while F. W. A. Argelander (Bonn, Germany) said it was 12–15′ long on the 6th. Moonlight began interfering thereafter, with Argelander reporting the tail was invisible on the 8th, 9th, 10th, and 17th, while A. C. Petersen (Altona, Germany) said the comet appeared faint and indistinct on the 15th, 16th, and 17th. With moonlight no longer an issue, J. E. B. Valz (Marseille, France) described the comet on November 21 as a difficult naked-eye object, which looked "like a star of the 7th magnitude in the seeker." He added that no tail or nucleus was visible, but there was an elongated central condensation. On the same evening, Argelander noted a nearly stellar condensation, which was distinctly elliptical, with the major axis directed towards the tail. Petersen simply noted the comet was "very nice and distinct." Valz reported no change in the comet's appearance on the 24th, and he noted it was a difficult object near the horizon on the 26th.

The comet was last detected on November 27.72, when M. Koller (Kremsmünster, Austria) gave the position as $\alpha = 19^h 14.3^m$, $\delta = -23° 16'$. This was Koller's only observation, as cloudy weather moved in following his receiving word of the comet's discovery. Argelander and Petersen said they did not see the comet after November 21, first because of overcast skies, and then because of the comet's closeness to the horizon.

The comet's short period of observation prevented anything other than a parabolic orbit from being calculated. The first such orbit was determined by Laugier, who used five positions obtained between October 28 and November 5, and computed a parabolic orbit with a perihelion date of December 16.3. Although he then pointed out a slight similarity to the orbit of the comet of 1301, he correctly linked comet 1P/Halley to the 1301 comet a few days later. Laugier's orbit proved very close to the comet's real path, although, a short time later, Petersen used positions from October 28 to November 8, and determined a more precise perihelion date of December 16.43. Later calculations by Valz, J. F. Encke, Laugier, C. K. L. Rümker, Petersen, and Argelander would only slightly improve the orbit.

Several astronomers have calculated definitive orbits for this comet. J. Kowalczyk (1873), K. Schwarzschild (1895), and R. J. Buckley (1979) each produced virtually identical parabolic orbits. Buckley also noted a slight hyperbolic trend in the motion, but preferred the parabolic orbit because of the short period of observation. His parabolic orbit is given below.

T	ω	Ω (2000.0)	i	q	e
1842 Dec. 16.4553 (TT)	240.4937	210.0085	106.4213	0.504685	1.0

ABSOLUTE MAGNITUDE: $H_{10} = 8.8$ (V1964)

FULL MOON: Oct. 19, Nov. 18, Dec. 17

SOURCES: P. A. E. Laugier, *CR*, **15** (1842 Oct. 31), p. 816; P. A. E. Laugier, *CR*, **15** (1842 Nov. 7), pp. 895–6; P. A. E. Laugier, *AN*, **20** (1842 Nov. 17), p. 79; P. A. E. Laugier, A. C. Petersen, and J. E. B. Valz, *CR*, **15** (1842 Nov. 21), pp. 948–9; P. A. E. Laugier, C. K. L. Rümker, A. C. Petersen, and J. F. Encke, *AN*, **20** (1842 Dec. 1), pp. 101–4; J. E. B. Valz, *CR*, **15** (1842 Dec. 12), pp. 1115–16; F. W. A. Argelander, A. C. Petersen, and J. E. B. Valz, *AN*, **20** (1843 Jan. 12), pp. 161–70; P. A. E. Laugier, *CR*, **16** (1843 Jan. 23), pp. 208–9; A. C. Petersen and M. Koller, *AN*, **20** (1843 Mar. 30), pp. 273–8, 281; P. A. E. Laugier, *AJS* (Series 1), **44** (1843 Jan.–Mar.), p. 211; J. Kowalczyk, *AN*, **81** (1873 Mar. 22), pp. 133–6; K. Schwarzschild, *AN*, **137** (1895 Feb. 21), pp. 177–90; V1964, p. 56; R. J. Buckley, *JBAA*, **89** (1979 Oct.), pp. 583–4.

C/1843 D1 (Great March Comet)

1843 I

Discovered: 1843 February 6.0 ($\Delta = 0.88$ AU, $r = 0.85$ AU, Elong. $= 54°$)
Last seen: 1843 April 19.76 ($\Delta = 1.93$ AU, $r = 1.50$ AU, Elong. $= 50°$)
Closest to the Earth: 1843 January 27 (0.8692 AU), 1843 March 5 (0.8420 AU)
Calculated path: SCL (Disc), CET (Feb. 11), AQR (Feb. 20), CET (Mar. 3), ERI (Mar. 17), ORI (Apr. 14)

The earliest observation of this comet occurred on the evening of 1843 February 5. This observation and another on February 11 were mentioned by J. F. Encke as having appeared in a newspaper. The story originated out of New York (USA). Although no details were given for the February 5 observation, that of February 11 placed the comet "in the vicinity of β Ceti." Encke examined the orbit of this comet and found it had been fairly close to β Ceti on the 11th, and he concluded that both observations were probably real. Calculations by the Author indicate the actual dates in universal time were probably February 6.0 and February 12.0. The comet was situated very low in the southwest at the end of astronomical twilight and, from New York, the head would have set about one hour later. Additional pre-perihelion observations are difficult to find in the literature of the time, but E. C. Herrick did report in the April–June 1843 issue of the *American Journal of Science*, "It appears quite probable that the train of this comet was seen in the evening before the perihelion passage, at Bermuda, Philadelphia, and Porto Rico, on the 19th, 23d and 26th of February." The Philadelphia observation was made by S. C. Walker (Central High School Observatory, Philadelphia, Pennsylvania, USA). According to B. Peirce (1844), Walker saw the tail on the evening of February 23 (February 24.0 UT) and eventually reported so in the Philadelphia *Gazette* during 1843 May.

The sun's glare hid the comet from view during the next few days, but the comet was brightening rapidly as it approached perihelion. One last observation was made before the comet passed behind the sun, as P. Ray (Concepción, Chile) independently discovered it in broad daylight on February 27.66. He estimated it was about 5' east of the sun's limb. The comet's nucleus passed behind the sun on February 27.87 and passed about 0.1° from the sun's center on February 27.89. The nucleus reappeared from behind the sun on February 27.91, right around the time it was passing perihelion. The comet slowly moved away from the sun until about 0.4° from the center on February 27.93 and then began moving back toward it. On February 27.97 the nucleus began transiting the face of the sun. It passed about 0.2° from the sun's center on February 27.99 and then ended its transit on February 28.02.

Unaware of the earlier sightings of this comet, numerous independent discoveries were reported in Italy in daylight around February 28.4. A. Colla (Parma) said numerous people saw the comet near the sun. He related one letter from an amateur astronomer in Collorno, which said the comet was seen on February 28.41 by blocking the sun with a wall. The amateur astronomer described the comet as a "very beautiful star", with a tail 4–5° long. G. B. Amici (Florence) said the comet was seen on February 28.47 by people in Bologna, one of whom wrote, "the mass, examined by an opera glass, [was] like a flame, badly defined, three times as long as it was wide, very luminous towards the sun, and a little smoky at the east."

Clouds are known to have prevailed over western Europe at this time, so the next daylight sightings were reported from North America. Beginning on February 28.52 "a large part of the adult population" of Waterbury, Connecticut, USA saw the comet through most of the day. G. L. Platt, M. C. Leavenworth, S. W. Hall, A. Blackman, and N. J. Buel gave particulars of these observations and said it first appeared not long after sunrise "east and below the sun." They described the comet as a round coma with a pale tail extending 2–3°, which was "melting away into the brilliant sky." The nucleus was detected with the naked eye and was distinctly round, "its light equal to that of the moon in midnight in a clear sky; and its apparent size about one eighth the area of the full moon." J. G. Clarke (Portland, Maine, USA) saw the comet on February 28.82. He determined that the nearest limb of the nucleus was situated 4° 06' 15" from the sun's farthest limb and the nucleus and tail appeared as well defined "as the moon on a clear day." He added that the comet looked like "a perfectly pure white cloud, without any variation, except a slight change near the head, just sufficient to distinguish the nucleus from the tail at that point." J. Bowring (Chihuahua, Mexico) said the comet was located 3° 53' 20" from the sun on February 28.97, with a tail extending about 34'. Further observations on the 28th were reported in newspapers worldwide, although many of these reports included no names, just locations from where the observations had been made and a few descriptions. For instance, an observer in Woodstock, Vermont, USA saw the comet in daylight on February 28.70 and said it looked like a small, white cloud about 3° long. He said a telescope revealed the tail divided near the nucleus into two distinct branches. The outer edge of each tail was convex and both measured 8–10° long. The Chinese also recorded that a large "broom star" was seen during the day.

The comet's tail extended into the evening sky on March 1, which prompted additional independent discoveries. Among some of the people who reported it were J. H. Kay (Royal Navy, Magnetical Observatory, Ross Bank, Tasmania) and W. Lloyd (Mauritius). A passenger aboard the *Lawrence* (en route from Sydney to Concepción) observed "a white streak of light, inclined at an angle of 40° to the horizon, and was imagined to be the zodiacal light." It was also seen from Pernambuco, Brazil. Additional reports came on March 2, most notably from J. C. Haile (Auckland, New Zealand) and P. P. King (Royal Navy, stationed at Port Stephens, New South Wales, Australia). King noted it was "producing great alarm among the natives." The Bishop of Australia made distinct notes about the comet's appearance on this evening and wrote, "my attention was drawn to the remarkable spectacle of a definite portion of the tail being deflected from the axis, or direction in which the general body of light continued to proceed. Perhaps, about one-sixth of the train might be thus drawn aside from that which may be termed the natural direction, so as to form therewith, at the point of separation, an angle which I should calculate to be about three degrees. . . ."

The most common aspect of the comet reported on March 3 was an apparent double tail, with the angle separating the two tails variously reported as between 10° and 15°. G. Maclean (Cape Coast Castle, Ghana) said clouds hid both extremities of the comet, but he noted that the visible tail was the same brightness throughout and never exceeded a width of 1°. P. Smyth (Royal Observatory, Cape of Good Hope, South Africa) described the nucleus as "a planetary disk, from which rays emerged in the direction of the tail." He added, "To the naked eye there appeared a double tail, about 25° in length, the two streamers making with each other an angle of about 15°, and proceeding from the head in perfectly straight lines. From the end of the forked tail, and on the north side of it, a streamer diverged at an angle of 6° or 7° towards the north, and reached a distance of upwards of 65° from the comet's head; a similar, though much fainter, streamer was thought to turn off south of the line of direction of the tail."

One of the few early indications of the comet's brightness came from Close (sailing on the *Ellenborough*) on March 4. He said the nucleus was equal to a star of magnitude 2–3 and noted, "it was occasionally brilliant enough to throw a strong light on the sea." King observed the nucleus with a refractor and described it as a "reddish stellar spot" with well-defined edges and about 1′ in diameter. The comet was 8° above the horizon. H. A. Cooper (Pernambuco, Brazil) described the comet's head "as particularly small, without any nebulosity, but of extreme brightness, of a golden hue." But observers were more interested in the tail than any other feature of this comet. Cooper said the tail was about 30° long and exhibited a brilliant silver color. Maclean said the tail was about 22° long. Observers in Bombay, India, saw the comet's tail and described it "as a long, straight beam of light streaming from the western horizon towards the zenith." This is interesting since Close noted the tail was considerably curved. Cooper said a line of a golden hue extended from the nucleus into the tail for about 4–5°, while Hopkins (on a voyage from the Cape of Good Hope to India) noted a dark line in the tail. F. W. Ludwig Leichhardt (Australia) remarked that the main tail was accompanied by a thin secondary tail "leaving the former from the under side at an acute angle and extending westward . . . It was many times longer than the main tail. . . . " Only a few observations were published for March 5. The most interesting include Smyth's estimation that the tail was about 35° long and that, "All the rays proceeding from the head were now of uniform brightness, excepting one bright streak, which could be traced along the tail." Maclean said an exceedingly small bright point, about the color of Venus, was first seen in his telescope and he said it "appeared to be a nucleus."

The comet was widely observed on March 6. Smyth said "the nucleus is the broadest part of that end of the comet; all rays come from the posterior side, and are pretty equal in brightness, with the exception of a narrow bright streak in the middle, which runs for 3° or 4° along the middle of the tail, and then verges to the north side." John Caldecott (Trivandrum Observatory, India) observed with a 7.5-foot focal length telescope and wrote, "The nucleus of the head presented rather a well-defined planet-like disc, the diameter of which I *estimated* to be about 12″, and that of the nebulosity surrounding it at about 45″." The tail was estimated as about 23° long by Kay, 27° long by Smyth, 36° long by Caldecott, about 43° long by Gilbert (St. Helena in South Atlantic Ocean), and about 50° long by a passenger aboard the *Lawrence* who said it was composed of "two streams of light, the outside edges being clear and well-defined." Kay added that the greatest width of the tail was 54′. Aside from a statement by John Belan (master of the British warship HMS *Albatross*) that the comet's head was reddish, observers on March 7 were more interested in the tail. The length was given as 26° by Kay, 27.4° by J. Burdwood (cruising off the western coast of Africa on the HM Sloop *Persian*), 32.5° by Haile, 34° by Maclean, 37.4° by Gilbert, and 43° long by the crew of the *Dublin*. Kay said the maximum width of the tail was 50′ and added, "A dark line commencing near the middle and extending to the end divided the tail into two portions." E. C. Herrick (New Haven, Connecticut, USA) said the tail was 1° wide where it was lost in the horizon haze and about 2° across at its upper extremity. He added, "the train shone with the distinctness and splendor of a bright auroral streamer." Maclean said, "Several stars were visible to the naked eye through the tail."

Despite the comet's continued brilliance, observations began declining in the coming evenings. Gilbert noted on March 8 that the tail color was similar to the rays of the moon and a passenger aboard the *Lawrence* said the nucleus was "as bright as stars of the third or fourth magnitude" on the 9th. Kay noted that no stellar point was visible in the nuclear condensation on the 11th. Aside from these observations, most observers were still interested in the tail length. Herrick said the tail was 43° long on the 8th, with the upper end of the tail spanning about 2.5°. He noted the tail was slightly curved. On the 9th, Haile estimated the length as 35.2°, while Kay said it was 39° long. Kay also said the maximum tail width was 76'. Haile estimated the tail length as 35.8° on the 10th. On the 11th, Kay said it extended 45°, while M. M. Franzini (Lisbon, Portugal) said it was about 36° long. Kay added that the maximum tail width was 76', while Franzini estimated it as 0.5°. Clerihew (India) said the comet exhibited two tails on the same date, with "the second being nearly twice as long as the old one, but fainter." The angle between the tails, as indicated by a drawing, was about 20°.

Observations on March 12 and 13 were not numerous, but some interesting descriptions were added to some of the standard fare. Edward Cooper (Nice, France) saw the comet for the first time on the 12th, after his servant had called attention to it. He described it as "a long white light near the western horizon which had somewhat the appearance of that kind of cloud commonly called cirrostratus. This I conceived it to be, although there were very few clouds in the sky at the time." Cooper confirmed this was "the tail of a very large comet" on the next evening. S. C. Walker and E. Otis Kendall (Central High School Observatory, Philadelphia, Pennsylvania, USA) saw the "nucleus" in a cometseeker and described it as a "well-defined disc larger than Jupiter in the same instrument" and said it was similar in brightness to ζ Ceti (magnitude 3.73 – SC2000.0). They also added that the 9-foot focal length refractor revealed this "nucleus" as "a nebulosity gradually condensed toward the centre, so that it was impossible to distinguish any nucleus." They also added that the tail extended between Rigel and Sirius, which implies a PA near 308°. Herrick wrote that the comet's appearance in the 13-cm refractor (55×) "was that of an indefinite globular body, somewhat elongated behind, with a concentration of light near or a little in advance of the centre, which at times seemed to consist of three faint stellar points." He added that the coma diameter was 3 to 4'. Kay said the tail was 42° long, with a maximum width of 80'. He noted a non-stellar nucleus. J. Caldecott gave the tail length as 45° and said the width was 33' at one-third its length and 60' at two-thirds. He also measured the diameter of the "bright part or disc of the head" with a parallel wire micrometer and determined it as

11″. He added that the nebulosity surrounding this nucleus was "about four times the diameter" of the disc.

The moon hampered observations a bit as it approached its full phase. E. Cooper saw the comet in "the strong light of the moon" on March 14 and said, "on sweeping down the tail, I discovered the nucleus with a little telescope the object-glass of which has a diameter of two english inches. It is stellar, and the coma was quite visible." Kay said the tail was 42° long. Gilbert saw the comet on the 15th and noted the tail "appeared to be much brighter."

The last exceptional day of observations came on March 17. With the moon about a day past full, it was again possible to see the comet in dark skies. The tail length was given as 30° by J. F. W. Herschel (Slough, England), 32.8° by Gilbert, 39–40° by D. F. J. Arago (Paris, France), and 43° by Maclean. John Grover (Pisa, Italy) estimated the tail width as about 40′ and noted the edges were "sharply and clearly defined." Herschel basically described the tail as a "vivid luminous streak" and said a slight curvature was suspected. He added that the head had the appearance of a star of magnitude 5. Maclean noted the nucleus was "very indistinct."

Observers continued to follow the comet with the naked eye during the remainder of March, but tail lengths mostly filled the few published reports. On the 18th, K. L. von Littrow (Vienna, Austria) said the tail was 40° long and 1° wide, while the crew of the *Dublin* said the tail was 42.3° long and Herrick said it was 34° long. On the 19th, Haile said the tail was extending 41.8°. Maclean noted, "the outline of the comet [was] very plainly marked. The bright spot or condensation in its head was distinctly perceptible to the naked eye." Tail lengths ranged from 40–47.5° on the 20th, 37–40° on the 22nd, 36–38° on the 23rd, and 25–39° on the 24th. In addition, Maclean said the tail was about 35° long on the 26th, while Kay reported it was 35° long on the 27th. Maclean reported on the difficulty of seeing the nucleus on the 22nd. Kay noted the comet was "evidently becoming more indistinct" on the 24th. Kay noted the tail was "much fainter in appearance" on the 27th. F. de Vico (Rome, Italy) said the nucleus was seen to scintillate on the 29th.

Observations quickly declined as April began. Franzini described the comet as very faint on the 2nd and added that the nucleus was no longer seen. Herrick last saw the comet with the naked eye on the 3rd and said it was "barely discernible." Caldecott turned his 7.5-foot focal length refractor toward the predicted position of the comet on April 6, but the comet was not detected. Smyth saw the comet on the 8th and remarked that the "comet was so excessively faint that it was in the field for many minutes before I could recognize it." Leichhardt last detected the comet on April 11 while at F. T. Rusden's station near Gwydir Falls (east of present day Narrabri in New South Wales, Australia). He said, "by straining my eyes, I could just make out the last faint glimmer of it. . . . "

The comet was last detected on April 19.76, when Smyth and T. Maclear (Cape of Good Hope, South Africa) gave the position as $\alpha = 5^h\ 02.0^m$, $\delta = -3°\ 29'$. Smyth said the comet was then "of the last degree of faintness. . . . "

The physical description alone prompted E. Cooper to write a letter on March 20 to the *Astronomische Nachrichten* suggesting that the comet might be identical to comets seen in 1668 and 1702. The orbital similarity to these comets was confirmed within the next few days, when E. Plantamour and Galle independently calculated the first orbits for C/1843 D1. Plantamour used positions from March 18 to March 21 and determined the perihelion date as 1843 February 27.97, while Galle took positions from March 20 to March 22 and determined the perihelion date as February 27.92. The general correctness of Galle's orbit was confirmed during the next several months by J. F. Encke, B. Valz, J. Nooney and J. Hadley, Walker and Kendall, B. Peirce, Plantamour, K. Knorre, P. A. E. Laugier and V. Mauvais, F. W. A. Argelander, F. W. Bessel, C. H. F. Peters, and F. B. G. Nicolai. But the question of identity with comets seen during the seventeenth and eighteenth century remained and some astronomers departed from the parabolic solution and published elliptical and hyperbolic solutions.

A lot of wishful thinking initially went into the elliptical solutions because of the similarity of both the orbit and appearance to previous comets. When Walker and Kendall published their parabolic orbit, they suggested the period was probably near 21.88 years in order to link it to comets seen in 1668 and 1689. Nicolai was more concerned with linking this comet to that seen in 1668 and suggested a period of 175 years. Laugier and Mauvais produced a similar orbit. T. Clausen wanted to link this comet to those seen in 1668 and 1689. He surmised that since most periodic comets moved in short-period orbits, then a solution of 6.36 years for the period seemed most likely.

Interestingly, while some astronomers were trying to force an elliptical orbit to the short observational arc, others found the positions were best represented by a hyperbolic orbit. Encke published the first hyperbolic orbit at the beginning of April and gave the eccentricity as 1.00021825. Although Walker had suggested with his colleague, Kendall, that the comet might be moving in an elliptical orbit, his computations using positions spanning the period of March 11 to April 10 revealed a hyperbolic orbit with an eccentricity of 1.00090495. About a month later, Walker and Kendall used what they considered were the 18 best positions from the period of March 20 to April 10 and determined an eccentricity of 1.0008560.

The question as to whether the comet moved in an elliptical or hyperbolic orbit remained unanswered for several years; however, during the period of 1849–1852, J. S. Hubbard published a series of papers in the *Astronomical Journal* that worked toward computing a definitive orbit. From the very beginning, Hubbard found that an elliptical solution fitted the positions best. Although his first papers offered orbits with periods ranging from 66 to 10567 years, a comparison with the rough positions determined by J. G. Clarke and Bowring on 1843 February 28 led him to believe the longer period was favored. A breakthrough came late in 1850 when Maclear published the observations made at the Cape of Good Hope. With precise positions now extending back to March 4, Hubbard's initial calculations revealed a period of 376 years. Hubbard continued to revise his calculations during 1851 and 1852. His final paper on the matter gave the period as 533 years. Hubbard also investigated the suggestion that the comet was previously observed in 1668, but he concluded, "the hypothesis of the identity of this comet with that of 1668 is not sustained." H. C. F. Kreutz (1895) reexamined the orbit of this comet as part of a series of papers he would ultimately publish on what would become known as the "Kreutz Sungrazing Comet Group." He concluded that an elliptical orbit with a period of 513 years fitted the positions best. This orbit is given below.

It should be added that this comet made a strong impression on the people of the time. Moncure Daniel Conway (1904) included a particularly interesting reference to this comet in his autobiography. He wrote, "But the greatest sensation was caused by the comet of 1843. There was a widespread panic, similar, it was said, to that caused by the meteors of 1832. Apprehending the approach of Judgement Day, crowds besieged the shop of Mr. Petty, our preaching tailor, invoking his prayers. Methodism reaped a harvest from the comet. The negroes, however, were not disturbed;—they were, I believe, always hoping to hear Gabriel's trump."

T	ω	Ω (2000.0)	i	q	e
1843 Feb. 27.9110 (UT)	82.6390	3.5272	144.3548	0.005 527	0.999 914

ABSOLUTE MAGNITUDE: $H_{10} = 4.9$ (V1964)

FULL MOON: Jan. 16, Feb. 14, Mar. 16, Apr. 14, May 13

SOURCES: G. L. Platt, M. C. Leavenworth, S. W. Hall, A. Blackman, N. J. Buel, J. G. Clarke, S. J. Parker, W. Mitchell, W. C. Bond, S. C. Walker, E. O. Kendall, and E. Loomis, *AJS* (Series 1), **44** (1843 Jan.–Mar.), pp. 412–17; D. F. J. Arago, *CR*, **16** (1843 Mar. 20), pp. 597–601; E. Plantamour, *CR*, **16** (1843 Mar. 27), pp. 608–9; *PMJS* (Series 3), **22** (1843 Apr.), pp. 323–4; P. A. E. Laugier and V. Mauvais, *CR*, **16** (1843 Apr. 3), pp. 640–3; J. E. B. Valz, J. F. Encke, P. A. E. Laugier, V. Mauvais, A. Colla, and A. Decous, *CR*, **16** (1843 Apr. 10), pp. 718–24; E. Cooper, J. F. W. Herschel, A. von Humboldt, J. G. Galle, K. L. von Littrow, E. Plantamour, J. F. Encke, F. W. Bessel, K. Kreil, and F. de Vico, *AN*, **20** (1843 Apr. 13), pp. 289–304; P. A. E. Laugier, V. Mauvais, and E. Plantamour, *CR*, **16** (1843 Apr. 17), pp. 781–2; F. B. G. Nicolai and F. W. A. Argelander, *AN*, **20** (1843 Apr. 22), pp. 313–16; P. A. E. Laugier, V. Mauvais, and J. E. B. Valz, *CR*, **16** (1843 Apr. 24), pp. 919–28; F. W. Bessel, C. H. F. Peters, E. Plantamour, K. Knorre, J. F. Encke, and F. B. G. Nicolai, *AN*, **20** (1843 May 11), pp. 337–52; G. B. Amici, *CR*, **16** (1843 May 15), pp. 1091–2; F. de Vico, *AN*, **20** (1843 May 18), p. 353; K. Kreil, *AN*, **20** (1843 May 20), pp. 369–72; S. C. Walker and E. O. Kendall, *AN*, **20** (1843 Jul. 8), pp. 385–96; J. Bowring, *CR*, **17** (1843 Jul. 10), pp. 84–5; M. M. Franzini, *AN*, **21** (1843 Aug. 3), pp. 59–64; S. C. Walker, E. O. Kendall, and E. C. Herrick, *AJS* (Series 1), **45** (1843 Apr.–Jun.), pp. 188–208; J. Taylor and J. G. Clarke, *AJS* (Series 1), **45** (1843 Apr.–Jun.), pp. 229–30; T. Clausen, *AN*, **21** (1843 Aug. 12), pp. 73–6; C. Darlu, *CR*, **17** (1843 Aug. 21), p. 362; S. C. Walker, *AN*, **21** (1843 Sep. 2), pp. 107–10; S. C. Walker, Tucker, J. C. Haile, Kay, P. P. King, W. S. Mackay, and W. Lloyd, *MNRAS*, **6** (1843 Nov. 10), pp. 2–9; J. Taylor and J. G. Clarke, *AN*, **21** (1843 Dec. 7), p. 175;

PRSL, **4** (1843), pp. 450–3, 456; B. Peirce, P. Ray, and S. C. Walker, *American Almanac.* Boston: James Munroe & Co. (1844), pp. 94–100; Clerihew, *AN,* **21** (1844 Jan. 6), pp. 199–202; J. Burdwood and J. G. Clarke, *MNRAS,* **6** (1844 Jan. 12), pp. 22–3; J. Calder, *AN,* **21** (1844 Feb. 8), p. 239; *AJS* (Series 1), **46** (1844 Apr.), p. 114; *PMJS* (Series 3), **24** (1844 Apr.), pp. 301–6; *PMJS* (Series 3), **24** (Supplement 1844), pp. 522–3; G. Maclean, *MNRAS,* **6** (1844 May 10), pp. 76–7; W. B. Clarke, *MNRAS,* **6** (1844 Jun. 14), p. 84; E. Loomis, *AN,* **22** (1844 Oct. 12), p. 205; G. Brand, *MNRAS,* **6** (1844 Dec. 13), p. 136; *PMJS* (Series 3), **26** (1845 Apr.), pp. 360–1; J. Caldecott, *MRAS,* **15** (1846), pp. 229–31; Georg Heinrich von Boguslawski, *Report on the comet of 1843.* London: R. and J. E. Taylor (1846), pp. 3–6; R. Main, J. H. Kay, and G. Maclean, *MRAS,* **16** (1847), pp. 23–53; J. S. Hubbard, *AJ,* **1** (1849 Dec. 13), pp. 10–13; J. S. Hubbard, *AJ,* **1** (1850 Jan. 7), p. 24; J. S. Hubbard, *AJ,* **1** (1850 Feb. 2), pp. 25–9; J. S. Hubbard, *AJ,* **1** (1850 May 7), pp. 57–60; J. S. Hubbard, *AJ,* **1** (1850 Dec. 23), pp. 153–4; T. Maclear, C. P. Smyth, and J. Gibbs, *MRAS,* **20** (1850–1), pp. 62–9; J. S. Hubbard, *AJ,* **2** (1851 Oct. 3), pp. 46–8; J. S. Hubbard, *AJ,* **2** (1851 Nov. 20), p. 57; J. S. Hubbard, *AJ,* **2** (1852 Jul. 10), pp. 153–6; H. C. F. Kreutz, *AN,* **139** (1895 Dec. 2), pp. 113–16; Moncure Daniel Conway, *Autobiography: Memoires and Experiences.* Boston: Houghton, Mifflin and Company (1904), p. 23; V1964, p. 56; *The Letters of F. W. Ludwig Leichhardt.* Volume 2. Translated by M. Aurousseau. Cambridge: University Printing House (1968), pp. 706, 717; HA1970, p. 86; personal correspondence from Arthur Beales (1996); SC2000.0, p. 37.

C/1843 J1 *Discovered:* 1843 May 3.08 ($\Delta = 1.69$ AU, $r = 1.62$ AU, Elong. $= 68°$)
(Mauvais) *Last seen:* 1843 October 2.26 ($\Delta = 1.65$ AU, $r = 2.51$ AU, Elong. $= 142°$)
  Closest to the Earth: 1843 August 10 (1.1523 AU)
1843 II *Calculated path:* PEG (Disc), PSC (Jul. 28), AQR (Aug. 14), SCL (Sep. 20)

F. V. Mauvais (Paris, France) discovered this comet on 1843 May 3.08, at a position of $\alpha = 21^h 43^m$, $\delta = +29° 27'$. He described it as a "small oval nebulosity of about 3' in diameter." Mauvais added that he "perceived a small brilliant point at the center." Mauvais confirmed his find on May 4.13.

F. W. Bessel (Königsberg, now Kaliningrad, Russia) described the comet as very pale on May 19. G. Santini (Padova, Italy) typically observed the comet under mostly unfavorable skies from May 24 to May 30. On June 1, he noted skies were very clear and commented that the comet was "much more dazzling" than when previously seen. M. Koller (Kremsmünster, Austria) said the comet was 3' in diameter on June 1 and 2, without a nucleus or tail. He said the comet was visible despite the full moon on the 12th, and was considerably faded by moonlight on June 17 and 18. F. W. A. Argelander (Bonn, Germany) described the comet as faint in moonlight during the period of June 15 to 17, but said the comet's position was still measurable because of its strong central condensation. During the period of June 6–30, Santini said the comet always presented the appearance of a large, faint nebulosity.

Argelander saw the comet on August 2, while observing with a 10-cm refractor. He last saw the comet on August 17, while his colleague, Bergius, followed it until August 31. C. K. L. Rümker (Hamburg, Germany) saw the comet for the final time on September 18.

The comet was last detected on October 2.26, when Elias Loomis (Hudson Observatory, Ravenna, Ohio, USA) gave the position as $\alpha = 23^h\ 02.5^m$, $\delta = -29° 21'$, as seen in his 10-cm refractor.

The first orbit was calculated by Mauvais using positions from May 5 to May 9. The resulting perihelion date was 1843 May 11.46. A few weeks later, Mauvais published a revision using positions up to June 3. The perihelion date was given as May 5.98. Additional parabolic orbits were published by W. C. Götze, H. Schlüter, J. R. Hind, Koller, A. Reslhuber, and Mauvais, which finally established the perihelion date as May 6.6.

The first hyperbolic orbit was published by Santini. He used positions from the period of May 5 to June 28, and determined the perihelion date as May 6.62 and the eccentricity as 1.0144067. Götze (1845) used positions from the period of May 4 to September 3 and determined the perihelion date as May 6.56 and the eccentricity as 1.0001798. R. J. Buckley (1979) refined the orbit of this comet using positions obtained between May 4 and September 3, and applied perturbations by Venus to Neptune. The resulting perihelion date was May 6.56, while the eccentricity was 1.000129. This orbit is given below.

T	ω	Ω (2000.0)	i	q	e
1843 May 6.5622 (TT)	124.2580	159.4425	52.7261	1.616419	1.000129

ABSOLUTE MAGNITUDE: $H_{10} = 4.2$ (V1964)

FULL MOON: Apr. 14, May 13, Jun. 12, Jul. 11, Aug. 10, Sep. 8, Oct. 8

SOURCES: F. V. Mauvais, *MNRAS*, **5** (1843 May 12), p. 289; F. V. Mauvais, *CR*, **16** (1843 May 15), pp. 1090–1; F. V. Mauvais, *AN*, **20** (1843 May 20), p. 369; F. V. Mauvais, *CR*, **16** (1843 Jun. 5), pp. 1207–8; F. V. Mauvais, C. K. L. Rümker, Schlüter, and F. W. Bessel, *AN*, **20** (1843 Jul. 8), pp. 385, 395–8; F. W. A. Argelander, F. V. Mauvais, M. Koller, H. Schlüter, F. W. Bessel, C. K. L. Rümker, and F. V. Mauvais, *AN*, **21** (1843 Aug. 3), pp. 35, 39–42, 49–54; G. Santini, *AN*, **21** (1843 Sep. 2), pp. 105–8; F. W. A. Argelander and G. Santini, *AN*, **21** (1843 Oct. 7), pp. 133–8; F. V. Mauvais, *CR*, **17** (1843 Oct. 23), pp. 886–8; C. K. L. Rümker, *MNRAS*, **6** (1843 Nov. 10), pp. 10–11; C. K. L. Rümker, *AN*, **21** (1844 Jan. 6), p. 193; J. R. Hind, *AN*, **21** (1844 Jan. 18), pp. 215–18; F. W. A. Argelander and Bergius, *AN*, **21** (1844 Feb. 8), p. 225; *PMJS* (Series 3), **24** (1844 Apr.), p. 306; W. C. Götze, *AN*, **21** (1844 Apr. 4), pp. 315–18; W. C. Götze, *MNRAS*, **6** (1844 Apr. 12), p. 68; M. Koller and A. Reslhuber, *AN*, **21** (1844 May 23), pp. 369–71; E. Loomis, *AN*, **22** (1844 Oct. 12), pp. 205–10; W. C. Götze, *AN*, **23** (1845 Apr. 12), pp. 67–72; V1964, p. 56; R. J. Buckley, *JBAA*, **89** (1979 Apr.), pp. 260, 262.

4P/1843 W1 *Discovered:* 1843 November 23.04 ($\Delta = 0.79$ AU, $r = 1.73$ AU, Elong. $= 154°$)
(Faye) *Last seen:* 1844 April 10.82 ($\Delta = 2.29$ AU, $r = 2.35$ AU, Elong. $= 81°$)
 Closest to the Earth: 1843 November 24 (0.7871 AU)
1843 III *Calculated path:* ORI (Disc), MON (Mar. 22)

H. A. E. A. Faye (Royal Observatory, Paris, France) discovered this comet on 1843 November 23.04, near γ Orionis. He estimated the position as α = 5^h 24.3^m, δ = +6° 56′ on November 23.11. The comet exhibited a distinct nucleus, which emitted faint indications of a tail extending 4′ away from the sun. Due to cloudy weather, Faye was not able to confirm his discovery until November 25.21.

The comet was moving away from the sun and Earth as observations began elsewhere. O. W. Struve (Pulkovo Observatory, Russia) said the comet was visible to the naked eye at the end of November. J. South (Kensington, England) saw the comet in an 30.2-cm refractor on November 30 and noted the nucleus was elongated towards the tail. The tail itself was later observed after moonset and estimated as 11′ long. Although the comet was faint, South said it was easily found in a 7.0-cm refractor. A. Reslhuber (Kremsmünster, Austria) saw the comet during the period spanning December 12–15 and noted a distinct nucleus and a tail 10–12′ long. Struve began carefully measuring the comet's position on December 15 and said the comet's brightness was similar to a star of magnitude 6 or 7 in a cometseeker, while the coma was 2′ across. T. Henderson (Edinburgh, Scotland) described the comet as "very faint" on the same evening. Struve saw the comet on December 19 and said the fan-like tail extended 16′. Struve said the nucleus was more distinct on the 25th than on previous nights.

Henderson said the comet was "barely visible" in a 9.1-cm refractor on January 10. He added, "It was seen only at times, according to the varying clearness of the sky." Struve saw it on the 24th and said it was very faint at a 10° altitude and with strong moonlight. G. B. Airy (Royal Observatory, Greenwich, England) found the comet excessively faint on February 19, 20, and 22. Struve found the comet very faint in the 38-cm refractor on the 19th. W. Hamilton and C. Thompson (Trinity College, Dublin, Ireland) last detected the comet on the 20th with a 13-cm refractor. Struve said the comet's light was "much weakened by moonlight" on March 21. He said the comet was very faint in the refractor on April 5.

The comet was last detected on April 10.82, when Struve gave the position as α = 6^h 47.0^m, δ = +11° 23′. He was using a 38-cm refractor and said the comet was moving within a few arc minutes of a 9th-magnitude star when first observed. Struve noted that by the time the comet had moved sufficiently past the star to enable position measurements, its light had faded because of haze near the horizon. Struve said cloudy weather was present during the next few nights and when clear skies returned on the evening of April 17 he could no longer find the comet.

The first parabolic orbit was calculated by Henderson. Using positions obtained on November 25, 29, and December 2, he determined the perihelion date as 1843 September 29.40. He added, "From the great distance of the comet, and slowness of its motion, small errors in the observations and quantities neglected in the computations have a considerable effect upon the elements." Henderson proved this later in the month, when positions up to December 15 revealed a perihelion date of September 9.34. Using positions covering a similar period of time, perihelion dates of September 7–11 were obtained by Faye, Funk and W. C. Götze, J. M. Agardh, B. Valz, and E. Plantamour, while a perihelion of August 28 was determined by J. G. Galle.

The first elliptical orbits were independently calculated by Henderson and J. C. Adams. Using precise positions obtained by J. Challis during the period of November 29 to December 16, both astronomers noticed unexpected large errors when they attempted a parabolic orbit. Henderson's subsequent elliptical orbit had a perihelion date of October 24.20 and a period of 6.58 years, while Adams' revealed a perihelion date of October 26.83 and a period of 6.39 years. Further elliptical orbits were determined by Henderson, A. C. Petersen, Agardh, J. R. Hind, F. W. A. Argelander, F. B. G. Nicolai, H. Goldschmidt, F. Carlini, O. W. Struve, M. V. Lyapunov, Le Jeune, G. Santini, U. J.-J. Le Verrier, and Faye. These gradually revealed the perihelion date was closer to October 18, while the period was around 7.4 years.

Later calculations using multiple apparitions and planetary perturbations were published by D. M. A. Möller (1865, 1872), A. Shdanow (1885), and B. G. Marsden and Z. Sekanina (1971). These revealed a perihelion date of October 17.64 and a period of 7.44 years. Marsden and Sekanina determined the nongravitational terms as $A_1=+0.601$ and $A_2=+0.01026$. The orbit of Marsden and Sekanina is given below.

T	ω	Ω (2000.0)	i	q	e
1843 Oct. 17.6437 (TT)	200.0107	211.7317	11.3596	1.692255	0.555816

ABSOLUTE MAGNITUDE: $H_{10} = 4.2$ (V1964)

FULL MOON: Nov. 7, Dec. 7, Jan. 5, Feb. 4, Mar. 4, Apr. 3, May 2

SOURCES: H. A. E. A. Faye, *CR*, **17** (1843 Nov. 27), p. 1248; T. Henderson and H. A. E. A. Faye, *MNRAS*, **6** (1843 Dec. 8), pp. 15–16; H. A. E. A. Faye, *CR*, **17** (1843 Dec. 11), p. 1308; H. A. E. A. Faye, Funck, W. C. Götze, J. M. Agardh, and E. Plantamour, *AN*, **21** (1844 Jan. 6), pp. 205–8; E. Plantamour and B. Valz, *CR*, **18** (1844 Jan. 8), pp. 56–8; T. Henderson, J. C. Adams, and W. Lassell, *MNRAS*, **6** (1844 Jan. 12), pp. 18–22; H. Goldschmidt, *CR*, **18** (1844 Jan. 15), pp. 96–7; J. F. Encke and J. G. Galle, *AN*, **21** (1844 Jan. 18), p. 223; H. A. E. A. Faye and H. Goldschmidt, *CR*, **18** (1844 Jan. 29), p. 186; F. W. A. Argelander, A. C. Petersen and J. M. Agardh, *AN*, **21** (1844 Feb. 8), pp. 225, 239; E. Plantamour, *CR*, **18** (1844 Feb. 19), pp. 309–10; W. Lassell, G. B. Airy, C. Thompson, T. Chevallier, and J. R. Hind, *MNRAS*, **6** (1844 Mar. 8), pp. 53–9; J. R. Hind, E. Plantamour, and H. Goldschmidt, *AN*, **21** (1844 Mar. 9), pp. 279–82; H. Goldschmidt, *CR*, **18** (1844 Mar. 25), pp. 528–31; H. A. E. A. Faye and J. South, *AJS* (Series 1), **46** (1844 Apr.), p. 210; C. Thompson, *MNRAS*, **6** (1844 Apr. 12), pp. 67–8; F. B. G. Nicolai, *AN*, **21** (1844 Apr. 18), pp. 325–8; F. Kaiser, Le Jeune, and G. Santini, *AN*, **21** (1844 Apr. 27), pp. 337–44; J. R. Hind, *MNRAS*, **6** (1844 May 10), p. 78; A. Reslhuber, *AN*, **21** (1844 May 23), p. 371; *PMJS* (Series 3), **24** (Supplement 1844), pp. 519–22; O. W. Struve, *AN*, **22** (1844 Jun. 15), pp. 1–16; O. W. Struve and M. V. Lyapunov,

AN, **22** (1844 Jun. 20), pp. 17–24; *PMJS* (Series 3), **25** (1844 Jul.), pp. 72–3; J. R. Hind, *AN*, **22** (1844 Jul. 6), p. 61; F. Carlini, *AN*, **22** (1844 Aug. 22), pp. 135–8; *PMJS* (Series 3), **25** (1844 Sep.), p. 224; E. Loomis, *AN*, **22** (1844 Oct. 12), p. 209; O. W. Struve, *AN*, **23** (1845 May 17), p. 121; U. J.-J. Le Verrier, *AN*, **23** (1845 Jun. 26), p. 195; W. Lassell, *MRAS*, **15** (1846), pp. 233–6; D. M. A. Möller, *AN*, **64** (1865 Apr. 10), pp. 145–51; D. M. A. Möller, *VJS*, **7** (1872), p. 96; A. Shdanow, *BASP*, **33** (1885), pp. 1–24; V1964, p. 56; B. G. Marsden and Z. Sekanina, *AJ*, **76** (1971 Dec.), pp. 1136–7.

C/1844 N1	*Discovered:* 1844 July 8.04 (Δ = 1.41 AU, r = 1.90 AU, Elong. = 102°)
(Mauvais)	*Last seen:* 1845 March 10.80 (Δ = 2.88 AU, r = 2.45 AU, Elong. = 55°)
	Closest to the Earth: 1844 July 17 (1.3878 AU), 1844 December 18 (1.0022 AU)
1844 II	*Calculated path:* HER (Disc), BOO (Jul. 15), CrB (Jul. 18), BOO (Jul. 19), VIR (Sep. 1), CRV (Oct. 19), HYA (Oct. 26), CEN (Nov. 11), HYA (Nov. 13), CEN (Nov. 14), VEL (Dec. 3), CAR (Dec. 5), VOL (Dec. 13), DOR (Dec. 21), PIC–DOR (Dec. 23), HOR (Jan. 3), ERI (Jan. 9), FOR (Jan. 13), ERI (Jan. 31)

F. V. Mauvais (Paris, France) discovered this comet on 1844 July 8.04, at a position of α = 16^h 30.0^m, δ = +46° 15′. He confirmed his find on July 8.96 and described the comet as rather bright, with a coma 3′ across, and a small bright nucleus. The comet was independently discovered by H. L. d'Arrest (University of Berlin, Germany) on July 10.00. He said the comet was quite bright and round, with an appreciable diameter as seen in a comet-seeker. The comet had reached its most northerly declination of +49° on June 25.

Although the comet was well observed, physical descriptions were few. E. Loomis (Hudson's Observatory, Ohio, USA) said the comet was visible in the 8-cm refractor of the transit circle on August 6 and 7. The comet was seen by J. M. Gilliss and J. H. C. Coffin (US Naval Observatory, Washington, DC) on August 17 and 18, while using the new 14-foot focal length refractor. It was described as 2′ across, with a nucleus of magnitude 9. They suspected a tail 6′ long on the 18th. F. B. G. Nicolai (Mannheim, Germany) last detected the comet on the evening of September 2. He said he could probably have continued observing the comet in twilight for a few more days thereafter, if not for clouds that accumulated each evening. The comet was last detected in evening twilight on September 6.81 by C. F. Gauss (Göttingen, Germany) and on September 8.79 by A. Reslhuber (Kremsmünster, Austria).

The comet steadily drew closer to the sun, reaching an elongation of 30° on September 12, 20° on September 21, and 10° on October 1. After attaining a minimum elongation of 6° on October 8, the comet began moving away from the sun. It reached an elongation of 10° on October 15 and 20° on October 24. The comet was recovered from morning twilight by W. Mann (Royal Observatory, Cape of Good Hope, South Africa) on October 28.11. He noted it was found "without difficulty soon after it rose. It appeared as a faint nebulous patch of light, nearly circular, about 1′ in diameter, and with a condensation of light near its centre."

Having passed perihelion during October, the comet changed little in brightness during November as it exited twilight and its distance from Earth decreased. Mann saw it on the 4th and described it as "a fine nebulous mass of light; the outline seemed to be of a parabolic figure, with a condensation of light in its focus. A faint trace of a tail was seen, about 4′ in length: its direction being *apparently* towards the sun." On the 9th, Mann noticed a distinct nucleus "about as bright as a star of the ninth magnitude," which was visible in the brightest part of the head. He estimated the length of the faint tail as 9′. Mann said the comet had become visible to the naked eye on November 11. F. W. Ludwig Leichhardt, who was on an expedition of exploration in Australia, saw the comet on the morning of November 28. Later that day he found a river in Queensland, Australia, which he named "Comet" River.

The comet began fading more rapidly after mid-December, as it was heading away from both the sun and Earth. On the 18th, it reached its most southerly declination of −66° and then began heading northward. T. Maclear (Royal Observatory, Cape of Good Hope, South Africa) said the comet became too faint to use an illuminated micrometer for position measurement after December 25. He was using a 3.8-foot focal length refractor. F. W. A. Argelander (Bonn, Germany) became the first European to recover the comet when he saw it on January 31. He described it as very faint in a somewhat hazy sky. Argelander saw the comet very well on February 4 and 8. A. Reslhuber (Kremsmünster, Austria) was able to see the comet intermittently on February 8. He added that a nucleus was visible, but no tail. Mann described the comet as extremely faint on February 16 and 26. Reslhuber described the comet as very faint on February 25, 28, and March 1.

The comet was last detected during March. Mann described it as "extremely faint" on March 2, 3, 4, 5, 6, and 10. The last date was also the final time the comet was seen, with Mann measuring the position as $\alpha = 3^h \, 03.2^m$, $\delta = -8° \, 37′$ on March 10.80.

Barely a week after his discovery of this comet, Mauvais calculated the first parabolic orbit. This indicated a perihelion date of 1844 October 15.26. J. E. B. Valz immediately suggested this comet might be identical to the comet of 1796, despite a sensible difference between the orbits of the two comets. By the end of the month, Mauvais was able to revise his calculations using positions he had obtained up to July 21. The resulting perihelion date was October 17.81. Later orbits by E. Plantamour, E. Cooper, A. Graham, Nicolai, G. Santini, Turazza, F. F. E. Brünnow, and J. R. Hind revealed the accuracy of Mauvais' second orbit.

A few years after the comet had faded from view, Plantamour (1847) determined ten Normal positions from the available observations and calculated an elliptical orbit with a perihelion date of October 17.84 and a period of about 102 050 years. F. E. Ross (1905) basically revised Planta-mour's work and determined a period of 98 153 years.

R. J. Buckley (1979) used 144 positions obtained between 1844 July 8 and 1845 February 11, as well as perturbations by Venus to Neptune, and computed an elliptical orbit with a perihelion date of October 17.85 and a period of about 208 855 years. This orbit is given below.

T	ω	Ω (2000.0)	i	q	e
1844 Oct. 17.8460 (TT)	211.2749	33.8480	131.4092	0.855401	0.999757

ABSOLUTE MAGNITUDE: $H_{10} = 4.9$ (V1964)

FULL MOON: Jun. 30, Jul. 29, Aug. 27, Sep. 26, Oct. 26, Nov. 24, Dec. 24, Jan. 23, Feb. 22, Mar. 23

SOURCES: F. V. Mauvais, H. L. d'Arrest, E. Plantamour, and J. E. B. Valz, *CR*, **19** (1844 Jul. 22), p. 239; F. V. Mauvais, *CR*, **19** (1844 Jul. 30), pp. 245–9; F. V. Mauvais and J. F. Encke, *AN*, **22** (1844 Aug. 1), p. 97; E. Plantamour, *CR*, **19** (1844 Aug. 26), pp. 415–18; *PMJS* (Series 3), **25** (1844 Sep.), p. 239; F. F. E. Brünnow, *AN*, **22** (1844 Sep. 5), p. 165; C. F. Gauss, E. Plantamour, and F. B. G. Nicolai, *AN*, **22** (1844 Oct. 3), pp. 189–94, 197; F. B. G. Nicolai, E. Cooper, A. Graham, J. M. Gilliss, J. H. C. Coffin, and A. Reslhuber, *AN*, **22** (1844 Oct. 12), pp. 201, 209–12; F. V. Mauvais, H. L. d'Arrest, and W. Lassell, *MNRAS*, **6** (1844 Dec. 13), pp. 129–32; T. Maclear and W. Mann, *MNRAS*, **6** (1845 Jan. 10), pp. 148–9; A. Reslhuber, *AN*, **22** (1845 Jan. 25), p. 357; T. Maclear, *MNRAS*, **6** (1845 Feb. 14), pp. 200–1; G. Santini and Turazza, *AN*, **23** (1845 Mar. 11), p. 15; F. B. G. Nicolai, *AN*, **23** (1845 Mar. 25), p. 21; *PMJS* (Series 3), **26** (1845 Apr.), pp. 358–9; T. Maclear, *MNRAS*, **6** (1845 Apr. 11), pp. 218–19; A. Reslhuber, *AN*, **23** (1845 Apr. 12), p. 67; T. Maclear, W. Lassell, J. R. Hind, and J. Glaisher, *MNRAS*, **6** (1845 May 9), pp. 231–4; J. F. Encke, T. Maclear, and W. Mann, *MNRAS*, **6** (1845 Jun. 13), pp. 249–51; T. Maclear and W. Mann, *AN*, **23** (1845 Jun. 14), p. 175; J. R. Hind, *AN*, **23** (1845 Jun. 26), p. 197; F. W. A. Argelander, *AN*, **23** (1845 Aug. 7), p. 237; A. Reslhuber, *AN*, **23** (1845 Aug. 23), p. 249; *AJS* (Series 1), **49** (1845 Oct.), p. 220; W. Lassell, *MRAS*, **15** (1846), pp. 240–1, 242; T. Maclear and W. Mann, *MRAS*, **15** (1846), pp. 244–50; E. Plantamour, *MSPG*, **11** (1847), p. 574; F. E. Ross, *EAN*, **2** (1905), p. 28; V1964, p. 56; *The Letters of F. W. Ludwig Leichhardt*. Volume 3. Translated by M. Aurousseau. Cambridge: University Printing House (1968), p. 1019; R. J. Buckley, *JBAA*, **89** (1979 Apr.), pp. 260, 262.

54P/1844 Q1 (de Vico–Swift– NEAT) 1844 I

Discovered: 1844 August 23.09 ($\Delta = 0.20$ AU, $r = 1.19$ AU, Elong. $= 158°$)
Last seen: 1844 December 31.80 ($\Delta = 1.13$ AU, $r = 1.82$ AU, Elong. $= 118°$)
Closest to the Earth: 1844 September 1 (0.1903 AU)
Calculated path: AQR (Disc), CET (Aug. 28), PSC (Oct. 23), ARI (Nov. 29)

F. de Vico (Rome, Italy) discovered this telescopic comet on 1844 August 23.09 at a position of $\alpha = 23^h 26.8^m$, $\delta = -23° 19'$. De Vico confirmed his find on August 24.09. Two independent discoveries were made during the first half of September. Melhop (Hamburg, Germany) found the comet with the naked-eye near β Ceti on September 6 and H. L. Smith (Cleveland, Ohio,

USA) saw the comet next to β Ceti on September 11.19, while observing with his 6-cm refractor. Smith said the nucleus was very bright and the tail was about 1° long.

This comet was widely observed by astronomers in Europe from September into November, but physical descriptions were not generally made. A. Reslhuber (Kremsmünster, Austria) saw the comet on September 14 and remarked on the bright nucleus. Reslhuber's next descriptions came on November 10 and 13, when he noted it was uncommonly faint, with a hardly noticeable nucleus.

The comet was very difficult to see in December and only a few observations were reported. Reslhuber looked for the comet on December 2 and found a small, dull spot very near the predicted position. J. R. Hind (London, England) and J. E. B. Valz (Marseille, France) independently continued their observations until December 6. The comet was last detected on December 31.80, when O. W. Struve (Pulkovo Observatory, Russia) saw it in a 38-cm refractor and gave the position as $\alpha = 2^h\ 25.6^m$, $\delta = +14°\ 44'$.

The earliest published parabolic orbit was by J. J. E. Goujon, who took positions obtained on September 2, 3, and 4, and determined the perihelion date as 1844 September 2.43. A later orbit by de Vico gave the perihelion date as August 31.04. It soon became obvious that a parabolic orbit could not represent the positions very well. On September 16, the *Comptes Rendus* published an elliptical orbit by H. A. E. A. Faye, which gave the perihelion date as September 3.09 and the period as 5.13 years. Faye published a revised orbit in the September 30 issue of that same publication. Using positions spanning September 2–19, he determined the perihelion date as September 3.01 and the period as 5.28 years. Additional refinements by Hind, H. Goldschmidt, Faye, F. B. G. Nicolai, and F. F. E. Brünnow (1846, 1859) would establish a perihelion date of September 2.98 and a period of 5.46 years.

Some astronomers believed this comet had been seen before. As early as 1844 September 9, the *Comptes Rendus* published a paper by P. A. E. Laugier and F. V. Mauvais stating that the orbit was very similar to that of a comet seen in 1585. In that paper, as well as in a follow-up on September 16, they even went so far as to suggest additional returns were observed in 1678, 1743, 1770, and 1819. Although the second and last comets were ultimately excluded, Laugier and Mauvais suggested an orbital period of 9.2 years. U. J.-J. Le Verrier added his expertise in the September 30 issue by conducting a thorough investigation into the orbital evolution of the 1770 comet. He examined the perturbations by Jupiter during the period of 1776–80 and concluded that the comet would return to perihelion in 1844. Le Verrier did note that the orbital planes of the 1770 comet and that of 1844 were not very similar. Eventually, it was realized that the 1678 comet was actually comet 6P/d'Arrest, while the 1770 comet was ejected by Jupiter into a much larger orbit with a period of over two centuries. No relationship to other pre-1844 comets has ever been proven.

The comet remained lost until it was accidentally rediscovered in 1894. Later studies of the orbit used multiple apparitions, with S. D. Shaporev (1978) indicating a perihelion date of September 2.97 and a period of 5.46 years. This orbit is given below.

T	ω	Ω (2000.0)	i	q	e
1844 Sep. 2.9748 (TT)	279.0523	65.6261	2.9202	1.186 379	0.617 410

ABSOLUTE MAGNITUDE: $H_{10} = 8.0$ (V1964)

FULL MOON: Jul. 29, Aug. 27, Sep. 26, Oct. 26, Nov. 24, Dec. 24, Jan. 23

SOURCES: F. de Vico, *CR*, **19** (1844 Sep. 2), p. 484; J. J. E. Goujon, *CR*, **19** (1844 Sep. 9), pp. 500–1; H. A. E. A. Faye, P. A. E. Laugier, and F. V. Mauvais, *CR*, **19** (1844 Sep. 16), pp. 557–62; H. A. E. A. Faye, *CR*, **19** (1844 Sep. 30), pp. 665–6; U. J.-J. Le Verrier, *CR*, **19** (1844 Sep. 30), pp. 666–70; H. L. Smith, *AJS* (Series 1), **47** (1844 Oct.), p. 419; F. de Vico, *AN*, **22** (1844 Oct. 3), p. 197; P. A. E. Laugier and F. V. Mauvais, *CR*, **19** (1844 Oct. 7), pp. 701–3; F. de Vico, *AN*, **22** (1844 Oct. 12), p. 213; H. A. E. A. Faye, *AN*, **22** (1844 Nov. 9), p. 247; F. B. G. Nicolai, *AN*, **22** (1844 Nov. 16), pp. 259–62; J. R. Hind, H. Goldschmidt, and H. L. Smith, *AN*, **22** (1844 Nov. 28), pp. 269–80; H. A. E. A. Faye, *CR*, **19** (1844 Dec. 9), pp. 1313–14; T. Dell, J. R. Hind, and R. Snow, *MN-RAS*, **6** (1844 Dec. 13), pp. 132–6; H. A. E. A. Faye, *AN*, **22** (1845 Jan. 21), p. 341; A. Reslhuber, *AN*, **22** (1845 Jan. 25), pp. 357–60; H. L. Smith, *AJS* (Series 1), **48** (1845 Apr.), pp. 219–20, 402; *PMJS* (Series 3), **26** (1845 Apr.), pp. 359–60; J. E. B. Valz, *AN*, **23** (1845 May 3), pp. 99–102; W. Lassell, and J. R. Hind, *MNRAS*, **6** (1845 May 9), p. 238; W. Lassell, *MRAS*, **15** (1846), pp. 237–8, 241; F. F. E. Brünnow, *AN*, **24** (1846 Jun. 25), pp. 165–80; O. W. Struve, *AN*, **25** (1847 Mar. 20), pp. 249–52; F. F. E. Brünnow, *Astronomical Notices*, No. 3 (1859), pp. 4–5; V1964, p. 56; S. D. Shaporev, *QJRAS*, **19** (1978), pp. 52–3, 57.

C/1844 Y1
(Great Comet)

1844 III

Discovered: 1844 December 16.9 ($\Delta = 1.16$ AU, $r = 0.27$ AU, Elong. $= 11°$)
Last seen: 1845 March 12.81 ($\Delta = 2.18$ AU, $r = 1.97$ AU, Elong. $= 64°$)
Closest to the Earth: 1845 January 6 (0.9553 AU)
Calculated path: SGR (Disc), MIC (Dec. 27), GRU (Jan. 1), PHE (Jan. 11), SCL (Jan. 16), FOR (Jan. 28), CET (Feb. 5), ERI (Feb. 11)

In the 1845 March 3 issue of the *Comptes Rendus*, D. F. J. Arago reported that this comet was discovered in twilight, low over the horizon in the evening sky from "English Guyana" (now Guyana, South America) on 1844 December 16. The probable universal time was December 16.9. No additional details were given. An independent discovery was made on December 18.8 by an observer at Green Point (Cape Town, South Africa). P. P. King (Port Stephens, New South Wales, Australia) independently discovered the comet on December 19.4. The first person to independently discover the comet and notify the proper authorities was E. Wilmot (Magnetic Observatory, Cape of Good Hope, South Africa) who first detected it on December 19.8. He said the tail was 3–4° long and noted that the head seemed to set below the horizon at the same time as Mercury. A few minutes later, Wilmot had informed T. Maclear (Royal Observatory, Cape of Good Hope, South Africa), who imperfectly detected the comet. He said the tail "was seen shooting above the edge of a nimbus or bank of clouds, but I doubted whether it might not be one of these beams that occasionally appear from behind a cloud with the setting sun." Sailors aboard the barque *Ceylon*, commanded by E. W. Beazley, first detected the comet on December 20 and J. C. Haile (Auckland, New Zealand) saw the comet on the same evening. R. Sheppard (Wellington, New Zealand) independently discovered the comet on December 22.

The comet was a couple of days past perihelion when first detected, but initially it did not show signs of fading because it was approaching Earth and steadily climbing out of evening twilight. J. Robinson (on a ship in the south Atlantic) saw the comet on December 23 and remarked that the "nucleus appeared like a star of the second magnitude, with an immense tail...." W. Mann (Royal Observatory, Cape of Good Hope, South Africa) said the tail was about 7° long on the 25th and seemed to run parallel to the equator. James Donald and W. Wilson (Georgetown, Guyana) independently discovered the comet about 5° above the southwest horizon on December 26, and Sheppard estimated the tail was 3–4° long that same evening. Mann described the tail as "bushy and slightly curved (sword-shape) towards the north" on December 29. He added that the tail measured 8° long. Maclear gave the tail length as 10° 30′ on the 30th and noted the nucleus was "pungently bright" in the 3.8-foot focal length refractor (43×). J. Caldecott (Trivandrum Observatory, India) saw the comet for the first time that same evening and said the comet was very bright for the short time it was seen before clouds moved in. He added that it "must have evidently been conspicuous for several evenings earlier at places having a clear sky." On December 31, W. H. Simms (Colombo, Ceylon, now Sri Lanka) first observed the comet and commented, "its nucleus was about as bright as a star of the third magnitude, and the tail about 15° long, its edges being sharp and clear, and the light very equally diffused between them."

During the first eight days of January, nearly every observer gave the tail length as between 8° and 10°. Sheppard said the tail was 1.5° wide at the end on the 1st and added that the tail was brightest along the center. Robinson commented that the tail was vertical on the 6th. The nucleus was also very noticeable during this time, with Sheppard estimating the naked-eye brightness as near 2.5. Andrew Lang (St Croix, Virgin Islands) said the nucleus appeared larger than Jupiter in a telescope on the 3rd. On January 6, Haile said the coma exhibited "a very minute stellar point" of about magnitude 5 or 6. He added that this nucleus was distinctly visible and was "very brilliant and sparkling." Caldecott said the nucleus appeared distinct on the 8th, like a 5th-magnitude star seen through a thin haze. Mann commented on the 2nd that the head was bright and "the general outline more sharp and clear than the great comet of March [1843]." G. Webbe (Nevis

Island in the West Indies) summarized the comet's appearance during the last days of this period. He wrote, "The nucleus is large, and rather suddenly condensed, but indicates nothing like a defined termination. The tail appears homogeneous, undivided, and straight, and at present seems to be about 10° in length." W. Pole (Elphinstone College, Bombay, India) also summarized the comet's appearance during this period. He wrote, "The nucleus resembled a star of the fifth or sixth magnitude, and the tail (which was straight and pointed directly from the sun) could be traced distinctly for 7 to 10 degrees. Its light was but faint, but the whole was visible to the naked eye, and formed a beautiful object in the south-west part of the heavens."

During the period of January 9–15, estimates of the comet's overall brightness were few. Caldecott did note that the comet "gradually diminished in length and brightness" following January 9, "both by reason of its own diminishing light and of the increasing moonlight." The tail length typically ranged from 6–7°. Caldecott noted on the 9th that the maximum width of the tail was about 1.25°, which occurred at about one-third of its length from the nucleus. Maclear noted "a faint ray of luminous matter" on the 11th. He estimated its length as about 1.25° and said it "was seen to extend from the anterior portion of the comet's head in a direction away to that of the tail. The breadth of this ray near the head was about 2′, increasing slightly towards the extremity. Its borders were comparatively well defined, and the light gradually diminished in intensity from that portion nearest the comet's head, until it became insensible. A dark space seemed to separate the head of the comet and ray." The coma was described by Haile as noticeably faint on the 10th, while on the 13th he said the coma resembled "a small nebulous mass of light, more condensed at the center than at the sides, but no distinct nucleus visible." The nucleus seemed to undergo rapid changes during this period. Haile described it as "still lucid and sparkling" on the 9th, while he noted it was only occasionally visible on the 11th. Donald and Wilson saw the nucleus on the 12th and said it was "not well defined, [and] appeared equal in size to a star of the fourth magnitude." Interestingly, King said the nucleus appeared unusually bright on the 13th.

The moon greatly affected observations of the comet during the second half of January. Donald and Wilson noted the comet's "appearance was so very indistinct" on January 16 and 17. Maclear noted the tail and anterior ray were both invisible from January 18 until January 27th because of moonlight. J. J. Waterston (Bombay, India) noted that after the full moon the comet could still be seen as a faint streak of light, until January 31 when it was only visible through a telescope. Maclear obtained an excellent observation of the comet on the 27th. With a comet-sweeper he said, "the northern borders of the tail and anterior ray appeared distinct and sharply defined, but the light became fainter towards the southern border of tail, and the corresponding border of the ray could no longer be traced. The whole presented a fan-like appearance." Using the 3.8-foot focal length refractor (43×), Maclear noted, "the anterior extension of the light could be traced as a distinct ray for about 5' from the comet's head, the luminous matter of the southern border becoming then diffused and scattered, while the northern border continued well defined. The actual connexion of the rays and head of the comet still could not be traced." Mann carefully measured the directions of the tail and the northern border of the luminous matter during the last days of January. On the 29th he said the tail extended toward PA 94.4°, while the luminous matter extended toward PA 301.1°. On the 30th he said the tail extended toward PA 92.8°, while the luminous matter extended toward PA 298.5°. On the 31st he said the tail extended toward PA 90.9°, while the luminous matter extended toward PA 301.8°.

The comet's fading took it below naked-eye visibility early in February. Caldecott remarked that the comet was no longer visible to the naked eye during the period of February 1–5, but the head was still visible in a telescope as a "blotch of light." Robinson saw the comet for the final time with the naked eye on the 2nd. Joseph Dayman (sailing from south of Australia to England) noted the comet was measured with a sextant on every possible evening "until its disappearance to the naked eye on the 6th of February." Beazley also reported sextant observations were made every favorable night to February 6. The comet also became visible for the first time in Europe, with independent discoveries being reported by A. Colla (Parma, Italy) on February 5 and C. H. F. Peters and E. Cooper (Naples, Italy) on February 7. Peters and Cooper said, "The new comet has no sharp distinct nucleus, and an oval shape, the long axis directed away from the sun, about 3' in diameter, diffuse." King said the comet was "only discernible as a small nebula" on February 4. Although Mann noted the tail was extending toward PA 94.8° on the 1st, while the northern border of the luminous matter extended toward PA 305.0°, this was the final time the tail was reported. Mann saw the comet on the 9th and described it as a nearly circular, bright nebulosity. The coma was about 3' across. He added that there was "no trace whatever of a tail or any other appendage." Peters and Cooper noted the comet was a few degrees west-southwest of comet C/1844 N1 (Mauvais) on the 7th, while Maclear reported both comets were within the same field of view of the comet-sweeper on the 9th. Mann said the comet was barely visible in bright moonlight on February 16 and 18, while the comet was hidden from view by the moonlight following the 18th until the 27th. A. C. Petersen (Altona, Germany) described the comet as very faint in a telescope and invisible in a cometseeker on the 24th, when the moon was just two days past full. Mann saw the comet on the 27th and described it as a faint nebulosity, about 2.25' across. He added that there was "no apparent condensation of light."

The comet was becoming a difficult object to see as March began. J. R. Hind (Regent's Park, England) observed the comet on the evening of March 3 and said it was "extremely faint, appearing as an oval mass of light, with a slight condensation of the cometic matter towards the centre." J. F. Encke

(Berlin, Germany) saw the comet on March 9, but was unable to find it thereafter. Peters described the comet as very faint on the 11th.

The comet was last seen on March 12.81, when Mann was barely able to detect it in a refractor of 3.8-foot focal length. In fact, the comet was so faint, extreme care had to be taken in the positional measurement so that α was acquired 20 minutes earlier than δ. The position was $\alpha = 3^h 49.9^m$, $\delta = -7° 22'$. A further attempt was made 24 hours later, but moonlight kept the comet hidden from view.

One of the first orbits was calculated by Caldecott, using positions from January 9, 14, and 19. The resulting perihelion date was 1844 December 13.85. As observations of the comet continued, several astronomers calculated revised parabolic orbits using positions from February and early March, including Peters, Petersen, Hind, and F. F. E. Brünnow and H. L. d'Arrest. The perihelion date was quickly narrowed down to December 14.19. Although G. P. Bond (1850) calculated a hyperbolic orbit with an eccentricity of 1.00035303, Z. Sekanina (1978) showed that the orbit was elliptical with a period of 6800 years.

T	ω	Ω (2000.0)	i	q	e
1844 Dec. 14.1914 (TT)	177.5055	120.5909	45.5651	0.250537	0.999302

ABSOLUTE MAGNITUDE: $H_{10} = 4.9$ (V1964)

FULL MOON: Nov. 24, Dec. 24, Jan. 23, Feb. 22, Mar. 23

SOURCES: D. F. J. Arago, *CR*, **20** (1845 Mar. 3), p. 575; G. Webbe, W. H. Simms, J. Robinson, J. J. Waterston, T. G. Taylor, J. Donald, and W. Wilson, *MNRAS*, **6** (1845 Mar. 14), pp. 206–10; C. H. F. Peters, E. Cooper, and A. C. Petersen, *AN*, **23** (1845 Mar. 25), pp. 17–22; *PMJS* (Series 3), **26** (1845 Mar.), p. 271; C. H. F. Peters, F. F. E. Brünnow, and H. L. d'Arrest, *AN*, **23** (1845 Apr. 1), pp. 43–6; T. Maclear, W. Mann, and J. Caldecott, *MNRAS*, **6** (1845 Apr. 11), pp. 213–18; E. O. Kendall, J. S. Hubbard, and F. Bradley, *AJS* (Series 1), **48** (1845 Apr.), pp. 402–3; T. Maclear, E. Wilmot, W. Pole, and J. R. Hind, *MNRAS*, **6** (1845 May 9), pp. 234–8; J. R. Hind, *AN*, **23** (1845 May 17), pp. 119–22; J. R. Hind, G. B. Airy, E. Wilmot, and G. Webbe, *AN*, **23** (1845 May 24), pp. 133, 141–4; C. P. Smyth, J. F. Encke, T. Maclear, W. Mann, J. Dayman, and E. W. Beazley, *MNRAS*, **6** (1845 Jun. 13), pp. 242–5, 249, 252–4; T. Maclear and W. Mann, *AN*, **23** (1845 Jun. 14), pp. 173–6; J. R. Hind and J. Caldecott, *AN*, **23** (1845 Jun. 21), pp. 177–82; J. R. Hind and J. C. Haile, *AN*, **23** (1845 Aug. 7), p. 227; J. R. Hind, P. P. King, and R. Sheppard, *AN*, **23** (1845 Aug. 23), p. 247; C. H. F. Peters, *AN*, **23** (1845 Sep. 20), pp. 301–4; A. Colla, *AJS* (Series 1), **49** (1845 Oct.), p. 220; T. Maclear and W. Mann, *MRAS*, **15** (1846), pp. 251–5; T. Maclear, *MNRAS*, **9** (1849 Apr. 13), pp. 130–3; G. P. Bond, *AJ*, **1** (1850 Jul. 26), pp. 97–103; V1964, p. 56; Z. Sekanina, *AJ*, **83** (1978 Jan.), p. 66.

C/1844 Y2
(d'Arrest)

1845 I

Discovered: 1844 December 28.80 ($\Delta = 0.80$ AU, $r = 0.93$ AU, Elong. $= 61°$)

Last seen: 1845 March 30.88 ($\Delta = 0.84$ AU, $r = 1.64$ AU, Elong. $= 126°$)

Closest to the Earth: 1844 October 18 (0.9968 AU), 1845 February 18 (0.2144 AU)

Calculated path: CYG (Disc), LYR (Jan. 9), CYG (Jan. 14), LYR (Jan. 16), DRA (Jan. 22), UMa (Feb. 12), CVn-UMa (Feb. 17), LMi (Feb. 21), LEO (Feb. 24), HYA (Mar. 4)

H. L. d'Arrest (Berlin, Germany) discovered this comet in the evening sky near 15 Cygni on 1844 December 28.80 at a position of $\alpha = 19^h$ 36.6^m, $\delta = +36°$ 19'. His colleague, J. F. Encke said the position was not considered accurate because the sky clouded after only one measurement. Encke added that the nebulosity was rather bright and was visible in a cometseeker.

A. C. Petersen (Altona, Germany) looked for the comet on January 3. He began with the position reported by d'Arrest for the 28th and noted no trace of a comet. He then began sweeping for the object and found a faint nebulosity north of the discovery position. A. Reslhuber (Kremsmünster, Austria) found the comet near the horizon on January 28 and described it as a large nebulosity, without a distinct nucleus, and no noticeable tail. The comet attained its most northerly declination of +62° on February 11.

The comet was last detected on March 30.88, when Encke described it as a "quite faint cloud." He gave the position as $\alpha = 9^h$ 02.9^m, $\delta = -11°$ 44'.

The first parabolic orbits were independently calculated by C. K. L. Rümker and F. W. A. Argelander, using positions from December 28, January 3, and 10. Rümker determined the perihelion date as 1845 January 8.74, while Argelander determined it as January 8.71. Rümker pointed out a similarity between this orbit and that computed for the comet of 1779. These orbits turned out to be very good representations of the comet's motion, with later calculations by M. L. G. Wichmann, H. A. E. A. Faye, d'Arrest, W. C. Götze, F. B. G. Nicolai, J. R. Hind, J. E. B. Valz, and J. Kowalczyk (1873) only slightly changing the perihelion date to January 8.66.

With observations spanning three months, nothing more than a parabolic orbit should be considered for this comet. Nevertheless, three astronomers have calculated hyperbolic and even elliptical orbits over the years. J. Sievers (1845) took positions spanning the period of December 28 to February 10 and determined an eccentricity of 1.0003323. A. W. Doberck (1874) took positions obtained between January 3 and March 12, applied perturbations by two planets, and computed parabolic and hyperbolic orbits. The parabolic orbit was little different than those calculated by earlier astronomers. The hyperbolic orbit had an eccentricity of 1.000247. R. J. Buckley (1979) used 69 positions obtained between January 9 and March 12, and computed a parabolic orbit with a perihelion date of January 8.66. No planetary perturbations were considered. Buckley said the comet was "indifferently observed" with poor accuracy at the ends of the observation arc. Although he supplied an elliptical orbit with a period of about 4.2 million years and a hyperbolic orbit with an eccentricity of 1.000783, he preferred the parabolic orbit, which is given below.

T	ω	Ω (2000.0)	i	q	e
1845 Jan. 8.6579 (TT)	114.5819	338.9135	46.8605	0.905204	1.0

ABSOLUTE MAGNITUDE: $H_{10} = 9.2$ (V1964)

FULL MOON: Dec. 24, Jan. 23, Feb. 22, Mar. 23, Apr. 22

SOURCES: H. L. d'Arrest, J. F. Encke, A. C. Petersen, and C. K. L. Rümker, AN, **22** (1845 Jan. 21), p. 343; F. W. A. Argelander, AN, **22** (1845 Feb. 22), p. 377; F. W. Bessel and M. L. G. Wichmann, AN, **23** (1845 Mar. 11), p. 5; C. K. L. Rümker and R. Snow, MNRAS, **6** (1845 Mar. 14), pp. 210–11; F. B. G. Nicolai, C. K. L. Rümker, H. A. E. A. Faye, and J. Sievers, AN, **23** (1845 Mar. 25), pp. 21–32; H. L. d'Arrest and C. K. L. Rümker, AJS (Series 1), **48** (1845 Apr.), p. 403; J. F. Encke and H. L. d'Arrest, AN, **23** (1845 Apr. 19), p. 81; J. E. B. Valz, AN, **23** (1845 May 3), p. 97; W. Lassell, C. K. L. Rümker, J. Glaisher, and J. B. Reade, MNRAS, **6** (1845 May 9), pp. 239–40; J. F. Encke, MNRAS, **6** (1845 Jun. 13), p. 249; W. C. Götze and F. B. G. Nicolai, AN, **23** (1845 Jun. 14), pp. 167–72; J. R. Hind, AN, **23** (1845 Jun. 26), p. 197; J. R. Hind, AN, **23** (1845 Aug. 7), pp. 225–8; A. Reslhuber, AN, **23** (1845 Aug. 23), p. 249; W. Lassell, MRAS, **15** (1846), pp. 239–40, 242; J. Kowalczyk, AN, **81** (1873 Mar. 22), pp. 136–43; A. W. Doberck, MNRAS, **35** (1874 Dec.), p. 104; A. W. Doberck, AN, **85** (1875 Mar. 13), pp. 205–8; V1964, p. 56; R. J. Buckley, JBAA, **89** (1979 Apr.), pp. 260, 262.

C/1845 D1 (de Vico)

1845 II

Discovered: 1845 February 25.93 ($\Delta = 0.65$ AU, $r = 1.50$ AU, Elong. $= 131°$)

Last seen: 1845 May 1.85 ($\Delta = 0.86$ AU, $r = 1.26$ AU, Elong. $= 85°$)

Closest to the Earth: 1845 March 21 (0.4736 AU)

Calculated path: UMa (Disc), LMi (Mar. 12), LEO (Mar. 17), CNC (Mar. 18), HYA (Apr. 1), MON (Apr. 7), PUP (Apr. 14)

F. de Vico (Rome, Italy) discovered this telescopic comet on 1845 February 25.93 at a position of $\alpha = 11^h 44.0^m$, $\delta = +55° 05'$. He confirmed his find on February 26.76. The comet was independently discovered by H. A. E. A. Faye (Paris, France) on March 7.02. He described it as a telescopic comet with a distinct central condensation, but no tail.

Other observers noted the distinct nuclear condensation throughout March and into April. C. A. F. Peters (Altona, Germany) noted it on March 14 and 15, and estimated it as about 15" across on the last date. During the period of March 29 to April 7, W. Lassell (Starfield Observatory, Liverpool, England) said the comet had "an evident nucleus."

A. Reslhuber (Kremsmünster, Austria) said moonlight weakened the comet "almost up to indiscernibility" on April 13. Reslhuber noted the comet was uncommonly faint on the 24th and attributed this in part to its nearness to the horizon and interference from twilight.

The comet was last detected on May 1.85, when J. E. B. Valz (Marseille, France) determined the position as $\alpha = 7^h 42.4^m$, $\delta = -25° 12'$.

The first parabolic orbit was calculated by Faye. Using positions spanning the period of February 25 to March 8, he determined the perihelion date as 1845 April 13.34. During early April additional orbits by J. Sievers, Faye, J. R. Hind, W. C. Götze, and Valz had determined perihelion dates within the range of April 21.48–21.56. But not all of the calculated orbits were parabolic. During April, T. Clausen took positions spanning the period of February 26 to March 26 and calculated an elliptical orbit with a perihelion date of April 23.50 and a period of 33.07 years. During September, C. Jelinek and K. Hornstein took a very short arc of March 13 to April 9 and determined a hyperbolic orbit with an eccentricity of 1.0039886. The most accurate orbit was calculated by A. Scheller (1902). Using 130 positions spanning the period of February 25 to April 23, as well as perturbations by four planets, he computed a parabolic orbit with a perihelion date of April 21.54. This orbit is given below.

T	ω	Ω (2000.0)	i	q	e
1845 Apr. 21.5404 (UT)	205.4516	349.2811	56.4038	1.254549	1.0

ABSOLUTE MAGNITUDE: $H_{10} = 7.5$ (V1964)

FULL MOON: Feb. 22, Mar. 23, Apr. 22, May 21

SOURCES: F. de Vico and H. A. E. A. Faye, *AN*, **23** (1845 Mar. 25), pp. 27–30; H. A. E. A. Faye and J. F. Encke, *MNRAS*, **6** (1845 Apr. 11), pp. 214–15; *PMJS* (Series 3), **26** (1845 Apr.), p. 361; J. Sievers, *AN*, **23** (1845 Apr. 12), p. 67; T. Clausen, *AN*, **23** (1845 Apr. 19), pp. 93–6; W. Lassell, J. Glaisher, J. B. Reade, and J. R. Hind, *MNRAS*, **6** (1845 May 9), pp. 238–9; J. R. Hind and W. C. Götze, *AN*, **23** (1845 May 17), pp. 119, 125–8; C. Jelinek and K. Hornstein, *AN*, **23** (1845 May 24), p. 143; J. F. Encke, *MNRAS*, **6** (1845 Jun. 13), p. 250; J. E. B. Valz, *AN*, **23** (1845 Jun. 14), p. 171; J. R. Hind, *AN*, **23** (1845 Jul. 26), p. 223; A. Reslhuber, *AN*, **23** (1845 Aug. 23), pp. 249–52; C. Jelinek and K. Hornstein, *AN*, **23** (1845 Sep. 18), pp. 277–80; C. H. F. Peters, *AN*, **23** (1845 Sep. 20), p. 303; H. A. E. A. Faye, *AJS* (Series 1), **49** (1845 Oct.), p. 220; W. Lassell, *MRAS*, **15** (1846), pp. 242–4; J. E. B. Valz, *AN*, **23** (1846 Jan. 10), p. 385; A. Scheller, *AN*, **157** (1902 Jan. 18), pp. 317–18; V1964, p. 56.

C/1845 L1 (Great June Comet)

1845 III

Discovered: 1845 June 2.10 ($\Delta = 0.85$ AU, $r = 0.42$ AU, Elong. $= 24°$)

Last seen: 1845 July 2.08 ($\Delta = 1.47$ AU, $r = 0.76$ AU, Elong. $= 29°$)

Closest to the Earth: 1845 June 6 (0.8059 AU)

Calculated path: PER (Disc), AUR (Jun. 7), LYN (Jun. 15), CNC (Jun. 22)

A. Colla (Paris, France) discovered this comet on June 2.10, "just below the head of Medusa" in Perseus. He said it was almost visible to the naked eye in twilight, but an easy object in a terrestrial telescope with a magnification of 25×. Colla said the nucleus was "very brilliant" and the tail extended about 1° towards the north-northeast. Many other independent discoveries were made. On the morning of June 2 (between 2 and 3 a.m.), an anonymous observer in the USA said the tail was very distinct and brilliant, with a length of about 1°. He added that the nucleus was about equal to Capella in brightness. E. B. Rice (aboard the British warship HMS *Rodney*) spotted the comet with the naked eye on June 3. G. P. Bond (Cambridge, Massachusetts, USA) found the comet on June 3.34, and was the first to give a position, which was $\alpha = 3^h\ 27.6^m$, $\delta = +38°\ 15'$. At about midnight on June 6, Richter (Berlin, Germany) saw the comet near Capella. He said it exhibited a tail, and shone at a brightness comparable to that of a 3rd-magnitude star. The comet was found with the naked eye by Raht (Komotau, now Chomutov, Czech Republic) and M. L. G. Wichmann (Königsberg, now Kaliningrad, Russia) on June 7. Many additional naked-eye discoveries were reported during the next few days.

Several months after the comet had faded from view, the 1845 October issue of the *American Journal of Science* reported that Bennett (on pilot boat *Aid*, at Norfolk, Virginia, USA) had seen the comet in Perseus at 3 a.m. on May 31. If this is an accurate observation, then with a universal time date of June 1.34, it would be the earliest observation of this comet.

The comet was closest to the sun and Earth on June 6 and began fading thereafter. J. F. Encke (Berlin, Germany) observed the comet with a magnification of 320× on June 7 and saw a "planet-like nucleus." He added that the tail extended about 1° in twilight. F. Schaub (Vienna, Austria) saw the comet on June 7 and 8, and noted a bright nucleus and a tail about 2.5° long. J. Bianchi (Modena, Italy) saw the comet in twilight on June 8 and noted it was as bright as Capella (magnitude 0.08 – SC2000.0) or Jupiter (magnitude about −2).

Several astronomers provided descriptions of the comet on June 11. E. Plantamour (Geneva, Switzerland) saw the comet with the naked eye and a telescope. He said it exhibited a very brilliant nucleus about 36" across and a tail 4–5° long. A. C. Petersen (Altona, Germany) said the comet was well observed with the naked eye. C. K. L. Rümker (Hamburg, Germany) said the nucleus was 10" across. A. Reslhuber (Kremsmünster, Austria) easily found the comet in twilight and moonlight. He said a telescope revealed an almost planetary nuclear condensation of about magnitude 2–3. Following the setting of the moon, he noted a narrow, fan-shaped tail about 3° long.

The comet steadily faded during the remainder of June. G. Santini (Padova, Italy) saw the comet with a telescope on the 14th and noted the bright nucleus, but it was not visible to the naked eye because of moonlight. Reslhuber described the comet as very faint on June 27. F. W. A. Argelander (Bonn, Germany) provided an interesting series of observations. He said the comet seemed brighter on the 11th than on 10th, but remarked on how the comet and particularly the tail had faded by the 12th. On the 15th, with some poor conditions, Argelander noted the head was again quite bright, though smaller than when previously seen, while the tail was fainter, but more fan-shaped. Argelander observed the comet with binoculars on the 16th and said it was comparable in brightness to a star of magnitude 3.

The comet was last detected on July 2.08, by J. H. C. Coffin (U.S. Naval Observatory, Washington, DC, USA) at a position of $\alpha = 8^h\ 51.4^m$, $\delta = +24°\ 29'$. Reslhuber was unable to find the comet on July 1 or 2.

The first parabolic orbit was calculated by B. Peirce. Using early positions obtained by Bond, he determined the perihelion date as 1845 June 6.04. This orbit was a good representation of the comet's motion, with the calculations of W. C. Götze, Funk, E. Mailly, F. B. G. Nicolai, E. Schubert, E. O. Kendall, J. Downes, J. R. Hind, A. Reslhuber, J. C. Houzeau, C. Jelinek and K. Hornstein, G. Santini, J. Bianchi, Encke, H. L. d'Arrest, and H. A. Peck (1904) increased the perihelion date to 6.18. Peck's orbit is given below.

An elliptical orbit was also independently determined by d'Arrest and Peck. D'Arrest determined the period as 250 years. He noted the similarity between the orbit of this comet and that of comet C/1596 N1 and suggested they might be one and the same. Peck, however, said the elliptical orbit would have a period of about 313 000 years and said the link to 1596 was not valid.

T	ω	Ω (2000.0)	i	q	e
1845 Jun. 6.1832 (UT)	75.7977	339.9729	131.0985	0.401077	1.0

ABSOLUTE MAGNITUDE: $H_{10} = 4.0$ (V1964)

FULL MOON: May 21, Jun. 19, Jul. 19

SOURCES: M. L. G. Wichmann, C. F. Gauss, J. Bianchi, Funk, C. K. L. Rümker, and J. R. Hind, *MNRAS*, **6** (1845 Jun., supplement), pp. 1–2; C. K. L. Rümker and R. Snow, *MNRAS*, **6** (1845 Jun. 13), pp. 254–5; Richter, A. Colla, A. C. Petersen, J. F. Encke, C. K. L. Rümker, W. C. Götze, Funk, and M. L. G. Wichmann, *AN*, **23** (1845 Jun. 26), pp. 197–200, 207; F. Schaub, J. R. Hind, E. Plantamour, F. B. G. Nicolai, E. Schubert, and M. L. G. Wichmann, *AN*, **23** (1845 Jul. 26), pp. 213, 217–20, 223; C. F. Gauss, J. R. Hind, H. L. d'Arrest, and F. W. A. Argelander, *AN*, **23** (1845 Aug. 7), pp. 225–8, 231–6; A. Reslhuber, J. C. Houzeau, C. Jelinek, and K. Hornstein, *AN*, **23** (1845 Aug. 23), pp. 251–6; G. Santini, E. B. Rice, K. Kreil, Raht, E. Mailly, *AN*, **23** (1845 Sep. 6), pp. 265–8, 271; Bennett, G. P. Bond, B. Peirce, E. O. Kendall, and J. Downes, *AJS* (Series 1), **49** (1845 Oct.), pp. 220–1; J. Bianchi, *AN*, **23** (1845 Oct. 2), pp. 309–14; H. L. d'Arrest, *AN*, **23** (1845 Oct. 30), pp. 349–52; J. S. Hubbard and J. H. C. Coffin, *AJ*, **1** (1850 Oct. 15), p. 134; H. A. Peck, *AJ*, **24** (1904 Feb. 1), pp. 17–25; V1964, p. 56; SC2000.0, p. 112.

2P/Encke

1845 IV

Recovered: 1845 July 5.33 ($\Delta = 1.67$ AU, $r = 0.89$ AU, Elong. $= 27°$)
Last seen: 1845 July 15.09 ($\Delta = 1.55$ AU, $r = 0.72$ AU, Elong. $= 22°$)
Closest to the Earth: 1845 August 31 (1.1840 AU)
Calculated path: AUR (Rec)

The prediction for this comet was published by J. F. Encke in the 1845 April 19 issue of the *Astronomische Nachrichten*. He began with the 1829 orbit, applied perturbations by Mercury to Saturn, and determined the perihelion date as 1845 August 10.13. The comet was recovered by S. C. Walker (Central High School Observatory, Philadelphia, Pennsylvania, USA) in morning twilight on 1845 July 5.33. He measured the position as $\alpha = 5^h 02.0^m$, $\delta = +29° 27'$ on July 5.34. Independent recoveries were made by F. de Vico (Rome, Italy) on July 10.09 and J. H. C. Coffin (Washington, DC, USA) on July 11.36. De Vico said the comet was difficult to see because of its low altitude and morning twilight. Coffin described the comet as nebulous and elliptical, with a diameter of 3–4″. He added that it was very faint in twilight.

This was not a particularly favorable apparition for this comet and its steadily decreasing solar elongation made it a very difficult object to follow. The comet was last observed on July 15.09, when de Vico measured the position as $\alpha = 6^h\, 04.6^m$, $\delta = +29°\, 27'$. He noted the comet was difficult to observe. A. Reslhuber (Kremsmünster, Austria) was unsuccessful at finding the comet at this return.

Minor refinements of this comet's orbit using positions from multiple apparitions, as well as planetary perturbations, have been published by numerous astronomers. Of most significance were the studies of Encke, F. E. von Asten (1877), and B. G. Marsden and Z. Sekanina (1974). These have typically revealed a perihelion date of August 10.12 and a period of 3.30 years. Marsden and Sekanina determined the nongravitational terms as $A_1 = +0.38$ and $A_2 = -0.03503$. The orbit of Marsden and Sekanina is given below.

T	ω	Ω (2000.0)	i	q	e
1845 Aug. 10.1164 (TT)	183.3941	336.5094	13.1472	0.338078	0.847444

ABSOLUTE MAGNITUDE: $H_{10} = 8.3$ (V1964)

FULL MOON: Jun. 19, Jul. 19

SOURCES: J. F. Encke, AN, **23** (1845 Apr. 19), pp. 81–8; A. Reslhuber and F. de Vico, AN, **23** (1845 Aug. 23), pp. 254–6; S. C. Walker, AN, **24** (1846 Jun. 11), p. 133; J. H. C. Coffin, AN, **24** (1846 Jun. 11), p. 145; J. S. Hubbard and J. H. C. Coffin, AJ, **1** (1850 Apr. 20), p. 56; J. H. C. Coffin and J. S. Hubbard, AJ, **1** (1850 Oct. 15), p. 134; F. E. von Asten, BASP, **22** (1877), p. 561; V1964, p. 56; B. G. Marsden and Z. Sekanina, AJ, **79** (1974 Mar.), pp. 415–16.

C/1846 B1
(de Vico)

1846 I

Discovered: 1846 January 24.91 ($\Delta = 0.81$ AU, $r = 1.48$ AU, Elong. $= 111°$)
Last seen: 1846 May 2.04 ($\Delta = 2.12$ AU, $r = 2.01$ AU, Elong. $= 70°$)
Closest to the Earth: 1846 January 18 (0.8038 AU)
Calculated path: ERI (Disc), TAU (Jan. 30), AUR (Feb. 28), LYN (Apr. 12)

F. de Vico (Rome, Italy) discovered this comet in the evening sky near 38 Eridani on 1846 January 24.91. The position was determined as $\alpha = 4^h\, 07.0^m$, $\delta = -7°\, 12'$. De Vico acquired additional positions on January 27, 30, and 31.

Moonlight prevented additional observations until February 14, at which time astronomers in both Berlin and Hamburg were able to measure the position. J. F. J. Schmidt (Bonn, Germany) said the comet was small and fairly bright on the 18th, with a coma 3′ across. J. South (Kensington Campdenhill Observatory, England) described it as "a comet's ghost" on February 26th, as seen in a 12-cm refractor. At the end of February, Schmidt noted the "brightness not very considerable," while the coma had increased to 8' across. A. Reslhuber (Kremsmünster, Austria) summarized the comet's general appearance during late February by noting it had a small nucleus and a short parabolic tail. F. W. A. Argelander (Bonn, Germany) said the comet was faint in the 13-cm refractor on April 1, 3, and 5, because of moonlight. He said the comet was exceptionally faint on April 27, 28, 29, and May 2. The last date was also the final time the comet was detected and Argelander gave the position as $\alpha = 7^h$ 43.0^m, $\delta = +48°$ 23' on May 2.04.

One of the earliest parabolic orbits was calculated by J. E. B. Valz. Using positions acquired through February, he determined the perihelion date as January 23.93. Similar orbits by J. R. Hind, A. Graham, K. L. von Littrow, Neumann, A. Reslhuber, and J. A. C. Oudemans (1847), established the perihelion date as around January 22.6.

C. Jelinek (1848) re-examined the orbit of this comet using 57 positions spanning the period of January 30 to May 2. He calculated both parabolic and elliptical orbits. The parabolic orbit was very similar to those calculated by other astronomers. The elliptical orbit had a perihelion date of January 22.59 and a period near 2721 years. The elliptical orbit was preferred and is given below.

T	ω	Ω (2000.0)	i	q	e
1846 Jan. 22.5939 (UT)	337.9898	113.2744	47.4256	1.480703	0.992403

ABSOLUTE MAGNITUDE: $H_{10} = 6$ (V1964)

FULL MOON: Jan. 12, Feb. 11, Mar. 13, Apr. 11, May 11

SOURCES: J. E. B. Valz, *CR*, **22** (1846 Mar. 9), p. 424; F. de Vico, and H. C. Schumacher, *AN*, **24** (1846 Mar. 16), pp. 25–8; J. R. Hind, A. Graham, and F. de Vico, *AN*, **24** (1846 Apr. 2), pp. 34–6, 43; J. South, *AN*, **24** (1846 Apr. 8), p. 63; K. L. von Littrow and Neumann, *AN*, **24** (1846 Jul. 2), pp. 187–90; J. F. J. Schmidt, *AN*, **24** (1846 Jul. 23), pp. 257–60; F. W. A. Argelander, *AN*, **24** (1846 Aug. 13), p. 293; J. A. C. Oudemans, *AN*, **25** (1847 Feb. 25), pp. 199–204; A. Reslhuber, *AN*, **25** (1847 Apr. 3), pp. 277–82; J. F. Encke, *AN*, **26** (1847 Jun. 14), p. 3; C. Jelinek, *CR*, **26** (1848 Feb. 28), pp. 280–1; V1964, p. 56.

3D/Biela *Recovered:* 1845 November 26.73 ($\Delta = 0.95$ AU, $r = 1.42$ AU, Elong. $= 95°$)

 Last seen: 1846 April 27.93 ($\Delta = 0.65$ AU, $r = 1.42$ AU, Elong. $= 116°$)

1846 II *Closest to the Earth:* 1846 March 20 (0.3711 AU)

 Calculated path: PEG (Rec), PSC (Dec. 12), CET (Jan. 21), ERI (Feb. 23), LEP (Mar. 15), CMa (Mar. 23), PUP (Apr. 1), HYA (Apr. 9)

G. Santini (1835) had predicted the comet would arrive at perihelion on 1839 July 23.51, but the comet was not well placed and no observations were made. Santini continued working on the orbit of 3D/Biela and predicted the next perihelion date as 1846 February 11.87. He provided an ephemeris for the period of 1845 November 23 to 1846 May 6. Using a 61-cm reflector, W. Lassell (Starfield Observatory, Liverpool, England) unsuccessfully searched for this comet around predicted positions during 1845 November, but it was recovered by F. de Vico (Rome, Italy) on 1845 November 26.73. J. G. Galle (Berlin, Germany) independently recovered this comet with a refractor on November 28.72, at a position of $\alpha = 22^h 27.3^m$, $\delta = +3° 40.5'$. He described it as a very faint nebulosity, with a faint point-like nucleus.

Although numerous positions were measured during December, clues to the comet's physical appearance are at a minimum. In fact, from Galle's simple comment that the comet was easily seen when the reticle was illuminated for measurement on December 21, it is only apparent that the comet was brightening.

A major discovery was made just before mid-January. M. F. Maury (Washington, DC, USA) described the comet as faint in moonlight on the 13th. Upon reobserving the comet on the 14th with the 23-cm refractor he noted the comet appeared double. He said a faint companion was located just over 1' north of the comet. The Author has adopted the modern nomenclature of referring to the southern comet as "comet A" and the northern comet as "comet B" for the remainder of the discussion of this comet. Maury said comet A was well seen, while comet B was too faint for accurate measurement. He added that comet B was one-eighth the magnitude of comet A and one-fourth the intensity. M. L. G. Wichmann (Königsberg, now Kaliningrad, Russia) independently discovered comet B on the 15th and described it as a faint nebula near the comet. Meanwhile, a few hours later, Maury said both comets were well observed and separated by 1.3'. He noted that averted vision showed a tail on each comet that extended roughly parallel to one another towards the northeast. Two earlier discoveries of comet B may exist. First, E. C. Herrick and F. Bradley (Yale College Observatory, New Haven, Connecticut, USA) observed the comet with a 13-cm refractor on December 30 and noticed it was "attended by a faint nebulous spot preceding, estimated to be rather more than a minute of space distant from its brightest point." Second, J. Challis (Cambridge, England) determined the precise position of the comet on five days during the period of December 1–30, but J. S. Hubbard later said that the positions from December 1 and 3 would better represent the probable position of comet B. Since, comet B was later shown to vary considerably in brightness, December observations cannot be ruled out.

After the announcement of a double comet, many observers turned their telescopes toward this comet. Maury said the comets were separated by 2.1' on the 19th. He added that comet A exhibited a tail extending a few arc minutes toward the northeast, while comet B's tail was shorter, but nearly parallel. Both comets were well condensed towards their centers. E. O. Kendall and S. C. Walker (Philadelphia, Pennsylvania, USA) first detected the companion on January 20, but noted it was so faint, the illumination of the measuring wires rendered it invisible. Challis and Maury independently observed the comet on the 23rd and each measured the separation between the two comets as 2.4′. Challis added that comet B was located toward PA 327.7°. During the remainder of January, numerous astronomers watched the two comets slowly separate from one another, with J. F. Encke (Berlin, Germany) reporting comet B was located 3.0′ from comet A toward PA 331.6° on the 30th. Several astronomers continued to provided general descriptions of the comet's appearance. Maury said comet B exhibited two tails on January 24, one extending parallel to the main comet's, while the second extended towards a point just south of comet A's nucleus. F. W. Bessel (Königsberg, now Kaliningrad, Russia) said comet A was 3–4 times brighter than comet B on the 26th and 27th. John Herschel saw the comet on January 28 and said the comet "was evidently double, consisting of two distinct nebulae a larger and a smaller, both round or nearly so, the preceding faint and small and not much brighter in the middle, the following nearly three times as bright and perhaps 1 1/2 times as large in diameter and a good deal brighter in the middle with an approach to a bright stellar point." He added that no tail was present, though the atmosphere was hazy. Maury observed a point-like nucleus within comet A on the 29th, and "caught glimpses of a point of condensed light" within comet B.

Measurements of the separation between the two comets dominated the published descriptions for February, with astronomers obtaining values in close agreement with one another. As the month began, the separation was given as 3.4′. This increased to 4.5′ by February 10, 6.1′ on February 20, and 8.2′ on February 28. Astronomers were also reporting that the brightnesses of the two comets were changing day to day, if not more frequently, while features such as the nuclei and tails also exhibited changes. On February 4 J. F. J. Schmidt (Bonn, Germany) said the companion was "considerably faint and small", while Maury and Walker noted the brightnesses of the two comets had become nearly equal. Maury said each comet contained "a decided stellar nucleus," and "tails reaching almost across the field and nearly parallel." Maury and Walker said the nucleus of comet A seemed to be flashing. On February 6, Schmidt said the comets were similar in size and brightness. On February 9, C. K. L. Rümker (Hamburg, Germany) said the comet appeared very faint because of moonlight. On February 10, Maury said comet B exhibited two-thirds more light than comet A, while Schmidt said the comets were similar in size and brightness. On February 12, Maury said comet B had one-third more light than the main comet. He added that comet B exhibited a "decidedly stellar" nucleus, while comet A exhibited a diffuse and fainter nucleus. On February 13, Maury said comet A appeared to possess two nuclei, while a greater magnification showed a possible third nucleus. Each of these nuclei seemed to possess a tail which extended towards comet B. Comet B exhibited a stellar nucleus, with a greater magnification showing a second nucleus. Each of these nuclei seemed to exhibit a tail which extended towards comet A. Comet A was larger, and each comet exhibited tails that extended towards the northeast. On February 17,

Maury observed the comets in twilight and noted comet B became visible first. He estimated it had one-third more light than comet A and exhibited "a diamond-like nucleus." On February 19, Maury and Walker began observing in twilight and noted comet A was twice as bright as comet B and exhibited a star-like nucleus. In dark skies, the comets' tails appeared and were described as parallel, and the "tail" extending from comet B to comet A was arched. Comet A's nucleus was diamond-like, with Maury noting a possible "fragment of a nucleus above...." Comet B's nucleus was not as distinct as on previous days. On February 21, Schmidt said comet B was only half as large as comet A, but similar in surface brightness. He added that the space between the comets was completely dark. On February 22, Maury said the comets were "pearly white," with comet B being one-half as bright as comet A. The nucleus of comet A was described as "a ragged condensation of light." Walker said the comets were "very white," with comet A appearing one-third brighter than comet B. He caught one or two glimpses of a diamond-like nucleus in comet A, and noted the main tails of both comets were white, parallel, and extended across the field of view. On February 23, Maury described the comets as "pearly white," with comet A appearing larger and comet B appearing more intense. Comet A seemed to exhibit two nuclei, with "An arch way of cometary matter [extending] between the two nuclei." The tail of comet A extended 45' and he said a second tail was visible that was inclined at an angle of 120° to the first. Maury also said, "a band of nebulous matter, a little arched, joins the two [comets]." On February 26, Schmidt said the comet was visible with the naked eye. He added that the large nebulosity appeared irregular in form, while the companion seemed fan-shaped. On February 27, Maury said comet A was twice as bright as comet B. Comet A was "pearly white," with a tripod tail. One tail extended toward comet B, the main tail extended in a direction nearly opposite to the sun, and the third tail extended in a direction away from comet B. Comet B appeared a "dirty reddish" color. Maury added, "Confused appearance about the nucleus of Biela [comet A] as though there were several nuclei."

Astronomers were still carefully measuring the distance between the two comets during March. The distance was generally given as 8.7' on March 1, 10.8' on March 10, and 13.4' on March 20. The comets also continued to exhibit change in brightness and appearance. On March 2, Lassell said, "The first or preceding comet was incomparably fainter than the second, perhaps not one-fourth the brightness. The second comet had a bright nucleus, the first scarcely any." On March 4, Maury said comet A was six times brighter than comet B. On March 5, Maury said comet A was not well observed, while comet B was "too faint for tolerable observations." On March 6, Maury said comet B was "exceedingly faint" with no condensation or nucleus. Schmidt said comet B was hardly discernible in the moonlight. It was about 3' across, while comet A was near 10' across. On March 9, Maury said comet A exhibited a main tail, as well as a second tail pointing directly east. Comet B was described as a "faint reddish muddy light with a bright spark towards the center." On March 11, Maury said comet B was one-twentieth as bright as comet A, with no stellar nucleus apparent in either comet. With the moon in the sky, no tails were visible. On March 12, Maury observed in moonlight and detected comet A, but not comet B. On March 15, Maury said comet B contained a "very faint reddish nucleus," but the whole comet disappeared after moonrise. He added, "Think I discover cometary fragments about Biela [comet A]. There appear to me to be three and they are more distinctly marked tonight than I have seen them before." Maury's assistant, E. Blunt thought there were five fragments. On March 18, Maury saw no fragments around comet A. He added that comet B was so faint it would have been invisible if comet A had not been used as a guide. He described comet B as "a very diffused mass of exceedingly faint nebulous matter." A. Reslhuber (Kremsmünster, Austria) saw the comet with the naked eye in moonlight during the period of March 20–30. On March 22, Maury said comet B was visible before comet A. Comet B was described as very faint and muddy, with "a shining point in a dim patch of light about its nucleus." On March 27, James South (Kensington, London) described the comet as extremely faint and irregular in shape, with no nucleus. Challis said that comet B "was so excessively faint as to be seen with the greatest difficulty." On March 30, Schmidt said comet B was extraordinarily small and faint, with a diameter no greater than 30". On March 31, Maury and Walker were unsuccessful in their search for comet B. Comet A exhibited two tails.

With the comet having passed close to Earth on March 20, its brightness seemed to fade rapidly. Walker last detected the comet with the 24.4-cm refractor on April 20, and Galle described the comet as very faint on the 24th. The comet was last detected on April 27.93, by F. W. A. Argelander (Bonn, Germany). The position was determined as $\alpha = 9^h\ 58.1^m$, $\delta = -11° 03'$.

Orbits for each comet have been calculated by several astronomers over the years. What was known as the primary or "A" comet was investigated by E. Plantamour (1846), Hubbard (1853, 1860), H. L. d'Arrest (1855), J. von Hepperger (1900, 1903), and B. G. Marsden and Z. Sekanina (1971), with the result being a perihelion date of February 11.49 and a period of 6.60 years. The secondary or "B" comet was investigated by the same group, with the result being a perihelion date of February 11.58 and a period of 6.60 years. Marsden's orbits are given below. The first orbit represents comet A, while the second is comet B.

T	ω	Ω (2000.0)	i	q	e
1846 Feb. 11.4942 (TT)	223.0580	248.1420	12.5753	0.856440	0.756599
1846 Feb. 11.5782 (TT)	223.0616	248.1381	12.5767	0.856460	0.756608

ABSOLUTE MAGNITUDE: $H_{10} = 8.0$ (V1964)

FULL MOON: Nov. 14, Dec. 13, Jan. 12, Feb. 11, Mar. 13, Apr. 11, May 11

SOURCES: G. Santini, AN, **12** (1835 Jan. 14), p. 117; G. Santini, AJS (Series 1), **28**

(1835 Jul.), p. 398; G. Santini, *AN*, **21** (1843 Dec. 7), p. 171; J. G. Galle, *AN*, **23** (1846 Jan. 10), pp. 379–82; F. de Vico, *AN*, **23** (1846 Jan. 10), pp. 387–90; J. F. Encke and J. G. Galle, *AN*, **23** (1846 Jan. 31), pp. 401–4; M. F. Maury, S. C. Walker, and G. Santini, *AJS* (Series 2), **1** (1846 Mar.), pp. 293–4; J. R. Hind, H. C. Schumacher, M. L. G. Wichmann, and C. K. L. Rümker, *AN*, **24** (1846 Mar. 16), pp. 19–28; F. de Vico, J. F. Encke, J. R. Hind, J. G. Galle, O. W. Struve, and J. Challis, *AJS* (Series 2), **1** (1846 May), pp. 446–7; J. R. Hind and W. Lassell, *AN*, **24** (1846 May 2), p. 67; J. G. Galle, *AN*, **24** (1846 May 21), p. 113; S. C. Walker, M. F. Maury, and J. S. Hubbard, *AN*, **24** (1846 Jun. 11), pp. 133–40; T. Brorsen, C. K. L. Rümker, and J. South, *AJS* (Series 2), **2** (1846 Jul.), p. 138; T. Maclear, *AN*, **24** (1846 Jul. 2), p. 181; M. L. G. Wichmann, M. F. Maury, and S. C. Walker, *AJS* (Series 2), **2** (1846 Nov.), pp. 435–8; E. Plantamour, *AN*, **25** (1846 Dec. 17), pp. 117–28; A. Reslhuber, *AN*, **25** (1847 Apr. 3), p. 277; J. F. Encke, *AN*, **26** (1847 Jun. 14), p. 1; T. Maclear, W. Mann, and H. Campion, *MNRAS*, **10** (1849 Nov. 9), pp. 8–15; M. F. Maury, J. H. C. Coffin, and S. C. Walker, *AJ*, **1** (1850 Oct. 15), pp. 135–6; J. S. Hubbard, *AJ*, **3** (1853 Apr. 19), pp. 57–64; J. S. Hubbard, *AJ*, **3** (1853 Apr. 25), pp. 65–8; J. S. Hubbard, *AJ*, **3** (1853 May 18), pp. 73–7, 79–80; J. S. Hubbard, *AJ*, **3** (1853 Jun. 27), pp. 89–94; J. F. J. Schmidt, *AN*, **37** (1853 Oct. 25), pp. 255–60; O. W. Struve, *AN*, **37** (1853 Nov. 1), p. 281; H. L. d'Arrest, *AN*, **39** (1855 Jan. 17), pp. 329–32; J. S. Hubbard, *AJ*, **6** (1860 Jul. 30), pp. 130–4; J. S. Hubbard, *AJ*, **6** (1860 Aug. 15), pp. 137–9; J. von Hepperger, *SAWW*, **109** Abt. IIa (1900), pp. 299–382; J. von Hepperger, *SAWW*, **112** Abt. IIa (1903), pp. 1329–76; V1964, p. 56; B. G. Marsden and Z. Sekanina, *AJ*, **76** (1971 Dec.), pp. 1138–41.

122P/1846 D1
(de Vico)

1846 IV

Discovered: 1846 February 20.76 ($\Delta = 1.08$ AU, $r = 0.72$ AU, Elong. $= 40°$)
Last seen: 1846 May 20.16 ($\Delta = 1.48$ AU, $r = 1.54$ AU, Elong. $= 138°$)
Closest to the Earth: 1846 February 15 (1.0730 AU)
Calculated path: CET (Disc), PSC (Feb. 26), AND (Mar. 21), CAS (Apr. 4), CEP (Apr. 29)

F. de Vico (Rome, Italy) discovered this comet close to 36 Ceti near the end of evening twilight on 1846 February 20.76. He gave the position as $\alpha = 0^h\ 57.9^m$, $\delta = -7°\ 29'$. De Vico described it as small, with a little tail, and noted a northward motion. News of this comet was slow to spread out of Rome and an independent discovery was made by G. P. Bond (Cambridge, Massachusetts, USA) on the evening of February 26. The comet was situated about 3.5° from the three-day-old moon on February 28.

The editor of the *Astronomische Nachrichten* (1846 Apr. 2) reported that when news of de Vico's discovery arrived on March 4 there was a suspicion that it was an earlier discovery of the comet found by T. Brorsen on February 26 (see 5D/1846 D2), since it was in the same area of the sky. Taking an orbit calculated by A. C. Petersen for Brorsen's comet and determining a position for February 20 revealed Brorsen's comet was several degrees away from the discovery position given by de Vico. Therefore, two comets had been found near each other within a few days. Interestingly, Brorsen independently discovered this comet on March 8.

The comet was widely observed during March. M. F. Maury (Washington, DC, USA) noted a tail 15′ long on the 11th. C. K. L. Rümker (Hamburg, Germany) noted the comet exhibited a large nucleus and considerable nebulosity on the 12th. He added that the tail was not seen with certainty because of moonlight and twilight. J. F. J. Schmidt (Bonn, Germany) said the coma was 3′ across on the 21st. He added that it was extremely bright and white, with a hardly distinguishable trace of tail. Schmidt noted that of the three comets then observable, one of which was visible to the naked eye, this was the brightest.

The comet continued to be observed during April and May, but physical descriptions were generally not provided. Moonlight prevented observations of the comet during the period of April 6–12 and May 6–18.

The comet was last observed on May 20.15 by Bond and on May 20.16 by S. C. Walker. Walker measured the position as $\alpha = 21^h\,08.8^m$, $\delta = +74°\,56′$.

The first parabolic orbits were calculated by J. J. E. Goujon and Bond. Goujon used positions spanning the period of February 20 to March 5 and determined the perihelion date as 1846 March 6.47. He noted a similarity to the orbit of the comet of 1707. Bond used positions spanning the period of February 26 to March 4 and determined the perihelion date as March 6.04. Additional calculations by F. Bradley and Bond indicated a perihelion date late on March 5.

Before the end of March, it was becoming obvious that this comet was moving in an elliptical orbit. B. Peirce calculated the first such orbit. Using positions spanning the period of February 26 to March 18, he determined the perihelion date as March 6.00 and the period as 94.91 years. Peirce said the period could be several years off, "but if the comet should be visible as late as June or July, I think it may be determined within a year." Later calculations would reveal a perihelion date of March 6.05, but there was still some uncertainty in the period, with H. Breen giving it as 75.66 years and J. R. Hind giving it as 55.43 years. After all of the positions had come in, additional attempts were made to pinpoint the period. A. J. van Deinse (1849, 1850) determined values of 72.74 and 73.25 years. J. von Hepperger (1887) determined a period of 75.71 years, with an uncertainty of three years.

This comet was virtually ignored until 1979, when R. J. Buckley and I. Hasegawa wrote papers mentioning it. Buckley took 98 positions obtained during the period of February 27 to May 20, applied perturbations by Venus to Neptune, and determined the period as 76.30 years. He suggested an uncertainty of ±1 year. Hasegawa was actually examining the motion of several ancient and medieval comets, when he found that one seen in 1391 had a path very consistent with what would be expected for de Vico's comet. He said the likely mean period between 1391 and 1846 was 75.8 years. Following the comet's accidental rediscovery in 1995, S. Nakano (1999) determined the period of the 1846 return as 76.06 years. This orbit is given below.

T	ω	Ω (2000.0)	i	q	e
1846 Mar. 6.0473 (UT)	12.9131	79.7003	85.1106	0.663779	0.963022

ABSOLUTE MAGNITUDE: $H_{10} = 7.2$ (V1964), 6.9 (Kronk)

FULL MOON: Feb. 11, Mar. 13, Apr. 11, May 11, Jun. 9

SOURCES: J. J. E. Goujon, *CR*, **22** (1846 Mar. 9), pp. 426–7; F. de Vico, A. C. Petersen, T. Brorsen, and C. K. L. Rümker, *AN*, **24** (1846 Apr. 2), pp. 40–6; G. P. Bond, W. C. Bond, B. Peirce, and F. Bradley, *AJS* (Series 2), **1** (1846 May), pp. 447–8; G. P. Bond and B. Peirce, *AN*, **24** (1846 May 16), p. 91; G. P. Bond, *AJS* (Series 2), **2** (1846 Jul.), p. 137; H. Breen, *AN*, **24** (1846 Jul. 2), p. 181; J. F. J. Schmidt, *AN*, **24** (1846 Jul. 23), p. 259; A. C. Petersen, *AN*, **24** (1846 Aug. 17), p. 327; J. R. Hind, *AN*, **24** (1846 Sep. 19), p. 381; C. K. L. Rümker, *AN*, **24** (1846 Sep. 19), pp. 389–92; J. F. Encke, *AN*, **26** (1847 Jun. 14), p. 3; A. J. van Deinse, *AN*, **29** (1849 Jul. 20), p. 129; A. J. van Deinse, *AN*, **30** (1850 May 16), p. 305; M. F. Maury and S. C. Walker, *AJ*, **1** (1850 Nov. 4), p. 137; M. F. Maury, *Washington Astronomical Observations: 1846*, **2** (1851), p. 334; J. von Hepperger, *SAWW*, **95** Abt. II (1887), pp. 870–912; J. von Hepperger, *AN*, **117** (1887 Aug. 13), p. 245; V1964, p. 56; R. J. Buckley, *JBAA*, **89** (1979 Apr.), pp. 260, 262; S. Nakano, *Nakano Note* No. 724 (1999 Nov. 19).

5D/1846 D2

(Brorsen)

1846 III

Discovered: 1846 February 26.81 ($\Delta = 0.67$ AU, $r = 0.65$ AU, Elong. $= 40°$)

Last seen: 1846 April 22.95 ($\Delta = 0.59$ AU, $r = 1.17$ AU, Elong. $= 91°$)

Closest to the Earth: 1846 March 27 (0.5211 AU)

Calculated path: PSC (Disc), AND (Mar. 10), CAS (Mar. 17), CEP (Mar. 29), DRA (Apr. 11)

T. Brorsen (Holstein, Germany) discovered this telescopic comet on 1846 February 26.81, near η Piscium. He estimated the position as α = 0^h 52.0^m, δ = $+14°$ 25′. A. C. Petersen (Altona, Germany) confirmed the discovery on February 28.80.

The comet had just passed perihelion when discovered and although it was widely observed, few physical descriptions were made. J. F. J. Schmidt (Bonn, Germany) provided the only useful details. He described the comet as "a formless white nebulosity" on March 9 and said the coma was 3–4′ across. Because the comet passed relatively close to Earth near the end of March, Schmidt noted the coma was 8–10′ across on March 25. M. L. G. Wichmann (Königsberg, now Kaliningrad, Russia) saw the comet on March 25 and 26, and described it as "a very indefinite, faint and diffuse nebulosity, without a nucleus." On April 3, Schmidt observed the comet despite the presence of a full moon.

The comet was last detected on April 22. F. W. A. Argelander (Bonn, Germany) saw the comet on April 22.01, while J. F. Encke (Berlin, Germany) found it on April 22.95. Argelander said the comet appeared very faint, partly because of an 8th-magnitude star in the same field of view. Encke described the comet as an uncommonly diffuse and faint nebulosity, which vanished even by the light of the faintest stars. Encke gave the position as $\alpha = 17^h\ 14.4^m$, $\delta = +70°\ 42'$.

Petersen took positions obtained by himself on February 28 and March 2, as well as a position obtained by C. K. L. Rümker on March 1, and calculated a parabolic orbit with a perihelion date of 1846 February 27.16. Additional parabolic orbits were determined by Funk, K. R. Powalky, and Petersen that pushed the perihelion date to around February 27.9; however, it was also becoming increasingly obvious that the comet was not moving in a parabolic orbit.

Elliptical orbits were soon to follow. J. F. Encke used positions obtained between February 28 and March 7, and computed an elliptical orbit with a perihelion date of February 24.49 and a period of 3.44 years. J. R. Hind used three positions obtained between February 28 and March 10, and computed an elliptical orbit with a perihelion date of February 25.83 and a period of 5.52 years. Hind's orbit proved closer to the truth and additional computations by J. J. E. Goujon, F. F. E. Brünnow (1849), Hind (1848,1849), and P. van Galen (1856) gradually refined the period to about 5.6 years. Hind also examined the probable close approach to Jupiter in 1842 and found the closest approach between the comet and this planet would have been 0.085 AU on May 19. No planetary perturbations were applied to obtain the details of this encounter. Hind added, "The similarity between the orbit of Brorsen's comet and those of the comets of 1532 and 1661, long supposed to be identical, may of itself be considered sufficient inducement to some one accustomed to these intricate computations [perturbations by Jupiter] to undertake the solution of the question with all possible accuracy."

Minor refinements of this comet's orbit using positions from multiple apparitions, as well as planetary perturbations, have been published by several astronomers. K. C. Bruhns (1868), B. G. Marsden and Z. Sekanina (1971), and W. Landgraf (1986) all established the perihelion date as February 25.87 and the period as 5.57 years. Marsden and Sekanina used 110 positions from the apparitions spanning 1846–73 and determined nongravitational terms of $A_1 = +0.32$ and $A_2 = -1.1545$. Landgraf took 102 positions from the apparition spanning 1846 to 1868 and determined nongravitational terms of $A_1 = +0.51$ and $A_2 = -0.0651$. Landgraf's orbit is given below.

T	ω	Ω (2000.0)	i	q	e
1846 Feb. 25.8685 (TT)	13.8141	104.8124	30.9143	0.650123	0.793070

ABSOLUTE MAGNITUDE: $H_{10} = 7.2–8.2$ (V1964)

FULL MOON: Feb. 11, Mar. 13, Apr. 11, May 11

SOURCES: T. Brorsen, A. C. Petersen, C. K. L. Rümker, Funk, and K. R. Powalky, *AN*, **24** (1846 Apr. 2), pp. 39–42; J. Encke, *AN*, **24** (1846 Apr. 2), p. 45; J. J. E. Goujon, *CR*, **22** (1846 Apr. 11), pp. 642–3; J. R. Hind, *AN*, **24** (1846 May 2), pp. 68–70; F. W. A. Argelander and J. F. Encke, *AN*, **24** (1846 May 21), pp. 111–14; T. Brorsen, and J. Encke, *AJS* (Series 2), **2** (1846 Jul.), p. 137; J. F. J. Schmidt, *AN*, **24** (1846 Jul. 23),

p. 259; F. W. A. Argelander, *AN*, **24** (1846 Aug. 13), p. 293; M. L. G. Wichmann, *AN*, **25** (1846 Dec. 10), p. 45; J. R. Hind, *CR*, **26** (1848 Jun. 5), pp. 604–5; J. R. Hind, *MNRAS*, **9** (1849 Jun. 8), pp. 203–4; F. F. E. Brünnow, *AN*, **29** (1849 Oct. 22), pp. 324–32; F. F. E. Brünnow, *AN*, **29** (1849 Nov. 12), pp. 377–9; P. van Galen, *AN*, **44** (1856 Oct. 17), pp. 311–20; K. C. Bruhns, *AN*, **71** (1868 Mar. 12), pp. 37–40; V1964, p. 56; B. G. Marsden and Z. Sekanina, *AJ*, **76** (1971 Dec.), pp. 1141–2; W. Landgraf, *QJRAS*, **27** (1986 Mar.), p. 116; personal correspondence from W. Landgraf (1993).

C/1846 J1 *Discovered:* 1846 May 1.01 ($\Delta = 0.37$ AU, $r = 0.97$ AU, Elong. $= 74°$)
(Brorsen) *Last seen:* 1846 June 15.90 ($\Delta = 1.52$ AU, $r = 0.67$ AU, Elong. $= 20°$)
  Closest to the Earth: 1846 May 6 (0.3022 AU)
1846 VII *Calculated path:* PEG (Disc), VUL–PEG (May 1), CYG (May 2), PEG–CYG–LAC (May 3), CAS (May 6), CAM (May 11), LYN (May 16), AUR (May 18), LYN (May 19), AUR (May 21), LYN (May 22), AUR (May 26), GEM (Jun. 9)

T. Brorsen (Holstein, Germany) discovered this comet on 1846 May 1.01, at a position of $\alpha = 21^h 18.4^m$, $\delta = +22.4°$. He said it resembled a large, round nebulosity. There was no tail, nor a recognizable nucleus. An independent discovery was made by M. L. G. Wichmann (Königsberg, now Kaliningrad, Russia) on May 1.96. Wichmann described the comet as a large and round nebulosity, with a quite bright central condensation. No tail was visible and the comet was easily seen in a cometseeker. Because of the slow communications across the Atlantic Ocean, an independent discovery was also made by G. P. Bond (Cambridge, Massachusetts, USA) on May 20.21. He simply described it as a telescopic comet.

The comet passed closest to Earth on May 6 and continued heading toward the sun. J. F. J. Schmidt (Bonn, Germany) saw the comet despite a full moon on May 8. Using a 5-foot focal length telescope at a magnification of $300\times$, he noted the comet was distinct, with no nucleus and a weak condensation. However, there were two branches of material extending from the coma on the side opposite the sun. F. Kaiser (Leiden, Netherlands) saw the comet with the naked eye after the moon had set on May 13 and 14. Schmidt detected weak traces of a tail on the 13th. On May 19 and 20, Schmidt said the comet was distinctly visible to the naked eye and appeared like a star of magnitude 6, while a telescope revealed weak traces of a tail. Because news of Brorsen's discovery did not arrive in the USA in a timely fashion, the large refractors of that country were not able to begin observations until after Bond's announcement had reached them. Schmidt saw traces of a tail on the 23rd and noted a thin tail about 20′ long on the 24th.

As June began, the comet was entering evening twilight, because of a steadily decreasing solar elongation. It was measured for the final time on June 12.91, when Kaiser gave the position as $\alpha = 6^h 57.9^m$, $\delta = +34° 02′$. Kaiser also saw the comet on June 13, 14, and 15. On the 15th, he noted the sky was extremely clear and he saw the comet "in one instant," very near the horizon and said it appeared so faint that he gave up hope of trying to find it thereafter. Because of the short time the comet was above the horizon at Leiden during this final observation, the probable universal time of the observation was June 15.90.

The first parabolic orbit was calculated by Petersen. Using positions from May 1, 2, and 3, he determined the perihelion date as 1846 June 5.75. He noted some similarity to the orbits of comets seen in 1701 and 1766. Additional calculations by H. L. d'Arrest, B. Peirce, F. F. E. Brünnow, J. R. Hind, H. Breen, and J. A. C. Oudemans revealed very similar orbits.

The first elliptical orbit was calculated by Wichmann. He used positions spanning the period of May 1 to June 5 and gave the perihelion date as June. 6.02 and the period as about 401 years. Oudemans followed a few weeks later with the determination that the perihelion date was June 5.98 and the period was about 500 years. The last investigation into this comet's orbit was conducted by A. Krause (1912). He used 170 positions obtained between May 2 and June 12, and computed a perihelion date of June 5.97 and a period of about 538 years. This orbit is given below.

T	ω	Ω (2000.0)	i	q	e
1846 Jun. 5.9710 (UT)	99.7253	263.9890	150.6810	0.633760	0.990414

ABSOLUTE MAGNITUDE: $H_{10} = 8.1$ (V1964)

FULL MOON: Apr. 11, May 11, Jun. 9, Jul. 8

SOURCES: T. Brorsen, A. C. Petersen, and M. L. G. Wichmann, *AN*, **24** (1846 May 16), pp. 97–100; M. L. G. Wichmann, *AN*, **24** (1846 May 21), p. 115; H. L. d'Arrest and F. F. E. Brünnow, *AN*, **24** (1846 Jun. 20), pp. 153–6; G. P. Bond and B. Peirce, *AJS* (Series 2), **2** (1846 Jul.), p. 138; J. A. C. Oudemans and J. R. Hind, *AN*, **24** (1846 Jul. 4), pp. 207–10, 212; M. L. G. Wichmann, *AN*, **24** (1846 Jul. 16), pp. 239–44; J. F. J. Schmidt, *AN*, **24** (1846 Jul. 23), p. 259; J. A. C. Oudemans, *AN*, **24** (1846 Aug. 13), p. 297; F. Kaiser, *AN*, **24** (1846 Aug. 17), pp. 309–24; H. Breen and F. Kaiser, *AN*, **24** (1846 Sep. 19), pp. 383, 389; J. F. J. Schmidt, *AN*, **25** (1847 Apr. 3), p. 287; M. L. G. Wichmann, *AN*, **29** (1849 Oct. 30), pp. 345–50; M. F. Maury and S. C. Walker, *AJ*, **1** (1850 Nov. 4), p. 137; A. Krause, *Definitive Bahnbestimmung des Kemeten 1846 VII*. Publication der Astronomischen Gesellschaft, No. 24 (1912), 35pp; V1964, p. 56.

80P/1846 M1
(Peters–Hartley)

1846 VI

Discovered: 1846 June 26.92 ($\Delta = 0.62$ AU, $r = 1.52$ AU, Elong. $= 135°$)
Last seen: 1846 July 23.9 ($\Delta = 0.84$ AU, $r = 1.60$ AU, Elong. $= 118°$)
Closest to the Earth: 1846 June 4 (0.5503 AU)
Calculated path: LIB (Disc), SCO (Jul. 19), OPH (Jul. 23)

C. H. F. Peters (Capodimonte Observatory, Naples, Italy) discovered this comet with the refractor on 1846 June 26.92 at a position of $\alpha = 15^h 07.4^m$, $\delta = -21° 39'$. Although he said the comet was then in Scorpius, the position indicates it was actually in Libra. Peters described the comet as very faint. He noted a nebula situated a degree away that was classified as Herschel VI, no. 19 (NGC 5897, mag. 8.6, size=12.6′), and said the comet was nearly equal in brightness, but more concentric and without a nucleus. Peters confirmed his discovery on June 27.87 and June 28.90.

The conditions for the observations were not the best. Peters acknowledged that interference from haze on the first night, the general faintness of the comet and the lack of a nucleus, the wide field of the refractor, and moonlight on the 28th, all prevented him from obtaining accurate positions. As a support for this statement, he added that each night he made several measurements that sometimes varied by an arc minute from one another. Nevertheless, he took the position from these first three evenings and calculated a rough parabolic orbit with a perihelion date of 1846 April 12.80. This orbit was sent to several observers in Germany in the hope that they might recover the comet after the moon had left the sky.

Peters obtained additional observations on June 29, 30, and July 1, with moonlight becoming more of a problem each night. After July 1 moonlight prevented him from seeing the comet for more than a week. Word had yet to reach other observatories, but the comet hunter F. de Vico (Rome, Italy) independently discovered the comet on July 2. Unfortunately, clouds hampered his searches during the following nights.

After the moon had left the sky Peters recovered the comet on July 11.86 and measured its position on every evening until the 19th. He last measured the comet's position on July 21.87, and gave it as $\alpha = 15^h\ 55.2^m$, $\delta = -8° 19′$. Peters saw the comet for the final time on July 23.9, when he noted it was too faint for measurements. His final attempts to find the comet were made on the evening of July 29 and on the morning of July 30.

Unfortunately, by the time the moon was gone, the German observatories were not able to locate the comet, with unsuccessful searches reported by observers in Altona, Berlin, Königsberg (now Kaliningrad, Russia), and Hamburg. As it turned out, Peters' orbit suffered some large errors because of the rough positions, with his perihelion date being about a month and a half too early. Although this orbit represented his first three rough positions well, it became increasingly inadequate shortly after July began. The resulting errors would have amounted to 1° by July 10, 2° by July 15, and 3° by July 20. Peters admitted that his first orbit was bad and somewhat in error by the time searches began.

A much more accurate orbit was calculated by Peters using his positions obtained up to July 18. The revised orbit had a perihelion date of May 31.00. Unfortunately, this orbit was not mailed until July 30 and its first publication was in the September 3 issue of the *Astronomische Nachrichten*.

H. L. d'Arrest (1846) computed two different sets of parabolic elements, one using Peters' first three positions and the other using three Normal positions covering the complete observational arc. The first was similar in many ways to Peters' original orbit, but with a perihelion date of April 22.10. The second orbit revealed a perihelion date of May 30.11 and definitely fitted the observations better, but large discrepancies were still present. D'Arrest had greater success when he computed an elliptical orbit with a perihelion date of June 1.60 and a period of 15.89 years.

Peters (1848) confirmed the elliptical nature of the orbit and determined a perihelion date of June 1.71 and a period of 12.85 years. J. R. Hind (1871), noting the discrepancy in the calculated periods for this comet, suggested that the only chance of recovering it would be by accidental recovery by one of the people "who occupy themselves in sweeping the heavens for telescopic comets."

Later attempts were made to determine the period of this comet more accurately. A. Berberich (1887) used 16 positions obtained between June 26 and July 21, and computed an elliptical orbit with a perihelion date of June 1.63 and a period of 13.38 years. He suggested the period was uncertain by about one year and added that the orbit indicated close approaches to Saturn were made in 1856 and 1883. R. J. Buckley (1979) used 15 positions obtained between June 26 and July 21, as well as perturbations by Venus to Neptune, and determined a perihelion date of June 1.66 and a period of 12.71 years. Buckley also suggested an uncertainty of at least a year in the period.

Interestingly, following the comet's accidental rediscovery in 1982, M. P. Candy and B. G. Marsden (1982) found it actually moved in a much shorter period than had been determined from the positions of 1846. They linked the apparitions of 1846 and 1982, and applied perturbations by five planets. The result was a perihelion date of June 3.68 and a period of 7.88 years.

T	ω	Ω (2000.0)	i	q	e
1846 Jun. 3.683 (UT)	340.917	263.412	28.904	1.49838	0.62166

ABSOLUTE MAGNITUDE: $H_{10} = 8.0$ (V1964), $H_{10} = 7.7$ (Kronk)

FULL MOON: Jun. 9, Jul. 8, Aug. 7

SOURCES: C. H. F. Peters, *AN*, **24** (1846 Jul. 30), p. 261; C. H. F. Peters and F. de Vico, *AN*, **24** (1846 Sep. 3), pp. 357–60; C. H. F. Peters and H. L. d'Arrest, *AN*, **24** (1846 Sep. 19), pp. 387–90; C. H. F. Peters and H. L. d'Arrest, *AJS* (Series 2), **3** (1847 Mar.), p. 282; C. H. F. Peters and H. L. d'Arrest, *AN*, **28** (1848 Dec. 25), p. 139; J. R. Hind, *MNRAS*, **31** (1871 May), pp. 218; A. Berberich, *AN*, **117** (1887 Aug. 20), pp. 249–52; V1964, p. 56; R. J. Buckley, *JBAA*, **89** (1979 Apr.), pp. 260, 262; B. G. Marsden and M. P. Candy, *CCO*, 5th ed. (1986), pp. 13, 49.

C/1846 O1
(de Vico–Hind)

1846 V

Discovered: 1846 July 29.88 ($\Delta = 1.71$ AU, $r = 1.64$ AU, Elong. $= 69°$)
Last seen: 1846 October 18.8 ($\Delta = 1.71$ AU, $r = 2.41$ AU, Elong. $= 123°$)
Closest to the Earth: 1846 February 28 (0.9566 AU), 1846 September 13 (1.2470 AU)
Calculated path: CAM (Disc), CAS (Aug. 2), AND (Sep. 8), LAC (Sep. 16), PEG (Sep. 23)

F. de Vico (Rome, Italy) discovered this comet on 1846 July 29.88. He described it as circular, with a distinct nucleus. An independent discovery was made by J. R. Hind (Regent's Park, London) on July 29.97. He described it as "a round nebulosity or very nearly so, with a minute stellar point near the centre, but its brightness was not sufficient to allow of my seeing it in the comet-sweeper. I found this object with our large refractor." Hind gave the position as $\alpha = 3^h\ 15.6^m$, $\delta = +60°\ 37'$ on July 30.00. He indicated that he had watched the comet for several hours as the night progressed and at one time had observed it pass over a 12th-magnitude star, "which was hardly affected by the intervention of the comet."

The comet was a couple of months past perihelion when discovered and was fading. Hind measured the position of the comet on August 4, but noted it was "not very good owing to the faintness of the comet." H. L. d'Arrest (Berlin, Germany) observed the comet in moonlight with a refractor on the 5th, but noted it was not discernible in a seeker. On August 12 and 14, F. W. A. Argelander (Bonn, Germany) described the comet as rather faint and small, but with a bright central condensation. On August 14 and 16, Hind simply described the comet as faint. M. L. G. Wichmann (Königsberg, now Kaliningrad, Russia) saw the comet on August 16, 23, and 25, and said it was extremely faint and hardly visible.

Argelander said the comet was very faint in September, especially during his final observations of September 25 and 26. The comet's position was determined for the final time on September 26.93, when Argelander and J. F. J. Schmidt gave it as $\alpha = 22^h\ 14.8^m$, $\delta = +30°\ 37'$. The comet was last observed on October 18.8, when Schmidt barely saw it in a heliometer.

The first parabolic orbit was calculated by Hind. He used three positions obtained on July 30 and 31, and determined the perihelion date as 1846 May 14.61. New orbits were published a few days later by Funk and K. R. Powalky. Using positions spanning the period of July 30 to August 3, they determined perihelion dates of July 31.06 and August 1.45, respectively. After a few more days J. A. C. Oudemans determined a perihelion date of July 28.96. During the last half of August it was obvious the perihelion date occurred in late May. H. Niebour used positions obtained up to August 13 and determined a perihelion date of May 21.61. Later orbits by Argelander, T. Brorsen, A. Graham, and H. C. Vogel (1868) established the perihelion date as May 28. Vogel's orbit is given below.

T	ω	Ω (2000.0)	i	q	e
1846 May 28.3959 (UT)	78.7516	163.4639	122.3771	1.375992	1.0

ABSOLUTE MAGNITUDE: $H_{10} = 6.2$ (V1964)

FULL MOON: Jul. 8, Aug. 7, Sep. 5, Oct. 4, Nov. 3

SOURCES: J. R. Hind, G. B. Airy, J. F. Encke, Funk, K. R. Powalky, and de Vico, *AN*, **24** (1846 Aug. 17), pp. 325–30; F. W. A. Argelander, J. F. J. Schmidt, and J. R. Hind, *AN*, **24** (1846 Sep. 19), p. 381; H. Niebour, *AN*, **24** (1846 Sep. 19), p. 393; F. W. A. Argelander, *AN*, **25** (1846 Oct. 22), p. 83; J. R. Hind, *AJS* (Series 2), **2** (1846 Nov.), p. 439; A. Graham, *MNRAS*, **7** (1846 Nov.), pp. 160–1; T. Brorsen, *AN*, **25** (1846 Nov. 5), p. 97; M. L. G.

Wichmann, *AN*, **25** (1846 Dec. 10), p. 45; J. R. Hind, F. de Vico, Funk, K. R. Powalky, and J. A. C. Oudemans, *AJS* (Series 2), **3** (1847 Mar.), p. 282; F. W. A. Argelander, *AN*, **25** (1847 Apr. 3), p. 289; F. W. A. Argelander, *AN*, **63** (1864 Dec. 22), p. 287; H. C. Vogel, *AN*, **71** (1868 Apr. 7), pp. 97–102; V1964, p. 56.

C/1846 S1 (de   *Discovered:* 1846 September 23.80 ($\Delta = 0.72$ AU, $r = 1.07$ AU, Elong. $= 75°$)
Vico)   *Last seen:* 1846 October 26.11 ($\Delta = 1.05$ AU, $r = 0.83$ AU, Elong. $= 48°$)
   Closest to the Earth: 1846 September 20 (0.7175 AU)
1846 VIII   *Calculated path:* UMa (Disc), CVn (Oct. 10)

F. de Vico (Rome, Italy) discovered this comet in Ursa Major on 1846 September 23.80. He measured the position and indicated the comet was at $\alpha = 8^h\ 35.2^m$, $\delta = +64°\ 15'$ on September 23.89.

H. C. Schumacher conjectured in the 1846 November 5 issue of the *Astronomische Nachrichten* that moonlight and bad weather had prevented de Vico from obtaining further observations. As a result, the comet was independently discovered by M. L. G. Wichmann (Königsberg, now Kaliningrad, Russia) on October 15.80. Wichmann simply described the comet as telescopic on both the 15th and 16th. Thereafter, the comet was seen by A. C. Petersen (Altona, Germany) on the 21st, and by C. K. L. Rümker (Hamburg, Germany) on October 21, 23, 24, and 26. Rümker's last observation was also the final time the comet was seen anywhere. He gave the position as $\alpha = 13^h\ 04.6^m$, $\delta = +34°\ 08'$ on October 26.11. Observations ended because of moonlight.

The first orbit was calculated by K. R. Powalky. Using positions spanning the period of October 15–21, he determined a parabolic orbit with a perihelion date of 1846 October 29.95. As November began, independent orbits were published by J. R. Hind and Powalky. Hind used positions from September 23 to October 21 and determined the perihelion date as October 30.25. Powalky took positions from September 23 to October 26 and determined the perihelion date as October 30.28. Although A. Quirling (1847) calculated an elliptical orbit with a perihelion date of October 30.42 and a period of about 1381 years, an investigation by S. Oppenheim (1890) showed that the parabolic orbit was preferred because of the short period of observation. Oppenheim's orbit is given below.

T	ω	Ω (2000.0)	i	q	e
1846 Oct. 30.2772 (UT)	93.9764	6.8367	49.7196	0.830669	1.0

ABSOLUTE MAGNITUDE: $H_{10} = 8$ (V1964)
FULL MOON: Sep. 5, Oct. 4, Nov. 3
SOURCES: F. de Vico, M. L. G. Wichmann, A. C. Petersen, C. K. L. Rümker, and K. R. Powalky, *AN*, **25** (1846 Nov. 5), pp. 93, 99; C. K. L. Rümker and K. R. Powalky, *AN*, **25** (1846 Nov. 5), p. 99; J. R. Hind, *AN*, **25** (1846 Nov. 26), p. 111; M. L. G. Wichmann, *AN*,

25 (1846 Dec. 10), p. 47; F. de Vico, *AJS* (Series 2), **3** (1847 Jan.), p. 132; A. Quirling, *AN*, **25** (1847 Mar. 20), pp. 253–6; S. Oppenheim, *AN*, **125** (1890 Jul. 12), p. 31; V1964, p. 56.

X/1846 U1 J. R. Hind (Bishop's Observatory, England) discovered this telescopic comet in Coma Berenices on 1846 October 19.17. He described it as faint, 2–2.5′ across, with a bright central condensation. Hind determined the position as $\alpha = 11^h\ 59^m 49.1^s$, $\delta = +14°\ 59'\ 32''$ on October 19.18 and $\alpha = 11^h\ 59^m 57.5^s$, $\delta = +14°\ 59'\ 08''$ on October 19.21. He commented, "The comet is much fainter than that of July 29 [*see C/1846 O1*], this might be partly owing to its low altitude and the morning twilight." Hind attempted to observe the comet on the morning of October 20, but the skies remained overcast. The comet was not seen again.

F. A. T. Winnecke (1881) tried to link comet 72P/1881 T1 to Hind's comet after early orbits revealed that the 1881 comet moved in a short-period orbit. However, modern calculations reveal comet 72P would have passed perihelion on 1847 March 6, which would have placed it far from Hind's positions.

FULL MOON: Oct. 4, Nov. 3
SOURCES: J. R. Hind, *AN*, **25** (1846 Nov. 5), pp. 93–6; J. R. Hind, *MRAS*, **16** (1847), pp. 299–300; J. R. Hind, *AJS* (Series 2), **3** (1847 Jan.), p. 132; J. R. Hind, *AN*, **25** (1847 Feb. 25), pp. 205–8; F. A. T. Winnecke, E. Hartwig, and L. Wutschichowsky, *AN*, **101** (1881 Nov. 26), p. 77.

C/1847 C1 *Discovered:* 1847 February 6.88 ($\Delta = 1.17$ AU, $r = 1.49$ AU, Elong. $= 87°$)
(Hind) *Last seen:* 1847 April 24.90 ($\Delta = 1.78$ AU, $r = 0.90$ AU, Elong. $= 20°$)
Closest to the Earth: 1847 March 23 (0.8372 AU)
1847 I *Calculated path:* CEP (Disc), CAS (Feb. 24), AND (Mar. 6), PSC (Mar. 24), CET (Mar. 30), PSC (Mar. 31), ARI (Apr. 11), TRI (Apr. 14), ARI (Apr. 15), TRI (Apr. 17)

This comet was discovered in the evening sky by J. R. Hind (Regent's Park, London, England) on 1847 February 6.88. He gave the position as $\alpha = 21^h\ 09.2^m$, $\delta = +71°\ 26'$ on February 6.89. Hind described the comet as "excessively faint, without nucleus, and scarcely presenting the central condensation which is usual in telescopic comets."

The discovery announcement did not spread quickly to other observatories and Hind remained the only observer of this comet for over two weeks. He considered the positions he measured during the period of February 6–10 as uncertain because of the lack of a condensation. J. F. J. Schmidt (Bonn, Germany) saw the comet with a seeker on the 13th and described it as a large, bright, centrally condensed nebulosity. Hind reported the comet was much brighter on the 15th and noted it occasionally exhibited a nucleus. The nucleus was even more distinct on the 19th, but was not central within the coma. He noted a possible short tail. C. K. L. Rümker (Hamburg, Germany) made his first observation on the morning of February 22, and continued measuring the comet's position in the evenings and mornings of the following days. Hind noted the nucleus was still eccentrically placed within the coma on the 24th. On the 27th, Hind said the nucleus was about magnitude 11–12 and was more central. There was an area of very condensed nebulosity surrounding it. Schmidt observed the comet that morning and suspected he saw it with the naked eye after the moon had set.

As March began the comet could still be seen in the morning before sunrise and in the evening after sunset. It was also steadily brightening as it continued to approach both the sun and Earth. It was independently discovered by G. P. Bond (Cambridge, Massachusetts, USA) on March 4.99, and was described as nearly visible with the naked eye. He noted a slight central condensation and suspected traces of a tail. Schmidt saw the comet with the naked eye on the 5th and said it looked like a star of magnitude 6. Bond said the comet was visible to the naked eye on the 8th and noted it was similar to a star of magnitude 5–6. Hind said the comet was distinctly seen in strong twilight on the 9th. Later that same night, Hind noted a tail 1° long that appeared as a "narrow ray of light proceeding directly from the centre of the comet." Schmidt said the comet was similar in brightness to σ Andromedae (magnitude 4.52 – SC2000.0) on March 15 and 16, while Hind said the comet appeared "brighter than a star of 4th magnitude" on the 16th. The tail steadily grew during this same period, with Schmidt reporting it as 12′ long on the 5th and 30′ long on the 8th. Schmidt and Hind independently gave lengths of 40′ and 30′, respectively, on the 9th, while Schmidt gave lengths of 50′ on the 10th and 56′ on the 11th. Bond said the tail was 2° or 3° long in the cometseeker on the 12th. Schmidt said the tail extended 1° 50′ on March 15, while Bond said it was 4° or 5° long. On the 16th, Schmidt and Hind independently gave lengths of 3° 33′ and 90′ or 100′, respectively. Bond said the coma was 4′ or 5′ across on the 8th.

The comet was easily observed as the second half of March began, but soon became a difficult object as it dropped into twilight. Schmidt saw the comet on the evening of the 17th and noted the coma was sharply defined and parabolic in shape. It exhibited a fan-shaped tail 4.0° long. The next evening the shape was unchanged, although the tail had increased to 4.3°. In addition, he noted the brightness was similar to π Andromedae (magnitude 4.36 – SC2000.0). Encke saw the comet at an altitude of only 5° on March 22, while Bond estimated the altitude as just 3° on the 24th. Schmidt observed the comet in bright morning twilight on the 23rd and said the bright portion of the coma was 1.5′ across, while the tail was no longer visible.

Following March 24, the comet was lost from view for a few days. It passed only 0.8° from the sun on March 30.01 and then began steadily moving away. Hind found the comet near the sun in broad daylight, while using a large refractor (40×) and a light green glass to protect his eye from the sun's light, on March 30.46. He said, "The nucleus . . . was round or nearly so, beautifully defined and planetary. Two short rays of light formed a divided tail, not more than 40" in length. At times I felt certain that the nucleus twinkled like a fixed star." He added that G. Bishop and others visiting the observatory saw the comet "with tolerable ease" at this time. Hind estimated the comet's position on March 30.56 and 30.58. On the last date he wrote, "For this observation a glass throwing a yellowish-brown tint over the field was employed, and for moments the comet was beautifully distinct, its bright round nucleus seemed about 8" in diameter and the smoky tails were apparent by glimpses." The comet attained an elongation of 2.5° on March 30.85 and then turned back toward the sun. It passed 0.7° away on March 31.95.

The comet was not observed again until April 22, when J. G. Galle (Berlin, Germany) found it at a low altitude, near the end of evening twilight. The comet was last detected on April 24.83 by Galle and on April 24.90 by A. Graham (Markree Observatory, Ireland). Graham's last measured position was $\alpha = 2^{\mathrm{h}} 25.7^{\mathrm{m}}$, $\delta = +32° 09'$.

The first parabolic orbit was calculated by Hind, using positions obtained during the period of February 7–9. It indicated a perihelion date of 1847 February 15.76. A few weeks later, Hind apologized for the "entirely erroneous" orbit, stating that the "extremely small arc" and "unavoidable errors in the observations of such a faint object" led to the inaccurate orbit. A much better representation of the orbit was obtained by Graham, when he used positions from February 6, 10, and 15, and determined the perihelion date as March 30.85. This proved to be only two hours later than the actual perihelion date. Later positions enabled more refined orbits to be calculated by H. L. d'Arrest, Graham, Hind, B. Peirce, Schmidt, Bond, K. Hornstein, F. Carlini, W. W. Boreham, and N. R. Pogson. Hind was the first to note the possibility that the comet would be observed in daylight at the end of March.

The first elliptical orbit was calculated by A. J. F. Yvon-Villarceau. Using positions from February to the beginning of March, he determined the perihelion date as March 30.82 and the period as 552 years. A short time later, A. Quirling used a slightly longer arc and determined the perihelion date as March 30.81 and the period as 608 years. Using positions spanning the period of February 6 to April 24, Graham was the first to show the period was over ten thousand years. Hornstein (1854, 1870) continued to improve the orbit. His last calculation applied perturbations from five planets and determined a perihelion date of March 30.78 and a period of about 10 219 years. Hornstein's orbit is given below.

T	ω	Ω (2000.0)	i	q	e
1847 Mar. 30.7844 (UT)	254.3565	23.8238	48.6637	0.042593	0.999910

ABSOLUTE MAGNITUDE: $H_{10} = 6.8$ (V1964)

FULL MOON: Jan. 31, Mar. 2, Mar. 31, Apr. 30

SOURCES: A. Graham, *CR*, **24** (1847 Mar. 8), pp. 337–8; J. R. Hind and C. K. L. Rümker, *AN*, **25** (1847 Mar. 11), pp. 239–42; J. R. Hind, *MNRAS*, **7** (1847 Mar. 12), pp. 247–9; A. J. F. Yvon-Villarceau, *CR*, **24** (1847 Mar. 15), pp. 448–9; J. R. Hind, *AN*, **25** (1847 Mar. 20), pp. 251–4; A. Graham, *CR*, **24** (1847 Mar. 22), pp. 499–500; J. R. Hind and A. J. F. Yvon-Villarceau, *CR*, **24** (1847 Mar. 29), pp. 563–5; J. F. J. Schmidt and J. R. Hind, *AN*, **25** (1847 Apr. 3), pp. 285–92; A. Quirling, *AN*, **25** (1847 Apr. 15), p. 301; J. R. Hind, J. F. J. Schmidt, H. L. d'Arrest, and J. F. Encke, *AN*, **25** (1847 Apr. 24), pp. 313–16, 319–22; G. P. Bond, B. Peirce, and J. R. Hind, *AJS* (Series 2), **3** (1847 May), p. 443; J. R. Hind and G. Bishop, *AN*, **25** (1847 May 1), pp. 331–4; W. C. Bond and B. Peirce, *AN*, **25** (1847 May 6), p. 355; W. C. Bond, A. Graham, and G. P. Bond, *MNRAS*, **7** (1847 May 14), pp. 273–4; K. Hornstein, *AN*, **25** (1847 May 20), p. 373; A. Graham, *CR*, **24** (1847 May 24), pp. 900–1; F. Carlini, *AN*, **26** (1847 Jun. 14), p. 9; N. R. Pogson, *MNRAS*, **8** (1848 Jun. 9), p. 181; J. F. Encke and J. G. Galle, *AN*, **26** (1847 Jun. 14), p. 5; K. Hornstein, *AN*, **26** (1847 Aug. 12), pp. 99–102; W. W. Boreham, *AN*, **26** (1847 Aug. 26), p. 143; K. Hornstein, *AN*, **38** (1854 Jun. 27), pp. 323–8; K. Hornstein, *SAWW*, **62** Abt. II (1870), pp. 244–60; K. Hornstein, *AN*, **77** (1871 May 20), p. 303; V1964, p. 56; SC2000.0, pp. 8, 14.

C/1847 J1 *Discovered:* 1847 May 7.85 ($\Delta = 1.83$ AU, $r = 2.14$ AU, Elong. $= 93°$)

(Colla) *Last seen:* 1847 December 30.80 ($\Delta = 3.21$ AU, $r = 3.17$ AU, Elong. $= 78°$)

Closest to the Earth: 1847 March 23 (1.3078 AU)

1847 II *Calculated path:* LMi (Disc), UMa (May 31), DRA (Oct. 13), CYG (Dec. 20)

A. Colla (Parma, Italy) discovered this telescopic comet in Leo Minor on 1847 May 7.85, at a position of $\alpha = 10^{\rm h} 04^{\rm m}$, $\delta = +36.5°$. He described it as very faint and wrote that it was "a small nebulosity, almost circular, and without tail, but with some indication of a bright point visible at intervals in the central part of the nebulosity." J. R. Hind remarked in the 1847 July 15 issue of the *Astronomische Nachrichten*, "It is really surprising that Prof. Colla should have been able to detect so faint an object in sweeping."

Although the comet was observed at numerous observatories from May 14, very few physical descriptions were obtained. K. L. von Littrow (Vienna, Austria) provided several observations that indirectly indicate the comet's brightness. Using a 15-cm refractor, he saw the comet occult a star of magnitude 7 or 8 on July 8. He noted that the comet grew faint, but always remained visible. On July 17, Littrow said the comet was very faint because of its nearness to a star of magnitude 7.4 and on the 21st he noted that the comet's position could not be measured because of moonlight. On August 8, the comet reached a minimum solar elongation of 43°. Littrow saw the comet on August 2 and 9. Littrow said it was difficult to measure the comet's position on September 3, because of the proximity of two stars of magnitude 11 or 12. Fortunately he was able to differentiate the comet's nucleus. Littrow saw the comet quite well on September 8. Littrow and K. Hornstein

(Vienna, Austria) saw the comet with the 15-cm refractor on September 10, 13, and October 11, and noted it was nearing the limit of visibility on the last date. They described the comet as small and slightly elongated. J. Challis (Cambridge, England) saw the comet on five occasions during the period spanning November 24 to December 8. He said it was of the last degree of faintness. Littrow saw the comet with a 15-cm refractor on November 26. J. G. Galle (Berlin, Germany) described the comet as a very small, faint nebulosity on November 28. W. Lassell (Starfield Observatory, Liverpool, England) was the only observer of the comet after December 8, thanks to a 61-cm telescope. He saw the comet on December 14 and 28.

The comet was last detected on December 30.80, when Lassell saw it at the limit of visibility in a 61-cm telescope. He said the micrometer could not be illuminated because of the "excessive faintness of the comet." The position was nevertheless given as $\alpha = 19^h 54.6^m$, $\delta = +53° 54'$, although it should be noted that α was determined 41 minutes prior to δ. Littrow (1848) examined the brightness of this comet and noted it should have been visible on 1848 March 1 with a 30-cm refractor.

From the publication of the very first orbit, it became obvious that this comet had a perihelion distance larger than 2 AU. Add to this the fact that the comet was discovered about a month and a half after it had made its closest approach to Earth, and you have a very slow-moving comet. This made it very difficult for the early orbits to pin down the comet's perihelion date. The first parabolic orbit was calculated by J. J. E. Goujon, who took positions obtained on May 13, 17, and 20, and determined the perihelion date as 1847 May 14.03. A short time later, H. Niebour and A. Quirling independently calculated orbits using identical positions from May 13, 18, and 23, and determined perihelion dates of June 18.59 and June 18.99, respectively. H. L. d'Arrest then published an orbit using positions from May 18, 21, and 24, which revealed a perihelion date of June 6.47. Even though d'Arrest's orbit would later prove to be only a day later than the actual perihelion date, the orbit was really no more reliable than the other calculated orbits. This is best exemplified by the fact that d'Arrest published a set of three orbits less than a week later that relied on a longer observational arc of May 13 to June 3, which indicated perihelion dates ranging from June 12.63 to June 14.02. Hind took positions from May 13 to May 30 and calculated a perihelion date of May 30.93. As July began, Goujon took positions spanning the period of May 13 to June 7 and calculated a perihelion date of June 6.08. Later orbits by Littrow and E. Gautier (1848) determined perihelion dates of June 5.26 and June 5.19, respectively.

The orbit has been re-examined on two other occasions and has been found to be hyperbolic. F. Engström (1882) used 54 positions obtained between May 13 and December 30, as well as perturbations by seven planets, and computed a parabolic orbit with a perihelion date of June 5.1925. He used the same positions to also compute a hyperbolic orbit with a perihelion date of June 5.2333 and an eccentricity of 1.0006549. R. J. Buckley (1979) used 37 positions obtained between May 13 and December 30, as well as perturbations by Venus to Neptune, and computed a hyperbolic orbit with a perihelion date of June 5.23 and an eccentricity of 1.000723. This orbit is below.

T	ω	Ω (2000.0)	i	q	e
1847 Jun. 5.2298 (TT)	32.3587	176.0849	100.4185	2.115749	1.000723

ABSOLUTE MAGNITUDE: $H_{10} = 4.8$ (V1964)

FULL MOON: Mar. 31, Apr. 30, May 30, Jun. 28, Jul. 27, Aug. 26, Sep. 24, Oct. 23, Nov. 22, Dec. 21, Jan. 20

SOURCES: A. Colla and J. R. Hind, *MNRAS*, **7** (1847 May 14), pp. 267–8; J. J. E. Goujon, *CR*, **24** (1847 May 24), p. 901; A. Colla, K. L. von Littrow, C. K. L. Rümker, H. Niebour, A. Quirling, and H. L. d'Arrest, *AN*, **25** (1847 Jun. 9), pp. 395–8; J. R. Hind, H. L. d'Arrest, and A. Quirling, *MNRAS*, **7** (1847 Jun. 11), pp. 288–90; H. L. d'Arrest, *AN*, **26** (1847 Jun. 29), pp. 37–42; J. J. E. Goujon and A. Colla, *CR*, **25** (1847 Jul. 5), pp. 31–2; J. R. Hind, *AN*, **26** (1847 Jul. 15), p. 49; K. L. von Littrow, *AN*, **26** (1847 Aug. 5), p. 83; K. L. von Littrow, *CR*, **25** (1847 Aug. 9), pp. 257–8; K. L. von Littrow, *AN*, **26** (1847 Aug. 12), pp. 101–4; K. L. von Littrow and K. Hornstein, *AN*, **26** (1847 Oct. 14), p. 195; A. Colla and H. L. d'Arrest, *AJS* (Series 2), **4** (1847 Nov.), p. 426; K. L. von Littrow and K. Hornstein, *MNRAS*, **8** (1847 Nov. 12), p. 12; J. Challis and W. Lassell, *MNRAS*, **8** (1847 Dec. 10), pp. 26–7; K. L. von Littrow and K. Hornstein, *AN*, **26** (1847 Dec. 16), pp. 309–16; W. Lassell, *MNRAS*, **8** (1848 Jan. 14), p. 48; W. Lassell, *CR*, **26** (1848 Jan. 17), p. 109; E. Gautier, *AN*, **27** (1848 Feb. 24), p. 33; K. L. von Littrow, *AN*, **27** (1848 Jun. 1), p. 219; J. G. Galle, *AN*, **29** (1849 Jul. 17), p. 125; F. Engström, *VJS*, **17** (1882), pp. 293–7; V1964, p. 56; R. J. Buckley, *JBAA*, **89** (1979 Apr.), pp. 260, 262–3.

C/1847 N1
(Mauvais)

1847 III

Discovered: 1847 July 4.95 ($\Delta = 1.83$ AU, $r = 1.83$ AU, Elong. $= 74°$)

Last seen: 1848 April 22.08 ($\Delta = 3.04$ AU, $r = 3.52$ AU, Elong. $= 111°$)

Closest to the Earth: 1847 July 8 (1.8276 AU), 1848 February 11 (1.9898 AU)

Calculated path: CEP (Disc), DRA (Jul. 8), UMi (Jul. 13), CAM (Jul. 18), UMi (Jul. 21), CAM (Jul. 24), UMi (Jul. 25), DRA (Aug. 2), UMa (Aug. 9), CVn (Aug. 25), COM (Oct. 13), CVn (Oct. 23), COM (Oct. 26), LEO (Feb. 9)

F. V. Mauvais (Paris, France) discovered this comet near the border of Cepheus and Ursa Minor on 1847 July 4.95. Although the sky was almost immediately obscured, he was able to reobserve the comet about 2 hours later and said the coma was 4–5′ across and slightly oblong, with a rather distinct nucleus. A position of $\alpha = 22^h 08.2^m$, $\delta = +80° 26′$ was given for July 5.06. The comet was independently discovered by G. P. Bond (Cambridge, Massachusetts, USA) on July 15.

The comet was well placed for observations as it passed just over 4° from the North Celestial Pole on July 12. J. F. J. Schmidt (Bonn, Germany) saw the comet on July 15 and noted a coma at least 5.2′ across and a fan-shaped tail. Schmidt's colleague F. W. A. Argelander said the tail appeared at least 8′ long in a seeker. On the following nights, Schmidt failed to see the tail, which he assumed was because of poor transparency, and by July 21 he noted the coma was only 2.6′ across.

Schmidt estimated the tail length as 8–10′ on August 1 and 2. He remarked that the comet became fainter toward the end of August, but was still rather well observed up to the middle of September. He last detected a trace of the tail on September 12.

The comet had become a relatively faint object as October began, mostly because of its increasing distance from both the sun and Earth. In addition, the elongation from the sun decreased to just within 40° on October 8. These two items were responsible for ending most observations. The only observatories that continued making observations into October were Cambridge (Massachusetts), Bonn, Kremsmünster, and Berlin. None of the first three continued beyond the 16th, but J. G. Galle (Berlin, Germany) continued to measure the comet's position until November 2.

The comet's distance from Earth increased to 2.64 AU by the end of October and then it began decreasing again as Earth made its way around the sun and began catching up to the slowly-moving comet. The result was that the comet's fading slowed dramatically and it moved to a higher altitude in the sky. Schmidt wrote that astronomers in Paris reported they had observed the comet in December. Schmidt then began his search on the evening of the 12th and found the comet early on the morning of December 13. He described it as extremely faint and at the limit of the 5-foot focal length refractor. Thereafter, other observatories began turning their large telescopes toward the comet again.

Astronomers in Vienna (Austria) found the comet in February, with F. Schaub locating the comet on February 4 and K. L. von Littrow observing it on February 5. Schaub also saw the comet on the 13th and said it was near the limit of a 15-cm refractor. He added that it appeared rather large, with an almost invisible nucleus that scintillated very noticeably. G. P. Bond (Cambridge, Massachusetts, USA) said the comet was found without difficulty on February 26, near the position predicted by B. Peirce's elements. He indicated it was similar in size and brightness to the "nebula" h843 (NGC 3599, magnitude 11.9, size 2.8′ – NGC2000.0). He noted that the comet was bright enough to allow red illumination of the wire micrometer and added that the stellar point was still barely visible and "a faint, diffused, nebulous light" extended from it for "some distance."

Littrow saw the comet on March 2 and said it was easy to observe, but "showed abrupt light changes." W. C. Bond (Cambridge, Massachusetts, USA) noted the comet seemed to scintillate in brightness on March 2 and 25. W. Lassell (Starfield Observatory, Liverpool, England) observed the comet during a lunar eclipse on March 19. Lassell said the comet was almost in the same field as a faint nebula on March 31, as seen with the 20-foot focal length refractor. He remarked that he "estimated the comet at about half the brightness of the nebula." The nebula is probably the galaxy NGC 3121, which is magnitude 13.9 and 1.7' in diameter. Lassell added, "The comet appears to have a *very minute* stellar nucleus surrounded by nebulosity."

Lassell said the comet had become a faint object by April 3, but thought it might be visible again after the moon left the sky. J. Challis (Cambridge, England) last saw the comet on April 6 while using the 29.5-cm refractor at 120× and 160×. He noted the comet was so faint, no illumination could be used with the micrometer. The comet was last seen on April 22.08, when W. C. Bond determined the position as $\alpha = 9^h\ 37.7^m$, $\delta = +12°\ 29'$. He described the comet as faint and said, "It exhibited to the last the star-like scintillations which have all along distinguished it."

The first parabolic orbit was calculated by Schmidt. Using positions from July 5, 9, and 16, he determined a perihelion date of August 12.67. As more positions became available, orbital calculations by Mauvais, A. Quirling and H. Niebour, H. L. d'Arrest, and Littrow (1848) eventually narrowed the perihelion date down to August 9.94. Although many more observations were obtained well into 1848, no follow-up orbits were calculated for several years. Finally, E. Gautier (1852) began his investigation of the orbit, using positions spanning the period of July 12 to April 2. He found the orbit was elliptical, with a perihelion date of August 9.84 and a period of about 44 000 years.

T	ω	Ω (2000.0)	i	q	e
1847 Aug. 9.8386 (UT)	91.5248	340.4045	96.5817	1.766058	0.998589

ABSOLUTE MAGNITUDE: $H_{10} = 5.3$ (V1964)

FULL MOON: Jun. 28, Jul. 27, Aug. 26, Sep. 24, Oct. 23, Nov. 22, Dec. 21, Jan. 20, Feb. 19, Mar. 19, Apr. 18

SOURCES: F. V. Mauvais, *CR*, **25** (1847 Jul. 5), pp. 5–6; F. V. Mauvais, *CR*, **25** (1847 Jul. 26), pp. 149–50; F. V. Mauvais, A. Quirling, and H. Niebour, *AN*, **26** (1847 Aug. 2), p. 79; K. Hornstein, J. F. J. Schmidt, and F. W. A. Argelander, *AN*, **26** (1847 Aug. 12), pp. 103, 109–12; J. R. Hind and W. W. Boreham, *AN*, **26** (1847 Aug. 26), p. 143; H. L. d'Arrest, *AN*, **26** (1847 Aug. 30), p. 147; G. P. Bond, *AN*, **26** (1847 Sep. 16), p. 167; J. F. J. Schmidt, *AN*, **26** (1847 Sep. 30), pp. 177–80; F. V. Mauvais, *AJS* (Series 2), **4** (1847 Nov.), p. 426; W. W. Boreham, *MNRAS*, **8** (1847 Nov. 12), p. 11; J. F. J. Schmidt, *AN*, **26** (1847 Nov. 15), pp. 259–62; F. V. Mauvais, *CR*, **25** (1847 Nov. 22), pp. 733–4; J. F. J. Schmidt, *AN*, **26** (1847 Dec. 16), p. 319; J. F. J. Schmidt, *AN*, **26** (1848 Jan. 22), p. 373; K. L. von Littrow and F. Schaub, *CR*, **26** (1848 Feb. 28), pp. 279–80; J. Challis and G. P. Bond, *MNRAS*, **8** (1848 Mar. 10), pp. 128–9; K. L. von Littrow, *AN*, **27** (1848 Mar. 23), pp. 109–12; K. L. von Littrow, *CR*, **26** (1848 Apr. 3), p. 414; W. Lassell, J. Challis, G. Bishop, and J. R. Hind, *MNRAS*, **8** (1848 Apr. 14), pp. 143–4; W. C. Bond, *CR*, **26** (1848 May 1), p. 485; W. Lassell, *AN*, **27** (1848 May 2), pp. 171–6; W. C. Bond, *MNRAS*, **9** (1848 Nov. 10), p. 10; J. G. Galle, *AN*, **29** (1849 Jul. 17), pp. 125–7; E. Gautier, *CR*, **35** (1852 Dec.), pp. 948–9; E. Gautier, *MNRAS*, **13** (1853 Jan. 14), p. 86; E. Gautier, *AN*, **36** (1853 Jan. 21), pp. 77–82; V1964, p. 57; NGC2000.0 (1988), p. 100.

23P/1847 O1
(Brorsen–
Metcalf)

1847 V

Discovered: 1847 July 20.98 ($\Delta = 0.77$ AU, $r = 1.20$ AU, Elong. $= 83°$)
Last seen: 1847 September 13.13 ($\Delta = 1.22$ AU, $r = 0.49$ AU, Elong. $= 23°$)
Closest to the Earth: 1847 August 6 (0.6503 AU)
Calculated path: TRI (Disc), ARI (Jul. 21), TRI (Jul. 22), PER (Jul. 27), AUR (Aug. 7), LYN (Aug. 20), CNC (Aug. 26), LEO (Sep. 4)

T. Brorsen (Altona, Germany) discovered this comet near α Arietis on 1847 July 20.98, at a position of $\alpha = 1^h 49.7^m$, $\delta = +26° 07'$. He described it as a weak nebulosity. Confirmations of the new telescopic comet were made by A. C. Petersen (Altona, Germany) on July 22.00, and C. K. L. Rümker (Hamburg, Germany) on July 22.02. Petersen said the comet was exceedingly faint. Rümker described it as a faint and undefined form, without a discernible nucleus.

The comet passed closest to Earth early in August and then passed perihelion in September. Petersen described it as exceedingly faint on July 25, while J. R. Hind (Bishop's Observatory, London, England) observed the comet on July 27, 28, and 30, and described it as extremely faint. J. F. J. Schmidt and F. W. A. Argelander (Bonn, Germany) searched for the comet during the last days of July, but were not able to locate it because of its faintness; however, M. L. G. Wichmann (Königsberg, now Kaliningrad, Russia) observed the comet with a heliometer during the period spanning July 26 to August 11, and this indicated a brightness greater than magnitude 7. The comet was independently discovered between β and θ Aurigae by K. G. Schweizer (Moscow, Russia) on August 11. He described it as round, without a tail. Schmidt simply noted the comet was very bright on August 11, while on the 12th he saw the first traces of a tail and said it was very faint with respect to the coma. Schmidt said the comet was very bright on the 15th, while the tail was 15′ long on the 18th. He also remarked that the comet grew brighter during the period between August 6 and August 19.

The comet was last seen by Rümker on September 13.13, at which time it was at $\alpha = 10^h 03.3^m$, $\delta = +16° 58'$.

The first parabolic orbits were calculated using positions obtained on July 21–25. Brorsen, H. Niebour, and A. Quirling determined perihelion dates of 1847 September 7.39, September 9.22, and September 9.24, respectively. K. R. Powalky used an additional position from the 26th and determined the perihelion date as September 8.99. As more positions became available, orbits by Brorsen, Niebour, and Schmidt indicated a perihelion date closer to September 9.6.

By mid-August it was becoming increasingly obvious that an elliptical orbit would fit the observations better. H. A. E. A. Faye was the first to offer such an orbit. Using positions spanning the period of July 21 to August 11, he determined the perihelion date as September 9.65 and the period as about 1439 years. Quirling and Niebour took positions from July 22 to August 24 and found a perihelion date of September 9.90 and a period of about 125 years. A short time later, H. L. d'Arrest determined the period as 28.09 years.

Following the comet's final observations, several astronomers worked on its orbit through the years. D'Arrest (1849) and B. A. Gould, Jr, (1850) were the first to tackle the problem. Ultimately, d'Arrest determined the period as 74.97 years, while Gould determined periods of 71.48 and 81.05 years. Interestingly, d'Arrest also computed a hyperbolic orbit with an eccentricity of 1.0412, but said his elliptical orbit was a better fit to the observations.

Following the comet's accidental rediscovery in 1919, similar elliptical orbits were computed by P. Duckert (1922), L. M. Belous (1975, 1985), B. G. Marsden and D. K. Yeomans (1978), and Yeomans (1986). While Duckert and Belous (1975) determined periods of just over 68 years, the three orbits calculated during the period of 1978–86 revealed a period of 73.1 years. The comet's recovery in 1989 proved the 73.1 year orbit to be correct. Yeomans' orbit is given below.

T	ω	Ω (2000.0)	i	q	e
1847 Sep. 10.0474 (TT)	129.3436	311.9757	19.1506	0.487918	0.972093

ABSOLUTE MAGNITUDE: $H_{10} = 9.6$ (V1964)

FULL MOON: Jun. 28, Jul. 27, Aug. 26, Sep. 24

SOURCES: T. Brorsen, A. C. Petersen, and C. K. L. Rümker, *AN*, **26** (1847 Aug. 5), p. 87; H. A. E. A. Faye, *CR*, **25** (1847 Aug. 16), pp. 265–6, 288; H. A. E. A. Faye, *CR*, **25** (1847 Aug. 23), p. 288; J. R. Hind, *AN*, **26** (1847 Aug. 26), p. 143; H. L. d'Arrest and M. L. G. Wichmann, *AN*, **26** (1847 Aug. 30), pp. 147–50; T. Brorsen, H. Niebour, A. Quirling, and K. R. Powalky, *AN*, **26** (1847 Aug. 30), pp. 155–8; J. F. J. Schmidt, A. Quirling, and H. Niebour, *AN*, **26** (1847 Sep. 30), pp. 177–80, 185; H. L. d'Arrest, *AN*, **26** (1847 Sep. 30), p. 191; *AJS* (Series 2), **4** (1847 Nov.), p. 426; H. A. E. A. Faye, *AJS* (Series 2), **5** (1848 Jan.), p. 134; M. L. G. Wichmann, *AN*, **27** (1848 Jan. 29), p. 7; H. L. d'Arrest, *AN*, **28** (1849 Feb. 10), pp. 219–22; T. Brorsen, C. K. L. Rümker, H. C. Schumacher, A. Quirling, H. Niebour, and H. L. d'Arrest, *AJ*, **1** (1850 Jun. 10), p. 80; B. A. Gould, Jr, J. R. Hind, and J. F. J. Schmidt, *AJ*, **1** (1850 Jun. 27), pp. 81–3; B. A. Gould, Jr, *AJ*, **1** (1850 Nov. 21), pp. 145–7; P. Duckert, *AN*, **215** (1922 Feb. 20), pp. 201–10; V1964, p. 57; L. M. Belous, *QJRAS*, **15** (1974), pp. 450–1, 459; B. G. Marsden and D. K. Yeomans, *QJRAS*, **19** (1978), pp. 80–1; L. M. Belous, *QJRAS*, **26** (1985), pp. 308–9; D. K. Yeomans, *QJRAS*, **27** (1986 Mar.), p. 116.

C/1847 Q1
(Schweizer)
1847 IV

Discovered: 1847 August 31.79 ($\Delta = 1.02$ AU, $r = 1.52$ AU, Elong. $= 96°$)
Last seen: 1847 November 28.72 ($\Delta = 2.65$ AU, $r = 2.12$ AU, Elong. $= 47°$)
Closest to the Earth: 1847 September 14 (0.9021 AU)
Calculated path: CAS (Disc), CEP (Sep. 5), CYG (Sep. 18), VUL (Oct. 5), SGE (Oct. 11), AQL (Oct. 17)

While observing the evening sky, K. G. Schweizer (Moscow, Russia) discovered this faint telescopic comet on 1847 August 31.79 near ε Cassiopeiae. He estimated the position as $\alpha = 2^h$, $\delta = +65.5°$, and said the comet was moving towards ψ Cassiopeiae.

Although the comet was well observed, as with many other comets of this time there were few published physical descriptions. Only a few really exist for the period between the discovery and final observations. O. W. Struve (Pulkovo Observatory, Russia) described the comet as faint, without a noticeable condensation on September 8. J. F. J. Schmidt (Bonn, Germany) said the comet was very faint and diffuse, with a diameter of 3′ on September 19 and 20. J. G. Galle (Berlin, Germany) described the comet as very faint on November 5. An indirect physical description can be inferred from the observations of A. C. Petersen (Altona, Germany). He measured the comet's position with a meridian circle on several occasions between September 11 and 27. Considering the apparent diffuseness, the magnitude must have then been between magnitude 6 and 7.

This comet was last detected on November 28.72, when Struve saw it with a 38-cm refractor. He gave the position as $\alpha = 19^h\ 11.9^m$, $\delta = -0°\ 47′$.

The first parabolic orbit was calculated by A. Quirling and H. Niebour. They took three positions from the period spanning September 11–14 and determined a perihelion date of 1847 August 1.72. By increasing the observation period to September 15, Petersen and J. R. Hind independently calculated orbits with perihelion dates of August 8.65 and August 8.74, respectively. Additional positions allowed refinements of the orbit by H. L. d'Arrest and N. R. Pogson.

The first orbit to use positions spanning the entire period of visibility was calculated by Struve and W. Döllen (1848). This was an elliptical solution with a perihelion date of August 8.69 and a period of about 228 years. Schweizer (1849) calculated an elliptical orbit with a period of about 14 000 years. He also calculated a parabolic solution with a perihelion date of August 9.85.

A definitive orbit was calculated by W. Schur (1876). He used 48 positions obtained between September 8 and November 28, and found a parabolic orbit with a perihelion date of August 9.85 fitted the positions best. This orbit is given below.

T	ω	Ω (2000.0)	i	q	e
1847 Aug. 9.8451 (UT)	55.4717	78.8836	147.3551	1.484824	1.0

ABSOLUTE MAGNITUDE: $H_{10} = 5.7$ (V1964)

FULL MOON: Aug. 26, Sep. 24, Oct. 23, Nov. 22, Dec. 21

SOURCES: K. G. Schweizer, A. Quirling, H. Niebour, and A. C. Petersen, *AN*, **26** (1847 Sep. 30), pp. 189–92; J. R. Hind and A. C. Petersen, *AN*, **26** (1847 Oct. 14), p. 207; K. G. Schweizer, J. R. Hind, A. Quirling, and H. Niebour, *MNRAS*, **7** (1847 Supplement), pp. 312–14; H. L. d'Arrest, *AN*, **26** (1847 Nov. 8), pp. 252–4; N. R. Pogson, *MNRAS*, **8** (1847 Nov. 12), p. 12; J. F. J. Schmidt, *AN*, **26** (1847 Nov. 15), p. 261; A. C. Petersen, *AJS* (Series 2), **5** (1848 Jan.), p. 135; O. W. Struve and W. Döllen, *AN*, **27** (1848 Jul. 31), pp. 321–6; J. G. Galle, *AN*, **29** (1849 Jul. 17), p. 127; K. G. Schweizer, *AN*, **29** (1849 Jul. 31), pp. 168–70; W. Schur, *AN*, **88** (1876 Aug. 15), pp. 217–20; V1964, p. 57.

C/1847 T1 *Discovered:* 1847 October 2.15 ($\Delta = 0.37$ AU, $r = 1.13$ AU, Elong. $= 100°$)
(Mitchell) *Last seen:* 1848 January 4.1 ($\Delta = 1.42$ AU, $r = 1.25$ AU, Elong. $= 59°$)
Closest to the Earth: 1847 October 12 (0.1914 AU)
1847 VI *Calculated path:* CEP (Disc), DRA (Oct. 4), HER (Oct. 10), OPH (Oct. 15), SCO (Oct. 17), LUP (Oct. 26), LIB (Nov. 20), SER (Dec. 18)

As a method of relaxing from her regular routine of observing, M. Mitchell (Nantucket, Massachusetts, USA) usually hunted for comets. On 1847 October 2.15, her numerous comet-sweeping sessions were finally rewarded when she found a diffuse, circular object a few degrees above Polaris. Mitchell was sure of its cometary nature, since this region of the sky had frequently been scrutinized by her 8-cm refractor in the past, but the clearness of the sky and faintness of the object caused her to wait until a confirmation could be obtained the next evening. On October 3.19, she found it had moved toward the northwest, thus confirming that a comet had been found, and gave the position as $\alpha = 0^h 11.2^m$, $\delta = +84° 22'$. Mitchell said the comet exhibited little condensation and no tail. The comet passed slightly less than $5°$ from the North Celestial Pole on October 3. This marked the first time that an American was credited with the original discovery of a comet. Mitchell also became the first woman to receive the comet medal from the King of Denmark, Christian VIII.

Several independent discoveries were made, namely by F. de Vico (Rome, Italy) on October 3.78, W. R. Dawes (Observatory at Camden Lodge, Cranbrook, England) on October 7.89, A. Colla (Parma, Italy) on October 8, and M. H. Rümker on October 11.85. Dawes made his discovery with the naked eye and said, "The comet appeared . . . as a hazy star of the 5th magnitude." Dawes added that the comet's rapid motion was quickly detected with a 15-cm refractor. Interestingly, it was later learned that G. P. Bond (Cambridge, Massachusetts, USA) had swept over the area of Mitchell's initial discovery on several occasions prior to the October 2 discovery, including October 1. The instrument used for these searches was a comet-sweeper, but nothing was noted. After considering Bond's sweeps, the numerous independent discoveries, and the comet's rapid brightening after the discovery, W. Mitchell concluded, "It is evident that its apparition even to the telescope was sudden."

The comet was found less than two weeks from its closest approach to Earth and its motion across the sky increased during most of the first half of October. Thereafter, it quickly moved into the evening twilight. W. C. Bond (Cambridge, Massachusetts, USA) saw the comet with the naked eye on the 8th, 10th, and 12th. On the 10th he noted a faint tail extending about $1.5°$. Dawes observed the comet with a 15-cm refractor on the 8th and said, "It was round, much condensed in the centre, but without any stellar nucleus." On the 11th, Dawes estimated the magnitude as 4 with the naked eye and added, "Its nebulosity extends over nearly 30′ in arc as seen" in a 15-cm refractor, and the coma was "nearly round, much condensed in the centre, but without stellar nucleus." Dawes also noted a star of 10th magnitude could be seen when in the center of the coma. K. L. von Littrow (Vienna, Austria) estimated the magnitude as 3–4 with the naked eye on the 12th. Dawes estimated the naked-eye magnitude as 4 on the 13th and again noted no stellar nucleus visible in a telescope. Because of the nearness to Earth, a large coma diameter of about 30′ was estimated by Littrow on the 13th and about 15′ was given by J. G. Galle (Berlin, Germany) on the 14th. As the comet edged closer to evening twilight, naked-eye observations were made by Galle on the 14th and 15th, Littrow on the 14th, and W. C. Bond on the 15th, 16th, and 19th. Galle estimated the magnitude as 4 on the 14th and 4 or 5 on the 15th. Littrow's observation was made in moonlight on the 14th. The comet was lost in the sun's glare after the 19th.

The comet remained hidden throughout November, but was finally recovered by de Vico on December 11.2 and F. Schaub (Vienna, Austria) on December 11.3. In both instances, the comet's low altitude, as well as haze near the horizon, prevented the measurement of its position. Galle described the comet as a small, but fairly bright nebulosity on the 14th, while J. F. J. Schmidt (Bonn, Germany) described the comet as a very small, faint nebulosity on the same morning. On December 17, Galle described the comet as diffuse and uncondensed.

The comet's position was measured for the final time on December 20.15, by Schaub, and on December 20.20, by G. F. W. Rümker, who gave the position as $\alpha = 15^h 19.7^m$, $\delta = -2° 05′$. The moon interfered thereafter, but after it had left the sky M. L. G. Wichmann (Königsberg, now Kaliningrad, Russia) made the final observation of the comet on January 4.1. He said it was then so faint that no position could be measured.

The first parabolic orbit was calculated by G. P. Bond. Using positions spanning the period of October 7–11, he determined the perihelion date as 1847 November 15.39. Using slightly longer observation arcs, M. Mitchell and K. R. Powalky independently determined perihelion dates of November 15.00 and November 14.90, respectively. A perihelion date of November 14.9 was confirmed in the coming weeks and months by B. Peirce, H. L. d'Arrest, H. Niebour, J. A. C. Oudemans, Burgersdyk, N. R. Pogson, Schaub, and G. F. W. Rümker (1848, 1857).

Rümker (1857) was the first person to point out that the comet's orbit was slightly hyperbolic. Using positions spanning October and December, he determined the perihelion date as November 14.90 and the eccentricity as 1.0001326. M. Palmer (1893) used 77 positions obtained between October 7 and December 20, as well as perturbations by four planets, and computed a hyperbolic orbit with a perihelion date of November 14.90 and an eccentricity of 1.0001727. Palmer's orbit is given below.

T	ω	Ω (2000.0)	i	q	e
1847 Nov. 14.8998 (UT)	276.6090	192.9665	108.1326	0.329024	1.000173

ABSOLUTE MAGNITUDE: $H_{10} = 7.3$ (V1964)

FULL MOON: Sep. 24, Oct. 23, Nov. 22, Dec. 21

SOURCES: G. F. W. Rümker, K. R. Powalky, F. de Vico, W. R. Dawes, A. C. Petersen, and H. L. d'Arrest, *AN*, **26** (1847 Nov. 8), pp. 245–50; M. Mitchell, F. de Vico, W. R. Dawes, K. L. Littrow, F. Schaub, W. C. Bond, H. L. d'Arrest, and G. P. Bond, *MNRAS*, **8** (1847 Nov. 12), pp. 9–11; H. Niebour and J. A. C. Oudemans, *AN*, **26** (1847 Nov. 15), pp. 259, 271; H. L. d'Arrest, Burgersdyk, G. P. Bond, B. Peirce, and F. Schaub, *AN*, **26** (1847 Dec. 9), pp. 275–8, 287–90; J. Challis, N. R. Pogson, and G. F. W. Rümker, *MNRAS*, **8** (1847 Dec. 10), p. 25; W. R. Dawes, K. L. von Littrow, and F. Schaub, *AN*, **26** (1847 Dec. 16), pp. 305–12; M. Mitchell, W. Mitchell, B. Peirce, G. P. Bond, and G. F. W. Rümker, *AJS* (Series 2), **5** (1848 Jan.), pp. 83–5, 135; G. F. W. Rümker, *AN*, **26** (1848 Jan. 10), p. 349; G. F. W. Rümker, *MNRAS*, **8** (1848 Jan. 14), pp. 47–8; N. R. Pogson, *AN*, **26** (1848 Jan. 17), p. 367; F. Schaub, *CR*, **26** (1848 Jan. 17), p. 109; J. F. J. Schmidt, *AN*, **26** (1848 Jan. 22), p. 375; M. L. G. Wichmann and F. Schaub, *AN*, **27** (1848 Jan. 29), pp. 11, 15; M. Mitchell, *MNRAS*, **8** (1848 Mar. 10), p. 130; F. de Vico, *AN*, **27** (1848 Jun. 10), p. 229; *AJS* (Series 3), **7** (1849 Jan.), p. 123; J. G. Galle, *AN*, **29** (1849 Jul. 17), p. 127; G. F. W. Rümker, *AN*, **45** (1857 Mar. 26), pp. 263–70; M. Palmer, *Transactions of the Astronomical Observatory of Yale University*, **1** (1893), pp. 187–207; V1964, p. 57.

C/1848 P1
(Petersen)

1848 I

Discovered: 1848 August 8.06 ($\Delta = 1.22$ AU, $r = 0.89$ AU, Elong. $= 46°$)

Last seen: 1848 August 27.16 ($\Delta = 0.85$ AU, $r = 0.48$ AU, Elong. $= 28°$)

Closest to the Earth: 1848 August 29 (0.8437 AU)

Calculated path: AUR (Disc), GEM (Aug. 14), CNC (Aug. 22)

A. C. Petersen (Altona, Germany) discovered this telescopic comet in Auriga on 1848 August 8.06. He wrote, "The comet was small, but pretty well defined, bright and easily observed." He gave the position as $\alpha = 6^h\ 15.0^m$, $\delta = +41°\ 19'$ on August 8.09. Petersen confirmed his discovery on August 11.03.

The comet steadily dropped deeper into morning twilight following discovery. J. F. J. Schmidt (Bonn, Germany) saw the comet on August 16 and said it was of moderate brightness in the 5-foot focal length refractor. M. L. G. Wichmann (Königsberg, now Kaliningrad, Russia) said the comet was visible in a heliometer on August 16 and 17. On August 20, J. R. Hind (London, England) said the comet was about 1.5' across, "with a very strong condensation of light, somewhat excentrically placed, on the side near the sun." Petersen noted the comet was still well observed on the 26th, and was visible in morning twilight after the disappearance of an 8th-magnitude star. The comet was last detected on August 27.16, when A. Reslhuber (Kremsmünster, Austria) gave the position as $\alpha = 8^h\ 27.6^m$, $\delta = +13°\ 23'$.

Several parabolic orbits for this poorly-observed comet were published after the comet was no longer visible. Perhaps the earliest orbit came from Schmidt, who took positions from August 8, 10, and 16, and determined the perihelion date as 1848 September 9.44. A few days later, Schmidt took positions from August 8, 16, and 22, and determined the perihelion date as September 8.55. This orbit proved quite accurate, as established by other orbits determined by A. Sonntag, A. Quirling and Sonntag, G. F. W. Rümker, and F. Bidschof (1887). The last astronomer used 30 positions obtained between August 8 and 27, and computed a perihelion date of September 8.55. This orbit is given below.

T	ω	Ω (2000.0)	i	q	e
1848 Sep. 8.5453 (UT)	260.9477	213.6499	95.5938	0.319947	1.0

ABSOLUTE MAGNITUDE: $H_{10} = 8.1$ (V1964)

FULL MOON: Jul. 16, Aug. 14, Sep. 13

SOURCES: A. C. Petersen, H. C. Schumacher, A. Sonntag, and G. F. W. Rümker, *MNRAS*, **8** (Supplement, 1848), pp. 207–8; A. C. Petersen, J. R. Hind, A. Sonntag, G. F. W. Rümker, and A. Quirling, *AN*, **27** (1848 Sep. 21), pp. 361–70; J. F. J. Schmidt and M. L. G. Wichmann, *AN*, **27** (1848 Oct. 12), pp. 369–72; *AJS* (Series 2), **6** (1848 Nov.), p. 437; H. C. Schumacher, *AJS* (Series 3), **7** (1849 Jan.), p. 122; A. C. Petersen, *MNRAS*, **9** (1849 Feb. 9), p. 68; F. Bidschof, *SAWW*, **96** Abt. II (1887), pp. 36–52; V1964, p. 57.

2P/Encke *Recovered:* 1848 August 28.28 ($\Delta = 1.20$ AU, $r = 1.68$ AU, Elong. $= 99°$)

Last seen: 1848 November 26.45 ($\Delta = 1.06$ AU, $r = 0.34$ AU, Elong. $= 19°$)

1848 II *Closest to the Earth:* 1848 October 20 (0.3703 AU)

Calculated path: PER (Rec), AUR (Sep. 22), LYN (Oct. 5), UMa (Oct. 12), COM (Oct. 26), VIR (Nov. 2), LIB (Nov. 18)

In the 1848 March 30 issue of the *Astronomische Nachrichten*, J. F. Encke reinvestigated the orbit of this comet and predicted it would next arrive at perihelion on 1848 November 26.62. The prediction included planetary perturbations, as well as "the effect of a resisting medium." He noted the comet would probably pass 0.038 AU from Mercury on November 22. The comet was recovered on 1848 August 28.28, when G. P. Bond (Cambridge, Massachusetts, USA) found it with a 38-cm refractor and gave the position as $\alpha = 3^h 19.5^m$, $\delta = +31° 57'$. He described it as "a misty patch of light, faint and without concentration. Its light is coarsely granulated, so that, were it not for its motion, it might be mistaken for a group of stars of the 21st magnitude." He added that it was 1–2' across. Bond had to wait nearly two days to confirm his recovery, but finally reobserved the comet on August 30.23 and 31.24. He said a condensation was only suspected. Bond added that on the 31st, "A slight elongation is suspected in the direction south-preceding, position 240°."

As September began, the comet was scrutinized by many of the larger telescopes in the world. J. R. Hind (London, England) saw it with an 18-cm refractor on the 4th and noted, "it is the faintest object that can be observed in a dark field." J. Challis (Cambridge, England) aimed the 29.5-cm refractor (120×) at the comet on September 3, 4, and 7. He said, "The comet had the appearance of an extremely faint nebulosity of very large diameter, and its point of maximum brightness was so difficult to fix upon" for measurement. J. F. J. Schmidt (Bonn, Germany) found the comet large, extremely faint, and indistinct on the 20th, with no condensation or tail. A. Colla (Parma, Italy) saw the comet with a 10.8-cm refractor on the 21st and wrote, "It appeared like a faint round nebulosity, the light condensing towards the centre, with traces of scintillation throughout the mass." He added that the comet was not seen in a 6.4-cm refractor. Schmidt said the coma was 6–7′ across on the 22nd. A. Sonntag (Altona, Germany) described it as faint and gave the diameter as 8′ on September 24. Hartwell (England) saw the comet with a 15-cm refractor on September 23 and 26, and wrote, "it was so faint as to require a 15-cm object glass and a practised eye. It appeared in a dull nebulous spot, about 3′ in diameter, rather oblong, a little condensed towards the centre, and shaded off indefinitely at the edge." Bond noted a "brush of light *towards* the sun" on the 27th and Schmidt said the comet had increased in brightness and size when seen on the 30th.

The comet continued to brighten during October as it approached both the Earth and sun. Schmidt found it shining brightly with a central condensation on the 2nd and noted on the 6th that the coma was 10′ across, while the "increase in intensity and brightness of the condensation was considerable." Bond found the comet "just visible to the naked eye" on the 9th. He added, "The brighter portion is very eccentrically situated with respect to the general mass. The fan-shaped brush of light is very evident on the side *towards* the sun, the angles of the sides opening by 75° or 80°. There is no other appendage which could be called a tail." Following interference from moonlight and the October 20 passage by Earth, the comet was seen by Schmidt on October 23. In a 12.7-cm refractor he noted the coma was 8′ across and appeared very bright, with the nucleus situated in the northern half of the coma. There was a slight tail-like extension. Bond saw the comet on the 28th and wrote, "The general mass of light is on the side of the nucleus, towards the sun; a faint ray, probably the commencement of the true tail, is thrown out on the side opposite to the sun."

Despite an increasing distance from Earth, the comet continued to brighten during November, as it dropped into morning twilight. Bond found the comet plainly visible to the naked eye on the 4th and said the tail was 1–2° long. Schmidt found the comet's light was intense and white, despite morning twilight on the 10th. He added that the coma was 3–4′ across, with considerable condensation and a tail. Bond saw the comet in strong twilight on the 14th and said the "comet shows an almost sparkling central point."

The comet was last detected on November 26.45, when Bond measured the position as $\alpha = 14^h 53.0^m$, $\delta = -16° 45′$. He said the comet was in morning twilight at an altitude of about 3°.

Minor refinements of this comet's orbit using positions from multiple apparitions, as well as planetary perturbations, have been published by numerous astronomers. Of most significance were the studies of F. E. von Asten (1877) and B. G. Marsden and Z. Sekanina (1974). These have typically revealed a perihelion date of November 26.58 and a period of 3.32 years. Marsden and Sekanina determined the nongravitational terms as $A_1 = +0.38$ and $A_2 = -0.03503$. The orbit of Marsden and Sekanina is given below.

T	ω	Ω (2000.0)	i	q	e
1848 Nov. 26.5805 (TT)	183.3956	336.5067	13.1641	0.336956	0.847838

ABSOLUTE MAGNITUDE: $H_{10} = 8.5$ (V1964)

FULL MOON: Aug. 14, Sep. 13, Oct. 12, Nov. 11, Dec. 10

SOURCES: J. F. Encke, *AN*, **27** (1848 Mar. 30), pp. 113–120; J. F. Encke, *MNRAS*, **8** (1848 Jun. 9), pp. 179–80; J. Challis, G. Bishop, and J. R. Hind, *MNRAS*, **8** (Supplement, 1848), pp. 208–9; J. R. Hind and A. Sonntag, *AN*, **27** (1848 Oct. 12), pp. 373, 387; G. P. Bond, *MNRAS*, **9** (1848 Nov. 10), pp. 10–11; J. Hartnup, *MNRAS*, **9** (1848 Dec. 8), pp. 28–9; M. L. G. Wichmann and J. F. J. Schmidt, *AN*, **28** (1849 Jan. 26), pp. 179–86; W. C. Bond, J. Challis, and J. G. Galle, *MNRAS*, **9** (1849 Feb. 9), pp. 68–9; W. C. Bond, *MNRAS*, **9** (1849 Mar. 9), pp. 106–7; G. P. Bond and W. C. Bond, *AN*, **31** (1850 Jul. 19), pp. 37–40; J. Ferguson and J. S. Hubbard, *AJ*, **1** (1850 Sep. 10), p. 117; F. E. von Asten, *BASP*, **22** (1877), p. 561; V1964, p. 57; B. G. Marsden and Z. Sekanina, *AJ*, **79** (1974 Mar.), pp. 415–16.

C/1848 U1
(Petersen)

1849 I

Discovered: 1848 October 26.81 ($\Delta = 1.38$ AU, $r = 1.70$ AU, Elong. $= 90°$)
Last seen: 1849 January 26.75 ($\Delta = 1.36$ AU, $r = 0.97$ AU, Elong. $= 45°$)
Closest to the Earth: 1848 December 8 (0.9344 AU)
Calculated path: DRA (Disc), CYG (Nov. 8), VUL (Dec. 3), PEG (Dec. 6), AQR (Dec. 22), SCL (Jan. 20)

A. C. Petersen (Altona, Germany) discovered this telescopic comet in Draco on 1848 October 26.81. The position was given as $\alpha = 18^h 17.7^m$, $\delta = +63° 15'$ on October 26.86. Petersen described the comet as rather bright, with a distinct nucleus. The comet's motion was confirmed by Petersen on October 26.95.

The comet was widely observed as November began. J. F. J. Schmidt (Bonn, Germany) said the coma was 3′ across and exhibited a considerably bright central condensation on the 1st. He reported a tail 5′ long on the 15th. J. R. Hind (London, England) observed the comet on the 19th and said, "The comet does not appear round in our telescope, but has a short fan-like tail about 8′ in length. There is no decided nucleus, but a sufficient condensation of the nebulous matter. . . . " Schmidt estimated the tail length as 6′ on the 20th and 8′ on the 25th. T. Chevallier and R. A. Thompson (Durham, England) noted the comet's nucleus was very close to a star on the 24th. They commented, "The star was seen as before, but the comet appeared much fainter when near it." The star was probably SAO 070259 (mag. 8.3). An independent discovery was made by G. P. Bond (Cambridge, Massachusetts, USA) on November 25.97. On November 25.99, W. C. Bond (Cambridge, Massachusetts, USA) said there was a "finely marked nucleus," which resembled a star of magnitude 9–10, as well as a coma 5′ in diameter and a tail 15–20′ long. W. C. Bond said the comet exhibited a well-defined central condensation on the 27th.

W. C. Bond saw the comet's nucleus pass within 1″ of a 12-magnitude star on December 1 and noted, "both appeared of the same magnitude, and formed a close double star, but were not in contact. . . . " Schmidt estimated the tail length as 8′ on the 2nd. Schmidt saw the comet pass almost centrally over a 9th-magnitude star on the 5th, but noted the star's light changed little. C. K. L. Rümker (Hamburg, Germany) saw the same event and said, "the star was seen distinctly *through* the comet." Schmidt found the tail barely visible in bright moonlight on the 7th, but estimated it as 7′ long in moonless skies on the 14th. He found the tail extending 10′ on the 18th and also reported the comet was barely visible to the naked eye. The tail began showing some complexity on the 19th, when W. C. Bond found it 2° long and noted traces of a second tail inclined 10–20° to the primary tail. Schmidt said the tail was 10′ long and distinctly fan-shaped on the 20th. He added that the coma displayed a nucleus that remained point-like even at a magnification of 200×. W. C. Bond saw the comet the same evening and said, "The breadth of the tail in its brightest part, at 20′ from the nucleus, is only about one minute of arc." Schmidt estimated the tail length as 18′ on December 21 and 30′ on the 22nd. He added on the last date that the tail was narrow and faint. Schmidt said the comet was barely visible to the naked eye on the 23rd and added it was not quite equal to a 6th-magnitude star. He noted the tail was 20′ long and the nucleus was even more distinct and remained so at higher magnifications. On December 24, Schmidt said the coma exhibited a parabolic shape and was 2–3′ across, but the nucleus was less sharp than on previous nights and the tail had decreased to 15′ in length. He reported the tail was 7′ long on the 25th and 26th, while it was 8′ long on the 28th.

As the new year began, the comet was becoming increasingly more difficult to observe as it moved southward. Schmidt saw a slight tail on January 2, but failed to detect a tail on the 15th when the comet was 10° above the horizon. He then described the comet as small, with a bright nucleus. That same night, J. Hartnup (Liverpool, England) said the comet was too faint to allow the illumination of the micrometer wires for a position measurement. Bond saw the comet at an altitude of only 8° on the 22nd.

The comet was last detected on January 26.75 by E. Plantamour (Geneva, Switzerland) at a position of $\alpha = 23^h\ 51.1^m$, $\delta = -29°\ 32′$.

The first parabolic orbits were calculated by J. F. Encke and G. F. W. Rümker. Encke used positions from October 28, 30, and November 2, and determined the perihelion date as 1849 January 20.15. G. F. W. Rümker took positions from October 26, 29, and November 2, and calculated a perihelion date of January 21.52. A couple of days later, Petersen took positions from October 26, 30, and November 4, and determined the perihelion date as January 20.55. Hind was the first person to determine the perihelion date within a couple of hours of what would prove to be the actual date. Using positions from October 26, November 8, and 19, he determined the perihelion date as January 19.66. Additional orbits by Petersen, A. Sonntag, A. Graham, Hind, N. R. Pogson, B. Peirce, T. Clausen, and H. L. d'Arrest refined the perihelion date to January 19.85.

Once observations ceased, astronomers suspected the orbit might not be a parabola. Petersen and Sonntag (1849) took positions spanning the period of October 26 to January 26, and computed a parabolic orbit with a perihelion date of January 19.85. They also found an acceptable hyperbolic solution with a perihelion date of January 19.85 and an eccentricity of 1.0000195. They said the hyperbolic orbit differed little from a parabola, and, after considering the probable errors, they considered the parabola best represented the comet's orbit. Using a similar span of positions, T. H. Safford (1849) computed an elliptical orbit with a perihelion date of January 19.85 and a period of about 383 000 years. R. J. Buckley (1979) used 127 positions obtained between October 26 and January 26, and determined a parabolic orbit with a perihelion date of January 19.85. He also determined an elliptical orbit with a period of about 73 000 years and a hyperbolic orbit with an eccentricity of 1.000415. Although he noted the elliptical and hyperbolic orbits actually fitted the observations slightly better than the parabola, he preferred the parabola given below.

T	ω	Ω (2000.0)	i	q	e
1849 Jan. 19.8534 (TT)	208.0066	217.3332	85.0400	0.959782	1.0

ABSOLUTE MAGNITUDE: $H_{10} = 4.6$ (V1964)

FULL MOON: Oct. 12, Nov. 11, Dec. 10, Jan. 8, Feb. 7

SOURCES: A. C. Petersen, H. C. Schumacher, J. F. Encke, C. K. L. Rümker, E. J. Cooper, A. Graham, A. Sonntag, J. R. Hind, and G. F. W. Rümker, *MNRAS*, **9** (1848 Nov. 10), pp. 11–13; A. C. Petersen, A. Sonntag, J. F. Encke, and G. F. W. Rümker, *AN*, **28** (1848 Nov. 20), pp. 59–64; J. Challis, C. K. L. Rümker, W. W. Boreham, W. C. Bond, H. L. d'Arrest, A. C. Petersen, A. Sonntag, and N. R. Pogson, *MNRAS*, **9** (1848 Dec. 8), pp. 25–8; J. R. Hind and H. L. d'Arrest, *AN*, **28** (1848 Dec. 25), pp. 137–9; G. P. Bond, *AJS* (Series 2), **7** (1849 Jan.), p. 123; J. Hartnup and C. K. L. Rümker, *MNRAS*, **9** (1849 Jan. 12), pp. 47–9; A. C. Petersen, *MNRAS*, **9** (1849 Feb. 9), p. 68; W. C. Bond, E. J. Cooper, A. Graham, T. Chevallier, R. A. Thompson, T. H. Safford, and H. L. d'Arrest, *MNRAS*, **9** (1849 Mar. 9), pp. 107–10; T. Clausen, A. Sonntag, and J. F. J. Schmidt, *AN*, **29** (1849 Jun. 15), pp. 37–46; E. Plantamour, *AN*, **29** (1849 Jul. 6), pp. 91–6; A. C. Petersen and A. Sonntag, *AN*, **29** (1849 Oct. 9), pp. 305–20; Bond, *AN*, **31** (1850 Jul. 19), p. 41; V1964, p. 57; R. J. Buckley, *JBAA*, **89** (1979 Apr.), pp. 260, 263.

C/1849 G1
(Schweizer)

1849 III

Discovered: 1849 April 11.79 ($\Delta = 0.44$ AU, $r = 1.35$ AU, Elong. $= 134°$)
Last seen: 1849 August 27.35 ($\Delta = 1.25$ AU, $r = 1.62$ AU, Elong. $= 91°$)
Closest to the Earth: 1849 April 28 (0.2050 AU)
Calculated path: CrB (Disc), BOO (Apr. 12), COM (Apr. 22), VIR (Apr. 25), LEO (Apr. 28), SEX (Apr. 30), HYA (May 3), PYX (May 7), PUP (May 10), CMa (May 18), COL (Jun. 16), LEP (Jul. 28), ERI (Aug. 23)

K. G. Schweizer (Moscow, Russia) discovered this comet near β Coronae Borealis on 1849 April 11.79. He described it as a small nebulosity, which was seen well in a cometseeker. Independent discoveries were made by G. P. Bond (Cambridge, Massachusetts, USA) on April 12.15 and A. Graham (Markree Observatory, Ireland) on April 14.99. Bond gave the position as $\alpha = 15^h\ 09.1^m$, $\delta = +28°\ 38'$ and said it exhibited an extensive coma, with a strong, stellar condensation, but no tail. Graham said it was "easily seen with an ordinary telescope. The nucleus tolerably bright; but badly defined. The coma much diffused."

Late on April 16, Graham and his colleague, E. J. Cooper, observed the comet with the 34-cm refractor and said it seemed more concentrated than on the 14th. There was also a distinct nucleus and a faint coma. On the evening of the 17th, with the same telescope, they noted the comet was considerably brighter, while the coma was much more extended, but still very faint. The nucleus was very distinct. Bond said the comet was visible to the naked eye on the 18th. J. Hartnup (Liverpool, England) saw the comet on April 26 and 28 and said it exhibited a well-defined nucleus about 2" across, which was "surrounded by a strong nebulous light." No tail was detected. On April 27 and 28, O. W. Struve (Pulkovo Observatory, Russia) said the comet was visible to the naked eye at low altitude, despite twilight and bright moonlight. Cooper and Graham said the nucleus was very sharp in the 34-cm refractor on April 28.

The comet was quickly moving southwestward as May began and was heading for evening twilight. Struve saw the comet with the naked eye on the 1st and said it equaled τ Leonis (magnitude 4.95 – SC2000.0) in brightness. He described the comet as round, with a well-defined central nucleus, but exhibiting no trace of a tail. A. Reslhuber (Kremsmünster, Austria) simply described the comet as uncommonly faint on the same evening, but attributed it to the low altitude. The comet was last detected before conjunction with the sun on May 9.85, by J. E. B. Valz (Marseille, France).

The comet was lost until Bond picked it up at low altitude in the morning sky with the 38-cm refractor on August 25.35. The comet was last detected on August 27 by J. Challis (Cambridge, England) and Bond. Challis saw the comet on August 27.14, while Bond found on August 27.35. Bond gave the position as $\alpha = 4^h\ 41.1^m$, $\delta = -27°\ 18'$.

The first parabolic orbit was calculated by B. Peirce. Taking three positions obtained on April 12 and 13, he determined the perihelion date as 1849 June 3.49. He immediately noted a similarity to the orbit of comet 1748

II. B. A. Gould used three positions obtained between April 12 and April 15, and computed a parabolic orbit with a perihelion date of June 8.12. He said, "The similarity of the orbit with that of the 2nd comet of 1748 is very striking, with the exception of the perihelion distance, which comes out very different. Could such a difference arise from any perturbation by the earth or Jupiter?" Additional calculations by A. Sonntag, K. T. R. Luther, H. L. d'Arrest, Graham, F. Hensel, Schweizer, and J. D. Runkle during the following weeks and months established the perihelion date as June 8.7.

Parabolic solutions eventually began to fail to represent the comet's motion. G. P. Bond computed a hyperbolic orbit with a perihelion date of June 7.25 and an eccentricity of 1.06703. Schweizer calculated a hyperbolic orbit with an eccentricity of 1.007066. D'Arrest (1850) computed an elliptical orbit with a perihelion date of June 8.70 and a period of about 8375 years. G. Respondek (1907) used 102 positions obtained between April 12 and August 27, as well as perturbations by two planets, and computed an elliptical orbit with a perihelion date of June 8.71 and a period of about 13557 years. Respondek's orbit is given below.

T	ω	Ω (2000.0)	i	q	e
1849 Jun. 8.7121 (UT)	236.5866	32.6348	66.9587	0.894360	0.998427

ABSOLUTE MAGNITUDE: $H_{10} = 7.4$ (V1964)

FULL MOON: Apr. 7, May 7, Jun. 5, Jul. 5, Aug. 4, Sep. 2

SOURCES: A. Graham and G. P. Bond, *MNRAS*, **9** (1849 Apr. 13), pp. 127–8; G. P. Bond, *AJS* (Series 2), **7** (1849 May), p. 449; K. G. Schweizer, W. C. Bond, O. Struve, J. Hartnup, J. G. Galle, F. Hensel, K. T. R. Luther, J. D. Runkle, and G. P. Bond, *MNRAS*, **9** (1849 May 11), pp. 162–5; K. G. Schweizer, G, Rümker, A. Sonntag, J. Breymann, A. Graham, J. J. E. Goujon, F. W. A. Argelander, K. T. R. Luther, H. L. d'Arrest, F. Hensel, G. P. Bond, and B. A. Gould, *AN*, **28** (1849 May 15), pp. 355–66; O. W. Struve, *AN*, **29** (1849 Jun. 8), p. 31; A. Sonntag, *AN*, **29** (1849 Jun. 15), p. 41; K. G. Schweizer, *AN*, **29** (1849 Jun. 23), p. 63; H. L. d'Arrest, *AN*, **29** (1849 Jul. 12), pp. 101–4; K. G. Schweizer, *AN*, **29** (1849 Jul. 17), pp. 121–4; A. Graham, *AN*, **29** (1849 Sep. 7), p. 283; E. J. Cooper and A. Graham, *MNRAS*, **9** (Supplement, 1849), pp. 222–3; A. Sonntag, *AJS* (Series 2), **8** (1849 Nov.), p. 428; G. P. Bond, *MNRAS*, **10** (1849 Nov. 9), p. 18; A. Reslhuber, *AN*, **29** (1849 Nov. 12), p. 381; G. P. Bond, *AN*, **30** (1849 Nov. 26), pp. 13–16; J. E. B. Valz, *AN*, **30** (1850 Jan. 10), p. 89; H. L. d'Arrest, *AN*, **30** (1850 Jan. 31), p. 115; K. G. Schweizer, G. P. Bond, and A. Graham, *MNRAS*, **10** (1850 Feb. 8), p. 90; H. L. d'Arrest, *AJ*, **1** (1850 Mar. 12), p. 37; G. Respondek, *Definitive Bahnbestimmung des Kometen 1849 III*. Dissertation: Göttingen (1907); V1964, p. 57; SC2000.0, p. 291.

C/1849 G2 *Discovered:* 1849 April 15.87 ($\Delta = 0.39$ AU, $r = 1.33$ AU, Elong. $= 142°$)
(Goujon) *Last seen:* 1849 September 22.91 ($\Delta = 1.76$ AU, $r = 2.12$ AU, Elong. $= 96°$)
    Closest to the Earth: 1849 April 23 (0.3519 AU)
1849 II *Calculated path:* ERI (Disc), CRT (Apr. 16), LEO (Apr. 22), UMa (May 3), DRA (Jun. 11), UMi (Jul. 5), DRA (Aug. 15)

J. J. E. Goujon (Paris, France) discovered this telescopic comet in Crater on 1849 April 15.87, and described it as a "large, circular nebulosity, without a tail." He added that the coma contained a bright nucleus. Goujon gave a position of $\alpha = 11^h\ 08.5^m$, $\delta = -25°\ 31'$ on April 15.95.

The comet passed closest to Earth a few days after discovery, but continued to approach the sun. A. Reslhuber (Kremsmünster, Austria) saw the comet on April 22, and said the coma was 5–6' across, "with a little nucleus somewhat excentrically situated." A. Graham (Markree Observatory, Ireland) observed with the 34-cm refractor on April 25 and noted a short tail extending away from the sun. Having earlier seen C/1849 G1 (Schweizer), he said C/1849 G2 (Goujon) was the brighter of the two, with a coma 3–4 times smaller.

Following a report on positions obtained during the period of April 23 to May 12, J. Hartnup (Liverpool, England) noted, "The light of the comet was strongly condensed towards the centre; but the nucleus was not so well defined, nor the coma so extensive." No tail was observed. Around mid-May, Reslhuber described the comet as diffuse and 5–6' across. He added that it exhibited an eccentric nucleus and a short fan-shaped tail. W. W. Boreham (Haverhill, Suffolk) listed precise positions for May 3–27 and noted, "The absence of any stellar nucleus in this comet renders it difficult to observe."

As June began, the comet was heading away from both the sun and Earth. M. V. Lyapunov (Kazan, Russia) observed with a refractor on the 16th and said the nucleus was similar to a star of magnitude 10, although he noted rapid variations in brightness. Graham said the comet was excessively faint in the 34-cm refractor on the 23rd. On June 24, Hartnup said the comet looked like a faint nebulous star, with a nearly stellar nucleus of magnitude 12–13 (refractor, 134×). The same night, W. Lassell (Liverpool, England) said, "The comet had a very minute stellar disc, and was easily observed." Graham said the comet was faint on the 27th, but better observed than on the 23rd.

The comet was not as well observed during the last few months as it slowly faded from view, despite the fact that it was well placed on July 21 when it passed about 17° from the North Celestial Pole. E. Plantamour (Geneva, Switzerland) said the comet was very faint on July 27. On August 21, Hartnup said a stellar nucleus was visible in the refractor, while the light of the coma was "pretty equally diffused around the nucleus." J. Challis (Cambridge, England) saw the comet with the 29.5-cm refractor on September 12, 17, and 19, and noted the comet was extremely faint. The two final observations of this comet were made less than one hour from one another, with Lassell seeing the comet on September 22.88 and J. G. Galle (Berlin, Germany) observing the comet on September 22.91. Galle gave the position as $\alpha = 18^h\ 43.8^m$, $\delta = +56°\ 32'$. Lassell searched for the comet on several occasions after the 22nd, but could no longer see it.

The first parabolic orbit was calculated by F. W. A. Argelander. He took positions from April 15, 18, and 20, and calculated a perihelion date of 1849

May 27.02. G. F. W. Rümker and J. Breymann took positions from April 15, 20, and 24, and calculated a perihelion date of 1849 May 27.16. The general correctness of these were confirmed by H. L. d'Arrest, Goujon, Plantamour, and G. D. E. Weyer (1849, 1850).

There have only been two attempts to calculate a definitive orbit. Weyer (1852) used 251 positions obtained between April 15 and September 22, and computed a hyperbolic orbit with a perihelion date of May 26.99 and an eccentricity of 1.0007079. No planetary perturbations were applied. R. J. Buckley (1979) used 150 positions obtained between April 15 and September 22, applied perturbations by Venus to Neptune, and computed a hyperbolic orbit with a perihelion date of May 26.99 and an eccentricity of 1.000940. This orbit is given below.

T	ω	Ω (2000.0)	i	q	e
1849 May 26.9923 (UT)	33.1655	204.6599	67.1526	1.159406	1.000940

ABSOLUTE MAGNITUDE: $H_{10} = 7.2$ (V1964)
FULL MOON: Apr. 7, May 7, Jun. 5, Jul. 5, Aug. 4, Sep. 2, Oct. 2
SOURCES: J. J. E. Goujon, J. R. Hind, W. W. Boreham, and J. Hartnup, *MNRAS*, **9** (1849 Apr. 13), p. 128; J. G. Galle, G. F. W. Rümker, J. Hartnup, H. L. d'Arrest, and G. D. E. Weyer, *MNRAS*, **9** (1849 May 11), pp. 160–2; G. F. W. Rümker, J. Breymann, F. W. A. Argelander, and H. L. d'Arrest, *AN*, **28** (1849 May 15), pp. 355–62; G. D. E. Weyer, *AN*, **29** (1849 May 31), pp. 13–16; E. Plantamour, *AN*, **29** (1849 Jun. 8), pp. 29–32; N. R. Pogson, J. Hartnup, W. Lassell, G. Bishop, J. R. Hind, G. F. W. Rümker, and W. W. Boreham, *MNRAS*, **9** (1849 Jun. 8), pp. 195–7; J. Hartnup, *AN*, **29** (1849 Jul. 20), p. 143; E. Plantamour and A. Graham, *AN*, **29** (1849 Sep. 7), p. 275, 285–8; M. V. Lyapunow, *AN*, **29** (1849 Oct. 9), p. 319; A. Reslhuber, *AN*, **29** (1849 Oct. 30), pp. 349–52; E. J. Cooper, A. Graham, J. Challis, W. Lassell, and J. Hartnup, *MNRAS*, **9** (Supplement, 1849), pp. 224–6; J. J. E. Goujon, *AJS* (Series 2), **8** (1849 Nov.), p. 429; W. Lassell, *MNRAS*, **10** (1849 Nov. 9), p. 8; G. D. E. Weyer, *AN*, **30** (1849 Dec. 27), pp. 75–8; J. J. E. Goujon, *MNRAS*, **10** (1850 Feb. 8), p. 90; G. D. E. Weyer, *AJ*, **1** (1850 Mar. 12), p. 36; J. J. E. Goujon and G. D. E. Weyer, *AN*, **30** (1850 Jun. 13), pp. 341–50; G. D. E. Weyer, *AN*, **30** (1850 Jun. 30), pp. 361–8; V1964, p. 57; R. J. Buckley, *JBAA*, **89** (1979 Apr.), pp. 261, 263.

C/1850 J1
(Petersen)

1850 I

Discovered: 1850 May 1.89 ($\Delta = 1.53$ AU, $r = 1.70$ AU, Elong. $= 81°$)
Last seen: 1850 October 16.12 ($\Delta = 1.75$ AU, $r = 1.71$ AU, Elong. $= 71°$)
Closest to the Earth: 1850 July 15 (0.4631 AU)
Calculated path: DRA (Disc), UMi (Jun. 9), DRA (Jun. 13), UMi (Jun. 16), DRA (Jun. 18), BOO (Jun. 27), CVn (Jul. 8), BOO-CVn (Jul. 9), BOO (Jul. 10), VIR (Jul. 17), HYA (Jul. 29), CEN (Aug. 2), MUS (Sep. 11), CHA (Oct. 5)

A. C. Petersen (Altona, Germany) discovered this comet on 1850 May 1.89, at a position of $\alpha = 19^h 25.0^m$, $\delta = +71° 10'$. He said the faint telescopic comet was badly defined and difficult to observe. Petersen remarked that the comet was only a few degrees from the spot where he discovered comet C/1848 U1. Petersen confirmed the discovery on May 2.88 and C. K. L. Rümker (Hamburg, Germany) confirmed it on May 2.90.

The comet approached both the sun and Earth during the next couple of months, steadily brightening and increasing in size. Petersen said the comet's central condensation appeared like a faint cluster of stars on May 8 and 9. J. F. J. Schmidt (Bonn, Germany) said the coma was 3′ across on the 8th, 9th, 10th, 11th, and 14th. He noted a nucleus of magnitude 10 on the last date. A. Reslhuber (Kremsmünster, Austria) said the comet was "somewhat bright", but weakly condensed on the 19th. He said it had become considerably brighter and larger by the 30th. The comet attained its most northerly declination of +74° on May 27. On May 29, J. R. Hind (London, England) said the coma was 80″ across. W. C. Bond (Cambridge, Massachusetts, USA) noted a decidedly stellar nucleus on the 30th. Schmidt estimated the nuclear magnitude as 9 to 10 on the 30th and 31st, and noted a trace of tail on the former date.

Schmidt again estimated the magnitude of the nucleus as 9–10 on June 1 and 2, and added that a tail was seen on the former date. J. Ferguson (National Observatory, Washington, DC) saw the comet with the 24.4-cm refractor on the 3rd and described it as about 1.5′ across and appearing like "a greyish white nebula, condensed at the center." On June 4, Ferguson said the comet was very faint, about 1.5′ across, and much less distinct than on the previous night, while Schmidt said the coma was 4.5′ across, and noted a tail 4′ long. Ferguson said the comet was very dim on June 5, with a centrally condensed coma about 1.5′ across. On June 7, Bond noted a bright stellar point in the center of the coma with the 38-cm refractor. Schmidt said the coma was 4.5′ across, with a discernible tail on the 8th and 9th. Although the coma was unchanged on the 10th, Schmidt did note that the tail was invisible. He added that the nuclear magnitude was 9–10. J. G. Galle (Berlin, Germany) said the comet appeared brighter with a more distinct nucleus on June 10 and 11 than when last seen at the end of May. On June 12, Ferguson said, "There seems a nucleus like a cluster of stars of the 11 magnitude. The central part of the comet was like white star dust." Schmidt saw the comet on the 12th and 13th and noted a nucleus of magnitude 9 to 10, but no trace of a tail. He added that the existence of the tail was doubtful on the 19th. Ferguson said the central condensation still looked like a star cluster on the 20th in the 24.4-cm refractor, with five or six of the stars being of magnitude 12–13. Schmidt saw the comet in moonlight on June 24 and 26. He estimated the nuclear magnitude as 9, but although the tail was distinctly visible on the former date, it was invisible on the latter. Reslhuber said the comet was easily visible in bright moonlight on the 25th. Schmidt saw the comet with the naked eye in moonlight on June 28 and said it appeared fainter than the Andromeda galaxy. He subsequently estimated the total magnitude as 6. Schmidt said the coma was 6.5′ across, the nucleus was magnitude 9, and the tail was 27′ long. On June 30, Schmidt estimated the nuclear magnitude as 9, and exhibited a tail 28′ long. R. C. Carrington (Durham, England) measured the coma's north–south length and determined a value of 2′ 06.3″.

The comet passed closest to Earth in mid-July. Schmidt compared the comet with ι Boötis on July 2 and estimated the magnitude as 6. He said the coma was 6.0′ across, the nucleus was magnitude 8–9, and the tail extended 1° 14′. Reslhuber said the coma was 5′ across on the 3rd, but said no tail was visible. He added that it was visible to the naked eye. Schmidt compared the comet to A Boötis (magnitude 4.81 – SC2000.0) on the 5th and estimated the magnitude as 6. He said the coma was 8.0′ across, the nucleus was magnitude 8–9, and the tail extended 1° 26′. Bond saw the comet the same night and said the nucleus was magnitude 11 in the 38-cm refractor. Schmidt compared the comet with A Boötis on the 6th and estimated the magnitude as 5. He estimated the nuclear magnitude as 8. Schmidt compared the comet with A Boötis and σ Boötis (magnitude 4.46 – SC2000.0) on the 8th and estimated the magnitude as 5. He said the coma was 10′ across, the nucleus was magnitude 8, and the tail extended 1° 29′. Bond said the nucleus was less compact on the 9th, while Schmidt said the tail was 1° 38′ long. Bond estimated the tail as 5° long on the 11th, while Schmidt estimated the nuclear magnitude as 8. Schmidt added that, despite excellent skies, the tail was hardly visible. Schmidt said the coma was 10′ across on the 13th, and exhibited a nucleus of magnitude 8–9. On July 14, W. Lassell (Liverpool, England) said, "The nucleus of the comet did not appear stellar, but more like a bright planetary nebula, surrounded by haze. Still it appeared to increase gradually in brightness to the centre. No obvious tail could be seen but the comet appeared to throw out two rays, one preceding and the other at about an angle of 45° north preceding." Schmidt also saw the comet that evening and compared the comet with υ Boötis (magnitude 4.06 – SC2000.0) and estimated the magnitude as 5.4. He said the coma was 10′ across, the nucleus was magnitude 8, and the tail was 37′ long. Galle saw the comet with the naked eye on the 14th and said it appeared like a nebulous star of magnitude 5–6. Examining the comet with a telescope, he said the weakest magnification revealed a stellar nucleus, while high magnification caused the nucleus to become diffuse. Ferguson said the condensation no longer appeared as a star cluster on the 15th, while Reslhuber noted that although the comet was near the moon, it was still well observed. On the 16th, Lassell saw a "slight increase in the brightness of the coma" on the preceding side.

Although the comet was still heading for perihelion, low altitude and twilight were beginning to take their toll on its appearance. Lassell said the comet appeared faint on the 17th and although he noted it seemed brighter on the 19th, he saw the comet for the final time on the 20th. Schmidt estimated the nuclear magnitude as 9 on the 21st and said no tail was visible. He found the nucleus to be between magnitude 9 and 10 on the 22nd, 23rd, and 24th. Reslhuber said the comet was faint near the horizon on the 23rd and 25th. The comet was seen for the final time in bright twilight by Bond on July 30.05.

The comet was lost in the sun's glare in late July and remained impossible to observe for over a month. On September 6.77, T. Maclear (Royal Observatory, Cape of Good Hope, South Africa) recovered the comet about 28.5° from the south celestial pole. He said, "the comet resembled a nearly circular nebula of about 5′ in diameter, with a considerable degree of diffused brightness in the centre."

The comet was last detected on October 16.12, when Maclear used the 8.5-foot focal length refractor and gave the position as $\alpha = 13^h\ 24.8^m$, $\delta = -79°\ 53′$. He said, "there was no increase of opacity towards the centre, nor anything to guide the judgement of the measurements other than the general form." Maclear said skies were cloudy on the mornings of October 17 and 18, and searches thereafter failed to reveal the comet.

The first parabolic orbits used positions obtained from May 2 to May 4. G. F. W. Rümker determined the perihelion date as 1850 June 12.40, A. Sonntag determined it as June 8.87, and R. Schumacher determined it as June 8.67. Although these orbits seemed very consistent, they missed the actual date of perihelion by over a month. A vast improvement came from R. Luther when he determined the next orbit using positions spanning May 2–7 and found a perihelion date of July 14.74. Similar orbits were determined shortly thereafter by G. F. W. Rümker, H. Niebour, and Olde using positions from May 2 to May 8, but these were all still at least a week early. The orbit was finally determined with fairly good accuracy when positions spanning the period of May 3–13 were used. J. S. Hubbard then determined the perihelion date as July 25.91, while Schumacher determined it as July 23.57. Very similar orbits were calculated by N. R. Pogson, A. J. F. Yvon-Villarceau, C. Mathieu, E. Plantamour, Sonntag, W. C. Götze, Hind, G. Weyer, J. Breen, S. C. Walker, G. P. Bond, Hubbard, Petersen, Schumacher, and H. L. d'Arrest.

The first elliptical orbit was calculated by Sonntag (1852). He used positions spanning the period of May 2 to July 28, and determined the perihelion date as July 24.02 and the period as about 23 million years. Sonntag also calculated a parabolic orbit, but noted that the elliptical fit the observations marginally better. Carrington (1853) was the first astronomer to add the positions obtained to October 16. He also added perturbations by six planets to his calculations and determined the period as about 29 000 years. R. J. Buckley (1979) also used positions spanning the entire period of visibility, as well as perturbations by Venus to Neptune, and determined the period as about 21 000 years. Buckley's orbit is given below.

T	ω	Ω (2000.0)	i	q	e
1850 Jul. 24.0286 (TT)	180.5495	94.9829	68.1848	1.081499	0.998599

ABSOLUTE MAGNITUDE: $H_{10} = 6.0$ (V1964)

FULL MOON: Apr. 26, May 25, Jun. 24, Jul. 24, Aug. 22, Sep. 21, Oct. 21

SOURCES: A. C. Petersen, *MNRAS*, **10** (1850 Apr. 12), p. 142; J. Challis, J. Hartnup, J. B. Reade, W. W. Boreham, A. C. Petersen, R. Schumacher, J. F. Encke, Wichmann, C. K. L. Rümker, N. Pogson, and J. Breen, *MNRAS*, **10** (1850 May 10), pp. 149–56;

A. C. Petersen, *CR*, **30** (1850 May 13), pp. 581–2; A. C. Petersen, G. F. W. Rümker, A. Sonntag, and R. Schumacher, *AN*, **30** (1850 May 16), p. 307; A. C. Petersen, *AJ*, **1** (1850 May 27), p. 72; R. Luther, Olde, R. Schumacher, G. F. W. Rümker, and H. Niebour, *AN*, **30** (1850 May 30), pp. 337–40; A. J. F. Yvon-Villarceau and C. Mathieu, *CR*, **30** (1850 Jun. 3), pp. 716–17; A. C. Petersen, A. Sonntag, C. K. L. Rümker, R. Schumacher, G. F. W. Rümker, H. Niebour, J. G. Galle, G. Weyer, and Olde, *AJ*, **1** (1850 Jun. 10), pp. 76–9; J. Hartnup, *AN*, **30** (1850 Jun. 13), p. 349; M. F. Maury, W. W. Boreham, R. C. Carrington, J. Hartnup, A. C. Petersen, and R. Schumacher, *MNRAS*, **10** (1850 Jun. 14), pp. 164–8; A. J. F. Yvon-Villarceau, *CR*, **30** (1850 Jun. 17), pp. 779–80; J. S. Hubbard, *AJ*, **1** (1850 Jun. 27), pp. 83–4; H. C. Schumacher and J. R. Hind, *AJ*, **1** (1850 Jun. 27), p. 87; E. Plantamour, *AN*, **30** (1850 Jun. 27), pp. 383–6; A. Sonntag and W. C. Götze, *AN*, **31** (1850 Jul. 2), p. 15; H. L. d'Arrest, *AN*, **31** (1850 Jul. 8), p. 17; H. L. d'Arrest, C. K. L. Rümker, G. Weyer, J. S. Hubbard, and S. C. Walker, *AJ*, **1** (1850 Jul. 17), pp. 92–3; J. Ferguson and W. C. Bond, *AJ*, **1** (1850 Jul. 17), pp. 94–6; J. Curley and W. C. Bond, *AJ*, **1** (1850 Jul. 26), pp. 103–4; J. R. Hind, *AN*, **31** (1850 Jul. 26), p. 61; J. R. Hind, A. C. Petersen, and R. Schumacher, *AN*, **31** (1850 Aug. 2), pp. 67–70, 79; J. G. Galle, *AN*, **31** (1850 Aug. 8), p. 81; A. Sonntag, W. C. Götze, H. L. d'Arrest, and W. C. Bond, *AJ*, **1** (1850 Aug. 9), pp. 109–11; W. Lassell, *AN*, **31** (1850 Aug. 14), p. 97; C. K. L. Rümker, *AJ*, **1** (1850 Sep. 10), pp. 113–14; J. Ferguson, *AJ*, **1** (1850 Oct. 15), pp. 132–3; A. J. F. Yvon-Villarceau, *AN*, **31** (1850 Oct. 20), pp. 225–30; J. F. J. Schmidt, *AN*, **31** (1850 Oct. 24), pp. 241–8; M. F. Maury, J. Ferguson, and A. Reslhuber, *AN*, **31** (1850 Oct. 31), pp. 257–60, 269–72; W. C. Bond, *MNRAS*, **11** (1850 Nov. 8), pp. 15–16; A. C. Petersen, *AN*, **31** (1850 Nov. 14), pp. 289–96; H. L. d'Arrest, *AJS* (Series 2), **11** (1851 Jan.), p. 130; E. Loomis, *AJ*, **1** (1851 Mar. 14), pp. 179–81; T. Maclear, *MRAS*, **21** (1851–2), pp. 136–40; F. Brünnow, *AJ*, **2** (1852 Jan. 23), p. 86; A. Sonntag, *AN*, **34** (1852 Mar. 15), pp. 69–84; A. Sonntag, *AN*, **34** (1852 Mar. 19), pp. 85–100; A. Sonntag, *AN*, **34** (1852 Apr. 23), pp. 165–78; R. C. Carrington, *MNRAS*, **13** (1853 May 13), pp. 218–19; T. Maclear, *AN*, **36** (1853 Mar. 15), pp. 181–6; R. C. Carrington, *AN*, **37** (1853 Jul. 19), pp. 41–8; V1964, p. 57; R. J. Buckley, *JBAA*, **89** (1979 Apr.), pp. 261, 263; SC2000.0, pp. 339, 348, 354.

C/1850 Q1 *Discovered:* 1850 August 30.14 ($\Delta = 0.63$ AU, $r = 1.20$ AU, Elong. $= 91°$)
(Bond) *Last seen:* 1850 November 14.42 ($\Delta = 1.45$ AU, $r = 0.80$ AU, Elong. $= 32°$)
Closest to the Earth: 1850 September 19 (0.4012 AU)
1850 II *Calculated path:* CAM (Disc), AUR (Sep. 7), LYN (Sep. 11), AUR-LYN (Sep. 12), AUR-LYN (Sep. 15), CNC (Sep. 19), LEO (Sep. 26), SEX (Sep. 29), HYA (Oct. 9), CRT (Oct. 12), CRV (Oct. 26), HYA (Nov. 6)

G. P. Bond (Cambridge, Massachusetts, USA) discovered this comet in Camelopardalis, about 10° north of α Persei, on 1850 August 30.14. He described it as a faint telescopic object, with "a very feeble concentration of light towards the center." He added that the brightest part of the comet was 1′ 30″ across.

The first ten days of September brought numerous observations, as well as independent discoveries as the comet's distance from the sun and Earth decreased. The comet attained its most northerly declination of +58° on the 1st. Bond noted on the 3rd that the comet had increased in size and brightness since its discovery. T. Brorsen (Senftenberg, Germany) independently discovered this comet on September 5.90 and described it as bright, with an ill-defined border, and no nucleus. Bond found the comet brighter and somewhat condensed on the 9th, while A. C. Petersen (Altona, Germany) described it as faint, without a distinct nucleus. C. Robertson (Markree Observatory, Ireland) was preparing the comet-seeking refractor for A. Graham when he independently discovered this nebulous object on September 10.04. Graham subsequently observed the object and said "it appeared like a cluster seen with low power: but there was a faint nebulosity which rendered it suspicious." After a while, Graham described it as, "A very diffused and faint nebulosity [which] almost fills the field of the 25-foot [focal length], comet power." No nucleus could be detected. F. V. Mauvais (Paris, France) independently discovered the comet on September 10.06 and described it as a white nebulosity, about 2–3' across, with no apparent condensation or tail.

The comet continued approaching the sun during the remainder of September, but it was closest to Earth on September 19. Petersen said the comet was much brighter on the 12th than on the 9th, and exhibited a weak nuclear condensation in the center of the coma. T. Clausen (Dorpat, now Tartu, Estonia) independently discovered the comet on the 13th. On September 14, J. Hartnup (Liverpool, England) said the coma was 1' 48" across, while Bond said the comet was "a much better object to observe than it has been on previous nights, though it has no nucleus." J. H. Mädler (Dorpat, now Tartu, Estonia) said the comet was rather bright in the 13-cm refractor on the 15th. He added that there was a faint nucleus, but no tail. Petersen found the comet brighter on the 16th than on the 12th, and said it exhibited a brighter, more distinct nucleus. Also on the 16th, A. Reslhuber (Kremsmünster, Austria) described the comet as round, without a noticeable nucleus and tail. J. Ferguson (Washington, DC, USA) described the comet as faint on the 17th and 18th. Petersen said the comet was beautifully seen on the 17th and exhibited a distinct central nucleus and a possible tail. On September 18, R. C. Carrington (Durham, England) said the comet was best seen with the Fraunhofer refractor. He noted, "I then thought it a little elongated in the north following direction: no nucleus was shown by this instrument." Petersen said the comet was not as beautiful and distinct on the 19th as on the 17th because of moonlight.

As October progressed, the comet moved toward morning twilight, which eventually hampered observations. Ferguson said the comet was "bright, much condensed, and nearly round" on the 5th. Reslhuber said the comet was bright and easily seen in twilight on the 9th and he was able to see the comet very near the horizon on the 10th. Bond obtained the final observations of the comet before it was lost in twilight, as he measured positions on October 14.41 and 15.41. Reslhuber looked for the comet in twilight on October 16, but was unsuccessful.

Following the perihelion passage, Bond recovered the comet at an altitude of 4° or 5° on October 28. He said, "The comet is bright enough to be observed with precision, though seen in strong twilight." Bond remained the only observer after perihelion. On November 7, he said the comet was fainter than on October 28, "but is not yet difficult to see." The comet was last detected on November 14.42, when Bond gave the position as $\alpha = 13^h \, 03.2^m$, $\delta = -24° \, 13'$.

The earliest parabolic orbit calculated for this comet was one by Bond using positions obtained from August 30 to September 3. The resulting perihelion date was 1850 October 19.87. T. H. Safford, Jr, and J. D. Runkle each used positions obtained during the period spanning August 30 to September 9 and both determined the perihelion date as October 19.84. Similar orbits were later calculated by H. Niebour, G. F. W. Rümker, Mauvais, E. Plantamour, A. Quirling, A. Graham, Runkle, Safford, H. L. d'Arrest, E. Vogel, J. E. B. Valz, J. Breen, and Reslhuber. A definitive parabolic orbit was calculated by G. Rechenberg (1894) that gave a perihelion date of October 19.84. R. J. Buckley (1979) used 50 positions obtained between August 30 and November 14, and computed a parabolic orbit with a perihelion date of October 19.84. An elliptical orbit with a period of about 46 000 years was also given, but his parabolic orbit is given below.

T	ω	$\Omega \, (2000.0)$	i	q	e
1850 Oct. 19.8374 (TT)	243.2029	208.1112	40.0621	0.565586	1.0

ABSOLUTE MAGNITUDE: $H_{10} = 7.9$ (V1964)

FULL MOON: Aug. 22, Sep. 21, Oct. 21, Nov. 19

SOURCES: G. P. Bond, *AJ*, **1** (1850 Sep. 10), p. 118; C. Robertson and A. Graham, *CR*, **31** (1850 Sep. 23), p. 452; *AN*, **31** (1850 Sep. 27), pp. 189–92; J. D. Runkle, and T. H. Safford, Jr, *AJ*, **1** (1850 Sep. 30), p. 128; F. V. Mauvais, A. Graham, H. Niebour, G. F. W. Rümker, E. Plantamour, W. C. Bond, and J. Hartnup, *AN*, **31** (1850 Oct. 10), pp. 209–14, 219; T. Brorsen, A. C. Petersen, A. Graham, and Rümker, *AJ*, **1** (1850 Oct. 15), p. 131–2; J. H. Mädler, A. C. Petersen, A. Quirling, and A. Graham, *AN*, **31** (1850 Oct. 20), pp. 229–36; J. E. B. Valz, J. D. Runkle, and T. H. Safford, Jr, *AN*, **31** (1850 Oct. 24), p. 253; J. Ferguson, W. C. Bond, and G. P. Bond, *AJ*, **1** (1850 Nov. 4), pp. 140–2; H. L. d'Arrest, E. Vogel, and J. E. B. Valz, *AN*, **31** (1850 Nov. 7), pp. 277, 285; G. P. Bond, R. C. Carrington, F. V. Mauvais, J. Hartnup, G. F. W. Rümker, and A. Graham, *MNRAS*, **11** (1850 Nov. 8), pp. 12–14; J. Breen, *AN*, **31** (1850 Nov. 14), p. 299; R. C. Carrington and A. Reslhuber, *AN*, **31** (1850 Dec. 3), pp. 321, 329; W. C. Bond and F. Bradley, *AJ*, **1** (1850 Dec. 23), pp. 155–6; T. H. Safford, Jr, *AJS* (Series 2), **11** (1851 Jan.), p. 130; W. C. Bond, *AN*, **31** (1851 Jan. 2), p. 357; R. C. Carrington, *MNRAS*, **11** (1851 Jan. 10), p. 63; F. Brünnow, *AJ*, **2** (1852 Jan. 23), p. 86; G. Rechenberg, *AN*, **135** (1894 Jun. 26), pp. 401–16; V1964, p. 57; R. J. Buckley, *JBAA*, **89** (1979 Apr.), pp. 261, 263.

4P/1850 W1 *Recovered:* 1850 November 28.86 ($\Delta = 2.07$ AU, $r = 2.07$ AU, Elong. $= 76°$)
(Faye) *Last seen:* 1851 March 4.82 ($\Delta = 2.48$ AU, $r = 1.72$ AU, Elong. $= 32°$)
Closest to the Earth: 1850 September 3 (1.6416 AU)
1851 I *Calculated path:* AQR (Rec), PSC (Jan. 7)

The prediction for this return was worked on by U. J.-J. Le Verrier (1845). Beginning with the 1843 orbit, he determined the next perihelion would come on 1851 April 4.00. Strafford calculated an ephemeris. The comet was recovered by J. Challis (Cambridge, England) with the 29.5-cm refractor on 1850 November 28.86. His assistant, J. Breen, then measured the position as $\alpha = 21^h\ 29.3^m$, $\delta = -7°\ 12'$, and the comet was found to be quite close to Strafford's ephemeris. Challis described the comet as "excessively faint" on that date and when confirmed on November 29.82, despite the sky being perfectly clear.

This was not a particularly well-placed apparition. The comet had passed closest to Earth nearly three months prior to its recovery, and although it was approaching perihelion, it remained a faint object and was not widely observed. Challis barely saw the comet in hazy skies on December 6, 7, and 25. W. C. Bond (Cambridge, Massachusetts, USA) observed with a 38-cm refractor on 1851 January 2 and 5, and described the comet as "a very faint object," which "when best seen . . . appeared slightly elongated in the direction of the sun." Challis continued to observe the comet with the 29.5-cm refractor (166×) and noted it was "extremely faint" on January 5, "very faint" with an occasionally visible nucleus on January 22, and so faint it was barely visible on January 24. O. W. Struve (Pulkovo Observatory, Russia) said the coma was only 25" across on the 24th. Challis said the comet was brighter on the 27th than it had been on the 24th and was easily seen. He saw "a slight tendency of the coma to the south-following direction." Struve remarked that the comet was brighter on February 1 than on January 24. Moonlight hampered observations early in February, with Challis finding the comet "excessively difficult" on the 6th and Struve describing it as "excessively faint" on the 7th. Challis noted that "a sparkling nucleus" was occasionally visible. After moonlight had ceased to be a problem, Struve estimated the coma diameter as 20" on February 21. Challis noted the comet was "scarcely perceptible" on the 26th and was "too near the horizon to be observed satisfactorily." Struve was only able to get glimpses of the comet on March 1 because of clouds, but he saw the comet very well on March 2.

The comet was last detected on March 4.74 and March 4.82 by Struve and Challis, respectively. Struve described the comet as very faint and measured the position as $\alpha = 0^h\ 59.5^m$, $\delta = +5°\ 42'$. Challis described it as "of the last degree of faintness on account of the zodiacal light: could scarcely be observed." He was not able to measure a complete position. Struve said bad weather and moonlight spoiled the nights immediately following March 4. He noted that when moonlight was no longer a problem the comet had "disappeared in the rays of the Sun."

Later calculations using multiple apparitions and planetary perturbations were published by D. M. A. Möller (1865, 1872), A. Shdanow (1885), and B. G. Marsden and Z. Sekanina (1971). These revealed a perihelion date of April 2.44 and a period of 7.46 years. Marsden and Sekanina determined the nongravitational terms as $A_1 = +0.601$ and $A_2 = +0.01026$. The orbit of Marsden and Sekanina is given below.

T	ω	Ω (2000.0)	i	q	e
1851 Apr. 2.4433 (TT)	200.1295	211.6545	11.3453	1.699909	0.554866

ABSOLUTE MAGNITUDE: $H_{10} = 5.5$ (V1964)

FULL MOON: Nov. 19, Dec. 19, Jan. 17, Feb. 16, Mar. 17

SOURCES: U. J.-J. Le Verrier, *AN*, **23** (1845 Jun. 26), p. 195; J. Challis and J. Breen, *MNRAS*, **11** (1850 Dec.), p. 38; U. J.-J. Le Verrier, *MNRAS*, **11** (1850 Dec.), p. 39; W. C. Bond, *MNRAS*, **11** (1851 Jan.), p. 63; W. C. Bond, *AN*, **32** (1851 Feb. 27), p. 63; J. Challis, *MNRAS*, **11** (1851 May), p. 158; J. Challis, *AN*, **32** (1851 Aug. 4), p. 391; O. W. Struve, U. J.-J. Le Verrier, and Strafford, *CR*, **34** (1852 Feb.), pp. 180–6; D. M. A. Möller, *AN*, **64** (1865 Apr. 10), pp. 145–51; D. M. A. Möller, *VJS*, **7** (1872), p. 96; A. Shdanow, *BASP*, **33** (1885), pp. 1–24; V1964, p. 57; B. G. Marsden and Z. Sekanina, *AJ*, **76** (1971 Dec.), pp. 1136–7.

6P/1851 M1 *Discovered:* 1851 June 28.01 ($\Delta = 0.71$ AU, $r = 1.18$ AU, Elong. $= 84°$)
(d'Arrest) *Last seen:* 1851 October 7.12 ($\Delta = 0.81$ AU, $r = 1.61$ AU, Elong. $= 125°$)
  Closest to the Earth: 1851 July 7 (0.7067 AU)
1851 II *Calculated path:* PSC (Disc), ARI (Jul. 13), CET (Jul. 25), TAU (Aug. 11), ERI (Sep. 16)

H. L. d'Arrest (Leipzig, Germany) discovered this comet on 1851 June 28.01, fairly low over the eastern horizon. He described it as very faint and quickly estimated the position as $\alpha = 0^h 31.3^m$, $\delta = +10° 32'$, before it was obscured by twilight. The comet was not found the next morning because of hazy skies, but it was located on June 30.02 and again described as very faint.

Although the comet was quickly confirmed at other observatories during the first days of July, it was moving away from both the sun and Earth after the 9th. Physical descriptions were initially rare, but A. Reslhuber (Kremsmünster, Austria) finally described the comet on the 29th when he said it was "pale, round, of moderate size, without nucleus and tail." J. Ferguson (Washington, DC, USA) described the comet as very faint and about 3' across on August 7 and 14. Reslhuber said it was extremely faint on August 21. R. C. Carrington (Durham Observatory, England) said the comet exhibited no nucleus during his observations of August 25, 31, September 2, and 7. Carrington said his attempts to find the comet with the refractor after September 7 were unsuccessful. M. L. G. Wichmann (Königsberg, now Kaliningrad, Russia) simply described the comet as very faint on September 22, 23, and 28. J. Challis (Cambridge, England) saw the comet on September 23 and said it was faint "and so much diffused that it was difficult to fix upon the brightest point." On October 1, Wichmann said the comet was especially faint, while Reslhuber said it was hardly discernible.

The comet was last detected on October 7.12, when J. G. Galle and R. Luther (Berlin, Germany) saw it, in a 38-cm refractor. They gave the position as $\alpha = 4^h$ 22.3^m, $\delta = -4°$ 49′. Challis said the moon and unfavorable weather prevented observations after October 5, and that his next attempt to locate the comet came on October 25. For that date he said, "it was only just visible in the large refractor...." The comet's faintness prevented the determination of a position.

The first published parabolic orbit was calculated by d'Arrest using positions obtained during the period of June 30 to July 7. His resulting perihelion date was 1851 July 6.82 and he noted that the orbit was similar to that of a comet seen in 1678. E. Vogel and G. F. W. Rümker took a similar set of positions and calculated a very similar orbit with a perihelion date of July 6.61. In early August, Vogel took three positions from the period of June 30 to July 24 and calculated a revised orbit with a perihelion date of July 9.89. He added that he suspected the orbit was elliptical.

The first elliptical orbit was calculated by Norman Pogson. He used three positions from the period of June 30 to July 25 and determined a perihelion date of July 10.50 and a period of 5.48 years. As the comet continued to be observed, astronomers were able to revise the perihelion date downward to around July 9.4, but there was still a fairly large variation in the period, with d'Arrest providing the lowest value of 7.05 years and A. J. F. Yvon-Villarceau providing the highest value of 8.00 years. Late in the comet's apparition and in the years that followed the period was established as about 6.4 years by d'Arrest, J. A. C. Oudemans, Yvon-Villarceau, L. R. Schulze, and G. Leveau.

Refinements of this comet's orbit using positions from multiple apparitions, as well as planetary perturbations, have been published by numerous astronomers. Some of the earliest were published by H. Lind (1859), Yvon-Villarceau (1859), and Schulze (1863, 1865), while the most recent was by B. G. Marsden, Z. Sekanina, and D. K. Yeomans (1973). These typically revealed a perihelion date of July 9.18 and a period of 6.39 years. A. W. Recht (1939) was the first person to recognize that the comet's motion was being altered by nongravitational forces. He wrote, "It seems clearly evident ... that some unknown force is operating to produce a consistent change in the dynamic elements of Comet d'Arrest," which causes the comet to "always [move] more slowly than predicted." Marsden, and coworkers determined the nongravitational terms as $A_1 = 0.00$ and $A_2 = +0.1038$ and applied them to their orbital computations. Because of a close approach to Jupiter in 1861 (0.35 AU), they said the value of A_1 was assumed to be 0.00. K. Kinoshita (2000) determined a very similar orbit, but with terms of $A_1 = +0.27$ and $A_2 = +0.0997$. His orbit is given below.

T	ω	Ω (2000.0)	i	q	e
1851 Jul. 9.1780 (TT)	174.5437	150.4598	13.9041	1.173319	0.659306

ABSOLUTE MAGNITUDE: $H_{10} = 9.5$ (V1964)

FULL MOON: Jun. 13, Jul. 13, Aug. 11, Sep. 10, Oct. 10

SOURCES: H. L. d'Arrest, G. F. W. Rümker, and E. Vogel, *MNRAS*, **11** (1851 Jun.), p. 170; H. L. d'Arrest, *AN*, **32** (1851 Jul. 7), p. 327; H. L. d'Arrest, *AN*, **32** (1851 Jul. 18), p. 341; E. Vogel, *AN*, **33** (1851 Aug. 22), p. 25; R. Luther, E. Vogel, G. F. W. Rümker, and H. L. d'Arrest, *AN*, **33** (1851 Aug. 28), pp. 33, 43–8; H. L. d'Arrest, E. Vogel, and G. F. W. Rümker, *AJ*, **2** (1851 Oct. 3), pp. 41–2; M. L. G. Wichmann, *AN*, **33** (1851 Oct. 21), pp. 173–6; J. Challis, *AN*, **33** (1851 Oct. 27), p. 181; R. C. Carrington, *AN*, **33** (1851 Nov. 27), p. 267; H. L. d'Arrest, J. Challis, R. C. Carrington, E. Vogel, and N. R. Pogson, *MNRAS*, **11** (Supplement, 1851), pp. 218–20; A. J. F. Yvon-Villarceau, *MNRAS*, **12** (1851 Nov. 14), pp. 17–18; F. W. A. Argelander and J. F. J. Schmidt, *MNRAS*, **12** (1851 Dec. 12), pp. 28–9; J. Challis and R. Luther, *AN*, **33** (1852 Jan. 9), pp. 325, 331; A. Reslhuber, *AN*, **33** (1852 Feb. 5), p. 401; H. L. d'Arrest, *MNRAS*, **12** (1852 Feb. 13), pp. 102–3; A. J. F. Yvon-Villarceau, *CR*, **35** (1852 Dec.), pp. 827–31; J. A. C. Oudemans, *CR*, **38** (1854 Jun. 19), pp. 1083–4; J. A. C. Oudemans, *AJ*, **5** (1857 Aug. 28), pp. 65–70; H. Lind, *AN*, **50** (1859 May 5), pp. 247–50; A. J. F. Yvon-Villarceau, *CR*, **48** (1859 May 9), pp. 924–7; L. R. Schulze, *AN*, **59** (1863 Feb. 21), pp. 189–92; L. R. Schulze, *AN*, **65** (1865 Sep. 12), pp. 163–8; G. Leveau and A. J. F. Yvon-Villarceau, *CR*, **81** (1875 Jul. 19), pp. 141–5; A. W. Recht, *AJ*, **48** (1939 Jul. 17), pp. 65–78; V1964, p. 57; B. G. Marsden and Z. Sekanina, *QJRAS*, **13** (1972), pp. 428–9; B. G. Marsden, Z, Sekanina, and D. K. Yeomans, *AJ*, **78** (1973 Mar.), p. 213; personal correspondence from K. Kinoshita (2000).

C/1851 P1
(Brorsen)

1851 III

Discovered: 1851 August 1.95 ($\Delta = 0.84$ AU, $r = 1.07$ AU, Elong. $= 70°$)

Last seen: 1851 September 30.80 ($\Delta = 0.53$ AU, $r = 1.15$ AU, Elong. $= 92°$)

Closest to the Earth: 1851 Jun. 17 (0.8489 AU), 1851 Oct. 11 (0.5090 AU)

Calculated path: CVn (Disc), BOO (Aug. 5), HER (Sep. 10), DRA (Sep. 11)

T. Brorsen (Senftenberg, Germany) discovered this telescopic comet on 1851 August 1.95. He described it as small, but rather bright. Brorsen determined a precise position of $\alpha = 13^h 55.0^m$, $\delta = +31° 27'$ on August 2.01.

Several astronomers began observations on August 4, including A. C. Petersen (Altona, Germany), G. F. W. Rümker (Hamburg, Germany), and R. Luther (Berlin, Germany). Petersen said the comet was visible "with great difficulty" in bad seeing, while Rümker described the comet's nucleus as "very distinct and fine." Rümker found the nucleus "very distinct and fine" on the 5th and 6th. On August 11, E. Plantamour (Geneva, Switzerland) and A. Reslhuber (Kremsmünster, Austria) found the comet very faint and difficult to see in moonlight. Rümker described the comet's nucleus as "very distinct and fine" during the period spanning August 19–29. Reslhuber said the comet was slightly brighter on the 20th than on the 11th, and was exhibiting a very faint nucleus. J. Ferguson (Washington, DC, USA) described the comet as very faint on August 27 and 28. Reslhuber described the comet as extremely faint on September 22, and noted that it occasionally vanished. R. C. Carrington (Durham Observatory, England) said no nucleus was detectable in the refractor during his observations covering August 19 to September 21, and the comet gradually became fainter during that period.

He said the comet was very difficult to see on the 21st and added that he saw the comet on two other occasions after the 21st, but was unable to measure positions. An apparently independent discovery was made by K. G. Schweizer (Moscow, Russia) on August 21.

The comet was last detected on September 30.80, when Brorsen gave the position as $\alpha = 18^{h} 24.5^{m}$, $\delta = +58° 28'$. Brorsen searched for the comet near the north celestial pole in early October, but failed to locate it in a 10-cm refractor.

The first parabolic orbit was calculated by G. F. W. Rümker using positions spanning the period of August 2 to 6. He determined the perihelion date as 1851 August 26.68. An additional orbit was calculated by C. W. Tuttle and Brorsen. Brorsen (1851) took positions obtained on August 1, August 26, and September 21, and calculated an elliptical orbit with a perihelion date of August 26.73 and a period of about 5543 years. R. Spitaler (1894) used 151 positions obtained between August 1 and September 30, and computed parabolic and elliptical orbits. The parabolic one had a perihelion date of August 26.75, while the elliptical one had a perihelion date of August 26.74 and a period of about 1.2 million years. Although the elliptical orbit fitted the positions best, Spitaler decided the parabolic one was best because of the short observational arc. This orbit is given below.

T	ω	Ω (2000.0)	i	q	e
1851 Aug. 26.7458 (UT)	87.2603	225.7722	38.2035	0.984753	1.0

ABSOLUTE MAGNITUDE: $H_{10} = 7.6$ (pre-perihelion) to 12 (post-perihelion) (V1964)
FULL MOON: Jul. 13, Aug. 11, Sep. 10, Oct. 10
SOURCES: T. Brorsen, K. G. Schweizer, A. C. Petersen, J. Hartnup, R. C. Carrington, J. Challis, and G. F. W. Rümker, *MNRAS*, **11** (1851, Supplement), pp. 220–2; T. Brorsen, A. C. Petersen, and G. F. W. Rümker, *AN*, **33** (1851 Aug. 22), p. 25; G. F. W. Rümker, *AN*, **33** (1851 Aug. 22), p. 31; R. Luther and E. Plantamour, *AN*, **33** (1851 Aug. 28), pp. 33, 45; T. Brorsen and J. Ferguson, *AJ*, **2** (1851 Oct. 3), pp. 42–3; G. F. W. Rümker, *AJ*, **2** (1851 Oct. 22), pp. 55–6; T. Brorsen, *AN*, **33** (1851 Nov. 14), p. 241; W. W. Boreham, *MNRAS*, **12** (1851 Nov. 14), pp. 18–19; R. C. Carrington, *AN*, **33** (1851 Nov. 27), p. 267; C. W. Tuttle, *AJS* (Series 2), **13** (1852 Jan.), p. 128; A. Reslhuber, *AN*, **33** (1852 Feb. 5), p. 403; T. Brorsen, *MNRAS*, **12** (1852 Feb. 13), p. 103; F. W. A. Argelander, *AN*, **34** (1852 Feb. 19), pp. 17–20; T. Brorsen, *AN*, **34** (1852 Jun. 26), p. 335; T. Brorsen, *AN*, **35** (1852 Sep. 21), pp. 169–74; R. Spitaler, *DAWW*, **61** (1894), pp. 323–46; V1964, p. 57.

C/1851 U1   *Discovered:* 1851 October 22.79 ($\Delta = 0.98$ AU, $r = 0.73$ AU, Elong. $= 43°$)
(Brorsen)   *Last seen:* 1851 November 21.73 ($\Delta = 1.46$ AU, $r = 1.39$ AU, Elong. $= 66°$)
Closest to the Earth: 1851 October 7 (0.8176 AU)
1851 IV   *Calculated path:* CVn (Disc), BOO (Oct. 26), HER (Nov. 20)

T. Brorsen (Senftenberg, Germany) discovered this comet in Canes Venatici on 1851 October 22.79. He said the coma exhibited a bright nucleus, and a brilliant tail extending more than 1°. He added that it looked similar to comet C/1847 C1. He estimated the position as $\alpha = 13^h 39.4^m$, $\delta = +32° 35.5'$ on October 22.80, not far from the position where he had found C/1851 P1 on August 1. Brorsen confirmed his find on October 23.18 and said the comet exhibited two tails, one pointing away from the sun, and the shorter pointing towards the sun.

The comet had already passed the sun and Earth when discovered and steadily faded during the next month. K. L. von Littrow (Vienna, Austria) said the comet was visible to the naked eye on October 24. With a refractor, he said the nucleus looked like a 6th-magnitude star. The primary tail extended 30', with a width of 5', away from the sun, while the other extended 8' towards the sun. J. Hartnup (Liverpool, England) saw the comet on November 12 and said it exhibited a nucleus that was steadily seen "generally as a very minute star, but occasionally, for intervals of a few seconds, it appeared quite bright, and had a rather large planetary disc. The tail was about 40' long and about 2' broad in the widest part." On November 14, F. W. A. Argelander (Bonn, Germany) said the comet was not visible in a comet-seeker, but was observed with a 5-foot focal length refractor. He said "the tail was a narrow ray, still distinctly recognized, the head however was difficult to observe." Hartnup said the comet was very faint on November 17.

The comet was last detected on November 21.73, when Littrow determined the position as $\alpha = 15^h 48.1^m$, $\delta = +46° 07'$. He said the nucleus was extremely weak, and the primary tail still pointed away from the sun and extended 32', while it was 3' wide. The sunward tail was barely visible.

Since this comet was only observed for one month, only parabolic orbits have been determined. The first came from Brorsen, using three positions from the period of October 23–29. The resulting perihelion date was 1851 October 1.57. E. Vogel and G. F. W. Rümker utilized positions through October 30 and determined the perihelion date as October 1.13. As soon as astronomers began applying the November positions the perihelion date was established as October 1.30. Among the astronomers calculating this orbit in 1851 were J. Breen, A. Sonntag, E. Schoenfeld and O. Lesser, and W. Klinkerfues. Later orbits using positions spanning the entire period of visibility were determined by P. Andries (1873) and Z. Sekanina (1978). Sekanina's orbit is given below.

T	ω	Ω (2000.0)	i	q	e
1851 Oct. 1.2962 (TT)	294.4471	46.4286	73.9862	0.142053	1.0

ABSOLUTE MAGNITUDE: $H_{10} = 6.0$ (V1964)

FULL MOON: Oct. 10, Nov. 8, Dec. 8

SOURCES: T. Brorsen, AN, 33 (1851 Oct. 30), p. 207; K. L. Littrow and K. T. R. Luther, AN, 33 (1851 Nov. 7), p. 223; T. Brorsen and K. Hornstein, AN, 33 (1851 Nov. 14), p. 243; T. Brorsen, J. Hartnup, and G. F. W. Rümker, MNRAS, 12 (1851 Nov. 14), pp. 1–2; E. Vogel and G. F. W. Rümker, AN, 33 (1851 Nov. 21), p. 257; T. Brorsen, AJ, 2 (1851

Dec. 9), p. 69; J. Breen, *AN*, **33** (1851 Dec. 12), p. 307; E. Vogel and G. F. W. Rümker, *AJ*, **2** (1852 Jan. 1), p. 79; K. T. R. Luther, K. L. Littrow, A. Sonntag, W. Klinkerfues, E. Schoenfeld, and O. L. Lesser, *AN*, **33** (1852 Jan. 9), pp. 333–40; J. Hartnup, *AN*, **34** (1852 Feb. 12), pp. 11–14; T. Brorsen, *MNRAS*, **12** (1852 Feb. 13), p. 103; F. W. A. Argelander, *AN*, **34** (1852 Feb. 19), pp. 17–20; P. Andries, *AN*, **81** (1873 Feb. 21), pp. 55–8; V1964, p. 57; Z. Sekanina, *QJRAS*, **19** (1978), pp. 80–1.

2P/Encke *Recovered:* 1852 January 9.75 ($\Delta = 1.55$ AU, $r = 1.35$ AU, Elong. $= 60°$)

Last seen: 1852 March 11.7 ($\Delta = 0.74$ AU, $r = 0.35$ AU, Elong. $= 16°$)

1852 I *Closest to the Earth:* 1852 March 20 (0.6463 AU)

Calculated path: PSC (Rec), PEG (Feb. 9), PSC (Feb. 10)

J. F. Encke (1851) examined the accuracy of the orbit for the 1848 apparition of this comet and then applied perturbations by Jupiter to bring it to this apparition. He predicted the comet would arrive at perihelion on 1852 March 15.29. The comet was recovered by E. Vogel (London, England) on 1852 January 9.75. He was then using a refractor of 11-foot focal length and gave the position as $\alpha = 23^h\ 01.8^m$, $\delta = +3°\ 48'$. Vogel described the comet as "a very faint, indistinct nebulosity." Independent recoveries were made by J. Hartnup (Liverpool, England) on January 11.81 and A. Reslhuber (Kremsmünster, Austria) on January 12.74. Hartnup described the comet as "a faint patch of nebulous light" between 1' and 2' across in his 22-cm refractor. Reslhuber simply noted the comet was very faint.

The comet was continually noted as a faint object for most of the remainder of January. R. C. Carrington (Durham, England) saw the comet in the refractor on January 12, 16, and 17. He reported the comet was slightly brighter, but still very faint on the 17th. J. Ferguson (Washington, DC, USA) said the comet was seen with difficulty with the refractor on the 13th and 14th. He described it as "a faint white nebula" on both nights although he added that it was seen more distinctly the second night. F. W. A. Argelander (Bonn, Germany) said it was very faint on the 16th and A. Graham (Markree Observatory, Ireland) said it was excessively faint without a nucleus in the 34-cm refractor on the 17th. Graham added, "there was an impression that whatever little condensation of light existed was not central, but northeast of the centre." On January 20, Reslhuber said the comet was "more recognizable, but still very faint," while Graham said the comet looked "like a very faint wispy cloud, rather more condensed toward the northern portion." Graham said the comet was extremely faint on the 23rd, although he noted its light was weakened by a nearby 9th-magnitude star. He commented, "I have never had a stronger impression of the very vapory nature of such bodies." K. L. von Littrow (Vienna, Austria), Reslhuber, and Graham all noted the comet was difficult to see on the 24th because of its proximity to stars of magnitude 8 and 9. Despite the nearby moon on the 25th, Hartnup said the comet was much brighter than on the 11th. Ferguson observed the comet on the 26th and 27th and found the condensation was 2' across and resembled "a cluster of stars of the 12th magnitude." The coma itself was 3' across and extended toward PA 70°. Reslhuber could not see the comet on the 31st because of bright moonlight.

The comet developed nicely during February as its distance from the sun and Earth decreased. Ferguson saw it on the 3rd, 4th, 8th, and 9th, and reported it completely unchanged since his late January observations. Reslhuber saw the comet on the 8th and noted it was "much brighter" than when previously seen. Graham observed on the 13th and said, "The alteration in the brightness of the comet is unexpectedly great. It is now a fine object. The light at least equal to that of a star 10th magnitude, and beautifully white. The appearance is that of a rich round nebula, with a concentration of light; but no nucleus. The faint nebulosity did not seem to extend so far in the NE direction as elsewhere." On February 15, Ferguson noted a bright condensation and said two rays were extending toward PA 300° and PA 90°. He said the condensation was still 2' across and added, "The light not nebulous and nearly as at first but fibrous." Graham said the comet was "beautifully shown" on the 21st and looked like, "a fine round nebula, uniformly surrounded by an atmosphere gradually fading off." On February 24, Reslhuber said the comet was detected in twilight and exhibited a good nucleus, but no tail. J. G. Galle (Breslau, now Wroclaw, Poland) saw the comet on the 24th, 25th, and 26th. He described it as well condensed and as bright as a 7th-magnitude star.

As the comet treked toward its close approaches to both the sun and Earth, it continued to brighten until finally lost in evening twilight. Ferguson saw it on the 4th and reported, "The comet... had a crescent appearance, the concave side near the sun. The condensation being nearest the southern point of the crescent...." He indicated that the condensation was about 2' across. F. Schaub (Trieste, Italy) saw the comet on the 7th, 8th, and 9th, and said its light was so strong that it was detected in a telescope prior to the appearance of a 7th-magnitude star. Also on the 7th, Reslhuber said the comet was bright, despite twilight, and might have been visible to the naked eye if not so close to the horizon. He added that a tail was discernible. Ferguson said the condensation was 2' across on the 8th and noted a "thin beam of light" extending 11' toward PA 325° 55'. He added that the tail was 30" wide. Graham saw an object near the comet's position that same evening and remarked, "but its light was so bright and concentrated that I had much doubt whether it was not a star." Since no star of comparable brightness was then near the comet's position, Graham must have seen the comet. Graham looked for the comet "in a tolerably clear sky" on the 10th, but found nothing.

The comet was last detected on March 11.00, by Ferguson and March 11.7, by C. Fearnley (Königsberg, now Kaliningrad, Russia). Ferguson gave the position as $\alpha = 0^h 25.8^m$, $\delta = +5° 31'$. He said the thin ray, which emanated from the condensation on the 8th, was no longer visible. Fearnley's observation was in bright twilight and no position was obtained. Reslhuber tried to find the comet on March 12 and 13, but nothing was found.

Hartnup and Carrington independently gave brief summaries of their observations. Hartnup remarked, "The light of the comet appeared to me much less diffused than it was in 1848–9. On favourable occasions, the appearance of the comet, with a low power, resembled that of a nebulous star.... This was more particularly the case on and after the 19th of February." Carrington noted, "On and after February 8 the comet was considerably brighter [than during previous observations], and exhibited a strong condensation of central light.... The whole extent of the coma was below two minutes of arc in all cases."

Minor refinements of this comet's orbit using positions from multiple apparitions, as well as planetary perturbations, have been published by numerous astronomers. Of most significance were the studies of F. E. von Asten (1877) and B. G. Marsden and Z. Sekanina (1974). These have typically revealed a perihelion date of March 15.20 and a period of 3.30 years. Marsden and Sekanina determined the nongravitational terms as $A_1 = +0.38$ and $A_2 = -0.03503$. The orbit of Marsden and Sekanina is given below.

T	ω	Ω (2000.0)	i	q	e
1852 Mar. 15.2020 (TT)	183.4287	336.4884	13.1492	0.337565	0.847586

ABSOLUTE MAGNITUDE: $H_{10} = 9.8$ (V1964)

FULL MOON: Jan. 7, Feb. 5, Mar. 6, Apr. 4

SOURCES: J. F. Encke, *AN*, **33** (1851 Nov. 21), pp. 245–50; J. Hartnup and A. Graham, *MNRAS*, **12** (1851 Dec. 12), pp. 27–8; J. Ferguson, *AJ*, **2** (1852 Jan. 23), p. 88; A. Graham, *CR*, **34** (1852 Feb.), pp. 179–80; E. Vogel and A. Reslhuber, *AN*, **33** (1852 Feb. 5), pp. 405–8; J. Hartnup, *AN*, **34** (1852 Feb. 12), pp. 11–14; F. W. A. Argelander, *AN*, **34** (Feb. 19, 1852), p. 17; J. J. E. Goujon and C. Mathieu, *CR*, **34** (1852 Mar.), pp. 363–4; E. J. Cooper and A. Graham, *AN*, **34** (1852 Mar. 2), p. 39; W. C. Bond, J. Hartnup, R. C. Carrington, and W. W. Boreham, *MNRAS*, **12** (1852 Mar. 12), pp. 134–40; F. Schaub, *AN*, **34** (1852 Apr. 16), p. 145; K. L. von Littrow, R. C. Carrington, and A. Reslhuber, *AN*, **34** (1852 Apr. 20), pp. 151, 155, 159–62; J. Ferguson, *AN*, **34** (1852 May 6), pp. 199–202; J. Ferguson, *AJ*, **2** (1852 May 13), p. 116; E. J. Cooper and A. Graham, *MNRAS*, **12** (1852 Jun. 11), pp. 204–6; C. Fearnley, *AN*, **34** (1852 Jun. 15), pp. 297–300; A. Graham, *CR*, **35** (1852 Aug.), pp. 258–60; E. J. Cooper, *AN*, **35** (1852 Aug. 10), pp. 33–6; J. G. Galle, *AN*, **36** (1853 Apr. 26), pp. 287–92; J. F. Encke, *AN*, **41** (1855 Jun. 16), pp. 113–20; F. E. von Asten, *BASP*, **22** (1877), p. 561; V1964, p. 57; B. G. Marsden and Z. Sekanina, *AJ*, **79** (1974 Mar.), pp. 415–16.

C/1852 K1   *Discovered:* 1852 May 16.08 ($\Delta = 0.83$ AU, $r = 1.02$ AU, Elong. = 66°)

(Chacornac)   *Last seen:* 1852 June 15.13 ($\Delta = 1.37$ AU, $r = 1.33$ AU, Elong. = 65°)

Closest to the Earth: 1852 May 16 (0.8298 AU)

1852 II   *Calculated path:* CEP (Disc), UMi (May 23), CAM (May 26), DRA (May 29), UMa (Jun. 2)

J. Chacornac (Marseille, France) discovered this comet on 1852 May 16.08 at a position of $\alpha = 22^h 34^m$, $\delta = +66° 00'$. He described it as small, very faint, and diffuse, with no tail or nucleus. Chacornac confirmed his discovery on May 17.12. Independent discoveries were made by A. C. Petersen (Altona, Germany) on May 17.93 and G. P. Bond (Harvard College Observatory, Massachusetts, USA) on May 19.24. Petersen said the comet was small, and "moving with considerable rapidity toward the north pole." Bond noted the comet was rather faint and round, with a diameter of about 2'.

Interestingly, there are two possible pre-discovery observations of this comet, but no positions are available for absolute confirmation. First, in a letter written on 1852 May 15 and published in the 1852 June issue of the *Compte Rendu*, J. E. B. Valz wrote that Chacornac "had taken the comet for a nebulosity on May 13, but it was not possible to find comparison stars [to measure the position] because of clouds and the moon." Second, in the 1852 July 24 issue of the *Astronomical Journal*, Petersen wrote that he had received a notice that K. G. Schweizer (Moscow, Russia) had found the comet on the evening of May 7, but clouds prevented a confirmation of the cometary nature. Schweizer traveled to St Petersburg on May 8 and he looked for the comet from Pulkovo that evening, but was unsuccessful because of strong twilight.

The comet had passed perihelion in late April and was closest to Earth on the day of discovery. Therefore, it steadily faded after discovery because of its increasing distances from the sun and Earth. Bond described the comet as faint, about 2' across, and without a marked condensation during the period of May 20–23 and on May 26. K. L. von Littrow (Vienna, Austria) said the comet was difficult to observe on the 22nd and was unable to find the comet on the 26th because of moonlight and poor conditions. The comet passed about 1° from the north celestial pole on May 25.

Only astronomers in Vienna and Cambridge (USA) were able to follow the comet in June. K. Hornstein (Vienna, Austria) noted the comet was very faint on June 5, 7, and 8. Bond said it was faint, without a marked condensation on the 6th. The comet was last detected on June 15.13, when Bond gave the position as $\alpha = 10^h 57.2^m$, $\delta = +56° 10'$. He said the observation was "made with much difficulty owing to the faintness of the comet. . . ."

The first parabolic orbit was calculated by A. Sonntag. He used three positions obtained at Altona between May 17 and 19, and determined a perihelion date of 1852 April 19.11. J. B. Bradford followed a couple of days later with an orbit using positions from May 19 to May 21. It revealed a perihelion date of April 20.17. Bradford's orbit proved to be within 2 hours of the comet's actual perihelion date. Later orbits by G. P. Bond, C. W. Tuttle, Sonntag, E. Vogel, and Valz were only slight improvements over Bradford's. Both Sonntag and Valz noted a similarity to comet D/1827 M1.

Following Valz' suggestion of a possible short-period orbit, E. Hartwig calculated an orbit in late July of 1852. What he arrived at was a hyperbolic orbit with a perihelion date of April 21.14 and an eccentricity of 1.0525041. Two decades later F. E. von Asten (1873) used 31 positions obtained between May 18 and June 15, and computed a parabolic orbit with a perihelion date of April 20.09, and a hyperbolic orbit with a perihelion date of April 20.58 and an eccentricity of 1.023850. No planetary perturbations were applied. Although the hyperbolic orbit was apparently confirmed, the short observational arc prompted the parabolic orbit to be preferred and it is given below.

T	ω	Ω (2000.0)	i	q	e
1852 Apr. 20.0873 (UT)	37.2061	319.2712	131.1243	0.905406	1.0

ABSOLUTE MAGNITUDE: $H_{10} = 9.8$ (V1964)

FULL MOON: May 3, Jun. 2, Jul. 1

SOURCES: J. Chacornac, A. C. Petersen, and A. Sonntag, *MNRAS*, **12** (1852 Apr. 7), p. 166; J. Chacornac, *CR*, **34** (1852 May), p. 804; G. P. Bond, *AJ*, **2** (1852 May 22), p. 123; A. C. Petersen and J. Chacornac, *AN*, **34** (1852 May 28), pp. 265–8; B. Valz, *CR*, **34** (1852 Jun.), p. 872; G. P. Bond, I. B. Bradford, C. W. Tuttle, and A. C. Petersen, *AJ*, **2** (1852 Jun. 4), p. 131; G. P. Bond, *MNRAS*, **12** (1852 Jun. 11), p. 203; A. C. Petersen, J. E. B. Valz, and J. Chacornac, *AJ*, **2** (1852 Jun. 15), pp. 142–4; G. P. Bond, K. L. von Littrow, M. L. G. Wichmann, K. Hornstein, A. Sonntag, and J. B. Bradford, *AN*, **34** (1852 Jun. 26), pp. 333, 345–8; E. Vogel, *AN*, **35** (1852 Jul. 19), p. 15; A. Sonntag, M. L. G. Wichmann, and K. G. Schweizer, *AJ*, **2** (1852 Jul. 24), pp. 167–8; W. C. Bond, *AJ*, **2** (1852 Aug. 10), p. 174; W. C. Bond and E. Hartwig, *AN*, **35** (1852 Aug. 19), pp. 49, 59; *AJS* (Series 2), **14** (1852 Nov.), p. 130; F. E. von Asten, *AN*, **81** (1873 Feb. 14), pp. 35–46; V1964, p. 57.

3D/Biela   *Recovered:* 1852 August 26.11 ($\Delta = 1.44$ AU, $r = 0.97$ AU, Elong. $= 42°$)

           Last seen: 1852 September 29.09 ($\Delta = 1.54$ AU, $r = 0.87$ AU, Elong. $= 32°$)

1852 III   *Closest to the Earth:* 1852 September 1 (1.4364 AU)

           Calculated path: GEM (Rec), CNC (Aug. 30), LEO (Sep. 15), SEX (Sep. 24)

The recovery of this comet began with G. Santini (1851). He took an orbit computed during the 1846 apparition, applied perturbations by Venus, Earth, Jupiter, and Saturn, and predicted the comet would next arrive at perihelion on 1852 September 29.18. An ephemeris was given covering the period of 1852 June 30 to 1852 September 30. Using this ephemeris, A. Secchi (Rome, Italy) recovered comet "A" in Gemini with a 15-cm refractor on 1852 August 26.11 and described it as small and very faint. He observed a central occultation of a 9th-magnitude star by the comet on August 26.12 and determined the comet's position as $\alpha = 7^h$ 29.5^m, $\delta = +21°$ 48'. He stated that the comet remained visible during the entire event. The position indicated the comet's perihelion date would be slightly less than six days earlier than Santini had predicted. This caused Secchi to wonder as to whether this was a new comet or "a portion of Biela." Based on the facts that this comet lay near the expected path of 3D/Biela, and that the hourly motion was consistent with what was expected for the comet, A. C. Petersen expressed his strong belief that this was 3D/Biela.

Comet "A" was the only portion of the comet observed for the next couple of weeks. It changed little in brightness during this period, having passed closest to Earth at the beginning of the month and with perihelion not coming until the 23rd. J. Breen (Cambridge, England) found it excessively faint and round on September 9.

Comet "B" was recovered by Secchi on September 16.13. He had not observed the comet for several days, however, and actually misidentified the two components, as he said the newly found portion followed the other part "at a distance of about two minutes of time, and was about half a degree farther south." In reality, this referred to comet "A". To really confuse the issue, comet "B" was brighter than comet "A" at that time. Secchi gave the position of comet "B" as $\alpha = 9^h\ 16.5^m$, $\delta = +11°\ 56'$. He described comet "A" as "very faint, without a nucleus, and of an elongated ovoid form, the apex being turned away from the sun." Comet "B" was described as "quite irregular and had two very faint streaks: it was more luminous in the centre, but without any nucleus."

The comet remained faint during the remainder of September and comets "A" and "B" took turns reigning in brightness. Breen said comet "A" was extremely faint and round on the 17th, while Secchi noted that only comet "A" was visible. Secchi noted that only comet "B" was visible on the 18th and 19th. O. W. Struve (Pulkovo Observatory, Russia) also saw only comet "B" on the 19th with the 38-cm refractor and noted it was at least 30" across. He added, "The nebulosity exhibited a considerable increase of brightness towards the centre, but there was no decided indication of a nucleus." Secchi found comet "B" the brightest on the 20th, but noted both comets seemed of equal brightness on the 21st. Struve also observed both comets on the 21st with the 38-cm refractor and found comet "B" slightly brighter than "A". He also noted comet "B" has a definite nucleus, while the nucleus of "A" was more diffuse. He added, "There is seen extending from [nucleus "A"] an emanation of bright nebulous matter in the direction of B." Struve also measured the distance between the two comets as 30' 35.1". Breen described comet "A" as bright, elongated, and about 50" across on the 22nd. He also noted a "bright stellar point in the centre." Struve saw both comets with the 38-cm refractor on September 24. He said comet "A" was fainter than "B" and the separation between the comets was 29' 43.7". He said the nebulosity of comet "A" was "not uniformly distributed about the point of maximum brightness" and said there was no trace of a nucleus. On the other hand, he said the nebulosity of comet "B" was symmetrical about the nucleus.

Comet "A" was last detected on September 26.08, by Struve with a 38-cm refractor. The position was determined as $\alpha = 10^h\ 05.4^m$, $\delta = +6°\ 12'$.

He said comet "A" was "decidedly fainter" than "B" and that "A" was round and 30" across, while "B" was "somewhat oblong" and 50–60" across. Struve added that comet "A" was situated 59' 15.1" from comet "B" in PA 117° 42.5'. In addition, Struve noted, "The brightest part of A is not situated in the centre of the nebulosity, but is turned *away* from B. The nucleus of B, on the other hand, is turned *towards* A." Comet "B" was last detected on September 29.09, by Struve with a 38-cm refractor. The position was determined as $\alpha = 10^h 17.2^m$, $\delta = +4° 46'$. He added, "There being bright moonlight and a strong twilight, "B" was recognizable only with great difficulty." J. R. Hind (Bishop's Observatory, London, England) searched for the comet again in strong twilight on the morning of October 12, but nothing was found.

Orbits for each comet have been calculated by several astronomers over the years. What was known as the comet "A" was investigated by H. L. d'Arrest (1855), J. S. Hubbard (1860), J. von Hepperger (1900, 1903), and B. G. Marsden and Z. Sekanina (1971), with the result being a perihelion date of September 23.54 and a period of 6.62 years. Comet "B" was investigated by the same group, with the result being a perihelion date of September 24.22 and a period of 6.62 years. Marsden's orbits are given below. The first orbit represents comet "A", while the second is comet "B". Hubbard (1854) said the motion of the two comets indicated 3D/Biela had split about 500 days before the 1846 apparition.

This was the final observed apparition of this comet. Although the next predicted return of May 1859 was unfavorable, that of 1866 was much better. J. Michez (1864) and Thomas Clausen (1865) both accepted the orbit from the 1852 return as accurate and provided predictions for the 1866 apparition, with the likely perihelion date coming around January 27. Searches were made by H. L. d'Arrest (Copenhagen, Denmark), Otto Struve (Pulkovo Observatory, Russia), E. Weiss (Vienna, Austria), K. C. Bruhns (Leipzig, Germany), J. F. J. Schmidt (Athens, Greece), Secchi (Rome, Italy), and Hind, but no trace was found. Although Michez (1872) believed there was little hope in finding the comet again, he did take his 1866 orbit, applied perturbations, and predicted the comet would next arrive at perihelion on 1872 October 6. Both he and Hind provided ephemerides, but the only well-documented search was that by Schmidt during the autumn of 1872. He used a refractor of 6-foot focal length, as well as two seekers, on the mornings of September 30, October 1, 4, and 11, but no trace of the comet was found. David Gill (Royal Observatory, Cape of Good Hope, South Africa) actually searched for this comet from late November to the first days of December during 1885, but again no trace was detected. It should be noted that several observers did report seeing nebulous objects along the predicted path of this comet during 1865 and 1872, and these are given in Appendix 1.

T	ω	Ω (2000.0)	i	q	e
1852 Sep. 23.5432 (TT)	223.1890	248.0070	12.5488	0.860594	0.755828
1852 Sep. 24.2212 (TT)	223.1912	248.0043	12.5500	0.860625	0.755879

ABSOLUTE MAGNITUDE: $H_{10} = 8.1$ (V1964)

FULL MOON: Jul. 31, Aug. 29, Sep. 28, Oct. 27

SOURCES: G. Santini, *MNRAS*, **11** (1851 May), pp. 155–7; A. Secchi and J. F. Encke, *MNRAS*, **12** (1852 Jun.), p. 203–4; A. Secchi, *CR*, **35** (1852 Sep.), pp. 334, 363–4; A. Secchi, *AN*, **35** (1852 Sep. 10), p. 89; A. Secchi, *AN*, **35** (1852 Oct. 1), p. 191; A. Secchi and A. C. Petersen, *AJ*, **3** (1852 Oct. 23), pp. 6–7; A. Secchi, *AN*, **35** (1852 Oct. 29), p. 251; J. Challis and A. Secchi, *MNRAS*, **13** (1852 Nov.), pp. 24–5; A. Secchi and J. S. Hubbard, *AJ*, **3** (1852 Nov. 20), pp. 13–14; J. Breen and J. Challis, *AN*, **35** (1852 Nov. 26), pp. 325–8; J. R. Hind, *AN*, **35** (1852 Dec. 24), p. 371; *AJS* (Series 2), **15** (1853 Jan.), p. 135; J. Challis, *AJ*, **3** (1853 Jan. 27), p. 27; O. W. Struve, *AN*, **37** (1853 Nov. 1), pp. 279–82; J. S. Hubbard, *AJ*, **4** (1854 Jul. 15), pp. 1–5; H. L. d'Arrest, *AN*, **39** (1855 Jan. 17), pp. 321–32; O. W. Struve, *MNRAS*, **16** (1856 Mar.), pp. 137–8; J. S. Hubbard, *AJ*, **5** (1858 Dec. 31), pp. 185–6; J. S. Hubbard, *MNRAS*, **19** (1859 Apr.), p. 229; J. S. Hubbard, *AJ*, **6** (1860 Aug. 15), pp. 140–1; J. Michez, *AN*, **63** (1864 Dec. 31), p. 297; T. Clausen, *BASP*, **8** (1865), pp. 57–62; O. W. Struve, *BASP*, **9** (1866), pp. 569–73; H. L. d'Arrest, *AN*, **66** (1866 Jan. 15), p. 109; A. Secchi, *AN*, **66** (1866 Feb. 12), p. 161; E. Weiss, *AN*, **66** (1866 Mar. 31), p. 265; K. C. Bruhns, *AN*, **67** (1866 Aug. 6), p. 253; J. Michez, *AN*, **79** (1872 Jun. 15), p. 331; J. R. Hind, *AN*, **80** (1872 Sep. 27), p. 137; J. R. Hind, *MNRAS*, **32** (1872 Supplementary Notice), p. 362; J. R. Hind, *MNRAS*, **33** (1873 Mar.), p. 322; J. F. J. Schmidt, *AN*, **82** (1873 Aug. 1), p. 89; D. Gill, *MNRAS*, **46** (1886 Jan.), p. 124; J. von Hepperger, *SAWW*, **109** Abt. IIa (1900), pp. 299–382; J. von Hepperger, *SAWW*, **112** Abt. IIa (1903), pp. 1329–76; V1964, p. 57; B. G. Marsden and Z. Sekanina, *AJ*, **76** (1971 Dec.), pp. 1138–41.

20D/1852 O1
(Westphal)

1852 IV

Discovered: 1852 July 24.97 ($\Delta = 1.12$ AU, $r = 1.70$ AU, Elong. $= 105°$)
Last seen: 1853 February 9.9 ($\Delta = 1.29$ AU, $r = 2.08$ AU, Elong. $= 131°$)
Closest to the Earth: 1852 September 13 (0.6048 AU)
Calculated path: CET (Disc), PSC (Jul. 26), ARI (Aug. 17), TRI (Aug. 20), AND (Aug. 27), PER (Sep. 4), AND-PER (Sep. 6), CAS (Sep. 11), CEP (Sep. 24), UMi (Oct. 4), CAM (Oct. 8), UMi (Oct. 17), DRA (Dec. 4), UMa (Jan. 19)

J. G. Westphal (Göttingen, Germany) discovered this comet on 1852 July 24.97 at a position of $\alpha = 1^h 11.7^m$, $\delta = +1° 04'$. He noted that a cometseeker showed it as a bright nebulosity several arc minutes across. Westphal confirmed his discovery on July 26.05. C. H. F. Peters (Constantinople, Turkey) independently discovered this comet on August 9.

Observations immediately following discovery were few because of moonlight. Nevertheless, F. F. E. Brünnow and G. F. W. Rümker (Berlin, Germany) described the comet as very faint on July 30, with morning twilight adding to the difficulty.

As August began, moonlight was still a problem and few, if any, observations seem to have been made until C. K. L. Rümker (Hamburg, Germany) and E. Schoenfeld (Bonn, Germany) determined positions on the 10th. Thereafter, the comet came under widespread observation. A. Reslhuber (Kremsmünster, Austria) described the comet as a faint nebulosity, with a bright nucleus on August 11. The distinct nucleus was a feature noted by all observers providing descriptions of the comet during August. Although Reslhuber said the coma was elliptical on August 14, no other observer reported this during the month, although H. L. d'Arrest and E. Hartwig (Leipzig, Germany) did note a tail-like extension toward PA 55° on the 17th, so Reslhuber's observation may have referred to a similar feature. No other mention was made of a tail until Reslhuber reported a distinct fan-shaped tail on the 27th. The coma was estimated by d'Arrest as 2.5' across in all of his observations made during the period of August 16–22, and Reslhuber reported it was 4–5' across on the 19th. From its relatively faint beginnings at the beginning of the month, the comet was reported as bright by Reslhuber on the 27th. Moonlight was again diminishing its brightness by the next night.

The comet continued to brighten during September, though how much is difficult to say. Schoenfeld said the comet was possibly of naked-eye brightness on the 5th, but no other observers reported naked-eye visibility until a month later, with some observers indicating the comet was still on the faint side early in the month. C. A. F. Peters (Königsberg, now Kaliningrad, Russia) described the comet as faint on September 4 and 6, as did Reslhuber on the 6th. Reslhuber also noted the comet was not easy to see on the 10th, when it was situated within the double cluster in Perseus. Showing signs of brightening as it headed for the sun, the comet was reported as bright by Reslhuber on the 23rd, despite strong moonlight. The tail was described early in the month. C. Fearnley (Christiania, now Oslo, Norway) said it extended 30' toward PA 225° on the 2nd, while Schoenfeld simply reported a distinct tail extending away from the sun on the 4th.

The comet had passed Earth during September, but was still heading toward the sun as October began. It was also visible throughout the night for most observers, and passed within 2° of the north celestial pole on October 5. Reslhuber first noted the comet was a naked-eye object on October 4 and said it was "easy to discern with naked eye" on the 7th; however, he reported it had faded by the 14th. The last naked-eye observation came from Schoenfeld on the 19th, when he noted it equaled the Andromeda galaxy in brightness. Moonlight began interfering thereafter and Reslhuber reported the comet was very faint on the 23rd. The tail was widely observed during the first half of the month, although it was generally diminishing as the comet steadily moved away from Earth. J. R. Hind (Bishop's Observatory, London, England) said it was 40' long on the 5th, Reslhuber said it was short and wide on the 7th, and Fearnley said it extended 5' toward PA 5–10° on the 9th. Fearnley noted the tail extended 9' toward PA 12° on the 11th. He then reported it extended toward PA 11° on the 12th and PA 25° on the 14th. The comet's most interesting day during this apparition was October 11. Hind described the comet as rather curious and then elaborated, "The nebulosity was extended in the usual direction of the tail and a small glimmering point was visible in the more condensed part situated near the boundary of the nebulosity toward the sun. From this point a ray of light shot out into the cometic matter forming a short tail: at moments this was very distinct and reminded me of some of the drawings of Halley's Comet with its luminous sector." Interestingly, Fearnley also reported a jet emanating from the nucleus and extending about 1' toward the tail.

The comet was moving away from both the sun and Earth as November began. All descriptions of it during the month indicate it was very faint, especially after moonlight began interfering from the 17th onward. Aside from the brightness, little else was reported about this comet other than positions, although Fearnley did note on the 12th that the coma was 3' across and exhibited a tail extending toward PA 30°.

The comet continued to fade during December and observations were obtained until the moon began interfering around mid-month. Reslhuber said the comet was seen with great difficulty on the 5th and was seen as a weak trace on the 10th. Other observers, such as C. A. F. Peters, typically reported the comet as very faint and diffuse. Fearnley said the comet was too faint to be measured on the 13th. The coma was 8' across on the 3rd, according to Fearnley, while the tail extended toward PA 0°. Fearnley also noted the tail extended toward PA 355–360° on the 6th.

As January began, only a handful of observers were continuing to follow the comet. C. A. F. Peters described the comet as very faint and diffuse on the 3rd and 7th, while N. R. Pogson (Radcliffe Observatory, Oxford, England) gave the same description on the 6th. Schoenfeld gave the position as $\alpha = 12^h\ 05.9^m$, $\delta = +69°\ 47'$ on January 11.75, which was the last position obtained anywhere. Following the period of moonlit evening skies, Schoenfeld looked for and found the comet on January 27, but the moon rose and prevented measurement of its position. Many days of bad weather followed and Schoenfeld found the comet again in the evening sky on February 9. Its large and very faint appearance prevented measurement of the position and no further observations were made.

The first parabolic orbit was calculated by A. Sonntag. He used three positions obtained during the period of July 26–30, and determined a perihelion date of 1852 October 14.58. Sonntag revised his calculations as additional observations were published. Using positions up to August 17, he determined the perihelion date as October 11.82, and, after acquiring positions up to September 7, he determined the perihelion date as October 11.80. Sonntag added in his discussion of the last orbit that it was evident the observations could not be adequately represented by a parabola. Other parabolic orbits were calculated by G. F. W. Rümker and B. Valz. G. F. W. Rümker pointed out the resemblance to the orbit of comet C/1793 S1. Although the orbital planes are somewhat similar, the difference in perihelion distance amounts to 0.25 AU and C/1793 S1 arrived at perihelion about two years after the date expected of Westphal's comet.

The first elliptical orbit was calculated by A. Marth using positions covering the period of July 28 to September 18. The resulting perihelion date was October 13.13 and the period was 67.8 years. Sonntag followed with an orbit using positions up to October 26. It had a perihelion date of October 13.32 and a period of 60.8 years. Marth then came back with an orbit using positions obtained up to November 14, which had a perihelion date of October 13.29 and a period of 58.35 years. After the comet was no longer visible, D. M. A. Möller (1858), Westphal (1859), and A. Hnatek (1910) determined orbits using positions spanning the entire period of visibility. The period was determined as 60.01 years by Möller, 60.53 years by Westphal, and 61.56 years by Hnatek.

Although predictions were made for the next return, searches proved fruitless until the comet was accidentally found in September 1913. Orbital computations thereafter revealed the perihelion date of the 1852 apparition had actually been October 13.22, while the period was 61.20 years. The last determinations of the orbit was by G. Schrutka and B. G. Marsden (1974), H. J. Carr (1974), L. M. Belous (1974), and D. K. Yeomans (1986). The orbit of Yeomans is given below.

T	ω	Ω (2000.0)	i	q	e
1852 Oct. 13.2187 (TT)	57.0313	348.2377	40.9373	1.249862	0.919521

ABSOLUTE MAGNITUDE: $H_{10} = 5.0$ (V1964)

FULL MOON: Jul. 31, Aug. 29, Sep. 28, Oct. 27, Nov. 26, Dec. 26, Jan. 25, Feb. 23

SOURCES: J. G. Westphal, A. C. Petersen, A. Sonntag, C. K. L. Rümker, and G. F. W. Rümker, *MNRAS*, **12** (1852 Jun. 11), pp. 201–3, 220; J. G. Westphal, A. Sonntag, and C. K. L. Rümker, *CR*, **35** (1852 Aug.), pp. 191–2; A. Sonntag, *CR*, **35** (1852 Aug.), p. 309; J. G. Westphal, C. K. L. Rümker, G. F. W. Rümker, A. Sonntag, and F. F. E. Brünnow, *AN*, **35** (1852 Aug. 10), pp. 43–6; J. G. Westphal, *AJ*, **2** (1852 Aug. 25), p. 183; B. Valz, *CR*, **35** (1852 Sep.), pp. 360–1, 436; A. C. Petersen, and A. Sonntag, *AJ*, **2** (1852 Sep. 18), pp. 188–9; A. Marth and H. L. d'Arrest, *AN*, **35** (1852 Oct. 1), p. 195; A. Sonntag, *AJ*, **3** (1852 Oct. 23), p. 5; C. K. L. Rümker, *AJ*, **3** (1852 Nov. 20), p. 15; A. Sonntag, *AN*, **35** (1852 Nov. 26), p. 321; J. R. Hind, C. A. F. Peters, and A. Marth, *AN*, **35** (1852 Dec. 24), pp. 371, 379; A. Sonntag, *AJ*, **3** (1853 Jan. 6), p. 23; J. G. Westphal, A. Secchi, A. Sonntag, A. Marth, A. Reslhuber, C. Fearnley, and J. R. Hind, *MNRAS*, **13** (1853 Jan. 14), pp. 81–2; C. K. L. Rümker, *AJ*, **3** (1853 Jan. 27), p. 25; A. Reslhuber, *AN*, **36** (1853 Jan. 28), pp. 89–94; C. Fearnley, *AN*, **36** (1853 Feb. 4), pp. 103–8; J. Ferguson, *AJ*, **3** (1853 Mar. 15), pp. 41–2; E. Schoenfeld, *AN*, **37** (1853 Sep. 13), pp. 141–4; C. A. F. Peters, *AN*, **37** (1853 Sep. 13), pp. 147–50; N. R. Pogson, *AN*, **37** (1853 Nov. 22), pp. 327–30; D. M. A. Möller, *AN*, **49** (1858 Dec. 31), p. 356; J. G. Westphal, *AN*, **50** (1859 Feb. 8), pp. 49–54; A. Hnatek, *AN*, **185** (1910 Sep. 5), pp. 345–88; V1964, p. 57; G. Schrutka, B. G. Marsden, H. J. Carr, and L. M. Belous, *QJRAS*, **15** (1974), pp. 450–1, 459; D. K. Yeomans, *QJRAS*, **27** (1986), p. 604.

C/1853 E1 *Discovered:* 1853 March 6.81 ($\Delta = 0.59$ AU, $r = 1.11$ AU, Elong. $= 84°$)

(Secchi) *Last seen:* 1853 April 18.88 ($\Delta = 1.99$ AU, $r = 1.39$ AU, Elong. $= 41°$)

Closest to the Earth: 1853 February 23 (0.3785 AU)

1853 I *Calculated path:* LEP (Disc), ERI (Mar. 7), ORI (Mar. 13), ERI (Mar. 15), ORI (Mar. 16), TAU (Mar. 19)

A. Secchi (Rome, Italy) discovered this comet in Lepus on 1853 March 6.81. He described it as 5′ across, with the brightest portion about 3′ across. Although a small, luminous point was present, Secchi noted it was not too distinct and several bright points may have been present. He determined the position as $\alpha = 4^h 52.8^m$, $\delta = -15°\ 51′$ on March 6.84. Independent discoveries were made by K. G. Schweizer (Moscow, Russia) on March 8.76, C. W. Tuttle (Harvard College Observatory, Cambridge, Massachusetts, USA) on March 9.05, and E. Hartwig (Leipzig, Germany) on March 10.80. Schweizer said the comet was about 8′ across, with no tail. Tuttle described the comet as round, centrally condensed, and about 3′ across. Hartwig simply noted the comet was quite large and fairly bright. The comet had attained its greatest southern declination of $-75°$ on February 20. The comet reached its greatest elongation of 95° on the day of perihelion.

On March 11, 15, 19, and 23, Tuttle noted the comet was round, centrally condensed, and about 3′ across. E. Schönfeld (Bonn, Germany) noted several bright points in the coma on March 13 and a few days thereafter. J. R. Hind (Bishop's Observatory, London, England) managed to observe the comet "in a hazy moonlit sky" on March 22, 23, and 25. A. Reslhuber (Kremsmünster, Austria) said the comet appeared as a faint nebulosity, about 2′ across, without a nucleus or tail on March 26 and 29. C. A. F. Peters (Königsberg, now Kaliningrad, Russia) described the comet as very faint on March 29. On March 30, S. Reedtz (Palsgaard, Denmark) said the coma was a blurred nebulosity about 3–4′ across. He added that there was a bright point north of the coma's center. C. K. L. Rümker (Hamburg, Germany) also saw the comet on March 30 and 31, as well as on April 5. He remarked that the comet rapidly decreased in brightness between the last days of March and April 5. For the last date he described the comet as fairly faint. Reslhuber said the comet was seen with extreme difficulty on April 7. Peters said the comet was very faint on the 10th.

The comet's position was apparently last measured on April 11.79 by Secchi, and April 11.84 by Peters. Peters indicated a position of $\alpha = 4^h 32.1^m$, $\delta = +13°\ 25′$ and said the comet was extraordinarily faint. The comet was last detected on April 18.88 by J. A. C. Oudemanns (Leiden, Netherlands), but he said clouds prevented a determination of the position. Oudemans tried to find the comet on April 28, but its faintness and location in bright twilight prevented an observation. Schönfeld failed to find the comet during searches on April 27, 28, and May 1.

The first parabolic orbit was calculated by Secchi using positions from March 6, 7, and 8. He determined the perihelion date as 1853 February 24.40. As further observations were obtained, calculations by K. C. Bruhns, K. L. von Littrow, Tuttle, E. Hartwig, J. D. Runkle, A. Marth, Reedtz, B. A. Gould, d'Arrest, and K. Hornstein increased the date to February 24.5. Several of these astronomers remarked on the similarity between the orbit of this comet and that seen in 1664, but after all of the observations were considered, it was obvious that they were two different comets. An elliptical orbit was also calculated by Hartwig, which had a perihelion date of February 24.50 and a period of 1215 years. Although it fitted the available positions better than the parabolic orbit, he considered the observation arc to have been too short to allow a reliable determination of anything other than a parabolic orbit. The last orbital calculation was by B. Cohn (1899). Using positions obtained between March 6 and April 11, he determined the perihelion date as February 24.52. Cohn's orbit is given below.

T	ω	Ω (2000.0)	i	q	e
1853 Feb. 24.5213 (UT)	275.9051	71.6574	159.7566	1.092195	1.0

ABSOLUTE MAGNITUDE: $H_{10} = 7.3$ (V1964)

FULL MOON: Feb. 23, Mar. 25, Apr. 23

SOURCES: A. Secchi, *CR*, **36** (1853 Mar.), p. 543; A. Secchi, K. G. Schweizer, J. Hartnup, C. W. Tuttle, and H. L. d'Arrest, *MNRAS*, **13** (1853 Mar. 11), pp. 162–4; W. C. Bond and C. W. Tuttle, *AJ*, **3** (1853 Mar. 15), p. 47; A. C. Petersen and E. Hartwig, *AN*, **36** (1853 Mar. 15), p. 195; E. Hartwig, K. G. Schweizer, A. Secchi and K. C. Bruhns, *AN*, **36** (1853 Mar. 25), pp. 205–8, 211; A. Secchi, *CR*, **36** (1853 Apr.), pp. 658–61; C. W. Tuttle, C. A. F. Peters, and G. F. W. Rümker, *AN*, **36** (1853 Apr. 5), p. 243; G. F. W. Rümker, J. R. Hind, R. Luther, A. C. Petersen, and A. Marth, *AN*, **36** (1853 Apr. 15), pp. 257–60; W. C. Bond, C. W. Tuttle, and J. Ferguson, *AJ*, **3** (1853 Apr. 19), pp. 60, 64; A. Secchi, K. Hornstein, C. A. F. Peters, and K. C. Bruhns, *AJ*, **3** (1853 Apr. 25), pp. 70–2; C. W. Tuttle and S. Reedtz, *AN*, **36** (1853 May 3), p. 303; G. F. W. Rümker and A. Marth, *AJ*, **3** (1853 May 18), p. 78; A. Reslhuber, *AN*, **36** (1853 May 24), p. 337; C. K. L. Rümker, *AJ*, **3** (1853 Jun. 10), pp. 81–2; E. Schönfeld, *AN*, **36** (1853 Jun. 14), p. 373; J. A. C. Oudemans, *AN*, **37** (1853 Aug. 2), p. 69; E. Hartwig and A. Secchi, *AN*, **37** (1853 Dec. 30), pp. 403–10; K. Hornstein, *AN*, **38** (1854 Mar. 28), p. 159; B. Cohn, *SAWW*, **108** Abt. II (1899), pp. 83–118; V1964, p. 57.

C/1853 G1

(Schweizer)

1853 II

Discovered: 1853 April 5.02 ($\Delta = 0.87$ AU, $r = 1.10$ AU, Elong. $= 72°$)

Last seen: 1853 June 11.74 ($\Delta = 1.47$ AU, $r = 1.07$ AU, Elong. $= 47°$)

Closest to the Earth: 1853 April 29 (0.0838 AU)

Calculated path: AQL (Disc), DEL (Apr. 6), EQU (Apr. 21), PEG (Apr. 23), PSC (Apr. 26), CET (Apr. 28), ERI (Apr. 29), LEP (May 1), CMa (May 4), PUP (Jun. 2)

K. G. Schweizer (Moscow, Russia) discovered this comet about 1.5° south of ρ Aquilae on 1853 April 5.02. He described it as "small and round, about 3' in diameter, and without a tail." There was also a faint nucleus visible from time to time. Schweizer gave the position as $\alpha = 20^h 03.3^m$, $\delta = +13° 04'$. Schweizer confirmed his discovery on April 5.98.

Because it was nearing its closest approach to Earth, the comet's daily motion had amounted to less than 1^m in α and $1'$ in δ during all of March. As April progressed it gradually picked up speed. G. F. W. Rümker (Hamburg, Germany) saw the comet on April 17 and said it appeared "tolerably distinct, with a nucleus and a short tail." K. C. Bruhns (Berlin, Germany) observed the comet in bright moonlight on the 20th, and noted a tail about 1.5' long, and a nucleus situated in the sunward half of the coma. He added that the comet seemed equal in brightness to a 6th-magnitude star. The comet was quickly moving toward morning twilight in an east-southeasterly direction which prevented further Northern Hemisphere observations after Bruhns saw it on April 25.

The comet passed 21° from the sun on April 29, the day it was closest to Earth, and its daily motion on the 28th and 29th was close to 24°. It burst into the skies of the Southern Hemisphere on April 30 and was independently discovered by several people. P. P. King (Paramatta, New South Wales, Australia) reported that a comet "suddenly made its appearance after sunset" on April 30.37. He added that it was "a beautiful object, the nucleus and coma of rather large dimensions, but much diffused in the hazy medium, for it was low in the western horizon." He indicated that the tail was 4° long. S. H. Wright (on the Royal Mail Steamer *Lady Jocelyn*, just off the southeast tip of Africa) found the comet on April 30.68 and said, "The nucleus of the comet was as large as a star of the 1st magnitude, and of a pale blue colour. The tail pointed towards Rigel, was about 6° in length, distended at the outer end about 20', and of a streaky appearance." W. W. Vine (on HMS *Waterwitch*, in the Atlantic Ocean not far from Luanda, Africa) found the comet on April 30.76 about 10° above the southwest horizon and said the tail was 3° 15' long and pointed toward Canopus. J. Parish (commanding HMS *Sharpshooter*, just north of Rio de Janeiro, Brazil) found the comet on April 30.93 about 9° above the horizon and estimated the tail length as 8–10°. J. S. Goodenough (on board HMS *Centaur*, near Buenos Aires, Argentina) discovered the comet on April 30.94 and said, "Its nucleus was very distinct from the coma, and shone as a star between the third and fourth magnitude. The coma extended (by estimation) about 2' round the nucleus; and the tail, pointing nearly direct to the zenith, had a length of about 4°, its width increasing but little." C. W. Moesta (Santiago, Chile) found the comet on May 1.04 and said its nearness to the horizon prevented a reliable estimate of the position.

As May began the comet was rapidly heading away from Earth, but still heading toward the sun. Southern Hemisphere observations were quite numerous. Independent discoveries were still being reported on May 1, with the observations of W. B. Edwards (Master on HM Brig *Penguin*, off the coast of South Africa) and A. G. Constable (a passenger on a ship sailing just off the northwest tip of New Zealand) being recorded in various journals. Edwards simply noted the tail was 5° long, while Constable said the comet was about magnitude 1, with a reddish hue, and a bright tail extending 4°. Bosquet (Port Louis, Mauritius) said the comet exhibited a nucleus of about magnitude 4, as well as a tail nearly 10° long. T. Maclear (Royal Observatory, Cape of Good Hope, South Africa) said the tail appeared 3.5–4° long and extended southward. He added that the nucleus was about magnitude 5 (he initially gave it as magnitude 4 in the *Monthly Notices*) and appeared fan-shaped when using a high power on the 8.5-foot focal length refractor. Vine said the comet was not as bright as on April 30, but still exhibited a tail 3° long. Wright said the nucleus had faded by a magnitude since the previous night, and the tail was no longer than 3°.

Observations continued to be numerous during the next few days. On May 2, Bosquet, Moesta, and Vine each noted the comet had faded and decreased in size since the previous night. Moesta added that the tail was 58′ long. Wright said the nucleus was no larger than a star of magnitude 3, while the tail had decreased to 1° 30′. King said the comet's form was "more defined, the coma more linear, and the nucleus had a stellar body of apparently half the size of Jupiter, and much resembling that planet as if seen through a mist." On May 3, Bosquet said the comet was of a faint and pale color, while the naked eye revealed a very pale nuclear condensation, with no bright points within it. He added that a telescope (60×) revealed a distinct nucleus. Goodenough reported that "the *distinct* part of the tail had diminished to 1° 30′ in length, the brightness of the nucleus having diminished to that of a star of the 6th mag." King said the stellar nucleus was no longer visible. The comet attained its most southerly declination of −14° on May 4. Bosquet then commented the comet had further diminished in size and brightness. "To the naked eye," he noted, "it seemed as a faint light like a cluster of nebulae, and had a very short tail. With a telescope the tail could be distinctly observed; its direction was slightly changed being a little more inclined towards the southward, and diverging more from the equator." King said the comet was "much more indistinct." Wright saw the comet on the 5th and said the tail partly surrounded the nucleus, extending 10′ westward and 45′ eastwards. The streaky appearance was gone, so that the tail appeared only as a pale light. On the 6th, Vine said the comet was "very indistinctly seen," while Wright said it was becoming very faint. Wright added that he saw it through a telescope and noted, "the nebulous part seemed completely to surround the nucleus in the form of an oval. The greatest diameter from E.S.E to W.N.W. about 1° in length. The lesser diameter about 30′." Vine said the comet was not bright enough for its position to be measured by a sextant on May 7, while Maclear noted the tail extended toward PA 118° 40′. Wright saw the comet with a telescope for several days following May 7. He said it gradually diminished in size, but retained "very nearly the same position in the heavens." On May 9, Maclear said the tail extended toward PA 121° 30′, while the nucleus was "remarkably bright," but not visible to the naked eye. He noted the nucleus was faded by the 11th, while the halo surrounding the head was more diffuse. Maclear found the tail extending toward PA 117° 30′ on the 12th. Vine saw the comet for the final time on May 14, while Goodenough saw the comet for the final time on the 17th, while using a telescope of 2-foot focal length and a magnification of 20×. Maclear continued measuring the position angle of the tail for several more days, with his final measure coming on the 17th when the tail extended toward PA 118° 45′. Maclear said the comet was faint in moonlight on the 20th, with a "barely distinguishable" tail. He said it was "very indistinct" as seen through thin clouds on the 23rd, and was very faint on the 29th.

Only a few descriptions were provided in June, before the comet was lost. On the 1st, Maclear said the comet's faintness was diminishing the precision of his position measurements. He found the comet very faint on the 3rd and reported that moonlight was diminishing the visibility of the comet on the 10th.

The comet was last detected on June 11.74, by Maclear. He determined the position as $\alpha = 7^h 23.5^m$, $\delta = -12° 33′$. Maclear said the next three nights were cloudy, and that his searches during the next clear nights were unsuccessful. Following the June 11 observation, the comet's daily motion dropped to less than 0.5^m in α and $0.5′$ in δ for the remainder of the month.

The first parabolic orbit was calculated by Bruhns. He used positions obtained between April 15 and 20, and determined the perihelion date as 1853 May 10.86. The orbit proved remarkably accurate and few attempts were made to revise the orbit until after the comet had faded from view.

The first elliptical orbits were published in 1857. J. N. Stockwell determined three Normal positions between the period of May 2 and June 8, and calculated a perihelion date of May 10.32 and a period of 646 years. He pointed out a similarity between this orbit and that of comet C/1844 N1, but added, "the elements I have given represent the observations too well to allow the supposition of a nine-year period." G. F. W. Rümker collected about 80 positions from the period spanning April 15 to June 11, applied the perturbations of Jupiter, and determined the perihelion date as May 10.33 and the period as 782 years.

T	ω	Ω (2000.0)	i	q	e
1853 May 10.3263 (UT)	199.2335	43.0218	122.1955	0.908693	0.989286

ABSOLUTE MAGNITUDE: $H_{10} = 6$ (V1964)

FULL MOON: Mar. 25, Apr. 23, May 22, Jun. 21

SOURCES: K. G. Schweizer, *MNRAS*, **13** (1853 Apr. 8), p. 185; K. G. Schweizer, *MNRAS*, **13** (1853 Apr. 8), p. 205; K. G. Schweizer, *AN*, **36** (1853 Apr. 15), supplement; K. C. Bruhns, *AN*, **36** (1853 May 3), pp. 297–9; K. G. Schweizer, *AJ*, **3** (1853 May 18), pp. 78–9; J. S. Goodenough, *MNRAS*, **13** (1853 Jun. 10), pp. 239–40; K. C. Bruhns, *AN*, **36** (1853 Jul. 1), p. 389; C. W. Moesta, *AJ*, **3** (1853 Jul. 11), p. 104; C. W. Moesta and J. S. Goodenough, *AJ*, **3** (1853 Sep. 3), pp. 117–18; *PRSL*, **6** (1853 Dec.), p. 375; J. S. Goodenough, W. W. Vine, T. Maclear, and S. H. Wright, *MNRAS*, **13** (Supplemental Notice, 1853), pp. 272–6; T. Maclear, A. G. Constable, Bosquet, J. Parish, P. P. King, *MNRAS*, **14** (1853 Nov. 11), pp. 1–11; J. Challis, *MNRAS*, **14** (1854 Jun. 9), p. 217; T. Maclear, *MNRAS*, **15** (1855 Jan. 12), pp. 73–9; T. Maclear, *AJ*, **4** (1855 Mar. 23), pp. 73–5; G. F. W. Rümker, *AN*, **45** (1857 Mar. 26), pp. 271–84; J. N. Stockwell, *AJ*, **5** (1857 Mar. 31), p. 36; V1964, p. 57.

C/1853 L1   *Discovered:* 1853 June 11.01 ($\Delta = 2.21$ AU, $r = 1.84$ AU, Elong. $= 55°$)
(Klinkerfues)   *Last seen:* 1854 January 12.0 ($\Delta = 1.97$ AU, $r = 2.58$ AU, Elong. $= 118°$)
Closest to the Earth: 1853 September 5 (0.7126 AU)
1853 III   *Calculated path:* UMa (Disc), LMi (Jun. 23), UMa (Aug. 2), LEO (Aug. 19), VIR-LEO (Aug. 31), CRT (Sep. 7), HYA (Sep. 10), ANT (Sep. 21), VEL (Oct. 27), PUP (Dec. 13)

W. Klinkerfues (Göttingen, Germany) discovered this comet on 1853 June 11.01 at a position of $\alpha = 9^h 31.9^m$, $\delta = +43° 28'$. He described it as small, with a tail 3–4′ long. Klinkerfues confirmed his find on June 11.95.

The comet was over two months from its closest approaches to the sun and Earth when discovered; however, despite its initial faintness, it was well observed during the remainder of June and throughout July. On June 17, K. Hornstein (Vienna, Austria) said the tail extended 2–3′ and was about 1.5′ wide. He added that the condensation gave the impression that it contained a conglomeration of bright nuclei. J. A. C. Oudemans (Leiden, Netherlands) described the comet as small and faint on the 21st, with a nucleus of magnitude 10 and a short tail. J. F. J. Schmidt (Olmütz, now Olomouc, Czech Republic) gave the nuclear magnitude as 8 on June 26 and 28, while the tail was described as between 2.5′ and 4′ long. A. Reslhuber (Kremsmünster, Austria) said the comet exhibited a short tail on the 28th, while Oudemans noted a distinct nucleus of magnitude 11. Oudemans said the comet appeared brighter on the 29th than on the 28th, while the tail extended toward PA 110°. Hornstein reported the tail was 6–7′ long on July 13, despite bright moonlight. On July 21, Oudemans said the tail was fainter and the coma brighter than when observed a couple of days earlier. The nucleus was eccentrically situated within the coma. Reslhuber described the comet as bright on the 24th, with a distinct nucleus and a short, fan-shaped tail. Hornstein said the tail was 9′ long on the 25th. Oudemans said the comet was brighter on the 27th than on the 21st and exhibited a distinct tail. He noted the tail extended toward PA 67.0° on the 30th. J. Ferguson (Washington, DC, USA) saw the comet on July 29 in misty skies. He said the comet would probably have been visible to the naked eye if the night had been clear, and added that the bright, condensed nucleus was about 8″ across. Schmidt continually gave the nuclear magnitude as between 7 and 8 throughout July, while the coma grew from 1′ to 2′ and the tail length increased from 4′ to 8.5′. Schmidt added that he noted a granulation within the coma beginning on July 25 that was visible at magnifications up to 300×.

The comet brightened rapidly during August. Naked-eye observations were reported by Schmidt on the 3rd, A. Secchi (Rome, Italy) on the 7th, and Oudemans on the 9th. Secchi estimated the magnitude as 6. The comet was easily observed in twilight with a telescope after the 13th and naked-eye observations were reported when the comet was in twilight after the 19th by Reslhuber, E. Plantamour (Geneva, Switzerland), and others. The magnitude of the nucleus was estimated by several astronomers. Schmidt gave it as 7 on the 3rd, 6.5 on the 5th and 9th, and 5.5 on the 11th and 12th. On the 13th, Schmidt gave it as 5, while Reslhuber estimated it as 4. Schmidt said it increased from 5 on the 14th to 3.5 on the 19th, while, on the latter date, Reslhuber said it was between 2 and 3. Schmidt made nuclear magnitude estimates of 3.5 on the 20th, 3 on the 21st and 22nd, 2.5 on the 24th, 2 on the 26th, about 1.5 on the 28th, and about 1 on the 30th. Reslhuber gave it as 1 on the 26th. Ferguson measured the diameter of the nuclear condensation as 10.5″ on the 13th, 11.4″ on the 22nd, 15.0″ on the 23rd, 11.5″ on the 27th, and 16.9″ on the 30th. Reslhuber said the nucleus seemed to flicker on the 22nd. According to an extensive series of coma measurements by Schmidt, the coma generally maintained a diameter of 1.6–1.7′ throughout August, with extremes of 1.24′ on the 22nd and 1.82′ on the 24th. Secchi reported jets were visible in the front of the coma on the 24th and 25th, and noted the coma seemed smaller on the 27th than on previous nights. Foster (Brussels, Belgium) described the telescopic appearance of the comet on the 28th as a "clear and well-defined globe . . . of a bright golden colour." The tail length underwent a rapid increase as August progressed, with numerous estimates coming from Reslhuber, Schmidt, and Secchi. According to Reslhuber, the tail was 4–5′ long on the 1st. Reslhuber and Schmidt indicated it had increased to between 1.8° and 2.5° by the 20th. Reslhuber and Schmidt gave lengths of 10–11.4° on the 26th, and Schmidt reported the longest tail length of 12.5° on the 28th. Oudemans measured the tail's position angle as it decreased from 76.6° on the 2nd to 57.1° on the 11th. J. Hartnup (Liverpool, England) described the tail on the 29th as consisting of "two curved rays of light which united at about 3° from the nucleus." Reslhuber described the tail as parabolic in shape on the 13th. In the August 22, 1853 issue of the *Compte Rendu*, D. F. J. Arago reported that he had received several letters relating to the appearance of a brilliant, naked-eye comet in the northern area of the sky on August 19 and 20. An independent discovery was made on the 23rd by W. Bradshaw (HM Troopship *Resistance*, near the Bahamas).

Astronomers prepared to search for the comet in daylight as it approached and passed the sun during the next few days. Schmidt was the first to see the comet in daylight with a telescope, when he found it with a 13-cm refractor on August 30.64 about 14.9° from the sun. Reslhuber saw the comet shortly after sunset on the 30th and said it displayed a bright nucleus, while Plantamour noted the comet was still visible to the naked eye. Although Reslhuber was unsuccessful in finding the comet in daylight on the 31st, Schmidt did locate it in a 3-foot focal length refractor on August 31.33 and in the 13-cm refractor on August 31.45. Schmidt said it was just 13.1° from the sun. Reslhuber tried to find the comet on September 1 a short time prior to sunset, but failed; however, he did locate the comet after sunset. Schmidt continued his daylight observations and noted the comet was 10.7° from the sun on September 1.72, 9.4° away on September 2.46, and 7.9° away on September 3.48. Hartnup found the comet less than an hour after local noon on September 3.54. He turned his 22-cm refractor (92×) to the predicted position "and on applying my eye to the telescope I immediately detected the comet almost in the centre of the field. Its appearance was planetary, the nucleus was round, well defined, of a whitish colour, and about 9" diameter. I could not detect the slightest appearance of a tail." The final daylight observation was made by Schmidt on September 4.49. He said the comet was then 7.8° from the sun.

An event occurred on the evening of September 2, which, although referred to as the aurora borealis, was argued by K. C. Bruhns (Berlin, Germany) in late September to have been the tail of this comet. At Liverpool (England), a "remarkable luminous appearance was observed . . . over that part of the horizon where the comet had set a couple of hours previous." At Durham (England), "a luminous streak like the tail of a comet was seen extending above the horizon in the same direction, and the surrounding part of the heavens was seen in a glare of light attributed to an aurora." On September 2.86, C. K. L. Rümker (Hamburg, Germany) saw an object resembling "the tail of a comet, slightly curved, with the concave side to the north." On September 2.88, at Rundhof (Germany), "a phenomenon resembling the tail of a comet was observed in the west-southwest, extending 45° above the horizon, inclining toward the south with the concave side southward, and spreading like a fan." This observer detected "a splendid aurora" to the north slightly more than one hour later. Bruhns computed the probable azimuth and inclination of the comet's tail to the horizon. He found a perfect agreement with the observed azimuth, but a different inclination. He pointed out, however, that "it has been frequently observed that the tails of comets are curved, and a similar curvature might have occasioned this deviation."

The comet's minimum elongation from the sun was about 8° on September 4 and it was lost in twilight for several nights thereafter. Because of its southern declination, the first observations after perihelion were made in the Southern Hemisphere. G. P. Heath (HMS *Calliope*, near Auckland, New Zealand) saw the comet distinctly on September 12.7, just a short time before sunrise. B. Drury (HMS *Pandora*, just north of New Zealand) saw the comet on September 12.73 and said, "to the naked eye the nucleus first appeared four times the size of Jupiter, of a pale, but considerable brightness, as if the light of Jupiter were spread over four times its surface. The tail was about 5° in length, making an angle with the vertical circle of 15° south of the zenith." T. Maclear (Royal Observatory, Cape of Good Hope, South Africa) recovered the comet and measured its position on September 13.16. He was using an 8.5-foot refractor and noted the comet was a bright orange color and exhibited a tail 6° or 7° long. Heath said the comet was colourless, with a tail 4° 45' long on the 14th. He added that the tail gradually widened further from the comet's head, with a maximum width of about 0.5°. C. W. Moesta (Santiago, Chile) independently discovered the comet in the morning sky on September 17 and said, "It is visible to the naked eye, its nucleus being very well defined, and as brilliant as a star of the 6–7 magnitude. Its tail is so faint that it is scarcely possible to determine its length." Drury saw the comet for the final time on the 25th and noted the brightness had considerably decreased since the 12th.

For a brief period during the latter half of September and early October Secchi resumed observations from the Northern Hemisphere. He simply noted it was a large nebulosity on the 27th. On the 29th he said it was visible to the naked eye, while a telescope revealed a rudimentary tail about 8′ long. Secchi again saw the comet with the naked eye on the 30th, when a telescope revealed a narrow tail about 1° long and 1–2′ wide, as well as a coma 5–6′ across. He found a tail about 2.75° long on October 1, and 2.5° long and 1.5′ wide on the 2nd.

From October 9 onward, observations were only made at Royal Observatory, Cape of Good Hope, South Africa, but no physical descriptions were provided. The comet attained its most southerly declination of −46° on December 7. The comet was last detected on 1854 January 10.10, when Maclear used the 8.5-foot refractor and gave the position as $\alpha = 6^{\rm h} 52.5^{\rm m}$, $\delta = -39° 58′$. Maclear added that the comet was also seen on the night of January 11/12, but it was "so faint that no observations worth recording could be made."

The first parabolic orbit was calculated by Bruhns using positions spanning the period of June 13–24. The resulting perihelion date was 1853 August 27.68. Bruhns later refined the orbit using positions spanning the period of June 17 to July 23 and determined the perihelion date as September 2.17. Further refinements by C. Mathieu, J. S. Hubbard, H. L. d'Arrest, and J. N. Stockwell (1856) only slightly improved the orbit. Although T. Krahl (1867) calculated a parabolic orbit very similar to those previously published, he also calculated a hyperbolic orbit with a perihelion date of September 2.21 and an eccentricity of 1.00026085. This hyperbolic orbit was confirmed by H. Büttner (1918), who also applied perturbations by eight planets. His result of a perihelion date of September 2.21 and an eccentricity of 1.00024641 is given below.

T	ω	Ω (2000.0)	i	q	e
1853 Sep. 2.2069 (UT)	170.4415	142.5660	61.5005	0.306839	1.000246

ABSOLUTE MAGNITUDE: $H_{10} = 4.8$ (V1964)

FULL MOON: May 22, Jun. 21, Jul. 20, Aug. 18, Sep. 17, Oct. 17, Nov. 15, Dec. 15, Jan. 14

SOURCES: W. Klinkerfues, *MNRAS*, **13** (1853 Jun. 10), p. 239; W. Klinkerfues, *AN*, **36** (1853 Jun. 14), p. 375; K. C. Bruhns, *AN*, **36** (1853 Jul. 1), p. 391; W. Klinkerfues, *AJ*, **3** (1853 Jul. 11), p. 104; K. C. Bruhns and C. K. L. Rümker, *AN*, **37** (1853 Aug. 9), pp. 85–8; D. F. J. Arago, *CR*, **37** (1853 Aug. 22), p. 293; V. Trettenero, *AN*, **37** (1853 Aug. 23), p. 93; J. Ferguson, K. C. Bruhns, and J. S. Hubbard, *AJ*, **3** (1853 Sep. 3), pp. 113–15, 118–19; C. Mathieu, *CR*, **37** (1853 Sep. 5), p. 412; K. Hornstein, *AN*, **37** (1853 Sep. 13), pp. 145–8; E. Plantamour, H. L. d'Arrest, and V. Trettenero, *AN*, **37** (1853 Sep. 27), pp. 189–92, 195; J. Hartnup and J. F. J. Schmidt, *AN*, **37** (1853 Oct. 4), pp. 209–18;

J. Ferguson, *AJ*, **3** (1853 Oct. 11), pp. 124–6; J. Ferguson, C. K. L. Rümker, and H. L. d'Arrest, *AN*, **37** (1853 Oct. 11), pp. 225–36; J. F. J. Schmidt, *AN*, **37** (1853 Oct. 18), pp. 237–50; W. Klinkerfues, K. C. Brühns, J. Hartnup, J. F. J. Schmidt, and H. L. d'Arrest, *MNRAS*, **13** (Supplemental Notice, 1853), pp. 276–7; AJS (Series 2), **16** (1853 Nov.), p. 431; A. Reslhuber, *AN*, **37** (1853 Nov. 8), pp. 291–6; Foster and J. Hartnup, *MNRAS*, **14** (1853 Nov. 11), pp. 11–12; J. A. C. Oudemans, *AN*, **37** (1853 Dec. 6), pp. 349–53; C. W. Moesta, *AJ*, **3** (1853 Dec. 8), p. 143; W. Bradshaw and C. K. L. Rümker, *MNRAS*, **14** (1853 Dec. 9), pp. 33–8; G. P. Heath and B. Drury, *MNRAS*, **14** (1854 Mar. 10), pp. 149–51; K. L. von Littrow, *AN*, **38** (1854 Mar. 13), p. 107; A. Secchi, *AN*, **38** (1854 Mar. 27), pp. 137–42; T. Maclear, *AJ*, **3** (1854 May 31), pp. 182–3; C. W. Moesta, *AJ*, **3** (1854 Jun. 15), p. 191; T. Maclear, *AJ*, **4** (1855 Nov. 5), pp. 121–5; J. N. Stockwell, *AJ*, **5** (1856 Nov. 26), pp. 1–6; T. Maclear, *MRAS*, **31** (1861–2), pp. 1–17; T. Krahl, *AN*, **70** (1867 Sep. 21), pp. 1–40; H. Büttner, *AN*, **207** (1918 Sep. 13), pp. 179–82; V1964, p. 57.

C/1853 R1   *Discovered:* 1853 September 12.01 ($\Delta = 1.20$ AU, $r = 1.04$ AU, Elong. $= 55°$)
(Bruhns)   *Last seen:* 1853 December 12.19 ($\Delta = 1.88$ AU, $r = 1.46$ AU, Elong. $= 50°$)
   Closest to the Earth: 1853 October 5 (0.6544 AU)
1853 IV   *Calculated path:* LYN (Disc), LMi (Sep. 19), LEO-LMi (Sep. 24), LEO (Sep. 27), LMi-LEO (Sep. 29), VIR (Oct. 5), CRV (Oct. 11), VIR (Oct. 12), BOO (Nov. 21), SER (Nov. 30)

K. C. Bruhns (Berlin, Germany) discovered this comet on 1853 September 12.01, and gave the position as $\alpha = 8^h\ 27.9^m$, $\delta = +44°\ 52'$. He described it as a "large faint nebulous comet, resembling a star cluster." It was easily visible through a cometseeker. Bruhns confirmed his find on September 12.09.

The comet approached the sun and Earth throughout September and was well observed, although only a few astronomers provided physical descriptions. Bruhns saw it on September 13 and 15, and noted there was no tail, while the nucleus looked like a star cluster with numerous points of light. On the 22nd, Bruhns said the comet was similar in brightness to a star of magnitude 6–7, and exhibited a nucleus, but no tail.

The comet was brightest during October, as it passed closest to Earth on the 5th and closest to the sun on the 17th. H. L. d'Arrest (Leipzig, Germany) estimated the magnitude as 5–6 on the 1st and said the coma was about 5' across, with a straight and narrow tail extending 3° 35' towards PA 47° 22'. On the same night, Bruhns said the coma contained a 4th-magnitude nucleus 0.8' across, while a tail extended 15' away from the sun. On October 3, d'Arrest estimated the magnitude as 5, A. Marth (Königsberg, now Kaliningrad, Russia) said it appeared about magnitude 4 to the naked eye, while Bruhns said the comet was similar in brightness to θ Leonis (magnitude 3.34 – SC2000.0). D'Arrest said the coma was about 5' across, with a straight and narrow tail extending towards PA 43° 15'. Marth said the coma exhibited a strong central condensation and the tail was 1° long. Bruhns said the tail extended 0.75–1°. On October 5, d'Arrest estimated the magnitude as 4–5. He added that the coma was about 5' across, with a straight and narrow tail extending 3° 50′ towards PA 41° 31′. Bruhns saw the comet in strong twilight on October 6.19, which marked the last time the comet was seen before its conjunction with the sun. He said the tail was distinctly visible.

The comet passed about 2° from the sun on October 24 and moved very slowly away from it as it faded. Bruhns recovered it on November 28.18 at an elongation of 37°. He described it as rather diffuse and faint. A. Colla (Parma, Italy) independently recovered the comet on the morning of November 29, while using a 10.2-cm refractor at 60×. He described it as "extremely faint, without nucleus, without trace of tail, exhibiting only, even with a magnifying power of 100, a very minute nebulosity of uniform brightness, with a feeble appearance of scintillation by intervals."

On December 1, Bruhns said the comet appeared diffuse, while d'Arrest described it as an extremely faint nebulosity because of its nearness to a star of magnitude 8–9. Colla observed the comet with a 10.2-cm refractor on the 3rd and wrote, "The nebulosity was more distinct, but there was no trace of nucleus, of tail, or of scintillation, even when employing a magnifying power of 100." The comet was last detected on December 12.19, when J. F. J. Schmidt (Olmütz, now Olomouc, Czech Republic) gave the position as $\alpha = 15^{\mathrm{h}}$ 19.0^{m}, $\delta = +18°$ 07′. He somewhat doubted the accuracy of the position because of the comet's faintness (5-foot focal length refractor, 40×).

The first parabolic orbit was calculated by Bruhns using positions obtained during the period of September 12–15. The resulting perihelion date was 1853 October 17.07. The orbit proved to be a very accurate representation, with the perihelion date just over 1 hour early. Later orbits by d'Arrest, Bruhns, and Hoffmann offered only minor alterations to Bruhns' initial parabolic solution. S. Alexander (Princeton, New Jersey, USA) suggested the comet of 1582 was a possible previous return of this comet, but this quickly proved impossible.

The comet ultimately proved to be moving in a hyperbolic orbit, as was first established by d'Arrest (1854). He used positions obtained between September 12 and December 12, and determined a perihelion date of October 17.11 and an eccentricity of 1.0012289. R. J. Buckley (1979) revised the orbit over a century later. He used 17 positions obtained between September 12 and December 11, applied perturbations by seven planets, and determined a perihelion date of October 17.12 and an eccentricity of 1.000664. This orbit is given below.

T	ω	Ω (2000.0)	i	q	e
1853 Oct. 17.1244 (TT)	277.8406	222.1158	118.9964	0.172863	1.000664

ABSOLUTE MAGNITUDE: $H_{10} = 7.0$ (V1964)

FULL MOON: Aug. 18, Sep. 17, Oct. 17, Nov. 15, Dec. 15

SOURCES: K. C. Bruhns, F. F. E. Brünnow, and Hoffmann, *AN*, **37** (1853 Sep. 20), p. 187; K. C. Bruhns, *AN*, **37** (1853 Sep. 27), p. 193; K. C. Bruhns, O. L. Lesser, and Hoffmann, *AN*, **37** (1853 Oct. 4), pp. 217–20; K. C. Bruhns, *AJ*, **3** (1853 Oct. 11), pp. 127–8;

H. L. d'Arrest, *AN*, **37** (1853 Oct. 11), pp. 231, 235; K. C. Bruhns and Hoffmann, *AN*, **37** (1853 Oct. 25), pp. 259–62; S. Alexander, *AJ*, **3** (1853 Nov. 1), pp. 130, 132; H. L. d'Arrest, A. Marth, and J. R. Hind, *AN*, **37** (1853 Nov. 1), pp. 275–80; K. C. Bruhns, *MNRAS*, **13** (Supplemental Notice, 1853), p. 277; K. C. Bruhns, *AN*, **38** (1854 Jan. 17), pp. 27–32; H. L. d'Arrest, *AN*, **38** (1854 Jan. 24), pp. 35–8; A. Colla, *MNRAS*, **14** (1854 Mar. 10), p. 166; J. F. J. Schmidt, *AN*, **38** (1854 Mar. 28), pp. 157–60; H. L. d'Arrest, *AN*, **38** (1854 Apr. 20), pp. 187–90; V1964, p. 57; R. J. Buckley, *JBAA*, **89** (1979 Oct.), p. 583–4; SC2000.0, p. 286.

C/1853 W1
(van Arsdale)
1854 I

Discovered: 1853 November 25.99 ($\Delta = 1.28$ AU, $r = 2.10$ AU, Elong. $= 135°$)
Last seen: 1854 March 1.79 ($\Delta = 2.78$ AU, $r = 2.15$ AU, Elong. $= 42°$)
Closest to the Earth: 1853 December 1 (1.2668 AU)
Calculated path: CAS (Disc), PER (Nov. 27), AND (Dec. 4), PSC (Dec. 18), CET (Feb. 9)

R. van Arsdale (Newark, New Jersey, USA) discovered this comet in Cassiopeia on 1853 November 25.99, and estimated the position as $\alpha = 2^h 07^m$, $\delta = +60° 12'$. He described it as "small, round, and bright" in a 16.2-cm refractor. Van Arsdale could not confirm the new object until December 1.08. An independent discovery was made by W. Klinkerfues (Göttingen, Germany) on December 3.14.

W. J. Förster (Bonn, Germany) said the comet was rather difficult to see on December 10–13, because of bright moonlight and unfavorable skies. Following the December full moon, C. K. L. Rümker (Hamburg, Germany) said the coma was moderately small and exhibited a planetary appearance. That same evening, Förster noted a considerable tail. K. C. Bruhns (Berlin, Germany) saw a short tail on the 25th. Förster estimated the nucleus as magnitude 8–9 on the 26th. K. L. von Littrow (Vienna, Austria) estimated the tail as 4′ long on the 30th and saw a tiny nucleus.

The comet passed perihelion during the first days of January and began fading as it headed away from both the sun and Earth. Littrow estimated the tail as 2′ long on the 2nd and then noted the comet was very faint in moonlight on the 3rd. After the January full moon, A. Reslhuber (Kremsmünster, Austria) described the comet as very faint on the 18th, 23rd, 24th, and 25th, while A. Secchi (Rome, Italy) said it was very faint from the 22nd to the 30th. J. F. J. Schmidt (Olmütz, now Olomouc, Czech Republic) saw the comet on January 23 and 25, and noted the comet was very small and faint, with a trace of a curved tail. Littrow described the comet as faint, with a tail about 1.5′ long on the 26th. Most observers no longer saw the comet after January, but K. C. Bruhns (Berlin) described the comet as a "not uncommonly faint" object on February 26.

The comet was last detected on 1854 March 1.79, by Förster, at a position of $\alpha = 1^h 38.3^m$, $\delta = -2° 31'$. He said the nucleus was seen with great difficulty. Bruhns failed to locate the comet in early March.

Klinkerfues determined the first orbit for this comet which was published on December 9. He computed the perihelion date as 1854 January 10.30. By mid-month, he used three positions obtained between December 3 and 12, and computed a parabolic orbit with a perihelion date of January 10.33. Bruhns used three positions obtained between December 3 and 17, and computed a parabolic orbit with a perihelion date of January 5.45. Additional calculations by Förster, J. A. C. Oudemans, A. Marth, Klinkerfues, and S. Alexander narrowed the perihelion date down to January 4.4. Rzepecki (1857) used 80 positions obtained between December 3 and March 1, and computed a parabolic orbit with a perihelion date of January 4.44. No planetary perturbations were applied. The comet might have been found sooner if it had not been for moonlight interference from the November 15 full moon.

T	ω	Ω (2000.0)	i	q	e
1854 Jan. 4.4351 (UT)	170.9202	229.0785	113.8757	2.044645	1.0

ABSOLUTE MAGNITUDE: $H_{10} = 4.3$ (V1964)

FULL MOON: Nov. 15, Dec. 15, Jan. 14, Feb. 13, Mar. 14

SOURCES: W. Klinkerfues, *AN*, **37** (1853 Dec. 6), p. 363; R. van Arsdale, *AJ*, **3** (1853 Dec. 8), p. 144; W. Klinkerfues, *MNRAS*, **14** (1853 Dec. 9), p. 33; W. Klinkerfues, *AN*, **37** (1853 Dec. 30), p. 409; C. K. L. Rümker and K. C. Bruhns, *AN*, **38** (1854 Jan. 10), pp. 11, 15; C. K. L. Rümker, *MNRAS*, **14** (1854 Jan. 13), pp. 65–6; K. C. Bruhns, *AN*, **38** (1854 Jan. 17), p. 31; B. A. Gould, R. van Arsdale, J. A. C. Oudemans, and A. Marth, *AN*, **38** (1854 Jan. 24), pp. 37–40, 45; W. J. Förster, *AN*, **38** (1854 Jan. 31), pp. 49, 54–6; S. Alexander, *AJ*, **3** (1854 Feb. 17), pp. 146–7, 150–1; R. van Arsdale, *AJS* (Series 2), **17** (1854 Mar.), p. 286; W. Klinkerfues, *AN*, **38** (1854 Mar. 4), p. 93; W. C. Bond, *MNRAS*, **14** (1854 Mar. 10), p. 167; K. L. von Littrow, *AN*, **38** (1854 Mar. 13), p. 109; W. C. Bond, *AJ*, **3** (1854 Mar. 16), pp. 159–60; A. Secchi, *AN*, **38** (1854 Apr. 20), p. 191; J. F. J. Schmidt, *AN*, **38** (1854 May 10), pp. 213–16; K. C. Bruhns, *AN*, **38** (1854 May 19), p. 269; A. Reslhuber, *AN*, **38** (1854 Jun. 27), p. 321; W. J. Förster, *AN*, **38** (1854 Jul. 4), p. 341; K. Hornstein, *AN*, **39** (1854 Nov. 20), p. 191; Rzepecki, *AN*, **47** (1857 Nov. 5), pp. 177–80; V1964, p. 57.

X/1854 F2 T. Brorsen (Senftenberg, Germany) discovered this object in the evening sky on 1854 March 16.81, and gave the position as $\alpha = 2^h 30.2^m$, $\delta = +1° 11'$. Shortly thereafter, the object dropped behind tree branches and no motion could be observed. It was not seen again. Upon observing the bright comet C/1854 F1 on April 1, K. C. Bruhns speculated that it might have been Brorsen's comet, but the brightness and motion of the former comet caused him to suspect they were not related. F. W. A. Argelander confirmed Bruhns' suspicion, when he determined new orbits for comet C/1854 F1 and found this comet to have been about 53° west of Brorsen's comet.

SOURCES: T. Brorsen, *AN*, **38** (1854 Mar. 27), p. 141; K. C. Bruhns, *AN*, **38** (1854 Apr. 12), p. 175; F. W. A. Argelander, *AN*, **38** (1854 Apr. 20), p. 185.

C/1854 F1 *Discovered:* 1854 March 23.25 ($\Delta = 1.01$ AU, $r = 0.28$ AU, Elong. $= 16°$)
(Great Comet) *Last seen:* 1854 April 28.59 ($\Delta = 1.38$ AU, $r = 0.98$ AU, Elong. $= 45°$)
Closest to the Earth: 1854 April 1 (0.8458 AU)
1854 II *Calculated path:* PEG (Disc), PSC (Mar. 25), ARI (Mar. 31), TAU (Apr. 8), ORI (Apr. 19)

A. de Menciaux (near Damazan, France) discovered this comet in the east on the morning of 1854 March 23.25. The comet was then very low over the horizon and he described it as brilliant and similar to comet C/1853 L1. The tail was nearly perpendicular to the horizon, with a slight inclination toward the north, and the comet's rising point was said to be more northerly than that of the sun. Menciaux immediately sent a letter to P. A. E. Laugier (Paris, France) announcing his find. The next morning, Laugier's colleagues C. Mathieu and E. Liouville looked for the comet, but found no trace. As it turned out the comet was heading westward and had moved closer to the horizon, with the tail shifting even more toward the north, the morning after Menciaux' discovery. It passed due north of the sun on March 26 and then entered the evening sky.

Independent discoveries were made from numerous locales during the period of March 26 to April 1. Dezautière (Decize, France) briefly noted the comet was seen on March 26.80. T. W. Webb wrote that one of the earliest observers was a 10-year-old boy in Ross, England. He said the boy spotted "a stranger in the sky" on March 28.81, which enabled several other people to see the comet before it "sank into a bank of clouds in the N.W. horizon." One of these people was W. H. Purchas (Ross, England), who told Webb that the nucleus was then as bright as β Arietis (magnitude 2.64 – SC2000.0), while the tail extended about 3° in the strong twilight. Purchas added that a 10-cm refractor revealed the convex side of the tail was very sharply defined. Another important independent discovery came from R. C. Carrington (Durham, England) on March 29.82. His careful observation led to the first precise position of $\alpha = 1^h 06.4^m$, $\delta = +19° 48'$ on March 29.83.

In addition to continued independent discoveries, numerous astronomers began routine observations during the final days of March. E. J. Lowe (Highfield House Observatory, near Nottingham, England) and J. R. Hind (Bishop's Observatory, London, England) both described the comet as large and brilliant, with a nucleus equal to a 2nd-magnitude star. Lowe added that the tail extended about 4° toward Polaris. Webb said his 9.4-cm refractor revealed the nucleus was several seconds across, while high magnifications revealed the light of the nucleus "was diffused into a hazy mass of considerable brightness, not much condensed towards the centre." He added, "the nucleus was encompassed by a narrow border of faint yellow light, forming a hemispherical cap or envelope on the side turned to the sun. . . . From this envelope rose the tail, which was very narrow in proportion to its length, widened very gradually, and exhibited a decided curvature. Its convex side seemed a little better defined than the other. . . . "

T. W. Burr (Highbury, England) saw the comet on the 30th and said the tail was 2° long. On the same date, Webb said his 9.4-cm refractor showed that the tail "ascended from the sides of the envelope in two apparent streams, very little denser, however, than the general mass, and soon becoming undistinguishable from it." A. Graham (Markree Castle, Ireland) said the nucleus was about magnitude 2, while the tail was about 3° long. Laugier, Mathieu, and Liouville observed the comet with a 10-cm refractor on the 31st. They noted the nucleus was visible in full twilight and was about 18" across. Laugier added that the edges of the tail seemed brighter than the tail's center.

The comet was closest to Earth on April 1 when K. C. Bruhns (Berlin, Germany) referred to it as a "magnificent naked eye comet." He estimated the magnitude as 2 and said the tail was 3° long. A. Reslhuber (Kremsmünster, Austria) estimated the nuclear magnitude was then 1 and described the tail as fan-shaped. Using a cometseeker he estimated the length as 4°. Laugier, Mathieu, and Liouville noted the nucleus was 18" across. K. Hornstein (Vienna, Austria) saw the comet on the 2nd and said the tail was about 2.5° long in his cometseeker, with the western side more distinct that the eastern side. He added that the nucleus was fairly sharp and the coma was oval-shaped. Hornstein commented that the comet looked similar to drawings of comet C/1811 F1 obtained during October 1811. During the period of April 1–2, U. J.-J. Le Verrier noted that drawings by E. Liais indicated a diffuse nucleus and a slightly curved tail. The tail was better defined on the southern side and was about 2° long. Burr observed the comet with a telescope on April 3, just 20 minutes after sunset. He said Saturn was as easy to see as the comet. Webb again studied the comet with his 9.4-cm refractor and with a magnification of 330× he said, "there was occasionally a suspicion of a strong luminous point in the centre of the diffused nebulous brightness; but this was seldom visible, and the air was too unsteady to admit of any certainty. With powers from 80× to 330×, the light of the nucleus appeared condensed or accumulated, though but very slightly, on the side next the sun." Webb reported the longest naked-eye tail length on April 4, noting it was over 5° long in moonlight. J. F. J. Schmidt (Olmütz, now Olomouc, Czech Republic) said the tail then extended toward PA 84.4°. He added that the nucleus was about magnitude 3. E. B. Powell (Madras, India) saw the comet for the first time on this date. He said his "attention was caught by a star-like object in the west, badly defined, and with a short, indistinct ray running upwards." On April 5, Schmidt said the nucleus was 3.72" across with a magnitude of 3. He added that the tail extended towards PA 86.5°. Laugier, Mathieu, and Liouville noted the nucleus was 19" across. Powell said a 10-cm refractor revealed a sharp outline to the envelope, while the comet's length was measured as 22'. He added that the nucleus was "pretty bright" and equaled a star of magnitude 3 or 4. Webb said the 9-cm refractor showed the nucleus was "becoming smaller and more definite." W. S. Jacob (Madras, India) said the nucleus was as bright as a star of magnitude 5 and was in the form of a crescent. He added that the outline of the envelope was "a fine parabola," while the extremity was "undefined." On April 6, Schmidt estimated the nuclear magnitude as 3–4, and said the tail extended towards PA 87.5°. Laugier, Mathieu, and Liouville noted the nucleus was 18" across. Powell said the 10-cm refractor revealed the nucleus as 8" across, with a ragged appearance, and equal to a star of magnitude 4. The tail was 30' long and seemed less broad than on the 5th. On April 7, Bruhns said the comet appeared appreciably fainter than when previously observed, and the very small nucleus exhibited an elliptical shape. J. S. Goodenough (on board HMS *Centaur*, Cape Verde islands) saw the comet at an altitude of just over 8°. Powell said clouds interfered with the observation, but he noted the nucleus seemed "somewhat larger" than on the 6th. Laugier, Mathieu, and Liouville noted the nucleus was 12" across. On April 8, Schmidt said the nucleus was 3.62" across, and the tail extended towards PA 87.2°. Laugier, Mathieu, and Liouville noted the nucleus was 16" across. Powell observed with a 10-cm refractor and said the comet appeared fainter because of moonlight and the outline of the envelope was "indiscernible." He added that the nucleus was similar in brightness to a star of magnitude 6. On April 9, Burr said the comet was not visible to the naked eye. He also noted that α Arietis became visible "some time before the comet could be discovered, although the star was in much the more unfavourable position with regard to the twilight." On April 10, Reslhuber said the comet continued to fade, while the tail became shorter and more pale. Schmidt said the comet was still visible to the naked eye, while a refractor showed a nucleus 3.49" across. Jacob observed with a refractor and said bright moonlight made the envelope "barely visible," although nearly 1° of tail was still detected. He added that the nucleus was "more concentrated" than when previously seen and appeared reddish "with ill-defined *horns*" toward the west.

With the comet rapidly fading and moonlight becoming a problem, some observers stopped following the comet, while others no longer provided physical descriptions. But a few continued to weather the increasingly adverse observing conditions. Schmidt continued to estimate the nuclear magnitude and measure the nuclear diameter. He gave them as magnitude 7 and 3.41" on April 11, 7 and 2.88" on the 12th, 7–8 and 4.32" on the 13th, 8 and 4.87" on the 14th, and 8 and 3.65" on the 15th. Thereafter, he estimated the nuclear magnitude as 8–9 on the 16th, 8 on the 17th, and 9 on the 19th. Reslhuber observed the comet in moonlight on April 14 and noted the tail was 12–15' long. He added that by sighting up the telescope tube he could barely see the comet with the naked eye. Powell observed with a 10-cm refractor on the 18th and said "the comet would not bear any illumination." Despite its faintness, he said the tail seemed about 1° long. Reslhuber described the comet as very faint on the 19th.

The comet was last detected on April 28.59, when Powell saw it with a 10-cm refractor and gave the position as $\alpha = 5^h \, 10.8^m$, $\delta = -2° \, 10'$. He said the comet "was so excessively faint that the observations were taken with the greatest difficulty. In fact, it was no more than a small, whitish, cloudy patch; still there was something like a faint star as a nucleus."

The first parabolic orbits were calculated by Carrington and Hind. They each used their own positions obtained on March 29, 30, and 31. Carrington's perihelion date was 1854 March 24.63, while Hind's was March 24.84. Several days later E. Schönfeld took positions from April 1, 2, and 3, and determined a perihelion date of March 24.47, while C. N. A. Krueger and W. J. Förster used positions from April 1 to April 5 to determine a perihelion date of March 24.50. These last two orbits proved quite accurate and no substantial improvement was obtained by later orbits determined by J. Challis, Reslhuber, C. W. Tuttle, Bruhns, J. Ferguson, Hind, Hornstein, A. Graham, F. W. A. Argelander, Mathieu, A. M. Nell, and Powell. Hind noted, "These elements have some slight resemblance to those of the second comet of 1799... but the similarity is not sufficiently great to justify a suspicion of identity." H. Oppenheim (1870, 1885) took about 200 positions obtained between 1854 March 29 and 1854 April 28, derived five Normal positions, and computed a parabolic orbit with a perihelion date of March 24.51. No planetary perturbations were applied. Oppenheim's orbit is given below.

Shortly, after the first orbits appeared, Bruhns suspected this comet might be identical to the object found by Brorsen on March 16 (see X/1854 F2), although he was somewhat uncertain because of the large differences in brightness and rate of motion. Argelander offered improved orbits for this comet shortly before mid-April and concluded this comet was not the same as Brorsen's object.

T	ω	Ω (2000.0)	i	q	e
1854 Mar. 24.5132 (UT)	101.6257	317.4957	97.4868	0.277064	1.0

ABSOLUTE MAGNITUDE: $H_{10} = 7.0$ (V1964)

FULL MOON: Mar. 14, Apr. 13, May 12

SOURCES: R. C. Carrington, E. J. Lowe, and J. R. Hind, *MNRAS*, **14** (1854 Mar.), pp. 152–3; J. Challis, R. C. Carrington, F. W. A. Argelander, Schönfeld, C. N. A. Krueger, W. J. Förster, P. A. E. Laugier, J. R. Hind, A. Graham, T. W. Burr, and J. C. Adams, *MNRAS*, **14** (1854 Apr.), pp. 174–81; J. S. Goodenough, *MNRAS*, **14** (1854 Apr.), pp. 187–8; U. J.-J. Le Verrier, E. Liais, P. A. E. Laugier, A. de Menciaux, C. Mathieu, and E. Liouville, *CR*, **38** (1854 Apr. 3), pp. 647–50; J. R. Hind and Dezautière, *CR*, **38** (1854 Apr. 3), p. 693; A. J. F. Yvon-Villarceau, J. Chacornac, A. Graham, F. W. A. Argelander, and T. Brorsen, *CR*, **38** (1854 Apr. 10), pp. 711–13; K. C. Bruhns and F. F. E. Brünnow, *AN*, **38** (1854 Apr. 12), p. 175; P. A. E. Laugier, C. Mathieu, E. Liouville, and K. Hornstein, *CR*, **38** (1854 Apr. 17), pp. 718–19, 749; A. J. F. Yvon-Villarceau, J. Chacornac, E. Cooper, and A. Graham, *CR*, **38** (1854 Apr. 24), pp. 783–4; E. Cooper and A. Graham, *MNRAS*, **14** (1854 May), p. 193; F. W. A. Argelander and K. Hornstein, *AN*, **38** (1854 Apr. 20), pp. 183–6; K. Hornstein and K. C. Bruhns, *AN*, **38** (1854 Apr. 29), pp. 201–4; C. N. A. Krueger and A. Graham, *CR*, **38** (1854 May 15), pp. 887, 890; J. Ferguson, *AJ*, **3** (1854 May 31), pp. 183–4; E. B. Powell and T. W. Webb, *MNRAS*, **14**

(1854 Jun.), pp. 218–24; A. M. Nell, *AN*, **38** (1854 Jun. 9), p. 311; C. Mathieu, *CR*, **38** (1854 Jun. 12), pp. 1064–6; W. C. Bond and C. W. Tuttle, *AJ*, **3** (1854 Jun. 15), p. 189; A. Reslhuber, *AN*, **38** (1854 Jun. 27), pp. 321–4; A. de Menciaux and J. Ferguson, *AJS* (Series 2), **18** (1854 Jul.), p. 138; J. F. J. Schmidt, *AN*, **38** (1854 Jul. 18), pp. 361–6; W. S. Jacob, *MNRAS*, **14** (Supplemental Notice, 1854), p. 246; E. B. Powell, *MNRAS*, **15** (1854 Nov. 10), p. 34; E. B. Powell, *MNRAS*, **15** (1854 Dec. 8), pp. 60–1; H. Oppenheim, *Bahnbestimmung von Comet II. des Jahres 1854*. Dissertation: Göttingen (1870), 39pp; J. G. Galle and H. Oppenheim, *AN*, **113** (1885 Nov. 30), p. 55; V1964, p. 57; SC2000.0, p. 38.

C/1854 L1   *Discovered:* 1854 June 5.06 ($\Delta = 1.19$ AU, $r = 0.75$ AU, Elong. $= 39°$)
(Klinkerfues)   *Last seen:* 1854 July 30.87 ($\Delta = 1.78$ AU, $r = 1.02$ AU, Elong. $= 28°$)
   Closest to the Earth: 1854 June 21 (0.9877 AU)
1854 III   *Calculated path:* TRI (Disc), AND (Jun. 7), PER (Jun. 10), CAM (Jun. 16), LYN (Jun. 25), CAM (Jun. 27), LYN (Jun. 28), UMa (Jul. 2), LMi (Jul. 13)

W. Klinkerfues (Göttingen, Germany) discovered this comet on 1854 June 5.06 at a position of $\alpha = 2^h 01.2^m$, $\delta = +32° 15'$. He confirmed his find on June 6.04. Klinkerfues noted, "Since the dawn was already rather bright, then the nature of this comet-like object remained somewhat doubtful [on the first night], and this doubt could only be completely lifted on the following night." The comet attained its greatest solar elongation of 39° on June 7.

This comet would pass closest to the sun and Earth on June 22 and June 21, respectively. F. W. A. Argelander (Bonn, Germany) remarked on the comet's rapidly increasing motion on June 11 and said it "is very bright, and could be seen very well in the finder... very near to the horizon, notwithstanding the light of the full moon." He added that it was centrally condensed, but exhibited neither a defined nucleus nor a tail. On June 16, K. Hornstein (Vienna, Austria) said the comet was about magnitude 6. He added that the coma was 5' across, while the tail was about 1° long and narrow, with a width near the coma of only 2–3'. J. Chacornac (Paris, France) described the comet as very brilliant on the 17th, with a tail 18–20' long. The tail was said to be brighter along the edges than in the middle. The next night Chacornac said the comet was not as well seen as on the previous morning, but the condensation appeared larger. The comet was observed with the naked eye on June 19 by J. F. J. Schmidt (Olmütz, now Olomouc, Czech Republic), A. Reslhuber (Kremsmünster, Austria), Chacornac and his colleague J. J. E. Goujon. Reslhuber then estimated the magnitude as 5 and added that the coma was 2' across, but contained no nucleus and exhibited a very fine jet extending about 0.5° away from the sun. Chacornac noted the nucleus was visible to the naked eye, while the tail was less visible than on the morning of the 17th. Reslhuber said the comet had brightened and the tail had increased in length between the 19th and 23rd, while Hornstein noted the tail had become brighter and wider between the 16th and 23rd. Schmidt saw the comet on the 23rd and estimated the magnitude as 5–6, the coma diameter as about 4.5', and the tail as 25' long. The comet attained its most northerly declination of +61° on June 25 and was then independently discovered by R. van Arsdale (Newark, New Jersey, USA). He described it as bright, with a tail about 1° long. On June 26, Reslhuber estimated the magnitude as 4, while Schmidt estimated it as 5. Schmidt indicated the coma was 4.4' across and the tail extended 27' toward PA 7° 57'. Hornstein saw the comet in a cometseeker on the 27th and estimated the magnitude as 5, while Reslhuber simply noted it was brighter than on the 26th. The tail length was estimated as 45' long by Hornstein and 1.5° long by Reslhuber. Schmidt estimated the magnitude as 5 on the 28th and indicated the coma was 4.6' across and the tail extended 25' toward PA 19° 54'. A. J. F. Yvon-Villarceau (Paris, France) then noted the nucleus was not very distinct. On June 30, Schmidt estimated the magnitude as 5. He said the tail extended 39' towards PA 30° 18'.

The comet moved away from both the sun and Earth as July progressed. On July 3, Schmidt estimated the magnitude as 5–6, while Reslhuber said the comet was the same brightness as on June 27. Schmidt said the tail was 20' long. Chacornac said the comet exhibited a very diffuse nucleus and seemed to contain several areas of condensation. Schmidt said the tail extended toward PA 52° 33' on the 6th. Reslhuber found the comet much fainter on the 8th, while Chacornac described the nucleus as very large and diffuse. Schmidt said the tail extended toward PA 85° 13' on the 10th. On July 11, Schmidt estimated the magnitude as 6–7. He indicated a coma diameter of 2.6' and said the tail was 7' long. Reslhuber said the decrease in brightness was appreciable by the 14th, and there was hardly more than a trace of tail. Schmidt then indicated a coma diameter of 2.4'. Reslhuber described the comet as large and fairly faint on the 16th, while on the 23rd he said it was very faint because of its low altitude. The comet was last detected on July 30.87, when Reslhuber gave the position as $\alpha = 10^h 38.2^m$, $\delta = +26° 19'$. He described it as very faint, but noted it was very near a star, and, thus, difficult to see.

The first parabolic orbit was calculated by Klinkerfues. He took three positions from the period of June 6–13 and determined the perihelion date as 1854 June 22.77. Later calculations by Klinkerfues, as well as F. A. T. Winnecke, C. F. Pape, Argelander, K. C. Bruhns, Reslhuber, B. Peirce, C. Mathieu, E. Liouville, J. A. C. Oudemans, and G. Santini, were very similar, although the perihelion date was found to be closer to June 22.5. Klinkerfues pointed out the similarity between the orbit of this comet and those of 961 and 1558. R. Keith took positions obtained during the period of June 11 to July 12 and calculated an elliptical orbit, but the period was about 18 260 years. A month later, Keith used positions from the period of June 6 to July 27 and noted a hyperbolic orbit fitted the positions best, with a resulting eccentricity of 1.000442. W. R. von Hillmayr (1902) took 227 positions spanning the period of June 6 to July 30 and found a parabolic orbit still fit the positions best. This orbit is given below.

T	ω	Ω (2000.0)	i	q	e
1854 Jun. 22.4970 (UT)	74.5689	349.6985	108.7032	0.648092	1.0

ABSOLUTE MAGNITUDE: $H_{10} = 6.4$ (V1964)

FULL MOON: May 12, Jun. 10, Jul. 10, Aug. 8

SOURCES: F. W. A. Argelander, W. Klinkerfues, and J. R. Hind, *MNRAS*, **14** (1854 Jun.), pp. 214–15; W. Klinkerfues, J. A. C. Oudemans, C. Mathieu, and E. Liouville, *CR*, **38** (1854 Jun. 19), pp. 1083, 1087; W. Klinkerfues and E. Schoenfeld, *AN*, **38** (1854 Jun. 27), p. 327; F. W. A. Argelander, C. Mathieu, E. Liouville, K. C. Bruhns, Hoffmann, W. Klinkerfues, F. A. T. Winnecke, and C. F. Gauss, *AN*, **38** (1854 Jul. 11), pp. 345, 349, 353–4; R. van Arsdale, W. Klinkerfues, F. W. A. Argelander, J. Ferguson, B. A. Gould, Jr, J. Winlock, and B. Peirce, *AJ*, **4** (1854 Jul. 15), pp. 5–7; J. Chacornac, J. J. E. Goujon, E. Liais, and A. J. F. Yvon-Villarceau, *CR*, **39** (1854 Jul. 17), pp. 158–9; J. A. C. Oudemans, *AN*, **38** (1854 Jul. 25), pp. 381–4; J. Ferguson, B. A. Gould, Jr, J. Winlock, and R. Keith, *AJ*, **4** (1854 Aug. 11), pp. 11–12, 14–15; A. Reslhuber, *AN*, **39** (1854 Aug. 25), pp. 43–6; R. Keith, *AJ*, **4** (1854 Sep. 19), p. 23; J. F. J. Schmidt, *AN*, **39** (1854 Sep. 29), pp. 103–8; V. Trettenero, *AN*, **39** (1854 Oct. 10), pp. 119–22; A. Reslhuber, *AN*, **39** (1854 Oct. 24), pp. 133–6; W. Klinkerfues, *AN*, **39** (1854 Nov. 10), p. 161; K. Hornstein, *AN*, **39** (1854 Nov. 20), p. 191; G. F. W. Rümker, *AJ*, **4** (1854 Dec. 5), p. 46; G. Santini, *AJS* (Series 2), **19** (1855 May), p. 447; F. A. T. Winnecke and C. F. Pape, *AN*, **42** (1855 Oct. 30), pp. 113–20; W. R. von Hillmayr, *DAWW*, **72** (1902), pp. 475–518; V1964, p. 57.

C/1854 R1
(Klinkerfues)

1854 IV

Discovered: 1854 September 11.83 ($\Delta = 0.84$ AU, $r = 1.17$ AU, Elong. $= 78°$)
Last seen: 1854 December 3.15 ($\Delta = 1.31$ AU, $r = 1.05$ AU, Elong. $= 52°$)
Closest to the Earth: 1854 September 16 (0.8374 AU)
Calculated path: CAM (Disc), UMa (Sep. 15), COM (Oct. 19), VIR (Nov. 5)

W. Klinkerfues (Göttingen, Germany) discovered this comet on 1854 September 11.83, at a position of $\alpha = 7^h\ 31.2^m$, $\delta = +75°\ 33'$. The comet was then independently discovered by K. C. Bruhns (Berlin, Germany) on September 12.79, R. van Arsdale (Newark, New Jersey, USA) on September 14.12, G. B. Donati (Florence, Italy) on September 18.85, M. Mitchell (Nantucket, Massachusetts, USA) on September 19, and M. M. Gussew (Wilna, now Vilnius, Lithuania) on September 21. Bruhns described it as a very fine nebulosity in the cometseeker, while a refractor showed a coma several arc minutes across with an indistinct appearance. The comet had attained a maximum declination of $+76°$ on September 10.

Although the comet was followed into November by astronomers in Europe and USA, descriptive observations were not plentiful and primarily came from A. Reslhuber (Kremsmünster, Austria). He described the comet as diffuse and 5–6' across on September 17, with a faint nucleus, but no tail. On the 18th and 22nd, he simply described it as very faint. U. J.-J. Le Verrier (Paris, France) reported for the 23rd that the comet was a small, very faint nebulosity, with no distinct nucleus or tail. Reslhuber battled moonlight on October 1 and 2, and noted the comet was observed with difficulty, but his observation before the moon rose on the 9th revealed the comet had increased in size. On October 20, Reslhuber said the comet was brighter than in September, but with a smaller coma and a sharp nucleus. He said the comet was very faint in moonlight on November 13. J. R. Hind (London, England) saw the comet on September 20 and 25, and wrote, "The brightest part appeared to me to follow the centre of the whole nebulosity, but the absence of a well-marked nucleus rendered the observations somewhat difficult and uncertain in an illuminated field."

The comet was last detected on December 3.15, when A. Colla (Parma, Italy) found it in the morning sky. The position was then near $\alpha = 13^h 02.8^m$, $\delta = -11° 20'$. Colla reported he had also found the comet on December 22 and 25, but these observations were later found to have been pre-discovery observations of comet C/1854 Y1. Interestingly, this comet passed only 0.8° from C/1854 Y1 on December 16.

Bruhns calculated one of the first parabolic orbits using positions obtained on September 12, 15, and 18. The resulting perihelion date was 1854 October 27.53. Similar orbits were determined by Reslhuber, Hind, B. A. Gould, Jr, F. W. Günther, F. A. T. Winnecke, C. F. Pape, and O. L. Lesser. Reslhuber pointed out the similarity between the orbit of this comet and that of the comet C/1844 Y2. Hind and Gould agreed, with Gould also noting a similarity to the orbit of comet C/1793 S1, although the perihelion distance of this comet was quite different.

The first elliptical orbit was calculated by Lesser, with the resulting perihelion date of October 28.01 and a period of about 1309 years. Very similar orbits were later calculated by C. G. T. Buschbaum and L. Steiner (1899) and R. J. Buckley (1979), with periods of 1089 years and 1286 years, respectively. Buckley's orbit is given below.

T	ω	Ω (2000.0)	i	q	e
1854 Oct. 28.0085 (TT)	129.8988	326.5159	40.9201	0.798762	0.993246

ABSOLUTE MAGNITUDE: $H_{10} = 8.3$ (V1964)

FULL MOON: Sep. 6, Oct. 6, Nov. 4, Dec. 4

SOURCES: K. C. Bruhns, *AN*, **39** (1854 Sep. 19), p. 95; U. J.-J. Le Verrier and G. B. Donati, *CR*, **39** (1854 Oct. 2), p. 646; J. Ferguson, *AJ*, **4** (1854 Oct. 7), pp. 25–6, 32; K. Hornstein, G. F. W. Rümker, G. B. Donati, and M. M. Gussew, *AN*, **39** (1854 Oct. 10), pp. 121–4; J. R. Hind, A. Reslhuber, A. M. Nell, F. A. T. Winnecke, and C. F. Pape, *AN*, **39** (1854 Oct. 24), pp. 131–8; K. C. Bruhns, *MNRAS*, **14** (Supplemental Notice, 1854), pp. 244–5; R. van Arsdale, *AJS* (Series 2), **18** (1854 Nov.), p. 430; B. A. Gould, Jr, J. Winlock, and J. Ferguson, *AJ*, **4** (1854 Nov. 10), pp. 36–9; W. Klinkerfues and F. W. Günther, *AN*, **39** (1854 Nov. 10), pp. 161–4; W. Klinkerfues, K. C. Bruhns, R. van Arsdale, M. Mitchell, G. B. Donati, and M. M. Gussew, *AJ*, **4** (1854 Dec. 5), p. 47; G. B. Donati, *CR*, **39** (1854 Dec. 26), p. 1218; A. Colla, *AN*, **39** (1855 Jan. 17), p. 333; F. W. Günther, *AN*, **41** (1855 Aug. 25), pp. 277–84; A. Reslhuber, *AN*, **43** (1856 Mar. 15), p. 39; O. L. Lesser, *AN*, **50**

(1859 Jun. 28), pp. 373–5; C. G. T. Buschbaum and L. Steiner, *AN*, **149** (1899 Jun. 2), pp. 341–4; V1964, p. 57; R. J. Buckley, *JBAA*, **89** (1979 Oct.), pp. 583–4.

C/1854 Y1	*Discovered:* 1854 December 22.18 ($\Delta = 1.55$ AU, $r = 1.36$ AU, Elong. $= 60°$)
(Winnecke–Dien)	*Last seen:* 1855 April 23.09 ($\Delta = 1.46$ AU, $r = 2.24$ AU, Elong. $= 129°$)
	Closest to the Earth: 1855 January 11 (1.5446 AU), 1855 April 20 (1.4619 AU)
1854 V	*Calculated path:* HYA (Disc), LIB (Jan. 12), SCO (Jan. 27), OPH (Feb. 22)

F. A. T. Winnecke (Berlin, Germany) discovered this comet on 1855 January 15.21 at a position of $\alpha = 15^h\ 01.3^m$, $\delta = -27°\ 11'$. He described it as a faint, granular appearing nebulosity. C. Dien (Imperial Observatory, Paris, France) independently discovered this comet on January 15.22. K. C. Bruhns (Berlin, Germany) confirmed Winnecke's find on January 16 and described the comet as rather faint, while Dien and U. J.-J. Le Verrier confirmed Dien's find on January 18 and described the comet as a very faint telescopic object which "presented many centers of light, and no tail."

Interestingly, although the above information was announced in journals published on January 24 and later, the January 17 issue of the *Astronomische Nachrichten* unknowingly published observations made in December by A. Colla (Parma, Italy). Colla was describing his final sightings of comet C/1854 R1 and told of his observations made on 1854 December 22.18 and December 25.18. For the former date, he gave a position of $\alpha = 13^h\ 50^m$, $\delta = -24°$, and described it as a "very faint circular nebulosity, a little more condensed within the center, without the slightest trace of the nucleus, nor a tail." Hazy skies prevented observations on the 23rd and 24th, but Colla saw the comet again on the 25th and noted it was "a little distance to the west of the star π" in Hydra. Colla tried to observe the comet again on the 26th, but was unsuccessful. He suggested the comet was then very near π Hydrae and that the star's brightness hid the comet. It was recognized that Colla's comet was an early sighting of comet C/1854 Y1 while the comet was still being observed. In fact, J. A. C. Oudemans (Leiden, Netherlands) even referred to the comet as "the Colla–Dien–Winnecke comet" in a paper published in the March 14 issue of the *Astronomische Nachrichten*.

Even though the comet was closest to Earth on April 20, its brightness did not benefit because it had already passed perihelion on December 16 and the closest distance from Earth was still nearly 1.5 AU. H. L. d'Arrest (Leipzig) said the coma was about 1.5′ across on January 22 and 24. A. Reslhuber (Kremsmünster, Austria) saw the comet at low altitude on the 24th and said it appeared very faint next to a 4th-magnitude star. In addition, no nucleus or tail was visible. Even though Reslhuber described the comet as very faint on the 25th, he said it was brighter on the 29th and exhibited a coma 4–5′ across, with a bright nucleus. G. B. Donati (Florence, Italy) described the comet as very faint on February 16. On February 18, Oudemans described the comet as a faint, round nebulosity, 1–2′ across, without a nucleus. Reslhuber said the comet was so faint it "could only be detected with extreme effort" on February 20. Reslhuber experienced bad weather from February 20 to March 22. On the morning of March 22, he unexpectedly found clear skies and looked for the comet. Although he found it, he noted it was very faint and required his eyes to be well dark adapted. Oudemans saw the comet on April 18. He said it appeared as a small, faint spot that became visible only after his eyes had become dark adapted. The comet was last detected on April 23.09, when Winnecke measured the position as $\alpha = 17^h 31.5^m$, $\delta = -21° 28'$.

The first parabolic orbit was calculated by Bruhns using positions from January 16 to January 19. He determined the perihelion date as 1854 December 18.20. Additional orbits by Reslhuber, Oudemans, Winnecke, B. Valz, and d'Arrest established the perihelion date as December 16.5. When the comet was no longer under observation, it became obvious that it was moving in a long-period orbit. Adam (1855) took 25 positions spanning the period of January 15 to April 19 and determined six Normal positions, which revealed a perihelion date of December 16.22 and a period of about 997 years. W. L. Elkin (1879) determined an almost identical orbit, using an arc spanning the period of January 16 to April 23. The orbit was revised by B. G. Marsden (1979). He took the best positions spanning the period of January 15 to April 19, and determined a perihelion date of December 16.31 and a period of 1956 years. This orbit is given below.

T	ω	Ω (2000.0)	i	q	e
1854 Dec. 16.3092 (TT)	287.0725	240.2563	14.1520	1.359463	0.991309

ABSOLUTE MAGNITUDE: $H_{10} = 6.9$ (V1964)

FULL MOON: Dec. 4, Jan. 3, Feb. 2, Mar. 3, Apr. 2, May 2

SOURCES: F. A. T. Winnecke, *MNRAS*, **15** (1855 Jan.), p. 90; A. Colla, *AN*, **39** (1855 Jan. 17), p. 333; F. A. T. Winnecke, K. C. Bruhns, and R. Luther, *AN*, **39** (1855 Jan. 24), p. 351; C. Dien, U. J.-J. Le Verrier, K. C. Bruhns, and H. L. d'Arrest, *AN*, **39** (1855 Jan. 31), pp. 353, 361–4; A. Reslhuber, *AN*, **40** (1855 Feb. 20), p. 183; C. Dien, U. J.-J. Le Verrier, F. A. T. Winnecke, and K. C. Bruhns, *AJ*, **4** (1855 Feb. 24), pp. 69–70; G. B. Donati, *MNRAS*, **15** (1855 Mar.), p. 159; J. A. C. Oudemans and F. A. T. Winnecke, *AN*, **40** (1855 Mar. 14), pp. 237–42; A. Reslhuber, *AN*, **40** (1855 Mar. 22), p. 263; C. Dien and F. A. T. Winnecke, *AJS* (Series 2), **19** (1855 May), p. 447; J. A. C. Oudemans, *AN*, **41** (1855 May 14), p. 15; A. Reslhuber, *AN*, **41** (1855 Jul. 2), p. 173; J. G. Galle, *AN*, **41** (1855 Aug. 25), p. 273; H. L. d'Arrest and Adam, *AN*, **41** (1855 Aug. 31), pp. 297–302; F. A. T. Winnecke, *AN*, **41** (1855 Sep. 20), pp. 381–4; W. L. Elkin, *AN*, **94** (1879 Jan. 22), pp. 73–8; V1964, p. 58; B. G. Marsden, *QJRAS*, **26** (1985), p. 91.

C/1855 G1 *Discovered:* 1855 April 11.85 ($\Delta = 1.34$ AU, $r = 2.32$ AU, Elong. = 163°)

(Schweizer) *Last seen:* 1855 June 5.91 ($\Delta = 2.44$ AU, $r = 2.58$ AU, Elong. = 86°)

Closest to the Earth: 1855 April 5 (1.3197 AU)

1855 I *Calculated path:* CRV (Disc), VIR (Apr. 18), CRT (Apr. 23), VIR (Apr. 24), LEO (May 2)

K. G. Schweizer (Moscow, Russia) discovered this telescopic comet on 1855 April 11.85 at a position of $\alpha = 12^h\ 18.7^m$, $\delta = -17°\ 20'$. He confirmed his discovery on April 14.88. Schweizer observed with an 8.0-cm refractor and described the comet as a small, faint nebulosity.

F. A. T. Winnecke (Berlin, Germany) saw the comet with the finder of the refractor he was using and said it was clearly visible. He estimated the coma diameter as 40″ in the refractor and noted a stellar nucleus. R. Schumacher (Altona, Germany) saw the comet in a 4-foot focal length refractor on May 9 and said the comet appeared very faint. A. Secchi (Rome, Italy) immediately found the comet with the aid of an ephemeris on May 18. He described it as a faint, oval nebulosity in the large refractor, but added that it nearly vanished when the wire of the micrometer was illuminated. Secchi also noted the comet was not visible in the finder.

The comet was last detected on June 5.91, when O. Lesser (Berlin, Germany) gave the position as $\alpha = 11^h\ 00.7^m$, $\delta = +14°\ 18'$.

The first parabolic orbit was calculated by Schweizer using positions from April 14, 16, and 19. The perihelion date was determined as 1855 January 24.54. The large perihelion distance made it difficult to pinpoint the perihelion date until the comet had been observed for over a month. Schumacher and J. A. C. Oudemans determined perihelion dates on February 9, while Winnecke determined it as February 6.23. B. Tiele (1859) used 21 positions obtained between April 14 and June 5, and computed a parabolic and two elliptical orbits. His best elliptical orbit had a perihelion date of February 5.55 and a period of 500 years. This orbit is given below.

T	ω	Ω (2000.0)	i	q	e
1855 Feb. 5.5474 (UT)	323.0929	191.7463	128.5764	2.193526	0.965185

ABSOLUTE MAGNITUDE: $H_{10} = 4.0$ (V1964)

FULL MOON: Apr. 2, May 2, May 31, Jun. 29

SOURCES: K. G. Schweizer, *MNRAS*, **15** (1855 Apr.), p. 174; K. G. Schweizer, *AN*, **40** (1855 May 3), p. 375; K. G. Schweizer, *AN*, **40** (1855 May 10), pp. 387–90; R. Schumacher, *AN*, **41** (1855 May 14), pp. 11–15; K. G. Schweizer, *AJ*, **4** (1855 May 26), p. 95; F. A. T. Winnecke, *AN*, **41** (1855 May 30), p. 63; F. A. T. Winnecke, *AN*, **41** (1855 Jun. 5), p. 79; J. A. C. Oudemans, *AN*, **41** (1855 Jun. 8), p. 95; A. Secchi and R. Schumacher, *AN*, **41** (1855 Jun. 11), p. 109; O. Lesser, *AN*, **41** (1855 Sep. 20), pp. 373–84; F. A. T. Winnecke, *MNRAS*, **15** (Supplemental Notice, 1855), p. 227; B. Tiele, *AN*, **52** (1859 Dec. 22), pp. 33–40; V1964, p. 58.

X/1855 K1   H. Goldschmidt (Paris, France) discovered this object while searching for periodic comet de Vico (now known as 54P/de Vico–Swift–NEAT). He estimated the position as $\alpha = 21^h\ 41.8^m$, $\delta = -15°\ 38'$ on 1855 May 17.08. Subsequent searches by Goldschmidt and other observers failed to detect the object and astronomers soon doubted that comet de Vico had actually been seen.

The comet's location near the ecliptic captured the attention of F. A. T. Winnecke (Bonn, Germany) in later years. Shortly after Tempel's comet of 1867 (9P/Tempel 1) was found to move in a short-period orbit, Winnecke did some calculations and found it would have passed perihelion in 1855. He said if the perihelion date was assumed to have been 1855 February 7.22, comet 9P/Tempel 1 would have been very close to the position of Goldschmidt's object. During a later investigation of the motion of 9P/Tempel 1, F. E. von Asten (1873) used positions from both the 1867 and 1873 apparitions to establish its motion. He said an earlier perihelion date would have occurred on 1856 February 2.04, which put the comet far from Earth and eliminated 9P/Tempel 1 as a candidate for Goldschmidt's object. Shortly after Denning's comet of 1881 was found to be moving in a short-period orbit, Winnecke (1881) took the orbit and found that a perihelion date of 1855 August 3 would put the comet within 3° of Goldschmidt's position. Denning's comet was unfortunately lost after the 1881 apparition, but was rediscovered in 1978 and was later given the designation of 72P/Denning–Fujikawa. K. Kinoshita (1998) investigated the comet's long-term motion and found that, although the comet did pass perihelion in 1855, the actual perihelion date was October 1.07.

SOURCES: H. Goldschmidt, *AN*, **41** (1855 Aug. 25), pp. 285–8; F. A. T. Winnecke, *AN*, **69** (1867 Jun. 20), pp. 205–8; F. E. von Asten, *AN*, **82** (1873 Nov. 3), pp. 273–6; F. A. T. Winnecke, E. Hartwig, and L. Wutschichowsky, *AN*, **101** (1881 Nov. 26), p. 77; personal correspondence from K. Kinoshita (1998).

C/1855 L1 *Discovered:* 1855 June 3.89 ($\Delta = 0.58$ AU, $r = 0.58$ AU, Elong. $= 28°$)
(Donati) *Last seen:* 1855 June 30.90 ($\Delta = 1.58$ AU, $r = 0.88$ AU, Elong. $= 31°$)
  Closest to the Earth: 1855 May 27 (0.4483 AU)
1855 II *Calculated path:* AUR (Disc), LYN (Jun. 6), GEM–LYN (Jun. 8), CNC (Jun. 15)

G. B. Donati (Florence, Italy) discovered this comet on 1855 June 3.89 at a position of $\alpha = 6^h 41.1^m$, $\delta = +36° 20'$. He said the comet exhibited no nucleus or tail, and seemed slightly fainter than the globular cluster M13 or NGC 6210. Independent discoveries were made by W. Klinkerfues (Göttingen, Germany) on June 4.91 and C. Dien (Paris, France) on June 4.91.

Observations were made at observatories in Germany, France, Italy, England, and the Netherlands, but few descriptions were published. K. C. Bruhns (Berlin, Germany) saw the comet on June 5 and said it was "of the brightness of the nebula in Aquarius." This nebula was probably NGC 7009, which is about magnitude 8. He added that the comet was about 1' across, but did not exhibit a tail. C. F. Pape (Göttingen, Germany) saw the comet on June 14 and noted that the comet had considerably faded.

The comet was last detected on June 30.90, when O. Lesser (Berlin, Germany) measured the position as $\alpha = 8^h 51.9^m$, $\delta = +30° 39'$.

The first parabolic orbit calculated for this comet was by Bruhns. Using positions obtained during the period of June 5–7, he determined the perihelion date as 1855 May 29.70. Although Donati determined a perihelion date of June 7.18 a few days later using a slightly longer arc, Bruhns' orbit was rather accurate, as proven by the later computations of V. Trettenero, L. R. Schulze, V. A. Puiseux, and Pape.

The first elliptical orbit was calculated by Donati using positions spanning the period of June 3–17. The resulting perihelion date was May 30.70 and the period was about 493 years. Donati was struck by the similarity between the orbit of this comet and that of comet C/1362 E1. He noted that, while the period was probably not accurate, it was still "extremely probable" that the two comets were identical. S. Alexander pointed out the variations in the available orbits for C/1362 E1, but said, "It will be observed that both of the orbits of 1362 differ less in several respects from that of 1855 II than those orbits differ among themselves." Later in the year, Schulze used positions covering the period of June 5–30 and calculated a period of 14.27 years, but he noted that positions determined around mid-June were not represented well and that a parabola was a better fit. Donati did not believe the last positions were reliable enough to determine a precise orbit and only chose those positions spanning June 3–17. His resulting period was about 493 years.

G. van Biesbroeck (1916) used 49 positions obtained between June 5 and June 19, as well as perturbations by Venus to Saturn, and computed an elliptical orbit with a perihelion date of May 30.65 and a period of about 252 years. He noted that the eccentricity and, consequently, the period, were very uncertain because of the short arc, with the possible period lying between 155 and 523 years. Van Biesbroeck considered the June 30 observation as unreliable, as neither this orbit nor a parabolic one represented it well. Van Biesbroeck also examined the suggestion that this comet was previously seen as the comet of 1362, but as the observations of the earlier comet were "not of sufficient number to permit the computation of an orbit," he concluded, "The identity of the comet of 1362 with that of 1855 cannot, in my opinion, be considered as in any way proved."

T	ω	Ω (2000.0)	i	q	e
1855 May 30.6540 (UT)	22.4883	262.2306	156.8707	0.567564	0.985780

ABSOLUTE MAGNITUDE: $H_{10} = 11.3$ (V1964)

FULL MOON: May 31, Jun. 29

SOURCES: W. Klinkerfues, *MNRAS*, **15** (1855 May), p. 203; G. B. Donati, *MNRAS*, **15** (1855 Jun.), pp. 209–10; W. Klinkerfues, *AN*, **41** (1855 Jun. 8), p. 95; K. C. Bruhns, *AN*, **41** (1855 Jun. 11), p. 111; C. Dien, V. A. Puiseux, W. Klinkerfues, and G. B. Donati, *CR*, **40** (1855 Jun. 11), pp. 1271–2; W. Klinkerfues, K. C. Bruhns, C. Dien, and G. B. Donati, *AN*, **41** (1855 Jun. 16), pp. 119–22, 127; G. B. Donati, *AN*, **41** (1855 Jun. 28), p. 157; W. Klinkerfues and C. F. Pape, *AN*, **41** (1855 Jul. 2), pp. 173–6; V. Trettenero,

AN, **41** (1855 Jul. 17), p. 205; W. Klinkerfues, C. Dien, G. B. Donati, and K. C. Bruhns, *AJ*, **4** (1855 Jul. 26), pp. 102–3; G. B. Donati, *AN*, **41** (1855 Aug. 25), p. 287; K. C. Bruhns, *AJS* (Series 2), **20** (1855 Sep.), p. 285; S. Alexander, *AJ*, **4** (1855 Sep. 5), p. 106; O. Lesser, *AN*, **41** (1855 Sep. 20), p. 383; G. B. Donati, *AN*, **42** (1855 Sep. 27), p. 63; J. Hartnup, *MNRAS*, **15** (Supplemental Notice, 1855), p. 225; K. C. Bruhns, *MNRAS*, **15** (Supplemental Notice, 1855), p. 231; G. B. Donati, *MNRAS*, **16** (1855 Nov.), pp. 14–16; L. R. Schulze, *AN*, **42** (1855 Dec. 8), pp. 199–202; G. van Biesbroeck, *AJ*, **29** (1916 Mar. 24), pp. 109–18; V1964, p. 58.

2P/Encke *Recovered:* 1855 July 12.77 ($\Delta = 0.96$ AU, $r = 0.45$ AU, Elong. $= 26°$)
Last seen: 1855 August 16.79 ($\Delta = 0.82$ AU, $r = 1.06$ AU, Elong. $= 69°$)
1855 III *Closest to the Earth:* 1855 August 3 (0.7374 AU)
Calculated path: CNC (Rec), LEO (Jul. 14), HYA-SEX (Jul. 16), LEO (Jul. 26), CRT (Jul. 27), CRV (Aug. 3), VIR (Aug. 10), HYA (Aug. 11)

During June of 1855, J. F. Encke took the orbit he had determined for the 1852 apparition, applied perturbations by Jupiter, and determined the perihelion date as 1855 July 1.69. T. Maclear (Royal Observatory, Cape of Good Hope, South Africa) recovered this comet on 1855 July 12.77 in evening twilight, about 3° or 4° above the horizon. He said the finder revealed the comet as a star of 6th magnitude, while the refractor revealed a "circular nebula with a bright centre." Maclear's colleague, W. Mann, measured the position as $\alpha = 9^h\ 11.0^m$, $\delta = +9°\ 57'$ on July 13.72. He said the coma was about 1.5′ across. Mann was the only observer of this comet through the remainder of the apparition.

The comet had passed perihelion nearly two weeks earlier, but was heading toward its closest approach to Earth, which acted to slow its rate of fading. The comet was seen at an altitude of 25° on July 16, when bright twilight illuminated the wire of the micrometer and the comparison star was still invisible. The 11th-magnitude comparison star became visible 25 minutes later. The comet was followed until it set behind "Lion Hill." The comet was observed through haze on the 17th, and the diameter was measured as 53″. Moonlight and a "foggy atmosphere" made observing difficult on July 22, and the comet was described as extremely faint. The comet was nearly invisible because of moonlight on July 24, 25, and 27. On July 28, the comet was so faint it was only seen using averted vision. Moonlight was no longer a factor on July 31 and the comet was at a higher altitude than during previous observations. Nevertheless, it had "greatly decreased in brightness" since the 16th.

Mann observed the comet occult the star BD-12° 3481 (magnitude 9.3) on August 1 and said the coma nearly disappeared. On August 7 the coma diameter was about 2.5′. Observations were continually interrupted by clouds on the 8th and "only faint glimpses" were seen. The most favorable observing conditions since the comet became visible occurred on August 9 and the coma diameter was then measured as about 2.25′. On August 11 it was difficult to see the comet because of its proximity to a 7th-magnitude star. The coma was measured as about 1.25′ on the 12th.

The comet was last detected on August 16.79, when Mann measured the position as $\alpha = 13^h\ 40.3^m$, $\delta = -25°\ 51′$. He said the coma was about 1.75′ across.

Minor refinements of this comet's orbit using positions from multiple apparitions, as well as planetary perturbations, have been published by numerous astronomers. Of most significance were the studies of F. E. von Asten (1877) and B. G. Marsden and Z. Sekanina (1974). These have typically revealed a perihelion date of July 1.54 and a period of 3.30 years. Marsden and Sekanina determined the nongravitational terms as $A_1 = +0.38$ and $A_2 = -0.03503$. The orbit of Marsden and Sekanina is given below.

T	ω	Ω (2000.0)	i	q	e
1855 Jul: 1.5386 (TT)	183.4119	336.4921	13.1529	0.337202	0.847719

ABSOLUTE MAGNITUDE: $H_{10} = 9.4$ (V1964)
FULL MOON: Jun. 29, Jul. 29, Aug. 27
SOURCES: J. F. Encke, *AN*, **41** (1855 Jun. 16), pp. 113–20; J. F. Encke, *AN*, **51** (1859 Aug. 12), pp. 81–90; T. Maclear, *MRAS*, **31** (1861–2), p. 2; T. Maclear and W. Mann, *MRAS*, **31** (1861–2), pp. 19–23; F. E. von Asten, *BASP*, **22** (1877), p. 561; V1964, p. 58; B. G. Marsden and Z. Sekanina, *AJ*, **79** (1974 Mar.), pp. 415–16.

C/1855 V1   *Discovered:* 1855 November 13.10 ($\Delta = 0.95$ AU, $r = 1.25$ AU, Elong. $= 80°$)
(Bruhns)   *Last seen:* 1856 January 3.79 ($\Delta = 1.11$ AU, $r = 1.37$ AU, Elong. $= 81°$)
Closest to the Earth: 1855 December 6 (0.2692 AU)
1855 IV   *Calculated path:* SEX (Disc), HYA (Nov. 20), CMi (Nov. 30), MON (Dec. 3), ORI (Dec. 4), TAU (Dec. 7), CET (Dec. 10), PSC (Dec. 14)

K. C. Bruhns (Berlin, Germany) discovered this comet on 1855 November 13.10 and said it looked "like a feeble nebula." He determined a position of $\alpha = 9^h\ 56.1^m$, $\delta = +2°\ 07′$ on November 13.19.

At discovery, the comet was heading toward both the sun and Earth. J. F. J. Schmidt (Olmütz, now Olomouc, Czech Republic) described the comet as a bright, large nebulosity, without a nucleus or tail, on November 20 and 21. On the 21st, both A. Reslhuber (Kremsmünster, Austria) and George Rümker (Hamburg, Germany) made observations after the moon had set and said the comet looked like a large, diffuse nebulosity. Reslhuber added that it was very faint, without a nucleus or a tail. Rümker observed the comet in moonlight on November 22 and 28, and said it was extremely faint.

Even though the comet had passed perihelion late in November, it continued approaching Earth early in December and became bright enough for Schmidt to see with the naked eye from December 2 to December 10. The comet was commonly described as large and diffuse during this period of time, with Schmidt estimating the coma as 15′ across on the 7th and Reslhuber noting that the coma exhibited no distinct outline on the 3rd. During most of this period no nucleus was visible, although there was a very prominent, large central condensation, which H. L. d'Arrest (Leipzig, Germany) measured as 224″ across on the 2nd and 253″ across on the 6th. Rümker first reported a nucleus on the 10th and he noted it had become more distinct by the 13th.

The comet should have faded after December 6, as its distances from the sun and Earth increased. As already noted, the comet did drop below naked-eye visibility after the 10th, although Reslhuber commented that the comet was a little brighter on the 13th than on the 10th. Whether this was real or related to sky conditions is uncertain, as no one else reported the event. Schmidt said the comet was an easy object in his refractor on the 11th and Reslhuber described the comet as faint on the 30th. Interestingly, W. Mitchell (Nantucket, Massachusetts, USA) independently discovered this comet on December 12.03.

As the new year began, Rümker said the comet was difficult to detect in poor skies on January 2, but could find no trace of the comet, despite good observing conditions, on the 3rd. The comet was last seen by d'Arrest on January 3.73 and F. A. T. Winnecke (Berlin, Germany) on January 3.79.

The first parabolic orbit was calculated by Rümker. He used three positions obtained between November 13 and 21, and determined a perihelion date of 1855 November 26.16. Similar orbits were later determined by Bruhns, d'Arrest, Adam, M. Hoek, Winnecke, and L. R. Schulze (1856). Even though Hoek (1856) determined an elliptical orbit with a perihelion date of November 25.88 and a period of about 9512 years, Schulze's parabolic orbit is given below.

T	ω	Ω (2000.0)	i	q	e
1855 Nov. 25.8927 (UT)	325.6188	53.6603	169.8235	1.230994	1.0

ABSOLUTE MAGNITUDE: $H_{10} = 8.1$ (V1964)

FULL MOON: Nov. 23, Dec. 23

SOURCES: K. C. Bruhns, *MNRAS*, **16** (1855 Nov.), p. 23; R. Luther, *AN*, **42** (1855 Nov. 29), pp. 187–92; J. Hartnup, *MNRAS*, **16** (1855 Dec.), p. 45; F. A. T. Winnecke, *AN*, **42** (1855 Dec. 8), pp. 205–8; H. L. d'Arrest, *AN*, **42** (1856 Jan. 3), pp. 237–40; K. C. Bruhns, *AJ*, **4** (1856 Jan. 9), pp. 139–42; J. F. J. Schmidt, *AN*, **42** (1856 Jan. 9), pp. 247–50; K. C. Bruhns, Adam, and H. L. d'Arrest, *AN*, **42** (1856 Feb. 19), pp. 359, 363–6; A. Reslhuber, *AN*, **43** (1856 Mar. 15), pp. 37–40; J. Ferguson, *AJ*, **4** (1856 Apr. 26), pp. 156–9; W. Mitchell, *AJS* (Series 2), **21** (1856 May), pp. 438–9; M. Hoek, *AN*, **44** (1856 Jul. 11), pp. 33–8; L. R. Schulze, *AN*, **44** (1856 Jul. 27), pp. 85–8; V1964, p. 58.

C/1857 D1   *Discovered:* 1857 February 23.16 ($\Delta = 1.54$ AU, $r = 0.93$ AU, Elong. $= 35°$)
(d'Arrest)   *Last seen:* 1857 May 2.83 ($\Delta = 1.48$ AU, $r = 1.10$ AU, Elong. $= 48°$)
              Closest to the Earth: 1857 March 30 (0.9991 AU)
1857 I   *Calculated path:* PEG (Disc), LAC (Mar. 10), AND (Mar. 13), CAS-AND (Mar. 24), PER (Apr. 1), AUR (Apr. 13), TAU (Apr. 19), ORI (Apr. 26)

This comet was discovered by H. L. d'Arrest (Leipzig, Germany) on 1857 February 23.16, during one of his routine searches for comets. He described it as rather bright, with a coma 1.5′ across and measured the position as $\alpha = 21^h 22.5^m$, $\delta = +22° 04′$. An independent discovery was made by R. van Arsdale (Newark, New Jersey, USA) on March 27.4. He described it as bright and resembling an unresolved globular cluster.

The comet was well observed during March as it approached both the sun and Earth. F. A. T. Winnecke (Bonn, Germany) said the coma was 3.0′ across on the 3rd, and he thought he occasionally saw a nucleus. The coma seemed more like 2.5′ across on the 4th, with a strong condensation. Although he saw no tail on either morning, he did note a distinct "beard" extending toward the sun on the 4th. Also on the 4th, A. Reslhuber (Kremsmünster, Austria) described the comet as round and 2′ across, with a small nucleus. Moonlight hampered observations for the next few days, although Winnecke was able to faintly see the comet on the 14th.

The comet's magnitude was estimated as 7 or 8 by C. Fearnley (Christiania, now Oslo, Norway) on March 17 and J. F. J. Schmidt (Olmütz, now Olomouc, Czech Republic) described the comet as bright, with considerable central condensation on that same date. Schmidt added that the radius of the coma on the sunward side of the condensation was 2.00′, while Winnecke gave a coma diameter of 2.50′. Reslhuber noted the comet was "brighter and more extensive" on the 18th than on previous mornings and Schmidt noted the comet was brighter on the 20th than when seen on the 17th. Also on the 20th, Winnecke measured the coma as 2.35′ across. An actual tail was reported by Schmidt on March 26, and he described it as broad and 4–5′ long. On the 28th, A. Secchi (Rome, Italy) also saw the tail and said it extended toward PA 196°.

The comet passed perihelion on March 21 and was closest to Earth on the 30th, but moonlight again interfered with observations. Schmidt saw the comet on the 30th and noted the tail was not well observed. He saw a bright central condensation on March 31 and a short trace of tail on April 1. Schmidt reported that the comet appeared fainter and tailless on the 3rd and Reslhuber said the comet was faintly seen in moonlight on the 8th. On April 11, Schmidt said the tail was more visible and the comet was brighter than on March 18, while Reslhuber remarked that the comet had faded since April 8. Schmidt said the coma's radius on the sunward side of the condensation was 2.80′. Schmidt said the comet was possibly visible to the naked eye on April 14, but only a trace of tail was visible in a telescope. He measured the coma's radius on the sunward side of the condensation as 2.50′.

Several factors combined to make the comet more difficult to view after mid-April. First, it was fading as it moved away from both the sun and Earth. Second, its position in the sky was becoming less favorable as it steadily neared the horizon. On the 17th, Schmidt had a clear horizon and noted a trace of tail was visible, but Winnecke was not so fortunate and reported the comet was faint because of haze near the horizon. Schmidt made a similar observation on the 18th. Winnecke saw the comet on the 19th and said, "the comet was very faint and difficult to observe because of the large zenithal distance." Schmidt said the comet was slightly fainter on the 19th and 20th than it had been on the 18th, but he did note that the tail was more distinct.

The comet was last detected on May 2.83, when G. B. Donati (Florence, Italy) determined the position as $\alpha = 6^h\ 02.2^m$, $\delta = +17°\ 26'$. Under very clear skies, but a nearly full moon, Reslhuber looked for the comet on May 8, but nothing was found. Reslhuber unsuccessfully tried to find the comet on May 15, but surmised it was then in strong twilight.

The two earliest parabolic orbits were by C. A. F. Peters and C. F. Pape. Both astronomers used positions obtained during the first three or four days the comet was visible and both obtained perihelion dates of 1857 March 14.55. After the first week of March, J. G. Galle determined a perihelion date of March 18.63, but it was mid-March before astronomers finally pinned down the perihelion date. At that time, Pape, W. Förster, and Winnecke determined dates of March 21.86, March 21.84, and March 21.80, respectively. Similar orbits were calculated in the coming weeks and months by J. C. Watson, E. Plantamour, V. Trettenero, d'Arrest, and Filippe Folque.

Later in 1857, L. R. Schulze determined a new parabolic orbit, as well as the first elliptical orbit. He used 48 positions obtained during the period of February 23 to April 26, and determined six Normal positions. The result was a parabolic orbit, with a perihelion date of March 21.8673. The elliptical orbit had a perihelion date of March 21.8636 and a period of 30 977 years. Although the elliptical orbit fitted the positions best, the difference was small, and Schulze felt the parabolic orbit better represented this comet.

The last investigation of this comet's orbit was made by M. Loewy (1859). He determined 12 Normal positions for the period between February 24 and May 2, and computed both parabolic and elliptical orbits. No planetary perturbations were considered for either one. The parabolic orbit had a perihelion date of March 21.8691, while the elliptical orbit had a perihelion date of March 21.8690 and an orbital period of about 8.3 million years. He said no noticeable deviation occurred from the parabolic orbit and considered it best represented the comet's motion. That orbit is given below.

T	ω	Ω (2000.0)	i	q	e
1857 Mar. 21.8691 (UT)	121.5663	315.1526	87.9479	0.772493	1.0

ABSOLUTE MAGNITUDE: $H_{10} = 7.1$ (V1964)

FULL MOON: Feb. 8, Mar. 10, Apr. 9, May 9

SOURCES: H. L. d'Arrest and C. F. Pape, *MNRAS*, **17** (1857 Mar.), pp. 160–1; H. L. d'Arrest and C. A. F. Peters, *AN*, **45** (1857 Mar. 4), p. 223; W. Förster, J. G. Galle, F. A. T. Winnecke, and C. F. Pape, *AN*, **45** (1857 Mar. 17), pp. 251–5; F. A. T. Winnecke, *AN*, **45** (1857 Mar. 26), p. 285; R. van Arsdale, *AJ*, **5** (1857 Mar. 31), p. 39; J. Challis, *MNRAS*, **17** (1857 Apr.), pp. 170–1; R. Hodgson, *MNRAS*, **17** (1857 Apr.), p. 171; E. Plantamour, *AN*, **45** (1857 Apr. 8), p. 331; V. Trettenero, *AN*, **45** (1857 Apr. 14), pp. 347–50; H. L. d'Arrest, R. van Arsdale, and C. F. Pape, *AJS* (Series 2), **23** (1857 May), p. 447; H. L. d'Arrest, *AN*, **46** (1857 May 18), p. 69; E. Plantamour, *AN*, **46** (1857 May 18), p. 75; J. F. J. Schmidt, *AN*, **46** (1857 May 28), pp. 107–12; J. Ferguson and J. C. Watson, *AJ*, **5** (1857 May 30), pp. 52, 54; A. Secchi and A. Reslhuber, *AN*, **46** (1857 Jun. 15), pp. 135–8; G. B. Donati, *AN*, **46** (1857 Jun. 30), pp. 215–18; A. Reslhuber and F. Folque, *AN*, **46** (1857 Jul. 17), pp. 283–6; L. R. Schulze, *AN*, **47** (1857 Sep. 25), pp. 83–8; W. Förster, *AN*, **47** (1857 Oct. 12), pp. 133, 142; F. A. T. Winnecke, *AN*, **47** (1858 Jan. 4), p. 293; C. Fearnley, *AN*, **47** (1858 Feb. 6), pp. 343–6; M. Loewy, *SAWW*, **35** (1859), pp. 392–409; V1964, p. 58.

5D/1857 F1 *Recovered:* 1857 March 18.80 ($\Delta = 1.17$ AU, $r = 0.65$ AU, Elong. $= 33°$)

(Brorsen) *Last seen:* 1857 June 22.94 ($\Delta = 1.13$ AU, $r = 1.55$ AU, Elong. $= 92°$)

Closest to the Earth: 1857 May 8 (0.7342 AU)

1857 II *Calculated path:* CET (Rec), ARI (Mar. 20), PER (Apr. 7), AUR (Apr. 23), CAM–AUR (Apr. 25), CAM (Apr. 28), LYN (May 2), CAM (May 6), UMa (May 10), CVn (Jun. 3), COM (Jun. 19), CVn (Jun. 23)

This comet had been predicted to return to perihelion on 1851 November, but attempts to observe it at Cambridge Observatory (England) and elsewhere were unsuccessful. P. van Galen took the orbit determined during the 1846 apparition and advanced it to predict the comet would next return to perihelion on 1857 June 26.27. K. C. Bruhns (Berlin, Germany) accidentally recovered this comet on 1857 March 18.80 and reported it as new. He described it as round and 2′ across. It exhibited a condensed nucleus, but no tail. A position of $\alpha = 2^h\ 03.3^m$, $\delta = +8°\ 19′$ was given by Bruhns and W. J. Förster on March 18.82.

The comet passed perihelion near the end of March, with several astronomers providing descriptions of it. G. F. W. Rümker (Hamburg, Germany) saw the comet on March 20 and said it looked like a round nebulosity about 1.5–2′ across, with a central condensation. On March 29, A. Reslhuber (Kremsmünster, Austria) said the comet was a round nebulosity about 1.5′ across, with a small nucleus, but no tail. A. Secchi (Rome, Italy) said the comet appeared diffuse. Reslhuber said the comet was difficult to detect because of thin cirrus clouds on the 31st.

Although having passed perihelion, the comet continued to approach Earth throughout April. J. F. J. Schmidt (Olmütz, now Olomouc, Czech Republic) said the comet equaled the brightness of C/1857 D1 on the 1st, but was smaller. He added that this comet was round, white, and tailless. F. A. T. Winnecke (Pulkovo Observatory, Russia) said it was very faint in twilight on the 2nd, while Reslhuber said it was faint because of moonlight and its low altitude on the 4th. Schmidt said the comet appeared smaller and fainter than C/1857 D1 on the 5th because of bright moonlight. J. Hartnup (Liverpool, England) saw the comet on April 7, 11, 14, and 16. He said the "nebulosity was round, and of about 1' 40" diameter. For about 15" from the centre it was very bright." Schmidt first saw the comet with the naked eye on April 8. On April 10, Reslhuber said the comet was observed in a very clear sky and found to be brighter than on previous days, 3' across, but with no tail. Schmidt observed the comet prior to moonrise on the 11th and said it was visible to the naked eye. He commented that a narrow tail extended 11' towards the point opposite to the sun. Schmidt added, "The middle of the coma is strongly condensed and a brilliant white, while the outer coma is very diffuse." He indicated a coma diameter of 3.0'. Schmidt again saw the comet with the naked eye on the 12th. He indicated the coma was 4.2' across on the 14th, with only a trace of tail visible. On April 17, Schmidt said the comet was a brilliant white and almost tailless, while Secchi said the coma was 3' across. On April 18, Schmidt indicated a coma diameter of 4.6', while Förster (Berlin, Germany) said the comet was about magnitude 8–9, with a trace of tail extending toward PA 250°. On April 19, H. L. d'Arrest (Leipzig, Germany) indicated a coma diameter of 3', and noted a strong central condensation of magnitude 7 or 8. C. H. F. Peters (Dudley Observatory, New York, USA) also saw a strong central condensation, and a possible stellar nucleus. He said the comet was similar in shape to C/1857 D1, but was brighter. Schmidt said the comet was a brilliant white on the 20th, with a prominent central condensation and a short tail. G. B. Donati (Florence, Italy) used a 28-cm refractor and observed the near-central occultation of the comet over a 12th-magnitude star on the 21st. He said the star became fainter and ill-defined, and actually vanished for about 30 seconds as the comet's central condensation passed over it.

The comet passed closest to Earth on May 8, but it was typically described as a faint object since it was over a month past perihelion. Schmidt said the comet was unexpectedly faint and diffuse on the 1st. C. Fearnley (Christiania, now Oslo, Norway) said the comet was extremely faint and about 3' across on the 9th. Förster noted the comet was very faint and vague when he observed it during the period of May 5–16. Reslhuber described it as faint on the 10th and Schmidt indicated a coma diameter of 3.8' on the 12th. On May 13, Winnecke described the comet as large and diffuse, without a central condensation. Several estimates of the coma diameter were made during the next few days. On the 14th, Schmidt gave it as 3.6', while d'Arrest said it was 2.5' across. D'Arrest noted no appreciable condensation. Schmidt indicated it was 3.9' across on the 15th and 2.7' on the 20th, while Winnecke gave it as 3.5' on the 18th, 3.0' on the 19th, and 4.5' on the 20th. Secchi noted the comet was very faint on the 17th and 22nd. Reslhuber also described the comet as "extremely faint" on the 20th. Schmidt continued to provide accurate measurements of the coma, noting it was 4.0' across on the 22nd and 23rd, and 3.2' across on the 25th. D'Arrest said the coma was a round, dull nebulosity of uniform brightness on the 21st, while he noted it was slightly brighter than Herschel's nebula h908 (aka NGC 3718, magnitude 10.5 – NGC2000.0) on the 22nd. Donati found the comet very faint in a 28-cm refractor on the 24th.

Förster said the comet was extraordinarily difficult to observe during June 18–22, both because of the comet's faintness and because it was in a region filled with faint stars. He and Bruhns saw the comet for the final time on June 22.94 and measured the position as $\alpha = 13^h\ 12.1^m$, $\delta = +29°\ 45'$.

From the three precise positions obtained between March 18 and 20, Bruhns computed a parabolic orbit with a perihelion date of 1857 March 25.45. C. F. Pape noted a strong similarity between this orbit and that computed for 5D/Brorsen. He said that if an assumed perihelion date of 1857 March 29.32 was substituted into van Galen's orbit, Bruhns' positions would be closely represented. Thus, Pape suggested comet Bruhns of 1857 was identical to 5D/Brorsen of 1846. At the beginning of April, Bruhns computed both parabolic and elliptical orbits for this comet. The elliptical orbit put the comet's identity with 5D/Brorsen beyond all doubt, with the resulting perihelion date being March 29.72. Additional orbits elliptical orbits were calculated by Pape, d'Arrest, and Bruhns.

Minor refinements of this comet's orbit using positions from multiple apparitions, as well as planetary perturbations, have been published by several astronomers. Bruhns (1868), B. G. Marsden and Z. Sekanina (1971), and W. Landgraf (1986) all established the perihelion date as March 29.75 and the period as 5.54 years. Marsden and Sekanina used 110 positions from the apparitions spanning 1846–73 and determined nongravitational terms of $A_1 = +0.32$ and $A_2 = -1.1545$. Landgraf took 102 positions from the apparitions spanning 1846–68 and determined nongravitational terms of $A_1 = +0.51$ and $A_2 = -0.0651$. Landgraf's orbit is given below.

T	ω	Ω (2000.0)	i	q	e
1857 Mar. 29.7462 (TT)	14.0333	103.7399	29.7971	0.620514	0.801707

ABSOLUTE MAGNITUDE: $H_{10} = 7.7$ (V1964)

FULL MOON: Mar. 10, Apr. 9, May 9, Jun. 7, Jul. 7

SOURCES: J. Breen, AN, **33** (1852 Jan. 9), p. 327; K. C. Bruhns, W. J. Förster, C. F. Pape, P. van Galen, and R. C. Carrington, MNRAS, **17** (1857 Mar.), pp. 161–3; K. C. Bruhns, AN, **45** (1857 Mar. 26), p. 285; J. Challis, J. Hartnup, and K. C. Bruhns, MNRAS, **17** (1857 Apr.), pp. 171–3; C. F. Pape, AN, **45** (1857 Apr. 2), pp. 317–20; G. F. W. Rümker, AN, **45** (1857 Apr. 8), p. 333; K. C. Bruhns, C. F. Pape, and P. van Galen, AJ, **5** (1857 Apr. 22), pp. 41, 48; H. L. d'Arrest and V. Trettenero, AN, **46** (1857 May 8), pp. 7–10; A. Secchi and A. Reslhuber, AN, **46** (1857 Jun. 15), pp. 133–8; J. F. J. Schmidt, AN, **46** (1857 Jun. 19), pp. 145–56; K. C. Bruhns, AN, **46** (1857 Jun. 26), pp. 187–92; G. B. Donati, AN, **46** (1857 Jun. 30), pp. 217–20; H. L. d'Arrest, AN, **46** (1857 Jul. 10), pp. 237–40; A. Secchi, *AN*, **47** (1857 Aug. 21), p. 13; W. J. Förster and K. C. Bruhns, *AN*, **47** (1857 Oct. 12), pp. 133, 142; C. Fearnley, *AN*, **47** (1858 Feb. 6), pp. 343–6; F. A. T. Winnecke, *AN*, **59** (1862 Dec. 27), pp. 67–70; K. C. Bruhns, *AN*, **71** (1868 Mar. 12), pp. 37–40; V1964, p. 58; B. G. Marsden and Z. Sekanina, *AJ*, **76** (1971 Dec.), pp. 1141–2; W. Landgraf, *QJRAS*, **27** (1986 Mar.), p. 116; personal correspondence from W. Landgraf (1993); NGC2000.0, p. 107.

C/1857 M1
(Klinkerfues)

1857 III

Discovered: 1857 June 23.02 ($\Delta = 1.24$ AU, $r = 0.76$ AU, Elong. $= 37°$)
Last seen: 1857 July 19.86 ($\Delta = 0.98$ AU, $r = 0.37$ AU, Elong. $= 21°$)
Closest to the Earth: 1857 July 11 (0.8765 AU)
Calculated path: PER (Disc), AUR (Jul. 1), LYN (Jul. 8), CNC (Jul. 17), LEO (Jul. 18)

W. Klinkerfues (Göttingen, Germany) discovered this comet on 1857 June 23.02, at a position of $\alpha = 3^h 26.1^m$, $\delta = +40° 06'$. Independent discoveries were made by C. Dien (Imperial Observatory, Paris, France) on June 24.03 and W. Habicht (Gotha, Germany) on June 25.05.

The comet was approaching both the sun and Earth throughout the remainder of June, but it was also approaching evening twilight. On June 24, H. L. d'Arrest (Leipzig, Germany) and G. F. W. Rümker (Hamburg, Germany) described the comet as bright. D'Arrest noted a strong central condensation, while Rümker simply added that it was small and round. The comet was unchanged when d'Arrest saw it on the 25th. F. A. T. Winnecke (Bonn, Germany) said the comet was 2′ across on the 26th and added, "Neither nucleus nor tail observed; the comet appeared somewhat fan-shaped." Winnecke said it was only 1.5′ across in bright twilight on the 27th. A. Reslhuber (Kremsmünster, Austria) saw the comet in rather strong twilight on the 28th and said it appeared bright with a short tail. Winnecke again found the comet 2′ across on the 29th. He also noted a sharp central condensation and an indication of a tail toward PA 285°.

The comet became a more difficult object to observe as July progressed. W. J. Förster (Berlin, Germany) saw the comet in his telescope's finder on the 2nd and said it was about equal to a star of magnitude 7 or 8. He added that a magnification of 214× revealed a good nucleus. That same night, C. A. F. Peters (Altona, Germany) saw the comet occult this same star and said the star vanished for a time. Förster said the comet was "decidedly brighter" on the 3rd and exhibited a narrow tail toward PA 309°. The comet attained its most northerly declination of +50° on the 6th. Reslhuber observed in bright moonlight and bright twilight on the 7th and said the comet was centrally condensed and pale. Förster said the tail extended toward PA 341° on the 8th. On July 10, Förster said the coma was 0.9′ across, with a nucleus that was situated closer to the tail than the coma's center. He said the tail extended 8′ in PA 8.8°. Förster measured the direction of the tail during his remaining observations, noting it extended toward PA 17.9° on the 12th, PA 27.5° on the 13th, and about PA 30° on the 14th. He said the tail was 10′ long on the 12th. Reslhuber said the comet had brightened by the 13th and was visible to the naked eye in twilight. That same evening, Winnecke estimated the magnitude as 6 and the coma as 0.4′ across, while the tail extended about 30′ toward PA 27.0°. Winnecke said the comet was about magnitude 6.5 on the 14th, with a distinct trace of tail. Reslhuber said the comet was seen faintly in poor skies on the 16th, and was detected faintly at low altitude on the 17th.

The comet was last detected on July 19.86, when E. Plantamour (Geneva, Switzerland) gave the position as $\alpha = 9^h 25.1^m$, $\delta = +27° 02′$. The comet had reached a minimum elongation from the sun on July 18 (21°), which was also the day of perihelion. Reslhuber could not find the comet on the 20th. Thereafter the comet began to fade as it moved away from both the sun and Earth. The geometry was such that the comet also stayed within 22° of the sun until August 18, at which time the comet should have faded about 5 magnitudes since the final observation.

The first parabolic orbit was calculated by Förster using positions spanning the period of June 24–26. The resulting perihelion date was 1857 July 18.57. This proved only 0.1 day late as shown by the orbits of Klinkerfues, R. Goltzsch, G. B. Donati, C. F. Pape, and R. König (1891). Although A. J. F. Yvon-Villarceau (1858) did calculate an elliptical orbit with a period of about 7028 years, the observational arc of less than one month puts great uncertainty on this orbit. Therefore, König's parabolic orbit is given below.

T	ω	Ω (2000.0)	i	q	e
1857 Jul. 18.4743 (UT)	134.0658	25.6942	121.0332	0.367535	1.0

ABSOLUTE MAGNITUDE: $H_{10} = 9.0$ (V1964)

FULL MOON: Jun. 7, Jul. 7, Aug. 5

SOURCES: W. Klinkerfues, *MNRAS*, **17** (1857 Jun.), p. 253; J. Challis, *MNRAS*, **17** (1857 Jun.), p. 253; C. F. Pape and W. J. Förster, *AJ*, **5** (1857 Jun. 30), p. 64; W. Klinkerfues, W. Habicht, C. F. Pape, W. J. Förster, and W. Klinkerfues, *AN*, **46** (1857 Jun. 30), pp. 219–22; G. B. Donati, *MNRAS*, **17** (1857 Jul.), p. 263; C. F. Pape, *MNRAS*, **17** (1857 Jul.), p. 263; H. L. d'Arrest, *AN*, **46** (1857 Jul. 10), p. 237; C. A. F. Peters, *AN*, **46** (1857 Jul. 13), p. 271; R. Goltzsch, *AN*, **46** (1857 Jul. 22), p. 303; E. Plantamour and A. Reslhuber, *AN*, **47** (1857 Aug. 21), pp. 7, 15; G. B. Donati, *AN*, **47** (1857 Sep. 25), pp. 81–4; F. A. T. Winnecke, *AN*, **47** (1858 Jan. 4), pp. 293–6; W. J. Förster, *AN*, **47** (1858 Jan. 20), pp. 305, 309; G. F. W. Rümker, *AN*, **48** (1858 Mar. 27), p. 81; A. J. F. Yvon-Villarceau, *CR*, **46** (1858 Jun. 7), pp. 1115–18; R. König, *SAWW*, **100** Abt. IIa (1891), pp. 21–70; R. König, *AN*, **128** (1891 Dec. 8), pp. 385–92; V1964, p. 58.

C/1857 O1 *Discovered:* 1857 July 26.20 ($\Delta = 0.65$ AU, $r = 0.94$ AU, Elong. = 64°)
(Peters) *Last seen:* 1857 October 22.41 ($\Delta = 1.46$ AU, $r = 1.32$ AU, Elong. = 61°)
Closest to the Earth: 1857 July 30 (0.6428 AU)
1857 IV *Calculated path:* CAM (Disc), AUR (Aug. 1), GEM (Aug. 17), CMi (Sep. 1), HYA (Sep. 11)

C. H. F. Peters (Dudley Observatory, New York, USA) discovered this comet with a cometseeker in the evening sky on 1857 July 26.20, at a position of $\alpha = 3^h\ 33.5^m$, $\delta = +59°\ 09'$. He described it as "exceedingly faint, and without visible nucleus." He confirmed his discovery on July 27.24. Peters proposed this comet be called the "Olcott Comet," to honor "the very beloved and esteemed name of the distinguished citizen who is identified with the history of the erection of this observatory." Independent discoveries were made by C. Dien (Imperial Observatory, Paris, France) on July 29.04 and W. Habicht (Gotha, Germany) on July 30.93. Habicht described it as fairly large, with a diameter of about 3', and extraordinarily faint. He mentioned no condensation, but said the coma was diffuse and of uniform brightness.

The comet passed closest to Earth on July 30 when K. C. Bruhns (Berlin, Germany) said it was fairly diffuse and about 1.5' across. He added that there was no tail. F. A. T. Winnecke (Bonn, Germany) said the comet was fainter than C/1857 D1 and about 2.7' across on August 1. He also noted that the condensation was not situated in the center of the coma, but in the sunward half. That same night, A. Secchi (Rome, Italy) described the comet as an irregular nebulosity, about 3' across, with an eccentric condensation. Secchi said the comet was elongated toward PA 117° on the 3rd. Winnecke said the comet was a fairly difficult object to see in moonlight on the 4th. J. C. Watson (Ann Arbor Observatory, Michigan, USA) noted the comet was extremely faint in moonlight on the 8th and 10th. He said it appeared "much diffused and ill-defined, without any marked concentration of light at the center." H. L. d'Arrest (Leipzig, Germany) saw the comet on the 15th, 16th, 23rd, and 26th. He simply noted it was bright. On August 21, W. J. Förster (Berlin, Germany) described the comet as very bright, with a coma about 1.5' across. He added that the nucleus was about magnitude 9.

Although observations continued after perihelion, there were few physical descriptions provided. Watson reported it was very difficult to measure the comet's position on September 24 and 26, because of its faintness.

The comet was last detected on October 22.41, when G. P. Bond (Harvard College Observatory, Cambridge, Massachusetts, USA) saw it with the 38-cm refractor and gave the position as $\alpha = 9^h\ 34.5^m$, $\delta = -16°\ 23'$. The observation was made with difficulty because of slight haze in the sky.

The first parabolic orbit was calculated by Peters using positions spanning the period of July 26–30. He determined the perihelion date as 1857 August 25.58. Additional orbits by L. R. Schulze, Bruhns, Winnecke, Peters, C. F. Pape, and Watson during the next couple of weeks narrowed down the perihelion date to between August 23.97 and 24.25. A more precise orbit seemed impossible because the positions indicated the comet was deviating from a parabola. At the beginning of September, Watson calculated an elliptical orbit with a perihelion date of August 24.54 and a period of 178 years. In the following weeks and years, the period was determined as 258 years by Peters (1858), 258 years by Pape (1858), 256 years and 243 years by

H. Lind (1858), and 235 years by D. M. A. Möller (1859). Möller's orbit is given below.

T	ω	Ω (2000.0)	i	q	e
1857 Aug. 24.4964 (UT)	180.9440	202.8313	32.7565	0.746843	0.980414

ABSOLUTE MAGNITUDE: $H_{10} = 9.1$ (V1964)
FULL MOON: Jul. 7, Aug. 5, Sep. 4, Oct. 3, Nov. 2
SOURCES: K. C. Bruhns, *MNRAS*, **17** (1857 Jul.), p. 265; C. H. F. Peters, *MNRAS*, **17** (1857 Jul.), p. 265; C. Dien and W. Habicht, *AN*, **46** (1857 Aug. 7), pp. 365–8; K. C. Bruhns, *AN*, **46** (1857 Aug. 11), p. 383; F. A. T. Winnecke, K. C. Bruhns, A. Secchi, and C. H. F. Peters, *AN*, **47** (1857 Aug. 21), pp. 7, 13; L. R. Schulze, *AN*, **47** (1857 Aug. 26), p. 17; C. H. F. Peters and J. C. Watson, *AJ*, **5** (1857 Aug. 28), pp. 71–2; J. C. Watson, *AJ*, **5** (1857 Sep. 14), p. 79; H. Lind, *AN*, **49** (1858 Sep. 25), p. 117; H. L. d'Arrest, *AN*, **47** (1857 Oct. 17), p. 147; F. A. T. Winnecke, C. H. F. Peters, and C. F. Pape, *AN*, **47** (1858 Jan. 4), pp. 295, 299–302; W. J. Förster, *AN*, **47** (1858 Jan. 20), pp. 305, 309; H. Lind and G. P. Bond, *AN*, **48** (1858 Mar. 18), pp. 73–6; J. C. Watson, *AJ*, **5** (1858 May 20), pp. 129–30, 133; V1964, p. 58; D. M. A Möller, *AN*, **49** (1859 Jan. 7), p. 357.

C/1857 Q1 *Discovered:* 1857 August 20.86 ($\Delta = 0.68$ AU, $r = 1.05$ AU, Elong. $= 74°$)
(Klinkerfues) *Last seen:* 1857 October 5.8 ($\Delta = 1.47$ AU, $r = 0.57$ AU, Elong. $= 15°$)
  Closest to the Earth: 1857 August 26 (0.6384 AU)
1857 V *Calculated path:* CAM (Disc), DRA (Aug. 24), UMa (Aug. 30), CVn (Sep. 3), BOO (Sep. 14), VIR (Sep. 30)

W. Klinkerfues (Göttingen, Germany) discovered this comet on 1857 August 20.86, at a position of $\alpha = 5^{\mathrm{h}}\ 21.5^{\mathrm{m}}$, $\delta = +77°\ 21'$.

Several astronomers provided physical descriptions during the remainder of August as the comet passed closest to Earth. K. C. Bruhns (Berlin, Germany) saw the comet on the 22nd and said it was very diffuse, about 2.5′ across, and twice as bright as comet C/1857 O1. C. F. Pape (Altona, Germany) saw the comet on August 23, 25, and 29. He noted it was about magnitude 8 on each evening, while the coma diameter changed from 4–2.5′, and then to 3′. He added that some condensation was present on the 29th. W. J. Förster (Berlin, Germany) said the coma was 3′ across on the 25th, with a bright nucleus eccentrically situated more toward PA 230° than in the center of the coma. A. Reslhuber (Kremsmünster, Austria) saw the comet with the naked eye, while a telescope revealed the comet as bright, round, 5–6′ across, without a nucleus or tail. Reslhuber noted traces of a tail on the 29th. H. L. d'Arrest (Leipzig, Germany) noted that during the first days after the discovery, the comet appeared faint and very diffuse, with a diameter of 5′.

Although the comet had passed Earth, it continued to draw closer to the sun throughout September. Pape estimated the magnitude on several occasions, indicating it was about magnitude 6 or 7 on the 3rd, about magnitude 4 on the 14th, 15th, 16th, 17th, and about magnitude 3 on the 22nd, 23rd, 24th, and 25th in twilight. Other estimates of the brightness were made by C. Fearnley (Christiania, now Oslo, Norway) and Reslhuber, with Fearnley estimating it as brighter than magnitude 6 on the 15th and brighter than magnitude 5 on the 17th, while Reslhuber estimated it as 5 on the 18th. Reslhuber said the comet was distinctly larger on September 7 than when last seen on August 29 and added that it was easily visible to the naked eye. J. Ferguson (Washington, DC, USA) estimated the tail as 25′ long on September 9. Förster saw the comet in the finder of his telescope on the 12th and said it appeared very bright and large, with a tail 1° long. On September 13, Fearnley said the coma was 6′ across, with a nucleus 8–10″ across. He also noted a distinct, perfectly straight tail. On September 15, Pape said the coma was about 2′ across, and the tail about 2.5° long and 8–10′ wide at the end. That same night, Förster observed with a cometseeker and said the coma was 3.5′ across and the tail extended 2° in PA 60–70°. He added that a distinct point-like nucleus was observed with the refractor and a magnification of 214×. G. F. W. Rümker (Hamburg, Germany) also saw the comet on the 15th and described it as a "very beautiful bright nebula." He added that a strong central condensation was present, as well as a tail 20–25′ long. F. A. T. Winnecke (Bonn, Germany) saw the comet on the 17th and noted the coma was 4.5′ across, with an exceptional condensation, but no nucleus. He added that the tail extended about 3° in PA 60.31°, and was about 2′ wide. That same evening, Förster said the tail extended toward PA 63.9°. Reslhuber also saw the comet on the 17th and noted it had brightened and exhibited a tail about 20′ long. It was also still an easy object for the naked eye. Ferguson said the faint tail was 40′ long on the 18th, while Reslhuber noted it was 2.5° long, with a coma 3–4′ across. On September 22, Förster and Bruhns said the tail was 4° long and the nucleus was visible with the naked eye. Reslhuber said the comet was brighter and larger on the 23rd than on the 18th. Winnecke said the tail extended in PA 62.48° on the 27th.

The comet passed perihelion as October began and was quickly dropping into evening twilight. Reslhuber said the comet was easy to see at low altitude and in bright twilight on the 1st. On the 3rd, Reslhuber noted the comet was visible in twilight when no stars could be seen nearby.

The comet's position was last measured on October 3.74 by Reslhuber, and on October 3.76 by Fearnley. Fearnley gave the position as $\alpha = 13^h\ 39.7^m$, $\delta = +4°\ 38′$. Reslhuber also saw the comet just before sunset on October 4.8 and 5.8, but no positions could be measured because of the lack of comparison stars. The comet passed about 3° from the sun on October 17.

The first parabolic orbit was calculated by Bruhns using positions from August 21, 22, and 23. He determined the perihelion date as 1857 October 1.32. He noted a slight resemblance to the orbits of comets C/1790 H1 and C/1825 K1. Additional orbits were calculated by Fearnley, Pape, and A. J. F. Yvon-Villarceau.

The first elliptical orbit was calculated by Yvon-Villarceau (1858). The resulting period was 1618 years. C. Linsser (1860) used 104 positions obtained between 1857 August 22 and October 3, and computed an elliptical orbit with a perihelion date of October 1.38 and a period of about 2462 years. This orbit is given below.

T	ω	Ω (2000.0)	i	q	e
1857 Oct. 1.3799 (UT)	124.8444	16.9550	123.9614	0.562898	0.996913

ABSOLUTE MAGNITUDE: $H_{10} = 4.9$ (V1964)

FULL MOON: Aug. 5, Sep. 4, Oct. 3, Nov. 2

SOURCES: K. C. Bruhns, *MNRAS*, **17** (1857 Jul.), pp. 265–6; W. Klinkerfues, *AN*, **47** (1857 Aug. 26), p. 27; K. C. Bruhns, *AN*, **47** (1857 Aug. 29), pp. 43–6; W. Klinkerfues, *AJ*, **5** (1857 Sep. 14), p. 80; A. J. F. Yvon-Villarceau, *CR*, **45** (1857 Sep. 14), pp. 378–9; C. Fearnley and C. F. Pape, *AN*, **47** (1857 Sep. 16), pp. 73–8; H. L. d'Arrest, *AN*, **47** (1857 Oct. 17), p. 147; J. Ferguson and C. F. Pape, *AJ*, **5** (1857 Oct. 24), pp. 85–8; C. F. Pape, *AN*, **47** (1857 Oct. 24), pp. 167–70; A. Reslhuber, *AN*, **47** (1857 Nov. 5), pp. 181–4; F. A. T. Winnecke, *AN*, **47** (1858 Jan. 4), p. 295; W. J. Förster, *AN*, **47** (1858 Jan. 20), pp. 305, 309; C. Fearnley, *AN*, **47** (1858 Feb. 6), pp. 339–44; G. F. W. Rümker, *AN*, **48** (1858 Mar. 27), pp. 81–4; J. Hartnup, *MNRAS*, **18** (1858 May), p. 230; C. Linsser, *AN*, **52** (1860 Jan. 18), pp. 97–106; V1964, p. 58; SC2000.0, p. 339.

C/1857 V1
(Donati-van Arsdale)

1857 VI

Discovered: 1857 November 10.76 ($\Delta = 0.66$ AU, $r = 1.02$ AU, Elong. $= 73°$)
Last seen: 1857 December 19.72 ($\Delta = 1.48$ AU, $r = 1.13$ AU, Elong. $= 50°$)
Closest to the Earth: 1857 November 14 (0.6488 AU)
Calculated path: DRA (Disc), HER-DRA (Nov. 14), LYR (Nov. 18), CYG (Nov. 24), VUL (Nov. 26), SGE (Nov. 30), AQL (Dec. 3), DEL (Dec. 4)

G. B. Donati (Florence, Italy) discovered this comet with a small refractor on 1857 November 10.76 at a position of $\alpha = 15^h 28.6^m$, $\delta = +55° 44'$. Donati confirmed his find on November 10.84. While using a 28-cm refractor, he described the comet as "very faint, and has no indication of nucleus or tail." An independent discovery was made by R. van Arsdale (Newark, New Jersey, USA) on November 11.02. He described it as a small, faint comet, and noted a motion of about 1' per hour in α, with a "slightly diminishing" δ. He added that the comet could be seen in a 10-cm cometseeker.

The comet was discovered a couple of days prior to its closest approach to Earth and about a week and a half before perihelion. J. Ferguson (Washington, DC, USA) saw the comet on several occasions with the 24.4-cm refractor during the period of November 13–18 and found it "very faint and shapeless." He noted that the comet seemed slightly brighter on the 18th than on the 13th. A. Secchi (Rome, Italy) reported the comet was faint during the period of November 15–21, but seemed brighter during the last few days. Also on November 18, G. F. W. Rümker (Hamburg, Germany) described the comet as a pale and diffuse nebulosity, A. J. F. Yvon-Villarceau

(Paris, France) described the comet as very faint, and A. J. G. F. von Auwers (Göttingen, Germany) said it was faint, slightly condensed, and about 5' across. A. Reslhuber (Kremsmünster, Austria) described the comet as very faint, without a nucleus or tail on the 19th and 20th. H. L. d'Arrest (Copenhagen, Denmark) said the comet was "so extraordinarily faint" on the 19th and 20th, that he doubted he would be able to see it again.

Two star occultations were apparently noted on November 19. F. A. T. Winnecke (Bonn, Germany) observed a near central occultation of a star of magnitude 12 or 13 by the comet. At one point it could not be discerned from the central condensation. W. J. Förster (Berlin, Germany) observed a central occultation of a star of magnitude 10 or 11 by the comet. He said the star remained steady during the occultation, but the comet actually disappeared for several minutes.

Winnecke saw the comet on December 4 with a heliometer and noted the comet was faint, 1.5–2' across, with no apparent condensation. Auwers said the comet was 2' across on the 8th, but appeared extraordinarily faint because of its position in the Milky Way. The comet was last detected on December 19.72, when Förster gave the position as $\alpha = 20^h\ 44.6^m$, $\delta = +2° 30'$.

The first parabolic orbit was calculated by Yvon-Villarceau using positions spanning the period of November 10–15. He determined the perihelion date as 1857 November 19.59. Additional orbits were soon published by C. F. Pape, d'Arrest, C. Struve, Winnecke, J. C. Watson, and C. W. Tuttle that narrowed the perihelion date down to November 19.56.

Auwers (1859) took 69 positions spanning the period of November 10 to December 19 and calculated both a parabolic and an elliptical orbit. The parabolic orbit had a perihelion date of November 19.56. The elliptical orbit had a perihelion date of November 19.57 and a period of about 6143 years. Because of the short period of observation, the parabolic orbit was favored and is given below.

T	ω	Ω (2000.0)	i	q	e
1857 Nov. 19.5641 (UT)	95.1021	141.3074	142.1557	1.009088	1.0

ABSOLUTE MAGNITUDE: $H_{10} = 9.9$ (V1964)

FULL MOON: Nov. 2, Dec. 1, Dec. 30

SOURCES: G. B. Donati, *MNRAS*, **18** (1857 Nov.), pp. 14–15; C. F. Pape, *MNRAS*, **18** (1857 Nov.), pp. 15–16; A. J. F. Yvon-Villarceau, *CR*, **45** (1857 Nov. 23), p. 898; G. B. Donati, F. A. T. Winnecke, and C. F. Pape, *AN*, **47** (1857 Nov. 24), pp. 217–20; H. L. d'Arrest and C. Struve, *AN*, **47** (1857 Dec. 3), pp. 231–4; R. van Arsdale and J. Ferguson, *AJ*, **5** (1857 Dec. 11), pp. 89, 95–6; A. Secchi and R. van Arsdale, *AN*, **47** (1857 Dec. 12), pp. 259, 263; F. A. T. Winnecke, *AN*, **47** (1858 Jan. 4), pp. 295–8; J. C. Watson and C. W. Tuttle, *AJ*, **5** (1858 Jan. 20), pp. 100–1; W. J. Förster, *AN*, **47** (1858 Jan. 20), pp. 307, 311; G. F. W. Rümker, *AN*, **48** (1858 Mar. 27), p. 83; A. Reslhuber, *AN*, **48** (1858 Apr. 21), p. 151; A. J. G. F. von Auwers, *AN*, **48** (1858 Jun. 24), p. 317; A. J. G. F. von Auwers, *AN*, **50** (1859 Mar. 16), pp. 115–22; V1964, p. 58.

6P/1857 X1 *Recovered:* 1857 December 5.82 ($\Delta = 1.70$ AU, $r = 1.17$ AU, Elong. $= 42°$)
(d'Arrest) *Last seen:* 1858 January 18.82 ($\Delta = 1.90$ AU, $r = 1.34$ AU, Elong. $= 41°$)
    Closest to the Earth: 1857 May 18 (1.6003 AU), 1857 November 25 (1.6936 AU)
1857 VII *Calculated path:* SGR (Rec), CAP (Dec. 7), AQR (Jan. 2)

J. A. C. Oudemans took the best orbit from the 1851 apparition, applied perturbations by Jupiter and Saturn, and determined a series of sweeping ephemerides for the 1857 return. The ephemerides covered the period of 1857 July 30 to 1858 January 26. For each day of this period, he gave five positions: one using the comet's orbit that he computed in 1857, and the other four derived by assuming variations in the mean daily motion of $+10''$, $+5''$, $-5''$, and $-10''$.

T. Maclear (Royal Observatory, Cape of Good Hope, South Africa) recovered this comet on 1857 December 5.82. His colleague, W. Mann, gave the position as $\alpha = 19^h 51.6^m$, $\delta = -21° 17'$. Mann considered the positions rough because of the comet's low altitude. Maclear confirmed the recovery on December 7.81.

The comet was only observed by Maclear and Mann. As it had passed perihelion during late November, it was steadily fading. Clouds allowed only brief glimpses of the comet on December 9 and 15. But very clear and calm skies allowed Mann to see the comet on the 20th, at which time he said moonlight "nearly obliterates the comet; it is detected with great difficulty." Mann said the comet was "barely visible" in moonlight on the 21st and he noted that only "The faintest trace of the comet visible" on the 24th.

Moonlight was too strong during the remainder of December and into the first days of January. Mann described the comet as "remarkably faint" on January 4, "barely visible" on the 7th, and "almost too faint for work" on the 10th. Additional observations were obtained by Mann on January 12, 13, 15, 16, and 17, during which he noted the comet was "of the last degree of faintness."

The comet was last detected on January 18.82, when Maclear measured the position as $\alpha = 22^h 54.7^m$, $\delta = -13° 54'$. Mann noted that the comet was "of the last degree of faintness, and the observations in consequence very difficult."

Maclear summarized the observations he made. He said, "The comet appeared to be a very faint nebulous body subtending a diameter of about 1.25'; and was observed generally under very unfavourable circumstances, owing to its low altitude (from $8°$ to $16°$), and the continued interruption caused by drifting clouds from the Table Mountain." He said the observations were made with an 8.5-foot focal length refractor ($90\times$).

Refinements of this comet's orbit using positions from multiple apparitions, as well as planetary perturbations, have been published by numerous astronomers. Some of the earliest were published by H. Lind (1859), A. J. F. Yvon-Villarceau (1859), and L. R. Schulze (1863, 1865), while the most recent were by A. W. Recht (1939), and B. G. Marsden, Z. Sekanina, and D. K. Yeomans (1972, 1973). These typically revealed a perihelion date of November 28.68 and a period of 6.38 years. Marsden, and coworkers determined the nongravitational terms as $A_1 = 0.00$ and $A_2 = +0.1038$ and applied them to their orbital computations. The orbit of Marsden and coworkers is given below.

T	ω	Ω (2000.0)	i	q	e
1857 Nov. 28.6831 (TT)	174.6290	150.4021	13.9170	1.169947	0.659871

ABSOLUTE MAGNITUDE: $H_{10} = 9.6$ (V1964)

FULL MOON: Dec. 1, Dec. 30, Jan. 29

SOURCES: J. A. C. Oudemans, *AJ*, **5** (1857 Aug. 28), pp. 65–70; T. Maclear and W. Mann, *MNRAS*, **19** (1858 Dec.), pp. 45–9; H. Lind, *AN*, **50** (1859 May 5), pp. 247–50; A. J. F. Yvon-Villarceau, *CR*, **48** (1859 May 9), pp. 924–7; L. R. Schulze, *AN*, **59** (1863 Feb. 21), pp. 189–92; L. R. Schulze, *AN*, **65** (1865 Sep. 12), pp. 163–8; G. Leveau and A. J. F. Yvon-Villarceau, *CR*, **81** (1875 Jul. 19), pp. 141–5; A. W. Recht, *AJ*, **48** (1939 Jul. 17), pp. 65–78; V1964, p. 58; B. G. Marsden and Z. Sekanina, *QJRAS*, **13** (1972), pp. 428–9; B. G. Marsden, Z. Sekanina, and D. K. Yeomans, *AJ*, **78** (1973 Mar.), pp. 213, 215.

8P/1858 A1
(Tuttle)

1858 I

Discovered: 1858 January 5.01 ($\Delta = 0.80$ AU, $r = 1.27$ AU, Elong. $= 90°$)
Last seen: 1858 March 24.02 ($\Delta = 1.09$ AU, $r = 1.11$ AU, Elong. $= 64°$)
Closest to the Earth: 1858 January 21 (0.7607 AU)
Calculated path: AND (Disc), PSC (Jan. 18), ARI (Feb. 1), PSC (Feb. 3), CET (Feb. 7), ERI (Feb. 18)

H. P. Tuttle (Harvard College Observatory, Cambridge, Massachusetts, USA) discovered this comet on 1858 January 5.01. He described it as "rather faint, but not so much so as to afford any difficulty in observing it with the great refractor." He gave the position as $\alpha = 23^h\ 42.5^m$, $\delta = +39°\ 46'$ on January 5.09. An independent discovery was made by K. C. Bruhns (Berlin, Germany) on January 11.89. He described it as "a very large diffuse object without nucleus and distinct border."

The comet was well observed during the remainder of January and passed closest to Earth during the second half of the month. F. A. T. Winnecke (Pulkovo Observatory, Russia) described the comet as faint and difficult to see on the 16th, with a coma diameter of 2.5'. J. C. Watson and F. F. E. Brünnow (Ann Arbor, Michigan, USA) observed with a 30-cm telescope on the 21st and 22nd and said the comet appeared very faint on account of moonlight. H. L. d'Arrest (Copenhagen, Denmark) saw the comet on the 23rd and 24th. He described it as large and hardly perceptible in bright moonlight. D'Arrest said the comet was only occasionally seen in moonlight on the 29th. A. Reslhuber (Kremsmünster, Austria) observed the comet for a short time before moonrise on the 30th, and described it as very dull, without a nucleus or tail. D'Arrest said the comet looked like a nebulous star as it occulted a 9th-magnitude star on the 31st.

Although past Earth, the comet continued to approach the sun during February and passed perihelion on the 24th. Reslhuber said it appeared brighter on February 2 than on January 30. Reslhuber described the comet as a bright, round nebulosity about 4′ across on February 3. Winnecke said the comet was faint and 1.8′ across on the 6th, with only slight central condensation. Reslhuber saw a faint nucleus on the 7th. G. F. W. Rümker (Hamburg, Germany) described the comet as large and diffuse on the 8th, but noted it was seen distinctly and was slightly condensed on the 10th. D'Arrest noted during the first half of February that the comet was noticeably condensed, but without a tail. On the 10th and 12th he said the coma was about 140″ across. A. J. G. F. von Auwers (Göttingen, Germany) described the comet as a bright, round nebulosity about 5–6′ across, with a strong central condensation. Rümker said the comet was an easy object on February 16, despite moonlight and bright northern lights. He also noted it had a granular appearance. Reslhuber and Auwers found the comet very faint in moonlight on the 19th, but Winnecke was able to estimate a diameter of 2.2′. Reslhuber said the comet was hardly visible in moonlight on the 20th, but he said it was still visible on the 27th despite the brightly illuminated field of view from the full moon. Before moonrise on the 28th, Reslhuber said the comet was better seen than on the previous evening.

The comet was moving away from both the sun and Earth during March. Reslhuber said the comet was exceptionally bright on the 2nd, with a short tail. Reslhuber found the comet faint because of hazy skies on the 3rd, but noticed an elliptical coma measuring about 3.5′ by 2.5′. J. F. J. Schmidt (Olmütz, now Olomouc, Czech Republic) saw the comet on March 3 and 4 with a 5-foot focal length refractor. He said it was easily seen with a trace of tail and possible granularity within the coma. Reslhuber said the comet was distinctly seen in very unfavorable skies on the 5th. Twilight and low altitude caused some interference with Reslhuber's observations during the next two weeks. He said it appeared faint with an occasionally visible nucleus on the 10th. He described it as faint, but centrally condensed on the 13th. Reslhuber was unable to find the comet on the 20th.

The comet was last detected on March 24.02, when G. P. Bond (Harvard College Observatory, Cambridge, Massachusetts, USA) saw it in the 38-cm refractor and gave the position as $\alpha = 4^h\ 12.4^m$, $\delta = -26° 19′$.

The first parabolic orbit was calculated by C. W. Tuttle using positions from January 5 to January 13. He determined the perihelion date as 1858 February 20.48. He noted "so strong a resemblance to [the orbit] of the second comet of 1790, that there would seem to be small reason for doubt of the identity of the comet." Similar orbits were calculated by J. C. Watson, Bruhns, and C. F. Pape, but where Watson and Pape agreed with Tuttle's suggestion of a strong resemblance to comet 1790 II, Bruhns said the orbit resembled comet 1785 I so closely that he considered identity beyond doubt. As the comet continued being observed, A. Hall and Watson independently published revisions that established a perihelion date around February 20.7. Watson noted that positions spanning January 5 to February 4 did not perfectly fit a parabolic orbit and reiterated his suggestion that the comet had been seen in 1790.

T. H. Safford published the first elliptical orbit. He took positions spanning the period of January 5 to February 7 and calculated a period of 13.63 years. During the remainder of February, elliptical orbits were calculated by Watson (13.96 years), Hall (13.71 years), Bruhns (14.81, 13.66, and 13.70 years), d'Arrest (14.56 years), H. K. F. K. Schjellerup (11.57 and 11.81 years), and Pape (13.97, 14.02, and 14.34 years). The perihelion date was then established as February 24.0. Further orbits were calculated during the remainder of the year by Hall and Bruhns, that narrowed down the period to about 13.7 years.

Several astronomers have considered multiple apparitions in the investigation of this comet's motion. The earliest such investigations were published by F. Tischler (1868) and J. Rahts (1885). The most recent was by B. G. Marsden and Z. Sekanina (1972). All determined the perihelion date as February 24.02 and the period as 13.74 years, with Marsden and Sekanina giving nongravitational terms of $A_1 = +0.32$ and $A_2 = +0.0131$.

T	ω	Ω (2000.0)	i	q	e
1858 Feb. 24.0181 (TT)	206.7770	271.0483	54.4065	1.025538	0.821207

ABSOLUTE MAGNITUDE: $H_{10} = 7.8$ (V1964)

FULL MOON: Dec. 30, Jan. 29, Feb. 27, Mar. 29

SOURCES: *MNRAS*, **18** (1858 Jan.), pp. 57–8; H. P. Tuttle, *AJ*, **5** (1858 Jan. 20), p. 101; K. C. Bruhns, *AN*, **47** (1858 Jan. 20), p. 317; C. F. Pape and H. K. F. K. Schjellerup, *AN*, **47** (1858 Jan. 30), pp. 327–32; H. P. Tuttle, *AN*, **47** (1858 Jan. 30), p. 335; K. C. Bruhns, *AN*, **47** (1858 Feb. 6), pp. 345–52; J. C. Watson and K. C. Bruhns, *AJ*, **5** (1858 Feb. 13), pp. 111–12; H. L. d'Arrest, *AN*, **47** (1858 Feb. 13), p. 367; C. F. Pape and K. C. Bruhns, *MNRAS*, **18** (1858 Mar.), p. 134; H. L. d'Arrest, H. K. F. K. Schjellerup, K. C. Bruhns, and C. F. Pape, *AN*, **48** (1858 Mar. 2), pp. 17–28; H. L. d'Arrest, *AN*, **48** (1858 Mar. 12), p. 45; T. H. Safford, J. C. Watson, and A. Hall, *AJ*, **5** (1858 Mar. 24), pp. 114, 119–20; A. Reslhuber, *AN*, **48** (1858 Apr. 21), pp. 151–4; J. Hartnup, *MNRAS*, **18** (1858 May), p. 230; H. P. Tuttle, K. C. Bruhns, and J. C. Watson, *AJS* (Series 2), **25** (1858 May), pp. 447–8; J. F. J. Schmidt, *AN*, **48** (1858 May 3), p. 183; K. C. Bruhns, *AN*, **48** (1858 May 18), p. 221; A. Hall, *AJ*, **5** (1858 Jun. 22), p. 138; G. F. W. Rümker and A. J. G. F. von Auwers, *AN*, **48** (1858 Jun. 24), pp. 311, 317–20; G. P. Bond, *AN*, **48** (1858 Jun. 28), p. 331; K. C. Bruhns, *AN*, **49** (1858 Aug. 14), pp. 33–40; K. C. Bruhns, *AJ*, **5** (1858 Nov. 2), pp. 170–4; *AN*, **59** (1862 Dec. 27), pp. 69–72; F. Tischler, *Dissertatio inauguralis.* Dissertation: Königsberg (1868), 32 pp.; J. Rahts, *AN*, **113** (1885 Dec. 22), pp. 169–206; V1964, p. 58; B. G. Marsden and Z. Sekanina, *QJRAS*, **13** (1972), pp. 427–9; B. G. Marsden, *CCO*, 12th ed. (1997), pp. 50–1.

7P/1858 E1
(Pons–Winnecke)

1858 II

Discovered: 1858 March 9.06 ($\Delta = 0.60$ AU, $r = 1.16$ AU, Elong. $= 90°$)
Last seen: 1858 June 23.44 ($\Delta = 1.38$ AU, $r = 1.14$ AU, Elong. $= 54°$)
Closest to the Earth: 1858 March 24 (0.5389 AU)
Calculated path: OPH (Disc), SER (Mar. 13), AQL (Mar. 19), AQR (Apr. 3), PSC (Apr. 23), CET (May 13), PSC (May 14), CET (Jun. 11),

F. A. T. Winnecke (Bonn, Germany) discovered this comet in the morning sky on 1858 March 9.06 at a position of $\alpha = 17^h\ 14.9^m$, $\delta = -1°\ 57'$. He confirmed the object was moving when it was seen on March 9.16. Winnecke described it as a pale and diffuse nebulosity, about 3′ across, with no trace of a nucleus.

The comet brightened slowly throughout most of March as it approached both the sun and Earth. The comet was near a 7th-magnitude star on the 12th, which K. T. R. Luther (Bilk, Germany) said caused the comet to appear small and faint. Winnecke said the comet was then uncommonly diffuse, about 2–3′ across, with no perceptible condensation. A. J. G. F. von Auwers (Göttingen, Germany) also saw the comet on the 12th and said it was faint and diffuse, with a diameter of about 4′, but displaying no nucleus. A. Reslhuber (Kremsmünster, Austria) saw the comet on March 19, 20, and 21, and described it as dull and diffuse, with a coma 2–3′ across, and a very small, faint nucleus. He added that the coma exhibited a blurred extension pointing away from the sun on the 20th. Winnecke gave the coma diameter as 3′ on the 19th and said the comet was considerably brighter than when previously seen, but exhibited little condensation. G. F. W. Rümker (Hamburg, Germany) said the coma was 2.5′ across on the 20th and contained a central condensation. Winnecke said the comet was 3–4′ across on the 21st and gradually brightened toward the middle. Winnecke simply noted the comet was faint on the 23rd. With bright moonlight interfering, Reslhuber found the comet faint on the 27th. Moonlight and morning twilight prompted Reslhuber to describe the comet as extraordinarily faint on the 29th.

The comet noticeably brightened during the first half of April, as it continued heading for perihelion. Reslhuber saw the comet through variably cloudy skies on the 6th and could see the comet had increased in brightness since March 29. Winnecke saw the comet on the 7th and also noted the comet was much brighter than when last seen in March. Winnecke found the comet "rather bright" on the 14th, while Reslhuber described it as a fairly bright, elliptical nebulosity, with no conspicuous nucleus on the 16th.

Although the comet was probably still brightening, it was becoming a difficult object for Northern Hemisphere observers during the second half of April and into May because of low altitude and morning twilight. Winnecke said the comet was extremely difficult to observe because of these circumstances on the 19th and 21st. J. Breen (Cambridge, England) simply noted the comet was "very faint and diffuse" on the 19th and 20th. Reslhuber found the comet fairly bright on the 20th and noted on the 27th that it was seen in twilight after the comparison stars were no longer visible. Reslhuber failed to find the comet on the 29th in fairly clear skies. The final Northern

Hemisphere observation was made by G. P. Bond (Harvard College Observatory, Cambridge, Massachusetts, USA) on May 3.

The comet was moving away from both the sun and Earth after May 2. It came under observation in the Southern Hemisphere on May 27, when C. W. Moesta (Santiago, Chile) described it as a diffuse nebulosity, with an intense central condensation. Moesta found the comet to be smaller on the 30th, although he observed in humid, hazy conditions. Moesta said the comet exhibited a small, fairly sharp nucleus on June 4. It was simply described as "fairly faint" by Moesta on June 19 and 22. The comet was last detected on June 23.44, when Moesta found it difficult to measure because of moonlight.

The first parabolic orbit was calculated by C. N. A. Krueger. He took positions from March 9 to 13 and calculated a perihelion date of 1858 April 23.21. Krueger immediately saw that the orbit was so similar to that of Pons' comet of June 1819 (see 7P/1819 L1) that he suggested the two were identical. V. Trettenero calculated a perihelion date of April 28.08. Shortly thereafter, Winnecke applied the orbit for the 1819 comet to the available positions and found an almost perfect agreement by assuming a perihelion date of May 2.47, and slightly increasing the inclination.

The first elliptical orbit calculated completely from positions obtained during 1858 was by Winnecke. Using positions from March 9 to April 7, he determined a perihelion date of May 2.99 and a period of 5.01 years. Similar elliptical orbits were calculated by Hänsel, H. Seeling, and others, but with a perihelion date of May 2.54 and a period of 5.55 years.

Later calculations using multiple apparitions and planetary perturbations were published by T. R. von Oppölzer (1870, 1880), J. F. Encke (1871), E. F. von Haerdtl (1888, 1889), and B. G. Marsden (1970, 1972). These revealed a perihelion date of May 2.54 and a period of 5.55 years. Oppölzer (1880) noted that the positions of 1858, 1869, and 1875 indicated a slight secular acceleration was present. This was later confirmed by B. G. Marsden (1970, 1972), when he determined nongravitational terms of $A_1 = +0.27$ and $A_2 = -0.0072$. This orbit is given below.

T	ω	Ω (2000.0)	i	q	e
1858 May 2.5396 (TT)	162.2032	115.4344	10.7939	0.768939	0.754845

ABSOLUTE MAGNITUDE: $H_{10} = 9.0$ (V1964)

FULL MOON: Feb. 27, Mar. 29, Apr. 28, May 27, Jun. 26

SOURCES: F. A. T. Winnecke, *MNRAS*, **18** (1858 Mar.), p. 134; F. A. T. Winnecke, K. T. R. Luther, and C. N. A. Krueger, *AN*, **48** (1858 Mar. 18), pp. 75–80; F. A. T. Winnecke, *MNRAS*, **18** (1858 Apr.), p. 165; F. A. T. Winnecke and F. W. A. Argelander, *AJ*, **5** (1858 Apr. 2), p. 127; V. Trettenero, *AN*, **48** (1858 Apr. 17), p. 139; F. A. T. Winnecke, *AN*, **48** (1858 Apr. 21), pp. 155–7; J. Challis and J. Breen, *AN*, **48** (1858 Jun. 14), p. 303; F. A. T. Winnecke, *AJ*, **5** (1858 Jun. 22), p. 138; A. J. G. F. von Auwers, *AN*, **48** (1858 Jun. 24), p. 319; G. P. Bond, *AN*, **48** (1858 Jun. 28), p. 331; F. A. T. Winnecke, *MNRAS*, **18** (1858 Jul.), pp. 319–20; J. C. Watson, *AJ*, **5** (1858 Jul. 30), pp. 145, 147; A. Reslhuber, *AN*, **49**

(1858 Sep. 1), pp. 65–7; C. W. Moesta, *AN*, **50** (1859 Mar. 16), p. 125; H. Seeling, *AN*, **55** (1861 Aug. 2), pp. 337–60; Hänsel, *AN*, **59** (1863 Mar. 19), p. 235; F. A. T. Winnecke, *AN*, **52** (1860 Mar. 29), pp. 307–14; G. F. W. Rümker, *AN*, **64** (1865 Feb. 19), p. 33; T. R. von Oppölzer, *SAWW*, **62** Abt. II (1870), pp. 655–75; J. F. Encke, *AN*, **77** (1871 May 30), p. 313; T. R. von Oppölzer, *AN*, **97** (1880 May 27), pp. 149–54; T. R. von Oppölzer, *AN*, **97** (1880 Jul. 30), pp. 337–42; E. F. von Haerdtl, *AN*, **120** (1888 Dec. 28), pp. 257–72; E. F. von Haerdtl, *DAWW*, **55** Abt. 2 (1889), pp. 220–7; E. F. von Haerdtl, *DAWW*, **56** Abt. 2 (1889), pp. 151–85; V1964, p. 58; B. G. Marsden, *AJ*, **75** (1970 Feb.), pp. 80–1; B. G. Marsden, *QJRAS*, **13** (1972), pp. 428–9; B. G. Marsden, *CCO*, 12th ed. (1997), pp. 48–9.

41P/1858 J1
(Tuttle–
Giacobini–
Kresák)
1858 III

Discovered: 1858 May 3.2 ($\Delta = 0.46$ AU, $r = 1.14$ AU, Elong. $= 94°$)
Last seen: 1858 June 2.17 ($\Delta = 0.54$ AU, $r = 1.20$ AU, Elong. $= 96°$)
Closest to the Earth: 1858 April 17 (0.4515 AU)
Calculated path: LMi (Disc), UMa (May 14), CVn (May 27)

H. P. Tuttle (Harvard College Observatory, Cambridge, Massachusetts, USA) discovered this comet in the evening sky on 1858 May 3.1, at a position of $\alpha = 9^h 47^m$, $\delta = +34° 40'$. He described it as a very faint object. Tuttle confirmed the find on May 4.15.

This comet was exclusively observed by astronomers at Harvard and J. C. Watson (Ann Arbor Observatory, Michigan, USA). Physical descriptions were virtually nonexistent, with Watson's comment that the comet "was exceedingly faint, and was observed with some difficulty" being the only one published during May and June. The comet was last detected on June 2.17, when Watson gave its position as $\alpha = 12^h 29.1^m$, $\delta = +37° 12'$.

Because of the short period of visibility, only two parabolic orbits were initially determined. Watson and A. Hall independently took positions spanning the period of May 4–13 and calculated perihelion dates of 1858 May 2.82 and May 2.54, respectively. Eight years later, Hall (1866) revised his computations using positions spanning the entire period of visibility and found a perihelion date of May 3.56. Interestingly, he commented, "if the excentricity is left indeterminate I have satisfied myself by trial that the resulting orbit would be slightly hyperbolic, and the residual errors although diminished still quite large."

The first elliptical orbit was calculated by L. Schulhof (1884). Using positions spanning the entire period of visibility, he determined the perihelion date as May 3.47 and the period as 6.61 years. Schulhof added that the period was likely between 5.8 and 7.5 years.

No further interest was shown toward this comet until W. H. Pickering (1914) noted the resemblance between its orbit and that of a comet found by M. Giacobini on 1907 June 1, which was observed for 14 days. Pickering indicated that Schulhof's period of 6.61 years did not allow an even number of returns to have occurred between 1858 and 1907, but, noting the range Schulhof provided, he suggested periods of either 6.13 years or 7.01 years were most likely. Pickering then conducted a statistical study and found that no comets then existed with periods between 6.00 and 6.39 years. With this knowledge, he threw out the 6.13-year period for the comet of Tuttle and Giacobini and predicted the comet would next return in 1914. The comet was not recovered.

A. C. D. Crommelin (1929) used eight positions obtained between May 4 and June 2, and computed an elliptical orbit with a perihelion date of May 3.45 and a period of 6.02 years. Crommelin studied the perturbations by Jupiter during the period of 1858 and 1907 and concluded that the most likely period for this comet was 5.44 years. He predicted the comet would have last returned to perihelion on 1928 November 17, but noted it was an unfavorable return.

The comet was not seen again until it was accidentally rediscovered by L. Kresák in 1951. Although several astronomers believed the comets seen in 1858, 1907, and 1951 were identical, the first astronomer to conclusively link the three observed apparitions of this comet was B. G. Marsden (1974). He took 55 positions obtained between 1858 and 1951, applied perturbations by all nine planets, as well as nongravitational terms, and determined a perihelion date of May 3.58 and a period of 5.35 years. He gave the nongravitational terms as $A_1 = +3.17$ and $A_2 = +0.0056$.

T	ω	Ω (2000.0)	i	q	e
1858 May 3.584 (UT)	25.775	177.810	18.899	1.14037	0.62718

ABSOLUTE MAGNITUDE: $H_{10} = 11.3$ (V1964)

FULL MOON: Apr. 28, May 27, Jun. 26

SOURCES: H. P. Tuttle, *MNRAS*, **18** (1858 May), p. 236; W. C. Bond and H. P. Tuttle, *AJ*, **5** (1858 May 20), p. 134; H. P. Tuttle and G. P. Bond, *AN*, **48** (1858 Jun. 8), p. 287; J. C. Watson, *AJ*, **5** (1858 Jun. 22), p. 141; H. P. Tuttle, A. Hall, and G. P. Bond, *AN*, **48** (1858 Jun. 28), p. 331; *AJS* (Series 2), **26** (1858 Jul.), p. 146; A. Hall, *MNRAS*, **18** (1858 Jul.), p. 322; J. C. Watson, *AJ*, **5** (1858 Jul. 30), p. 147; J. C. Watson, *AN*, **49** (1858 Sep. 25), p. 119; A. Hall, *AN*, **66** (1866 Jan. 27), pp. 137–40; L. Schulhof, *BA*, **1** (1884), pp. 171–5; L. Schulhof, *AN*, **108** (1884 May 13), p. 425; W. H. Pickering, *AN*, **198** (1914 Aug. 1), p. 471; A. C. D. Crommelin, *MNRAS*, **89** (1929 Feb.), pp. 361–4; V1964, p. 58; B. G. Marsden, *QJRAS*, **15** (1974), pp. 450–1.

C/1858 K1   *Discovered:* 1858 May 22.04 ($\Delta = 1.09$ AU, $r = 0.64$ AU, Elong. $= 35°$)
(Bruhns)   *Last seen:* 1858 June 20.86 ($\Delta = 1.40$ AU, $r = 0.64$ AU, Elong. $= 25°$)
  Closest to the Earth: 1858 May 29 (1.0468 AU)
1858 IV   *Calculated path:* AND (Disc), PER (May 26), AUR (Jun. 3), LYN (Jun. 13), AUR (Jun. 14), LYN (Jun. 18)

K. C. Bruhns (Berlin, Germany) discovered this comet in Andromeda on 1858 May 22.04. He gave the position as $\alpha = 1^h 36.2^m$, $\delta = +39° 58'$ on May 22.06. Bruhns said the comet was "easily observed," and was 3–4' across.

The comet attained its most northerly declination of +52° on June 3, at which time K. L. von Littrow (Vienna, Austria) said the comet was similar in brightness to a star of magnitude 7. He added that the centrally condensed coma was about 2' across, while the tail extended about 0.5°. A. Reslhuber (Kremsmünster, Austria) said the comet was bright on the 5th. He added that the coma was of a fairly uniform brightness across the middle, and the tail was 30' long. That same evening, W. J. Förster (Berlin, Germany) said the tail extended toward PA 7° 53'. G. F. W. Rümker (Hamburg, Germany) said the tail extended about 20' toward PA 28° on June 8. Förster saw the comet twice that day. He said that early in the morning the tail extended toward PA 16° 12', while late that evening it extended toward PA 25° 23'. On June 13, Reslhuber said the comet's brightness was rather considerable, while Förster said the tail extended toward PA 39° 02'. E. Luther (Königsberg, now Kaliningrad, Russia) saw the comet with a heliometer on June 14, which indicated a magnitude brighter than 7. Around mid-June, H. L. d'Arrest (Copenhagen, Denmark) estimated the comet's magnitude as 7–8. Reslhuber said the comet was observed low over the horizon on the 16th, while he noted on the 19th that the comet was observed despite the sky not being very clear. The comet was last detected on June 20.86, when V. Trettenero (Padova, Italy) gave the position as $\alpha = 7^h 36.2^m$, $\delta = +36° 22'$.

A later final observation was published by J. G. Galle (1894) and S. K. Vsekhsvyatskij (1964). Given as July 15, the observation was credited to observers at Marseille Observatory. Galle even provided a reference of *Comptes Rendus*, volume 47, page 306. Upon checking this source, the Author found an observation from Marseille for July 15 and noted that the writer expressed his surprise that the comet had not been seen elsewhere since June 20. After providing a precise position, the writer then began a discussion of comet C/1858 L1 (Donati). Although the June 20 comment would seem to indicate that the observation refers to C/1858 K1, the position matches that of comet C/1858 L1, so that a date later than June 20 seems unlikely. Additional factors to consider would be the comet's brightness, which would have been fading after early June because of the comet's increasing distance from the sun and Earth, and the comet's elongation from the sun, which would have decreased to only 22° by July 15.

The first parabolic orbits were calculated by Bruhns and M. F. Karlinski. Using positions from May 22, 24, and 27, they determined the perihelion date as 1858 June 5.68 and June 5.71, respectively. Similar orbits were calculated by d'Arrest and M. Loewy. A. J. G. F. von Auwers (1859) used about 90 positions obtained between May 22 and June 20, and computed a parabolic orbit with a perihelion date of June 5.80. No planetary perturbations were applied. Auwers' orbit is given below.

T	ω	Ω (2000.0)	i	q	e
1858 Jun. 5.7956 (UT)	98.8587	326.9503	99.9713	0.544261	1.0

ABSOLUTE MAGNITUDE: $H_{10} = 8.6$ (V1964)

FULL MOON: Apr. 28, May 27, Jun. 26

SOURCES: K. C. Bruhns, *MNRAS*, **18** (1858 May), p. 257; K. C. Bruhns, *AN*, **48** (1858 May 26), p. 239; K. C. Bruhns and M. F. Karlinski, *AN*, **48** (1858 Jun. 8), pp. 285–8; K. C. Bruhns, *AJ*, **5** (1858 Jun. 22), p. 140; K. C. Bruhns, *MNRAS*, **18** (1858 Jul.), p. 322; H. L. d'Arrest, K. L. von Littrow, and M. Loewy, *AN*, **48** (1858 Jul. 8), pp. 353–6, 361–4; A. Reslhuber, *AN*, **49** (1858 Sep. 1), p. 67; W. J. Förster, *AN*, **49** (1858 Oct. 19), p. 165; E. Luther, *AN*, **50** (1859 Feb. 15), p. 71; K. L. von Littrow, *AN*, **50** (1859 Feb. 25), p. 83; M. Hoek, *AN*, **50** (1859 Mar. 29), pp. 155–8; A. J. G. F. von Auwers and V. Trettenero, *AN*, **51** (1859 Aug. 25), pp. 113–24; G. F. W. Rümker, *AN*, **64** (1865 Feb. 19), p. 33; G1894, p. 237; V1964, pp. 58, 202.

4P/Faye *Recovered:* 1858 September 8.0 ($\Delta = 1.50$ AU, $r = 1.69$ AU, Elong. $= 82°$)

Last seen: 1858 October 17.12 ($\Delta = 1.24$ AU, $r = 1.73$ AU, Elong. $= 101°$)

1858 V *Closest to the Earth:* 1858 December 15 (1.0212 AU)

Calculated path: TAU (Rec), ORI (Sep. 12), GEM (Sep. 28)

K. C. Bruhns (Berlin, Germany) recovered this comet with a 23-cm refractor on 1858 September 8.0, while sweeping using J. R. Hind's ephemeris. He described it as "a very faint nebulous spot," but could not measure its position because the sky quickly grew cloudy. Bruhns and W. J. Förster confirmed the recovery on September 9.04, and determined the position as $\alpha = 5^h 31.5^m$, $\delta = +19° 23'$. Bruhns added that it was as difficult to see as a 13th-magnitude star.

Very few physical descriptions were obtained of this comet. J. Challis (Cambridge, England) saw the comet on September 16, 17, October 9, 10, and 12. He commented that positions "were taken with difficulty on account of the faintness of the comet." Bruhns saw the comet on September 16 and said it resembled an 11th-magnitude star. On October 7, Bruhns said the comet was barely visible.

The comet was last detected on October 17.12, when Bruhns measured the position as $\alpha = 6^h 51.3^m$, $\delta = +12° 59'$. He said the comet was then so faint that it was detected only with the greatest effort.

Following the comet's recovery, Bruhns used an orbit computed by U. J.-J. Le Verrier and published in the 1845 June 26 issue of the *Astronomische Nachrichten*, and applied corrections attributed to precession. He then varied the longitude of the ascending node and the date of perihelion until the new observations were best represented. The result was a perihelion date of 1858 September 13.10.

Later calculations using multiple apparitions and planetary perturbations were published by D. M. A. Möller (1865, 1872), A. Shdanow (1885), and B. G. Marsden and Z. Sekanina (1971). These revealed a perihelion date of September 13.39 and a period of 7.45 years. Marsden and Sekanina determined the nongravitational terms as $A_1 = +0.601$ and $A_2 = +0.01026$. The orbit of Marsden and Sekanina is given below.

T	ω	Ω (2000.0)	i	q	e
1858 Sep. 13.3850 (TT)	200.1228	211.6912	11.3555	1.694100	0.555725

ABSOLUTE MAGNITUDE: $H_{10} = 8.1$ (V1964)

FULL MOON: Aug. 24, Sep. 23, Oct. 22

SOURCES: J. F. Encke, *MNRAS*, **18** (1858 Jul.), pp. 324–5; K. C. Bruhns, *AN*, **49** (1858 Sep. 15), pp. 107–10; *AJ*, **5** (1858 Nov. 2), p. 169; J. Challis, *AN*, **50** (1859 May 5), pp. 241–6; K. C. Bruhns, *AN*, **52** (1860 Jan. 11), pp. 81–8; D. M. A. Möller, *AN*, **57** (1862 May 6), pp. 215–24; D. M. A. Möller, *AN*, **64** (1865 Apr. 10), pp. 145–52; D. M. A. Möller, *VJS*, **7** (1872), p. 96; A. Shdanow, *BASP*, **33** (1885), pp. 1–24; V1964, p. 58; B. G. Marsden and Z. Sekanina, *AJ*, **76** (1971 Dec.), pp. 1136–7.

C/1858 L1
(Donati)

1858 VI

Discovered: 1858 June 2.89 ($\Delta = 2.47$ AU, $r = 2.23$ AU, Elong. $= 64°$)

Last seen: 1859 March 4.81 ($\Delta = 3.14$ AU, $r = 2.71$ AU, Elong. $= 56°$)

Closest to the Earth: 1858 October 10 (0.5378 AU)

Calculated path: LEO (Disc), LMi (Jul. 30), UMa (Sep. 4), CVn (Sep. 22), COM (Sep. 29), CVn–COM (Oct. 1), BOO (Oct. 2), VIR (Oct. 9), SER (Oct. 10), LIB (Oct. 12), SCO (Oct. 14), OPH (Oct. 16), SCO (Oct. 22), CrA (Oct. 30), TEL (Nov. 9), IND (Dec. 29), TUC (Feb. 17)

G. B. Donati (Florence, Italy) discovered this comet on 1858 June 2.89 at a position of $\alpha = 9^h 24.1^m$, $\delta = +23° 55'$. He described it as a very faint, small nebulosity of uniform brightness and about 3′ across. He confirmed the discovery on June 7.90.

Donati observed the comet every evening during the period of June 7–11 and described the comet as very faint. K. C. Bruhns and W. J. Förster (Berlin, Germany) provided the only other physical descriptions of this comet during June. Unfortunately, because of the 52° latitude of the observatory, twilight was always a problem and observations on June 13–16 revealed the comet was "very faint."

Word was slow arriving in the USA and two independent discoveries were reported. H. M. Parkhurst (Perth Am'boy, New Jersey, USA) found the comet on June 30.1 and M. Mitchell (Nantucket, Massachusetts, USA) found it on July 7.1. Parkhurst was not able to confirm the comet until July 2.08. Mitchell found the comet very low in the northwest while searching for comets in the evening sky. G. M. Searle (Dudley Observatory, New York, USA) confirmed Parkhurst's "discovery" on July 8 and described it as very diffuse, without a definite nucleus. He informed the observatory's director, B. A. Gould, Jr, but Gould, who was also the editor of the *Astronomical Journal*, recognized this comet as that found by Donati because of a notice he had received a few days earlier.

The comet was still sitting in evening twilight for many observers as August began, and actually reached a minimum solar elongation of 18° on the 10th. But its decreasing distances from both the sun and Earth caused it to rapidly brighten throughout July and into August. A. Reslhuber (Kremsmünster, Austria) said the comet was clearly visible in bright evening twilight on the August 5th. He noted a coma 2' across, but saw neither a nucleus nor a tail. J. C. Watson (Ann Arbor Observatory, Michigan, USA) saw the comet on the 15th and said it "appeared as bright as a star of the 4.5 magnitude." J. Breen (Cambridge, England) provided two testimonies as to the comet's increasing brightness. First, on August 19, he said the comet was easy to see with the 7-cm finder through dense cloud. Second, on August 23, he said the comet was visible despite a nearly full moon. Bruhns reported the first naked-eye observation on the 28th, with Reslhuber and R. Hodgson (Eversley, England) noting naked-eye visibility on the 30th. G. F. W. Rümker (Hamburg, Germany) saw a tail extending 35', with a width of no more than 5' on the 29th, while Reslhuber noted the tail was about 1/2° long on the 30th.

The comet was widely seen with the naked eye throughout September, and it continued to brighten. Numerous very rough magnitude estimates were made with the naked eye, with Bruhns estimating 3–4 on the 2nd, Searle estimating 3 on the 5th, and Breen and Bruhns estimating 2–3 on the 7th. J. Chacornac (Marseille, France) said the comet's naked-eye appearance was similar to that of ε Ursae Majoris on the 10th, indicating a brightness near 1.8, and Bruhns said it was similar to α Ursae Majoris on the 15th, which also indicated a value near 1.8. Bruhns noted the comet was near α Leonis in brightness on the 17th, indicating a magnitude near 1.4. Interestingly, estimates of the total magnitude were virtually nonexistent thereafter, as the comet's tail and nucleus began dominating the scene. Estimates of the tail length varied widely depending on the instrument used. Observations by Rümker, Reslhuber, and Förster on the 1st indicated the tail extended 30–50' toward PA 5°. Donati estimated it as 2° long on the 3rd, and Rümker found it 4° long in a cometseeker on the 9th. Winnecke said the tail extended toward PA 357° on the 11th, while Förster and Bruhns indicated the tail extended 5–6° toward PA 353° on the 15th. Although the tail length continued to increase thereafter, the rotation of the tail reversed direction after the 15th because of the changing Earth–comet–sun geometry. Bruhns and Winnecke indicated the tail extended 7° toward PA 355° on the 16th, T. W. Burr (Highbury, England) and Johan Henrich Mädler (Dorpat, now Tartu, Estonia) indicated the tail extended 8° toward PA 359° on the 21st, Mädler, Winnecke, and Förster indicated the tail extended 10° toward PA 4° on the 25th, and numerous astronomers reported a tail 18° to 25° long on the 30th, which extended toward PA 17°. The tail was described as "brush-like" by Burr on the 8th, "striped and slightly curved toward the left" by S. H. Schwabe (Dessau, Germany) on the 11th, and "saber-shaped" by Reslhuber on the 16th. Although the nucleus was visible throughout the month, Förster's observations on the 15th were an interesting sign of things to come, as he noticed a "very distinct emanation" toward PA 80° and a distinct bright ray toward PA 50°. On the 19th, Chacornac said the nucleus had a planetary appearance, although the side facing the sun looked nebulous. C. A. F. Peters and C. F. Pape (Altona, Germany) scrutinized the nucleus on the 20th and said the nucleus was fuzzy on the sunward side and looked as though material was flowing in the coma and then bending around to form the tail. Schwabe saw the comet on the 21st and said the nucleus exhibited several distinctly curved beams that turned back into the tail. By the 23rd, Chacornac was reporting four semicircular hoods on the sunward side of the nucleus, with the nearest hood to the nucleus being brightest. These hoods were well observed to the end of the month. Astronomer Royal Sir George Biddell Airy took a doubly refracting prism and unsilvered glass and viewed the comet with the naked eye on September 27. He said that on doing so the tail nearly disappeared, indicating the light was polarized in the plane of the comet's tail. He tried the same technique on the comet's head, as viewed through the Sheepshanks telescope, and noticed "feeble signs of polarisation."

The date of 1858 September 27 was a landmark in the field of cometary astronomy when a portrait artist named W. Usherwood (Walton Common, England) obtained the first photograph of a comet. Being impressed with the sight of comet Donati, Usherwood took an $f/2.4$ focal ratio portrait lens and obtained a 7-second exposure which revealed the bright region surrounding the nucleus, as well as a part of the comet's tail. Interestingly, September 28 was another landmark as G. P. Bond (Harvard College Observatory, Cambridge, Massachusetts, USA) obtained the first photograph of a comet through a telescope. Bond used the $f/15$ refractor at the observatory and obtained an exposure of 6 minutes. This photograph revealed only the bright region surrounding the nucleus.

The comet had passed perihelion on the final day of September, and approached Earth until October 10. During the first ten days of the month, the comet remained an impressive sight. Some astronomers attempted to find the comet in broad daylight with the use of telescopes, but no one succeeded. On the 1st, several observers indicated the tail extending 21–27° toward PA 19°, with some even noting a straight, very thin tail extending about 3° along the convex side of the main tail. In addition, A. J. G. F. von Auwers (Göttingen, Germany) noted a wisp or fan extending somewhat sunward toward PA 218°. The main tail continued displaying a curved shape and lengthened with each passing day. Observers estimated it extended 28–35° toward PA 31° on the 4th, 27–35° toward PA 40° on the 6th, and 30–41° toward PA 59° on the 9th. The width of the widest portion was typically given as 4–8°. The thin, straight tail continued to be observed by Auwers and quickly grew in length. He said it was 30° long on the 3rd, 32° on the 4th, and 40° on the 6th. Interestingly, less than perfect seeing usually prevented an observation of this tail. Auwers' "wisp or fan" continued to be observed. It extended toward PA 216° on the 2nd, PA 191° on the 4th, PA 233° on the 5th, and PA 228° on the 8th. The region around the nucleus continually drew the attention of observers. Reslhuber noted on the 3rd that a hood nearly touching the nucleus on the 1st had separated from it by the 3rd. Observations of 6–7 luminous hoods were not uncommon, with the brightest usually closest to the nucleus. The nucleus itself only appeared stellar during the early days of this period, but was most frequently described as oblong by observers with large telescopes. F. A. T. Winnecke (Bonn, Germany) began reporting a secondary nucleus on the 7th, when he noted it was situated 1 diameter of the primary nucleus away in PA 303.1°. He said this secondary nucleus gradually moved away from the primary and was situated at 1.5 diameters toward PA 306.2° on the 8th, and 2.25 diameters toward PA 306.1° on the 9th. Although no other astronomers reported this nucleus, several did note a gradual lengthening of the nucleus. October 10 was also the date when the comet first became visible to observers in the Southern Hemisphere, and it was seen by Bernhard von Wüllerstorf and Robert Müller (aboard the frigate *Novara* in the southern Pacific Ocean) and T. Maclear (Royal Observatory, Cape of Good Hope, South Africa). Wüllerstorf and Müller estimated the comet's brightness as about equal to α Lyrae, or magnitude 0. Maclear said clouds prevented observations of this comet prior to this evening, but a clear view was obtained on this night for a few minutes, while the tail continued to remain visible afterwards through breaks in the clouds.

The comet began moving away from both the sun and Earth after October 10, and it started to fade more rapidly. What's more, the comet's steadily decreasing declination meant the major observatories of Europe and the USA would lose the comet in only nine days. The dust tail was still estimated as 40–43° long on the 11th, with a maximum width of 10–16°. It was shooting out of the coma toward PA 74°. It was still nearly 33° long on the 15th, 20° long on the 16th, and no more than 5° long on the 17th. There were three reasons for the rapid decline of the tail. First, the comet was heading away from both the sun and Earth. Second, moonlight had become a factor. Third, observers in Europe and the USA were seeing the comet drop lower over the southwestern horizon and, during the period of October 7–19, the tail direction changed from a position perpendicular to the horizon to a position nearly parallel with it. The tail was said to have been composed of multiple branches on the 11th, but it had basically become featureless by the 19th, primarily because of the same three factors mentioned above. There were no reports of the long narrow tail during this period and observations of the "wisp or fan" extending from the sunward side of the coma ceased after the 14th. The region around the nucleus remained an interesting area. Chacornac was still detecting 7–8 hoods during the period of October 12–15, but he obtained no further observations thereafter. Although other observers continued to report these hoods, the features rarely exceeded two in number. Winnecke again reported the appearance of a secondary nucleus and said it was situated toward PA 305.2°. This was the final observation of the secondary nucleus. The overall appearance of the comet on October 15 and 16 was described by Challis (Cambridge, England) as like "an inverted comma," especially in strong twilight. Bruhns, J. Ferguson (Washington, DC, USA), and J. A. C. Oudemans last saw the comet on the 19th, with Bruhns noting the comet was very near the horizon.

During the remainder of October the comet steadily faded and was only visible to observers in the Southern Hemisphere and the low latitudes of the Northern Hemisphere. The only estimate of the brightness came from Wüllerstorf and Müller on the 25th, when they said the magnitude was about 3.7. Rough measures of the tail length were made in moonlight early in this period, with Wüllerstorf and Müller indicating it was 5° long on the 21st and Maclear noting it was about 4° long on the 22nd. After the moon had left the sky, the tail was still on the short side, with Eyre B. Powell (Madras, India) estimating it as 4° or 5° on the 27th and J. W. Callow (commanding the *Charles* in the southern Atlantic Ocean) indicating a similar length on the 28th. The region around the nucleus was again scrutinized. On the 23rd, W. Mann (Royal Observatory, Cape of Good Hope, South Africa) noted, "The comet's head when viewed with the ordinary observing power appears to consist of two concentric luminous envelopes nearly surrounding the nucleus on the side next the sun." He was using a telescope of 8.5-foot focal length and also noted a jet extending a short distance from the nucleus. Powell observed the comet with a 7-cm refractor on the 26th and saw one hood still present. When Mann reobserved the comet on the 29th he noted the jet was gone and a "somewhat bright condensation of light about 10″ in diameter [was] enveloping the nucleus." Mann added that the coma was 4.3′ across on the 29th and 2.6′ across on the 31st. The comet passed only 2° from Venus on October 20.

The comet's rapid fading continued throughout November. Although Wüllerstorf and Müller estimated the magnitude as 3.7 on the 1st, which implied no change since October 25, other observers indicated otherwise. Captain James Clarke (aboard the ship *Marco Polo* in the south Atlantic Ocean) noted on November 5 that the comet was much fainter than when he first saw it on October 11. He did not see the comet again. Wüllerstorf and Müller saw the comet with the naked eye on the 9th, but failed to find it on the 11th. Meanwhile, Callow said the comet was "seen very indistinctly" with the naked eye on the 11th, but last saw the comet in a 6.4-cm refractor on the 16th. Callow provided the last description of the tail on November 4, when he said it looked "like a double bow." The hoods and jets described during previous weeks were apparently no longer present, and the only descriptive observation that referred to the coma was a diameter measurement of 2.9′ on the 5th.

From December onward, C. W. Moesta (National Observatory, Santiago, Chile) and Mann were the only observers of this comet, with Mann providing occasional descriptions. He said the comet was very faint in moonlight on December 14 and was a faint nebulous body about 90″ across on the 23rd, with slight central condensation, but no tail. Mann said the coma was 98″ across on December 30. During the period of January 14–19, he said the comet was "wholly obliterated" by moonlight, while he noted the coma was 60" across on January 24. After describing the comet as exceedingly faint on February 5, Mann was still able to measure the coma as 54" across on February 26.

The comet was last seen by Moesta on March 2.03, while Mann's final observation came on March 4.81. Mann gave the position as $\alpha = 22^h 32.1^m$, $\delta = -62° 05'$.

The comet was noteworthy enough to warrant being included in the diaries of nonastronomers. James Hector, then leading a branch expedition in northern Saskatchewan, Canada as part of the John Palliser expedition, first observed this comet in the evening sky on September 11. Captain McClintock (aboard the Arctic discovery yacht *Fox* near Boothia Peninsula, Northwest Territories, Canada) first saw the comet on September 14. David Livingston (traveling down the Shire River near present day Vila Fontes, Mozambique, in Africa) wrote on October 14, "Observed a comet this evening. The kroomen say they had seen it three nights before and Matengo, my Makololo, had seen it the night before. It is a fine one, the tail a little bent." In addition, sometime during the period of October 7 to November 5, a large "broom-star" was recorded in China.

T. W. Webb (Hardwick Parsonage, Herefordshire, England) summarized his observations of this comet in an article in the November 1858 issue of the *Monthly Notices of the Royal Astronomical Society*. Concerning the nucleus he said, "This appeared invariably circular, several seconds in diameter, of uniform light throughout; its definition decreasing with the increase of the power employed." He noted that the color was a "clear yellow." With respect to the coma, he said, "This was of considerable extent, much fainter than the nucleus, but brighter than the exterior haze and train. Instead of being a hemispherical cap, serving as a base to the tail... its outline was distinctly continued through an arc of much more than 180°, round to the edges of the central darkness of the train." With respect to the two streams he said, "The two streams which formed the tail were for a long time unequal in breadth, but were never observed to change sides so as to indicate rotation: the antecedent branch showed greater fullness and density near its origin...." He added, "During the whole time the angle at which they came off from the nucleus underwent a steady increase: the antecedent stream was always far better defined than the other at it exterior edge. These two streams were connected round the sunny side of the envelope or photosphere, by a border of the same material, much narrower and fainter than the envelope." With respect to the tail, he noted, "This gave some intimation of its structure even to the naked eye; in the telescope its central darkness was very conspicuous; and as it advanced in its course the sides of the hollow paraboloid became more thin, and the angle included by them increased."

Admiral William Henry Smyth (England) compared this comet with C/1811 F1 (Great Comet) during November of 1858. Although he found the comet of 1811 to have had several naked-eye features that were striking he said, "But recollect that in these remarks, I mean nothing disrespectful to the Donati. On the contrary, with those exceptions, it is one of the most beautiful objects I have ever seen in the heavens. The head is certainly not so fully pronounced as in that of 1811; but greatly its physical interest is increased by segments of light and a dark hollow, giving the aspect a resemblance to the gaslight called a bat's-wing...."

Challis made some general comments about the comet. He said, "In the course of the observations my attention was especially directed to the following particulars, respecting which I can speak with confidence. The brightness contiguous to the nucleus preponderated on the right side (as seen in the telescope) till October 2; and on October 9 the excess had passed to the left side. The excess of brightness of the right-hand stream of the tail above that of the other attained its maximum about October 2; after which there was a gradual diminution, till, on October 11, 15, and 16, the two streams were not sensibly unequal. The dark band separating the two portions of the tail was of uniform width and definite boundary on September 30 and October 2; and in proportion as the boundaries afterwards became indefinite, and the intervening space was gradually filled with luminosity, the angular divergence of the two streams also increased."

The comet was quite far from both the sun and Earth when discovered and its subsequent motion was so slow that it moved less than 4° during its first month of visibility. This made the early orbital calculations somewhat more uncertain than usual. Bruhns was the first to calculate a parabolic orbit. He used positions from June 7 to June 16 and determined a perihelion date of 1858 September 6.26. Orbits by astronomers like Searle and C. W. Tuttle followed. They used positions up to around mid-July and were able to push the perihelion date into the second half of September. Interestingly, it was Donati himself who was the first to produce a parabolic orbit that had a perihelion date within one day of what the final value would prove to be. He determined it as September 29.71. Later parabolic orbits by Watson, Bruhns, Winnecke, S. Stampfer, J. F. Tennant, W. S. Mackay, Powell, and M. Loewy would improve little on Donati's calculations.

By October, astronomers were beginning to note that the positions no longer fitted a parabolic solution. The first elliptical orbits were independently calculated by Bruhns and Watson at the beginning of the month. Both astronomers determined the perihelion date as September 30.45, but Bruhns gave the period as 2101 years, while Watson gave it as 2415 years. Later calculations by S. Newcomb, Stampfer, Searle, Brünnow, and Loewy confirmed the general accuracy of the orbit, although the period was revised downward to near 2000 years. Interestingly, an earlier elliptical orbit was calculated by A. J. F. Yvon-Villarceau. His solution was published in the 1858 August 16 issue of the *Comptes Rendus* and gave a perihelion date of September 21.90 and a period of 69 years. Yvon-Villarceau said he preferred the parabolic solution.

From the wealth of positions gathered during this apparition, E. von Asten (1865) and G. W. Hill (1865) calculated definitive orbits for this comet. Hill began with nearly 1000 positions obtained between June 7 and March 4, determined 16 Normal positions, and applied perturbations by five planets. The result was a perihelion date of September 30.46 and a period of about 1950 years. Asten began with nearly as many positions, determined ten Normal positions spanning the period of June 19 to February 18, and also applied perturbations by five planets. The result was a perihelion date of September 30.47 and a period of about 1880 years. Since Hill's orbit used more positions spanning a longer arc, it is the one given below.

T	ω	Ω (2000.0)	i	q	e
1858 Sep. 30.4645 (UT)	129.1144	167.3044	116.9512	0.578469	0.996295

ABSOLUTE MAGNITUDE: $H_{10} = 3.3$ (V1964)

FULL MOON: May 27, Jun. 26, Jul. 26, Aug. 24, Sep. 23, Oct. 22, Nov. 21, Dec. 20, 1859 Jan. 19, Feb. 17, Mar. 18

SOURCES: G. B. Donati, *MNRAS*, **18** (1858 Jun.), p. 271; K. C. Bruhns, *MNRAS*, **18** (1858 Jun.), p. 272; G. B. Donati, W. J. Förster, and K. C. Bruhns, *AN*, **48** (1858 Jun. 28), p. 333; G. B. Donati, *MNRAS*, **18** (1858 Jul.), pp. 322–3; G. B. Donati, *AN*, **48** (1858 Jul. 8), pp. 355–8; H. M. Parkhurst, G. M. Searle, G. B. Donati, and J. Ferguson, *AJ*, **5** (1858 Jul. 30), pp. 148–50, 152; A. J. F. Yvon-Villarceau, *CR*, **47** (1858 Aug. 16), pp. 306–7; H. M. Parkhurst, M. Mitchell, G. B. Donati, and K. C. Bruhns, *AN*, **49** (1858 Aug. 20), pp. 55–60; S. Stampfer, *AN*, **49** (1858 Sep. 15), pp. 101–4; C. W. Tuttle, J. C. Watson, and J. Ferguson, *AJ*, **5** (1858 Sep. 21), pp. 154–6, 158–9; J. C. Watson and C. F. Pape, *AN*, **49** (1858 Sep. 25), p. 119, 127; M. Loewy and K. C. Bruhns, *AN*, **49** (1858 Oct. 6), pp. 133–8; J. C. Watson, J. Ferguson, and W. C. Taylor, *AJ*, **5** (1858 Oct. 16), pp. 165–7; S. Stampfer, *AN*, **49** (1858 Oct. 19), p. 173; M. Loewy, *AN*, **49** (1858 Oct. 26), pp. 177–80; S. H. Schwabe, *AN*, **49** (1858 Oct. 29), pp. 205–8; G. B. Airy, R. Main, W. Christy, J. Challis, W. Lassell, T. W. Webb, J. Slatter, J. C. Haile, T. W. Burr, W. Gray, F. Selby, W. H. Smyth, W. de la Rue, and W. R. Grove, *MNRAS*, **19** (1858 Nov.), pp. 11–9; W. C. Bond and J. C. Watson, *AJS* (Series 2), **26** (1858 Nov.), pp. 433–4; D. Trowbridge and L. Swift, *AJ*, **5** (1858 Nov. 2), p. 176; J. H. Mädler and A. J. G. F. von Auwers, *AN*, **49** (1858 Nov. 20), pp. 225–38; J. Hartnup, *MNRAS*, **19** (1858 Dec.), pp. 54–6; E. B. Powell, R. L. J. Ellery, and G. M. Balfour, *MNRAS*, **19** (1858 Dec.), pp. 62–6; A. Reslhuber, J. Hartnup, and E. Heis, *AN*, **49** (1858 Dec. 10), pp. 257–72; S. Newcomb, J. Ferguson, Chacornac, G. B. Donati, G. M. Searle, S. H. Schwabe, C. A. F. Peters, C. F. Pape, and T. M. Logan, *AJ*, **5** (1858 Dec. 20), pp. 178, 180, 182–4; G. M. Searle, *AJ*, **5** (1858 Dec. 31), pp. 188–90; C. F. Pape, *AN*, **49** (1858 Dec. 31), pp. 309–54; W. Lassell, *MNRAS*, **19** (1859 Jan.), pp. 79–80; R. Hodgson and W. R. Dawes, *MNRAS*, **19** (1859 Jan.), pp. 86–91; J. Ferguson, *AN*, **49** (1859 Jan. 7), pp. 363–6; J. G. Galle, *AN*, **50** (1859 Feb. 2), pp. 37–42; E. Luther, *AN*, **50** (1859 Feb. 15), pp. 71–6; J. F. Tennant, *MNRAS*, **19** (1859 Mar.), p. 185; J. A. C. Oudemans, *AN*, **50** (1859 Mar. 8), pp. 107–10; M. Hoek, *AN*, **50** (1859 Mar. 29), pp. 155–60; T. Maclear, *MNRAS*, **19** (1859 Apr.), pp. 229–30; B. von Wüllerstorf and R. Müller, *AN*, **50** (1859 Apr. 22), pp. 211–18; J. Challis, *AN*, **50** (1859 May 5), pp. 241–6; F. A. T. Winnecke, *AN*, **50** (1859 May 27), pp. 305–18;

N. M. Edmondson, *MNRAS*, **19** (1859 Jul.), pp. 305–6; K. C. Bruhns and W. J. Förster, *AN*, **51** (1859 Aug. 6), pp. 65–74; J. W. Callow, *MNRAS*, **20** (1859 Dec.), pp. 49–50; Trivett, *MNRAS*, **20** (1859 Dec.), p. 50; McClintock, *MNRAS*, **20** (1859 Dec.), p. 51; J. Clarke, *MNRAS*, **20** (1859 Dec.), p. 51; T. Maclear, W. Mann, and G. W. H. Maclear, *MRAS*, **29** (1859–60), pp. 59–77; F. A. T. Winnecke, *BASP*, **1** (1860), pp. 1–15; C. W. Moesta, *AJ*, **6** (1860 Apr. 27), pp. 100–3; T. Maclear, *MNRAS*, **20** (1860 Jul.), pp. 334–5; W. S. Mackay, *MNRAS*, **22** (1862 Mar.), p. 160; G. P. Bond, *AN*, **58** (1862 Aug. 8), pp. 81–6; G. P. Bond, *AN*, **60** (1863 Jun. 3), pp. 49–60; G. W. Hill and E. von Asten, *AN*, **64** (1865 Apr. 27), pp. 185–92; *The Zambezi Expedition of David Livingston (1858–1893)*. Edited by J. P. R. Wallis. London: Chatto & Windus Limited (1956), p. 49; V1964, p. 58; *The Papers of the Palliser Expedition*. Edited by Irene M. Spry. Great Britain: Robert MacLehose and Company Limited (1968), p. 320; HA1970, p. 86.

C/1858 R1   *Discovered:* 1858 September 6.37 ($\Delta = 1.17$ AU, $r = 1.52$ AU, Elong. $= 88°$)
(Tuttle)   *Last seen:* 1858 November 10.76 ($\Delta = 1.42$ AU, $r = 1.48$ AU, Elong. $= 73°$)
  Closest to the Earth: 1858 October 5 (0.4903 AU)
1858 VII   *Calculated path:* AUR (Disc), PER (Sep. 6), AND (Sep. 23), PER (Sep. 25), AND (Sep. 26), PEG (Oct. 4), EQU–AQR (Oct. 16), CAP (Oct. 29)

H. P. Tuttle (Harvard College Observatory, Cambridge, Massachusetts, USA) discovered this comet in the evening sky on 1858 September 6.37 at a position of $\alpha = 4^h 41.0^m$, $\delta = +44° 47'$. He confirmed his find on September 8.26.

Although the comet steadily brightened during September, no physical descriptions were made until October. C. F. Pape (Altona, Germany) saw the comet on October 3 and described it as "a bright, fairly large nebulosity with a nuclear condensation. . . . " This nuclear condensation was estimated as magnitude 8 by G. F. W. Rümker (Hamburg, Germany) on the 4th. A. J. G. F. von Auwers (Göttingen, Germany) saw the comet rather distinctly with the naked eye on October 6 and 8. K. Hornstein (Vienna, Austria) saw the comet on the 7th and described it as "a large, round, very diffuse nebulosity of about 6' diameter; it is fairly bright with a small nuclear condensation in the middle. . . . " On October 8, both Rümker and A. Reslhuber (Kremsmünster, Austria) independently estimated the coma as 4–5' across. Rümker noted the comet appeared diffuse, while Reslhuber described the comet as round, with a weak nucleus, but no tail. M. Hoek (Leiden, Netherlands) found the comet very faint on the 16th. Rümker saw the comet on the 26th and 27th, and said the coma was over 4' across and that the "core" was no longer visible.

The comet rapidly faded after passing perihelion in mid-October and many observers were no longer following it by the time November began. On November 9, Reslhuber described the comet as "very faint", while E. Weiss (Vienna, Austria) described it as "extremely faint." The comet was last detected by Reslhuber on November 10.73 and by Hornstein on November 10.76. Reslhuber then described the comet as very faint, and

Hornstein described it as extremely faint. Hornstein gave the position as $\alpha = 20^h\ 15.1^m$, $\delta = -19°\ 08'$.

The first parabolic orbit was calculated by C. W. Tuttle. He used three positions spanning the period of September 6–10 and determined the perihelion date as 1858 October 17.54. During the next few months, calculations by Auwers, H. L. d'Arrest, and Weiss (1858, 1859) established the perihelion date as early on October 13. Weiss took 58 positions obtained throughout the comet's period of visibility and determined six Normal positions. He determined a parabolic orbit with a perihelion date of October 13.31 and an elliptical orbit with a perihelion date of October 13.33 and a period of about 6004 years. Weiss' parabolic orbit is given below.

T	ω	Ω (2000.0)	i	q	e
1858 Oct. 13.3102 (UT)	155.5411	161.7449	158.7051	1.427002	1.0

ABSOLUTE MAGNITUDE: $H_{10} = 5.9$ (V1964)

FULL MOON: Aug. 24, Sep. 23, Oct. 22, Nov. 21

SOURCES: H. P. Tuttle, *MNRAS*, **18** (1858 Jul.), p. 303; H. P. Tuttle and C. W. Tuttle, *AJ*, **5** (1858 Sep. 21), p. 154; H. P. Tuttle, A. Hall, and C. F. Pape, *AN*, **49** (1858 Oct. 6), p. 141; J. Ferguson, *AJ*, **5** (1858 Oct. 16), pp. 166–7; C. F. Pape, *AN*, **49** (1858 Oct. 26), pp. 183–6; A. J. G. F. von Auwers, *AN*, **49** (1858 Oct. 29), p. 205; H. P. Tuttle and C. W. Tuttle, *AJS* (Series 2), **26** (1858 Nov.), p. 434; H. L. d'Arrest, *AN*, **49** (1858 Nov. 11), p. 221; A. Reslhuber, *AN*, **49** (1858 Dec. 10), pp. 263–6; J. Ferguson and J. F. Encke, *AJ*, **5** (1858 Dec. 20), pp. 180, 182; J. Ferguson, *AN*, **49** (1859 Jan. 7), p. 365; K. Hornstein and E. Weiss, *AN*, **50** (1859 Feb. 8), p. 57; E. Luther, *AN*, **50** (1859 Feb. 15), p. 73; M. Hoek, *AN*, **50** (1859 Apr. 5), p. 167; E. Weiss, *AN*, **50** (1859 Apr. 30), pp. 231–6; J. Challis, *AN*, **50** (1859 May 5), pp. 241–6; J. Challis, *AJ*, **6** (1859 May 23), pp. 25–6; W. Förster and K. C. Bruhns, *AN*, **51** (1859 Aug. 6), pp. 65–70; N. R. Pogson, *MNRAS*, **20** (1859 Nov.), pp. 5–6; G. F. W. Rümker, *AN*, **64** (1865 Feb. 19), p. 37; V1964, p. 58.

2P/Encke

1858 VIII

Recovered: 1858 August 8.02 ($\Delta = 1.46$ AU, $r = 1.44$ AU, Elong. $= 68°$)

Last seen: 1858 October 8.18 ($\Delta = 1.05$ AU, $r = 0.44$ AU, Elong. $= 25°$)

Closest to the Earth: 1858 September 20 (0.9037 AU)

Calculated path: PER (Rec), AUR (Aug. 11), GEM (Sep. 1), LYN (Sep. 10), GEM (Sep. 11), CNC (Sep. 12), LEO (Sep. 20), VIR (Oct. 8)

Using the observations of this comet obtained by T. Maclear in 1855, and the perturbations by Jupiter computed by K. R. Powalky for the period 1855–8, J. F. Encke predicted this comet would next arrive at perihelion on 1858 October 18.86. Powalky subsequently computed an ephemeris. Encke continued to refer to this comet as "Pons' comet." W. J. Förster (Berlin, Germany) recovered this comet on 1858 August 8.02, at a position of $\alpha = 4^h\ 12.7^m$, $\delta = +31°\ 25'$. Encke had surmised that this would be the evening "we could entertain any hope" of finding the comet for the first time. Förster found it as soon as his eyes had become adapted to the dark. He described it as excessively faint and 1' across. His colleague, K. C. Bruhns, said the comet was similar in brightness to a star of magnitude 12–13. Förster confirmed the recovery on August 10.05.

The comet was steadily approaching both the sun and Earth when recovered. R. Hodgson (Eversley, England) saw it in the morning sky on August 14 and said, "It had a filmy aspect, and appeared larger, perhaps, than *Jupiter*, but not 8' or 10', as it is said to be. It was, however, too low to see it fairly." A. Reslhuber (Kremsmünster, Austria) said the comet was exceptionally faint on the 17th. The comet attained its greatest northern declination of +35° on August 31.

Although numerous positions were measured during September, physical descriptions do not adequately indicate how the comet evolved. E. Luther (Königsberg, now Kaliningrad, Russia) said the comet appeared as "a faint nebulosity, without definite boundaries" for the entire period of September 8–30. Bruhns observed with a cometseeker on the 10th and noted the comet looked like an 8th-magnitude star. He added that the coma was 2' across. That same evening, Reslhuber described the comet as a bright, round nebulosity, about 2' across, with a central condensation, but no tail. G. F. W. Rümker (Hamburg, Germany) said the comet was 3–4' across on the 13th, with a noticeable central condensation. On the 17th, Rümker noted the nucleus was not at the center of the coma. Förster said the coma was 1.2' across on September 23, while Reslhuber noted it was not an easy object to see in moonlight. In a letter written by Encke to G. B. Airy, the British Astronomer Royal that was dated 1858 September 11, Encke said, "it is now so bright that it cannot escape notice even with inferior instruments. The form of the comet offers no determinate point of condensation."

The comet was still heading toward perihelion as October began and was steadily dropping into twilight. Bruhns saw it with the naked eye on the 2nd and said it looked like a 6th-magnitude star. He added that the coma was 0.5' across, with a nucleus and traces of a tail. The comet was last detected on October 8.18, when Bruhns gave the position as $\alpha = 11^h\,31.4^m$, $\delta = +8°\,03'$.

During September 1858 Encke again discussed his "resisting medium" hypothesis with respect to this comet. From an analysis of the observations obtained during the comet's returns of 1819–48, he showed how there continued to be a steady decrease in the comet's period with each return that could not be accounted for by considering the gravitational pulls of the planets.

Minor refinements of this comet's orbit using positions from multiple apparitions, as well as planetary perturbations, have been published by numerous astronomers. Of most significance were the studies of Encke (1859), F. E. von Asten (1877), and B. G. Marsden and Z. Sekanina (1974). These have typically revealed a perihelion date of October 18.87 and a period of 3.30 years. Marsden and Sekanina determined the nongravitational terms as $A_1 = -0.78$ and $A_2 = -0.03170$. The orbit of Marsden and Sekanina is given below.

T	ω	Ω (2000.0)	i	q	e
1858 Oct. 18.8658 (TT)	183.4475	336.4761	13.0894	0.340769	0.846410

ABSOLUTE MAGNITUDE: $H_{10} = 8.7$ (V1964)

FULL MOON: Jul. 26, Aug. 24, Sep. 23, Oct. 22

SOURCES: J. F. Encke, *MNRAS*, **18** (1858 Jul.), pp. 306–11, 324–5; R. Hodgson, *MNRAS*, **18** (1858 Jul.), p. 311; J. F. Encke and W. J. Förster, *AN*, **49** (1858 Aug. 14), pp. 45–8; J. F. Encke, T. Maclear, and K. R. Powalky, *AJ*, **5** (1858 Sep. 21), pp. 154–5; J. Ferguson, *AJ*, **5** (1858 Oct. 16), pp. 166–7; A. Reslhuber, *AN*, **49** (1858 Dec. 10), p. 263; J. Ferguson, *AN*, **49** (1859 Jan. 7), p. 365; K. Hornstein, *AN*, **50** (1859 Feb. 8), p. 57; E. Luther, *AN*, **50** (1859 Feb. 15), pp. 73–6; M. Hoek, *AN*, **50** (1859 Apr. 5), p. 169; J. Challis, *AN*, **50** (1859 May 5), p. 243; J. F. Encke, W. J. Förster, and K. C. Bruhns, *AN*, **51** (1859 Aug. 12), pp. 81–90; G. F. W. Rümker, *AN*, **64** (1865 Feb. 19), p. 39; F. E. von Asten, *BASP*, **22** (1877), p. 561; V1964, p. 58; B. G. Marsden and Z. Sekanina, *AJ*, **79** (1974 Mar.), pp. 415–16.

C/1859 G1
(Tempel)

1859

Discovered: 1859 April 2.81 ($\Delta = 0.87$ AU, $r = 1.45$ AU, Elong. $= 102°$)

Last seen: 1859 July 1.30 ($\Delta = 1.25$ AU, $r = 0.96$ AU, Elong. $= 49°$)

Closest to the Earth: 1859 April 24 (0.7134 AU)

Calculated path: UMi (Disc), DRA (Apr. 6), CAM (Apr. 9), DRA (Apr. 10), CAM (Apr. 14), UMa (Apr. 16), CAM (Apr. 18), LYN (Apr. 24), AUR (Apr. 29), TAU (May 12), ORI (May 21), TAU (May 28), PER (Jun. 13)

E. W. L. Tempel (Venice, Italy) discovered this comet on 1859 April 2.81 at a position of $\alpha = 14^h 30^m$, $\delta = +71°$. Moonlight briefly interfered and then several European astronomers began their observations. C. A. F. Peters (Altona, Germany) said the coma was 4′ across, with a weak central condensation on April 18. He added that it was a "difficult object in moonlight." A. Reslhuber (Kremsmünster, Austria) saw the comet on the 20th and described it as "a faint nebulosity without nucleus and tail, about 2′ to 3′ diameter."

News of the discovery did not reach the USA for several weeks. Subsequently, independent discoveries were made around the time the comet was passing closest to Earth by J. C. Watson (Ann Arbor, Michigan, USA) on April 24.11, H. P. Tuttle (Harvard College Observatory, Cambridge, Massachusetts, USA) on April 28.10, and J. Ferguson (Washington, DC, USA) on April 29.1. Watson said the comet was "quite bright, and seemed to be considerably elongated in a direction opposite from the sun." Tuttle described the comet as bright, with a tail 10′ long in a hazy sky, and moving southward at a rate of 2° daily.

The comet continued its trek toward perihelion during late April and into May. Reslhuber noted on April 27 that the comet was a little brighter than it had been on the 20th. Also on the 27th, Rümker said the comet was quite bright and exhibited a noticeable condensation. Reslhuber said the comet exhibited a short, faint tail on the 29th. Rümker noted a distinct nucleus and a slight tail on the 30th. Tuttle described the comet as bright on May 3 and noted a faint tail about 20′ long. On the same evening, C. Fearnley (Christiania, now Oslo, Norway) said the comet was 5′ across, with a very distinct nucleus, and a tail 8′ long. G. F. W. Rümker (Hamburg, Germany) said the comet was quite bright with a tail on May 5. The comet's final pre-perihelion observation came on May 17.8, when Reslhuber noted the comet was visible in twilight, but no comparison star could be detected in its vicinity.

The comet passed slightly less than 11° from the sun on May 30. The only observations obtained after perihelion were by G. P. Bond (Harvard Observatory, Cambridge, Massachusetts, USA) on June 27, 28, and July 1. He said the comet was seen with much difficulty because it was then very faint and diffuse. When the comet was last seen on July 1.30, Bond measured the position as $\alpha = 3^h\ 13.7^m$, $\delta = +53°\ 51′$.

The first parabolic orbit was calculated by V. Trettenero, using positions from April 7, 10, and 14. The resulting perihelion date was 1859 May 30.40. Later orbits by C. F. Pape, T. H. Safford, A. Hall, M. Loewy, Trettenero, A. J. G. F. von Auwers, S. Stampfer, and S. Hertzsprung (1860) determined perihelion dates of May 29.73.

A definitive orbit was calculated by O. Klein (1911). He determined five Normal positions spanning the period of April 16 to May 14 and calculated a parabolic orbit with a perihelion date of May 29.73. He pointed out that the residuals of Bond's June and July positions were high enough to indicate they were not reliable. He then took the same Normal positions and calculated a hyperbolic orbit with a perihelion date of May 29.73 and an eccentricity of 1.0000291. Because of the relatively short arc the hyperbolic orbit is based on, Klein's parabolic orbit is given below.

T	ω	Ω (2000.0)	i	q	e
1859 May 29.7261 (UT)	282.0049	359.3146	95.4894	0.201032	1.0

ABSOLUTE MAGNITUDE: $H_{10} = 7.0$ (V1964)

FULL MOON: Mar. 18, Apr. 17, May 16, Jun. 15, Jul. 15

SOURCES: C. F. Pape, *MNRAS*, **19** (1859 Apr.), pp. 227–8; E. W. L. Tempel, C. A. F. Peters, and C. F. Pape, *AN*, **50** (1859 Apr. 22), pp. 221–4; M. Loewy and V. Trettenero, *AN*, **50** (1859 Apr. 30), p. 239; J. C. Watson, H. P. Tuttle, T. H. Safford, and A. Hall, *AJ*, **6** (1859 May 10), pp. 23–4; A. J. G. F. von Auwers, *AN*, **50** (1859 May 17), p. 287; J. Ferguson, *AJ*, **6** (1859 May 23), p. 32; J. C. Watson and S. Stampfer, *AN*, **50** (1859 May 23), pp. 295–8; G. P. Bond, *MNRAS*, **19** (1859 Jun.), pp. 296–7; H. P. Tuttle, J. C. Watson, J. Ferguson, and T. H. Safford, *AN*, **50** (1859 Jun. 4), pp. 329–32; E. W. L. Tempel, *AJS* (Series 2), **28** (1859 Jul.), p. 153; A. Reslhuber, *AN*, **51** (1859 Jul. 22), p. 25; J. Ferguson, *AJ*, **6** (1859 Sep. 15), p. 49; G. P. Bond, *AN*, **51** (1859 Oct. 15), p. 275; C. Fearnley, *AN*, **52** (1860 Mar. 21), pp. 281–4; S. Hertzsprung, *AN*, **53** (1860 May 26), pp. 149–52; G. F. W. Rümker, *AN*, **64** (1865 Feb. 19), p. 39; O. Klein, *AN*, **188** (1911 Jul. 10), p. 385; V1964, p. 58.

C/1860 D1 (Liais)   *Discovered:* 1860 February 27.01 ($\Delta = 0.77$ AU, $r = 1.21$ AU, Elong. $= 85°$)
Last seen: 1860 March 14.00 ($\Delta = 1.09$ AU, $r = 1.27$ AU, Elong. $= 74°$)

1860 I   *Closest to the Earth:* 1860 February 2 (0.4423 AU)
Calculated path: DOR (Disc), RET–DOR (Mar. 3), RET (Mar. 7), DOR (Mar. 9)

Emmanuel Liais (Olinda, Brazil) discovered this faint double comet near μ Doradus on February 27.01. He was using a 7.6-cm refractor and measured the position as $\alpha = 5^h\ 04.8^m$, $\delta = -61°\ 58'$. He had established that it was moving toward the northwest by February 27.02. Liais said, "The larger portion preceded the smaller and was elongated along the radius vector of the sun. It is true also that the side towards the sun was the most brilliant, and that towards its extremity appeared a small luminous point comparable to a star of 9th magnitude. The intensity of the comet is very weak and the observations difficult when light illuminated the field." Liais said the main comet was about 1′ wide and 3–3.5′ long, while the smaller comet was nearly circular and about 29″ across.

Liais remained the only observer of this comet. He next observed it on March 1 and 4, and noted the small comet was hardly visible on the last date. His attempts to locate the comets in moonlight on the 6th were unsuccessful. Liais found the small comet much more brilliant on the 10th than on the 4th and described it as circular and 34–42″ in diameter. He noted that the leading part of the nebulosity was the brighter area. The larger nebulosity had lengthened since the 4th and had a width 2–3 times greater than the small nebulosity. Although the stellar point had vanished, a very notable condensation was still present. Liais saw the comet for a short time on March 11, but was unable to measure a position before clouds moved in. He did note that the larger comet had changed little in appearance, except the condensation had decreased. Meanwhile, the smaller comet was still round, but had faded from the previous night. Liais again saw the comet on the 12th. He said the larger comet was similar in form, but with larger dimensions and a weaker condensation. The smaller comet was difficult to distinguish and only occasionally seen.

The larger comet was last detected on March 14.00, when Liais determined its position as $\alpha = 4^h\ 02.6^m$, $\delta = -54°\ 41'$. He said it was very faint, about 2′ 45″ across, and exhibited only slight ellipticity. Liais considered the diameter estimate as "dubious" because the edges of the coma were not well defined. Only a weak condensation was present. He was unable to find any trace of the smaller comet. Liais attempted to see the comet in the days that followed, but a number of cloudy and rainy days passed before the skies cleared enough for another search. Liais could then no longer see the comet.

On the final date, Liais made another type of observation of this comet. He used a prism to produce a second image of the comet and then attached a diaphragm to the opening of the telescope to decrease the amount of light coming in. Liais then rotated the prism to check for polarization. The result was that the prism image did change in brightness from the regular image, which Liais concluded indicated "a notable polarization...."

Liais' earliest announcement was a letter written to the *Astronomische Nachrichten* on March 8 that was published in the April 14 issue. C. F. Pape used positions spanning the period of February 27 to March 4 and calculated a parabolic orbit with a perihelion date of 1860 February 17.23. Liais sent a report containing observations up to March 14 that was published in the June 11 issue of the *Comptes Rendus*. He included an orbit that he had determined from the entire observation arc that indicated a perihelion date of February 17.17.

C. F. Pechüle (1868) used 17 positions obtained between February 27 and March 14 to generate three Normal positions, and computed a parabolic orbit with a perihelion date of February 17.12. For the second comet, he used three positions obtained between February 28 and March 10, and computed a parabolic orbit with a perihelion date of February 17.17. No planetary perturbations were applied. These orbits are given below, with the first orbit representing comet A, and the second orbit comet B.

T	ω	Ω (2000.0)	i	q	e
1860 Feb. 17.1249 (UT)	209.7599	326.0179	79.6819	1.198875	1.0
1860 Feb. 17.1697 (UT)	209.6880	326.0119	79.6193	1.198174	1.0

ABSOLUTE MAGNITUDE: $H_{10} = 7$ (V1964)

FULL MOON: Feb. 7, Mar. 7, Apr. 5

SOURCES: E. Liais and C. F. Pape, *AN*, **52** (1860 Apr. 14), pp. 379–82; E. Liais, *CR*, **50** (1860 Apr. 16), p. 763; E. Liais and C. F. Pape, *AJ*, **6** (1860 May 31), pp. 107–8; E. Liais, *CR*, **50** (1860 Jun. 11), pp. 1089–93; E. Liais, *MNRAS*, **20** (1860 Jul.), p. 336; E. Liais, *CR*, **51** (1860 Jul. 9), pp. 65–7; C. F. Pechüle, *AN*, **72** (1868 Oct. 23), pp. 235–8; V1964, p. 58.

C/1860 H1 *Discovered:* 1860 April 17.92 ($\Delta = 2.08$ AU, $r = 1.45$ AU, Elong. $= 40°$)
(Rümker) *Last seen:* 1860 June 12.15 ($\Delta = 2.58$ AU, $r = 1.93$ AU, Elong. $= 40°$)
Closest to the Earth: 1859 October 29 (1.9414 AU), 1860 March 13 (2.0087 AU)
1860 II *Calculated path:* PER (Disc), CAM (Apr. 27), AUR-CAM (May 12), LYN (May 23)

G. F. W. Rümker (Hamburg, Germany) discovered this comet on 1860 April 17.92 at a position of $\alpha = 2^h\ 46.3^m$, $\delta = +48°\ 29'$. He confirmed the find on April 18.88, and added that the comet was very faint. The comet was at its most southerly declination of $+48°$ on the 17th. It was already past its closest distances from the sun and Earth.

Rümker found the comet at the limit of visibility because of clouds on April 20 and 21, while he found it easier to see on the 22nd. He said moonlight greatly weakened the light of the comet on the 25th, and this might explain why C. A. F. Peters (Altona, Germany) noted the comet was too faint to be seen after the 25th. C. N. A. Krueger (Bonn, Germany) saw the comet with a heliometer on the 26th and described it as very faint and 1.3′ across. A. J. G. F. von Auwers (Königsberg, now Kaliningrad, Russia) said the comet was seen with great difficulty on the 28th because of moonlight. Auwers said the comet was hardly visible during the period of April 23 to May 20 because of haze near the horizon. He nevertheless noted it was always round, about 2′ across, with a strong condensation.

A. Reslhuber (Kremsmünster, Austria) saw the comet in bright moonlight on May 6, with the 7-foot focal length refractor. He described it as extraordinarily faint, without a nucleus. R. Luther (Bilk, Germany) said the comet was difficult to detect on the 7th because of its faintness. Reslhuber said the comet grew fainter as it neared a 7th-magnitude star on the 9th, while it appeared "very faint" when situated between two 10th-magnitude stars on the 10th. On May 17, Reslhuber said the comet had noticeably diminished in brightness compared with previous nights. On the 18th, Reslhuber said the comet's light was "very suppressed" by a nearby star of magnitude 9 or 10. The comet attained its most northerly declination of +56° on May 19. A. Sonntag (Dudley Observatory, New York, USA) said the comet was "very faint" on the 21st. Reslhuber said it required a great effort to see the comet in the 7-foot focal length refractor on the 23rd. Sonntag simply described it as "very faint" on the 23rd and 24th, while he could not find the comet on the 25th "on account of its low altitude and the increased moonlight." Reslhuber was only able to occasionally glimpse the comet on the 26th.

As June began, the comet's faintness and low altitude were making it a challenge to follow. Reslhuber said the comet could not be seen on the 4th and 5th because of its advancement into twilight. The comet was last observed on June 12.15, when G. P. Bond (Cambridge, Massachusetts, USA) determined the position as $\alpha = 7^h\ 47.0^m$, $\delta = +53°\ 13′$.

The first parabolic orbit was calculated by Rümker. Using positions spanning the period of April 17–20, he determined the perihelion date as 1860 March 1.65. Later orbits by Krueger, B. Tiele, G. V. Schiaparelli, T. H. Safford, Jr, A. Murmann, H. Romberg, and H. Seeling, eventually established the perihelion date as March 6.1. A definitive orbit was calculated by J. A. H. Gyldén (1863) using 55 positions spanning the whole period of visibility. The resulting perihelion date was March 6.07. No planetary perturbations were applied. This orbit is given below.

T	ω	Ω (2000.0)	i	q	e
1860 Mar. 6.0655 (UT)	41.2172	10.8260	48.2359	1.306664	1.0

ABSOLUTE MAGNITUDE: $H_{10} = 5.2$ (V1964)

FULL MOON: Apr. 5, May 5, Jun. 3

SOURCES: O. W. Struve and G. V. Schiaparelli, *BASP*, **2** (1860), pp. 255–7; G. F. W. Rümker and C. A. F. Peters, *AN*, **53** (1860 May 1), pp. 91–4; A. J. G. F. von Auwers,

AN, **53** (1860 May 18), p. 139; A. Murmann, *AN*, **53** (1860 May 26), p. 153; G. F. W. Rümker, C. N. A. Krueger, and B. Tiele, *AJ*, **6** (1860 May 31), p. 108; H. Romberg, *AN*, **53** (1860 Jun. 4), p. 175; T. H. Safford, Jr, and A. Sonntag, *Astronomical Notices*, No. 19 (1860 Jun. 9), pp. 149–50; R. Luther, *AN*, **53** (1860 Jun. 18), p. 255; A. Reslhuber, *AN*, **53** (1860 Jul. 13), pp. 313–16; A. J. G. F. von Auwers, *AN*, **53** (1860 Aug. 17), pp. 353–6; H. Seeling, *MNRAS*, **20** (1860 Supplement), pp. 357–8; G. P. Bond, *AN*, **55** (1861 May 12), pp. 199–204; J. A. H. Gyldén, *BASP*, **6** (1863), pp. 363–74; G. F. W. Rümker, *AN*, **64** (1865 Feb. 19), p. 39; V1964, p. 58.

C/1860 M1
(Great Comet)

1860 III

Discovered: 1860 June 18.9 ($\Delta = 1.03$ AU, $r = 0.30$ AU, Elong. $= 17°$)

Last seen: 1860 October 18.81 ($\Delta = 2.95$ AU, $r = 2.48$ AU, Elong. $= 53°$)

Closest to the Earth: 1860 July 11 (0.4570 AU)

Calculated path: AUR (Disc), LYN (Jun. 27), CNC (Jul. 3), LEO (Jul. 6), SEX (Jul. 12), LEO (Jul. 13), CRT (Jul. 16), CRV (Jul. 20), HYA (Jul. 25), CEN (Jul. 28), LUP (Aug. 21), NOR (Sep. 20)

This comet suddenly appeared in the evening sky as a bright naked-eye object and was subsequently discovered by many people around the world. The editor of the *Astronomical Journal* estimated that 50–100 people independently discovered this comet in the USA alone.

The earliest observation appears to come from Italy, at least according to a vague reference by A. Secchi (Rome, Italy) in the 1860 October 22 issue of the *Astronomische Nachrichten*. He said the comet was seen on June 18, but gave no other details. The comet would not have been above the horizon for long on that evening, so a more precise date would likely be June 18.9. Secchi also said that Camorri (Senigallia, Italy) saw the comet the next evening.

Some of the more significant independent discoveries were made by Caswell from the deck of the steamship *Arabia* on June 20, Gronemann (The Netherlands) on June 22, H. P. Tuttle (Harvard College Observatory, Cambridge, Massachusetts, USA) on June 22.07, and F. A. T. Winnecke (Bonn, Germany) on June 23.94. Tuttle measured the first position of $\alpha = 6^h\ 29.2^m$, $\delta = +41°\ 57'$ on June 22.11. He said the comet was visible to the naked eye, with a tail 6° or 7° long and 30′ wide at a point 5° from the nucleus. His colleague, G. P. Bond, said the nucleus was as bright as a star of magnitude 4. Winnecke said he noticed "an unfamiliar 4th-magnitude star low in the northwest." A telescope revealed its true nature. The center of the tail extended toward PA 21.07°, while the width was determined as 4.5′ at a distance of 5′ from the nucleus, and 7′ at a distance of 10′. In addition, the left-hand side of the tail was very bright, while the right-hand side was very faint. The nucleus was considerably lengthened in a direction perpendicular to the tail axis.

Because of the numerous independent discoveries word spread quickly and observations generally began around June 22. On that date, J. M. Stothard (Dublin, Ireland) remarked, "it was much brighter and more readily visible to the naked eye than Donati's Comet during the first few nights of its appearance." Dezautière (Nièvre, France) said the nucleus was perfectly visible as a star of magnitude 5. It also exhibited a tail that extended 2° or 3° toward the celestial north pole. Stothard saw the comet again on the 23rd and noted a distinct tail in his 7.6-cm refractor.

June 24 was an exceptional day for observations, with the comet then attaining its most northerly declination of +42°. J. F. J. Schmidt (Athens, Greece) distinctly saw the comet in the northwestern sky in twilight with the naked eye and said the comet's head was reddish and similar in brightness to a 3rd-magnitude star. The straight and thin tail was 7–10° long in twilight, but nearly 20° long in dark skies. He noted that after the nucleus had set below the horizon, the tail remained visible for nearly an hour. Winnecke said the comet could be seen in twilight soon after sunet. About an hour after sunset, the comet was examined with a telescope, and Winnecke said the nucleus was very eccentrically situated in the coma, with very little nebulosity being seen in the direction toward the sun. The nucleus was about 7" across and yellowish. The tail contained a dark, parabolically-shaped area in the center, which became visible while the sky was still bright. After the sky had darkened further, the length of the tail was estimated as 15°. At 9° from the head the tail width was estimated as 1°. Colonel De Rottenburg (Gibraltar) said, "It was very easily seen by the naked eye, the nucleus was a little less brilliant than Castor. Its altitude above the western horizon about equal to that of Venus at the same time." He turned his 11.4-cm refractor to the comet and noted, "It has a bifid tail, very like that of the year 1846. . . . I used powers of 26, 50, and 100 on the comet; the nucleus has a very sensible disk. It bore powers up to 100 very well; one portion of the tail is much longer than the other, the south preceding being the longer. . . ." De Rottenburg added, "With 100 power the nucleus was situated within the nebulosity, and the nebulosity was more arched and prominent on the south preceding part. It was first seen by a gentleman here on Saturday evening 23d June." Edwin Clarke (Observatory House, Forest Hill, England) saw the comet in the evening with the naked eye and then worked as quickly as possible to mount his 10.2-cm refractor outdoors for a better look. Unfortunately, he had "a brilliant view . . . for a minute or two only, when clouds again covered it. . . . " He added, "the nucleus was very sharply defined, having a large distinct disk, the bruch was vertical nearly below the head. It was but slightly divergent and of great length. It was a very striking object to the naked eye, and of great brilliancy."

The comet continued to be well observed during the remainder of June. Schmidt distinctly saw the comet with a small telescope in twilight 43 minutes after sunset on the 25th, and observed it with the naked eye 7 minutes later and said it was similar in brightness to a 3rd-magnitude star. He said the "very compact nucleus" was bright, sharp, and stood out from a tail consisting of two distinct arms. The tail was 15° long, and appeared brighter on the left-hand side (as seen through a telescope). That same evening, T. Spratt

(HMS *Medina*, near Retuno, Crete) said the comet remained visible to the naked eye for the 20 minutes between the end of twilight and the rising of the moon. He also noted a tail 15° long and said it was slightly curved. W. J. Förster (Berlin, Germany) estimated the brightness of the nucleus as magnitude 3 on the 26th and said two tails extended from the coma, separated by a dark region. He gave the tail length as 15°, while H. Dembowski (Milan, Italy) said it was closer to 20° and was shaped like a feather. Winnecke also saw the comet on the 26th, while visiting Brussels, Belgium, and commented that it was quite remarkable in twilight. Schmidt said the comet was seen with the naked eye in twilight about 39 minutes after sunset on the 27th. He estimated the total magnitude as 1–2. Bond and Dembowski gave descriptions for that same evening. Bond said, "the nucleus was surrounded on its sunward side by a faint dark halo, 1" or 2" broad, distant 18" at its apex. Within it, in the south preceding quadrant of the envelope, was an irregular clouded mass, while the nucleus itself presented the appearance of a streak of very bright nebulosity, curiously elongated in the direction of right ascension." He added that the nucleus was "spindle-shaped, the longer axis 10" or 15"; the shorter 3"." Dembowski observed the nucleus with a very small refractor and noted the nucleus "had the strange appearance of a parabolic crescent." Schmidt saw the comet with the naked eye about 51 minutes after sunset on the 28th. He estimated the magnitude as 2–3, while Secchi said the nucleus was about magnitude 2. Schmidt gave the tail length as 15°, while Secchi said it extended toward PA 43°. Schmidt estimated the total magnitude as 3 on the 29th and 4 on the 30th, while Dembowski estimated the nucleus as magnitude 3 on the 30th. Schmidt estimated the tail length as 6° on the 29th and 11° on the 30th, while Secchi said it extended toward PA 22°.

The comet was still heading toward a close approach to Earth as July began. Schmidt estimated the magnitude as 4–5 in moonlight on the 1st, while Winnecke saw it with the naked eye while visiting Cranford, England. Winnecke added that the 33-cm telescope revealed a distinct central condensation, but no tail. Schmidt estimated the magnitude, in moonlit skies, as 4 on the 2nd and 5 on the 3rd. Interestingly, with less moonlight, Schmidt estimated the magnitude as 3–4 on the 5th, while E. Liais (Rio de Janeiro, Brazil) gave it as 2. The tail length was given as 9° by Schmidt and 15° by Liais. W. Scott (Sydney, New South Wales, Australia) said "a large comet" was reported from different parts of Australia soon after sunset on the 6th. That same evening, Schmidt saw the comet with the naked eye about 35 minutes after sunset. He gave the total magnitude as 2–3, the nuclear magnitude as 4.5, and the tail length as 8–9°. R. L. J. Ellery (Williamstown, Victoria, Australia) described the comet as very bright, with a very nebulous head on the 7th. He noted a tail over 4° long. Schmidt estimated the magnitude as 3–4 on the 8th, while the tail length was given as 6° by Schmidt and about 5° by Liais.

The comet was closest to Earth on July 11 and was well observed for a couple of days before and after. Many observations were made on the

9th. Schmidt estimated the magnitude as 3–4 and said the tail was 6° long. Scott described the comet as brilliant and said the tail extended upwards at an angle of 60° to the horizon. W. Mann (Royal Observatory, Cape of Good Hope, South Africa) observed the comet with the naked eye and said the tail was 4° long. Using the refractor of 8.5-foot focal length he said, "The comet appears as a diffused mass of light enveloping a nucleus of 3" or 4" in diameter." Schmidt estimated the magnitude as 3 on the 11th and 3–4 on the 12th, while he gave the tail length as 1° on the 11th and 0.5° on the 12th. Schmidt added that the coma was 9' across on the 11th. Secchi saw the comet at a very low altitude on the 12th and said it was easily visible even though no stars were seen nearby.

During the remainder of July the comet became a more difficult object for Northern Hemisphere observers as it neared and then dropped below the southern horizon. Schmidt's last physical description came on the 13th, when he said it was still easily visible to the naked eye, but was nearly tail-less. Scott said it was still a naked-eye object on the 18th. Mann observed the comet with an 8.5-foot focal length refractor on the 23rd. He estimated the coma diameter as 3' and said the nuclear condensation was about magnitude 7.5. Scott measured the comet's position on several occasions between July 12 and 18, inclusive, and made some general comments about its appearance. He said, "Throughout the whole series of observations I have been unable to distinguish any marked peculiarity in its appearance. There was a decided brilliant nucleus, but not sufficiently well defined to form a good subject for observation, around which the light appeared diffused very uniformly. In consequence of the want of definition of the nucleus I consider all the transits to be liable to errors of at least one second. . . . "

As the comet headed away from both the sun and Earth, the number of observations steadily declined. Mann estimated the coma diameter as 3.5' on August 6 and said the tail was no longer visible. Mann said the nuclear condensation was about magnitude 8 on the 15th and he gave its diameter as about 5". Mann added that the nucleus was "situated nearer to the *preceding* portion of the nebulous image of the comet." Scott described the comet as faint and diffuse on August 17 and 18, while Mann found it exceedingly faint on the 28th and 31st. Mann described the comet as very faint on September 22 and October 5. He gave the comet's diameter as about 30" on the last date. Mann said moonlight made it very difficult to see the comet on October 17.

The comet was last detected on October 18.81, when Mann used a refractor of 8.5-foot focal length to determine the position as $\alpha = 16^h\ 15.9^m$, $\delta = -51°\ 21'$. Mann said moonlight was present and the comet was detected with difficulty.

The first parabolic orbits were independently calculated by H. P. Tuttle and T. H. Safford, using positions spanning the period of June 21–25. The perihelion date was determined as 1860 June 16.27 by Tuttle and June 16.18 by Safford. Additional orbits were calculated by A. J. G. F. von Auwers,

C. W. Tuttle, K. R. Powalky, M. Loewy, G. M. Searle, Liais, and C. G. Moesta (1861) which established the perihelion date as June 16.6. Although Liais gave an elliptical orbit in August, with a period of 1089 years, Auwers (1868) calculated a definitive orbit using positions obtained between 1860 June 22 and October 18, and concluded that a parabolic orbit fitted the observations best. He gave a perihelion date of June 16.56. This orbit is given below. Calculations reveal the comet had passed less than 4° from the sun on June 4.

T	ω	Ω (2000.0)	i	q	e
1860 Jun. 16.5610 (UT)	76.8841	86.6271	79.3233	0.292885	1.0

ABSOLUTE MAGNITUDE: $H_{10} = 5.8$ (V1964)

FULL MOON: Jun. 3, Jul. 3, Aug. 1, Aug. 31, Sep. 30, Oct. 29

SOURCES: E. Brunner and Dezautière, *CR*, **50** (1860 Jun. 25), pp. 1200–1; A. C. Petersen and F. A. T. Winnecke, *AN*, **53** (1860 Jun. 29), p. 287; J. M. Stothard, *MNRAS*, **20** (1860 Jul.), p. 344; G. De Rottenburg, *MNRAS*, **20** (1860 Jul.), pp. 344–5; E. Clarke, *MNRAS*, **20** (1860 Jul.), p. 345; M. Hoek, Gronemann, G. B. Donati, W. J. Förster, and A. J. G. F. von Auwers, *AN*, **53** (1860 Jul. 6), p. 299; J. F. J. Schmidt, K. Hornstein, M. Loewy, and K. R. Powalky, *AN*, **53** (1860 Jul. 13), pp. 315–20; J. Ferguson, C. W. Tuttle, and G. M. Searle, *AJ*, **6** (1860 Jul. 17), pp. 127–8; H. P. Tuttle, T. H. Safford, V. Trettenero, and J. F. J. Schmidt, *AN*, **53** (1860 Jul. 28), pp. 325, 331; H. Dembowski, *AN*, **53** (1860 Aug. 8), p. 343; J. Ferguson, *AJ*, **6** (1860 Aug. 15), p. 143; J. F. J. Schmidt, *AN*, **53** (1860 Aug. 17), pp. 359–68; E. Liais, *CR*, **51** (1860 Aug. 20), pp. 301–2; H. P. Tuttle, H. Seeling, and E. Liais, *AN*, **54** (1860 Sep. 5), pp. 5–10; E. Liais, *AN*, **54** (1860 Sep. 20), p. 45; E. Liais, *CR*, **51** (1860 Sep. 24), pp. 503–4; E. Liais, *AN*, **54** (1860 Oct. 10), p. 91; A. Secchi and Camorri, *AN*, **54** (1860 Oct. 22), pp. 113–16; G. P. Bond, W. Scott, R. L. J. Ellery, and T. Spratt, *MNRAS*, **20** (1860 Supplement), pp. 358–63; E. Liais, *AJ*, **6** (1860 Nov. 13), p. 164; W. Scott, *MNRAS*, **21** (1860 Dec.), pp. 46–8; F. A. T. Winnecke, *BASP*, **3** (1861), pp. 110–12; C. G. Moesta, *MNRAS*, **21** (1861 Apr.), pp. 186–9; T. Maclear and W. Mann, *MRAS*, **31** (1861–62), pp. 29–35; C. W. Moesta, *MNRAS*, **26** (1865 Dec.), pp. 59–60; A. Fischer, *AN*, **67** (1866 Aug. 17), pp. 273–300; A. Fischer, *AN*, **68** (1867 Jan. 16), p. 239; A. J. G. F. von Auwers, *VJS*, **3** (1868), pp. 117–26; A. J. G. F. von Auwers, *MNRAS*, **28** (1868 Feb.), pp. 99–100; V1964, p. 58.

C/1860 U1 *Discovered:* 1860 October 24.17 ($\Delta = 0.45$ AU, $r = 0.93$ AU, Elong. = 69°)

(Tempel) *Last seen:* 1860 October 26.14 ($\Delta = 0.46$ AU, $r = 0.96$ AU, Elong. = 72°)

Closest to the Earth: 1860 October 8 (0.3907 AU)

1860 IV *Calculated path:* LEO (Disc), LMi (Oct. 24)

This comet was barely observed as it moved across the sky. E. W. L. Tempel (Marseille, France) discovered this small comet on 1860 October 24.17 at a position of $\alpha = 10^h 04^m 38^s$, $\delta = +28°\ 27'$. He confirmed his find on October 25.11. The comet was last detected on October 26.14, when A. J. F. Yvon-Villarceau (Paris, France) estimated its position as $\alpha = 10^h 05^m 18^s$, $\delta = +31°\ 26'$. The editor of the *Astronomische Nachrichten* wrote in the 1860

November 1 issue, "The bright moonlight has so far made it impossible to find the comet" at Altona, Germany. Tempel looked for the comet again on the nights of November 7 and 8, but it was not found.

The first orbit was calculated by J. E. B. Valz (1861). He used the three available positions and determined the perihelion date as 1860 September 16.66. Later orbits were calculated by J. Kowalczyk (1869, 1870) and T. R. von Oppölzer (1869), with Kowalczyk giving a perihelion date of September 22.81 and Oppölzer indicating September 21.59. Kowalczyk's 1870 orbit is given below. It indicated the comet would have been at its brightest (by about 1.5 magnitudes) one month prior to discovery, but near its minimum elongation of 34° from the sun. It moved rapidly northward as it approached Earth, with the magnitude slowly fading and the elongation from the sun increasing.

Interestingly, H. P. Tuttle (Cambridge, Massachusetts, USA) announced that a comet had been seen on November 14 near Polaris. Tempel remarked that the available orbits indicated C/1860 U1 would have passed near such a position several days after the 14th. Nevertheless, Valz did not doubt this was Tempel's comet and he computed a parabolic orbit with a perihelion date of 1860 September 28.78 and a perihelion distance of 0.9537 AU. Kowalczyk (1870) said it was unlikely that Tuttle's object referred to C/1860 U1.

T	ω	Ω (2000.0)	i	q	e
1860 Sep. 22.8119 (UT)	311.9768	46.7741	32.2096	0.682653	1.0

ABSOLUTE MAGNITUDE: $H_{10} = 9.5$ (V1964)

FULL MOON: Sep. 30, Oct. 29

SOURCES: *MNRAS*, **21** (1860 Nov.), p. 39; E. W. L. Tempel and A. J. F. Yvon-Villarceau, *AN*, **54** (1860 Nov. 1), p. 143; *AJ*, **6** (1860 Nov. 13), p. 166; E. W. L. Tempel; *AJS* (Series 2), **31** (1861 Jan.), p. 136; E. W. L. Tempel and J. E. B. Valz, *AN*, **54** (1861 Jan. 10), p. 285; E. W. L. Tempel, H. P. Tuttle, and J. E. B. Valz, *AN*, **55** (1861 Mar. 30), p. 79; J. Kowalczyk, *AN*, **73** (1869 Jan. 16), p. 81; T. R. von Oppölzer, *AN*, **73** (1869 Feb. 16), p. 189; J. Kowalczyk, *AN*, **75** (1870 Jan. 12), p. 165; V1964, p. 58.

C/1861 G1 *Discovered:* 1861 April 5.17 ($\Delta = 0.83$ AU, $r = 1.37$ AU, Elong. $= 97°$)
(Thatcher) *Last seen:* 1861 September 7.09 ($\Delta = 1.58$ AU, $r = 1.81$ AU, Elong. $= 85°$)
Closest to the Earth: 1861 May 5 (0.3356 AU)

1861 I *Calculated path:* DRA (Disc), UMi (Apr. 19), DRA (Apr. 24), UMa (Apr. 28), LMi (May 5), LEO (May 7), CNC (May 8), HYA (May 16), MON (May 24), PUP (May 26), CMa (Jun. 22), PUP (Jul. 16), COL (Aug. 16), PUP (Aug. 25), PIC (Sep. 5)

A. E. Thatcher (New York, New York, USA) discovered this comet on 1861 April 5.17. He was then using an 11-cm refractor and estimated the comet's position as $\alpha = 17^h\ 33^m$, $\delta = +56°$. He communicated the news to Henry Fitz (New York, New York, USA), the maker of his telescope. Fitz searched with his 19.1-cm refractor and was able to confirm the comet on April 7.1. He described the comet as nearly circular, 2' across, and slightly condensed at the center.

The comet was heading toward both the sun and Earth when discovered, and was rather well observed during the remainder of April as it steadily brightened. J. Ferguson (Washington, DC, USA) described the centrally condensed comet as circular and about 2' across on the 11th and 12th. C. H. F. Peters (Hamilton College Observatory, Clinton, New York, USA) observed the comet on the 18th and noted, "There was a blurred nucleus or rather a condensation of light of about 20" diameter. The coma, nearly globular, extends over more than 10' in diameter; both nucleus and coma show some swelling in the direction of the sun." Thatcher observed the comet with an 11-cm refractor (30×) on the 20th and said it appeared "as bright as the cluster in the sword-handle of Perseus, or the nebula of Cancer, to the unassisted eye; and with a diameter of 3'." He added that the comet's brightness "remained nearly the same from April 9 to 20." Peters said bright moonlight did not prevent the comet's observation on the 25th. He added, "it has neither a sharply defined nor a bright nucleus, though good concentration well adapted for pointing." G. W. Hough (Dudley Observatory, New York, USA) said the comet was "very difficult to observe" in strong moonlight on the 26th. Rutherford saw the comet with the naked eye and added that a telescope revealed a round coma about 6' across. The comet was independently discovered by C. W. Baeker (Berlin, Germany) with the naked eye on the 28th. W. J. Förster (Berlin, Germany) confirmed Baeker's find on the 30th when he observed the comet with the naked eye. He estimated it as magnitude 4 or 5. That same night, Peters said the comet was visible to the naked eye as a nebulous star. With a magnification of 90× he found the tail extending 3° and appearing as a "pale narrow streak" with a width of only 5'. The coma was 14' or 15' across. He added, "Remarkable is the radiating distribution of light in the coma, especially distinct farther away from the center and on the circumference."

The comet passed closest to Earth on May 5 and several descriptions were published leading up to that date. Parkin (Regent's Park, London, England) saw the comet with the naked eye on the 1st. C. F. Pape (Altona, Germany) reported it was between magnitude 3 and 4 on the 4th, while S. Gorton and Gaunt (Clapton, England) reported it was visible to the naked eye, "appearing like an indistinct star" of magnitude 2 or 3. Gorton and Gaunt continued, "It was sufficiently bright to be well seen in the field [of an 8.6-cm refractor] under an illumination which extinguished stars apparently of the 6th magnitude. In the dark field a tail was readily to be traced, extending in the [south following] quadrant to a distance of about a degree, as far as I could judge." On that same date, E. Kayser (Danzig, now Gdansk, Poland) described the comet as round, with a thin, straight tail about 3.5° long. Pape still noted the comet was between magnitude 3 and 4 on the 5th. That same evening, Peters said the coma had a "radiating structure." Peters also said the tail structure was remarkable and seemed "as if originating from behind the head, and appears pretty bright in its first track; then it is narrowing and in the same time fading away, so as to be scarcely visible at the distance of about three quarters of a degree from the head; beyond, however, the light again increases, and the tail spreads as a narrow fan, exhibiting a bright phosphorescence at about two degrees distance. The entire length of the tail may be estimated at three degrees."

During the remainder of May the comet moved away from Earth, but was still heading toward perihelion. E. Schönfeld (Mannheim, Germany) noted a trace of tail on the 6th and 11th. Kayser said only a weak trace of the tail was present on the 7th. A. Reslhuber (Kremsmünster, Austria) wrote about his observation of the 9th, "The comet appears as a vast nebulosity about one-third degree in diameter with a faint hazy nucleus." He added that there was no tail, and the comet was visible to the naked eye. Förster also saw the comet on the 9th and said it exhibited a faint tail that appeared about 1° long in the finder. He indicated the tail extended toward PA 96°. On May 10, Reslhuber described the comet as fairly round, while C. Fearnley (Christiania, now Oslo, Norway) saw the comet with the naked eye in bright twilight. In darker skies, Fearnley noted a bright coma 15–20′ across, without a distinct tail. A. Secchi (Rome, Italy) saw the comet with the naked eye on the 10th and 12th. He also noted a rudimentary tail. J. F. J. Schmidt (Athens, Greece) saw the comet in moonlight on the 16th and noted it was at the limit of naked-eye visibility. He said his telescope's finder revealed a coma 7–8′ across, but no tail. Schönfeld saw the comet in twilight and moonlight on the 17th. He described it as 4′ across, with a good central condensation. Schmidt saw only a trace of a condensation on the 19th. Peters found the comet in strong moonlight and at a low altitude on the 23rd and described it as "faint and ill defined." Schmidt said the central condensation was brighter and sharper on the 24th and 25th than it had been on the 19th. The comet was last seen in the Northern Hemisphere on the 28th, when Schmidt saw it in bright twilight, but it was first detected in the Southern Hemisphere by C. W. Moesta (Santiago, Chile) on July 31.

T. Maclear and W. Mann (Royal Observatory, Cape of Good Hope, South Africa) first observed this comet on August 15. They used an 8.5-foot focal length refractor, with a magnification of 126×, and described the comet as "a faint, round nebulous body, about 2′ in diameter, and without any decided central condensation of light." Unfavorable weather moved in for several days, and they next searched for the comet on the morning of the 19th. They said the comet was not visible until after moonset, at which time it was very faint because of the beginning of twilight. On September 1, they said the comet was extremely faint and became invisible when the moon rose.

The comet was last detected on September 7.09, when Mann measured the position as $\alpha = 5^h\,56.7^m$, $\delta = -47°\,07'$. He said the position was "not very precise, owing to the extreme faintness of the Comet."

The first parabolic orbit was calculated by T. H. Safford, using positions spanning the period of April 11–19. He determined the perihelion date as 1861 June 4.72. Additional orbits were calculated by J. M. Gillis, Förster, F. Tietjen, A. J. G. F. von Auwers, Safford, T. R. von Oppölzer, and Pape, which ultimately revealed a perihelion date near June 3.6.

The first elliptical orbit was calculated by Pape. He used positions spanning the period of April 11 to May 18 and determined the perihelion date as June 3.69 and the period as 1849 years. Oppölzer (1862, 1864) worked toward the computation of a definitive orbit. His 1864 study began with 187 positions from the period spanning April 11 to September 7, and resulted in a perihelion date of June 3.89 and a period of 415 years. Perturbations by Jupiter and Saturn were applied and this orbit is given below.

During 1866 and 1867, several astronomers had suggested and proven the links between some comets and meteor showers. E. Weiss (1867) calculated the probable close encounters between Earth and comet orbits and found that the orbit of C/1861 G1 passed within 0.002 AU of Earth on April 20. He searched through various publications for evidence of meteor activity around this date and found several references to observed meteor showers. A month later, J. G. Galle mathematically confirmed the link between C/1861 G1 and the Lyrids and successfully traced the history of the shower back to March 16 of 687 BC.

T	ω	Ω (2000.0)	i	q	e
1861 Jun. 3.8899 (UT)	213.4496	31.8674	79.7733	0.920700	0.983465

ABSOLUTE MAGNITUDE: $H_{10} = 5.5$ (V1964)

FULL MOON: Mar. 26, Apr. 24, May, 24, Jun. 22, Jul. 21, Aug. 20, Sep. 19

SOURCES: A. E. Thatcher and J. Ferguson, *MNRAS*, **21** (1861 Apr.), p. 192; T. H. Safford, S. Gorton, and Gaunt, *MNRAS*, **21** (1861 May), p. 216; G. P. Bond, J. Ferguson, C. W. Baeker, and W. J. Förster, *AN*, **55** (1861 May 5), p. 189; C. F. Pape and A. J. G. F. von Auwers, *AN*, **55** (1861 May 12), p. 205; W. J. Förster and F. Tietjen, *AN*, **55** (1861 May 18), pp. 217–20; E. Schönfeld and C. F. Pape, *AN*, **55** (1861 Jun. 10), pp. 251–6; J. F. J. Schmidt, *AN*, **55** (1861 Jun. 18), p. 257; A. Reslhuber, *AN*, **55** (1861 Jun. 25), p. 283; A. E. Thatcher, H. Fitz, Rutherford, Parkin, and T. H. Safford, *AJS* (Series 2), **32** (1861 Jul.), p. 134; N. M. Edmondson, C. F. Pape, G. W. Hough, and G. P. Bond, *MNRAS*, **21** (1861 Jul.), pp. 240–1; T. H. Safford, *AN*, **55** (1861 Jul. 2), p. 299; A. Secchi, *AN*, **56** (1861 Sep. 11), pp. 67–70; C. Fearnley, *AN*, **56** (1861 Oct. 12), p. 139; T. Maclear and W. Mann, *MRAS*, **31** (1861–62), pp. 41–2; T. R. von Oppölzer, *AN*, **56** (1862 Jan. 11), pp. 369–74; E. Kayser, *AN*, **57** (1862 Feb. 1), p. 21; W. J. Förster, *AN*, **57** (1862 Apr. 12), pp. 177, 182; T. R. von Oppölzer, *AN*, **58** (1862 Jul. 15), pp. 5–8; *AN*, **60** (1863 Jun. 30), pp. 113–16; T. R. von Oppölzer, *AN*, **62** (1864 Jun. 3), pp. 177–88; E. Weiss, *AN*, **68** (1867 Mar. 9), pp. 381–4; J. G. Galle, *AN*, **69** (1867 Apr. 2), pp. 33–6; V1964, p. 58.

C/1861 J1
(Great Comet, also
Tebbutt)

1861 II

Discovered: 1861 May 13.37 ($\Delta = 1.27$ AU, $r = 0.99$ AU, Elong. $= 50°$)
Last seen: 1862 April 30.90 ($\Delta = 4.71$ AU, $r = 4.45$ AU, Elong. $= 68°$)
Closest to the Earth: 1861 June 30 (0.1326 AU)
Calculated path: ERI (Disc), ORI (Jun. 25), TAU (Jun. 28), AUR (Jun. 29), LYN (Jul. 1), UMa (Jul. 2), DRA (Jul. 6), UMa (Jul. 7), BOO (Jul. 14), HER (Sep. 10), LYR (Dec. 12), DRA (Dec. 23), CYG (Jan. 11), CEP (Feb. 18)

The Australian farmer and self-taught astronomer J. Tebbutt (Windsor, New South Wales, Australia) discovered this comet on 1861 May 13.37, at a position of $\alpha = 3^h 54.2^m$, $\delta = -30° 44'$. He had been searching for comets in the westen sky when he found this nebulous object. Tebbutt believed this might be a new comet, but even though the object was not found in a catalog of nebulae that Tebbutt had available, he still wanted to make sure the object was moving. Hazy skies the next morning prevented a reliable confirmation, but clear skies on the evening of May 14 seemed to indicate the object had not moved and Tebbutt concluded it was probably a nebula. Wanting to confirm this conclusion, Tebbutt had to wait until May 21 for the next clear evening. Upon focusing the telescope on the star the object was compared with on the 14th, he found the "nebula" had shifted its position and was therefore a comet. He communicated news of his discovery to W. Scott (Sydney, New South Wales, Australia) on the 21st and to the *Sydney Morning Herald* of May 25. Scott managed to see the comet on May 22, and on the 27th he noted, it "was just visible after sunset to the naked eye."

The comet was heading towards close approaches to both the sun and Earth as June began; however, despite a perihelion around mid-month, it would be the very close approach to Earth at the end of the month that would make this comet truly spectacular. G. W. H. White (Royal Observatory, Cape of Good Hope, South Africa) saw the comet near the eastern horizon in the morning twilight of June 3. He said the nucleus was about magnitude 2 or 3, while the tail was 3° long and was inclined toward the south pole. R. L. J. Ellery (Williamstown, Victoria, Australia) made the first of several observations with the 5-foot focal length refractor (50×) on the 4th. He said the nucleus was "badly defined and nebulous", with a tail extending about 4° toward the south. Scott found the comet "plainly visible to the naked eye" on the 8th. He added, "The comet is now sufficiently brilliant to be seen without a telescope 40 minutes before sunrise; its tail extends 18° in a direction 15° W. of South, one narrow stream of light extending twice as far as the rest of the tail. The nucleus is distinct and round, presenting no remarkable features." G. W. H. Maclear (Royal Observatory, Cape of Good Hope, South Africa) also saw the comet on the 8th, but through stratus clouds. Nevertheless, he said the comet displayed a bright, diffuse nucleus. E. Liais (Rio de Janeiro, Brazil) observed the nucleus on the 12th and gave its brightness as magnitude 2 or 3 and its diameter as 21.5". He also noted a tail 40° long. Ellery also saw the comet that night and said the nucleus was "bright and planetary," while an apparently higher power revealed it as fan-shaped. He added that the tail was faint and 7° long. Ellery saw the comet again on the 14th and found the nucleus "much brighter" and "distinctly fan-shaped with stellar point." He added that the tail extended 22° toward the south-southwest. Maclear estimated the tail length as 36° on the 19th. Ellery said the nucleus was magnitude 2 on the 20th and very distinctly fan-shaped. He added that the tail was double, with the western or main tail extending over 40° and the eastern tail extending about 5°. They were separated by an angle of 34°. The eastern tail was also slightly curved toward the east. Maclear saw the comet for the final time just before sunrise on the 28th. He noted that the sky was "foggy" and no stars were then visible.

The comet was well observed as it passed closest to Earth on June 30. The total brightness was estimated as "not as bright as Jupiter" (fainter than magnitude −2) by J. F. J. Schmidt (Athens, Greece), while the nuclear magnitude was estimated as 1 by T. Brorsen (Senftenberg, Germany), but "intermediate" between Venus and Jupiter (between −4 and −2) according to T. W. Webb (Hardwick Parsonage, Herefordshire, England). He added that the comet had a golden hue to it. H. Goldschmidt (Paris, France) said the nucleus was 4″ across, while K. G. Schweizer (Moscow, Russia) measured it as 3.07″. Webb observed with a 14-cm refractor and estimated it as 2″, but admitted that he probably underrated it. He added that it was "a fine luminous disc" with "a very ill-terminated, but still definite, limb." Webb also observed the comet at 27× and described the sight "as though a number of light, hazy clouds were floating around a miniature full moon." He was describing six "luminous veils" located within the coma, the brightest of which was nearest the nucleus, while the faintest was farthest away. J. G. Galle (Breslau, now Wroclaw, Poland) said the nucleus was "extremely bright and distinct," while R. Main (Radcliffe Observatory, Oxford, England) said it was elliptical, with the major axis "directed nearly towards the Sun." Main added, "A stream of light went off from the upper apparent part of the nucleus, and turned round towards the apparent west in the shape of a sickle. Another but fainter stream was seen on the apparent east side of the first stream, also turning round towards the west." The coma was described as a parabolic curve by Galle, with Schmidt estimating its diameter as 60–70′. The tail was very impressive and contained a number of rays. Although Galle estimated the length as 30–40° and Goldschmidt said it was 35° long and 3–4° wide, other observers noted a much longer length. Some of these extreme lengths were 122° by Schmidt, 90° by Brorsen, at least 90° by Webb, and "considerably longer" than 43° by Main. George Williams (Liverpool, England) observed a tail ray extending through Boötis into Ursa Major, as well as a somewhat brighter ray extending into Cassiopeia. He suspected they might have been clouds, but noted that both pointed towards the comet's nucleus. Webb and his wife noticed a faint ray "of perfectly similar character to the tail, stretching under the square of Ursa Major, about 3° or 3.5° broad . . . and traceable about half way from the latter star to

Arcturus: it pointed to the Comet, but in the twilight no connexion could be made out." Webb added that about 20 minutes later it had become brighter. He concluded it was probably a "cirrus cloud, brought up by the N. W. wind". W. C. Burder (Clifton, England) saw the comet in the morning twilight and said it was "as bright as Capella. It had no tail that I could see, but it had a bright nucleus surrounded by a nebulous haze." He said it disappeared in the brightening twilight. J. M. Stothard (Dublin, Ireland) had just completed an observation of Jupiter in the evening sky when, "I was very much surprised at seeing a remarkable-looking nebulosity in the full glare of the twilight, in the north-west part of the heavens, which twilight was then so great as to render invisible all stars in those regions nearly as far as the zenith. I soon perceived in the object a distinct and planetary-looking nucleus, which was well seen in an opera-glass, and shone out like the disk of a planet seen with a low power with a dense halo round it." Stothard added, "As the twilight faded, the tail came out; but, although I watched until midnight, the night was not dark enough to show it well." Tidmarsh (Downside College, Bath, England) saw the comet early in the evening in strong twilight. He said, "The nucleus is very large and bright, and was distinctly visible to the naked eye...." Although Chinese accounts simply note that a large "broom-star" was seen in the northwest sometime during the period of June 8 to July 7, it was probably seen on June 30, or perhaps during the first days of July.

Extensive observations continued on July 1. H. Dembowski (Gallarate, Italy) said the comet was comparable in brightness to Jupiter (about magnitude −2), while A. Secchi (Rome, Italy) said the nucleus was comparable to Saturn (about magnitude 1). The diameter of the nucleus was given as 2" by O. W. Struve (Pulkovo Observatory, Russia) and 2.74" by Schweizer, while Schmidt said the coma was 40' across. Goldschmidt said the tail was 45° long, while Secchi noted it extended 118°, Schmidt said it was 114°, and Dembowski said it was 90° long. J. M. Gilliss (US Naval Observatory, Washington, DC, USA) noted a bank of clouds from which there was a "pulsating light extending towards the zenith." He initially thought it was an aurora. Dembowski said the nucleus was well defined on the side opposite the sun and very diffuse in the other direction, while there was a slight indication of its being elongated north to south. Dembowski also noticed three envelopes near the nucleus, with the two closest ones described as well defined, and the outer one as rather diffuse. Tidmarsh said the comet was at its lower transit at about 0:10 UT "and the huge tail then extended at least sixty or seventy degrees across the heavens. It could be distinctly traced across Polaris, and considerably beyond, nearly in the direction of Vega."

Still another day of extensive observations came on July 2. K. L. von Littrow (Vienna, Austria) estimated the nuclear magnitude as 1 and its diameter as a few arc seconds. K. Hornstein (Gratz, Austria) said the nucleus was very small and estimated it as about an arc second across. He added that it appeared white and about as bright as a 1st magnitude star. Webb said the nucleus seemed smaller and less distinguished than on the 30th, although he admitted it seemed equally bright. The six veils seen on the 30th were again present, but had changed considerably, with some combining together. Webb also noted the luminous sector fanned out between 190° and 315°. Main said the nucleus still appeared elliptical. He noted a sickle-shaped stream of light extending westward from the nucleus for a short distance and then turning rapidly away from the sun. The second stream of light seen on the 30th was still visible but was not as well defined as when previously seen. Dembowski again noted envelopes near the nucleus and said they were unchanged since the previous night. The tail length was estimated as 97° by Littrow, 107° by Schmidt, 100° by Dembowski, nearly 90° by Hornstein, about 80° by Webb, and at least 70° by Main. Gilliss was again greeted by mostly cloudy skies, but, although the beam of light was in a slightly different location, he still could not see the coma or nucleus and thought the spectacle was an aurora.

Observations were still extensive on July 3. Goldschmidt estimated the diameter of the nucleus as 4″. Dembowski said the nucleus seemed more oblong, with a possible very diffuse fan extending towards PA 120°. Webb's 14-cm refractor showed the nucleus as a "sharp and vivid point" at 110×, but a "small, dull, and ill-defined" object at 460×. Gilliss said the nucleus "appeared as a small planetary disc of only a few seconds diameter. From this emanated towards the sun a luminous sector, or fan-shaped head, terminated by a well-defined convex line." Schmidt said the coma was 40′ across. C. F. Pape (Altona, Germany) said the tail extended toward PA 57° 40′. Tail lengths were estimated as 100° by Dembowski, 93° by Schmidt, 75° by Goldschmidt, 80–85° by Gilliss, 70° by G. F. W. Rümker (Hamburg, Germany), and about 60° by Main. Goldschmidt added that the tail was 5° wide at a point 20° from the nucleus. Gilliss said that while the tail was long and straight, a shorter, western branch was also seen which was slightly curved. It measured 8–10° in length. With Gilliss finally experiencing clear skies, he noted "The constancy of the light near the nucleus was interrupted by flashings or pulsations, closely resembling those of the aurora."

There were indications that the spectacle was beginning to subside on July 4. Main remarked that the comet was "rapidly becoming fainter . . . but the brilliancy of the nucleus is still very great." Littrow estimated the magnitude of the nucleus as 1. Gilliss said the nucleus was 11.2″ across in a telescope, though "evidently elongated in a line perpendicular to the direction of motion." Webb said the nucleus "seems to stand at the vertex of a wide but very ill-marked parabola. . . . " C. H. F. Peters (Hamilton College Observatory, Clinton, New York, USA) observed with a 34-cm refractor and said the nucleus was stellar and cast a shadow that extended 20″ or 30″. He added that the nucleus "is surrounded by fine envelopes, of which four are counted, the innermost much brighter than the others." He noted that within the first envelope the region was "filled with jets or flames, magnificently radiating from the nucleus, and bending before reaching the limb [of the first envelope]." Gilliss said the luminous sector seen the previous night was now 101" long and its boundary "was parabolic rather than elliptical, but there was in its western half an irregularity as though a segment had been cut from that wing." Webb said his 14-cm refractor showed the sector was about 1.5' long and fanned out between PA 207° and 340°. F. Tietjen (Berlin, Germany) said the brightest jet extended toward PA 243.7°, while the middle of the sector was pointing toward PA 261.0°. Main said a telescope revealed the nucleus was pear-shaped, with two curved streams of light still flowing from it. The lower stream was sickle-shaped, but not as well defined as when seen previously. The upper arc was also ill defined. The space between the sectors spanned 100° of a circle. Schmidt estimated the coma diameter as 36' and Webb estimated it as 20'. Several observers were seeing two tails. The long tail was estimated as about 40° long by Main, 60–65° by A. Quirling (Radcliffe Observatory, England), 84° by Schmidt, 82° long by Littrow, and about 96.5° long by Peters. It was described as nearly straight by Webb. The short tail was described as 30° long and 8° wide by Littrow, and 24° long and 1.5° wide by Gilliss. Gilliss also noted the long tail's width never exceeded 4°, while its eastern edge was more sharply defined than its western edge.

Fewer physical descriptions were being given by July 5th, but these still indicated the tail was of great length. The main tail was estimated as 63° long by Littrow, 85° long by Schmidt, and 45° long by Quirling, with Quirling also indicating the greatest width was 10°. The second tail was estimated as wide and 30° long by Littrow. Schmidt estimated the coma diameter as 50'. Webb swept rapidly across the tail with his comet eyepiece and noted a slightly darker region extended from the nucleus for a short distance into the tail. Activity was still visible within the coma. Main said the nucleus was about magnitude 1. He added that the two streams of light "now pass symmetrically on each side of the nucleus." Peters gave the nuclear diameter as 5.7" or possibly smaller and said the 34-cm refractor showed one bright inner envelope and two faint outer ones. In addition, he noted "many fine jets streaming out from the nucleus, part of them recurving to the right, others to the left."

This comet was independently discovered by David Livingston on July 6, who was then traveling down the Shire River near present day Blantyre, Malawi, in Africa. He noted "a large comet in Ursa Major" and estimated the tail length as 23°. Littrow estimated the main tail as 59° long and about 2.5° wide, while the second tail was about 30° long and distinctly fainter since the 5th. Schmidt gave the tail length as 74°. Webb said the tail seemed slightly turned to the left again. Peters said, "The secondary tail is quite bright, and as wide as the principal tail, branching off from it to the west under an acute angle." He added that the 34-cm refractor revealed a nucleus 3.8" across. Fearnley estimated the magnitude as 1–2. He said a telescope revealed a stellar nucleus from which a bright fan of light extended toward PA 260°. Tietjen said the middle of the sector was directed toward PA 272.8°.

The last big day of observations for this comet came on July 7. John Kirk, who was traveling independent of Livingston down the Shire River, in Africa wrote, "This night we got sight of a splendid comet in the Great Bear moving rapidly from the sun." Heis said the brightness equalled γ Ursae Majoris (magnitude 2.44 – SC2000.0). Dembowski estimated the tail length as 30°, while Gilliss said it was no more than 25° long and 3° wide. Peters agreed that the tail seemed to have decreased in brightness at its end, but was wider and still visible over 30°. Rümker said it was 25° long. Dembowski said the fan extending from the nucleus was less definite than on the 3rd. Gilliss commented that the luminous sector was "much smaller and fainter, and for the greater part of the time could scarcely be discerned at all as distinct from the general mass of light." Peters said the 34-cm refractor indicated the outer envelopes were no longer visible, while the inner envelope seemed to be undulating before his eyes. Tietjen said the middle of the sector was directed toward PA 270.5°.

Gilliss said the nucleus was brighter than ε Ursae Majoris (magnitude 1.77 – SC2000.0) on July 8 and added that, although well condensed, it was not stellar. The tail was estimated as 31° long by Littrow, 57° long by Schmidt, 14° or 15° by Webb, and 27° by Quirling, with Quirling noting that the second tail had disappeared. Webb said the tail appeared streaked to the naked eye. Webb specifically noted that the veils were no longer visible in his 14-cm refractor. Fearnley estimated the nuclear condensation as 1–2" across. On the 9th Gilliss said the nucleus appeared "planetary" and was about equal in brightness to ζ Ursae Majoris (magnitude 2.27 – SC2000.0). Schmidt estimated the coma diameter as 34'. Gilliss added that the luminous sector was more distinct than on the previous evening, and "the marked brush or rays of light diverging from the nucleus extended across the general mottled surface of the sector...." Quirling estimated the tail as 20–24° long. Main said there was a bright envelope near the nucleus and a very faint, diffuse one just beyond it. Schmidt estimated the tail length as 48° on the 10th and Webb estimated the coma diameter as 18'. Webb added that the comet appeared distinctly white to the naked-eye, while his comet eyepiece revealed it tinged with bluish-green or greenish-blue. On July 11 Heis said the brightness was nearly the same as that of α₂ Canum Venaticorum (magnitude 2.90 – SC2000.0), while Fearnley said it was slightly fainter than η and ζ Ursae Majoris (η is magnitude 1.86 and ζ is magnitude 2.27 – SC2000.0). Schmidt reported the coma was 28' across. The brightness was fairly well determined on July 12, with Schmidt estimating it as 3 and E. Heis (Münster, Germany) noting it was between that of γ Ursae Majoris (magnitude 2.44 – SC2000.0) and γ Ursae Minoris (magnitude 3.05 – SC2000.0). The main tail was estimated as over 30° long by Littrow and 33° long by Schmidt, while the wide tail was estimated as 21° long by Littrow. Schmidt added that the coma was 26' across. Tietjen said the middle of the sector was directed toward PA 201.4°. Littrow estimated the tail length as 27° on the 13th, while Peters said it was no longer than 35–40°. Peters said the main and secondary tails were separated by a "narrow dark straight line." On the 14th Littrow estimated the tail length as 19°, while Quirling estimated is as 15° or 16°. Quirling added that the comet seemed to have grown "considerably smaller" since previous nights, with the head not seeming brighter than a star of 3rd magnitude. He also noted that both borders of the tail were well defined, although the northern edge could be traced farther than the southern one. Tietjen said the middle of the sector was directed toward PA 193.5°. Littrow reported the tail was 18° long on July 15. Webb then noted that the nucleus was "extremely dim and nebulous" when seen with the 14-cm refractor at 460×, while at lower powers the coma appeared 13' across. He added that the nucleus had a greenish-yellow color.

Descriptive observations were sparse during the second half of July. On the 16th Heis said the brightness nearly equalled that of ι Draconis (magnitude 3.29 – SC2000.0), while Quirling and Rümker estimated tail lengths as 12° or 13° and 6°, respectively. On the 17th Heis said the brightness was between those of ι Draconis and α Draconis (magnitude 3.65 – SC2000.0) and on the 24th he said the brightness was between those of θ Boötis (magnitude 4.05 – SC2000.0) and ι Boötis (magnitude 4.75 – SC2000.0). Fearnley said the coma was 15' across and contained a very distinct nucleus. While the comet faded the nucleus was beginning to change as well. E. Schönfeld (Mannheim, Germany) said the nucleus appeared a little diffuse on the 18th, while Peters noted on the 24th, "The nucleus now appears much less stellar than before, rather as a blurred surface, of 8" in diameter, though this measure is little reliable." He added that the outline of the envelope was no longer visible, though the moon was nearly full.

The comet faded as the year progressed, but where measured positions were plentiful, physical descriptions decreased as each month passed. Fearnley said the tail was 2° long in the cometseeker on August 6, while a larger telescope revealed a very distinct nucleus of magnitude 8. Heis said the comet was "three steps fainter" than 44 Boötis (magnitude 4.76 – SC2000.0) on August 10. Peters looked at the comet with a 34-cm refractor on the 12th and wrote, "The nucleus has become small and is rather dim. . . ." Schönfeld estimated the coma diameter as 5' across on the 13th and 4' across on the 14th. He added that the nucleus was not centrally placed on the former date. Fearnley also saw the comet on the 14th and said the tail was 2.5° long. Heis saw the comet with the naked eye on the 15th. Schönfeld said the comet was 5' across on September 2, while the nucleus was magnitude 9 on the 3rd. Schönfeld said moonlight reduced the coma to 2.5' on the 12th. Peters saw the comet in moonlight with the 34-cm refractor and noted its faintness only allowed a slight illumination of the wires used for measuring its position. K. C. Bruhns (Leipzig, Germany) said the comet appeared rather faint in the 6-foot focal length refractor on October 1, 4, and 8.

Schönfeld said the nucleus was magnitude 11 on the 5th, while the coma was 40″ across on the 9th. Peters described the comet as "dull" in the 34-cm refractor on the 11th, while the nuclear condensation had become more stellar on the 19th. Schönfeld described the comet as round, 2′ across, with a nucleus of magnitude 11 on the 13th, while it was 3′ across, with a faint, eccentrically situated nucleus on the 25th. On November 4, Schönfeld said the coma was fairly faint, with an indistinct nucleus of about magnitude 11–12. He measured the coma diameter as 1.3′ on November 5, 1.5′ on the 21st, and 0.7′ on the 28th. Schönfeld also noted a condensation of magnitude 12–13 on the 25th and said the condensation was weak on the 28th. Peters again observed the comet with the 34-cm refractor on the 22nd and simply noted the nucleus was very small. On December 22, Schönfeld said the comet was fairly bright, round, and 20″ across, while the well-condensed nucleus was about magnitude 11.

As the new year began, the number of observers had considerably decreased, and few physical descriptions were published. Peters saw the comet with the 34-cm refractor on January 3 and 5. On the latter date he noted that moonlight and strong winds made the comet difficult to see. Schmidt said the comet could only be seen with difficulty using averted vision on February 3 and 6. F. A. T. Winnecke (Pulkovo Observatory, Russia) found the comet still reasonably visible in a dark field on March 20 and 22. He estimated the diameter as 0.5′. Winnecke also noted that the comet was oblong on the 20th, with the north-following part the brightest. Struve said the coma was 40″ across on April 16. He noted, "Its light is, even in its present faint state, not quite uniform, but shows distinct traces of concentration." The comet appears to have been last detected on April 30.90, when Struve and Winnecke gave the position as $\alpha = 22^h\ 35.2^m$, $\delta = +76°\ 57′$. A position was reported by Struve for May 1.93, but he himself noted in the 1868 issue of the *Mémoires de l'Académie Impériale des Sciences de St. Pétersbourg* that the position was over 2′ in error when compared to positions predicted by a reliable orbit. Struve suggested a faint star had been measured instead of the comet.

F. Abbott (Hobart Town, Tasmania) gave an interesting summary of the comet's appearance during June. He said it was first seen in Hobart Town on the 4th and by the 20th it was "appearing every clear morning in the south-eastern hemisphere, and in the evening near the south-western horizon." He then went on to say, "From its slow motion and general appearance it may continue for some time to be an object of interest in the southern sky. The tail, which is quite straight and about 10 degrees in length, points in the direction of the star Achernar, the nucleus forming nearly a right angle with that star and Canopus. The nucleus of the comet, measured with a Cavallo's micrometer in an eye-piece of 50, was 33″ in diameter; the coma or nebulosity surrounding the nucleus, as seen with a comet eye-piece of 27 in a 5-feet telescope, was 30 minutes in diameter. The breadth from nebulosity at the apex of the tail 50 minutes, where it becomes very diffuse."

Schmidt (1869) tabulated the tail lengths from numerous observers and noted a periodic fluctuation. Fitting a curve to the lengths brought him to conclude that the period was 25.6 days. He also took measurements of the coma diameter and noted a fluctuation which fitted a period of 25.4 days. Comparing the times of the maximum and minimum size of both the tail and coma revealed similar dates, which prompted him to conclude both were being influenced by a similar factor.

The first parabolic orbit was calculated by Tebbutt, using positions obtained on May 24, June 3, and 11. He determined the perihelion date as 1861 June 14.23. He noted "the near approach of Earth to the comet's tail for June 29, and also that the comet itself would probably become visible in full daylight." During the next few weeks and months, several astronomers, including A. Murmann, H. Seeling, L. Hopff, M. Loewy, J. R. Hind, Schweitzer, A. Hall, J. S. Hubbard, H. P. Tuttle, Hawkins, Pape, W. S. Mackay, and O. M. Mitchell established the perihelion date as June 12.0. Interestingly, using positions from the period of July 3–18, Hubbard found it impossible to represent the comet's motion by a parabola. Instead, he obtained a satisfactory fit with a hyperbolic orbit. The result was a perihelion date of June 12.57 and an eccentrcity of 1.0265470.

By September, astronomers were discovering that the comet's true orbit was elliptical. A. J. G. F. von Auwers was the first to recognize this and gave possible periods of 656 and 601 years. Aditional elliptical orbits were calculated by E. Fergola (1797 years), J. Michez (181 Years), H. Seeling (420 years), and T. H. Safford (395 years). After the comet was no longer visible, new orbits were calculated using positions spanning the entire period of visibility. Sluzki determined the period as about 400 years. A Savitch (1863) determined it as about 422 years. H. C. F. Kreutz (1880) determined it as about 409 years, and his orbit is given below.

T	ω	Ω (2000.0)	i	q	e
1861 Jun. 12.0068 (UT)	330.0841	280.9099	85.4424	0.822384	0.985070

ABSOLUTE MAGNITUDE: $H_{10} = 3.9$ (V1964)

FULL MOON: Apr. 24, May 24, Jun. 22, Jul. 21, Aug. 20, Sep. 19, Oct. 18, Nov. 17, Dec. 17, Jan. 16, Feb. 14, Mar. 16, Apr. 14, May 13

SOURCES: Burder, *MNRAS*, **21** (1861 Jul.), p. 242; J. M. Stothard, *MNRAS*, **21** (1861 Jul.), pp. 242–3; Tidmarsh, *MNRAS*, **21** (1861 Jul.), p. 243; W. J. Carpenter, *MNRAS*, **21** (1861 Jul.), p. 243; J. G. Galle, F. A. T. Winnecke, C. F. Pape, L. Hopff, Schweitzer, and A. Secchi, *AN*, **55** (1861 Jul. 12), pp. 305–12; M. Loewy and J. R. Hind, *CR*, **53** (1861 Jul. 15), p. 80; E. Liais, *AN*, **55** (1861 Jul. 23), supplement; K. L. von Littrow, A. Murmann, E. Liais, and H. Seeling, *AN*, **55** (1861 Aug. 2), pp. 361–8; J. F. J. Schmidt, H. Dembowski, and H. Goldschmidt, *AN*, **55** (1861 Aug. 14), pp. 369–76; R. L. J. Ellery and E. J. White, *AN*, **56** (1861 Aug. 31), p. 53; T. H. Safford, A. Hall, G. P. Bond, J. S. Hubbard, and M. C. Stevens, *AJS* (Series 2), **32** (1861 Sep.), pp. 252–66; E. Heis and A. J. G. F. von Auwers, *AN*, **56** (1861 Sep. 11), pp. 71, 77–80; A. J. G. F. von Auwers and J. Michez, *AN*, **56** (1861 Sep. 23), pp. 93–6; W. Scott, Hawkins, G. W. Hough,

O. M. Mitchell, J. Ferguson, and F. Abbott, *MNRAS*, **21** (Supplemental Notice, 1861), pp. 253–61; C. Fearnley, *AN*, **56** (1861 Oct. 12), pp. 139–44; J. M. Gilliss, *AJS* (Series 2), **32** (1861 Nov.), pp. 305–11; A. L. Mansell, N. M. Edmondson, F. J. Krabbe, T. Maclear, and G. N. Hough, *MNRAS*, **22** (1861 Nov.), pp. 15–19; T. H. Safford and H. P. Tuttle, *AN*, **56** (1861 Nov. 30), pp. 269–72; J. Hartnup, *MNRAS*, **22** (1861 Dec.), pp. 46–7; R. Main and A. Quirling, *MNRAS*, **22** (1861 Dec.), pp. 50–8; R. Main, *MNRAS*, **22** (1862 Jan.), pp. 93–4; E. Schönfeld, *AN*, **56** (1862 Jan. 2), pp. 363–6; H. Seeling, *AN*, **57** (1862 Feb. 10), p. 41; E. Fergola, *AN*, **57** (1862 Feb. 28), p. 93; E. Schönfeld, *AN*, **57** (1862 Mar. 8), pp. 97–106; W. S. Mackay, *MNRAS*, **22** (1862 Mar. 14), pp. 159–62; J. F. J. Schmidt, *AN*, **57** (1862 Apr. 3), p. 161; O. W. Struve, *MNRAS*, **22** (1862 Apr.), pp. 244–6; W. J. Förster and F. Tietjen, *AN*, **57** (1862 Apr. 12), pp. 177–80, 183; F. A. T. Winnecke, *AN*, **57** (1862 Apr. 25), pp. 205–8; J. C. Adams, and J. Challis, *MNRAS*, **22** (1862 May), pp. 267–72; T. W. Webb, *MNRAS*, **22** (Supplemental Notice, 1862), pp. 302–14; K. G. Schweizer and Sluzki, *AN*, **58** (1862 Sep. 18), pp. 195–200; G. W. H. Maclear, *MRAS*, **32** (1862–3), pp. 5–7; A. Savitch, *BASP*, **6** (1863), pp. 102–11; K. Hornstein, *AN*, **59** (1863 Jan. 26), pp. 137–40; J. Tebbutt, Jr, *AN*, **59** (1863 Jan. 31), p. 157; K. C. Bruhns, *AN*, **60** (1863 Jun. 18), pp. 83, 92; C. H. F. Peters, *AN*, **60** (1863 Jun. 30), pp. 117–24; G. F. W. Rümker, *AN*, **64** (1865 Feb. 19), pp. 39–42; O. W. Struve, *MASP*, **12** (1868), pp. 1–46; J. F. J. Schmidt, *AN*, **73** (1869 Mar. 11), pp. 241–60; J. Tebbutt, *MNRAS*, **38** (1878 Jun.), p. 412; J. Tebbutt, *AN*, **93** (1878 Aug. 14), p. 47; H. C. F. Kreutz, *Untersuchungen über die Bahn des grossen Kometen von 1861 (1861 II)*. Dissertation: Bonn (1880), 155 pp.; *The Zambezi Expedition of David Livingston (1858–1893)*. Edited by J. P. R. Wallis, London: Chatto & Windus Limited (1956), p. 181; V1964, p. 59; *The Zambesi Journal and Letters of Dr. John Kirk (1858–63)*. Edited by Reginald Foskett, Edinburgh: Oliver & Boyd (1965), p. 349; HA1970, pp. 86–7; W. Orchiston, *IAJ*, **25** (1998), pp. 167–78; SC2000.0, pp. 299, 320, 321, 330, 338, 344, 348, 351, 365, 371, 373.

C/1861 Y1   *Discovered:* 1861 December 29.32 ($\Delta = 0.68$ AU, $r = 0.93$ AU, Elong. $= 65°$)
(Tuttle)   *Last seen:* 1862 February 2.82 ($\Delta = 0.82$ AU, $r = 1.32$ AU, Elong. $= 93°$)
   Closest to the Earth: 1862 January 13 (0.3848 AU)
1861 III   *Calculated path:* VIR (Disc), BOO (Jan. 4), DRA (Jan. 14), UMi (Jan. 17), DRA (Jan. 18), UMi (Jan. 19), DRA (Jan. 20), CEP (Jan. 22), CAS (Jan. 27)

H. P. Tuttle (Harvard College Observatory, Cambridge, Massachusetts, USA) discovered this telescopic comet on 1861 December 29.32. He gave the position as $\alpha = 14^h$ 12.9^m, $\delta = -5°$ 13$'$ on December 29.47. An independent discovery was made by F. A. T. Winnecke (Pulkovo Observatory, Russia) on 1862 January 9.01. Winnecke described it as bright, 3–4$'$ across, with a strong central condensation.

C. A. F. Peters (Altona, Germany) observed the comet on January 18 and 19. He described it as a weak nebulosity, about 2$'$ across, without a visible nucleus. The comet attained it greatest northern declination of $+81°$ on the 22nd. E. Schönfeld (Mannheim, Germany) described the comet as diffuse and difficult to observe on January 25 and 26.

The comet was last detected on February 2.82, when F. Tietjen (Berlin, Germany) gave the position as $\alpha = 1^h$ 11.9^m, $\delta = +64°$ 42$'$. Schönfeld said he was not able to find the comet in February. J. F. J. Schmidt (Athens, Greece) searched for the comet on 1862 February 22, using an ephemeris by C. F. Pape, but no trace was found.

The first parabolic orbit was calculated by T. H. Safford. He used positions from December 29, 31, and January 2 and determined the perihelion date as 1861 December 7.70. This was quite close to the comet's true orbit, which was refined by Tuttle, Winnecke, Tietjen, Pape, A. Hall (1862), V. Fuss (1865), and M. Noether (1867).

T	ω	Ω (2000.0)	i	q	e
1861 Dec. 7.6744 (UT)	331.5973	147.0373	138.0019	0.839027	1.0

ABSOLUTE MAGNITUDE: $H_{10} = 9.3$ (V1964)

FULL MOON: Dec. 17, Jan. 16, Feb. 14

SOURCES: G. P. Bond, H. P. Tuttle, and T. H. Safford, *MNRAS*, **22** (1862 Jan.), pp. 94–5; F. A. T. Winnecke and C. A. F. Peters, *AN*, **57** (1862 Jan. 23), pp. 9–12; H. P. Tuttle, T. H. Safford, F. Tietjen, and C. F. Pape, *AN*, **57** (1862 Feb. 1), pp. 29–32; H. P. Tuttle, *AN*, **57** (1862 Mar. 21), p. 131; E. Schönfeld, *AN*, **57** (1862 May 10), p. 239; J. F. J. Schmidt, *AN*, **57** (1862 Apr. 3), p. 163; A. Hall, *AN*, **58** (1862 Jul. 24), p. 29; F. A. T. Winnecke, *BASP*, **5** (1863), pp. 59–61; K. C. Bruhns, *AN*, **60** (1863 Jun. 18), p. 83; V. Fuss, *BASP*, **8** (1865), pp. 50–7; M. Noether, *AN*, **69** (1867 May 6), pp. 103–8; V1964, p. 59.

2P/Encke

1862 I

Recovered: 1861 October 4.90 ($\Delta = 1.09$ AU, $r = 2.07$ AU, Elong. $= 164°$)

Last seen: 1862 March 13.14 ($\Delta = 1.37$ AU, $r = 0.86$ AU, Elong. $= 38°$)

Closest to the Earth: 1862 January 31 (0.6247 AU)

Calculated path: PSC (Rec), PEG (Oct. 9), AQR (Jan. 10), CAP (Jan. 30), AQR (Jan. 31), CAP (Feb. 3)

J. F. Encke (1861) took the comet's orbit for the 1858 apparition, applied perturbations by Jupiter, and predicted the comet would next arrive at perihelion on 1862 February 6.67. He provided a daily ephemeris for the period of 1861 October 3 to 1862 January 1. Encke continued to refer to this comet as "the comet of Pons." W. J. Förster (Berlin, Germany) recovered this comet on 1861 October 4.90 at a position of $\alpha = 0^h 19.1^m$, $\delta = +18° 49'$. He said it was extremely difficult to see.

Several astronomers immediately began their observations of this faint comet. Förster continued his observations on October 8, 26, and 31, and simply described the comet as diffuse and difficult to measure. A. Reslhuber (Kremsmünster, Austria) saw a weak trace of the comet on October 11. E. Schönfeld (Mannheim, Germany) said the comet appeared too faint for measurement on October 24. The comet was also independently recovered by G. P. Bond (Harvard College Observatory, Cambridge, Massachusetts, USA) on 1861 October 24.99, while using a 38-cm refractor. He described it as faint and diffuse, but noted, "The Comet would probably have been found several weeks earlier, if the Ephemeris had come to hand in season to escape the interference of the Moonlight in the early part of the month." J. F. J. Schmidt (Athens, Greece) first began looking for the comet on October 4. Nothing convincing was detected on that evening, or on the 7th. He believed he found the comet on October 8.78, and measured a position, which ultimately proved to be about 2′ in error. Another apparent observation on the 10th proved to have been a nebula when the area was rescanned on the 11th. Searches on October 25 and November 8 again proved unsuccessful.

The comet steadily brightened during November and December as it approached the sun and Earth. Förster said the comet was distinctly seen on November 4. Schmidt said the comet was 2.5′ across and centrally condensed on the 21st, while very clear skies on the 23rd allowed Schmidt to see a 5.5′ coma. On the last date he also noted an appreciable condensation, but no nucleus. F. A. T. Winnecke and O. W. Struve (Pulkovo Observatory, Russia) said the comet was 3′ in diameter on November 25. Schmidt said the comet was visible in a small cometseeker on the 28th, while his larger telescope revealed a coma 5.8′ across, but still no nucleus. Schönfeld reported he saw the comet on several occasions during this month, but each time faint stars within the coma made it impossible to measure the comet's position. Winnecke and Struve found the comet 3–4′ across on December 1, with an inner condensation 30–40″ across. On the 2nd, they estimated the coma diameter as 4–5′ and noted the comet was visible in the refractor's finder. Schönfeld described the comet as very diffuse and 3′ across on December 2 and 4, while Schmidt said it was 4.9′ across on the 3rd, with a hardly noticeable condensation. Schmidt estimated the diameter as 4′ in moonlight on December 5. Several days passed with no physical descriptions because of moonlight, but Schmidt noted on the 18th that the comet was visible in a cometseeker despite the moonlight. He also noticed a central condensation. Winnecke and Struve found the condensation was oval on the 19th and noted a faint protrusion extending 6′ toward PA 235–295°. Schmidt determined the coma diameter as 6.2′ on the 21st and 7.1′ on the 22nd. He noticed it was bright and well condensed on both evenings, but saw a bright, wide, fan-shaped nebulosity on the sunward side of the nucleus on the last date. Reslhuber said the comet was rather bright on the 21st, but was still without a nucleus or tail. Schönfeld also said the comet was bright and well condensed on December 22 and 25. F. Tietjen (Berlin, Germany) and K. C. Bruhns (Leipzig, Germany) gave the coma diameter as 2.5′ to 3′ on the 23rd, with Bruhns adding that the nucleus was just north of the coma's center. Bruhns found the coma elongated on the 25th, with the larger axis about 3′ long. Winnecke noted three envelopes were visible in the heliometer. Tietjen said the coma was 1.9′ across on the 30th, while he and Förster said a protrusion extended toward PA 250–270°.

The comet continued to brighten throughout January. T. Maclear (Royal Observatory, Cape of Good Hope, South Africa) observed with the 8.5-foot focal length refractor on the 3rd and measured the coma as 2′ 29.2″ across. Schmidt described the comet as bright on the 5th and noted a nearly pure white central condensation. Tietjen gave the coma diameter as 1.5' on the 6th and 1.2' on the 10th, while Bruhns said it was about 1.25' across on the 10th. Tietjen said the protrusion extended toward PA 260° on the 15th and PA 250° on the 18th. He added that the coma was 2.4' across on the last date. Bruhns said the nucleus was similar to a star of magnitude 6 or 7 on January 20, while the coma was 1.5' across. Schönfeld described the comet as very bright and strongly condensed in evening twilight on the same date. Schönfeld saw the comet in bright twilight on the 25th and noted a short tail.

The comet remained lost in the sun's glare for a few weeks, passing just 2.5° from the sun on February 2. W. Scott (New South Wales, Australia) saw the comet with an 18-cm refractor on February 23 and 24, and described it as "very indistinct and ill defined." Maclear said the outer coma was 1' 46" across on the 3rd, while the inner coma was 47" across. On the 8th, he described the comet as very faint, with a diameter of 2' 20".

The comet was last detected on March 13.14, when Maclear measured the position as $\alpha = 21^h 15.9^m$, $\delta = -23° 50'$.

Minor refinements of this comet's orbit using positions from multiple apparitions, as well as planetary perturbations, have been published by numerous astronomers. Of most significance were the studies of F. E. von Asten (1877) and B. G. Marsden and Z. Sekanina (1974). These have typically revealed a perihelion date of February 6.75 and a period of 3.30 years. Marsden and Sekanina determined the nongravitational terms as $A_1 = -0.78$ and $A_2 = -0.03170$. The orbit of Marsden and Sekanina is given below.

T	ω	Ω (2000.0)	i	q	e
1862 Feb. 6.7469 (TT)	183.4705	336.4656	13.1005	0.339950	0.846739

ABSOLUTE MAGNITUDE: $H_{10} = 8.5 - -9.7$ (V1964)

FULL MOON: Oct. 18, Nov. 17, Dec. 17, Jan. 16, Feb. 14, Mar. 16

SOURCES: J. F. Encke, *MNRAS*, **21** (Supplemental Notice, 1861), pp. 248–53; J. F. Encke, *AN*, **56** (1861 Sep. 23), pp. 83–90; W. J. Förster, *AN*, **56** (1861 Nov. 20), p. 231; G. P. Bond, *AN*, **56** (1861 Nov. 30), p. 269; J. F. J. Schmidt, *AN*, **56** (1861 Dec. 12), p. 315; A. Reslhuber, *AN*, **57** (1862 Mar. 29), p. 153; J. F. J. Schmidt, *AN*, **57** (1862 Apr. 3), pp. 161–4; J. Hartnup, *MNRAS*, **22** (1862 Apr.), p. 238; F. A. T. Winnecke and O. W. Struve, *AN*, **57** (1862 Apr. 25), pp. 203–6; W. Scott, *MNRAS*, **22** (1862 May), pp. 272–3; E. Schönfeld, *AN*, **57** (1862 May 10), p. 239; T. Maclear, *MRAS*, **32** (1862–3), pp. 1–3; F. Tietjen, *AN*, **60** (1863 Jun. 12), pp. 73–6; K. C. Bruhns, *AN*, **60** (1863 Jun. 18), pp. 83, 92; F. E. von Asten, *BASP*, **22** (1877), p. 562; V1964, p. 59; B. G. Marsden and Z. Sekanina, *AJ*, **79** (1974 Mar.), pp. 415–16.

C/1862 N1 *Discovered:* 1862 July 2.87 ($\Delta = 0.12$ AU, $r = 1.00$ AU, Elong. = 78°)
(Schmidt) *Last seen:* 1862 July 31.11 ($\Delta = 1.06$ AU, $r = 1.18$ AU, Elong. = 69°)
Closest to the Earth: 1862 July 4 (0.0982 AU)
1862 II *Calculated path:* CAS (Disc), CEP (Jul. 3), DRA-UMi (Jul. 4), DRA-BOO (Jul. 5), CVn (Jul. 6), BOO (Jul. 8), COM (Jul. 10), VIR (Jul. 11)

J. F. J. Schmidt (Athens, Greece) discovered this telescopic comet on 1862 July 2.87. He gave the position as $\alpha = 23^h 56.8^m$, $\delta = +56°\ 23'$ on July 2.93 and described the comet as tailless, with a condensed coma 22' across. Schmidt indicated that to the naked eye the comet was about half the brightness of the Andromeda Galaxy. Independent discoveries were made by E. W. L. Tempel (Marseille, France) on July 2.99, S. Simons (Dudley Observatory, New York, USA) on July 4.16, and G. P. Bond (Harvard College Observatory, Cambridge, Massachusetts, USA) on July 4.25. Tempel said the comet was as large and bright as a star of magnitude 4–5. Bond was actually looking for comet C/1861 J1, which was supposed to be near the north celestial pole, when he found this comet. He said it was plainly visible to the naked eye, while the telescope revealed "a small but decided nucleus of the 14th magnitude and 0.3 arcsec in diameter, which could be observed with precision."

The comet had passed perihelion on June 22, while on July 4 it passed closest to Earth and also attained it most northerly declination of +78°. Schmidt distinctly saw the comet with the naked eye on July 4 and 6. With a telescope he estimated the coma diameter as 31–34' on the 4th and 32' on the 6th. Also on the 4th, Schmidt saw a nucleus of magnitude 11–12 and a faint trace of a tail extending about 1/2°. G. W. Hough (Dudley Observatory, New York, USA) observed with a refractor on the 4th and described the comet "as a faint nebula somewhat condensed at the center. The envelope was equally distributed on all sides." Hough added that with the absence of moonlight the comet was just visible to the naked eye. Schmidt last saw the comet with the naked eye on July 7 in bright moonlight.

Moonlight prevented observers giving physical descriptions of the comet for several days, but Schmidt resumed on July 14 by measuring the coma diameter as 9.7'. In fact, Schmidt continued measuring the coma diameter on every clear night until it was too faint to be observed. He measured it as 7.9' on the 16th, 7.5' on the 17th, 6.0' on the 19th, 4.0' on the 21st, 6.0' on the 22nd, 4.5' on the 24th, 2' on the 29th, and 1.5' on the 30th. Schmidt added on the 30th that the observation was "very difficult because of the great faintness of the uncondensed nebulosity." C. H. F. Peters (Hamilton College Observatory, Clinton, New York, USA) noted on the 26th that "no distinct nucleus was visible under very clear skies." The comet was last detected by Bond on July 31.09 and by Peters on July 31.11. Peters gave the position as $\alpha = 13^h 14.9^m$, $\delta = +1°\ 19'$.

The first parabolic orbits were independently calculated by A. Hall and H. P. Tuttle using positions obtained during the period of July 4–6. Hall determined the perihelion date as 1862 June 22.53, while Tuttle determined it as June 24.30. Hall's orbit turned out to be a very close representation, as later proved by the computations of E. Weiss, H. Seeling, and V. Cerulli (1888). Cerulli's orbit is given below.

T	ω	Ω (2000.0)	i	q	e
1862 Jun. 22.5296 (UT)	27.1664	328.4371	172.1091	0.981328	1.0

ABSOLUTE MAGNITUDE: $H_{10} = 9.4$ (V1964)

FULL MOON: Jun. 12, Jul. 11, Aug. 9

SOURCES: E. W. L. Tempel, *CR*, **55** (1862 Jul. 7), p. 46; E. W. L. Tempel, *AN*, **58** (1862 Jul. 15), p. 15; J. F. J. Schmidt and H. Seeling, *AN*, **58** (1862 Jul. 24), pp. 29–32; G. P. Bond, A. Hall, H. P. Tuttle, and E. Weiss, *AN*, **58** (1862 Aug. 8), pp. 88–90, 93; J. F. J. Schmidt, *AN*, **58** (1862 Aug. 14), pp. 97–102; S. Simons and J. F. J. Schmidt, *AN*, **58** (1862 Aug. 19), pp. 115–22; H. Seeling, *AN*, **58** (1862 Aug. 25), p. 141; C. H. F. Peters, *AN*, **58** (1862 Oct. 24), p. 271; E. Weiss, *MNRAS*, **22** (Supplemental Notice, 1862), pp. 314–15; G. P. Bond, *AN*, **60** (1863 May 30), p. 35; V. Cerulli, *AN*, **118** (1888 Jan. 5), pp. 193–204; V1964, p. 59.

109P/1862 O1 *Discovered:* 1862 July 16.1 ($\Delta = 1.50$ AU, $r = 1.16$ AU, Elong. $= 51°$)
(Swift–Tuttle) *Last seen:* 1862 October 31.8 ($\Delta = 1.93$ AU, $r = 1.49$ AU, Elong. $= 49°$)
  Closest to the Earth: 1862 August 30 (0.3417 AU)
1862 III *Calculated path:* CAM (Disc), DRA (Aug. 14), CAM (Aug. 15), UMi (Aug. 19), DRA (Aug. 23), BOO-DRA-BOO (Aug. 26), CrB (Aug. 29), SER (Aug. 31), OPH (Sep. 6), SCO (Sep. 7), ARA (Oct. 5)

L. Swift (Marathon, New York, USA) discovered this comet on 1862 July 16.1, following over three years of searching for comets. He was then examining the northern sky with his 11.4-cm refractor and described the comet as a somewhat bright telescopic object. He confirmed it on July 17.11, and indicated a position of $\alpha = 5^h\ 19.7^m$, $\delta = +67°\ 08'$. Interestingly, Swift did not report the comet right away, because he thought he was observing comet C/1862 N1, which had been found on July 2. Without knowledge of Swift's observation, H. P. Tuttle (Harvard College Observatory, Cambridge, Massachusetts, USA) independently discovered this comet on July 19.19. He confirmed his find on July 19.20 and noted the comet was heading northward. Following Tuttle's announcement, Swift immediately realized the comet seen on July 16 was not C/1862 N1 and made his announcement to get credit for his first comet discovery. Independent discoveries were made by T. Simons (Dudley Observatory, New York, USA) on July 19.30, A. Pacinotti and C. Toussaint (Florence, Italy) on July 22.95, N. K. Rosa (Rome, Italy) on July 25.96, and H. K. F. K. Schjellerup (Copenhagen, Denmark) on July 26/27. Simons remarked, "When first seen it appeared as a nebula considerably condensed at the centre, the light being intense enough to be easily observed when the wires of the micrometer were illuminated." Rosa said the comet was round and centrally condensed in the finder, while the refractor revealed it was 3–4' across, with a well-defined central condensation. Pacinotti and Toussaint said the comet exhibited a faint trace of tail, as well as a distinct nucleus at high magnifications. Schjellerup described it as a bright nebulosity with a very slow movement.

The comet was approaching the sun and Earth when discovered and became widely observed during the remaining days of July. G. V. Schiaparelli (Milan, Italy) examined the area on the 25th and found a "beautiful comet."

He said the centrally condensed coma was about magnitude 6 and measured about 8′ across. A narrow tail extended over 1.5° toward PA 312°. He noted it was visible, with difficulty, to the naked eye. The comet was also seen that night with the naked eye by Tuttle, Pacinotti, and Toussaint. Tuttle said it resembled the brightness of M13 (magnitude 5.9 – NGC2000.0), while Pacinotti and Toussaint said it was about magnitude 4. Simons saw the comet on the 26th and noted the "light was more concentrated on one side, showing that the tail was already in process of formation." On the 27th, Schjellerup said the nucleus equaled a star of magnitude 7 and noted that a magnification of 226× revealed a distinct extension in the direction of the sun, while the surrounding nebulosity was 3′ across. Rosa saw the comet that same evening and said the comet exhibited no tail, but was slightly visible to the naked eye. Pacinotti and Toussaint estimated the naked-eye brightness as 4 on the 28th, 29th, and 30th. K. C. Bruhns (Leipzig, Germany) saw the comet with the naked eye on the 31st, but noted with a telescope that the coma was about 1.5′ across and a tail extended about 1°.

The comet steadily brightened and increased in size as August progressed, as it passed both the sun and Earth late in the month. It remained a naked-eye object throughout the month and was around its greatest brightness by month's end. It was also well poised for Northern Hemisphere observers as it passed its most northerly declination of +82° on the 15th. Among the Northern Hemisphere observers were the Chinese who later noted the comet was seen on the 19th, 20th, and 21st in the northwest. There were three different sets of magnitude estimates made: coma brightness, condensation brightness, and brightness of the nucleus. Unfortunately, depending on whether the estimates were made with the naked eye, binoculars, or telescopes, the observers did not correctly identify which feature was being estimated. The obvious estimates of the coma brightness were few, but are as follows: Pacinotti and Toussaint estimated magnitude 4 on the 1st and 2nd, Schiaparelli gave it as 4 on the 8th, Winnecke said it was slightly fainter than δ Ursae Majoris (magnitude 3.31 – SC2000.0) on the 15th, Chambers said it "decidedly surpasses" α Coronae Borealis (magnitude 2.23 –SC2000.0) on the 31st. Estimates of the magnitude of the condensation were quite numerous, but the observers sometimes attributed them to the total brightness and sometimes to the nucleus. It is certainly true that at maximum brightness the magnitude estimates of the condensation were very similar to the estimates of the coma brightness. Some of the estimates are as follows: E. Weiss (Vienna, Austria) estimated a magnitude of 6 on the 1st and 2nd, G. J. Chambers (Kensington, London, England) estimated it as 6 on the 3rd, F. A. T. Winnecke (Pulkovo Observatory, Russia) said it was fainter than 24 Camelopardalis (magnitude 6.05 – SC2000.0) on the 5th, slightly fainter than that star on the 6th, and about equal to or slightly brighter than that star on the 7th, Chambers gave it as 4.5 on the 15th and 3 on the 22nd, J. Hartnup (Liverpool, England) gave it as 5 on the 23rd, A. Reslhuber (Kemsmünster, Austria) said it was about 2 or 3 on the 26th.

Only Winnecke and J. F. J. Schmidt (Athens, Greece) provided magnitude estimates of the nucleus. Winnecke gave estimates of 9–10 on the 4th, 10 on the 7th, 10 on the 14th, 10 on the 20th, 10 on the 24th, and 9 on the 27th. Schmidt gave estimates of 7–8 on the 13th, 8 on the 14th, 8–9 on the 15th and 17th, 8 on the 18th and 19th, 7–8 on the 20th and 21st, 9 on the 22nd, 7–8 on the 23rd, 8 on the 24th, 11 on the 25th, 7–8 on the 26th and 27th, 10–11 on the 28th, 6–7 on the 29th, 8 on the 30th, and 11 on the 31st. Schmidt added that these magnitude estimates indicated that, on the average, maximum light occurred every 2.691±0.269 days, while minimum light occurred every 2.711 ± 0.284 days. Both Schmidt and Winnecke, as well as other observers, reported the nucleus remained well defined and nearly stellar throughout the month. This made another feature very noticeable. Chambers observed with his 8-cm refractor on the 3rd and noted a "jet or fan of light" extended from the nucleus toward the sun. On the 18th, Chambers wrote, "The luminous sector or fan spreading from the nucleus, on the side nearest the sun, is very distinct tonight." That same night, G. Knott (Woodcroft Observatory, Cuckfield, England) saw "a curious ray or jet proceed from [the nucleus], having a position-angle of 280° or 285°." Reslhuber also noted a "bushy extension" from the nucleus on the 18th. Chambers said the "angular size of the fan has diminished" on the 22nd, despite the comet's steadily increasing brightness. On the 24th, Chambers wrote, "The sector has largely increased in amplitude; it now covers fully 120° of a circle." Knott said the jet was curved on the 25th. Chambers wrote on the 27th, "the jet is smaller but in line with the tail which seems to indicate the existence of some motion of rotation." Bruhns also noted a "fan-shaped sector" on the 27th and said it was brightest near the nucleus. Bruhns said the sector was indistinct on the 28th, but easier to see on the 29th. Knott saw an "extraordinary change in the form of the jet" on the 29th and gave the position angle as 279.0°. Knott said the jet was back to its usual form on the 30th and gave the position angle as 225.6°. Schmidt commented that the jets seen emanating from the nucleus would vary their position angle by as much as 100° in three days. Interestingly, although astronomers were interested in what was happening inside the coma, few were interested in the coma itself, but it underwent a large change as a result of the close approach to Earth at the end of the month. Actual descriptions of the coma's shape are few, with Weiss describing it as fan-shaped on the 1st and 2nd, Chambers saying it was globular on the 3rd, and Reslhuber describing it as irregularly shaped on the 21st and 28th. In addition, Chambers remarked that the coma became more strongly condensed as the month progressed. Measurements of the coma diameter were made by few astronomers. After F. Tietjen (Berlin, Germany) estimated it as 4′ across on the 1st, and Bruhns said it was 1.5–2′ across on the 2nd, Schmidt provided estimates through the remainder of the month. He gave the diameter as 16′ on the 14th, 12′ on the 15th, 14′ on the 17th, 16′ on the 18th, 19′ on the 19th, 21′ on the 20th, 22′ on the 21st, 24′ on the 22nd, 27′ on the 23rd, 29′ on the 24th, 33′ on the 25th, 31′ on the 26th, 34′ on the 27th and 28th, 30′ on the 29th and 30th, and 29′ on the 31st. Certainly the most impressive feature of the month was the tail. It was a telescopic object as the month began, with Weiss estimating it as 0.5° long on the 1st and 2nd, and Bruhns noting it was 1° long on the 1st and 1.5° long on the 2nd. Tietjen said the tail extended toward PA 316.9° on the 1st and 311.5° on the 2nd. Chambers said the tail was 2° long on the 3rd, while Tietjen said the tail extended 1.5° toward PA 310.9° on the 4th. Schiaparelli estimated the tail length as 5° or 6° on the 8th. Although many other tail length estimates followed, Schmidt provided the most consistent series, while Tietjen continued to measure the position angle. Their combined numbers were as follows: 1.0° in PA 7.5° on the 16th, 1.5° in PA 21.7° on the 17th, 8.5° in PA 49.2° on the 20th, 8.5° in PA 64.4° on the 21st, 11.0° in PA 69.7° on the 23rd, 16.0° in PA 73.5° on the 24th, 15.0° in PA 81.4° on the 25th, 23.0° in PA 79.7° on the 26th, 20.0° in PA 78.9° on the 28th, and 21.5° in PA 85.8° on the 29th. Tietjen did not provide measurements of the position angle during the last two days of the month, but Schmidt reported rapidly decreasing tail lengths of 17° on the 30th and 9° on the 31st. Schmidt's length estimates were reflected in the estimates of others, especially Knott, who's maximum tail length of 20° came on the 27th. Bruhns reported the tail was only 1.5° wide at the end on the 30th. Many descriptions of the tail were also published. Chambers reported two tails on the 18th, noting the right-hand branch was bright and straight, while the left-hand branch was fainter and ill defined. Reslhuber described it as fan-shaped on the 21st. Chambers wrote on the 22nd, "The tail has a curious aspect; its bifidity differs from any thing I ever saw before either naturally or pictorially; a long narrow ray stretches from the nucleus for 5° or 6° and on the Eastern side there is another and much fainter strip of luminous matter, slightly inclined to the former but not reaching more than 2° from the nucleus." During the next few days, Chambers reported the two tails changed places. After noting they had "nearly coalesced" on the 25th, he wrote on the 27th, "The two branches of the tail have partly changed places; whereas the long ray was on the right (E.) a few nights ago, forming a well defined margin, with the faint branch on the left, there is now nebulous matter on both sides of the main ray, but as the original faint branch is narrower it would seem that the long ray had shifted its position a little." For several days near the end of August, A. Murmann (Vienna, Austria) detected some polarization in the light of the comet.

The comet had traveled far enough southward to enable observations in the Southern Hemisphere as September began. In fact, J. Tebbutt, Jr (Windsor, New South Wales, Australia) independently discovered this comet on the 1st, not yet having received a notification of its prior discovery. With an 8-cm telescope, he noted the comet's nucleus was badly defined and did not admit accurate determinations of position. The comet continued to be visible to the naked eye through most of the month, but only two apparent magnitude estimates were made of the coma. They came from T. Maclear (Royal Observatory, Cape of Good Hope, South Africa) on the 11th and 12th.

He estimated it was about magnitude 3 or 4 on the 11th and about magnitude 4 on the 12th. Some indications of the brightness come from the observations of F. Abbott (Hobart, Tasmania). He said the comet was distinctly visible to the naked eye on the 6th despite a 13-day-old moon. Abbott also noted the comet was just visible to the naked eye when pointed out on the 21st, but was a telescopic object on the 22nd. The only apparent estimate of the condensation brightness came from W. Chimmo (Canna Island, Hebrides, Scotland) on the 1st, when he called it the nucleus and said it was about magnitude 3 or 4 in his telescope. Magnitude estimations of the nucleus were made by Schmidt and Maclear, with Schmidt estimating magnitude 7 on the 1st and 9–10 on the 2nd, while Maclear said it was 9th-magnitude on the 9th. Chimmo said the nucleus was "surrounded by considerable nebulosity of an irregular oval, perhaps paraboloidal." Schmidt did provide a good series of coma diameter estimates until the comet had moved below his horizon. He gave the diameter as 12' on the 11th, 14' on the 12th and 13th, 12' on the 14th and 15th, 14' on the 16th, and 13' on the 17th. Only two other coma diameter estimates were made in September and that was 40' across on the 11th by Abbott and about 8' across by Maclear on the 12th. The lack of coma diameter estimates early in the month is probably because of moonlight. Maclear said the coma had become more diffuse on the 9th and Hartnup made a similar comment on the 10th. Maclear described the coma as globular on the 19th and noted, "the south side [is] brighter than the north." The prominent tail of August was much diminished by moonlight as September began, at least for everyone except Tuttle. Where Schmidt gave the tail length as 4.0° on the 1st and Chambers said it was 2° long on the 3rd, Tuttle said the comet's greatest brightness and greatest tail length came on the 3rd. He wrote, "I was able to trace the tail as far as thirty degrees from the nucleus. Beyond ten degrees from the nucleus the tail was so faint that it would escape the notice of one not accustomed to celestial observation." Other tail length estimates were 1.5° by Maclear on the 9th, 5° by Abbott on the 11th, 2° by Maclear on the 11th and 12th, and 2.0° by Schmidt on the 12th. Maclear added that the widest portion of the tail was 16' or 17'. Tietjen provided two additional measurements of the tail's position angle. He said it extended toward 89.5° on the 2nd and 91.6° on the 3rd. Chimmo said the comet's most attractive feature "was the well-defined limb of the south or underneath portion of the tail." Abbott described the tail as curved on the 11th and Maclear also noted it was "slightly concave towards the north" on that date. Maclear said the tail was not easily traced in the finder on the 22nd. Maclear provided an interesting description of the comet on the 11th. He said the view through a telescope revealed "the head somewhat resembles a cone of light enveloped in loose cometic matter, the brighter face towards the south; the nucleus at the apex of the cone is not so bright or distinct as it appeared to be on the 9th. The loose matter of the head seems to extend beyond the general outline of the north border of the tail."

The comet was too faint for everyone but the South African astronomers as October began. On the 2nd, Maclear said the comet was "very much reduced in size and brightness" as a result of moonlight. In addition, no tail was visible. The head was about 3' across, "and the light is more condensed towards its southern border." The comet's position was determined for the final time on October 27.81, when Mann measured it as $\alpha = 16^h\ 51.0^m$, $\delta = -52°\ 49'$. Maclear saw the comet on October 31.8, but no position could be determined because of the proximity to several faint stars, as well as the comet's extreme faintness because of moonlight. Maclear also said he searched in vain for the comet on November 6 when the moon was low.

The first parabolic orbit was calculated by Tuttle using positions from July 19, 22, and 26. The resulting perihelion date was 1862 August 22.22. Additional parabolic orbits were calculated by G. W. Hough, K. Hornstein, Bruhns, A. Secchi, J. R. Hind, G. V. Schiaparelli, R. Engelmann, J. Calandrelli, A. Hall, T. R. Oppölzer, H. Seeling, and Tebbutt, with the perihelion date generally falling within the range of August 23.5–24.0. By early September, it was realized that the difficulty in pinpointing the perihelion date was because the orbit was elliptical. The first such orbit was calculated by S. Stampfer, with a resulting period of 114 years. Additional periods were calculated by Winnecke (427 years), Oppölzer (122, 123, and 124 years) and T. H. Safford (142 years). The first definitive orbit was calculated by F. Hayn (1889). From positions spanning the entire period of visibility, he determined the period as 120 years, although he said it could lie anywhere between 117.4 and 121.9 years. These calculations were generally confirmed by B. G. Marsden and Z. Sekanina (1973), when they also determined the period as 120 years.

Interest in this comet was heightened a few years later when G. V. Schiaparelli (1867) compared its orbits with that determined for the Perseid meteor shower. Because of the striking agreement he concluded (according to the February 1867 issue of the *Monthly Notices of the Royal Astronomical Society*) that the comet "is nothing more than a very large meteor of the August system."

Although astronomers believed the orbit was well determined, W. T. Lynn (1902) discussed this comet and made an interesting comment at the end of his article. He wrote, "The period of the third comet of 1862 . . . is probably about 125 years in length; it is possible that the second comet of 1737 may be identical with it, but the orbit of the latter . . . is very uncertain." This comment was included in a paper by B. G. Marsden (1973) that tried to establish when the comet was likely to return by looking for previous apparitions. He said the 120-year period favored a comet seen by Wargentin early in February of 1750, although Kegler's comet of July 1737 had a more similar orbit. Following the comet's accidental rediscovery late in 1992, Marsden (1992) wrote that it "confirms the suggestion that Kegler's 1737 observations were indeed of P/Swift–Tuttle." A paper published by Marsden, *et al.* (1993) revealed the period for this apparition was 132 years and identified a problem with the South African positions of 1862 October. This orbit is given below.

T	ω	Ω (2000.0)	i	q	e
1862 Aug. 23.4229 (UT)	152.7737	139.3714	113.5664	0.962658	0.962798

ABSOLUTE MAGNITUDE: $H_{10} = 4.0$ (V1964)

FULL MOON: Jul. 11, Aug. 9, Sep. 8, Oct. 7, Nov. 6

SOURCES: G. V. Schiaparelli, H. K. F. K. Schjellerup, and H. Seeling, *AN*, **58** (1862 Aug. 4), pp. 77–80; K. C. Bruhns, Rosa, A. Pacinotti, and C. Toussaint, *AN*, **58** (1862 Aug. 8), pp. 93–6; H. P. Tuttle, K. Hornstein, J. R. Hind, and J. Chacornac, *CR*, **55** (1862 Aug. 11), pp. 291–3; K. Hornstein and H. K. F. K. Schjellerup, *AN*, **58** (1862 Aug. 14), pp. 109–12; H. P. Tuttle, A. Pacinotti, C. Toussaint. G. V. Schiaparelli, and R. Engelmann, *AN*, **58** (1862 Aug. 19), pp. 115–20; A. Secchi, *AN*, **58** (1862 Aug. 25), p. 143; T. Simons and G. W. Hough, *AJS* (Series 2), **34** (1862 Sep.), p. 294; A. Hall and S. Stampfer, *AN*, **58** (1862 Sep. 18), p. 203; T. R. von Oppölzer and J. F. J. Schmidt, *AN*, **58** (1862 Oct. 14), pp. 249–54; F. A. T. Winnecke and J. F. J. Schmidt, *AN*, **58** (1862 Oct. 24), pp. 261–70; K. Hornstein, C. B. Chalmers, and W. Chimmo, *MNRAS*, **22** (Supplemental Notice, 1862), p. 314–16; R. Main, J. Hartnup, G. Knott, and F. Abbott, *MNRAS*, **23** (1862 Nov.), pp. 28–32; T. H. Safford, G. J. Chambers, L. Swift, T. Simons, and G. W. Hough, *AN*, **59** (1862 Nov. 29), pp. 25–32; J. F. J. Schmidt, *AN*, **59** (1862 Dec. 9), pp. 33–42; T. R. von Oppölzer, *AN*, **59** (1862 Dec. 19), pp. 49–58; A. Secchi and J. Calandrelli, *AN*, **59** (1862 Dec. 27), pp. 71–6; T. Maclear, W. Mann, and G. Maclear, *MRAS*, **32** (1862–3), pp. 193–7; H. Romberg, J. Tebbutt, Jr, and N. M. Edmondson *MNRAS*, **23** (1863 Jan.), pp. 92–8; J. Tebbutt, Jr, *AN*, **59** (1863 Jan. 31), pp. 157–60; H. P. Tuttle, *AN*, **59** (1863 Feb. 21), pp. 187–90; A. Reslhuber, *AN*, **60** (1863 May 18), pp. 23–6; K. C. Bruhns, *AN*, **60** (1863 Jun. 18), pp. 85, 93; F. Tietjen, *AN*, **60** (1863 Aug. 24), pp. 257–62; A. Murmann, *AN*, **60** (1863 Oct. 2), pp. 365–7; F. A. T. Winnecke, *MASP*, **7**, no. 7 (1864), pp. 1–39; G. V. Schiaparelli, *MNRAS*, **27** (1867 Feb.), p. 134; G. V Schiaparelli and T. R. von Oppölzer, *AN*, **68** (1867 Feb. 20), pp. 331–4; T. R. von Oppölzer, *AN*, **69** (1867 May 1), pp. 81–8; G. V. Schiaparelli, *MNRAS*, **35** (1875 Feb.), pp. 222–3; F. Hayn, *AN*, **123** (1889 Nov. 2), p. 112; W. T. Lynn, *The Observatory*, **25** (1902 Aug.), pp. 304–5; L. Swift, *Rochester History*, **9** (1947 Jan.), p. 6; V1964, p. 59; HA1970, p. 87; B. G. Marsden and Z. Sekanina, *AJ*, **78** (1973 Sep.), pp. 654–8; B. G. Marsden, *IAUC*, No. 5620 (1992 Sep. 27); B. G. Marsden, G. V. Williams, G. W. Kronk, and W. G. Waddington, *Icarus*, **105** (1993 Oct.), pp. 420–6; SC2000.0, pp. 125, 299, 307, 376, 475.

C/1862 W1	*Discovered:* 1862 November 28.22 ($\Delta = 1.62$ AU, $r = 0.99$ AU, Elong. $= 35°$)
(Respighi)	*Last seen:* 1863 February 20.76 ($\Delta = 1.82$ AU, $r = 1.27$ AU, Elong. $= 42°$)
	Closest to the Earth: 1863 January 6 (0.6291 AU)
1862 IV	*Calculated path:* VIR (Disc), HYA (Dec. 11), CEN (Dec. 17), LUP (Dec. 24), CIR (Dec. 29), TrA (Jan. 1), APS (Jan. 3), OCT (Jan. 6), IND (Jan. 10), TUC (Jan. 12), PHE (Jan. 16), SCL (Jan. 25), CET (Feb. 7)

L. Respighi (Bologna, Italy) discovered this comet on 1862 November 28.22, at a position of $\alpha = 13^h 56.0^m$, $\delta = -11° 35'$. He described the comet as a centrally condensed nebulosity 3′ across, with a nucleus. The comet was independently discovered by K. C. Bruhns (Leipzig, Germany) on December 2.23. He said the comet was then in morning twilight and was rather bright, with a coma 1–1.5′ across. Bruhns' discovery was made one day after he discovered C/1862 X1.

Although the comet's position was measured frequently during the next three months, physical descriptions were virtually nonexistent. A. Secchi (Rome, Italy) described the comet as round and centrally condensed on December 6 and 10. Bruhns reported the comet was faint because of its low altitude on December 17. The comet reached its greatest southern declination of −79° on January 7. The comet was last detected on February 18 and 20, when Bruhns said it was found with great difficulty in evening twilight. He gave the position as $\alpha = 1^{\rm h}\,06.6^{\rm m}$, $\delta = -17°\,25'$ on February 20.76.

The first parabolic orbits were calculated by Bruhns and W. J. Förster, using positions obtained on December 2, 3, and 5. Bruhns determined the perihelion date as 1862 December 31.66, while Förster gave it as December 29.30. Using positions spanning the period of November 28 to December 10, Respighi determined the perihelion date as December 28.35. R. Engelmann took positions from December 2 to December 16 and calculated the perihelion date as December 28.65. After the comet was no longer visible, T. Krahl took positions from December 3 to February 19 and determined the perihelion date as December 28.67. This orbit is given below.

T	ω	Ω (2000.0)	i	q	e
1862 Dec. 28.6741 (UT)	230.5765	357.6950	137.5412	0.803238	1.0

ABSOLUTE MAGNITUDE: $H_{10} = 6.5$ (V1964)

FULL MOON: Nov. 6, Dec. 6, Jan. 5, Feb. 3, Mar. 5

SOURCES: K. C. Bruhns, *MNRAS*, **23** (1862 Dec.), pp. 77–8; K. C. Bruhns, *AN*, **59** (1862 Dec. 9), p. 47; L. Respighi, K. C. Bruhns, and W. J. Förster, *AN*, **59** (1862 Dec. 19), pp. 57–64; L. Respighi and A. Secchi, *AN*, **59** (1863 Jan. 6), pp. 91–6; R. Engelmann, *AN*, **59** (1863 Jan. 19), p. 127; K. C. Bruhns, *AN*, **60** (1863 Jun. 18), pp. 85–7, 93; T. Krahl, *AN*, **65** (1865 Jul. 24), pp. 59–62; V1964, p. 59.

C/1862 X1
(Bruhns)

1863 I

Discovered: 1862 December 1.15 ($\Delta = 1.05$ AU, $r = 1.41$ AU, Elong. $= 88°$)
Last seen: 1863 March 13.18 ($\Delta = 1.69$ AU, $r = 1.06$ AU, Elong. $= 36°$)
Closest to the Earth: 1863 January 1 (0.4462 AU)
Calculated path: SEX (Disc), LEO (Dec. 4), VIR (Dec. 15), COM (Dec. 21), BOO (Dec. 27), CVn-BOO (Dec. 29), CrB (Jan. 3), HER (Jan. 6), LYR (Jan. 13), CYG (Jan. 21), VUL (Jan. 30), DEL (Feb. 20), PEG (Feb. 26)

K. C. Bruhns (Leipzig, Germany) discovered this comet with a 6-foot focal length refractor on 1862 December 1.15. He determined the position as

$\alpha = 10^h 32.5^m$, $\delta = -3° 03'$ on December 1.17. Bruhns confirmed his find on December 2.17 and described the comet as a very faint and diffuse nebulosity about 2' across.

The only other physical descriptions obtained during December were from Bruhns. He described the comet as fairly bright on December 3, while he found it slightly brighter, but still difficult to observe on the 16th.

Moonlight interfered early in January, but numerous descriptions came forth during the second half of the month. E. Weiss (Vienna, Austria) said the comet appeared diffuse on the 20th, as it passed by a small star. Adolph (Königsberg, now Kaliningrad, Russia) described the comet as very faint, diffuse, and about 2' across on the 22nd. He added that the comet was difficult to observe at an altitude of about 7° and said no distinct nucleus was seen. Bruhns said the coma was about 1' across and contained a good nucleus on the 25th, while on the 26th he commented that it was not as well observed, but displayed a nucleus 11" across. On January 27, M. F. Karlinski (Krakow, Poland) observed with a refractor and saw a fine nucleus in the middle of the coma. Adolph said the comet appeared brighter but smaller on the 28th than on the 22nd. He added that the coma was centrally condensed. He said the comet was much fainter on the 29th than on the previous night. The comet had attained it most northerly declination of +40° on January 8.

The comet passed perihelion on February 3, at which time H. Romberg saw a faint, but distinct nucleus. Although the comet was moving away from both the sun and Earth thereafter, it showed signs of brightening as February progressed, perhaps indicating an outburst. Bruhns said the coma was over 1' across and contained a very bright nucleus on the 10th. Adolph also saw the comet that morning, but on the morning of the 11th he said the comet appeared fainter and more diffuse. Bruhns and Adolph found the comet rather bright and easy to see on the 14th. Bruhns also noted a sharply defined nucleus and a coma elongated away from the sun, with this longest axis measuring 1.25' across. Bruhns said the coma's elliptical shape was more pronounced on the 18th, and was 1.5' long and 1' wide. He added that the nucleus was at the focus of the ellipse. Weiss said the comet was "unexpectedly bright" on the 19th, and exhibited traces of a tail. On February 22, Adolph said, "The right side (in telescope) of the comet appeared sharply defined." There was also a broad, fan-shaped tail extending from the left side.

As March began, the comet's brightness had notably decreased. Adolph found the comet very faint and hard to see on the 1st, although he added it was involved in morning twilight. Bruhns saw the comet on the 2nd and said it was noticeably fainter than when last observed on February 22, and appeared diffuse and 1.5' across. The comet was last detected on March 13.18, when Bruhns observed with the 12-foot focal length telescope and determined the position as $\alpha = 21^h 20.7^m$, $\delta = +12° 31'$. He described it as hardly noticeable.

R. Engelmann calculated the first parabolic orbit. He used positions obtained by Bruhns on December 1, 2, and 3, and determined the perihelion date as 1863 February 2.17. Additional orbits during the following weeks and months by F. Tietjen, Engelmann, and Romberg established the perihelion date as February 3.99.

The first astronomer to investigate the orbit using positions spanning the entire period of visibility was Engelmann. He began with 67 positions and calculated two parabolic orbits and one elliptical one. The parabolic orbits were little different from those calculated by Engelmann and others during the comet's period of visibility. The elliptical orbit had a period of about 1.8 million years. A. von Flotow (1902) was the next investigator of this orbit. He took 120 positions, as well as perturbations by five planets, and computed parabolic and hyperbolic orbits. The hyperbolic orbit had an eccentricity of 1.000049. Flotow's hyperbolic orbit is given below.

T	ω	Ω (2000.0)	i	q	e
1863 Feb. 3.9902 (UT)	74.4657	118.8387	85.3553	0.794764	1.000049

ABSOLUTE MAGNITUDE: $H_{10} = 8.4$ (V1964)

FULL MOON: Nov. 6, Dec. 6, Jan. 5, Feb. 3, Mar. 5, Apr. 4

SOURCES: K. C. Bruhns, *MNRAS*, **23** (1862 Dec.), pp. 76–8; K. C. Bruhns, *AN*, **59** (1862 Dec. 9), p. 47; R. Engelmann, K. C. Bruhns, *AN*, **59** (1862 Dec. 19), p. 59; R. Engelmann, *AN*, **59** (1863 Jan. 6), p. 93; F. Tietjen, *AN*, **59** (1863 Jan. 10), p. 111; F. Tietjen, R. Engelmann, and H. Romberg, *AN*, **59** (1863 Feb. 21), pp. 181, 185–7; H. Romberg, *MNRAS*, **23** (1863 Apr. 10), pp. 198–9; *AJS* (Series 2), **35** (1863 May), p. 461; K. C. Bruhns, *AN*, **60** (1863 Jun. 18), pp. 85–7, 94; R. Engelmann, *AN*, **60** (1863 Jul. 14), pp. 145–52; A. von Flotow, *AN*, **160** (1902 Nov. 17), pp. 217–31; V1964, p. 59.

C/1863 G1
(Klinkerfues)

1863 II

Discovered: 1863 April 12.12 ($\Delta = 0.81$ AU, $r = 1.07$ AU, Elong. $= 72°$)
Last seen: 1863 November 15.1 ($\Delta = 3.18$ AU, $r = 3.31$ AU, Elong. $= 89°$)
Closest to the Earth: 1863 April 26 (0.6529 AU)
Calculated path: AQR (Disc), DEL (Apr. 14), VUL (Apr. 21), CYG (Apr. 24), DRA (May 5), UMi (May 15), CAM (May 22), UMi (May 23), CAM (May 24), DRA (May 29), UMa (Jun. 9), DRA (Nov. 7)

W. Klinkerfues (Göttingen, Germany) discovered this comet in Aquarius on 1863 April 12.12, at a position of $\alpha = 20^h 36.2^m$, $\delta = -2° 51'$. The comet was independently discovered by G. B. Donati (Florence, Italy) on April 15.12, and he noted it exhibited a tail about a degree long.

The comet had passed perihelion about a week before the discovery and would pass closest to Earth near the end of April. Physical descriptions were not plentiful. J. G. Galle (Breslau, now Wroclaw, Poland) saw it on the 16th and described it as a bright, but not well-defined nebulosity, exhibiting a granular structure. He added that there were traces of a tail. R. Hodgson

(Grove House, England) said the comet was about 4' across with a bright condensed nucleus on the 21st.

Hodgson said the comet was less bright on May 2 than on April 21. On May 4, Hodgson said the comet was "fast diminishing in brightness." J. Hartnup (Liverpool, England) noted the nucleus was stellar on the 5th and wrote, "Nebulosity pretty equally diffused." The comet reached its most northerly declination of $+80°$ on May 21. In a letter written on May 6, H. Romberg (Leyton, England) said the comet appeared as "a round nebula strongly condensed in the middle, and having a very faint nucleus." He added that it was easily seen in moonlight.

G. F. W. Rümker (Hamburg, Germany) saw the comet at the limit of visibility on August 10 and 12. H. L. d'Arrest (Copenhagen, Denmark) observed the comet with a refractor on August 16 and 19. He described it as round with a diameter of about 30". The comet was brighter in the middle and contained a nucleus equal to a star of about magnitude 11 or 12.

H. Schulz (Uppsala, Sweden) saw the comet with a refractor on September 2, 8, and 14. He described it as similar to a nebula of class 2 or 3 (referring to Herschel's classes of nebulae), but with a distinct nucleus visible from time to time. He tried to see the comet again on the 26th, but was prevented by haze and moonlight. D'Arrest said the comet was very well visible on September 4 and October 1, between which bad weather interfered. He said that, had moonlight not become a factor, the comet could possibly have been followed for another 2 or 3 weeks.

No further descriptions were provided, but the comet continued to be followed during the remainder of October by astronomers at Leipzig and Pulkovo. Using the 38-cm refractor, the Pulkovo astronomers were the only ones to follow the comet into November. It was last detected on November 15.1, when O. W. Struve indicated a position of $\alpha = 12^h\ 49.3^m$, $\delta = +66°\ 09'$.

The first parabolic orbit was calculated by R. Engelmann using positions from April 14, 16, and 18. He determined the perihelion date as April 5.41. This orbit proved to be fairly accurate, as shown by later parabolic orbits by F. Tietjen, Romberg, von Raschkoff, T. R. von Oppölzer, and J. Frischauf (1864). Frischauf's orbit was the only one to use positions spanning the entire period of visibility.

R. J. Buckley (1979) began his investigation of this comet's orbit with 55 positions obtained over a 6-month observational arc. He determined both a parabolic orbit and an elliptical orbit with perihelion dates of April 5.41 and April 5.39, respectively. The elliptical orbit had a period of about 21 069 years. Noting that the ellipse fitted the observations best, he removed six rather discordant positions and redetermined the elliptical orbit. The result was a perihelion date of April 5.40 and a period of about 45 257 years. This last elliptical orbit is given below.

T	ω	Ω (2000.0)	i	q	e
1863 Apr. 5.3961 (TT)	3.9669	253.1658	112.6210	1.068038	0.999159

ABSOLUTE MAGNITUDE: $H_{10} = 5$ (V1964)

FULL MOON: Apr. 4, May 3, Jun. 1, Jul. 1, Jul. 30, Aug. 28, Sep. 27, Oct. 26, Nov. 25

SOURCES: W. Klinkerfues, E. Weiss, R. Engelmann, H. Romberg, G. B. Donati, J. G. Galle, F. Tietjen, and R. Engelmann, *AN*, **59** (1863 Apr. 30), pp. 273–82; H. Romberg, *MNRAS*, **23** (1863 May), pp. 225–6; R. Hodgson, *MNRAS*, **23** (1863 May), p. 227; H. Romberg, *AN*, **60** (1863 Jun. 12), p. 71; J. Ferguson and A. Hall, *AN*, **60** (1863 Jun. 30), p. 123; M. Loewy, *AJS* (Series 2), **36** (1863 Jul.), p. 143; T. R. von Oppölzer, *AN*, **60** (1863 Aug. 13), pp. 227–32; H. L. d'Arrest, *AN*, **60** (1863 Aug. 29), p. 285; von Raschkoff, *AN*, **60** (1863 Sep. 24), p. 345; A. Reslhuber and G. Strasser, *AN*, **60** (1863 Oct. 9), pp. 379–82; H. L. d'Arrest, *AN*, **61** (1863 Dec. 15), p. 187; J. Frischauf, *SAWW*, **49** Abt. II (1864), pp. 345–50; H. Schultz, *AN*, **61** (1864 Feb. 4), pp. 327–30; G. F. W. Rümker, *AN*, **62** (1864 Apr. 19), p. 111; J. Frischauf, *AN*, **62** (1864 Aug. 4), pp. 341–4; J. Hartnup, *MNRAS*, **25** (1865 Jan. 13), p. 64; V1964, p. 59; R. J. Buckley, *JBAA*, **89** (1979 Oct.), pp. 583–4.

C/1863 G2 *Discovered:* 1863 April 13.07 ($\Delta = 0.86$ AU, $r = 0.65$ AU, Elong. $= 40°$)
(Respighi) *Last seen:* 1863 June 2.03 ($\Delta = 1.91$ AU, $r = 1.06$ AU, Elong. $= 23°$)
  Closest to the Earth: 1863 April 9 (0.8474 AU)
1863 III *Calculated path:* PEG (Disc), AND (Apr. 19), PER (May 6), AND (May 7), PER (May 12)

L. Respighi (Bologna, Italy) discovered this comet on 1863 April 13.07 at a position of $\alpha = 22^h 43^m$, $\delta = +18°$, and confirmed it was a comet on April 15.12. He said the comet presented a distinct nucleus similar in brightness to a star of magnitude 6, with a tail 40′ long. Independent discoveries were made by C. W. Baeker (Nauen, Germany) on April 14.1 and F. A. T. Winnecke (Pulkovo Observatory, Russia) on April 17.03. Winnecke described the comet as bright, with a 3° long tail.

Upon discovery this comet was moving almost due north and had actually been rising around the same time as the sun during the last half of March. Interestingly, the comet had passed just 0.7° from C/1863 G1 on March 17, at a time when they were situated at a declination of $-32°$, about 50° southwest of the sun. They were only above the Northern Hemisphere horizon during daylight, but they should have been easy Southern Hemisphere objects, being about 30° above the eastern horizon at the beginning of morning twilight.

The comet had already passed closest to Earth when discovered, but was still two weeks from perihelion. W. J. Förster (Berlin, Germany) noted it was visible to the naked eye on April 16, while a short tail was seen in a telescope. On April 17, E. W. L. Tempel (Marseille, France) observed the comet not far from 61 Pegasi. He said the comet was as bright as a star of magnitude 4 or 5, and the tail was very bright with a length of about 1°. H. Romberg (Leyton, England) said, "for the week following April 19 it was a very brilliant little object with low powers on the refractor." He also estimated it as about magnitude 5, with a very sharp nucleus. A. Reslhuber (Kremsmünster, Austria) saw the comet on April 21 and estimated it was about magnitude 3, with a planet-like nucleus and a fan-like tail about a degree long. M. F. Karlinski (Krakow, Poland) said the comet was well seen on the 22nd and shone like a star of magnitude 4–5. Reslhuber said the nucleus was small, but brighter on the 23rd, while the tail was brighter and "beautifully parabolic" in form. On April 27, R. Hodgson (Grove House, England) said the comet was, "A beautiful object, very brilliant, and a miniature of Donati's Comet; decided stellar nucleus, and tail of at least a degree in length; visible to the naked eye."

The comet was moving away from both the sun and Earth during May and June. Reslhuber had experienced cloudy skies since April 23, but upon seeing the comet again on May 6 he noted it had considerably decreased in brightness. He added that the nucleus was small, but well seen, while only traces of the tail remained. Hodgson said the comet was much less brilliant on May 8 than on April 27, and was not visible to the naked eye. The comet reached its most northerly declination of +48° on May 12. Reslhuber saw the comet on the 9th and then noted it was considerably fainter on the 12th. C. H. F. Peters (Hamilton College Observatory, Clinton, New York, USA) described the comet as 10′–12′ across on the 18th, and said the tail was about 3° long. The comet was last detected on June 2.03 when Romberg gave the position as $\alpha = 4^h\,08.4^m$, $\delta = +44°\,27'$.

The first orbit was calculated by Respighi using positions from April 15, 17, and 19. The result was a parabolic orbit with a perihelion date of 1863 April 21.42. Slight refinements came from the later calculations of A. J. G. F. von Auwers, M. F. Karlinski, Romberg, J. A. H. Gyldén, J. Frischauf, and J. E. B. Valz, with the perihelion date being established as April 21.36. Several years passed before a definitive orbit was calculated. G. Ericsson (1888) began with 175 positions obtained during the period of April 15 to June 2 and ultimately calculated an elliptical orbit with a perihelion date of April 21.36 and a period of about 17 740 years. This orbit is given below.

T	ω	Ω (2000.0)	i	q	e
1863 Apr. 21.3647 (UT)	55.5926	252.0822	85.4962	0.628781	0.999076

ABSOLUTE MAGNITUDE: $H_{10} = 6.8$ (V1964)

FULL MOON: Apr. 4, May 3, Jun. 1, Jul. 1

SOURCES: L. Respighi, C. W. Baeker, F. A. T. Winnecke, and E. W. L. Tempel, AN, **59** (1863 Apr. 30), pp. 273–8; H. Romberg, R. Hodgson, and M. F. Karlinski, MNRAS, **23** (1863 May), pp. 225–7; L. Respighi, M. F. Karlinski, and A. J. G. F. von Auwers, AN, **60** (1863 May 14), pp. 5, 9–12, 15; H. Romberg, AN, **60** (1863 Jun. 12), p. 71; J. A. H. Gyldén and J. Frischauf, AN, **60** (1863 Jun. 26), pp. 109–12; J. E. B. Valz, AJS (Series 2), **36** (1863 Jul.), p. 14; C. H. F. Peters, AN, **60** (1863 Sep. 19), pp. 331–4; A. Reslhuber, AN, **61** (1863 Oct. 27), p. 21; H. Schultz, AN, **61** (1864 Feb. 4), pp. 327–30; H. Romberg, *Astronomical Observations Taken During the Years 1862–64, at the Private Observatory of Joseph Gurney Barclay, Esq, FRAS. Leyton, Essex.* Volume 1, London: Williams and Norgate (1865), pp. 75, 102; G. Ericsson, AN, **118** (1888 Mar. 13), pp. 353–60; V1964, p. 59.

C/1863 T1   *Discovered:* 1863 October 10.1 ($\Delta = 2.04$ AU, $r = 1.76$ AU, Elong. $= 59°$)

(Baeker)   *Last seen:* 1864 April 14.11 ($\Delta = 2.06$ AU, $r = 2.01$ AU, Elong. $= 72°$)

Closest to the Earth: 1863 December 2 (1.3021 AU)

1863 VI   *Calculated path:* LEO (Disc), LMi (Oct. 13), UMa (Oct. 28), CVn (Nov. 11), BOO (Nov. 27), HER (Dec. 9), LYR (Jan. 5), VUL (Jan. 24), SGE (Feb. 16), DEL (Mar. 9)

C. W. Baeker (Nauen, Germany) discovered this comet on 1863 October 10.1, at a position of $\alpha = 9^h 36^m$, $\delta = +29°$. He described it as nebulous. K. C. Bruhns (Leipzig, Germany) observed the comet on October 12 and described it as a diffuse nebulosity, with a not quite central condensation. E. W. L. Tempel (Marseille, France) independently discovered the comet on October 14.1.

Although this comet was approaching both the sun and Earth, its distances from these bodies were never closer than 1.3 AU, so there was little change in the comet's brightness. J. F. J. Schmidt (Athens, Greece) first found the comet on November 7 and said it was easily seen in the finder, with a coma 5′ across. Schmidt observed the comet with a 17-cm refractor on the 13th and noted a nucleus of magnitude 12, while the tail was 8′ long. He said the coma was 3.1′ across. On the 14th Schmidt indicated the coma was between 3.1′ and 3.8′ across. Schmidt saw the comet again on the 17th and said it was 4.8′ across and contained a nucleus of magnitude 11. Schmidt added that during November he occasionally noted a tail up to 30′ long. H. Schultz (Uppsala, Sweden) was hindered by bad weather following the comet's discovery announcement, but was finally able to see it for the first time on November 8. As he observed it throughout November, he noted the nucleus was never very small and initially appeared about magnitude 11, but was around 10 or 11 near the end of the month.

H. L. d'Arrest (Copenhagen, Denmark) saw the comet on December 2 and noted the nucleus did not appear brighter than magnitude 10. H. Romberg (Leyton, England) saw the comet on December 2 and 6. He described it as an oblong nebula, with a strong central condensation. He noted, "Its diameter is about 1′, and it shines now as a star of the 8th mag." Schultz saw the comet on several occasions during the period of December 6–29. He said the nucleus was always between magnitude 9 and 10, while the coma was rather faint, with vague edges, and between 2′ and 3′ across. C. N. A. Krueger (Helsingfors, now Helsinki, Finland) measured the tail's position angle on several occasions during December, giving values of 327.3° on the 3rd, 324.3° on the 9th, 339.5° on the 13th, 342.5° on the 28th, and 348.4° on the 31st.

Although the comet continued to be followed into 1864, physical descriptions were obviously not a priority to astronomers. Krueger continued to measure the tail's position angle and gave values of 345.7° on January 1, 347.0° on the 5th, and 353.0° on the 27th. Schmidt said the coma was about 3.1′ across on March 8. Interestingly, the comet passed only 3.4° from

C/1863 V1 on January 1. The comet was last detected on April 14.11, when R. Engelmann (Leipzig, Germany) gave the position as $\alpha = 20^h\ 33.2^m$, $\delta = +10°\ 47'$.

Romberg took positions obtained on October 12, 15, and 17 and computed a parabolic orbit with a perihelion date of 1863 December 28.21. Refinements came from d'Arrest, S. Stampfer, T. R. von Oppölzer, and Engelmann and the perihelion date was established as December 29.67. P. G. Rosén (1866) took positions spanning the entire period of visibility and determined a hyperbolic orbit with an eccentricity of 1.0006499. The hyperbolic orbit was confirmed by F. H. Julius (1867), with a resulting eccentricity of 1.0009055. Julius' orbit is below.

T	ω	Ω (2000.0)	i	q	e
1863 Dec. 29.6647 (UT)	78.1144	106.9360	83.3159	1.313149	1.000906

ABSOLUTE MAGNITUDE: $H_{10} = 4.2$ (V1964)
FULL MOON: Sep. 27, Oct. 26, Nov. 25, Dec. 25, Jan. 23, Feb. 22, Mar. 23, Apr. 22
SOURCES: C. W. Baeker and K. C. Bruhns, AN, **61** (1863 Oct. 21), p. 15; H. Romberg, MNRAS, **24** (1863 Nov.), pp. 29–30; H. Romberg, MNRAS, **24** (1863 Dec.), pp. 44–5; H. L. d'Arrest, AN, **61** (1863 Dec. 15), p. 189; S. Stampfer, T. R. von Oppölzer, and R. Engelmann, AN, **61** (1863 Dec. 18), pp. 201–6; H. Schultz and J. F. J. Schmidt, AN, **61** (1864 Feb. 4), pp. 325–30; H. Romberg, AJS (Series 2), **37** (1864 Mar.), p. 293; C. N. A. Krueger, AN, **62** (1864 Mar. 24), p. 41; J. F. J. Schmidt, AN, **62** (1864 Jul. 31), p. 327; P. G. Rosén, AN, **68** (1866 Nov. 30), pp. 145–60; P. G. Rosén and F. H. Julius, VJS, **2** (1867), pp. 181–7; F. H. Julius, AN, **69** (1867 Mar. 19), p. 6; V1964, p. 59.

C/1863 V1 *Discovered:* 1863 November 5.19 ($\Delta = 0.81$ AU, $r = 0.71$ AU, Elong. $= 45°$)
(Tempel) *Last seen:* 1864 February 10.14 ($\Delta = 2.04$ AU, $r = 1.80$ AU, Elong. $= 62°$)
Closest to the Earth: 1863 November 17 (0.7315 AU)
1863 IV *Calculated path:* CRT (Disc), VIR (Nov. 7), COM-BOO (Nov. 21), CrB (Dec. 3), HER (Dec. 14), LYR (Jan. 11)

E. W. L. Tempel (Marseille, France) discovered this comet on 1863 November 5.19 near θ Crateris after spending several hours testing a telescope by observing double stars and other objects. Noting that bright moonlight and morning twilight were beginning to interfere, he quickly plotted the comet's position and on November 5.21 estimated it as $\alpha = 11^h\ 32.9^m$, $\delta = -10°$. Tempel said the tail was 2° long and the nucleus had the brightness of a 4th-magnitude star. He confirmed the discovery on November 6.21.

The comet was well observed during November as it passed closest to both the sun and Earth. K. C. Bruhns (Leipzig, Germany) said the comet was visible to the naked eye near Venus on the 10th and said the cometseeker revealed a very bright nucleus and a tail about 1° long. J. F. J. Schmidt (Athens, Greece) independently discovered the comet on November 13.13. He said it was then visible to the naked eye and was about magnitude 4. The tail was ultimately measured as 4.0° long in the finder. He confirmed the discovery on November 14.11. The comet was then estimated as magnitude 4, with a coma 6.2' across and a tail 3.2° long. A nucleus of magnitude 7 or 8 was also seen. In the days following Schmidt's discovery, he said the finder showed tail lengths of 4.5° on the 15th, 7.2° on the 16th, 6.1° on the 18th, and 10.6° on the 23rd. In addition, he measured the coma diameter as 6.4' on the 18th. The comet's total magnitude was estimated as 4 on each date except for the 23rd, when it was either 4 or 5. An 8th-magnitude nucleus was seen on the 18th. On November 20, R. Engelmann (Leipzig, Germany) and G. F. W. Rümker (Hamburg, Germany) independently observed the comet. Engelmann said the sharp nucleus was 5"–6" across, while the tail was 3° long. Rümker estimated the nuclear magnitude as 3–4 and said the tail was 2.5–3° long in the finder. Rümker saw the comet again on the 21st and said the nucleus was magnitude 3, while the tail was 4–4.5° long.

The comet was moving away from the sun and Earth during the remainder of its apparition. Schmidt said the tail appeared only 2.0° long on November 26 and December 4, while the total magnitude was given as 5. Rümker estimated the nuclear magnitude as 3 on December 1, while moonlight reduced the tail length to 0.5°. H. L. d'Arrest (Copenhagen, Denmark) said a magnification of 100× revealed a nucleus of magnitude 6 on the 2nd, while a magnification of 356× showed the nucleus as a nebulous feature with a very strong central condensation. H. Romberg (Leyton, England) saw the comet on December 3 and said it "appeared round, about 2' in diameter. Its well-defined nucleus was as bright as a star of mag. 7. The tail appeared very faint, through clouds." C. N. A. Krueger (Helsingfors, now Helsinki, Finland) found the tail's position angle virtually unchanged during the period of December 3 to January 5, as it only extended into the range of 338.3–346.9°. The comet passed only 3.4° from C/1863 T1 on January 1. The comet was last detected on 1864 February 10.14, when Bruhns gave the position as $\alpha = 19^h 02.1^m$, $\delta = +36° 16'$.

The first parabolic orbit was calculated by Romberg using positions spanning the period of November 10–18. The resulting perihelion date was 1863 November 10.00. A few days later, Engelmann took positions from November 10 to November 20 and determined the perihelion date as November 9.99. Both of these orbits were excellent representations of the comet's motion, although refinements by G. B. Donati, J. Michez, T. R. von Oppölzer, Rümker, and S. Stampfer established the perihelion date as November 9.97.

A definitive orbit was calculated by A. Svedstrup (1887). He began with about 150 positions spanning the entire period of visibility, and calculated an elliptical orbit with a perihelion date of November 9.98 and a period of about 18 367 years. R. J. Buckley (1979) began his investigation of this comet's orbit with 76 positions and determined both parabolic and elliptical orbits. The elliptical orbit also had a period of about 22 063 years. Noting that the ellipse fitted the observations best, he removed 18 rather discordant positions and redetermined the elliptical orbit, which resulted in a period of about 15 819 years. This last elliptical orbit is given below.

T	ω	Ω (2000.0)	i	q	e
1863 Nov. 9.9718 (TT)	357.2129	99.3938	78.0817	0.706413	0.998879

ABSOLUTE MAGNITUDE: $H_{10} = 5.7$ (V1964)
FULL MOON: Oct. 26, Nov. 25, Dec. 25, Jan. 23, Feb. 22
SOURCES: E. W. L. Tempel and K. C. Bruhns, *AN*, **61** (1863 Nov. 13), p. 95; E. W. L. Tempel, *AN*, **61** (1863 Nov. 25), p. 125; H. Romberg, *MNRAS*, **24** (1863 Dec.), pp. 44–5; H. Romberg, R. Engelmann, and G. F. W. Rümker, *AN*, **61** (1863 Dec. 1), pp. 137–42; J. F. J. Schmidt, G. B. Donati, J. Michez, and T. R. von Oppölzer, *AN*, **61** (1863 Dec. 8), pp. 163–6, 171–4; H. L. d'Arrest and G. F. W. Rümker, *AN*, **61** (1863 Dec. 15), pp. 189–92; S. Stampfer, *AN*, **61** (1863 Dec. 18), pp. 201–4; T. R. von Oppölzer, *AN*, **61** (1864 Jan. 9), pp. 245–50; J. F. J. Schmidt, *AN*, **61** (1864 Feb. 4), pp. 323–6; H. Romberg, *AJS* (Series 2), **37** (1864 Mar.), pp. 292–3; C. N. A. Krueger, *AN*, **62** (1864 Mar. 24), p. 42; J. F. J. Schmidt, *AN*, **62** (1864 Jul. 31), pp. 327–30; J. Hartnup, *MNRAS*, **25** (1865 Jan.), p. 64; A. Svedstrup, *AN*, **117** (1887 Aug. 13), pp. 221–42; V1964, p. 59; R. J. Buckley, *JBAA*, **89** (1979 Oct.), pp. 584–5.

C/1863 Y1 *Discovered:* 1863 December 28.75 ($\Delta = 0.83$ AU, $r = 0.77$ AU, Elong. $= 49°$)
(Respighi) *Last seen:* 1864 March 1.85 ($\Delta = 0.79$ AU, $r = 1.41$ AU, Elong. $= 104°$)
  Closest to the Earth: 1864 January 31 (0.1826 AU)
1863 V *Calculated path:* HER (Disc), LYR (Dec. 28), CYG (Jan. 12), LAC (Jan. 25), CEP-LAC (Jan. 26), CAS (Jan. 27), AND (Jan. 30), PER-AND (Jan. 31), PER (Feb. 1), TAU (Feb. 3), ORI (Feb. 8), MON (Feb. 27)

L. Respighi (Bologna, Italy) discovered this comet on 1863 December 28.75 at a position of $\alpha = 18^h 49.4^m$, $\delta = +25° 58'$. He confirmed the discovery on December 29.23. The comet was independently discovered by C. W. Baeker (Nauen, Germany) on January 1. K. C. Bruhns (Leipzig, Germany) confirmed Baeker's find on January 4.25. He described it as a "rather bright, round nebulosity, with a short tail." Bruhns later noted the tail was only seen in a cometseeker operating with a magnification of 18×, while a 12-foot focal length refractor (192×) revealed a nebulous mass about 1′ across. He added that, in the seeker, the comet gave the "same impression as a star of 6 or 7 magnitude."

Additional independent discoveries were made by M. F. Karlinski (Krakow, Poland) on January 9.74 and J. C. Watson (Ann Arbor, Michigan, USA) on January 10.00. Karlinski said the coma was 3–4′ across and exhibited a nucleus of magnitude 8 and a tail about 30′ long. Watson described it as large and bright, with a tail about 1.5° long and a strong central condensation.

Bruhns said the cometseeker showed a tail about 0.5° long on January 10 and it was still visible on the 15th despite moonlight. C. N. A. Krueger

(Helsingfors, now Helsinki, Finland) said the tail extended toward PA 7.9° on the 20th, PA 40.8° on the 27th, PA 49.5° to 50.2° on the 28th, and PA 53.9° on the 29th. J. Hartnup (Liverpool, England) said the nucleus appeared "elongated and stellar" on January 28. He added, "Tail, as seen through the finder, about one degree long." Bruhns said the refractor was used on the 30th and revealed "the nucleus was not round," while a "sector" was visible in the direction of the tail. The tail was then extending toward PA 55°.

Bruhns saw the comet with the refractor on February 1 and said the tail was then extended toward PA 75°. The nucleus was sharp and the "sector" was still visible. The seeker revealed a tail length of about 1.5°. Karlinski described the comet as very faint on February 11.

The comet was last detected on March 1.85, when H. Romberg (Leyton, England) gave the position as $\alpha = 5^h 55.6^m$, $\delta = -10° 29'$. He placed this observation under the designation "1863 IV" instead of "1863 V". J. F. J. Schmidt (Athens, Greece) tried to find the comet on March 4, but was unsuccessful.

The first parabolic orbit was calculated by S. Stampfer using positions spanning the period of December 28 to January 6. He determined the perihelion date as 1863 December 29.14. As further positions came forth, refinements were published by F. Peters, R. Engelmann, E. Weiss, Watson, J. Michez, and Stampfer, so that the perihelion date was established as December 28.26.

Several astronomers explored the possibility that the comet was moving in an elliptical orbit. Although Weiss initially gave a parabolic orbit, he suggested a period of 53.3 years based on a similarity to the orbit of the comet of 1810. Watson added that the orbital elements "almost exactly resemble those of the comet of 1810, so that there can be very little doubt of the identity of the two comets." Early in February, Engelmann said the 53-year period was unlikely, while F. Tietjen claimed to have confirmed it. Later in February, Michez calculated an orbit with a period of 108 years. The question of whether the comet moved in an elliptical orbit or not had to await a dissertation by W. Valentiner (1869). He took positions spanning the entire period of observation, established nine Normal positions, and calculated a parabolic orbit with a perihelion date of December 28.26. This orbit is given below.

T	ω	Ω (2000.0)	i	q	e
1863 Dec. 28.2627 (UT)	115.6692	306.6287	64.4906	0.771492	1.0

ABSOLUTE MAGNITUDE: $H_{10} = 8.2$ (V1964)

FULL MOON: Dec. 25, Jan. 23, Feb. 22, Mar. 23

SOURCES: C. W. Baeker and L. Respighi, *AN*, **61** (1864 Jan. 9), p. 255; M. F. Karlinski, E. Weiss, and S. Stampfer, *AN*, **61** (1864 Jan. 16), pp. 283–6; S. Stampfer and F. Peters, *AN*, **61** (1864 Jan. 23), pp. 301–4; J. C. Watson, *AN*, **61** (1864 Feb. 4), p. 327; E. Weiss, *AN*, **61** (1864 Feb. 11), pp. 347–50; K. C. Bruhns, *AN*, **61** (1864 Feb. 18), pp. 355–64;

M. F. Karlinski, *AN*, **61** (1864 Feb. 23), p. 383; E. Weiss and J. C. Watson, *AJS* (Series 2), **37** (1864 Mar.), p. 293–4; L. Respighi, E. Weiss, and J. Michez, *MNRAS*, **24** (1864 Mar.), pp. 124–7; J. Michez, *AN*, **62** (1864 Mar. 8), p. 13; C. N. A. Krueger, *AN*, **62** (1864 Mar. 24), p. 41; J. F. J. Schmidt, *AN*, **62** (1864 Jul. 31), p. 329; H. Romberg, *Astronomical Observations Taken During the Years 1862–64, at the Private Observatory of Joseph Gurney Barclay, Esq, FRAS. Leyton, Essex.* Volume 1. London: Williams and Norgate (1865), pp. 76–7, 103; J. Hartnup, *MNRAS*, **25** (1865 Jan.), p. 64; W. Valentiner, *Determinatio orbitae cometae V, anni MDCCCLXIII: dissertatio inauguralis.* Dissertation: Berlin (1869), 24pp; V1964, p. 59.

C/1864 N1 *Discovered:* 1864 July 5.07 ($\Delta = 1.37$ AU, $r = 1.17$ AU, Elong. $= 56°$)
(Tempel) *Last seen:* 1864 October 5.01 ($\Delta = 2.06$ AU, $r = 1.26$ AU, Elong. $= 27°$)
Closest to the Earth: 1864 August 8 (0.0963 AU)
1864 II *Calculated path:* ARI (Disc), TAU (Jul. 27), AUR (Aug. 4), GEM (Aug. 6), CNC (Aug. 7), LEO (Aug. 8), VIR (Aug. 10), LIB (Aug. 29)

E. W. L. Tempel (Marseille, France) discovered this comet on 1864 July 5.07. The position was roughly compared with 54 Arietis, which resulted in $\alpha = 2^h 57^m$ and $\delta = +18° 28'$. He confirmed the discovery on July 6.09. Independent discoveries were made by L. Respighi (Bologna, Italy) on July 6 and by M. F. Karlinski (Krakow, Poland) on July 11. Karlinski described it as a telescopic comet of moderate brightness.

Confirmations came from several astronomers. G. V. Schiaparelli (Brera Observatory, Milan, Italy) checked out the report from Respighi and found the comet on July 9.08. He said it appeared round, with an ill-defined nucleus of magnitude 8 or 9. The diameter of the coma was given as 3′ or 4′, but no tail was seen. Tempel's announcement brought confirmations on July 10.02 by A. C. Petersen (Altona, Germany) and on July 10.06 by K. C. Bruhns and R. Engelmann (Leipzig, Germany). Petersen said the comet was seen in bright twilight, which made it appear very faint.

The comet was well observed during the remainder of July and physical descriptions were quite numerous during the last days of the month. G. F. W. Rümker (Hamburg, Germany) found the comet "surprisingly bright" on the 28th, despite the moon being only about 9° away. He said the coma was 5–6′ across. That same night, J. F. J. Schmidt (Athens, Greece) saw the comet with the naked eye and estimated the brightness as magnitude 4–5. He said the coma was 10.5′ across and the tail was 0.3° long. Bruhns saw the comet on July 29 and said it was rather bright with an off-center condensation. The coma was elongated away from the sun. Engelmann saw it on the 30th and said the coma was round and 5′ across, while the unsharp nucleus was somewhat below right of the coma's center in the 12-foot focal length refractor. Schmidt estimated the comet's naked-eye magnitude as 4 on the 31st. He gave a coma diameter of 14.3′ and a tail length of 1.1°.

As August began, the comet was heading for a very close approach to Earth. Engelmann and Schmidt saw the comet with the naked eye on the

1st, with Schmidt estimating its brightness as 3–4. Engelmann said the coma was 6' across, while Schmidt found it 18' in diameter. Although Engelmann did not see a tail, Schmidt said it was 2.22° long, perhaps indicating it was a rather faint feature. On the same night, Rümker noted the coma contained an inner coma with a distinct nucleus. Schmidt estimated the naked-eye brightness as 3–4 on August 2 and said the finder of his refractor revealed a coma 19' across and a tail 3.65° long. Schmidt said the naked-eye brightness was 3 on the 3rd, while the finder revealed a coma 27' across and a tail 8.1° long. Bruhns saw the comet on the 4th and said the coma was 10' across, with a diffuse nucleus eccentrically situated within it. Engelmann said a refractor revealed the comet as a large, diffuse, nebulous mass, with a diffuse nucleus on the 5th. He added that a tail 3° long was seen in the finder. Schmidt also saw the comet on the 5th and estimated the naked-eye magnitude as 3. He said the finder revealed a coma 27' across, which exhibited a central condensation of magnitude 10 and a tail 9.1° long. Bruhns saw the comet on the 6th and said the coma filled the entire 15' field of view of the refractor, while the tail was brighter than on previous nights. A magnification of 144× showed a sharp nucleus, while 288× revealed it was slightly diffuse, but with a jet directed away from the sun. That same night, Schmidt said the naked-eye magnitude was 2–3, while he also noted a central condensation of magnitude 9–10, a coma 32' across, and a tail 11.5° long. On the 7th, Schmidt estimated the naked-eye magnitude as 2–3, with the finder revealing a central condensation of magnitude 9, a coma diameter of 47', and a tail length of 14.0°. Schmidt added that he suspected that there was a very faint nebulous light protruding from the tip of the tail that extended about 3 times longer than the easily visible tail. The comet passed less than nine million miles from Earth on August 8 and moved from the morning into the evening sky. No observations appear to have been made on the 8th as the comet remained in daylight and twilight when above the horizon and was below the northern horizon at night. The comet reached its most northerly declination of +33° on August 7. It passed about 12° from the sun on August 8.

A very important event happened during early August, when G. B. Donati (Florence, Italy) became the first person to apply a prism to the light of the comet. He said, "The spectrum of this Comet resembles the spectra of the metals; in fact, the dark portions are broader than those which are more luminous, and we may say these spectra are composed of three bright lines." This was the first time the spectrum of a comet was observed and the three lines were later identified as the Swan bands of carbon. Interestingly, W. Huggins (London, England), who would become the most diligent observer of cometary spectra of the next two decades, made several unsuccessful attempts to observe this comet's spectra, but was always hampered by weather and the comet's position.

Schmidt found the comet on August 9.79 and estimated the naked-eye magnitude as 3, but he was unable to measure a position. Schmidt again saw the comet on the 10th and estimated the naked-eye magnitude as 2–3, while the finder revealed a tail 2.0° long. On the same date, H. P. Tuttle (US Ironclad *Catskill*, off the coast of Charleston, South Carolina, USA) looked at the comet through a telescope and noted a bright nucleus and a very faint tail. J. Tebbutt (Windsor, New South Wales, Australia) said the comet was first detected in Australia by E. Quaife (Windsor, New South Wales) on August 10.35. Schmidt estimated the naked-eye magnitude as 3–4 on the 11th and said the finder revealed a central condensation of magnitude 9. C. W. Moesta (Santiago, Chile) independently discovered the comet with the naked eye on August 11.99, although he correctly suspected it might have been the comet seen by northern observatories during July. He said its appearance was like that of a 2nd-magnitude star. The telescope revealed an unsharp nucleus with a diameter of about 30″. This was surrounded by a nebulous mass of about 1° diameter. Schmidt saw the comet on the 12th with the finder and estimated the magnitude of the central condensation as 8–9. On August 13, Schmidt estimated the naked-eye magnitude as 4, while the finder revealed a central condensation of magnitude 8. Schmidt said the naked-eye magnitude was 4 on the 14th. Engelmann saw the comet on the 15th and said the comet appeared as a diffuse, nebulous mass about 3′ across. The faint nucleus was estimated as magnitude 10. That same evening, Schmidt gave the naked-eye magnitude as 4. Schmidt gave the naked-eye magnitude as 4–5 on the 16th. W. Mann (Royal Observatory, Cape of Good Hope, South Africa) saw the comet with the 8.5-foot focal length refractor on the 18th and said it appeared as a circular nebulosity with a diameter of about 7′. There was a central condensation, but no nucleus or tail. Schmidt provided the majority of the physical descriptions during the remainder of August. He said the naked-eye brightness was magnitude 4–5 during the period spanning August 19–25, magnitude 5 on the 26th, 27th, and 30th, and magnitude 5–6 on the 28th and 29th, and 31st. He also obtained a series of coma diameters and tail lengths with the finder. He gave the coma diameter as 6.82′ on the 23rd, 6.74′ on the 24th, 6.22′ on the 25th, 5.40′ on the 26th, 5.26′ on the 27th, 4.78′ on the 29th, 4.80′ on the 30th, and 4.52′ on the 31st. He gave the tail length as 2.0° on the 21st, 2.5° on the 22nd, 2.0° on the 24th, 1.0° on the 27th, 29th, and 30th, and 0.75° on the 31st. The magnitude of the condensation was given as 9 on the 23rd, 24th, 25th, 26th, 29th, 8 on the 30th, and 8–9 on the 31st. From the Southern Hemisphere, Mann gave the coma diameter as 3′ on the 27th, while Moesta gave the tail length as 40′ on the 29th and 20′ on the 31st. Moesta added that the comet was hardly visible to the naked eye on the 29th.

The comet steadily faded during September as it moved away from both the sun and Earth. Schmidt once again provided the majority of the physical descriptions, although the comet was moving closer to the horizon and becoming more favorable for Southern Hemisphere observers. Schmidt gave the naked-eye brightness as magnitude 5–6 on the 1st and 6 on the 2nd. The last date was the final time he saw the comet with the naked eye. Schmidt used the finder to measure the coma diameter as 4.50′ on the 1st, 4.20′ on the 2nd, 4.02′ on the 3rd, and 4.18′ on the 4th. Schmidt said the finder revealed a tail length of 0.75° on the 1st and 0.7° on the 4th. He said the central condensation was magnitude 8–9 on the 1st, 9 on the 2nd and 3rd, and 8–9 on the 4th. Schmidt made his final precise measurement of the comet's position on September 4. Schmidt said the comet was a difficult object on the 12th and 13th because of its altitude of 5–10°. He added that the low altitude and moonlight prevented observations on the 14th, 15th, and 16th. Observations were made on September 17 and 18 by Schmidt. On the 18th he said the comet was about 2.5′ in diameter, with a central condensation of magnitude 11. Although positions were determined on September 19–22, he indicated that these were inferior to his earlier positions because of the comet's low altitude. Other physical descriptions made during September came from Moesta, Tebbutt, and Mann in the Southern Hemisphere. Moesta said the tail was 20′ long in a telescope on the 1st, but noted the tail was no longer visible on the 9th. Mann said the comet appeared faint in the bright moonlight of September 13, and described it as excessively faint in moonlight on the 14th, 16th, and 19th. Moesta also saw the comet in moonlight on the 13th and 14th. Tebbutt observed the comet with his 8.3-cm telescope on September 19, 20, and 25. Precise positions were obtained each evening, although he considered the last position as "not so good . . . owing to the faintness of the Comet." Mann last saw the comet on the 26th, but said clouds prevented measurement of the position. Tebbutt obtained a single position measurement on the 27th, which he said was "not reliable", and a "serious illness" thereafter prevented him from seeing the comet again.

The comet was last detected on October 5.01, when Moesta gave the position as $\alpha = 14^h\ 26.5^m$, $\delta = -15°\ 12'$. Because of the difficulty in observing the comet, he suggested the position was imperfect. The α was actually measured about 2 minutes before δ.

The first parabolic orbit was calculated by O. Lesser using positions from July 9, 12, and 14. The resulting perihelion date was 1864 August 12.34. Additional orbits by F. Tietjen, M. F. Karlinski, S. Stampfer, G. Celoria, Schiaparelli, A. Graham, Tebbutt, J. Frischauf, Moesta, and T. R. von Oppölzer (1869) eventually established the perihelion date at August 16.0.

J. Kowalczyk (1866) first suggested the comet was moving in an elliptical orbit. He gave the perihelion date as August 16.07 and the period as about 4754 years. A later revision by Kowalczyk (1870) took 206 positions spanning the entire period of visibility and determined a perihelion date of August 16.08 and a period of about 3933 years. This last orbit is given below.

T	ω	Ω (2000.0)	i	q	e
1864 Aug. 16.0766 (UT)	151.5939	97.6776	178.1768	0.909291	0.996351

ABSOLUTE MAGNITUDE: $H_{10} = 6.2$ (V1964)

FULL MOON: Jul. 19, Aug. 17, Sep. 15, Oct. 15

SOURCES: E. W. L. Tempel, A. C. Petersen, and C. N. A. Krueger, *AN*, **62** (1864 Jul. 14), p. 287; M. F. Karlinski, L. Respighi, G. V. Schiaparelli, and O. Lesser, *AN*, **62** (1864 Jul. 21), pp. 301–4; F. Tietjen, M. F. Karlinski, S. Stampfer, and G. Celoria, *AN*, **62** (1864 Aug. 4), pp. 349–52; G. Celoria, *AN*, **62** (1864 Aug. 13), p. 363; R. Engelmann and K. C. Bruhns, *AN*, **62** (1864 Aug. 28), pp. 381–4; A. Graham, *AN*, **63** (1864 Sep. 5), p. 31; J. F. J. Schmidt, *AN*, **63** (1864 Oct. 14), pp. 123–6; J. Frischauf, *AN*, **63** (1864 Oct. 21), p. 143; H. P. Tuttle, *AJS* (Series 2), **38** (1864 Nov.), p. 432; C. W. Moesta, *AN*, **63** (1864 Nov. 2), p. 157; H. Cacciatore, *AN*, **63** (1864 Nov. 9), p. 175; J. Tebbutt, Jr., *AN*, **63** (1864 Nov. 30), pp. 237–40; A. Graham, *MNRAS*, **24** (Supplemental Notice, 1864), pp. 221–2; J. Tebbutt, *MNRAS*, **25** (1864 Dec.), pp. 43–5; C. W. Moesta, *AN*, **63** (1865 Jan. 24), pp. 359–62; G. B. Donati, *MNRAS*, **25** (1865 Feb.), p. 114; G. F. W. Rümker, *AN*, **64** (1865 Feb. 19), p. 43; J. Tebbutt, *MNRAS*, **25** (1865 Apr.), p. 194; J. Frischauf and J. Kowalczyk, *AN*, **65** (1865 Sep. 2), pp. 145–52; W. Mann, *MRAS*, **34** (1866), pp. 45–50; W. Huggins, *PRSL*, **15** (1866 Jan. 18), p. 5; J. Kowalczyk, *AN*, **66** (1866 Mar. 31), pp. 261–4; J. F. J. Schmidt, *AN*, **72** (1868 August 11), pp. 75–80; T. R. von Oppölzer, *AN*, **73** (1869 Jan. 4), pp. 55–8; J. Kowalczyk, *AN*, **75** (1870 Jan. 12), pp. 161–4; V1964, p. 59.

C/1864 O1
(Donati–
Toussaint)

1864 III

Discovered: 1864 July 23.89 ($\Delta = 1.76$ AU, $r = 1.63$ AU, Elong. $= 65°$)
Last seen: 1865 February 25.04 ($\Delta = 2.87$ AU, $r = 2.33$ AU, Elong. $= 48°$)
Closest to the Earth: 1864 June 25 (1.6655 AU), 1864 December 3 (1.1311 AU)
Calculated path: COM (Disc), VIR (Aug. 4), CRV (Sep. 16), HYA (Oct. 7), CEN (Oct. 22), MUS (Nov. 18), CAR (Nov. 20), CHA (Nov. 25), MEN (Dec. 1), HYI (Dec. 8), HOR (Dec. 14), HYI (Dec. 15), ERI (Dec. 20), PHE (Dec. 25), SCL (Jan. 4), CET (Jan. 22)

G. B. Donati and C. Toussaint (Florence, Italy) discovered this comet on 1864 July 23.89 and estimated a position of $\alpha = 13^h 06^m$, $\delta = +21° 15'$. It was described as very faint with a diameter of about 2′. Donati confirmed the find on July 27.89.

The comet's perihelion was still three months away, but the comet would never approach close enough to Earth to become a bright object. G. V. Schiaparelli (Brera Observatory, Milan, Italy) observed the comet on July 30 and said it was very difficult to see because of the proximity of the twilight. Donati said the comet exhibited a central luminous point and a tail 15′ long on August 3. R. Engelmann (Leipzig, Germany) saw the comet with a 12-foot focal length refractor on August 4 and noted a star-like nucleus, but added that the comet appeared small and faint because of an unclear sky. E. Weiss (Vienna, Austria) described the comet as faint, with a stellar nucleus on the 5th. T. R. von Oppölzer (Vienna, Austria) saw the comet on the 6th and wrote, "The comet is not very faint, resembles a nebula of the first class of moderate brightness and exhibits a good central condensation." The statement "nebula of the first class" refers to terminology used in W. Herschel's eighteenth century catalog of nebulae. Oppölzer's description indicates a likely brightness around magnitude 9. Engelmann said the comet was seen with much trouble on the 13th because of an altitude of only 5°. Engelmann last saw the comet on the August 15.8, but the low altitude prevented a position measurement.

The comet's motion took it into the skies of the Southern Hemisphere, where it was recovered on August 30.98 by C. W. Moesta (Santiago, Chile), who was using a refractor and low magnification. He confirmed the recovery on August 31.99. He noted a tail 20′ long. Moesta saw the comet on September 8 and noted an indefinite nebulous mass with a weak condensation, but no trace of tail. The comet passed less than 16° from the sun on September 24. Although Moesta continued to observe the comet during October, November, and December, he did not provide further physical descriptions. The comet reached its most southerly declination of −82° on December 3 and then turned northward.

The comet reappeared in the skies of the Northern Hemisphere during January of 1865. J. F. J. Schmidt (Athens, Greece) recovered it on January 19.74 and described it as very faint, with a curved tail, and a sharp nucleus of magnitude 11. He indicated a coma diameter of 66″, while the tail was 7′ long. Schmidt indicated the coma was 122″ across on the 21st and 116″ across on the 22nd. On the former date he also noted a nucleus of magnitude 11–12. Schmidt said the nucleus was still magnitude 11–12 on the 26th, but was magnitude 12 on the 27th. The tail on the latter date was 5.5′ long. Schmidt failed to see the comet in very clear skies on the 29th and attributed this to the comet's position within the zodiacal light. Schmidt saw the comet for the final time on the 30th. Engelmann said he believed he saw this comet on February 13, but noted it was too weak to measure.

The comet was last detected on February 25.04, by C. H. F. Peters (Clinton, New York, USA). He gave the position as $\alpha = 1^h 49.6^m$, $\delta = -9° 40′$ and noted "it seemed as if the comet had a very small stellar nucleus." Interestingly, Peters doubted the observation a few days later, when he reobserved the area and saw a "possible object" in the position measured for the comet on the 25th. The matter seemed closed until early 1881, when Peters mapped all of the nebulous objects in the area and found that none were within 5′ of the comet's position on 1865 February 25. He then took an orbit determined by E. von Asten and calculated where the comet would have been at the exact moment of his observation. The difference between the calculated and observed positions was $\Delta\alpha = -0.5^s$, $\Delta\delta = +0.2′$. In the 1865 May 27 issue of the *Astronomische Nachrichten*, Schmidt said he used Engelmann's ephemeris but could not find the comet on February 20 and March 20.

The first parabolic orbit was calculated by C. N. A. Krueger using positions obtained on July 28, 31, and August 2. He determined the perihelion date as 1864 October 11.82. Similar orbits were calculated by Engelmann, Asten, Donati, G. Celoria, Toussaint, F. Tietjen, and Oppölzer, which established the perihelion date as October 11.90.

Definitive orbits were calculated by Asten (1866) and J. F. Schröter (1906) which indicated the comet was moving in an elliptical orbit. Both astronomers used positions spanning the entire period of observation, with Asten determining a perihelion date of October 11.90 and a period of about 2.8 million years, and Schröter determining a perihelion date of October 11.89 and a period of about 55 242 years. Schröter's orbit is given below.

T	ω	Ω (2000.0)	i	q	e
1864 Oct. 11.8921 (UT)	232.4593	33.6659	109.7123	0.931212	0.999358

ABSOLUTE MAGNITUDE: $H_{10} = 5.2$ (V1964)

FULL MOON: Jul. 19, Aug. 17, Sep. 15, Oct. 15, Nov. 13, Dec. 13, Jan. 11, Feb. 10, Mar. 12

SOURCES: G. B. Donati, C. Toussaint, G. V. Schiaparelli, T. R. von Oppölzer, C. N. A. Krueger, and E. Weiss, *AN*, **62** (1864 Aug. 13), pp. 363–8; G. Celoria, G. B. Donati, F. Tietjen, T. R. von Oppölzer, and R. Engelmann, *AN*, **62** (1864 Aug. 28), pp. 375–84; C. N. A. Krueger, *MNRAS*, **24** (Supplemental Notice, 1864), pp. 221–2; C. Toussaint, *AN*, **63** (1864 Oct. 4), p. 95; T. R. von Oppölzer and E. von Asten, *AN*, **63** (1864 Dec. 8), pp. 249–56; E. von Asten, *AN*, **63** (1865 Jan. 7), pp. 323–6; C. W. Moesta, *AN*, **63** (1865 Jan. 24), pp. 359–62; J. F. J. Schmidt, *AN*, **64** (1865 Feb. 24), p. 63; R. Engelmann, *AN*, **64** (1865 Feb. 24), p. 63; J. F. J. Schmidt, *AN*, **64** (1865 Apr. 10), pp. 157–60; R. Engelmann, *AN*, **64** (1865 May 19), p. 248; J. F. J. Schmidt, *AN*, **64** (1865 May 27), p. 272; E. von Asten, *AN*, **66** (1866 Jan. 20), pp. 121–4; C. H. F. Peters, *AN*, **99** (1881 Mar. 12), p. 203; J. F. Schröter, *AJB*, **7** (1906), pp. 149–50, 176–7; V1964, p. 59.

C/1864 R1
(Donati)

1864 I

Discovered: 1864 September 10.13 ($\Delta = 1.79$ AU, $r = 1.09$ AU, Elong. = 32°)
Last seen: 1864 October 20.9 ($\Delta = 1.59$ AU, $r = 1.71$ AU, Elong. = 79°)
Closest to the Earth: 1864 June 21 (0.7749 AU)
Calculated path: LMi (Disc), UMa (Sep. 18)

G. B. Donati (Florence, Italy) discovered this comet on 1864 September 10.13 at a position of $\alpha = 10^h\ 23.3^m$, $\delta = +35°\ 21'$. He described it as very faint. Donati reobserved the comet on September 11.12 and determined it was moving northward.

This comet passed perihelion about a month and a half before it was discovered, but was poorly placed for observing because of its nearness to the sun. In fact, it passed just 10° from the sun on August 13. More interesting was the fact that the comet had passed closest to Earth on June 21 and was not seen by anyone around that time. It was then south of the sun and could have been visible to observers in the Southern Hemisphere as it was then near the limit of naked-eye visibility, but no observations were made.

The comet remained a faint object as it moved away from the sun and Earth. Consequently, few physical descriptions were made. G. V. Schiaparelli (Milan, Italy) said the comet was hardly visible on September 14 and its position was measured with extreme difficulty. R. Engelmann (Leipzig, Germany) saw the comet on September 15 and said it appeared round, about 1.5′ across, but without a distinct condensation. A. Reslhuber

(Kremsmünster, Austria) observed the comet on September 28, October 1, 4, and 9, and said the comet was so faint it was difficult to measure its position.

The comet's fading light made measurements more and more difficult as October progressed. The position was measured for the final time on October 11.05, when Engelmann gave it as $\alpha = 10^h 48.1^m$, $\delta = +54° 03'$. He noted that the observing conditions were not the best, and the comet was extremely faint and difficult to see. Reslhuber was not able to look for the comet for over a week after his October 9 observation because of clouds and moonlight. He recorded the final observations of this comet on October 19 and 20, although he noted it was not then seen with certainty in the evening sky. No positions were measured and no times were given as to when the comet was seen. Since the comet was circumpolar, the Author has guessed that the comet was seen on October 20.9.

The first parabolic orbit was calculated by Donati, using positions obtained on September 10, 11, and 12. He determined the perihelion date as 1864 July 30.17. Additional orbits were calculated by G. Celoria, W. Valentiner, J. Frischauf (1866), and J. Kowalczyk (1869), which established the perihelion date as July 28.31. Kowalczyk's orbit is given below and indicates the comet reached its most northerly declination of +88.9° on November 19.

T	ω	Ω (2000.0)	i	q	e
1864 Jul. 28.3117 (UT)	346.0948	176.8804	134.9821	0.626106	1.0

ABSOLUTE MAGNITUDE: $H_{10} = 7.0$ (V1964)

FULL MOON: Aug. 17, Sep. 15, Oct. 15, Nov. 13

SOURCES: G. B. Donati, R. Engelmann, G. V. Schiaparelli, and G. Celoria, AN, **63** (1864 Sep. 27), pp. 77, 79; G. B. Donati, AN, **63** (1864 Oct. 4), p. 95; G. B. Donati, AN, **63** (1864 Oct. 14), p. 127; W. Valentiner, AN, **63** (1864 Oct. 21), p. 143; W. Valentiner, MNRAS, **25** (1864 Nov.), p. 31; R. Engelmann, AN, **63** (1864 Nov. 2), p. 159; A. Reslhuber, AN, **63** (1864 Nov. 9), p. 175; J. Frischauf, AN, **68** (1866 Nov. 12), p. 111; J. Kowalczyk, AN, **73** (1869 Jan. 16), pp. 81–4; V1964, p. 59.

C/1864 X1   *Discovered:* 1864 December 15.71 ($\Delta = 1.50$ AU, $r = 0.78$ AU, Elong. $= 28°$)
(Baeker)   *Last seen:* 1865 February 25.76 ($\Delta = 2.06$ AU, $r = 1.41$ AU, Elong. $= 38°$)
    Closest to the Earth: 1865 January 1 (1.4245 AU)
1864 IV   *Calculated path:* AQL (Disc), AQR (Jan. 1), PSC (Feb. 5), CET (Feb. 5)

C. W. Baeker (Nauen, Germany) discovered this comet on 1864 December 15.71 at a position of $\alpha = 18^h 32^m$, $\delta = -1°$. It was then in the evening sky and low over the western horizon. He added that the brightness equaled a star of magnitude 7. F. Tietjen (Berlin, Germany) received a telegram announcing the discovery on the 16th, but bad weather prevented a confirmation until December 18.72.

E. Weiss (Vienna, Austria) said the comet was "quite bright" despite its low altitude on December 23 and he even noticed traces of a tail. He was uncertain whether a nucleus was present. On December 29, K. C. Bruhns (Leipzig, Germany) said the comet "was as bright as a star of magnitude 6." He added that the coma was 1' across and exhibited a "good nucleus." On December 30, Bruhns said the comet appeared brighter than on the previous evening and noted the beginnings of a tail. He said the 12-foot focal length telescope revealed a nucleus 15" across at a magnification of 144×.

H. L. d'Arrest (Copenhagen, Denmark) saw the comet on January 11, 13, 16, and 25, with some interference from twilight. He said the comet always appeared large, very bright, and round, with a strong central condensation. The beginnings of a short, bright tail were suspected. J. F. J. Schmidt (Athens, Greece) obtained precise positions of this comet on nine occasions during the period of 1865 January 7–26, inclusive. On the 22nd, his telescopic observation revealed a "new, somewhat brighter tail" had developed to the left of the normal tail with the angular separation being 115°.

Bruhns and Engelmann saw the comet on February 5, 6, and 7. They noted it was far weaker than around mid-January and was not easy to see. Schmidt obtained a precise position on February 20. He said the comet was still bright and distinct. A. Murmann (Vienna, Austria) last saw the comet in his 15-cm refractor on the 24th.

The comet was last detected on February 25.75 by F. Tischler (Königsberg, now Kaliningrad, Russia) and on February 25.76 by A. Reslhuber (Kremsmünster, Austria). Reslhuber gave the position as $\alpha = 1^h\ 07.2^m$, $\delta = -5°\ 34'$. Schmidt searched in vain for the comet on March 20. He said his ephemeris only went up to March 5.

The first parabolic orbit was calculated by C. F. W. Peters using positions obtained on December 18, 23, and 28. The resulting perihelion date was 1864 December 23.20. Additional orbits were calculated by Tietjen, Tischler, A. Hall, and J. Kowalczyk (1869), with the perihelion date eventually being narrowed down to December 22.95. Kowalczyk's orbit is given below.

Interestingly, C/1864 N1 (Tempel) passed only 0.5° from C/1864 X1 (Baeker) on 1864 August 11.9 – this was about four months before the discovery of C/1864 X1. Should C/1864 N1 have helped astronomers find C/1864 X1 at this time? Probably not. Based on the absolute magnitude value below, C/1864 X1 was then probably near magnitude 11 – certainly bright enough to be seen in telescopes of that time. But there were two factors against an earlier discovery. First, C/1864 N1 was still moving at a rate of about 4° per day, so the amount of time spent near C/1864 X1 was very limited. Second, the location low in the western sky would have allowed a dark-sky observing window of less than 50 minutes for observatories in Europe and North America.

T	ω	Ω (2000.0)	i	q	e
1864 Dec. 22.9510 (UT)	118.4521	205.1122	48.8623	0.770731	1.0

ABSOLUTE MAGNITUDE: $H_{10} = 5.2$ (V1964)

FULL MOON: Dec. 13, Jan. 11, Feb. 10, Mar. 12

SOURCES: F. Tietjen, C. W. Baeker, and E. Weiss, *AN*, **63** (1864 Dec. 31), pp. 299–304; C. F. W. Peters and K. C. Bruhns, *AN*, **63** (1865 Jan. 4), p. 319; F. Tietjen, *AN*, **63** (1865 Jan. 7), p. 335; F. Tischler, *AN*, **63** (1865 Jan. 15), p. 351; C. W. Baeker and C. F. W. Peters, *MNRAS*, **25** (1865 Feb.), p. 147; F. Tietjen, *AN*, **64** (1865 Feb. 4), p. 13; J. F. J. Schmidt, *AN*, **64** (1865 Feb. 24), p. 64; H. L. d'Arrest, *AN*, **64** (1865 Mar. 7), p. 73; A. Hall and H. L. d'Arrest, *AN*, **64** (1865 Mar. 29), pp. 121–4, 127; R. Engelmann and K. C. Bruhns, *AN*, **64** (1865 May 19), pp. 241–4, 247; J. F. J. Schmidt, *AN*, **64** (1865 May 27), p. 271; A. Reslhuber, *AN*, **65** (1865 Jul. 11), p. 31; A. Murmann, *AN*, **68** (1867 Jan. 30), p. 261; F. Tischler, *AN*, **69** (1867 Aug. 25), p. 357; J. Kowalczyk, *AN*, **73** (1869 Jan. 16), pp. 83–90; V1964, p. 59.

C/1864 Y1 *Discovered:* 1864 December 31.23 ($\Delta = 1.04$ AU, $r = 1.12$ AU, Elong. $= 67°$)
(Bruhns) *Last seen:* 1865 January 29.90 ($\Delta = 0.31$ AU, $r = 1.23$ AU, Elong. $= 137°$)
Closest to the Earth: 1865 January 26 (0.2916 AU)
1864 V *Calculated path:* VIR (Disc), CRV (Jan. 16), CRT (Jan. 20), HYA (Jan. 21), ANT (Jan. 25), PYX (Jan. 26), PUP (Jan. 27), CMa (Jan. 29)

K. C. Bruhns (Leipzig, Germany) discovered this comet on 1864 December 31.23 in the morning sky at a position of $\alpha = 14^h 01.9^m$, $\delta = -13° 06'$. He said the comet appeared as a diffuse nebulosity about 2′ across.

Although the comet had just passed perihelion, it approached Earth throughout most of January and should have brightened as it moved southwestward. Although several positions were obtained, there were few physical descriptions and the comet remained a faint object. F. Tietjen (Berlin, Germany) described the comet as very diffuse on January 4, but attributed this partly to the fact that ice had formed on the telescope's main objective. On January 22, J. F. J. Schmidt (Athens, Greece) wrote, "Under difficult circumstances I observed this very faint comet once..." Bruhns also saw the comet on January 22 and said a hazy sky made it extremely faint and difficult to see. R. Engelmann (Leipzig, Germany) described the comet as very faint on January 22 and 26. The comet was last detected on January 29.90, when Bruhns and Engelmann gave its position as $\alpha = 7^h 23.5^m$, $\delta = -19° 39'$.

The first parabolic orbit was calculated by Bruhns, using positions from December 31, January 2, and 3. He determined the perihelion date as 1864 December 27.83. Additional orbits were calculated by W. Valentiner, Engelmann, and Bruhns, and the perihelion date was eventually established as December 28.2.

F. Wesely (1909) calculated a definitive orbit using positions spanning the entire period of visibility. He determined a parabolic orbit with a perihelion date of December 28.22. He also gave a hyperbolic orbit with an eccentricity of 1.00006, but preferred the parabolic one and this is given below.

T	ω	Ω (2000.0)	i	q	e
1864 Dec. 28.2188 (UT)	178.4992	342.7789	162.8931	1.114634	1.0

ABSOLUTE MAGNITUDE: $H_{10} = 9$–10 (V1964)

FULL MOON: Dec. 13, Jan. 11, Feb. 10

SOURCES: K. C. Bruhns, *AN*, **63** (1865 Jan. 4), p. 319; F. Tietjen, *AN*, **63** (1865 Jan. 7), p. 335; K. C. Bruhns, W. Valentiner, and R. Engelmann, *AN*, **63** (1865 Jan. 24), p. 367; K. C. Bruhns, *MNRAS*, **25** (1865 Feb.), p. 147; R. Engelmann, *AN*, **64** (1865 Feb. 4), pp. 13–16; J. F. J. Schmidt, *AN*, **64** (1865 Feb. 24), p. 64; R. Engelmann and K. C. Bruhns, *AN*, **64** (1865 May 19), pp. 243, 248; J. F. J. Schmidt, *AN*, **64** (1865 May 27), p. 271; R. Engelmann and W. Valentiner, *AN*, **68** (1866 Nov. 19), p. 119; F. Wesely, *DAWW*, **84** (1909), pp. 641–654; V1964, p. 59.

C/1865 B1 (Great Southern Comet)

1865 I

Discovered: 1865 January 17.43 ($\Delta = 0.96$ AU, $r = 0.19$ AU, Elong. $= 11°$)

Last seen: 1865 May 2.78 ($\Delta = 2.55$ AU, $r = 2.47$ AU, Elong. $= 74°$)

Closest to the Earth: 1865 January 15 (0.9439 AU)

Calculated path: MIC (Disc), GRU (Jan. 25), PHE (Feb. 15), ERI (Mar. 15), HOR (Mar. 29), RET (Apr. 24)

F. Abbott (Hobart, Tasmania) discovered this comet on 1865 January 17.43. He said, "it was a distinct object low down in the south-west horizon, alternately disappearing and reappearing amongst dark cumuli clouds. It has a fine white nucleus, with a straight tail 10 or 12 degrees long." He later estimated the diameter of the nucleus as 14".

Several independent discoveries were made during the next couple of nights. C. W. Moesta (Santiago, Chile) said he first became aware of the comet on January 18 after receiving a report from Colina, Chile. Although no date and time were given for the Colina observation, it is probably safe to assume it occurred around January 18.0. R. L. J. Ellery (Melbourne, Australia) saw a "large Comet . . . in the south-west horizon" on January 18.46. He added that the tail was convex towards the south pole. A report also came from Green Point, South Africa, where the tail was seen in the evening sky on January 18.79, with Table Mountain blocking the head. C. C. Copsey (São João del Rei, Brazil) saw the comet on January 21 and said "my attention was called to a bright streak of light of some breadth in the S.W., near one of the stars of the constellation Toucan. Though it was visible only for a short time on account of the unfavourable state of the heavens and its small elevation above the horizon, I became convinced that it was a comet's tail; and this impression was verified by subsequent observation, and also by information I have received of its having been seen on that night from other parts of Brazil."

The comet had just passed perihelion and its closest distance to Earth when discovered, but remained a naked-eye object through the remainder of January. Although no total magnitude estimates were made, there were a couple of observations that indicate the comet's brightness, although these may refer more to the nucleus or the coma condensation. Moesta saw the comet with the naked eye in strong evening twilight on January 20 and said it became visible before any stars in that area of the sky were seen. Abbott saw the comet with the naked eye on the 22nd in twilight and between clouds. He said the comet was seen as soon as Fomalhaut (magnitude 1.16 – SC2000.0) and a few minutes before α and β Gruis (α is magnitude 1.74 and β is magnitude 2.11 – SC2000.0). In addition, Ellery noted the comet was "not nearly so bright" as C/1858 L1 (Donati). The only other magnitude estimate was made by Copsey. Although he said the estimate referred to the "nucleus", it probably referred to the condensation within the coma. He gave the magnitude as 3 on the 26th. He saw the nucleus at a low altitude on the 21st and said it was not round and exhibited an "effusion" from it, while on the 22nd he noted jets extending from the nucleus. Moesta saw the nucleus in hazy skies and noted the jets were not as strongly developed. Moesta said the nucleus was stellar on the 24th and even sharper on the 25th, with a diameter of 5.3". Moesta measured the diameter of the nucleus as 4.8" on the 26th. Abbott described the nucleus as "planetary" on the 28th. W. Mann (Royal Observatory, Cape of Good Hope, South Africa) saw the nucleus in an 8.5-foot focal length refractor on the 30th and said it was slightly elongated in a direction at right angles to the axis of the tail. The only estimate of the coma diameter came on January 31, when Moesta indicated it was 4' across. He described it as round and diffuse. The tail was well seen and, as is apparent from the independent discoveries above, was what initially caught the attention of observers. The length was given as 15–16° by Ellery on the 20th, 25° by Moesta on the 21st, 10–12° by Abbott on the 22nd, 7° by Ellery on the 24th, 26° by Copsey on the 26th, 14° by Abbott and 18° by Mann and Moesta on the 28th, 17° by Mann and 12° by Tebbutt on the 30th, and 17° by Mann and Moesta on the 31st. Abbott indicated the tail was pointing toward PA 130° on the 22nd. There were several good descriptions of the tail. Moesta found it straight, narrow, and conical on the 20th. Moesta said the tail was sharply bordered on the southern side and diffuse on the northern edge when seen on the 21st. He added that the width was 10' near the coma and 1.5° at the end. Moesta found the southern edge of the tail strongly curved on the 23rd and noted the lateral tail was broader than on the previous night. Moesta said the main tail was more strongly curved on the 24th and noticeably fainter on the 25th. Copsey said the tail was perfectly straight on the 26th and 28th, and gradually widened from the nucleus. He added that it was "well-defined up to its extremity." That same evening, Moesta said a telescope revealed the right-hand side of the tail was more strongly curved than the left-hand side and very narrow near the coma. Abbott said the tail was "bushy" on the 28th and slightly curved. Also on the 28th, Moesta said the tail was 1° wide at a distance of 12° from the coma. Mann said the tail was "almost perfectly straight" on the 30th and "presents the appearance of a long tapering brush." He added that no division was noticed, while both borders were equally well defined. That same night, J. Tebbutt, Jr (Windsor, New South Wales, Australia) said the tail was "slightly curved, being convex on the western side." In the May 1865 issue of the *Monthly Notices of the Royal Astronomical Society*, Ellery reported, "For several days after perihelion it presented a very imposing appearance, with a tail almost perfectly straight 150° long." This was an error as Ellery later gave the length as 15–16°.

Moonlight was interfering with observations as February began and most astronomers were not able to see the comet with the naked eye for several evenings after the 6th. Tebbutt said the comet was faintly visible to the naked eye on the 14th, just before moonrise, and reported naked-eye observations of the tail through the 22nd. The nuclear condensation was no longer the well-defined feature seen in January. Tebbutt said it was ill defined on February 8 and indistinct on the 9th. Abbott remarked that the "planetary appearance" of January was absent on February 14. Although Tebbutt noted the nucleus was more distinct on the 15th, he reported an indistinct condensation on the 22nd, 23rd, and 24th. Moonlight also hampered observations of the tail early in the month. The length was given as 5° by Tebbutt on the 3rd, 4° by Abbott and 4° or 5° by Tebbutt on the 4th, 1° by Abbott, 4° or 5° by Mann, and 2° or 3° by Tebbutt on the 14th, 3° or 4° by Tebbutt on the 15th, 10° by Tebbutt on the 17th, 1° 20' by Mann on the 21st, 2° by Tebbutt on the 23rd, 33' by Mann on the 26th, and 22' by Mann on the 28th. Tebbutt (1887) remarked that he sketched the tail on the 17th and believed the great length was due to "the withdrawal of the moon and an unusual clearness of the sky...." He indicated that the tail then extended toward PA 130°. Tebbutt noted the tail was "extremely faint to the naked eye" on the 22nd.

As with February, March began with the moon interfering with observations. Tebbutt said the comet was "excessively faint" with the moon at first quarter on the 4th, while Mann simply noted the comet was very faint in moonlight on the 5th. Mann's 8.5-foot focal length refractor was barely able to detect the comet on the 8th, while Tebbutt noted the comet was "scarcely distinguishable" on the 9th. The comet's normal fading was not helping observers and Tebbutt failed to find the comet *before moonrise* on the 14th, despite "beautifully clear" skies. Tebbutt did see it on the 16th and 17th and noted it was "excessively faint." Mann also saw the comet on the 16th and said it was "almost entirely invisible" after moonrise in the 8.5-foot focal length refractor. Mann described the comet as a faint nebulosity about 45" across, with a tail 3–4' long on the 17th. On March 18, Mann described the comet as "very faint and frequently obliterated by vapour," while Tebbutt used his 8.3-cm refractor and said, "Comet of the last degree of faintness, and could only be seen by looking obliquely into the telescope." On March 20, Mann said the comet's diameter was about 30", while the tail was faintly traced for about 4'. That same evening, Tebbutt said the comet was "seen only by oblique vision" in his 8.3-cm refractor. Tebbutt said the comet was "barely distinguishable" in the 8.3-cm refractor on the 28th despite very clear and dark skies. Mann saw the comet on the 29th in the 8.5-foot focal length refractor and described it as a "a very faint nebula" about 25–30" across, with a "very faint ray of light or tail" extending 2' or 3' in a direction opposite to that of the sun.

During the final month of visibility, Mann was the only observer to continue following the comet, thanks to an 8.5-foot focal length refractor. In moonlight on April 3 he said the comet was very faint and no tail was visible. Mann said the comet was extremely faint on the 21st. The comet was last detected on May 2.78, when he determined the position as $\alpha = 3^h 54.7^m$, $\delta = -56° 57'$. Mann noted the moon was then seven days old and "almost overpowers the comet; it is difficult at times to assure oneself of its presence in the field."

The first parabolic orbit was calculated by E. J. White using positions obtained on January 20, 22, and 24. He determined the perihelion date as 1865 January 14.73. Additional orbits were calculated by Moesta, Tebbutt, and F. Körber (1887), which eventually established the perihelion date as January 14.83. Körber's orbit is given below. Moesta had initially suggested the comet was a return of C/1843 D1, while Tebbutt noted rough similarities to C/1826 U1, as well as a comet seen in −371, for which A. G. Pingré (1783) had calculated a rough orbit. None of these suggestions proved possible.

T	ω	Ω (2000.0)	i	q	e
1865 Jan. 14.8253 (UT)	111.7175	254.8252	92.4945	0.025844	1.0

ABSOLUTE MAGNITUDE: $H_{10} = 3.8$ (V1964)

FULL MOON: Jan. 11, Feb. 10, Mar. 12, Apr. 11, May 10

SOURCES: A. G. Pingré, *Cometographie ou Traité Historique et Théorique des Cometes*, Paris (1783), Volume 1, pp. 259–63; C. C. Copsey, *MNRAS*, **25** (1865 Mar.), pp. 174–5; C. W. Moesta, *AN*, **64** (1865 Mar. 26), pp. 109–12; J. Tebbutt, Jr, R. L. J. Ellery, and F. Abbott, *MNRAS*, **25** (1865 Apr.), pp. 195–9; R. L. J. Ellery and E. J. White, *MNRAS*, **25** (1865 May), p. 219; R. L. J. Ellery, E. J. White, and J. Tebbutt, Jr, *AN*, **64** (1865 May 5), pp. 219–24; J. Tebbutt, Jr, *AN*, **64** (1865 May 27), pp. 269–72; J. Tebbutt, Jr, *MNRAS*, **25** (1865 Jun.), pp. 257–60; J. Tebbutt, Jr, *AN*, **65** (1865 Jul. 31), p. 79; J. Tebbutt, Jr, *MNRAS*, **25** (Supplemental Notice, 1865), p. 271; R. L. J. Ellery, *MNRAS*, **26** (1865 Nov.), pp. 30–2; J. Tebbutt, Jr, *MNRAS*, **26** (1866 Jan.), p. 84; T. Maclear and W. Mann, *MRAS*, **34** (1866), pp. 35–43; J. Tebbutt, Jr, R. L. J. Ellery, and F. Koerber, *AN*, **117** (1887 Oct. 11), p. 385; F. Körber, *Ueber den Cometen 1865 I*. Dissertation: Breslau (1887), 60pp; V1964, p. 59; *VA*, **35** (1992), pp. 324–5; SC2000.0, pp. 563, 576, 582.

2P/Encke *Recovered:* 1865 January 25.8 ($\Delta = 2.60$ AU, $r = 2.05$ AU, Elong. $= 46°$)

Last seen: 1865 July 23.8 ($\Delta = 0.44$ AU, $r = 1.22$ AU, Elong. $= 107°$)

1865 II *Closest to the Earth:* 1865 July 7 (0.2972 AU)

Calculated path: PSC (Rec), ARI (Apr. 10), TAU (May 6), ORI (May 29), GEM (May 30), ORI (May 31), GEM (Jun. 2), CMi (Jun. 12), MON (Jun. 22), HYA (Jun. 25), PYX (Jul. 1), HYA (Jul. 2), ANT (Jul. 3), CEN (Jul. 10), LUP (Jul. 21)

R. Farley took the orbit calculated by J. F. Encke for this comet's 1862 apparition, applied approximate perturbations by Venus to Saturn, and predicted a perihelion date of 1865 May 28.07. Farley and J. R. Hind independently calculated ephemerides and, although the comet was not favorably placed for observation prior to perihelion, it was recovered by H. L. d'Arrest (Copenhagen, Denmark) on 1865 January 25.8. D'Arrest was using Hind's ephemeris and found a "faint glow" very near the predicted position. The comet was then very near the western horizon. Bad weather followed and d'Arrest did not see the comet again during this apparition. An independent recovery was made by K. C. Bruhns and R. Engelmann (Leipzig, Germany) on February 13.77. Although they noted it was then "too weak for a precise measurement," they did give a position of $\alpha = 23^{\mathrm{h}}\,45.6^{\mathrm{m}}$, $\delta = +5° 19'$.

The comet steadily moved into evening twilight and was not observed again until after perihelion. J. F. J. Schmidt (Athens, Greece) said he attempted no observations at the beginning of 1865 because the ephemeris beginning with January did not arrive until the end of March. The comet passed about 9° from the sun on April 28 and passed perihelion on May 28.4.

Following its perihelion passage, the comet was first seen in Australia on June 24.37 by John Tebbutt, Jr (Windsor, New South Wales, Australia). He said, "I found it after a short search, but at a considerable distance from its calculated place. It was about two minutes in diameter, faint, and without the slightest condensation of light in the centre." He saw the comet again on the 29th. G. R. Smalley (Government Observatory, Sydney, Australia) first saw the comet on June 28, while using a 17.8-cm refractor. He obtained additional observations on June 29, 30, and July 1, and described the comet as "very faint throughout." Smalley noted that he never saw a nucleus, condensation, or tail. He summarized by saying the comet was "a mere faint patch of haze."

T. Maclear and W. Mann (Royal Observatory, Cape of Good Hope, South Africa) were using an 8.5-foot focal length refractor and had searched for the comet with the aid of an ephemeris on several occasions, but an unfavorable western sky had prevented success until June 24.73. On that date it was an easy object in the telescope's finder. It was described as about 2.5' in diameter and "appeared to be somewhat elongated in a direction parallel to the equator." The comet appeared extremely faint because of moonlight and haze on June 28, and it was seen with great difficulty because of the moonlight on the 29th.

Beginning on July 11, Maclear and Mann noted it became more difficult to determine the comet's position as it had "entered a region bordering the

Milky Way, where the sky is generally thickly studded with minute stars; and as the Comet presented no sensible condensation of light, stars of the 11th or 12th magnitude shining through the faint nebulous matter rendered it almost impossible to form a just estimate of the centre of the mass." On the 15th they said it was about 2.25′ in diameter and still visible in the finder. It was no longer visible in the finder on the 21st.

The comet was last detected on July 23.8, when Maclear noted that "micrometrical observations were quite impracticable, owing to the multitude of small stars in the field." Maclear made an attempt to find the comet on July 26, with the moon above the horizon, but no trace was found.

Minor refinements of this comet's orbit using positions from multiple apparitions, as well as planetary perturbations, have been published by numerous astronomers. Of most significance were the studies of E. Becker and F. E. von Asten (1868), Asten (1875, 1877) and B. G. Marsden and Z. Sekanina (1974). These have typically revealed a perihelion date of May 28.42 and a period of 3.30 years. Marsden and Sekanina determined the nongravitational terms as $A_1 = -0.78$ and $A_2 = -0.03170$. The orbit of Marsden and Sekanina is given below.

T	ω	Ω (2000.0)	i	q	e
1865 May 28.4221 (TT)	183.4907	336.4561	13.0810	0.340970	0.846316

ABSOLUTE MAGNITUDE: $H_{10} = 8 - 10$ (V1964)

FULL MOON: Jan. 11, Feb. 10, Mar. 12, Apr. 11, May 10, Jun. 9, Jul. 8, Aug. 7

SOURCES: R. Farley, *MNRAS*, **25** (1865 Jan.), pp. 64–5; H. L. d'Arrest, *AN*, **64** (1865 Mar. 7), p. 73; R. Engelmann and C. Bruhns, *AN*, **64** (1865 May 19), p. 248; J. F. J. Schmidt, *AN*, **64** (1865 May 27), p. 272; J. Tebbutt, Jr, *AN*, **65** (1865 Oct. 6), p. 237; J. Tebbutt, Jr, *MNRAS*, **26** (1865 Nov.), pp. 29–30; G. R. Smalley, *MNRAS*, **26** (1865 Dec.), p. 63; T. Maclear and W. Mann, *MRAS*, **35** (1867), pp. 17–20; E. Becker and F. E. von Asten, *AN*, **71** (1868 May 8), p. 179; F. E. von Asten, *BASP*, **20** (1875), p. 344; F. E. von Asten, *BASP*, **22** (1877), p. 562; V1964, p. 59; B. G. Marsden and Z. Sekanina, *AJ*, **79** (1974 Mar.), pp. 415–16.

4P/Faye   *Recovered:* 1865 August 23.01 ($\Delta = 1.36$ AU, $r = 2.34$ AU, Elong. $= 162°$)

    Last seen: 1866 March 17.77 ($\Delta = 2.25$ AU, $r = 1.71$ AU, Elong. $= 45°$)

1866 II   *Closest to the Earth:* 1865 September 20 (1.2649 AU)

    Calculated path: PEG (Rec), AQR (Sep. 21), PSC (Dec. 10), CET (Jan. 21), PSC (Jan. 27), CET (Feb. 28), ARI (Mar. 13)

D. M. A. Möller (1865) used positions obtained during the first three apparitions of this comet, determined 15 Normal positions for the period of 1843 November 30 to 1858 October 9, and predicted the comet would next arrive at perihelion on 1866 February 14.52. He provided an ephemeris that covered the period of 1865 July 1 through November 22. Using this ephemeris,

T. N. Thiele (Copenhagen, Denmark) recovered the comet with a refractor on 1865 August 23.01. He determined its position as $\alpha = 22^h 02.9^m$, $\delta = +5° 49'$. Thiele described the comet as like a nebula of the third class, with a centrally condensed coma 25" across and exhibiting a small, distinct nucleus.

The comet was well observed during the next couple of months. R. Engelmann (Leipzig, Germany) said it appeared faint in the reasonably clear air and exhibited a faint condensation on August 26. He indicated the coma was about 0.75' across. C. H. F. Peters (Hamilton College Observatory, Clinton, New York, USA) noted the comet was so faint on that same night that very favorable skies were necessary to see it. The sky was clearer on the 27th for Engelmann and he said the comet was easily seen. He also noted a stellar nucleus of magnitude 12.5. On August 28, Englemann said the comet appeared fainter than on the previous night, but noted that "small stars seemed very disturbed" indicating poor observing conditions. J. F. J. Schmidt (Athens, Greece) saw the comet with averted vision through a 17-cm Reinfelder refractor on September 11 and 12. He described it as an "extremely faint nebulosity." A. Secchi (Rome, Italy) obtained precise positions on the 16th and 17th, and then stopped observing the comet because of its faintness. On the 18th Engelmann simply described it as faint, while on the 23rd he said it was very faint and not as bright as in August, but again indicated somewhat poor observing conditions. He did note that the comet was 40" across. On the 25th Engelmann said the comet was faint, "but as bright as on September 23." Engelmann saw the comet for the final time on September 27, when he noted it was not seen with certainty, and indicated all stars near the comet's position appeared slightly nebulous. V. Fuss (Pulkovo Observatory, Russia) also saw the comet on the 27th and noted it was faint because of clouds. He added that the comet's position was difficult to measure early on because the comparison star was in the coma. K. C. Bruhns (Leipzig, Germany) managed to determine precise positions of the comet on November 13 and 14. On the 13th he said the comet was 45" across and only faintly seen during sharp seeing. On the 14th he noted the comet was close to a star of magnitude 9 and 10, which hampered the accuracy of his measured position.

Observations seemed to have ceased after mid-November, but resumed in December following the "discovery" of a comet by Secchi, while he was trying to recover Biela's comet. He found a diffuse object on December 9 and announced his find after confirming it had moved on the 10th. G. B. Donati (Florence, Italy) began observing Secchi's comet on the 11th and continued observing until the 21st. He described the comet's appearance during this period as "of extraordinary faintness." Secchi observed the comet again on December 12, and then sent a letter to the *Astronomische Nachrichten* on the 13th stating that the "new" comet was none other than Faye's periodic comet. Donati, however, was preoccupied with the idea that Secchi's comet might be Biela's comet. On December 22, he sent a letter to the *Monthly Notices of the Royal Astronomical Society* giving his precise positions, as well as a parabolic orbit with a perihelion date of 1866 January 20.82, based on Secchi's December 9 position and his own of December 13 and 19. He commented, "Although the elements are only a rough approximation, we see that the orbit is quite different from that of the Comet of Biela, especially in the perihelion distance and in the position of the orbit in its plane. But the new Comet moves in a plane not very different from that of the Comet of Biela, and passes its perihelion nearly at the same time. These coincidences and the circumstance of not finding the Comet of Biela at its place may be the result of accident, or may have a physical reason." In a letter written to the *Monthly Notices of the Royal Astronomical Society* on December 27, Donati acknowledged that Secchi's comet was none other than periodic comet Faye.

H. L. d'Arrest (Copenhagen, Denmark) observed this comet with a refractor on 1866 January 12. He said it was "still seen with ease" and he hoped "to be able to still pursue it a rather long time." Fuss measured the comet's position on the 14th, but noted it was not a good observation. On January 18, he said the comet appeared brighter than when formerly seen on the 14th, with a distinct nucleus and a small coma. The comet passed 0.1° from Neptune on January 26.

Fuss was the only person to see the comet after January, but his observations are not even mentioned in later comet catalogs. J. G. Galle's *Cometenbahnen* (1894) says the final observation of this comet came on January 12, S. K. Vsekhsvyatskij's *Physical Characteristics of Comets* (1964) gave the final observation as January 19, and B. G. Marsden's *Catalogue of Cometary Orbits* (1993) agreed with Vsekhsvyatskij. Nevertheless, Fuss continued to measure the comet's position precisely and gave occasional descriptions. On February 13, he said the comet was seen reasonably well, while on the 17th it was "very faint." Although he said the moon interfered with the observation on the 18th, he still managed to determine a precise position. The comet became more challenging in March as its slow motion was causing it to slowly sink in the evening sky. Fuss saw it on the 7th at the end of evening twilight and noted the comet was then at a low altitude and hard to see. He described it as "very faint" on the 12th. The comet was last detected on March 17.77, when Fuss measured the position as $\alpha = 2^h\ 45.6^m$, $\delta = +10°\ 31'$. He was using the Pulkovo refractor and said the comet disappeared in the twilight thereafter.

Later calculations using multiple apparitions and planetary perturbations were published by D. M. A. Möller (1872), A. Shdanow (1885), and B. G. Marsden and Z. Sekanina (1971). These revealed a perihelion date of February 14.48 and a period of 7.41 years. Marsden and Sekanina determined the nongravitational terms as $A_1 = +0.560$ and $A_2 = +0.0608$. The orbit of Marsden and Sekanina is given below.

T	ω	Ω (2000.0)	i	q	e
1866 Feb. 14.4757 (TT)	200.1865	211.6250	11.3553	1.682119	0.557598

ABSOLUTE MAGNITUDE: $H_{10} = 6.4$ (V1964)

FULL MOON: Aug. 7, Sep. 5, Oct. 4, Nov. 3, Dec. 2, Jan. 1, Jan. 30, Mar. 1, Mar. 31

SOURCES: D. M. A. Möller, *AN*, **64** (1865 Apr. 10), pp. 145–58; H. L. d'Arrest and T. N. Thiele, *AN*, **65** (1865 Aug. 28), p. 143; J. F. J. Schmidt and C. H. F. Peters, *AN*, **65** (1865 Oct. 21), pp. 267–70; G. B. Donati, *MNRAS*, **26** (1865 Dec.), p. 68; A. Secchi, *AN*, **66** (1865 Dec. 30), p. 75; G. B. Donati, *MNRAS*, **26** (1866 Jan.), pp. 81–2; H. L. d'Arrest, *AN*, **66** (1866 Jan. 27), p. 137; C. Bruhns and R. Engelmann, *AN*, **67** (1866 Aug. 6), pp. 243, 249; V. Fuss, *BASP*, **12** (1867), pp. 108–14; D. M. A. Möller, *VJS*, **7** (1872), p. 97; A. Shdanow, *BASP*, **33** (1885), pp. 1–24; V1964, p. 59; B. G. Marsden and Z. Sekanina, *AJ*, **76** (1971 Dec.), pp. 1136–7; B. G. Marsden, *CCO*, 8th ed. (1993), p. 41.

55P/1865 Y1
(Tempel–Tuttle)

1866 I

Discovered: 1865 December 19.82 ($\Delta = 0.21$ AU, $r = 1.05$ AU, Elong. $= 102°$)
Last seen: 1866 February 9.74 ($\Delta = 1.77$ AU, $r = 1.09$ AU, Elong. $= 33°$)
Closest to the Earth: 1865 December 21 (0.1947 AU)
Calculated path: UMi (Disc), DRA (Dec. 20), CEP (Dec. 21), LAC (Dec. 23), AND (Dec. 24), PEG (Dec. 26), PSC (Jan. 4), AQR (Jan. 24)

E. W. L. Tempel (Marseille, France) discovered this comet on 1865 December 19.82 near β Ursae Minoris. He described it as a round, centrally condensed nebulosity measuring about 12′ across. There was also a tail extending 0.5°. News of the comet had not yet reached the USA, when H. P. Tuttle (Harvard College Observatory, Cambridge, Massachusetts, USA) independently discovered this comet on January 9. The comet attained its most northerly declination of +78° on December 20.

The comet passed closest to Earth on December 21 and was then seen by T. R. von Oppölzer (Josephstadt, Austria). He said the comet was not very bright and appeared very diffuse. The coma was oval and about 5′ across. It also exhibited a very faint, eccentric nucleus of about magnitude 12. The comet remained a diffuse object during the remainder of December. K. C. Bruhns (Leipzig, Germany) saw the comet on the 22nd and said it was difficult to observe and about 3′ across. On the 26th, Bruhns said the coma was about 2′ across and exhibited no condensation at a magnification of 144×.

The comet passed perihelion just before mid-January and was observed by several observatories during the month. H. C. Vogel (Leipzig, Germany) said the comet appeared very bright on the 4th and exhibited a centrally condensed coma about 9′ across. He said the central condensation was brighter on the 6th, while the coma was only 8′ across. F. Tischler (Königsberg, now Kaliningrad, Russia) determined precise positions on January 7, 9, 10, and 11. He said the comet resembled a diffuse nebulosity, with a central condensation, and added that there was no certain nucleus. C. H. F. Peters (Hamilton College Observatory, Clinton, New York, USA) described the comet as round, about 1′ across, without a nucleus on the 15th. J. F. J. Schmidt (Athens, Greece) read of the comet in newspapers as early as

January 5, but was unable to find it until the ephemerides of Bruhns arrived. His first observation came on January 14. On this and the next few days, Schmidt noted the comet was "rather well condensed" with a coma 2.75' across and a nucleus of magnitude 12. Moonlight began to interfere by January 23, but Schmidt noted the comet was still "rather easily seen." Bruhns was dealing with foggy conditions on this evening and said the comet appeared very faint. Vogel said the comet appeared extremely faint on the 28th.

Schmidt next saw the comet on February 3 and 8. On the last date he said the comet's low altitude of 12.5° and the zodiacal light caused it to appear "extremely weakened" to the point that averted vision had to be used to recognize it. Oppölzer also saw the comet on the 8th, but barely, as it was sitting within a brighter section of the zodiacal light. The comet was last detected on February 9.74, when Oppölzer measured the position as $\alpha = 23^h 44.7^m$, $\delta = -7° 13'$. He said the comet was almost too faint to observe.

W. Huggins (London, England) and A. Secchi (Rome, Italy) managed to detect the spectrum of this comet. Secchi detected three lines on January 8, of which one was very bright and the other two were too faint to fix their positions. Huggins looked at the comet on January 9 and detected a continuous spectrum, which he surmized came from the reflected light of the sun. He also detected one other bright line, which he said matched the brightest lines in the spectra of nebulae. He wrote, "This line may perhaps be interpreted as an indication that cometary matter consists chiefly of nitrogen, or of a more elementary substance existing in nitrogen."

The first orbit was calculated by Bruhns using positions from December 22, 23, and 25. This parabolic solution revealed a perihelion date of 1866 January 10.05. Additional orbits were calculated by C. F. Pechüle and C. H. F. Peters during January. By the end of the month, Pechüle noted the positions were no longer being adequately represented by a parabolic orbit and he published an elliptical orbit on January 30, which had a perihelion date of January 11.68 and a period of 31.91 years. About two weeks later three astronomers confirmed the elliptical nature, with the period being given as 10.76 years by H. L. d'Arrest, 29.77 years by Oppölzer, and 53.15 years by Pechüle. It was later proven by Oppolzer (1867) that Pechüle's original elliptical orbit was closest to the truth. Oppölzer then gave the period as 33.18 years. The most recent computation was by S. Nakano (1999), when he used positions from the apparitions of 1866, 1965, and 1998, and determined the period as 33.22 years.

G. V. Schiaparelli (1867) compared the orbit of this comet with the Leonid meteor stream and noted a strong similarity. In his paper published in the 1867 February 20 issue of the *Astronomische Nachrichten* he asked, "Are the meteors to be considered as swarms of minute comets, or as the products of the dissolution of large comets?"

T	ω	Ω (2000.0)	i	q	e
1866 Jan. 11.6274 (UT)	170.9048	233.2477	162.6891	0.976575	0.906050

ABSOLUTE MAGNITUDE: $H_{10} = 8.0$ (V1964)

FULL MOON: Dec. 2, Jan. 1, Jan. 30, Mar. 1

SOURCES: E. W. L. Tempel, K. C. Bruhns, and T. R. von Oppölzer, *AN*, **66** (1865 Dec. 30), p. 75–8; J. A. H. Gyldén, *BASP*, **10** (1866), pp. 1–4; H. L. d'Arrest and C. F. Pechüle, *AN*, **66** (1866 Jan. 15), p. 109; W. Huggins, *PRSL*, **15** (1866 Jan. 18), pp. 5–7; H. L. d'Arrest, C. F. Pechüle, and T. R. von Oppölzer, *AN*, **66** (1866 Jan. 27), pp. 137–40; A. Secchi, *CR*, **62** (1866 Jan. 29), p. 210; C. F. Pechüle, *AN*, **68** (1867 Jan. 30), pp. 267–70; C. H. F. Peters, T. R. von Oppölzer, H. L. d'Arrest, and C. F. Pechüle, *AN*, **66** (1866 Feb. 12), pp. 167–74; T. R. von Oppölzer, *AN*, **66** (1866 Mar. 5), p. 221; J. F. J. Schmidt, *AN*, **66** (1866 Mar. 14), p. 231; T. R. von Oppölzer, *AN*, **66** (1866 Mar. 22), pp. 249–52; F. Tischler, *AN*, **67** (1866 May 18), p. 11; C. Bruhns and H. C. Vogel, *AN*, **67** (1866 Aug. 6), pp. 243, 251; T. R. von Oppölzer, *AN*, **68** (1867 Jan. 28), pp. 241–50; G. V Schiaparelli, *AN*, **68** (1867 Feb. 20), p. 331; T. R. von Oppölzer, *AN*, **68** (1867 Feb. 20), p. 333; G. V. Schiaparelli, *MNRAS*, **27** (1867 Apr.), pp. 246–7; F. Bidschof, *AN*, **144** (1897 Oct. 14), pp. 299–304; V1964, p. 59; S. Nakano, *Nakano Note* No. 722A (1999 Nov. 17).

38P/1867 B1 *Discovered:* 1867 January 22.9 ($\Delta = 1.12$ AU, $r = 1.58$ AU, Elong. = 96°)

(Stephan– *Last seen:* 1867 April 4.11 ($\Delta = 1.83$ AU, $r = 1.82$ AU, Elong. = 73°)

Oterma) *Closest to the Earth:* 1866 November 25 (0.9187 AU)

Calculated path: ARI (Disc), TAU (Feb. 15), PER (Mar. 6), AUR (Mar. 10)

1867 I

A case of mistaken identity caused J. E. Coggia (Marseilles, France) to lose his first comet discovery. He was searching the sky on the night of 1867 January 22.9, when he found what he thought was an uncataloged nebula. The sky clouded up almost immediately and remained completely cloudy until the night of January 24, when E. J. M. Stephan (Marseille, France) checked on the nebula through a brief break in the clouds and saw that it had moved. Stephan was able to confirm this was a comet on January 25.86 and gave the position as $\alpha = 2^h\ 33.9^m$, $\delta = +15°\ 34'$. Stephan said the comet was rather brillant, round, with a very marked nucleus. The initial announcements did not mention Coggia's name and the comet was named after Stephan.

E. W. L. Tempel (Marseille, France) independently discovered this comet on January 28.86 near π Arietis. The sky clouded over before movement could be detected, but he found the comet near 40 Arietis on January 29.86. Tempel said J. E. B. Valz (Marseille, France) had left a message for him indicating that Stephan had discovered the comet on January 25. Valz told Tempel, "It is rather brilliant, generally round in appearance, with a nucleus, and a vague tail." Tempel said that although the positions agreed with that of his comet, the description was not the same, as his comet was "hardly 3' in diameter and is very faint." As proof of the comet's faintness he said it had passed over an 8th-magnitude star on January 29.94 and was invisible for several minutes.

The comet was already moving away from both the sun and Earth when discovered, and the comet remained a faint object for observers. E. Weiss (Vienna, Austria) observed with a 15-cm refractor on February 4 and said the comet looked like a very faint, diffuse nebula, without a distinct nucleus. He added that within the coma "several fixed star-like points flashed." K. C. Bruhns (Leipzig, Germany) said the comet was already very faint on the 4th at a magnification of 96×, while Bruhns and his colleague H. C. Vogel described it as very faint at a magnification of 192× on the 5th. Vogel found the comet faint, but elongated toward PA 75° on the 6th, and noted a small, distinct nucleus. T. R. von Oppölzer (Vienna, Austria) simply noted the comet was quite faint on the 7th. The comet was seen in bright moonlight by Vogel on the 9th and he gave the coma diameter as 1.5'. Vogel said the comet was extremely faint and diffuse on the 21st. J. F. J. Schmidt (Athens, Greece) said the comet was small and faint on the 23rd, with a coma 2.0' across and a sharp, central condensation of magnitude 12.

As March began, few astronomers were still observing the comet, but J. Winlock (Cambridge, Massachusetts, USA) continued to measure it using a 38-cm refractor. He made observations on March 27, 29, 30, 31, April 1, and 3, and continually described it as very faint. The comet was last seen on April 4.11, when Winlock saw it with the 38-cm refractor at a position of $\alpha = 5^h 46.3^m$, $\delta = +35° 35'$. He described it as very faint, but suggested observations could have continued if not for the increasing moonlight.

The first parabolic orbit was calculated by W. Valentiner. Using positions from the period spanning January 25 to February 4, he determined the perihelion date as 1867 January 28.84. During the next couple of weeks, the perihelion date was determined as January 20.17 by Oppölzer and January 19.27 by H. C. Vogel. During the last days of the comet's appearance, G. M. Searle took positions spanning the period of January 27 to March 29 and noted that only an elliptical orbit represented the positions well. He determined the perihelion date as January 20.36 and the period as 33.62 years.

L. Becker (1891) determined a definitive orbit. He began with 63 positions spanning the period of January 25 to April 4, and ultimately found a perihelion date of January 20.71 and a period of 40.1 years. Becker recognized that there was some uncertainty in the period because of the relatively short period of visibility. He concluded that the comet would likely return sometime between 1902 and 1912. He provided an ephemeris giving rough positions for every month based on five different perihelion dates. Unfortunately, the comet was not recovered then and was not seen again until accidentally found by L. Oterma in 1942.

The most significant studies using multiple apparitions and planetary perturbations were published by M. Y. Shmakova (1971) and D. K. Yeomans (1986). These revealed a perihelion date of January 20.65 and a period of

37.12 years. Yeomans calculated nongravitational terms of $A_1 = +0.17$ and $A_2 = -0.0030$, which were listed by B. G. Marsden (1986). Yeomans' orbit is given below.

T	ω	Ω (2000.0)	i	q	e
1867 Jan. 20.6477 (TT)	357.5370	80.3159	18.2037	1.575367	0.858435

ABSOLUTE MAGNITUDE: $H_{10} = 7.2$ (V1964)

FULL MOON: Jan. 20, Feb. 18, Mar. 20, Apr. 18

SOURCES: E. J. M. Stephan, *CR*, **64** (1867 Jan. 28), pp. 151–2; *MNRAS*, **27** (1867 Feb.), p. 134; E. W. L. Tempel, J. E. B. Valz, and W. Valentiner, *AN*, **68** (1867 Feb. 9), pp. 301–4; T. R. von Oppölzer, *AN*, **68** (1867 Feb. 20), p. 335; T. R. von Oppölzer, *AN*, **68** (1867 Mar. 2), p. 363; H. C. Vogel, *AN*, **68** (1867 Mar. 9), p. 379; T. R. von Oppölzer, *MNRAS*, **27** (1867 Apr.), p. 255; H. C. Vogel, J. F. J. Schmidt, J. Winlock, and G. M. Searle, *AN*, **69** (1867 May 6), pp. 103, 109–12; J. E. Coggia, *VJS*, **3** (1868), p. 210; E. Weiss, *AN*, **75** (1870 Mar. 3), p. 273; J. R. Hind, *MNRAS*, **31** (1871 May), pp. 214–15; J. Winlock and E. C. Pickering, *Annals of the Astronomical Observatory of Harvard College*, **13** (1882), p. 176; L. Becker, *MNRAS*, **51** (1891 Jun.), pp. 475–94; V1964, p. 59; M. Y. Shmakova, *QJRAS*, **12** (1971), pp. 266–7, 272; B. G. Marsden, *CCO*, 5th ed. (1986), p. 66; D. K. Yeomans, *QJRAS*, **27** (1986), p. 604.

9P/1867 G1
(Tempel 1)

1867 II

Discovered: 1867 April 3.90 ($\Delta = 0.71$ AU, $r = 1.64$ AU, Elong. $= 147°$)
Last seen: 1867 August 27.8 ($\Delta = 1.30$ AU, $r = 1.81$ AU, Elong. $= 103°$)
Closest to the Earth: 1867 May 15 (0.5683 AU)
Calculated path: LIB (Disc), SER (Apr. 6), LIB (May 15), SCO (Jul. 24), OPH (Aug. 5)

E. W. L. Tempel (Marseille, France) discovered this faint, diffuse comet in Libra on 1867 April 3.90, at a position of $\alpha = 15^h 03.0^m$, $\delta = -2° 27'$. He noted, "The comet had an apparent diameter of 4'–5' and was in the middle, where several little stars were seen pulsating, only a little condensed." The comet had reached a minimum declination of $-3.5°$ on March 3.

The comet approached both the sun and Earth during the remainder of April. On April 13, H. C. Vogel (Leipzig, Germany) described the comet as "rather bright," with a distinct nucleus. The coma was elongated in PA 180°, and measured 75" by 40". W. J. Förster and E. Becker (Berlin, Germany) found the comet easy to see in the refractor on April 13 and 21, despite moonlight. They noted a distinct nucleus. The comet attained its most northerly declination of $-2°$ on the 21st and K. C. Bruhns (Leipzig, Germany) noted it looked like a round nebulosity. He added that it was well seen despite moonlight. G. F. W. Rümker (Hamburg, Germany) obtained several observations of this comet with a 5-foot focal length refractor. On the 24th, he described the comet as fairly bright and 2–3' across. He added that the nucleus was eccentrically situated within the coma. V. Fuss (Pulkovo Observatory, Russia) said the tail extended 2' toward PA 215° on the 27th. E. Weiss

(Vienna, Austria) observed the comet with a 15-cm refractor on the 30th and described it as a rather large, very diffuse nebulosity, exhibiting an eccentrically situated condensation with a starlike nucleus.

The comet passed closest to Earth around mid-May and reached perihelion about a week later. Vogel said the comet was bright, with a very bright star-like nucleus on May 1. He added that the coma was about 4' across and there was an extension in PA 210°. J. F. J. Schmidt (Athens, Greece) noted the comet was well condensed and contained a star-like nucleus of magnitude 11 on the 3rd. The nucleus was eccentrically situated with the coma. C. G. Talmage (Leyton, England) saw the comet in the morning sky on the 4th. He said a 25.4-cm refractor (250×) seemed to reveal the nucleus had "a division across its centre." That evening he used the same telescope at 130× and remarked, "the comet was exceedingly faint through clouds." Several astronomers described the comet on May 6. Vogel wrote, "Comet very bright with star-like nucleus." He added that there was an extension toward PA 230°. R. Engelmann (Leipzig, Germany) found the comet bright, but "rather small," and possibly elliptical. He initially noted the nucleus equaled a star of magnitude 9.7, but later revised it to 10.5. Rümker said the coma was 3'–4' across and very diffuse. In addition, there was an eccentrically situated, faint condensation. Weiss observed the comet with a 15-cm refractor and described it as a rather large, very diffuse nebulosity, exhibiting an eccentrically situated condensation with a starlike nucleus. The comet attained a maximum solar elongation of 165° on May 7. Engelmann and Weiss independently saw the comet that night. Engelmann noted the comet was bright, with a coma about 1.5' across and contained a nucleus of magnitude 10.5. He added that the coma was elongated in PA 220°. Weiss described it as a rather large, very diffuse nebulosity, exhibiting an eccentrically situated condensation with a star-like nucleus. Engelmann said the comet was bright, with a nucleus of magnitude 10.5 on the 8th. Rümker said the comet appeared very faint in moonlight on the 11th. Tempel saw the comet on the 19th and noted it was larger and more diffuse than when he last saw it in late April. Bruhns said the comet was rather bright on the 25th, with a coma 1.5' across. He added that there was a nucleus eccentrically situated toward the south. Engelmann described the comet as rather large and faint on the 29th. He noted a nucleus of magnitude 11. That same night, Bruhns said the comet was rather bright and elongated toward PA 120°. He noted a nucleus of magnitude 11.5. Engelmann found the comet rather faint and round on the 30th, with a nucleus of magnitude 11.

The comet was moving away from the sun and Earth during June and July. Talmage found it "exceedingly faint" on the 1st. That same night Engelmann noted it was rather faint, large, and round, with a nucleus of magnitude 11. He estimated the nuclear magnitude as 11.5 on the 2nd. Although Engelmann found the comet faint, large, and round, on the 5th, Bruhns described it as "rather bright." The nuclear magnitude was given as 11.5 by Engelmann and 12 by Bruhns. Bruhns also estimated the coma diameter as

2′. Bruhns found the comet bright, with a distinct nucleus on the 9th, despite moonlight. Additional observations by Bruhns in strong moonlight on the 19th, 20th, and 21st, revealed a faint, centrally condensed object. Without moonlight, Bruhns found the comet easy to detect on the 22nd and noted a nucleus of magnitude 12. Bruhns described the comet as fairly bright on the 24th and easy to see with a distinct nucleus on the 27th. On July 4, Bruhns observed despite some clouds and noted the comet was at the limit of his telescope. J. Winlock (Harvard College, Cambridge, Massachusetts, USA) last detected the comet with the 38-cm refractor on August 1. Because of the comet's southerly declination and faintness, Schmidt was the only observer able to follow the comet thereafter. Using an ephemeris computed by Bruhns, he found the comet with a 17-cm Reinfelder refractor on the 18th and described it as an extremely faint nebulosity about 2′ across. Schmidt measured the comet's position for the final time on August 21.78, giving it as $\alpha = 16^h\,48.1^m$, $\delta = -23° \, 20′$. He observed the comet on the 22nd and 23rd, but clouds prevented measurements of the position. From August 24 to August 26, Schmidt could only see the comet intermittently. The comet was last detected on August 27.8, but its faintness prevented a measure of its position.

W. Huggins (1867) observed this comet from London, England with a telescope on May 4 and 8, and noted, "the comet appeared to consist of a slightly oval coma surrounding a minute and not very bright nucleus." He then applied a spectroscope to the telescope and noted, "a continuous spectrum was formed by the light of the coma. I was unable, on account of the faintness of the nucleus, to distinguish with certainty the spectrum of its light from the broad spectrum of the coma on which it appeared projected. Once or twice I suspected the presence of two or three bright lines, but of this observation I was not certain." Huggins added, "The prismatic observation of this faint object, though imperfect, appears to show that this small comet is probably similar in physical structure to Comet I, 1866."

C. F. W. Peters calculated the first parabolic orbits. The first used positions from April 3, 12, and 21, and produced a perihelion date of 1867 March 7.40. He considered the orbit unreliable and calculated a second orbit using positions from April 12, 21, and 25. The resulting perihelion date was February 28.35.

The first elliptical orbit was calculated by Bruhns using positions spanning the period of April 13 to May 1. He determined a perihelion date of May 17.24 and the period as 5.74 years. Additional orbits by E. Becker, Bruhns, G. M. Searle, and A. J. Sandberg (1869) eventually established the perihelion date as May 24.2 and the period as 5.7 years. The most recent analysis of this comet's motion over multiple apparitions came from G. Schrutka (1974). Using 394 positions from the apparitions of 1867, 1873, 1879, and 1972, he applied perturbations by Venus to Pluto and determined the perihelion date as May 24.22 and the period as 5.65 years. This orbit is given below.

While the comet was still under observation, F. A. T. Winnecke realized that Tempel's comet would have passed perihelion during 1855, and calculated the path of 9P/Tempel 1 across the sky in that year. He found that if a perihelion date of 1855 February 7.22 was assumed, then P/Tempel 1 would have passed very close to the position of the "comet" seen by Goldschmidt. Winnecke assumed the mystery had been solved. F. E. von Asten (1873) recalculated the past orbital elements of 9P/Tempel 1, including perturbations by Jupiter. He found a perihelion date of 1856 February 2.04, and said the comet would not only have been a fairly great distance from Goldschmidt's object, but it would have been much fainter, with a distance from Earth of about 3.2 AU.

T	ω	Ω (2000.0)	i	q	e
1867 May 24.2174 (TT)	134.9855	102.9168	6.3897	1.562104	0.507708

ABSOLUTE MAGNITUDE: $H_{10} = 8.4$ (V1964)

FULL MOON: Mar. 20, Apr. 18, May 18, Jun. 17, Jul. 16, Aug. 15, Sep. 14

SOURCES: E. W. L. Tempel, *AN*, **69** (1867 Apr. 12), p. 63; W. J. Förster and E. Becker, *AN*, **69** (1867 Apr. 23), p. 79; E. W. L. Tempel and C. G. Talmage, *MNRAS*, **27** (1867 May), p. 275; G. F. W. Rümker and C. F. W. Peters, *AN*, **69** (1867 May 1), pp. 93–6; K. C. Bruhns, H. C. Vogel, C. G. Talmage, and R. Engelmann, *AN*, **69** (1867 May 22), pp. 141–4; E. Becker, *AN*, **69** (1867 May 27), p. 149; W. Huggins, *MNRAS*, **27** (1867 Jun.), p. 288; E. Becker and F. A. T. Winnecke, *AN*, **69** (1867 Jun. 20), pp. 203–8; K. C. Bruhns, *AN*, **69** (1867 Jul. 22), pp. 285–8; R. Engelmann and J. F. J. Schmidt, *AN*, **69** (1867 Aug. 1), pp. 299–302; J. F. J. Schmidt, *AN*, **69** (1867 Aug. 7), pp. 317–20; E. W. L. Tempel, *AN*, **69** (1867 Aug. 25), pp. 365–8; J. Winlock and G. M. Searle, *AN*, **70** (1867 Sep. 21), p. 45; J. F. J. Schmidt, *AN*, **70** (1867 Sep. 28), p. 63; H. L. d'Arrest, *AN*, **70** (1867 Oct. 14), p. 93; K. C. Bruhns, *AN*, **70** (1867 Oct. 28), pp. 119–24; C. G. Talmage, *MNRAS*, **27** (Supplemental Notice, 1867), p. 317–8; W. Huggins, *MNRAS*, **28** (1868 Feb.), p. 93; V. Fuss, *AN*, **71** (1868 May 26), pp. 253–6; A. J. Sandberg, *AN*, **73** (1869 Jan. 12), p. 77; A. J. Sandberg and G. F. W. Rümker, *AN*, **74** (1869 Jun. 8), pp. 103, 109; E. Weiss, *AN*, **75** (1870 Mar. 3), p. 273; F. E. von Asten, *AN*, **82** (1873 Nov. 3), pp. 273–6; V1964, p. 59; G. Schrutka, *QJRAS*, **15** (1974), pp. 450–1, 459.

C/1867 S1 *Discovered:* 1867 September 26.86 ($\Delta = 1.10$ AU, $r = 1.09$ AU, Elong. $= 62°$)
(Baeker– *Last seen:* 1867 November 1.24 ($\Delta = 1.13$ AU, $r = 0.38$ AU, Elong. $= 19°$)
Winnecke) *Closest to the Earth:* 1867 October 14 (0.9002 AU)
Calculated path: UMa (Disc), CVn (Oct. 8), BOO (Oct. 18), VIR (Oct. 29), SER
1867 III (Nov. 1)

C. W. Baeker (Nauen, Germany) discovered this comet on 1867 September 26.86. It was independently discovered by F. A. T. Winnecke (Honigesstein on Rhein, Germany) about four hours later.

The comet was well observed during October. H. L. d'Arrest (Copenhagen, Denmark) saw the comet on October 3, 4, and 5. He described it as a "beautiful, bright nebulosity of 3' to 4' in diameter, with a strong central condensation." He added that a broad, tail-like extension began in the direction of the sun. E. Weiss (Vienna, Austria) observed the comet with a 15-cm refractor on the 6th and said it appeared "uncommonly faint" because of moonlight. T. R. von Oppölzer (Vienna, Austria) simply noted the comet was faint on the 14th and 17th. K. C. Bruhns and H. C. Vogel (Leipzig, Germany) said the comet was very bright on the 17th, with a distinct, bright nucleus measuring 35" across. A tail extended 10' toward PA 346°. Bruhns and Vogel said the comet appeared very bright on the 21st and 22nd, while they noted it was very bright with a sharp nuclear condensation on the 23rd. On the last date, they noted the tail extended 15' toward PA 355°. C. N. A. Krueger (Helsinki, Finland) said the tail extended 5' toward PA 5.8° on the 25th. V. Fuss (Pulkovo Observatory, Russia) said the comet was difficult to measure because of its diffuse nucleus and low altitude on the 25th and 29th. G. L. Tupman (Portsmouth, England) saw the comet on October 27 and said it was too faint to withstand illumination of the micrometer wires within the eyepiece. He noted, "The nucleus of the comet appeared equal to a star of the 8.0 or 8.1 magnitude by comparison with six stars. Coma faint, no tail." Tupman said the comet had become too faint for illumination of the micrometer wires on the 27th and 28th. On October 29, Krueger said the tail extended 5' toward PA 14.7°. J. F. J. Schmidt (Athens, Greece) did not receive word of the comet's September discovery until October 11, and an ephemeris did not arrive until October 25. He managed to see the comet on October 25, 26, and 27, but at a very low altitude of 3–5°. On the 25th he said the tail was perhaps 10' in length, while the coma was 4–5' in diameter and contained a stellar nucleus of magnitude 7. Schmidt thought his chances of seeing the comet on October 28 and 29 would have been quite good, but clouds and rain prevented any further observations after the 27th.

As the comet dropped into twilight, its position was measured for the final time on October 31.66, when Krueger determined it as $\alpha = 15^h 02.2^m$, $\delta = +3° 04'$. He said the comet's low altitude and the bright twilight prevented the tail from being seen. Bruhns and Vogel saw the comet on November 1.24, but bright twilight made it impossible to measure a position.

The first orbits were calculated by T. R. von Oppolzer and T. Wolff. Oppolzer used positions from the period spanning September 30 to October 6 and determined the perihelion date as 1867 November 7.52. Wolff used positions from the period spanning September 27 to October 5 and determined the perihelion date as November 7.54. As additional observations were obtained, more refined orbits were calculated by C. F. Pechüle, F. Tietjen, and Oppolzer. The perihelion date was then narrowed down to November 7.46. The most extensive study of this comet was conducted by P. Broch (1888). He took positions covering the period of Sepember 30 to October 28 and derived both parabolic and elliptical orbits. For the parabolic orbit he determined the perihelion date as November 7.46. For the elliptical orbit he determined the perihelion date as November 7.46 and the period as about 37 128 years. Broch remarked on the similarity between this orbit and that of comet C/1785 E1, but added that both comets were likely of very long period. Because of the short period of observation, the parabolic orbit is given below.

T	ω	Ω (2000.0)	i	q	e
1867 Nov. 7.4608 (UT)	148.6408	66.8417	96.5735	0.330359	1.0

ABSOLUTE MAGNITUDE: $H_{10} = 7.9$ (V1964)

FULL MOON: Sep. 14, Oct. 13, Nov. 12

SOURCES: H. L. d'Arrest, C. F. Pechüle, T. R. von Oppölzer, T. Wolff, *AN*, **70** (1867 Oct. 14), pp. 93–6; T. R. von Oppölzer, *AN*, **70** (1867 Oct. 28), p. 125; C. W. Baeker, F. A. T. Winnecke, and F. Tietjen, *MNRAS*, **28** (1867 Nov.), p. 15; K. C. Bruhns, H. C. Vogel, and C. N. A. Krueger, *AN*, **70** (1867 Nov. 22), p. 189; G. L. Tupman, *AN*, **70** (1867 Nov. 30), p. 205; J. F. J. Schmidt, *AN*, **70** (1867 Dec. 10), pp. 215–18; C. W. Baeker and F. A. T. Winnecke, *VJS*, **3** (1868), pp. 212–13; C. W. Baeker, *MNRAS*, **28** (1868 Feb.), p. 98; V. Fuss, *AN*, **71** (1868 May 26), pp. 253–6; T. R. von Oppölzer, *AN*, **73** (1869 Jan. 4), p. 57; E. Weiss, *AN*, **75** (1870 Mar. 3), p. 273; P. Broch, *SAWW*, **97** Abt. IIa (1888), pp. 1477–1504; P. Broch, *AN*, **121** (1889 Jun. 5), pp. 353–8; V1964, p. 59.

5D/Brorsen

1868 I

Recovered: 1868 March 22.82 ($\Delta = 1.56$ AU, $r = 0.78$ AU, Elong. $= 25°$)

Last seen: 1868 June 23.86 ($\Delta = 1.11$ AU, $r = 1.32$ AU, Elong. $= 76°$)

Closest to the Earth: 1868 May 25 (0.9107 AU)

Calculated path: CET (Rec), PSC-CET (Mar. 25), ARI (Apr. 3), TAU-ARI (Apr. 13), TAU (Apr. 14), PER (Apr. 25), AUR (Apr. 27), LYN (May 15), UMa (May 25), CVn (Jun. 22), COM (Jun. 23)

K. C. Bruhns took the orbits for the returns of 1846 and 1857, applied perturbations by Jupiter, and predicted this comet would next pass perihelion on 1868 April 18.95. There were three independent recoverers of this comet. E. W. L. Tempel (Marseille, France) first saw it on 1868 March 22.82, followed by J. F. J. Schmidt (Athens, Greece) on April 11.78 and Bruhns (Leipzig, Germany) on April 12.83. Tempel indicated it was then at a very low altitude near the end of evening twilight. Schmidt said the coma was 120″ across and contained a nuclear condensation of magnitude 8–9. There was also a tail 5′ long.

The comet passed perihelion around the middle of April and was well observed. During most of the month, Schmidt provided most of the physical descriptions. He gave coma diameter estimates of 128″ on the 14th, 140″ on the 15th, 110″ on the 17th, and 194″ on the 25th. Schmidt also estimated the magnitude of the nucleus on several occasions, but noted it was always within the range of magnitude 7–8. He also supplied tail lengths of 20′ on the 14th and 15th, 15′ on the 17th, 25′ on the 18th, 30′ on the 20th, and 15′

on the 25th. Engelmann described the comet as bright and round, with a central condensation on the 15th. He estimated the coma diameter as 1' and said the nucleus was similar in appearance to a star of magnitude 9. G. F. W. Rümker (Hamburg, Germany) first saw the comet on April 21. He was using a 24.4-cm refractor, but noted that clouds made his measurements somewhat uncertain. Bruhns saw the comet on April 23 and said it appeared rather bright and was 1.7' across. He also noted a distinct nuclear condensation 30" across.

Having passed perihelion during April, the comet continued to approach Earth and was nearest near the end of May. It attained its most northerly declination of +50° on the 24th. Schmidt's observations indicate the comet continued to develop during May. The coma was more than 200" across when seen on the 11th, and had surpassed a diameter of 300" by the 16th. The largest coma diameters were recorded by Schmidt on May 22 and 23 with values of 378" and 362", respectively. Meanwhile, the nucleus was brightest on May 11 and 12, when Schmidt gave the magnitude as 7. It had faded to 9 when seen on the 22nd and 23rd. The longest tail length was given as 40' on May 9. It quickly declined to 10–15' on the 11th and 12th, then to 7' on the 14th, and finally to 5' on the 16th. No tail was noted thereafter. H. L. d'Arrest (Copenhagen, Denmark) also obtained several observations during May. He saw the comet on May 2 and said the brightness of the central condensation was similar to a star of magnitude 8 or 9. He said there was little change on the 13th. D'Arrest added that the central condensation seemed to consist of three or four bright points. On May 17, d'Arrest said the center of the "bright comet" passed just 26" south of a 7th-magnitude star, so that the coma covered it. He added that during the occultation the comet and star looked like a "magnificent nebula of the fourth class." This classification referred to W. Herschel's famous list of nebulous objects and indicates a planetary nebula. D'Arrest's last sightings of the comet were on May 25, 27, and 28, with the "bright summer dawn" making it a challenging object to see. He added that at the end of May, "I always looked for it in vain." H. C. Vogel (Leipzig, Germany) and Rümker also provided several descriptions. Vogel said the comet was rather bright on the 2nd, with a centrally condensed, not quite round coma measuring about 3' across. On May 5, Vogel described the comet as very bright and well observed, despite some clouds and a nearby moon. Vogel saw a tail extending 8–10' toward PA 75° on the 8th. He added that the bright inner coma of the comet, which was 50" in diameter, contained several condensations. Vogel said it was rather bright with a tail on the 14th. He added that there was no sharp nucleus, but there were several bright spots in the southern part of the coma. Rümker said that on both May 20 and 27, the "coma of the comet seemed to expand ever more and fade," with the comet "quite faint" on the last date. Rümker said moonlight and weather prevented any further observations. Vogel last detected the comet on May 29 and said it appeared faint in the bright northern sky. He estimated the coma diameter as 5'.

Schmidt's observations during June were primarily limited to coma diameters, as the tail was never seen and his final estimate of the magnitude of the nucleus was given as 12 on the 8th. His measurements of the coma diameter fell within the range of 266–274″ on June 8, 9, and 10. On the 11th he found the coma 302″ across, and this dropped to 280″ by the 12th. His final measurements came on June 18 and 19, when the coma diameter was 256″ and 266″, respectively.

The comet was last seen on June 23.86, when Schmidt gave the position as $\alpha = 12^h \ 06.6^m$, $\delta = +33° \ 31'$. He noted the comet was large and diffuse, without a nucleus. With very clear skies on the evening of July 10, Schmidt seached for the comet in the area predicted by Bruhns' ephemeris, but no trace of it was found.

Schmidt was interested in trying to find a reason for the apparent fluctuation in his measurements of the radius of the coma. He noted that a curve could be fitted that indicated a period of about 23.5 days. He also noted that a similar investigation of the coma variations of C/1861 J1 had revealed a period of 25.4 days. He encouraged further investigations into the coma diameters of comets in order to determine whether or not the size fluctuates because of the rotation of the sun.

A. Secchi (1868) and W. Huggins (1868) examined the spectrum of this comet on several nights during the first half of May 1868. In addition to a faint continuous spectrum, three bright bands were seen, the positions of which they noted matched those seen by G. B. Donati in comet C/1864 N1. Huggins said the length of the bands "shows that they are not due alone to the stellar nucleus, but are produced by the light of the brighter portions of the coma." He said the middle band was the brightest. Huggins compared these bands to the bands of magnesium, sodium, hydrogen, and nitrogen. Although the brightest band was "nearly in the position" with "the double line of the spectrum of nitrogen," it was "in a small degree less refrangible than the line of nitrogen." He said the difference could not be explained by the comet's motion. He concluded, "The positions of the bands in this comet would seem to indicate a chemical constitution different from that of the nebulae, which give a spectrum of bright lines."

Minor refinements of this comet's orbit using positions from multiple apparitions, as well as planetary perturbations, have been published by several astronomers, including L. R. Schulze (1873, 1878), B. G. Marsden and Z. Sekanina (1971), W. Landgraf (1986), and S. Nakano (1997). These have typically revealed a perihelion date of April 17.92 and a period of 5.48 years. Nongravitational terms were determined by Marsden and Sekanina, as well as Landgraf. These were fairly comparable, with Landgraf's being $A_1 = +0.51$ and $A_2 = -0.0651$. Nakano's orbit is given below.

T	ω	Ω (2000.0)	i	q	e
1868 Apr. 17.9204 (TT)	14.8323	103.0207	29.3674	0.596978	0.807912

ABSOLUTE MAGNITUDE: $H_{10} = 9$ (V1964)

FULL MOON: Mar. 8, Apr. 7, May 6, Jun. 5, Jul. 4

SOURCES: K. C. Bruhns, *MNRAS*, **28** (1868 Mar. 13), pp. 158–60; K. C. Bruhns, *MNRAS*, **28** (1868 May), pp. 221–3; K. C. Bruhns and E. W. L. Tempel, *AN*, **71** (1868 May 8), pp. 185–6; J. F. J. Schmidt, *AN*, **71** (1868 May 11), p. 207; A. Secchi, *CR*, **66** (1868 May 11), pp. 881–4; M. Huggins, *PRSL*, **16** (1868 May 14), pp. 386–9; H. L. d'Arrest and T. R. von Oppölzer, *AN*, **71** (1868 Jun. 3), pp. 267–70; B. F. Sands, A. Hall, and H. L. d'Arrest, *AN*, **72** (1868 Jul. 31), pp. 45–8; J. F. J. Schmidt, *AN*, **72** (1868 Aug. 5), pp. 53–6; J. F. J. Schmidt and G. F. W. Rümker, *AN*, **72** (1868 Aug. 11), pp. 65–72; J. F. J. Schmidt, *AN*, **72** (1868 Sep. 4), p. 123; K. C. Bruhns, R. Engelmann, and H. C. Vogel, *AN*, **72** (1868 Nov. 10), pp. 279–81; L. R. Schulze, *AN*, **82** (1873 Sep. 6), pp. 173–84; L. R. Schulze, *AN*, **93** (1878 Sep. 27), pp. 177–88; V1964, p. 59; B. G. Marsden and Z. Sekanina, *AJ*, **76** (1971 Dec.), pp. 1141–2; W. Landgraf, *QJRAS*, **27** (1986 Mar.), p. 116; B. G. Marsden and W. Landgraf, *CCO*, 12th ed. (1997), pp. 48–9; S. Nakano, *Nakano Note* No. 663 (1997 May 5).

C/1868 L1
(Winnecke)

1868 II

Discovered: 1868 June 13.98 ($\Delta = 1.02$ AU, $r = 0.65$ AU, Elong. $= 37°$)

Last seen: 1868 July 17.80 ($\Delta = 0.97$ AU, $r = 0.74$ AU, Elong. $= 44°$)

Closest to the Earth: 1868 July 1 (0.6202 AU)

Calculated path: PER (Disc), CAM (Jun. 19), AUR (Jun. 22), CAM (Jun. 23), LYN (Jun. 25), UMa-LYN (Jul. 1), LMi (Jul. 4), LEO (Jul. 5), SEX (Jul. 17)

This comet was discovered by F. A. T. Winnecke (Karlsruhe, Germany) on 1868 June 13.98, and the position was estimated as $\alpha = 3^h 07.3^m$, $\delta = +47° 18'$. He confirmed this was a comet on June 14.03 and noted a short tail was visible in the finder, while the 11-cm refractor revealed a strong, but tiny nucleus. An independent confirmation came from K. C. Bruhns and H. C. Vogel (Leipzig, Germany) on June 14.93. Bruhns said the nucleus was 0.4' across, while the tail extended toward PA 298°. Vogel described it as bright and 45" across, with a tail extending 2.5' toward PA 300°. He added that the coma was remarkably sharp on the north-following side.

The comet approached the sun and Earth through most of the remainder of June, with perihelion coming near the end of the month. R. Engelmann (Leipzig, Germany) saw the comet in twilight on the 16th and described it as rather faint, with a stellar nucleus. On the 17th, Winnecke said the comet was "visible to the naked eye as a star of the 5th magnitude." He also noted a tail 1–2° long in the cometseeker. Vogel said the comet was visible to the naked eye on the 18th, while the cometseeker revealed a tail 2° long and the telescope (92×) revealed a tail extending 17' toward PA 296°. Engelmann noted the nucleus appeared granular on the 18th. G. F. W. Rümker (Hamburg, Germany) observed the comet with the 24.4-cm refractor on the morning of June 19 and said it appeared "extraordinarily bright" with a distinct nucleus of magnitude 8.5. There was also a "concentric, surrounding nebulous mass" about 22.5" across, which was surrounded by an "eccentric, fine coma 90″ across.″ Using a thread micrometer, Rümker measured the nucleus as 5.86″ across. Late in the evening of the 19th, Rümker measured the nucleus as 6.35″ across. That same evening, R. Copeland (Göttingen, Germany) and Vogel independently obtained observations. Copeland said the comet was visible to the naked eye, while a cometseeker revealed a tail about 1° long. Vogel said the comet was situated in a bright field. With the telescope (92×), the coma appeared oval with the largest dimension being 4′. There was also a large condensation measuring 25″ within the coma and the tail was 25′ long. The cometseeker revealed a tail extending 2.5°. Vogel said the cometseeker revealed a tail 3° long on the 20th. On June 21, Rümker measured the nucleus as 5.31″ across. He added that no tail was noted at the end of June because the comet's entrance into the western evening sky enabled it only to be observed in twilight. Also on the 21st, Vogel said the comet appeared unchanged since the 19th. The coma was still oval with the largest width being 3.5′ and contained a condensation 30″ across. The cometseeker revealed a tail 3.0° long. W. Huggins (London, England) described the comet as nearly circular on the 22nd, with a "nearly round spot of light" in the center and a tail nearly 1° long. A. Secchi (Rome, Italy) saw the comet that morning and said it was small and about magnitude 6. The comet attained its most northerly declination of +56° on the 24th. Vogel noted the comet's brightness had "very much decreased" on the 26th, while the tail was extending toward PA 300°. Engelmann did not notice the tail on the 22nd and said the nucleus was still granular in appearance. J. F. J. Schmidt (Athens, Greece) said the comet was bulb-shaped on the 27th, with a coma 5′ across, a nucleus of magnitude 8, and a tail 60′ long. He said the nucleus was magnitude 7–8 through the remainder of June.

The comet was moving away from the sun as July began and it passed closest to Earth on the 1st. That same night Rümker said the comet was "still quite bright" in the 24.4-cm refractor, but was "already considerably fainter" than during his first observations. He added that the nucleus was surrounded by an inner coma 30–40″ across and an outer coma 1.5′ in diameter. Vogel saw the comet with the telescope on the 4th and noted it was "still quite bright." He added that the tail extended 2° toward PA 350°. Schmidt said the nucleus was about magnitude 7–8 on July 2 and 4, while the coma was 4.4′ across on the 6th. Schmidt saw the comet on July 7 and said the coma was 4.3′ across, with a nucleus of magnitude 7–8, and a tail 45′ long. On July 8, Vogel said the tail extended toward PA 0°, while Schmidt said the coma was 4.1′ across, with a nucleus of magnitude 7–8, and a tail 30′ long. Schmidt continued to observe the comet during the period of July 9–16, during which time the nucleus remained between magnitude 7 and 8. The coma and tail were last measured on the 14th, at which time they were 5.2′ across and 5′ long, respectively. The tail steadily decreased during the period.

The comet was last seen on July 17.80, when Schmidt gave the position as $\alpha = 10^h\ 40.7^m$, $\delta = +7°\ 02′$. He said the nucleus was magnitude 8.

Huggins (1868, 1869) obtained spectroscopic observations of this comet on June 22, 23, and 25. He noticed three bright bands and wrote, "The author found this cometic spectrum to agree exactly with a form of the spectrum of carbon which he had observed and measured in 1864." This was the first time a constituent of comets had been identified. Secchi also saw the comet on the 22nd, but noted a solar spectrum.

The first parabolic orbit was calculated by Winnecke using positions from June 14, 16, and 17. It indicated a perihelion date of 1868 June 26.39. A slightly better orbit was determined by F. Tietjen on June 19. He used positions obtained on June 15–19 and determined a perihelion date of June 26.70. Additional orbits were calculated by W. Plummer, C. N. J. Börgen, Copeland, and M. F. Karlinski, which eventually established the perihelion date as June 26.98.

T	ω	Ω (2000.0)	i	q	e
1868 Jun. 26.9763 (UT)	126.6431	54.1124	131.5581	0.578577	1.0

ABSOLUTE MAGNITUDE: $H_{10} = 7.6$ (V1964)
FULL MOON: Jun. 5, Jul. 4, Aug. 3
SOURCES: F. A. T. Winnecke, *MNRAS*, **28** (1868 Jun.), p. 243; F. A. T. Winnecke, *CR*, **66** (1868 Jun. 15), p. 1207; W. Huggins, *PRSL*, **16** (1868 Jun. 18), pp. 481–2; F. A. T. Winnecke and K. C. Bruhns, *AN*, **71** (1868 Jun. 19), p. 317; F. A. T. Winnecke, *CR*, **66** (1868 Jun. 22), pp. 1230–1; F. A. T. Winnecke, F. Tietjen, R. Engelmann, C. N. J. Börgen, and R. Copeland, *AN*, **71** (1868 Jun. 25), pp. 333–6; A. Secchi, *CR*, **66** (1868 Jun. 29), pp. 1299–1302; H. C. Vogel, *AN*, **71** (1868 Jun. 30), p. 351; C. N. J. Börgen, R. Copeland, W. Huggins, and R. Engelmann, *AN*, **71** (1868 Jul. 15), pp. 381–4; W. Plummer, *AN*, **72** (1868 Aug. 5), p. 63; G. F. W. Rümker, *AN*, **72** (1868 Aug. 11), pp. 69, 73; J. F. J. Schmidt, *AN*, **72** (1868 Sep. 4), pp. 117–24; K. C. Bruhns and H. C. Vogel, *AN*, **72** (1868 Nov. 10), pp. 281–3; W. Huggins, *MNRAS*, **29** (1869 Feb.), p. 169; G. F. W. Rümker, *AN*, **74** (1869 Jun. 8), pp. 109–10; M. F. Karlinski, *Annuaire pour l'An 1885. Publié par le Bureau des Longitudes. Avec des Notices scientifiques* (1884), p. 214; V1964, p. 59.

2P/Encke

1868 III

Recovered: 1868 July 17.04 ($\Delta = 1.70$ AU, $r = 1.28$ AU, Elong. $= 48°$)
Last seen: 1868 September 4.14 ($\Delta = 1.26$ AU, $r = 0.44$ AU, Elong. $= 19°$)
Closest to the Earth: 1868 August 27 (1.2331 AU)
Calculated path: TAU (Rec), AUR (Jul. 22), GEM (Aug. 11), CNC (Aug. 22), LEO (Aug. 31)

E. Becker and F. E. von Asten (1868) began with an orbit for the 1862 return, applied perturbations by Jupiter for the period of 1862–68, and predicted the comet would arrive at perihelion on 1868 September 15.18. The comet was recovered by F. A. T. Winnecke (Karlsruhe, Germany) on 1868 July 17.04. Unfortunately, clouds prevented a determination of the position. He described it as a very faint, diffuse nebulosity about 1.5′ across. Winnecke was able to measure the position as $\alpha = 4^h\ 18.0^m$, $\delta = +29°\ 40'$ on July 18.06. The comet must have been especially faint around this time as H. L. d'Arrest (Copenhagen, Denmark) failed to find the comet on a few mornings prior to July 19, and J. F. J. Schmidt (Athens, Greece) failed to locate it on July 20.

The comet was observed by several astronomers during the remainder of July. Winnecke found the comet extremely faint in unfavorable skies on the 20th. Schmidt said he searched for the comet on July 21 under very clear skies, but did not see it with certainty. On July 28, he said there was "hardly a noticeable trace of condensation toward the center." K. C. Bruhns and H. C. Vogel (Leipzig, Germany) first saw the comet on July 23, but noted it was only barely visible because of hazy skies. They found it easier to see on the 25th and described it as very faint and round, with a coma 3′ in diameter that was very slightly brighter in the middle. W. J. Förster (Berlin, Germany) first saw the comet on July 25 and said it appeared rather faint and was difficult to see because of morning twilight. Vogel said the comet was very faint and round on July 26, with a coma 3′ across. D'Arrest also saw the comet on the 26th. Using the large refractor he described it as extraordinarily faint and about 45″ across.

The comet continued to brighten throughout August and passed closest to Earth near the end of the month. Vogel said the comet was rather bright and easy to see on the 14th. He noted the coma was oval with the longest axis extending toward PA 40°. A bright stellar condensation was located in the south preceding side of the coma. Vogel found the comet considerably brighter on the 16th, and although no nucleus was seen, he did note several points of condensation somewhat southward from the center of the coma. Vogel detected three envelopes on the 21st. He said the one closest to the center of the coma was round and 50″ in diameter. The second was eccentrically situated with respect to the first and was found with a diameter of 2.7′. The third was described as tail-like and very faint, and was situated in PA 45°. Vogel found the comet very bright, with a nuclear condensation 20–30″ across, which equaled a star of magnitude 7. Schmidt said that during the period of August 21 to 30, the coma was small, but contained a strong condensation. Meanwhile, the cometseeker revealed that the comet looked like a nebulous star of magnitude 7–8. Magnifications of 200× and 300× failed to reveal a true nucleus. He carefully measured the coma diameter as 1.87′ on the 27th, 1.72′ on the 29th, and 1.72′ on the 30th. Schmidt first recognized a tail on August 25 and estimated its length as 4′. He described it as narrow and straight. He further estimated its length as 10′ on the 29th and 9′ on the 30th.

The comet was last detected on September 4.14 by Vogel at a position of $\alpha = 9^h\ 48.4^m$, $\delta = +17°\ 26'$. Vogel said the field of view was bright and the nucleus was as bright as a 7th-magnitude star.

Minor refinements of this comet's orbit using positions from multiple apparitions, as well as planetary perturbations, have been published by numerous astronomers. Of most significance were the studies of S. P. von

Glasenapp (1871), F. E. von Asten (1875, 1877), and B. G. Marsden and Z. Sekanina (1974). These have typically revealed a perihelion date of September 15.12 and a period of 3.29 years. Marsden and Sekanina determined the nongravitational terms as $A_1 = -0.78$ and $A_2 = -0.03170$. The orbit of Marsden and Sekanina is given below.

T	ω	Ω (2000.0)	i	q	e
1868 Sep. 15.1216 (TT)	183.6462	336.3879	13.1279	0.333602	0.849163

ABSOLUTE MAGNITUDE: $H_{10} = 9.0$ (V1964)

FULL MOON: Jul. 4, Aug. 3, Sep. 2, Oct. 1

SOURCES: E. Becker and F. E. von Asten, *AN*, **71** (1868 May 8), p. 179; F. A. T. Winnecke, H. L. d'Arrest, K. C. Bruhns, H. C. Vogel, and W. J. Förster, *AN*, **72** (1868 Jul. 31), pp. 45–8; H. L. d'Arrest, *AN*, **72** (1868 Aug. 5), p. 63; J. F. J. Schmidt, *AN*, **72** (1868 Sep. 4), p. 123; K. C. Bruhns and H. C. Vogel, *AN*, **72** (1868 Nov. 10), pp. 281–4; J. F. J. Schmidt, *AN*, **72** (1868 Nov. 24), pp. 321–8; S. P. von Glasenapp, *AN*, **78** (1871 Aug. 4), pp. 87–96; F. E. von Asten, *BASP*, **20** (1875), p. 344; F. E. von Asten, *BASP*, **22** (1877), p. 562; V1964, p. 59; B. G. Marsden and Z. Sekanina, *AJ*, **79** (1974 Mar.), pp. 415–16.

7P/1869 G1
(Pons–Winnecke)

1869 I = 1869a

Recovered: 1869 April 9.99 ($\Delta = 0.65$ AU, $r = 1.46$ AU, Elong. $= 123°$)
Last seen: 1869 October 12.99 ($\Delta = 0.76$ AU, $r = 1.71$ AU, Elong. $= 155°$)
Closest to the Earth: 1869 July 6 (0.2341 AU)
Calculated path: LMi (Rec), LYN (Jun. 11), GEM (Jun. 27), AUR (Jul. 4), GEM (Jul. 5), TAU (Jul. 9), ORI-TAU (Jul. 10), ORI (Jul. 16), TAU (Jul. 22), ERI (Aug. 10), CET (Sep. 27)

The comet was missed during the 1863 apparition, despite a prediction by H. Seeling that it would pass perihelion on November 23.69. The problem was probably a combination of the comet's poor placement and the fact that Seeling had not included perturbations by any of the planets. C. Linsser (1869) began with the 1858 orbit, applied perturbations by Jupiter up to 1869, and determined the perihelion date as July 4.11. He later revised his prediction and concluded the comet would arrive at perihelion on June 30.46. F. A. T. Winnecke (Karlsruhe, Germany) recovered this comet with a 5-foot focal length telescope on 1869 April 9.99 and gave the position as $\alpha = 10^h 32.1^m$, $\delta = +33° 57'$. He confirmed motion on April 10.06 and wrote, "This stranger will prove doubtless the expected Comet 1858 II [7P/1858 E1], in the track of which it is situated." Winnecke described it as faint, but centrally condensed, and 6–8' in diameter. O. W. Struve (Pulkovo Observatory, Russia) had searched for the comet with the 38-cm refractor on several occasions during March and into April, before Winnecke's recovery. He said Winnecke's excellent eye and telescope recovered this object, while the smaller field of view of the larger Pulkovo refractor actually hampered his own search attempts. Struve also noted that the comet's path during

March took it through eastern Leo, which he noted was a region containing numerous nebulosities.

During the period of April 10–12, E. Schönfeld (Mannheim, Germany) saw the comet despite unfavorable skies; however, he said it was very diffuse. During the period of April 11–14, T. Wolff (Bonn, Germany) said the comet was exceptionally faint when seen using a heliometer. On April 14, Winnecke commented that the comet appeared distinctly brighter and was 8' across, while K. C. Bruhns and H. C. Vogel (Leipzig, Germany) said the comet was faint because of a hazy sky. Bruhns and Vogel also noted it was large and round, with very diffuse edges. G. F. W. Rümker and F. R. Helmert (Hamburg, Germany) described the comet as very faint on the 15th.

The comet was widely observed during the last days of April. On the 28th, Vogel estimated the diameter of the coma as 4'. He added that the comet was round and gradually became brighter in the middle. D. M. A. Möller (Lund, Sweden) said the comet appeared very faint and large, with two light spots within the coma. The preceding spot was the brightest. Rümker and Helmert examined the comet with a magnification of 180× and said it was a round nebulosity over 2' across, with a strong central condensation 40" across. A nucleus was occasionally suspected. On the 29th, E. Weiss (Vienna, Austria) saw the comet with a 15-cm refractor and said it was a large and diffuse nebulosity, with an "eccentric lying condensation of granular appearance." Vogel said the air was very transparent, which made the comet appear considerably brighter than on the 28th. The coma was 3.5' across, which was somewhat brighter in the center. Vogel added that a well-defined round condensation about 20" across was at the center and occasionally exhibited a fine stellar point. Möller said the comet seemed more difficult to observe than on the 28th. He also still noted two points of light within the coma, with the brighter point following the fainter one. Rümker and Helmert said two nuclei were seen very close together. H. Wortham (Royston, England) saw the comet with a 15-cm refractor (50×) and said it was "distinctly seen as a faint nebulous patch of some little size, appearing occasionally to brighten somewhat to a centre. . . . " On the 30th, Vogel said the comet was unchanged since the 29th, although the center was brighter and a stellar point of magnitude 13 was at the center of the coma. Schönfeld said the comet was 5' across and looked like a diffuse nebulosity with a slightly brighter center. Möller said a magnification of 250× revealed four or five points of light within the coma. A lesser magnification simply showed a central condensation.

The comet continued to approach the sun and Earth during May. Weiss saw the comet in the refractor on the 1st and said it was still a diffuse nebulosity, with an eccentrically situated condensation. Schönfeld saw a very weak nucleus on the 2nd and 5th. Also on the 5th, Vogel said the comet was rather faint and about 4' across. It was gradually brighter towards the center, with the bright area being about 45" across and exhibiting a granular appearance. J. F. J. Schmidt (Athens, Greece) described the comet as very faint and weakly condensed on the 8th, but added it was visible in the refractor's finder. He measured the coma diameter as 5.01'. Schmidt noted a condensation, but no nucleus on the 11th, and gave the coma diameter as 5.40'. S. J. Perry (Stonyhurst College Observatory, England) and Vogel independently saw the comet the same night. Perry saw "a slight condensation towards the centre" with his 20-cm refractor (100×), while Vogel said the coma was centrally condensed, about 4' across, and exhibited a weak nebulosity extending toward PA 338°. Descriptions came from Schmidt, Schönfeld, and Perry on the 12th. Schmidt said the magnitude of the central condensation was 12 and noted that some nebulosity was extending away from the sun. The coma diameter was 6.20'. Schönfeld said, "I believed to have noticed several obvious nuclei." Perry said there was "a slight condensation towards the centre, but no decided nucleus." Schmidt said the central condensation was about magnitude 11–12, while the nebulosity seemed granular. He measured the coma diameter as 6.05'. N. C. Dunér (Lund, Sweden) and Vogel saw the comet on the 14th. Dunér simply noted the comet was very large and faint, while Vogel saw a tail-like extension toward PA 330° and a stellar nucleus near the coma's center. Dunér said the comet was brighter and more condensed on the 15th, but fainter on the 16th. On May 15 and 16, Rümker and Helmert estimated the magnitude of the nucleus as 11–12. Schönfeld noted a very faint nucleus on the 17th. Moonlight interfered for several days and the next physical descriptions came from Schmidt, Vogel, and Weiss on the 28th. Schmidt gave the magnitude of the central condensation as 11–12. He added that on the sunward side of the coma, the nebulosity was closer to the condensation and broader. He measured the coma diameter as 6.54', with the preceding half being 3.65' across. Vogel saw the comet through cirrus clouds and gave the coma diameter as 2.5–3'. He also saw a very small central condensation. Weiss said the 15-cm refractor revealed the comet was rather bright, with traces of a tail. Schmidt also saw the comet on the 30th and said it had changed little since the 28th, with the exception that the central condensation was brighter. He measured the coma as 8.02' in diameter and said the preceding half was 4.36' across.

Although the comet continued to approach the sun and Earth during June, it was also moving toward evening twilight and became temporarily lost after mid-month. Schmidt gave the magnitude of the central condensation as 10–11 on the 1st and measured the coma diameter as 8.47'. He added that the preceding half of the coma was 4.68' across. Vogel also saw the comet on this night in poor sky conditions and said it appeared granular. On the 2nd, Vogel observed with "quite transparent" skies and said the comet was bright, about 2.5' across, and exhibited a slight central condensation. He added that the comet more resembled a star cluster than a nebula. Rümker and Helmert also saw the comet this night and said it was very large, with the bright central part measuring about 2.5' across. Schmidt said the comet's appearance on the 3rd was similar to that of the 1st, except the central condensation was magnitude 10 and the coma was 8.82' across. He added that the preceding half of the coma was 5.29′ across. Schmidt said the comet was still basically unchanged on the 9th, except that the central condensation was magnitude 9–10. He added that the coma diameter was 8.62′, with the preceding half being 4.99′. Vogel and Weiss saw the comet on the 12th. Vogel observed in poor conditions, but said the coma was 2′ across and "completely had the appearance of a star cluster." Weiss observed with a 15-cm refractor and said the comet seemed to possess a double nucleus, while traces of a tail were visible. On June 6, 7, and 12, Rümker and Helmert said the diameter of the bright central part was 1–1.4′. Schmidt said the comet's appearance was still unchanged on the 13th, except that the condensation was magnitude 9.

Following June 13, observations ceased at most locations as evening twilight allowed only low-altitude observations. Vogel saw the comet on June 21, despite the bright field of view, and Schmidt saw the comet at an altitude of 8° or 9° on June 25 and 26. Schmidt said the nucleus was magnitude 8 and the diameter of the coma was only 30″. The comet passed perihelion on June 30, passed 7° from the sun on July 3, and then passed closest to Earth on July 6. Its movement out of the sun's glare was slow and its decreasing brightness aided in making it difficult to find. It was finally recovered on August 8.07 by Wolff and F. W. A. Argelander (Bonn, Germany) and on August 8.09 by F. Tietjen (Berlin, Germany). Bruhns said the comet was large and as bright as the nebulae in Ophiuchus on the 11th. He added that the condensation was 1′ across. Weiss found the comet with a 15-cm refractor on the 12th, but it disappeared in the morning twilight before a position could be obtained. On August 13, Weiss and T. R. von Oppölzer said the comet "had a granular appearance, and in a bright halo of diffuse light, a beautiful, eccentric lying star-like nucleus." Rümker saw the comet through clouds on August 16, and on the 19th he estimated it was about magnitude 9.

Bruhns said the comet was several minutes across and very diffuse on September 1. Vogel said the comet was rather bright and 3′ across on the 4th. He added that the strongest condensation was on the south-following side of the comet's center. Möller said the comet was easy to observe on the 5th. Vogel said the comet was rather bright and 3.5′ across on the 6th. He added that the edges of the comet were diffuse, and the rather strong condensation was on the south-following side of the comet's center. Möller also saw the comet on the 6th and said the coma contained two points of light within the same hour angle, with the more southerly being brightest. Vogel said the comet was 4′ across with a bright center on the 9th, while Dunér said the comet was large and diffuse. On September 10, Dunér said the comet was faint and condensed.

Vogel said the comet was very faint, round, and 2.5′ across on October 7. It also gradually brightened towards the center. On October 9, Vogel said the sky was very transparent and the comet appeared rather faint, with little condensation in the middle. Vogel said the comet was extremely faint on the

11th, with a slight central condensation. The comet was last seen on October 12.99, when Weiss determined the position as $\alpha = 2^h\ 05.7^m$, $\delta = -13°\ 58'$. He observed with a 15-cm refractor and described the comet as an uncommonly faint, pale, diffuse nebulosity, in which a "star-like nucleus occasionally flashes."

A few weeks after the comet was recovered, Linsser (1869) took several positions and corrected his earlier prediction to reveal a perihelion date of June 30.44. Later calculations using multiple apparitions and planetary perturbations were published by Oppölzer (1870, 1880), J. F. Encke (1871), E. F. von Haerdtl (1888, 1889), and B. G. Marsden (1970, 1972). These revealed a perihelion date of June 30.44 and a period of 5.59 years. Oppölzer (1880) noted that the positions of 1858, 1869, and 1875 indicated a slight secular acceleration was present. This was later confirmed by B. G. Marsden (1970, 1972), when he determined nongravitational terms of $A_1 = +0.27$ and $A_2 = -0.0072$. This orbit is given below.

T	ω	Ω (2000.0)	i	q	e
1869 Jun. 30.4419 (TT)	162.4566	115.2939	10.7970	0.781520	0.751927

ABSOLUTE MAGNITUDE: $H_{10} = 9.6$ (V1964)

FULL MOON: Mar. 27, Apr. 26, May 25, Jun. 24, Jul. 23, Aug. 22, Sep. 20, Oct. 20

SOURCES: J. R. Hind, *MNRAS*, **23** (1863 Jun.), pp. 255–7; C. Linsser, *BASP*, **13** (1869), pp. 454–8; O. W. Struve and F. A. T. Winnecke, *BASP*, **14** (1869), pp. 248–50; F. A. T. Winnecke, *AN*, **73** (1869 Apr. 10), p. 383; F. A. T. Winnecke, K. C. Bruhns, F. W. A. Argelander, and T. Wolff, *AN*, **74** (1869 Apr. 22), pp. 13–16; F. A. T. Winnecke, H. Wortham, and S. J. Perry, *MNRAS*, **29** (1869 May), p. 299; E. Weiss and C. Linsser, *AN*, **74** (1869 May 10), p. 43; G. F. W. Rümker and F. R. Helmert, *AN*, **74** (1869 Jun. 8), p. 105; J. F. J. Schmidt, K. C. Bruhns, H. C. Vogel, E. Schönfeld, F. W. A. Argelander, and T. Wolff, *AN*, **74** (1869 Aug. 2), pp. 227–38; G. F. W. Rümker and F. R. Helmert, *AN*, **74** (1869 Sep. 3), pp. 285–8; F. W. A. Argelander and T. Wolff, *AN*, **75** (1869 Nov. 1), pp. 29–32; H. C. Vogel, *AN*, **75** (1869 Nov. 16), p. 61; T. R. von Oppölzer, *SAWW*, **62** Abt. II (1870), pp. 655–75; K. C. Bruhns and H. C. Vogel, *AN*, **75** (1870 Jan. 27), p. 197; D. M. A. Möller and N. C. Dunér, *AN*, **75** (1870 Jan. 27), p. 201; E. Weiss and T. R. von Oppölzer, *AN*, **75** (1870 Mar. 3), pp. 273–6; J. F. Encke, *AN*, **77** (1871 May 30), p. 313; F. Tietjen, *AN*, **81** (1873 Jun. 4), p. 345; T. R. von Oppölzer, *AN*, **97** (1880 May 27), pp. 149–54; T. R. von Oppölzer, *AN*, **97** (1880 Jul. 30), pp. 337–42; E. F. von Haerdtl, *AN*, **120** (1888 Dec. 28), pp. 257–72; E. F. von Haerdtl, *DAWW*, **55** Abt. II (1889), pp. 230–43; E. F. von Haerdtl, *DAWW*, **56** Abt. II (1889), pp. 151–85; V1964, p. 60; B. G. Marsden, *AJ*, **75** (1970 Feb.), pp. 80–1; B. G. Marsden, *QJRAS*, **13** (1972), pp. 428–9; B. G. Marsden, *CCO*, 12th ed. (1997), pp. 48–9.

C/1869 T1	*Discovered:* 1869 October 12.2 ($\Delta = 1.82$ AU, $r = 1.23$ AU, Elong. $= 40°$)
(Tempel)	*Last seen:* 1869 November 13.16 ($\Delta = 1.35$ AU, $r = 1.34$ AU, Elong. $= 68°$)
	Closest to the Earth: 1869 November 29 (1.2537 AU)
1869 II = 1869b	*Calculated path:* SEX (Disc), HYA (Oct. 27), ANT (Nov. 12)

E. W. L. Tempel (Marseille, France) discovered this comet on 1869 October 12.2, at a position of $\alpha = 10^h\ 34.0^m$, $\delta = +2°\ 10'$. He simply described it as a faint comet.

The comet had just passed perihelion when discovered, and, although it was still heading toward its closest passage by Earth, the distance was too great to prevent the comet from fading. E. Weiss (Vienna, Austria) observed with his 15-cm refractor throughout this apparition and on October 13 described the comet as rather bright and round, with a central, nebulous condensation. He noted that after morning twilight began, a stellar nucleus became visible within the condensation. On the 14th, Weiss described the comet as a faint nebulosity under mostly cloudy conditions. H. C. Vogel (Leipzig, Germany) said the comet was rather bright and round, with a coma 1.2' across and a central stellar nucleus of magnitude 11. Weiss described the comet as round and 1.5' across on the 28th, with a stellar nucleus. On November 1, Weiss saw the comet with the 15-cm refractor and said it was rather bright, "with a pretty, star-like nucleus." Of this observation on November 1, 2, and 8, E. J. M. Stephan (Marseille, France) wrote, "The Comet is round, without a tail but with a quite apparent point of condensation; it is rather faint and is quickly moving Southward, which I fear will end observations."

The comet was seen for the final time on November 13.16, when Weiss measured the position as $\alpha = 10^h\ 16.6^m$, $\delta = -29°\ 35'$. He observed the comet with a 15-cm refractor and said that, despite the comet's low altitude, it was still rather bright, with an eccentric, stellar nucleus.

The first parabolic orbit was calculated by J. R. Hind. Using positions obtained up to October 18, he determined the perihelion date as 1869 October 8.94. The actual perihelion date was quickly isolated to October 10 by G. Leveau, H. C. Vogel, H. Oppenheim, and T. R. von Oppölzer. Unfortunately, since the comet was only observed for one month, only a parabolic orbit could be calculated and this continued to be revised during the next few years by W. Grünert (1870), A. Seydler (1871), J. Kowalczyk (1873), and A. W. Doberck (1874). Doberck's orbit is given below.

T	ω	Ω (2000.0)	i	q	e
1869 Oct. 10.3497 (UT)	188.1940	313.3275	111.6854	1.230731	1.0

ABSOLUTE MAGNITUDE: $H_{10} = 5.1$ (V1964)

FULL MOON: Sep. 20, Oct. 20, Nov. 19

SOURCES: E. W. L. Tempel, *AN*, **74** (1869 Oct. 15), p. 383; J. R. Hind, *MNRAS*, **30** (1869 Nov.), p. 27; E. Weiss and H. C. Vogel, *AN*, **75** (1869 Nov. 1), p. 31; H. C. Vogel and T. R. von Oppölzer, *AN*, **75** (1869 Nov. 16), p. 63; H. Oppenheim and E. J. M. Stephan, *AN*, **75** (1869 Nov. 22), pp. 75–8; G. Leveau, *AN*, **75** (1869 Dec. 10), p. 109; G. Leveau, *MNRAS*, **30** (1870 Jan.), pp. 74–5; K. C. Bruhns and H. C. Vogel, *AN*, **75** (1870 Jan. 27), p. 197; E. Weiss and W. Grünert, *AN*, **75** (1870 Mar. 3), p. 275; A. Seydler, *SAWW*, **63** Abt. II (1871), pp. 421–2; J. Kowalczyk, *AN*, **81** (1873 Mar. 22), p. 143; A. W. Doberck, *MNRAS*, **34** (1874 Jun.), p. 426; V1964, p. 60.

**11P/1869 W1
(Tempel-Swift–
LINEAR)**

1869 III = 1869c

Discovered: 1869 November 27.83 ($\Delta = 0.26$ AU, $r = 1.07$ AU, Elong. $= 102°$)
Last seen: 1870 January 4.07 ($\Delta = 0.34$ AU, $r = 1.22$ AU, Elong. $= 128°$)
Closest to the Earth: 1869 December 7 (0.2485 AU)
Calculated path: PEG (Disc), AND (Dec. 10), PSC (Dec. 15), TRI (Dec. 21),
ARI (Dec. 23)

E. W. L. Tempel (Marseille, France) discovered this comet on 1869 November 27.83 at a position of $\alpha = 22^h 45.0^m$, $\delta = +14° 16'$. The comet was already about a week past perihelion. Physical descriptions were provided by H. C. Vogel (Leipzig, Germany) and E. Weiss (Vienna, Austria) on the 29th. Vogel described the comet as rather faint, large, and diffuse. He said the refractor was set to a magnification of 96× and revealed a coma 3' across, while the wider field of view of the cometseeker revealed a coma 6' across. Weiss observed with a 15-cm refractor and described the comet as round, pale, very diffuse, and about 5' across. He added that it was only weakly condensed.

The comet was fairly well observed during December and passed closest to Earth on the 7th. E. Schönfeld (Mannheim, Germany) reported the comet was diffuse with a weak central condensation on the 1st, 5th, 7th, and 8th. He added on the 1st that the coma was definitely more than 2' across and possibly 2.5' across. Vogel said the refractor revealed the comet as very faint and very large on the 7th, with only slight central condensation. He noted an extension toward PA 300°. Vogel and Weiss each saw the comet on the 8th. Vogel said the comet was very faint, about 5' across, and exhibited very little condensation. Weiss saw the comet with a 15-cm refractor and said it was quite large and diffuse. The brightest portion of the comet was eccentrically situated to the east of the coma's center. Independent observations were made by F. Tischler (Königsberg, now Kaliningrad, Russia) and Vogel on the 9th. Tischler said the comet was uncommonly faint and diffuse, only being visible after his eye had dark adapted. Vogel noted the comet was very faint, round, and difficult to see in moonlight. Vogel said the comet was observed with difficulty in moonlight on the 10th. After the moon had left the sky, Vogel said the comet was "noticed only with effort" on the 21st. On December 29, K. C. Bruhns (Leipzig, Germany) said the comet was not seen with certainty, because of its close proximity to an 8th-magnitude star. He said the comet was very faint in the refractor (96×) on the 30th. G. Strasser (Kremsmünster, Austria) said the comet was uncommonly faint during the period spanning December 7–31. Schönfeld saw the comet on the evenings of December 30 and January 1. He said only a very weak trace was visible, and it was more doubtful on the 1st.

The final observations of this comet were made with the 38-cm refractor at Harvard College Observatory, USA on 1870 January 1.12 and January 4.07. On the last date the position was measured as $\alpha = 3^h 10.4^m$, $\delta = +26° 30'$. The comet was described as faint, without a nucleus, and about 2' across.

The first parabolic orbit was calculated by B. Tiele. Using positions spanning the period of November 29 to December 7, he determined the perihelion date as 1869 November 20.82. Similar orbits were calculated by L. Schulhof, T. R. von Oppölzer, Bruhns, and K. L. von Littrow.

The first elliptical orbit was calculated by L. Schulhof and J. F. Bossert (1880), following the accidental recovery of this comet by Swift in 1880. They determined the perihelion date as November 19.30 and the period as 5.48 years. K. Zelbr (1880) and S. C. Chandler, Jr (1880) made similar assumptions that the period was either 5.5 or 11 years within the next couple of weeks. Both favored the 5.5-year period. Orbits using multiple apparitions were determined by Bossert (1886) and B. G. Marsden (1971). They found perihelion dates of November 19.31 and a period of 5.48 years. Marsden determined the nongravitational terms as $A_1 = +0.04$ and $A_2 = -0.0459$. Marsden's orbit is given below.

T	ω	Ω (2000.0)	i	q	e
1869 Nov. 19.3053 (TT)	106.0659	298.7378	5.4085	1.063102	0.658121

ABSOLUTE MAGNITUDE: $H_{10} = 11.4$ (V1964)

FULL MOON: Nov. 19, Dec. 18, Jan. 17

SOURCES: E. W. L. Tempel, *AN*, **75** (1869 Nov. 30), p. 95; K. C. Bruhns, H. C. Vogel, K. L. von Littrow, and E. Weiss, *AN*, **75** (1869 Dec. 10), p. 109; B. Tiele, *AN*, **75** (1869 Dec. 16), p. 127; T. R. von Oppölzer, *AN*, **75** (1869 Dec. 23), p. 143; L. Schulhof, *MNRAS*, **30** (1870 Jan.), p. 75; K. C. Bruhns and K. L. von Littrow, *AN*, **75** (1870 Jan. 19), pp. 181–4; K. C. Bruhns and H. C. Vogel, *AN*, **75** (1870 Jan. 27), pp. 197–200; E. Weiss and G. Strasser, *AN*, **75** (1870 Mar. 3), pp. 275–8, 283; F. Tischler, *AN*, **76** (1870 May 27), pp. 51, 56; E. Schönfeld, *AN*, **76** (1870 Jul. 9), p. 123; L. Schulhof and J. F. Bossert, *AN*, **99** (1880 Dec. 23), pp. 11–16; K. Zelbr, *AN*, **99** (1880 Dec. 28), pp. 17–20; S. C. Chandler, Jr, *AN*, **99** (1880 Dec. 31), p. 45; E. C. Pickering, *AN*, **99** (1881 Jan. 27), p. 95; J. Winlock and E. C. Pickering, *Annals of the Astronomical Observatory of Harvard College*, **13** (1882), p. 176; J. F. Bossert, *BA*, **3** (1886 Feb.), pp. 65–78; V1964, p. 60; B. G. Marsden and Z. Sekanina, *AJ*, **76** (1971 Dec.), pp. 1142–3; B. G. Marsden, *CCO*, 12th ed. (1997), pp. 50–1.

C/1870 K1
(Winnecke–
Tempel)

1870 I = 1870a

Discovered: 1870 May 30.04 ($\Delta = 1.67$ AU, $r = 1.27$ AU, Elong. $= 49°$)

Last seen: 1870 July 10.05 ($\Delta = 0.52$ AU, $r = 1.01$ AU, Elong. $= 74°$)

Closest to the Earth: 1870 July 18 (0.4166 AU)

Calculated path: PSC (Disc), ARI (Jun. 28), PSC (Jul. 2), CET (Jul. 4), ERI (Jul. 10)

F. A. T. Winnecke (Karlsruhe, Germany) discovered this comet on 1870 May 30.04 at a position of $\alpha = 0^h\ 49.0^m$, $\delta = +29°\ 06'$. He said it was fairly bright, round, about 2.5′ across, with a strong central condensation. E. W. L. Tempel (Marseille, France) independently discovered the comet on May 30.06. The comet was confirmed on May 31 by H. C. Vogel (Leipzig, Germany),

E. Weiss (Vienna, Austria), and Winnecke. Vogel described it as rather bright and 2' across, with a strong central condensation. Weiss said it was a round, diffuse nebulosity, with a rather bright, nearly stellar nucleus.

The comet was widely observed during June as its distances from the sun and Earth decreased. S. J. Perry (Stonyhurst College Observatory, England) picked up the comet at low power almost immediately on the 7th even though twilight and moonlight distinctly illuminated the field of view. He added that the comet was not far above the eastern horizon. Vogel described the comet as rather faint, with a nucleus on the 15th because of moonlight, but reported it was bright and large on the 16th, with a central condensation. Vogel also found the comet bright and 2–3' across on the 17th and 19th. He added that it exhibited a tail toward PA 270° on the last date. F. R. Helmert (Hamburg, Germany) said the comet was quite bright in a refractor on the 20th and noted a granular structure in the coma. In addition, he noted an envelope that was convex with a diameter of 20". Vogel described the comet as "bright with a strong, nuclear condensation" on the 22nd and saw an extension toward PA 270°. J. F. J. Schmidt (Athens, Greece) began observing the comet before the end of June and provided extensive details of the comet's appearance. On June 26, he indicated the coma was 3.36' across and exhibited a central condensation of magnitude 7–8. There was also a short, wide tail 4.0' long. On June 30, Schmidt indicated the coma was 3.86' across and exhibited a central condensation of magnitude 7–8. There was also a short, wide tail 5.2' long.

The comet passed closest to the sun and Earth around the middle of July, but moonlight and the comet's rapid movement toward the skies of the Southern Hemisphere ultimately ended observations. Schmidt said the coma was 4.22' across on the 3rd and exhibited a central condensation of magnitude 7–8. There was also a short, wide tail 5.6' long. J. Michez (Bologna, Italy) observed the comet with a 6-cm refractor on July 7, 8, and 9, and described it as a small, faint nebulous object with a distinct nucleus. He added it was then about 14–16° above the horizon. On July 7, Schmidt indicated the coma was 4.94' across and exhibited a central condensation of magnitude 7–8. There was also a short, wide tail 9.9' long. On July 8, Schmidt indicated the coma was 5.66' across and exhibited a central condensation of magnitude 7–8. There was also a short, wide tail 9.4' long.

This comet was last detected on 1870 July 10.05, by Schmidt. He determined the position as $\alpha = 2^h 33.5^m$, $\delta = -6° 18'$. Schmidt also indicated the coma was 5.52' across and exhibited a central condensation of magnitude 7–8. There was also a short, wide tail 20' long. Schmidt said he looked for the comet in full moonlight on July 13.06, but did not find it. He commented, "I would have surely still observed it, if the ephemeris had continued to precalculate the place of the comet one more week."

C. J. E. Wolf and G. A. P. Rayet were able to detect the spectrum of this comet on several occasions, beginning during the first days of June. They reported a continuous spectrum with the three luminous bands of carbon.

The first parabolic orbit was calculated by E. Becker, using positions from May 30, June 3, and 5. He determined the perihelion date as 1870 July 14.81. Using positions obtained through June 6, independent calculations were made by Winnecke and H. Oppenheim which gave perihelion dates of July 13.37 and July 14.40, respectively. Following the comet's apparition, A. Seydler (1871) and J. Dreyer (1872) independently calculated parabolic orbits with perihelion dates of July 14.58. Dreyer's orbit is given below.

T	ω	Ω (2000.0)	i	q	e
1870 Jul. 14.5822 (UT)	198.2279	143.5674	121.7835	1.008692	1.0

ABSOLUTE MAGNITUDE: $H_{10} = 6.2$ (V1964)
FULL MOON: May 15, Jun. 13, Jul. 12
SOURCES: F. A. T. Winnecke and S. J. Perry, *MNRAS*, **30** (1870 Jun.), pp. 208–9; F. A. T. Winnecke and H. C. Vogel, *AN*, **76** (1870 Jun. 4), p. 79; E. Becker and H. Oppenheim, *AN*, **76** (1870 Jun. 17), pp. 93–6; E. Weiss, E. W. L. Tempel, and F. A. T. Winnecke, *AN*, **76** (1870 Jun. 24), pp. 109–12; C. J. E. Wolf and G. A. P. Rayet, *CR*, **71** (1870 Jul. 4), p. 49; G. F. W. Rümker and F. R. Helmert, *AN*, **76** (1870 Aug. 20), p. 221; J. F. J. Schmidt, P. Tacchini, and J. Michez, *AN*, **76** (1870 Sep. 15), pp. 281–4; K. C. Bruhns and H. C. Vogel, *AN*, **76** (1870 Oct. 27), p. 355; A. Seydler, *SAWW*, **64** Abt. II (1871 Jul.), pp. 153–4; G. F. W. Rümker, *AN*, **77** (1871 Apr. 18), p. 247; J. Dreyer, *AN*, **80** (1872 Nov. 9), pp. 219–22; V1964, p. 60.

C/1870 Q1
(Coggia)

1870 II = 1870b

Discovered: 1870 August 28.98 ($\Delta = 1.22$ AU, $r = 1.82$ AU, Elong. $= 109°$)
Last seen: 1870 December 25.8 ($\Delta = 2.53$ AU, $r = 2.31$ AU, Elong. $= 66°$)
Closest to the Earth: 1870 September 27 (0.8847 AU)
Calculated path: CET (Disc), ARI (Sep. 5), PSC (Sep. 19), AND (Sep. 27), PEG (Oct. 3), CYG (Oct. 30), VUL (Nov. 20), CYG (Dec. 7)

J. E. Coggia (Marseille, France) discovered this comet on 1870 August 28.98, at a position of $\alpha = 3^h 07.1^m$, $\delta = +5° 35'$. As seen through the 80-cm reflector, the comet was described as rather bright, round, with a diameter of 2'. There was also a nucleus at the center of the coma. Confirmation came on August 30.15, when E. J. M. Stephan (Marseille, France) determined a precise position. Interestingly, Coggia sent the first position and an incorrect daily motion of -24^s in α and $-1' 20''$ in δ to several observatories. Astronomers at Vienna received his telegram on September 1 and unsuccessfully searched for the comet on the night of September 1/2, under hazy skies. T. R. von Oppölzer searched a wider area on September 2 and found the comet at a position of $\alpha = 2^h 54.0^m$, $\delta = +8° 27'$. He gave a correct daily motion of $-2^m 36^s$ in α and $+34'$ in δ.

The comet passed perihelion at the beginning of September and passed closest to Earth near the end. K. C. Bruhns (Leipzig, Germany) saw the comet in the morning sky of the 5th and described it as fairly round, about 1' across, with an eccentric condensation. H. C. Vogel (Leipzig, Germany) saw the comet in the late evening hours of the 5th and described it as "quite bright with a central condensation." He estimated the diameter as 2'. Vogel found the comet faint in moonlight on the 9th. Vogel said the comet was quite bright and 1.5' across on the 19th, with a central condensation. On September 20, Vogel and Albrecht found the comet quite bright and 2.7' across, with a distinct nucleus, while J. J. Plummer (Durham Observatory, England) said the comet was "sufficiently bright to be easily observed." J. F. J. Schmidt (Athens, Greece) measured the coma diameter as 7.34' on the 21st. H. Gedmüyden (Christiania, now Oslo, Norway) said the comet was easy to see on the 21st and 23rd and exhibited a distinct nucleus. G. F. W. Rümker (Hamburg, Germany) described the comet as a very large oval, without a tail on the 24th. He also noted a rather sharp condensation. Rümker found the comet slightly fainter on the 25th than on the 24th. He saw the comet on the 26th at a magnification of 260× and noted a small stellar nucleus of magnitude 10.5. Gedmüyden said the comet was easy to see and exhibited a distinct nucleus on the 28th. Rümker and C. F. Pechüle said a magnification of 120× resolved the condensation into several bright points on the 29th. Schmidt measured the coma diameter as 7.65' on the 29th and noted it remained an easy object to measure both because of its brightness and substantial central condensation. He also saw a nucleus of magnitude 12, which remained distinct even at a magnification of 300×. Although Schmidt said the coma was oval, he was never sure that a tail was recognizable. On September 30, Rümker and Pechüle said the comet was very bright, but with no tail. A magnification of 90× revealed several bright points in the condensation.

The comet was moving away from both the sun and Earth as October began. J. Joynson (Waterloo, England) described the nucleus as "tolerably bright" on the 1st. Joynson found the comet fainter on the 2nd, while Plummer described it as "still tolerably bright but more diffused." Gedmüyden said the comet was easy to see and exhibited a distinct nucleus on the 3rd, while Joynson said the nucleus was very faint in a moonlit sky. On October 4, Joynson said the nucleus was barely visible in a moonlit sky. Strong moonlight was present during the next few days, but descriptive details again became available on October 12, when Rümker noted the comet was very faint as a result of moonlight and clouds. Gedmüyden also saw the comet on this evening and said it was fainter than when previously seen, but still exhibited a nucleus. On the 17th, Rümker and Pechüle said the comet was very large with several bright points, one of which was particularly bright and stellar. They again noted it was very large on the 18th and added that the nucleus was distinct at 120×, rather distinct at 180×, and no longer visible at 260×. On October 26, Plummer said the comet was "much fainter than when last observed, but not smaller." Gedmüyden said the comet was faint, with almost no trace of a nucleus on the 29th.

Moonlight again interfered as November began. Rümker and Pechüle said the comet was rather faint because of moonlight on the 2nd. No observations were made during the period of November 9–12. Plummer tried to find the comet on November 11 and 14, but it could not be seen. Schmidt saw the comet on the 15th and noted it was difficult to measure the position because of its faintness and weakly condensed coma. He measured the coma diameter as 2.98'. Schmidt said the comet continued to be visible in the refractor's finder on the 17th, 19th, and 25th. He measured the coma diameter on those dates as 3.40', 2.93', and 2.88', respectively. Rümker and Pechüle said the comet was still quite bright on the 24th, although fainter than when previously seen. They added that a faint, stellar nucleus was visible in the center of the nebulosity. On November 30, Rümker and Pechüle said the comet was rather faint because of moonlight.

Rümker and Pechüle again noted the comet was rather faint because of moonlight on December 1. During the period of December 2–10, no observations were made because of moonlight and the comet's decreasing brightness. Schmidt saw the comet every evening during the period of December 16–20. He said the comet was faint, with little condensation, although he noted it was easier to see than 6P/d'Arrest during that time. Rümker and Pechüle said the comet was difficult to see on the 21st and very large, faint, and diffuse on the 22nd.

The position of the comet was measured for the final time on December 23.85, when Pechüle determined it as $\alpha = 21^h\ 02.1^m$, $\delta = +28°\ 55'$. Rümker and Pechüle said it was then very faint. The comet was last seen on December 25.8, when Rümker and Pechüle noted it was difficult to see and said several small stars shining through the coma prevented a position measurement. Rümker said moonlight and bad weather followed. By the time searches could be resumed in January, the comet could not be found.

The first parabolic orbit came from Oppölzer. He took positions from August 29, September 2, and 5, and determined a perihelion date of 1870 September 4.29. The orbit was steadily refined during the next few months by J. R. Hind, H. Seeliger, J. Palisa, and T. N. Thiele. J. Gerst (1872) was the first person to utilize positions spanning the comet's entire period of visibility and determined the perihelion date as September 2.70. A. Schobloch (1896) and R. J. Buckley (1979) began their studies of the comet's orbit with 311 positions. Schobloch determined a perihelion date of September 2.69, while Buckley gave it as September 2.67. Although Buckley noted a hyperbolic orbit fitted the positions slightly better than the parabola, he preferred the parabola because of the comet's large perihelion distance, and this is given below.

T	ω	Ω (2000.0)	i	q	e
1870 Sep. 2.6676 (TT)	354.9389	14.7590	99.3580	1.816602	1.0

ABSOLUTE MAGNITUDE: $H_{10} = 3.8$ (V1964)

FULL MOON: Aug. 11, Sep. 9, Oct. 9, Nov. 8, Dec. 8, Jan. 6

SOURCES: J. E. Coggia, *CR*, **71** (1870 Aug. 29), p. 405; J. E. Coggia, *AN*, **76** (1870 Sep. 7), p. 255; E. J. M. Stephan, J. E. Coggia, R. Luther, K. C. Bruhns, and T. R. von Oppölzer, *AN*, **76** (1870 Sep. 15), pp. 285–8; J. Palisa, *AN*, **76** (1870 Oct. 12), pp. 333–6; J. R. Hind, *AN*, **76** (1870 Oct. 19), p. 341; K. C. Bruhns, H. C. Vogel, Albrecht, and H. Seeliger, *AN*, **76** (1870 Oct. 27), p. 355; S. J. Perry, *MNRAS*, **31** (1870 Nov.), pp. 28–9; T. N. Thiele, *AN*, **77** (1870 Nov. 30), p. 23; J. Joynson, *MNRAS*, **31** (1870 Dec.), pp. 43–4; J. F. J. Schmidt, *AN*, **77** (1870 Dec. 23), pp. 69–72; T. N. Thiele, *MNRAS*, **31** (1870 Jan.), p. 87; J. F. J. Schmidt, *AN*, **77** (1871 Feb. 14), p. 143; J. J. Plummer, *AN*, **77** (1871 Feb. 21), pp. 155–8; G. F. W. Rümker and C. F. Pechüle, *AN*, **77** (1871 Apr. 18), pp. 241–4; H. Gedmüyden, *AN*, **77** (1871 May 30), pp. 311–14; J. Gerst, *AN*, **80** (1872 Nov. 21), p. 237; A. Schobloch, *AN*, **141** (1896 Oct. 16), pp. 377–402; V1964, p. 60; R. J. Buckley, *JBAA*, **89** (1979 Oct.), pp. 584–5.

6P/d'Arrest   *Recovered:* 1870 August 31.87 ($\Delta = 0.81$ AU, $r = 1.31$ AU, Elong. $= 91°$)
Last seen: 1870 December 20.71 ($\Delta = 1.55$ AU, $r = 1.64$ AU, Elong. $= 77°$)
1870 III = 1870c   *Closest to the Earth:* 1870 August 17 (0.8085 AU)
Calculated path: OPH (Rec), SER (Sep. 11), OPH (Sep. 12), SGR (Sep. 19), CAP (Oct. 24), PsA (Nov. 12), AQR (Nov. 21)

G. Leveau (1869) took an orbit computed for 1863 and applied perturbations from Mars, Jupiter, and Saturn in order to determine this comet's motion for the period of 1863–70. The resulting perihelion date was 1870 September 23.06. He provided a daily ephemeris for the period of 1870 January 31 to 1871 January 2. In a letter dated 1870 June 8 and published in the *Monthly Notice of the Royal Astronomical Society*, S. J. Perry (Stonyhurst College Observatory, England) said, "I have several times swept in vain for D'Arrest's comet." No dates were given. F. A. T. Winnecke (Karlsruhe, Germany) unsuccessfully searched for this comet on several occasions during July and August of 1870, but was finally successful on August 31.87. He then described it as a very faint nebulosity and gave the position as $\alpha = 16^h 38.0^m$, $\delta = -10° 40'$, which was quite close to the ephemeris published by Leveau.

Although the comet was already past its closest approach to Earth, it was heading toward a late September perihelion. Winnecke observed it with the refractor on September 16 and described it as a pale mark of 1–2′ diameter. Winnecke described the comet as 2′ across on the 19th, and faint with a diameter of 2–3′ on the 20th. G. F. W. Rümker (Hamburg, Germany), C. F. Pechüle (Hamburg, Germany), and Winnecke continued to provide descriptions during the remainder of September, with rough estimates of the coma diameter typically within the range of 2–3′. They also generally noted it was faint and diffuse. Winnecke noted "several flashing central condensations" on the 22nd, and "several bright points" on the 23rd. Rümker saw a faint central condensation on the 27th. The comet was only visible at a low altitude as the month came to an end.

The comet was moving away from both the sun and Earth as October began. J. F. J. Schmidt (Athens, Greece) saw the comet for the first time on the 10th and he began making an extensive series of coma diameter measurements beginning of the 13th. He expressed some uncertainty in these measurements as the comet remained diffuse, with little condensation, but also consistently noted the preceding half of the comet was larger than the following half. He gave the diameter as 3.74' on the 13th, 3.30' on the 14th, 3.82' on the 16th, 3.45' on the 17th, 3.38' on the 20th, 3.24' on the 21st, 3.51' on the 23rd, 3.72' on the 24th, and 3.96' on the 26th. Rümker saw the comet at low altitude on the 18th and said it seemed no fainter than when previously seen. Rümker and Pechüle described the comet as rather faint on the 21st. The comet attained its most southerly declination of −28° on October 30.

Schmidt's more southerly latitude enabled him to continue making valuable observations of the comet during the remainder of its apparition. Consequently, he was the only astronomer to continue providing physical descriptions. He again noted the comet remained diffuse, with little condensation, and his extensive coma measurements again showed the preceding half was more extensive than the following half. He gave the diameter as 3.04' on the November 10, 3.31' on the 12th, 3.24' on the 15th, 3.25' on the 16th, 2.92' on the 17th, 3.21' on the 18th, 3.09' on the 19th, and 3.06' on the 20th. During December, Schmidt noted a trace of condensation on the 11th, and suspected a nucleus of magnitude 13 on the 12th. He again indicated the preceding half of the coma was more extensive that the following half. He measured the coma diameter as 2.85' on the 11th, 2.97' on the 12th, 2.82' on the 13th, and 2.71' on the 16th.

The comet was last seen on December 20.71, when Schmidt measured the position as roughly $\alpha = 23^h 29^m$, $\delta = -18°$ 26'. A slightly later observation was reported. Rümker and Pechüle continually had their south and southwestern horizon affected by haze during November and the first half of December. On December 23.76, they determined the position of the comet as $\alpha = 23^h 36.9^m$, $\delta = -17°$ 35', but noted it did not agree well with the Athens observation of the 20th. In fact, the position was several arc minutes from the predicted position and probably does not refer to the comet.

Minor refinements of this comet's orbit using positions from multiple apparitions, as well as planetary perturbations, have been published by numerous astronomers. Of most significance were the studies of G. Leveau (1883), A. W. Recht (1939), and B. G. Marsden and Z. Sekanina (1972). These have typically revealed a perihelion date of September 23.17 and a period of 6.57 years. Marsden and Sekanina determined the nongravitational terms as $A_1 = -0.50$ and $A_2 = +0.0960$. Their orbit is given below.

T	ω	Ω (2000.0)	i	q	e
1870 Sep. 23.1677 (TT)	172.2622	148.2079	15.6457	1.280249	0.634925

ABSOLUTE MAGNITUDE: $H_{10} = 8.2$ (V1964)

FULL MOON: Aug. 11, Sep. 9, Oct. 9, Nov. 8, Dec. 8, Jan. 6

SOURCES: G. Leveau, *AN*, **74** (1869 Sep. 24), pp. 329–36; S. J. Perry, *MNRAS*, **30** (1870 Jun.), p. 209; F. A. T. Winnecke, *AN*, **76** (1870 Sep. 15), p. 287; F. A. T. Winnecke, *AN*, **76** (1870 Oct. 12), pp. 331–4; J. F. J. Schmidt, *AN*, **77** (1870 Dec. 23), pp. 65–70; J. F. J. Schmidt, *AN*, **77** (1871 Feb. 14), p. 141; G. F. W. Rümker and C. F. Pechüle, *AN*, **77** (1871 Apr. 18), p. 245; G. Leveau, *AN*, **105** (1883 Mar. 14), pp. 19–22; A. W. Recht, *AJ*, **48** (1939 Jul. 17), pp. 65–78; V1964, p. 60; B. G. Marsden and Z. Sekanina, *QJRAS*, **13** (1972), pp. 428–9; B. G. Marsden, Z. Sekanina, and D. K. Yeomans, *AJ*, **78** (1973 Mar.), pp. 213, 215; B. G. Marsden and Z. Sekanina, *CCO*, 12th ed. (1997), pp. 48–9.

C/1870 W1
(Winnecke)

1870 IV = 1870d

Discovered: 1870 November 24.16 ($\Delta = 0.53$ AU, $r = 0.77$ AU, Elong. $= 51°$)
Last seen: 1870 December 1.23 ($\Delta = 0.42$ AU, $r = 0.63$ AU, Elong. $= 25°$)
Closest to the Earth: 1870 December 1 (0.4148 AU)
Calculated path: VIR (Disc), LIB (Nov. 30)

F. A. T. Winnecke (Karlsruhe, Germany) discovered this comet on 1870 November 24.16, at a position of $\alpha = 12^{\text{h}} 41.6^{\text{m}}$, $\delta = -3° 30'$. He described it as small and bright, with a diameter of 2'. He confirmed the comet on November 24.22. Winnecke then revised the diameter estimate as 2.5' and said the coma was round, with a strong central condensation.

The comet was moving towards the sun and Earth, but unfortunately dived into twilight after about a week of visibility. Every observer who managed to see the comet in November reported it as bright and round, with a strong, nearly central condensation. E. Weiss and J. Palisa (Vienna, Austria) said the comet was 3–4' across on the 25th, with no distinct nucleus. K. C. Bruhns (Leipzig, Germany) said the comet was about 2' across on the 27th. G. F. W. Rümker and C. F. Pechüle (Hamburg, Germany) said the comet was 2' across on the 29th.

The comet was last seen on December 1.23, when Rümker measured the position as $\alpha = 15^{\text{h}} 09.4^{\text{m}}$, $\delta = -5° 55'$. He said the comet was extraordinarily faint and at a low altitude in bright twilight. J. F. J. Schmidt (Athens, Greece) received notice of the comet from both Karlsruhe and Vienna, and even had Winnecke's ephemeris to December 6 in his hands by December 8. With the comet predicted to move into the evening sky, he began searching for it on December 9 with three different size telescopes offering different fields of view and different limiting magnitudes. Nothing was found. Additional unsuccessful searches with the same instruments were made on every clear night from December 10 to December 23. On the final date the altitude of the search area was too low and, for this reason, no further searches were made.

Using positions he had measured for November 24, 25, and 26, Winnecke calculated a parabolic orbit with a perihelion date of 1870 December 20.30. He also provided an ephemeris for the period of November 26 to

December 6, inclusive. Winnecke added, "The brightness of the comet will be substantially greater with its appearance in the evening sky than now." Palisa and L. Schulhof took positions from November 24, 25, and 27, and gave the perihelion date as December 20.38. D. M. A. Möller and N. C. Dunér determined the perihelion date as December 20.38 using positions from November 24, 27, and December 1. Schulhof (1875) calculated a definitive orbit by taking 14 positions spanning the whole period of visibility and determining four Normal positions. The result was a perihelion date of December 20.38. Schulhof's orbit is given below.

T	ω	Ω (2000.0)	i	q	e
1870 Dec. 20.3761 (UT)	90.6273	96.5866	147.2703	0.389262	1.0

ABSOLUTE MAGNITUDE: $H_{10} = 11.1$ (V1964)

FULL MOON: Nov. 8, Dec. 8

SOURCES: F. A. T. Winnecke, E. Weiss, J. Palisa, and K. C. Bruhns, *AN*, **77** (1870 Nov. 30), pp. 29–32; F. A. T. Winnecke, *MNRAS*, **31** (1870 Dec.), p. 46; J. Palisa, L. Schulhof, and G. F. W. Rümker, *AN*, **77** (1870 Dec. 8), p. 47; D. M. A. Möller and N. C. Dunér, *AN*, **77** (1870 Dec. 17), p. 61; J. F. J. Schmidt, *AN*, **77** (1871 Feb. 14), p. 143; G. F. W. Rümker and C. F. Pechüle, *AN*, **77** (1871 Apr. 18), p. 247; L. Schulhof, *AN*, **85** (1875 May 5), p. 323; V1964, p. 60.

C/1871 G1
(Winnecke)

1871 I = 1871a

Discovered: 1871 April 7.84 ($\Delta = 1.82$ AU, $r = 1.40$ AU, Elong. $= 50°$)
Last seen: 1871 August 5.79 ($\Delta = 1.13$ AU, $r = 1.27$ AU, Elong. $= 73°$)
Closest to the Earth: 1871 July 26 (1.0898 AU)
Calculated path: PER (Disc), AUR (May 9), TAU (May 17), ORI (Jun. 1), MON (Jun. 21), CMa (Jun. 28), PUP (Jul. 13), VEL (Jul. 24)

F. A. T. Winnecke (Karlsruhe, Germany) discovered this comet on 1871 April 7.84 at a position of $\alpha = 2^h 27.0^m$, $\delta = +53° 55'$. He described it as small and pale. Notices went out to several German observatories and confirmations came the next night. Independent discoveries were made by A. L. N. Borrelly (Marseille, France) on April 13.90 and Lewis Swift (Marathon, New York, USA) on April 16.1. Borrelly said, "It is rather beautiful and presents a nucleus, with a tail directed from the south towards north."

The independent discoveries came as a result of the discovery news being slow to leave Germany; however, several German observatories confirmed Winnecke's find a day later. G. F. W. Rümker and C. F. Pechüle (Hamburg, Germany) saw the comet on April 8.85 and said it showed a bright nucleus and exhibited a tail extending 4' in PA 11.3°. They noted bright spots within the coma just below the nucleus. Additional confirmations came from B. Tiele (Bonn, Germany) on April 8.88 and from K. C. Bruhns and C. N. J. Börgen (Leipzig, Germany) on April 8.97.

The comet was heading toward both the sun and Earth during April and May. Rümker and Pechüle said the comet seemed fainter on the 9th because of the bright northern lights. The comet was well seen on the 10th. H. C. Vogel (Bothkamp, Germany) described the comet as rather bright, with a distinct nucleus. He also noted a tail extending 4' toward PA 8°. Bruhns and Börgen said the comet had a distinct nucleus and an extension about 1.5' long that was either a tail or a portion of the badly defined coma. Rümker and Pechüle said the coma was 1' across and exhibited a tail extending 3' in PA 11.5°. J. R. Hind (Bishop's Observatory, Twickenham, England) said, "There was an evident extension of the nebulosity on the side opposite to the Sun, as if a tail might be expected on the comet's nearer approach. At present it will not be observed without a good telescope." Winnecke saw a tail on the 11th. Bruhns and Börgen said the extension to the coma was clearly perceptible as about 2' long in the refractor (144×) on the 14th, and was about 4' long in the large cometseeker (24×). D. M. A. Möller and E. A. Wijkander (Lund, Sweden) measured the tail position angle as 16.2° and 17.2°, respectively, on the 20th. Möller added, "The comet showed 2 nuclei; position angle of the connecting line = 31.1°...." Möller said the northern lights caused the comet to appear indistinct on the 23rd. J. F. J. Schmidt (Athens, Greece) said the comet was very bright and several minutes in diameter on the 24th. Schmidt also saw the comet on April 26 and 27, despite moonlight and a hazy sky. He said the nucleus was sharp and about magnitude 8. Wijkander said moonlight caused the comet to appear fainter on the 27th than on the 20th. He said the tail extended toward PA 27.0°. Wijkander said the tail extended toward PA 23.5° on the 28th and PA 20.5° on the 30th.

The moon was nearly full on May 1 and Schmidt said the nucleus was about magnitude 11. Wijkander said the tail extended toward PA 30.8° on the 3rd. Schmidt noted a nucleus of magnitude 8 and a tail 8' long on the 6th. He also measured the coma diameter as 2.2'. Rümker and Pechüle also saw the comet on the same night and said the tail was indistinct, but extending toward PA 21°. S. J. Perry (Stonyhurst College Observatory, England) wrote, following his May 8 observation, "the twilight and the nearness of the Sun rendered the object much less distinct than it had previously been, though it was still much more conspicuous than the comets of 1870." Several observers provided descriptions on the 9th. Schmidt estimated the nucleus as magnitude 8 and said the tail was 10' long. The coma diameter perpendicular to the tail was 2.1'. Rümker and Pechüle said the tail was faint, but was extending toward PA 17°. N. C. Dunér (Lund, Sweden) said the tail extended toward PA 12.7°. On May 10, Schmidt said the nucleus was magnitude 8 and the tail was 7' long. Schmidt said the nucleus was magnitude 8–9, while the coma was 2.1' across. No tail was visible. Observations were becoming more difficult during the next few days as the comet drew closer to the sun. Perry determined positions on May 9 and 10, and but still managed to see the comet again on May 11 and 12. Although he said measurements of the position were obtained on the last two dates, "the nearness of the comet to the horizon and the passing clouds rendered the results less trustworthy." Rümker said the comet was low in bright twilight on the 14th, which resulted in the comet and comparison star appearing very faint. Rümker said the comet was extremely faint in bright twilight on the 15th. Schmidt gave the coma diameter as 2.0′ on the 16th and noted a nucleus of magnitude 8. No tail was visible. Rümker said sky conditions were good on the 17th, but the comet could not be seen.

E. W. L. Tempel (Milan, Italy) and A. Hall (Washington, DC, USA) provided some general descriptions of the comet for April and May. Tempel saw the comet on several occasions during the period of April 16 to May 8, inclusive. He said the comet was faint and initially displayed a tail. Later, he described the comet as large and diffuse, but after the moon had entered the sky it became "remarkably smaller, more condensed, and brighter than in former times, equalling a star of magnitude 6–7." He added that the comet's low altitude prevented further observations after May 8. Hall saw the comet on several occasions during the period of April 21 to May 11, inclusive, and said, "This comet had a well-defined nucleus and could be observed with ease."

The comet was last detected on August 5.79 by observers at the Cape of Good Hope, South Africa. The position was measured as $\alpha = 10^h\ 11.9^m$, $\delta = -53°\ 54'$. These observations did not come to light until Hall sent them to J. Holetschek in 1874.

The spectrum of the comet was observed by Vogel, W. Huggins, and A. Secchi. Vogel saw the comet on April 11, 14, and 22. He provided precise measurements of two bright, somewhat blurred lines that were observed on the last two dates. The centers of these two lines were at 5570 Å and 5110 Å, with the second being brighter. Vogel also said, "The very bright nucleus of the comet produced, like in former times, a continuous spectrum...." Huggins saw the comet on April 13 and May 2. He saw the three bright bands of carbon and measured the center of the brightest band as 5100 Å. He wrote, "It would appear that this comet is similar in constitution to the comets which I examined in 1868." Secchi observed the comet on April 18 and saw only the brightest band.

The first orbits were independently calculated by Winnecke and E. Weiss using positions from April 7, 9, and 11. Winnecke determined the perihelion date as 1871 June 13.45, while Weiss determined it as June 14.29. Very similar orbits were calculated before the end of April by Weiss, F. Tietjen, J. H. Hind, and Hall, with the last two astronomers determining the perihelion date as June 11. Using positions obtained to May 11, Hall revised his calculations and determined a perihelion date of June 11.09. Holetschek (1873) determined parabolic orbits that differed little from Hall's last orbit, but in 1874 he calculated an elliptical orbit which used positions from the period of April 7 to August 5. It established the perihelion date as June 11.10 and the period as about 5188 years. The elliptical orbit is given below.

T	ω	Ω (2000.0)	i	q	e
1871 Jun. 11.0999 (UT)	222.5068	281.1119	87.6035	0.654300	0.997814

ABSOLUTE MAGNITUDE: $H_{10} = 5.3$ (V1964)

FULL MOON: Apr. 5, May 4, Jun. 3, Jul. 2, Jul. 31, Aug. 30

SOURCES: F. A. T. Winnecke and J. R. Hind, *MNRAS*, **31** (1871 Apr.), pp. 198–9; F. A. T. Winnecke, B. Tiele, K. C. Bruhns, C. N. J. Börgen, H. Vogel, and E. Weiss, *AN*, **77** (1871 Apr. 18), pp. 249–50, 253–6; E. Weiss, A. L. N. Borrelly, and E. J. M. Stephan, *AN*, **77** (1871 Apr. 25), pp. 267–70; J. R. Hind, *MNRAS*, **31** (1871 May), p. 218; K. C. Bruhns, C. N. J. Börgen, H. Vogel, and F. Tietjen, *AN*, **77** (1871 May 6), pp. 285–8; A. Secchi, *AN*, **77** (1871 May 20), p. 300; W. Huggins, *PRSL*, **19** (1871 May 25), pp. 490–1; A. Hall, *AN*, **77** (1871 May 30), p. 319; S. J. Perry, *MNRAS*, **31** (1871 Jun.), pp. 246–7; J. Palisa, *AN*, **77** (1871 Jul. 1), pp. 375–8; J. F. J. Schmidt, *AN*, **78** (1871 Jul. 25), pp. 55–8; G. F. W. Rümker and C. F. Pechüle, *AN*, **78** (1871 Aug. 9), pp. 109–12; E. W. L. Tempel, *AN*, **78** (1871 Aug. 28), pp. 153–6; D. M. A. Möller, N. C. Dunér, and E. A. Wijkander, *AN*, **78** (1871 Oct. 13), pp. 227–34; G. F. W. Rümker and C. F. Pechüle, *AN*, **79** (1872 Mar. 7), p. 93; A. Hall, *AN*, **79** (1872 Mar. 13), pp. 97, 111; L. Schulhof, *AN*, **79** (1872 Jun. 24), p. 345; A. Hall, *AN*, **80** (1872 Jul. 29), pp. 29–32; J. Holetschek, *AN*, **82** (1873 Nov. 11), p. 301; J. Holetschek and A. Hall, *AN*, **84** (1874 Oct. 20), pp. 323–30; V1964, p. 60.

C/1871 L1   *Discovered:* 1871 June 14.93 ($\Delta = 1.40$ AU, $r = 1.29$ AU, Elong. $= 62°$)
(Tempel)   *Last seen:* 1871 September 20.89 ($\Delta = 1.02$ AU, $r = 1.40$ AU, Elong. $= 87°$)
   Closest to the Earth: 1871 April 28 (0.9109 AU), 1871 October 17 (0.7624 AU)
1871 II = 1871b   *Calculated path:* UMa (Disc), CAM (Aug. 25), LYN (Sep. 10), CAM (Sep. 20)

E. W. L. Tempel (Milan, Italy) discovered this comet on 1871 June 14.93 and described it as a faint, diffuse object, with a diameter of 3–4'. He precisely measured its position as $\alpha = 10^h\ 27.3^m$, $\delta = +57°\ 05'$ on June 14.99. Tempel obtained another precise position on June 15.92, which confirmed this was a comet.

The comet was well observed during the remainder of June. K. C. Bruhns (Leipzig, Germany) described it as faint on the 20th, while G. F. W. Rümker and C. F. Pechüle (Hamburg, Germany) said it was "rather faint" in bright twilight on the 24th. Rümker and Pechüle added, "It is rather large and diffuse with a very fine stellar nucleus." Tempel summarized the comet's appearance during the period of June 14–22, by noting that in clear skies the comet was diffuse with a nucleus. He added that when the comet was placed under the thread of the micrometer he could easily see that the center was speckled or consisted of several stars.

The comet passed perihelion near the end of July. On July 4, J. F. J. Schmidt (Athens, Greece) measured the coma diameter as 2.3' and estimated the brightness of the central region as magnitude 12–13. L. Schulhof (Vienna, Austria) said the comet appeared faint because of moonlight. Schmidt continued to measure the coma diameter and estimate the brightness of the central condensation throughout the month. He gave the coma diameter as 2.2′ on the 6th, 2.7′ on the 10th, 2.9′ on the 12th, 2.7′ on the 14th, 2.6′ on the 17th, 2.6′ on the 19th, 2.5′ on the 20th. He gave the magnitude of the condensation as 11–12 on the 6th, 12 on the 10th, 12–13 on the 12th, 11–12 on the 14th, 12–13 on the 17th, 12 on the 19th, 12–13 on the 20th. Pechüle saw a rather distinct condensation on the 17th, and specifically noted that several bright points were observed. This was an interesting comment because Schmidt noted that during his observations of July, there frequently seemed to be different nuclei of magnitude 12 and 13 present. On July 21 and 22, A. Hall (US Naval Observatory, Washington, DC, USA) saw the comet and said, "This comet was faint and difficult to observe on both days." C. N. J. Börgen (Leipzig, Germany) said the coma was about 1.5–2′ across, with a central condensation on the 22nd. Börgen and Pechüle independently noted the comet was extremely faint because of moonlight on the 27th. Pechüle said the comet was faint because of moonlight on the 30th.

Although the comet had already passed perihelion, it would approach Earth during the next few months, which acted to slow its fading. Pechüle said the comet was faint in moonlight on August 2, but he found it "quite bright" with a central condensation on the 9th and noted an extension toward PA 330°. Schmidt continued to measure the coma and estimate the brightness of the central condensation during most of the month. He measured the coma as 2.1′ on the 4th, 2.2′ on the 5th, 2.8′ on the 9th, 2.7′ on the 11th, and 2.6′ on the 16th. He estimated the magnitude of the condensation as 12 on the 4th, 12–13 on the 5th, 12 on the 9th and 11th, and 12–13 on the 16th. On August 27, Börgen said the comet was extremely diffuse and vague. On September 3, Pechüle said the comet was "unexpectedly faint and diffuse," while Schmidt measured the coma diameter as 2.0′ and estimated the magnitude of the central condensation as 13. Schulhof saw the comet for the final time on the 7th and said it was too faint to measure the position. The comet attained its most northerly declination of +61° on September 10, at which time Rümker described it as faint and diffuse. Rümker found the comet very faint on the 11th. Around the middle of September, Schulhof conducted searches for the comet, but all were in vain.

The comet was last detected on September 20.89, when Rümker determined the position as $\alpha = 6^h\ 03.0^m$, $\delta = +60°\ 22′$. He said this observation was made more difficult by the close proximity of several faint stars.

The first parabolic orbit was calculated by E. Weiss and L. Schulhof using positions from June 14, 17, and 22. They determined the perihelion date as 1871 August 2.70. Pechüle followed a short time later with an orbit based on positions from June 16, 20, and 24, which indicated a perihelion date of August 2.34. Pechüle and Schulhof revised their orbits during the next few weeks and found the perihelion date was July 27. Definitive orbits were calculated by Schulhof (1875) and P. N. Cramer (1875). Using 181 positions spanning the entire period of visibility, Schulhof determined eight Normal positions, and calculated a parabolic orbit with a perihelion date of 1871 July 27.52. Cramer took 185 positions, derived ten Normal positions, applied perturbations by seven planets, and calculated parabolic and hyperbolic orbits. The parabolic orbit had a perihelion date of July 27.53. The hyperbolic had a perihelion date of July 27.53 and an eccentricity of 1.0000243. Because Cramer applied perturbations and since he considered the parabolic orbit satisfactory because of the three-month arc, this orbit is given below.

T	ω	Ω (2000.0)	i	q	e
1871 Jul. 27.5344 (UT)	96.3220	213.7030	101.9782	1.083363	1.0

ABSOLUTE MAGNITUDE: $H_{10} = 6.5$ (V1964)

FULL MOON: Jun. 3, Jul. 2, Jul. 31, Aug. 30, Sep. 28

SOURCES: E. W. L. Tempel, F. A. T. Winnecke, E. Weiss, L. Schulhof, G. F. W. Rümker, and C. F. Pechüle, *AN*, **77** (1871 Jul. 1), pp. 375–82; C. F. Pechüle, *AN*, **78** (1871 Jul. 29), p. 79; E. W. L. Tempel and G. V. Schiaparelli, *AN*, **78** (1871 Aug. 28), pp. 153–6; L. Schulhof, *AN*, **78** (1871 Sep. 12), pp. 173–6; J. F. J. Schmidt, *AN*, **78** (1871 Nov. 9), pp. 277–80; K. C. Bruhns and C. N. J. Börgen, *AN*, **79** (1872 Feb. 14), pp. 41–4; G. F. W. Rümker and C. F. Pechüle, *AN*, **79** (1872 Mar. 7), pp. 85–8; A. Hall, *AN*, **79** (1872 Mar. 13), pp. 97, 111; L. Schulhof, *AN*, **79** (1872 Jun. 24), pp. 345–8; P. N. Cramer, *Berekening van de loopbaan der komeet II. 1871*. Dissertation: Leiden (1875), 36 pp.; V1964, p. 60.

8P/1871 T1 *Recovered:* 1871 October 13.17 ($\Delta = 1.09$ AU, $r = 1.27$ AU, Elong. $= 75°$)
(Tuttle) *Last seen:* 1872 January 30.95 ($\Delta = 1.33$ AU, $r = 1.36$ AU, Elong. $= 70°$)
  Closest to the Earth: 1871 November 23 (0.6875 AU)
1871 III = 1871d *Calculated path:* LYN (Rec), UMa (Oct. 13), LYN (Oct. 15), LMi (Oct. 18), LEO (Oct. 24), SEX (Nov. 12), HYA (Nov. 22), CRT (Nov. 24), HYA (Nov. 29), CEN (Dec. 5), CRU (Dec. 20), MUS (Dec. 29), APS (Jan. 11)

An orbit determined by F. Tischler was published in the 1871 April 18 issue of the *Astronomische Nachrichten*. The orbit had been found among the papers of Tischler and included partial determinations of the effects of planetary perturbations. The predicted date of perihelion passage was given as 1871 November 30.96. During May of 1871, J. R. Hind computed an ephemeris for the period of 1871 September 1 to October 19, inclusive. Hind added, "The comet's track in the heavens would, therefore, appear to be very favourable for observation in these latitudes."

The comet was recovered by A. L. N. Borrelly (Marseille, France) on 1871 October 13.17. He described it as "rather faint" and measured the position as $\alpha = 9^h\ 09.7^m$, $\delta = +44°\ 16'$. Independent recoveries were made by F. A. T. Winnecke (Karlsruhe, Germany) on October 16.02 and by H. P. Tuttle (US Naval Observatory, Washington, DC, USA) on October 23.38. Winnecke said the comet was "rather bright" with a diameter of 2–3′ and exhibiting a "very fine nucleus." He said the nucleus seemed to occasionally brighten. Tuttle simply noted it had a well-defined nucleus. Although numerous observers began to follow the comet during the remainder of October, only a handful of additional physical descriptions were made. G. F. W. Rümker and C. F. Pechüle (Hamburg, Germany) said the comet was reasonably bright on the 18th, while Pechüle found it very faint on the 19th.

The comet made a moderately close approach to Earth during the second half of November, and continued to approach the sun throughout the month. A. Hall (US Naval Observatory, Washington, DC, USA) said the comet exhibited a well-defined nucleus on November 8, 9, 12, and 17. Rümker and Pechüle described the comet as "rather bright" on the 10th and added that the comet was extended toward PA 235°, "in which direction a granular appearing core is situated." Pechüle said the diameter was 2' on the 13th and noted the very condensed region was shifted slightly north of center. Rümker estimated the coma diameter as 2.5' on the 14th and said a noticeable condensation was somewhat north of center. That same night, Winnecke said the comet appeared like a centrally condensed nebulosity about 3.3' across.

The comet passed perihelion on December 2 and was moving away from both the sun and Earth thereafter. Few physical descriptions were provided from December onward. Rümker saw the comet in the morning sky on December 1 and said he could not measure its position because of its low altitude and bright moonlight. Winnecke said the comet was 2.5' across on the 3rd and did not exhibit a substantial condensation. He added that he was surprised to find it so faint.

This comet was last detected on January 30.95, by observers at the Royal Observatory, Cape of Good Hope, South Africa. The position was given as $\alpha = 15^h\ 41.1^m$, $\delta = -78°\ 14'$, although it should be noted that δ was determined 12 minutes before α. On this date it was also just 3.9° from C/1871 V1.

Several astronomers have considered multiple apparitions in the investigation of this comet's motion. The earliest such investigations were published by J. Rahts (1885, 1894). The most recent was by B. G. Marsden and Z. Sekanina (1972). All determined the perihelion date as December 2.30 and the period as 13.82 years, with Marsden and Sekanina giving nongravitational terms of $A_1 = +0.32$ and $A_2 = +0.0131$.

T	ω	Ω (2000.0)	i	q	e
1871 Dec. 2.2976 (TT)	206.7765	271.1140	54.2819	1.030109	0.821103

ABSOLUTE MAGNITUDE: $H_{10} = 8.0$ (V1964)

FULL MOON: Sep. 28, Oct. 28, Nov. 27, Dec. 26, Jan. 25, Feb. 24

SOURCES: F. Tischler, *AN*, **77** (1871 Apr. 18), p. 255; J. R. Hind and F. Tischler, *MNRAS*, **31** (1871 May), pp. 215–16; E. J. M. Stephan, A. L. N. Borrelly, C. A. F. Peters, and F. A. T. Winnecke, *AN*, **78** (1871 Oct. 24), p. 255; J. R. Hind, *MNRAS*, **32** (1871 Nov.), p. 28; G. F. W. Rümker, C. F. Pechüle, E. J. M. Stephan, and J. R. Hind, *AN*, **78** (1871

Nov. 9), pp. 281–6; J. R. Hind, *MNRAS*, **32** (1871 Dec.), pp. 67–8; K. C. Bruhns and C. N. J. Börgen, *AN*, **79** (1872 Feb. 14), p. 43; G. F. W. Rümker and C. F. Pechüle, *AN*, **79** (1872 Mar. 7), p. 91; A. Hall and H. P. Tuttle, *AN*, **79** (1872 Mar. 13), pp. 97, 111; F. A. T. Winnecke, *AN*, **94** (1879 Feb. 6), p. 117; J. Rahts, *AN*, **113** (1885 Dec. 22), pp. 169–206; J. Rahts, *AN*, **136** (1894 Aug. 2), pp. 65–8; V1964, p. 60; B. G. Marsden and Z. Sekanina, *QJRAS*, **13** (1972), pp. 427–9; B. G. Marsden, *CCO*, 12th ed. (1997), pp. 50–1.

C/1871 V1
(Tempel)

1871 IV = 1871e

Discovered: 1871 November 3.79 ($\Delta = 1.27$ AU, $r = 1.15$ AU, Elong. $= 59°$)
Last seen: 1872 February 21.36 ($\Delta = 0.67$ AU, $r = 1.37$ AU, Elong. $= 110°$)
Closest to the Earth: 1871 October 12 (1.1940 AU), 1872 February 20 (0.6665 AU)
Calculated path: SCT (Disc), SGR (Nov. 10), CrA (Dec. 5), TEL (Dec. 16), PAV (Dec. 31), APS (Jan. 21), OCT (Jan. 31), CHA (Feb. 6), VOL (Feb. 11), CAR (Feb. 16), PUP (Feb. 21)

E. W. L. Tempel (Milan, Italy) discovered this comet near the open cluster M26 on 1871 November 3.77. He simply described it as small and faint. He measured the position as $\alpha = 18^h$ 37.9^m, $\delta = -9°$ 15′ on November 3.79. Tempel and his colleague G. V. Schiaparelli measured another position on November 4.75, confirming the object was a comet.

The comet had actually passed 1.19 AU from Earth around mid-October and was temporarily moving away from Earth during November, but it was approaching perihelion. K. C. Bruhns and C. N. J. Börgen (Leipzig, Germany) said the comet was faint on November 5, with a diameter of 1–2′. The comet was also seen by F. A. T. Winnecke (Karlsruhe, Germany) and K. L. von Littrow (Vienna, Austria) that same night. Winnecke said the comet was round, centrally condensed, and 2.5′ across, while Littrow said it was centrally condensed, round, faint, and 2–3′ across. On November 6, G. F. W. Rümker and C. F. Pechüle (Hamburg, Germany) described the comet as well condensed, about a minute in diameter, and fairly bright. J. F. J. Schmidt (Athens, Greece) also saw the comet on the 6th and said it was faint and not easily recognizable in the finder, although he noted a well-defined condensation and a nucleus whose magnitude was about 12. Börgen found the centrally condensed coma 2–3′ across on the 8th. F. W. A. Argelander (Bonn, Germany) said the comet appeared very faint on the 10th. Rümker and Pechüle said the comet was faint on the 14th, but noted it occasionally seemed to brighten. Schmidt said the comet was easily seen on the 15th, despite being 10–15° from the moon. Rümker and Pechüle said the comet was fairly easy to see on the 17th and 18th, despite the low altitude and moonlight.

The comet had moved almost due south since its discovery and dropped below the horizon for Northern Hemisphere observers during the second half of November. After passing 27° from the sun on December 16, the comet appeared to Southern Hemisphere observers around mid-January. B. A. Gould (Argentine National Observatory, Cordoba, Argentina) saw the comet on the evening of January 18 and said, "It appeared in the telescope as a nearly circular nebula, somewhat condensed toward the center, but without a definite nucleus, and not less than 2' in diameter." Gould described it as "exceedingly filmy" on the 20th and added that a 10th-magnitude star was 30" from the center of the coma and "scarcely obscured by an appreciable amount." Moonlight began to interfere by the next evening, as Gould said the comet was scarcely distinguishable and that no nucleus was perceptible. Gould could not find the comet in bright moonlight on the 22nd. During the short period between the end of evening twilight and moonrise on the 27th, Gould wrote, "The comet still appeared not less than 2' in diameter, & without perceptible nucleus or outline." The comet passed 3.9° from 8P/Tuttle on the 30th and Gould described it as "ill defined, but more distinct than on previous nights, the moon being absent. Apparently nearly circular, & with a slight tendency to granulation toward the center."

The comet attained its most southerly declination of −84° on February 3. Gould continued to provide physical descriptions during February as the comet faded from view. He indicated the nuclear condensation was about magnitude 12 on the 2nd and 4th. On February 6, Gould said the comet appeared "more spread and dilute" than when previously seen. He estimated the coma as 2.5' across. On the 12th, he said, "Comet only visible after the eye had been kept for several minutes in perfect darkness." By the 13th, Gould wrote, "Comet very wide of predicted place and found with difficulty just before moonset."

The comet was last detected on 1872 February 21.36, when Gould measured the position as $\alpha = 7^h 55.7^m$, $\delta = -51° 56'$. He remarked, "No fainter object could be recognized with this telescope of 28-cm aperture." L. Schulhof (Vienna, Austria) searched for the comet during the first days of March, but never found it. He noted that it had then moved back into the Milky Way.

The first parabolic orbit was calculated by J. R. Hind using positions spanning the period of November 5–10. He determined the perihelion date as 1871 December 20.79. Additional orbits were calculated by C. F. W. Peters and Schulhof. The first elliptical orbit was calculated by A. Lindhagen (1885). He used positions spanning the entire observational arc and determined a perihelion date of December 20.88 and a period of about 2690 years. I. Lagarde (1898) used the same arc and determined a perihelion date of December 20.87 and a period of 2048 years. His orbit is given below.

T	ω	Ω (2000.0)	i	q	e
1871 Dec. 20.8722 (UT)	242.8935	148.9290	98.2992	0.691268	0.995714

ABSOLUTE MAGNITUDE: $H_{10} = 8.0$ (V1964)

FULL MOON: Oct. 28, Nov. 27, Dec. 26, Jan. 25, Feb. 24

SOURCES: E. W. L. Tempel, K. C. Bruhns, C. N. J. Börgen, F. A. T. Winnecke, K. L. von Littrow, G. F. W. Rümker, and C. F. Pechüle, *AN*, **78** (1871 Nov. 9), pp. 285–8; E. W. L. Tempel, G. V. Schiaparelli, C. F. W. Peters, and F. W. A. Argelander, *AN*, **78** (1871 Nov. 17), p. 303; C. F. W. Peters, *AN*, **78** (1871 Nov. 24), p. 319; J. R. Hind, W. E. Plummer, T. R. von Oppölzer, and L. Schulhof, *MNRAS*, **32** (1871 Dec.), pp. 66–8; J. F. J. Schmidt, *AN*, **78** (1871 Dec. 5), p. 331; L. Schulhof, *AN*, **78** (1872 Jan. 3), p. 383; K. C. Bruhns and C. N. J. Börgen, *AN*, **79** (1872 Feb. 14), pp. 44–6; G. F. W. Rümker and C. F. Pechüle, *AN*, **79** (1872 Mar. 7), p. 93; L. Schulhof, *AN*, **79** (1872 Jun. 24), p. 347; B. A. Gould, *AN*, **94** (1879 Feb. 6), pp. 117–22; A. Lindhagen, *AN*, **111** (1885 Feb. 2), p. 111; I. Lagarde, *BA*, **15** (1898), p. 129; V1964, p. 60.

2P/Encke

1871 V = 1871c

Recovered: 1871 September 18.9 ($\Delta = 1.03$ AU, $r = 1.86$ AU, Elong. $= 131°$)
Last seen: 1871 December 10.70 ($\Delta = 0.45$ AU, $r = 0.58$ AU, Elong. $= 18°$)
Closest to the Earth: 1871 November 16 (0.3121 AU)
Calculated path: TRI (Rec), AND (Oct. 6), LAC (Nov. 1), PEG (Nov. 6), CYG (Nov. 8), VUL (Nov. 13), SGE (Nov. 19), AQL (Nov. 22), SER (Nov. 30), OPH (Dec. 4), SER (Dec. 5), OPH (Dec. 8)

Two predictions were made for the return of this comet. J. R. Hind took W. J. Förster's orbit for the 1868 return and advanced it, without the addition of planetary perturbations, and determined the upcoming perihelion date as 1871 December 29.5. S. P. von Glasenapp (1871) began with the 1868 orbit of F. E. von Asten and E. Becker, and applied corrections based on observations made during 1868. Glasenapp then applied perturbations by Jupiter and predicted the comet would pass perihelion on 1871 December 29.09.

Two independent recoveries were made near mid-September. E. J. M. Stephan (Marseille, France) found a very faint object on September 18.9, but it was visible "during too short a time for me to form a conviction." He said the state of the sky did not allow a further search until his observation on October 9. He added that he discovered seven new nebulae in the course of searching for this comet. F. A. T. Winnecke (Karlsruhe, Germany) began searching for 2P/Encke during the first days of September, but continually met with failure. On September 19.90 he spotted a very faint nebulous mass at a position of $\alpha = 2^h\ 00.2^m$, $\delta = +30°\ 25'$. This position indicated Glasenapp's orbit required a correction of $+0.2$ day. A search on September 20, with hazy skies, failed to reveal anything at the above position or in the ephemeris position. Rainy weather and moonlit skies followed and no further searches could be conducted until October 4. On that evening, Winnecke confirmed the comet. He described it as a nebulous mass, with a coma diameter of 6–8'.

There were also a couple of suspected recoveries which later proved erroneous. J. F. J. Schmidt (Athens, Greece) made his first attempt to find this comet on 1871 September 12. He was using Glasenapp's ephemeris and occasionally noticed an extremely faint nebulosity that was visible in both a refractor and a seeker. He estimated the position as $\alpha = 2^h\ 05.9^m$, $\delta = +28°\ 23'$, but considered the observation as doubtful. Indeed, the position was several degrees from where it should have been, based on later observations. Hind and W. E. Plummer (Twickenham, England) spotted "an extremely faint nebulous object" and measured its position on September 22.96 and 23.01. Although they suspected it was 2P/Encke, this proved to be incorrect.

The comet was widely observed during October, but as most astronomers were observing to measure a precise position, mostly large refractors were being used. From October 4 to October 20, observations by N. C. Dunér and D. M. A. Möller (Lund, Sweden), Schmidt, Hind, Stephan, G. F. W. Rümker and C. F. Pechüle (Hamburg, Germany), K. C. Bruhns and C. N. J. Börgen (Leipzig, Germany), W. Huggins (London, England), and John J. Plummer (Durham Observatory, England) indicated the comet was large and faint. But Schmidt also commented that the comet was not hard to detect in his telescope's finder (8×) on the 4th, 5th, and 6th, so a telescope with a wider field of view indicated the comet was not as faint as others indicated. Although Stephan reported a central condensation on the 9th and 10th, everyone else reported that either the comet was very diffuse, or exhibited no condensation, or they failed to mention the condensation at all. Stephan measured the coma diameter as 4′ on the 9th and 10th, and noted the comet appeared as a milky nebulosity. On the 18th, Bruhns and Börgen said the coma was 4′ across, while H. C. Key (Stretton Rectory, Hereford, England) noted the coma was elliptical in a 46-cm reflector. Once the moon had left the sky, Rümker and Pechüle reported the comet was distinctly brighter on October 31.

The comet was still approaching the sun and Earth during the first half of November. Astronomers began reporting a nucleus early in November. Rümker and Pechüle first noticed a strong condensation on the 4th and Key saw the nucleus for the first time on the 7th in his 46-cm reflector. The nucleus was not centered within the coma: Börgen commented it was shifted toward the east side on the 6th and 8th, while Plummer and L. Schulhof (Vienna, Austria) independently confirmed this on the 8th and 10th, resepctively. Although several astronomers provided coma diameter estimates, most used large refractors and were seeing only the brighter inner portion. Schmidt continued to use his telescope's 8× finder. He measured the coma diameter as 12.3′ on the 1st, 15.0′ on the 3rd, 15.2′ on the 4th, 14.6′ on the 7th, 17.1′ on the 11th, 14.3′ on the 14th, and 14.6′ on the 15th. As indicated by other astronomers, Schmidt reported the coma was elliptical as seen in his refractor at 40× and added that the nucleus was distinctly off center. On the 4th, he said 7.71′ of the coma preceded the nucleus, while 2.05′ followed it. On the 15th, Schmidt said 7.38′ of the coma preceded the nucleus, while 1.67′ followed it. Pechüle said the long axis was directed toward PA 225° 06′ on the 13th. Throughout this period, astronomers provided some interesting descriptions of the comet. Rümker and Pechüle said the comet fanned out toward PA 240° 03′ on the 4th. Huggins reported a strong condensation shifted east of center, while the more condensed part of the comet was fan-shaped with a "tolerably defined contour" on the east side that resembled a parabolic curve. A much fainter nebulosity surrounded this brighter portion. J. Carpenter (Royal Observatory, Greenwich, England) said the condensation was "somewhat fan-shaped" on the 9th, but he noted with the 32-cm refractor, "I was able to make out a considerable extension of the illumination beyond the bright fan-shaped condensation, but on one side (the spreading side) only. On the opposite side this diffused illumination appeared to be cut off nearly in a straight line immediately behind (following) the apex of the fan." H. W. Hollis (Keele, England) saw the comet with a 20-cm refractor (120×) on the 12th and said, "My attention was particularly arrested by the sharp definition of the two edges of the fan-shaped nebulosity; and I could not help asking myself whether the true form of the body were not in all probability a hollow cone. I say a hollow cone, because I distinctly saw by averted vision a slight condensation of the light towards both edges, particularly the one running N. and S., on the following, or eastern, side of the comet, whereas a solid cone would appear brightest in the middle and grow fainter to the edges." Schulhof said the comet was elliptical or leaf-shaped on the 15th, while Schmidt noted there was a fan-shaped extension from the nucleus that extended toward the sun, while the area immediately behind the nucleus was dark.

The comet passed closest to Earth on November 16, but continued its trek toward the sun. The comet also continued to brighten. Where Plummer noted how the comet was "rendered nearly invisible" when it passed over an 8th-magnitude star on the 17th, Schmidt reported it was visible to the naked eye on the 29th and estimated its magnitude as between 6 and 7. On the 30th, Schmidt said the comet was not visible to the naked eye, even though stars of magnitude 6 were seen. Schmidt was still providing the best measures of the coma while using his telescope's 8× finder. He measured it as 11.1' across on the 29th and 11.0' across on the 30th. On the last date, Schmidt also measured the elliptical coma with his 17-cm refractor (40×) and noted that 6.73' preceded the nucleus, while 2.37' followed it. Schmidt saw the first traces of tail on the 29th, but could not see the tail on the 30th. H. Schultz (Uppsala, Sweden) noted strong parabolic jets radiating in the direction of the comet's movement on the 19th, while the coma exhibited a granular appearance. Although he noted the jets were more indistinct on the 20th, he said they were again very noticeable on the 30th.

Although the comet continued to approach the sun in December, it was also moving into evening twilight. Chinese astronomers reported a "broom star" visible to the naked eye around this time. A. Hall (US Naval Observatory, Washington, DC) saw the comet with the naked eye on December 1, 2, and 5. Schmidt said the comet was easily seen with the naked eye on the 2nd as a small star, with little nebulosity, shining at magnitude 5.2 or 5.3. Schmidt said the comet's low altitude prevented a naked-eye observation on the 6th, although it was still seen in the 8× finder. Hall said the comet was still visible to the naked eye on the 7th. Schmidt continued with his careful measurements of the coma diameter with the 8× finder. He gave it as 9.4′ on the 2nd, 8.3′ on the 6th, and 7.4′ on the 7th. He also continued to measure the long axis of the elliptical portion with his refractor (40×). For the 2nd, he said 4.32′ preceded the nucleus, while 2.88′ followed it. On the 7th, he said 3.52′ preceded the nucleus, while 1.71′ followed it. The tail made a reappearance as the comet drew closer to the sun. Rümker and Pechüle said it was narrow and 10′ long during the period of December 4–6. Börgen used a 13-cm cometseeker on the 5th and noted a rather faint, thin tail extending 2° away from the sun. Schmidt noted a tail 15′ long in the 8× finder on the 7th, despite the comet's low altitude. Schmidt said a bright halo was surrounding the nucleus on the 2nd and noted it had not been previously noted at various magnifications. He added that, on this occasion, even a magnification of 300× continued to show dense nebulosity surrounding the nucleus. Schmidt was only able to see the comet for about 3 minutes during a brief break in the clouds on the 3rd. He noted the halo was gone, but commented that the central region of the coma was very bright and strongly condensed. On the 6th, Rümker and Pechüle noted two very weak jets extending from the nucleus toward the sun and then curving backwards. Schmidt saw a nucleus of magnitude 8 or 9 on the 7th. The comet was last detected on December 10.70, when Rümker and Pechüle measured the position as $\alpha = 17^h 42.9^m$, $\delta = -6° 24'$. The comet passed 12° from the sun on December 17.

The spectrum was examined by Huggins on November 8, 9, 12, 13, and 14, and C. A. Young (Princeton College Observatory, New Jersey, USA) during early December. Young detected the three carbon bands at 5575 Å, 5174.5 Å, and 4702 Å. Huggins saw the brightest of the carbon bands at about 5160 Å on the first nights, and suspected the two fainter ones. On the last night, Huggins distinctly saw all three bands, with the fainter ones at 5632 Å and 4735 Å.

Refinements of this comet's orbit using positions from multiple apparitions, as well as planetary perturbations, have been published by numerous astronomers. Of most significance were the studies of Asten (1875, 1877), J. O. Backlund (1884, 1886), and B. G. Marsden and Z. Sekanina (1974). These have typically revealed a perihelion date of December 29.30 and a period of 3.29 years. Marsden and Sekanina determined the nongravitational terms as $A_1 = -0.78$ and $A_2 = -0.03170$. The orbit of Marsden and Sekanina is given below.

T	ω	Ω (2000.0)	i	q	e
1871 Dec. 29.3036 (TT)	183.6331	336.3907	13.1390	0.332985	0.849412

ABSOLUTE MAGNITUDE: $H_{10} = 8.8$ (V1964)
FULL MOON: Sep. 28, Oct. 28, Nov. 27, Dec. 26

sources: S. P. von Glasenapp, *BASP*, **16** (1871), pp. 321–32; J. R. Hind, *MNRAS*, **31** (1871 May), p. 216; S. P. von Glasenapp, *AN*, **78** (1871 Aug. 4), pp. 87–96; J. R. Hind and W. E. Plummer, *AN*, **78** (1871 Oct. 9), p. 223; E. J. M. Stephan, *CR*, **73** (1871 Oct. 9), p. 889; A. Winnecke, *AN*, **78** (1871 Oct. 13), p. 235; J. R. Hind, D. M. A. Möller, N. C. Dunér, and E. J. M. Stephan, *AN*, **78** (1871 Oct. 24), p. 253–6; J. Carpenter and J. R. Hind, *MNRAS*, **32** (1871 Nov.), pp. 25–8; G. F. W. Rümker, C. F. Pechüle, E. J. M. Stephan, S. P. von Glasenapp and J. R. Hind, *AN*, **78** (1871 Nov. 9), pp. 281–6; W. Huggins, *PRSL*, **20** (1871 Nov. 23), pp. 45–7; H. W. Hollis, *MNRAS*, **32** (1871 Dec.), pp. 65–6; W. Huggins, *AN*, **78** (1871 Dec. 19), p. 357; W. Huggins, *PRSL*, **20** (1871 Dec. 21), pp. 87–9; J. F. J. Schmidt, *AN*, **79** (1872 Jan. 24), pp. 17–24; K. C. Bruhns and C. N. J. Börgen, *AN*, **79** (1872 Feb. 14), pp. 44–6; H. C. Key, *MNRAS*, **32** (1872 Mar.), pp. 217–18; G. F. W. Rümker and C. F. Pechüle, *AN*, **79** (1872 Mar. 7), pp. 89–92; A. Hall, *AN*, **79** (1872 Mar. 13), pp. 97, 111; J. R. Hind and W. E. Plummer, *MNRAS*, **32** (1872 Apr.), p. 249; H. Schultz, *AN*, **79** (1872 May 17), p. 249; L. Schulhof, *AN*, **79** (1872 Jun. 24), p. 347; J. J. Plummer, *AN*, **80** (1872 Jul. 16), pp. 13–16; F. E. von Asten, *BASP*, **20** (1875), p. 344; F. E. von Asten, *BASP*, **22** (1877), pp. 568–9; C. A. Young, *The Observatory*, **3** (1879 Jun.), pp. 56–7; HA1970, p. 87; J. O. Backlund, *MASP*, **32** No. 3, 7th Series (1884), p. 36; J. O. Backlund, *MASP*, **34** No. 8, 7th Series (1886), p. 38; V1964, p. 60; B. G. Marsden and Z. Sekanina, *AJ*, **79** (1974 Mar.), pp. 415–16.

X/1871 Y1   E. W. L. Tempel (Milan, Italy) found this object near β Cygni on 1871 December 29.73. It was described as a faint, diffuse nebula. Because of the rich star fields, it took Tempel several minutes to find the region on charts, but he eventually gave the position as α = 19^h 51^m 15^s, δ = +29° 56'. He noted that a nebula listed in Herschel's general catalog of nebulae was located at α = 19^h 51^m 22^s, δ = +28° 58' and suspected that the 1° difference in δ might indicate a misprint in Herschel's catalog. On the other hand, he thought the object might be a new comet. Unfortunately, because of the amount of time taken to determine the position, the moon had risen when he resumed observing and the region had moved to very near the horizon. Cloudy skies prevailed on the 30th, but a period of partly cloudy skies on the 31st allowed Tempel to find the position measured on the 29th was empty, while Herschel's nebula was in the appropriate place. No searching was possible because of increasing clouds. Although the night of January 1 was cloudy, Tempel did have very clear skies on the 2nd; however, a search was unsuccessful. A search in less favorable conditions on the 3rd was also in vain. Tempel wrote, "Its appearance was exactly like the Encke Comet on October 15." K. C. Bruhns wrote in the 1872 issue of *Vierteljahrsschrift Astronomische Gesellschaft* that Tempel probably made a 10° error in recording α and misidentified M56. On the other hand, Bruhns wrote that δ = +39° 56', which was 10° in error from what Tempel recorded, made the M56 suggestion invalid.

sources: K. C. Bruhns, *VJS*, **7** (1872), p. 98; E. W. L. Tempel, *AN*, **78** (1872 Jan. 3), p. 383; E. W. L. Tempel, *AN*, **80** (1872 Jul. 29), pp. 27–9.

X/1872 X1   This is one of the most interesting of the "uncertain" objects in this catalog. It began with a great display of meteors on the night of 1872 November 27. As astronomers quickly attempted to understand the display, W. Klinkerfues sent a telegram to N. R. Pogson (Madras Observatory, India) that read, "Biela touched Earth on 27th: search near Theta Centauri." Pogson experienced clouds and rain on the mornings of December 1 and 2, but there were breaks in the clouds on the morning of the 3rd. He began searching for the comet with a 20-cm refractor (99×) around 4 a.m. (local time), which was when it was expected to rise above the horizon. At 5:15 a.m. (December 3.00 UT) he spotted an object that he described as "Circular, bright, with a decided nucleus, but no tail, and about 45″ in diameter." Pogson made four comparisons with an anonymous star and noted the object moved 2.5^s in α in four minutes, which he said confirmed it was Biela. This observation occurred in twilight, and Pogson was able to obtain a position of $\alpha = 14^h\ 07^m 12.66^s$, $\delta = -34°\ 45'\ 21.1″$ on December 3.0070. Pogson said he obtained a better observation of the comet the next morning, making ten comparisons with three different stars. The result was a position of $\alpha = 14^h\ 21^m 55.11^s$, $\delta = -35°\ 04'\ 37.5″$ on December 3.995 and a position of $\alpha = 14^h\ 22^m 02.72^s$, $\delta = -35°\ 04'\ 42.3″$ on December 4.003. He described the comet as "Circular; diameter 75″; bright nucleus; a faint but distinct tail, 8′ in length and spreading, a position angle from nucleus about 280°." Pogson (1874) later revised the tail length to 7.4′ and said it extended toward PA 260°. Pogson sent numerous letters out on December 5 and 6, one of which went to Klinkerfues and another to G. B. Airy, the English Astronomer Royal. In the letter to Klinkerfues he added for December 4, "I was too anxious to secure one good place of the one in hand to look for the other comet, and the fourth morning was cloudy and rainy. At a guess I should describe it as three times as bright as the cluster 80 Messier, in the field with R, S, and T Scorpii" (globular cluster; magnitude 7.2; diameter 8.9′).

Klinkerfues took the information provided by Pogson and, by incorporating the orbit for the 1852 apparition of Biela, he determined that the comet's distance from Earth was only 0.073 AU on December 3 and 0.088 AU on December 4. He concluded, "An examination of the foregoing figures leaves scarcely a doubt as to the identity of the comet observed by Mr. Pogson with one of the two heads of Biela's comet, and at the same time they show that an unusually small minimal distance from the Earth occurred in the course of November 27th."

G. L. Tupman (Portsmouth, England) was one of the first to recognize problems with the supposed identity of Pogson's comet with that of Biela. In the January 1873 issue of the *Monthly Notices* he said Klinkerfues' "happy and bold idea that the comet itself had touched the Earth" could not place Biela in the positions indicated by Pogson. In particular, he noted the comets would have been about 12 weeks behind their predicted orbital positions and Pogson's positions indicate a reduction of over 7° in the orbital inclination. Further examination by Tupman indicated the 12-week late arrival could not be explained by planetary perturbations, but the required inclination difference could be accounted for if Pogson saw the secondary comet of Biela on the 3rd and the primary comet on the 4th. He strengthened his argument in a more elaborate paper published in the March 1873 issue of the *Monthly Notices*. In particular, he determined the orbit of the meteors seen on the night of November 27 and found a likely perihelion date of December 27.4, whereas the predicted perihelion date for Biela had been October 6.9.

Five parabolic orbits have been calculated for this object, based on the three positions measured by Pogson. They indicate that identity with Biela's comet was unlikely. K. C. Bruhns (1875) took the three positions measured by Pogson and calculated two parabolic orbits. The first had a perihelion date of 1872 December 15.88, while the second had a perihelion date of December 15.18. H. C. F. Kreutz (1886, 1902) calculated a total of three parabolic orbits with perihelion dates ranging from December 15.18 to 16.62, which were fairly similar to Bruhns' first orbit. Kreutz' 1902 orbit is given below.

W. H. S. Monck (1892) favored Tupman's idea that a different comet was seen each night by Pogson. He suggested the two nebulous objects seen by J. Buckingham (see Appendix) on 1865 November 9 might be identical to Pogson's object and suggested a possible return in 1893, which did not happen.

T	ω	Ω (2000.0)	i	q	e
1872 Dec. 16.616 (TT)	63.43	48.76	148.44	0.0637	1.0

ABSOLUTE MAGNITUDE: $H_{10} = 6.3$ (V1964)

FULL MOON: Nov. 15, Dec. 14

SOURCES: N. R. Pogson, *MNRAS*, **33** (1872 Dec.), p. 116; W. Klinkerfues and G. L. Tupman, *MNRAS*, **33** (1873 Jan.), pp. 128–30; C. F. W. Peters, *AN*, **80** (1873 Jan. 7), p. 335; W. Klinkerfues, *AN*, **80** (1873 Jan. 16), p. 349; J. Holetschek and T. R. von Oppölzer, *AN*, **80** (1873 Jan. 31), pp. 379–82; G. L. Tupman and J. R. Hind, *MNRAS*, **33** (1873 Mar.), pp. 317–23; T. R. von Oppölzer, *AN*, **81** (1873 May 13), pp. 281–8; N. R. Pogson, *AN*, **84** (1874 Aug. 11), pp. 183–6; K. C. Bruhns, *VJS*, **10** (1875), pp. 8–16; K. C. Bruhns, *AN*, **86** (1875 Sep. 10), pp. 219–24; H. C. F. Kreutz, *AN*, **114** (1886 Mar. 18), p. 73; W. H. S. Monck, N. R. Pogson, J. Buckingham, and C. G. Talmage, *PASP*, **4** (1892 Nov. 26), pp. 253–4; H. C. F. Kreutz, *AJB*, **3** (1902), pp. 190–1; V1964, p. 60.

9P/1873 G1 *Recovered:* 1873 April 4.13 ($\Delta = 1.01$ AU, $r = 1.80$ AU, Elong. $= 127°$)

(Tempel 1) *Last seen:* 1873 July 1.97 ($\Delta = 0.90$ AU, $r = 1.83$ AU, Elong. $= 145°$)

 Closest to the Earth: 1873 May 26 (0.7654 AU)

1873 I = 1873a *Calculated path:* OPH (Rec), SCO (Jun. 4), OPH (Jun. 9), SCO (Jun. 17)

The recovery of this comet began with the calculations of several astronomers. H. Seeliger and F. E. von Asten began with A. J. Sandberg's orbit for the 1867 apparition and each carefully determined the perturbations by

Jupiter. Both astronomers noted the approach to within 0.32 AU of Jupiter early in 1870, which lengthened the period by a few months. Seeliger's prediction gave a perihelion date of 1873 May 9.41, while Asten gave it as May 9.25. Two additional predictions were published by W. E. Plummer and A. W. Doberck. Plummer began with Searle's orbit for the 1867 apparition. He calculated rough perturbations by Jupiter, and determined the perihelion date as May 5.5. Doberck actually applied Plummer's rough perturbation calculations to Sandberg's orbit, and determined the perihelion date as May 21. E. J. M. Stephan (Marseille, France) recovered this comet using Seeliger's ephemeris on 1873 April 4.13, at a position of $\alpha = 16^h 26.4^m$, $\delta = -10°39'$. He described the comet as excessively faint in a large telescope. Stephan caught a brief glimpse of the comet on the 5th under poor observing conditions.

The comet passed perihelion during the first half of May and passed closest to Earth during the second half. Stephan said the comet was still very faint on May 2, but brighter than on April 4. On May 16, C. F. Pechüle (Hamburg, Germany) described the comet as faint with a noticeable condensation. W. H. M. Christie and H. Carpenter (Royal Observatory, Greenwich, England) saw the comet on May 20 and 23. They wrote, "The comet was slightly oblong in the direction of parallel, about 40" in diameter, and had a centrical nucleus, appearing like a blurred star of the twelfth or thirteenth magnitude. It was examined with eye-pieces of various powers, and was distinctly and steadily visible with all." Stephan saw the comet on May 21 and 23. He wrote, "The light of the comet is always faint; but it very noticeably increased" since May 2. C. H. F. Peters (Clinton, New York, USA) saw the comet without difficulty when he first looked for it on May 24. On May 28, G. F. W. Rümker (Hamburg, Germany) described the comet as very faint and small, with a central condensation. The comet reached a maximum solar elongation of 176° on the 29th.

Rümker said the comet was very faint in twilight on June 2. During the period of June 12–23, J. F. J. Schmidt (Athens, Greece) described the comet as 1–1.5' across with a very slight central condensation in the middle. He said it was difficult to measure the position on the 23rd, because the comet was in the midst of many 13th-magnitude stars. Stephan obtained observations on June 24, 26, and 29.

The comet was last detected on July 1.97 at a position of $\alpha = 16^h 14.1^m$, $\delta = -23°07'$. The observation was not published in any of the major astronomy periodicals of the time. The Author acquired this information from J. Caplan (2002), astronomer at Marseille Observatory, who found this observation during a search through the archives. Although no astronomer's name was part of this record, the Author believes it was probably made by Stephan, just as were all of the other observations of this comet.

Observations by C. L. F. André and B. Baillaud (Paris, France) were reported to have been made on August 1 by K. C. Bruhns (1875), J. G. Galle (1894), and S. K. Vsekhsvyatskij (1964); however, the Author has checked the sources and found that these observations actually belonged to comet 10P/1873 N1 (Tempel 2).

J. R. Hind used the positions obtained by Stephan on April 4 and May 2, as well as those he himself obtained on May 22 and computed an elliptical orbit with a perihelion date of May 10.24. Additional elliptical orbits were calculated by Sandberg (1875) and R. Gautier (1879, 1885). The most recent analysis of this comet's motion over multiple apparitions came from G. Schrutka (1974). Using 394 positions from the apparitions of 1867, 1873, 1879, and 1972, he applied perturbations by Venus to Pluto and determined the perihelion date as May 10.28 and the period as 5.98 years. This orbit is given below.

T	ω	Ω (2000.0)	i	q	e
1873 May 10.2804 (TT)	159.4083	80.4078	9.7664	1.771159	0.462625

ABSOLUTE MAGNITUDE: $H_{10} = 9.2$ (V1964)

FULL MOON: Apr. 12, May 12, Jun. 10, Jul. 10

SOURCES: F. E. von Asten, *BASP*, **18** (1873), pp. 557–63; W. E. Plummer and H. Seeliger, *MNRAS*, **33** (1873 Mar.), pp. 325–8; W. E. Plummer, *AN*, **81** (1873 Mar. 12), pp. 65–8; H. Seeliger, *AN*, **81** (1873 Mar. 26), pp. 145–8; E. J. M. Stephan, *MNRAS*, **33** (1873 Apr.), p. 410; A. W. Doberck, *AN*, **81** (1873 Apr. 7), p. 189; E. J. M. Stephan, *CR*, **76** (1873 Apr. 7), p. 875; E. J. M. Stephan, *AN*, **81** (1873 Apr. 18), p. 223; J. R. Hind and E. J. M. Stephan, *MNRAS*, **33** (1873 May), p. 459; E. J. M. Stephan, *CR*, **76** (1873 May 26), p. 1291; J. R. Hind, W. H. M. Christie, and H. Carpenter, *MNRAS*, **33** (1873 Jun.), pp. 498–500; J. R. Hind, *CR*, **76** (1873 Jun. 2), p. 1344; E. von Asten, *AN*, **81** (1873 Jun. 4), p. 337; E. J. M. Stephan, *AN*, **81** (1873 Jun. 14), p. 379; C. H. F. Peters, *AN*, **82** (1873 Jul. 1), p. 43; J. F. J. Schmidt, *AN*, **82** (1873 Aug. 1), pp. 89–94; C. L. F. André and B. Baillaud, *BIOP* (1873 Nov. 1); K. C. Bruhns, *VJS*, **10** (1875), p. 17; A. J. Sandberg, *AN*, **85** (1875 Apr. 29), p. 309; G. F. W. Rümker and C. F. Pechüle, *AN*, **86** (1875 Jul. 26), pp. 87–90; R. Gautier, *AN*, **94** (1879 Feb. 15), p. 157; R. Gautier, *AN*, **111** (1885 Mar. 26), pp. 241–6; G1894, p. 262; V1964, pp. 60, 236; G. Schrutka, *QJRAS*, **15** (1974), pp. 450–1, 459; personal correspondence from J. Caplan (2002).

10P/1873 N1
(Tempel 2)

1873 II = 1873b

Discovered: 1873 July 4.04 ($\Delta = 0.70$ AU, $r = 1.35$ AU, Elong. $= 102°$)
Last seen: 1873 October 20.99 ($\Delta = 0.89$ AU, $r = 1.82$ AU, Elong. $= 150°$)
Closest to the Earth: 1873 August 16 (0.6431 AU)
Calculated path: PSC (Disc), CET (Jul. 8)

E. W. L. Tempel (Brera Observatory, Milan, Italy) discovered this comet on 1873 July 4.04 at a position of $\alpha = 0^h\ 07.5^m$, $\delta = -4°\ 34'$. He simply described it as faint, with a slow southeastern motion. No further observations were possible on the 4th, but Tempel again saw the comet on the 5th. He remarked that it was brighter than expected and exhibited a coma diameter of 5'.

The comet was already past perihelion when discovered, but was continuing to approach Earth as July progressed. On July 7, L. Schulhof (Vienna, Austria) described the comet as large and bright, with three stellar condensations, the brightest of which was about magnitude 11. On the 18th, Schulhof described the comet as large and diffuse, with many stellar condensations, and an apparent short tail-like extension. Schulhof noted the comet appeared brighter on the 19th. He also pointed out that the left-most condensation was stronger. Tempel reported that moonlight caused the comet to appear faint on the 18th and 19th. On July 22, K. C. Bruhns and C. N. J. Börgen (Leipzig, Germany) described the comet as somewhat oblong, with an eccentric condensation, a granular appearance, and a diameter of 2'. Schulhof said the comet appeared "rather bright" on the 23rd and 26th. He also noted two separate, condensed nuclei on the 23rd and three distinct nuclei of magnitude 11 on the 26th. On July 27, Schulhof said the comet was fainter and less detailed than on the previous night.

The comet passed closest to Earth around mid-August. On the 1st, Bruhns said the comet was about 2' across, with a rather distinct, but eccentrically situated nucleus. Schulhof found the comet very faint in moonlight on the 17th. On August 18, C. F. Pechüle (Hamburg, Germany) said the comet was faint with a short tail toward the northeast. Börgen said the comet appeared faint on the 21st. He added that the "nucleus or condensation (was) hardly noticeable." R. Engelmann was visiting Lund Observatory in Sweden on the 26th and described the comet as rather faint, granular, and possibly elliptical in PA 0.0°. He also noted that a stellar point of magnitude 11.7 was within the coma. On August 29, Pechüle simply described the comet as faint.

The comet was moving away from both the sun and Earth as September began. On the 2nd, Börgen said the comet was faint and the condensation was not seen. On September 3, Schulhof said the comet appeared faint and rather large, with a very diffuse boundary. There was also a bright, large nucleus of magnitude 12. G. F. W. Rümker (Hamburg, Germany) saw the comet on the 19th and wrote, "Comet very pale, about 1.5' in diameter, with a weak central condensation." Schulhof found the comet faint on the 20th. Rümker reported the comet was brighter on the 25th than it had been on the 19th and 20th. He added that it appeared oblong with a weak condensation. On September 29, Rümker said the comet was very faint, with a faint central condensation.

Only a handful of observers measured the comet's position during October, and even fewer reported physical descriptions. Tempel did not measure any further positions of the comet after August, although he did see it up to October 2, when his telescope revealed it as large and faint. The comet was last seen on October 20, when W. E. Plummer and J. R. Hind measured positions within four minutes of one another at George Bishop's Observatory at Twickenham, England. Both observers "placed no reliance on their observations" because of the extreme faintness of the comet. Nevertheless, on October 20.99, Hind's final measurement of the comet's position revealed $\alpha = 1^h 22.2^m$, $\delta = -19° 56'$.

The first orbit was calculated by Schulhof and J. Holetschek using positions from July 8, 17, and 23. This was a parabolic orbit with a perihelion date of 1873 July 2.11. After a few additional observations were obtained, Börgen took positions spanning the period of July 22–28 and calculated the perihelion date as June 24.95. By the end of July, astronomers found the positions were no longer being represented by a parabolic orbit. Bruhns and Hind independently took positions spanning July and calculated elliptical orbits. Bruhns determined a perihelion date of June 25.99 and a period of 5.47 years, while Hind determined a perihelion date of June 26.22 and a period of 6.11 years. Additional orbits were determined by Börgen, Plummer, and Schulhof, which ultimately revealed a perihelion date of June 25.8 and a period of 5.1 years.

Minor refinements of this comet's orbit using positions from multiple apparitions, as well as planetary perturbations, have been published by numerous astronomers. The earliest came from Schulhof (1883, 1894), while the most recent were by B. G. Marsden and Z. Sekanina (1971) and Sekanina (1985). These have typically revealed a perihelion date of June 25.84 and a period of 5.21 years. Sekanina determined the nongravitational terms as $A_1 = +0.08$ and $A_2 = +0.0021$ and his orbit is given below.

T	ω	Ω (2000.0)	i	q	e
1873 Jun. 25.8358 (TT)	185.2126	122.6591	12.7478	1.344165	0.552593

ABSOLUTE MAGNITUDE: $H_{10} = 8.5$ (V1964)

FULL MOON: Jun. 10, Jul. 10, Aug. 8, Sep. 6, Oct. 6, Nov. 4

SOURCES: K. C. Bruhns, *MNRAS*, **33** (Supplemental Notice, 1873), p. 581; E. W. L. Tempel, K. C. Bruhns, and C. N. J. Börgen, *AN*, **82** (1873 Jul. 26), p. 79; L. Schulhof, J. Holetschek, and C. N. J. Börgen, *AN*, **82** (1873 Aug. 6), p. 111; E. W. L. Tempel and L. Schulhof, *AN*, **82** (1873 Aug. 11), pp. 113–16, 125; J. R. Hind, *AN*, **82** (1873 Aug. 16), p. 135; C. N. J. Börgen, *AN*, **82** (1873 Aug. 30), pp. 149–52; L. Schulhof and C. F. Pechüle, *AN*, **82** (1873 Sep. 6), pp. 187–92; J. C. Watson, *AN*, **82** (1873 Oct. 8), p. 242; E. W. L. Tempel, *AN*, **82** (1873 Oct. 18), p. 269; E. J. M. Stephan, A. L. N. Borrelly, and J. E. Coggia, *MNRAS*, **34** (1873 Nov.), p. 42; G. F. W. Rümker, *AN*, **82** (1873 Nov. 3), p. 273; W. E. Plummer, *MNRAS*, **34** (1873 Dec.), p. 76; D. M. A. Möller and R. Engelmann, *AN*, **83** (1874 Jan. 20), pp. 37, 41; L. Schulhof, *AN*, **83** (1874 Mar. 29), pp. 183–6; K. C. Bruhns and C. N. J. Börgen, *AN*, **84** (1874 June 10), p. 39; G. F. W. Rümker and C. F. Pechüle, *AN*, **86** (1875 Jul. 26), p. 89; J. Palisa, *AN*, **92** (1878 Jul. 20), pp. 375–80; L. Schulhof, *Annuaire pour l'An 1884. Publié par le Bureau des Longitudes. Avec des Notices Scientifiques* (1883), p. 229; L. Schulhof, *BA*, **11** (1894 Jun.), pp. 254–6; V1964, p. 60; B. G. Marsden and Z. Sekanina, *AJ*, **76** (1971 Dec.), pp. 1137–8; Z. Sekanina, *QJRAS*, **26** (1985), p. 114; B. G. Marsden and Z. Sekanina, *CCO*, 1st ed. (1972), pp. 50–1.

4P/Faye   *Recovered:* 1873 September 4.16 ($\Delta = 2.07$ AU, $r = 1.75$ AU, Elong. $= 57°$)

       Last seen: 1873 December 24.30 ($\Delta = 1.50$ AU, $r = 2.24$ AU, Elong. $= 128°$)

1873 III = 1873f   *Closest to the Earth:* 1874 January 12 (1.4675 AU)

       Calculated path: GEM (Rec), CMi (Sep. 18), CNC (Sep. 25), HYA (Oct. 22)

D. M. A. Möller took the previous orbit of this comet and advanced it to this apparition, with the resulting perihelion date being 1873 July 18.99. E. J. M. Stephan (Marseille, France) recovered this comet on 1873 September 4.16, at a position of $\alpha = 7^h\ 00.8^m$, $\delta = +15°\ 47'$. He commented, "The comet is excessively faint and very small, but with a small well-defined nucleus."

This apparition was not a well-observed one. No observations were made during the remainder of September or throughout October. On November 29, Stephan saw the comet "under unsatisfactory circumstances" and described it as "excessively faint, and only visible by glimpses." He saw the comet under better circumstances on December 1.

The comet was last detected on December 24.30, when C. H. F. Peters (Litchfield Observatory, Clinton, New York, USA) measured the position as $\alpha = 9^h\ 18.3^m$, $\delta = -2°\ 15'$. K. C. Bruhns wrote in the 1875 volume of *Vierteljahrsschrift Astronomische Gesellschaft* that unsuccessful searches for this comet were made at observatories in Copenhagen, Leipzig, and Pulkovo during this apparition.

Later calculations using multiple apparitions and planetary perturbations were published by A. Shdanow (1885) and B. G. Marsden and Z. Sekanina (1971). These revealed a perihelion date of July 18.98 and a period of 7.41 years. Marsden and Sekanina determined the nongravitational terms as $A_1 = +0.560$ and $A_2 = +0.0608$. The orbit of Marsden and Sekanina is given below.

T	ω	Ω (2000.0)	i	q	e
1873 Jul. 18.9847 (TT)	200.3449	211.5150	11.3505	1.682547	0.557381

ABSOLUTE MAGNITUDE: $H_{10} = 7.4$ (V1964)

FULL MOON: Sep. 6, Oct. 6, Nov. 4, Dec. 4, Jan. 2

SOURCES: D. M. A. Möller, *AN*, **80** (1873 Jan. 16), p. 337; E. J. M. Stephan, *CR*, **77** (1873 Sep. 8), p. 606; E. J. M. Stephan, *AN*, **82** (1873 Sep. 27), pp. 215–18; E. J. M. Stephan, *MNRAS*, **34** (1873 Nov.), p. 42; E. J. M. Stephan, *CR*, **77** (1873 Dec. 8), pp. 1364–5; E. J. M. Stephan, *AN*, **82** (1873 Dec. 13), p. 383; E. J. M. Stephan, *MNRAS*, **34** (1874 Jan.), p. 126; C. H. F. Peters, *MNRAS*, **34** (1874 Mar.), p. 271; H. L. d'Arrest, *AN*, **84** (1874 Jul. 7), p. 140; K. C. Bruhns, *VJS*, **10** (1875), p. 23; A. Shdanow, *BASP*, **33** (1885), pp. 1–24; V1964, p. 60; B. G. Marsden and Z. Sekanina, *AJ*, **76** (1971 Dec.), pp. 1136–7.

C/1873 Q1
(Borrelly)
1873 IV = 1873c

Discovered: 1873 August 20.99 ($\Delta = 1.30$ AU, $r = 0.89$ AU, Elong. $= 43°$)
Last seen: 1873 September 21.41 ($\Delta = 0.87$ AU, $r = 0.82$ AU, Elong. $= 51°$)
Closest to the Earth: 1873 September 21 (0.8713 AU)
Calculated path: LYN (Disc), GEM (Aug. 24), CNC (Sep. 4), HYA (Sep. 12)

A. L. N. Borrelly (Marseille, France) discovered this comet on 1873 August 20.99 at a position of $\alpha = 7^h\ 27^m$, $\delta = +38°\ 45'$. He simply described it as bright. Borrelly confirmed the comet on August 21.14 and said the motion was toward the south at nearly 1° per day. His colleague E. J. M. Stephan saw the comet on August 22.11 and described it as rather brilliant, round, with an almost central condensation. The comet had passed 41° from the sun on July 30.

The comet was well observed during the remainder of August as it approached both the sun and Earth. On August 23, E. Weiss (Vienna, Austria) said the comet was rather bright and appeared elliptical. That same night, C. F. Pechüle (Hamburg, Germany) described the comet as round and gradually condensed toward the middle. C. N. J. Börgen (Leipzig, Germany) described the comet as very bright on the 24th. He said the coma was 3' across and added that a magnification of 145× revealed a strong central condensation. L. Schulhof (Vienna, Austria) said the comet was large and bright on the 25th, with a nucleus of 9th magnitude. Pechüle noted that the comet was round and 2' across on the 23rd, 28th, and 29th. He also reported a central condensation. J. F. J. Schmidt (Athens, Greece) reported the comet was visible in the finder of his telescope on August 27.

The comet passed perihelion during the first half of September and passed closest to Earth during the second half. For September 2, E. J. M. Stephan (Marseille, France) wrote, "The comet is excessively faint and diffuse in appearance, with a trace of condensation...." Schulhof saw the comet on the 3rd and noted, "Although not very clear, a faint tail was perceived." Schmidt was unable to find the comet on the 13th, even though, as he noted, it should have been brighter and still favorably situated. The comet passed 2.7° from comet 5D/Brorsen on the 13th.

The final observations of this comet were made by G. Strasser (Kremsmünster, Austria) on September 21.15 and A. Hall (US Naval Observatory, Washington, DC) on September 21.41. Hall gave the position as $\alpha = 8^h\ 35.8^m$, $\delta = -13°\ 19'$ and noted the comet was then in twilight.

Several astronomers examined the spectrum of this comet. C. J. E. Wolf and G. A. P. Rayet detected a continuous spectrum on August 22, with "two luminous bands, one in the green, and the other in the blue." H. C. Vogel examined it on September 1 and detected all three carbon bands.

The first parabolic orbit was calculated by E. Weiss using positions from August 22, 23, and 24. He determined the perihelion date as 1873 September 11.99. Later orbits by C. F. W. Peters, W. E. Plummer, and Weiss eventually established the perihelion date as September 11.2. A few years passed before this comet captured the interest of another astronomer. R. Gautier (1877, 1878) calculated parabolic orbits with perihelion dates of September 11.30, but he also noted that "the data make it tolerably certain that the orbit is elliptic." His last elliptical orbit had a perihelion date of September 11.28 and a period of about 3375 years. This orbit is given below.

T	ω	Ω (2000.0)	i	q	e
1873 Sep. 11.2837 (UT)	193.7729	232.3621	95.9662	0.794061	0.996471

ABSOLUTE MAGNITUDE: $H_{10} = 6.7$ (V1964)

FULL MOON: Aug. 8, Sep. 6, Oct. 6

SOURCES: A. L. N. Borrelly, C. J. E. Wolf, and G. A. P. Rayet, *MNRAS*, **33** (Supplemental Notice, 1873), pp. 581–2; A. L. N. Borrelly, C. J. E. Wolf, and G. A. P. Rayet, *CR*, **77** (1873 Aug. 25), pp. 528–9; A. L. N. Borrelly, C. N. J. Börgen, and E. Weiss, *AN*, **82** (1873 Aug. 30), p. 155; A. L. N. Borrelly and E. J. M. Stephan, *CR*, **77** (1873 Sep. 1), pp. 563–4; E. Weiss, C. F. W. Peters, C. F. Pechüle, E. J. M. Stephan, and A. L. N. Borrelly, *AN*, **82** (1873 Sep. 6), pp. 187, 191; E. J. M. Stephan, *AN*, **82** (1873 Sep. 12), p. 193; H. C. Vogel, *AN*, **82** (1873 Sep. 27), p. 217; J. F. J. Schmidt and E. W. L. Tempel, *AN*, **82** (1873 Oct. 18), pp. 265–72; W. E. Plummer, E. J. M. Stephan, and A. L. N. Borrelly, *MNRAS*, **34** (1873 Nov.), pp. 41–2; E. Weiss, *AN*, **82** (1873 Nov. 20), pp. 305–8; D. M. A. Möller, A. Lindstedt, and G. Strasser, *AN*, **83** (1874 Jan. 20), pp. 37, 43–8; L. Schulhof, *AN*, **83** (1874 Mar. 29), pp. 183–6; A. Hall, *AN*, **84** (1874 Jun. 8), p. 17; C. F. Pechüle, *AN*, **86** (1875 Jul. 26), p. 89; R. Gautier, *AN*, **91** (1877 Nov. 8), pp. 49–58; R. Gautier, *The Observatory*, **1** (1878 Mar.), p. 374; R. Gautier, *AN*, **92** (1878 Apr. 2), p. 71; A. Hall, *AN*, **92** (1878 Jul. 16), p. 365; V1964, p. 60.

C/1873 Q2
(Henry)

1873 V = 1873d

Discovered: 1873 August 23.95 ($\Delta = 1.12$ AU, $r = 1.02$ AU, Elong. $= 57°$)

Last seen: 1873 December 18.21 ($\Delta = 1.98$ AU, $r = 1.69$ AU, Elong. $= 58°$)

Closest to the Earth: 1873 September 17 (0.4884 AU)

Calculated path: LYN (Disc), UMa (Aug. 31), LMi (Sep. 10), UMa-LMi (Sep. 11), LEO (Sep. 15), VIR (Sep. 18), CRV (Sep. 24), HYA (Sep. 29), VIR (Nov. 2)

Paul Henry (Paris, France) discovered this comet on 1873 August 23.95 at a position of $\alpha = 7^h\ 27^m$, $\delta = +59°\ 30'$. It was described as round, centrally condensed, and almost visible to the naked eye. The diameter was about 3–4'. Henry also said the comet was moving rapidly towards the east.

The comet was well observed during the remainder of August as it approached both the sun and Earth. A. Hall (US Naval Observatory, Washington, DC, USA) described it as a bright telescopic object on the 26th. G. A. P. Rayet and C. L. F. André examined the comet at moderately high powers on the 26th and 27th. They noted the coma was about 6' across and exhibited "a very visible condensation of light in the centre; there was no trace of a nucleus, or of successive envelopes." The condensation was similar to a star of magnitude 7. They added that the comet resembled M13 "as seen through a telescope not sufficiently powerful to resolve the cluster into stars." On August 27, 29, and 31, C. F. Pechüle (Hamburg, Germany) said the comet was "very brilliant with a star-like nucleus situated to the east side." He gave the coma diameter as 4'. Interestingly, Pechüle added that the "nebulous mass appeared distinctly marbled." E. W. L. Tempel (Brera Observatory, Milan, Italy) did not see a tail in a telescope on the 28th. The comet was especially well observed on the 29th, with descriptions coming from several astronomers. Rayet and André said the coma diameter was "much increased" and nearly 8′ across. There was also a rather broad tail extending 25′ and directed away from the sun. They added, "The head of the comet had preserved its roundness, and the brightness of the central nucleus had increased to that of a star of the sixth magnitude. The tail, less luminous near the nebulosity of the head, became brighter afterwards, and disappeared gradually at the extremity." Tempel noted the small, narrow tail was about a degree long. He also said the coma was round, sharply bordered, and 8–10′ across, with a strong central condensation. Tempel added, "The tail did not begin directly from the head, but there was a gap of several minutes and it began as a fine point of one minute width, extended and ended broad and diffuse." D. M. A. Möller, Anderson, and A. Lindstedt (Lund, Sweden) said the tail extended toward PA 321.3°, 313.9°, and 315.4°, respectively. L. Schulhof (Vienna, Austria) said the comet was very bright, large, round, and a little more condensed than comet C/1873 Q1. On August 31, Hall described the comet as a bright telescopic object. Möller said the tail extended toward PA 316.8°, while Lindstedt said it extended toward PA 317.8°.

The comet was closest to Earth around mid-September and was lost in the sun's glare shortly thereafter. On the 1st, Schulhof saw a faint tail 10′ long, while C. N. J. Börgen (Leipzig, Germany) said the comet was very bright, about 3′ across, with a central condensation. Although Hall reported the tail was 15–20′ long on the 3rd, Tempel said it was more than 3° long. He noted that it "began 2′ to 3′ outside the coma" and was 2′ wide about 15′ from the coma and at least 16–20′ wide at the end. Interestingly, Tempel added that he saw a "distinct flickering in the tail . . . as if waves of light went off and on." H. C. Vogel (Bothkamp, Germany) said the coma was 5′ or 6′ across with a strong, central condensation on the 4th. He reported the tail was 0.8° long, while Möller said it was directed toward PA 324.2° and Lindstedt said it extended toward PA 319.4°. A tail length of 2° was also provided by Rayet and André on that morning, while the coma was estimated as 8′ or 9′ across. R. Engelmann (Leipzig, Germany) said the comet was faint on the 6th, while Vogel noted the nucleus was 8–10″ across. Engelmann found the comet rather bright and round, with a granular appearance on the 7th. On September 8, Schulhof noted an elliptical nucleus and a fan-shaped tail 1–2′ long, with its major axis directed toward PA 100°. Engelmann described the comet as very bright and round on the 9th. He added that the coma was 2′ across and granular in appearance, with a nucleus of magnitude 7 or 8. Schulhof saw the comet with the naked eye on the 10th. He added that a telescope revealed a sharp, distinct nucleus "eccentric toward the left," with a faint tail 20′ long. That same evening, Engelmann saw the comet at an altitude of 3.3° and described it as rather faint. N. C. Dunér (Lund, Sweden) said the comet appeared elliptical on the 13th, with the major axis directed toward PA 56°. G. F. W. Rümker (Hamburg, Germany) described the comet as very faint at low altitude on the 13th. On September 17.15, Schulhof was only able to make one position measurement before the sky became too bright for further observations.

The comet passed 1.2° from the sun on September 21 and attained its most southerly declination of −27° on October 8. It remained lost in the sun's glare until November 29.17, when J. Palisa (Pula, Yugoslavia) measured the position of a nebulous object he found while searching for this comet. He said, "In the absence of a complete nebula catalog I cannot decide whether the object observed by me is the Henry comet." The object was within a few arc seconds of the predicted position. The comet was last seen on December 18.21, when Palisa gave the position as $\alpha = 13^h\ 41.1^m$, $\delta = -13°\ 02'$.

The spectrum was examined by Rayet and André on August 26 and 27, and by Vogel on September 4, 6, and 11. The three carbon bands were reported each time. Rayet and André said there was no trace of the continuous spectrum.

The first parabolic orbit was calculated by E. Weiss using positions spanning the period of August 27–30. He determined the perihelion date as 1873 October 2.33. Additional orbits by O. Stone, Möller, Dunér, L. de Ball, W. E. Plummer, W. Fabritius, Weiss, and A. Zielinsky eventually established the perihelion date as October 2.27. A definitive orbit was calculated by H. C. F. Kreutz (1894). Beginning with positions spanning the entire period of visibility, he applied the perturbations of Mars, Jupiter, and Saturn, and determined the elliptical orbit below.

T	ω	Ω (2000.0)	i	q	e
1873 Oct. 2.2672 (UT)	233.7544	178.4959	121.4625	0.384913	0.999730

ABSOLUTE MAGNITUDE: $H_{10} = 6.4$ (V1964)

FULL MOON: Aug. 8, Sep. 6, Oct. 6, Nov. 4, Dec. 4, Jan. 2

SOURCES: P. Henry, G. A. P. Rayet, and C. L. F. André, *MNRAS*, **33** (Supplemental Notice, 1873), pp. 581–2; Paul Henry, *CR*, **77** (1873 Aug. 25), p. 528; G. A. P. Rayet and C. L. F. André, *CR*, **77** (1873 Sep. 1), pp. 564–6; L. Schulhof and C. F. Pechüle, *AN*, **82** (1873 Sep. 6), pp. 189–92; E. Weiss, Paul Henry, D. M. A. Möller, and N. C. Dunér, *AN*, **82** (1873 Sep. 12), pp. 193, 199; G. A. P. Rayet and C. L. F. André, *CR*, **77** (1873 Sep. 15), pp. 638–9; H. C. Vogel, R. Engelmann, and L. de Ball, *AN*, **82** (1873 Sep. 27), pp. 217–22, 239; B. F. Sands, A. Hall, and O. Stone, *AN*, **82** (1873 Oct. 8), p. 243; E. W. L. Tempel, *AN*, **82** (1873 Oct. 18), pp. 269–72; W. E. Plummer and E. J. M. Stephan, *MNRAS*, **34** (1873 Nov.), pp. 40–2; W. Fabritius, *AN*, **82** (1873 Nov. 3), p. 283; H. C. Vogel, *AN*, **82** (1873 Nov. 11), p. 297; E. Weiss, *AN*, **82** (1873 Nov. 20), pp. 305–8; J. Palisa, *AN*, **82** (1873 Dec. 13), p. 379; D. M. A. Möller, Anderson, N. C. Dunér, and A. Lindstedt, *AN*, **83** (1874 Jan. 20), pp. 39–42; E. Weiss and A. Zielinsky, *AN*, **83** (1874 Jan. 28), pp. 49, 53–6; J. Palisa, *AN*, **83** (1874 Mar. 5), pp. 131–6; L. Schulhof, *AN*, **83** (1874 Mar. 29), pp. 183–6; K. C. Bruhns and C. N. J. Börgen, *AN*, **84** (1874 June 10), p. 39; J. J. Plummer, *AN*, **84** (1874 Jun. 16), p. 79; G. F. W. Rümker and C. F. Pechüle, *AN*, **86** (1875 Jul. 26), p. 91; H. C. F. Kreutz, *Publication der Sternwarte in Kiel*, **9** (1894), pp. 21–37; V1964, p. 60.

5D/Brorsen

1873 VI = 1873e

Recovered: 1873 September 1.16 ($\Delta = 1.02$ AU, $r = 0.96$ AU, Elong. $= 56°$)
Last seen: 1873 October 27.22 ($\Delta = 1.44$ AU, $r = 0.68$ AU, Elong. $= 24°$)
Closest to the Earth: 1873 September 10 (1.0037 AU)
Calculated path: MON (Rec), CMi (Sep. 8), HYA (Sep. 11), SEX (Sep. 24), LEO (Oct. 2), VIR (Oct. 11)

The prediction of this comet's return came from J. R. Hind (1873), although L. R. Schulze (1873) also published a prediction around the time of the comet's recovery. Hind noted that the comet would experience no appreciable perturbations from Jupiter or Saturn during the interval between its 1868 apparition and the next return in 1873. Consequently, Hind determined the perihelion date as 1873 October 11.5. Schulze predicted a date of October 10.73. The comet was recovered by E. J. M. Stephan (Marseille, France) on 1873 September 1.16, using an ephemeris by Plummer that had been provided by Hind. The comet was then about 10° above the horizon in the east-southeastern sky, just minutes before the beginning of morning twilight. No position was obtained. On September 2.15, Stephan gave the position as $\alpha = 7^h 06.5^m$, $\delta = -1° 55'$. He described the comet as "an ovoid nebulosity, diffuse, of excessive faintness, with a hardly detectable trace of central condensation; the observation was difficult."

Although the comet was well observed during September and October, only E. W. L. Tempel (Milan, Italy) provided a physical description. He saw the comet on September 20 and 22, as well as on October 1, 3, and 4. Tempel described it as appearing like a nebulous star about 1' across, that looked like the second companion to the Andromeda galaxy. He noted the brightness was that of a star of magnitude 7 or 8. The comet was closest to Earth in the first half of September. On October 10, the comet passed perihelion and attained its most northerly declination of $+7°$.

The comet was last seen on October 27.22, when W. E. Plummer (Twickenham, England) gave the position as $\alpha = 13^h 02.4^m$, $\delta = +5° 30'$. He described the comet as round, about 1.25' across, "with considerable condensations towards the centre."

Not long after the comet had been recovered, Plummer took the Marseille position of September 1 and used it to correct the predicted orbit. The result was a perihelion date of October 10.97 and a period of 5.47 years. Very similar results, derived from positions spanning multiple apparitions, were later obtained by Schulze (1878), E. A. Lamp (1892), P. H. Harzer (1894), B. G. Marsden and Z. Sekanina (1971), W. Landgraf (1986), and S. Nakano (1997). Nongravitational terms were determined by Marsden and Sekanina, as well as Landgraf. They were fairly comparable, with Landgraf's being $A_1 = +1.27$ and $A_2 = +0.1343$. Nakano's orbit is given below.

T	ω	Ω (2000.0)	i	q	e
1873 Oct. 10.9803 (TT)	14.8637	102.9964	29.4035	0.593761	0.808804

ABSOLUTE MAGNITUDE: $H_{10} = 9.2$ (V1964)

FULL MOON: Aug. 8, Sep. 6, Oct. 6, Nov. 4

SOURCES: J. R. Hind, *MNRAS*, **33** (1873 Mar.), pp. 324–5; L. R. Schulze, *AN*, **82** (1873 Sep. 6), pp. 173–84; E. J. M. Stephan, *CR*, **77** (1873 Sep. 8), pp. 605–6; E. J. M. Stephan, *AN*, **82** (1873 Sep. 12), p. 193; W. E. Plummer, *AN*, **82** (1873 Oct. 8), pp. 241–4; E. W. L. Tempel, *AN*, **82** (1873 Oct. 18), pp. 269–72; G. Bishop and W. E. Plummer, E. J. M. Stephan, and J. E. Coggia, *MNRAS*, **34** (1873 Nov.), p. 40, 42; E. W. L. Tempel, *Pubblicazioni del Reale Osservatorio di Brera in Milano*, No. 5 (1874), p. 7; L. R. Schulze, *AN*, **93** (1878 Sep. 27), pp. 177–88; E. A. Lamp, *Publication der Sternwarte in Kiel*, **7** (1892), pp. 31–59; P. H. Harzer, *Publication der Sternwarte in Kiel*, **9** no. 2 (1894), pp. 21–37; V1964, p. 60; B. G. Marsden and Z. Sekanina, *AJ*, **76** (1971 Dec.), pp. 1141–2; W. Landgraf, *QJRAS*, **27** (1986 Mar.), p. 116; B. G. Marsden and W. Landgraf, *CCO*, 12th ed. (1997), pp. 48–9; S. Nakano, *Nakano Note* No. 663 (1997 May 5).

27P/1873 V1 *Discovered:* 1873 November 10.84 ($\Delta = 0.24$ AU, $r = 0.85$ AU, Elong. $= 49°$)
(Crommelin) *Last seen:* 1873 November 16.73 ($\Delta = 0.22$ AU, $r = 0.80$ AU, Elong. $= 29°$)
Closest to the Earth: 1873 November 17 (0.2183 AU)
1873 VII = 1873g *Calculated path:* HER (Disc)

J. E. Coggia (Marseille, France) discovered this comet on 1873 November 10.84 and measured the position as roughly $\alpha = 16^h 23.0^m$, $\delta = +27° 26'$. He described it as faint and noted movement toward the southwest. He reobserved the comet on November 11.78 and described it as faint, with a central condensation. An independent discovery was made by F. A. T. Winnecke (Strasbourg, France) on November 11.73. He said the comet appeared as a pale disk 3′ in diameter and was surrounded by a pale glow. Winnecke confirmed his find on November 12.76 and said the comet was uniformly bright with a diameter of 6′.

Observations were obtained at other observatories, but they were few. G. F. W. Rümker and C. F. Pechüle (Hamburg, Germany) measured positions on November 12, while Rümker obtained additional positions on the 13th and 14th. For the 12th, Rümker wrote, "The comet is about 4–5′ across, round and very diffuse." He added, "In consequence of its strongly granular appearance it can only be observed with difficulty." Rümker added that the comet was very faint and diffuse on the 13th and "reasonably bright" on the 14th. K. C. Bruhns and C. N. J. Börgen (Leipzig, Germany) saw the comet on November 12 at a magnification of 145× and said it "resembled a 3–4′ diameter nebula, with a slight condensation on the sunward side." They added that it had a granular texture. E. Weiss and L. Schulhof (Vienna, Austria) measured precise positions on November 12 and 13.

The comet was last detected on November 16.73, when Winnecke measured the position as $\alpha = 15^h 31.5^m$, $\delta = +10° 07'$. The comet passed 20° from the sun on November 21. Although J. R. Hind had provided an ephemeris for Southern Hemisphere observers that covered the period of December 16 to January 1, no additional observations were acquired.

The earliest parabolic orbits were independently calculated by E. Weiss and J. R. Hind. Both astronomers used positions from November 11, 12, and 13, with Weiss determining a perihelion date of 1873 December 4.60 and Hind giving it as November 30.78. Both astronomers noted the similarity between this comet's orbit and that of Pons' comet of February 1818. Additional orbits were calculated by W. Fabritius and Weiss (1874), which indicated a perihelion date of December 1.7. F. W. A. Argelander commented on the similarity of Fabritius' orbit and that of the comet of February 1818. Weiss found that by assigning a perihelion date of 1818 February 7.10 to his orbit, the first and last rough positions of Pons' 1818 comet were represented within 15' or 16'. He determined two hypothetical elliptical orbits, the first had a perihelion date of December 2.09 and a period of 55.82 years, while the second had a perihelion date of December 3.44 and a period of 6.98 years. Two years later, Weiss published a revised set of possible elliptical orbits indicating possible periods of 55.82, 18.61, and 6.20 years.

Schulhof spent the most time trying to link the short observation arcs of comets C/1818 D1 and C/1873 V1. In 1885, he suggested periods of 7, 9.3, or 55.8 years were most likely. He said the 7-year period would cause the comet to pass 0.18 AU from Jupiter during March of 1841. The 9.3-year period would raise the possibility that Pons' comet seen during September 1808 could be an early appearance. If the 55.8-year period were true, then comet C/1457 A1 could be an earlier apparition. Schulhof (1886) determined more precise orbits with periods of 6.20 years and 55.82 years. He published additional investigations in 1887 and 1892, but the 1818 and 1873 observations were just too rough and he never derived the correct period.

The comet was accidentally discovered again in 1928 and, for a while was known by the name of Pons–Coggia–Winnecke–Forbes. However, it was astronomer A. C. D. Crommelin (1936) who established the true orbit of this comet and successfully linked the apparitions of 1818, 1873, and 1928, so the comet is now named after him. One of the most recent orbits calculated for this comet was determined by D. K. Yeomans and P. W. Chodas (1986). They used 279 positions obtained during the apparitions spanning 1818–1984, and determined the period as 28.1 years. This orbit is given below.

T	ω	Ω (2000.0)	i	q	e
1873 Dec. 2.4190 (TT)	196.0245	251.5880	28.7941	0.747466	0.919076

ABSOLUTE MAGNITUDE: $H_{10} = 11.6$ (V1964)

FULL MOON: Nov. 4, Dec. 4

SOURCES: E. Weiss and J. R. Hind, *MNRAS*, **34** (1873 Nov.), pp. 47–8; J. E. Coggia, F. A. T. Winnecke, E. J. M. Stephan, G. F. W. Rümker, C. F. Pechüle, E. Weiss, and L. Schulhof, *AN*, **82** (1873 Nov. 20), pp. 315–20; J. R. Hind, *MNRAS*, **34** (1873 Dec.), p. 83; K. C. Bruhns, C. N. J. Börgen, W. Fabritius, and F. W. A. Argelander, *AN*, **82** (1873 Dec. 13), pp. 369, 381; E. Weiss, *AN*, **83** (1874 Jan. 9), pp. 5–10; L. Schulhof, *AN*, **83** (1874 Mar. 29), p. 185; K. C. Bruhns and C. N. J. Börgen, *AN*, **84** (1874 Jun. 10), p. 39; G. F. W.

Rümker and C. F. Pechüle, *AN*, **86** (1875 Jul. 26), p. 91; E. Weiss, *AN*, **87** (1876 Jan. 19), pp. 121–4; F. A. T. Winnecke, *AN*, **91** (1878 Jan. 10), pp. 249–52; L. Schulhof, *AN*, **113** (1885 Dec. 15), p. 143; L. Schulhof, *BA*, **3** (1886 Mar.), pp. 125–34; L. Schulhof, *BA*, **4** (1887 Feb.), pp. 51–4; L. Schulhof, *BA*, **9** (1892 Mar.), pp. 118–21; A. C. D. Crommelin and H. A. Kobold, *AN*, **261** (1936 Nov. 10), pp. 83–6; V1964, p. 60; D. K. Yeomans and P. W. Chodas, *QJRAS*, **27** (1986), p. 604.

C/1874 D1	*Discovered:* 1874 February 21.14 ($\Delta = 0.76$ AU, $r = 0.69$ AU, Elong. $= 44°$)
(Winnecke)	*Last seen:* 1874 February 26.17 ($\Delta = 0.74$ AU, $r = 0.54$ AU, Elong. $= 33°$)
	Closest to the Earth: 1874 February 26 (0.7427 AU)
1874 I = 1874a	*Calculated path:* VUL (Disc), PEG (Feb. 25)

F. A. T. Winnecke (Strasbourg, France) discovered this "very small comet" around 1874 February 21.14. He estimated a rough position of $\alpha = 20^h 35.1^m$, $\delta = +26° 05'$ on February 21.15, and a short time later noted that the coma was 2′ across. C. F. Pechüle (Hamburg, Germany) confirmed the discovery on February 22.19.

This comet moved quickly into morning twilight after its discovery and remained visible for only 6 days. Winnecke observed it under very clear skies on February 23 and noticed a very faint tail extending 2′. He added that the nucleus frequently flashed into view like an 11th-magnitude star. L. Schulhof saw the comet on the 24th and described it as rather bright and round, with a diameter of 2′. There was also a central condensation. Schulhof noted that the comet had brightened by the next morning and said the central condensation was nearly stellar.

The comet was seen for the final time by J. Palisa (Pula, Yugoslavia) on February 26.15 and by Schulhof on February 26.17. Schulhof measured the position as $\alpha = 21^h 20.1^m$, $\delta = +18° 03'$ and said the comet was larger, more diffuse, and fainter than on the previous morning. Winnecke searched in vain for the comet on April 6 and 7.

The short period of visibility has only allowed the calculation of parabolic orbits. The first was by W. Schur, who took positions from February 21, 22, and 23, and determined the perihelion date as 1874 March 10.28. Additional orbits were calculated by Schur, L. Schulhof, and A. Wittstein (1879) which determined the perihelion date as March 10.43. Wittstein's best orbit is given below.

T	ω	Ω (2000.0)	*i*	*q*	*e*
1874 Mar.10.4346 (UT)	269.5099	32.0539	58.8932	0.044568	1.0

ABSOLUTE MAGNITUDE: $H_{10} = 11.0$ (V1964)

FULL MOON: Feb. 1, Mar. 3

SOURCES: F. A. T. Winnecke and L. Schulhof, *MNRAS*, **34** (1874 Mar.), p. 272; F. A. T. Winnecke, C. F. Pechüle, and W. Schur, *AN*, **83** (1874 Mar. 5), pp. 141–4; L. Schulhof, *AN*, **83** (1874 Mar. 11), p. 147; F. A. T. Winnecke, *VJS*, **10** (1875), pp. 183–4; L. Schulhof,

AN, **85** (1875 Apr. 29), p. 285; C. F. Pechüle, *AN*, **86** (1875 Jul. 26), p. 91; A. Wittstein, F. A. T. Winnecke, J. Palisa, and L. Schulhof, *AN*, **94** (1879 Feb. 28), pp. 193–200; V1964, p. 60.

**C/1874 G1
(Winnecke)**

1874 II = 1874b

Discovered: 1874 April 12.12 ($\Delta = 1.01$ AU, $r = 1.03$ AU, Elong. $= 61°$)
Last seen: 1874 June 17.92 ($\Delta = 1.53$ AU, $r = 1.82$ AU, Elong. $= 89°$)
Closest to the Earth: 1874 May 7 (0.5369 AU)
Calculated path: AQR (Disc), EQU–DEL (Apr. 21), AQL (Apr. 28), SGE (Apr. 29), VUL (May 1), LYR (May 4), HER (May 8), BOO (May 17), CVn (May 27)

F. A. T. Winnecke (Strasbourg, France) discovered this comet on 1874 April 12.12, at a position of $\alpha = 21^h 23.1^m$, $\delta = -6° 56'$. He said the comet appeared bright and round, with a coma 4' across that was strongly condensed in the center. Independent discoveries were made by A. L. N. Borrelly (Marseille, France) on April 16.1 and E. W. L. Tempel (Milan, Italy) on April 19.06.

The comet had already passed perihelion, but was still heading for a moderately close approach to Earth. It was well observed during the remainder of April. K. C. Bruhns (Leipzig, Germany) saw the comet on the 18th and noted that its brightness resembled that of the nebula in Aquarius, although it was larger. At 192× the diameter was estimated as 4'. On April 22, L. Schulhof (Vienna, Austria) said the comet was very bright, circular, and about 5' across, while G. F. W. Rümker (Hamburg, Germany) said the comet was rather large, with a condensation 1.5' across. Rümker also noted the coma extended toward the sun. Schulhof said the comet was nearly circular on the 23rd, with a "rather elliptical" condensation. On April 24, J. F. J. Schmidt (Athens, Greece) observed the comet in the morning sky after the moon had set. With the finder (8×) he noted the comet was tailless, round, well-condensed, and about 7' across. He added that it appeared larger than M15, but was slightly fainter. Following moonset on the morning of April 25, Schmidt said the comet was visible to the naked eye with difficulty. The 17-cm Reinfelder refractor revealed a faint condensation, with a nucleus of magnitude 11. He indicated the coma diameter was 8.82'. Schulhof found the nucleus large and irregular on the 25th. C. F. Pechüle (Hamburg, Germany) said the comet was rather faint in morning twilight on the 27th. On April 28, Rümker said the coma was probably 6–8' across in the bright moonlight, with a pale condensation about 50" across.

The comet passed closest to Earth during the first half of May and moved away from both the sun and Earth thereafter. On May 6, Schulhof said the comet was fainter and more diffuse than at the end of April, while Schmidt said the comet was bright and condensed, with a nucleus of magnitude 12. Schmidt indicated the coma was 8.66' across. Schmidt found the comet was an easy target in the finder on May 7, 8, and 9. H. C. Vogel (Berlin, Germany) said the coma was 7–8' across on May 7 and 10. Schmidt described the comet as large and faint in the 17-cm Reinfelder refractor on the 12th. He gave the diameter as 6.66' and said it was not easy to see with the naked eye. Schmidt added that the comet was fainter and less condensed than M13. On May 13, Schmidt said the comet was condensed, with a nucleus of magnitude 12 or 13. The coma was 9.44' across. Schmidt said the comet maintained this appearance in the seeker during the period of May 15–19. Rümker said the comet was still large on the 14th, but had decreased in brightness. On the 18th, Schmidt said the 17-cm Reinfelder refractor revealed a coma 9.88' across, with a nucleus of magnitude 12.5. The comet attained its most northerly declination of +43° on the 19th and Rümker described it as very diffuse and strongly granular, with a noticeable condensation in the preceding section. G. Koch (Leipzig, Germany) said the comet was very faint and without a nucleus on the 20th, while he described it as faint and diffuse on the 21st. On May 25, Rümker observed the comet in bright moonlight and said that it was pale and diffuse, with occasional stellar points in the preceding section.

As June began, and with the moon no longer hampering observations, it was obvious the comet had faded considerably. On the 3rd, Schulhof described it as faint, while Schmidt saw it near the zenith with the finder on the 4th and noted it was very difficult to see. Schulhof also described the comet as very faint on the 7th. During the period of June 9–13, Schmidt observed with the refractor and said the comet was very faint, with little condensation and no nucleus. His measurements of the coma diameter indicated values of 4.82' on the 9th, 4.72' on the 10th, 5.76' on the 11th, and 5.20' on the 12th. On June 8 and 10, Rümker said the comet was at the limit of visibility in twilight. Bruhns said the comet was extremely diffuse on the 12th.

The comet was seen for the final time on June 17.92, when Schulhof gave the position as $\alpha = 12^h\ 44.5^m$, $\delta = +32°\ 35'$. He said the comet was very faint. Schulhof failed to find the comet on June 10, because conditions were not good and the comet should have been near a star of magnitude 9.5.

The spectrum was observed by Vogel on May 7 and 10. He detected the three bright cometary bands caused by C_2.

The first parabolic orbits were calculated by W. Schur and E. Weiss. Schur took positions from the period spanning April 13–20 and determined a perihelion date as 1874 March 14.50, while Weiss took positions from April 13, 18, and 21, and detemined a perihelion date of March 14.46. A definitive orbit was calculated by G. Burggraf (1904). It established the perihelion date as March 14.44. Although Burggraf also gave an ellipse with a period of about 13 000 years, he preferred the parabolic orbit and this is given below.

T	ω	Ω (2000.0)	i	q	e
1874 Mar.14.4355 (UT)	331.7167	275.8486	148.4122	0.885773	1.0

ABSOLUTE MAGNITUDE: $H_{10} = 5.0$ (V1964)
FULL MOON: Apr. 1, May 1, May 31, Jun. 29

SOURCES: F. A. T. Winnecke, *AN*, **83** (1874 Apr. 24), p. 263; K. C. Bruhns, *AN*, **83** (1874 Apr. 29), p. 285; F. A. T. Winnecke and E. Weiss, *MNRAS*, **34** (1874 May), pp. 361–2; W. Schur, F. A. T. Winnecke, E. W. L. Tempel, and E. Weiss, *AN*, **83** (1874 May 6), pp. 293–8; J. F. J. Schmidt, *AN*, **84** (1874 Jul. 14), pp. 157–60; H. C. Vogel, *AN*, **85** (1874 Dec. 22), p. 17; K. C. Bruhns and G. Koch, *AN*, **85** (1875 Feb. 8), pp. 87–90, 101; L. Schulhof, *AN*, **85** (1875 Apr. 29), pp. 285–8, 290; G. F. W. Rümker and C. F. Pechüle, *AN*, **86** (1875 Jul. 26), pp. 91–6; G. Burggraf, *SAWW*, **113** Abt. IIa (1904), pp. 97–182; V1964, p. 60.

C/1874 H1	*Discovered:* 1874 April 17.82 ($\Delta = 1.64$ AU, $r = 1.67$ AU, Elong. $= 75°$)
(Coggia)	*Last seen:* 1874 October 19.12 ($\Delta = 1.79$ AU, $r = 1.94$ AU, Elong. $= 83°$)
	Closest to the Earth: 1874 July 23 (0.2905 AU)
1874 III = 1874c	*Calculated path:* CAM (Disc), LYN (Jul. 7), GEM (Jul. 18), CNC (Jul. 21), CMi (Jul. 22), MON (Jul. 24), PUP (Jul. 27), VEL (Aug. 6), CAR (Aug. 12), VOL (Sep. 2), MEN (Oct. 5)

J. E. Coggia (Marseille, France) discovered this comet on 1874 April 17.82 and estimated the position as $\alpha = 6^h 28.1^m$, $\delta = +69° 58'$. He described it as faint with a nucleus and noted a slow southwestward motion. He confirmed this object was a comet when he noticed it had moved on April 18.03.

The comet's motion was quite slow during the initial weeks following discovery, with the closest approaches to Earth and sun not occurring until July. The comet was also well observed during the remainder of April. F. A. T. Winnecke (Strasbourg, France) found the comet rather bright and 3–4' across on the 19th. He also noted a very faint nucleus. On April 20, E. W. L. Tempel (Arcetri, Italy) said the comet was 3' across and only slightly brighter in the center, while L. Schulhof (Vienna, Austria) said the comet exhibited a strong, stellar condensation, which seemed granular with three nuclei sometimes seen. On the 21st, Schulhof said the comet was rather faint because of moonlight and fog, but appeared very diffuse, with a large, round nucleus. G. F. W. Rümker (Hamburg, Germany) said the comet appeared round and compact, with a diameter of about 50". Tempel noted the comet was slightly brighter on the 24th than on the 20th. On April 28, Rümker said, "The comet was very beautiful despite the bright moonlight, it had become considerably brighter and shows a clearly recognizable, 1 minute large, slightly elongated condensation with a star-like nucleus."

Although the comet steadily brightened and developed during May, there were no direct estimates of the comet's total brightness. J. F. J. Schmidt (Athens, Greece) did report that the comet was "very insignificant" in the finder scope on the 3rd, while other astronomers simply noted it was bright in their telescopes during the rest of the month. The only actual brightness estimations were of the stellar nucleus. Schmidt used a 17-cm Reinfelder refractor at a magnification of 50× and gave the nuclear brightness as 10.5 on the 3rd, 10.0 on the 4th, 9.5 on the 6th and 13th, 9.0 on the 14th, and 8.5 on the 18th. Tempel also estimated the brightness of the nucleus. He said it was 9 on the 14th, 9 or 10 on the 16th, and 8 on the 18th. C. F. Pechüle (Hamburg, Germany) said the sharply marked nucleus-like condensation was about 40″ across on the 5th. Schulhof said the nucleus was above and to the left of the coma's center on the 6th. Schmidt provided the only indications of the coma size and these were actually measurements of the preceding half only, which do not represent the actual radius of the coma. He gave this as 1.80′ on the 3rd, 2.07′ on the 4th, 2.73′ on the 6th, 2.81′ on the 13th, 2.90′ on the 14th, and 3.54′ on the 18th. Pechüle described the coma as very delicate on the 5th. Tempel said the coma was smaller and more condensed on the 14th than on previous nights. The first appearance of the tail was reported by Schmidt on May 13, when he gave the length as 0.10°. He followed with measures of 0.08° on the 14th, 0.12° on the 17th, and 0.27° on the 18th. Tempel gave the length as 15′ on the 18th and 18′ on the 24th. He also gave a length of 10′ on the 28th, but noted that a nearby 5th-magnitude star had probably prevented the fainter portion from being seen. Tempel reported the tail was fan-like on the 18th, less fan-like on the 24th, and no longer fan-like on the 29th. Pechüle said the tail extended toward PA 35° on May 16.

The comet was widely observed during June. Schmidt reported the first naked-eye observation on June 1 and estimated the comet's magnitude as 6.5. He continued estimating the naked-eye magnitude throughout the month, giving values of 6.5 on the 2nd, 6.3 on the 3rd, 6.1 on the 4th and 5th, 6.0 on the 6th and 7th, 5.7 on the 8th, 5.3 on the 9th, 5.5 on the 10th, 5.2 on the 11th, 12th, and 13th, 5.1 on the 14th, 5.0 on the 15th, 4.8 on the 16th, 4.6 on the 17th, 4.5 on the 18th, 19th, and 20th, 4.3 on the 21st, 4.2 on the 22nd, 4.0 on the 24th, 27th, and 28th, and 3.7 on the 29th. The only actual estimate of the full coma diameter was 4–5′ by K. C. Bruhns (Leipzig, Germany) on the 11th, but Schmidt provided a fine series of measurements of the preceding half of the coma through most of the month. He gave values of 2.84′ on the 1st, 3.20′ on the 4th, 3.50′ on the 6th, 2.89′ on the 8th, 3.76′ on the 9th, 3.62′ on the 10th, 3.56′ on the 11th, 3.64′ on the 12th, 3.36′ on the 13th, and 3.46′ on the 17th. Schmidt provided a good series of magnitude estimates of what he referred to as the "nucleus" during the month. He was using a refractor at a magnification of 50×, which indicates he was likely seeing the well-defined condensation. Schmidt gave estimates of 8.0 on the 1st, 8.0 on the 4th, 7.5 on the 6th, 7.7 on the 8th and 9th, 8.0 on the 10th and 11th, 7.8 on the 12th, 7.7 on the 13th, 7.5 on the 17th, 7.7 on the 18th, 7.0 on the 20th, 7.2 on the 22nd, 6.8 on the 24th, and 7.0 on the 27th. Tempel made a couple of estimates using his telescope at a magnification of 50×. He simply noted the magnitude as 8 on the 1st and "brighter than 8" on the 4th. Schmidt did provide a few magnitude estimates of the "nucleus" using a higher magnification. He gave it as 10.0 on the 1st and 11th, 9.0 on the 17th, 9.0 on the 18th, 8.0 on the 20th, 8.5 on the 22nd, 8.0 on the 24th, and 9.0 on the 27th. Some astronomers obtained measurements of the diameter of the nucleus during the month.

The diameter was given as 3.4″ by Rümker on the 11th, 1.5″ by Schmidt on the 17th, 7.9″ by Bruhns on the 22nd, about 5″ by W. H. M. Christie (Royal Observatory, Greenwich, England) and 8″ by Bruhns on the 23rd, and 6″ by G. Koch (Leipzig, Germany) on the 30th. Bruhns found the nucleus elliptical on the 24th and said the north–south axis was 6.5″ across, while the west–east axis was 5″ across. Rümker described the nucleus as "somewhat irregular" on the 27th and gave the north–south axis as 3.2″ across. There were some interesting observations of the nuclear region. Tempel noted on the 4th that magnifications of 60× and 300× revealed the condensation was composed of individual pulsating stars. Schulhof said the nucleus was eccentrically situated within the condensation on the 6th. J. Dreyer (Copenhagen, Denmark) saw the comet with the 28-cm refractor on the 20th and said the light in front of the nucleus was like the top of a fountain. The tail developed nicely during June, with Schmidt providing the most extensive series of measurements. Using a refractor, he gave the length as 0.47° on the 1st, 0.70° on the 2nd, 0.85° on the 3rd, 0.80° on the 4th, 0.85° on the 5th, 1.00° on the 6th and 7th, 1.10° on the 8th, 1.20° on the 9th, 1.25° on the 10th, 1.35° on the 11th, 1.25° on the 12th, 1.40° on the 13th, 1.60° on the 14th, 1.75° on the 15th, 2.00° on the 16th, 1.95° on the 17th, 2.20° on the 18th, 2.30° on the 19th, 2.25° on the 20th, 2.20° on the 21st, 2.15° on the 22nd, and 2.50° on the 24th. Other estimates of the tail length came from Tempel, Schulhof, and Bruhns, and very closely reflected those of Schmidt. Descriptions of the tail's appearance came from several people during the month. Tempel said it was narrow, with sharply defined edges on the 1st. Schulhof said the end was slightly curved to the left on the 3rd. Schulhof said the tail was broader on the 5th than on the 3rd. Tempel noted the tail contained large waves of brighter areas, which he said became clearer with his telescope's 12× finder on the 8th. Schulhof said the tail was parabolic in shape on the 16th. C. Abbe (Washington, DC, USA) said the tail's south-following edge was concave. Rümker said the tail was broad in a telescope on the 21st and "displayed a darker region down the center." Abbe said the tail appeared "somewhat smoother and more condensed" on the 23rd and he suspected a second tail "opposite to the main tail." Bruhns said the tail extended toward PA 30° on the 24th. Abbe said no second tail was seen on the 25th, but "the outer boundary of tail [was] very bright producing a resemblance to streamers." He added that the tail was concave on the north-preceding side, which was the brighter side. On June 26, Abbe said the north-preceding side of the tail was straight, while the south-following side was slightly concave.

This comet was most spectacular during the first half of July, with perihelion coming early in the month. The Chinese reported a "broom-star" was visible in the northwest on the 1st. The comet was an easy naked-eye object throughout the period, with Schmidt once again producing the most extensive series of magnitudes. He gave the naked-eye magnitude as 3.3 on the 1st, 3.2 on the 2nd, 3.1 on the 3rd, 3.0 on the 4th and 5th, 2.9 on the 6th, 2.7 on the 7th, 2.5 on the 8th, 2.0 on the 9th, 1.9 on the 10th,

1.6 on the 11th, 1.5 on the 12th, and 1.5 on the 13th. Interestingly, there are indications that Schmidt's magnitude estimates may have actually referred to the comet's strong condensation, as several magnitude estimates of the "nucleus" by J. Holetschek (Vienna, Austria), Bruhns, and Christie were comparable, if not slightly brighter than Schmidt's estimates. There were no naked-eye estimates of the comet's magnitude after the 14th. Despite these apparent magnitude estimates of the "nucleus," Schmidt was also providing his own nuclear magnitude estimates using a telescope at a magnification of 50× and "high power." Although it is doubtful the actual nucleus was being observed, Schmidt's high magnification estimates of the magnitude were 7.5 on the 2nd, 4th, and 6th, 7.0 on the 8th, 7.5 on the 10th, 7.0 on the 12th, 6.5 on the 13th, and 7.0 on the 14th. The diameter of the nucleus was measured by astronomers on several occasions during the month and remained surprisingly constant, with Bruhns, Koch, Christie, and A. Lindstedt (Hamburg, Germany) almost always finding it between 4" and 6". Schmidt's 17-cm Reinfelder refractor revealed a much smaller nucleus, as he gave measurements of 1.49" on the 6th, 1.29" on the 8th, 1.12" on the 12th, and 1.19" on the 15th.

The nucleus and its immediate region presented some interesting characteristics during July. On the 2nd, Bruhns said it was very sharply defined on the side opposite the sun, while a fan was emitted on the sunward side. He added that the fan extended 20–25" toward PA 100°. Bruhns reported the fan was slightly brighter on the 6th and extended 20" toward PA 90°. G. H. With (Hereford, England) observed the comet with a 21.6-cm reflector (125×) on the 8th and said, "The nucleus was small and bright, with a fan-shaped jet of about 130° ± issuing from it." That same night Bruhns said the fan extended 15" toward PA 110°. Schmidt noted an envelope situated 53" from the nucleus on the 6th. Lindstedt noted two envelopes on the 8th, while the nucleus was slightly elliptical. Bruhns said the fan extended 25" toward PA 140° on the 9th, while Lindstedt said it was 32" long. Bruhns added that an envelope was located about 1' from the nucleus. Bruhns again noted the nucleus was sharply defined on the side away from the sun on the 10th. He added that the fan was 20" long. With observed with a 31.1-cm reflector (90×–300×) on the 11th and said, "A singular change had taken place in the shape of the jet proceeding from the nucleus. A spiral tendency appeared to have been developed in it; the fibrous structure was most distinct." On July 12, R. S. Newall (Gateshead, England) said, "The nucleus was very bright, with a disk tolerably well defined. In front of the nucleus was a fan-shaped light, which seemed to arise from the overlapping, or duplication of the two tails." That same night, Bruhns said the fan extended 20" toward PA 120°, while Schmidt saw envelopes 37.2" and 43.7" from the nucleus. On the 13th, Dreyer said the nucleus "was pretty sharply defined and formed the centre of a small sector of light, which faded away gradually." On the same night, Bruhns said the fan was 30" long, while an envelope was 90" from the nucleus. With saw the comet with a 21.6-cm reflector on the 14th and wrote, "The jet had resumed its fan-like shape, but was now enriched with a broad, well-defined border of condensed light. The envelope extended only a very small distance beyond the arc of the fan, which included, as on July 8, about 135°." With continued by adding, "But above the primary envelope, and quite separate from it, were three faintly luminous, flocculent clouds of cometary matter symmetrically arranged and well defined in shape, though fading into blackness at the edges." Schmidt continued providing a precise indication of the coma size during this period, although he was continuing to measure only the preceding radius. He gave values of 3.79' on the 4th, 3.66' on the 6th, 4.01' on the 8th, 4.39' on the 10th, 4.46' on the 12th, and 4.03' on the 13th. Christie provided an interesting description of the coma on the 8th. After observing with a 21.6-cm reflector (125×), he wrote, "The envelope and coma were sharply defined at their outer edges; the central axis of the coma appeared almost black, with a fairly defined boundary." The tail was extensively observed during the first half of July. Although remarkably similar estimates of the length were made by Schmidt, Tempel, Bruhns, and Christie, it was Schmidt who provided the most continuous series. He gave the length as 6.2° on the 1st, 8.5° on the 2nd, 10.0° on the 3rd, 11.0° on the 4th, 13.0° on the 5th, 14.0° on the 6th, 15.1° on the 7th, 17.0° on the 8th, 18.5° on the 9th, 20.5° on the 10th, 22.0° on the 11th, 25.0° on the 12th, 29.0° on the 13th, 36.0° on the 14th, 41.5° on the 15th, and 47.2° on the 16th. Schulhof said the tail was "perfectly parabolic" in a telescope on the 3rd. To the naked eye, observers generally found the tail to be perfectly straight and narrow, with Abbe, Christie, and Tempel independently noting the width of the tail at its end was usually between 1° and 2°. Abbe noted the borders of the tail were very straight, but ill defined on the 3rd, while Schulhof noted a "sharp, dark strip" running down the center of the tail. Abbe noted the tail had a rosy tinge not far from the nucleus on the 6th. Lindstedt said the preceding part of the tail was less bright than the following part on the 9th. Bruhns found the tail was plainly darker in the middle than toward the edges on the 10th. Dreyer observed with the 28-cm refractor on the 13th and wrote, "Two luminous tails streamed out from the foremost part of the coma, turning back left and right. They were decidedly brighter than the region round the nucleus. A faint parabolic arc of light was seen in front of these tails." On the 14th, Newell noted, "The two streams of the tail have separated and become shorter, leaving a wider dark space between, while from the two corners of the fan proceed two anthenæ, which appear to be projections of the inner sides of the tails, and preceding these is still a luminous cloud." Abbe noted the tail was slightly concave on the south-following side on the 15th. Dreyer said the tails seemed to stream directly from the nucleus on the 16th.

The comet was moving rapidly southward during the second half of July, as it passed closest to Earth. In addition, its low altitude and involvement in twilight made it difficult to observe anything but the tail. Bruhns reported the nucleus was no longer visible on the 17th. He and Schmidt estimated tail lengths of 36° and 54.0°, respectively. Abbe gave the tail length as 30° on the 18th, but he noted that a friend observing with him followed the tail for about 60°. Schmidt then reported a tail length of 55.9°. On the 19th, the tail was estimated as 45° long by Abbe and 48° long by Bruhns. The comet passed only 2.2° from the sun on the 20th. Bruhns then indicated a tail length of over 70°, while Schmidt measured it as 63.3°. The tail length was given as 60° by Abbe and 65.8° by Schmidt on the 21st. Schmidt measured the length as 64.6° on the 22nd. On July 22.07, E. L. Trouvelot (Medford, Massachusetts, USA) noticed a "bright ray of light darting from the northwest horizon, which he at first took for an auroral streamer." For more than an hour "the tail kindled and extinguished more than a hundred times, being sometimes so bright that in spite of the moon (then in her first quarter) it could be distinguished in all its contours to its very extremity, and at other times so faint that not a trace could be seen." Although Abbe said the tail was no longer visible on the 23rd, Schmidt became the final Northern Hemisphere astronomer to see the comet when he saw a trace of the tail near ν and ξ Ursae Majoris.

The first documented observation in the Southern Hemisphere was by Andrew A. Anderson, who was traveling by ox-waggon in South Africa a short distance from Barkly. He first saw the comet on July 27.18, when it was in the morning sky at an altitude of 15°. He logged observations until August 8, and said the comet appeared very bright. When moonlight became a factor, the tail length was decreased. R. L. J. Ellery (Melbourne, Australia) first saw the comet on July 27.87, although he reported it had been seen on the 25th at some locations. Ellery noted, "It has been very bright, and the nucleus very stellar, so that full illumination of the wires of a fine filar micrometer could be borne throughout the observations sent."

John Tebbutt (Windor, New South Wales, Australia) first saw the comet on August 1, and noted it was still a conspicuous naked-eye object. On the same date, E. J. Stone (Cape of Good Hope, South Africa) said the comet "appeared of about the brightness of a third magnitude star." By August 8, Ellery noted the comet "had much diminished in brightness." Virtually not a day went by that the comet was not observed by at least one observatory. The observatories providing positions included those at Windsor, Melbourne, and Cordoba (Argentina). Likewise, the comet continued to be observed by the same observatories throughout September.

Tebbutt made his final observations on October 2, 6, and 7. He described the comet as "excessively faint and much diffused" during that period. The comet was last observed by Ellery on October 6. J. M. Thome (Argentine National Observatory, Cordoba, Argentina) last detected the comet on October 19.12, when he gave the position as $\alpha = 6^h 39.0^m$, $\delta = -77° 43'$.

Numerous spectroscopic observations were obtained. Observations by G. A. P. Rayet during the second half of May revealed a continuous spectrum. Although a continuous spectrum was also detected by N. de Konkoly (O'Gyalla Observatory, Hungary) during June and July, and by Christie during July, A. Secchi (Rome, Italy) and H. L. d'Arrest (Copenhagen, Denmark) began reporting the three cometary bands around the middle of June. H. C. Vogel (Berlin, Germany) observed the comet on several occasions during the period of May 6 to June 22. The brightest cometary band was seen on every occasion, while all three bands were seen during the period of June 4–15. W. Huggins (London, England) saw the three cometary bands on July 7 and 8.

The polarization of the comet was examined by several astronomers as July progressed. Christie first noted polarization on the 4th and noted the tail and coma were "partially polarised in a plane through the axis of the tail," although he added, "the polarisation of the tail appeared to be slightly greater than that of the coma." A. C. Ranyard (England) first noted polarization on the 7th and said it became more and more conspicuous in the tail through the 15th. W. Zenker (Charlottenburg, Germany) said polarization in the coma became more noticeable beginning on July 14.

The first parabolic orbit was calculated by J. Holetschek, who took positions spanning the period of April 17–22. He determined a perihelion date of 1874 June 16.18. The comet was still beyond the orbit of Mars and was moving slowly. Subsequently, the perihelion date shifted greatly once a longer observational arc became available, beginning with W. E. Plummer, who took positions from the period of April 17 to May 9 and determined the perihelion date of July 6.88. Additional orbits by A. Svedstrup, W. Fabritius, and J. R. Hind established the perihelion date as July 9.4.

By the end of July, astronomers noted, the comet's path could no longer be represented by a parabolic orbit. L. Schulhof took positions from April 17 to July 13 and determined an elliptical orbit with a perihelion date of July 9.36 and a period of about 12 175 years. M. Geelmuyden later determined the period as about 10 445 years. The orbit was revisited a few years later by J. Seyboth (1879) and J. von Hepperger (1882). Seyboth gave a period of about 5711 years, while Hepperger gave it as about 13 708 years. Hepperger's orbit is given below.

T	ω	Ω (2000.0)	i	q	e
1874 Jul. 9.3583 (UT)	152.3803	120.4948	66.3438	0.675782	0.998820

ABSOLUTE MAGNITUDE: $H_{10} = 5.7$ (V1964)

FULL MOON: April 1, May 1, May 31, Jun. 29, Jul. 29, Aug. 27, Sep. 25, Oct. 25

SOURCES: J. E. Coggia, AN, **83** (1874 Apr. 24), p. 263; J. E. Coggia and W. E. Plummer, MNRAS, **34** (1874 May), p. 362; F. A. T. Winnecke, J. Holetschek, F. Tietjen, C. F. Pechüle, and C. F. W. Peters, AN, **83** (1874 May 6), pp. 295–6, 299; A. Svedstrup, AN, **84** (1874 Jun. 10), p. 45; J. R. Hind, Nature, **10** (1874 Jun. 25), p. 149; H. L. d'Arrest, AN, **84** (1874 Jul. 7), pp. 137–40; J. R. Hind, BIOP (1874 Jul. 22); L. Schulhof, W. Zenker, and N. de Konkoly, AN, **84** (1874 Aug. 4), pp. 169, 173; W. Fabritius, AN, **84** (1874 Aug. 28), p. 209; M. Geelmuyden and C. Fearnley, AN, **84** (1874 Sep. 28), pp. 261–4; A. C. Ranyard, E. J. Stone, and W. H. M. Christie, MNRAS, **34** (1874, Supplementary Notice), pp. 489–92; J. Ellery, J. Tebbutt, and A. A. Anderson, MNRAS, **35** (1874

Nov.), pp. 57–61; C. Abbe, *AN*, **84** (1874 Nov. 15), pp. 353–66; J. M. Wilson and G. M. Seabroke, *MNRAS*, **35** (1874 Dec.), pp. 83–4; H. C. Vogel and J. Holetschek, *AN*, **85** (1874 Dec. 22), pp. 17–34, 37; J. Tebbutt, *MNRAS*, **35** (1875 Jan.), pp. 110–12; W. Huggins, *PRSL*, **23** (1875 Jan. 7), pp. 154–9; J. Kowalczyk, *AN*, **85** (1875 Jan. 16), p. 61; K. C. Bruhns and G. Koch, *AN*, **85** (1875 Feb. 8), pp. 89, 101–4; J. Tebbutt, *MNRAS*, **35** (1875 Mar.), p. 313; E. W. L. Tempel, *AN*, **85** (1875 Mar. 8), pp. 177–90; L. Schulhof, *AN*, **85** (1875 Apr. 29), pp. 287, 291; G. F. W. Rümker, C. F. Pechüle, A. Lindstedt, and J. Tebbutt, *AN*, **86** (1875 Jul. 26), pp. 95–100, 119–22; J. Tebbutt, *MNRAS*, **35** (1875, Supplementary Notice), pp. 406–7; J. F. J. Schmidt, *AN*, **87** (1875 Dec. 3), pp. 33–48; G. H. With and R. S. Newall, *MNRAS*, **36** (1876 Mar.), pp. 277–9; A. C. Ranyard, *MNRAS*, **36** (1876 Mar.), pp. 279–80; J. Dreyer, *MNRAS*, **36** (1876 May), pp. 339–40; F. A. Bredikhin, *AN*, **88** (1876 Aug. 15), pp. 219–22; E. L. Trouvelot, *The Observatory*, **2** (1878 Aug.), p. 126; J. M. Thome, *AN*, **94** (1879 Feb. 25), pp. 177–86; J. Seyboth, *AN*, **95** (1879 Jun. 7), p. 79; J. von Hepperger and G. A. P. Rayet, *SAWW*, **86** Abt. II (1882), pp. 451–510; V1964, p. 60; HA1970, p. 87; SC2000.0, pp. 69, 112, 216, 282, 496.

C/1874 O1 *Discovered:* 1874 July 26.07 ($\Delta = 0.60$ AU, $r = 1.13$ AU, Elong. $= 84°$)
(Borrelly) *Last seen:* 1874 October 21.10 ($\Delta = 0.69$ AU, $r = 1.35$ AU, Elong. $= 105°$)
Closest to the Earth: 1874 July 2 (0.5312 AU), 1874 October 30 (0.6775 AU)
1874 V = 1874d *Calculated path:* DRA (Disc), UMi (Aug. 4), DRA (Aug. 23), UMa (Sep. 9), CAM (Oct. 7), LYN (Oct. 14)

A. L. N. Borrelly (Marseille, France) discovered this comet on 1874 July 26.07 at an estimated position of $\alpha = 15^h 52.3^m$, $\delta = +59° 32'$. He described it as moderately bright. E. J. M. Stephan (Marseille, France) confirmed the discovery on July 26.93. He described the comet as 3–5′ across, with an eccentrically-placed trace of condensation.

This comet experienced two moderately close approaches to Earth during its apparition. The first occurred over three weeks before the discovery, while the second occurred after the final observation. What this meant was that the comet was moving away from Earth and approaching the sun through the remainder of July and most of August. F. A. T. Winnecke (Strasbourg, France) described the comet as 5′ across with a central condensation on July 30. K. C. Bruhns (Leipzig, Germany) said the comet was moderately bright, very diffuse, and over 2′ across on August 2. L. Schulhof (Vienna, Austria) said the comet was bright and large on the 3rd, with an irregular border and a granular-looking condensation. On August 4, A. Lindstedt (Hamburg, Germany) said the comet was very diffuse, without a recognizable condensation. G. Koch (Leipzig, Germany) said the comet was 2′ across on the 9th. On August 10, Schulhof said the comet was noticeably fainter and smaller than when last seen, except for the bright nucleus. There were indications that several fainter nuclei were occasionally flashing from within the main nucleus. Bruhns said a magnification of 192× revealed the coma was 1.5′ across on the 16th, with a condensation eccentrically situated north of center. The coma also had a granular appearance. On August 30,

Bruhns noted that, despite rather bright moonlight, the comet was still about 1.5′ across and centrally condensed. The comet attained its most northerly declination of +74° on August 31.

As September began, the comet was moving away from the sun, but had begun to approach the Earth again. On September 2, Schulhof said the comet was brighter than when previously seen. On the 3rd, he said the comet was circular, with a sharp, distinct nucleus, while on the 7th, he reported the comet had increased in size and brightness since the 3rd. Koch found the comet faint and diffuse on the 18th. For the 21st, J. Holetschek (Vienna, Austria) wrote, "The comet appeared as an expanded, roundish nebulosity, without a clearly perceptible nucleus." Holetschek said the comet was seen with difficulty on the 23rd. On September 26 and 29, Lindstedt noted the comet was extraordinarily faint in moonlight and "unfavorable air."

Despite still approaching Earth, the comet's increasing distance from the sun was taking a toll on its brightness as October progressed. Holetschek was unable to find it on the 1st, despite a very clear sky. Schulhof reported it was large, diffuse, and very faint in the refractor on the 9th. Another observation by Schulhof on the 12th, in foggy conditions, revealed the comet at the limit of visibility. Lindstedt was barely able to see it on the 13th, while Schulhof reported it was extraordinarily faint on the 19th. Schulhof summarized the comet's appearance by noting that it "showed clear fluctuations in its brightness."

The comet was last detected on October 21.10, when Lindstedt observed with a refractor and gave the position as $\alpha = 6^h 58.5^m$, $\delta = +56° 08′$. He said the comet was barely visible.

The first parabolic orbit was calculated by J. Holetschek using positions from July 27, 31, and August 3. He determined the perihelion date as 1874 August 27.18. About a month later, J. R. Hind gave the perihelion date as August 27.34.

The first elliptical orbit was calculated by A. Grützmacher using positions from July 28, September 1, and 16. He determined the perihelion date as August 27.35 and the period as about 45 638 years. Definitive orbits were calculated by L. Gruber and I. Kurländer (1877) and G. Gruss (1878), which gave a perihelion date of August 27.34. Gruber and Kurländer determined the period as about 19 849 years, while Gruss gave it as about 24 368 years. Gruss' orbit is given below.

T	ω	Ω (2000.0)	i	q	e
1874 Aug. 27.3436 (UT)	92.6123	253.2797	41.8266	0.982649	0.998831

ABSOLUTE MAGNITUDE: $H_{10} = 8.2–10.5$ (V1964)

FULL MOON: Jun. 29, Jul. 29, Aug. 27, Sep. 25, Oct. 25

SOURCES: A. L. N. Borrelly, *CR*, **79** (1874 Jul. 27), p. 218; A. L. N. Borrelly, *AN*, **84** (1874 Aug. 4), p. 173; F. A. T. Winnecke, K. C. Bruhns, J. Holetschek, and A. L. N. Borrelly, *AN*, **84** (1874 Aug. 11), p. 191; A. L. N. Borrelly, E. J. M. Stephan, and

A. Gromadzki, *AN*, **84** (1874 Aug. 28), pp. 215–18; J. R. Hind, *BIOP* (1874 Sep. 17); A. Grützmacher, *AN*, **84** (1874 Oct. 20), p. 323; A. Grützmacher, *AN*, **84** (1874 Nov. 9), p. 345; J. Holetschek, *AN*, **85** (1874 Dec. 22), p. 37; K. C. Bruhns and G. Koch, *AN*, **85** (1875 Feb. 8), pp. 89–92, 104; L. Schulhof, *AN*, **85** (1875 Apr. 29), pp. 289–92; A. Lindstedt, *AN*, **86** (1875 Jul. 26), pp. 101–4; K. C. Bruhns and G. Koch, *AN*, **86** (1875 Aug. 29), p. 165; L. Gruber and I. Kurländer, *AN*, **91** (1877 Nov. 10), pp. 77–80; G. Gruss, *SAWW*, **78** Abt. II (1878), p. 276; V1964, p. 60.

C/1874 Q1	*Discovered:* 1874 August 19.99 ($\Delta = 1.53$ AU, $r = 1.74$ AU, Elong. $= 84°$)
(Coggia)	*Last seen:* 1874 November 15.09 ($\Delta = 1.37$ AU, $r = 2.27$ AU, Elong. $= 147°$)
	Closest to the Earth: 1874 October 19 (1.2931 AU)
1874 IV = 1874e	*Calculated path:* TAU (Disc), ORI (Sep. 23), ERI (Nov. 8)

J. E. Coggia (Marseille, France) discovered this comet on 1874 August 19.99 at a position of $\alpha = 3^h 57.7^m$, $\delta = +27° 06'$. He described the comet as faint, round, and without a condensation.

The comet was about a month past perihelion when discovered and although it would make its closest approach to Earth shortly after mid-October, it remained a faint object and physical descriptions were not plentiful. L. Schulhof (Vienna, Austria) described the comet as faint on August 21 and 22. He added that the coma was 2–3′ across, while the nucleus was about magnitude 12. In a paper published in the *Astronomische Nachrichten* in 1875, Schulhof added that there was a central condensation on the 21st and the nucleus was eccentrically situated to the right of center on the 22nd.

Few observations were made during September and October. A. Lindstedt (Hamburg, Germany) said the comet was very faint in the 26-cm refractor on September 4, although he noticed the sky was unfavorable. The comet was described as very faint by Schulhof on the 15th and G. Koch (Leipzig, Germany) on the 17th. Lindstedt said the comet was seen with difficulty in the refractor on the 20th. On October 9, Lindstedt said the comet was easy to see in the 26-cm refractor. Schulhof said the comet was very faint with a nucleus of magnitude 12 on the 10th. No observations were apparently made during the period of October 23–30.

The comet was last detected on November 15.09, when J. J. Plummer (Orwell Park Observatory, England) gave the position as $\alpha = 4^h 49.9^m$, $\delta = -6° 10'$ using a 25-cm refractor.

Plummer observed the comet with a 25-cm reflector from August 24 until November 15. He wrote, "The Comet though faint was easily seen throughout the whole period of the observations, except when atmospheric circumstances were unfavorable, and during the first month possessed a sensible condensation of light towards the centre. Subsequently it became a hazy patch of light, but was observed without much difficulty until the middle of November when the return of moonlight interrupted the observations. It was not looked for in December as its diminished brilliancy and southerly declination rendered the successful observation of it highly improbable."

The first parabolic orbit was calculated by Schulhof using positions from August 20, 21, and 22. The result was a perihelion date of 1874 July 5.63. Later orbits by Schulhof, J. R. Hind, and J. Holetschek gave perihelion dates as July 17 and 18.

The first elliptical orbit was calculated by Holetschek (1881). He took positions from August 22, September 18, October 10, and November 10, and determined a perihelion date of July 18.18 and a period of about 299 years. Holetschek (1882) revised his computations using positions spanning the entire period of visibility. He applied perturbations from Venus to Saturn, and gave the perihelion date as July 18.20 and the period as about 306 years. This orbit is given below.

T	ω	Ω (2000.0)	i	q	e
1874 Jul. 18.1995 (UT)	149.5838	217.6272	34.1268	1.687978	0.962831

ABSOLUTE MAGNITUDE: $H_{10} = 6.2$ (V1964)

FULL MOON: Jul. 29, Aug. 27, Sep. 25, Oct. 25, Nov. 23

SOURCES: G. Koch, J. E. Coggia, and E. J. M. Stephan, *AN*, **84** (1874 Aug. 28), pp. 221–4; J. R. Hind, *BIOP* (1874 Sep. 17); L. Schulhof, F. A. T. Winnecke, E. W. L. Tempel, W. Klinkerfues, and J. Holetschek, *AN*, **84** (1874 Sep. 28), pp. 261, 269; K. C. Bruhns and G. Koch, *AN*, **85** (1875 Feb. 8), pp. 91, 104; J. J. Plummer and L. Schulhof, *AN*, **85** (1875 Apr. 29), pp. 277–82, 289–92; A. Lindstedt, *AN*, **86** (1875 Jul. 26), p. 103; K. C. Bruhns and G. Koch, *AN*, **86** (1875 Aug. 29), p. 165; J. Holetschek, *AN*, **100** (1881 Jul. 16), p. 109; J. Holetschek, *SAWW*, **86** Abt. II (1882), pp. 1098–1124; V1964, p. 60.

C/1874 X1
(Borrelly)

1874 VI = 1874f

Discovered: 1874 December 7.15 ($\Delta = 1.29$ AU, $r = 1.17$ AU, Elong. $= 60°$)

Last seen: 1875 January 7.78 ($\Delta = 1.29$ AU, $r = 1.69$ AU, Elong. $= 94°$)

Closest to the Earth: 1874 September 28 (1.1052 AU), 1874 December 24 (1.2555 AU)

Calculated path: CrB (Disc), HER (Dec. 11), DRA (Dec. 22)

A. L. N. Borrelly (Marseille, France) discovered this comet on 1874 December 7.15 at a position of $\alpha = 15^h 59.8^m$, $\delta = +36° 07'$. He described it as rather bright, with a diameter of 3', and moving toward the north-northeast. E. J. M. Stephan (Marseille, France) confirmed motion when he saw the comet on December 7.76.

The comet was discovered about a month and a half after it had passed perihelion and over two months after its closest approach to Earth. Interestingly, the comet was heading for a second, lesser, approach to Earth when found, but this hardly slowed the comet's fading. G. F. W. Rümker and A. Lindstedt (Hamburg, Germany) said the comet appeared faint in the 26-cm refractor on December 10, at first because of its low altitude and then because of unfavorable skies. F. A. T. Winnecke (Strasbourg, France) said the comet was 4′ across on the 11th, without a distinct central condensation. On December 18, L. Schulhof (Vienna, Austria) said the condensation was eccentrically situated to the right and below the center of the coma. K. C. Bruhns (Leipzig, Germany) saw the comet on 1875 January 4 and described it as very faint and diffuse, with a diameter of about 2′.

The comet was last detected on 1875 January 7.78, when Bruhns gave the position as $\alpha = 16^h 51.7^m$, $\delta = +68° 47′$. He described the comet as very diffuse. Schulhof said he looked for the comet on January 18, but failed to find it.

The first parabolic orbits were independently calculated by G. Koch and L. J. Gruey, using positions spanning the period of December 7–10. Koch determined a perihelion date of 1874 November 5.02, while Gruey determined it as October 20.48. Revisions by J. Holetschek (1875, 1879) established the perihelion date as October 19.44. The 1879 orbit is given below.

T	ω	Ω (2000.0)	i	q	e
1874 Oct. 19.4429 (UT)	16.2682	283.7162	99.2205	0.508226	1.0

ABSOLUTE MAGNITUDE: $H_{10} = 7.6$ (V1964)

FULL MOON: Nov. 23, Dec. 23, Jan. 21

SOURCES: E. J. M. Stephan, A. L. N. Borrelly, and F. A. T. Winnecke, AN, **85** (1874 Dec. 22), pp. 43–6; J. Holetschek, K. C. Bruhns, and G. Koch, AN, **85** (1875 Jan. 16), pp. 53–6, 61; E. J. M. Stephan, A. L. N. Borrelly, and L. J. Gruey, CR, **80** (1875 Feb. 1), pp. 313–14; L. Schulhof, AN, **85** (1875 Apr. 29), pp. 289–92; G. F. W. Rümker and A. Lindstedt, AN, **86** (1875 Jul. 26), p. 105; K. C. Bruhns, AN, **86** (1875 Aug. 29), pp. 165, 169; J. Holetschek, AN, **94** (1879 Feb. 25), pp. 185–90; V1964, p. 61.

7P/Pons–Winnecke

1875 I = 1875b

Recovered: 1875 February 2.22 ($\Delta = 1.39$ AU, $r = 1.03$ AU, Elong. $= 48°$)
Last seen: 1875 February 17.43 ($\Delta = 1.36$ AU, $r = 0.91$ AU, Elong. $= 42°$)
Closest to the Earth: 1875 February 15 (1.3569 AU)
Calculated path: SER (Rec), SGR (Feb. 6)

T. R. von Oppölzer (1874) took the orbit calculated for this comet's 1869 apparition and applied perturbations by Jupiter and Saturn. The result was a prediction that the comet would next pass perihelion on 1875 March 12.68. A. L. N. Borrelly (Marseille, France) recovered this comet on 1875 February 2.22, at a position of $\alpha = 17^h 42.8^m$, $\delta = -15° 29′$. He described it as faint, rather extended, and diffuse. The position indicated Oppölzer's prediction was less than two hours late.

E. W. L. Tempel (Arcetri, Italy) was using the telescope of Amici on February 9. He found this comet and described it as "surprisingly large and rather bright, with several pulsating nuclei." G. V. Schiaparelli (Brera

Observatory, Milan, Italy) found the comet on the 14th and 15th with a 20-cm telescope, but noted it was difficult to see because of haze near the horizon.

The final observations of this comet were made by Schiaparelli on February 16.20 and by Willson (Harvard College Observatory, Cambridge, Massachusetts, USA) on February 16.42 and February 17.43. Willson gave the position as $\alpha = 19^h\ 09.4^m$, $\delta = -16°\ 30'$ on the last date. Schiaparelli found the comet easier to see in the 20-cm telescope than on the 14th and 15th, because it was better condensed toward the center. Willson said further observations were impossible.

Later calculations using multiple apparitions and planetary perturbations were published by T. R. von Oppölzer (1880), E. F. von Haerdtl (1888, 1889), and B. G. Marsden (1970, 1972). These revealed a perihelion date of March 12.60 and a period of 5.73 years. Oppölzer (1880) noted that the positions of 1858, 1869, and 1875 indicated a slight secular acceleration was present. This was later confirmed by Marsden (1970, 1972), when he determined nongravitational terms of $A_1 = +0.27$ and $A_2 = -0.0072$. This orbit is given below.

T	ω	Ω (2000.0)	i	q	e
1875 Mar. 12.5979 (TT)	165.2122	113.1617	11.2763	0.829006	0.740992

ABSOLUTE MAGNITUDE: $H_{10} = 7.6$ (V1964)

FULL MOON: Jan. 21, Feb.20

SOURCES: T. R. von Oppölzer, *AN*, **84** (1874 Nov. 23), pp. 373–8; A. L. N. Borrelly, *CR*, **80** (1875 Feb. 1), p. 315; A. L. N. Borrelly and T. R. von Oppölzer, *AN*, **85** (1875 Feb. 20), p. 159; G. V. Schiaparelli and E. W. L. Tempel, *AN*, **85** (1875 Mar. 13), pp. 201–4; T. R. von Oppölzer, *AN*, **97** (1880 May 27), pp. 149–54; T. R. von Oppölzer, *AN*, **97** (1880 Jul. 30), pp. 337–42; E. F. von Haerdtl, *AN*, **120** (1888 Dec. 28), pp. 257–72; E. F. von Haerdtl and Willson, *DAWW*, **55** (1889), pp. 247–50; E. F. von Haerdtl, *DAWW*, **56** Abt. II (1889), pp. 151–85; V1964, p. 61; B. G. Marsden, *AJ*, **75** (1970 Feb.), p. 80; B. G. Marsden, *QJRAS*, **13** (1972), pp. 428–9; B. G. Marsden, *CCO*, 12th ed. (1997), pp. 48–9.

2P/Encke

1875 II = 1875a

Recovered: 1875 January 26.98 ($\Delta = 1.98$ AU, $r = 1.51$ AU, Elong. $= 47°$)

Last seen: 1875 May 17.84 ($\Delta = 0.57$ AU, $r = 0.85$ AU, Elong. $= 57°$)

Closest to the Earth: 1875 May 4 (0.5496 AU)

Calculated path: PSC (Rec), ARI (Mar. 26), CET (Apr. 18), PSC (Apr. 24), CET (Apr. 25)

The prediction for this apparition came from F. E. von Asten (1875). Although published in 1875, it was completed in October of 1874 and was distributed to several observatories. Asten took positions from the 1865, 1868, and 1871 apparitions, applied perturbations by Venus to Saturn, and predicted the comet would next arrive at perihelion on 1875 April 13.57. The comet was recovered by E. S. Holden and H. P. Tuttle (US Naval Observatory, Washington, DC, USA), while using the 66-cm refractor on 1875 January 26.98. They described it as extremely faint and barely detectable. It was noted that the position was only slightly different from the prediction by Asten. An independent recovery was made by E. J. M. Stephan (Marseille, France) on January 27.79. He said the observations were extremely difficult with the comet appearing as a small milky spot, with an undefined boundary and no condensation.

The comet approached the sun and Earth during the next couple of months. On February 26, J. Palisa (Pula, Yugoslavia) saw a nebulosity, but, because of the lack of reliable sources, was uncertain whether it was this comet. He described it as faint, but well seen, with a possible nucleus. The position was almost exactly where 2P/Encke should have been. K. C. Bruhns (Leipzig, Germany) observed the comet at a magnification of $144\times$ on March 2 and said it appeared extremely faint, although he attributed this to the zodiacal light. On the 3rd, Bruhns noted the comet was "considerably brighter than yesterday, diameter 2', nucleus only faintly visible." Bruhns again found the centrally condensed coma 2' across on the 14th. N. de Konkoly (O'Gyalla Observatory, Hungary) found the comet very bright on March 22 and noted it had a noncentral nucleus of magnitude 12. Konkoly saw a tail on the 26th that was as long as the coma was wide. F. A. Bredikhin (Moscow, Russia) said the comet remained faint through March 24, while his observations on March 27 and 30 revealed the comet was near magnitude 7, with traces of a tail. C. F. W. Peters (Kiel, Germany) described the comet as quite bright despite an altitude of $5°$. Bredikhin last detected the comet on April 10.

Following perihelion the comet appeared to astronomers in the Southern Hemisphere. J. Tebbutt (Windsor, New South Wales, Australia) wrote a letter to the *Astronomische Nachrichten* dated 1875 May 13. Published in the July 26 issue, it said, "A comet, which I supposed to be Encke's, was discovered by me" in the morning sky on May 7.80. He described it as "extremely faint and diffused" in the 11-cm refractor, "with scarcely any perceptible condensation of light." He reobserved it on May 8, 9, 10, 11, and 14, with the 14th being his final observation. Tebbutt added that no ephemeris for comet Encke had been received and that the "discovery" had been telegraphed to Sydney and Melbourne. E. J. White (Melbourne, Australia) had made several attempts to observe the comet, but had continually been foiled by bad weather. On May 17.81 he found the comet in the morning sky with an 11-cm refractor and described it as "an excessively faint nebulous object, without nucleus, compact shape, but slight elongated." He followed the comet for over half an hour and last detected it when it "was of the last degree of faintness... owing to a slight haze, and the presence of twilight" on May 17.84. Moonlight interfered thereafter and no further observations were made.

Konkoly examined the comet's spectrum on March 22 and noted the three cometary bands at $5610\,\text{Å}$, $5160\,\text{Å}$, $4751\,\text{Å}$. The continuous spectrum was not detected.

Refinements of this comet's orbit using positions from multiple apparitions, as well as planetary perturbations, have been published by numerous astronomers. Of most significance were the studies of Asten (1877), J. O. Backlund (1884, 1886), and B. G. Marsden and Z. Sekanina (1974). These have typically revealed a perihelion date of April 13.49 and a period of 3.29 years. Marsden and Sekanina determined the nongravitational terms as $A_1 = +0.07$ and $A_2 = -0.02657$. The orbit of Marsden and Sekanina is given below.

T	ω	Ω (2000.0)	i	q	e
1875 Apr. 13.4901 (TT)	183.6486	336.3831	13.1372	0.332964	0.849373

ABSOLUTE MAGNITUDE: $H_{10} = 9.9$ (V1964)

FULL MOON: Jan. 21, Feb.20, Mar. 22, Apr. 20, May 20

SOURCES: F. E. von Asten, *BASP*, **20** (1875), pp. 340–66; E. J. M. Stephan, *CR*, **80** (1875 Feb. 1), pp. 314–15; E. S. Holden and H. P. Tuttle, *AN*, **85** (1875 Feb. 20), p. 157; J. Palisa and C. F. W. Peters, *AN*, **85** (1875 Mar. 13), p. 207; N. de Konkoly, *AN*, **85** (1875 Apr. 29), pp. 317–20; C. F. W. Peters, *AN*, **85** (1875 May 5), p. 331; F. E. von Asten and F. A. Bredikhin, *AN*, **85** (1875 May 25), pp. 337–54, 365–8; F. A. Bredikhin, *AN*, **85** (1875 May 31), p. 383; F. A. Bredikhin, *AN*, **86** (1875 Jun. 10), pp. 9–12; J. Tebbutt, *AN*, **86** (1875 Jul. 26), p. 122; K. C. Bruhns and G. Koch, *AN*, **86** (1875 Aug. 29), pp. 167–70; E. J. White, *AN*, **86** (1875 Aug. 30), p. 191; J. Tebbutt, *AN*, **86** (1875 Sep. 10), p. 223; J. Tebbutt, *MNRAS*, **35** (1875, Supplementary Notice), p. 406; F. A. Bredikhin, *MNRAS*, **36** (1876 Feb.), pp. 191–2; F. E. von Asten, *BASP*, **22** (1877), pp. 568–9; J. O. Backlund, *MASP*, **32** No. 3, 7th Series (1884), p. 36; J. O. Backlund, *AN*, **106** (1883 Sep. 14), pp. 289–94; J. O. Backlund, *MASP*, **34** No. 8, 7th Series (1886), p. 38; F. A. Bredikhin, *Annales de l'Observatoire de Moscou*, **2**, Series 2 (1889), pp. 27–8; V1964, p. 61; B. G. Marsden and Z. Sekanina, *AJ*, **79** (1974 Mar.), pp. 415–16.

C/1877 C1 *Discovered:* 1877 February 9.14 ($\Delta = 0.45$ AU, $r = 0.90$ AU, Elong. $= 65°$)
(Borrelly) *Last seen:* 1877 April 3.92 ($\Delta = 1.70$ AU, $r = 1.54$ AU, Elong. $= 63°$)
Closest to the Earth: 1877 February 18 (0.2789 AU)
1877 I = 1877a *Calculated path:* OPH (Disc), HER (Feb. 13), DRA (Feb. 18), CEP (Feb. 22), CAS (Feb. 24), CAM (Mar. 1), PER (Mar. 14), AUR (Apr. 3)

The discovery of this comet broke a long period of cometless skies. Not only had there been no comet discoveries or even recoveries during 1876, but no comet had been under observation for nearly 21 months.

A. L. N. Borrelly (Marseille, France) discovered this comet on 1877 February 9.14, at a position of $\alpha = 17^h\ 13.3^m$, $\delta = -1°\ 37'$. It was described as bright, round, and 3.5' across, with a well-defined nucleus. He had confirmed a rapid motion by February 9.19. C. F. Pechüle (Copenhagen,

Denmark) independently discovered the comet on February 10.20 and described it as bright, round, with a central condensation. He confirmed it on February 11.15 and did not receive notice of Borrelly's discovery until the 15th.

The comet had passed perihelion three weeks before its discovery, but it was heading toward a close approach to Earth. J. F. J. Schmidt (Athens, Greece) said the comet was easily seen with the naked eye during his observations of February 11–19, although he noted it "was not as bright as a star of magnitude 6." Paul Henry (Paris, France) saw the comet with the naked eye on the 16th and 17th. He said it was like a faint nebulosity, only slightly brighter than M13. J. Palisa (Pula, Yugoslavia) said the comet was very well seen with the naked eye at magnitude 5 or 6 on the 19th. The coma was generally described as circular, although E. S. Holden (US Naval Observatory, Washington, DC, USA) noted a hint of a longer axis oriented along the north–south line on the 10th. As a result of the close approach to Earth, estimates of the coma diameter varied greatly during the month because of the wide range of telescopes used. Schmidt obtained the largest estimates while using the telescope's 8× finder. He said the coma was about 20′ across on the 11th, 15.5′ on the 16th, and 19.5′ on the 23rd. Schmidt said the comet never became larger than the "Andromeda Nebula." Additional estimates were 3–4′ by H. Geelmuyden (Christiania, now Oslo, Norway) on the 13th, 6′ by K. C. Bruhns and B. Peter (Leipzig, Germany), as well as F. A. T. Winnnecke (Strasbourg, France) on the 16th, 9′ by G. L. Tupman (Royal Observatory, Greenwich, England) on the 16th, 10′ by Henry on the 16th and 17th, and 8′ or 9′ by Tupman on the 20th. The comet displayed some condensation throughout the month, with the strength depending on the type of instrument used to observe it. Small telescopes, including finders, revealed a strong condensation, while larger telescopes revealed very little condensation. The same was true for the "nucleus", with only the smaller telescopes revealing it as stellar. At no time during the month was a tail reported. There were a few particularly interesting observations during February. Peter said the comet appeared less condensed, but larger on the 17th than on the 16th. A. Secchi (Rome, Italy) wrote for the 17th, "The nucleus of the comet seemed to be composed of a number of small patches of light arranged in a curve, and surrounded by a vast excentric indeterminable nebulosity, greatly diffused towards the south." In moonlight on the 28th, Bruhns described the comet as a faint, diffuse spot. The comet attained its most northerly declination of +78° on February 23.

The comet was moving away from both the sun and Earth as March began. The only real estimate of its total brightness came from Peter on the 17th when he said the comet resembled a star of magnitude 10. Otherwise, the brightness can only be implied from several vague descriptions. Peter observed the comet at a magnification of 144× on the 1st and said the coma filled the largest part of the visual field of view with the appearance of a faint nebula. Bruhns and Peter said the comet was very faint and diffuse on the 8th. Peter said the comet was reasonably visible in morning twilight on the 10th, but was diffuse and without a border. R. Copeland (Dun Echt Observatory, Scotland) said the comet was very faint on the 19th. C. Pritchard (Oxford University, England) saw the comet with "considerable difficulty" with a 31-cm refractor on the 30th. Schmidt continued his estimates of the coma diameter using the telescope's 8× finder. He gave a value of 9.5' on the 1st, 12.0' on the 4th, 11.1' on the 5th, 9.1' on the 7th, 8.5' on the 8th, 7.5' on the 10th, and 8–9' on the 12th. Schmidt said the 17-cm Reinfelder refractor revealed a coma diameter of 5.53' on the 12th. Additional coma diameters from other astronomers were estimated using larger telescopes. Pechüle said it was 80–90" across on the 15th. Peter said it was 1.5–2.0' across on the 17th.

The comet was last seen during the first days of April. Pechüle found it on the 1st and 2nd, and noted it was more difficult to see on the latter date because of poorer sky conditions. The comet was last detected on April 3.92, when Pechüle measured its position as $\alpha = 4^h 42^m 05.18^s$, $\delta = +45° 58'$ 10.5". He saw it with a large refractor and said it was observed with great difficulty because of somewhat poor sky conditions. Pritchard observed with a 31-cm refractor on April 4 and said the comet "was too faint to be certainly seen...."

The spectrum of this comet was observed on several occasions. The three usual cometary bands attributed to the carbon molecule C_2 were detected by Holden on the mornings of February 10 and 11, by A. Secchi on February 17, by W. Harkness on February 14 and 18, and by N. de Konkoly on March 1 and 2. De Konkoly also used an instrument to detect polarization of the comet's light, but noted none.

The first parabolic orbits were independently calculated by J. Holetschek and Pechüle using positions from February 9, 11, and 13. Holetschek determined the perihelion date as 1877 January 19.69, while Pechüle gave it as January 19.70. Additional orbits were calculated by H. Oppenheim, E. Hartwig, A. N. Skinner, and J. R. Hind, which eventually established the perihelion date as January 19.68. A definitive orbit was calculated by A. Thraen (1881) and this is given below.

T	ω	Ω (2000.0)	i	q	e
1877 Jan. 19.6786 (UT)	347.1611	188.9608	152.8948	0.807441	1.0

ABSOLUTE MAGNITUDE: $H_{10} = 8.2$ (V1964)

FULL MOON: Jan. 29, Feb. 27, Mar. 29, Apr. 27

SOURCES: A. L. N. Borrelly, *MNRAS*, **37** (1877 Feb.), p. 189; A. L. N. Borrelly, C. F. Pechüle, and N. C. Dunér, *CR*, **84** (1877 Feb. 12), p. 285; V. Knorre, K. C. Bruhns, and P. Henry, *CR*, **84** (1877 Feb. 19), p. 336; A. L. N. Borrelly and C. F. Pechüle, *AN*, **89** (1877 Feb. 20), p. 93; J. Holetschek, K. C. Bruhns, C. F. Pechüle, and J. Holetschek, *AN*, **89** (1877 Feb. 28), p. 111; A. Secchi, *CR*, **84** (1877 Mar. 5), p. 427; H. Oppenheim, F. A. T. Winnecke, and E. Hartwig, *AN*, **89** (1877 Mar. 27), pp. 129, 133; A. N. Skinner, *AJS* (Series 3), **13** (1877 Apr.), pp. 323–4; C. Pritchard, *MNRAS*, **37** (1877 Apr.), p. 360;

G. L. Tupman and A. Secchi, *The Observatory*, **1** (1877 Apr.), p. 19; G. L. Tupman and N. de Konkoly, *AN*, **89** (1877 Apr. 7), pp. 169–72; C. F. W. Peters, *AN*, **89** (1877 Apr. 11), p. 179; J. F. J. Schmidt, *AN*, **89** (1877 Apr. 25), pp. 217–20; N. de Konkoly, *The Observatory*, **1** (1877 May), p. 58; K. C. Bruhns and B. Peter, *AN*, **89** (1877 May 5), pp. 253–6; G. L. Tupman, *AN*, **89** (1877 May 15), p. 287; H. Geelmuyden, *AN*, **90** (1877 Jun. 28), p. 17; C. F. Pechüle, *AN*, **90** (1877 Aug. 16), pp. 153–6; E. S. Holden and W. Harkness, *AN*, **90** (1877 Aug. 20), pp. 169–72; J. R. Hind, *Nature*, **16** (1877 Sep. 6), pp. 398–9; K. C. Bruhns and Hartwig, *VJS*, **13** (1878), pp. 185–6; G. F. W. Rümker and G. Koch, *AN*, **92** (1878 May 29), pp. 257–60; J. Palisa, *AN*, **92** (1878 Jul. 20), pp. 375–80; A. Thraen, *AN*, **100** (1881 Aug. 30), pp. 231–8; R. Copeland, *MNRAS*, **46** (1886, Supplemental Notice), pp. 489, 493; V1964, p. 61.

C/1877 G1 *Discovered:* 1877 April 6.09 ($\Delta = 1.42$ AU, $r = 0.97$ AU, Elong. $= 43°$)
(Winnecke) *Last seen:* 1877 July 13.81 ($\Delta = 2.49$ AU, $r = 1.71$ AU, Elong. $= 32°$)
  Closest to the Earth: 1877 May 4 (0.9870 AU)
1877 II = 1877b *Calculated path:* PEG (Disc), LAC (Apr. 20), CAS (Apr. 29), CEP (May 1), CAS (May 2), CEP (May 3), CAS (May 8), CEP (May 10), CAM (May 15), UMa (May 23), LYN (Jun. 24), LMi (Jun. 28), LEO (Jul. 13)

F. A. T. Winnecke (Strasbourg, France) discovered this comet on 1877 April 6.09 at a position of $\alpha = 22^{h}\ 07.8^{m}$, $\delta = +15°\ 00'$. He said there was a nucleus. In a letter published in the April 11 issue of the *Astronomische Nachrichten* he wrote, "The comet certainly shows a strange, very weak, double tail. The longer tail extended over a degree and is casually turned away from the sun, while the other, shorter one forms an angle of about 95° with this." An independent discovery was made by E. E. Block (Odessa, Ukraine) on April 11.07. He described it as "rather bright," with a possible tail about 30′ long. Twilight soon ended his observation of this comet, but he acquired more extensive observations the next morning, which confirmed its motion. For the 12th he said, "The comet has a bright nucleus, approximately equal to a star of 7th magnitude." Interestingly, Block had found C/1877 G2 less than 7 hours earlier.

The comet was approaching both the sun and Earth when discovered, but passed perihelion shortly after mid-April. C. F. Pechüle (Copenhagen, Denmark) said the nucleus was located in the eastern half of the coma on April 7, while C. H. F. Peters (Litchfield Observatory, Clinton, New York, USA) described the comet as "a brilliant object, with good concentration; the coma fading out into a short and rather wide, but indistinct tail to the north preceding side." On the 8th, C. Pritchard (Oxford University, England) saw the comet and said it "was tolerably conspicuous, with considerable condensation, approaching to a planetary nucleus, the whole nebularity, in early twilight, having a diameter of about a minute and a half." That same night, the comet was seen by K. C. Bruhns (Leipzig, Germany) and E. S. Holden (US Naval Observatory, Washington, DC, USA). Bruhns said the comet looked like a star of magnitude 7, with a distinct nucleus. He added that the coma was 4′ across, while a tail was pointing away from the sun. Holden observed with a 24-cm refractor and described the comet as like "a nebulous star" in the hazy sky. There was no tail, but a "decided elongated nucleus." A strong condensation was reported by J. F. J. Schmidt (Athens, Greece), Bruhns, and Peters on the 9th. Schmidt gave the naked-eye magnitude as 6.7 and the condensation magnitude as 7. He added that the coma was 5.5′ across and the tail was 0.33° long. Bruhns said the coma was 3′ across, with a nucleus about 4″ across. Peters estimated the tail was about 1° long, while the condensation was "a few seconds" across. He added, "it seems as if the comet is about to form an envelope." B. Peter (Leipzig, Germany) described the comet as oval on the 11th, with an eccentrically situated condensation. The nucleus was not too distinct. Pritchard said the comet was unchanged in shape and brightness on the 12th and was "distinctly visible" in a 5-cm telescope. That same night R. Copeland and J. G. Lohse (Dun Echt Observatory, Scotland) said the comet had a narrow tail about 18′ long, while the nucleus was magnitude 6.5. On April 13, H. Geelmuyden (Christiania, now Oslo, Norway) described the comet as "very bright, with a distinct nucleus, the preceding part is elongated." Copeland and Lohse said the tail extended 31′ toward PA 260° on the 14th. Geelmuyden said the comet was visible to the naked eye on the 15th, while a telescope revealed the elongation on the preceding side was very distinct. Bruhns and Peter said the area surrounding the nucleus appeared granular on the 15th, so that the nucleus seemed divided. That same night Schmidt determined the naked-eye magnitude as 6.5, while the seeker revealed a condensation of magnitude 7 and a tail 0.50° long. On April 16, the comet was seen by Geelmuyden, Pechüle, Bruhns, and Peter. Geelmuyden said the preceding side of the coma extended 6.5′ from the nucleus, while the following side extended only 2′. Pechüle said the tail was clearly seen in the cometseeker, but not in the refractor. Bruhns and Peter said the area around the nucleus was similar in appearance to the 15th, but less granular. On April 17, Schmidt determined the naked-eye magnitude as 6.2, while the seeker revealed a coma 6.0′ across with a condensation of magnitude 7 and a tail extending 0.60° toward PA 268°. Geelmuyden said the nucleus was more distinct on the 18th than on the 16th, and sometimes looked like a small star. G. L. Tupman (Royal Observatory, Greenwich, England) said the comet was 3.5′ across on the 20th and about magnitude 6.5. On April 23, Holden said the comet was "just visible to the naked eye," while the refractor (175×) revealed a nucleus 4.99″ across and a coma 1.8′ across. He said the diameter of the nucleus was 3.74″ on the 24th. He added that a lower power failed to reveal the two tails reported by Winnecke. Winnecke wrote a letter to the *Astronomische Nachrichten* dated April 16 that was published in the April 25 issue. He said, "Comet 1877b is very strange by its tail development. In the morning with almost always hazy skies, I could not decide whether we have it with two narrow tails, which are almost perpendicularly to each other, or whether it is a large, fan-like tail opened over 90° ;

the branch which is almost situated in the parallel was yesterday night at 2° long. To the naked eye the comet appeared like a star of 5 or 6 magnitude." On April 26, W. Valentiner (Mannheim, Germany) reported that a tail was barely visible in bright moonlight and low magnification. Schmidt said the finder revealed a condensation of magnitude 7.5 on the 29th. On April 30, Schmidt said the finder revealed a coma 7.5′ across with a condensation of magnitude 7 and a tail extending 0.50° toward PA 304°.

Although the comet was moving away from the sun as May began, it was approaching Earth and passed closest on the 4th. Numerous naked-eye magnitudes were given during the month, although there is a chance they refer more to the brightness of the condensation. Schmidt provided the longest series, giving it as 6.5 on the 3rd, 6.4 on the 5th, 6.3 on the 6th, 6.5 on the 7th, 6.4 on the 12th, 6.3 on the 13th, 7.0 on the 28th, and 6.8 on the 29th. Bruhns and Peter provided a few estimates, giving values of 5–6 on the 1st, and 5 on the 7th and 11th. R. Engelmann (Kissingen, Germany) determined the magnitude using binoculars and gave values of 6.4 on the 6th, 6.5 on the 9th, 6.2 on the 10th, 6.4 on the 11th and 13th, 6.8 on the 14th, and 7.0 on the 15th. Astronomers said the nucleus was distinct throughout the month, with Schmidt estimating its brightness in a 17-cm refractor as 9 on the 6th, 8 on the 9th, 9 on the 12th and 20th, 7–8 on the 29th, and 8 on the 31st. Copeland and Lohse observed with a 38-cm refractor on the 3rd, at a magnification of 430× and wrote, "the nucleus is fairly defined as a very small nebula in the midst of the coma, very suddenly exceedingly much brighter in the middle." That same evening Bruhns and Peter said the nucleus was eccentric and granular. G. Koch (Hamburg, Germany) gave the diameter of the nucleus as 55″ on the 7th. Geelmuyden said the nucleus looked like a star on the 20th. The coma was well observed by Schmidt and Engelmann. Schmidt measured it with the finder and gave the diameter as 10.0′ on the 1st, 9.8′ on the 3rd, 9.4′ on the 5th, 10.5′ on the 7th, 8.5′ on the 11th, 8.6′ on the 12th, 9.7′ on the 13th, 9.3′ on the 14th, 8.6′ on the 15th, 9.7′ on the 16th, 6.5′ on the 18th, 10.7′ on the 19th, and 8.1′ on the 31st. Engelmann made his observations with a 12.2-cm cometseeker and gave the diameter as 5.5′ on the 9th, 4′ on the 11th, 4.1′ on the 13th, 4.7′ on the 14th, 4.5′ on the 15th, and 5.5′ on the 17th. Peters noted for the 3rd, "the comet looks as if consisting of discrete particles, making the same impression upon the eye as a cluster of fine stars at the point of being resolved." On the same night, Bruhns and Peter said there was less nebulosity to the right of the nucleus. The tails were generally well observed. Valentiner said they were separated by an angle of 100° on the 2nd. On May 3, Copeland and Lohse said the main tail was 56′ long, while a "lateral tail" was also visible. Schmidt began a fine series of measurements of the main tail at this time, as viewed through the finder. He said it extended 0.50° on the 3rd, 1.60° on the 5th, 1.00° on the 6th, 1.40° toward PA 302° on the 7th, 1.30° toward PA 330° on the 10th, 1.00° toward PA 332° on the 11th, 1.80° toward PA 338° on the 12th, 1.70° toward PA 351° on the 13th, 1.40° toward PA 11° on the 14th, 1.20° toward

PA 35° on the 15th, 1.30° toward PA 53° on the 16th, 1.20° toward PA 42° on the 18th, 0.50° toward PA 24° on the 18th, 0.40° on the 29th, and 0.33° on the 31st. Engelmann consistently estimated the lengths of both the main and secondary tails while observing with the 12.2-cm cometseeker. He gave the lengths as 48' and 20' on the 6th, 67' and 27' on the 10th, 48' and 14' on the 11th, 54' and 27' on the 13th, 61' and 27' on the 14th, 54' and 41' on the 15th, and 54' and 27' on the 17th. Bruhns and Peter referred to the tail as "broad" on the 3rd and they said the secondary tail was hardly noticeable on the 11th. The comet passed about 10° from the North Celestial Pole on May 14.

The comet was continuing to move away from the sun and Earth in June. The only total magnitude estimate for the month was from Schmidt on the 3rd, when he said it was 7.0 to the naked eye. As with Schmidt's estimates during May, this may have referred to the condensation. Schmidt also continued to provide some magnitude estimates of the "nucleus" while using the refractor. He gave values of 9 on the 5th, 9.5 on the 12th, and 11.5 on the 28th. Bruhns and Peter said the nucleus was about magnitude 12 or 13 on the 12th, "difficult to see" on the 14th, "only suspected" on the 17th, and definitely visible on the 19th. Although Schmidt obtained several measurements of the coma diameter, some were made with the finder and others with the refractor. Using a finder, he said the diameter was 9.0' on the 1st and 7.5' on the 2nd. The refractor revealed a diameter of 5.7' on the 5th, 6.84' on the 12th, and 4.82' on the 28th. Bruhns and Peter said the coma was only 0.5–0.7' across on the 14th, but this probably referred to the central condensation. Bruhns and Peter noted on the 3rd that the distinct nucleus was surrounded by a granular appearance. The only measurements of the tail length came from Schmidt. Using a finder he said it extended 0.33° toward PA 66° on the 1st, 0.40° toward PA 80° on the 3rd, 0.25° toward PA 28° on the 5th, 0.25° toward PA 34° on the 7th, 0.25° on the 11th, and 0.15° on the 12th. Schmidt said his refractor revealed a tail 0.11° long on the 28th. Bruhns and Peter looked for the comet on June 30 and the following nights, but were not successful in finding it.

The only physical descriptions reported during July came from Schmidt and these were only made with the refractor. On the 2nd, he noted a nucleus of magnitude 11 and gave the coma diameter as 4.8'. The tail was 0.16° long. On July 4, Schmidt saw a nucleus of magnitude 11.5 and indicated a coma diameter of 4.0'. Schmidt said the nucleus was magnitude 11 on the 6th, while the coma was 4.46' across. The tail was 0.09° long. Schmidt saw a nucleus of magnitude 11 on the 10th.

This comet was last detected on July 13.81, when Schmidt measured the position as $\alpha = 9^h\ 48.3^m$, $\delta = +33°\ 13'$. He said the refractor revealed the nucleus was magnitude 11.5.

The spectrum of this comet was obtained on several nights during April and May by J. L. Lindsay, Copeland, and Lohse (Dun Echt Observatory, Scotland), W. H. M. Christie and E. W. Maunder (Royal Observatoy,

Greenwich, England), M. C. Wolff (Paris, France), F. A. Bredikhin (Moscow, Russia), H. C. Vogel (Potsdam, Germany), and A. Secchi (Rome, Italy). The three carbon bands, now attributed to the molecule C_2, were detected in the coma by each observer, while the nucleus displayed a continuous spectrum.

The first parabolic orbit was calculated by E. Hartwig using positions from April 6, 7, and 8. He determined the perihelion date as 1877 April 18.64. Winnecke noted, "The analogy of these elements with those of the Comets 1827 II and 1852 II is great, and the circumstance of the intervals being nearly equal gives it a certain significance." Additional orbits were calculated by Pritchard, N. C. Dunér, A. Lindstedt, E. F. von d'Sande Bakhuyzen, J. C. Kapteyn, C. W. Plath, G. Becka, J. R. Hind, and D. Kirkwood which eventually revealed a perihelion date early on April 18. Kirkwood agreed that the orbit of this comet was similar to those of 1827 II and 1852 II, but added, "it seems impossible to reconcile the changes of perihelion and inclination with the theory of identity."

A definitive orbit was calculated by C. W. Plath (1878) using 280 positions spanning the entire period of visibility. He determined both a parabolic and an elliptical orbit. The parabolic orbit had a perihelion date of April 18.17. The elliptical orbit had a perihelion date of April 18.16 and a period of about 19 765 years. The elliptical orbit is given below.

T	ω	Ω (2000.0)	i	q	e
1877 Apr. 18.1562 (UT)	63.1204	318.3341	121.1548	0.949980	0.998701

ABSOLUTE MAGNITUDE: $H_{10} = 5.7$ (V1964)

FULL MOON: Mar. 29, Apr. 27, May 27, Jun. 25, Jul. 25

SOURCES: F. A. T. Winnecke, E. Hartwig, W. Schur, and C. Pritchard, *MNRAS*, **37** (1877 Apr.), pp. 359–61; C. Pritchard and F. A. T. Winnecke, *The Observatory*, **1** (1877 Apr.), pp. 20–1, 27; F. A. T. Winnecke, *AN*, **89** (1877 Apr. 11), p. 179; C. F. W. Peters, W. Klinkerfues, and V. Knorre, *AN*, **89** (1877 Apr. 19), p. 205; K. C. Bruhns, F. A. T. Winnecke, E. E. Block, N. C. Dunér, and A. Lindstedt, *AN*, **89** (1877 Apr. 25), pp. 217, 221–4; M. C. Wolff, *CR*, **84** (1877 Apr. 30), pp. 929–30; M. C. Wolff, *The Observatory*, **1** (1877 May), p. 58; J. R. Hind, *Nature*, **16** (1877 May 3), p. 15; G. Koch, E. F. von d'Sande Bakhuyzen, and J. C. Kapteyn, *AN*, **89** (1877 May 5), pp. 247, 251; C. W. Plath, *AN*, **89** (1877 May 11), pp. 267–70; G. L. Tupman, *AN*, **89** (1877 May 15), p. 287; J. L. Lindsay, R. Copeland, and J. G. Lohse, *MNRAS*, **37** (1877 Jun.), pp. 430–2; A. Secchi, *CR*, **84** (1877 Jun. 4), pp. 1289–92; H. Geelmuyden, *AN*, **90** (1877 Jun. 28), pp. 17–20; D. Kirkwood, *AN*, **90** (1877 Jul. 21), p. 87; J. F. J. Schmidt, *AN*, **90** (1877 Aug. 14), pp. 139–44; C. F. Pechüle and W. Upton, *AN*, **90** (1877 Aug. 16), pp. 155–8; C. H. F. Peters, *AN*, **90** (1877 Aug. 27), p. 193; R. Engelmann, *AN*, **90** (1877 Aug. 30), pp. 209–18; K. C. Bruhns, B. Peter, and E. S. Holden, *AN*, **90** (1877 Sep. 24), pp. 323–30, 335; G. Strasser, *AN*, **90** (1877 Sep. 27), pp. 347–50; W. H. M. Christie and E. W. Maunder, *MNRAS*, **37** (1877, Supplemental Notice), pp. 469–70; Deichmüller, *AN*, **91** (1877 Nov. 10), pp. 73–8; W. Valentiner, *AN*, **91** (1877 Nov. 14), pp. 89–92; K. C. Bruhns and C. W. Plath, *VJS*, **13** (1878), pp. 186–8; G. F. W. Rümker and G. Koch, *AN*, **92** (1878 May 29), pp. 257–60; J. Palisa, *AN*, **93** (1878 Aug. 14), p. 33; A. Palisa, *AN*, **94** (1879 Feb. 28), p. 203; H. C.

Vogel, *AN*, **96** (1880 Jan. 1), p. 189; G. Becka, *AN*, **101** (1882 Feb. 2), p. 205; R. Copeland and J. G. Lohse, *MNRAS*, **46** (1886, Supplemental Notice), pp. 489, 493–4; V1964, p. 61.

C/1877 G2
(Swift–Borrelly–
Block)

1877 III = 1877c

Discovered: 1877 April 10.8 (Δ = 1.49 AU, r = 1.05 AU, Elong. = 44°)
Last seen: 1877 June 4.98 (Δ = 1.60 AU, r = 1.20 AU, Elong. = 48°)
Closest to the Earth: 1877 May 5 (1.3087 AU)
Calculated path: CAS (Disc), CAM (Apr. 26), LYN (May 10), CNC (Jun. 2)

This comet was discovered by E. E. Block (Odessa, Ukraine) on 1877 April 10.8. He gave the position as roughly $\alpha = 0^h 36^m$, $\delta = +51°$ 53, but at that time he thought this was one of Herschel's faint nebulae and did not announce the discovery. Independent discoveries were made by L. Swift (Rochester, New York, USA) on April 11 and A. L. N. Borrelly (Marseille, France) on April 14.88. Swift simply described the comet as faint, while Borrelly said it was round and bright, with a small nucleus. Block realized on April 16 that the nebula was really a comet and announced his find at that time.

The comet was approaching both the sun and Earth when discovered, and passed perihelion near the end of April. Several astronomers observed the comet on April 15. C. F. Pechüle (Copenhagen, Denmark) described it as "small, quite bright, and very condensed in the center," with a diameter of 50–55″. K. C. Bruhns (Leipzig, Germany) described the comet as round and rather faint, with a diameter of about 1.5′. F. A. T. Winnecke (Strasbourg, France) said the comet was about 1.5′ across, oval, without a stellar nucleus. On the 16th, B. Peter (Leipzig, Germany) said the round comet exhibited diffuse edges, as well as a small condensation. R. Copeland and J. G. Lohse (Dun Echt Observatory, Scotland) saw the comet with a 38-cm refractor on the 19th and said it was round, about 1.5′ across, and "much brighter in the middle." H. Seeliger (Bonn, Germany) said the comet was faint in a 5-foot focal length refractor on both the 19th and 20th, but he detected a barely perceptible nucleus on the last date. H. Geelmuyden (Christiania, now Oslo, Norway) said the comet was faint, with a trace of a nucleus on the 20th. G. Koch (Hamburg, Germany) simply noted the comet was faint in moonlight on the 21st. On April 24, Geelmuyden said the observation was difficult because of moonlight, but he did notice a trace of a nucleus. During the period of April 19–26, M. C. Wolff (Paris, France) described the comet as a small nebulosity, with a weak central condensation. On the 29th, Geelmuyden said the comet "was perhaps a little brighter than when previously seen." He added that the nucleus was still not distinct. Lohse said the refractor revealed the comet was round and about 57″ across on the 30th. He added that it was "gradually brighter in the middle."

The comet was moving away from the sun as May began and it passed closest to Earth on the 5th. Geelmuyden said the comet was faint and the nucleus indistinct on the 2nd. Peter said the comet was 2' across on the 3rd, with diffuse edges and a nucleus. Geelmuyden said the comet seemed slightly brighter on the 4th than on the 2nd, but the nucleus was still indistinct and difficult to see. H. Seeliger described the comet as rather faint in his refractor, without a perceptible nucleus. On May 6, Geelmuyden said, "The comet had a good nucleus today; undoubtedly it would have been good to observe on a less bright sky." He said the comet was diffuse, with a very indistinct nucleus. On May 7, he said the comet was 1.5' across, with some condensation. Peter found the comet fainter and more diffuse on the 10th, with no visible nucleus. He said the comet was 1.5' across on the 11th and had the brightness of a 10th-magnitude star. He also noted an eccentrically situated condensation. Peter found the comet more diffuse on the 13th and, although there was a condensation, he said the nucleus was questionable. On May 30, C. H. F. Peters (Litchfield Observatory, Clinton, New York, USA) described the comet as "a rather faint nebulous mass, without much concentration.

The final observations came at the beginning of June. Bruhns saw the comet with the 30-cm telescope on the 3rd and described it as an extremely fine nebulous mass. The comet was last detected on June 4.98, by W. E. Plummer (Oxford University, England). He was using a 31-cm refractor and measured the position as $\alpha = 8^h 30.2^m$, $\delta = +31° 08'$.

J. L. Lindsay (Dun Echt Observatory, Scotland) examined the spectrum of this comet on the evening of May 5 and said it consisted of three bright lines. The centers of these lines were measured as 5282 Å, 5079 Å, and 4676 Å. He added that the former line "is very close to E of the Solar Spectrum. There we have to deal with a comet closely allied to Brorsen's Comet of 1868 and Comet I. 1871."

The first parabolic orbit was calculated by J. Holetschek, who took positions from April 14, 15, and 16, and determined the perihelion date as 1877 April 28.66. The orbit was fairly close to the truth, as computations by C. F. W. Peters, G. Celoria, J. R. Hind, C. W. Plath, C. Pritchard, R. Poenisch, and K. Zelbr, as well as a revision by Holetschek, varied by only a day or so in the perihelion date. Holetschek initially noted that the orbit was similar to that of C/1762 K1, but he later said a link was not possible.

The first person to calculate a nonparabolic orbit was Holetschek. He used positions obtained during the period of April 15 to May 30, and determined an elliptical orbit with a perihelion date of April 27.30 and a period of about 28 246 years. Interestingly, J. W. Nichol (1878) took 60 positions covering the same period of time as Holetschek's computations, determined five Normal positions, and calculated a hyperbolic orbit with a perihelion date of April 27.33 and an eccentricity of 1.0003446.

A definitive orbit was calculated by R. Poenisch (1886). He took 173 positions, determined six Normal positions, and produced both parabolic and elliptical orbits. The parabolic orbit had a perihelion date of April 27.32, while the elliptical orbit had a perihelion date of April 27.31. The period of the elliptical orbit was about 10 718 years and it was a better fit to the positions. It is given below.

T	ω	Ω (2000.0)	i	q	e
1877 Apr. 27.3058 (UT)	116.7721	347.7984	77.1916	1.009053	0.997923

ABSOLUTE MAGNITUDE: $H_{10} = 6.7$–9.2 (V1964)

FULL MOON: Mar. 29, Apr. 27, May 27, Jun. 25

SOURCES: L. Swift and A. L. N. Borrelly, *The Observatory*, **1** (1877 Apr.), p. 27; A. L. N. Borrelly, *CR*, **84** (1877 Apr. 16), pp. 759–60; A. L. N. Borrelly and V. Knorre, *AN*, **89** (1977 Apr. 19), p. 205; L. Swift, K. C. Bruhns, C. F. W. Peters, F. A. T. Winnecke, and W. Valentiner, *AN*, **89** (1877 Apr. 25), pp. 209, 217, 221; J. R. Hind, *Nature*, **15** (1877 Apr. 26), pp. 549–50; M. C. Wolff, *CR*, **84** (1877 Apr. 30), pp. 929–31; J. Holetschek, E. E. Block, V. Knorre, G. V. Schiaparelli, and T. R. von Oppölzer, *AN*, **89** (1877 May 1), p. 233; G. Koch and G. Celoria, *AN*, **89** (1877 May 5), pp. 247, 255; C. W. Plath, *AN*, **89** (1877 May 11), p. 267; C. F. W. Peters, *AN*, **89** (1877 May 15), p. 287; J. L. Lindsay, *MNRAS*, **37** (1877 Jun.), p. 432; C. F. Pechüle, *AN*, **90** (1877 Aug. 16), p. 155; C. H. F. Peters, *AN*, **90** (1877 Aug. 27), p. 193; K. C. Bruhns and B. Peter, *AN*, **90** (1877 Sep. 12), pp. 269–72; G. Strasser, *AN*, **90** (1877 Sep. 27), pp. 347–50; Deichmüller and H. Seeliger, *AN*, **91** (1877 Nov. 10), pp. 73–8; J. Holetschek, *AN*, **91** (1877 Dec. 19), pp. 163–6; K. Zelbr, *SAWW*, **78** Abt. II (1878), p. 976; K. C. Bruhns, *VJS*, **13** (1878), pp. 188–90; J. Holetschek, *The Observatory*, **1** (1878 Mar.), pp. 374–5; G. F. W. Rümker and G. Koch, *AN*, **92** (1878 May 29), p. 259; J. W. Nichol, *AN*, **93** (1878 Aug. 14), pp. 37–42; R. Poenisch, *AN*, **100** (1881 Jun. 17), pp. 59–64; R. Poenisch, W. E. Plummer, and C. A. Jenkins, *AN*, **115** (1886 Sep. 20), pp. 161–90; R. Copeland and J. G. Lohse, *MNRAS*, **46** (1886, Supplemental Notice), pp. 490, 494; V1964, p. 61.

6P/d'Arrest   *Recovered:* 1877 June 13.8 ($\Delta = 1.61$ AU, $r = 1.38$ AU, Elong. $= 58°$)

Last seen: 1877 September 11.01 ($\Delta = 1.47$ AU, $r = 1.90$ AU, Elong. $= 98°$)

1877 IV = 1877d   *Closest to the Earth:* 1877 October 20 (1.3949 AU)

Calculated path: PSC (Rec), CET (Jun. 21), TAU (Jul. 20), ORI (Sep. 5)

G. Leveau (1876) investigated the orbit of this comet using positions obtained during the apparitions of 1851, 1857, and 1870, as well as perturbations produced by Mars, Jupiter, and Saturn. He predicted the comet would next arrive at perihelion on 1877 May 10.83. C. Pritchard (Oxford University, England) wrote a letter to the *Monthly Notices* on April 13 that said, "D'Arrest's Comet has here been looked for in vain, in the places assigned for it in Leveau's Ephemeris. The twilight and the Moon may render observations of this interesting body very difficult for some time to come." The comet was recovered by E. J. White (Melbourne, Australia) on 1877 June 13.8 at a position of about $\alpha = 1^h 38^m$, $\delta = +5° 16'$. According to the Melbourne Observatory annual report (*Monthly Notices*, 1878 February) the comet was observed on five mornings during June. Independent recoveries were made by J. E. Coggia (Marseille, France) on July 10.10 and E. W. L.

Tempel (Arcetri, Italy) on July 10.05. Coggia said the comet was very faint, while Tempel described it as 3′ across and diffuse, with a brightness resembling a Herschel nebula of class I or II. The resulting positions were very close to the prediction of Leveau.

The comet was not particularly well placed, having passed perihelion around mid-May and did not pass particularly close to Earth. It therefore remained a faint and difficult object to follow. Tempel said the sky was not as pure on July 11, and he noted the comet was only 1.5′ across and displayed several stars at the center. One of these "stars" remained steadily visible, while the others pulsated in and out of view. J. F. J. Schmidt (Athens, Greece) said the comet was extraordinarily faint on the 14th. K. C. Bruhns and B. Peter (Leipzig, Germany) saw the comet on August 7 and described it as a rather faint and diffuse spot measuring 1.5–2.0′ across. G. V. Schiaparelli (Milan, Italy) saw the comet on July 11, 18, August 8 and 13 with a 20-cm refractor and noted, "The comet appeared like a simple spot of very faint and indeterminate light, with a diameter of 1.5′ or 2′."

The comet was last detected on September 11.01, when Schmidt gave the position as $\alpha = 4^{\mathrm{h}}\ 42.4^{\mathrm{m}}$, $\delta = +2°\ 44′$. He was using a 17-cm Reinfelder refractor and described it as one of the faintest and smallest comets he had ever seen.

Minor refinements of this comet's orbit using positions from multiple apparitions, as well as planetary perturbations, have been published by numerous astronomers. Of most significance were the studies of G. Leveau (1883), A. W. Recht (1939), and B. G. Marsden and Z. Sekanina (1972). These have typically revealed a perihelion date of May 10.95 and a period of 6.66 years. Marsden and Sekanina determined the nongravitational terms as $A_1 = -0.50$ and $A_2 = +0.0960$. Their orbit is given below.

T	ω	Ω (2000.0)	i	q	e
1877 May 10.9545 (TT)	173.0022	147.8093	15.7109	1.318010	0.627784

ABSOLUTE MAGNITUDE: $H_{10} = 8.8$ (V1964)

FULL MOON: Jun. 25, Jul. 25, Aug. 23, Sep. 22

SOURCES: *MNRAS*, **36** (1876 Feb.), pp. 193–4; G. Leveau, *CR*, **82** (1876 Mar. 13), pp. 624–6; G. Leveau, *CR*, **83** (1876 Aug. 28), pp. 508–10; G. Leveau, *MNRAS*, **37** (1877 Feb.), pp. 189–90; C. Pritchard, *MNRAS*, **37** (1877 Apr.), p. 360; E. W. L. Tempel and G. Leveau, *The Observatory*, **1** (1877 Jul.), p. 122; J. E. Coggia, *CR*, **85** (1877 Jul. 16), p. 131; J. E. Coggia, E. W. L. Tempel, and G. Leveau, *AN*, **90** (1877 Jul. 21), pp. 87–92; J. E. Coggia and E. W. L. Tempel, *The Observatory*, **1** (1877 Aug.), p. 160; J. F. J. Schmidt, *AN*, **90** (1877 Aug. 20), pp. 165, 173; E. W. L. Tempel and J. F. J. Schmidt, *AN*, **90** (1877 Aug. 26), p. 191; G. V. Schiaparelli, *AN*, **90** (1877 Aug. 30), p. 223; J. F. J. Schmidt, *AN*, **90** (1877 Oct. 1), p. 367; K. C. Bruhns, *VJS*, **13** (1878), p. 190; K. C. Bruhns and B. Peter, *AN*, **91** (1878 Jan. 29), pp. 299–302; E. J. White, *MNRAS*, **38** (1878 Feb.), p. 188; G. Leveau, *AN*, **105** (1883 Mar. 14), pp. 19–22; A. W. Recht, *AJ*, **48** (1939 Jul. 17), pp. 65–78; V1964, p. 61; B. G. Marsden and Z. Sekanina, *QJRAS*, **13** (1972), pp. 428–9; B. G. Marsden, Z. Sekanina, and D. K. Yeomans, *AJ*, **78** (1973 Mar.), pp. 213, 215; B. G. Marsden and Z. Sekanina, *CCO*, 12th ed. (1997), pp. 48–9; personal correspondence from Wayne Orchiston (2002).

C/1877 R1
(Coggia)

1877 VI = 1877e

Discovered: 1877 September 14.18 (Δ = 1.85 AU, r = 1.58 AU, Elong. = 58°)
Last seen: 1877 December 10.88 (Δ = 1.12 AU, r = 1.99 AU, Elong. = 141°)
Closest to the Earth: 1877 November 19 (0.9031 AU)
Calculated path: UMa (Disc), LYN (Sep. 19), GEM (Oct. 20), ORI (Nov. 11), ERI (Nov. 30)

J. E. Coggia (Marseille, France) discovered this comet on 1877 September 14.18 at a position of $\alpha = 8^h 32.7^m$, $\delta = +48°$ 30'. He simply described it as faint.

The comet had just passed perihelion when discovered, but was still heading toward its closest approach to Earth. K. C. Bruhns and B. Peter (Leipzig, Germany) described the comet as rather faint, with a trace of nucleus on September 17, and gave the diameter as 0.7'. On September 19, F. A. T. Winnecke (Strasbourg, France) described the comet as moderately bright, about 1.5' across, with a central condensation. That same night, Bruhns and Peter said the comet appeared faint and diffuse in the 20-cm refractor. On October 5, Bruhns and Peter noted a slight condensation, but no nucleus. They said the coma was 0.5' across on the 6th. G. Koch (Hamburg, Germany) also saw the comet on the 6th and said it was extremely faint and diffuse. On October 10, G. F. W. Rümker (Hamburg, Germany) said the comet was very faint, with a round coma about 1.3' across. He noted there was also a "fine star-like condensation in the center." Bruhns and Peter found the coma to be 1.5' across in a refractor on the 15th and noted a nucleus of about magnitude 12.

The comet passed closest to Earth shortly after mid-November and began to fade more rapidly thereafter. On November 29, Bruhns and Peter said the coma diameter was 1.5–2' across and contained a strong condensation, with a distinct nucleus. They were unable to find the comet with the refractor on December 7. The comet was last detected on December 10.88, when J. Palisa (Pula, Yugoslavia) gave the position as $\alpha = 4^h 00.3^m$, $\delta = -11°$ 08'.

The first parabolic orbits were independently calculated by J. Holetschek and J. R. Hind using positions obtained over less than one week. Holetschek determined a perihelion date of 1877 September 28.40, while Hind determined it as September 6.91. Later orbits calculated by E. Hartwig and W. E. Plummer indicated a perihelion date of September 11. A definitive orbit was calculated by R. Larssén (1886). Using positions spanning the period of September 17 to December 3, he determined the orbit below.

T	ω	Ω (2000.0)	i	q	e
1877 Sep. 11.7182 (UT)	143.2049	252.7104	102.2274	1.575904	1.0

ABSOLUTE MAGNITUDE: $H_{10} = 6.3$ (V1964)

FULL MOON: Sep. 22, Oct. 22, Nov. 20, Dec. 20

SOURCES: J. E. Coggia, *AN*, **90** (1877 Sep. 19), p. 303; K. C. Bruhns and B. Peter, *AN*, **90** (1877 Sep. 21), p. 319; F. A. T. Winnecke, *AN*, **90** (1877 Sep. 24), p. 329; J. Holetschek, *AN*, **90** (1877 Sep. 27), pp. 349–52; J. R. Hind, *Nature*, **16** (1877 Sep. 27), pp. 460–1; E. Hartwig, *The Observatory*, **1** (1877 Oct.), p. 229; F. A. T. Winnecke, *AN*, **91** (1877 Oct. 17), p. 15; E. Hartwig, *AN*, **91** (1877 Oct. 27), pp. 31–2; C. Pritchard and W. Plummer, *MNRAS*, **38** (1877 Nov.), p. 56; W. E. Plummer, *AN*, **91** (1877 Nov. 14), p. 91; K. C. Bruhns and W. Plummer, *VJS*, **13** (1878), pp. 191–2; K. C. Bruhns and B. Peter, *AN*, **91** (1878 Jan. 29), pp. 299–302; G. F. W. Rümker and G. Koch, *AN*, **92** (1878 May 29), p. 259; A. Palisa, *AN*, **94** (1879 Feb. 28), p. 203; R. Copeland and J. G. Lohse, *MNRAS*, **46** (1886, Supplemental Notice), p. 490; R. Larssén, *AN*, **116** (1886 Dec. 8), pp. 23–6; V1964, p. 61.

C/1877 T1
(Tempel)

1877 V = 1877f

Discovered: 1877 October 2.86 ($\Delta = 0.88$ AU, $r = 1.86$ AU, Elong. $= 162°$)

Last seen: 1877 October 31.8 ($\Delta = 1.61$ AU, $r = 2.20$ AU, Elong. $= 113°$)

Closest to the Earth: 1877 March 24 (0.9913 AU), 1877 September 16 (0.7212 AU)

Calculated path: CET (Disc), AQR (Oct. 3), PsA (Oct. 25)

E. W. L. Tempel (Arcetri, Italy) discovered this comet on 1877 October 2.86 at a position of $\alpha = 23^{\mathrm{h}} 51.0^{\mathrm{m}}$, $\delta = -10° 19'$. He described it as small, bright, with a tail, and also noted a daily motion of $-70'$ in α and $+63'$ in δ. G. V. Schiaparelli (Observatoire Royal de Bréra, Milan, Italy) first saw the comet low over the horizon in the morning sky on October 3.09. He noted it was "veiled considerably by the vapor of the horizon."

The comet came under immediate observation throughout Europe. Schiaparelli noted a well-condensed nucleus of magnitude 10.5–11 on October 4 and 7. He also reported a coma $3'$ across and a tail extending 4–5' toward PA 30°. K. C. Bruhns and B. Peter (Leipzig) estimated the magnitude as 10 on the 5th and described it as small, with an eccentric nucleus. F. A. T. Winnecke (Strasbourg, France) saw the comet on the 6th and said the coma was $0.4'$ across and exhibited a distinct nucleus of magnitude 11. He added that a faint tail extended $4'$ toward PA 24.0°. Bruhns and Peter saw the comet that same night and said the coma was $1.5'$ across, while the tail was $5'$ long. On October 8, J. F. J. Schmidt (Athens) said the coma was $2.7'$ across and exhibited a 12th-magnitude nucleus. Schiaparelli saw the comet on the 9th and said the coma was $3'$ across, while the nucleus was still around magnitude 10.5–11. The tail extended $4.4'$ toward PA 25.0°. He added, "The tail was so rare and so badly defined, that I dared not establish its width or form." Bruhns and Peter saw the comet with a 20-cm refractor (192×) and described it as faint, 1.0–1.5' across, with a distinct nucleus. Tempel also saw the comet and said it was unchanged from the 2nd.

The comet showed signs of some changes beginning on October 11. G. Koch (Hamburg, Germany) said the comet was very faint, while Schmidt said the comet was difficult to see in the finder. Bruhns and Peter said the 20-cm refractor revealed a fainter nucleus of magnitude 11.5 and a tail 2–3′ long. Schiaparelli said the coma was 3′ across on the 12th and he noted it was "irregular and extremely rare." He added that the tail was no longer visible. Bruhns and Peter said the comet was very difficult to see in the 20-cm refractor. On October 13, Schmidt said the comet was very faint in the refractor, but showed a diameter of 2.7′ and a stellar nucleus of magnitude 12. Schiaparelli last saw the comet on the 13th and 14th. He again noted the faintness and irregularity of the coma.

The final precise positions of this comet came from several astronomers on October 14, with measurements coming from Bruhns and Peters on October 14.90, Schiaparelli on October 14.95, and Greenwich observers on October 14.97. Bruhns and Peter observed with the 20-cm refractor and said the comet appeared as a faint, diffuse spot, with a condensation. The Greenwich observers measured the position as $\alpha = 23^h 13.8^m$, $\delta = -20° 46′$.

Following the interference of moonlight, the comet was next seen on October 26, when Tempel described it as a diffuse, round nebulosity, without a condensation or tail. Schiaparelli tried to find the comet at the end of October, following the full moon, but he said, "The vapor of our southern horizon ruined all the attempts that I made." The comet was last detected on October 30.8 and 31.8 by Tempel, but there was no convenient comparison star to determine the position. Tempel again looked for the comet on November 1, but despite clear skies, no trace was found.

E. M. Pittich (1969) believed this comet is a strong candidate for a pre-discovery outburst in brightness. He noted that if no outburst had occurred then conditions should have enabled the comet to have been found 198 days earlier, of which 83 days would have been favorable with no moon and a northern sky location.

The first parabolic orbit was calculated by J. Holetschek and A. Palisa. They took positions obtained during the first days following the discovery and determined a perihelion date of 1877 June 26.62. The next orbit came from F. K. Ginzel. He took positions obtained on October 2, 7, and 13, and determined a perihelion date of June 27.56. G. Gruss (1882) took positions from the period spanning October 3–14 and determined the perihelion date as June 27.57. This orbit is given below.

T	ω	Ω (2000.0)	i	q	e
1877 Jun. 27.5703 (UT)	103.2436	185.9981	115.7261	1.070451	1.0

ABSOLUTE MAGNITUDE: $H_{10} = 6.0$ (V1964)

FULL MOON: Sep. 22, Oct. 22, Nov. 20

SOURCES: J. Holetschek and A. Palisa, *The Observatory*, **1** (1877 Oct.), pp. 229–30; E. W. L. Tempel, *AN*, **90** (1877 Oct. 10), p. 381; C. F. W. Peters, K. C. Bruhns, B. Peter, and F. A. T. Winnecke, *AN*, **91** (1877 Oct. 17), p. 15; J. Holetschek and A. Palisa, *AN*,

91 (1877 Oct. 27), pp. 29–32; J. F. J. Schmidt, *AN*, **91** (1877 Nov. 8), p. 63; F. K. Ginzel, *AN*, **91** (1877 Dec. 4), p. 143; G. V. Schiaparelli, *AN*, **91** (1877 Dec. 19), p. 167; K. C. Bruhns and F. K. Ginzel, *VJS*, **13** (1878), p. 191; K. C. Bruhns and B. Peter, *AN*, **91** (1878 Jan. 29), pp. 299–304; G. F. W. Rümker and G. Koch, *AN*, **92** (1878 May 29), p. 259; J. Palisa, *AN*, **93** (1878 Aug. 14), p. 33; E. W. L. Tempel, *AN*, **93** (1878 Aug. 17), p. 49; A. Palisa, *AN*, **94** (1879 Feb. 28), p. 203; G. Gruss, *SAWW*, **85** Abt. II (1882), pp. 98–104; V1964, p. 61; E. M. Pittich, *BAC*, **20** (1969), p. 258.

C/1878 N1	*Discovered:* 1878 July 7.30 ($\Delta = 0.50$ AU, $r = 1.41$ AU, Elong. $= 133°$)
(Swift)	*Last seen:* 1878 July 24.11 ($\Delta = 0.54$ AU, $r = 1.39$ AU, Elong. $= 124°$)
	Closest to the Earth: 1878 July 13 (0.4734 AU)
1878 I = 1878a	*Calculated path:* HER (Disc), OPH (Jul. 9), SCO (Jul. 22), OPH–SCO (Jul. 23)

L. Swift (Rochester, New York, USA) discovered this comet on 1878 July 7.30 at a position of $\alpha = 17^h 40.0^m$, $\delta = +18° 00'$. He described it as faint and large, with no tail or nucleus. C. H. F. Peters (Dudley Observatory, New York, USA) confirmed the comet on July 8.23. He said it did not show a nucleus and that a magnification of $80\times$ revealed it was quite large and bright.

Peters was the only observer thereafter. After measuring its position on July 11, moonlight completely blocked it from view on July 14–16. He again found the comet on July 20 and noted it was rather bright until the moon rose. His final observation came on July 24.11, when the position was measured as $\alpha = 16^h 12^m 41.6^s$, $\delta = -21° 18' 17''$. Peters said it was then very faint, but suggested this was because it was low over the southern horizon.

The first parabolic orbit was calculated by J. Holetschek during August of 1878. He took positions measured by Peters on July 8, 20, and 24, and determined a perihelion date of 1878 July 21.73. He provided an ephemeris for the period of August 8 to September 17. Peters (1879) took his four positions and determined a perihelion date of July 21.22. H. Büttner (1880) took the same four positions and determined a perihelion date of July 21.19. This orbit is given below.

There are two reasons why this comet was not observed in Europe. The September 1878 issue of *The Observatory* stated, "The comet was not observed at any European observatory, although carefully searched for – a circumstance which is probably due to a misunderstanding as to the date of discovery. . . . " Swift had given the discovery date as 2 a.m. on July 7, which was not the generally accepted astronomical form of the date and time. With the astronomical day considered as beginning at noon, astronomers took this to mean 14 hours on July 7, which actually moved the discovery date forward 24 hours. In the 1879 issue of the *Vierteljahrsschrift*, K. C. Bruhns pointed out another problem with the discovery announcement. He said it stated the comet was traveling slowly, while in reality it was initially moving at more than 2° per day.

T	ω	Ω (2000.0)	i	q	e
1878 Jul. 21.1907 (UT)	177.5898	103.9645	78.1761	1.391967	1.0

ABSOLUTE MAGNITUDE: $H_{10} = 8.5$ (V1964)

FULL MOON: Jun. 14, Jul. 14, Aug. 13

SOURCES: L. Swift, *AN*, **92** (1878 Jul. 11), p. 349; L. Swift, *CR*, **87** (1878 Jul. 15), p. 104; L. Swift, *The Observatory*, **2** (1878 Aug.), p. 130; C. H. F. Peters and J. Holetschek, *AN*, **93** (1878 Aug. 29), p. 71; J. Holetschek, *The Observatory*, **2** (1878 Sep.), p. 163; K. C. Bruhns, *VJS*, **14** (1879), p. 95; C. H. F. Peters, *AN*, **95** (1879 May 8), p. 21; H. Büttner, *AN*, **97** (1880 Jul. 8), p. 277; V1964, p. 61.

2P/Encke

1878 II = 1878c

Recovered: 1878 August 3.35 ($\Delta = 1.18$ AU, $r = 0.39$ AU, Elong. $= 19°$)

Last seen: 1878 September 7.01 ($\Delta = 1.13$ AU, $r = 1.00$ AU, Elong. $= 55°$)

Closest to the Earth: 1878 August 21 (1.0295 AU)

Calculated path: LEO (Rec), SEX (Aug. 4), LEO (Aug. 9), VIR (Aug. 15), CRT (Aug. 16), VIR (Aug. 17), CRV (Aug. 21), VIR (Aug. 26), HYA (Sep. 3), LIB (Sep. 7)

F. E. von Asten (1878) considered the perturbations by Mercury through Saturn, as well as the effects of a resisting medium, for the period of 1874–1878 and calculated this apparition's perihelion date as 1878 July 26.62. J. Tebbutt (Windsor, New South Wales, Australia) recovered the comet with an 11-cm refractor on 1878 August 3.35, with the help of the ephemeris provided by Asten. He said, "It presented the appearance of a small round nebula gradually condensed towards the centre, and was within a few minutes of arc of the position assigned to it in the Ephemeris." It was about 2′ in diameter. Tebbutt only had time to measure a rough position of $\alpha = 10^h\ 00.5^m$, $\delta = +8.5°$, as the comet was near the horizon and setting. The following evening was cloudy, but E. W. L. Tempel (Arcetri, Italy) confirmed this was the expected comet on August 5.35. He said Asten's ephemeris required a correction of $+17^s$ in α and $-3′$ in δ. B. A. Gould (Argentine National Observatory, Cordoba, Argentina) independently recovered the comet on August 3.97. The comet was then at an altitude of $3°$ or $4°$ and in evening twilight. Gould said the comet was very conspicuous in his telescope and resembled "a small and dense white cloud, nearly circular, but of indefined outline, and without indication of any tail." The comet had passed $2°$ from the sun on July 20.

The comet had already passed perihelion and its closest approach to Earth during the second half of August was not particularly close. J. M. Thome (Argentine National Observatory, Cordoba, Argentina) said the comet was circular and about magnitude 8 on August 10. Tebbutt said the comet appeared "excessively faint" on August 11, 12, and 13, because of increased moonlight. Thome said the coma diameter was about 1′ on the 17th. Tebbutt provided a precise position for the final time on August 17, but continued to observe the comet until the 28th, when it was excessively faint in the 11-cm refractor. At the end of this comet's apparition, Tebbutt remarked that, overall, "I found it very difficult to observe the comet owing to the absence of a nucleus or other well defined condensation."

This comet was last detected on September 7.01, when Thome gave the position as $\alpha = 14^h 12.8^m$, $\delta = -23° 55'$. He said it was very faint and difficult to see when placed near the measuring threads of the telescope.

Refinements of this comet's orbit using positions from multiple apparitions, as well as planetary perturbations, have been published by numerous astronomers. Of most significance were the studies of J. O. Backlund (1884, 1886) and B. G. Marsden and Z. Sekanina (1974). These have typically revealed a perihelion date of July 26.67 and a period of 3.29 years. Marsden and Sekanina determined the nongravitational terms as $A_1 = +0.07$ and $A_2 = -0.02657$. The orbit of Marsden and Sekanina is given below.

T	ω	Ω (2000.0)	i	q	e
1878 Jul. 26.6677 (TT)	183.6437	336.3796	13.1253	0.333513	0.849144

ABSOLUTE MAGNITUDE: $H_{10} = 10.1$ (V1964)

FULL MOON: Jul. 14, Aug. 13, Sep. 11

SOURCES: F. E. von Asten, *AN*, **92** (1878 May 16), pp. 193–200; J. Tebbutt, *AN*, **93** (1878 Oct. 9), p. 223; J. Tebbutt, *MNRAS*, **39** (1878 Nov.), pp. 75–6; B. A. Gould and J. M. Thome, *AN*, **93** (1878 Nov. 12), p. 329; J. Tebbutt and B. A. Gould, *The Observatory*, **2** (1878 Dec.), p. 276; K. C. Bruhns, *VJS*, **14** (1879), pp. 96–7; J. Tebbutt, *AN*, **94** (1879 Jan. 22), pp. 71–4; B. A. Gould, *MNRAS*, **39** (1879 Feb.), p. 265; J. Tebbutt, *MNRAS*, **39** (1879 Mar.), pp. 321–2; J. O. Backlund, *MASP*, **32** No. 3, 7th Series (1884), p. 36; J. O. Backlund, *MASP*, **34** No. 8, 7th Series (1886), p. 38; V1964, p. 61; B. G. Marsden and Z. Sekanina, *AJ*, **79** (1974 Mar.), pp. 415–16.

10P/1878 O1
(Tempel 2)

1878 III = 1878b

Recovered: 1878 July 19.86 ($\Delta = 0.74$ AU, $r = 1.45$ AU, Elong. $= 110°$)
Last seen: 1878 December 21.8 ($\Delta = 1.77$ AU, $r = 1.74$ AU, Elong. $= 72°$)
Closest to the Earth: 1878 June 29 (0.7232 AU)
Calculated path: LIB (Rec), SCO (Aug. 13), OPH (Aug. 24), SGR (Sep. 17), MIC (Oct. 28), PsA (Nov. 14), AQR (Nov. 25)

L. Schulhof (1878) predicted this comet would next return to perihelion on 1878 September 2.00. E. W. L. Tempel (Arcetri, Italy) had been searching for this comet since 1878 May 3, but finally managed to recover it on July 19.86. He gave the position as $\alpha = 15^h 16.7^m$, $\delta = -4° 15'$ and described it as very large, with a bright condensation. An independent recovery was made by F. A. T. Winnecke (Strasbourg, France) on July 20.89. He described the comet as rather faint, 2–3' across, with a central condensation.

Although the comet was already three weeks past its closest approach to Earth, it was still heading toward perihelion. Prosper Henry (Paris, France) described the comet as very faint, round, and about 3' across on July 23. He added that there was no apparent nucleus. J. F. J. Schmidt (Athens, Greece) said the comet was seen with difficulty in the large finder of his 17-cm Reinfelder refractor on July 26, 27, and 29, and he described it as extremely faint, 3' across, and without a nucleus. W. E. Plummer (Oxford University, England) observed the comet with the 31.1-cm Grubb refractor on July 27, 28, 31, and August 1. He described it as resembling "a faint round nebula" with a diameter of about 1'. There was a "very slight" condensation. Tempel saw the comet again on July 30 with a 27-cm refractor and noted it was "pretty large, like a nebula, of 3' to 4' diameter. The south preceding side (on its apparent path) is brighter and shows several small nuclei, the following side is diffused, so that it has the oval form and appearance of Encke's comet, though on the whole much smaller." Tempel saw the comet again on August 16, 17, and 18 and said the appearance was little changed from July 30. On the 18th he noted the comet "could be very well seen" with the 6.5-cm cometseeker of his observatory. Schmidt said a slight condensation was seen on the eastern side of the coma on August 20, 21, and 29.

The comet passed perihelion early in September and was moving away from both the sun and Earth thereafter. Schmidt saw the comet on September 16 and 27. He said it was easily visible, about 2' across and well condensed in the middle. The comet passed near the star cluster M55 on October 15 and Schmidt said the comet appeared fainter. The comet attained its most southerly declination of $-32°$ on October 16. Schmidt said the comet was "still reasonably bright and almost 2' across" on October 26, but a precise position could not be measured because a 10th-magnitude star was within the coma.

The comet continued to be viewed at various observatories during November and into December, but no physical descriptions were provided as the comet grew fainter.

The position of the comet was measured for the final time on December 18.76, when Tempel gave it as $\alpha = 23^h 03.2^m$, $\delta = -19° 16'$. Tempel also saw the comet for the final time on December 21.8. He described it as extremely faint and large. Although a suitable comparison star was near the comet, clouds moved in before the measurements were complete. Cloudy and foggy conditions remained for several days thereafter. Tempel used Schulhof's ephemeris to search for the comet on January 12 and 14, but saw nothing of a nebulous nature.

The comet remained a rather faint object throughout its apparition, causing problems for some astronomers. J. Tebbutt (Windsor, New South Wales, Australia) wrote to the *Astronomische Nachrichten* on August 13 and said, "I may state that I have on several occasions sought for Tempel's comet II 1873, in the positions indicated in *Nature* of March 21st last, but in vain." L. Boss (Dudley Observatory, New York, USA) noted that his observatory was unsuccessful in finding Tempel 2, but did observe 25 nebulae, of which four "could not be found in any of the Catalogues in the Observatory library." They were using the 33-cm refractor.

Using positions obtained early in this apparition, Schulhof corrected his prediction and determined the perihelion date as September 7.74. Minor refinements of this comet's orbit using positions from multiple apparitions, as well as planetary perturbations, have been published by numerous astronomers. The earliest came from Schulhof (1883, 1894), while the most recent were by B. G. Marsden and Z. Sekanina (1971) and Sekanina (1985). These have typically revealed a perihelion date of September 7.77 and a period of 5.20 years. Sekanina determined the nongravitational terms as $A_1 = +0.08$ and $A_2 = +0.0021$ and his orbit is given below.

T	ω	Ω (2000.0)	i	q	e
1878 Sep. 7.7721 (TT)	185.1903	122.6562	12.7594	1.339700	0.553687

ABSOLUTE MAGNITUDE: $H_{10} = 9.4$ (V1964)

FULL MOON: Jul. 14, Aug. 13, Sep. 11, Oct. 11, Nov. 10, Dec. 9, Jan. 8

SOURCES: L. Schulhof, *CR*, **86** (1878 May 6), pp. 1124–5; L. Schulhof, *The Observatory*, **2** (1878 Jun.), p. 72; L. Schulhof, *AN*, **92** (1878 Jul. 11), p. 351; E. W. L. Tempel and F. A. T. Winnecke, *CR*, **87** (1878 Jul. 22), pp. 156–7; E. W. L. Tempel, *AN*, **93** (1878 Jul. 24), p. 15; Pr. Henry, *CR*, **87** (1878 Jul. 29), p. 201; E. W. L. Tempel, *The Observatory*, **2** (1878 Aug.), p. 131; F. A. T. Winnecke, *AN*, **93** (1878 Aug. 9), p. 31; L. Schulhof, *AN*, **93** (1878 Aug. 29), pp. 71–4; J. F. J. Schmidt and E. W. L. Tempel, *AN*, **93** (1878 Sep. 4), pp. 107, 111; J. F. J. Schmidt, *AN*, **93** (1878 Sep. 22), pp. 165–8; E. W. L. Tempel, *The Observatory*, **2** (1878 Oct.), pp. 191–2; J. Tebbutt, *AN*, **93** (1878 Oct. 9), p. 223; C. Pritchard, W. E. Plummer, and J. Tebbutt, *MNRAS*, **39** (1878 Nov.), pp. 75–6; J. F. J. Schmidt, *AN*, **93** (1878 Nov. 1), pp. 301–4; E. W. L. Tempel, *AN*, **93** (1878 Nov. 12), pp. 333–6; J. F. J. Schmidt, *AN*, **93** (1878 Nov. 29), p. 367; K. C. Bruhns, *VJS*, **14** (1879), p. 97; E. W. L. Tempel, *AN*, **94** (1879 Jan. 26), pp. 91–4; E. W. L. Tempel, *AN*, **94** (1879 Feb. 6), p. 127; E. W. L. Tempel, *AN*, **94** (1879 Feb. 8), pp. 139–41; L. Boss, *The Observatory*, **2** (1879 Mar.), p. 392; L. Schulhof, *Annuaire pour l'An 1884. Publié par le Bureau des Longitudes. Avec des Notices Scientifiques* (1883), p. 229; L. Schulhof, *BA*, **11** (1894 Jun.), pp. 254–6; V1964, p. 61; B. G. Marsden and Z. Sekanina, *AJ*, **76** (1971 Dec.), pp. 1137–8; Z. Sekanina, *QJRAS*, **26** (1985), p. 114; B. G. Marsden and Z. Sekanina, *CCO*, 12th ed. (1972), pp. 50–1.

5D/Brorsen

1879 I = 1879a

Recovered: 1879 January 14.75 ($\Delta = 1.92$ AU, $r = 1.43$ AU, Elong. $= 45°$)

Last seen: 1879 May 23.93 ($\Delta = 0.73$ AU, $r = 1.14$ AU, Elong. $= 80°$)

Closest to the Earth: 1879 May 10 (0.6881 AU)

Calculated path: SCL (Rec), AQR (Jan. 24), CET (Feb. 3), PSC (Mar. 13), CET (Mar. 20), ARI (Mar. 22), PER (Apr. 8), AUR–CAM (Apr. 25), AUR (Apr. 27), CAM (Apr. 28), LYN (May 3), CAM (May 4), UMa (May 10)

L. R. Schulze (1878) revised the orbit calculated for this comet's 1868 and 1873 returns, and then advanced it forward to the 1879 apparition. The resulting prediction for the perihelion date was 1879 March 30.58. E. W. L. Tempel (Arcetri, Italy) recovered this comet near the horizon on 1879 January

14.75, northeast of the 10th-magnitude galaxy NGC 7507. He indicated a position of $\alpha = 23^h\ 06^m$, $\delta = -29°$. Tempel said the comet was smaller, but brighter than this object.

The comet's distances from the sun and Earth decreased after its recovery, but physical descriptions were nonexistent until the second half of February. J. Tebbutt (Windsor, New South Wales, Australia) found "a faint nebulous object" with his 11-cm refractor on February 22. Although he was uncertain if this was the comet, he noted it was close to the position predicted by Schulze. He said, "It presented the appearance of an elliptic nebula, the major axis being nearly coincident with a parallel of declination. Unfortunately, the object was only visible in the fading twilight for about ten minutes, it having disappeared behind the walls of the new Observatory now in course of erection." Tebbutt succeeded in seeing the comet again on the evening of the 23rd. Although conditions were very similar to the previous night, he did succeed in establishing a rough position with respect to the star 17 Ceti. H. C. Russell (Sydney, New South Wales, Australia) described the comet as faint on the 26th and noted a diameter of less than 1'. He said there was no nucleus and little condensation. That same night, Tebbutt said there was no condensation.

The comet continued to approach the sun and Earth throughout March, and actually passed perihelion on the final day of the month. Tempel said the comet was as bright as a star of magnitude 8 on March 9. Tebbutt saw no condensation in his 11-cm refractor on the 11th. The comet reached a minimum solar elongation of 32° on March 14 and G. Strasser (Kremsmünster, Austria) said it was bright, with an easy to observe nucleus. N. de Konkoly (O'Gyalla Observatory, Hungary) said the comet was bright with a strong condensation on the 18th. He added that it was visible in the finder of the 15-cm refractor. On the 19th, K. C. Bruhns and B. Peter (Leipzig, Germany) said the refractor of 12-foot focal length revealed a coma 1.5' across, with a central condensation of magnitude 8. They said the cometseeker revealed a coma diameter of 4'. J. F. J. Schmidt (Athens, Greece) saw the comet through his telescope's finder on the 23rd and estimated the magnitude of the central condensation as 7.5. Schmidt said the 17-cm Reinfelder refractor revealed a central condensation of magnitude 7.7 or 7.8 on the 24th. He indicated a coma diameter of 2.56' and noted a "very insignificant tail" 3' in length. On the same night, C. F. Pechüle (Copenhagen, Denmark) observed with the "great refractor" and described the comet as circular, brilliant, and very condensed at the center. He indicated that the coma was about 1.5' across. Pechüle added that a tail was only suspected toward PA 70°. On March 25, Peter said the comet was 4' across, while the nucleus was 5" across. J. J. Plummer (Orwell Park Observatory, England) simply described the comet as bright on the 29th.

The comet was moving away from the sun as April began, but was still closing in on Earth. Schmidt gave the magnitude as 7.7 in the finder on the 7th, 7 on the 8th, and 7.5 on the 9th. He noted a tail 0.5° long on the 8th, but only suspected it on the 9th. J. Franz (Königsberg, now Kaliningrad, Russia) also saw the comet on the 9th. He said a very faint tail extended toward PA 62°. Schmidt said the comet very condensed on the 10th, but did not see a stellar nucleus. He noted the coma was round with a diameter of about 4.5', while the tail extended 7'. On April 11, Plummer said the comet was bright, with some indication of a tail. R. Copeland and J. G. Lohse (Dun Echt Observatory, Scotland) described the tail as "ray-like" on the 12th, with a length of about 25'. They said the nucleus was highly condensed, while the coma was "obviously elongated at right angles to the direction of the tail, its greatest diameter being 5'." Schmidt gave the magnitude as 6.9 or 7.0 on the 14th and 7 on the 15th. He added that a trace of tail was visible in the finder. Copeland said the tail extended 10' toward PA 57.8° on the 16th and 13' toward PA 51.9° on the 18th. Schmidt also saw the comet on the 18th and said the refractor revealed a very strong condensation of magnitude 8, while measurements indicated a coma diameter of 4.82'. No tail was visible in the seeker. Schmidt said the finder revealed a magnitude of 7 on the 19th. Although the finder did not then reveal a tail, Schmidt did estimate it as 10' long on the 21st. Peter said the coma was 3' across on the 25th and he noted a diffuse, granular condensation on that date and on the 26th. Heidorn (Göttingen, Germany) said the comet exhibited a noticeable condensation on the 25th. Peter said the coma was about 4' across, with a distinct nucleus on the 27th.

The comet passed moderately close to Earth during the first half of May and moved away from both the sun and Earth thereafter. Schmidt and Pechüle both reported the comet was difficult to see in moonlight on the 1st. On the 8th, just before moonrise, Schmidt said the comet was insignificant in the finder, while Pechüle noted the coma and condensation were 4.5' across and 2.3', repectively. Schmidt said the comet was well seen in the finder on the 9th and 11th. Pechüle found the comet was round and very condensed in the middle on the 10th, while he noted the condensation had become less definite on the 10th and 12th. Peter said the comet was very faint on the 11th. He noted the coma was 4' across, while the condensation was magnitude 11 on both the 11th and 13th. Schmidt said the comet was well seen in the finder on the 14th, with a weakly condensed coma 6–7' across. Schmidt said the comet was still recognizable in the finder on the 15th, while the refractor revealed a coma diameter of 5.1' and a nucleus of magnitude 11.7–12. Heidorn said the comet exhibited a noticeable condensation on the 18th. On the 19th, Plummer said the comet was "much fainter, though still large, exceeding 2' in diameter. There is still some condensation towards the centre." The comet appeared very weak to Schmidt on the 20th and he noted little condensation and no nucleus. The coma was about 5.7' across. That same night, Peter described the comet as a large, faint nebulosity.

The comet was last detected on May 23.93, when Bruhns and Peter gave the position as $\alpha = 11^h 01.5^m$, $\delta = +57° 56'$. They said the comet was difficult to see, although there was still a slight central condensation.

The spectrum of this comet was observed during the period of March–May by F. A. Bredikhin, de Konkoly, W. H. M. Christie and E. W. Maunder (Royal Observatory, Greenwich, England), Copeland and Lohse, and C. A. Young (Princeton College Observatory, New Jersey, USA). In every case, the three usual cometary bands, which are now attributed to carbon molecule C_2, were detected.

This was the fifth and final observed apparition of this comet. P. H. Harzer (1880) used three positions obtained at Leipzig to correct the prediction made by Schulze. The resulting perihelion date was March 31.03, while the period was 5.46 years. Virtually identical results, derived from positions spanning multiple apparitions, were later obtained by E. A. Lamp (1892), P. M. Harzer (1894), B. G. Marsden and Z. Sekanina (1971), W. Landgraf (1986), and S. Nakano (1997). Nongravitational terms were determined by Marsden and Sekanina, as well as Landgraf. Their results were fairly comparable, with Landgraf's being $A_1 = +1.27$ and $A_2 = +0.1343$. Nakano's orbit is given below.

Several predictions for the comet's return were published. The 1884 July 24 issue of *Nature* contained an anonymous article, which may have been written by J. R. Hind, that noted the perturbations between 1879 and 1884 "will not have been very material." The author subsequently gave the perihelion date as 1884 September 15.0. C. Trépied (Alger, now al-Jazâ'ir, Algeria) wrote of his unsuccessful search. Nakano later showed that perihelion probably came on September 13.25. E. A. Lamp (1889) predicted the comet would next arrive at perihelion on 1890 February 24.60, but searches by Tebbutt, W. R. Brooks (Smith Observatory, Geneva, New York, USA), L. Swift (Warner Observatory, Rochester, New York, USA), J. Bauschinger (Munich, Germany), and E. E. Barnard (Lick Observatory, California, USA) revealed nothing.

T	ω	Ω (2000.0)	i	q	e
1879 Mar. 31.0340 (TT)	14.9466	102.9677	29.3818	0.589843	0.809795

ABSOLUTE MAGNITUDE: $H_{10} = 9.3$ (V1964)

FULL MOON: Jan. 8, Feb. 7, Mar. 8, Apr. 6, May 6, Jun. 4

SOURCES: J. F. W. Herschel, *PTRSL*, **154** (1864), p. 134; L. R. Schulze, *AN*, **93** (1878 Sep. 27), pp. 177–88; E. W. L. Tempel, *The Observatory*, **2** (1879 Feb.), p. 355; E. W. L. Tempel, *AN*, **94** (1879 Feb. 8), pp. 139–42; G. Strasser, S. Ferrari, and E. W. L. Tempel, *AN*, **94** (1879 Mar. 24), p. 287; G. Strasser, *AN*, **94** (1879 Mar. 29), p. 303; L. Schulhof, L. R. Schulze, G. Strasser, and E. W. L. Tempel, *The Observatory*, **2** (1879 Apr.), pp. 418, 425–6; E. W. L. Tempel and N. de Konkoly, *AN*, **94** (1879 Apr. 16), p. 335; W. H. M. Christie, E. W. Maunder, J. L. Lindsay, R. Copeland, J. G. Lohse, J. Tebbutt, and H. C. Russell, *MNRAS*, **39** (1879 May), pp. 428–31; L. R. Schulze, *The Observatory*, **3** (1879 May), p. 30; J. Tebbutt, *MNRAS*, **39** (1879 Jun.), p. 486; C. A. Young, *The Observatory*, **3** (1879 Jun.), pp. 56–7; J. F. J. Schmidt, *AN*, **95** (1879 Jul. 3), pp. 153–6; K. C. Bruhns and B. Peter, *AN*, **95** (1879 Sep. 10), pp. 303–8; Heidorn, *AN*, **96** (1879 Oct. 14), p. 13; K. C. Bruhns and P. H. Harzer, *VJS*, **15** (1880), pp. 3–5; J. J. Plummer, *AN*, **96** (1880 Mar. 9), pp. 329–34; C. F. Pechüle, *AN*, **98** (1880 Oct. 7), pp. 153–6; J. Franz, *AN*, **100** (1881 Sep. 3), pp. 247, 251; *Nature*, **30** (1884 Jul. 24), pp. 300–1; E. Weiss and K. Zelbr, *AN*, **109** (1884 Aug. 11), p. 255; C. Trépied, *AN*, **109** (1884 Sep. 17), p. 349; ARSI, part 1 (1886), p. 384; E. A. Lamp, *AN*, **123** (1889 Nov. 7), pp. 75–8; J. Tebbutt, *MNRAS*, **50** (1890 Feb.), p. 216; W. R. Brooks and L. Swift, *AN*, **124** (1890 Apr. 18), pp. 221–3; W. R. Brooks, *The Observatory*, **13** (1890 May), p. 188; J. Tebbutt and J. Bauschinger, *AN*, **124** (1890 May 21), p. 265; E. E. Barnard, *AN*, **125** (1890 Jul. 17), p. 43; E. A. Lamp, *Publication der Sternwarte in Kiel*, **7** (1892), pp. 31–59; P. H. Harzer, *Publication der Sternwarte in Kiel*, **9** no. 2 (1894), pp. 21–37; V1964, p. 61; B. G. Marsden and Z. Sekanina, *AJ*, **76** (1971 Dec.), pp. 1141–2; W. Landgraf, *QJRAS*, **27** (1986 Mar.), p. 116; B. G. Marsden and W. Landgraf, *CCO*, 12th ed. (1997), pp. 48–9; S. Nakano, *Nakano Note* No. 663 (1997 May 5).

C/1879 M1	*Discovered:* 1879 June 17.26 ($\Delta = 1.70$ AU, $r = 1.26$ AU, Elong. $= 47°$)
(Swift)	*Last seen:* 1879 August 24.19 ($\Delta = 2.13$ AU, $r = 2.12$ AU, Elong. $= 75°$)
	Closest to the Earth: 1879 February 22 (1.5959 AU), 1879 July 4 (1.6526 AU)
1879 II = 1879c	*Calculated path:* CAS (Disc), CEP (Jul. 3), UMi (Jul. 13), DRA (Aug. 8), BOO (Aug. 24)

L. Swift (Rochester, New York, USA) discovered this comet on 1879 June 17.26 at a position of $\alpha = 2^h 30^m$, $\delta = +58°$. He described it as bright, with a short tail, and noted a daily motion of a "little over one degree of north."

This comet did not come particularly close to Earth and had actually been closest nearly four months prior to its discovery. It was also about a month and a half past perihelion. It was, nevertheless, well observed during the remainder of June. F. A. T. Winnecke (Strasbourg, France) described the comet as bright on the 21st, with a coma 3′ across. There was also a trace of a faint tail. On the 23rd, B. Peter (Leipzig, Germany) said the nucleus equaled a star of magnitude 12. L. Boss (Dudley Observatory, New York, USA) said the comet was easily seen in the 33-cm refractor on the 25th and there was a slight suspicion of a tail. Heidorn (Göttingen, Germany) found the comet faint and round on the 28th. Heidorn noted that on the evenings following the 28th of June the comet was only visible for a short time between clouds. Nevertheless, he said there was a clear trace of a tail in the refractor of 6-foot focal length.

The comet made another approach to Earth early in July, although this approach was more distant that the one in February. Boss said moonlight made the comet a difficult object to see in the refractor on the 1st. Boss said the refractor revealed a coma about 1′ across on the 9th. On July 14, Peter said the comet was rather bright and condensed. The comet passed just under 2′ from the north celestial pole on July 15. Peter said there was a "planetary condensation" about 15″ across and about magnitude 10.5 on the 18th, while the coma was 1′ across. On July 19, Peter said a distinct, point-like nucleus was situated within a condensation 0.2′ across. The coma was 2.0′ across. Peter said the comet was 2′ across and condensed on the 20th, while a distinct, point-like nucleus was seen on the 22nd. On August 12, Peter said the comet appeared more diffuse. He also noted the central condensation was 0.7′ across and equaled a star of magnitude 12 or 13. Peter said the comet was 1.5′ across on August 17.

This comet was last detected on August 24.19, when O. C. Wendell (Harvard College Observatory, Cambridge, Massachusetts, USA) gave the position as $\alpha = 15^h\ 32.2^m$, $\delta = +53°\ 18′$. He was using the 38-cm refractor.

The first parabolic orbit was calculated by K. Zelbr, using positions from June 21, 24, and 28. The resulting perihelion date was 1879 April 27.49. A short time later, F. Küstner took positions spanning the period of June 21 to July 2 and determined the perihelion date as April 27.80. Additional orbits by Leitsmann, T. H. Safford, J. Franz, and A. Abetti (1880) eventually established the perihelion date as April 27.9. The last investigation of this comet's orbit came from V. F. K. Kremser (1883, 1884). Beginning with positions spanning the period of June 21 to August 24, he calculated the orbit below.

T	ω	Ω (2000.0)	i	q	e
1879 Apr. 27.9225 (UT)	3.7528	47.4545	107.0439	0.896547	1.0

ABSOLUTE MAGNITUDE: $H_{10} = 4.5$ (V1964)

FULL MOON: Jun. 4, Jul. 3, Aug. 2, Aug. 31

SOURCES: L. Swift, *AN*, **95** (1879 Jun. 24), p. 127; F. A. T. Winnecke, *AN*, **95** (1879 Jun. 30), p. 143; G. L. Tupman, *The Observatory*, **3** (1879 Jul.), p. 101; K. Zelbr and F. Küstner, *AN*, **95** (1879 Jul. 14), pp. 187–90; L. Boss, *AN*, **95** (1879 Aug. 13), p. 233; Leitsmann and T. H. Safford, *AN*, **95** (1879 Aug. 24), p. 269; J. Franz, *AN*, **95** (1879 Sep. 4), p. 299; Heidorn, *AN*, **96** (1879 Oct. 14), p. 13; O. C. Wendell, *AN*, **96** (1879 Oct. 24), pp. 21–4; K. C. Bruhns and B. Peter, *AN*, **96** (1879 Dec. 11), pp. 141–4; K. C. Bruhns and J. Franz, *VJS*, **15** (1880), pp. 6–7; A. Abetti, *AN*, **98** (1880 Sep. 2), pp. 49–54; J. Franz, *AN*, **100** (1881 Sep. 3), p. 247; V. F. K. Kremser, *Die Bahn des zweiten Comet von 1879*. Dissertation: Breslau (1883), 43pp; V. F. K. Kremser, *AN*, **108** (1884 Feb. 25), p. 101; V1964, p. 61.

9P/Tempel 1

1879 III = 1879b

Recovered: 1879 April 25.07 ($\Delta = 0.88$ AU, $r = 1.77$ AU, Elong. $= 140°$)
Last seen: 1879 July 8.9 ($\Delta = 0.94$ AU, $r = 1.85$ AU, Elong. $= 143°$)
Closest to the Earth: 1879 May 28 (0.7695 AU)
Calculated path: OPH (Rec), SCO (Jul. 4)

The recovery of this comet began with R. Gautier (1879). He took his orbit for the 1873 return, applied perturbations by Jupiter, and predicted the comet would return to perihelion on 1879 May 11.40. He supplied an ephemeris for the period of 1879 February 11 to June 14. He also determined ephemerides for assumed perihelion dates of May 7.40 and May 14.40. E. W. L. Tempel (Arcetri, Italy) received a sweeping ephemeris for this comet at the beginning of February, but was met with several periods of bad weather. Nevertheless, during the clear nights of February 18 and March 27, he was able to search and was confident the comet was not yet visible. Tempel found a small, faint object on April 25.07 at a position of $\alpha = 16^h 51.0^m$, $\delta = -13° 32'$. Although it was fairly near the ephemeris made using Gautier's assumed May 7.40 perihelion date, Tempel regarded it as a nebula of the III class (according to Herschel's descriptions). On April 25.97, he found the object had moved exactly in accordance with the ephemeris.

The comet passed perihelion early in May and closest to Earth near the end of the month. B. Peter (Leipzig, Germany) observed with a 12-foot focal length telescope on May 15 and described the comet as a delicate nebulosity 2' across, with a small condensation just south of the coma's center. He saw the comet again on the 19th and noticed a distinct, star-like nucleus of magnitude 12 or 13. On the 24th, 25th, and 26th, L. Cruls (Rio de Janeiro, Brazil) described the comet as excessively faint and round, with a slight central condensation. A. A. Common (Ealing, England) saw the comet with a 46-cm reflector on the 25th and said, "it appeared rather faint, with an apparent nucleus that seemed considerably out of centre" toward PA 40°. Clouds moved in and Common said there was a possibility the "nucleus" was a star.

The comet was moving away from both the sun and Earth during June. K. C. Bruhns (Leipzig, Germany) observed with a 12-foot focal length telescope on the 1st and said the coma was 1' across. The comet reached its greatest solar elongation of 177° on the 2nd. Common again saw the comet with the large reflector on the 10th and said, "it appeared fainter than before, round and diffused, about 30" in diameter." J. M. Thome (Cordoba, Argentina) found the comet with a 28-cm refractor on the 22nd. He described it as very faint and circular, with a diameter of 1.5–2'. He added that there was "a perceptible condensation toward the center" and the brightness "was comparable with that of a star of the 13th magnitude." Thome again saw the comet on the 27th and noted it "could not be seen with the faintest illumination of the threads until at least half an hour after it was found."

The comet was last detected on July 7.90 and July 8.9 by Tempel. He gave the position on the first night as $\alpha = 16^h 29.0^m$, $\delta = -25° 12'$, but was not able to measure a position on the second night. Tempel wrote, "The comet was in the proximity of many stars on both evenings so that it was not possible to determine which star belonged to the faint nebulosity of the comet."

Gautier took positions obtained by Tempel during the period spanning April 25 to May 13 and corrected the perihelion date to May 7.48, but, in 1885, he took observations from the 1873 and 1879 apparitions and published a perihelion date of May 7.62. Nevertheless, despite three consecutive well-observed returns, this was the final time the comet would be seen for 87 years. The comet passed 0.55 AU from Jupiter during October of 1881 and Gautier (1885) predicted it would next pass perihelion on 1885

September 26.23. The comet was not found. J. R. Hind (1891) predicted the comet would next pass perihelion on 1892 March 30.0 but careful searches in 1892 by R. Spitaler (Vienna, Austria) on May 24 and 25, and J. Tebbutt (Windsor, New South Wales, Australia) on June 19 failed to locate it. Gautier (1898) suggested the comet was not found in 1885 and 1892 because of the increased perihelion distance caused by the 1881 encounter with Jupiter. He subsequently provided a prediction that the comet would next arrive at perihelion on 1898 October 4.47. His search ephemeris also covered perihelion dates eight days before and after his optimum prediction. C. D. Perrine (Lick Observatory, California, USA) searched for this comet on several occasions during the summer of 1898. He noted that Gautier's ephemeris was not received until after the comet's opposition so that "the available time in which the comet's place was favorably situated for observing with the large telescope was very much restricted." Perrine said the 91-cm refractor was used for sweeping on June 11, 18, and July 9, 10, and 18, and he noted that the entire region indicated by Gautier's ephemeris was swept over twice.

The comet was finally recovered in 1972 as a result of new predictions. The most recent analysis of this comet's motion over multiple apparitions came from G. Schrutka (1974). Using 394 positions from the apparitions of 1867, 1873, 1879, and 1972, he applied perturbations by Venus to Pluto and determined the perihelion date as May 7.60 and the period as 5.98 years. This orbit is given below.

T	ω	Ω (2000.0)	i	q	e
1879 May 7.6034 (TT)	159.5783	80.3640	9.7675	1.771126	0.462546

ABSOLUTE MAGNITUDE: $H_{10} = 10.4$ (V1964), $H_{10} = 9.6$ (Kronk)

FULL MOON: Apr. 6, May 6, Jun. 4, Jul. 3, Aug. 2

Sources: R. Gautier, *AN*, **94** (1879 Feb. 15), pp. 157–60; E. W. L. Tempel, *AN*, **95** (1879 May 24), p. 45; R. Gautier, *AN*, **95** (1879 Jun. 7), pp. 77–80; L. Cruls, *AN*, **95** (1879 Jun. 30), p. 141; E. W. L. Tempel, R. Gautier, and A. A. Common, *The Observatory*, **3** (1879 Jul.), pp. 90–1; E. W. L. Tempel, *AN*, **95** (1879 Jul. 24), pp. 199–202; K. C. Bruhns and B. Peter, *AN*, **95** (1879 Sep. 17), p. 333; E. W. L. Tempel, *AN*, **96** (1879 Nov. 8), pp. 61–4; K. C. Bruhns and R. Gautier, *VJS*, **15** (1880), pp. 5–6; B. A. Gould, *AN*, **97** (1880 Jul. 8), p. 287; R. Gautier, *AN*, **111** (1885 Jan. 27), p. 75; R. Gautier, *AN*, **111** (1885 Mar. 26), pp. 241–6; W. E. Plummer and R. Gautier, *MNRAS*, **46** (1886 Feb.), p. 230; J. R. Hind, *The Observatory*, **14** (1891 Aug.), p. 290; J. Tebbutt, *MNRAS*, **53** (1892 Dec.), p. 70; R. Spitaler, *AN*, **131** (1893 Jan. 10), p. 390; J. Tebbutt, *AN*, **131** (1893 Jan. 12), p. 405; R. Gautier, *AN*, **146** (1898 May 9), pp. 177–84; C. D. Perrine, *AN*, **148** (1899 Feb. 10), p. 285; V1964, p. 61; G. Schrutka, *QJRAS*, **15** (1974), pp. 450–1, 459.

C/1879 Q1 *Discovered:* 1879 August 21.10 ($\Delta = 1.90$ AU, $r = 1.25$ AU, Elong. = 36°)

(Palisa) *Last seen:* 1879 October 22.74 ($\Delta = 1.77$ AU, $r = 1.04$ AU, Elong. = 30°)

Closest to the Earth: 1879 September 25 (1.5759 AU)

1879 V = 1879d *Calculated path:* UMa (Disc), CVn (Sep. 8), BOO (Sep. 26), SER (Oct. 14)

A. Palisa (Pula, Yugoslavia) discovered this comet on 1879 August 21.10. He described it as a rather large nebulosity, and could not find it in the nebula catalogs of Herschel or d'Arrest. He measured its distance from a nearby star and found a difference in α of 52^m. About 45 minutes later he redetermined the difference as 56^m, but considered the difference uncertain because of increasing morning twilight. He confirmed the motion on August 21.90 and established this was a comet. He gave the position as $\alpha = 10^h\ 02.4^m$, $\delta = +49°\ 07'$ and described it as "round, small, but bright."

The comet was moving toward both the sun and Earth. E. C. Pickering and O. C. Wendell (Harvard College Observatory, Cambridge, Massachusetts, USA) determined the nuclear magnitude as 9.1 on August 24, and K. C. Bruhns and B. Peter (Leipzig, Germany) said the comet was very bright and 2' across on the 26th.

The comet was well observed during September. J. F. J. Schmidt (Athens, Greece) measured the coma and estimated the nuclear magnitude on several occasions during September. He said the coma was 3.66' across on the 6th, 5.10' on the 9th, 5.38' on the 13th, and 5.12' on the 17th. He added that most of the nebulosity was extending toward the sun. Schmidt gave the magnitude of the "nucleus" as 8.0 on the 6th, 8.5 on the 9th, and 8.3 on the 13th and 17th. P. Tacchini (Rome, Italy) said the refractor revealed a small, faint nucleus on the 3rd. On September 6, 7, and 13, H. A. Kobold (Göttingen, Germany) observed with a refractor of 6-foot focal length and said the comet was bright and round. A distinct nucleus was seen on the 13th. Tacchini measured the coma as 2.0' on the 7th. Wendell gave the nuclear magnitude as 8.6 on the 9th and 7.3 on the 10th. E. Millosevich (Rome, Italy) noted the comet passed over a 10th-magnitude star on the 10th, but did not diminish the star's brightness. H. C. Vogel (Potsdam, Germany) said a 29-cm Schröder refractor revealed a short, tail-like extension. He also noted the comet looked similar to a star cluster. R. Copeland (Dun Echt Observatory, Scotland) said the comet was 3' across on the 23rd, but added that the 38-cm refractor (307×) did not reveal a nucleus.

The comet had passed closest to Earth in late September and passed perihelion during the first days of October. Tacchini measured the coma diameter as 1.8' on the 3rd and 1.7' on the 7th. On October 4, N. de Konkoly (O'Gyalla Observatory, Hungary) described the comet as "rather large, very diffuse, and quite faint." On both the 4th and 6th, Konkoly noted the comet had a granular appearance in the 15-cm refractor and exhibited a central condensation. On October 18, Schmidt indicated a coma diameter of 3.74' and noted a nucleus of magnitude 9.0. Copeland described the comet as bright and round on the 19th. Although he added that there was a gradual brightening toward the middle, no nucleus was visible. J. G. Lohse (Dun Echt Observatory, Scotland) saw the comet very low over the horizon on the 20th. The comet was last detected on October 22.74 by Tacchini, before being lost in evening twilight. He gave the position as $\alpha = 15^h\ 33.3^m$, $\delta = +3°\ 09'$.

The spectrum of this comet was also measured. Pickering and Wendell independently found the spectrum was continuous on August 24. The three bright carbon bands were detected by Vogel on September 13 and 15, Copeland and Lohse on September 23 and October 10, and Konkoly on October 4 and 6.

The first parabolic orbit was calculated by K. Zelbr using positions from August 21, 22, and 24. He determined the perihelion date as 1879 September 25.93. Zelbr revised his orbit as September began. Using positions from August 21, 24, and 26, he determined the perihelion date as October 4.97. Additional orbits by Copeland, Lohse, Zelbr, J. R. Hind, H. Leitzmann, and Palisa (1880) established that perihelion occurred early on October 5.1. A definitive orbit by K. Laves (1895) gave the perihelion date as October 5.13. Laves' orbit is given below.

T	ω	Ω (2000.0)	i	q	e
1879 Oct.5.1250 (UT)	115.4560	88.8686	77.1301	0.989608	1.0

ABSOLUTE MAGNITUDE: $H_{10} = 5.5$ (V1964)

FULL MOON: Aug. 2, Aug. 31, Sep. 30, Oct. 30

SOURCES: A. Palisa, *AN*, **95** (1879 Aug. 24), p. 269; K. Zelbr, *AN*, **95** (1879 Sep. 4), pp. 299–302; R. Copeland and J. G. Lohse, *AN*, **95** (1879 Sep. 10), pp. 315–18; K. Zelbr, *AN*, **95** (1879 Sep. 20), p. 349; H. Leitzmann, *AN*, **95** (1879 Sep. 29), p. 367; J. R. Hind, *Nature*, **20** (1879 Oct. 2), pp. 533–4; H. A. Kobold and K. Zelbr, *AN*, **96** (1879 Oct. 14), pp. 13–16; J. L. Lindsay, R. Copeland, and J. G. Lohse, *MNRAS*, **40** (1879 Nov.), pp. 23–5; N. de Konkoly, *AN*, **96** (1879 Nov. 7), pp. 39–42; J. F. J. Schmidt, *AN*, **96** (1879 Nov. 24), pp. 91–4; P. Tacchini and E. Millosevich, *MNRAS*, **40** (1879 Dec.), pp. 72–3; A. Palisa, *SAWW*, **81** Abt. II (1880), pp. 564–75; K. C. Bruhns and K. Zelbr, *VJS*, **15** (1880), pp. 8–9; H. C. Vogel, *AN*, **96** (1880 Jan. 1), p. 189; K. C. Bruhns and B. Peter, *AN*, **96** (1880 Feb. 5), p. 255; J. Franz, *AN*, **100** (1881 Sep. 3), p. 247; K. Laves, *AJ*, **16** (1895 Dec. 31), pp. 9–16; E. C. Pickering and O. C. Wendell, *Annals of the Astronomical Observatory of Harvard College*, **33** (1900), pp. 150, 153; V1964, p. 61.

C/1879 Q2	*Discovered:* 1879 August 24.92 ($\Delta = 1.13$ AU, $r = 0.99$ AU, Elong. $= 55°$)
(Hartwig)	*Last seen:* 1879 September 18.81 ($\Delta = 1.54$ AU, $r = 1.05$ AU, Elong. $= 43°$)
	Closest to the Earth: 1879 August 18 (1.1049 AU)
1879 IV = 1879e	*Calculated path:* UMa (Disc), CVn (Aug. 30), BOO (Sep. 15)

E. Hartwig (Strasbourg, France) discovered this comet with a 16.1-cm comet-seeker on 1879 August 24.92 at a position of $\alpha = 12^h 19.3^m$, $\delta = +61° 02'$. He described it as rather faint and 1' across, with movement toward the southeast.

The comet had already passed closest to Earth and passed perihelion before the end of August. Observations were not plentiful. K. C. Bruhns and B. Peter (Leipzig, Germany) said the comet was faint and diffuse on August 26. On August 26 and 29, Hartwig said the comet was 2' across, with a central condensation that appeared granular. On September 9, E. Millosevich (Rome, Italy) observed with a refractor and described the comet as weak, with an oval shape. Bruhns and Peter said the comet was 4' across on the 13th. Millosevich said the comet was a difficult object to see on the 14th.

This comet was last observed on September 18.81, when J. Franz (Königsberg, now Kaliningrad, Russia) gave the position as $\alpha = 14^h\ 06.5^m$, $\delta = +27°\ 11'$. He described it as faint and small. J. F. J. Schmidt (Athens, Greece) was not able to find this comet. He said he failed to find it after the first messages and a new ephemeris did not arrive until mid-October. October 18 was his first clear night and his search for the comet was without success.

The first parabolic orbit was calculated by Hartwig using positions from August 25, 26, and 29. He calculated a perihelion date of 1879 August 26.93. Hartwig revised his orbit a couple of weeks later, using positions spanning 16 days. The result was a perihelion date of August 29.74. Millosevich (1884, 1888) extensively worked on the orbit for this comet. His final orbit had a perihelion date of August 29.78 and is given below.

T	ω	Ω (2000.0)	i	q	e
1879 Aug. 29.7787 (UT)	84.2634	34.1187	107.7629	0.991480	1.0

ABSOLUTE MAGNITUDE: $H_{10} = 10.5$ (V1964)

FULL MOON: Aug. 2, Aug. 31, Sep. 30

SOURCES: E. Hartwig and B. Peter, *AN*, **95** (1879 Sep. 2), p. 285; E. Hartwig, *AN*, **95** (1879 Sep. 10), p. 315; E. Hartwig, *AN*, **95** (1879 Sep. 20), p. 335; J. F. J. Schmidt, *AN*, **96** (1879 Nov. 24), p. 93; P. Tacchini and E. Millosevich, *MNRAS*, **40** (1879 Dec.), p. 74; K. C. Bruhns and E. Hartwig, *VJS*, **15** (1880), pp. 7–8; B. Peter, *AN*, **96** (1880 Feb. 5), p. 255; J. Franz, *AN*, **100** (1881 Sep. 3), p. 249; J. Franz, *AN*, **105** (1883 Apr. 25), pp. 181–4; E. Millosevich, *Memorie della Società degli Spettroscopisti Italiani*, **13** (1884), p. 27; E. Millosevich, *Memorie della Società degli Spettroscopisti Italiani*, **17** (1888), p. 55; V1964, p. 61.

C/1880 C1 (Great Southern Comet)

1880 I = 1880a

Discovered: 1880 February 1.4 ($\Delta = 0.77$ AU, $r = 0.28$ AU, Elong. $= 12°$)

Last seen: 1880 February 20.05 ($\Delta = 0.80$ AU, $r = 0.88$ AU, Elong. $= 58°$)

Closest to the Earth: 1880 January 3 (0.6100 AU), 1880 February 9 (0.6754 AU)

Calculated path: PsA (Disc), SCL (Feb. 7), FOR (Feb. 17)

The actual discoverer of this comet will probably never be known because of second-hand or worse reports. But these same reports do provide enough details to guarantee their authenticity and indicate the comet was discovered in the evening sky on 1880 February 1.

H. C. Russell (Sydney, New South Wales, Australia) said the earliest notice he received of the comet was from a gentleman living "in the northern part of this colony." He said that on February 1 the person was shooting quail after sunset, when he looked to the southwest and "was surprised by a bright streak of light, stretching from the horizon towards the South Pole. It was a striking object from its brightness, and about three degrees wide." Russell said the gentleman saw it again on February 2. Although only a rough location is given, it would seem likely that the date was February 1.4. Charles Todd (Adelaide Observatory, New South Wales, Australia) said he first saw the comet on the evening of February 2, but notes in his report that the comet "appears to have been seen on Sunday evening, February 1." If this observation was made in the general region of the observatory, then the likely date was February 1.47. There is always the chance that both observations refer to the same person.

First-hand accounts of the sighting of this comet on February 1 came from crew members of HMS *Garnet*. Two of them wrote letters describing observations on February 1.99, which were published in the March 1880 issue of the *Monthly Notices*. Lieutenant B. E. W. Gwynne said "after a long twilight, the comet's tail was observed in a S.S.W. direction, having an altitude of about 20° above the horizon." He added that the tail brightness was "about that of the Milky Way on a clear frosty night. . . ." The Reverend S. S. O. Morris said the nucleus of the comet was below the horizon and that "the altitude of the extremity of the tail visible to the naked eye was about 30°. The tail lay across the constellations of Grus and Piscis Australis in direction parallel to the line of direction of the Milky Way." Although no position could be determined, Morris estimated that $\alpha = 21^h$. Morris continued by noting, "There was a slight arch in the tail concave towards the south pole, convex towards Fomalhaut, which lay to the west of the comet."

During the days that followed, independent discoveries were made around the world. These include the following: William Bone (Castlemaine, Victoria, Australia) on February 2.41, B. A. Gould (Argentine National Observatory, Cordoba, Argentina) on February 3.03, and John Tebbutt (Windsor, New South Wales, Australia) on February 3.44.

The comet had just passed perihelion on January 28 and was still close to the sun when it was discovered. Because of this the head and nucleus remained below the horizon for the first few days, so only descriptions of the tail are available. Having seen the comet on the 1st, Morris noted the tail was reaching a higher altitude on the 2nd. On the same evening, R. L. J. Ellery (Melbourne, Australia) said the tail nearly reached β Gruis, which would indicate a length of about 20°. Todd wrote that the tail "appeared as a narrow whitish auroral streak . . . the upper portion curving somewhat sharply to the south." L. A. Eddie (Grahamstown, South Africa) described the tail as 1° wide and reaching an altitude of 20°. He added, "It had a decided curvature concave to the south, and shone with a light of a pale straw colour about equal in brightness to that of the Milky Way, and far more brilliant than on any subsequent evening." Bone said he "noticed a peculiar band of light near the position of the Sun, which had recently set.

At first I thought it was a shoot of light from one of the foundry furnaces, but its stability soon convinced me it could only be the tail of a Comet. As the light of day waned, this was quite apparent. The nucleus was below the horizon, and must have been very close to the Sun. The tail extended obliquely upwards for about 35°...."

The comet was probably observed most on February 3 as more southern observatories got word of it. Gould described it as a streak of light and noted, "It showed no marked difference in brightness through its whole length, but seemed to taper to a point at each end, being about a degree broad in the middle." Tebbutt said, "it extended about 25° above the horizon, the head being either behind a bench of clouds on the horizon or really below the horizon." Morris wrote, "We watched carefully for the first appearance of the comet on the evening of the 3rd, but, although the tail, now of immense dimensions, soon became visible, the nucleus had not yet become visible." Todd first noticed the tail in strong twilight and said it extended to an altitude of 27° once the skies darkened. Russell fought clouds to see the comet, but finally estimated the tail was visible for at least 30° and was never wider than about 1.3°.

The apparent first sighting of the comet's head and nucleus came on February 4. Bone said he distinctly saw the nucleus near the horizon for about 10 minutes before a building blocked his view. He noted, "I could clearly see a 'vacuole' and the reflected envelopes, similar to the drawings of Donati's Comet." The tail was reported as 40° long by Eddie, Gould, and E. Liais (Rio de Janeiro, Brazil), with Liais indicating that it extended toward PA 342°. Although Eddie said the tail had lost its curvature, David Gill (Cape of Good Hope, South Africa) described it as narrow and curved. Gill also indicated a length of nearly 50°.

More sightings of the comet's head were reported on February 5. Eddie saw a faint nucleus "equal in size to [M57] and resembling 47 Toucani." M57 is oval, with dimensions of 1.1' by 2.5', while 47 Toucani is a globular cluster composed of thousands of stars, which are more concentrated toward the center. Eddie added that the nucleus had a pale yellow color. Gould reported he saw "what must have been the head of the comet. Through the haze and twilight it appeared like a coarse, ill-defined, mass of dull light 2' or 3' in diameter, and without any visible nucleus." Gould said the tail was more than 40° long, Eddie said 50°, and Gill indicated nearly 60°.

An important turning point for this comet came on February 6. Until that date there were contradictory reports as to whether the comet was becoming brighter or fainter, but the overall consensus on the 6th was that the comet was fading. Gill's fine series of drawings of the tail from the Cape of Good Hope indicated the tail was longest on this day, with a length of about 75°. Both Todd and Gill still noted some curvature to the tail, where other astronomers reported straight tails. This was probably because Todd and Gill were still seeing longer tails than the others. This was the last night Todd would note any curvature.

Todd simply referred to the tail as a "thin nebulous streak" on February 7, while Gill described it as narrow and not as curved as on previous evenings. The tail was reported as 38° long by Eddie and about 60° by Gill. Interestingly, where Eddie said he watched the nucleus approach γ Sculptoris with his 7.6-cm refractor, Gill remarked, "I certainly saw no nucleus, but others up country say it was seen."

The nucleus was finally becoming more apparent by February 8 and 9, but the fading comet was beginning to bow to sky conditions in some locations. Gill finally detected the nucleus on the 8th and described it as nebulous in an opera-glass. Although he again saw it with the opera-glass on the 9th, Gill remarked that the faintness would force him to stay with the observatory telescope to acquire further observations. Eddie said it took longer to spot the comet on the 8th than on previous evenings and Liais indicated the tail extended about 50° toward PA 340°. Gwynn had followed the comet with a "spyglass" since the 1st, but noted on the 8th that "it was difficult to trace the direction of the tail, although the night was favourably dark."

Descriptive observations became sparse for this comet after February 9 as the comet faded and became smaller. Two naked-eye observations on the 12th best illustrate the change in appearance, as Eddie noted it appeared as a "slight hazy light" and Ellery said the tail had almost disappeared. Russell noted on the 13th that the "comet had faded so much that it was difficult to see any part of it with the naked eye." But a telescope revealed a very faint nucleus measuring about 1' wide. Russell added that the telescope revealed the tail was still about 15° long and measured about 30" wide at the coma and 15' wide at its widest part. Tebbutt failed to see a nucleus or condensation in his telescope on the 14th. The comet was described as "excessively faint" by Ellery and E. J. White (Melbourne, Australia) on the 16th, and both astronomers said it was "barely visible" on the 17th.

The comet was last detected on February 20.05, when Gould gave the position as $\alpha = 2^h 09.6^m$, $\delta = -28° 18'$. He was observing with the 28.5-cm refractor and he noted, "it could only be recognized as a scarcely perceptible whiteness." Gould looked for the comet on the evening of the 21st, but said "it could not be detected notwithstanding the ephemeris was good and the comet must have been in the field."

Although no northern hemisphere observations have been reported for this comet in the past, there is a probable observation from China reported in the Chinese text *Chhing Chhao Hsü Wên Hsien Thung Khao* (1910). This text is a continuation of the historical records of the Ch'ing Dynasty. It notes that Chinese astronomers observed a large "broom-star" comet sometime during the period of January 12 to February 9. It was "seen in the evening at the SW." The capital of the Ch'ing Dynasty was Peking (now Beijing). Although the comet would probably not have been seen in that city, southern cities of the Dynasty, such as Shanghai could have seen the comet sometime during the period of February 7–9.

Orbits for this comet were not quick to come, because of the inability to measure the head or nucleus. Although the first reliable position came on February 7, and the last came on the 20th, most of the precise positions were obtained during the period of February 9–17. Among the first parabolic orbits published was one by Gould. He used positions obtained on February 7, 10, and 13, and determined the perihelion date as 1880 January 28.13. He also noted that the perihelion distance of 0.00524 AU, as well as the other orbital parameters, bore a strong resemblance to that of comet C/1843 D1, and he suspected the two comets were identical. Gould continued to revise the orbit as more positions became available. One of his last parabolic orbits used positions from February 7, 13, and 19, and revealed a perihelion date of January 28.12. Gould continued to suggest this comet was identical to C/1843 D1, with a period of between three and four decades. Other astronomers providing similar parabolic orbits included J. R. Hind, R. Copeland, H. T. Carpenter, Tebbutt, H. Oppenheim, and H. C. F. Kreutz (1886).

Interestingly, Gould was not the only astronomer who suspected the brilliant comets of 1843 and 1880 were identical. Although E. Weiss determined a parabolic orbit, he suggested a period of 36.9 years. He looked through various sources and suggested several objects as possible previous returns, including comets seen in −371 and 1106. M. W. Meyer (1880, 1882) similarly gave orbits with periods of 36.91 and 36.84 years. Overall, there was little debate on the matter, as the orbits were just so similar. This would all change with the appearance of the Great September Comet of 1882 (see C/1882 R1).

T	ω	Ω (2000.0)	i	q	e
1880 Jan. 28.1187 (UT)	86.3112	7.8552	144.6761	0.005487	1.0

ABSOLUTE MAGNITUDE: $H_{10} = 7.1$–8.9 (V1964)

FULL MOON: Jan. 27, Feb. 26

SOURCES: E. Weiss, *SAWW*, **82** Abt. II (1880), pp. 95–114; B. A. Gould, *AN*, **96** (1880 Feb. 12), p. 271; B. Gwynne, S. S. O. Morris, R. L. J. Ellery, C. Todd, L. A. Eddie, and D. Gill, *MNRAS*, **40** (1880 Mar.), pp. 295–301; B. A. Gould, *AN*, **96** (1880 Mar. 23), pp. 363–6; E. Liais, R. Copeland, and H. T. Carpenter, *AN*, **96** (1880 Mar. 27), pp. 379–82; R. L. J. Ellery, E. J. White, and H. C. Russell, *MNRAS*, **40** (1880 Apr.), pp. 377–9; B. A. Gould, *AN*, **97** (1880 Apr. 8), pp. 43–6; B. A. Gould and E. Weiss, *AN*, **97** (1880 Apr. 17), pp. 57–64; J. R. Hind and B. A. Gould, *Nature*, **21** (1880 Apr. 22), pp. 597–8; B. A. Gould, *AJS* (Series 3), **19** (1880 May), pp. 396–402; J. Tebbutt and H. Oppenheim, *AN*, **97** (1880 May 1), p. 75; M. W. Meyer, *AN*, **97** (1880 Jun. 7), p. 185; B. A. Gould, *Nature*, **22** (1880 Jul. 8), p. 231; M. W. Meyer, *AN*, **97** (1880 Jul. 30), pp. 343–6; J. Tebbutt, *AN*, **98** (1880 Oct. 7), p. 155; D. Gill and W. H. Finlay, *MNRAS*, **40** (1880, Supplementary Notice), pp. 623–7; W. Bone, *MNRAS*, **41** (1880 Dec.), pp. 85–8; F. Deichmüller, *VJS*, **17** (1882), pp. 140–3; B. A. Gould, *AN*, **101** (1882 Feb. 20), pp. 253–6; M. W. Meyer, *AN*, **102** (1882 May 23), pp. 83–90; H. C. F. Kreutz, *AN*, **114** (1886 Mar. 18), p. 73; V1964, p. 61; HA1970, p. 87.

C/1880 G1
(Schaeberle)

1880 II = 1880b

Discovered: 1880 April 7.19 ($\Delta = 1.98$ AU, $r = 2.11$ AU, Elong. $= 83°$)
Last seen: 1880 September 12.09 ($\Delta = 2.20$ AU, $r = 2.03$ AU, Elong. $= 67°$)
Closest to the Earth: 1880 November 5 (1.5950 AU)
Calculated path: CAM (Disc), LYN (May 10), AUR (May 27), LYN (Jun. 3), AUR (Jun. 6), GEM (Jul. 23)

J. M. Schaeberle (Ann Arbor, Michigan, USA) discovered this comet on 1880 April 7.19 at a position of $\alpha = 7^h 20.0^m$, $\delta = +84° 25'$. He said the sky had been cloudy earlier in the evening, but after it cleared he found the interloper while sweeping with a 20-cm reflector. Schaeberle described the comet as 3' across, with a tail. The comet had passed about 2.5° from the north celestial pole on April 1.

This comet never came particularly close to Earth, with its nearest approach coming nearly two months after the final observation. It was, however discovered almost three months prior to perihelion and the comet was well observed. On April 9, C. A. Young (Princeton, New Jersey, USA) observed with the 24-cm refractor and said the comet had a bright stellar nucleus of magnitude 12 and a brush-shaped tail 3' or 4' long. On April 8, 12, and 16, Paul and Prosper Henry (Paris, France) described the comet as small and rather condensed, with a nucleus and a faint tail 3–4' long. Young said the refractor revealed the comet was much brighter and larger on the 13th than on the 9th. B. Peter (Leipzig, Germany) observed with a 22-cm refractor. He said the comet exhibited a tail 3' long on the 15th, with a sharp, stellar nucleus of magnitude 10. He noted a bright nucleus and tail on the 16th and a stellar nucleus of magnitude 10 on the 17th, with a trace of tail. On the 17th and 19th, G. Bigourdan (Paris, France) observed a distinct nucleus, which looked like a star of magnitude 11. The tail was about 3' long. On April 19, P. Tacchini (Rome, Italy) saw the comet with a refractor and said it was difficult because of the moon, but showed a coma 40" across, with the suspicion of a nucleus. That same night, Peter said the nucleus was like a star of magnitude 10. Moonlight prevented the tail from being seen and revealed a coma diameter of only 0.5'. C. F. Pechüle (Copenhagen, Denmark) observed with a 27-cm refractor and said the comet was rather bright on the 20th, with a "quasi-nucleus" and a tail extending about 1.5' toward PA 60°. J. Franz (Königsberg, now Kaliningrad, Russia) said the comet was very faint in moonlight on the 20th and 21st, and suggested it was fainter than stars of magnitude 11. Young said the comet was faint because of moonlight on the 21st and 22nd. Peter described the comet as round and faint on the 22nd, but noted that no nucleus or tail was seen in moonlight. On April 23, J. G. Lohse (Dun Echt Observatory, Scotland) observed in bright moonlight with the 38-cm refractor. He said the tail was seen at magnifications of 229× and 442×, and was "a little broader than the head" and extended toward PA 58.7°. He also noted the nucleus was still not visible at 442×. Lohse said the comet was faint, with an unrecognizable nucleus on the 26th. He noted that there was a broad tail. That same night, Bigourdan said the nucleus was magnitude 12. On the 27th, Franz noted a small tail extending toward PA 49.1°, while Pechüle said the tail extended 4' toward PA 50°. Tacchini said the tail was about 4.8' long on the 28th, while the nucleus seemed fainter than a star of magnitude 10–11 that was seen through the tail. Tacchini tried to observe the spectrum on this date, but detected nothing. On the 28th and 29th, Peter said the comet displayed a nucleus and a fan-shaped tail. The tail was 2' long and 0.5' wide. On April 29, Young said the comet was "faint and nebulous," while the tail was not seen very distinctly. H. A. Kobold and Heidorn (Göttingen, Germany) said the comet was bright, and exhibited a tail about 2' long. There was also a distinct nucleus of magnitude 11. Franz said the tail was 1–2' long, with a slight widening at the end. On April 30, Kobold and Heidorn said the comet appeared rather faint, with a distinct tail. Peter said the tail was 4' long.

The comet continued to move toward the sun and Earth during May. Kobold said the refractor revealed the comet as faint, with a tail 2' long on the 1st. That same night, F. F. Tisserand (Paris, France) and Bigourdan noted the tail extended toward PA 50.7° and 52.6°, respectively. R. Copeland (Dun Echt Observatory, Scotland) observed with the 38-cm refractor (229×) on the 2nd. He noted a nearly stellar nucleus of about magnitude 13.5, as well as a tail extending about 4' toward PA 48.7°. On the 3rd, Tisserand and Bigourdan respectively said the tail extended toward PA 48.2° and 54.7°. Young estimated the magnitude as 10 on the 4th, but saw no nucleus in the refractor. The tail was described as 6' long and 2' wide, and seemed slightly concave on the south-following side. Tacchini said the tail was 2.4' long on the 5th and was initially hard to see as it passed through a group of 10th-magnitude stars. On the same night, Copeland said the tail extended 7' toward PA 51.2°, while a stellar nucleus was visible in the refractor. Copeland found the tail 6' long on the 8th. G. F. W. Rümker (Hamburg, Germany) said the nucleus was almost stellar in a 24-cm refractor on the 9th. Bigourdan estimated the magnitude of the nucleus as 13 on the 11th, and added that the tail extended toward PA 47.1°. Tacchini noted hardly a trace of the nucleus on the 13th, and noted the tail was 2.4' long. G. A. P. Rayet (Bordeaux, France) said the comet was very faint and diffuse on May 10, 13, and 14. J. J. Plummer (Orwell Park Observatory, England) described the comet as faint on the 13th. On the 14th, Lohse said the sharp nucleus had the brightness of a 12th-magnitude star in the 38-cm refractor. The finder revealed the comet as an "exceedingly faint object." On the 27th, the comet was seen by Tacchini and Peter. Tacchini said the coma was 2' across, with a hardly detectable brightening in the center. Peter said the comet was extraordinarily faint in the 22-cm refractor. M. W. Meyer (Geneva, Switzerland) said the comet was extremely faint on the 30th, with a nucleus-like condensation of magnitude 14–15. The tail was about 1' long. In a letter written on May 28 and published in the June 10 issue of the *Astronomische Nachrichten*, Copeland and Lohse noted, "In this latitude the midnight twilight has now become so strong as to obliterate every trace of the comet."

The comet had become very difficult to observe as June began and the last two positions were measured by I. Y. Kortazzi (Nicolajew, now Mykolayiv, Ukraine) with a 23-cm refractor on June 7.88 and 8.86. The comet passed perihelion on July 2 and then passed 18° from the sun on July 7. The comet was recovered on September 4.14, when Plummer located it in the morning sky. On that date, as well as September 5 and 10, he described the comet as "very faint, having apparently less than its theoretical brilliancy." The comet was last detected on September 11.12 and 12.09 by Bigourdan. Although no physical descriptions were published, he gave α as 6^h 54.8^m on the first date and δ as $+20°$ 24' on the last.

The first parabolic orbit was calculated by J. Holetschek and K. Zelbr using positions obtained on April 10, 11, and 13. The resulting perihelion date was 1880 June 12.23. Additional orbits using positions obtained into May were published by T. H. Safford, Schaeberle, J. R. Hind, Copeland, Lohse, Martin, and Bigourdan. These revealed a perihelion date around July 2.

Definitive orbits have been calculated by J. Mayer (1881) and J. Polak (1913). Mayer determined a parabolic orbit with a perihelion date of July 2.24. Polak determined a parabolic orbit with a perihelion date of July 2.23, as well as a hyperbolic orbit with a perihelion date of July 2.20 and an eccentricity of 1.0008409. Polak's hyperbolic orbit is given below.

T	ω	Ω (2000.0)	i	q	e
1880 Jul. 2.2020 (UT)	145.1661	258.9234	123.0568	1.814416	1.000841

ABSOLUTE MAGNITUDE: $H_{10} = 2.7$ (V1964)

FULL MOON: Mar. 26, Apr. 24, May 24, Jun. 22, Jul. 21, Aug. 20, Sep. 18

SOURCES: G. Bigourdan and Pa. and Pr. Henry, CR, **90** (1880 Apr.19), pp. 911–12; J. Holetschek and K. Zelbr, MNRAS, **40** (1880 Apr.), p. 379; J. M. Schaeberle, AN, **97** (1880 Apr. 17), p. 63; G. A. P. Rayet, CR, **90** (1880 May 17), p. 1153; R. Copeland, J. G. Lohse, and J. R. Hind, MNRAS, **40** (1880 May), pp. 438–9; J. Holetschek, K. Zelbr, P. Tacchini, E. Millosevich, and B. Peter, AN, **97** (1880 May 1), pp. 77–80; P. Tacchini and E. Millosevich, AN, **97** (1880 May 27), p. 157; J. M. Schaeberle, AJS (Series 3), **19** (1880 Jun.), p. 494; T. H. Safford and G. Bigourdan, MNRAS, **40** (1880 Jun.), pp. 558–9; R. Copeland, J. G. Lohse, and Martin, AN, **97** (1880 Jun. 10), pp. 221–4, 235; C. A. Young, AN, **97** (1880 Jun. 19), p. 237; J. M. Schaeberle, AN, **97** (1880 Jul. 1), pp. 265–8; P. Tacchini and E. Millosevich, AN, **97** (1880 Jul. 8), pp. 279–82; G. Bigourdan and F. F. Tisserand, CR, **91** (1880 Jul. 12), pp. 71–5; I. Y. Kortazzi, AN, **97** (1880 Jul. 22), p. 335; M. W. Meyer, AN, **98** (1880 Sep. 11), pp. 91–4; G. Bigourdan, CR, **91** (1880 Sep. 13), pp. 483–4; H. A. Kobold and Heidorn, AN, **98** (1880 Oct. 16), p. 205; K. C. Bruhns and B. Peter, AN, **100** (1881 Jun. 29), pp. 65, 69; J. Franz, AN, **100** (1881 Sep. 3), pp. 249–51; J. Mayer, AN, **100** (1881 Oct. 25), p. 383; F. Deichmüller, VJS, **17** (1882), p. 143; J. J. Plummer, AN, **101** (1882 Jan. 16), pp. 163, 167; C. F. Pechüle, AN, **103** (1882 Nov. 15), p. 337; J. Franz, AN, **105** (1883 Apr. 25), pp. 181–4; G. F. W. Rümker, AN, **120** (1888 Oct. 19), pp. 81–4; J. Polak and I. Y. Kortazzi, AN, **193** (1913 Feb. 5), pp. 393–404; V1964, p. 61.

X/1880 P1 L. Swift (Rochester, New York, USA) discovered a faint "nebulous object elongated in the direction of the sun" in the morning sky on 1880 August 11. It was then situated about 1° from NGC 3682, at a position of $\alpha = 11^h 28^m$, $\delta = +68°$. As NGC 3682 was very familiar to Swift, the nebulous intruder was instantly believed to be a new comet. Discovered with a power of 25×, the object was further observed at powers of 36× and 72×. After an hour, the sky clouded over and, as no motion was detected, Swift decided not to announce the object until he could confirm it. The skies remained cloudy for the next five days, although a short period of partly cloudy conditions on August 16 allowed Swift to note that the object was absent from the vicinity of NGC 3682. On the morning of August 17, the sky cleared and, following the setting of the moon, a search was conducted until daylight. Nothing unusual was detected. An announcement was then made, but no additional observations were obtained.

SOURCES: L. Swift, *AN*, **98** (1880 Aug. 26), p. 47; L. Swift, *AN*, **98** (1880 Sep. 11), p. 95; F. Deichmüller, *VJS*, **17** (1882), p. 146.

C/1880 S1 *Discovered:* 1880 September 29.80 ($\Delta = 0.53$ AU, $r = 0.69$ AU, Elong. $= 40°$)
(Hartwig) *Last seen:* 1880 November 30.74 ($\Delta = 2.34$ AU, $r = 1.83$ AU, Elong. $= 47°$)
Closest to the Earth: 1880 September 27 (0.5176 AU)
1880 III = 1880d *Calculated path:* BOO (Disc), CrB (Oct. 3), SER (Oct. 5), HER (Oct. 6), OPH (Oct. 21), AQL (Nov. 12)

E. Hartwig (Strasbourg, France) discovered this comet on 1880 September 29.80, at a position of $\alpha = 14^h 08.2^m$, $\delta = +29° 45'$. He said it was bright with a tail extending 50' toward PA 39°. H. A. Kobold (O'Gyalla Observatory, Hungary) confirmed the find on September 30.78. He said the tail was slightly curved and seemed to glisten. Hartwig also saw the comet again on the 30th and said the tail extended toward PA 43°. An independent discovery was made by M. Harrington (Ann Arbor, Michigan) on October 1.21, but the sky clouded up before he was able to secure an accurate position. He noted that the large refractor showed a well-defined, planet-like nucleus.

The comet had passed perihelion earlier in September and had been closest to Earth about two days prior to discovery. Several astronomers, including J. F. J. Schmidt (Athens, Greece) and W. Upton (US Naval Observatory, Washington, DC, USA) reported the comet was faintly visible to the naked eye early in October, with Schmidt giving magnitude estimates of 6.2 on the 1st and 3rd, 6.3 on the 5th, and 6.5 on the 7th and 8th. Although K. C. Bruhns (Leipzig, Germany) gave the coma diameter as 3' on the 3rd, it was Schmidt who provided the most measurements during this period. He indicated the diameter was 7.5' on the 1st, 17.4' on the 3rd, 10.1' on the 4th, 10.9' on the 5th, 11.4' on the 6th, 12.7' on the 7th, and 16.2' on the 8th. A bright nucleus was reported by virtually all astronomers, with G. Bigourdan

(Paris, France) saying it was 9" across on the 1st, and Bruhns giving it as 0.7' across on the 3rd. Both measurements indicate this was a strong condensation, rather than the actual nucleus. Nevertheless, some rather consistent magnitude estimates were made. Schmidt gave estimates of 7 on the 1st, 7.5 on the 3rd, and 7.7 on the 5th. At Harvard College Observatory (Cambridge, Massachusetts, USA), A. Searle and E. C. Pickering gave estimates of 7.6 and 7.7, respectively, on October 6, and 7.8 and 7.2, respectively, on the 7th. Also on the 7th, their colleague, O. C. Wendell, said the nucleus was between magnitude 7.7 and 8.1. The tail was widely reported and was described as slightly curved by Kobold on the 1st, "narrow" by Bruhns on the 3rd, and "broad" by Kobold on the 6th. On the 1st, Kobold said the tail extended 1° 51' toward PA 49° 0.5', while Hartwig said the tail extended about 1.7° toward PA 46° and Bigourdan said the tail extended 1° toward PA 42.9°. Schmidt said the finder of the refractor revealed a tail extending 1.50°. On the 2nd, Kobold said the tail extended toward PA 45° 53'. Bruhns gave the tail length as 1° on the 3rd, while Schmidt said it was then 2.00° long in the finder. Schmidt followed with tail lengths of 2.55° on the 4th, 3.20° on the 5th and 6th, 2.20° on the 7th, and 1.20° on the 8th. Hartwig noted the tail extended toward PA 61° on the 8th. The tail must have had a fairly low surface brightness as it was not detected at low altitude by C. Pritchard (Oxford, England) on the 5th or by J. Franz (Königsberg, now Kaliningrad, Russia) on the 8th.

Although no longer visible to the naked eye after October 8, the comet was still well observed during the remainder of October, although moonlight was a problem early on. There were no estimates of the total brightness during this period, but there were several significant descriptions. In weak moonlight on the 9th, Kobold said the comet was bright. Franz said it was very faint in moonlight on the 11th. Hartwig said it was difficult to see in bright moonlight on the 15th. P. Tacchini and E. Millosevich (Rome, Italy) said it was faint in moonlight on the 16th. Although M. W. Meyer (Geneva, Switzerland) said it was very easily seen in the 25.4-cm telescope in "bad air, clouds, and moonlight" on the 16th, A. J. G. F. von Auwers (Berlin, Germany) found it very faint and difficult to observe with the 15-cm refractor in moonlight on the 17th. J. J. Plummer (Orwell Park Observatory, England) and Pritchard independently noted the comet was faint because of the full moon on the 18th. Upton said it was very faint in the 24-cm refractor on the 19th. With moonlight no longer a problem, Hartwig reported it was easier to see on the 24th than on the 15th. Kobold found it bright, broad, and round on the 25th. Tacchini and Millosevich said it was faint on the 27th. Auwers said it was "rather bright" on the 30th. Although several estimates of the coma diameter were given, they really should be divided into two groups: those made by Schmidt and those made by everyone else. Schmidt indicated a diameter of 17.00' on the 9th, 11.70' on the 21st, 11.20' on the 22nd, 8.90' on the 23rd, 10.00' on the 24th, 9.58' on the 28th, 8.96' on the 29th, and 8.52' on the 30th. Other measurements of the coma diameter included: 2.3' by Tacchini and 4′ by Meyer on the 9th, 2.5′ by Tacchini on the 10th, and 3′ by Auwers on the 30th. Although few descriptions of the coma are available, most astronomers agreed that it was round. Although astronomers again reported observations of the nucleus, it would seem these still referred to the condensation. The diameter of this "nucleus" was given as 10″ by Meyer on the 9th and 50″ by Hartwig on the 24th. Schmidt estimated the magnitude of the nucleus on several occasions, giving it as 9 on the 9th and 10th, 10 on the 21st, 11 on the 22nd, 23rd, and 24th, 11.5 on the 28th, 11 on the 29th, and 12 on the 30th. Auwers said the comet contained a very gradually condensed, diffuse nucleus of magnitude 11 or 12 on the 30th. Interestingly, Meyer observed with the 25.4-cm telescope on the 9th and reported a hood by the nucleus. He noted the nucleus was very diffuse on the 24th and 25th. J. G. Lohse (Dun Echt Observatory, Scotland) wrote for the 25th, "There is no real nucleus, but the central part appears granulated." The tail again seemed to have a relatively low surface brightness, as most astronomers had difficulty seeing it early on because of moonlight. On the 9th, Tacchini said there was an indication of a tail, while Kobold reported it as insignificant because of moonlight. On the other hand, Schmidt's transparent skies allowed him to see the tail extending 1.20° in the finder on the same date. Hartwig reported the tail extended toward PA 67° on the 10th. Using the finder, Schmidt followed with measurements of 0.37° on the 21st, and 0.33° on the 22nd and 23rd. On the 25th, Lohse said the tail extended 15′ toward PA 62.6°, while Hartwig said it extended 16′ toward PA 62°. Schmidt gave the tail length as 0.33° on the 28th, 29th, and 30th. Auwers said the tail extended toward PA 60–65° on the 30th. Franz was another astronomer who failed to see the tail near the end of the month, although on the 25th and 30th he noted a possible extension toward the sun.

The comet was still well observed as November began, but it was becoming quite faint. Most observatories ended their observations of this comet once the moon entered the sky. Tacchini and Millosevich found the comet faint, with a condensation but no tail on the 1st. G. F. W. Rümker (Hamburg, Germany) saw the comet with a 24-cm refractor on the 2nd and described it as large, round, with a strong central condensation. That same night, Kobold reported the comet was very faint, round, with hardly a condensation, while Strasser noted it was round. Auwers observed with a 15-cm refractor on the 2nd and 3rd, and found the comet faint, with a diffuse nucleus of magnitude 12. Plummer said the comet was faint and difficult to measure on the 1st, 3rd, 4th, and 20th. H. S. Pritchett (Morrison Observatory, Glasgow, Missouri, USA) observed with a 31-cm refractor on the 2nd, 3rd, and 7th, and wrote, "The comet was very faint on each day of observation and without any appearance of condensation at center." On the 4th, Schmidt observed with the 17-cm Reinfelder refractor and indicated a coma diameter of 8.26′ and a nucleus of magnitude 12, while the finder revealed a tail 0.17° long. B. Peter (Leipzig, Germany) saw the comet on the same night with a 22-cm refractor and described it as very faint, somewhat elongated, with a point-like condensation. He added that the coma was about 1.5′ across. Tacchini and Millosevich said the comet appeared faint on the 5th. Pritchett saw the comet for the final time on November 9, but said it was very faint in moonlight. His attempts to find the comet with a 31-cm refractor after the November moon were unsuccessful. Franz could not find the comet because of moonlight. Auwers was not able to see the comet in a 15-cm refractor on November 20. Plummer said the comet was so faint on the 22nd and 25th that the resulting positions were uncertain. Bigourdan said the comet was extremely faint and without a condensation on the 22nd.

E. W. L. Tempel's (Arcetri, Italy) final observations of this comet came on November 28.75, 29.73, and 30.74. Precise positions were difficult to obtain on the 28th and 29th, because the comet's nebulosity spread over several small stars. Although the comet was "probably isolated from stars" on the 30th, the nucleus was so small and indefinite that good measurements were again difficult. Nevertheless, the position on the final date was given as $\alpha = 19^h 04.8^m$, $\delta = +6° 19′$. Although Tempel suggested he would be able to easily follow the comet for several days with the 27-cm refractor, no such observations were reported.

Several astronomers examined the spectrum of this comet during late September and the first half of October. Many reported only a continuous spectrum from the nucleus, while the coma displayed the usual three carbon bands. N. de Konkoly's (O' Gyalla Observatory, Hungary) observations were the most extensive and during the period of September 30 to October 5 he detected four bands, the average measurements of which were 5609 Å, 5492 Å, 5169 Å, and 4859 Å.

The first parabolic orbits was independently calculated by K. Zelbr and E. Hartwig using positions from September 29, 30, and October 1. Zelbr determined the perihelion date as 1880 September 7.71, while Hartwig gave it as 1880 September 7.42. Although both were excellent first orbits, Hartwig's would later prove to have missed the actual perihelion date by only about 28 minutes. Additional parabolic orbits were calculated by J. R. Hind, H. Oppenheim, Meyer, Zelbr, C. F. W. Peters, and Upton. F. A. T. Winnecke expressed his opinion that it was highly probable that this comet was seen in 1506. Although Hind said he could only see this as a possibility, Winnecke came back with the suggestion that the comet was seen in 1382, 1444, 1506, and 1569.

Before the end of October it had become obvious that the comet's positions could not be represented by a parabola. W. Schur and Hartwig took positions spanning the period of September 29 to October 24 and calculated an elliptical orbit with a perihelion date of September 7.05 and a period of 62.3 years. L. Schulhof and J. F. Bossert calculated an elliptical orbit near the end of November, with a perihelion date of September 7.39 and a period of 1280 years. They admitted this orbit still did not fit the positions adequately, and, near the end of December, they gave an equally uncertain period of 63 years.

459

T. Molien (1883) calculated a definitive orbit for this comet. Taking positions spanning the entire period of visibility, he ultimately found that a hyperbolic orbit with a perihelion date of September 7.45 and an eccentricity of 1.0007659 fitted the observations best. However, since the comet was only observed for two months, he preferred the parabolic orbit given below.

T	ω	Ω (2000.0)	i	q	e
1880 Sep. 7.4351 (UT)	323.1288	47.0063	141.9119	0.354634	1.0

ABSOLUTE MAGNITUDE: $H_{10} = 7.7$ (V1964)

FULL MOON: Sep. 18, Oct. 18, Nov. 16, Dec. 16

SOURCES: E. Hartwig, *AN*, **98** (1880 Oct. 6), p. 143; K. C. Bruhns, *AN*, **98** (1880 Oct. 7), p. 159; K. Zelbr and E. Hartwig, *AN*, **98** (1880 Oct. 9), p. 175; G. Bigourdan, *CR*, **91** (1880 Oct. 11), p. 610; H. Oppenheim, *AN*, **98** (1880 Oct. 13), p. 191; F. A. T. Winnecke and J. R. Hind, *Nature*, **22** (1880 Oct. 14), p. 569; M. W. Meyer, *AN*, **98** (1880 Oct. 20), p. 223; P. Tacchini and K. Zelbr, *AN*, **98** (1880 Oct. 23), pp. 235–40; C. F. W. Peters, *AN*, **98** (1880 Oct. 28), p. 255; C. Pritchard and W. H. M. Christie, *MNRAS*, **41** (1880 Nov.), pp. 49–50, 52–3; F. A. Bredikhin, *AN*, **98** (1880 Nov. 2), p. 271; E. W. L. Tempel, *AN*, **98** (1880 Nov. 11), pp. 299–302; W. Upton, N. de Konkoly, and H. A. Kobold, *AN*, **98** (1880 Nov. 19), pp. 311–18; E. E. Block, *AN*, **98** (1880 Nov. 29), p. 329; P. Tacchini and E. Millosevich, *AN*, **98** (1880 Dec. 6), pp. 343–6; G. Bigourdan, L. Schulhof, and J. F. Bossert, *CR*, **91** (1880 Dec. 6), pp. 918–21; M. W. Meyer, *AN*, **98** (1880 Dec. 9), pp. 363–6; J. Palisa and G. Strasser, *AN*, **98** (1880 Dec. 16), pp. 369–78; L. Schulhof and J. F. Bossert, *AN*, **99** (1880 Dec. 23), p. 15; L. Schulhof and J. F. Bossert, *CR*, **91** (1880 Dec. 27), pp. 1051–2; E. W. L. Tempel, *AN*, **99** (1880 Dec. 28), pp. 21–4; J. F. J. Schmidt, *AN*, **99** (1881 Feb. 2), pp. 103–8; J. M. Schaeberle and M. Harrington, *AN*, **99** (1881 Feb. 3), p. 127; H. S. Pritchett, *AN*, **99** (1881 Feb. 11), p. 143; K. C. Bruhns and B. Peter, *AN*, **100** (1881 Jun. 29), pp. 65, 70; A. J. G. F. von Auwers, *AN*, **100** (1881 Jul. 16), p. 103; F. Deichmüller, *VJS*, **17** (1882), pp. 143–5; J. J. Plummer, *AN*, **101** (1882 Jan. 16), pp. 163–8; J. Franz, *AN*, **105** (1883 Apr. 25), pp. 181–5; T. Molien, *AN*, **105** (1883 Jun. 4), pp. 353–62; E. Hartwig, *AN*, **108** (1884 Apr. 5), p. 267; H. S. Pritchett, *Publications of the Morrison Observatory*, No. 1 (1885), p. 101; J. G. Lohse, *MNRAS*, **46** (1886, Supplemental Notice), pp. 490, 494; G. F. W. Rümker, *AN*, **120** (1888 Oct. 19), pp. 81, 85; *Annals of the Astronomical Observatory of Harvard College*, **33** (1900), pp. 150, 153; V1964, p. 61.

11P/1880 T1 (Tempel–Swift– LINEAR)

1880 IV = 1880e

Discovered: 1880 October 11.1 ($\Delta = 0.21$ AU, $r = 1.13$ AU, Elong. $= 127°$)

Last seen: 1881 January 26.88 ($\Delta = 0.55$ AU, $r = 1.47$ AU, Elong. $= 147°$)

Closest to the Earth: 1880 November 17 (0.1277 AU)

Calculated path: PEG (Disc), LAC (Nov. 1), AND (Nov. 9), CAS (Nov. 13), PER (Nov. 23), AUR (Dec. 11), GEM (Jan. 8)

L. Swift (Rochester, New York, USA) discovered this comet on 1880 October 11.1 at a position of $\alpha = 21^{\text{h}} 30^{\text{m}}$, $\delta = +18°$. Telegrams were sent to the proper authorities and E. E. Barnard (Nashville, Tennessee, USA) described the comet as large, diffuse, and perfectly transparent on October 22 and 23, while he noted it was visible in a 3.2-cm finder on the 25th. Unfortunately, the telegram that went to G. B. Airy, the Astronomer Royal of England, said the motion was slow and "probably north-west." In reality, the comet was heading north-northeastward at over half a degree per day, so that European observers were looking in the wrong spot for the comet. What made matters even worse was that the comet was heading for a close approach to Earth, which meant its speed across the sky was increasing. The first European observation was actually an independent discovery by J. G. Lohse (Dun Echt Observatory, Scotland) on November 7.97, with bad weather preventing an exact measurement of the position until November 8.15. Soon after the measurement, Lohse said the comet "disappeared in the rays" of a 5th-magnitude star and was not seen again that night because of approaching daylight.

The comet passed perihelion around the time of Lohse's discovery, and passed rather close to Earth on November 17. Despite this close approach, it remained a faint object, with Lohse reporting he was unable to find it with the 38-cm refractor on November 16, 17, and 18, because of strong moonlight. Once the moon was gone, observations were easier, with C. N. J. Börgen (Wilhelmshaven, Germany) observing the comet with a 10.2-cm Steinheil refractor and B. Peter (Leipzig, Germany) finding it with a 22-cm refractor on the 18th. Börgen described it as very faint and diffuse, while Peter said it was faint, round, and without a nucleus. Lohse said the comet was "fairly well seen" on the 19th, even though the moon was several degrees above the horizon. He added that the coma seemed "a little elongated on the following side" and was centrally condensed. By the 20th, Lohse said the comet was "perfectly round" and "considerably brighter in the middle." Lohse reported on the 24th that the comet was "apparently brighter and more extended" than on the 20th. Also on the 24th, H. A. Kobold (O'Gyalla Observatory, Hungary) said the comet was wide and diffuse, without a trace of condensation. On the 25th, Lohse again scrutinized the comet with the large refractor, but this time with a magnification of 229×, and wrote, "the texture of the comet is uneven as compared with a nebula. It has no nucleus, although an area in the centre 20" in diameter is brighter than the rest." He noted it was "brighter and larger" in the finder, which had an aperture of 9.5-cm, and displayed "no appreciable deviation from the circular form." That same night, Peter said the comet was diffuse and irregular in shape, as seen in the 22-cm refractor. Lohse said the comet was "decidedly somewhat brighter" on the 27th than on the 25th. He added that it remained round and hazy looking. Kobold described the comet as faint, wide, and without a condensation on the 28th, while Peter said it was an irregular nebulosity, with its extreme boundaries indicating a diameter of 4–5'. On the 29th, Peter observed with the 22-cm refractor at 72× and said the comet filled almost half the field of view. At 192× there was a small faint nebulosity with a condensation and an occasionally flashing granulation. That same night, A. J. G. F. von Auwers (Berlin, Germany) said the comet was rather bright, round, and 3′ across in the 15-cm refractor. He also noted a strong central condensation of magnitude 11, but no nucleus. On the 30th, J. F. J. Schmidt (Athens, Greece) measured the coma as 9.45′ across, while Peter gave the diameter as 5–6′. Peter said the comet was irregular in shape and exhibited a condensation and nucleus.

The comet was moving away from both the sun and Earth during the remainder of its apparition. Despite it faintness, it was well observed during December. On the 1st, Schmidt gave the coma diameter as 15.47′, while Peter gave it as 4′. Peter also noted a bright condensation and a distinct nucleus of magnitude 11 in the 22-cm refractor. E. W. L. Tempel (Arcetri, Italy) also reported a bright nucleus. Auwers described the comet as rather faint in the 15-cm refractor on the 1st and 2nd. On the 2nd, Lohse said the comet had not noticeably changed since November 29. He switched to observing with the 15.4-cm Simms refractor and noted, "the comet appears very large and tolerably bright, but without a nucleus." The coma was 10.5′ across and "is so faint as only to be visible with very low powers." Schmidt provided several measures of the coma diameter around this time. He gave it as 13.26′ on the 2nd, 13.88′ on the 3rd, and 13.83′ on the 4th. Also on the 4th, E. E. Block (Odessa, Ukraine) observed at a magnification of 180× and noted the nucleus was 5″ wide and 10″ long. Lohse remarked that the comet was "still tolerably bright, considering the already strong moonlight" on the 10th. Moonlight hampered observations for the next several nights, with the next physical description not coming until the 21st when Peter observed under moonlit skies and described the comet as a very faint, small, and condensed nebulosity. Auwers found the comet hardly visible in the 15-cm refractor on the 23rd. On the 25th and 31st, E. C. Pickering (Harvard College Observatory, Cambridge, Massachusetts, USA) described the comet as faint and 2′ across, but with no nucleus. Auwers said the comet was no longer visible in his refractor on the 26th. With respect to his December observations, Schmidt said the comet "was round, hardly condensed, perhaps without a nucleus," although he occasionally noted a star-like object of magnitude 12.5 within the coma that he suspected was not a real star.

Only observatories with large telescopes continued their observations into January. Pickering said the comet was faint and 2′ across in the 38-cm refractor on the 4th. He added that no nucleus was visible. A 38-cm refractor was also used by Lohse on the 5th, when he noted, "The comet is exceedingly faint and very difficult to measure." He added, "The condensation in the middle is very insignificant." For the 7th, Lohse wrote, "The comet is so exceedingly faint as only to be recognizable in a perfectly dark field." F. A. T. Winnecke (Strasbourg, France) also saw the comet on the 7th with a 46-cm refractor. He described it as a roundish, centrally condensed, nebulosity, with a diameter of 1′. Winnecke provided the only descriptions during the remainder of the apparition. He said it appeared as a faint nebulosity about 1.5′ across on the 8th. It was still 1.5′ across when he saw it on the 21st, although he added that the center was only slightly brighter than the surrounding nebulosity. On the 22nd, Winnecke said the comet was 1' across, without a definite condensation.

The comet was last seen on January 26.88, when Winnecke measured the position as $\alpha = 6^h\ 18.2^m$, $\delta = +23° 45'$. He said the 46-cm refractor revealed the comet as a small, nebulous spot, which was easily visible when the nearby comparison star was covered by the wire of the micrometer.

Because the initial telegram to Europe sent those astronomers searching in the wrong direction, astronomers in the USA were the first to link this comet to that found by E. W. L. Tempel in 1869. S. C. Chandler Jr calculated the first orbit using positions obtained on October 22, 26, and 29. His resulting parabolic orbit gave a perihelion date of 1880 November 8.43. He immediately noted a similarity to the orbit of Tempel's comet and wrote that Swift's comet "will in all probability prove to be periodic, the resemblance of the elements to those of III. 1869 appearing remarkable." L. Boss came to a similar conclusion a day or so later, with his orbit indicating a perihelion date of November 8.68. Shortly before mid-November, enough observations had been obtained in Europe to allow the computation of a parabolic orbit. One of the first was determined by Lohse and R. Copeland, who gave the perihelion date as November 7.41 and showed that Lohse's comet was identical to Swift's October comet. In addition, they noted the strong similarity to the orbit of 11P/1869 W1. Additional parabolic orbits were calculated by W. Upton, K. Zelbr, J. von Hepperger, and H. Oppenheim.

Because the link between the 1869 and 1880 comets appeared so likely, L. Schulhof and J. F. Bossert calculated two hypothetical elliptical orbits, which were published in the 1880 December 6 issue of the *Compte Rendu des Séances de l'Académie des Sciences*. The first assumed a period of 5.5 years and gave the perihelion date as November 8.50, while the second assumed a period of 11 years and gave the perihelion date as November 8.43. Schulhof and Bossert published a new paper in the 1880 December 23 issue of the *Astronomische Nachrichten*, which took positions from October 25 to December 5 and revealed a perihelion date of November 8.50 and a period of 5.45 years. Additional orbits based solely on the positions from the 1880 apparition were published later. E. Frisby (1881) gave the period as 5.96 years. Upton (1881) gave the period as 5.99 years. W. Beebe and A. W. Phillips (1889) gave periods of 5.38 and 5.44 years. Orbits using multiple apparitions were determined by Bossert (1886) and B. G. Marsden (1971). They found a perihelion date of November 8.50 and a period of 5.50 years. Marsden determined the nongravitational terms as $A_1 = +0.04$ and $A_2 = -0.0459$. His orbit is given below.

T	ω	Ω (2000.0)	i	q	e
1880 Nov. 8.4970 (TT)	106.0852	298.6602	5.4065	1.067202	0.657333

ABSOLUTE MAGNITUDE: $H_{10} = 12.2$ (V1964)

FULL MOON: Sep. 18, Oct. 18, Nov. 16, Dec. 16, Jan. 15, Feb. 14

SOURCES: L. Swift, *AN*, **98** (1880 Oct. 16), p. 207; L. Swift, *The Observatory*, **3** (1880 Nov.), p. 621; J. G. Lohse, *AN*, **98** (1880 Nov. 11), insert; S. C. Chandler, Jr, R. Copeland, and J. G. Lohse, *AN*, **98** (1880 Nov. 19), p. 319; L. Swift and W. Upton, *Science*, **1** (1880 Nov. 20), p. 258–9; R. Copeland, J. G. Lohse, E. E. Barnard, K. Zelbr, J. von Hepperger, S. C. Chandler, Jr, G. Bigourdan, F. A. T. Winnecke, E. E. Barnard, and H. Oppenheim, *AN*, **98** (1880 Nov. 29), pp. 325–32; L. Swift, J. G. Lohse, S. C. Chandler, Jr, K. Zelbr, J. von Hepperger, and K. C. Bruhns, *The Observatory*, **3** (1880 Dec.), pp. 662–3; C. N. J. Börgen, *AN*, **98** (1880 Dec. 6), p. 351; L. Schulhof and J. F. Bossert, *CR*, **91** (1880 Dec. 6), p. 921; L. Boss and W. Bellamy, *AN*, **98** (1880 Dec. 16), pp. 377–80; L. Schulhof and J. F. Bossert, *AN*, **99** (1880 Dec. 23), pp. 11–16; J. von Hepperger, E. W. L. Tempel, and J. F. J. Schmidt, *AN*, **99** (1880 Dec. 28), pp. 19–26; J. G. Lohse and R. Copeland, *MNRAS*, **41** (1881 Jan.), pp. 158–61; H. A. Kobold, *AN*, **99** (1881 Jan. 11), p. 61; E. C. Pickering, *AN*, **99** (1881 Jan. 27), p. 95; E. Frisby, *AN*, **99** (1881 Feb. 2), p. 111; W. Upton, *AN*, **99** (1881 Feb 26), p. 171; E. E. Block, *AN*, **99** (1881 Mar. 22), p. 209; K. C. Bruhns and B. Peter, *AN*, **100** (1881 Jun. 29), pp. 65, 70; A. J. G. F. von Auwers, *AN*, **100** (1881 Jul. 16), p. 103; J. Franz, *AN*, **100** (1881 Sep. 3), p. 249; F. Deichmüller, *VJS*, **17** (1882), pp. 145–6; O. C. Wendell, *AN*, **101** (1882 Feb. 27), p. 299; E. Hartwig, *AN*, **108** (1884 Apr. 5), p. 269; J. F. Bossert, *BA*, **3** (1886 Feb.), pp. 65–78; F. A. T. Winnecke, *AN*, **114** (1886 May 6), pp. 229, 233; W. Beebe and A. W. Phillips, *AJ*, **9** (1889 Dec. 18), pp. 113–17; W. Beebe and A. W. Phillips, *AJ*, **9** (1889 Dec. 26), pp. 121–4; V1964, p. 61; B. G. Marsden and Z. Sekanina, *AJ*, **76** (1971 Dec.), pp. 1142–3; B. G. Marsden, *CCO*, 12th ed. (1997), pp. 50–1.

C/1880 Y1 *Discovered:* 1880 December 16.75 (Δ = 1.57 AU, r = 1.00 AU, Elong. = 38°)

(Pechüle) *Last seen:* 1881 March 31.85 (Δ = 3.25 AU, r = 2.50 AU, Elong. = 35°)

Closest to the Earth: 1880 December 3 (1.5410 AU)

1880 V = 1880f *Calculated path:* AQL (Disc), SGE (Dec. 16), VUL (Jan. 4), PEG (Jan. 20), CYG (Jan. 22), PEG (Jan. 25), AND (Feb. 19)

C. F. Pechüle (Copenhagen, Denmark) discovered this comet during a lunar eclipse on 1880 December 16.75 at a position of $\alpha = 18^h 49.0^m$, $\delta = +10° 31'$. He said the comet was round, with a diameter of 1' and a condensation. Pechüle confirmed the find on December 17.70.

The comet had passed perihelion over a month before the discovery and had been closest to Earth nearly two weeks before the discovery. On December 17, G. F. W. Rümker (Hamburg, Germany) said that despite the moon and hazy skies, the comet was seen in the 24-cm refractor as very bright with a large nucleus. H. A. Kobold (O'Gyalla Observatory, Hungary) observed the comet with a 15-cm refractor on the 20th and said it was bright and diffuse, without a nucleus. There was possibly a short tail toward the northeast. Kobold estimated the total magnitude as 7.5 on the 22nd and described the comet as round and centrally condensed. That same night, A. J. G. F. von Auwers (Berlin, Germany) saw it with the 15-cm refractor and said the comet was bright, round, and 4' across, with a nucleus of magnitude 10. K. C. Bruhns (Leipzig, Germany) saw the comet with the

22-cm refractor on the 23rd and described it as rather bright and centrally condensed. On the 26th, Kobold said the comet was bright, broad, and about magnitude 8, with a strong condensation, while Bruhns said the coma was 2–3' across. E. Millosevich (Rome, Italy) saw the comet on the 26th and 27th, and said it was easily visible in the 7.6-cm finder. He added that the refractor revealed a coma 1.5' across, as well as a nucleus and a trace of tail. Kobold described the comet as broad, with a rather strong condensation on the 28th. Observations were made by J. F. J. Schmidt (Athens, Greece) and Bruhns on the 30th. Schmidt saw the comet with the 17-cm Reinfelder refractor and said the preceding coma radius was 4.83', while the following radius was 2.17'. A stellar nucleus had a magnitude of 9. Bruhns said the comet was very granular in the 22-cm refractor, with a central condensation. Schmidt and Pechüle saw the comet on the 31st. Schmidt saw it with the refractor and said the preceding coma radius was 6.27', while the following radius was 2.48'. Pechüle said a "quasi-nucleus" was faintly visible in the 27-cm refractor.

As January began, Kobold estimated the comet's magnitude as 8 and added that the 15-cm refractor revealed a round coma about 1' 45" across, but no significant condensation. Bruhns and his colleague B. Peter saw the comet with the 22-cm refractor on the 3rd and said the coma was 1.5' across and exhibited a distinct, disk-like condensation. Schmidt said his refractor revealed a nucleus of magnitude 8.5 on the 4th. Kobold and Peter saw the comet on the 6th. Kobold said the round coma was 1' 15" across and exhibited a central condensation of magnitude 8.5, while Peter said the coma was 2' across and contained a fine, point-like nucleus. Weak moonlight was present on the 8th when observations were made by Kobold and Peter. Kobold said the round coma was 1' 18" across, with a strong condensation. Peter said there was a strong coma and a condensation of magnitude 11. From January 6–9, G. Strasser (Kremsmünster, Austria) said the comet was always round, with a distinct nucleus. Kobold estimated the comet's magnitude as 8.5 on the 14th, but said it was difficult to see because of the nearly full moon. From January 16 to January 24, Strasser said the comet appeared fainter, but the nucleus was still visible. Schmidt saw the comet on the 19th and said the refractor revealed the preceding coma radius was 3.73', while the following radius was 1.82'. The nucleus was magnitude 10.5. During the period of January 23–26, Auwers said the comet was bright and varied in diameter from 2' to 3' across, as seen in the 15-cm refractor. Kobold gave the comet's magnitude as 9.5 on the 23rd and measured the coma as 1' 18" across. That same night, Schmidt said the preceding coma radius was 5.02', while the following radius was 1.72'. The nucleus was magnitude 11. Peter said the coma was 1.5' across on the 24th and exhibited a distinct nucleus. Although Strasser said the nucleus was no longer visible on the 25th, Peter said it was distinctly point-like in the 22-cm refractor. On the 26th, Peter said the faint coma was 1.5–2.0' across, with a condensation and a nucleus. B. von Engelhardt (Dresden, Germany) said the comet appeared very faint in a 30.5-cm refractor on the 27th. J. J. Plummer (Orwell Park Observatory, England) said the comet was "faint owing to light of aurora" on the 31st.

The comet continued its slow fading during February. Peter said the faint coma was 1.5–2.0′ across on the 3rd, with a condensation and nucleus. He then reported on the 4th that the comet appeared diffuse. On the 5th and 7th, Engelhardt said the comet appeared very faint. Peter said the faint coma was 2.5–3.0′ across on the 7th and still contained a point-like nucleus. Following the interference from moonlight, Schmidt found the comet with the refractor on the 17th and said the preceding coma radius was 2.35′, while the following radius was 1.01′. He reported that there was no nucleus. Kobold said the comet was faint on the 21st and added that "a very weak point" would occasionally appear within the condensation. He reported it was faint and condensed on the 23rd. Auwers estimated the coma diameter as 1.5′ on the 23rd and 1′ on the 24th. On February 25, Schmidt said the preceding coma radius was 3.05′. Pechüle described the comet as faint in his 27-cm refractor on the 27th.

Few observations were attempted as March began. Engelhardt reported the comet was at the limit of the 30.5-cm refractor on the 2nd and 3rd. Schmidt failed to locate it on the 2nd in the 17-cm Reinfelder refractor, although he noted some moonlight was present. After the moon had left the sky, E. W. L. Tempel (Arcetri, Italy) saw the comet for the final time on March 17. He noted the comet was faint, but still easy to see, and added that observations ceased because of the inconvenient placement in the sky.

The comet was last detected on March 30.84 and March 31.85, when G. Bigourdan (Paris, France) described it as extremely faint in very clear skies. On the last date he gave the position as $\alpha = 0^h 54.3^m$, $\delta = +39° 48′$.

The spectrum was recorded by N. de Konkoly (O'Gyalla Observatory, Hungary) on December 26. He detected the three bright carbon bands at 5603 Å, 5163 Å, and 4763 Å.

The earliest parabolic orbits were determined before the end of December. H. Oppenheim took positions from December 16, 19, and 21, and determined the perihelion date as 1880 November 9.69. This orbit represented the comet's motion fairly well, but the later orbits by S. C. Chandler, Jr, L. F. A. Ambronn, J. Holetschek, Oppenheim, and Bigourdan revealed a perihelion date of November 9.92. Nearly 100 years later, G. Schrutka (1978) re-examined this comet's orbit. He began with 299 positions spanning the entire period of visibility, applied perturbations by all nine planets, and determined the orbit below.

T	ω	Ω (2000.0)	i	q	e
1880 Nov. 9.9200 (UT)	11.6877	251.0495	60.7007	0.659736	1.0

ABSOLUTE MAGNITUDE: $H_{10} = 5.8$ (V1964)

FULL MOON: Dec. 16, Jan. 15, Feb. 14, Mar. 15, Apr. 14

SOURCES: C. F. Pechüle, *AN*, **99** (1880 Dec. 23), p. 15; H. Oppenheim, *AN*, **99** (1880 Dec. 31), p. 47; E. Millosevich, *AN*, **99** (1881 Jan. 11), p. 63; G. Bigourdan, *CR*, **92** (1881 Jan. 17), pp. 117–18; J. Holetschek and H. Oppenheim, *AN*, **99** (1881 Jan. 22), pp. 75–80; G. Bigourdan, *CR*, **92** (1881 Jan. 24), p. 172; N. de Konkoly and L. F. A. Ambronn, *AN*, **99** (1881 Jan. 27), pp. 93–6; S. C. Chandler, Jr and A. Abetti, *AN*, **99** (1881 Feb. 2), pp. 109–12; P. Tacchini and E. Millosevich, *AN*, **99** (1881 Feb. 11), pp. 137–40; H. A. Kobold, G. Strasser, and B. von Engelhardt, *AN*, **99** (1881 Mar. 22), pp. 217–20, 223; J. F. J. Schmidt, *AN*, **99** (1881 Mar. 25), pp. 253–6; G. Bigourdan, *CR*, **92** (1881 May 2), pp. 1045–7; E. W. L. Tempel, *AN*, **99** (1881 May 7), p. 351; K. C. Bruhns and B. Peter, *AN*, **100** (1881 Jun. 29), pp. 67, 71; A. J. G. F. von Auwers, *AN*, **100** (1881 Jul. 16), pp. 105–8; F. Deichmüller, *VJS*, **17** (1882), pp. 146–7; J. J. Plummer, *AN*, **101** (1882 Jan. 16), pp. 165–8; C. F. Pechüle, *AN*, **103** (1882 Nov. 15), p. 339; W. Schur, *AN*, **114** (1886 Mar. 23), p. 81; G. F. W. Rümker, *AN*, **120** (1888 Oct. 19), pp. 81, 85; V1964, p. 61; G. Schrutka, *QJRAS*, **19** (1978), pp. 80–1, 88.

X/1880 Y2
1880g

W. F. Cooper (Sheffield, England) reported that he found a fairly bright comet just below ε and ζ Piscium in the evening sky on 1880 December 21.88, while he was observing Saturn. The position was estimated as $\alpha = 1^h 05^m$, $\delta = +6°$. Cloudy weather prevented observations during the next couple of days, but he finally obtained additional observations on December 24.92 ($\alpha = 1^h 24^m$, $\delta = +2.5°$) and December 25.79 ($\alpha = 1^h 29^m$, $\delta = +2°$). There was no apparent change in brightness, which was described as nearly equal to ξ Piscium (magnitude 4.86 – SC2000.0). The comet's diameter was given as about 20″. No further observations were reported by Cooper.

Once word of the discovery got out, searches were made by a few astronomers, the most notable of whom was C. H. F. Peters (Clinton, New York, USA). He was hampered by clouds until December 29 and began his search at that time with an outside temperature near −12°F. He carefully searched over several degrees around the position extrapolated from Cooper's positions, but all that was found was a faint nebula, which turned out to be number 574 in J. F. W. Herschel's *General Catalogue of Nebulae and Clusters of Stars*.

A parabolic orbit was calculated by H. Oppenheim (1881) using the three rough positions. It revealed a perihelion date of 1880 November 9.34 and is given below. This orbit revealed two puzzling aspects of this cometary apparition. First, the comet was apparently less than 0.09 AU from Earth on December 15. Oppenheim said the comet should then have been 11 times brighter than at discovery (magnitude 1–2) and added "it is strange that the comet was not seen since it could have been detected with the naked eye." The second puzzle involved the fact that the comet should have been fading rapidly at the time of discovery, so that at the time of Cooper's last observation, the comet would have been at least one magnitude fainter than at discovery–in strong contrast to Cooper's report that the comet did not change in brightness.

Despite the apparent discrepancies between Cooper's observations and Oppenheim's predictions, Oppenheim's ephemeris for the days following discovery show that the failure of observers to confirm Cooper's comet should not be considered as unexpected. The comet's rapid motion would actually have placed it 10° or more from the region searched by most observers. In addition, most searches began around December 29, at which time the comet should have been at least 2 magnitudes fainter than at discovery.

T	ω	Ω (2000.0)	i	q	e
1880 Nov. 9.3443 (UT)	73.5567	257.5983	129.1950	0.386687	1.0

ABSOLUTE MAGNITUDE: $H_{10} = 7.4$ (V1964)

SOURCES: J. F. W. Herschel, *PTRSL*, **154** (1864), p. 56; W. F. Cooper, *AN*, **99** (1880 Dec. 31), p. 47; W. F. Cooper, *The Observatory*, **4** (1881 Jan.), p. 30; C. H. F. Peters, *AN*, **99** (1881 Feb. 11), p. 141; H. Oppenheim, *AN*, **100** (1881 Jun. 29), p. 73; F. Deichmüller, *VJS*, **17** (1882), pp. 146–7; V1964, p. 61; SC2000.0, p. 25.

4P/Faye

1881 I = 1880c

Recovered: 1880 August 2.92 ($\Delta = 1.52$ AU, $r = 2.35$ AU, Elong. $= 135°$)
Last seen: 1881 March 30.82 ($\Delta = 2.33$ AU, $r = 1.85$ AU, Elong. $= 50°$)
Closest to the Earth: 1880 October 4 (1.0905 AU)
Calculated path: PEG (Rec), PSC (Sep. 18), CET (Dec. 31), PSC (Jan. 16), CET (Feb. 12), ARI (Mar. 7), TAU (Mar. 15)

D. M. A. Möller (1880) took the orbit for the 1873 apparition and predicted the comet would next arrive at perihelion on 1881 January 23.17. A. A. Common (Ealing, England) began searching for this comet with a reflector of 3-foot focal length on 1880 June 11. He was hampered by several periods of unfavorable weather, but finally recovered it on August 2.92. Common found the comet very near the position predicted by Möller and described it as "faint, round, and a little brighter towards the middle." The night of the 4th was not a good one for observing and Common was only able to confirm that the place where the comet had been on the 2nd was now empty. An object he thought was the comet would turn out to be a nebula when next checked on the 8th. The comet was finally detected again on the 8th, at which time the position was roughly measured as $\alpha = 23^h \, 16.0^m$, $\delta = +10° \, 47'$. Independent recoveries were made by E. W. L. Tempel (Arcetri, Italy) on August 11.9 and H. A. Kobold (O'Gyalla Observatory, Hungary) on September 6.88. Tempel said it was then too faint to measure the position, but, after a period of bad weather and moonlight, he was finally able to locate it again on August 25.84. Using a 27-cm refractor (112×), Tempel said the coma was remarkably small, hardly 30" in diameter, although there was a distinct nucleus. The comet was heading toward both the sun and Earth when discovered.

On September 8, G. F. W. Rümker (Hamburg, Germany) said the comet was very faint and about 30″ in diameter in a 24-cm refractor. B. Peter (Leipzig, Germany) said the nucleus was like a star of magnitude 12.5 on the 10th, while the condensed coma was 1.5′ across in a 22-cm refractor. G. Bigourdan (Paris, France) said the comet looked like a star of magnitude 13 on the 11th, with no nucleus or tail. P. Tacchini and E. Millosevich (Rome, Italy) saw the comet with the 23-cm refractor on the 24th and described it as an irregular nebulosity with a nucleus of magnitude 13. On September 28, Peter said the comet was diffuse with a central condensation.

The comet was closest to Earth on October 4 and continued to approach the sun for the rest of the year. Peter said the comet was faint with an excellent, point-like nucleus on the 9th. On the 22nd, Tempel said the coma was 1′ across. Tacchini and Millosevich said the comet was very faint on the 25th, while M. W. Meyer (Geneva, Switzerland) said the comet was very faint in the 25.4-cm telescope, because of poor transparency, but it still displayed a distinct nucleus. J. F. J. Schmidt (Athens, Greece) saw the comet with a 12.25-cm refractor on October 30, but was not able to measure a position because he could not sufficiently differentiate the comet from the fine stars in the field. He did note the coma was 1.5–1.7′ across, while the moderately condensed center was magnitude 12.8–13. On October 30 and November 1, Meyer said the comet generally had the same appearance as on the 25th, although the nucleus was estimated as magnitude 14. Rümker said the comet was extraordinarily faint in a 24-cm refractor on the 2nd and 4th. Peter said the comet was a faint, irregular, but condensed nebulosity about 0.5′ across on the 3rd. Peter noted the comet was small, faint, with slight condensation on the 4th. C. Pritchard (Oxford University, England) obtained numerous precise positions of this comet during September, October and into November. He noted, "The comet has been so faint during its present apparition that the observations have been made with considerable difficulty. In moonlight it has been impossible to see it with the 31-cm refractor."

The comet passed perihelion during the second half of January, but few descriptions were provided through the remainder of its apparition. F. A. T. Winnecke (Strasbourg, France) provided the majority of these observations, beginning on January 22, when he saw the comet with the 46-cm refractor and said it was quite bright, with a stellar center, and about 1.5′ across. On March 2, Winnecke said the round coma was noticeably condensed toward the center, with a diameter of 1′. He described it as centrally condensed and easily visible on the 17th, with a diameter of about 40″ across. Winnecke estimated the coma diameter as 30″ on the 19th. G. A. Hill (US Naval Observatory, Washington, DC, USA) said the comet was easily observed with the 66-cm refractor on March 25th and the nucleus appeared small and sharp.

The comet was last detected on 1881 March 28.82 and 30.82. Winnecke gave the position as $\alpha = 3^h\,51.9^m$, $\delta = +12°\,41′$ on the first date, but could not measure it on the last. Also on the 30th, Winnecke described the comet as a small, pale spot about 20″ in diameter in the 46-cm refractor.

Later calculations using multiple apparitions were published by A. Shdanow (1885) and B. G. Marsden and Z. Sekanina (1971). These revealed a perihelion date of January 23.16 and a period of 7.56 years. Marsden and Sekanina determined the nongravitational terms as $A_1 = +0.601$ and $A_2 = +0.01026$. This orbit is given below.

T	ω	Ω (2000.0)	i	q	e
1881 Jan. 23.1579 (TT)	201.1685	211.3127	11.3157	1.738122	0.548965

ABSOLUTE MAGNITUDE: $H_{10} = 7.4$ (V1964)

FULL MOON: Jul. 21, Aug. 20, Sep. 18, Oct. 18, Nov. 16, Dec. 16, Jan. 15, Feb. 14, Mar. 15, Apr. 14

SOURCES: D. M. A. Möller, *BAJ for 1882*, **107** (1880), p. 138; A. A. Common and D. M. A. Möller, *The Observatory*, **3** (1880 Sep.), p. 551; H. A. Kobold, D. M. A. Möller, and N. C. Dunér, *AN*, **98** (1880 Sep. 11), p. 95; G. Bigourdan, *CR*, **91** (1880 Sep. 13), pp. 483–4; P. Tacchini and E. Millosevich, *AN*, **98** (1880 Oct. 23), p. 235; C. Pritchard, *MNRAS*, **41** (1880 Nov.), pp. 51–2; E. W. L. Tempel, *AN*, **98** (1880 Nov. 11), pp. 299–302; P. Tacchini and E. Millosevich, *AN*, **98** (1880 Dec. 6), pp. 343–6; M. W. Meyer, *AN*, **98** (1880 Dec. 9), pp. 363–6; J. F. J. Schmidt, *AN*, **99** (1881 Feb. 2), p. 103; K. C. Bruhns and B. Peter, *AN*, **100** (1881 Jun. 29), pp. 65, 69; G. A. Hill, *AN*, **100** (1881 Sep. 15), p. 273; C. F. Pechüle, *AN*, **103** (1882 Nov. 15), p. 337; A. Shdanow, *BASP*, **33** (1885), pp. 1–24; F. A. T. Winnecke, *AN*, **114** (1886 May 6), pp. 229, 233; G. F. W. Rümker, *AN*, **120** (1888 Oct. 19), p. 81, 85–6; V1964, p. 61; B. G. Marsden and Z. Sekanina, *AJ*, **76** (1971 Dec.), pp. 1136–7.

C/1881 J1 (Swift)

1881 II = 1881a

Discovered: 1881 May 1.32 ($\Delta = 1.11$ AU, $r = 0.73$ AU, Elong. $= 40°$)

Last seen: 1881 May 12.10 ($\Delta = 0.99$ AU, $r = 0.62$ AU, Elong. $= 36°$)

Closest to the Earth: 1881 May 26 (0.9231 AU)

Calculated path: AND (Disc), PSC (May 12)

L. Swift (Rochester, New York, USA) discovered this comet on 1881 May 1.32, and roughly estimated the position as $\alpha = 0^h\ 00^m$, $\delta = +37°$. He described it as bright and well defined, with a slow movement toward the south.

The comet passed closest to both the sun and Earth during the second half of May, but it was lost before that because of moonlight and low altitude. A. L. N. Borrelly (Marseille, France) saw the comet on May 3 and said it was brilliant, roughly circular, and about 2′ across. There was no nucleus. On the 7th, E. Hartwig (Strasbourg, France) said the comet was not distinct because of cirrus clouds. B. Peter (Leipzig, Germany) observed in moonlight and some twilight on the 11th and described the comet as round and faint with a diffuse condensation. Observations were also made in Vienna (Austria) during the period of May 3–9 by E. Weiss, J. Palisa, and K. Zelbr.

The comet was last detected on May 12.06 by Peter, and on May 12.10 by Borrelly. Peter observed in moonlight and some twilight and described the comet as round and faint, with a diffuse condensation. Borrelly gave the position as $\alpha = 0^h\ 43.1^m$, $\delta = +25°\ 26'$.

The first parabolic orbit was calculated by G. Bigourdan using positions from May 3, 5, and 7. The resulting perihelion date was 1881 May 21.55. Additional orbits by E. E. Block, H. Oppenheim, Zelbr, Bigourdan, G. Gruss (1883), and H. C. F. Kreutz (1902) eventually established the perihelion date as May 20.94. Gruss also calculated an elliptical orbit with an assumed semi-major axis of 1000 AU, but it left larger errors than his parabolic solutions. Kreutz' orbit is given below.

T	ω	Ω (2000.0)	i	q	e
1881 May 20.9409 (UT)	173.8057	128.1073	77.9336	0.591090	1.0

ABSOLUTE MAGNITUDE: $H_{10} = 8.4$ (V1964)

FULL MOON: Apr. 14, May 13

SOURCES: L. Swift, *AN*, **99** (1881 May 7), p. 351; G. Bigourdan, *CR*, **92** (1881 May 9), pp. 1100–1; E. E. Block and H. Oppenheim, *AN*, **99** (1881 May 21), pp. 381–4; G. Bigourdan, *CR*, **92** (1881 May 30), pp. 1272–4; K. Zelbr, *AN*, **100** (1881 Jun. 4), p. 15; A. L. N. Borrelly, *AN*, **100** (1881 Jul. 26), p. 117; B. Peter, *AN*, **100** (1881 Aug. 13), p. 175; *AAOHC*, **13** (1882), pp. 178–9; O. C. Wendell, *AN*, **101** (1882 Feb. 27), p. 301; L. Swift, *SM*, **1** (1882 Mar.), p. 10; E. Weiss, J. Palisa, and Zelbr, *AN*, **102** (1882 Jul. 18), p. 277; G. Gruss, *AN*, **105** (1883 May 28), pp. 315–18; E. Hartwig, *AN*, **108** (1884 Apr. 5), p. 269; H. C. F. Kreutz, *AN*, **160** (1902 Nov. 8), pp. 185–90; V1964, p. 61.

X/1881 J2 On the morning of 1881 May 12, E. E. Barnard (Nashville, Tennessee, USA) discovered a very faint comet in the same field with α Pegasi. A steady watch between 3 a.m. and 4 a.m. revealed no motion and a position of $\alpha = 22^h\ 59.3^m$, $\delta = +14°\ 24'$ was obtained using a ring-micrometer. Barnard found the comet missing from its previous position when he observed on May 13, but just as dawn was beginning, it was found very close to α Pegasi to the north. It was only possible to observe the comet when a ring-micrometer was used to block the star's light. A position of $\alpha = 22^h\ 58.9^m$, $\delta = +14°\ 36'$ was obtained and a telegram was sent to L. Swift.

Independent searches by Barnard and Swift on the morning of the 14th revealed no trace of the comet, and, in the following days, O. C. Wendell (Cambridge, Massachusetts, USA), C. H. F. Peters (Hamilton College Observatory, Clinton, New York, USA), and others also failed to locate the comet.

There are two possible explanations for why the comet was lost. First, the comet's obviously slow motion would have made it very difficult to lose, unless it faded very rapidly. Secondly, the comet's proximity to α Pegasi 2 (a 2.5-magnitude star) may allow it to be identified as a ghost image of that star. Barnard discovered his first official comet on 1881 September 18, and he went on to discover a total of 15. He gained the reputation of being a careful observer.

SOURCE: C. N. A. Krueger, O. C. Wendell, and C. H. F. Peters, *AN*, **100** (1881 Jul. 16), p. 111; E. E. Barnard and L. Swift, *AN*, **100** (1881 Jul. 26), p. 127; E. E. Barnard and L. Swift, *AN*, **100** (1881), p. 127; E. E. Barnard, *SM*, **1** (1882 May), p. 33.

C/1881 K1 (Great Comet)

1881 III = 1881b

Discovered: 1881 May 22.37 ($\Delta = 0.77$ AU, $r = 0.89$ AU, Elong. $= 58°$)
Last seen: 1882 February 15.13 ($\Delta = 3.83$ AU, $r = 3.68$ AU, Elong. $= 73°$)
Closest to the Earth: 1881 June 20 (0.2825 AU)
Calculated path: CAE (Disc), COL (May 26), LEP (Jun. 3), ORI (Jun. 11), TAU (Jun. 18), AUR (Jun. 20), CAM (Jun. 26), DRA (Jun. 13), CAM (Jun. 14), UMi (Aug. 7), DRA (Sep. 22), CEP (Nov. 20), CAS (Jan. 8)

J. Tebbutt (Windsor, New South Wales, Australia) was "scanning the western sky . . . with the unassisted eye" on May 22, when he "detected a hazy looking object just below the constellation Columba, which from my familiarity with that part of the heavens I regarded as new." Examination with a "small marine telescope" revealed three objects: a 4.5-magnitude star, a 5.5-magnitude star, and the head of a comet. A position of $\alpha = 4^h 58.3^m$, $\delta = -35° 40'$ was given on 1881 May 22.37. He noted a tail was only seen with optical aid. He subsequently sent notice of the discovery to R. L. J. Ellery (Melbourne, Australia). Tebbutt noted the tail was visible to the naked eye on the 23rd and was about $2°$ long on the 25th. Ellery confirmed Tebbutt's find on May 23.36. He said it was "moderately large, easily visible to the naked eye and with a bright stellar nucleus equal to a star of the 5th magnitude."

Additional independent discoveries were immediately reported by B. A. Gould and W. G. Davis (Cordoba, Argentina) on May 26.0 and L. A. Eddie (Grahamstown, South Africa) on May 27.72. Gould and Davis were walking to the observatory when Davis pointed out "a bright star of peculiar appearance in the constellation Columba, and on closer inspection he recognized a tail-like appendage." Gould then examined the object with an opera glass and said it was "a comet with a large and brilliant head and a tail, which, although faint, was clearly discernible for a length of several degrees." Gould said a refractor revealed a nearly straight, faint tail extending about $14°$, but not exceeding $12'$ in width. The nucleus was about magnitude 4. Eddie wrote, "I discerned a large comet about $10°$ or $12°$ above the south-western horizon on the boundaries of the constellations Columba and Caelum, forming almost an isosceles triangle with o Columbae and γ Caeli, about $4°$ north preceding o Columbae. This comet appeared to the naked eye like a good-sized planetary nebula or nebulous star, with a faint stream of light extending upwards from it towards the south-east at an angle of about 45° to the horizon, and passing midway between the two stars above-named." Eddie said he was away from his observatory, but did have an opera glass handy. Through this instrument he said the nucleus "appeared of a rich golden colour." The tail was described as "a faint yellow or straw colour, about 4' in breadth, from 6–8° in length, and with little if any curvature." Eddie added that his wife told him that she had seen a comet with two tails on the previous evening, and he surmised this was the same comet that she had seen.

The comet was heading toward close approaches to the sun and Earth when discovered. It remained visible exclusively to Southern Hemisphere observers during the remainder of May. Gould estimated the nucleus was about 1" across, while the coma was nearly 7' in diameter on the 28th. He added that the nucleus had an irregular appearance, "its diameter being decidedly greater in the direction at right angles to the tail, and its outline ill-marked but with a depression at the point corresponding to the apex, as though partially bifid." D. W. Barker and others aboard the ship *Superb* (sailing from southern Africa toward South America) first saw the comet southwest of Canopus and Sirius on the 29th. Barker said the nucleus was "comparable with β Columbae (magnitude 3.12 – SC2000.0), perhaps a shade brighter, round and well defined." He also noted the tail was about 6° long. That same night, Gould said the nucleus had a diameter of 40" or 50". E. C. Markwick (Pretoria, South Africa) and Eddie independently saw the comet on the 30th. Markwick's observation was another independent discovery as he wrote, "my attention was attracted by a nebulous-looking star in the S.W. quarter of the sky, and not 20° above the horizon. In a few moments I could detect a faint tail. . . . " Markwick turned a 4-cm refractor toward the comet and "saw at once that it was a very pretty comet, with a brilliant head. No distinct nucleus could then be seen, but the central brilliancy was gradually softened off all round with a woolly light. The tail seemed to start back in a straight line opposite to the Sun's direction, and faded away rapidly, although with the naked eye I should estimate it might be traced some 4° or 5°, if not more. It was inclined to the horizon at an angle of between 30° and 40°." Eddie only saw the comet for a few minutes through breaks in the clouds. He said an 8-cm refractor "revealed a small but brilliant nucleus, condensed at the centre, and surrounded by an extensive coma of moderate brilliancy, even to its outer edge." He continued, "The graduation of light from nucleus to coma was sudden, though unbroken by any dark interval. The tail appeared broad, scattered and ragged, but very faint, and soon fading away to invisibility." According to the Chinese text *Shih Êrh Chhao Tung Hua Lu*, which was written by Chiang Liang-Chhi and others, Chinese astronomers saw a "broom star" comet in the northeast on the 30th. Eddie and Markwick again provided fine descriptions on May 31. Eddie said it had "sensibly increased in brightness, and shone as a star of the second magnitude. The tail had also increased in length and visibility, and could be traced to about 12° or 13° from the nucleus." Once again, Eddie turned the 8-cm refractor toward the comet and said, "the coma seemed much broader than on the previous evening, and was fully 6' in diameter. The width of the tail had increased to 12–15' at its widest part." Markwick said the comet was basically unchanged, except for a slight increase in brightness. He added, "In intrinsic brightness I estimated it between β and γ Leporis, or between 4th and 5th mag., but owing to there being quite a disk visible to the naked eye, the head appeared more like a 2nd mag. star" (β is magnitude 2.84 and γ is magnitude 3.60 – SC2000.0).

The comet passed closest to the sun on June 16 and closest to Earth on June 20. Most importantly, as it approached Earth, it actually passed almost exactly between our planet and the sun. The result was that the comet was lost in the sun's glare for a couple of weeks after June 7. There were no estimates of the comet's total magnitude between June 1 and 7, but Barker made a couple of estimates of the nuclear magnitude. Barker said the nucleus was as bright as α Columbae (magnitude 2.64 – SC2000.0) on the 2nd and slightly brighter on the 3rd. B. E. W. Gwynne (on board HMS *Garnet*, in the Strait of Magellan, Chile) noted on the 1st and 2nd that, "The nucleus presented the appearance of a planet (such as Jupiter) with its brilliancy "dulled" by the nebulous matter that slightly precedes as well as surrounds it and also forms its tail." Eddie provided the most important observations of the nuclear region at this time with his 24-cm reflector. For June 1st, he wrote, "The nucleus appeared to be inclosed in a cometic envelope or series of envelopes of parabolic form. It was not, however, situated in the focus of this curve, as was the case with the comets [C/1858 L1 (Donati) and C/1874 H1 (Coggia)], but was on one side of this point, but about the breadth of its own diameter on the side towards the Sun. The immediate focus was occupied by a condensed stream of coma, issuing, as it were, from the nucleus. This stream was bounded on its exterior edge by a darker interval, also partaking of the parabolic form, but not extending over the nucleus. There was another dark rift of similar form around the cometic envelope near the interior limits of the comet; beyond this a faint light fading away to invisibility. There was also a dark interval immediately behind the nucleus, which gave it a sharp definition on its exterior edge. The tail appeared like the continuation of this luminous parabolic envelope, widening out as its distance from the focus increased, and fading away to invisibility as the luminous matter of which it is composed became more and more rare." For June 2, he wrote, "it was immediately apparent that the comet had undergone a vast change since I observed it on the previous evening, and presented quite a new character. The nucleus was still out of the focus of the parabola, but the dense stream of cometic matter which appeared to issue from it laterally on the evening before had entirely disappeared, as had also the dark interval surrounding it; but streams of light seemed now to issue in a spiral form from the exterior edge of the nucleus, and to be beaten back again towards the tail after they had reached the outer limits of the coma, and were beaten back on that side that was farthest from the nucleus. There was still a dark interval on the interior edge of the nucleus. A great portion of the detail which existed on the previous evening had disappeared, and these spiral streams issuing from the nucleus constituted the principal feature on this occasion. The nucleus appeared slightly larger but less condensed; the graduation of light from nucleus to coma was less sudden, and the coma above the nucleus somewhat brighter." Eddie noted little change on the 4th, although he did say the nucleus was closer to the center of the parabola and the spiral streams were less prominent. For the 4th, Eddie wrote, "the spiral character of the jet from the nucleus could no longer be detected." On the 5th, Eddie said, "the nucleus and upward jet of cometary vapour slightly inclined, and bent over towards the opposite side from where the nucleus was situated." Although no estimates of the coma diameter were made, the tail was widely observed, despite increasing moonlight. The length was given as 8.5° by Tebbutt on the 1st, 4° by Gwynne on the 1st and 2nd, 8° by Eddie and Barker on the 2nd, 9° by Eddie and Barker on the 3rd, 11° by Barker on the 4th, 14° by Eddie on the 5th, and 5° or more by Markwick on the 6th. The tail gave the naked-eye appearance of a "beautiful, dull rocket" on the 1st, according to Markwick. Eddie said it was very straight on the 5th and "shorter, but broader" on the 7th.

The comet passed only 7° from the sun on June 19 and continued its northward motion into the skies of the Northern Hemisphere. C. F. W. Peters (Kiel, Germany) spotted the comet on June 22.90 and noted it was about 1st magnitude, with an easily perceptible tail. As in early June, there were no estimates of the comet's total magnitude, but several astronomers did estimate the magnitude of the nearly stellar nucleus. It was given as 2 by F. A. T. Winnecke (Strasbourg, France) on the 24th, 1.5 by A. L. N. Borrelly (Marseille, France) and 1 by W. Noble (Forest Lodge, Maresfield, England) on the 25th, 2.2 by F. Schwab (Frankfurt, Germany) on the 29th, and about 2 by T. W. Backhouse (Tyndrum, Scotland), 2.1 by Schwab, and 3 by Noble on the 30th. Although several estimates were made of the diameter of the nucleus, H. A. Kobold (O'Gyalla Observatory, Hungary) obtained some consistent measurements with his 15-cm refractor. He gave the diameter as 3.96" on the 24th, 3.58" on the 25th, 3.09" on the 26th, 3.15" on the 27th, 2.98" on the 28th, and 4.33" on the 29th. Several astronomers examined the nucleus and its surrounding region with various sized telescopes. Noble observed with an 11-cm refractor (42×) on the 24th and wrote, "From the brilliant nucleus there extended towards the N. and E. a line of light into the tail. This seemingly expanded into a fan-shaped aigrette to the S. and E. of the nucleus. A small aigrette of light also issued from the nucleus to the south and west of it." On the 25th, Noble wrote, "a most remarkable change had taken place in the appendages of the nucleus." He continued, "Instead of appearing with any jets or line issuing from it, it was surrounded by two sectors of light placed unsymmetrically round it.... The nucleus itself seemed to be connected with the inner concave edge of the interior sector by a kind of radiating hazy luminosity." Viewing the comet about

30 minutes later, Noble said, "the strange arrangement of sectors ... which earlier seemed to bound the Comet's outline, as it were, were now at some distance inside the coma." Borrelly noted three jets emanating from the nucleus on the 27th, while W. J. Förster (Berlin, Germany) said a tuft-like "effusion" extended 46.3" toward PA 142.3°. Nobel wrote after observing the comet on the 28th, "The nucleus had certainly diminished, and the sectoral arrangement surrounding it had changed notably in form. It gave the idea of spouting from the nucleus, but it was very fuzzy and ill-defined...." Also on the 28th, Förster said, the main jet of the "effusion" extended near PA 180° or "perhaps some degrees over it." He added, "The whole jet is much more diffuse than yesterday, also the fan is more diffuse and smaller, but near the same position as yesterday." Förster's colleague, A. J. G. F. von Auwers estimated that the brightest jet extended toward PA 190°. On the 29th, Borrelly saw "several brilliant jets emanating from the nucleus toward the sun." Noble added that two sectors of light were seen emanating from the nucleus, with the preceding one being the brighter. He said they gave the impression of a "sea-gull in flight." F. Küstner (Berlin, Germany) said the "effusion" was sharply limited and extending toward PA 210°. On the 30th, Nobel said the nucleus appeared more stellar than before and saw, "a hazy aigrette of light, bounded on the [south following] side by a dark sector. Faint traces existed of the line of light which was such a feature on the night of the 24th, but it was very short." Küstner described the "effusion" as "very remarkable" and on June 30.90 he measured the PA of the middle as 157.2°, the right-hand side as 131.7°, and the left-hand side as 172.9°. Auwers measured the PA of the middle as 163.0°. Some measurements of the coma were obtained during this period, but most interesting was that some observers were detecting both an inner and an outer coma. N. de Konkoly (Ostend, Belgium) first noticed this on the 26th. Kobold measured the inner coma as 32.48" and the outer one as 106.73" on the 27th, while H. Pomerantzeff (Tashkent, Russia, now Uzbekistan) gave the outer coma as 1' 43". Kobold gave the inner one as 15.40" and the outer one as 55.86" on the 28th, while Pomerantzeff gave the outer one as 1' 28". Kobold gave the inner one as 33.77" and the outer as 85.48" on the 29th. The tail was very well observed during this period. On the 24th, H. J. H. Groneman (Gröningen, Holland) gave the length as 10°, while J. J. Plummer (Orwell Park Observatory, England) said it was 5–6° in evening twilight. On the 25th, Borrelly gave the tail length as 20°, while Kobold said it extended toward PA 351° 50'. On the 26th, Borrelly said the tail had diminished in length and brightness, while Kobold said it extended toward PA 351° 05'. On the 27th, the length was given as 15° by Borrelly and 11° by Pomerantzeff, while W. Schur (Strasbourg, France) and Kobold said it extended toward PA 349° and 353° 16', respectively. On the 28th, Pomerantzeff gave the length as 12°, while Kobold said it extended toward PA 353° 50'. On the 29th, the length was given as 12° by Borrelly and 14° by Schwab, while Kobold said it extended toward PA 345° 34'. On the 30th, the length was given as 5.5° by

Backhouse and 13.5° by Schwab, while Schur said it extended toward PA 350°. Borrelly noted the eastern portion of the tail was brighter and longer on the 27th, and Backhouse suspected some curvature on the 30th. H. Draper (New York, USA) obtained a 2 hour 45 minute photographic exposure on the 24th and said the tail was 10° long. The Chinese text *Chhing Chhao Hsü Wên Hsien Thung Khao*, which was written by Liu Chin-Tsao, notes that Chinese astronomers observed a large "broom-star" comet sometime during the summer toward the northwest. It was described as purple. P. J. C. Janssen (Meudon, France) was the first person to show what photography could really do for comets. On June 30 he photographed this comet through a 50-cm telescope and revealed a radial structure within the tail, as well as several streamers that emanated from around the comet's head and extended into the tail. Up to this time comet photographs had not been of a very high quality and revealed few details.

The comet was moving away from both the sun and Earth as July began, but it was still widely observed during the first half of the month as an easy naked-eye object. T. W. Backhouse (Newhaven, England) provided a few magnitude estimates. He said it was equal to ε Ursae Majoris (magnitude 1.77 – SC2000.0) and brighter than α Ursae Majoris (magnitude 1.79 – SC2000.0) on July 2. He noted it equaled α Ursae Minoris (magnitude 2.02 – SC2000.0) and η Ursae Majoris (magnitude 1.86 – SC2000.0) on the 4th. He said it was "scarcely fainter" than α Ursae Minoris and equal to ζ Ursae Majoris (magnitude 2.27 – SC2000.0) on the 5th. Backhouse (Darlington, England) said he believed the head was between α and γ Ursae Majoris in brightness (magnitude 2.44 – SC2000.0) on the 8th. Schwab reported several magnitude estimates of the "nucleus", which probably referred to the central condensation. He gave estimates of 2.7 on the 1st, 2.9 on the 3rd, 3.0 on the 4th, 5th, and 6th, 3.2 on the 10th, 3.4 on the 12th, 3.25 on the 13th, and 3.4 on the 14th. An indirect indication also came from B. Peter (Leipzig, Germany), who wrote that the comet "disappeared almost suddenly" about 34 minutes before sunrise, at which time it was located about 37° above the northeast horizon. Other magnitude estimates of the "nucleus" were 8 by F. W. Fabritius (Kiev, Ukraine) on the 8th and 5 by Groneman on the 14th. Pomerantzeff reported the nucleus was oblong on the 3rd, with a minor axis measuring 9.1", and then gave a diameter of 4.6" on the 4th. Plummer said the nucleus was "bright and starlike" on the 4th. Kobold measured the diameter of the nucleus as 3.91" on the 5th. On the 7th, O. Stone and H. C. Wilson (Cincinnati, Ohio, USA) reported that the 28.6-cm refractor revealed two nuclei separated by 6" and nearly equal in size. Although they reported their observations of the nuclei spanned several hours, Pomerantzeff later noted that he saw only a single nucleus 7.0" across in his 15-cm refractor on that night. On the 8th, A. Hall, A. N. Skinner, and M. Rock (US Naval Observatory, Washington, DC, USA) observed the comet with the 66.0-cm refractor at magnifications of 175×, 400×, and 900×, and noted the nucleus maintained a "solid stellar appearance."

Kobold said the nucleus was 4.31" across on the 13th. Küstner continued to watch the "effusion" that extended from the nucleus through a 23-cm refractor. He said it was broad on the 1st and measured the position angle of the right side as 105.6°, the center as 178.3°, and the left side as 259.7°. He said it was similar in general appearance on the 2nd, but added that the right-hand side of the fan was sharper and brighter, extending toward PA 140°, while the left-hand side was very diffuse and extended toward PA 250°. Küstner described it as a narrow bundle on the 4th and said it extended toward PA 140.0°. On the 5th, it was again fan-like, with Küstner measuring the position angle of the right-hand side as 115.8°, the center as 176.5°, and the left-hand side as 227.8°. He described it as a strongly curved bundle on the 6th and noted the right-hand side was brightest. His measurements indicated the more prominent line extended toward PA 152.3°, while the convex and sharp right-hand side extended toward PA 137° and the concave and diffuse left-hand side extended toward PA 205°. On the 7th, Küstner said the brightest part of the fan-like emission extended toward PA 151.8°, while the bright, sharp right-hand side extended toward PA 122.2° and the faint, diffuse left-hand side extended toward PA 208.4°. He said the "effusion" extended toward PA 210° on the 11th. Pomerantzeff continued to provide measurements of the coma diameter. He gave it as 1' 22" on the 3rd, 1' 14" on the 4th, and 1' 18" on the 7th. Kobold gave the diameter of the inner coma as 17.34" on the 5th. The tail was widely observed and very consistent estimates of the length came from Backhouse, Schwab, Pomerantzeff, and Groneman. Backhouse's estimates provide a more continuous record and were 13° on the 1st, 17.5° on the 2nd, 20.5° on the 3rd, 17° on the 4th, 16° on the 5th, and 8.3° on the 8th. He provided no additional estimates during the remainder of this period, but Schwab continued his observations. He gave the length as 6–8° on the 10th, 2° in moonlight on the 12th and 13th, and 4–5° in moonlight on the 14th. Kobold said the tail extended toward PA 5° 56' on the 5th. Backhouse indicated the tail was about 2.5° wide on the 2nd, 3.8° wide at 12.3° from the nucleus on the 4th, and 4.5° wide at 11° from the nucleus on the 5th. He also reported it was "probably curved" on the 1st, "slightly curved" on the 2nd, and on the 4th its brightness at 6° from the nucleus was "as bright as the brightest parts of the Milky Way in Cygnus, which are in rather darker sky." The Chinese text *Ta Chhing Li Chhao Shih Lu* notes that on July 4 "an edict was issued to the Grand Secretariat saying that a 'broom-star' comet was seen in the north for the past few days." An imperial decree was subsequently "written to the palace and court officers ordering them to perform their respective duties conscientiously."

Fewer descriptions were published for the remainder of July and the comet attained its most northerly declination of +82° on the 17th. Backhouse continued to make occasional estimates of the total brightness. On the 15th while in North Berwick (Scotland), he said the head equalled ε Cassiopeiae (magnitude 3.38 – SC2000.0). On the 21st, when back in Sunderland, he said the head was "rather brighter" than κ Draconis (magnitude

3.87 – SC2000.0), and about half way between ε Cassiopeiae (magnitude 3.38 – SC2000.0) and ζ Ursae Minoris (magnitude 4.32 – SC2000.0). On the 28th, Backhouse said the comet's head was "rather brighter" than δ Ursae Minoris (magnitude 4.36 – SC2000.0), "slightly brighter" than λ Draconis (magnitude 3.84 – SC2000.0), "considerably fainter" than κ Draconis (magnitude 3.87 – SC2000.0), and two-thirds the brightest range from κ Draconis to δ Ursae Minoris. Schwab also made several estimates of the total magnitude, giving magnitudes of 3.45 on the 15th, 3.3 on the 16th, 3.4 on the 17th, 3.7 on the 18th, and 4.75 on the 28th. The nucleus was not scrutinized as much as before, but a few interesting observations were provided by Pomerantzeff, A. Abetti (Padova, Italy), and O. C. Wendall (Harvard College Observatory, Cambridge, Massachusetts, USA). Pomerantzeff measured the diameter as 10.2″ on the 16th and 10.5″ on the 17th. Abetti described the nucleus as a luminous point of magnitude 9 on the 20th, but he noted it was irregular in shape on the 24th. Also on the 24th, Wendell reported the nucleus was elongated, with an apparent "notch", which he said indicated a possible division. Measurements indicated the nucleus was 10.9″ long, with the elongation extending toward PA 226°. The nucleus was 5.4″ in width. Wendall said the nucleus had regained its normal appearance when he next saw it on July 26. Abetti said the nucleus appeared as a disk rather than a luminous point on the 27th. The only measurements of the coma diameter came from Pomerantzeff, who gave it as 1′ 00″ on the 16th and 43″ on the 17th. Backhouse, Schwab, and Pomerantzeff provided the only estimates of the tail length during this period, with Backhouse again providing the most continuous series. Backhouse gave the length as 8.5° on the 15th, 11° on the 21st, 9.5° on the 22nd, and 9.5° on the 28th. He added that it was 3° wide at the end on the 21st.

Total magnitude estimates were made by Backhouse and Schwab during August. Backhouse's were again made with the naked eye. He reported the comet's head was "slightly brighter" than 5 Ursae Minoris (magnitude 4.25 – SC2000.0) or ε Ursae Minoris (magnitude 4.23 – SC2000.0) and "rather fainter" than δ Ursae Minoris (magnitude 4.36 – SC2000.0) on the 1st. On the 5th, he said the head was "scarcely" equal to 4 Ursae Minoris (magnitude 4.82 – SC2000.0) and "considerably brighter" than 5 Ursae Minoris (magnitude 4.25 – SC2000.0). Schwab was using a telescope and reported magnitudes of 6.0 on the 7th, 6.1 on the 9th, 6.5 on the 13th, 6.7 on the 16th and 20th, 6.9 on the 28th, and 7.0 on the 29th and 30th. It is probable that Schwab's magnitudes were referring to the central condensation or inner coma, especially as Backhouse said he saw the comet with the naked eye in dark skies on the 18th, as well as later on in September. There were no observations of the nucleus or coma, but Backhouse, Schwab, and C. F. Pechüle (Copenhagen, Denmark) reported information on the tail. Backhouse gave naked-eye tail lengths of 7.5° on the 2nd, 6.5° on the 5th, and 1° on the 18th. Schwab noted only a trace of tail in moonlight on the 7th, 9th, and 13th. With the moonlight no longer an issue, Schwab still reported that only a trace of tail was visible during the period spanning August 16–28. Schwab did note a tail 0.3–0.4° long on the 29th. Pechüle saw a tail 0.75° long through the finder of the 27-cm refractor on the 16th. He said the tail extended toward PA 82° ± 2°.

The only probable estimates of the comet's total magnitude for September came from Backhouse, but they were few. On both September 14 and 19, he noted that a pair of 3.5 × 38 binoculars revealed the comet's head was "considerably fainter" than 20 Ursae Minoris (magnitude 6.4 – SC2000.0). Moonlight was interfering on the first date. Backhouse specifically noted the comet was not visible to the naked eye because of moonlight on the 14th. In dark skies, Backhouse reported he could not see the comet with the naked eye on the 16th and 19th. Interestingly, Backhouse said he saw the comet with the naked eye while visiting in Scarborough (England) on the 28th. He said it was "visible together with a much fainter star." Schwab also provided several magnitude estimates of the comet's head, but these may have referred more to the central condensation. He said it was magnitude 8.2 on the 14th, 8.3 on the 17th, 8.25 on the 20th, 8.15 on the 24th and 25th, 8.20 on the 28th, and 8.35 on the 30th. Several estimates of the coma diameter were obtained by Backhouse while observing with a 10.8-cm refractor (20×). He gave the diameter as 8′ on the 14th, 7.5′ on the 16th, and 15′ on the 19th. J. F. J. Schmidt (Athens, Greece) said the coma was 5–6′ across on the 28th. Several estimates of the tail length were made by various astronomers, using small telescopes, finders, and binoculars. In moonlight, Backhouse gave the length as 40′ on the 7th and 25′ on the 14th. He added that the tail was 8′ wide on the 14th. In dark skies the length was given as 40′ by Backhouse on the 16th, 10–15′ by Tempel on the 17th, and 10′ by Backhouse on the 19th. On September 25, Schwab noted a possible trace of tail.

Physical descriptions were not plentiful during the remainder of this comet's apparition. The only descriptions published for October were a continuation of Schwab's total brightness estimates. He gave the magnitude as 8.6 on the 2nd, 8.8 on the 14th and 15th, 8.7 on the 16th, 8.8 on the 17th, 9.0 on the 18th, and 8.85 on the 19th. Peter described the comet as a faint nebulosity on November 9, with a condensation and one or more nuclei. The coma was irregular in shape and elongated in an east–west direction. On November 22, Peter said the very faint coma was 3–4′ across, with a slight condensation and one or more nuclei. No tail was seen in the 12-foot focal length refractor. Schmidt said the comet was faint and about 5′ across on November 25, with a difficult to determine center. It was invisible before the moon had set. Backhouse observed the comet on several occasions from October 30 to November 24 using the 10.8-cm refractor (20× and 38×). He always found it "very faint, but rather large; generally 6′ to 8′ in diameter." He said it was too faint to be observed "shortly after the last date mentioned." On December 10, 11, and 20, Schmidt said the extremely faint comet was 2–3′ across. He noted it was impossible to measure its position. Peter described the comet as a very faint and expanded nebulosity with some granulation on the 10th. Observations during January were primarily limited to observers using large refractors, most notably E. Frisby (US Naval Observatory, Washington DC, USA), who was using a 66-cm refractor, and Wendell, who was using a 38-cm. Winnecke observed the comet with the 46-cm refractor on February 11 and 12. On the former date he said the comet was 20″ in diameter, without a distinct nucleus.

The comet was last detected on 1882 February 15.13, when Wendell observed with the 38-cm refractor and gave the position as $\alpha = 0^h$ 15.9^m, $\delta = +55° 05'$. He said it was very faint and "doubtful."

The spectrum was observed by several astronomers. Backhouse observed the spectrum on several occasions beginning on July 2 and ending on September 16. He was always able to see the continuous spectrum from the nucleus, which was bright at first, but quite faint in September. He also observed "the three usual bright cometary bands," as well as a much fainter one in the violet part of the spectrum. Müller and Kempf (Potsdam, Germany) saw only traces of the comet bands on June 24, but noted a steady brightening so that all three were visible on June 30. They last observed the spectrum on July 6 and noted bands at 5540 Å, 5090 Å, and 4698 Å. W. Huggins obtained a one-hour photographic exposure of the comet's spectrum on June 24 and noted a continuous spectrum, as well as "a pair of bright lines . . . in the ultra violet region, which appear to belong to the spectrum of carbon (in some form)." He measured the wavelength of these lines as 3870 Å and 3890 Å. These lines were confirmed by a one-and-a-half hour exposure on the 25th. This marked the first time a comet's spectrum had been photographed. Other spectra were obtained during the period of late June to early July by W. H. M. Christie and E. W. Maunder (Royal Observatory, Greenwich, England), L. Thollon, and Konkoly (O'Gyalla Observatory, Hungary).

Additional details of this comet come from F. A. Bredikhin and Konkoly. Bredikhin's examination of the observations led him to conclude the comet exhibited both a type I and a type II tail. Konkoly observed with a Savart polariscope on June 26 and noticed lines appearing within the coma, which were more intense on the western side of the tail and faint on the opposite side.

The first parabolic orbit was calculated by Gould using positions from May 26, 28, and 30. He determined the perihelion date as 1881 June 18.87. He noted it was "essentially the same orbit that Bessel found for the great comet of 1807, for which the most probable elliptic elements indicated a return after 1540 years, allowance being made for the effect of planetary perturbations. We thus have a new case analogous to that of the comet of February 1880; inasmuch as we are brought to the alternative of supposing on the one hand, that the true length of the major axis is not more than the sixth or seventh part of that which has been deduced with all possible care from good observations, or on the other hand, that more than one large comet is traversing the same orbit." Additional parabolic orbits were published during the next few months by R. L. J. Ellery, E. J. White, W. Fabritius, J. M. Thome, Tebbutt, C. F. W. Peters, G. Bigourdan, M. W. Meyer, W. L. Elkin, J. R. Hind, A. Graham, H. T. Vivian, S. C. Chandler, Jr, O. C. Wendell, T. Wittram, L. Weinek, J. Rahts, F. Contarino, F. Angelitti, A. Lindstedt, H. Romberg, H. Oppenheim, Frisby, F. Deichmüller, K. Zelbr, Abetti, and V. Ventosa. Deichmüller noted that a parabola perfectly fitted the observations and said this should settle the question about this comet being identical to that seen in 1807.

The first elliptical orbit was calculated by N. C. Dunér and F. Engström using positions spanning the period of May 23 to August 13. They determined the perihelion date as June 16.94 and the period as about 2385 years. Using positions up to September 3, these astronomers later revised their elliptical orbit and gave the perihelion date as June 16.94 and the period as about 2954 years. At the end of October, J. F. Bossert took positions obtained up to September 29 and also gave the period as about 2954 years. A definitive orbit was calculated by J. K. R. Riem (1896) using positions spanning the entire period of visibility. The period was given as about 2439 years. This orbit is given below.

T	ω	Ω (2000.0)	i	q	e
1881 Jun. 16.9407 (UT)	354.2350	272.6308	63.4253	0.734558	0.995946

ABSOLUTE MAGNITUDE: $H_{10} = 4.1$ (V1964)

FULL MOON: May 13, Jun. 12, Jul. 11, Aug. 9, Sep. 8, Oct. 7, Nov. 6, Dec. 5, Jan. 4, Feb. 3

SOURCES: P. J. C. Janssen, *VJS*, **16** (1881), pp. 308–11; B. A. Gould, *AN*, **100** (1881 Jun. 4), p. 15; L. Cruls, *AN*, **100** (1881 Jun. 17), p. 53; C. N. A. Krueger, C. F. W. Peters, G. F. W. Rümker, B. Peter, C. N. J. Börgen, and F. A. T. Winnecke, *AN*, **100** (1881 Jun. 29), pp. 75–80; W. Fabritius, *AN*, **100** (1881 Jul. 6), p. 95; L. Thollon, *CR*, **93** (1881 Jul. 11), pp. 37–9; W. L. Elkin and J. R. Hind, *Nature*, **24** (1881 Jul. 14), p. 248; B. A. Gould, F. A. T. Winnecke, W. Schur, and M. W. Meyer, *AN*, **100** (1881 Jul. 16), pp. 107–12; G. Bigourdan, *CR*, **93** (1881 Jul. 25), pp. 197–8; B. A. Gould, J. M. Thome, H. Oppenheim, J. Tebbutt, R. L. J. Ellery, S. C. Chandler, Jr, O. C. Wendell, T. Wittram, L. Weinek, J. Rahts, F. Contarino, F. Angelitti, A. Lindstedt, H. Romberg, V. Vendosa, La Cruz, and Jimenez, *AN*, **100** (1881 Jul. 26), pp. 113, 119–26; W. Huggins, *AN*, **100** (1881 Jul. 29), p. 143; B. A. Gould, *Nature*, **24** (1881 Aug. 11), p. 342; F. Deichmüller, J. Tebbutt, and E. Frisby, *AN*, **100** (1881 Aug. 13), pp. 171–6; E. J. White, *AN*, **100** (1881 Aug. 17), p. 187; K. Zelbr, *AN*, **100** (1881 Aug. 22), pp. 203–6; N. C. Dunér and F. Engström, *AN*, **100** (1881 Aug. 26), p. 217; N. de Konkoly, *The Observatory*, **4** (1881 Sep.), pp. 257–60; F. Deichmüller, *AN*, **100** (1881 Sep. 3), p. 255; J. Kowalczyk and E. Becker, *AN*, **100** (1881 Sep. 8), p. 271; N. C. Dunér and F. Engström, *AN*, **100** (1881 Sep. 15), p. 283; A. Abetti, H. C. Vogel, Müller, Kempf, A. Belopolsky, and A. Socoloff, *AN*, **100** (1881 Sep. 30), pp. 293–304; J. Tebbutt, *AN*, **100** (1881 Oct. 7), p. 335; A. L. N. Borrelly and H. A. Kobold, *AN*, **100** (1881 Oct. 15), pp. 337–44; E. W. L. Tempel and V. Ventosa, *AN*, **100** (1881 Oct. 25), pp. 373–6; J. F. Bossert, *CR*, **93** (1881 Oct. 31), pp. 659–60; R. L. J. Ellery, E. J. White, L. A. Eddie, B. E. W. Gwynne, E. C. Markwick, A. Graham, and J. Tebbutt, *MNRAS*, **41** (1881, Supplementary Notice), pp. 432–43;

W. H. M. Christie, E. W. Maunder, S. J. Perry, H. T. Vivian, C. Todd, W. Noble, I. Maling, S. S. O. Morris, and B. E. W. Gwynne, *MNRAS*, **42** (1881 Nov.), pp. 14–17, 43–50; W. Huggins, *PRSL*, **33** (1881 Nov. 17), pp. 1–3; F. W. Fabritius, *AN*, **101** (1881 Nov. 21), pp. 55–8; H. J. H. Groneman, *AN*, **101** (1881 Nov. 26), pp. 69–72; W. F. Denning, *MNRAS*, **42** (1881 Dec.), pp. 78–9; H. Pomerantzeff, J. R. Eastmann, A. N. Skinner, M. Rock, and W. C. Winlock, *AN*, **101** (1881 Dec. 5), pp. 89–92; J. Winlock, E. C. Pickering, and O. C. Wendell, *AAOHC*, **13** (1882), pp. 175, 178–81; F. A. Bredikhin, *AN*, **101** (1882 Jan. 16), p. 161; F. Schwab, *AN*, **101** (1882 Jan. 20), pp. 189–92; P. Andries, P. Tacchini, and E. Millosevich, *AN*, **101** (1882 Feb. 10), pp. 231, 235–8; J. F. J. Schmidt, *AN*, **101** (1882 Feb. 20), pp. 249–52; J. F. J. Schmidt, *AN*, **101** (1882 Feb. 27), p. 299; D. W. Barker, *MNRAS*, **42** (1882 Mar.), p. 266; P. Henry, L. Swift, W. Doberck, H. Draper, N. de Konkoly, and O. C. Wendell, *SM*, **1** (1882 Mar.), pp. 10–13; B. Peter, *AN*, **101** (1882 Mar. 7), p. 335; J. J. Plummer and B. Peter, *AN*, **101** (1882 Mar. 24), pp. 371–6, 379–84; E. Frisby, A. Hall, A. N. Skinner, and M. Rock, *AN*, **102** (1882 May 31), p. 105; J. F. Parson, *MNRAS*, **42** (1882 Jun.), pp. 421–2; J. Palisa and K. Zelbr, *AN*, **102** (1882 Jul. 18), p. 277; F. Küstner, W. J. Förster, and A. J. G. F. von Auwers, *AN*, **103** (1882 Aug. 17), pp. 29–32; O. Stone and H. C. Wilson, *SM*, **1** (1882 Oct.), pp. 163–4; D. Gill and W. L. Elkin, *MNRAS*, **43** (1882 Nov.), pp. 7–9; C. F. Pechüle, *AN*, **103** (1882 Nov. 15), pp. 339–42; W. Schur, *AN*, **104** (1883 Feb. 12), pp. 305–12; L. de Ball, *AN*, **108** (1884 Mar. 13), p. 157; H. Romberg, *AN*, **113** (1885 Dec. 15), p. 145; F. A. T. Winnecke, *AN*, **114** (1886 May 6), pp. 229, 233; J. K. R. Riem, *Ueber eine frühere Erscheinung des Kometen 1881 III Tebbut*. Göttingen: E. A. Huth (1896), 26 pp.; V1964, p. 61; HA1970, p. 87; *Griffith Observer*, **48** (1984 Jul.), p. 11; W. Orchiston, *The Observatory*, **111** (1991 Dec.), pp. 313–14; SC2000.0, pp. 38, 50, 117, 123, 126, 129, 282, 292, 299, 314, 320, 330, 338, 345, 352, 380, 391, 403, 423.

C/1881 N1   *Discovered:* 1881 July 14.33 ($\Delta = 1.79$ AU, $r = 1.03$ AU, Elong. $= 29°$)
(Schaeberle)   *Last seen:* 1881 October 21.42 ($\Delta = 2.15$ AU, $r = 1.33$ AU, Elong. $= 26°$)
   Closest to the Earth: 1881 August 25 (0.5771 AU)
1881 IV = 1881c   *Calculated path:* AUR (Disc), LYN (Aug. 4), UMa (Aug. 14), CVn (Aug. 24), COM (Aug. 25), CVn-COM (Aug. 26), VIR (Aug. 31), BOO (Sep. 1), VIR (Sep. 3), LIB (Sep. 12), HYA (Oct. 3), CEN (Oct. 17), LUP (Oct. 18)

J. M. Schaeberle (Ann Arbor, Michigan, USA) discovered this comet on 1881 July 14.33, at a position of $\alpha = 5^h 45.0^m$, $\delta = +38° 37'$. He said it exhibited a stellar nucleus within a coma less than 2′ across.

The comet was approaching both the sun and Earth when discovered. On July 20, J. E. Coggia (Marseille, France) said the comet was "very beautiful, with a bright center." Although there was no distinct nucleus, the 26-cm refractor (130×) revealed several bright points. A. Abetti (Padova, Italy) described the comet as a tiny, round spot of uniform appearance on the 21st. He added that no tail was present. On the 22nd, E. W. L. Tempel (Arcetri, Italy) estimated the brightness as 6 and said the 27-cm refractor revealed a very diffuse tail about 1° long. Abetti still said no tail was visible. M. W. Meyer (Geneva, Switzerland) said the comet was bright and 3′ across in the 25.4-cm telescope, but reported no tail or distinct nucleus. That same night,

J. Franz (Königsberg, now Kaliningrad, Russia) said the comet was bright, round, but without a tail. It was not seen with the naked eye. Abetti suspected a tail on the 25th, while Meyer reported a trace of tail and a distinct star-like nucleus. T. W. Backhouse (Sunderland, England) said the comet was seen with the naked eye in weak twilight on July 27. He also noted a straight tail extending about 1.25°. Using a small pair of opera glasses, with a magnification of 2.5×, he said the tail extended 2.7° and was 0.7° wide. He also estimated the brightness of the head as equal to that of 55 Aurigae (magnitude 5.02 – SC2000.0) and slightly fainter than that of 56 Aurigae (magnitude 5.22 – SC2000.0). Meyer also saw the comet with the naked eye that night and said the 25.4-cm telescope revealed a strong condensation, while the finder showed a tail about 0.5° long. Another description came from G. Cacciatore (Palermo, Italy), who noted a faint nucleus, round inner coma, and a faint tail in the 25-cm refractor. On the 28th, Franz noted the comet had developed a parabolic shape, while Coggia noted a round, well-defined nucleus of magnitude 10. The length of the faint tail was given as 4′ by Coggia and 7′ by Meyer. Backhouse, Coggia, and Franz all reported the comet was visible to the naked eye on the 29th, with Backhouse estimating the brightness equaled that of 21 Lyncis (magnitude 4.64 – SC2000.0). Cacciatore also reported the comet and tail appeared brighter. Backhouse added that the naked-eye tail length was 5.8°. Cacciatore reported a more pronounced nucleus and a better developed tail on July 30. G. V. Schiaparelli (Brera Observatory, Milan, Italy) added that the coma was 2′ across. On the 31st, Abetti saw the tail with the naked eye, while Backhouse reported it was slightly curved, with the concave side to the south.

The comet was a naked-eye object throughout August, as it passed closest to both the sun and Earth during the second half of the month. Indications of how bright the comet was becoming came from Abetti and Backhouse. Abetti noted the comet was seen with the naked eye despite foggy skies on August 1, and Backhouse said it was just visible to the naked eye on the 9th despite a full moon. A few estimates or indications of the comet's total magnitude were made during the month. Backhouse said it was "considerably brighter" than 21 Lyncis (magnitude 4.64 – SC2000.0) on the 5th and 9th, while it was "rather fainter" than κ Ursae Majoris (magnitude 3.60 – SC2000.0) on the 9th. Backhouse said it was "rather fainter" than θ Ursae Majoris (magnitude 3.17 – SC2000.0) on the 14th. On the 16th, Meyer said it was magnitude 2 or 3, while Franz estimated it as 3. Backhouse said it was "rather brighter" than θ Ursae Majoris on the 18th, while F. Schwab (Frankfurt, Germany) estimated the magnitude as 4.05. Backhouse noted it was "considerably brighter" than θ Ursae Majoris (magnitude 3.17 – SC2000.0) or ψ Ursae Majoris (magnitude 3.01 – SC2000.0) on the 20th. Schwab gave the magnitude as 3.62 on the 20th, 3.45 on the 22nd, and 3.4 on the 25th. Backhouse said it was "much brighter" than β Comae Berenices (magnitude 4.26 – SC2000.0) on the 27th and not quite as bright as η Boötis (magnitude 2.68 – SC2000.0) on the 28th. Schwab gave magnitudes of 3.63 on the 28th, 4.10 on the 29th, and 4.36 on the 30th. Many observers reported a "well-defined" and even "stellar" nucleus during the month, but there were a few exceptions. Coggia reported the nucleus was oval and diffuse in the 26-cm refractor (130×) on the 2nd. J. J. Plummer (Orwell Park Observatory, England) said it was ill defined on the 6th and elongated north and south on the 11th. Several estimates of the nuclear magnitude were made as well. It was given as 7 by Schiaparelli on the 2nd, 5 by Abetti on the 16th and 25th, 7 by Abetti on the 27th, and 8 by Abetti on the 29th. A few observers noted features in the region near the nucleus. C. F. Pechüle (Copenhagen, Denmark) said the 27-cm refractor revealed several "effusions" extending from the nucleus into the tail on the 17th. Backhouse wrote on the 18th, "There is an indefinite projection from the nucleus towards [south-preceding], to a distance of 12′ or 15′ from it." W. Noble (Forest Lodge, Maresfield, England) observed with an 11-cm refractor (115×) and wrote, "I saw that the nucleus proper was an exceedingly minute point, a considerable distance within the head. Nothing whatever in the shape of a sector, jet, or aigrette of light was visible." Numerous measurements of the coma diameter were made. Some of these were 4′ by Cacciatore on the 1st, 3–4′ by Schiaparelli on the 2nd, 6′ by Cacciatore on the 5th, 6′ by Schiaparelli on the 18th, 8′ by Schiaparelli on the 24th, 9′ by Schiaparelli on the 25th, 10′ by Schiaparelli on the 26th, and 5′ by Pechüle on the 29th. A much smaller or inner coma was reported by some observers. Cacciatore gave its size as 45″ across on the 5th, while Pomerantzeff gave it as 40″ across on the 29th. The tail was the most watched feature of the comet, but where many only gave lengths as seen through telescopes, some observers indicated the tail was much longer to the naked eye or binoculars. Backhouse gave a very good series of tail length estimates. He said it was 6.8° long and strongly curved on the 4th, 0.9° long with a full moon on the 9th, 2° long in moonlight on 10th, 4.75° long and slightly curved on the 14th, 8.5° long and 1.5° wide on the 18th, 10.3° long and 1.5° wide on the 19th, 8° long and 1.5° wide on the 20th, and 2.8° long and 0.3° wide on the 28th. Backhouse did not estimate tail lengths during the period of August 20–28; however, tail lengths were then reported as 4° 15′ by F. Terby (Leuven, Belgium) on the 22nd, 7–8° by Pechüle on the 23rd, 5° 25′ by Terby on the 24th, 5–6° by Schiaparelli on the 24th and 8° by Schiaparelli on the 25th. Pechüle carefully measured the direction in which the tail was pointing on several occasions. He gave the direction as PA 352.6° ± 1.0° on the 17th, PA 17.6° ± 1.2° on the morning of the 22nd and PA 29.9° ± 1.0° on the evening of the 22nd, PA 38° on the 23rd, and PA 71.0° ± 0.3° on the 29th. Pechüle added that the eastern edge of the tail was sharper than the western edge on the 17th and 22nd. On the 18th, Backhouse observed with an 11-cm refractor (20×) and wrote, "From near the nucleus to a distance of a degree from it, the brightest part of the Comet is a narrow ray, almost straight, only 1′ or 2′ wide. Preceding this the tail is very faint, except near the nucleus, and at a distance of 1° or 1.5° from the nucleus fades out altogether, the [preceding] edge of the tail being, beyond that, almost a continuation of the narrow ray."

News of the comet was very slow to reach observatories in the Southern Hemisphere. Consequently, several independent discoveries were reported. M. Clarkson (Kimberly, South Africa) first saw the comet on the evening of August 28.7 and acquired a couple of additional observations before sending a telegram to Dun Echt Observatory in Scotland. J. Tebbutt (Windsor, New South Wales, Australia) was sweeping for comets with his 8.3-cm refractor when he spotted " a small nebulous object" on September 17.41. He checked the available catalogs of nebulae, but found nothing in the position of this object. After an hour he noted that it had slightly shifted its position. The comet was oval, "with a very gradual condensation towards the centre," and was so faint it would not withstand the illumination of the filar micrometer. He observed it the next evening and noted the comet had "moved considerably east and south of its former position." Tebbutt said there was no nucleus or tail.

Meanwhile, the comet's southward motion made September the final month of observations for Northern Hemisphere observers. B. Peter (Leipzig, Germany) described the comet as about 9th magnitude on the 5th, with a condensation and nucleus. Tempel saw the comet on the 9th and said it still appeared bright in the 10-cm refractor. B. A. Gould (Cordoba, Argentina) photographed the comet on the 10th and noted it was still visible to the naked eye on the 14th. Tebbutt reported a "great change" in the comet on the 19th, when he said, "It was the only evening on which there was anything like a central condensation with tolerably defined limits." He also noticed "rudiments of a tail." Tebbutt said the comet "became rapidly fainter" thereafter.

The final observations were made in October. W. L. Elkin (Royal Observatory, Cape of Good Hope, South Africa) described the comet as faint in the 8.5-foot focal length refractor on the 2nd. Wiggin (Cordoba, Argentina) also reported it as faint on the 7th. Tebbutt last saw it on October 14 and 15 and noted "it could be observed only by looking obliquely into the telescope." D. Gill (Royal Observatory, Cape of Good Hope, South Africa) said the comet was only barely visible in the 8.5-foot focal length refractor on the 16th.

The final observations of this comet were made by R. L. J. Ellery (Melbourne, Australia). On October 19.42, he gave the comet's position as $\alpha = 14^h 56.2^m$, $\delta = -30° 01'$. This was the last time the position could be measured. Ellery's final observation came on October 21.42, but no position could be obtained. The comet's angular distance from the sun was continuing to decrease.

The spectrum was observed by several astronomers. Backhouse used a spectroscope on his 11-cm refractor on July 27 and August 19. On the 27th he said the spectrum "consists of the three usual bright bands" with a faint continuous spectrum. On the 19th he noted the spectrum was the same, but brighter, with the "three bright bands by far the most conspicuous part." On August 21, Noble used a spectroscope on an 11-cm refractor and saw "a very bright and distinct spectrum of carbon... the three bright lines being crossed by a horizontal continuous one, presumably arising from the minute nucleus." Reports were also submitted by H. C. Vogel and Cacciatore.

Noble viewed the comet with a Nicol's prism in the 11-cm refractor on August 24 and said "the whole coma exhibited distinct traces of polarization."

The first parabolic orbit was calculated by J. von Hepperger using positions from July 19 to July 21. He determined the perihelion date as 1881 August 12.10. Using positions spanning the period of July 19–24, Hepperger revised the perihelion date to August 22.69. Additional orbits were calculated by G. Bigourdan, M. W. Meyer, H. T. Vivian, H. Oppenheim, O. Stone, and Hepperger. The perihelion date was eventually determined as August 22.81.

A definitive orbit was calculated by C. Stechert (1884) using positions spanning the entire period of visibility. He determined both a parabolic and a hyperbolic orbit. The parabolic was very similar to those calculated by other astronomers, while the hyperbolic had a perihelion date of August 22.81 and an eccentricity of 1.0001243. The hyperbolic orbit is given below.

T	ω	Ω (2000.0)	i	q	e
1881 Aug. 22.8062 (UT)	122.1446	98.7221	140.2181	0.633536	1.000124

ABSOLUTE MAGNITUDE: $H_{10} = 6.3$ (V1964)

FULL MOON: Jul. 11, Aug. 9, Sep. 8, Oct. 7, Nov. 6

SOURCES: J. von Hepperger, C. Schrader, V. Knorre, and C. F. W. Peters, *AN*, **100** (1881 Jul. 26), p. 127; J. von Hepperger, *AN*, **100** (1881 Jul. 29), p. 143; G. Bigourdan, *CR*, **93** (1881 Aug. 1), pp. 258–9; M. W. Meyer and C. Schrader, *AN*, **100** (1881 Aug. 13), p. 173; H. Oppenheim and O. Stone, *AN*, **100** (1881 Aug. 17), pp. 189–92; J. von Hepperger and E. W. L. Tempel, *AN*, **100** (1881 Aug. 26), p. 217, 221; B. Peter, *AN*, **100** (1881 Sep. 3), p. 255; M. Clarkson, *AN*, **100** (1881 Sep. 8), p. 271; A. Abetti and H. C. Vogel, *AN*, **100** (1881 Sep. 30), pp. 297–304; J. E. Coggia and H. A. Kobold, *AN*, **100** (1881 Oct. 15), pp. 337–40, 343; B. Peter, *AN*, **100** (1881 Oct. 20), p. 367; E. W. L. Tempel, *AN*, **100** (1881 Oct. 25), p. 373; T. W. Backhouse, W. Noble, and I. Maling, *MNRAS*, **42** (1881 Nov.), pp. 40–3, 48–50; J. Tebbutt, *AN*, **101** (1881 Nov. 11), p. 31; W. F. Denning, *MNRAS*, **42** (1881 Dec.), pp. 78–9; H. Pomerantzeff, *AN*, **101** (1881 Dec. 5), p. 91; B. A. Gould, J. M. Thome, W. G. Davis, Wiggin, and R. L. J. Ellery, *AN*, **101** (1882 Jan. 2), pp. 139–44; F. Schwab, *AN*, **101** (1882 Jan. 20), p. 191; F. Terby, *AN*, **101** (1882 Feb. 4), p. 223; J. J. Plummer and G. Cacciatore, *AN*, **101** (1882 Feb. 10), pp. 233–6; J. Tebbutt, *AN*, **101** (1882 Feb. 27), pp. 301–4; J. Tebbutt, *MNRAS*, **42** (1882 Mar.), pp. 263–5; J. M. Schaeberle, *SM*, **1** (1882 Mar.), p. 13; M. W. Meyer, *AN*, **101** (1882 Mar. 24), pp. 375–80; E. Weiss, J. Palisa, K. Zelbr, and J. von Hepperger, *AN*, **102** (1882 Jul. 18), p. 277; D. Gill, W. L. Elkin, and H. T. Vivian, *MNRAS*, **43** (1882 Nov.), pp. 7–12, 33; C. F. Pechüle, *AN*, **103** (1882 Nov. 15), p. 341; G. V. Schiaparelli, *AN*, **105** (1883 Mar. 14), pp. 25–8; J. C. Adams, *MNRAS*, **43** (1883 Apr.), pp. 318–19; J. Franz, *AN*, **105** (1883 Apr. 25), pp. 181–6; C. Stechert, *AN*, **108** (1884 Mar. 20), pp. 177–232; C. Stechert, *AN*, **108** (1884 May 13), p. 437; F. A. T. Winnecke, *AN*, **114** (1886 May 6), p. 231; V1964, p. 62; SC2000.0, pp. 158–9, 182, 234, 247, 284, 326, 341.

C/1881 S1   *Discovered:* 1881 September 18.04 ($\Delta = 1.10$ AU, $r = 0.46$ AU, Elong. $= 24°$)
(Barnard)   *Last seen:* 1881 October 28.41 ($\Delta = 1.60$ AU, $r = 1.09$ AU, Elong. $= 42°$)
    Closest to the Earth: 1881 August 30 (0.8429 AU)
1881 VI = 1881e   *Calculated path:* VIR (Disc), BOO (Sep. 24), CVn (Oct. 28)

E. E. Barnard (Nashville, Tennessee, USA) discovered this comet on 1881 September 18.04 in the course of a regular sweep for comets first begun during May. He was using a 13-cm refractor. The comet was then at a low altitude and set before a position could be determined, although he noted it was about 2° west of ζ Virginis. Barnard described the comet as "round, brightly condensed and about 2′ in diameter." Barnard saw the comet again on the 19th and suspected a faint tail, but once again it was seen for only a short time and no position was obtained. Notifications were sent to Harvard College Observatory and to L. Swift (Rochester, New York, USA). On the 20th, the comet was again seen by Barnard, but it was then too far north of ζ Virginis to allow a position measurement using his ring-micrometer. Barnard went to Vanderbilt University on the 21st and saw O. H. Landreth. Using a 15-cm refractor, Landreth found the comet and gave its position as $\alpha = 13^h\ 28.0^m$, $\delta = +3°\ 47'$ on September 21.04. The initial announcement from the Smithsonian Institution to observatories in the USA and Europe contained an erroneous position of $\alpha = 7^h\ 46^m$, $\delta = +13°\ 28'$. A correction was sent the next day and both the incorrect and correct positions appeared in the September 30 issue of the *Astronomische Nachrichten*.

The comet was moving away from both the sun and Earth, and, as Barnard later noted, it remained a difficult object to see because of its continued close proximity to the horizon. On September 29, Barnard and Landreth noted a "short, faint, hazy tail" which extended 6′ or 8′ toward a southeasterly direction. On October 3, E. Hartwig (Strasbourg, France) saw the comet through a large finder and said the magnitude was slightly greater than 8, while the coma was round and 2′ across. Meanwhile, the large refractor showed the comet as a bright nebulosity, without a nucleus. E. Millosevich (Rome, Italy) noted a nucleus on the 4th. F. A. T. Winnecke (Strasbourg, France) described the comet as well seen in bright moonlight with the 46-cm refractor on the 6th. He added that the coma was about 1.5′ across with a very faint, star-like nucleus. J. F. J. Schmidt (Athens, Greece) said the comet was too faint to measure with the 17-cm Reinfelder refractor on the 8th and he was unable to find the comet thereafter. On the 11th, B. Peter (Leipzig, Germany) said the comet was visible in weak evening twilight. After darkness, it was described as bright, with a central condensation and a distinct nucleus." That same night, Barnard said the comet was "quite faint." G. F. W. Rümker (Hamburg, Germany) described the comet as very faint and of average size on the 18th. Millosevich said the comet was extremely faint on the 21st and noted it was impossible to measure its position in the 23-cm refractor.

The comet was last detected on October 28.41, using the 38-cm refractor at Harvard College Observatory. The position was given as $\alpha = 13^h\ 51.3^m$, $\delta = +28°\ 41'$.

The first parabolic orbit was calculated by K. Zelbr using positions from September 20, 25, and October 2. He determined the perihelion date as 1881 September 14.68. Additional orbits were calculated by H. Oppenheim, Zelbr, S. C. Chandler, Jr, and Millosevich (1882). Millosevich's orbit is given below.

T	ω	Ω (2000.0)	i	q	e
1881 Sep. 14.8653 (UT)	6.2860	275.8195	112.8160	0.449200	1.0

ABSOLUTE MAGNITUDE: $H_{10} = 8.3$ (V1964)

FULL MOON: Sep. 8, Oct. 7, Nov. 6

SOURCES: O. H. Landreth, *AN*, **100** (1881 Sep. 30), p. 303; E. E. Barnard, *The Observatory*, **4** (1881 Oct.), p. 309; E. Hartwig, *AN*, **100** (1881 Oct. 7), p. 335; K. Zelbr, E. Millsevich, and B. Peter, *AN*, **100** (1881 Oct. 15), pp. 347–52; G. F. W. Rümker and H. Oppenheim, *AN*, **100** (1881 Oct. 25), pp. 377, 381; K. Zelbr, *The Observatory*, **4** (1881 Nov.), p. 340; E. Millosevich, F. A. T. Winnecke, and A. J. G. F. von Auwers, *AN*, **101** (1881 Nov. 7), pp. 13–16; S. C. Chandler, Jr and B. Peter, *AN*, **101** (1881 Nov. 21), pp. 57–62; J. Winlock and E. C. Pickering, *Annals of the Astronomical Observatory of Harvard College*, **13** (1882), p. 182; J. F. J. Schmidt, *AN*, **101** (1882 Feb. 27), p. 295; E. E. Barnard, *SM*, **1** (1882 Mar.), p. 14; E. E. Barnard, *SM*, **1** (1882 May), pp. 33–4; E. E. Barnard, *AN*, **102** (1882 Jun. 6), p. 155; E. Millosevich, *AN*, **102** (1882 Jul. 6), p. 269; J. Palisa, *AN*, **102** (1882 Jul. 18), p. 277; F. A. T. Winnecke, *AN*, **114** (1886 May 6), pp. 231–4; V1964, p. 62.

72P/1881 T1 (Denning–Fujikawa)

1881 V = 1881f

Discovered: 1881 October 4.13 ($\Delta = 0.76$ AU, $r = 0.81$ AU, Elong. $= 52°$)

Last seen: 1881 November 25.13 ($\Delta = 1.07$ AU, $r = 1.40$ AU, Elong. $= 86°$)

Closest to the Earth: 1881 August 5 (0.1157 AU)

Calculated path: LEO (Disc)

W. F. Denning (Bristol, England) discovered this comet in the morning sky on 1881 October 4.13 and estimated the position as $\alpha = 9^h\ 22^m$, $\delta = +16°$. The daily motion was given as 30' eastward and the comet was described as "a small bright nebula, round, and much brighter in the middle." An independent discovery was made by W. R. Brooks (Phelps, New York, USA) with a 13-cm reflector. In the *Siderial Messenger* of March 1882, it was said that Brooks made the observation on October 4.33 "but that clouds obscured it before a position could be obtained."

The comet had already passed closest to the sun and Earth when discovered. J. E. Coggia (Marseille, France) said the comet could only be seen in the morning sky for a few minutes after the moon had set on the 6th. J. G. Lohse (Dun Echt Observatory, Scotland) described the comet as very small and fairly bright on the 10th. F. A. T. Winnecke (Strasbourg, France) observed with the 46-cm refractor (208×) on the 19th and said the comet appeared about the brightness of a nebula of class I, with a distinct, eccentric condensation. The diameter was about 40″. Winnecke described the comet as round, diffuse, centrally condensed, and about 1′ across on the 20th. J. F. J. Schmidt (Athens, Greece) could not find the comet in the morning sky with the 17-cm Reinfelder refractor on October 20 and 21; however, he did locate it with the same telescope on the 28th and described it as an extremely faint nebulosity. On the 29th, Winnecke said the comet was faintly seen during poor seeing, but appeared oblong during moments of good seeing, with a major axis of about 2′. E. Hartwig (Strasbourg, France) saw the comet very easily in the finder of the 15-cm refractor on this morning.

The only additional physical descriptions came from E. J. M. Stephan (Marseille, France) and Schmidt. Stephan said the comet was very faint with slight condensation in an 80-cm telescope on November 3. Schmidt said the comet appeared as an extremely faint nebulosity in the 17-cm refractor on November 4. The comet was last seen on November 25.13, when Winnecke gave the position as $\alpha = 10^h\ 44.4^m$, $\delta = +15°\ 03′$.

The first orbit was calculated by H. Oppenheim using positions obtained on October 6, 10, and 12. It was parabolic with a perihelion date of 1881 September 12.96. A few days later, L. Schulhof took positions spanning the period of October 6–19 and determined the perihelion date as September 13.89, while J. Palisa took a similar set of positions and found a perihelion date of September 13.13. Schulhof noted that this orbit left generally large errors between the observed and calculated positions. He then found that an elliptical orbit with a perihelion date of September 13.54 and a period of 7.74 years fitted the positions much better. In the following weeks the period was determined as 8.34 years by S. C. Chandler, Jr, 8.45 years by Schulhof, 8.41 years by E. Hartwig and L. Wutschichowsky, 9.11 years by E. E. Block, and 8.83 years by Hartwig. Block suggest this comet was a return of D/1819 W1.

B. Matthiessen (1889) was the first person to utilize positions spanning the entire period of visibility. In the 1889 June 5 issue of the *Astronomische Nachrichten*, he applied perturbations by the planets Mercury to Saturn, and determined the perihelion date as September 13.81 and the period as 8.69 years. He continued his analysis in the 1889 December 19 issue and formerly predicted the comet would return to perihelion on 1890 May 19.33. The comet was not found and it remained lost until accidentally rediscovered by S. Fujikawa in 1978. B. G. Marsden (1978) approximately linked the apparitions of 1881 and 1978. For this apparition, he determined the perihelion date as September 13.75 and the period as

8.71 years. Perturbations by Jupiter to Pluto were included. Interestingly, R. J. Buckley (1979) wrote a paper that recomputed the orbits of several nineteenth century comets, including this one. The paper had been accepted for publication and was undergoing the review process when Fujikawa rediscovered the comet. Buckley had used 23 positions obtained during the period of October 10 to November 25, applied perturbations by Venus to Neptune, and determined the perihelion date as September 13.77 and the period as 8.71 years. K. Muraoka (1995) determined the orbit below and gave nongravitational terms of $A_1 = -0.01$ and $A_2 = +0.0240$.

T	ω	Ω (2000.0)	i	q	e
1881 Sep. 13.8301 (UT)	312.6627	67.4826	6.8613	0.725411	0.828668

ABSOLUTE MAGNITUDE: $H_{10} = 9.0$ (V1964)

FULL MOON: Oct. 7, Nov. 6, Dec. 5

SOURCES: W. F. Denning and C. N. A. Krueger, *AN*, **100** (1881 Oct. 7), p. 335; J. E. Coggia, R. Copeland, and J. G. Lohse, *AN*, **100** (1881 Oct. 15), p. 349; H. Oppenheim, *AN*, **100** (1881 Oct. 20), p. 367; F. A. T. Winnecke, E. Hartwig, and L. Wutschichowsky, *MNRAS*, **42** (1881 Nov.), pp. 45–6; W. F. Denning, L. Schulhof, and J. Palisa, *The Observatory*, **4** (1881 Nov.), pp. 331, 339–40; L. Schulhof, F. A. T. Winnecke, and E. Hartwig, *AN*, **101** (1881 Nov. 7), pp. 13–16; E. J. M. Stephan and L. Schulhof, *CR*, **93** (1881 Nov. 7), pp. 676, 693–4; E. Hartwig and L. Wutschichowsky, *AN*, **101** (1881 Nov. 11), pp. 29–32; E. E. Block, *AN*, **101** (1881 Nov. 21), pp. 61–4; F. A. T. Winnecke and E. Hartwig, *AN*, **101** (1881 Nov. 26), p. 77; S. C. Chandler Jr, *AN*, **101** (1881 Dec. 5), p. 93; O. C. Wendell and S. C. Chandler, Jr, *AN*, **101** (1882 Feb. 10), p. 231; J. F. J. Schmidt, *AN*, **101** (1882 Feb. 27), p. 295; W. F. Denning and W. R. Brooks, *SM*, **1** (1882 Mar.), p. 14; F. A. T. Winnecke, *AN*, **114** (1886 May 6), pp. 231–4; B. Matthiessen, *AN*, **121** (1889 Jun. 5), pp. 359–66; B. Matthiessen, *AN*, **123** (1889 Dec. 19), p. 221; V1964, p. 62; B. G. Marsden, *IAUC*, No. 3300 (1978 Nov. 6); R. J. Buckley, *JBAA*, **89** (1979 Apr.), pp. 261, 263; B. G. Marsden and K. Muraoka, *CCO*, 10th ed. (1995), pp. 64–5.

2P/Encke

1881 VII = 1881d

Recovered: 1881 August 21.00 ($\Delta = 1.32$ AU, $r = 1.64$ AU, Elong. $= 88°$)

Last seen: 1881 November 12.21 ($\Delta = 1.03$ AU, $r = 0.36$ AU, Elong. $= 20°$)

Closest to the Earth: 1881 October 11 (0.5404 AU)

Calculated path: PER (Rec), AUR (Sep. 7), LYN (Sep. 28), LMi (Oct. 8), LEO (Oct. 17), COM (Oct. 23), VIR (Oct. 24)

O. Backlund (1881) examined the orbit of this comet and predicted it would next arrive at perihelion on 1881 November 15.57. The comet was recovered by E. Hartwig (Strasbourg, France) on 1881 August 21.00. He had been searching with a 15-cm cometseeker when he found a weak nebulosity very near Backlund's ephemeris position. Although he noted it "seemed to be completely evident in that instrument," Hartwig failed to see it when the 46-cm refractor was directed toward it because of developing fog. The next clear morning did not come until August 26, when Hartwig noted the comet had a diameter of 4'. He gave the position as $\alpha = 3^h 54.6^m$, $\delta = +33° 12'$ on August 26.08. His colleague F. A. T. Winnecke also looked at the comet on the morning of the 26th. With the 15-cm cometseeker he said the coma was 4' across, while the 46-cm refractor (208×) revealed only the brighter, central part of the coma, which was 1' across. Independent recoveries were reported by B. Peter (Leipzig, Germany) on the night of August 21.0, E. W. L. Tempel (Arcetri, Italy) on August 22.0, Otto Struve (Pulkovo Observatory, Russia) on August 24.87, and A. A. Common (Ealing, England) on August 27. Peter was unable to measure a position because he could not illuminate the thread of the micrometer. Tempel was using a 20-cm refractor and was not able to obtain a measurement of the position. He described the comet as large and very diffuse, but was unable to differentiate a nucleus. After making additional observations on August 27, 28, and 29, he wrote that the tail had "a marvelous hook form." Struve was assisted by E. E. Block, who was visiting from Odessa. Struve said the comet was close to the limit of the 38-cm refractor and appeared as a very faint nebula standing out against the very dark sky. Both Struve and Block noted movement before the transparency of the sky deteriorated, but no position was measured. Common described the comet as very faint and about 2' across. He added that it was no brighter than a star of magnitude 10.

The comet moved toward the sun and Earth during September. On the 5th, M. W. Meyer (Geneva, Switzerland) said the comet appeared very diffuse, but was not faint in a 25.4-cm telescope. J. F. J. Schmidt (Athens, Greece) found the comet easily in the finder of the 17-cm Reinfelder refractor on the 19th, while the appearance in the refractor was that of a large, very faint nebula, without much condensation. He indicated a coma diameter of 6.68'. Schmidt also indicated a coma diameter of 7.64' on the 21st and 7.76' on the 22nd. G. F. W. Rümker (Hamburg, Germany) observed the comet with a 24-cm refractor on the 23rd and said it was very large, diffuse, and without a noticeable nucleus. The comet attained its most northerly declination of +43° on the 27th.

The comet made a moderately close approach to Earth just before mid-October, but continued to approach perihelion. On October 1, 7, and 11, Peter noted the comet "appeared as a large, vague nebulosity, about 5' in diameter, with little condensation." Meyer said the comet was very easy to see in the 25.4-cm telescope on the 6th. He said the faint condensation appeared eccentric, but there was hardly a nucleus. After the moon was no longer a problem, Meyer saw the comet on the 17th and described it as large and bright, with an eccentrically situated condensation. On the 18th, Winnecke said the 46-cm refractor revealed a nucleus from which a fan extended about 59". Winnecke said the fan extended 104" toward PA 108.5° on the 19th. Schmidt found the comet large and bright in the finder on October 20 and added that he might have seen it with the naked eye. He said the refractor revealed a very condensed coma, and he indicated the diameter was 5.54'. No nucleus was visible. Schmidt also wrote that the comet's shape was similar to what he had observed during October of 1848. That same morning, Winnecke said the 46-cm refractor revealed the sector was 123″ long, with the center line extending toward PA 129.8°. Schmidt measured the northern coma radius as 4.5′ on the 21st, while the radius along the east–west axis was 2.95′. Schmidt said the comet was not seen with the naked eye on the 28th. The refractor revealed a distinct, narrow tail 5–7′ long, and Schmidt indicated the coma diameter was 4.4′. He again noted that no nucleus was visible. On 1882 March 4, E. E. Barnard wrote of an interesting observation in a letter to the *Astronomische Nachrichten*. Observing on the morning of October 21, he watched the comet pass centrally over a 9th-magnitude star. The comet was then "very bright and round, with moderate condensation" and was plainly visible in his 3.2-cm finder. Except for the difference in contrast, the star never appeared to diminished in brightness and always stayed "remarkably distinct." He said the sharpness of the star gave him the unusual impression that the star was passing in front of the comet.

The comet was quickly moving into morning twilight as it headed toward perihelion. On November 4, Schmidt said the finder revealed the comet as an indistinct star of magnitude 5. He noted that the refractor showed the comet as a very small, white nebulosity, and indicated a coma diameter of about 2.3′. The tail was 7′ long. P. Tacchini (Rome, Italy) said the comet was visible to the naked eye with great difficulty because of some haze and its low altitude on the 6th. With a telescope he described it as bright, with a globular shape, and a slightly eccentric condensation. He added that a diffuse nucleus was seen. Winnecke reported the comet was visible in the finder (30×) on the 7th and was similar in brightness to a star of magnitude 6 or 7. Through the 46-cm refractor, he gave the coma diameter as 50″ on both the 7th and 9th, but noted the comet had a slightly different appearance on each of these dates. He said it was round and centrally condensed on the 7th, but bright, diffuse, and almost homogenous on the 9th. On the 10th, Winnecke described the comet as a "rather homogeneously illuminated disk."

The comet was last detected on November 12.21, when E. Millosevich (Rome, Italy) gave the position as $\alpha = 13^h\ 52.8^m$, $\delta = -10°\ 00'$ while using the 23-cm refractor. He said the comet was then bright and at the limit of naked-eye visibility. Schmidt tried to find the comet on November 13.18, but the morning twilight was already too bright. Following perihelion, J. Tebbutt (Windsor, New South Wales, Australia) looked for the comet. He wrote in the March 1882 issue of the *Monthly Notices* that "with the help of M. Backlund's Ephemeris . . . I have sought for Encke's Comet, but in vain. When the comet was sufficiently south, and removed from the morning twilight, it was considerably beyond the grasp of my telescope." Tebbutt was using an 8.3-cm refractor.

Refinements of this comet's orbit using positions from multiple apparitions, as well as planetary perturbations, have been published by numerous astronomers. Of most significance were the early studies of J. O. Backlund

(1884, 1886) and the later study of B. G. Marsden and Z. Sekanina (1974). These have typically revealed a perihelion date of November 15.80 and a period of 3.31 years. Marsden and Sekanina determined the nongravitational terms as $A_1 = +0.07$ and $A_2 = -0.02657$. The orbit of Marsden and Sekanina is given below.

T	ω	Ω (2000.0)	i	q	e
1881 Nov. 15.8034 (TT)	183.9014	336.2470	12.8973	0.343434	0.845347

ABSOLUTE MAGNITUDE: $H_{10} = 9.8$ (V1964)

FULL MOON: Aug. 9, Sep. 8, Oct. 7, Nov. 6

SOURCES: O. Backlund, *The Observatory*, **4** (1881 Jul.), pp. 217–18; O. Backlund, *AN*, **100** (1881 Jul. 16), p. 111; O. Struve, E. E. Block, E. Hartwig, and F. A. T. Winnecke, *AN*, **100** (1881 Aug. 30), p. 239; B. Peter and E. W. L. Tempel, *AN*, **100** (1881 Sep. 3), p. 255; E. Millosevich, *AN*, **100** (1881 Oct. 15), p. 345; E. W. L. Tempel, *AN*, **100** (1881 Oct. 25), p. 371; B. Peter and L. Weinek, *AN*, **101** (1881 Nov. 21), pp. 59–62; P. Tacchini, *AN*, **101** (1881 Nov. 26), p. 79; E. Millosevich, *AN*, **101** (1882 Jan. 2), p. 141; J. F. J. Schmidt, *AN*, **101** (1882 Feb. 27), pp. 295–300; J. Tebbutt, *MNRAS*, **42** (1882 Mar.), pp. 263–4; *SM*, **1** (1882 Mar.), pp. 13–14; M. W. Meyer, *AN*, **101** (1882 Mar. 24), pp. 375–80; E. E. Barnard, *AN*, **102** (1882 Jun. 15), p. 207; J. Palisa, *AN*, **102** (1882 Jul. 18), p. 277; J. O. Backlund, *MASP*, **32** No. 3, 7th Series (1884), p. 36; J. O. Backlund, *MASP*, **34** No. 8, 7th Series (1886), p. 38; F. A. T. Winnecke, *AN*, **114** (1886 May 6), pp. 231–4; G. F. W. Rümker, *AN*, **120** (1888 Oct. 19), pp. 81, 86; V1964, p. 62; B. G. Marsden and Z. Sekanina, *AJ*, **79** (1974 Mar.), pp. 415–16.

C/1881 W1   *Discovered:* 1881 November 17.2 ($\Delta = 1.20$ AU, $r = 1.93$ AU, Elong. $= 123°$)

(Swift)   *Last seen:* 1882 January 12.78 ($\Delta = 2.13$ AU, $r = 2.03$ AU, Elong. $= 71°$)

   Closest to the Earth: 1881 November 22 (1.1840 AU)

1881 VIII = 1881g   *Calculated path:* CAS (Disc), AND (Dec. 6), PEG (Dec. 20)

This comet was discovered by L. Swift (Rochester, New York, USA) on 1881 November 17.2. The comet was confirmed by O. C. Wendell (Harvard College Observatory, Cambridge, Massachusetts, USA) on November 18.26, when he gave the position as $\alpha = 1^h 57.2^m$, $\delta = +74° 04'$.

The comet passed closest to both the sun and Earth during the days following its discovery. F. A. T. Winnecke (Strasbourg, France) saw the comet with a 46-cm refractor (208×) on November 25 and 26, and said it was about 1' across. Using the finder of the refractor he said the coma appeared 4' across and was centrally condensed. P. Tacchini (Rome, Italy) said the comet appeared faint on November 26. G. Bigourdan (Paris, France) saw the comet on the 27th and described it as a faint nebulosity, without a tail, but with a slight brightening toward the center. He added that it was similar in brightness to a 12th-magnitude star.

The comet was moving away from both the sun and Earth for the remainder of the comet's apparition. J. F. J. Schmidt (Athens, Greece) observed the comet on December 10–13 and 20–22 with the 17-cm Reinfelder refractor and the Hertel refractor. He said it was faint, over 2′ across, and with little condensation. More precise measurements were later published and are as follows: December 10, 2.58′ diameter, 11th, 3.93′ diameter, 13th, 4.17′ diameter, 20th, 3.60′ radius, and 21st, 3.20′ radius. J. Palisa (Vienna, Austria) described the comet as faint, with a distinct nucleus on the 12th. B. von Engelhardt (Dresden, Germany) saw the comet with a 30-cm refractor and said it was faint, but remained visible as it passed a star of magnitude 7. J. Franz (Königsberg, now Kaliningrad, Russia) saw the comet on December 19 and said it appeared very faint and diffuse. B. Peter (Leipzig, Germany) said the comet was very faint and oblong, with some condensation on the 22nd.

As the moon entered the sky during late December, the comet was too faint for observations, with Palisa and R. Copeland and J. G. Lohse (Dun Echt Observatory, Scotland) independently detecting it for the final time on the 25th. Palisa provided an ephemeris for the period of January 6–24 to aid searches made after the full moon. The comet was recovered by Bigourdan on January 7, while Palisa located it with the 30.1-cm refractor on the 10th. The comet was last detected on January 12.78, when Palisa gave the position as $\alpha = 23^h\ 38.9^m$, $\delta = +17°\ 57′$.

As would later be shown, the comet's far northerly position and large distance from the sun and Earth made early orbits difficult to determine. The first parabolic orbit was calculated by S. C. Chandler, Jr, using positions obtained by Wendell on November 18, 20, and 21. The resulting perihelion date was 1882 February 5.85. Chandler noted the orbit resembled that of the comet C/1791 X1. Palisa took positions from November 23, 25, and 27, and calculated a perihelion date of 1881 December 9.31. Palisa and H. Oppenheim independently took positions obtained up to December 12 and published what would later prove to be very close representations of the orbit in the 1881 December 24 issue of *Astronomische Nachrichten*. The perihelion dates were given as November 20.50 and 20.20, respectively. An additional, very similar orbit was calculated by Bigourdan a few days later.

Two definitive orbits have been calculated. S. Oppenheim (1885) and K. G. Olsson (1886) both took positions spanning the entire period of visibility and calculated parabolic and elliptical orbits. Both astronomers said the elliptical orbits fitted the positions best. Oppenheim actually calculated two elliptical orbits, one with a period of 968 years and the other with a period of 2740 years. He showed how the longer period orbit fitted the positions best. Olsson's elliptical orbit had a period of 612 years, but Oppenheim's longer period orbit was still a better fit and it is given below.

T	ω	Ω (2000.0)	i	q	e
1881 Nov. 20.2418 (UT)	117.9487	183.0392	144.8016	1.924973	0.990169

ABSOLUTE MAGNITUDE: $H_{10} = 4.3$ (V1964)

FULL MOON: Nov. 6, Dec. 5, Jan. 4

SOURCES: R. Copeland, S. C. Chandler, and O. C. Wendell, *AN*, **101** (1881 Nov. 26), p. 79; G. Bigourdan, *CR*, **93** (1881 Nov. 28), p. 889; S. C. Chandler, *The Observatory*, **4** (1881 Dec.), p. 365; F. A. T. Winnecke and P. Tacchini, *AN*, **101** (1881 Dec. 5), p. 95; J. Palisa, L. Swift, O. C. Wendell, and H. Oppenheim, *AN*, **101** (1881 Dec. 24), pp. 123–8; G. Bigourdan, *CR*, **93** (1881 Dec. 26), pp. 1122–3; B. von Engelhardt and J. Palisa, *AN*, **101** (1882 Jan. 2), pp. 139, 143; O. C. Wendell, P. Tacchini, and E. Millosevich, *AN*, **101** (1882 Feb. 10), pp. 231, 237; J. F. J. Schmidt, *AN*, **101** (1882 Feb. 20), p. 253; G. Bigourdan, *CR*, **94** (1882 Feb. 27), pp. 573–4; *SM*, **1** (1882 Mar.), p. 14; B. Peter, *AN*, **101** (1882 Mar. 7), p. 335; J. F. J. Schmidt, *AN*, **102** (1882 Jun. 15), p. 205; J. Palisa, *AN*, **102** (1882 Jul. 18), p. 279; E. Hartwig, *AN*, **108** (1884 Apr. 5), p. 269; S. Oppenheim, *SAWW*, **92** Abt. II (1885), pp. 232–56; J. Franz, *AN*, **111** (1885 Jan. 31), pp. 89–94; S. Oppenheim, *AN*, **113** (1885 Nov. 30), pp. 49–56; K. G. Olsson, *AN*, **114** (1886 Apr. 21), pp. 201–6; F. A. T. Winnecke, *AN*, **114** (1886 May 6), pp. 231–4; V1964, p. 62.

C/1882 F1 (Wells)   *Discovered:* 1882 March 18.37 ($\Delta = 1.79$ AU, $r = 2.06$ AU, Elong. $= 90°$)

Last seen: 1882 August 16.75 ($\Delta = 2.38$ AU, $r = 1.75$ AU, Elong. $= 41°$)

1882 I = 1882a   *Closest to the Earth:* 1882 May 21 (0.8900 AU), 1882 June 18 (0.9473 AU)

Calculated path: HER (Disc), LYR (Mar. 29), DRA (Apr. 10), CYG (Apr. 18), DRA (Apr. 19), CEP (May 2), CAS (May 13), CAM (May 22), PER (May 29), AUR (Jun. 6), TAU (Jun. 9), ORI (Jun. 13), GEM (Jun. 14), CNC (Jun. 21), LEO (Jun. 29), VIR (Jul. 25)

C. S. Wells (Dudley Observatory, New York, USA) discovered this comet on March 18.37 and described it as small and bright, with a nucleus and a very narrow tail about 5′ long. He gave the position as $\alpha = 17^h 51.7^m$, $\delta = +32° 31′$ on March 18.41. The discovery of this comet has an interesting story behind it. Shortly after the completion of the rebuilding of Dudley Observatory, a group of people were visiting its director, L. Boss. According to J. Ashbrook (1959), "someone remarked that comets were being discovered at other institutions, and that Albany should not be left behind. In a joking way, Boss turned to his assistant and said, 'You see, Mr. Wells, you must discover a comet.'" Wells' discovery occurred a week later.

The comet changed little during the remainder of March. Boss noted on the 20th that it appeared "like a great comet in miniature." The only estimate of the total magnitude came from Boss on the 20th, when he said it was near 8. There were more estimates of the nuclear magnitude, with Boss estimating it as 10 on the 20th, J. Ritchie (Boston, Massachusetts, USA) estimating it as 8 on the 20th, E. Block (Odessa, Ukraine) estimating it as 10 on the 22nd and 23rd, and L. Weinek (Leipzig, Germany) determining it as 10.0 on the 24th and 9.7 on the 30th. The coma remained small, with estimates between 0.5′ and 1′ coming from Boss, Ritchie, Block, and Weinek. Although most observers reported the tail as between 4′ and 9′ long, Ritchie estimated it as 30′ long on the 20th and Block estimated it as 20′ on the 22nd and 23rd. E. Weiss (Währing, Austria) said the tail was directed toward PA

262° on the 21st and PA 269° on the 26th, while Weinek noted it was directed toward PA 264° on the 24th and PA 262° on the 30th.

The comet was very well observed during April as its distance from the sun and Earth decreased. The only estimates of the total magnitude came from R. Engelmann (Leipzig, Germany). His estimates using a 5.5-cm finder (25×) were 9.5 on the 7th, 9.0 on the 9th, 8.5 on the 12th and 14th, 8.0 on the 22nd, and 8.5 on the 27th. The comet was described as faint on the 30th by W. G. Thackeray and H. P. Hollis (Royal Observatory, Greenwich, England), and G. F. W. Rümker (Hamburg, Germany). Although numerous estimates of the nuclear magnitude were made, on any given day these estimates would generally range between 9 and 10, depending on the size of telescope being used. Ultimately, Engelmann provided the most consistent series with his 20-cm refractor (155×). After noting that the nucleus occasionally flashed into view on the 2nd and was not brighter than 11th magnitude, he gave estimates of 10.3 on the 5th, 10.0 on the 6th, 10.5 on the 7th, 10.7 on the 8th, 10.5 on the 9th, 10.3 on the 12th, 10.0 on the 13th and 14th, 10.5 on the 16th, 9.5 on the 22nd, and 9.8 on the 27th. Engelmann, B. von Engelhardt (Dresden, Germany), and Rümker reported the nucleus was star-like throughout the month. Rümker was observing with a 26-cm refractor. The coma appeared to be quite small at the beginning of the month, with Engelmann and Weinek independently estimating diameters of about 0.5′ on the 2nd. Thereafter, the diameter was given as 0.5–0.7′ by Engelmann and Weinek on the 5th, 0.7–0.8′ by Engelmann and Weinek on the 6th, 0.9′ by Weinek on the 7th, 0.7′ by Engelmann on the 8th, 0.7–0.9′ by Engelmann and Weinek on the 9th, 1.1′ by Engelmann on the 12th, 1.2′ by Engelmann on the 16th, and 1.5′ by Engelmann on the 22nd. Engelmann noted the comet was strongly condensed and round on the 2nd, with a center that looked granular. There were extensive measurements of the tail, with Engelmann, Weinek, Weiss, and C. F. Pechüle (Copenhagen, Denmark) giving both tail lengths and position angles. Engelmann provided the best series with the help of a refractor (70×). He said the tail extended 6.4′ toward PA 262.6° on the 2nd, 9.1′ toward PA 262.0° on the 5th, 12.7′ toward PA 261.5° on the 6th, 15.4′ toward 261.7° on the 7th, 16.4′ toward 260.8° on the 8th, 23.6′ toward 259.0° on the 9th, 22.7′ toward 258.7° on the 11th, 25.5′ toward 260.0° on the 13th, 16.4′ toward 260.9° on the 14th, 23.7′ toward 259.4° on the 16th, and 36.0′ toward 260.1° on the 22nd. Pechüle added that the tail was narrow and 1′ wide on the 17th.

The comet passed closest to Earth during the last half of May, and was still heading toward perihelion thereafter. Several observers provided estimates of the total magnitude during the month. Using a 5.5-cm finder (25×), Engelmann gave values of 8.0 on the 10th, 7.3 on the 12th, 7.0 on the 13th, 6.7 on the 17th, 7.3 on the 22nd, 6.7 on the 23rd, about 5 on the 28th, and 5.0 on the 29th and 30th. During the last days of May, several other astronomers were providing telescopic total magnitude estimates. Some of these estimates were 7.5 by W. H. Robinson (Radcliffe Observatory, Oxford,

England) on the 24th, 7–7.5 by F. A. Bellamy (Radcliffe Observatory, Oxford, England) on the 25th, 6–7 by Robinson on the 26th, 5 by A. M. W. Downing (Royal Observatory, Greenwich, England) on the 27th, 5 or 6 by Robinson on the 29th, and 4.5 by Robinson on the 31st. Robinson noted the comet and tail were seen with the naked eye on May 30 and 31. On the last date, he said the naked eye revealed the comet was "identical in brightness" to δ Persei (magnitude 3.01 – SC2000.0). A very interesting magnitude estimate was made on May 11 by G. Knott (Knowles Lodge, Cuckfield, England). He said he threw the telescope out of focus and noted the head of the comet was equal to a star of magnitude 7.2. A short time later, he was observing with an opera-glass and threw it out of focus. He noted the comet was slighter brighter than the magnitude 7.2 star, but "decidedly less" than a star of magnitude 6.3. The nucleus continued to be described as a "bright point" and "nearly stellar" throughout the month. Although a number of magnitude estimates were made, it was Engelmann who again provided the longest, most consistent values using his refractor (155×). He gave the magnitude as 9.8 on the 7th, 8.7 on the 10th, 8.0 on the 12th, 8.5 on the 13th, 7.9 on the 17th, 8.2 on the 21st, 8.1 on the 22nd, and 8.2 on the 23rd. The only magnitude estimate made after the 23rd came from W. Wickham (Radcliffe Observatory, Oxford), who said it equaled a star of magnitude 6.3 on the 27th. In general, nothing unusual was reported in the region of the nucleus, except on the 19th, when F. Lakits noted a faint "effusion" was visible. Although a very small coma diameter of 12″ was given by G. W. Hough (Dearborn Observatory, Evanston, Illinois, USA) on May 2, he was using a 47-cm refractor at a magnification of 400×, and so probably lost the outer portion. With a refractor at 70×, Engelmann gave the coma diameter as 1.8′ on the 10th, 2.2′ on the 12th, 1.8′ on the 13th, 2.0′ on the 17th, 2.2′ on the 22nd, and 2.1′ on the 23rd. Numerous meaurements of the tail were provided throughout May by Engelmann, Lakits, A. Abetti (Padova, Italy), V. Knorre (Berlin, Germany), Armsby (Carleton College, Northfield, Minnesota, USA), and H. A. Kobold (O'Gyalla Observatory, Hungary). As in April, Engelmann provided the longest, most consistent series with his refractor (70×). He said the tail extended 22′ toward PA 286.6° on the 7th, 41.8′ toward PA 297.7° on the 10th, 36′ toward PA 309.1° on the 12th, 36′ toward PA 313.9° on the 13th, 36′ toward PA 331.6° on the 17th, toward PA 344.5° on the 21st, 45′ toward PA 345.9° on the 22nd, toward PA 351.1° on the 23rd, toward PA 353.1° on the 28th, toward PA 354.5° on the 29th, and about 20′ toward PA 355.2° on the 30th. Armsby saw the comet with a 21-cm refractor (50×) on the 17th and said the tail was nearly 1° long. The comet attained its most northerly declination of +75° on May 11.

The comet was heading southward as June began and quickly brightened as it neared the sun. Abetti gave the nuclear magnitude as 3 on the 1st, while Knorre said the tail extended toward PA 355.0°. On the 2nd, the nuclear magnitude was given as 4.5 by Engelmann and 3 by Weiss. Engelmann added that the tail extended toward PA 359.6°. Boss observed the comet with difficulty using the transit circle about 19 minutes before noon on June 6. E. W. Maunder (Royal Observatory, Greenwich, England) observed with a telescope on the 8th and noted, "a distinct planetary disk of an orange color, nearly as deep and vivid as that of Mars." He added that the comet continued to be seen in the finder until 4 minutes after sunrise. J. F. J. Schmidt (Athens, Greece) observed the comet in broad daylight with a 15-cm refractor and weak magnification at just after 3 p.m. local time on the 10th. He said the sky was very clear, but the comet was difficult to see, as it was 2.8° from the edge of the sun. He noted it was 5–6" in diameter and exhibited no tail or protrusions from the nucleus.

The comet passed perihelion on June 11 and, as seen from Earth, it passed just 2.6° from the sun. Thereafter, it turned eastward and finally became visible to Southern Hemisphere observers. Maunder observed with a 32.4-cm refractor in daylight on the 11th. Although a light-red glass was necessary to cut the sky brightness at a magnification of 60×, he noted that no such filter was needed at magnifications of 220× and 310×. Maunder said the comet "could be easily seen as a dull yellow stellar point of light, the disk being no larger than that of a star. It was judged to be not so bright as Capella, which had been seen previously, making every allowance for the much brighter background on which the comet was seen." Boss observed the comet around noon that same day and said the "true nucleus" was seen using a magnification of 180×. He described it as "a perfectly round and sharply defined disk of moderate illumination." Boss estimated the diameter of the disk as 0.75". The coma was "quite uniform and faint" with a diameter of about 10". Boss again saw the comet around noon on the 12th and said it was a very easy object to see. He noted, "The nebulosity had increased in brightness." W. H. Finlay (Royal Observatory, Cape of Good Hope, South Africa) saw the comet on the 14th with an 18-cm refractor and said, "the comet had a large diffused head without any marked condensation, and a tail about 2° long in the twilight. The centre of the diffused head was observed throughout, but on two occasions, when the definition was better than usual, a bright stellar nucleus was seen in the preceding part of the head." J. Tebbutt (Windsor, New South Wales, Australia) saw the comet for the first time on the 15th, but noted that twilight prevented him seeing any stars to allow a position determination. Finlay observed with the refractor on the 17th and said, "the tail was considerably curved towards the south, and the northern side and edge were much brighter and better defined than the southern." L. G. Puckle (on board a ship near Singapore, Malaysia) wrote for June 23, "we observed a large comet a little to the southward of the planet Venus." His sextant revealed the nucleus was at an altitude of 12.0°. He added, "The nucleus was well defined and bright, like a star of the second magnitude, the tail spreading out like a fan to the extent of about two and a half degrees of altitude, as visible with the naked eye. The tail stretched upwards from the horizon, with a slight curve in it towards the southward, at the upper end." Puckle said they failed to find the comet, even with the aid of a telescope on the following nights.

The comet remained at a relatively low altitude throughout July and into August and physical descriptions were not plentiful. Schmidt saw the comet with the naked eye on the 1st, when at a low altitude in bright twilight. Finlay described the comet as faint through clouds in the 18-cm refractor on July 14. Schmidt simply noted the comet was faint in moonlight on the 20th. Schmidt saw the comet each night during the period of August 2–5, and noted it was very faint, with a slight central condensation, and about 3.5' across. On August 8 and 14, Finlay saw the comet with the 18-cm refractor and said it was "extremely faint." The comet was last seen on August 16.75, when Finlay gave the position as $\alpha = 12^h 25.1^m$, $\delta = +3° 01'$. Moonlight prevented observations thereafter.

The spectrum of this comet was observed at several observatories. Maunder reported a continuous spectrum was seen on April 22, 24, May 11, 14, and 21. On June 1, a bright line had developed in the yellow around 5903 Å, which he noted was "coincident with the D lines of sodium." This line was described as of "extraordinary brilliancy" on June 8 and the wavelength was measured as 5894.5 Å. The sodium line was also reported by N. C. Dunér (Lund, Sweden) on June 3, with two measurements using a 24.5-cm refractor and a spectroscope revealing the wavelength was around 5895 Å or 5894 Å, and by F. A. Bredikhin (Moscow, Russia) on June 4 and 6, with the measurement indicating a wavelength of 5892 Å.

The first parabolic orbits were determined within a few days of the discovery and resulted in a large range of perihelion dates. The earliest orbit came from Boss on March 22. He took positions obtained at Dudley Observatory on March 18, 20, and 22, and calculated a perihelion date of 1882 June 16.26. A. Thraen also calculated an orbit based on a five-day arc. Using positions obtained at Kiel on March 21, 23, and 25, he determined a perihelion date of June 28.85. Using slightly longer observation arcs that reached nearly the end of March, H. Oppenheim determined the perihelion date as June 17.08 and H. C. F. Kreutz determined it as June 8.89. Kreutz also pointed out that the comet would become a bright object near the sun in June. The first person to determine a perihelion date within a few hours of what would later prove to be the actual date was H. V. Egbert. Using positions obtained on March 20, 24, and 30, he obtained a perihelion date of June 10.81. Numerous orbits were calculated thereafter, many using positions spanning three weeks and more, and these slightly revised the perihelion date upwards to somewhere within the first hour of June 11. Some of the astronomers who produced these orbits were: A. Graham, O. C. Wendell, E. A. Lamp, Thraen, G. Bigourdan, E. Frisby, Kreutz, Oppenheim, J. R. Hind, and Wells.

By June it was becoming obvious that there was some deviation from a parabola and astronomers began calculating elliptical orbits. Thraen (1882) used positions obtained up to June 2 and determined a period of about

35 000 years. S. Wolyncewicz (1882) used positions obtained up to July 10 and determined a period of about 859 000 years. F. J. Parsons (1883) used positions obtained up to July 17 and determined a period of about 400 000 years. A definitive orbit came from E. von Rebeur-Paschwitz (1887). Using positions spanning the entire period of visibility, he calculated parabolic, elliptical, and hyperbolic orbits. The elliptical orbit had a perihelion date of June 11.03 and a period of about 1.2 million years. There were actually two hyperbolic orbits, one based on the pre-perihelion positions, and the other the post-perihelion positions. The hyperbolic solution actually fitted the positions best, but von Rebeur-Paschwitz believed the elliptical orbit was the best overall solution, and that orbit is given below.

T	ω	Ω (2000.0)	i	q	e
1882 Jun. 11.0296 (UT)	208.9853	206.5908	73.7978	0.060763	0.999995

ABSOLUTE MAGNITUDE: $H_{10} = 4.1$ (V1964)

FULL MOON: Mar. 5, Apr. 3, May 3, Jun. 1, Jul. 1, Jul. 30, Aug. 28

SOURCES: J. Ritchie and E. A. Lamp, *AN*, **101** (1882 Mar. 24), p. 383; J. G. Lohse, L. Weinek, L. de Ball, E. Block, P. Tacchini, M. W. Meyer, N. C. Dunér, H. Oppenheim, and H. C. F. Kreutz, *AN* (Special Edition), **102** (1882 Apr. 6), pp. 1–3; G. Bigourdan, *CR*, **94** (1882 Apr. 17), pp. 1101–4; E. Block, N. C. Dunér, H. Oppenheim, A. Thraen, R. Engelmann, and L. Weinek, *AN*, **102** (1882 Apr. 21), pp. 29–32, 37, 41–3; E. W. Maunder and A. Graham, *MNRAS*, **42** (1882 May), p. 351–2; C. S. Wells and H. V. Egbert, *SM*, **1** (1882 May), p. 36; E. Luther, B. von Engelhardt, C. S. Wells, L. Boss, H. V. Egbert, and E. A. Lamp, *AN*, **102** (1882 May 1), pp. 57–63; J. R. Hind, *Nature*, **26** (1882 May 4), p. 18; A. Thraen, C. S. Wells, E. Frisby, and H. C. F. Kreutz, *AN*, **102** (1882 May 12), pp. 75–8; J. R. Hind, *Nature*, **26** (1882 May 18), p. 68; E. W. Maunder, N. C. Dunér, and F. C. Penrose, *MNRAS*, **42** (1882 Jun.), pp. 410–13; G. W. Hough, Armsby, and E. A. Lamp, *SM*, **1** (1882 Jun.), p. 96–8; J. R. Hind, *Nature*, **26** (1882 Jun. 1), p. 114; I. Y. Kortazzi, *AN*, **102** (1882 Jun. 6), p. 157; E. A. Lamp and F. A. Bredikhin, *AN*, **102** (1882 Jun. 15), p. 207; J. F. J. Schmidt, *AN*, **102** (1882 Jun. 21), p. 223; A. Thraen, C. S. Wells, L. Boss, A. Belopolsky, and A. Socoloff, *AN*, **102** (1882 Jul. 6), pp. 265–70; J. F. J. Schmidt, *AN*, **102** (1882 Jul. 31), pp. 353–64; W. Huggins and L. Boss, *SM*, **1** (1882 Aug.), pp. 114–16, 127; T. Zona, *AN*, **102** (1882 Aug. 7), pp. 381–4; J. F. J. Schmidt, *AN*, **103** (1882 Sep. 22), pp. 131–6; J. F. J. Schmidt, *AN*, **103** (1882 Sep. 25), p. 157; R. Engelmann, *AN*, **103** (1882 Oct. 28), pp. 275–80; E. J. Stone, W. Wickham, W. H. Robinson, F. A. Bellamy, and L. G. Puckle, *MNRAS*, **42** (1882, Supplementary Notice), pp. 444–6; D. Gill, W. H. Finlay, W. G. Thackeray, H. P. Hollis, and A. M. W. Downing, *MNRAS*, **43** (1882 Nov.), pp. 8, 13–18, 26–7; F. Deichmüller and F. Lakits, *AN*, **103** (1882 Nov. 1), pp. 299–304; C. F. Pechüle, *AN*, **103** (1882 Nov. 15), p. 343; J. Tebbutt, *MNRAS*, **43** (1882 Dec.), pp. 58–61; S. Wolyncewicz, *AN*, **104** (1883 Dec. 12), pp. 57–60; A. Abetti, *AN*, **104** (1883 Jan. 2), pp. 121–8; O. C. Wendell, *AN*, **104** (1883 Feb. 7), p. 287; H. P. Hollis, *MNRAS*, **43** (1883 Mar.), pp. 288–9; H. A. Kobold, *AN*, **105** (1883 Apr. 2), pp. 89–91; E. Weiss, *AN*, **105** (1883 Jun. 11), pp. 379, 385; F. J. Parsons, *MNRAS*, **44** (1883 Nov.), pp. 10–12; V. Knorre, *AN*, **111** (1885 Mar. 17), pp. 209–26; E. von Rebeur-Paschwitz, *AN*, **117** (1887 Sep. 3), pp. 281–8; G. F. W. Rümker, *AN*, **120** (1888 Oct. 19), pp. 83, 86; G. Knott, *The Observatory*, **16** (1893 Feb.), p. 113; J. Ashbrook, *ST*, **18** (1959 Apr.), p. 312; V1964, p. 62; SC2000.0, p. 76.

X/1882 K1 Astronomers gathered in the region of Sawhâj, Egypt around mid-May 1882 as a total eclipse of the sun was predicted to occur on the 17th. About the middle of totality, a luminous streak was observed near the sun.

Some of the earliest reports of naked-eye sightings of this comet at totality came from C. Trépied and P. Tacchini. Trépied observed from the town of Akhmîm. He first noted the object to the right of the sun at a zenithal angle of nearly 90°. He referred to it as "in evident discordance with the rest of the corona." Tacchini reported "an isolated plume" west of the sun and near the corona. He measured the position of its nucleus as $\alpha = 3^h 35^m 16^s$, $\delta = +18° 30' 17''$ on May 17.267. Although Tacchini immediately knew this was a comet, Trépied said the idea of this object being a comet did not cross his mind until he saw the first photographs of the eclipse an hour later. He said, "The brightness of the comet appeared to one to be of the same order as that at the exterior parts of the corona."

A. Schuster and W. de W. Abney (members of an English team) took photographs during the eclipse. They wrote, "The nucleus is exceedingly well and sharply defined, the tail is somewhat curved; it did not point toward the sun's centre, but in a direction nearly tangential to the limb. The extent of the tail was roughly two-thirds of a solar diameter." One of the photographs showed the comet's nucleus situated slightly more than one solar diameter from the sun's limb. The position for 1882 May 17.267 was about $\alpha = 3^h 34.7^m$, $\delta = +18° 35'$. Schuster and Abney investigated numerous photographs from other observers. They noted "a slight but progressive change in the comet's position." Since the comet was measured with respect to the moon's limb, they suggested this movement was due to the motion of the moon; however, even after this was taken into account, there still remained a change in the comet's distance from the moon's center which they suggested was "probably in part due to the proper motion of the comet, which in that case must have moved away from the sun during the eclipse."

A special correspondent for the London *Daily News* summarized the eclipse experience in the August 1882 issue of the *Sidereal Messenger*. He said, "while the sky darkened and assumed a leaden hue . . . the great silence gave way, and from the river and palm-shaded slope arose a shout of wonder and fear, which reached its climax at the moment of the sun's disappearance, nor ceased then, for in addition to the horror of an eclipse . . . there appeared in the heavens on the right of the sun an unmistakable scimetar. The eclipse had indeed revealed the existence of a new comet."

According to the 1885 October 9 issue of *Science*, the various eclipse parties met after the eclipse and jointly agreed to name the comet "Tewfik . . . in recognition of the Khedive's generous hospitality."

Although an orbit cannot be calculated directly from the positions, B. G. Marsden (1967, 1989) has determined orbits based on the assumption that this comet was a member of the sungrazing family. In 1967, he wrote, "The tail was curved much as would be expected for a comet very close to the sun and rapidly approaching perihelion." He gave the probable perihelion date as 1882 May 17.5. This was revised to May 17.46 in 1989 and Marsden conjectured that the comet was associated with comet C/1880 C1. Marsden's hypothetical orbit is given below.

T	ω	Ω (2000.0)	i	q	e
1882 May 17.463 (UT)	86.16	7.69	144.50	0.00534	1.0

FULL MOON: May 3, Jun. 1
SOURCES: *SM*, **1** (1882 Jun.), p. 99; C. Trépied, *Nature*, **26** (1882 Jun. 29), p. 210; H. C. F. Kreutz, *AN*, **102** (1882 Jul. 6), p. 271; *SM*, **1** (1882 Aug.), p. 129; P. Tacchini, *CR*, **95** (1882 Nov. 13), pp. 897–8; W. de W. Abney and A. Schuster, *Science*, **6**, (1885 Oct. 9) p. 313; W. de W. Abney, A. Schuster, and Baillie, *PTRSL*, **175** (Part 1) (1884), pp. 261–262, C1889, p. 587; B. G. Marsden, *AJ*, **72** (1967 Nov.), pp. 1177; B. G. Marsden, *AJ*, **98** (1989 Dec.), pp. 2306–21.

C/1882 R1 (Great September Comet)

1882 II = 1882b

Discovered: 1882 September 1.2 (Δ = 1.37 AU, *r* = 0.71 AU, Elong. = 30°)
Last seen: 1883 June 1.97 (Δ = 5.16 AU, *r* = 4.42 AU, Elong. = 39°)
Closest to the Earth: 1882 September 17 (0.9773 AU)
Calculated path: SEX (Disc), LEO (Sep. 15), VIR (Sep. 17), LEO (Sep. 18), SEX (Sep. 29), HYA (Oct. 10), ANT (Nov. 14), PYX (Nov. 18), PUP (Dec. 10), CMa (Dec. 29), LEP (Feb. 3), MON (Mar. 24)

This ranks as one of the most spectacular comets of the nineteenth century. It became visible to observers in the Southern Hemisphere and the southern latitudes of the Northern Hemisphere in the morning sky early in September of 1882. The earliest observations are not first-hand accounts. J. G. Galle (1894) wrote that the comet was seen on 1882 September 1 at the Gulf of Guinea and at the Cape of Good Hope. The likely times in Universal Time (UT) were September 1.2 for both observations. W. T. Lynn (1903) reported the earliest observation was made in Auckland, New Zealand, on 1882 September 3. The likely time in UT was September 2.7. B. A. Gould (Cordoba, Argentina) reported in the 1883 January 5 issue of the *Astronomische Nachrichten* that he was informed of the comet on September 6, and that the informant had seen it in the morning sky on September 5. The likely time in UT was September 5.4. The informant claimed the comet was as bright as Venus with a brilliant tail. Gould added, "Inquiry showed that it had been seen for several days by employees of the railroad and other persons whose duties required them to rise before daylight." Another observation was reported in the October 1882 issue of the *Sidereal Messenger*, which simply noted that the comet was seen from Panama on September 6. This comet was observed on September 6.8 by members of the crew of the steamer *Caraki*. It was subsequently reported to H. C. Russell (Sydney, Australia) on September 7 by Springwell, the chief officer of the steamer.

The first astronomer to observe the comet was W. H. Finlay (Cape of Good Hope, South Africa,) on September 8.16, which was also an independent discovery. He was then on his way home from Royal Observatory after observing an occultation of 5 Cancri by the moon. He said the comet was a conspicuous object with a large head, a nucleus of magnitude 3, and a tail about a degree long. He returned to the observatory to make observations before sunrise. David Gill of the same observatory sent a notice to the English Astronomer Royal. It was later discovered that Finlay made an error reading off the declination circle and misidentified the comparison star. This led to an erroneous motion in declination that was included in Gill's letter. After discovering the error, Finlay determined a new position of $\alpha = 9^h\ 31.7^m$, $\delta = -1°\ 00'$ for September 8.17.

Additional independent discoveries were made by J. Tebbutt (Windsor, New South Wales, Australia) on September 8, Joseph Reed (on board HMS *Triumph*, just south of the Cape Verde Islands) on September 10, and L. Cruls (Rio de Janeiro, Brazil) on September 12. Tebbutt said the nucleus was "large and brilliant", while the tail was about 3–4° long. Cruls said the comet was visible to the naked eye and surmised it was probably the expected comet Pons of 1812.

W. L. Elkin (Royal Observatory, Cape of Good Hope, South Africa), having been alerted by Finlay, first saw the comet on the morning of September 9. He said, "It then appeared to the unassisted eye about as bright as a star of the third or fourth magnitude, with a straight tail about 2.5° in length. The colour of the comet's light struck me as remarkably white, perhaps contrasting it from recollection with comet Wells, which was of a brilliant golden hue." He said the coma was 40–50" across, while the well-defined nucleus was 10" or 15" in diameter and strongly condensed in the center. Elkin gave the position angle of the tail as 253.1° and noted, "The southern edge of the tail was sharply defined for a considerable distance, but the northern one faded away some 12–15' from the head."

The comet was heading toward perihelion when discovered and steadily moved toward the sun and morning twilight in the days that followed; however, instead of becoming lost in the dawn sky, the comet continued to be followed. Reed said the crew of his ship saw the comet on September 12 and 13. He noted, "the comet was visible for only a few minutes before sunrise; the twilight prevented my determining the length of the tail, but it appeared to extend through an arc of two or two and a half degrees. The whole of the coma is very brilliant, the nucleus surrounded by a still brighter ring; the tail was not curved." On September 13, L. A. Eddie (Grahamstown, South Africa) saw the comet shortly after it had risen above the horizon. He said that in the strong twilight it appeared "as a brilliant but narrow band of ruddy light, terminating in a very bright nucleus, equalling Jupiter in brilliancy and apparent size." He added that the tail was straight and about 12° long. On September 15, Eddie said the nucleus shone with a light equal to Jupiter's. On September 16, Eddie said his 24-cm telescope revealed "the nucleus appeared less sharply defined on its preceding boundary, and the breadth of the coma was greater on the northern side than on the southern. The tail seemed to spread out for a short distance behind the comet, and it was darker in the centre, as if split open." Gould said the comet was visible in the finder throughout the day.

The comet's increase in brightness was so great that it became easily visible in broad daylight for more than two days. On September 16.98, Tebbutt saw the comet about 4° west of the sun "and moving fast towards that luminary. The head and the tail for about a third of a degree were well seen." On September 17.17, Eddie saw the comet rise about 14 minutes before the sun and said it had further increased in size and brightness. The tail was then noted as about 8' long. He said the comet remained visible throughout the day and continued to show an overall length of about 1°. He noted, "So apparent was it to the naked eye, that one had but to look in the direction of the Sun when it could immediately be seen without any searching." He also looked at it with his 24-cm telescope and said, "The nucleus appeared as a solid globe with a white light surpassing that of Venus when viewed in the daytime. The coma and tail for a short distance behind the head were also very brilliant. There was but little coma preceding the nucleus, so that the nucleus appeared as situated at almost the extremity of the tail. The coma was bounded on the margins by a denser stream of light than that composing its interior portion, and the northern side was narrower and brighter than the southern, but the southern extended further behind the nucleus than the northern." J. P. y Ferrer said the inhabitants of Reus, Spain were astronished by the appearance of this comet on September 17.34. He noted that the comet was 1.5° west of the sun and was so brilliant it could be seen through light clouds. Eddie also noted that the comet continued to approach the sun throughout the day so that by September 17.51 there was some difficulty in being able to spot it. Because of approaching clouds, Eddie last saw the comet on September 17.61 when his 8-cm refractor (50×) revealed it about 14' from the sun's edge. On September 17.45, A. A. Common (Ealing, England) independently discovered the comet with a 15-cm helioscope during a routine search for comets near the sun, which he began shortly after the announcement of the eclipse comet of 1882 May 17. He said the nucleus was "large, bright, and quite round, with a diameter of about 45".... The tail was then very bright." On September 17.62, Gould said the comet was easily found in full daylight, although a "shade-glass" had to be used because of the comet's proximity to the sun. On September 17.64, Gould said the comet and sun were in the same field of view. Finlay and Elkin said the comet was visible throughout the day at the Cape of Good Hope and they made a large number of measurements using the great Indian theodolite. Finlay began watching the comet in the afternoon using a 15-cm refractor (110×) equiped with a neutral-tint wedge and noted the comet was rapidly approaching the sun's limb. Two measurements with the micrometer revealed the comet's disk was 4" across. The comet and the sun's limb were both in the same field of view on September 17.6430. Finlay commented, "The silvery light of the comet presented a striking contrast to the reddish-yellow of the Sun; the tail could only be traced to a very short distance now." Elkin said "I actually observed it to disappear among the undulations of the Sun's limb" on September 17.6506. Finlay finally lost sight of the comet about 8.5 seconds later "when the Sun's limb was boiling all about it. I fancied I caught a glimpse of it 3 s later, but was not sure. I then examined the Sun's disk very carefully, but could not see the slightest trace of the comet." Gould tried to find the comet on September 17.68, but failed, as it was then transiting the sun. He noted, "although I carefully scrutinized [the sun] and especially the preceding limb as it traversed the field of the meridian-circle, no token of the comet could be seen, nor could it be found during the afternoon."

As the comet approached perihelion it began transiting the sun on September 17.65. The transit ended on September 17.69, after 1 hour and 17 minutes. The comet passed perihelion on September 17.72, and by September 17.74 it reached a maximum solar elongation of 27' and then began heading towards an occultation by the sun. The occultation began on September 17.79, and ended on September 17.87, after 1 hour and 58 minutes.

Not long after the comet exited from behind the sun, it again came under observation. On September 18.06, Tebbutt said, "I could plainly distinguish it without optical aid simply by screening the eye from the Sun's direct rays. It was then less than a degree west of the Sun's western limb, and moving west, having obviously shortly before passed its perihelion." D. Gill was at Simons Bay, South Africa, on September 18.19 and said, "I was astonished at the brilliancy of the comet as it rose behind the mountains on the eastern side of False Bay. The Sun rose a few minutes afterwards, but to my intense surprise the comet seemed in no way dimmed in brightness, but becoming instead whiter and sharper in form as it rose above the mists of the horizon. I left Simons Bay and hurried back to the observatory, pointing out the comet in broad daylight to the friends I met by the way. It was only necessary to shade the eye from direct sunlight with the hand at arm's-length to see the comet with its brilliant white nucleus, and dense white, sharply-bordered tail of quite 1/2° in length." Gill estimated the nucleus was about magnitude 1. On September 18.48, L. Thollon (Nice, France) saw the comet 3° west of the sun and noted it was very brilliant. About 20' of tail was visible to the naked eye. It was like a half-ellipse in shape and exhibited a prominent nucleus. Gould reported the comet was again visible throughout the day on the 18th. Early that day he said, "its brilliancy attracted popular attention throughout the country, and the 'blazing star near the sun' was the one topic of remark." He noted a small telescope revealed a "brilliant nebulous mass." On September 18.78, H. C. Wilson (Cincinnati Observatory, Ohio, USA) saw the comet with the naked eye in daylight. Eddie said the comet was visible on September 19.26 in daylight about 4° northwest of the sun and "now moving off in almost the same direction as it approached the Sun."

Word of this comet appeared in the morning papers in the USA on September 19. The skies were cloudy in New Jersey during the early hours of daylight, but C. A. Young (Princeton, New Jersey, USA) said, "the comet suddenly became easily visible to the naked eye" on September 19.71. He said the glare of the sun on the object glass of the 58-cm refractor was so bad that only the nucleus could be seen, but "the comet was beautifully seen in the 13-cm finder, which was screened from the glare by the tube of the great telescope. The nucleus was diffuse, not stellar, (magnifying power about 75); the first envelope was pretty bright and well defined, extending out on each side to form the tail, and the second envelope was easily visible, though rather faint. The interesting feature, however, was the pair of eccentric arcs connecting the two envelopes: they were not conspicuous, but I think there is no doubt as to their reality."

The comet was only visible in morning twilight for the next few days, but as it moved further away from the sun the tail gradually became longer. Eddie saw the comet in bright twilight on September 20 and said it was 5° long and about 1.5° wide, but it was on the 22nd that the worldwide astronomical community really began to study this comet. E. E. Barnard (Nashville, Tennessee, USA) said the very bright, straight, narrow tail was about 12° long and was slightly convex toward the south. He also noted that the comet remained visible to the naked eye until about 15 minutes after sunrise. With a telescope, Barnard remarked, "A bright outline passes completely around the head, forming a parabolic curve of yellowish light which streaming backward traces the boundary of the tail." E. Millosevich (Rome, Italy) saw the comet about 20 minutes after noon with a telescope and said it appeared like snowy wool that was unequally illuminated. J. M. Schaeberle (Ann Arbor, Michigan, USA) saw the comet shortly before noon with the 15-cm refractor and said the nucleus was sharply defined and elongated, with a length of 11.9″ and a width of 3.8″.

During the remainder of September, astronomers primarily described the tail, although a few kept tabs on the region around the nucleus. By the 24th, observers in the Southern Hemisphere were estimating tail lengths of 15–25°, while the width remained near 1°. The tail was not as well situated for Northern Hemisphere observers and was typically estimated as 5–10° long. A. Riccò (Palermo, Italy) said the tail was concave toward the north. C. W. Pritchett (Morrison Observatory, Glasgow, Missouri) said the nucleus was "planet-like, and apparently about the size of Uranus." On the 25th, R. L. J. Ellery (Melbourne, Australia) noted the comet was "markedly less bright," while G. F. Parson (sailing south of the equator on the *Earnock* and traveling eastward toward Melbourne, Australia) reported "The body of the comet apparently equal to a star of the 1st magnitude." A. J. G. F. von Auwers (sailing on the SS *Theben*, near St Vincent) estimated the brightness as closer to magnitude 0. Barnard looked at the comet with a 13-cm Byrne refractor

(78×) on the 27th and noted the nucleus was elongated in the direction of the tail. By the 30th, O. C. Wendell (Harvard College Observatory, Cambridge, Massachusetts, USA), Riccò, and Tebbutt were also noting the elongation of the nucleus.

The first half of October was an interesting time for the comet, as telescopes began revealing things happening both inside and outside the coma. Total magnitude estimates were not plentiful, but did indicate the comet was fading, despite finally becoming visible in dark skies. On the 1st, Barker estimated the magnitude as about 0.5, while E. E. Markwick (Pietermaritzburg, South Africa) noted its brightness equaled a star of the 1st magnitude. On the 6th, L. Weinek (Gohlis, Germany) estimated its brightness as close to magnitude 1.8. Most observers were reporting the comet was a distinct white color early in this period, while Barnard reported a "pearly hue" on the 15th. The tail length stayed remarkably consistent at between 15° and 20° throughout and it remained quite narrow. A dark streak was also reported extending down the center of the tail for a short distance from the coma, which had vanished by the 8th. As this period began, observers examining the region within the inner coma were still reporting the nucleus as elongated, but this began to change on October 2. On that date, Pritchett said the nucleus had become egg-shaped, with "the wider end towards the sun." On the 3rd, F. Terby (Leuven, Belgium) and Eddie independently reported two nuclei, with Eddie using a 24-cm reflector and noting they were distinctly ellipsoidal. On the 4th, Eddie wrote, "The preceding nucleus was larger and brighter than the other, and they resembled in shape two grains of rice placed end to end." On the 5th, both Barnard and Wilson independently reported the presence of three nuclei. Wilson was observing with a 28-cm refractor and said the nuclei were "in a row, nearly parallel with the right (north) side of the tail." During the next few days observers with small telescopes continued to report one very elongated nucleus, while those with large telescopes saw two or three nuclei. On the 6th, Pritchett said the nucleus presented three centers of light "situated in a right line, extending parallel to the axis of the tail." He added that they "seemed to float in a cloud of yellow dust." C. N. A. Krueger said the two brightest nuclei were separated by 13" on the 8th. Finlay indicated they were 22" apart on the 10th. On the 11th, Eddie said the nuclei "have greatly altered in shape since I first detected the division, and they have also somewhat considerably opened out. The preceding nucleus is now much smaller and condensed into a very bright starlike point, from which spiral streams seem to emanate. The hindmost nucleus has become greatly elongated, and now possesses as it were two centres of condensation, and appears to be fast approaching to a further division. This elongated nucleus resembles a dumb-bell with the sphere towards the tail more oval than the other." On the 15th, Eddie obtained another interesting observation of the nuclear region. With his 24-cm reflector he noted one completely distinct nucleus which "resembled in colour the electric light," and two more nuclei within a "bar of light." When he increased the magnification to 100× the condensations within the bar of light "seemed again doubled, so that the whole nucleus resembled a string of five ill-defined luminous beads."

During the second half of October, most astronomers were concentrating on the nucleus, but other observations were made. The only estimate of the comet's brightness came from Markwick on the 24th, when he said it was about magnitude 1. The tail length continued to fall in the range of 15–20° throughout the period, except when moonlight interfered, at which time only 10° of tail was visible. The tail axis was pointing toward PA 276° on October 19 and PA 282° on the 31st. Although the tail continued to be fairly narrow and convex toward the south, it fanned out at the end and Barnard commented that it appeared "swallow tailed." The tail's curvature began decreasing near the end of the month. The nucleus was still seen by many observers as strongly elongated, but larger telescopes continued to show up to six distinct nuclei, which Schaeberle described on the 18th as "like very small beads strung on a thread of worsted." I. Y. Kortazzi (Nicolajew, now Mykolayiv, Ukraine) said the major axis of the nucleus was directed toward PA 286° on the 19th and 293° on the 31st. Eddie monitored the nuclei very closely nearly every night with his 24-cm reflector and noted that the order of brightest to faintest changed from one night to the next. Other astronomers gave similar descriptions. On the sunward side of the nucleus five or six hoods were being observed. C. H. F. Peters (Hamilton College Observatory, Clinton, New York, USA) reported that the innermost hood actually joined "the bright linear light of the nuclear region."

During the whole of October a very unusual phenomenon presented it-self to all observers who either examined the area around the comet's head with a telescope or had the good fortune to see the comet under extremely dark skies. Markwick reported on October 5 that when he looked at the comet with a 7-cm refractor, he noted, "South, preceding the comet's head, at this time were seen, about 1.5° distant, two wisps or pieces of nebulous-looking light. Whether they had anything to do with the comet I cannot say; but I can trace no nebula as being in this position, and moreover I have not been able since to recover them." Unaware of Markwick's observation, J. F. J. Schmidt (Athens, Greece) saw an object about 4° southwest of the comet on October 8 and actually reported it as a new comet in the 1882 October 12 issue of the *Astronomische Nachrichten*. Schmidt reported further positions for the object on October 10 and 11, which allowed H. Oppenheim and K. Zelbr to independently calculate parabolic orbits with perihelion dates of September 24.90 and 25.11, respectively. The remaining orbital el-ements were somewhat similar to those of the large comet, including the very small perihelion distance; however, although Oppenheim produced an ephemeris for the object, no further observations were made. Interest-ingly, E. Hartwig (Strasbourg, France) also saw the object on October 10. He noted a large nebula southwest of the bright comet, "which looked like a comet with a bright nucleus and a fan-shaped tail." He could not find the object on October 13. Barnard contributed observations of the unusual phenomenon on the 14th. He swept the 13-cm Byrne refractor around the comet and to the south he saw "a large distinct cometary mass, fully 15' in diameter. A similar object but less bright was seen close beside this, their edges touching – apparently a double comet – and on the opposite side of the first object was a third fainter mass, the three almost in line, east and west." He shifted the telescope southeastward and noted several more objects, "one very elongated in form." He said, "There were, at least, six or eight of those objects near one another within about 6° south by west of the large comet's head." He added that each object had the appearance of "distinct telescopic comets with very slightly brighter centers." Barnard was so impressed with the observation that he had his wife come out and confirm they were there. Interestingly, a few other astronomers reported, and even drew an extremely faint light which had seemed to almost surround this comet since mid-September and which became most extensive during the period of October 6–17, and eventually faded from view in early November. The earliest observation seems to come from C. Grover, an observer with the British Expedition traveling to Brisbane (Australia) to observe the December transit of Venus. Grover saw the comet on the morning of September 14 from the deck of the steamer and wrote, "The Comet seems enclosed in a large and faint envelope, which, singular to say, does not take the curved figure of the Tail, but shows a straight outline, and projects in front on either side of the nucleus, ending in two points...." He said the region between these two points "looks quite black." Another "cometary mass" was seen by W. R. Brooks (Phelps, New York, USA) on October 21. It was about 2° long and was located about 8° east of the comet. The portion of this vast cloud of light extending outward from the sunward side of the coma became very prominent during the period of October 16–24. Barnard and Auwers indicated lengths of 4–6° on the 16th and 17th, but the most interesting aspect of this "sunward tail" was that its sides were straight and parallel with one another. E. W. L. Tempel (Arcetri, Italy) commented on the 18th that the appearance in an opera glass was like that of a tube as "the sides were bright and nebulous, while the axial line was dark." Astronomers reported this feature had noticeably faded by the 20th and had disappeared by the 27th. No similar phenomenon of this nature seems to have ever been reported for any comet, but it is almost certain that the "comets" reported by Schmidt, Hartwig, and Barnard were probably condensations within this envelope, which might represent some of the dust released during the comet's previous perihelion passage nearly eight centuries earlier or might be the dust released by other members of the vast sungrazing comet family of which this comet is a member.

Observations are conflicting as to how fast the comet faded during November, with B. J. Hopkins (London, England) estimating the magnitude of the *nucleus* as 2 on the 9th and Markwick saying the comet was "superior in brightness" to a 5th-magnitude star on the 21st. But estimates of the tail length seemed to imply no change from October, with most estimates falling within the range of 17–20° throughout the month until moonlight began interfering after the 23rd. Kortazzi said the tail was directed toward PA 288° on the 8th. Hopkins actually said the tail was 30° long on the 15th and looked like the Greek letter γ. He said the southern half was brighter and straighter than the northern half, with the northern half curving upwards at the end. The tail was also displaying the dark streak noted during October by many people, but no observations of it were made after the 9th. Many astronomers were still reporting the nucleus as very elongated, while some referred to it as a "line of condensation." Common measured its dimensions as about 100" long and 11" wide on the 2nd, while Riccò said the nucleus was about 123" long and 12.7" wide on the 15th. Finlay, W. Schur (Strasbourg, France), and others who observed an oblong nucleus said it was more diffuse than during October. Eddie continued to observe the multiple nuclei with his 24-cm refractor, but said the varying brightnesses of the nuclei caused some to disappear. Eddie reported four nuclei on the 1st, six on the 2nd, and five on the 10th and 17th. On the last date he remarked that the sixth nucleus "has now quite disappeared." He added, "These extremely tiny points of light, situated as it were in a nebulous mist, and revealing themselves at intervals on very favourable nights only, by their twinkling do indeed present a wonderful phenomenon." N. de Konkoly (O'Gyalla Observatory, Hungary) made an interesting comment about the nucleus on the 2nd. He said, "A radiation from the nucleus, which large comets generally show, did not exist...." This most likely refers to the fact that no observers reported jets of material extending from the nucleus during this apparition of the comet.

The comet remained visible to the naked eye throughout the month of December, but the only estimate of the brightness was made by Markwick on the 1st, when he said the head was as bright as a 5th-magnitude star. The tail was showing its persistence at the beginning of the month, when Markwick and Eddie noted naked-eye lengths of 15° and 12°, respectively, in moonlight on the 1st. Several observers reported tail lengths of 8–15° after the moon had left the sky, but Markwick, who made the most extensive series of tail length estimates during the month, said the tail was completely invisible on the 22nd when moonlight again began interfering. Eddie said the hook at the end of the tail was still plainly visible on the 2nd. Markwick and others reported the tail was straight from December 10 onward. Astronomers were still keeping an eye on the region around the nucleus, but many were beginning to discover there was nothing more to see. Tebbutt could no longer see a nucleus in the 11-cm refractor on December 1, 2, and 8, and Markwick said his 7-cm refractor revealed no sign of the nucleus on the 8th. Eddie observed three nuclei on December 1 and 4, with his 24-cm reflector, but on the last date he said the nucleus "resembles a string of luminous beads almost hidden by a luminous gaseous veil stretched over them." Eddie described the nucleus as "oval and granulous" on the 5th, and on the 9th he said, "I could no longer detect the points." Interestingly, as the nucleus and comet faded, larger telescopes were turned toward the comet and were able to continue observations of the nuclear region. Wilson saw two nuclear condensations in a 71-cm refractor on the 12th, and J. M. Thome (Cordoba, Argentina) noted two condensations on the 25th.

The comet continued to remain a naked-eye object throughout January of 1883. The only estimate of the brightness was made by Auwers on the 30th, when he gave the magnitude as 6. On the other hand, Pritchett's comment on the 14th that the comet "was surprisingly plain to the naked eye" certainly indicates the ease of observability. Schmidt provided a nice series of naked-eye tail length estimates during the month, which indicate the tail extended 11° on the 3rd, 8° on the 10th, 5.5° on the 28th, and over 3° on the 31st. No further descriptive information was reported about the tail, other than E. A. Lamp (Kiel, Germany) stating that it extended toward about PA 211° on the 12th and 245° on the 31st. Many astronomers saw little or no trace of the nucleus, but a few were still seeing the long bar, which Riccò estimated was 185" long on the 4th. Common turned a 3-foot focal length reflector toward the comet on January 27 and saw all five remaining nuclei still situated in a straight line. He said the second and third nuclei were the brightest at magnitude 11, while the first and fourth were "one-third as bright" and the fifth nucleus, which was situated closest to the tail, was the faintest.

The comet's fading took it to the borderline of naked-eye visibility during February. Auwers estimated its magnitude as 5–6 on the 1st and 6 on the 7th. Meanwhile, Schmidt reported the comet was difficult to see with the naked eye on the the 7th and that there was only a doubtful trace visible to the naked eye on the 8th. Gould said he saw the comet with the naked eye on the 11th, while sailing across the Atlantic from Rio de Janeiro toward London; he commented, "had the sky continued clear, it would have remained visible to the naked eye for several days longer." Tebbutt noted the comet was "just visible to the unassisted eye" on the 12th. Estimates of the tail length had dropped considerably, with Schmidt estimating it as 2° to the naked eye on the 5th, Auwers estimating it as 4.5° long and 0.8° wide on the 7th, and Gould estimating it as 6° long on the 11th. In addition, Lamp said the tail was directed toward PA 255°. Observations of the nuclei were at a minimum. Schmidt said two nuclei were visible on the 3rd and estimated both magnitudes as 12.7. Thereafter, he suspected the nucleus was double on the 7th and 8th. Schur also saw two nuclei on the 5th and 11th. Both Schur and Riccò continued to indicate the nuclei or elongated condensation closely matched the tail direction. Common noted that the third nucleus was brightest on the 5th and 24th. On the last date he said the line of nuclei was concave on the north side and added that the fifth nucleus (that closest to the tail) was the faintest. Interestingly, Common added, "I strongly suspected another nucleus following the last in order, and in such a position that it would be symmetrically placed. This, however, may have been a star."

The comet was usually described as faint by observers during March, although Gould said he saw it for the last time with the naked eye on the 8th. No tail was reported by any observer, although Gill noted the coma was elliptical on the 6th, with dimensions of 3.5' by 1.5'. Schmidt said he possibly saw two nuclei of magnitude 12.5 in his refractor on the 8th.

Few descriptive details of the comet became available during April and May. Wendell saw the comet under less than perfect seeing conditions on April 7 and said it appeared faint, with "three points of light seen." Schmidt described the comet as rather faint on the 24th. Thome described it as "an irregular, exceedingly thin white cloud" on May 26.

The comet was last detected on June 1.97, when Thome gave the position as $\alpha = 6^h 32.9^m$, $\delta = -5° 21'$. He said he could then only see "an excessively faint whiteness." Gould commented, "The comet was finally lost to view, not so much from want of intrinsic brightness, as in consequence of its lowness in the West at nightfall."

Several astronomers examined the spectrum of this comet. Thollon observed the comet on September 18.48 and said the spectrum was very bright, with the lines of sodium being its leading characteristic. He also noted that the nucleus included a brilliant continuous spectrum. A sodium line was also noted in the days following perihelion by Russell, Millosevich, and P. Tacchini (Rome, Italy). Thollon's next observation of the spectrum came on October 9, at which time he noted the sodium lines were gone, leaving only the carbon bands. Konkoly observed the spectrum on November 2 and found the nucleus was displaying a "very intense continuous spectrum," while the coma displayed "a tolerably bright comet spectrum, characterized by the hydrocarbon bands." Upon moving to a 24-cm refractor, he measured the wavelengths of the bands as 5620 Å, 5147 Å, 5026 Å, and 4720 Å. There was also a suspected band in the red, which he could not measure. Konkoly noted the band at 5147 Å was the brightest. E. von Gothard (Herény, Hungary) obtained numerous measurements of the spectrum during the period of November 2–19. The same three distinct lines were measured on each occasion and the average wavelengths of the lines were 5619 Å, 5153 Å, and 4710 Å. The line at 5153 Å was the brightest, while that at 5619 Å was the faintest. Riccò said the three hydrocarbon bands were stronger during the period of October 4–26, while the continuous spectrum of the nucleus became fainter.

In a way, this comet inspired the first photographic star catalog. Gill borrowed a 6-cm aperture portrait camera and attached it to the 15-cm refractor. The camera was not shooting through the telescope, but was pointing directly at the sky and simply riding piggyback on the telescope to enable it to track the stars. Gill obtained six images between October 20 and November 15, with exposures ranging from 30 minutes to 2 hours and 20 minutes. Although he was impressed with the comet on the photographs, he was also impressed by the number of stars. Gill was inspired by this and began work on the *Cape Photographic Durchmusterung*, the first photographic star catalog, which gave positions and brightnesses for 454 875 stars.

The comet was also seen by Chinese astronomers/astrologers. The Chinese text *Ta Chhing Li Chhao Shih Lu* noted that "in the second third of this month" a "broom-star" was seen in the southeast. This comet was thought to have been the reappearance of the bright comet of 1881 and its reappearance was thought "due to frequent mistakes committed by those employed in the administration; the hardship of the village folks had not been adequately presented to the throne. Order is now given to make a thorough investigation."

The Chinese text *Chhing Chhao Hsü Wên Hsien Thung Khao* noted that Chinese astronomers discovered a "broom-star" on October 5. It was situated in the southeast. There is an additional statement that the comet of 1881 "went out of sight" sometime during the period of 1882 August 14 to 1882 September 11. Since comet C/1881 K1 was not observed by large telescopes after 1882 February 15, the Author suggests the statement of a comet seen in either August or September refers to another pre-perihelion observation of comet C/1882 R1.

Interestingly, A. V. Nursinga Row (Daba Gardens, Vizagapatam, India) said, "Our Hindu astronomers predicted the appearance of a comet in the southern hemisphere in their printed calendar for the present year. As to the identity, they gave no other particulars except that it would possess a bright copper colour like the rising moon, and a long tail. The name given by them to their predicted comet is 'Silpacam'."

The first parabolic orbit was calculated by E. J. White. He used positions obtained at Melbourne Observatory on September 9, 13, and 17, and determined a perihelion date of September 17.67. He added, "The exceedingly small perihelion distance, as well as the other elements of this comet, exhibit a decided similarity to that of 1843." This orbit later proved to have been just over 1 hour from the true perihelion date. Later parabolic orbits were calculated during the next few weeks by Finlay, Elkin, E. Frisby, A. N. Skinner, R. Gautier, H. Oppenheim, J. R. Hind, Weiss, Zelbr, S. C. Chandler, and O. C. Wendell. Most of these only slightly revised White's initial orbit and supported the suggestion that the comet's orbit resembled that of the comet of 1843, as well as the comet of 1880.

By the end of October, it had become obvious that the comet was moving in an elliptical orbit. Chandler and Wendell determined three Normal positions covering the period of September 20 to October 20, and calculated a perihelion date of September 17.72 and a period of about 4070 years. About a month later, Frisby calculated an orbit with a very similar elements, but with a period of about 794 years. He said the orbit showed "a considerable resemblance to Comet I, B.C. 371; and it may be its third return, a very brilliant comet having been seen in full daylight A.D. 363." As it turned out, Frisby's orbit was quite accurate. During the next few weeks and months the period was determined as 269 years by Gautier, 4070 years by Chandler, 843 years by H. C. F. Kreutz, 39 and 1000 years by Elkin, 1372 years by J. Tatlock, Jr, 993 and 712 years by J. Morrison, and Fabritius.

Kreutz (1888) revised his orbit for the main nucleus, which is today known as C/1882 R1–B, and determined a period of about 772 years. But in 1891, Kreutz realized he had enough positions to determine orbits for four of the nuclei. The resulting periods were about 671 years for C/1882 R1–A, about 772 years for C/1882 R1–B, about 875 years for C/1882 R1–C, and about 955 years for C/1882 R1–D. Although L. Hufnagel (1919) did determine an orbit for C/1882 R1–B, which indicated a period of about 761 years, the orbits determined by Kreutz are given below. The first orbit represents comet A, the second is B, the third is C, and the fourth is D.

T	ω	Ω (2000.0)	i	q	e
1882 Sep. 17.7241 (UT)	69.5842	347.6561	142.0111	0.007750	0.999899
1882 Sep. 17.7241 (UT)	69.5873	347.6558	142.0109	0.007750	0.999908
1882 Sep. 17.7241 (UT)	69.5840	347.6541	142.0106	0.007751	0.999915
1882 Sep. 17.7241 (UT)	69.5809	347.6508	142.0092	0.007749	0.999920

ABSOLUTE MAGNITUDE: $H_{10} = 0.8$ (V1964)

FULL MOON: Aug. 28, Sep. 27, Oct. 26, Nov. 25, Dec. 24, Jan. 23, Feb. 21, Mar. 23, Apr. 22, May 22, Jun. 20

SOURCES: K. Zelbr, *SAWW*, **86** Abt. II (1882), pp. 1090–7; R. Copeland and L. Cruls, *AN*, **103** (1882 Sep. 17), p. 127; H. Oppenheim and L. Thollon, *AN*, **103** (1882 Sep. 25), p. 159; L. Thollon, Gouy, J. P. y Ferrer, and W. de Fonvielle, *CR*, **95** (1882 Sep. 25), pp. 555–60; E. Frisby, A. N. Skinner, and C. A. Young, *SM*, **1** (1882 Oct.), pp. 165–7; C. N. A. Krueger and H. Oppenheim, *AN*, **103** (1882 Oct. 3), pp. 173–6; F. Terby and C. F. W. Peters, *AN*, **103** (1882 Oct. 10), p. 205; J. F. J. Schmidt, G. Cacciatore, A. Riccò, S. C. Chandler, Jr, O. C. Wendell, W. H. Finlay, and C. N. A. Krueger, *AN*, **103** (1882 Oct. 12), p. 209, 219–24; J. R. Hind, *Nature*, **26** (1882 Oct. 12), pp. 582–3; K. Zelbr, *AN*, **103** (1882 Oct. 16), pp. 235–8; H. Oppenheim, E. Frisby, A. N. Skinner, E. Millosevich, P. Tacchini, and E. Weiss, *AN*, **103** (1882 Oct. 23), pp. 251–6, 267–72; L. Weinek, A. Auerbach, H. C. Vogel, A. Riccò, H. Oppenheim, and K. Zelbr, *AN*, **103** (1882 Oct. 28), pp. 273, 279–86; D. Gill, W. H. Finlay, W. L. Elkin F. C. Penrose, W. G. Thackeray, A. Pead, A. M. W. Downing, R. L. J. Ellery, E. J. White, H. C. Russell, Springwell, J. Tebbutt, Jr, A. V. Nursinga Row, and G. Pochrane, *MNRAS*, **43** (1882 Nov.), pp. 19–25, 27–34; C. S. Hastings and C. A. Young, *SM*, **1** (1882 Nov.), pp. 171–4, 187–9; J. F. J. Schmidt, J. Tebbutt, Jr, E. Millosevich, and E. W. L. Tempel, *AN*, **103** (1882 Nov. 8), pp. 305–14, 317; B. von Engelhardt, J. Palisa, E. Frisby, A. Riccò, and W. Schur, *AN*, **103** (1882 Nov. 10), pp. 327, 331–4; S. C. Chandler, Jr, R. L. J. Ellery, and C. L. F. André, *AN*, **103** (1882 Nov. 15), pp. 347–50; A. Riccò and G. Cacciatore, *AN*, **103** (1882 Nov. 20), pp. 363–6; A. Graham, D. Gill, J. T. Stevenson, N. de Konkoly, J. Reed, T. P. Parry, W. G. Lettsom, C. Grover, I. Y. Kortazzi, and J. Tebbutt, Jr, *MNRAS*, **43** (1882 Dec.), pp. 50–62; L. Cruls, *CR*, **95** (1882 Dec. 18), pp. 1270–1; J. F. J. Schmidt and A. N. Skinner, *AN*, **104** (1882 Dec. 20), pp. 89–94; A. M. W. Downing, C. L. Prince, G. F. Parson, D. W. Barker, B. J. Hopkins, and F. C. Penrose, *MNRAS*, **43** (1883 Jan.), pp. 84–91; B. A. Gould, *AN*, **104** (1883 Jan. 5), pp. 129–32; H. C. F. Kreutz and E. Frisby,

AN, **104** (1883 Jan. 12), pp. 157–60; A. A. Common, *MNRAS*, **43** (1883 Feb.), p. 197; E. E. Barnard and W. R. Brooks, *AN*, **104** (1883 Feb. 5), pp. 267–72; W. L. Elkin, *AN*, **104** (1883 Feb. 7), p. 281; Jedrzejewicz, W. Schur, and E. A. Lamp, *AN*, **104** (1883 Feb. 12), pp. 315–20; J. F. J. Schmidt, *AN*, **104** (1883 Feb. 27), pp. 361–6; L. A. Eddie, *MNRAS*, **43** (1883 Mar.), pp. 289–97; A. J. G. F. von Auwers, *AN*, **105** (1883 Mar. 10), pp. 3–10; E. A. Lamp, *AN*, **105** (1883 Mar. 14), p. 25; J. F. J. Schmidt, A. Riccò, and E. Frisby, *AN*, **105** (1883 Mar. 19), pp. 33–40; W. H. Finlay. D. Gill, and H. C. Wilson, *AN*, **105** (1883 Mar. 29), pp. 71–6; W. H. Finlay, D. Gill, W. L. Elkin, G. Strahan, E. W. L. Tempel, and E. E. Markwick, *MNRAS*, **43** (1883 Apr.), pp. 319–25; R. H. Tucker, Jr, *AN*, **105** (1883 Apr. 2), p. 93; J. Tebbutt, *AN*, **105** (1883 Apr. 9), pp. 117–20; B. A. Gould and J. M. Thome, *AN*, **105** (1883 Apr. 25), pp. 185–92; A. A. Common and J. Tebbutt, Jr, *MNRAS*, **43** (1883 May), pp. 382–94; O. H. Landreth, *AN*, **105** (1883 May 9), p. 237; J. Tebbutt and B. von Engelhardt, *AN*, **105** (1883 May 16), pp. 263, 267–70; J. Tatlock, Jr and E. de Gothard *MNRAS*, **43** (1883 Jun.), pp. 419–20, 424; J. F. J. Schmidt C. W. Pritchett, and R. Gautier, *AN*, **105** (1883 Jun. 4), pp. 341–50, 363–6; J. R. Eastman and J. Tatlock, Jr., *AN*, **106** (1883 Jun. 19), pp. 7–14; E. W. L. Tempel, *AN*, **106** (1883 Jun. 23), pp. 27–9; E. Hartwig, *AN*, **106** (1883 Aug. 31), pp. 225–34; B. A. Gould and J. M. Thome, *AN*, **106** (1883 Sep. 12), pp. 273–82; C. H. F. Peters, *AN*, **107** (1883 Oct. 31), pp. 85–8; J. Morrison, *MNRAS*, **44** (1883 Dec.), pp. 49–55; G. H. Willis, *MNRAS*, **44** (1884 Jan.), p. 86; W. Schur, *AN*, **108** (1884 Apr. 29), pp. 389–92; E. Weiss, *AN*, **109** (1884 Jul. 14), pp. 161–72; C. W. Pritchett, *Publications of the Morrison Observatory*, No. 1 (1885), pp. 102–4; W. Schur, *AN*, **114** (1886 Mar. 23), pp. 81–4; H. C. F. Kreutz, *Publication der Sternwarte in Kiel*, No. III (1888), p. 107; J. M. Schaeberle, *AN*, **119** (1888 Apr. 30), pp. 85–8; H. C. F. Kreutz, *AJ*, **8** (1889 Apr. 15), p. 184; H. C. F. Kreutz, *Publication der Sternwarte in Kiel*, No. VI (1891), pp. 35–51; G1894, p. 283; *Annals of the Astronomical Observatory of Harvard College*, **33** (1900), pp. 153–4; J. Pidoux and A. A. Common, *AN*, **162** (1903 Jul. 22), pp. 353–6; W. T. Lynn, *The Observatory*, **26** (1903 Aug.), pp. 326–7; L. Hufnagel, *AN*, **209** (1919 Jul. 1), pp. 17–22; V1964, p. 62; HA1970, pp. 87–8; A. H. Benners, personal correspondence from G. Linden (2000); SC2000.0, p. 131.

C/1882 R2
(Barnard)

1882 III = 1882c

Discovered: 1882 September 14.37 ($\Delta = 1.53$ AU, $r = 1.40$ AU, Elong. $= 63°$)
Last seen: 1882 December 8.48 ($\Delta = 1.67$ AU, $r = 1.05$ AU, Elong. $= 36°$)
Closest to the Earth: 1882 October 22 (0.8999 AU)
Calculated path: GEM (Disc), CMi (Sep. 19), MON (Sep. 30), HYA (Oct. 5), PUP (Oct. 8), HYA (Oct. 10), PYX (Oct. 11), VEL (Oct. 21), CAR (Nov. 1), CEN (Nov. 4), CRU (Nov. 7), MUS (Nov. 8), CIR (Nov. 15), TrA (Nov. 21),

E. E. Barnard (Nashville, Tennessee, USA) was conducting one of his regular searches for comets when he found this one on 1882 September 14.37. He gave the position as $\alpha = 7^h\ 17.6^m$, $\delta = +16°\ 45'$ on September 14.43. Barnard said it was 2′ across, but distinct enough to be seen in his 13-cm telescope. He estimated the brightness as "tenth magnitude or fainter." There was also a slight central condensation, but no tail. Barnard confirmed his find the next morning and described it as "round without a trace of tail."

The comet was closing in on both the sun and Earth when discovered. On September 17, J. G. Lohse (Dun Echt Observatory, Scotland) observed it with the 38-cm refractor (229×) and noted, "The comet is fairly bright, gradually much brighter in the middle and about 2′ in diameter." C. N. A. Krueger (Kiel, Germany) also saw the comet on the 17th. He said hazy skies made the comet difficult to see and noted it was drowned out when the wires of the micrometer were illuminated. Krueger next observed the comet on the 20th and said it was faint enough to disappear when the threads of the micrometer were illuminated. T. Zona (Palermo, Italy) described it as a "faint object without a sensible nucleus" on the 24th. H. C. Wilson (Cincinnati, Ohio, USA) described the comet as very faint in the 28-cm refractor on the 25th, with a very small, but bright point in the center. There was no perceptible tail. L. Weinek (Gohlis, Germany) observed with a 9.5-cm Steinheil refractor on the 26th and said the comet was easily seen in bright moonlight.

The comet was closest to Earth during the second half of October. Early in the month, moonlight created a slight problem. J. F. J. Schmidt (Athens, Greece) saw the comet with the 17-cm Reinfelder refractor on the 5th and 6th, but said moonlight prevented it from being seen in the finder. B. von Engelhardt (Dresden, Germany) said the comet was rather bright in the 30-cm refractor on the 6th, and noted a bright condensation and a trace of tail, but no nucleus. Krueger said the comet was very faint on the 8th. Schmidt described the comet as very faint and about 8–9′ long on the 11th. Wilson said the comet was "very faint in haze near the horizon" in the 28-cm refractor on the 12th. Nevertheless, he did note a slight central condensation. On the 13th, Wilson said the comet was "strongly condensed in the center" with no sign of a tail. Barnard said the comet was easily visible in a 3.2-cm finder on the 16th and had "a smudge of tail pointing away from the sun." Schmidt observed on the 18th and said the coma was strongly condensed, but exhibited no nucleus. He estimated the brightness of the condensation as 8. He added that the comet was easily seen in the finder when near the horizon. This was the final Northern Hemisphere observation.

The comet attained its most southerly declination of $-66°$ on November 16. Southern Hemisphere observers were basically unaware of the comet, because news had not yet reached them. J. Tebbutt (Windsor, New South Wales, Australia) received word on November 30, and found it that night in the evening sky. Clouds prevented measurement of the position. W. H. Finlay was stationed at Aberdeen Road (South Africa) for the transit of Venus and first saw this comet with a 15-cm refractor on December 3. Further observations were obtained on December 5 and 7, and he wrote, "The comet presented the appearance of a small round diffused mass, slightly more condensed towards the centre."

The comet was last detected on December 8.48, when Tebbutt gave the position as $\alpha = 16^h 50.2^m$, $\delta = -59° 03′$. He noted the comet was "hardly distinguishable" in his 8.3-cm refractor.

The first parabolic orbit was calculated by H. Oppenheim using positions from September 15, 17, and 20. He gave the perihelion date as 1882

November 13.41. Very similar orbits were also calculated by H. Büttner, K. Zelbr, and J. R. Hind during the following weeks, with the final perihelion date being given as around November 13.5. S. Wolyncewicz (1883) took positions spanning the period of September 18 to November 11 and calculated an elliptical orbit with a period of about 43 606 years. A definitive orbit was calculated by L. de Ball (1896). He calculated a parabolic orbit with a perihelion date of November 13.48, as well as a hyperbolic orbit with a perihelion date of November 13.42 and an eccentricity of 1.00007. Because of the relative short period of visibility, the parabolic orbit is given below.

T	ω	Ω (2000.0)	i	q	e
1882 Nov. 13.4754 (UT)	254.2882	250.7703	96.1464	0.955567	1.0

ABSOLUTE MAGNITUDE: $H_{10} = 7.4$ (V1964)

FULL MOON: Aug. 28, Sep. 27, Oct. 26, Nov. 25, Dec. 24

SOURCES: C. N. A. Krueger and H. Oppenheim, *AN*, **103** (1882 Sep. 17), p. 127; H. Oppenheim and C. N. A. Krueger, *AN*, **103** (1882 Sep. 22), p. 143; J. Franz, *AN*, **103** (1882 Sep. 25), p. 159; E. E. Barnard, *SM*, **1** (1882 Oct.), p. 168; H. Büttner, *AN*, **103** (1882 Oct. 3), p. 171; T. Zona and C. N. A. Krueger, *AN*, **103** (1882 Oct. 12), pp. 221–4; K. Zelbr and E. Millosevich, *AN*, **103** (1882 Oct. 23), pp. 253–6; J. R. Hind, *Nature*, **26** (1882 Oct. 26), p. 635; L. Weinek and A. Auerbach, *AN*, **103** (1882 Oct. 28), p. 273; E. W. Maunder, *MNRAS*, **43** (1882 Nov.), p. 29; J. F. J. Schmidt, E. Millosevich, and P. Tacchini, *AN*, **103** (1882 Nov. 8), pp. 309, 317; B. von Engelhardt, *AN*, **103** (1882 Nov. 10), p. 327; T. Zona, *AN*, **103** (1882 Nov. 20), p. 365; I. Y. Kortazzi, *AN*, **104** (1882 Dec. 12), p. 61; S. Wolyncewicz, *AN*, **104** (1883 Jan. 25), pp. 217–22; E. E. Barnard, *AN*, **104** (1883 Feb. 7), p. 287; J. Tebbutt, *AN*, **104** (1883 Feb. 12), p. 317; E. Weiss and H. C. Wilson, *AN*, **105** (1883 Jun. 11), pp. 379, 387; W. H. Finlay, *MNRAS*, **44** (1884 Jan.), p. 87; J. G. Lohse, *MNRAS*, **46** (1886, Supplemental Notice), pp. 490, 494; L. de Ball, *Publikationen der Kuffner-Sternwarte*, **4** (1896), p. B65; V1964, p. 62.

C/1883 D1
(Brooks–Swift)

1883 I = 1883a

Discovered: 1883 February 24.02 ($\Delta = 1.17$ AU, $r = 0.77$ AU, Elong. $= 40°$)
Last seen: 1883 April 24.79 ($\Delta = 2.01$ AU, $r = 1.40$ AU, Elong. $= 40°$)
Closest to the Earth: 1883 March 1 (1.1555 AU)
Calculated path: PEG (Disc), AND (Feb. 28), PSC (Mar. 7), TRI (Mar. 11), ARI (Mar. 19), TAU (Mar. 29), ORI (Apr. 18)

W. R. Brooks (Phelps, New York, USA) discovered this comet with a 23.5-cm telescope on 1883 February 24.02. About 15 minutes later, on February 24.03, L. Swift (Rochester, New York, USA) independently discovered the comet and estimated the position as $\alpha = 22^h 50^m$, $\delta = +28° 00'$. This was barely three months after the completion of Warner Observatory. The comet had already passed perihelion.

Numerous observations were acquired during the remainder of February, and all observers reported the comet was bright, with a strong condensation and a straight, slender tail. The coma diameter was estimated as 3.5′ by H. C. Vogel (Potsdam, Germany) and 3–4′ by E. W. L. Tempel (Arcetri, Italy) on the 26th, and 3′ across by G. V. Schiaparelli (Brera Observatory, Milan, Italy) on the 28th. The magnitude of the "nucleus" was estimated as 8 by E. Millosevich (Rome, Italy) and 7 by Schiaparelli on the 28th. The tail length was given as 20′ by J. Franz (Königsberg, now Kaliningrad, Russia) on the 25th, 4′ by Vogel on the 26th, 1° 10′ by O. H. Landreth (Vanderbilt University Observatory, Tennessee, USA) on the 27th, and 18′ by E. J. M. Stephan (Marseille, France) on the 28th. Millosevich said the tail extended northward on the 28th, while Schiaparelli said it extended toward PA 20°. Stephan added that the comet was very beautiful and exhibited a granular condensation.

The comet made a moderately close approach to Earth as March began and attained its most northerly declination of +32° on the 4th. Although moving away from both the sun and Earth thereafter, it was well observed, with astronomers frequently describing it as round, with a central condensation, and a stellar nucleus. No estimates of the total magnitude were made; however, during the period spanning March 1–8, R. Engelmann (Leipzig, Germany), B. von Engelhardt (Dresden, Germany), G. Bigourdan (Paris, France), and H. C. Wilson (Cincinnati, Ohio, USA) all gave estimates of the central condensation of between magnitude 6 and 7. Wilson reported the comet was brighter on the 9th than on the 8th. The coma diameter was given as 1.7′ by Engelmann and at least 2′ by Engelhardt on the 1st, 3′ by Schiaparelli on the 3rd, 1.7′ by G. Meyer (Göttingen, Germany) on the 3rd and 4th, 2.5′ by Bigourdan on the 4th, 3.7′ by J. F. J. Schmidt (Athens, Greece) on the 8th, 3.5′ by Schmidt on the 11th, 2′ by Schiaparelli on the 24th, 5.4′ by Schmidt on the 25th, and 5.0′ by Schmidt on the 26th and 29th. Numerous estimates were made of the nuclear magnitude, although most were by Schmidt. Engelmann gave it as 10 on the 1st. Using a 5-foot focal length refractor, Schmidt then gave estimates of 8.7 on the 8th, 8 on the 11th and 12th, 8.5 on the 13th, 9 on the 14th, 9.5 on the 16th and 19th, 9 on the 25th, and 9.5 on the 26th and 29th. The tail was described as faint, straight, and narrow, before it faded from view. On the 1st, B. Peter (Leipzig, Germany) gave the length as 20′, while Engelmann said it extended toward PA 10° and Schiaparelli said it extended toward PA 22°. On the 2nd, F. Gonnessiat (Lyon, France) said it extended 13′ toward PA 11°, while H. A. Kobold (O'Gyalla Observatory, Hungary) said it extended toward PA 25.6°. Schiaparelli said the tail extended 40′ toward PA 25° on the 3rd, while Wilson noted it was scarcely visible. Wilson said the tail was very faint on the 8th. S. C. Chandler, Jr (Harvard College Observatory, Cambridge, Massachusetts, USA) watched the comet pass almost centrally over an 8th-magnitude star on March 6. Observing with a 40-cm refractor he noted "no appreciable effect either of apparent augmentation or diminution of the star's light during occultation."

The comet was last seen during April. Although well observed early in the month, moonlight temporarily halted observations around mid-month.

Schmidt continued to provide estimates of the coma diameter and nuclear magnitude. He gave the coma diameter as 5.5′ on the 1st, 5.7′ on the 3rd, and 6.6′ on the 4th. His estimates of the nuclear magnitude were 9.2 on the 1st, 9.5 on the 3rd, and 10 on the 4th. J. Lamp (Bothkamp, Germany) said the nucleus of the comet passed over a star of magnitude 9–10 on the 1st, but the star did not fade in brightness. On April 8 and 9, C. W. Pritchett (Morrison Observatory, Glasgow, Missouri, USA) said, "there were two centres of condensation, very near together."

The comet's position was last determined on April 15.80, when Millosevich gave it as $\alpha = 4^h\ 28.8^m$, $\delta = +14°\ 46′$. He said the observation was difficult because of moonlight. The comet was last detected on April 24.79, when Schmidt said it was extremely faint, about 0.5′ across, and without a condensation. He was not able to measure a position.

The three bright cometary bands were seen in the spectrum by Vogel, N. de Konkoly (O'Gyalla Observatory, Hungary), A. Riccò (Palermo, Italy), and E. von Gothard (Herény Observatory, Hungary) during February and March. Konkoly wrote for March 3, "I saw in the spectrum only three bright bands, and with a widely-opened slit a very faint continuous spectrum...." He measured the wavelengths of the three bands as 5599 Å, 5156 Å, and 4702 Å. Gothard saw similar bands on March 2 and 4, but noted a faint fourth band on the last date.

The first parabolic orbits were independently calculated by Meyer and J. von Hepperger using positions from February 24, 25, and 26. Meyer determined the perihelion date as 1883 February 20.35, while Hepperger determined it as February 20.66. Additional orbits by H. Oppenheim, Hepperger, A. Graham, R. Bryant, H. Büttner, A. Berberich, Chandler, O. C. Wendell, and M. McNeill eventually established the perihelion date as February 19.45. Interestingly, Wendell later took 41 positions spanning 41 days and found that a parabola did not satisfy the observations. He then calculated an elliptical orbit with a perihelion date of February 19.44 and a period of about 23 949 years.

E. Hellebrand (1906) set out to determine a definitive orbit for this comet. He ultimately calculated parabolic, hyperbolic, and elliptical orbits. The parabolic one had a perihelion date of February 19.45. The hyperbolic orbit had a perihelion date of February 19.46 and an eccentricity of 1.0003439. The elliptical orbit had a perihelion date of February 19.45 and an *assumed* period of about 31 622 years. The elliptical orbit was the worst fit, while the hyperbolic was the best. Since the positions spanned less than two months, the parabolic one was preferred and is given below.

T	ω	Ω (2000.0)	i	q	e
1883 Feb. 19.4533 (UT)	110.8946	279.7735	78.0648	0.760062	1.0

ABSOLUTE MAGNITUDE: $H_{10} = 7.0$ (V1964)

FULL MOON: Feb. 21, Mar. 23, Apr. 22, May 22

SOURCES: L. Swift, W. R. Brooks, J. Lamp, C. N. A. Krueger, M. W. Meyer, and W. Schur, *AN*, **104** (1883 Feb. 27), pp. 365–8; L. Swift, *SM*, **2** (1883 Mar.), p. 31; R. Engelmann, B. Peter, M. W. Meyer, H. C. Vogel, F. Engström, and J. von Hepperger, *AN*, **104** (1883 Mar. 5), pp. 381–4; E. J. M. Stephan, G. Bigourdan, and F. Gonnessiat, *CR*, **96** (1883 Mar. 5), p. 612, 632–3; B. von Engelhardt, H. Oppenheim, and J. von Hepperger, *AN*, **105** (1883 Mar. 10), pp. 13–16; B. von Engelhardt, L. Weinek, E. W. L. Tempel, and A. Riccò, *AN*, **105** (1883 Mar. 14), pp. 29–32; F. Gonnessiat, J. Franz, and H. Büttner, *AN*, **105** (1883 Mar. 19), pp. 43, 47; L. Swift, N. de Konkoly, H. Oppenheim, and A. Berberich, *AN*, **105** (1883 Mar. 29), pp. 71–9; J. C. Adams, A. Graham, W. G. Thackeray, and N. de Konkoly, *MNRAS*, **43** (1883 Apr.), pp. 326–9; H. A. Kobold, *AN*, **105** (1883 Apr. 2), pp. 89–91; B. von Engelhardt, G. Meyer, S. C. Chandler, Jr, and O. C. Wendell, *AN*, **105** (1883 Apr. 9), pp. 125–8; E. von Gothard and A. Berberich, *AN*, **105** (1883 Apr. 14), pp. 135, 143; O. H. Landreth and E. Millosevich, *AN*, **105** (1883 May 9), pp. 237, 247–50; B. von Engelhardt, *AN*, **105** (1883 May 16), p. 267; J. F. J. Schmidt, *AN*, **105** (1883 May 28), pp. 317–20; C. W. Pritchett, *AN*, **105** (1883 May 31), pp. 331–4; G. V. Schiaparelli, *AN*, **105** (1883 Jun. 4), pp. 349–52; M. McNeill, *AN*, **105** (1883 Jun. 11), p. 389; S. C. Chandler, Jr and A. Donner, *AN*, **106** (1883 Jun. 19), pp. 9–14; E. W. L. Tempel, *AN*, **106** (1883 Jun. 23), p. 27; J. Lamp, *AN*, **106** (1883 Aug. 4), p. 143; R. Bryant, *MNRAS*, **44** (1884 Jan.), p. 88; F. Gonnessiat, *AN*, **107** (1884 Jan. 8), p. 367; H. C. Wilson, *AN*, **108** (1884 Mar. 25), pp. 245–8; E. Hartwig, *AN*, **108** (1884 Apr. 5), p. 269; W. Schur, *AN*, **108** (1884 Apr. 29), p. 391; C. W. Pritchett, *Publications of the Morrison Observatory*, No. 1 (1885), p. 104; ARSI, part 1 (1886), p. 383; O. C. Wendell, *SM*, **5** (1886 Mar.), p. 92; W. Schur, *AN*, **114** (1886 Mar. 23), pp. 83, 89; B. Peter, *AN*, **115** (1886 Oct. 7), pp. 235–8; E. Hellebrand, *AN*, **171** (1906 May 14), pp. 145–8; L. Swift, *Rochester History*, **9** (1947 Jan.), pp. 13–14; V1964, p. 62.

12P/1883 R1 *Discovered:* 1883 September 2.1 ($\Delta = 2.37$ AU, $r = 2.44$ AU, Elong. $= 82°$)
(Pons–Brooks) *Last seen:* 1884 June 2.3 ($\Delta = 1.76$ AU, $r = 2.22$ AU, Elong. $= 103°$)
  Closest to the Earth: 1884 January 9 (0.6344 AU)
1884 I = 1883b *Calculated path:* DRA (Disc), HER (Nov. 13), DRA (Nov. 26), LYR (Nov. 27), CYG (Dec. 9), LYR (Dec. 10), CYG (Dec. 11), VUL (Dec. 29), PEG (Dec. 30), PSC (Jan. 10), AQR (Jan. 16), CET (Jan. 22), SCL (Jan. 30), PHE (Feb. 16), ERI (Mar. 6), HOR (Mar. 17), RET (Apr. 6), DOR (Apr. 24), VOL (May 11), CAR (May 31)

The prediction of the return of this comet was not easy since it had only been observed for barely two months in 1812. L. Schulhof and J. F. Bossert (1882) worked to improve the 1812 orbit and make a viable prediction for the comet's return. After working diligently on the calculations they concluded the comet would most likely pass perihelion on 1884 September 4.14 and that the most probable period was 73.18 years; however, they added that the uncertainty was still ±4.5 years. Searches for the comet had not yet commenced when W. R. Brooks (Phelps, New York, USA) announced the discovery of a comet on 1883 September 2.1. Around the middle of September, C. F. W. Peters (Kiel, Germany) suggested this was the return of Pons' comet of 1812, indicating the comet had arrived at perihelion about

7 months earlier than predicted. The comet was over four months from its closest approaches to the sun and Earth.

On September 4, O. C. Wendell (Harvard College Observatory, Cambridge, Massachusetts, USA) described the comet as circular and less than 1' across. It was estimated as 10th magnitude. Wendell added that there was a well-defined nucleus, but no tail. G. Bigourdan (Paris, France) saw the comet on the 5th and described it as a faint, round nebulosity of magnitude 12. E. W. Maunder (Royal Observatory, Greenwich, England) observed the comet with a 33-cm telescope on the 6th and said, "The comet was a very diffused and exceedingly faint object, becoming more and more difficult as it got lower." That same night, J. Franz (Königsberg, now Kaliningrad, Russia) noted a stellar nucleus, while C. A. Young (Halsted Observatory, Princeton, New Jersey, USA) described the comet as very faint in the 58-cm refractor, with a slight central condensation. Young added that the coma was about 1.5–2' across, with no definite outline or tail. E. W. L. Tempel (Arcetri, Italy) said the comet was faint on the 7th, but with a "bright, speckled nucleus." G. V. Schiaparelli (Reale Osservatorio di Brera a Milano, Italy) also saw the comet on the 7th and said it was 3' across, with a central condensation. Descriptions were published for the 9th by E. Hartwig (Strasbourg, France), H. Struve (Pulkovo Observatory, Russia), Schiaparelli, and Bigourdan. Hartwig described the comet as round and centrally condensed, with a diameter of 1'. He added that a nucleus was occasionally seen. Schiaparelli noted a nucleus of magnitude 12–13. Struve saw the comet with a 38-cm refractor and noted the nucleus was not brighter than magnitude 11. Bigourdan said a weak magnification revealed a coma 40" across, while a magnification of 500× showed a badly-defined nucleus. On September 10, E. Millosevich (Rome, Italy) said the comet was very faint in the 23-cm refractor. Franz noted the stellar nucleus of the 6th was absent on the 12th. On September 19, Millosevich said the comet was hardly visible in the refractor.

The comet experienced an outburst on September 22 and 23. S. C. Chandler, Jr (Harvard College Observatory, Cambridge, Massachusetts, USA) wrote that on September 22.07 the comet was very faint and diffuse, with a faint central condensation of about magnitude 11. With the 15.9-cm refractor, he noted the coma was round and about 1.5' across. Schiaparelli said that on September 22.83 the comet was 3' across, with a nucleus of magnitude 13. A. Abetti (Padova, Italy) described the comet on September 22.86 as a shining nucleus surrounded by a nebulosity in the 18.7-cm refractor (122×). For September 22.99, Chandler wrote, "There is a most surprising change in the appearance of the comet since last night." He added, "it is a bright starlike object surrounded by a faint nebulosity." The nebulosity was about 0.75' across. Chandler estimated the brightness of the nucleus as 8.5. O. Struve (Pulkovo Observatory, Russia) noted that on September 23.76 the comet had brightened considerably since the previous night. The 10-cm finder (25×) revealed a nucleus of about magnitude 8 or 9, but little trace of a coma. The 38-cm refractor (210×) revealed a "shining, round, rather sharply defined nebulous mass, with a much weaker tail." The coma was 32" across and a nucleus could occasionally be seen. Schiaparelli said the comet "had grown in splendor in an extraordinary way" on September 23.82. He said it looked like a star of magnitude 8, and was surrounded by a thin nebulosity 1–1.5′ across. There was a notable central condensation. Bigourdan said that on September 23.83 the comet was about magnitude 8. R. Engelmann (Leipzig, Germany) described the comet on September 23.87 as very bright in the 19.1-cm refractor (110×), with a granular nuclear condensation of magnitude 10. This condensation measured 37" long and 25" wide, with the major axis directed toward PA 223°. He added that the finder (25×) revealed the comet's head as magnitude 8.5. On September 23.89, G. F. W. Rümker (Hamburg, Germany) noted the comet was "already quite bright," with a strong condensation and a diameter of 45". The condensation was situated "somewhat north" of the coma's center. Abetti said that on September 23.94 he was amazed at the change in the comet from the previous evening. He noted it was round, with a diameter of 40" in the 18.7-cm refractor (122×), and gave the magnitude as 7.

News of the outburst spread quickly and the comet was widely observed on the 24th. Bigourdan reported a diameter of nearly 2′. W. T. Sampson (US Naval Observatory, Washington, DC, USA) observed with the 24.4-cm refractor and said, "The light was a bright yellow, without showing any decided nucleus." Rümker said the comet was very bright with a strong condensation. Chandler wrote, "The physical appearance has again greatly changed. The nucleus now appears spread out into a confused bright disc with ill-defined edges. . . . " He said the "bright disc" was about 0.5′ across and was surrounded by a nebulous envelope about 1.5′ across. Chandler estimated the magnitude of the comet as 8. Franz said the comet was noticeably brighter than on the 22nd. O. Struve said the diameter of the bright part of the coma was 68" in the 38-cm refractor.

Extensive observations were also made on the 25th. H. G. van de Sande Bakhuyzen (Leiden, Netherlands) described the comet as 2.5′ across with a central condensation, but no nucleus. Schiaparelli said the coma was 3′ across. Abetti said the comet was still round, but had grown to about 1.5′ in diameter. He noted it was an opaque, cloudy white in the refractor (122×). Sampson said the comet's appearance had not changed since the previous evening. C. W. Pritchett (Morrison Observatory, Glasgow, Missouri, USA) said the central condensation was round and "surprisingly bright." O. Struve said the bright part of the coma was 98" across and exhibited a central nucleus.

On September 26, Maunder observed with a 33-cm telescope and wrote, "Comet fairly bright; round; no tail; central condensation." Measures of the apparent diameter produced a mean value of 2′ 21.0". Chandler said the comet had "spread out into a confused disc" about 2′ across and slightly elongated south-preceding and north-following. A faint nucleus or condensation was no brighter than magnitude 11. Abetti said the comet seemed slightly elongated in the refractor (122×). Young said the coma was 2′ 46″ across and exhibited a faint streak toward PA 61°, which he surmised was the beginning of a tail. Pritchett said there was a great change in the comet's appearance as it now appeared like an "expanded nebulous mass." O. Struve said the bright part of the nebulosity was 128″ across. In addition, he noted a distinct central nucleus, as well as a faint point of light north following the center.

Although numerous observations were made during the remainder of September, they all basically indicated the same thing – the comet was continuing to enlarge, as well as grow fainter and more diffuse. Sampson wrote for the 27th, "It was much fainter and the light had lost its yellow color." Abetti noted on the 30th that the comet had noticeably diminished in brightness since the 26th. The comet diameter was given as 2.5′ and 2.7′ by Chandler and O. Struve, respectively, on the 27th. On the 28th, the diameter was given as 3′ by B. von Engelhardt (Dresden, Germany), 3.2′ by H. Pomerantzeff (Taschkent, Uzbekistan), and about 4′ by Engelmann. On the 29th, the diameter was given as 5′ by Hartwig, 6′ by L. Weinek (Prague, Czech Republic), and 7–10′ by Franz. Schiaparelli noted it was 3–4′ across on the 30th. The magnitude of the nucleus was given as 10–11 by Bigourdan on the 27th, 11–12 by Engelmann and 11 by J. B. Messerschmitt (Leipzig, Germany) on the 28th, and 11 by both Schiaparelli and Abetti on the 30th. O. Struve observed with the 38-cm refractor on the 27th and noted a sharp nucleus, as well as a fainter nucleus situated 25″ away toward PA 61°. Engelhardt said the nucleus was situated north of center on the 28th and added that the northern half of the coma was brighter than the southern half. Hartwig said the nucleus was no longer symmetrically situated within the coma on the 29th, while Millosevich noted several dots were located in the position of the nucleus. Hartwig saw the beginning of a tail on the 29th and noted it extended toward PA 55°. Millosevich also saw a tail on the same date. He gave its length as 15′ and said it gave the impression of being detached from the comet's head. Abetti saw a long, wide tail in the 18.7-cm refractor (122×) on the 30th.

As October began, both Engelhardt and Engelmann noted the comet was not as bright as on September 28. Young reported it was fainter on the 6th than on the 4th. On October 6, Bigourdan said the comet's appearance resembled that of September 5. Thereafter, the comet seems to have resumed its normal brightening, with F. Küstner (Hamburg, Germany) estimating the total magnitude as 9 on the 21st and O. Struve estimating it as 8–9 in the 10-cm finder on the 29th. Engelhardt described the nucleus as "rather large" on the 1st, but Young managed to spot a stellar nucleus in the 58-cm refractor on the 2nd. Schiaparelli gave the nuclear magnitude as 12 on the 5th and 7th, while Bigourdan indicated a similar magnitude on the 6th. On the 7th, Engelmann estimated the nuclear magnitude as 11, while Messerschmitt gave it as 10–11. Struve said the nucleus was magnitude 10 on the 10th and 19th. Engelhardt said the nucleus was only occasionally seen in the 30-cm refractor on the 21st. The nuclear magnitude was estimated as 11 by Abetti and 11.5 by Engelmann on the 23rd. On the 25th, Abetti said the 18.7-cm refractor (122×) revealed a nucleus of magnitude 11, while Young said the 58.4-cm refractor revealed a nucleus of magnitude 9.5. Schiaparelli gave the nuclear magnitude as 11 on the 26th. Abetti estimated it as 10 on the 27th and 29th. Young noted the nucleus "throws off small brush towards the sun" on the 3rd and 4th, as seen in the 58.4-cm refractor. Numerous estimates of the coma diameter were made, with larger diameters coming from observers using small telescopes. The diameter was given as 5′ by Engelhardt and 5.5′ by Engelmann on the 2nd, 5′ by Engelmann on the 7th, 3′ by Engelmann on the 23rd, 2.8′ by Pomerantzeff on the 24th, and 2.4′ by Pomerantzeff and 3′ by Schiaparelli on the 26th. The short tail was well observed during the month. On the 2nd, Engelhardt described the tail as fan-shaped, Engelmann said it extended toward PA 65°, and Young noted it extended 8′ toward PA 57°. Young said the tail was brighter on the southern edge on the 3rd and extended toward PA 66°. On the 4th, Young said the tail extended 5′ toward PA 59°. Messerschmitt said the tail was seen during favorable moments of seeing on the 16th, and was directed toward the northwest. Engelmann said the tail extended 4′ toward PA 80° on the 23rd. T. W. Backhouse (West Hendon, England) said the tail extended 20′ toward PA 75° on the 25th, while Young noted the comet was tailless.

The comet continued its normal brightening during November, with the total magnitude being given as 8 by H. Struve on the 11th and 6–7 by Bigourdan on the 19th. Several astronomers reported naked-eye observations beginning on November 20 and continuing until the end of the month. Although numerous observers gave magnitude estimates of the nucleus during the month, most were probably giving estimates of the comet's condensation. Schiaparelli was the only observer to provide a consistent series of observations of a stellar nucleus for the entire month. He gave magnitude estimates of 12 on the 5th, 10–11 on the 11th, 10 on the 16th, 8 on the 18th, and 10 on the 25th and 28th. Overall, the coma diameter seemed smaller during the first half of November than in October, but did increase throughout the month. The diameter was given as 2′ by Schiaparelli on the 5th, 1′ by Schiaparelli on the 11th, 2′ by Schiaparelli on the 16th, 3′ by Schiaparelli on the 18th, 1′ by Engelhardt and 6′ by G. A. P. Rayet (Bordeaux, France) on the 24th, 3.5′ by Weinek on the 25th, 4′ or 5′ by Maunder and 11′ by Hartwig on the 27th, 8′ by Schiaparelli on the 28th, 5′ by Messerschmitt, 6′ by Engelhardt and 7′ by Rayet on the 29th, and 3.1′ by Pomerantzeff on the 30th. The only observation of a tail during the first half of November was reported by Young on the 6th. He wrote, "A faint brush, like an incipient tail, at 50° position-angle; very faint and indistinct." During the second half of the month, there were numerous observations, which included a wide range of lengths. Hartwig provided the most consistent, although short series using a small telescope. He said it extended 40′ toward PA 39° on the 18th, 45′ on the 24th, and 42′ toward PA 34° on the 27th. Most other observers reported a length of less than 4' during this period. Some astronomers noted an unusual shape to the coma during the latter half of the month. Bigourdan said the coma was not symmetrical around the nucleus on the 19th and said it extended toward PA 110–140° and was brightest toward PA 280–290°. Young observed with a 58.4-cm refractor on the 29th and wrote, "Comet flattened out and better defined on side next sun, the major axis of the nebulosity being at right angles to the sun's direction and about twice the minor axis; tail straight and pointing from sun." He added that the comet's shape was like the letter "T".

The comet was a naked-eye object throughout December and maintained a stellar nucleus, as well as a tail that was steadily increasing in length. Total magnitude estimates were not made early in the month, mostly because of moonlight. During the second half of the month, the total magnitude was given as 3 by Weinek on the 18th, 5 by Franz on the 20th, 6 by Abetti on the 22nd, brighter than ζ Cygni (magnitude 3.20 – SC2000.0) by C. Braun (Kalocsa, Hungary) on the 26th, 3 by Engelhardt and 5.5 by Engelmann on the 29th, 5.5 by Engelmann on the 30th, and 3 by Braun on the 31st. In addition, G. Gruss (Prague, Czech Republic) saw the comet on the 29th and said it was brighter than ζ Cygni (magnitude 3.20 – SC2000.0) and fainter than ε Cygni (magnitude 2.46 – SC2000.0). Estimates of the nuclear magnitude were rather inconsistent between astronomers, but everyone seemed to indicate there was little change during the month. Schiaparelli once again provided estimates that spanned the entire month and these were 8 on the 4th and 6th, and 9 on the 15th and 31st. Abetti also provided a good series, but was probably referring more to the central condensation, especially since he gave the diameter as 30" on the 6th. He gave the "nucleus" brightness as 7 on the 8th, 8 on the 22nd, and 7 on the 24th, 26th, and 27th. Engelmann observed the comet with a 19.1-cm refractor (110× and 215×) on December 29 and saw a nucleus of magnitude 7, as well as a second nucleus of magnitude 10 situated about 13" away toward PA 280°. On the 30th, Engelmann said the main nucleus was magnitude 7.5, while the second nucleus was situated about 15" away toward PA 220°. In addition, Braun said the nucleus was less sharp on the 30th, while Abetti noted it was large and diffuse on the 31st. H. C. Vogel (Potsdam, Germany) said a magnification of 250× on the 29th revealed a distinct nucleus exhibiting a fan-shaped extension protruding 25–30" toward the region between PA 170° and 310°. Estimates of the coma diameter also varied considerably depending on the observing instrument being used. Schiaparelli's observations seem to reflect those of several astronomers, such as Gruss, Weinek, Engelhardt, Messerschmitt and Engelmann, but span most of the month. He gave the coma diameter as 6' on the 6th and 15th, 5–6' on the 19th, and 8' on the 24th. The tail was extensively observed by Hartwig, Weinek, Messerschmitt, Abetti, Müller, Pomerantzeff, H. Struve, Engelmann, Gruss, H. J. A. Perrotin (Nice, France), E. A. Lamp (Kiel, Germany), and F. Gonnessiat (Lyon, France), but it was Schiaparelli who, once again, obtained a month-long series of measurements. Early in the month, Schiaparelli reported two tails were visible. On the 4th he said the major tail extended 1.2° toward PA 34° and the minor tail extended 0.4° toward PA 330°. On the 6th, he said the major tail extended 1.0° toward PA 31°, while the minor tail extended 0.4° toward PA 335°. On the 15th, he said the major tail extended 1.4° toward PA 30°, while the minor tail extended 0.8° toward PA 0°. He then said the tail extended 2.8° toward PA 30° on the 18th, 1.0° toward PA 33° on the 19th, 4.6° toward PA 35° on the 24th, and 5.1° toward PA 39° on the 31st.

The comet seems to have undergone a minor outburst as the new year began. Müller saw the comet steadily brighten more than half a magnitude and then fade to near its original brightness in the course of about an hour and a half. He first observed the comet on January 1.70 and said the nucleus was not large and was extraordinarily diffuse. Upon next observing the comet on January 1.769 he said he was "surprised by the changed appearance of the comet. In place of the diffuse nucleus was an almost perfectly point-like star of about the brightness of a star of magnitude 7." He made his first photometric comparison of the nucleus with nearby stars on January 1.775 and gave the magnitude as 7.53. It then brightened to 7.35 by January 1.784, 6.97 by January 1.796, and 6.89 by January 1.802. The comet then faded to 7.03 by January 1.816, brightened to 7.00 by January 1.823, then steadily faded until the last measurement of 7.33 on January 1.844.

January was the best month for observing this comet as it passed closest to Earth during the first half of the month and passed perihelion near the end. The comet was a naked-eye object throughout the month and numerous naked-eye magnitude estimates were made of the coma, of which some were reported as of the "nucleus". Among these magnitude estimates were 3 by Gruss, 2–3 by Abetti, and "brighter than 3" by Braun on the 1st, 4 by Engelmann and 3–4 by Franz on the 3rd, 4 by A. L. N. Borrelly (Marseille, France) on the 4th, and 3 by Abetti on the 10th. Following interference from moonlight, there were estimates of 2 by Weinek on the 13th and 15th, 2 by G. F. Parson (on board the *Earnock*, sailing in the southern Atlantic Ocean) and 3 by Borrelly on the 15th, 2.5 by Parson and Young on the 17th, 3 by Gruss on the 22nd, and 4.3 by D. W. Barker (on board the *Superb*, sailing in the southern Pacific Ocean) on the 29th. Several observers noted the comet was fading after the 26th. A. V. Nursinga Row (Vizagapatam, India) reported on the 15th, "A comet was in our western sky . . . sufficiently conspicuous to the naked eye to attract general attention." He indicated he thought it was the comet Pons–Brooks. The nucleus was basically described as stellar during the entire month and numerous magnitude estimates were made. Some of these were 7 by Engelmann on the 1st, 6.5 by Engelmann and 8 by Abetti, Müller, and H. Struve on the 2nd, 7 by Engelmann and Abetti and 6.5 by Young on the 3rd, 7–8 by Seeliger on the 27th, and 7–8 by Abetti on the 29th. Engelmann continued to follow the secondary nucleus. He said it was 5″ away toward PA 230° on the 1st, 17″ away toward PA 205° on the 2nd, 22″ away toward PA 190° on the 3rd. He added that only a trace of this secondary nucleus was visible on the 3rd and estimated its magnitude as 9–10. W. G. Davis (Cordoba, Argentina) reported the nucleus appeared divided on the 23rd and said the south-preceding portion was brighter. Davis failed to find the secondary nucleus thereafter. Interestingly, where Engelmann reported a secondary nucleus on several occasions, other observers were reporting jets and fan-shaped features extending from the nucleus. Vogel said a magnification of 250× revealed a fan-shaped feature protruding from the nucleus on the 2nd. It was roughly 28″ long and spanned the region between PA 188° and 307°. H. Struve also noted a fan-shaped feature on the 2nd and said it extended between PA 192° and 262°. Young said the 58.4-cm refractor revealed a reddish nucleus with four or five "brushes" on the 3rd. He said each brush extended 20–30″ long and was directed toward the sun. Young noted three curved jets within the head on the 4th. H. Struve said the fan-shaped feature extended from the nucleus toward PA 200–280° on the 5th. On January 11 and 13, Pritchett said the once stellar nucleus "presented a very changed appearance. Two curving jets, similar to short-horns, proceeded from it." He noted the brighter was northward, while the fainter was southward. The jets were very conspicuous on the 11th and "a little fainter" on the 13th. Pritchett said the jets had disappeared by the 14th. Pomerantzeff said the nucleus was 9.7″ across and exhibited jets extending toward the sun on the 15th. Gonnessiat said the nucleus was "lengthened transversely and equipped with two brushes opposed to the tail" on the 16th. Also on the 16th, Young noted several brushes directed toward the sun. On the 17th, Young saw "a bright and pretty well-defined sector on east side of nucleus, nearly at right angles to the general trend of the tail." The coma was not a prominent feature of the comet even when it was at its brightest, with Braun describing it as a fine veil on the 13th. The coma diameter was estimated as 7′ by Engelmann on the 1st and 3rd, 10′ across by Abetti on the 2nd and 3rd, 1.5′ by Davis on the 12th, more than 2′ by Davis on the 16th, and 2.5′ by Abetti on the 29th. The tail was widely observed. Some of the longer tail lengths given were 4.8° toward PA 45° by Schiaparelli on the 1st, 4° by Young on the 3rd, 4.5° toward PA 58.2° by Pomerantzeff and 7° by Borrelly on the 15th, 8° toward PA 59° by Gonnessiat on the 16th, 6° by Young on the 17th, 9° by Parson on the 21st, 4° by A. S. Thomson (Canary Islands) on the 22nd, 4.7° toward PA 69° by Hartwig on the 24th, and 4° by J. P. Holdich (on board the *British Envoy*, sailing within a couple of degrees of the southern fortieth parallel heading from the Atlantic Ocean across the Indian Ocean). Young noted the tail was "cut out a little on the north side" near the head on the 3rd. Gonnessiat noted the main tail was straight, while a smaller tail 4–5° long diverged from it. Young said the tail was better defined on the southern side than on the northern on the 17th. Parson said the tail seemed more distinct on the 19th and 20th, and brighter than usual on the 23rd. W. H. Finlay (Royal Observatory,

Cape of Good Hope, South Africa) said the tail was faint and straight on the 24th.

The comet was moving away from both the sun and Earth as February began and was only visible from the Southern Hemisphere observatories. Finlay noted a stellar nucleus on the 4th and said the "head" was "much more diffused" on the 6th. Also on the 6th, Davis said the comet had slightly faded. On the 9th, Finlay wrote, "Comet's head very diffused; takes about 1^s to transit the wire." Davis said the comet was still easily visible to the naked eye on the 15th. On the 16th, Barker (now in the southern Atlantic Ocean) said the comet was much fainter, with the tail no longer visible to the naked eye. Holdich said the tail was hardly visible and the nucleus was only seen when the moon was clouded over. By the 18th, Davis noted the comet had "greatly diminished in brightness." Holdich said the tail was "barely visible, except with glasses" on the 20th. Parson (now sailing in the southern Indian Ocean) and J. Tebbutt (Windsor, New South Wales, Australia) both noted the comet was not visible to the naked eye. Parson added that it was still visible " by the aid of an ordinary ship's telescope. . . . " On the 25th, Holdich wrote, "Comet getting very indistinct to naked eye, and through glasses getting more the appearance of a nebulous cloud with bright centre." That same night, Finlay wrote, "A very faint tail visible in the finder; position-angle approximately 134°." Finlay said the tail extended toward PA 114° on the 26th and added that "traces of another very faint tail could just be seen" in about PA 153°. On February 28, Davis said the coma was 1' across, but less condensed in the center.

On March 2, Finlay said he had occasional glimpses of a stellar nucleus. On the 3rd, Parson said the comet was "distinguishable through telescope, but too faint for sextant observations," while Davis said the comet appeared slightly more condensed in the center than when seen previously. On March 11 and 13, Davis said the comet was very faint because of moonlight. On the 21st, Finlay said the "head" was "bright, but ill-defined, surrounded by nebulous mass." He added that there was still a "faint trace of very short tail."

On April 1, Tebbutt observed the comet in spite of moonlight and said the comet was excessively faint in his 11-cm refractor. Davis found the comet very faint on the 3rd and "exceedingly faint" on the 7th. For the 13th, Finlay wrote, "An 11th-magnitude star near the Comet very troublesome. The Comet was almost centrally over this star earlier in the evening, and the star then was brighter than when the Comet had passed away from it. The moonlight was, however, increasing rapidly towards the end of the measures." Davis reported the comet seemed brighter on the 19th and added that the coma was slightly condensed at the center and about 1' across. Tebbutt said the comet was "of the last degree of faintness" in his 11-cm refractor on the 21st, while Pett (Royal Observatory, Cape of Good Hope, South Africa) described the comet as "exceedingly faint." Davis said the comet was "very faint and ill-defined" on the 27th.

Most observers were no longer able to follow the comet during May as it continued to fade and Davis provided the only physical descriptions. He wrote that the comet was "exceedingly faint, a mere blur without visible nucleus" on the 11th. On the 18th, he said the comet was "without form or nucleus." For the 22nd, Davis wrote that the comet was faint, but "somewhat brighter than for several nights past. The sky however was exceptionally clear." He found the comet barely visible on the 23rd and 26th, but managed to acquire the last measurement of the comet's position on May 26.98, which was $\alpha = 8^h\ 34.3^m$, $\delta = -68°\ 22'$. The comet had attained its most southerly declination of $-70°$ on May 14. The comet was observed for the final time on June 2.3, by A. S. Atkinson (Nelson, New Zealand), but it was then too faint to measure.

The spectrum of this comet was measured by many astronomers, including C. Trépied, L. Thollon, C. F. Pechüle, Perrotin, Maunder, Rayet, Young, and Vogel, making it the most studied spectrum up to this time. The most diligent observers were N. de Konkoly and R. Kövesligethy (O'Gyalla Observatory, Hungary), who obtained numerous measurements from November into January. They continually noted the three bright cometary bands and said the middle one was always brightest, while the first band was always faintest. Measurements of the first band fell within the range of 5630–5585 Å. Measurements of the middle band fell within the range of 5190–5112 Å. Measurements of the last band fell within the range of 4784–4714 Å. Interestingly, Konkoly saw six bands on December 29 and on January 1, but clouds prevented measurements on both occasions.

The first parabolic orbit was calculated by Palisa using positions from September 3, 7, and 10. He determined a perihelion date of 1884 February 3.95. Peters took positions obtained on September 4, 7, and 11, and calculated a parabolic orbit with a perihelion date of February 7.71. He added that the orbit was similar to that of the comet seen by Pons in 1812. Shortly thereafter, Hartwig and Palisa independently noted the same similarity. Although additional parabolic orbits were calculated by J. Seyboth and H. Oppenheim within the next few days, astronomers were quickly convinced this was a return of Pons' 1812 comet.

The first elliptical orbit was calculated by Schulhof and Bossert, using positions obtained up to September 8 to correct their earlier prediction. They determined the perihelion date as January 26.29 and the period as 71.48 years. An orbit was also calculated by J. Morrison, using positions from October 11, December 27, and January 22. He found a period of 69.57 years.

The comet was next recovered in 1954. Thereafter, several astronomers used multiple apparitions to firmly determine the orbit for the 1884 apparition. The most recent orbits were calculated by P. Herget and H. J. Carr (1972), D. K. Yeomans (1985, 1986), and K. Kinoshita (2000). These all indicated an orbital period close to 71.7 years. Kinoshita determined nongravitational terms of $A_1 = -0.20$ and $A_2 = -0.0270$. This orbit is given below.

T	ω	Ω (2000.0)	i	q	e
1884 Jan. 26.2161 (TT)	199.1757	255.7745	74.0403	0.775729	0.955057

ABSOLUTE MAGNITUDE: H_{10} = 4.9 (V1964)

FULL MOON: Aug. 18, Sep. 16, Oct. 16, Nov. 14, Dec. 14, Jan. 12, Feb. 11, Mar. 11, Apr. 10, May 10, Jun. 8

SOURCES: L. Schulhof and J. F. Bossert, *Nature*, **26** (1882 Oct. 26), p. 635; L. Schulhof and J. F. Bossert, *AN*, **103** (1882 Nov. 1), pp. 289–98; L. Schulhof, J. F. Bossert, and W. E. Plummer, *MNRAS*, **43** (1883 Feb.), pp. 209–10; O. C. Wendell and C. N. A. Krueger, *AN*, **106** (1883 Sep. 5), p. 251; G. Bigourdan, *CR*, **97** (1883 Sep. 10), p. 636; E. W. L. Tempel, *AN*, **106** (1883 Sep. 12), p. 269; E. Hartwig and J. Palisa, *AN*, **106** (1883 Sep. 14), p. 303; L. Schulhof and J. F. Bossert, *CR*, **97** (1883 Sep. 17), pp. 662–4; E. Hartwig, J. Palisa, and C. F. W. Peters, *AN*, **106** (1883 Sep. 21), pp. 331–6; G. Bigourdan, *CR*, **97** (1883 Sep. 24), pp. 701–4; G. F. W. Rümker, J. Seyboth, L. Schulhof, and J. F. Bossert, *AN*, **106** (1883 Sep. 28), pp. 377, 383; W. R. Brooks, *SM*, **2** (1883 Oct.), pp. 220–23; H. Oppenheim and H. G. van de Sande Bakhuyzen, *AN*, **107** (1883 Oct. 3), pp. 13–16; G. Bigourdan, *CR*, **97** (1883 Oct. 8), pp. 794–7; E. Millosevich and E. Hartwig, *AN*, **107** (1883 Oct. 9), p. 27; J. Ritchie, Jr, E. Millosevich, G. F. W. Rümker, and B. von Engelhardt, *AN*, **107** (1883 Oct. 15), pp. 43–6; L. Weinek, *AN*, **107** (1883 Oct. 31), p. 89; A. M. W. Downing and E. W. Maunder, *MNRAS*, **44** (1883 Nov.), pp. 13–14; T. W. Backhouse, *Nature*, **29** (1883 Nov. 1), p. 7; C. G. Talmage, *Nature*, **29** (1883 Nov. 8), p. 45; S. C. Chandler, Jr, and G. V. Schiaparelli, *AN*, **107** (1883 Nov. 19), pp. 131–4, 139–44; G. Bigourdan, *CR*, **97** (1883 Nov. 19), pp. 1126–7; A. Abetti, L. Schulhof, and J. F. Bossert, *AN*, **107** (1883 Nov. 28), pp. 223–8; E. Frisby, W. T. Sampson, A. M. W. Downing, W. G. Thackeray, H. P. Hollis, and E. W. Maunder, *MNRAS*, **44** (1883 Dec.), pp. 56–63; B. von Engelhardt, P. Tacchini, E. Millosevich, and H. Pomerantzeff, *AN*, **107** (1883 Dec. 1), pp. 253–8; S. C. Chandler, Jr, *AN*, **107** (1883 Dec. 4), p. 275; Pa. Henry, Pr. Henry, and G. A. P. Rayet, *CR*, **97** (1883 Dec. 10), pp. 1352–3; F. Gonnessiat, *CR*, **97** (1883 Dec. 24), p. 1469; H. J. A. Perrotin and C. Trépied, *CR*, **97** (1883 Dec. 31), pp. 1539–41; O. Struve, H. Struve, and J. Seyboth, *BASP*, **29** (1884), pp. 229–36; C. Trépied, L. Thollon, and E. L. Trouvelot, *CR*, **98** (1884 Jan. 7), pp. 32–5; F. Gonnessiat and E. A. Lamp, *AN*, **107** (1884 Jan. 8), pp. 367–70; G. Müller, *AN*, **107** (1884 Jan. 14), pp. 381–4; L. Schulhof and J. F. Bossert, *AN*, **108** (1884 Jan. 21), p. 15; F. Gonnessiat, *CR*, **98** (1884 Jan. 21), p. 133; H. C. Vogel, W. T. Sampson, E. Frisby, A. Abetti, L. Schulhof, and J. F. Bossert, *AN*, **108** (1884 Jan. 28), pp. 21–30; E. L. Trouvelot, *CR*, **98** (1884 Jan. 28), pp. 207–9; R. Engelmann, *AN*, **108** (1884 Feb. 4), pp. 45–8; F. Küstner, *AN*, **108** (1884 Feb. 8), p. 61; H. J. A. Perrotin, L. Thollon, and G. A. P. Rayet, *CR*, **98** (1884 Feb. 11), pp. 344–9; L. Weinek, G. Gruss, P. Tacchini, and E. Millosevich, *AN*, **108** (1884 Feb. 14), pp. 71–80; F. Gonnessiat, E. Becker, A. Abetti, and C. A. Schultz-Steinheil, *AN*, **108** (1884 Feb. 25), pp. 103–12; A. L. N. Borrelly and C. Lamey, *CR*, **98** (1884 Feb. 25), pp. 495–8; N. de Konkoly, R. Kövesligethy, W. T. Sampson, E. Frisby, A. V. Nursinga Row, W. L. Rosseter, and S. S. O. Morris, *MNRAS*, **44** (1884 Mar.), pp. 251–6; J. Franz, H. Pomerantzeff, and J. B. Messerschmitt, *AN*, **108** (1884 Mar. 13), pp. 149–58; G. Müller and N. von Konkoly, *AN*, **108** (1884 Mar. 19), pp. 161–72; W. T. Sampson and E. Frisby, *AN*, **108** (1884 Mar. 25), p. 247; E. F. Sawyer and R. L. J. Ellery, *The Observatory*, **7** (1884 Apr.), pp. 113–16; E. Hartwig, T. Wolff, and R. L. J. Ellery, *AN*, **108** (1884 Apr. 5), pp. 265–8, 271–4, 277; C. A. Young and B. von Engelhardt, *AN*, **108** (1884 Apr. 17), pp. 305–10; H. Seeliger and G. Cacciatore, *AN*, **108** (1884 Apr. 25), pp. 373–6; J. Morrison, D. W. Barker, and G. F. Parson, *MNRAS*, **44** (1884 May), pp. 368–73; C. W. Pritchett, *AN*, **108** (1884 May 8), pp. 415–24; C. Braun, *AN*, **109** (1884 Jun. 12), p. 61; G. V. Schiaparelli, *AN*, **109** (1884 Jun. 17), pp. 65–78; H. Struve and O. Struve, *AN*, **109** (1884 Sep. 24), pp. 369–86; A. S. Atkinson, *The Observatory*, **7** (1884 Oct.), p. 306; J. Tebbutt, *AN*, **110** (1884 Oct. 31), pp. 135–40; J. Tebbutt and J. P. Holdich, *MNRAS*, **44** (1884, Supplementary Notice), pp. 445–9, 464; A. B. Biggs, *MNRAS*, **45** (1884 Dec.), pp. 116–17; C. W. Pritchett, *Publications of the Morrison Observatory*, No. 1 (1885), p. 104; B. A. Gould and W. G. Davis, *AN*, **111** (1885 Jan. 15), pp. 23–32; W. H. Finlay and Pett, *AN*, **112** (1885 Jul. 13), pp. 141–4; W. H. Finlay, *MNRAS*, **45** (1885, Supplementary Notice), pp. 471–6; C. F. Pechüle, *AN*, **113** (1886 Feb. 11), pp. 361–4; T. Bruhns, *AN*, **122** (1889 Jul. 25), pp. 121–38; V1964, p. 62; P. Herget and H. J. Carr, *QJRAS*, **13** (1972), pp. 428–9, 434; D. K. Yeomans, *QJRAS*, **26** (1985), p. 91; D. K. Yeomans, *QJRAS*, **27** (1986), p. 604; personal correspondence from K. Kinoshita (2000); SC2000.0, pp. 20, 525, 539, 585.

X/1883 Y1 Details of this comet first surfaced in the 1884 April and May issues of *The Observatory*, as well as the 1884 May 1 issue of *Nature*. In these publications, J. Tebbutt mentioned a comet seen in the morning sky from Tasmania on 1883 December 25 and December 27. The comet remained a mystery until nearly a century later, when W. Orchiston (1983) examined the letters of some of the key players and reported the following details.

Although the comet was initially reported to have been found by N. C. Sharland (New Norfolk, Tasmania), Tebbutt's correspondence with various individuals, including Sharland, revealed this had been in error. His quest for correct information eventually brought him in contact with meteorogist J. Shortt (Hobart Town, Tasmania). Shortt received letters from two men on 1884 February 11. The first was from Parsons (Kingston, Tasmania) who said Henry Clevers, an employee on his farm, had seen a comet at daybreak on Christmas Day. Clevers later visited Shortt and noted the comet was "about 8 or 10 degrees above the horizon, bearing about East, the tail extending downward E45N and of about 2° to 3° in length." Clevers said he watched the comet for some time. The probable time in Universal Time was December 24.7. The second letter came from L. C. Thirlwall (Kingston, Tasmania) who said he first saw the comet on the morning of December 27. He noted a tail and estimated the position as due east. The probable time in Universal Time was December 26.77. Both observers noted it was comparable in appearance to comet C/1882 R1.

Although Tebbutt had written in the 1884 April issue of *The Observatory* that he suspected this comet was identical to C/1884 A1 before any orbit had been calculated, he announced in the 1884 May issue that a link was impossible. H. C. F. Kreutz came to the same conclusion in the 1884 May 1 issue of the *Astronomische Nachrichten*. In addition, Kreutz said the comet's likely position on December 24 was $\alpha = 14.8^h$, $\delta = -6°$, while it was likely near $\alpha = 16.5^h$, $\delta = 0°$ on December 26. I. Hasegawa (1980) estimated the comet's probable magnitude as 0.

SOURCE: J. Tebbutt, *The Observatory*, **7** (1884 Apr.), pp. 116–17; J. Tebbutt, *Nature*, **29** (1884 Apr. 24), p. 606; J. Tebbutt, *The Observatory*, **7** (1884 May), pp. 140–1; *Nature*, **30** (1884 May 1), p. 21; H. C. F. Kreutz, J. Tebbutt, and J. Shortt, *AN*, **108** (1884 May 8), p. 423; J. Tebbutt, *AN*, **109** (1884 Aug. 11), p. 253; H1980, p. 90; W. Orchiston, *Proceedings of the Astronomical Society of Australia*, **5** (1983), pp. 282–4.

C/1884 A1   *Discovered:* 1884 January 7.5 ($\Delta = 0.74$ AU, $r = 0.49$ AU, Elong. $= 28°$)

(Ross)   *Last seen:* 1884 February 19.5 ($\Delta = 1.92$ AU, $r = 1.36$ AU, Elong. $= 42°$)

     Closest to the Earth: 1884 January 1 (0.6464 AU)

1883 II = 1884a   *Calculated path:* MIC (Disc), GRU (Jan. 8), PHE (Jan. 24)

D. Ross (Elsternwick, near Melbourne, Australia) was searching for 12P/1883 R1 (Pons–Brooks) on the evening of 1884 January 7. After looking around the predicted position of the comet for a while, he began sweeping the sky with his telescope and on January 7.5 he found a nebulous object and immediately suspected it was a comet. This was confirmed within the next couple of days and Ross reported his find to Melbourne Observatory on January 11. R. L. J. Ellery (Melbourne Australia) confirmed the comet on the evening of January 11 and gave the position as $\alpha = 22^h\ 03.6^m$, $\delta = -40°\ 07'$ on January 12.47. He described the comet as "a faint nebulous object with an ill-defined central condensation and distinct, but very small tail-like projection, pointing east. The light at the brightest part was about equal to a star of the 10th magnitude."

The comet was already moving away from both the sun and Earth at discovery. L. Cruls (Rio de Janeiro, Brazil) described the centrally condensed comet as circular, with a faint tail on January 17. He estimated the tail as 1.5° long on the 18th. J. Tebbutt (Windsor, New South Wales, Australia) said the comet "was then just beyond the reach of unassisted vision" on the 19th. His 11-cm telescope revealed "a round nebula, about two minutes of arc in diameter, with a bright condensation in the centre. There was no definite nucleus . . . , but a short tail could be distinguished." On the 21st, N. R. Pogson (Madras, India) described the comet as round and 90" in diameter, with a "decided condensation towards the centre." The comet attained its most southerly declination of $-42°$ on the 24th. Ellery described the comet as a "simple roundish nebulous mass with an undefined central condensation" on the 29th.

The comet was becoming quite faint by the time February began. A. B. Biggs (Launcestown, Tasmania) described the comet as nebulous on the 1st and said there was no definite point to measure from. Tebbutt saw the comet with "extreme difficulty" in his 11-cm telescope on the 2nd. Ellery said the comet's "rapidly diminishing brightness and the increasing moonlight" made it a very difficult object to see in a 20-cm refractor on the 4th.

The comet was last detected on February 19.5 by Ellery. He gave the position as $\alpha = 0^h\ 25.8^m$, $\delta = -40°\ 56'$, and said, "The comet was then so faint as to be seen with difficulty, and no glimpse of it has been obtained since, owing to continual cloudy evenings."

The first parabolic orbit was calculated by Ellery. He took positions from January 12, 18, and 28, and determined the perihelion date as 1883 December 25.88. A similar orbit was calculated by Tebbutt, when he used positions spanning the period of January 19–28 and determined a perihelion date of December 25.80. H. Oppenheim used a slightly longer arc than Tebbutt, but obtained essentially the same orbit.

The orbit became the basis for a series of arguments within the pages of the *Monthly Notices*. R. Bryant (1885) took Tebbutt's positions exclusively and calculated two elliptical orbits. Both had a perihelion date of December 25.63, while the resulting periods were 86.92 and 94.07 years. J. F. Tennant (1886) took all of the available positions and determined Normal positions for January 17, 27, and 30. Although he determined an elliptical orbit with a perihelion date of December 25.82 and a period of about 1986 years, he said a parabolic orbit with a perihelion date of December 25.82 fitted the positions best. Bryant (1887) said Tennant's orbit did not represent the January 27 position very well and believed this confirmed his earlier conclusion that the comet was moving in an elliptical orbit. This time, however, he considered other observations besides Tebbutt's. His conclusion was that the perihelion date was December 25.73 and the period was 414 years. Tennant (1887) rechecked his computations following the publication of Bryant's paper. He once again concluded that a parabolic orbit with a perihelion date of December 25.80 still represented the positions best. He added, "that the difference of the results at which Mr. Bryant and I have finally arrived is due to the modes in which we have respectively deduced our normal places from observation."

The question of whether the comet moved in a parabolic or elliptical orbit was revisited over two decades later. E. Moravi (1910) took all available positions and determined five Normal positions. He concluded that an elliptical orbit with a perihelion date of December 25.61 and a period of 64.63 years fitted the positions best. Although this orbit satisfied astronomers for many years, B. G. Marsden and Z. Sekanina (1972) called Moravi's orbit "completely fictitious." Using 15 positions covering the period spanning January 19 to February 4, they calculated the parabolic orbit given below.

T	ω	Ω (2000.0)	i	q	e
1883 Dec. 25.7926 (TT)	138.5968	265.9923	114.9770	0.309607	1.0

ABSOLUTE MAGNITUDE: $H_{10} = 7.0$ (V1964)

FULL MOON: Dec. 14, Jan. 12, Feb. 11, Mar. 11

SOURCES: R. L. J. Ellery, *AN*, **108** (1884 Jan. 21), p. 7; D. Ross, *The Observatory*, **7** (1884 Feb.), p. 53; N. R. Pogson, *MNRAS*, **44** (1884 Mar.), pp. 256–7; R. L. J. Ellery and D. Ross, *AN*, **108** (1884 Mar. 19), p. 175; R. L. J. Ellery and H. C. F. Kreutz, *AN*, **108**

(1884 Mar. 31), p. 263; J. Tebbutt and R. L. J. Ellery, *MNRAS*, **44** (1884 Apr.), pp. 283–5; R. L. J. Ellery, D. Ross, and J. Tebbutt, *The Observatory*, **7** (1884 Apr.), pp. 115–17; H. Oppenheim, H. C. F. Kreutz, and J. Tebbutt, *AN*, **108** (1884 Apr. 5), pp. 275–8; J. Tebbutt, *AN*, **108** (1884 Apr. 17), pp. 309–12; J. Tebbutt, *AN*, **108** (1884 Apr. 25), p. 375; J. Tebbutt, *MNRAS*, **44** (1884 May), p. 377; J. Tebbutt, *The Observatory*, **7** (1884 May), pp. 140–1; R. L. J. Ellery, *SM*, **3** (1884 May), p. 124; L. Cruls, *AN*, **109** (1884 Jun. 21), p. 111; R. L. J. Ellery, *AN*, **109** (1884 Jun. 25), p. 127; J. Tebbutt, *AN*, **109** (1884 Aug. 11), pp. 251–4; J. Tebbutt, *MNRAS*, **44** (1884, Supplementary Notice), p. 444; A. B. Biggs, *MNRAS*, **45** (1884 Dec.), pp. 116–17; R. Bryant, *MNRAS*, **45** (1885 Jun.), pp. 428–31; *ARSI*, part 1 (1886), p. 383; J. F. Tennant, *MNRAS*, **47** (1886 Nov.), pp. 24–6; R. Bryant, *MNRAS*, **47** (1887 May), pp. 434–7; J. F. Tennant, *MNRAS*, **47** (1887 Jun.), p. 520; J. F. Tennant, *MNRAS*, **47** (1887, Supplementary Notice), pp. 554–5; E. Moravi, *AN*, **184** (1910 Apr. 4), pp. 129–32; V1964, p. 62; B. G. Marsden and Z. Sekanina, *CCO*, 1st ed. (1972), pp. 19, 44; B. G. Marsden, *QJRAS*, **13** (1972), pp. 428–9.

D/1884 O1
(Barnard 1)

1884 II = 1884b

Discovered: 1884 July 17.14 ($\Delta = 0.44$ AU, $r = 1.33$ AU, Elong. $= 127°$)

Last seen: 1884 November 20.79 ($\Delta = 1.19$ AU, $r = 1.66$ AU, Elong. $= 99°$)

Closest to the Earth: 1884 July 24 (0.4394 AU)

Calculated path: LUP (Disc), SCO (Jul. 22), SGR (Aug. 25), CAP (Sep. 26), AQR (Nov. 1)

E. E. Barnard (Nashville, Tennessee, USA) discovered this comet while sweeping for comets in the southwestern sky on 1884 July 17.14. He immediately noted it as a "suspicious-looking object" and after consulting available catalogs of nebulae, he thought it a probable comet and gave the position as $\alpha = 15^h\ 50.7^m$, $\delta = -37°\ 10'$. The next evening was cloudy, but the evening of the 19th was clear for a short time and he saw the comet "for a few seconds only...." He wrote, "It was located precisely the same [as two nights ago] with respect to a star... the declination was the same. I at once concluded it was a nebula, but thought it brighter" than when seen previously. The sky was again cloudy on the 20th, but Barnard found the comet through breaks in the clouds and noted it had moved. The motion was confirmed on the 21st and Barnard noted the comet appeared fainter. He added, "It is large for a telescopic comet, gradually a little brighter in the middle."

The comet quickly came under observation at observatories in the Southern Hemisphere and in the more southerly latitudes of the Northern Hemisphere. In addition, although it was not realised until near the end of July, the comet passed rather close to Earth on July 24. C. Trépied (Alger, now al-Jazâ'ir, Algeria) described the comet as centrally condensed on the 23rd, but did not see a tail. He added that the nucleus was like a star of magnitude 11.5. W. H. Finlay (Royal Observatory, Cape of Good Hope, South Africa) and J. Tebbutt (Windsor, New South Wales, Australia) independently described the comet as exceedingly faint on the 24th. The comet attained its most southerly declination of $-37°$ on the 28th and Finlay described it as a "small diffused nebulous mass." He said this was the only evening he definitely saw the nucleus and, even then, they were only glimpses with an 18-cm refractor. On July 30, Finlay said the comet still appeared as a "small diffused nebulous mass" in bright moonlight. He added, "the point of greatest condensation (in which the nucleus had been seen) was slightly towards the following part of the mass."

The comet passed perihelion around the middle of August. Finlay observed with the 18-cm refractor on the 1st and 7th, and said the comet was very faint in bright moonlight. E. Millosevich (Rome, Italy) saw the comet with a 23-cm refractor on August 9, 10, and 12. He noted a slight central condensation and an irregular cloudiness toward the south. On August 10 and 11, E. W. L. Tempel (Arcetri, Italy) said the comet was very faint in his small refractor, while a 10-cm refractor revealed a large, diffuse nebulosity about 5' across, with only slight condensation. H. Pomerantzeff (Tashkent, Russia, now Uzbekistan) described the comet as very faint on the 10th. E. Neison (Natal Observatory, Durban, South Africa) observed with a 20-cm refractor on the 12th and described the comet as a "round, hazy body with a hazy nucleus, and a faint tail-like extension." On the 13th, Finlay said the comet was faint. J. A. Perrotin (Nice, France) observed the comet from August 15 until August 18 and said it had an ill-defined coma about 1.5' across, "presenting bright granulations in the center." Neison said the comet was very faint on the 16th, with no distinct nucleus. He noted there was a "rounded preceding edge and faint tail." Trépied found the comet very faint on the 17th. On the 18th, Neison said the comet was a "faint nebulous mass, without definite edges or nucleus." There was "slight evidence of tail on following side." Finlay said the comet was very faint in an 18-cm refractor on the 19th. With the same telescope on the 20th, Finlay described the comet as "exceedingly faint, scarcely possible to observe at all." Neison said the comet was very faint and about 2' in diameter on the 21st. He added that there were "two tiny points of light in the centre resembling 15-magnitude stars, which they probably were." Tebbutt saw the comet for the final time on the 22nd. In his 11-cm refractor he said it "was seen with the greatest difficulty."

Moonlight interfered with observations as September began. On the 12th, W. Schur (Strasbourg, France) described the comet as an oblong nebulous mass in a 46-cm refractor, which was difficult to observe because of haze near the horizon. Neison said the comet was about 2' across with a central condensation on the 13th. B. von Engelhardt (Dresden, Germany) said it was a faint nebulosity about 1' across in the 30-cm refractor on the 14th, with no nucleus. On the 15th, Finlay described the comet as small and diffuse, but noted a "very striking" increase in brightness; however, Engelhardt said the comet was fainter than on the 14th and appeared oblong, with a mottled appearance. On the 16th, Schur found the comet extremely faint in a 46-cm refractor, while Neison observed with a 20-cm telescope and wrote that the "comet seemed to glitter in the condensed centre as if there were one or two nuclei, or possibly very small stars, shining through the comet, as a number of faint stars were visible near." He gave the coma diameter as 2.5′ and said the condensation was 0.9′ across. Neison also noted a "slight trace of tail." Finlay again reported the comet was unusually bright on the 17th, while Schur described the comet as a circular, nebulous mass, with a central condensation in which an occasional point of light would appear. Perrotin and his colleague A. H. P. Charlois saw the comet on several occasions between the 9th and 17th. They noted the coma was 2.5′ in diameter and exhibited a granulated nucleus 10–12″ across. A thin jet protruded from the nucleus toward PA 50°. On September 18, Neison said the comet was "nearly round and brighter than before." He gave the coma diameter as 3.5′ and noted a "decided nucleus." A. A. Common (Ealing, England) reported the coma was 4′ across on the 22nd. F. Gonnessiat (Lyon, France) said the coma was ill defined on the 23rd, with only a faint central condensation. That same night, C. L. Prince (Scarborough, England) saw the comet with a 17-cm refractor and described it as "an exceedingly faint nebulous object without central condensation or nucleus of any kind."

The comet was becoming a challenging object for many astronomers during October. Moonlight again interfered during the first days of the month. G. V. Schiaparelli (Milan, Italy) said the comet was difficult to see in a 20-cm telescope on the 11th. Finlay described the comet as faint on the 15th, very faint on the 16th, and "excessively faint" on the 20th and 24th. On the 18th, V. Knorre (Berlin, Germany) said the comet was an extremely faint, diffuse nebulosity in a 24.4-cm refractor. Schiaparelli said the comet was so faint on the 19th, that it could not be measured in a 20-cm telescope. Schur observed with a 46-cm refractor and described the comet as a round, nebulous mass on the 19th and a diffuse, nebulous mass, without a nucleus on the 21st.

The comet could not be found by Neison with a 20-cm refractor on November 4, while Perrotin reported it was near the limit of visibility in a 38-cm refractor on the 7th and 8th. Tempel looked for the comet on the 8th, but was unsuccessful at first. Then he noted an 11th-magnitude star with a hint of nebulosity surrounding it and realized the comet was over the star. Barnard managed to see the comet with the 13-cm telescope in the evening sky on November 6–12. For the final date he wrote, "it was *extremely faint*, a mere breath of the dimmest haze." Following a night of cloudy skies, Barnard was not able to see the comet on the 14th. C. F. Pechüle (Copenhagen, Denmark) reported that he could not find the comet on November 10. Tempel quickly found the comet on the 15th, but poorer seeing conditions on the 16th hid it from view despite his persistent searches.

The comet was last detected on November 20.79, when Perrotin gave the position as $\alpha = 22^h\ 38.4^m$, $\delta = -7° \ 18′$. The comet's appearance in the 38-cm refractor was described as a barely visible, round, whitish spot, with a diameter of 2–3′. Pechüle said he could not find the comet on November 21.

The only report of a spectrum was by Perrotin for the evenings of September 16, 17, and 18. He said two of the normal three cometary bands were easily visible, while the third was occasionally suspected.

The first parabolic orbit was calculated by E. Weiss using positions from July 17, 23, and 26. The perihelion date was determined as 1884 August 15.71. Additional orbits by S. C. Chandler, Jr, Weiss, H. Oppenheim, C. Stechert, and G. Ravené quickly established the perihelion date as August 18. Weiss noted that the orbit was somewhat similar to that of the short-period comet found by F. de Vico in 1844, which had been lost ever since (see 54P/1844 Q1).

By late September, astronomers began to realize that this comet was moving in a short-period orbit. The first such orbit was calculated by A. Berberich using positions spanning the period of July 23 to September 12. The resulting perihelion date was August 16.98 and the period was 5.50 years. He said a link to de Vico's comet seemed unlikely because that comet would not have been due to return in 1884 and Barnard's comet is a much fainter object. Additional orbits were calculated by J. Morrison, Finlay, E. Frisby, H. V. Egbert, and Berberich (1885, 1889), which established a period of about 5.4 years. Berberich provided predictions for this comet's next two apparitions. By taking his orbit for 1884, he simply added the period of 5.40 years and determined perihelion dates of 1890 January 10.33 and early May of 1895. Berberich noted the 1890 return was not favorable, as the comet was opposite the sun from Earth when at perihelion, but that the return of 1895 was better. Berberich (1894) revised his last prediction and indicated a perihelion date of 1895 June 3.96. No observations were obtained at either return. R. J. Buckley (1979) used 179 positions obtained between July 24 and November 8, applied perturbations by Venus to Neptune, and computed a perihelion date of August 16.97 and a period of 5.38 years. He suggested the period was uncertain by 2 or 3 days and noted the comet was well observed and "unaccountably not seen since." This orbit is given below.

T	ω	Ω (2000.0)	i	q	e
1884 Aug. 16.9680 (TT)	301.0508	6.7602	5.4701	1.279482	0.583252

ABSOLUTE MAGNITUDE: $H_{10} = 8.9$ (V1964)

FULL MOON: Jul. 8, Aug. 6, Sep. 5, Oct. 4, Nov. 3, Dec. 2

SOURCES: E. E. Barnard, *AN*, **109** (1884 Jul. 21), p. 207; C. Trépied, E. Frisby, C. N. A. Krueger, and S. C. Chandler, Jr, *AN*, **109** (1884 Aug. 1), pp. 221–4; C. Trépied, *CR*, **99** (1884 Aug. 4), pp. 228–9; E. Weiss, S. C. Chandler, Jr, and H. Oppenheim, *AN*, **109** (1884 Aug. 5), p. 237; E. Weiss and K. Zelbr, *AN*, **109** (1884 Aug. 11), p. 253; E. Millosevich and C. Stechert, *AN*, **109** (1884 Aug. 18), p. 271; J. A. Perrotin, *CR*, **99** (1884 Aug. 18), p. 321; J. A. Perrotin, *CR*, **99** (1884 Aug. 25), p. 367; E. W. L. Tempel, *AN*, **109** (1884 Sep. 2), p. 303; C. Trépied and J. A. Perrotin, *AN*, **109** (1884 Sep. 4), p. 317; J. Palisa and E. E. Barnard, *AN*, **109** (1884 Sep. 17), p. 351; G. Ravené and A. Berberich, *AN*, **109** (1884 Sep. 22), p. 365; W. Schur, *AN*, **109** (1884 Sep. 24), p. 385; B. von Engelhardt, *AN*, **110** (1884 Sep. 29), p. 9; J. A. Perrotin and A. H. P. Charlois,

CR, **99** (1884 Sep. 29), pp. 533–4; A. A. Common and C. L. Prince, *The Observatory*, **7** (1884 Oct.), pp. 305–6; H. Pomerantzeff and F. Gonnessiat, *AN*, **110** (1884 Oct. 11), p. 39; W. H. Finlay, J. Morrison, and J. Tebbutt, *MNRAS*, **45** (1884 Nov.), pp. 45–54; W. H. Finlay and W. Schur, *AN*, **110** (1884 Nov. 14), pp. 201–6; J. A. Perrotin, *CR*, **99** (1884 Dec. 1), pp. 959–60; G. V. Schiaparelli, *AN*, **110** (1884 Dec. 4), p. 253; E. W. L. Tempel and J. A. Perrotin, *AN*, **110** (1884 Dec. 17), pp. 285–8; C. W. Pritchett, *Publications of the Morrison Observatory*, No. 1 (1885), p. 104; A. Berberich, *AN*, **111** (1885 Jan. 12), p. 15; E. Neison, *MNRAS*, **45** (1885 Mar.), pp. 328–30; H. V. Egbert, *AN*, **111** (1885 Mar. 30), p. 267; E. E. Barnard, *The Observatory*, **8** (1885 Apr.), pp. 123–4; C. W. Pritchett, *AN*, **111** (1885 May 29), pp. 395–8; W. H. Finlay, *AN*, **112** (1885 Aug. 7), p. 187; W. H. Finlay, *MNRAS*, **45** (1885, Supplementary Notice), pp. 476–7; C. F. Pechüle, *AN*, **113** (1886 Feb. 11), pp. 361–4; W. Schur, *AN*, **114** (1886 Mar. 23), pp. 83, 89; V. Knorre, *AN*, **117** (1887 Aug. 3), pp. 169, 196; A. Berberich, C. Trépied, and G. V. Schiaparelli, *AN*, **123** (1889 Dec. 4), pp. 145–76; A. Berberich, *AN*, **136** (1894 Oct. 29), p. 333; A. Berberich, *PA*, **2** (1894 Dec.), p. 189; *PA*, **3** (1895 Sep.), p. 38; V1964, p. 62; R. J. Buckley, *JBAA*, **89** (1979 Apr.), pp. 261, 263.

14P/1884 S1   *Discovered:* 1884 September 17.9 ($\Delta = 0.82$ AU, $r = 1.69$ AU, Elong. $= 135°$)
(Wolf)   *Last seen:* 1885 April 7.04 ($\Delta = 2.57$ AU, $r = 2.08$ AU, Elong. $= 50°$)
   Closest to the Earth: 1884 October 2 (0.8028 AU)
1884 III = 1884c   *Calculated path:* VUL (Disc), PEG (Sep. 21), AQR (Nov. 4), PSC (Nov. 17), AQR (Nov. 26), PSC (Dec. 12), CET (Dec. 24), TAU (Mar. 7)

M. F. J. C. Wolf (Heidelberg, Germany) discovered this comet on 1884 September 17.9 at a probable position of $\alpha = 21^h\ 14.3^m$, $\delta = +23°\ 39'$. Confirming observations on September 18 and 19 revealed the comet was moving southeastward. W. Schur (Strasbourg, France) confirmed the comet with a 46-cm refractor on September 20.93. He noted it was a bright nebulous mass, with a central condensation, but with no tail. R. Copeland (Dun Echt Observatory, Scotland) independently discovered this comet on September 23.00 while he was sweeping the sky for "remarkable spectra." Using a 15-cm refractor, with an object-glass prism, he was mainly looking in the Milky Way and had found some nebulae and interesting stars when comet Wolf's interesting spectra showed up. Another 12 hours passed before news of Wolf's discovery reached Dun Echt Observatory. The comet was approaching both the sun and Earth.

The comet was well observed during the remainder of September. Several astronomers obtained physical descriptions on the 21st. B. von Engelhardt (Dresden, Germany) said it was strongly condensed in the 30-cm refractor, with a nucleus of magnitude 8. He added that the condensation was 12.6″ across. The coma itself was 2′ across in the finder. P. Tacchini (Rome, Italy) said the comet exhibited a brilliant nucleus. I. Benko von Boinik (Pula, Yugoslavia) said the coma was 2′ across and exhibited a nucleus of magnitude 8.5. There was also a trace of tail toward PA 50°. E. W. L. Tempel (Arcetri, Italy) said the comet was smaller than Barnard's comet (D/1884 O1), but brighter. Hermann Struve (Pulkovo Observatory, Russia) said the 38-cm refractor revealed a nebulosity about 1' across, with a stellar condensation. The finder of the refractor revealed the nucleus to be only slightly fainter than magnitude 9.5. On the 23rd, Schur said the 46-cm refractor revealed a bright nucleus surrounded by a nebulous mass that extended upward. G. L. Tupman (Harrow, England) observed with the 11-cm refractor (66×) and said the comet was 2' across, with no tail on the 24th. A stellar nucleus had a diameter of 2" or 3" and was about magnitude 9.5. That same night, L. Weinek (Prague, Czech Republic) said the comet was rather round and bright, with a diameter of 1.5'. A distinct nucleus shone like a star of magnitude 9.5. Another flurry of descriptions were published for the 25th. J. Power (Royal Observatory, Greenwich, England) saw the comet with a transit circle and described it as a "faint object, diffused at edges but slightly condensed towards centre." Copeland observed with a 38-cm refractor and said the comet was about 3' across and elongated toward the north-preceding side. He added that the nucleus was not in the middle. G. Bigourdan (Paris, France) said the coma was 2.5' across, with a nucleus of magnitude 9–10. The coma was more developed toward PA 130°. H. J. A. Perrotin saw the comet on September 24th, 25th, and 26th. He estimated the total magnitude as 7–8 and noted a well-defined, circular nucleus about 8" across. The tail extended 1' toward PA 165°. On the 26th, W. Winkler (Gohlis, Germany) said the coma was round and about 40" across, while the nucleus shone at magnitude 9–10. On the 27th, T. Lewis (Royal Observatory, Greenwich, England) observed the comet with a transit circle and said it appeared as an "extremely faint patch of light, brighter towards centre." That same night, Copeland observed with a 38-cm refractor (229×) and said the greatest condensation was north of and preceding the coma center, while D. Wierzbicki (Krakow, Poland) simply described the comet as faint. During the period of September 25–28, F. Courty (Bordeaux, France) said the nucleus was about magnitude 9, while the coma was unsymmetrical. A. M. W. Downing (Royal Observatory, Greenwich, England) saw the comet with a transit circle on the 28th and said it appeared "fairly bright in dark field with nucleus." He also looked at it in the 17-cm refractor and said it was "very faint but with stellar nucleus." That same night, Engelhardt said the comet was about 0.5' across in the rather bright moonlight. During the period of September 25–28, E. L. A. Périgaud (Paris, France) said the condensation was about magnitude 10. On September 29, Winkler said the nucleus was about magnitude 9 or 10.

The comet was closest to Earth at the beginning of October, but was still heading toward perihelion. On October 1, Struve said the comet was basically unchanged in appearance from previous observations, although the nucleus was less sharp. Tupman noted the coma was 2' in diameter in the 11-cm refractor (66×) on the 4th, while the nucleus was 2" across and shone at magnitude 10. No tail was present. H. Seeliger (Munich, Germany) said the nucleus was distinct on the 11th. Struve reported the comet was a little brighter on the 12th than on the 1st. On the 14th, A. Kammermann (Geneva, Switzerland) said the comet was somewhat irregular in shape and exhibited a nucleus of magnitude 10. He noted it was visible in the 5-cm finder. Courty estimated the nuclear magnitude as 9 on the 13th and 16th. Struve said the comet was somewhat brighter on the 17th than on the 12th. On October 19, Seeliger said the nucleus was rather distinct and about magnitude 9, while Schur said the 46-cm refractor revealed a round coma about 1′ across, with a distinct nucleus of magnitude 12–13. On October 20, J. Lamp (Bothkamp, Germany) said the comet was faint, with a diameter of 0.4′. Winkler estimated the total magnitude as 9 and described the comet as a round nebulosity with a central condensation. Schur said the comet was a nebulous mass on the 21st, with a stellar nucleus and a trace of tail. On the 22nd, Tupman said the comet was large and faint in the 11-cm refractor (66×), with a distinct nucleus of magnitude 10. That same night, Schur described the comet as very bright in the 46-cm refractor, with a nucleus of magnitude 10. On the 23rd, Weinek noted the coma was 2.0′ across, with a nucleus of magnitude 9.7, while Seeliger said the comet was quite faint and the nucleus was less distinct. Courty described the comet as extremely faint and difficult to see on October 30 and 31, because of moonlight.

The comet was moving away from Earth as November began, but passed perihelion just after mid-month. On November 3, W. H. Finlay (Royal Observatory, Cape of Good Hope, South Africa) described the comet as very faint. Schur said the stellar nucleus was magnitude 11 on the 4th. Seeliger said the nucleus was distinct, with a magnitude of 9 on the 6th. On the 7th, H. P. Hollis (Royal Observatory, Greenwich, England) observed with a 61-cm reflector and described the comet as very faint as the moon was rising. That same night, Tupman said moonlight was present, but the comet seemed brighter in his 11-cm refractor (66×). On the 8th, Downing observed with the 61-cm reflector and said the comet was "bright with nucleus when sky was clear. A small star seen through the nebulous envelope." That same night, Tupman said the comet was bright and 4′ or 5′ across, with a "sub-stellar" nucleus of magnitude 11, while Winkler said the comet was about magnitude 10. Where Winkler said there was no nucleus, Struve said a distinct nucleus was visible in the 38-cm refractor. On November 10, A. Müller (Stonyhurst College Observatory, England) observed with a 20-cm telescope and said the comet was well defined. On November 11, Engelhardt said the nucleus was bright and granular in appearance, and preceded the center of the coma by a small distance. He said the whole comet was similar in brightness to a magnitude 7 star. Although Müller said the comet was fainter on the 13th than on the 10th, Struve said the comet seemed somewhat brighter on the 16th than on the 9th. Since similar observations were not reported elsewhere, these discrepancies might have been due to atmospheric conditions rather than an indication of fluctuations in brightness. Using an 18.7-cm refractor on November 13, 14, and 20, A. Abetti (Padova, Italy) said the nucleus was magnitude 11 and was surrounded by a coma no larger than 1.5' across. Tupman gave the comet's diameter as 4' on the 19th and noted "the coma is not uniformly luminous." He added that the central condensation seemed brighter than when previously seen and was 4" or 5" across. On the 20th, Müller said the nucleus appeared fainter than on previous evenings. On the 21st, Struve said the nucleus had become less sharp and the brightness had slightly faded.

The comet moved away from both the sun and Earth during the remainder of the apparition. On December 9, Müller said the nucleus was "very poorly defined," Tupman said the coma was 5' across, and Finlay said the nucleus equaled a star of the 11th-magnitude. On the 10th, Engelhardt said the comet was rather bright and oval, with a diffuse nucleus, Weinek said the coma was 1.3' across, with a distinct nucleus, and Schur said the comet was rather faint, with a nucleus of magnitude 13. Müller could not see the nucleus in his 20-cm telescope on the 14th. On December 18 and 19, Winkler said the comet was rather bright, round, and centrally condensed. The comet attained its most southerly declination of −6° on December 24.

The comet was "very much fainter" when observed with an 11-cm refractor by Tupman on January 7, while Schur described it as faint, with a distinct, stellar nucleus in a 46-cm refractor. On January 9, Engelhardt said the comet was rather bright, with an indistinct nucleus in a 30-cm refractor. Finlay and Wierzbicki both said the comet was faint on the 10th. On the 17th and 18th, J. Lamp (Bothkamp, Germany) said the comet still showed a distinct central condensation. On the 18th, Schur said the comet was faint, but the nucleus was not well observed in the 46-cm refractor. On the 19th, Winkler said the comet was about magnitude 11 or 12, while Schur said the nucleus was about magnitude 15. Finlay and Schur both said the comet was very faint in moonlight on the 21st.

On February 4, Schur said the 46-cm refractor revealed the comet was very faint, without a distinct nucleus. Tupman observed with an 11-cm refractor (66×) on the 5th and said the comet was faint and difficult to see. It had a diameter of 1.5–2' and a stellar nucleus was occasionally seen. Schur found the comet quite faint on the 8th and noted a nucleus of magnitude 14. Engelhardt said the comet was rather faint in a 30-cm refractor on the 12th and exhibited a nucleus of magnitude 14. Winkler saw the comet on February 12 and 13, and described it as a faint nebulosity, with indefinite borders and no clear nucleus. On the 14th, Schur said the comet was faint.

During the last few weeks that the comet remained visible only observers with larger telescopes were able to follow it. Schur described the comet as very faint and diffuse in a 46-cm refractor on March 2. E. A. Lamp (Kiel, Germany) described the comet as very faint and only slightly brighter toward the center on the 12th, while Kammermann said the comet was very faint and round in a 25.4-cm telescope, and did not exhibit a noticeable central condensation. Schur said the comet was very faint on April 1 and extremely faint on April 4. The comet was last detected on 1885 April 7.04, when C. A. Young (Halsted Observatory, Princeton, New Jersey, USA) gave the position as $\alpha = 4^h\ 26.5^m$, $\delta = +3°\ 22'$.

Müller gave an overview of his observations of this comet. He said, "During the whole period of the observations the comet has been a sufficiently distinguishable object, consisting of a round luminous patch, having a diameter of from 1' to 1.7', with a nucleus situated somewhat towards the N.W. of the centre. This bright patch sometimes presented the appearance of having luminous projections, emanating from the side opposite to the nucleus. This was especially the case on November 10 and 14 when they assumed the appearance of a V."

As noted earlier, Copeland independently discovered this comet on September 23 while sweeping the sky for "remarkable spectra" with a 15-cm refractor and an object-glass prism. He simply noted it was a "gaseous body." Perrotin also observed the spectrum around the period of September 24–26. He said the nucleus gave a bright continuous spectrum, as well as the three cometary bands. Tacchini probably detected the carbon bands on the 21st.

The first parabolic orbit was calculated by S. C. Chandler, Jr., using positions from September 20, 21, and 22. He gave the perihelion date as 1884 November 25.53. Shortly thereafter, A. Berberich and H. Oppenheim independently calculated orbits using positions spanning the period of September 20–24. Berberich gave the perihelion date as November 24.64, while Oppenheim gave it as November 24.73.

The first elliptical orbit was calculated by C. N. A. Krueger using positions spanning the period of September 20 to October 1. It had a perihelion date of November 18.40 and a period of 6.55 years. Additional orbits were calculated during the next couple of months by Chandler, Wendell, K. Zelbr, F. Gonnessiat, Krueger, and A. Thraen, which narrowed the period down to 6.77 years. Krueger noted that the comet's ascending node lay very close to the orbit of Jupiter and calculated the probable perihelion dates going back to 1864 and forward to 1891. He noted the comet passed very close to Jupiter during the first half of 1875. R. Lehmann-Filhés wanted to determine further details of the Jupiter encounter. He began with Krueger's orbit and meticulously integrated it backwards. He found that on 1875 May 25 the comet was so influenced by Jupiter that it was moving in a strongly hyperbolic orbit. Prior to this encounter, the comet had passed perihelion on 1868 September 25.06 and had a period of 9.91 years.

The first multiple apparition orbit came from Thraen (1891). Using positions from the 1884–5 and 1891 apparitions, he applied perturbations by Earth, Jupiter, and Saturn, and determined the perihelion date as November 18.29 and the period as 6.77 years. Very similar results were later obtained by M. Kamienski (1922, 1959) and D. K. Yeomans (1978). Yeomans (1975) determined the nongravitational terms as $A_1 = +0.139$ and $A_2 = -0.0081$ and his orbit is given below.

T	ω	Ω (2000.0)	i	q	e
1884 Nov. 18.2894 (TT)	172.6905	208.0024	25.2482	1.571967	0.560906

ABSOLUTE MAGNITUDE: $H_{10} = 6.2$ (V1964)

FULL MOON: Sep. 5, Oct. 4, Nov. 3, Dec. 2, Jan. 1, Jan. 30, Mar. 1, Mar. 30, Apr. 29

SOURCES: W. Schur, *AN*, **109** (1884 Sep. 22), p. 367; W. Schur, C. N. A. Krueger, C. Schrader, F. Engström, B. Peter, and B. von Engelhardt, *AN*, **109** (1884 Sep. 24), pp. 385–8; R. Copeland, P. Tacchini, J. Lamp, C. N. A. Krueger, V. Knorre, F. Engström, C. Schrader, S. C. Chandler, A. Berberich, and H. Oppenheim, *AN*, **110** (1884 Sep. 29), pp. 9–16; G. Bigourdan, E. L. A. Périgaud, and F. Courty, *CR*, **99** (1884 Sep. 29), pp. 535–7; H. J. A. Perrotin, *CR*, **99** (1884 Oct. 6), pp. 564–5; B. von Engelhardt, I. B. von Boinik, E. W. L. Tempel, and C. N. A. Krueger, *AN*, **110** (1884 Oct. 11), pp. 41–8; L. Weinek, *AN*, **110** (1884 Oct. 15), p. 77; F. Courty, *CR*, **99** (1884 Oct. 20), p. 641; W. Winkler and K. Zelbr, *AN*, **110** (1884 Oct. 21), pp. 107, 111; J. Lamp, S. C. Chandler, Jr, and O. C. Wendell, *AN*, **110** (1884 Oct. 31), p. 143; A. M. W. Downing, J. Power, T. Lewis, and H. P. Hollis, *MNRAS*, **45** (1884 Nov.), pp. 55–7; A. Thraen, *AN*, **110** (1884 Nov. 8), p. 191; F. Gonnessiat and F. Courty, *CR*, **99** (1884 Nov. 10), pp. 774–5; C. N. A. Krueger, *AN*, **110** (1884 Nov. 14), p. 207; R. Copeland, *MNRAS*, **45** (1884 Dec.), pp. 90–1; W. Winkler and R. Lehmann-Filhés, *AN*, **110** (1884 Dec. 4), pp. 253–6; A. Thraen, *AN*, **110** (1884 Dec. 17), p. 287; A. Müller, *MNRAS*, **45** (1885 Jan.), pp. 156–8; W. Schur, *AN*, **111** (1885 Jan. 12), p. 15; A. Kammermann, *AN*, **111** (1885 Jan. 27), p. 77; B. von Engelhardt, *AN*, **111** (1885 Feb. 16), p. 157; W. Winkler, *AN*, **111** (1885 Feb. 27), p. 191; J. Lamp and E. A. Lamp, *AN*, **111** (1885 Mar. 17), p. 235; B. von Engelhardt, A. Kammermann, and C. W. Pritchett, *AN*, **111** (1885 May 29), pp. 391–8; G. L. Tupman, *MNRAS*, **45** (1885 Jun.), pp. 402–4; W. Winkler and C. A. Young, *AN*, **112** (1885 Jun. 5), p. 31; W. H. Finlay, *AN*, **112** (1885 Aug. 22), pp. 257–60; A. Abetti, *AN*, **112** (1885 Sep. 18), p. 305; L. Weinek and H. Seeliger, *AN*, **112** (1885 Oct. 1), pp. 365–70; W. H. Finlay, *MNRAS*, **45** (1885, Supplementary Notice), pp. 478–80; E. Frisby, Wierzbicki, and A. Donner, *AN*, **113** (1886 Jan. 11), pp. 257–62; C. F. Pechüle, *AN*, **113** (1886 Feb. 11), p. 361; W. Schur, *AN*, **114** (1886 Mar. 23), pp. 83–6, 89–91; H. Struve, *AN*, **115** (1886 Jul. 28), p. 39; B. Peter, *AN*, **115** (1886 Oct. 7), p. 235; R. Copeland, *MNRAS*, **46** (1886, Supplemental Notice), pp. 491, 494; P. Baracchi, *AN*, **116** (1887 Feb. 3), pp. 145–50; A. Thraen, *AN*, **117** (1887 Jun. 16), pp. 65–98; V. Knorre, *AN*, **117** (1887 Aug. 3), pp. 169, 197; A. Berberich, *AN*, **117** (1887 Aug. 20), pp. 251–4; G. L. Tupman, *MNRAS*, **47** (1887, Supplemental Notice), p. 545; A. Thraen, *AN*, **127** (1891 Jun. 20), pp. 315–20; A. Thraen, *AN*, **128** (1891 Dec. 22), pp. 421–4; M. Kamienski, *AOAT*, **5** No. 5 (1922), pp. 1–5; M. Kamienski, *AA*, **9** (1959), p. 58; V1964, p. 62; D. K. Yeomans, *PASP*, **87** (1975 Aug.), pp. 635–7; D. K. Yeomans, *QJRAS*, **19** (1978), pp. 52–3, 57.

2P/Encke

1885 I = 1884d

Recovered: 1884 December 13.74 ($\Delta = 1.42$ AU, $r = 1.61$ AU, Elong. = 82°)

Last seen: 1885 April 23.38 ($\Delta = 1.11$ AU, $r = 1.06$ AU, Elong. = 60°)

Closest to the Earth: 1885 March 10 (0.6480 AU)

Calculated path: PEG (Rec), PSC (Dec. 16), AQR (Mar. 8)

J. O. Backlund (1884) provided a prediction for the 1885 apparition of this comet, which gave the perihelion date as 1885 March 8.15. He provided an ephemeris for the period of 1884 December 1 to 1885 January 31. E. W. L. Tempel (Arcetri, Italy) recovered this comet on 1884 December 13.74 at a position of $\alpha = 22^h\ 44.4^m$, $\delta = +3°\ 44'$. He described it as a "nebulous glow" and noted it was "very faint and without a bright center."

The comet was not seen again until it was observed by C. Trépied (Alger, now al-Jazâ'ir, Algeria) on January 2. It was then described as very faint. Several astronomers saw the comet on the 3rd. Tempel noted the comet was larger and brighter than when seen on December 13. He added that no condensation or nucleus was present and said the finder revealed the comet as a faint nebulosity. Trépied again noted the comet was very faint. W. Schur (Strasbourg, France) observed with a 46-cm refractor and said it appeared large and nebulous, but without a nucleus. On the 7th, B. von Engelhardt (Dresden, Germany) described the comet as an extremely faint nebulosity about 1.5′ across, without a condensation. Tempel observed it on the 5th and 8th and said it was bright and "shows a granular centre, and was also seen with the finder as a small faint nebula." That same night, Engelhardt said the comet was fainter than on the previous evening. On the 9th, A. Kammermann (Geneva, Switzerland) said the comet was 45″ across, without a notable condensation, while Schur said the comet was quite faint. Kammermann said the comet was easier to see on the 13th than on the 9th. Schur said the comet was very faint in the 46-cm refractor on the 14th. Engelhardt said the comet was very faint on the 15th, but added that it was brighter than on the 7th and 8th. H. Seeliger (Munich, Germany) described the comet as faint in the 27-cm refractor on the 16th. F. Gonnessiat (Lyon, France) described the comet as a very faint nebulosity, with a faint central condensation on the 16th and 18th. Schur said the comet was very faint, with a nearby moon on the 20th, extremely faint in moonlight on the 21st, and rather bright on the 31st.

The comet continued to approach the sun and Earth during February and into March. It passed perihelion on March 8 and made a moderately close approach to Earth on March 10. On February 2, Seeliger saw the comet and said it was much brighter than when seen two weeks earlier, with a rather distinct nucleus. Engelhardt simply described it as bright in the 30-cm refractor on the 2nd and 6th. Schur said it was very bright, with a strong condensation on the 4th. Kammermann observed with a 25.4-cm telescope on the 6th and described the comet as bright, with a strong condensation situated just southwest of the center of the coma. Schur found it very bright on the 8th. On February 12, E. E. Barnard (Vanderbilt University, Nashville, Tennessee, USA) observed the comet with a 13-cm telescope and noted a faint tail extending about 10′. That same night, A. Donner (Helsingfors, now Helsinki, Finland) noted a trace of tail and said the condensation was not in the center of the coma, while Engelhardt said the comet was very bright, with a coma 2′ in diameter. Engelhardt added that it had a granular appearance and the bright condensation was situated just north of center. Tempel noted an extremely faint tail on the 12th and 13th. On the

14th, Barnard observed with the 15-cm refractor (90×) and described the comet as "small and brightly condensed; the point of greatest brightness was not central, but very slightly following the centre in the direction of the tail." He added that the tail was "very slender, faint, and straight, and extended about 14' or 15'." That same night, H. Pomerantzeff (Tashkent, Russia, now Uzbekistan) said the comet looked like a round nebulosity about 0.8' across, with central condensation. Pomerantzeff said the coma was 0.7' across on the 16th. Barnard said the comet's appearance was generally the same on the 17th as on the 14th, except that the tail was "a little longer and slightly more distinct." E. L. A. Périgaud (Paris, France) saw the comet on the 21st and described it as a round nebulosity of magnitude 9. On February 27, Schur noted a central condensation. Schur said he could hardly see the comet in the 46-cm refractor on March 2, because of the comet's low altitude.

The comet was recovered in the morning sky with a 28-cm refractor (60×) by J. M. Thome (National Argentine Observatory, Cordoba, Argentina) on March 28.38 and 29.38. Thome also observed the comet on April 14, 18, 20, 21, 22, and 23, and said it was of the "last degree of faintness, being mainly an irregular whiteness." Thome's final observation was also the last time the comet was detected and he gave the position as $\alpha = 22^h 39.3^m$, $\delta = -19° 41'$ on April 23.38.

Refinements of this comet's orbit using positions from multiple apparitions, as well as planetary perturbations, have been published by numerous astronomers. Of most significance were the studies of Backlund (1886) and B. G. Marsden and Z. Sekanina (1974). These have typically revealed a perihelion date of March 8.14 and a period of 3.31 years. Marsden and Sekanina determined the nongravitational terms as $A_1 = -0.14$ and $A_2 = -0.02263$. The orbit of Marsden and Sekanina is given below.

T	ω	Ω (2000.0)	i	q	e
1885 Mar. 8.1371 (TT)	183.9047	336.2397	12.9147	7 0.342372	0.845756

ABSOLUTE MAGNITUDE: $H_{10} = 9.7$ (V1964)

FULL MOON: Dec. 2, Jan. 1, Jan. 30, Mar. 1, Mar. 30, Apr. 29

SOURCES: J. O. Backlund, *BASP*, **29** (1884), pp. 497–504; J. O. Backlund, *BA*, **1** (1884 Nov.), pp. 536–8; E. W. L. Tempel, *AN*, **110** (1884 Dec. 17), p. 283; E. W. L. Tempel, *The Observatory*, **8** (1885 Jan.), p. 30; W. Schur, *AN*, **110** (1885 Jan. 4), p. 399; C. Trépied, *AN*, **110** (1885 Jan. 6), p. 415; E. W. L. Tempel, *AN*, **111** (1885 Jan. 12), p. 9; A. Kammermann, *AN*, **111** (1885 Jan. 26), p. 57; F. Gonnessiat, *AN*, **111** (1885 Jan. 31), pp. 89–94; E. W. L. Tempel and E. E. Barnard, *The Observatory*, **8** (1885 Feb.), pp. 61–2; B. von Engelhardt, *AN*, **111** (1885 Feb. 16), p. 157; E. L. A. Périgaud, *CR*, **100** (1885 Mar. 9), p. 730; E. W. L. Tempel, *AN*, **111** (1885 Mar. 26), p. 247; E. E. Barnard, *The Observatory*, **8** (1885 Apr.), pp. 122–3; A. Donner and H. Pomerantzeff, *AN*, **111** (1885 May 7), p. 335; B. von Engelhardt and A. Kammermann, *AN*, **111** (1885 May 29), pp. 391–4; J. M. Thome and H. Seeliger, *AN*, **112** (1885 Oct. 1), pp. 365–70; J. O. Backlund, *MASP*, **34** No. 8, 7th Series (1886), p. 38; E. Frisby and Wierzbicki, *AN*, **113** (1886 Jan. 11), pp. 257–62; C. F. Pechüle, *AN,* **113** (1886 Feb. 11), p. 361; W. Schur, *AN,* **114** (1886 Mar. 23), pp. 85, 91; V. Knorre, *AN,* **117** (1887 Aug. 3), pp. 169, 197; V1964, p. 62; B. G. Marsden and Z. Sekanina, *AJ,* **79** (1974 Mar.), pp. 415–16.

X/1885 G1   While searching for Tempel's comet on 1885 April 6, L. Swift (Warner Observatory, Rochester, New York, USA) came across a nebula that he had not seen before. Since he was cataloging the nebulae he found, he measured its position as $\alpha = 11^h\,54^m40^s$, $\delta = +20°\,02'\,35''$. It was described as pretty faint, pretty large, and round. Although the evening of April 7 was cloudy, the evening of the 8th was clear, but Swift could not find an object in the position noted on the 6th. He made a "most determined effort" to relocate the object on the 8th, as well as on the 10th and 13th, but nothing was found. Swift also alerted Harvard College Observatory to the potential comet, but their attempts to locate it were also unsuccessful. Although it was labelled number 19 in his first catalog of nebulae, Swift later suggested it be struck out because of his certainty that it was a comet.

SOURCES: L. Swift, *AN,* **112** (1885 Sep. 24), p. 313; L. Swift, *AN,* **113** (1886 Jan. 26), p. 305.

C/1885 N1   *Discovered:* 1885 July 8.28 ($\Delta = 1.60$ AU, $r = 2.53$ AU, Elong. $= 148°$)
(Barnard)   *Last seen:* 1885 September 4.06 ($\Delta = 2.35$ AU, $r = 2.53$ AU, Elong. $= 88°$)
      Closest to the Earth: 1885 June 30 (1.5904 AU)
1885 II = 1885a   *Calculated path:* OPH (Disc), SCO (Aug. 16)

E. E. Barnard (Nashville, Tennessee, USA) discovered this faint comet with a 13-cm refractor on 1885 July 8.28, near NGC 4301. In fact, because sky conditions were not the best, he was not absolutely sure the diffuse object he was looking at was not NGC 4301, although he suspected that that object was following the comet and south of it. On July 8.35 he confirmed he had found a comet and gave the position as $\alpha = 17^h\,21.4^m$, $\delta = -4°\,57'$. Barnard described it as small and dim on July 9, and noted a very small, ill-defined nucleus in the 15-cm refractor. The comet was about a week past its closest approach to Earth and was still a month from perihelion.

   On July 10, Barnard noted, "an indefinite brightening in the middle to a tiny nucleus." A. Kammermann (Geneva, Switzerland) said it was round, with a central condensation on the 11th. A. H. P. Charlois (Nice, France) saw the comet on the 11th and noted a nucleus of magnitude 10.5. The nucleus was surrounded by a weak nebulosity "of rather confused form," which was about 1.5' across. On the 11th and 13th, E. W. L. Tempel (Arcetri, Italy) said the comet was round and 1' across. Observations came from B. von Engelhardt (Dresden, Germany), E. Millosevich (Rome, Italy), and W. Schur (Strasbourg, France) on the 12th. Schur said the 46-cm refractor revealed a very faint nebulous object, with neither a nucleus nor a tail. Engelhardt said the comet was rather bright and round in the 30.5-cm refractor. It was 0.5′ across and exhibited a central condensation. He added that the 12.7-cm finder (20×) revealed the comet as a cloudy star. Millosevich said the comet was faint, with a nucleus of magnitude 11. He noted the nucleus was a little eccentric toward the preceding half of the coma. E. W. Maunder (Royal Observatory, Greenwich, England) said the comet was "an exceedingly faint object" seen with great difficulty in the refractor on the 14th. It also appeared slightly elongated and exhibited no central condensation. Maunder indicated that the coma was between 1.5′ and 1.75′ across. Barnard observed under hazy skies on the 15th, but wrote, "There seems to be considerable but indefinite condensation, probably excentric." G. L. Tupman (Harrow, England) observed the comet with a 47.0-cm reflector on the 15th and 17th. He said the sky was hazy on both nights and the comet appeared very faint. Tupman added that the coma was 30–40″ in diameter and exhibited a central condensation. On the 16th, Schur said he looked at the comet with the 46-cm refractor (208×) and noted it was 1′ across, with a bright condensation and a star-like nucleus of magnitude 11. At a magnification of 408× he noted an uncondensed nebula. J. G. Lohse (Wiggleworth's Observatory, Scarborough, England) said the comet was fainter on the 17th than on the 14th. He described it as round and gradually brighter toward the middle, although there was "no real nucleus." Also on the 17th, J. Tebbutt (Windsor, New South Wales, Australia) detected the comet in his 11-cm telescope "as a hazy speck of the last degree of faintness." G. A. P. Rayet (Bordeaux, France) observed the comet on several occasions during the period of July 11–20 with a 36-cm refractor. He described it as about 0.5′ across, with a central condensation of magnitude 10 to 11. C. A. Young (Halsted Observatory, Princeton, New Jersey, USA) observed with a 58-cm telescope (114×) on July 11, 12, 13, 14, 16, 17, 18, 19, and 30. He summarized that the comet was about 0.75′ across, somewhat elongated and "much condensed in the centre," although there was no stellar nucleus. A faint "slightly fan-shaped" tail extended about 2.5–4′ toward PA 35°. No observations appear to have been made during the period of July 21–29, because of moonlight. On July 29, Barnard noted, "Occasionally indefinite brightening near centre."

The comet passed perihelion early in August. Schur found it very faint on the 1st. On the 5th, Tempel said the comet had a distinct decrease in brightness, while Schur described the comet as an indefinite nebulous mass, with a nucleus. Millosevich said the comet was faint on the 8th, with a nucleus of magnitude 11 or 12. That same night, Miranda (Imperial Observatory, Rio de Janeiro, Brazil) said the comet was hardly visible in the 25-cm refractor. Millosevich noted the comet was faint in the 25-cm refractor on the 12th. G. Aguello (Palermo, Italy) observed the comet with the refractor from July 30 to August 12. He said the comet had become fainter by the time the last observations were made. Barnard said the comet was exceptionally faint in the 15-cm refractor during the period of August 4–13.

The comet was last detected on September 4.06, when F. P. Leavenworth (Leander McCormick Observatory, Virginia, USA) measured the position as $\alpha = 16^h 26.9^m$, $\delta = -30° 58'$ with the 66-cm refractor.

The spectrum was observed by some astronomers. Young saw it on July 12 and 13 using a spectroscope connected to the 58-cm refractor. He said the spectrum was "almost continuous, the usual cometary bands being visible only as three slight intensifications of brightness." L. Thollon and H. J. A. Perrotin saw a faint continuous spectrum on July 13 and 15. They added that the cometary bands were only occasionally seen.

The first orbit was calculated by J. Holetschek. He used positions obtained on July 10–13, and calculated a parabolic orbit with a perihelion date of 1885 September 25.62. Additional orbits were calculated by E. A. Lamp, Charlois, H. Oppenheim, Holetschek, A. Hall, Jr, H. V. Egbert, Thome, and A. Berberich. These revealed that the comet's slow motion was primarily due to its rather large perihelion distance of 2.5 AU and that the perihelion date was actually around August 6. Near the end of August, Lamp published an orbit that indicated the comet moved on an elliptical orbit with a period of about 8700 years.

A. Berberich (1890) published a definitive orbit of this comet. Beginning with 221 positions obtained during the period of July 8 to August 17, he calculated both a parabolic and a hyperbolic orbit. The parabolic orbit had a perihelion date of August 6.18, while the hyperbolic orbit had a perihelion date of August 6.04 and an eccentricity of 1.0028519. Although Berberich found the hyperbolic orbit represented the positions better than the parabolic orbit, he said the parabolic orbit still fitted the positions in an acceptable manner, and, considering the short period of observation, he chose the parabolic orbit to represent this comet's motion. The parabolic orbit is given below.

T	ω	Ω (2000.0)	i	q	e
1885 Aug. 6.1818 (UT)	178.5163	93.8993	80.6239	2.506732	1.0

ABSOLUTE MAGNITUDE: $H_{10} = 5.8$ (V1964)

FULL MOON: Jun. 27, Jul. 27, Aug. 25, Sep. 24

SOURCES: W. Schur and G. Cacciatore, *AN*, **112** (1885 Jul. 13), p. 139; B. von Engelhardt, E. Millosevich, A. Kammermann. E. W. L. Tempel, W. Schur, and J. Holetschek, *AN*, **112**, (1885 Jul. 18), pp. 155–8; A. H. P. Charlois, L. Thollon, and H. J. A. Perrotin, *CR*, **101** (1885 Jul. 20), pp. 231–2; G. A. P. Rayet and A. H. P. Charlois, *CR*, **101** (1885 Jul. 27), pp. 301–3; E. A. Lamp, *AN*, **112** (1885 Jul. 31), p. 175; G. L. Tupman, *AN*, **112** (1885 Aug. 7), p. 189; E. Millosevich, H. Oppenheim, J. Holetschek, and E. A. Lamp, *AN*, **112** (1885 Aug. 22), pp. 259–64; G. Aguello, E. W. L. Tempel, and A. Hall, Jr, *AN*, **112** (1885 Sep. 14), p. 291; E. E. Barnard and H. V. Egbert, *AN*, **112** (1885 Oct. 1), pp. 371–4; E. W. Maunder, *AN*, **112** (1885 Oct. 8), p. 405; C. A. Young, *AN*, **113** (1885 Oct. 31), pp. 27–30; G. L. Tupman, *MNRAS*, **45** (1885, Supplementary Notice), p. 481; E. Frisby, A. Hall, E. Millosevich, and J. Tebbutt, *AN*, **113** (1886 Jan. 11), pp. 257, 261–4; O. Stone and F. P. Leavenworth, *AN*, **113** (1886 Feb. 22), pp. 393–6; L. Cruls, Miranda,

H. Morize, and Duarte, *CR*, **102** (1886 Feb. 22), pp. 404–5; J. M. Thome, *AN*, **114** (1886 Mar. 6), pp. 33–6; W. Schur, *AN*, **114** (1886 Mar. 23), pp. 87, 92; L. Cruls, Miranda, Morize, and Duarte, *AN*, **114** (1886 Mar. 30), p. 123; A. Berberich, *AN*, **114** (1886 Apr. 8), p. 157; J. G. Lohse, *MNRAS*, **47** (1886 Nov.), pp. 28–9; P. Baracchi, *AN*, **116** (1887 Feb. 3), pp. 145–50; V. Knorre, *AN*, **117** (1887 Aug. 3), p. 169; A. Berberich, *AN*, **123** (1890 Feb. 10), pp. 385–406; V1964, p. 62.

X/1885 Q1 L. Swift (Warner Observatory, Rochester, New York, USA) was cataloging nebulae during his routine comet searches when he found this object on 1885 August 20. He determined the position as $\alpha = 4^h\, 05^m\, 12^s$, $\delta = +27° \, 24' \, 30''$. Swift described the object as very faint and resembling a comet. He tried to verify objects he found as soon as possible, but on this occasion moonlight and clouds prevented this until September 6, at which time nothing could be found. Swift wrote, "Am certain of its place, and of its configuration with 4 stars. Have examined the place three times and am certain of its absence. Seeing on one occasion as good as when discovered." Although this was labeled number 29 in Swift's 2nd catalog of nebulae, he suggested it be "struck out" since it was probably a comet.

SOURCES: L. Swift, *AN*, **113** (1886 Jan. 26), pp. 307–10.

C/1885 R1 *Discovered:* 1885 September 1.1 ($\Delta = 1.11$ AU, $r = 0.86$ AU, Elong. $= 47°$)
(Brooks) *Last seen:* 1885 October 5.88 ($\Delta = 1.01$ AU, $r = 1.28$ AU, Elong. $= 79°$)
  Closest to the Earth: 1885 September 28 (1.0024 AU)
1885 III = 1885c *Calculated path:* CVn (Disc), BOO (Sep. 6), HER (Sep. 22)

W. R. Brooks (Phelps, New York, USA) discovered this comet in the evening sky on 1885 September 1.1. E. C. Pickering (Harvard College Observatory, Cambridge, Massachusetts, USA) confirmed the comet on September 3.20 and described it as round, about 2′ across, and about magnitude 9. There was some condensation, but no tail. He roughly measured the position as $\alpha = 13^h\, 42.0^m$, $\delta = +36° \, 37'$. A. A. Common (Ealing, England) independently discovered the comet on September 4.

The comet had already passed perihelion, but a moderately close approach to Earth was to occur in late September. On September 4, E. W. Maunder (Royal Observatory, Greenwich, England) saw the comet with a 33-cm refractor and said it was circular, with a central condensation, and very diffuse around the edges. His colleague, H. H. Turner, said the comet was "exceedingly faint" and diffuse in a 17-cm refractor, with no nucleus or tail. That same night, J. Winlock (US Naval Observatory, Washington, DC, USA) observed with a 24-cm refractor and described the comet as a "nebulous mass, about 1.6′ in diameter without nucleus or tail." On the 5th and 9th, E. Millosevich (Roman College Observatory, Rome, Italy) said the comet was faint, round, and 2′ across. G. L. Tupman (Harrow, England) observed with a 47-cm reflector (60×) on the 6th and said the comet had no nucleus. E. A. Lamp (Kiel, Germany) and Turner also saw the comet that night. Lamp said the comet was faint and somewhat brighter toward the center, while Turner said the comet "appeared a little brighter than on September 4," but was still very faint. He added that it was brighter in the center. On the 7th, Winlock said the comet was "excessively faint" and actually fainter than on the 4th, while H. Pomerantzeff (Tashkent, Russia, now Uzbekistan) said it was very faint, with no apparent condensation. Winlock also said the coma was 3.9′ across. On the 8th, G. F. W. Rümker (Hamburg, Germany) and Lamp both noted the comet was diffuse, without a nucleus. That same night, C. A. Young and M. McNeill (Halsted Observatory, Princeton, New Jersey, USA) said the comet was 2′ in diameter in the 58-cm refractor, with a bright center, but no nucleus. On the 9th, Tupman noted the comet seemed "a little brighter" than on the 6th; however, Maunder noted the comet was very faint, diffused, and more difficult to see than on the 4th. A. M. W. Downing (Royal Observatory, Greenwich, England) simply described the comet as a "very faint nebulous mass" in the 17-cm refractor on the 9th. H. Battermann (Berlin, Germany) saw the comet on the 10th and said he only suspected the comet as a nebulous mass about 2′ across. There was no condensation. Winlock saw the comet through breaks in the clouds with a 24-cm refractor on the 10th and 11th. He described it as a "faint white spot" about 1.2′ across. J. E. Coggia (Marseille, France) observed with a 25.8-cm telescope on the 11th and 16th, and said the comet "presented the aspect of a rather extended nebulosity, of granular appearance, without a trace of a nucleus or a tail." B. von Engelhardt (Dresden, Germany) said the comet was not faint in the 30-cm refractor on the 12th, but was diffuse without a condensation. The coma was slightly elongated with a diameter of about 2′. Tupman said the coma was 3′ across, as seen in the 47-cm reflector (120×) on the 13th, but noted no nucleus. E. W. L. Tempel (Arcetri, Italy) also saw the comet on the 13th and said it was similar to a class II nebula, but more diffuse and irregular in shape. Engelhardt said the comet was faint in moonlight on the 15th, with a coma about 2.5′ across. Young and McNeill observed with the 58-cm refractor on the 17th and said the comet was very diffuse, with "no trace of condensation." The comet attained its most northerly declination of +43° on September 25.

The comet was last detected on October 5.88, when L. Becker (Dun Echt Observatory, Scotland) gave the position as $\alpha = 17^h\ 29.6^m$, $\delta = +41°\ 36'$. He observed the comet with a 38-cm refractor and described it as very faint and 3′ in diameter.

The first parabolic orbit was calculated by J. Holetschek using positions obtained during the period of September 3–7. The resulting perihelion date was 1885 August 10.19. This would prove to be about a half day off. Later orbits were calculated by Lamp, R. Radau, H. Oppenheim, and A. Berberich (1886).

Definitive orbits were calculated by W. W. Campbell (1888), J. Gallenmüller (1892), and D. Klumpke (1896). Campbell determined an elliptical orbit with a period of about 496 years. Gallenmüller calculated both a parabolic and an elliptical orbit. He found the elliptical one represented the positions far better than did the parabolic one and gave the period as about 403 years. Klumpke calculated an elliptical orbit with a period of about 274 years and this is given below.

T	ω	Ω (2000.0)	i	q	e
1885 Aug. 10.6574 (UT)	42.8501	206.3673	59.0970	0.749149	0.982263

ABSOLUTE MAGNITUDE: $H_{10} = 9.5$ (V1964)

FULL MOON: Aug. 25, Sep. 24, Oct. 23

SOURCES: R. Radau, *BA*, **2** (1885 Aug.), p. 451; W. R. Brooks and E. C. Pickering, *AN*, **112** (1885 Sep. 5), p. 247; G. L. Tupman, G. F. W. Rümker, E. A. Lamp, and J. Holetschek, *AN*, **112** (1885 Sep. 14), p. 293; E. Millosevich and E. A. Lamp, *AN*, **112** (1885 Sep. 18), pp. 309–12; H. Battermann, G. L. Tupman, E. W. L. Tempel, and H. Oppenheim, *AN*, **112** (1885 Sep. 24), pp. 325–8; E. E. Barnard and B. von Engelhardt, *AN*, **112** (1885 Oct. 1), p. 375; E. W. Maunder, H. H. Turner, and A. M. W. Downing, *AN*, **112** (1885 Oct. 8), p. 405; L. Swift, *AN*, **113** (1885 Nov. 30), p. 63; L. Becker, *MNRAS*, **46** (1885 Dec.), p. 59; J. E. Coggia, *BA*, **3** (1886 Jan.), pp. 27–8; G. L. Tupman, *MNRAS*, **46** (1886 Jan.), pp. 123–4; J. Winlock, E. Frisby, C. A. Young, M. McNeill, H. Pomerantzeff, and E. Millosevich, *AN*, **113** (1886 Jan. 11), pp. 257–60, 263; A. A. Common, *MNRAS*, **46** (1886 Feb.), p. 217; A. Berberich, *AN*, **114** (1886 Mar. 6), p. 43; E. Stuyvaert and M. L. Niesten, *AN*, **114** (1886 Mar. 23), p. 91; G. L. Tupman, *MNRAS*, **47** (1887, Supplemental Notice), p. 545; W. W. Campbell, *AN*, **120** (1888 Oct. 4), pp. 49–60; J. Gallenmüller, *AN*, **130** (1892 Sep. 17), pp. 345–64; D. Klumpke, *BA*, **13** (1896 Sep.), pp. 329–37; V1964, p. 62.

8P/Tuttle

1885 IV = 1885b

Recovered: 1885 August 9.13 ($\Delta = 1.91$ AU, $r = 1.15$ AU, Elong. $= 30°$)

Last seen: 1885 September 17.40 ($\Delta = 1.70$ AU, $r = 1.03$ AU, Elong. $= 34°$)

Closest to the Earth: 1885 September 23 (1.6941 AU)

Calculated path: GEM (Rec), CNC (Aug. 19), HYA (Sep. 7)

The prediction of this comet's return came from J. Rahts (1885). In a paper published in the 1885 July 18 issue of the *Astronomische Nachrichten*, he took positions from the 1858 and 1871 apparitions, applied perturbations by Mercury to Uranus, and predicted the next perihelion date would occur on 1885 September 11.64. H. J. A. Perrotin and A. H. P. Charlois (Nice, France) recovered this comet with a refractor on 1885 August 9.13 and gave the position as $\alpha = 7^{\text{h}} 20.4^{\text{m}}$, $\delta = +28° 37'$. The observation was difficult because of the comet's faintness and low altitude. The coma was about 2′ across and was without a condensation.

Although Perrotin and Charlois measured the comet's position on every clear night up to August 23rd, they obtained only one physical description after the recovery. On August 11, with exceptional skies, they noted the coma was elongated.

The comet was last detected on September 17.40, when O. Stone (Leander McCormick Observatory, Virginia, USA) gave the position as $\alpha = 9^h 26.1^m$, $\delta = -1° 42'$, while using the 66-cm refractor.

Some astronomers have considered multiple apparitions in the investigation of this comet's motion. One of the earliest such investigations was published by Rahts (1885, 1894). The most recent was by B. G. Marsden and Z. Sekanina (1972). All determined the perihelion date as September 11.80 and the period as 13.76 years, with Marsden and Sekanina giving nongravitational terms of $A_1 = +0.32$ and $A_2 = +0.0131$.

T	ω	Ω (2000.0)	i	q	e
1885 Sep. 11.7988 (TT)	206.7661	271.2471	54.3303	1.024713	0.821542

ABSOLUTE MAGNITUDE: $H_{10} = 8.5$ (V1964)

FULL MOON: Jul. 27, Aug. 25, Sep. 24

SOURCES: J. Rahts, *AN*, **112** (1885 Jul. 18), p. 159; H. J. A. Perrotin, *AN*, **112** (1885 Aug. 22), p. 263; H. J. A. Perrotin and A. H. P. Charlois, *AN*, **113** (1885 Oct. 31), p. 29; J. Rahts, *AN*, **113** (1885 Dec. 22), pp. 169–206; O. Stone, *AN*, **113** (1886 Feb. 22), pp. 393–6; J. Rahts, *AN*, **136** (1894 Aug. 2), pp. 65–8; V1964, p. 62; B. G. Marsden and Z. Sekanina, *QJRAS*, **13** (1972), pp. 427–9; B. G. Marsden, *CCO*, 12th ed. (1997), pp. 50–1.

C/1885 X1   *Discovered:* 1885 December 1.88 ($\Delta = 1.57$ AU, $r = 2.29$ AU, Elong. $= 126°$)
(Fabry)   *Last seen:* 1886 July 30.77 ($\Delta = 2.28$ AU, $r = 2.15$ AU, Elong. $= 69°$)
   Closest to the Earth: 1886 May 1 (0.1980 AU)
1886 I = 1885d   *Calculated path:* PSC (Disc), PEG (Dec. 15), AND (Mar. 18). TRI (Apr. 24), ARI (Apr. 27), TAU (Apr. 29), ERI (May 2), ORI-ERI (May 3), ORI-LEP (May 4), CMa (May 7), PUP (May 14), VEL (May 27)

L. Fabry (Paris, France) discovered this comet on 1885 December 1.88, and one of his colleagues, M. Loewy, was able to immediately determine a position of $\alpha = 0^h 39.1^m$, $\delta = +21° 02'$. The comet was described as round, about magnitude 12, with a diameter of about 1′ and exhibiting a central, stellar nucleus. It was moving through Pisces at a rate of 0.6° per day toward the west. G. Bigourdan, another colleague, saw the comet on December 1.95. Later calculations revealed the comet had reached a declination of +21° on November 16 before proceeding in a west-southwest direction. It had moved to within 1.5639 AU of Earth on November 28, before Earth's orbit began taking it away from the comet.

The comet was observed at numerous European observatories on December 2. In general, astronomers indicated the comet was diffuse, tailless, and about 1.5′ in diameter, with a central star-like nucleus of magnitude

12–13. In the days that followed many more astronomers turned their telescopes toward this comet and made some interesting, detailed descriptions. E. W. L. Tempel (Arcetri, Italy) saw the comet on the 4th and said its small, faint, nebulous coma resembled a nebula of class II and III (*these classes were determined by W. Herschel and represented faint and very faint nebulae, respectively, or objects as faint as magnitude 10–12*). On the same date, F. Laschober (Pula, Yugoslavia) observed with a 15-cm refractor and said the comet had a compact, elliptical form with diffuse edges and a central condensation with a distinct boundary. On the 5th, B. von Engelhardt (Dresden, Germany) observed with a 30.5-cm telescope and said the comet was rather bright and exhibited a good distinct star-like nucleus of magnitude 11. The brightest part of the coma was 15" across, while the entire coma appeared 2' across in a 12.7-cm seeker. R. Copeland (Dun Echt Observatory, Scotland) observed with a 38-cm telescope on December 7 and said the comet was 40" in diameter and was "brighter towards the preceding side." During the period of December 8–10 astronomers generally described the comet as bright, round, and 1' across. The nucleus was about magnitude 11.

The moon was full on December 21, and no observations seem to have been made on this date or on the previous day. In addition, observations were sparse during the period of December 13–22, most likely because many observers did not wish to fight moonlight for a glimpse of the comet. Meanwhile, the comet entered Pegasus on December 15 and then began to slowly work its way northward.

Observations picked up during the last nine days of December, and most observers generally reported the comet as very faint, with a condensation, and a faint nucleus. G. L. Tupman (Harrow, England) said the comet was "very much brighter" than when previously observed and was 2' across on December 23. J. G. Lohse (Wigglesworth Observatory, Scarborough, England) saw the comet with a 39-cm refractor on December 24 and said it was faint and round, with a stellar nucleus. On December 27 Tupman said the comet appeared bright and exhibited a "sub-stellar nucleus," J. Bauschinger (Munich, Germany), observing with a 12.7-cm refractor, described it as bright, with a distinct nucleus of magnitude 11, and A. M. W. Downing (Royal Observatory, Greenwich, England), observing with a 17-cm refractor, described it as faint, circular, with a very faint nucleus. L. Weinek (Prague, Czech Republic) observed on the 28th with a 16.3-cm refractor (74×) and described the comet as bright with a distinct nucleus of magnitude 11. W. Luther (Hamburg, Germany) saw the comet with a 25-cm refractor on the 29th and said the nucleus was magnitude 11. The only coma diameter estimates during these last days of 1885 came from Fabry on the 23rd and Lohse on the 24th, when they found it about 2' across.

With the comet steadily approaching the sun as 1886 began, the comet should have brightened slowly because its distance from the Earth was increasing. Unfortunately, there were no physical descriptions reported during the first half of January, despite numerous positional measurements from around the world. The moon was full on January 20, and its light was apparently responsible for the sparse observations during the period of January 12–21. Nevertheless, Bauschinger observed the comet with his 12.7-cm refractor on the 18th, and described it as faint and difficult to see because of moonlight. On the 21st he obtained another observation with the same instrument and described the comet as a bright, large, centrally condensed nebulosity without a tail. On the 22nd Engelhardt reported a star-like nucleus visible in his 30.5-cm telescope, while Bauschinger's 12.7-cm refractor indicated a distinct nucleus of magnitude 8–9. The distance between the comet and Earth increased to 1.7361 AU by January 26, and began decreasing thereafter. On that same date, Tupman estimated the comet was perhaps magnitude 6, with a coma between 3' and 5' across. Although a bright central condensation was present, Tupman reported no stellar component.

The comet's distances from the sun and Earth steadily decreased during February and its brightness had become great enough so that not even the full moon of the 18th hid it from view. Bauschinger continued to view the comet in his 12.7-cm refractor, but curiously indicated little change during the month. His observations on the 3rd, 25th, and 27th indicated the comet was bright and circular, with a nucleus of magnitude 8. J. Holetschek (Vienna, Austria) saw the comet on February 3 and determined the coma diameter as 1.8'. E. von Gothard (Astrophysical Observatory, Herény, Hungary) observed on the 25th and found the comet shaped like a parabola, with the beginnings of a tail.

Notable changes in the comet's appearance were reported during March. A short tail was seen by Weinek while observing the comet with a 9.8-cm refractor (48×) on March 1. He described it as extending about 3.75' toward PA 30°. Weinek added that the coma was about 2.3' in diameter with a total magnitude of 7.5. A. Kammermann (Geneva, Switzerland) observed the comet on March 4, with a 25.4-cm telescope, and reported the coma was parabolic in shape, with a short tail. W. Luther (Hamburg, Germany) also reported the parabolic coma on the 5th and added that it was bright and symmetrical. A very bright nucleus was also reported by observers early in the month. Weinek estimated it as magnitude 7.5 on the 8th, V. Knorre (Berlin, Germany) estimated it as magnitude 6–7 on the 9th, and Bauschinger estimated it as magnitude 7.5 on the 12th. Holetschek observed the comet with his 15-cm telescope on March 10 and 30 and estimated the total magnitude as 7.0 and 6.0, respectively. On the 11th, T. Lewis (Royal Observatory, Greenwich, England) used the 17-cm refractor and said the comet was very bright, with "a slight condensation to right of centre." Tupman reported the comet was seen at a low altitude on March 9 and 31. He described it as faint on the 9th and bright on the 31st. F. Porro (Torino, Italy) described the comet as beautiful and bright on the 31st, with a fan-shaped tail.

April could have been a spectacular month for this comet since it passed perihelion on the 6th and approached Earth during the entire month;

however, it was continually very near the horizon around the time twilight began and its full splendor must have suffered. Fortunately, there were a number of observers who braved the unfavorable observing geometry and obtained numerous observations. On the 1st, Engelhardt observed the comet in evening twilight with a 30.5-cm telescope. He estimated the coma as 2′ across, and the condensation as 0.5′ across. Luther saw the comet on the 2nd with a 25-cm refractor and said the nucleus was about magnitude 7.5, while only a slight tail was visible. Porro saw the comet with a refractor on the 3rd and noted a bright, well-defined nucleus from which he noted a trace of a luminous jet. Holetschek observed with a 15-cm telescope (26×) on the 6th and estimated the magnitude as 6.0. He also noted the coma was 3.5′ across, while the tail was three-fourths of a degree long. Tupman simply described the comet as "bright" on the same date and noted the altitude was then only 5.5°. G. A. P. Rayet (Bordeaux, France) saw the comet on the 7th and said the coma was 3.1′ across, with a bright circular inner coma 18″ in diameter, and a diffuse central nucleus about 4″ across. He added that the tail was 15′ long. E. E. Barnard (Vanderbilt University Observatory) saw the comet easily with the naked eye on April 8. Examining it with a 15-cm refractor, he described the comet as "remarkably beautiful," with a bright, star-like, yellowish nucleus and a long, slender, straight tail extending toward PA 314.17°. On the same date, Holetschek estimated the magnitude as 5.5, while the coma was 2.7′ across and the tail was 1.5° long. Tupman saw the comet on the 10th and compared the brightness of its head to ι Andromedae (magnitude 4.29 – SC2000.0) and 15 Andromedae (magnitude 5.59 – SC2000.0). He said, "It was a little brighter than 15 and much inferior to ι . . . from which I infer that the Comet was equal to a star of 5 or 5.1 magnitude." Tupman noted that the tail was conspicuous and more than 3° long, and added that it was narrow and straight, and was "a little broader at the extremity than at the head." Bauschinger noted it was then visible to the naked eye, while observations with a 12.7-cm refractor revealed a nucleus of magnitude 5 and a bright tail 1° in length. Barnard also saw it with his naked eye and noted a tail about 2° long. With a 15-cm refractor he traced the tail for 5° toward PA 315.30°. Porro estimated the magnitude of the nucleus as 3 on the 11th. Rayet saw the comet again on the 13th and said the nucleus was slightly more diffuse, with a diameter of about 7″, while the coma was 4.8′ across.

The comet attained its most northerly declination of +40° on April 13. Up to this time the comet was heading westward, but after the 13th it made a quick turn around and began heading southeastward. The full moon of April 18 did nothing to discourage observations, with Holetschek making a naked-eye observation on the 19th and estimating the magnitude as 3. Rayet saw the comet on the 21st and said the sky was still brightly lit by the moon. He noted the nucleus was "very brilliant and slightly yellowish," with a diameter little changed from the 13th. The nucleus was still located within a bright inner coma, which seemed brighter toward the former part of the comet. On the 22nd Holetschek again made a naked-eye observation and estimated the magnitude as 2, while H. Geelmuyden (Christiania, now Oslo, Norway) said the tail was 1.3° long. Bauschinger found the comet easily visible to the naked eye near β Andromedae on April 23. He also observed the comet with a 27-cm refractor and said the nucleus was very bright and magnitude 3 or 4. On the 24th and 25th naked-eye observations were made by Knorre, who estimated it as 3rd magnitude, M. Schnauder (Leipzig, Germany), who found it between magnitude 3 and 4, and F. Gonnessiat (Lyon, France), who estimated it as 3. Gonnessiat also observed the comet with a 15-cm telescope and said the nucleus was magnitude 4. On the 25th, Gonnessiat said the tail extended 5° toward PA 335°.

The comet was last detected in morning twilight in the Northern Hemisphere on April 26.12, when C. F. Pechüle (Copenhagen, Denmark) saw it with the naked eye as a star-like point with a narrow tail 10° long, and on April 27.14, when Holetschek saw it about 20 minutes before sunrise. The comet was then about 14° from the sun. The angular distance between the comet and sun continued to decrease to a minimum of 9.4° on the 28th. Holetschek searched for the comet on the evening of the 27th and on the morning of the 28th, but saw no trace of it.

The comet passed closest to Earth on May 1 and was first recovered in evening twilight on May 1.72 by both David Gill (Royal Observatory, Cape of Good Hope, South Africa) and L. A. Eddie (Grahamstown, South Africa). L. Cruls (Rio de Janeiro, Brazil) saw the comet on May 2 and reported the nucleus as magnitude 5, while on the 13th he described the nucleus as diffuse. It seems possible that Cruls was describing the condensation rather than the nucleus during the first half of May, as several other astronomers typically described the nucleus as star-like with a magnitude of 9–9.5. Despite this, however, Cruls still made some valuable observations during May. He reported a tail length of 5–6° on the 2nd and estimated the coma diameter as 20′ on the 2nd and 10′ on the 13th. Eddie also saw the comet on the 2nd. It was first detected with optical aid in twilight about 18° above the western horizon. Eddie said the comet was very conspicuous, with a coma 15′ across and a perfectly-straight, sharply-defined tail extending 9° and widening to about 1.5° at the end. He added that the comet "shone with a pale straw-coloured light." The telescope also revealed a bright, well-condensed nucleus from which "luminous gas-jets" were seen extending toward the sun for a short distance before bending back and streaming in the opposite direction. Observations were made every day during May, including the days surrounding the full moon of the 18th. Eddie also obtained an excellent view of the comet on the 4th. He first spotted it with the naked eye in twilight. In dark skies the tail was over 6° long and "seemed slightly curved, with the convex and more sharply defined margin on the south." In his 24-cm Calver telescope, Eddie noted the nucleus "appeared very bright and condensed, and sharply defined, and of a ruddy brown colour, surrounded by a pale yellow coma of hyperbolic form, with its vertex almost pointed, and spreading out considerably in a lateral direction, so as to give the appearance of its possessing two luminous wings." Eddie noted a tail length of only 2° on the 7th. W. H. Finlay (Royal Observatory, Cape of Good Hope, South Africa) observed the comet with the 18-cm refractor on the 4th and noted a stellar nucleus of magnitude 9. He made further observations on the 12th and 13th and noted a stellar nucleus of magnitude 9.5. H. C. Russell (Sydney, Australia) saw the comet on the 14th with a 29.2-cm refractor and said the nucleus looked like a 9th-magnitude star. Technically, the final Northern Hemisphere observations were actually made on May 5 and 7, by Captain William Randall. He was sailing on the ship *Earl of Shaftesbury* and was in the eastern Atlantic Ocean just a few degrees of latitude above the equator. On both evenings he measured the comet's position from the stars Sirius and Bellatrix with a sextant.

The comet faded rapidly during June as its distances from the sun and Earth continued to increase. Russell obtained several observations with a 29.2-cm refractor and reported both the comet and nucleus as "faint" on several occasions. A similar observation was also made by Finlay on the 22nd, while using a 17.8-cm refractor. Moonlight also affected comet observations. With the moon full on June 16, observations virtually ceased during the period of June 13–19.

Several astronomers began losing the comet during the first days of July. Interestingly, Russell reported the comet was "looked for but not seen" with his 29.2-cm refractor on July 3.4, but noted a star which "looked hazy," and suggested the star may have been shining through the comet's coma. He gave the star's position as $\alpha = 9^h 29^m 09^s$, $\delta = -44° 03' 03''$, and the Author notes this is quite close to a 10th-magnitude star. The Author's calculations reveal the comet virtually passed over this star on July 3.45. Moonlight had also become a major factor. Finlay saw the comet on the 6th and 8th, but noted it appeared very faint because of moonlight. Following the 8th, observations stopped as the moon grew brighter each day. It was full on July 16 and observations finally resumed on the 19th. After this date only two astronomers were still observing the comet: Finlay, at the Cape, and Jefferson, at Cordoba (Argentina).

Finlay saw the comet on several occasions during July using the 17.8-cm refractor. On the 5th he noted the comet was near a 10.5-magnitude star and appeared very faint. On the 6th and 8th he said the comet appeared very faint because of moonlight. He described the comet as small and faint on July 26. The comet was last detected on July 30.77, when Finlay found it with a 17.8-cm refractor and gave its position as $\alpha = 10^h 06.8^m$, $\delta = -48° 21'$. He said the comet appeared small and faint.

Several astronomers studied the spectrum of this comet. In particular, G. Müller (Potsdam Astrophysical Observatory, Germany) wrote that the continuous spectrum dominated. N. de Konkoly (O'Gyalla Observatory, Hungary) observed a faint continuous spectrum on December 9. Gothard noted a continuous spectrum on March 10, but by April 26 he noted the three bright cometary bands at 5612 Å, 5164 Å, and 4725 Å. Rayet and

C. Trépied (Alger, now al-Jazâ'ir, Algeria) also noted the three cometary bands were present during April.

Some astronomers began checking comets for a phase effect similar to that seen in asteroids. Müller remarked that his observations indicated no effect of phase. He suggested that this may be due to "a variation in the inherent light of the comet as it approaches the sun and earth, or we may assume that the nucleus is made up of discrete particles by which the phase phenomena must to a great extent be modified."

It should be noted that Chinese astronomers also observed this comet. The Chinese text *Ch'ing Ch'ao Hsü Wên Hsien Thung Khao*, which was written by Liu Chin-Tsao early in the twentieth century, noted that a "broom-star" appeared sometime during the period of 1886 May 4 to June 1.

The first parabolic orbit was calculated by H. Oppenheim using positions from December 1, 4, and 7. He determined the perihelion date as 1886 March 10.26. During the remainder of December, astronomers indicated their difficulty in pinpointing the perihelion date when some rather discordant dates were published. Some examples were March 30.73 by L. Schulhof, April 24.74 by S. Oppenheim, April 20.80 by Schulhof, and April 11.02 by S. Oppenheim. In the 1886 January 5 issue of the *Astronomische Nachrichten*, H. Oppenheim published a new orbit using positions obtained up to December 26, which gave a perihelion date of April 4.64, which was less than two days from what would ultimately prove to be the actual date. Additional orbits were calculated by H. Oppenheim, S. Oppenheim, L. Estes, A. Svedstrup, and A. Donner.

Using positions spanning the period of December 8 to June 7, J. Morrison (1887) was the first astronomer to realize the comet was moving in a hyperbolic orbit. He gave the perihelion date as April 6.45 and the eccentricity as 1.00047857. The orbit was improved by E. Redlich (1911), using positions obtained through June 22. Definitive orbits using positions spanning the entire period of visibility have been calculated by B. G. Marsden (1978) and S. Nakano (2002). Marsden applied perturbations by all nine planets, while Nakano applied perturbations by Mercury through Neptune, as well as the minor planets Ceres, Pallas, and Vesta. Marsden determined nongravitational terms of $A_1 = +2.6$ and $A_2 = -0.2$, and gave the eccentricity as 1.000272. Nakano determined nongravitational terms of $A_1 = +2.560$, and $A_2 = -0.35341$, and gave the eccentricity as 1.0002740. Nakano's orbit is given below.

T	ω	Ω (2000.0)	i	q	e
1886 Apr. 6.4542 (TT)	126.5878	37.9630	82.6283	0.642363	1.000274

ABSOLUTE MAGNITUDE: $H_{10} = 5.2$ (V1964)

FULL MOON: Nov. 22, Dec. 21, Jan. 20, Feb. 18, Mar. 20, Apr. 18, May 18, Jun. 16, Jul. 16, Aug. 14

SOURCES: L. Fabry and M. Loewy, *BA*, **2** (1885 Dec.), pp. 564–5; R. Copeland, *MNRAS*,

46 (1885 Dec.), p. 59; E. A. B. Mouchez and C. F. Pechüle, *AN*, **113** (1885 Dec. 5), p. 119; H. Oppenheim, *AN*, **113** (1885 Dec. 11), p. 135; A. Abetti, L. de Ball, N. C. Dunér, E. Millosevich, A. Kammermann, E. W. L. Tempel, W. Luther, B. von Engelhardt, J. Lamp, V. Knorre, E. A. Lamp, and L. Schulhof, *AN*, **113** (1885 Dec. 15), pp. 147–52; S. Oppenheim, *AN*, **113** (1885 Dec. 21), p. 167; J. G. Lohse and L. Schulhof, *AN*, **113** (1885 Dec. 22), pp. 205–8; F. Laschober, E. Millosevich, V. Cerulli, B. von Engelhardt, W. Winkler, J. Lamp, V. Knorre, W. Luther, and S. Oppenheim, *AN*, **113** (1885 Dec. 30), pp. 235–8; R. Copeland, L. Fabry, and H. Oppenheim, *AR*, **24** (1886 Jan.), pp. 22–3; A. L. N. Borrelly, J. E. Coggia, and V. A. Lebeuf, *BA*, **3** (1886 Jan.), pp. 27–9; A. M. W. Downing, *MNRAS*, **46** (1886 Jan.), pp. 125–6; F. Gonnessiat, *CR*, **102** (1886 Jan. 4), pp. 39–40; H. Oppenheim, *AN*, **113** (1886 Jan. 5), p. 255; S. Oppenheim and V. A. Lebeuf, *CR*, **102** (1886 Jan. 25), pp. 197–9; E. Weiss, *AN*, **113** (1886 Jan. 26), p. 317; J. Bauschinger, *AN*, **113** (1886 Jan. 30), p. 347; G. A. P. Rayet and F. Courty, *CR*, **102** (1886 Feb. 8), pp. 305–6; J. Lamp, H. P. Hollis, and W. Luther, *AN*, **113** (1886 Feb. 11), pp. 363–6; H. Oppenheim and S. Oppenheim, *AN*, **113** (1886 Feb. 13), pp. 381–4; N. de Konkoly and H. Oppenheim, *AN*, **113** (1886 Feb. 22), pp. 391, 397; H. H. Turner, *MNRAS*, **46** (1886 Mar.), p. 303; C. Trépied and V. A. Lebeuf, *CR*, **102** (1886 Mar. 1), pp. 492–3; J. Bauschinger, L. Estes, and A. Svedstrup, *AN*, **114** (1886 Mar. 6), pp. 43–6; F. Courty, *CR*, **102** (1886 Mar. 8), p. 544; V. A. Lebeuf, *CR*, **102** (1886 Mar. 15), pp. 596–7; A. Donner, *AN*, **114** (1886 Mar. 16), p. 63; C. Wagner and F. Schwab, *AN*, **114** (1886 Mar. 18), p. 77; E. Stuyvaert, *AN*, **114** (1886 Mar. 23), p. 91; C. Trépied, *CR*, **102** (1886 Mar. 29), pp. 731–2; E. A. Lamp and E. W. L. Tempel, *AN*, **114** (1886 Mar. 30), pp. 123–6; T. Lewis, *MNRAS*, **46** (1886 Apr.), pp. 348–9; A. Svedstrup and S. Oppenheim, *AN*, **114** (1886 Apr. 8), pp. 157–60; J. Lamp, G. L. Tupman, and S. Oppenheim, *AN*, **114** (1886 Apr. 16), p. 171; B. von Engelhardt, *AN*, **114** (1886 Apr. 21), pp. 205–8; G. A. P. Rayet, *CR*, **102** (1886 Apr. 27), p. 970; H. H. Turner, *MNRAS*, **46** (1886 May), pp. 399–401; C. Trépied, *CR*, **102** (1886 May 3), pp. 1009–10; J. Bauschinger, *AN*, **114** (1886 May 18), p. 315; F. Gonnessiat, *BA*, **3** (1886 May), pp. 236–7; C. Trépied, *BA*, **3** (1886 Jun.), p. 293; L. Cruls, *CR*, **102** (1886 Jun. 15), pp. 1364–5; L. A. Eddie, *MNRAS*, **46** (1886 Jun.), pp. 455–7; J. H. Price, *MNRAS*, **46** (1886 Jun.), pp. 457–8; G. Müller, *AN*, **114** (1886 Jun. 5), pp. 363–6; H. Geelmuyden, *AN*, **114** (1886 Jun. 15), p. 379; J. Holetschek, *AN*, **115** (1886 Jul. 28), pp. 41–4; C. W. Pritchett, *AN*, **115** (1886 Aug. 31), p. 107; H. C. Russell, *MNRAS*, **46** (1886, Supplementary Notice), pp. 495–6; J. Campbell, *MNRAS*, **46** (1886, Supplementary Notice), p. 498; D. S. Cromarty, *MNRAS*, **46** (1886, Supplementary Notice), p. 499; B. Peter, and M. Schnauder, *AN*, **115** (1886 Oct. 7), pp. 235–8; J. G. Lohse, *MNRAS*, **47** (1886 Nov.), pp. 28–9; E. E. Barnard and F. Porro, *AN*, **115** (1886 Nov. 19), pp. 323–6, 331–4; C. F. Pechüle, *AN*, **115** (1886 Nov. 25), pp. 385, 389; F. Schwab and J. Tebbutt, *AN*, **115** (1886 Nov. 25), pp. 391–6; L. Weinek, *AN*, **116** (1886 Dec. 22), p. 57; W. Randall, *MNRAS*, **47** (1887 Jan.), p. 117; E. von Gothard, *AN*, **116** (1887 Jan. 22), p. 121; R. L. J. Ellery and P. Baracchi, *AN*, **116** (1887 Feb. 3), pp. 145–50; W. H. Finlay and D. Gill, *MNRAS*, **47** (1887 Mar.), pp. 277–9; J. M. Schaeberle, *AN*, **114** (1886 Mar. 6), p. 43; W. H. Finlay and D. Gill, *AN*, **116** (1887 Apr. 17), pp. 305, 315; S. Oppenheim and J. Palisa, *AN*, **116** (1887 Apr. 23), pp. 345–8; J. Morrison, *MNRAS*, **47** (1887 May), pp. 437–8; V. Knorre, *AN*, **117** (1887 Aug. 3), pp. 171, 197; P. H. Harzer and M. Nyrèn, *AN*, **117** (1887 Aug. 20), pp. 253–6; G. L. Tupman, *MNRAS*, **47** (1887, Supplementary Notice), pp. 546–7; W. Luther, *AN*, **120** (1888 Nov. 3), pp. 113, 131; E. Redlich, *AN*, **187** (1911 Feb. 27), pp. 193–300; V1964, p. 62; HA1970, p. 88; B. G. Marsden, Z. Sekanina, and E. Everhart, *AJ*, **83** (1978 Jan.),

p. 66; B. G. Marsden, *CCO*, 3rd ed. (1979), pp. 19, 47; SC2000.0, pp. 595, 596; S. Nakano, *Nakano Note* No. 861 (2002 May 10).

C/1885 X2
(Barnard)

1886 II = 1885e

Discovered: 1885 December 4.13 ($\Delta = 1.76$ AU, $r = 2.72$ AU, Elong. $= 161°$)
Last seen: 1886 July 26.85 ($\Delta = 1.66$ AU, $r = 1.76$ AU, Elong. $= 78°$)
Closest to the Earth: 1886 May 27 (0.3351 AU)
Calculated path: TAU (Disc), CET (Dec. 29), ARI (Jan. 11), TRI (Mar. 15), AND (Apr. 12), TRI (May 12), ARI (May 17), TAU (May 23), ERI (May 25), LEP (May 30), COL (Jun. 1), PUP (Jun. 9), VEL-CAR (Jun.21), VEL (Jun. 23)

E. E. Barnard (Nashville, Tennessee, USA) discovered this comet with a 15-cm refractor on 1885 December 4.13. The position was estimated as $\alpha = 4^h\ 21.9^m$, $\delta = +4°\ 45'$. He described the round comet as "faintish with indefinite brightening near the middle." Barnard said motion was soon detected.

The comet was discovered about 5 months prior to perihelion and nearly 6 months prior to its closest approach to Earth. Nevertheless, it was well observed during the remainder of December. B. von Engelhardt (Dresden, Germany) observed with a 30-cm refractor on the 5th and described the comet as rather bright, but somewhat fainter than comet C/1885 X1. It was 0.5' across, but although there was no real nucleus, a central condensation was present. For that same night, Barnard observed with a 15-cm refractor and noted, "A very small and difficult star-like nucleus in the preceding part of nebulosity." C. A. Schultz-Steinheil (Uppsala, Sweden) described the comet as very faint with a nucleus on the 6th, while J. G. Lohse (Wigglesworth Observatory, Scarborough, England) noted the comet was faint and small in his 39-cm refractor. Lohse added, "The coma extends farther on the following side than in other directions. The central bright part appears granulated, and equals in brightness about an 11th mag. Star. The diameter of the comet is 4 to 5 seconds in time." On the 7th, Engelhardt remarked that the comet was "moderately bright, with a central condensation," while Barnard suspected a short, faint tail. J. Lamp (Bothkamp, Germany) said the comet was brighter than comet C/1885 X1 on the 8th. That same night, R. Copeland (Dun Echt Observatory, Scotland) observed with a 38-cm refractor and said the comet was 1' across and "considerably brighter in the middle." On the 9th, L. Weinek (Prague, Czech Republic) estimated the nuclear magnitude as 11. E. W. L. Tempel (Arcetri, Italy) also noted the comet was brighter than comet C/1885 X1 when he observed on December 10. He added that the small nebulosity was similar to a nebula of class II with a small star in the middle. That same night, Lohse said the comet was "decidedly more extended on the following side, and about the same brightness" as on December 6. On the 11th, Barnard said the comet was considerably brighter than during previous observations and a small nucleus was in the preceding part of the nebulosity. There was also a short, faint tail. Engelhardt said the comet was brighter on the 12th than on the 5th and 7th, and exhibited a well-defined condensation located in the preceding half of the coma. That same night, J. Bauschinger (Munich, Germany) observed with a 13-cm refractor and said the comet was small and faint, with a trace of an 11th-magnitude nucleus. On the 15th, H. H. Turner (Royal Observatory, Greenwich, England) observed with a 17-cm refractor and said the comet was a small, round object, with no nucleus or tail. A. L. N. Borrelly (Marseille, France) observed with a 25.8-cm telescope on the 16th and said the comet was "of an irregular shape, 2′ in extent, little brilliant, with no nucleus."

Moonlight seriously hampered observations for several days, and then Barnard described the comet as "pretty dim" on the 26th, with a very faint diffused tail extending about 1′ in the following direction. There was also a stellar nucleus in the preceding part of the nebulosity. On the 27th, A. M. W. Downing (Royal Observatory, Greenwich, England) observed with a 17-cm refractor and described the comet simply as faint, with a nucleus, while G. L. Tupman (Harrow, England) said the comet was "pretty bright," but not as bright as C/1885 X1. N. de Konkoly (O'Gyalla Observatory, Hungary) added that the comet was "quite faint, rather circular, with a bright nucleus." Tupman gave the diameter as 90″ and noted a "sub-stellar nucleus." Engelhardt said the comet was rather bright in a 30-cm refractor on the 28th and noted a condensation and traces of a fan-shaped tail. That same night, Lohse made a particularly difficult observation. Noting it was occasionally snowing and storming, he observed the comet through breaks in the clouds with a 39-cm refractor and said it was barely visible. He detected that it was "elongated on the following side" and said the nucleus was brighter than C/1885 X1. Tempel saw a small tail 15′ long on the 31st.

On 1886 January 2, Lohse saw the comet in a hazy sky with his 39-cm refractor and noted it appeared "very faint." On the 5th, J. Palisa (Vienna, Austria) said the nucleus was about magnitude 11 and was very eccentrically situated within the coma. Moonlight again prevented physical descriptions, until Bauschinger observed the comet with a 13-cm refractor on January 21 and detected a faint nebulosity without a nucleus in moonlight. On the 22nd, Bauschinger said the comet was well seen with a distinct nucleus of magnitude 9. W. Luther (Hamburg, Germany) observed under hazy skies on the 24th and said the comet appeared diffuse. On the 28th, Tupman said the coma was 90″ across and extended toward PA 120° or 130°.

On February 1, Tupman said the coma was 90″ across and exhibited a possibly stellar nucleus of magnitude 9.5 or 10. Barnard said the comet was very diffuse, very dim, and vague, with a very faint extension just north of the following direction. Bauschinger said the comet appeared as a faint, round nebulosity, with a nucleus of magnitude 9–10 on the 3rd. Bauschinger said the comet was very poorly observed because of moonlight on the 8th, while E. von Gothard (Astrophysical Observatory, Herény, Hungary) said the comet was poorly defined with a diffuse nucleus. Bauschinger described the comet as a feeble nebulosity without a nucleus in a 13-cm refractor on the 21st. On the 23rd, Barnard said the comet was brighter than when previously seen, with a tail about one-fourth of a degree long. Bauschinger said the comet was faint on the 24th, with a hardly visible nucleus.

On March 5, Barnard said the comet was "pretty much brighter" with a tail at least one-half of a degree long and a nucleus that appeared as a tiny star. On the 7th, Turner said the comet was fainter than comet C/1885 X1, with a suspected nucleus. Engelhardt said the comet was moderately bright, condensed, and 1' in diameter on the 8th. Tupman described it as 2–3' across on the 9th, while V. Knorre (Berlin, Germany) said the comet appeared in the finder of the 24.4-cm refractor as about magnitude 9. He also noted the faint beginning of a tail. On March 9, 10, and 11, Bauschinger described the comet as round, with a diffuse nucleus. Barnard said the comet was easily seen in the 4-cm finder on the 23rd. He noted a short tail 4–5' long, with a faint extension that was near 0.5° long. On the 24th, Engelhardt said the comet was bright, condensed, and 3' in diameter. That same night, Barnard said the comet was bright, with a very distinct, broad tail about 10' long. L. Weinek (Prague, Czech Republic) said the comet was bright, with a round coma 1.9' across and a nucleus of magnitude 8.5, while, Luther said the nucleus was magnitude 9. Bauschinger described the comet as bright, with a nucleus of magnitude 9 on the 25th. J. Holetschek (Vienna, Austria) observed with a 15-cm telescope ($26\times$) and estimated the total magnitude as 8.5 on the 28th and 29th, 8.0 on the 30th, and 7.5 on the 31st. He added that the coma was 1.7' across on the 30th. Weinek observed on the 28th and said the comet was quite bright, with a coma 2.5' across. The nucleus was magnitude 8.5.

The comet was well observed during April as it was nearing its closest approach to the sun and Earth, but it was dipping into evening twilight and was generally observed at a low altitude. Toward the end of the month it was visible in the morning sky. A fine series of magnitude estimates was made by Holetschek using a 15-cm telescope ($26\times$). He gave it as 7.5 on the 1st, 7.3 on the 2nd, 7.7 on the 3rd, 7.2 on the 7th, 7.0 on the 8th, 7.5 on the 11th, 6.7 on the 18th, 6.2 on the 22nd, and 5.5 on the 27th. Other magnitude estimates were 5 by Tupman on the 3rd and 7 by Engelhardt on the 22nd. Barnard said the comet was brighter than M31 (magnitude 3.5 – NGC2000.0) on the 30th. The nucleus was not consistently observed during the month. Engelhardt reported it as a stellar condensation on the 1st. On the 2nd, Barnard was not certain he was seeing a nucleus in a 15-cm refractor, while F. Porro (Torino, Italy) estimated it as magnitude 7–8. E. A. Lamp (Kiel, Germany) reported a nucleus on the 7th. Barnard reported an ill-defined nucleus on the 8th. Engelhardt said a bright condensation, but no nucleus was seen in the 30-cm refractor on the 22nd. Bauschinger said the finder of the 13-cm refractor revealed a diffuse nucleus of about magnitude 6. The coma diameter was given as at least 3′ by Tupman on the 3rd, 1.5′ by Holetschek on the 7th, and 1.7′ by Holetschek on the 22nd. The tail was described as fan-shaped and 2′ long on the 1st, while Barnard described it as faint and about 0.5° long on the 2nd. Barnard reported the tail was rather faint on the 8th and invisible at low altitude on the 14th. Engelhardt said the tail was short and fan-shaped on the 22nd. Knorre observed with a 24.4-cm refractor on the 24th and said the tail was elliptical in shape, but not over 2.5′ long. Bauschinger reported an extremely faint, short tail on the 26th. On April 30, Barnard saw a faint, slender tail extending at least 2°.

The comet passed perihelion early in May and passed closest to Earth near the end of the month. Shortly before mid-month, it began to move rapidly southward. The comet attained its most northerly declination of +41° on the 1st. On that same morning, Barnard said he suspected he saw it with the naked eye, while the 15-cm refractor revealed the head and tail were "remarkably slender and straight", but there was no decided nucleus. H. Geelmuyden (Christiania, now Oslo, Norway) said binoculars revealed the comet's head was nearly as bright as τ Andromedae (magnitude 4.94 – SC2000.0), while Barnard said the comet was small, round, and bright, although the tail was lost in dawn. E. A. Lamp said the condensation was eccentrically situated toward the south on the 4th, while the tail was straight. Barnard noted the tail was longer and broader on the 4th than on the 2nd. He wrote, "Along the axis, near head, the tail is much brighter as if a ray of light shot backward from the nucleus." Barnard also noted "faint indications of stellar nucleus." Holetschek gave the magnitude as 5.0 on the 8th and also noted a coma 2.7′ across and a tail about 2.5° long. That same morning, Barnard said the comet was visible to the naked eye as "an indefinite hazy spot," while the tail was brightest near the following edge and at least 3° long. It was also straighter and broader than when previously seen and extended toward PA 312.35°. The nucleus was bright, but not stellar. Also on the 8th, Gothard said the comet was quite bright and visible to the naked eye despite morning twilight. The nucleus was yellow, while the tail was narrow and straight. Barnard said the comet was "dimly seen with the naked eye" on the 13th. The tail was again longer and broader than when previously seen and extended toward PA 300.1°. Among Northern Hemisphere observers, the comet was last seen by Holetschek on May 17.11 and by Barnard on May 17.37, when the solar elongation was 19°. Barnard said it was bright in the 15-cm refractor, with faint traces of a tail. The first person in the Southern Hemisphere to see the comet was W. H. Finlay (Royal Observatory, Cape of Good Hope, South Africa), who first saw it on May 29.70. J. Tebbutt (Windsor, New South Wales, Australia) said the comet was plainly visible to the naked eye on May 31.

As the comet moved away from both the sun and Earth, only a handful of observers followed it from the Southern Hemisphere. Tebbutt said it was plainly visible to the naked eye on June 3, while an 11-cm refractor revealed a slight condensation on the 11th. On the 13th, Tebbutt said the comet nearly became invisible when it passed close to a star of magnitude 8. Finlay saw the comet on June 15, but noted that moonlight almost overpowered it. On June 27 and 29, Tebbutt said the comet was excessively faint in an 11-cm refractor.

Tebbutt reported the comet was "of the last degree of faintness" in his 11-cm refractor on July 1. Finlay obtained several observations during the remainder of the month with an 18-cm refractor. On the 5th he said the comet was faint, with very bad definition, while on the 6th he noted it was faint in moonlight. On the 8th, Finlay said the comet was difficult to see because it was, "Large, faint and diffused." On the 20th the comet appeared "Very faint, and so mixed up with two stars that it was almost impossible to fix its position." On the 24th, he said the comet was "exceedingly faint." The comet was last detected on July 26.85, when Finlay gave the position as $\alpha = 9^h 53.7^m$, $\delta = -56° 11'$. He described the comet as "an exceedingly faint and ill-defined patch of light."

Gothard checked the spectrum of the comet on April 3. He saw the three cometary bands at 5648 Å, 5168 Å, and 4735 Å, with the middle band being very sharply defined and bright. Only a weak spectrum was noticed by de Konkoly on December 27.

The first parabolic orbit was calculated by H. Oppenheim using positions from December 5, 7, and 10. He determined the perihelion date as 1886 May 14.61. Similar orbits were independently calculated by C. N. A. Krueger and J. von Hepperger a short time later. Both Krueger and Hepperger remarked on the similarity between this orbit and that of the comet C/1785 E1. Hepperger revised his calculations a few days later and even determined a hypothetical elliptical orbit with a period of 101 years, based on the assumption that this was C/1785 E1. Krueger followed a short time later with a parabolic orbit and said the suggestion that this was a return of C/1785 E1 was not valid.

The first hyperbolic orbit was calculated by J. Morrison using positions spanning the period of December 12 to March 1. He determined the perihelion date as May 3.78 and the eccentricity as 1.0006711. About four months later, A. Thraen calculated a very similar orbit, but with an eccentricity of 1.000398. Thraen (1893) revised his calculations and gave the eccentricity as 1.00022867. This orbit is given below.

T	ω	Ω (2000.0)	i	q	e
1886 May 3.7868 (UT)	119.6224	69.9096	84.4385	0.479270	1.000229

ABSOLUTE MAGNITUDE: $H_{10} = 6.6$ (V1964)

FULL MOON: Nov. 22, Dec. 21, Jan. 20, Feb. 18, Mar. 20, Apr. 18, May 18, Jun. 16, Jul. 16, Aug. 14

SOURCES: R. Copeland, *MNRAS*, **46** (1885 Dec.), p. 59; E. E. Barnard, *AN*, **113** (1885 Dec. 5), p. 119; B. von Engelhardt, C. A. Schultz-Steinheil, F. Engström, E. A. Lamp,

V. Knorre, and J. Lamp, *AN*, **113** (1885 Dec. 11), p. 135; W. Luther, E. W. L. Tempel, V. Knorre, J. Lamp, E. A. Lamp, H. Oppenheim, C. N. A. Krueger, and J. von Hepperger, *AN*, **113** (1885 Dec. 15), pp. 149–52; B. von Engelhardt, E. Millosevich, V. Cerulli, L. Weinek, and W. Luther, *AN*, **113** (1885 Dec. 22), p. 207; J. von Hepperger and C. N. A. Krueger, *AN*, **113** (1885 Dec. 30), p. 237; E. E. Barnard and H. Oppenheim, *AR*, **24** (1886 Jan.), pp. 22–3; A. L. N. Borrelly and J. E. Coggia, *BA*, **3** (1886 Jan.), pp. 27–8; A. M. W. Downing and H. H. Turner, *MNRAS*, **46** (1886 Jan.), pp. 125–6; B. von Engelhardt and H. Oppenheim, *AN*, **113** (1886 Jan. 5), pp. 253–6; C. N. A. Krueger, *AN*, **113** (1886 Jan. 21), p. 303; E. Weiss and J. von Hepperger, *AN*, **113** (1886 Jan. 26), p. 317; J. Bauschinger, *AN*, **113** (1886 Jan. 30), p. 347; F. Courty, *CR*, **102** (1886 Feb. 8), p. 306; J. Lamp, H. P. Hollis, W. Luther, and C. N. A. Krueger, *AN*, **113** (1886 Feb. 11), pp. 363–8; J. von Hepperger, *AN*, **113** (1886 Feb. 13), pp. 381–4; N. de Konkoly and B. von Engelhardt, *AN*, **113** (1886 Feb. 22), pp. 391, 395; H. H. Turner, *MNRAS*, **46** (1886 Mar.), p. 303; J. Bauschinger, *AN*, **114** (1886 Mar. 6), p. 43; F. Courty, *CR*, **102** (1886 Mar. 8), p. 543; F. Schwab, *AN*, **114** (1886 Mar. 18), p. 77; E. Stuyvaert, *AN*, **114** (1886 Mar. 23), p. 91; C. Trépied, *CR*, **102** (1886 Mar. 29), pp. 731–2; E. W. L. Tempel, *AN*, **114** (1886 Mar. 30), p. 125; H. H. Turner and H. P. Hollis, *MNRAS*, **46** (1886 Apr.), pp. 348–9; J. Morrison, *The Observatory*, **9** (1886 Apr.), pp. 157–8; B. von Engelhardt, *AN*, **114** (1886 Apr. 21), pp. 205–8; H. P. Hollis, T. Lewis, and H. H. Turner, *MNRAS*, **46** (1886 May), pp. 399–401; H. Geelmuyden and B. von Engelhardt, *AN*, **114** (1886 Jun. 15), pp. 379–82; J. Holetschek and J. Bauschinger, *AN*, **115** (1886 Jul. 28), pp. 41–4, 47; A. Thraen, *AN*, **115** (1886 Aug. 17), p. 79; E. A. Lamp and C. W. Pritchett, *AN*, **115** (1886 Aug. 31), p. 107; H. C. Russell and J. A. Pollock, *MNRAS*, **46** (1886, Supplementary Notice), p. 497; B. Peter and Hahn, *AN*, **115** (1886 Oct. 7), p. 235; J. G. Lohse, *MNRAS*, **47** (1886 Nov.), pp. 28–9; E. E. Barnard and F. Porro, *AN*, **115** (1886 Nov. 19), pp. 323–6, 331–4; C. F. Pechüle and F. Schwab, *AN*, **115** (1886 Nov. 25), pp. 387–92; L. Weinek, *AN*, **116** (1886 Dec. 22), p. 57; E. von Gothard and J. Tebbutt, *AN*, **116** (1887 Jan. 22), pp. 121–6; R. L. J. Ellery, E. J. White, Pringle, and P. Baracchi, *AN*, **116** (1887 Feb. 3), pp. 147–50; W. H. Finlay, *MNRAS*, **47** (1887 Mar.), pp. 281–3; W. H. Finlay, *AN*, **116** (1887 Apr. 17), p. 307, 315–16; J. Palisa and S. Oppenheim, *AN*, **116** (1887 Apr. 23), pp. 345–7; V. Knorre, *AN*, **117** (1887 Aug. 3), pp. 171, 197; P. H. Harzer and M. Nyrèn, *AN*, **117** (1887 Aug. 20), p. 255; G. L. Tupman, *MNRAS*, **47** (1887, Supplementary Notice), pp. 547–8; W. Luther, *AN*, **120** (1888 Nov. 3), pp. 113–16, 131; A. Thraen, *AN*, **132** (1893 May 1), pp. 241–84; A. Thraen, *The Observatory*, **16** (1893 Jun.), p. 241; A. Thraen, *AN*, **136** (1894 Aug. 29), pp. 133–8; V1964, p. 62; NGC2000.0 (1988), pp. 3, 5; SC2000.0, p. 34.

C/1885 Y1   *Discovered:* 1885 December 26.99 ($\Delta = 1.87$ AU, $r = 1.19$ AU, Elong. $= 34°$)
(Brooks)   *Last seen:* 1886 March 1.81 ($\Delta = 2.54$ AU, $r = 1.84$ AU, Elong. $= 36°$)
    Closest to the Earth: 1885 December 12 (1.8461 AU)
1885 V = 1885f   *Calculated path:* AQL (Disc), DEL (Jan. 2), EQU (Jan. 11), DEL-PEG (Jan. 14), EQU-PEG (Jan. 15)

W. R. Brooks (Phelps, New York, USA) discovered this comet around 1885 December 26.99, when it was about 2.5° south of β Aquilae and very near the horizon. It was confirmed by E. C. Pickering (Harvard college Observatory, Cambridge, Massachusetts, USA) on December 28.04, when he gave the position as $\alpha = 19^h\ 55.7^m$, $\delta = +4°\ 08'$. Pickering gave the daily motion as roughly $+2°\ 45'$ in α and $+2°$ in δ. E. E. Barnard (Nashville, Tennessee, USA) independently discovered the comet in the evening sky on December 28.04. He described it as "pretty bright, round and much condensed, but no tail was visible." Barnard informed L. Swift and Harvard College Observatory of his find, but Swift responded by telegram a few hours later telling him that Brooks had already reported this comet. Pickering next saw the comet on December 28.97 and described it as circular and 3' across, with a magnitude of 9. He added that there was a "strong eccentric condensation," but no tail. The comet was already a month past perihelion and about two weeks past its closest approach to Earth.

On December 30, Barnard said the comet had an excessively faint, short tail, as seen in a 15-cm refractor (120×). B. von Engelhardt (Dresden, Germany) described the comet as rather bright in a 13-cm finder, while the 30-cm refractor revealed a centrally condensed coma 2.5' across. E. A. Lamp (Kiel, Germany) described the comet as "rather bright and elliptical, with the large axis directed north to south." W. Luther (Hamburg, Germany) said the comet appeared as a round nebulosity in the 25-cm refractor, with a nucleus of magnitude 10. On December 31, Barnard said the comet was less bright in the 15-cm refractor.

On the first day of the new year, C. Wagner (Kremsmünster, Austria) described the nucleus as rather bright, while Barnard said the entire comet was still "pretty bright," although fainter than when previously seen. Barnard also noted a very faint, diffuse tail and a small, ill-defined nucleus in the 15-cm refractor. A. Abetti (Padova, Italy) saw the comet during the period of January 1–4 and noted a faint nebulosity about 1' across that exhibited neither a nucleus or tail. G. Bigourdan (Paris, France) described the comet as a faint nebulosity about 2.5' across on January 5, with a small, faint nucleus that was slightly eccentric. No tail was seen. On January 6, J. Palisa (Vienna, Austria) described the comet as a faint, diffuse, nebulosity, about 2' in diameter, and with an eccentrically situated nucleus. J. Lamp (Bothkamp, Germany) observed under hazy skies on the 24th and described the comet as very faint. On the 22nd, Engelhardt observed with the 30-cm refractor and said the comet was very faint, diffuse, and without a condensation. On February 6, Barnard said the comet was faint and difficult to observe in the 15-cm refractor. He added that there was no definite condensation.

The comet was last detected on March 1.81, when C. F. Pechüle (Copenhagen, Denmark) gave the position as $\alpha = 23^h\ 43.2^m$, $\delta = +26°\ 54'$. Pechüle described the comet as extremely faint in the 27-cm refractor. L. Becker (Earl of Crawford's Observatory, Dunecht, Scotland) failed to find the comet on March 9 and 11, even though the ephemeris indicated it should still have been visible. Pechüle said the comet was invisible in the refractor on March 25.

The first parabolic orbit was calculated by Palisa using positions from December 28, 30, and January 1. He determined the perihelion date as

1885 November 28.71. Additional orbits by H. Oppenheim, J. Müller, J. Hackenberg (1889), and A. Berberich (1890) rolled the perihelion date back to November 26. F. Cohn (1894) calculated a definitive orbit, which gave a perihelion date of November 26.01. It is given below.

This comet was frequently referred to as "Brooks 2" in the *Astronomische Nachrichten* not long after its discovery. This was only to distinguish it from comet C/1885 R1 (Brooks) and is quite distinct from the short-period comet 16P/1889 N1 (Brooks 2).

T	ω	Ω (2000.0)	i	q	e
1885 Nov. 26.0099 (UT)	35.6003	263.8088	42.4420	1.079629	1.0

ABSOLUTE MAGNITUDE: $H_{10} = 6.3$ (V1964)

FULL MOON: Dec. 21, Jan. 20, Feb. 18, Mar. 20

SOURCES: E. C. Pickering, *AN*, **113** (1885 Dec. 30), p. 239; W. Luther, B. von Engelhardt, and E. A. Lamp, *AN*, **113** (1886 Jan. 5), p. 255; B. von Engelhardt, C. Wagner, E. A. Lamp, J. Palisa, and H. Oppenheim, *AN*, **113** (1886 Jan. 11), p. 271; G. Bigourdan and A. H. P. Charlois, *CR*, **102** (1886 Jan. 11), pp. 100–1; H. Oppenheim, *AN*, **113** (1886 Jan. 21), p. 303; H. Oppenheim, *AN*, **113** (1886 Jan. 26), p. 319; G. A. P. Rayet and F. Courty, *CR*, **102** (1886 Feb. 8), p. 307; J. Lamp, W. Luther, and H. Oppenheim, *AN*, **113** (1886 Feb. 11), pp. 363–8; J. Müller and B. von Engelhardt, *AN*, **113** (1886 Feb. 22), pp. 389, 395; A. Abetti, *AN*, **113** (1886 Feb. 22), p. 397; E. E. Barnard, *SM*, **5** (1886 Mar.), p. 89; G. A. P. Rayet, *CR*, **102** (1886 Mar. 8), p. 543; E. Stuyvaert, *AN*, **114** (1886 Mar. 23), p. 91; A. Berberich, *AN*, **124** (1890 Mar. 29), p. 149; E. W. L. Tempel, *AN*, **114** (1886 Mar. 30), p. 125; E. Millosevich, *AN*, **114** (1886 Apr. 21), p. 205; E. E. Barnard, *AN*, **115** (1886 Nov. 19), pp. 323, 327; C. F. Pechüle, *AN*, **115** (1886 Nov. 25), pp. 387–90; L. Becker, *MNRAS*, **47** (1887 Feb.), p. 159; J. Palisa, *AN*, **116** (1887 Apr. 23), pp. 345–7; V. Knorre, *AN*, **117** (1887 Aug. 3), p. 169; W. Luther, *AN*, **120** (1888 Nov. 3), pp. 113, 131; J. Hackenberg, *AN*, **121** (1889 Jun. 5), p. 365; F. Cohn, *AN*, **135** (1894 Mar. 12), pp. 17–32; V1964, p. 62.

C/1886 H1
(Brooks)

1886 V = 1886a

Discovered: 1886 April 28.00 ($\Delta = 1.36$ AU, $r = 1.10$ AU, Elong. $= 53°$)
Last seen: 1886 July 30.73 ($\Delta = 1.95$ AU, $r = 1.33$ AU, Elong. $= 39°$)
Closest to the Earth: 1886 June 16 (1.1046 AU)
Calculated path: CAS (Disc), PER (May 9), AND (May 15), PER (May 18), TAU (May 31), ORI (Jun. 10), MON (Jun. 20), HYA (Jul. 2)

W. R. Brooks (Phelps, New York, USA) discovered this comet in the evening sky on 1886 April 28.00 at a position of $\alpha = 0^h\ 15.0^m$, $\delta = +62°\ 00'$. He said the comet was moving slowly southeastward. Confirming observations were made by J. Palisa (Vienna, Austria) on April 29.84 and H. H. Turner (Royal Observatory, Greenwich, England) on April 29.98. Turner was using a 17-cm refractor and described the comet as fairly bright, with no nucleus.

The comet was approaching both the sun and Earth. As a result, it steadily brightened during May. The only two total magnitude estimates came from G. L. Tupman (Harrow, England). Using an 11-cm refractor, he estimated it as 8 or 9 (depending on the source) on May 6 and 5–6 when very near the horizon on May 20. Although the central condensation was frequently noted, few observers took note of the stellar nucleus. Several observers made nuclear magnitude estimates during the month. W. Luther (Hamburg, Germany) observed with a 25-cm refractor and gave it as 11 on the 3rd and 6th, and 7 on the 21st. J. Bauschinger (Munich, Germany) observed with a 27-cm refractor and gave it as 8 on the 19th and 20th, and 6–7 on the 22nd, 23rd, and 26th. The nucleus had not been reported before G. Bigourdan (Paris, France) suspected it near the center of the coma on May 2. B. von Engelhardt (Dresden, Germany) said the condensation was eccentrically situated within the northern portion of the coma on the 3rd and 5th. Also reporting a non-central nucleus during the period of May 1–12, were C. F. Pechüle (Copenhagen, Denmark) and E. J. M. Stephan (Marseille, France). Engelhardt noted on the 5th that the nucleus sometimes seemed composed of multiple bright points, while on the 9th he reported a granular nucleus which contained several bright points. The coma was generally reported as fairly round during the entire month, with the exceptions of Bauschinger who noted it seemed oval on the 2nd, and M. Schnauder (Leipzig, Germany), who said an oblong shape was suspected on the 3rd. Some of the diameter estimates were about 2′ by Bigourdan on the 1st, 1′ 40″ by Bauschinger and 1.5′ by Schnauder on the 2nd, 2′ by Engelhardt and E. W. L. Tempel (Arcetri, Italy) on the 3rd, and 2′ by Tupman on the 6th. During the period of May 1–6, Stephan estimated the coma diameter as about 1.5′. F. Gonnessiat (Lyon, France) estimated it as about 2′ on several occasions between May 2 and 8. Pechüle gave it as about 1′ during the period of May 7–12. Bauschinger gave the coma diameter as 2′ 20″ on May 22, 23, and 26. L. Weinek (Prague, Czech Republic) estimated it as 1′ on the 22nd in hazy skies. The comet became more difficult to observe as the month progressed as a result of its decreasing elongation from the sun. Palisa noted on the 29th that it was well observed in a 30-cm telescope in spite of bright twilight. The final Northern Hemisphere observation was made by Tupman on May 30.

The comet passed 12° from the sun on June 6 and remained lost in the sun's glare until early July. J. Tebbutt (Windsor, New South Wales, Australia) spotted the comet with his 11-cm refractor on the 3rd and described it as 1.5′ across, with a slight condensation. Tebbutt said it was 1.5′ across on the 4th, with a slight condensation. W. H. Finlay (Royal Observatory, Cape of Good Hope, South Africa) observed the comet with an 18-cm refractor on the 5th and described it as a "bright circular mass" about 45″ across which "gradually condensed towards the centre." Tebbutt found the comet "hardly distinguishable in the 11-cm refractor on the 11th, 12th, and 21st. Finlay was the sole observer for the remainder of the month and continually used the

18-cm refractor. He found the comet faint in bright moonlight on the 14th and described it as a "diffused mass without any particular condensation" on the 20th. The comet appeared as a "rather faint diffused patch of light" on the 24th, while on the 29th, Finlay described it as "faint and ill-defined. A difficult object to observe." The comet was last detected on July 30.73, when Finlay gave the position as $\alpha = 10^h\ 18.0^m$, $\delta = -12°\ 13'$. He simply described the comet as "very faint."

The first parabolic orbits was independently calculated by H. C. F. Kreutz and J. Holetschek using positions spanning the period of April 29 to May 3. Kreutz gave the perihelion date as 1886 June 7.42, while Holetschek gave it as June 6.96. Additional orbits were calculated by A. Berberich, H. V. Egbert, V. A. Lebeuf, C. N. A. Krueger, H. Oppenheim, and J. Müller, which eventually established the perihelion date as June 7.9.

Definitive orbits have been calculated by D. Klumpke-Roberts (1897), G. Bucht (1908), and B. G. Marsden (1978). Each of these orbits gave the perihelion date as about June 7.89, while the period was given as about 745 years by Klumpke-Roberts, about 771 years by Bucht, and about 768 years by Marsden. Marsden's orbit is given below.

T	ω	Ω (2000.0)	i	q	e
1886 Jun. 7.8854 (UT)	201.2872	194.2040	87.6617	0.269804	0.996783

ABSOLUTE MAGNITUDE: $H_{10} = 7.5$ (V1964)

FULL MOON: Apr. 18, May 18, Jun. 16, Jul. 16, Aug. 14

SOURCES: V. A. Lebeuf, *BA*, **3** (1886 May), p. 237; H. H. Turner, A. M. W. Downing, and T. Lewis, *MNRAS*, **46** (1886 May), pp. 400–1; W. R. Brooks and J. Ritchie, *AN*, **114** (1886 May 1), p. 223; G. Bigourdan, *CR*, **102** (1886 May 3), pp. 1008–9; H. C. F. Kreutz, *AN*, **114** (1886 May 6), p. 239; B. von Engelhardt, *AN*, **114** (1886 May 7), p. 287; F. Gonnessiat, *CR*, **102** (1886 May 10), pp. 1052–3; J. Palisa, E. W. L. Tempel, B. von Engelhardt, G. L. Tupman, V. Knorre, J. Holetschek, A. Berberich, and H. V. Egbert, *AN*, **114** (1886 May 13), pp. 299–304; V. A. Lebeuf and C. Rambaud, *CR*, **102** (1886 May 17), pp. 1096–7; W. Luther, E. A. Lamp, H. Oppenheim, and J. Müller, *AN*, **114** (1886 May 18), pp. 317–20; A. H. P. Charlois, *CR*, **102** (1886 May 24), pp. 1149–50; E. A. Lamp, W. Luther, V. Knorre, V. A. Lebeuf, C. N. A. Krueger, and A. Berberich, *AN*, **114** (1886 May 26), pp. 329–34; E. J. M. Stephan, *BA*, **3** (1886 Jun.), pp. 275–6; H. P. Hollis, *MNRAS*, **46** (1886 Jun.), p. 459; H. Oppenheim, *AN*, **114** (1886 Jun. 5), p. 367; B. von Engelhardt and J. Bauschinger, *AN*, **114** (1886 Jun. 15), pp. 379–82; W. R. Brooks, *AN*, **114** (1886 Jul. 1), p. 413; M. Schnauder, *AN*, **115** (1886 Oct. 7), pp. 235–8; C. F. Pechüle, *AN*, **115** (1886 Nov. 25), pp. 387–90; G. Gruss, *AN*, **116** (1886 Dec. 22), p. 57; J. Tebbutt, *AN*, **116** (1887 Jan. 22), pp. 123–6; W. H. Finlay, *MNRAS*, **47** (1887 Mar.), pp. 284–5; W. H. Finlay, *AN*, **116** (1887 Apr. 17), pp. 307, 315; J. Palisa and H. Oppenheim, *AN*, **116** (1887 Apr. 23), p. 347; H. V. Egbert, *AJ*, **7** (1887 Jun. 6), pp. 97, 99; V. Knorre, *AN*, **117** (1887 Aug. 3), pp. 171, 197; G. L. Tupman, *MNRAS*, **47** (1887, Supplementary Notice), p. 548; W. Luther, *AN*, **120** (1888 Nov. 3), pp. 115, 131; D. Klumpke-Roberts, *BA*, **14** (1897 Aug.), pp. 305–7; G. Bucht, *AMAF*, **5** (1909 Feb. 9), pp. 1–45; V1964, p. 62; B. G. Marsden, *AJ*, **83** (1978 Jan.), p. 66; B. G. Marsden, *CCO*, 3rd ed. (1979), pp. 19, 47.

C/1886 J1
(Brooks)

1886 III = 1886b

Discovered: 1886 May 1.00 ($\Delta = 1.01$ AU, $r = 0.85$ AU, Elong. $= 49°$)
Last seen: 1886 June 3.9 ($\Delta = 1.30$ AU, $r = 1.00$ AU, Elong. $= 49°$)
Closest to the Earth: 1886 May 9 (0.9716 AU)
Calculated path: PEG (Disc), AND (May 6), CAS (May 13), CAM (May 30)

This comet was discovered by W. R. Brooks (Phelps, New York, USA) on 1886 May 1.00 at a position of $\alpha = 23^h 00.0^m$, $\delta = +21° 00'$. It was his second comet discovery in four days. Brooks described it as having a small but bright and stellar head, and a conspicuous tail. He telegraphed a discovery announcement within two hours.

On May 3, E. A. Lamp (Kiel, Germany) said the comet was very bright in bright twilight, with a notable tail and a yellowish nucleus, while E. J. M. Stephan (Marseille, France) observed with a refractor (128×) and said the comet exhibited a nucleus of magnitude 7–8, as well as a "frayed" tail 10–12° long. There were numerous observers on May 4, the day the comet passed perihelion. Lamp said the comet seemed fainter and dull white in color. B. von Engelhardt (Dresden, Germany) observed with the 46-cm refractor and said the comet was very bright in the bright field of view. There was a disk-shaped nucleus from which a "bright linear discharge" extended 2' toward the tail axis. The tail was 8' long and very bright. It appeared narrow at first and then fanned out and curved northward. He added that the tail split at a distance of 6' from the nucleus. E. Millosevich (Rome, Italy) said the comet had a stellar nucleus and a tail extending 8' toward PA 265°. M. Schnauder (Leipzig, Germany) observed with a 30-cm refractor and said the comet was bright and very elongated, with a sharply defined coma. Although he noted no nucleus, there was a tail 10' long. A. Kammermann (Geneva, Switzerland) said the tail was bright and 15' long in the 25.4-cm telescope. G. Celoria (Milan, Italy) observed with a 20-cm refractor and said the tail was bright. It began thin, but slightly fanned out and displayed an "elegant bend."

The comet was moving away from the sun on May 5, but still approaching Earth. Lamp said the distinct yellowish nucleus was again visible in twilight even though observing conditions were worse than on the 4th. Kammermann said the bright tail extended toward PA 270°. W. Luther (Hamburg, Germany) observed with a 25-cm refractor and said the comet was bright, with a nucleus of magnitude 9. He added that the tail was curved, with the concave side pointing southward. G. Bigourdan (Paris, France) said the comet exhibited a very small, brilliant head. The tail extended 10–12° and displayed two bright edges separated by about 10°. The tail extended toward PA 252° for the first 3', but then curved. F. Gonnessiat (Lyon, France) observed with a 15-cm refractor and noted the coma was elongated toward PA 258°. There was also a tail about 12' long.

The comet was well observed throughout the remainder of May and passed 0.97 AU from Earth on the 9th. On May 6, V. Knorre (Berlin, Germany) observed with a 24.4-cm refractor and said the tail was 5–6' long. Bigourdan observed the central occultation of a 9.5-magnitude star by the comet. The star was seen to vanish, while the comet did not change its appearance. He added that the comet looked like a miniature version of Donati's comet. On the same day Stephan said its general appearance was unchanged since the 3rd, although the comet seemed slightly fainter. On May 7, Engelhardt found the tail about 40' long and nearly straight. He added that about 8' from the nucleus there was a faint secondary tail bending towards the south. J. Palisa (Vienna, Austria) observed with a 30-cm telescope and said the comet looked like a great comet in miniature, with a bright nucleus and a tail extending 11' toward PA 268.8°. The tail left the coma in the region between 262.0° and 272.6°, and turned southward. Stephan said the comet seemed slightly fainter than on the previous morning. On the 9th, Engelhardt said the comet was not as bright as on the 7th. He noted the tail was narrow, the nucleus was disk-shaped, and the color was whitish-blue. C. F. Pechüle (Copenhagen, Denmark) observed with a 27-cm refractor from May 3 to May 12. He said a nebulous band was noted extending from a very faint nucleus to a diffuse nucleus, from which a 10' long tail curved toward the south. He reported the tail extended toward PA 283° on the 12th. Luther said the comet appeared as an extremely faint nebulosity on the 19th, without a trace of condensation. On the 21st, Knorre wrote, "The comet appeared, perhaps by the effect of the moon, only as a barely visible ghost from 5–10' long." That same night the comet was also observed by E. W. L. Tempel (Arcetri, Italy) and G. Gruss (Prague, Czech Republic). Tempel observed with a 20-cm telescope (70×) and said the comet appeared as a spindle-shaped nebulosity 12' long and about 1.5' wide. He added that the head looked like a nebulous spot. Gruss saw a distinct tail about 28' long and 2.5' wide extending toward PA 40°. On the 22nd, Engelhardt said there was no nucleus or condensation, but a narrow tail did extend 16' toward PA 289.7°. Celoria saw the comet with a 20-cm refractor on the evenings of May 21–24. He noted the comet had suddenly decreased in brightness, and appeared oblong with no trace of a nucleus.

The comet's position was last measured on June 2.94, when H. A. Kobold (Strasbourg, France) gave it as $\alpha = 4^h 23.3^m$, $\delta = +71° 34'$. The comet was then near its most northerly declination. The comet was last seen on June 3.9, when Celoria saw it with a 20-cm refractor. He said it was barely detected, so that no position could be measured.

The first parabolic orbit was calculated by E. A. Lamp. He used positions obtained on May 3, 4, and 5, and determined the perihelion date as 1886 May 4.59. This orbit would prove to be within 12 hours of the actual perihelion date. Numerous orbits were computed during the next couple of weeks by O. C. Wendell, H. Oppenheim, R. Spitaler, A. Berberich, E. Frisby, and Celoria (1887), which produced a more accurate representation of the comet's motion. Celoria's was the last for several years. His orbit used positions spanning the period of May 5–23 and indicated a perihelion date of May 4.94. Kobold (1909) revealed the comet's orbit had hyperbolic characteristics, but still relied on the parabolic orbit to best represent the comet's motion because of the short period of observability. His hyperbolic orbits had perihelion dates of May 5.02 and eccentricities of about 1.013, while the parabolic orbit had a perihelion date of May 4.96.

E.. Weiss wrote in the 1886 June 22 issue of the *Astronomische Nachrichten* that the comet's ascending node passed 0.075 AU from the orbit of Earth on July 9. He suggested a meteor shower might occur from a radiant of $\alpha = 1.3^h$, $\delta = -42°$. No activity was ever reported from this radiant.

T	ω	Ω (2000.0)	i	q	e
1886 May 4.951 (UT)	38.559	289.365	100.218	0.84199	1.0

ABSOLUTE MAGNITUDE: $H_{10} = 4.9$ (V1964)

FULL MOON: Apr. 18, May 18, Jun. 16

SOURCES: W. R. Brooks, *AN*, **114** (1886 May 1), p. 223; E. A. Lamp, *AN*, **114** (1886 May 6), p. 237; B. von Engelhardt and E. A. Lamp, *AN*, **114** (1886 May 7), p. 287; G. Bigourdan and F. Gonnessiat, *CR*, **102** (1886 May 10), pp. 1051–3; E. Millosevich, W. Luther, R. Spitaler, H. Oppenheim, A. Berberich, and E. Frisby, *AN*, **114** (1886 May 13), pp. 301–4; C. Rambaud, *CR*, **102** (1886 May 17), pp. 1096–7; B. von Engelhardt, V. Knorre, and E. A. Lamp, *AN*, **114** (1886 May 18), p. 317; A. H. P. Charlois, *CR*, **102** (1886 May 24), pp. 1149–50; V. Knorre and E. W. L. Tempel, *AN*, **114** (1886 May 26), p. 331; E. J. M. Stephan, *BA*, **3** (1886 Jun.), pp. 275–6; B. von Engelhardt, *AN*, **114** (1886 Jun. 15), pp. 379–82; E. Weiss, *AN*, **114** (1886 Jun. 22), p. 399; W. R. Brooks and O. C. Wendell, *AN*, **114** (1886 Jul. 1), pp. 413–16; M. Schnauder and Parsons, *AN*, **115** (1886 Oct. 7), p. 237; C. F. Pechüle, *AN*, **115** (1886 Nov. 25), pp. 387, 391; G. Gruss, *AN*, **116** (1886 Dec. 22), p. 57; J. Palisa and S. Oppenheim, *AN*, **116** (1887 Apr. 23), pp. 345–7; G. Celoria, *AN*, **117** (1887 May 9), p. 9; A. Kammermann, *AN*, **117** (1887 Jun. 23), p. 107; V. Knorre, *AN*, **117** (1887 Aug. 3), pp. 171, 197; W. Luther, *AN*, **120** (1888 Nov. 3), pp. 115, 132; H. A. Kobold, *AN*, **151** (1899 Dec. 28), pp. 161–6; H. A. Kobold, *AN*, **182** (1909 Aug. 13), pp. 33–42; V1964, p. 62.

D/1886 K1
(Brooks 1)

1886 IV = 1886c

Discovered: 1886 May 23.2 ($\Delta = 0.57$ AU, $r = 1.34$ AU, Elong. $= 112°$)
Last seen: 1886 July 3.90 ($\Delta = 0.71$ AU, $r = 1.36$ AU, Elong. $= 102°$)
Closest to the Earth: 1886 May 2 (0.5576 AU)
Calculated path: VIR (Disc)

W. R. Brooks (Phelps, New York, USA) discovered this comet in the evening sky on 1886 May 23.2. He described it as a large, nearly round, and feebly luminous spot with a slight condensation occasionally visible. Although he immediately telegraphed an announcement, he noted that the comet was "suspected." Brooks confirmed the find on May 24.13 and gave the position as $\alpha = 11^h 55.2^m$, $\delta = +8° 55'$.

The comet had already passed closest to Earth, but was still approaching the sun. On May 25, E. Millosevich (Rome, Italy) said the comet was faint and difficult to observe. E. W. L. Tempel (Arcetri, Italy) said the comet resembled a round nebulosity of class I or II and was 2′ in diameter, with a speckled condensation. E. E. Barnard (Nashville, Tennessee, USA) said the comet was "faintish" and pretty large in a 15-cm refractor. On May 25 and 26, H. A. Kobold (Strasbourg, France) said the comet was a faint and round nebulosity about 1′ across. He added that a point-like condensation was occasionally visible. J. Palisa (Vienna, Austria) observed with a 30.1-cm refractor on the 26th and said the comet was faint and 2′ across. He then noted it was very faint and difficult to see in the same telescope on the 28th. Palisa said the comet was hard to see in the 76-cm refractor on the 30th. F. Gonnessiat (Lyon, France) observed with a 15-cm refractor and said the comet's brightness noticeably decreased during the period of May 25–30. On the 31st, Gonnessiat described the comet as a faint luminous spot about 1′ across, with a central condensation.

Gonnessiat found the comet even fainter on June 1 than on May 31. C. Trépied (Alger, now al-Jazâ′ir, Algeria) saw the comet with a 50-cm telescope on June 2 and 3, and described it as an ill-defined nebulosity, with hardly a trace of a central condensation. Palisa was now observing the comet with the 69-cm refractor and described it as extremely faint on the 3rd. Interestingly, Barnard continued his observations with smaller telescopes than Palisa. He observed with a 15-cm refractor on the 25th and wrote, "Not very faint. Dim and large." On the 29th, Barnard observed with a 13-cm refractor (40×) and said the comet was pretty faint.

The comet was last detected during the first days of July. Barnard described it as not very faint and still large in the 13-cm refractor (40×) on the 2nd. The final observations of this comet were obtained by H. C. Russell (Sydney Observatory, Australia) on July 3.46, by Tempel on July 3.89, and by astronomers at Nice on July 3.90. Russell simply described the comet as "very faint." The Nice astronomers gave the position as $\alpha = 13^h 18.0^m$, $\delta = -16° 18′$.

The first parabolic orbit was calculated by S. Oppenheim using positions from May 25, 28, and 30. The resulting perihelion date was 1886 June 3.37. H. Oppenheim took positions from May 26, 28, and 31, and gave a perihelion date of June 2.27. Although S. Oppenheim published a revision in June that gave a perihelion date of June 5.09, his next orbit used positions obtained up to July 1 and revealed an elliptical orbit with a perihelion date of June 7.28 and a period of 9.05 years. J. R. Hind used a similar set of positions a short time later and determined a perihelion date of June 7.07 and a period of 6.30 years.

A definitive orbit was calculated by S. Oppenheim (1891), which indicated a period of 5.60 years. He predicted the comet would next arrive at perihelion on 1892 January 13, but added that there was probably little chance of seeing the comet at that return, unless perihelion occurred later than predicted. Although he provided ephemerides based on perihelion dates spanning March 1 to July 29, the comet was not found. R. J. Buckley (1979) published a revised orbit, using 39 positions spanning the entire period of visibility. Although he found a period of 5.44 years, he noted that the positions were badly spread across the releatively short observation arc, thus making a likely uncertainty of 1–2 months. Buckley's orbit is given below.

T	ω	Ω (2000.0)	i	q	e
1886 Jun. 7.2122 (TT)	176.8756	55.1391	12.6710	1.325273	0.571388

ABSOLUTE MAGNITUDE: $H_{10} = 8.9$ (V1964)

FULL MOON: May 18, Jun. 16, Jul. 16

SOURCES: W. R. Brooks, *AN*, **114** (1886 May 26), p. 335; A. H. P. Charlois, *CR*, **102** (1886 May 31), p. 1230; H. V. Egbert, E. Millosevich, E. W. L. Tempel, J. Palisa, H. A. Kobold, S. Oppenheim, and H. Oppenheim, *AN*, **114** (1886 Jun. 5), pp. 365–8; F. Gonnessiat, *CR*, **102** (1886 Jun. 7), pp. 1303–4; C. Trépied, *CR*, **102** (1886 Jun. 21), pp. 1438–9; J. Palisa and S. Oppenheim, *AN*, **114** (1886 Jun. 22), p. 399; W. R. Brooks, *AN*, **114** (1886 Jul. 1), p. 413; E. W. L. Tempel and S. Oppenheim, *AN*, **115** (1886 Jul. 28), p. 47; J. R. Hind, *CR*, **103** (1886 Aug. 23), pp. 427–8; H. C. Russell, *MNRAS*, **46** (1886, Supplementary Notice), p. 497; E. E. Barnard, *AN*, **115** (1886 Nov. 19), pp. 323, 328; Pringle, *AN*, **116** (1887 Feb. 3), pp. 145–50; J. Palisa, *AN*, **116** (1887 Apr. 23), p. 347; S. Oppenheim, *AN*, **128** (1891 Nov. 3), pp. 297–306; S. Oppenheim, *The Observatory*, **14** (1891 Dec.), pp. 427–8; V1964, p. 62; R. J. Buckley, *JBAA*, **89** (1979 Apr.), pp. 261, 263.

7P/Pons–Winnecke

1886 VI = 1886d

Recovered: 1886 August 19.79 ($\Delta = 1.17$ AU, $r = 0.92$ AU, Elong. $= 49°$)

Last seen: 1886 November 29.83 ($\Delta = 1.39$ AU, $r = 1.51$ AU, Elong. $= 77°$)

Closest to the Earth: 1886 October 4 (0.9040 AU)

Calculated path: VIR (Rec), LIB (Sep. 6), SCO (Sep. 24), SGR (Oct. 14), CrA (Oct. 22), SGR (Oct. 25), MIC (Nov. 6), PsA (Nov. 17)

Its short duration of visibility in 1875 and the fact that this comet was missed at the apparition of 1880 made it crucial that it be recovered at this apparition. Early in 1886, A. Palisa took positions from the apparitions of 1858, 1869, and 1875, and calculated a perihelion date of 1886 September 16.94. E. A. Lamp provided an ephemeris for the optimum perihelion date, as well as for 4 days before and after. W. H. Finlay (Royal Observatory, Cape of Good Hope, South Africa) recovered the comet with an 18-cm refractor on August 19.79, and gave a position of $\alpha = 13^h 07.1^m$, $\delta = -0° 34'$. Another observation on August 20.74 confirmed this object was the expected comet. It was about 5.5° from the expected position, indicating it had passed perihelion 12 days earlier than Palisa's prediction. Finlay described it as diffuse, circular, with a diameter of 1' and a magnitude of about 10. There was some central condensation, but no tail.

The comet was nearing perihelion when recovered. T. Zona (Palermo, Italy) said the comet was faint and difficult to see on August 28, with a trace of a nucleus. H. C. Russell (Sydney, New South Wales, Australia) observed the comet exclusively with the 29-cm refractor. On the 2nd, with the moon nearby, Russell described the comet as faint and diffuse, with no nucleus. Finlay said the comet was faint in the 18-cm refractor on the 4th because of moonlight. Russell observed in strong moonlight on the 8th and said the comet was "much more condensed than on Sept. 2." He added that the coma was about one-fourth of an arc minute across. On September 9, Russell said the comet was "much brighter even with increased moonlight; a suspicion of a nucleus." Russell said the comet was "extremely faint" because of haze on the 10th, but appeared brighter with a suspicion of a nucleus on the 14th. Russell said the comet appeared "much brighter" on the 15th, with a coma about 1' across. E. E. Barnard (Nashville, Tennessee, USA) accidentally found the comet on the 17th during one of his routine comet searches. He described it as round, with an average brightness of between magnitude 9 and 10. Finlay said the comet was well condensed toward the center on the 18th, with "scarcely a stellar nucleus." That same night, Barnard said the the comet was brighter in the middle and exhibited a small, faint nucleus. Barnard said the comet was "brightish with glimpses of a stellar nucleus" on the 19th. Russell said the comet appeared brighter on the 20th and was "condensed in the centre, with a very faint nucleus." On September 21, Russell said the comet was bright, with a central condensation. A possible very faint nucleus was occasionally seen. The comet passed over a 9th-magnitude star, but the star did not diminish in brightness. Barnard said the comet was about 1' across, with a magnitude of about 9.5–10. The coma was round and condensed into a small ill-defined nucleus. Russell said the comet was not as bright on the 22nd as on the 21st. Barnard described the comet as "brightish" on the 23rd and said it was white, with a coma 1' across and an ill-defined nucleus of magnitude 9 or less. Russell glimpsed the nucleus on the 24th. He found the comet faint on the 28th, with a diffuse coma and a faint nucleus. He said it appeared even fainter on the 29th.

The comet passed Earth early in October. Barnard described the comet as of "average size" and "faintish" on the 17th. The comet attained its most southerly declination of $-37°$ on the 25th. Finlay and J. Tebbutt (Windsor, New South Wales, Australia) saw the comet on several occasions during October, but provided no physical descriptions. Tebbutt last saw the comet on the 29th.

Finlay and Barnard were the only observers providing physical descriptions during November. Finlay found it faint in moonlight on the 2nd. Barnard said the comet was not very faint on the 16th, but was difficult to see because of a nearby 7th-magnitude star. Finlay simply described the comet as "faint" in the 18-cm refractor on November 17, 19, 23, 25, and 26. Barnard noted the comet was faint and large when seen on November 19 and 26. The comet was last detected on November 29.83, when Finlay measured the position as $\alpha = 22^h\ 09.1^m$, $\delta = -28°\ 10'$. He simply described it as faint.

Barnard was upset about this apparition. He wrote in the 1887 September 20 issue of the *Astronomische Nachrichten* "that no ephemeris of this comet was published during this apparition, for I have seen none. If such had been published more observations of the comet could have been made. I mistook it several times for a new comet having swept it up whilst comet seeking, and remained in considerable suspense, not knowing certainly that it was Winnecke's comet."

Later calculations using multiple apparitions and planetary perturbations were published by E. F. von Haerdtl (1888, 1889), B. G. Marsden (1970), and Marsden and Z. Sekanina (1972). These revealed a perihelion date of September 4.89 and a period of 5.82 years. Marsden, Sekanina, and D. K. Yeomans (1973) determined nongravitational terms of $A_1 = -0.01$ and $A_2 = -0.0021$. This orbit is given below.

T	ω	Ω (2000.0)	i	q	e
1886 Sep. 4.8864 (TT)	172.0920	105.6074	14.5220	0.885499	0.726178

ABSOLUTE MAGNITUDE: $H_{10} = 9.2$ (V1964)

FULL MOON: Aug. 14, Sep. 13, Oct. 13, Nov. 11, Dec. 11

SOURCES: T. R. von Oppölzer, *AN*, **97** (1880 Jul. 30), pp. 337–42; A. Palisa and E. A. Lamp, *AN*, **114** (1886 Mar. 30), p. 127; E. A. Lamp, *AN*, **114** (1886 May 26), p. 335; W. H. Finlay, *AN*, **115** (1886 Aug. 31), p. 111; T. Zona, *AN*, **115** (1886 Sep. 7), p. 143; E. E. Barnard, *AN*, **115** (1886 Nov. 19), p. 328; H. C. Russell, *MNRAS*, **47** (1886 Dec.), pp. 67–8; W. H. Finlay and E. A. Lamp, *MNRAS*, **47** (1887 Feb.), p. 166; W. H. Finlay, *MNRAS*, **47** (1887 Mar.), pp. 286–9; J. Tebbutt, *MNRAS*, **47** (1887 Mar.), pp. 293–301; W. H. Finlay, *AN*, **116** (1887 Apr. 17), pp. 307–10, 316; E. E. Barnard, *AN*, **117** (1887 Sep. 20), p. 337; E. F. von Haerdtl, *AN*, **120** (1888 Dec. 28), pp. 257–72; E. F. von Haerdtl, *DAWW*, **55** Abt. II (1889), pp. 250–69; E. F. von Haerdtl, *DAWW*, **56** Abt. II (1889), pp. 151–85; V1964, p. 62; B. G. Marsden, *AJ*, **75** (1970 Feb.), pp. 80–1; B. G. Marsden and Z. Sekanina, *QJRAS*, **13** (1972), pp. 428–9; B. G. Marsden, Z. Sekanina, and D. K. Yeomans, *AJ*, **78** (1973 Mar.), pp. 214, 216.

15P/1886 S1
(Finlay)

1886 VII = 1886e

Discovered: 1886 September 26.83 ($\Delta = 1.14$ AU, $r = 1.28$ AU, Elong. = 73°)

Last seen: 1887 April 12.9 ($\Delta = 2.31$ AU, $r = 2.05$ AU, Elong. = 62°)

Closest to the Earth: 1886 July 18 (1.1252 AU), 1886 December 16 (0.8224 AU)

Calculated path: OPH (Disc), SGR (Oct. 9), CAP (Nov. 16), AQR (Dec. 9), PSC (Dec. 27), ARI (Jan. 25), TAU (Feb. 22)

W. H. Finlay (Royal Observatory, Cape of Good Hope, South Africa) discovered this comet with an 18-cm refractor on 1886 September 26.83, and gave the position as $\alpha = 17^h 02.0^m$, $\delta = -26° 04'$. He described it as round, 1' across, and "very slightly more condensed towards the centre." He added that the comet was faint and about magnitude 11, but exhibited no tail. He was able to determine additional positions on September 27.80 and 27.85, but noted the comet was "completely overpowered" by an 11th-magnitude star after the first position.

On September 29, E. Millosevich (Rome, Italy) estimated the coma diameter as 2' across, and noted a faint nucleus. A. Kammermann (Geneva, Switzerland) observed the comet with a 25-cm refractor and described it as round, with an eccentrically placed condensation. The coma was about 1' across. On September 30, E. E. Barnard (Vanderbilt University, Nashville, Tennessee, USA) estimated the total magnitude as 10.5–11. He added that the coma was round and exhibited no nucleus. H. C. Russell (Sydney, New South Wales, Australia) simply described the comet as faint in the 29-cm refractor.

On October 5, Russell said the comet appeared faint in moonlight. John Tebbutt (Windsor, New South Wales, Australia) observed with a 20-cm refractor on the 8th and noted the comet was "hardly distinguishable." Russell said the comet appeared brighter on the 8th than on the 7th. Using the 29-cm refractor, Russell said the comet appeared very faint in very hazy skies on the 11th and faint on the 12th. The comet attained its most southerly declination of –27° on the 15th. On the 22nd, Tebbutt noted the comet exhibited a "faint" condensation. On the 28th, Tebbutt said the comet exhibited a "very slight" condensation. J. Palisa (Vienna, Austria) observed with the 30-cm refractor on the 29th and described the comet as a rather bright, diffuse nebulosity some 2.5' across. He added that there was a condensation, but no actual nucleus. On October 31, J. M. Thome (Cordoba, Argentina) described the comet as "faint in clouds."

On November 5, Thome watched as the comet passed nearly centrally over a star of magnitude 11.5 without diminishing its brightness. On the 7th, Tebbutt wrote, "Considering the moon's presence the comet was brighter than expected. Slight condensation which could be pretty well observed." On the 15th, Tebbutt said the comet exhibited a "marked" condensation, while Thome watched the comet pass nearly centrally over a star of magnitude 11.5 and noticed no decrease in light. On the 16th, Tebbutt noted the comet's "condensation not quite so distinct as on 15th." On November 18, L. Boss (Dudley Observatory, New York, USA) described the comet as "bright, with a strong nuclear condensation," while J. I. Plummer (Orwell Park Observatory, England) observed with the 25-cm refractor and described the comet as "fairly bright but without marked nucleus." On the 19th, Boss described the comet as bright and 2–3' in diameter. He estimated the magnitude of the central condensation as 9.5. That same night, A. Abetti (Padova, Italy) described the comet as a small, round, nebulous spot, about 1' across, and gave the magnitude as 9. On the 25th, Finlay said, "the central condensation seemed somewhat elongated in position angle 165° ±." Tebbutt noted the comet "gradually condensed towards centre" on the 26th. On the 27th, B. von Engelhardt (Dresden, Germany) described the comet as bright, diffuse, and round, with a diameter of 2'. The coma was condensed and there was a nucleus. That same night, Tebbutt said the comet was still slightly condensed, "but notwithstanding a clear sky and an absent moon the comet was obviously considerably fainter." J. Holetschek (Vienna, Austria) observed the comet with a 15-cm refractor on the 28th and found it 2′ across, with a central condensation of magnitude 10. On November 30, Plummer said the comet was faint in moonlight.

On December 1, Tebbutt used a 20-cm refractor and saw the comet was only 2–3° from the moon. He commented, "Comet hardly distinguishable owing to haze and moonlight." On the 3rd, E. A. Lamp (Kiel, Germany) observed with a refractor and described the comet as extremely faint. Engelhardt observed on the 5th, in spite of bright moonlight and haze. He found the comet rather bright and diffuse, with a star-like nucleus of magnitude 11. On the 9th, Thome described it as "very faint," and watched as the comet passed nearly centrally over a star of magnitude 8 without diminishing its brightness. Thome described the comet as "very faint and ill defined" on the 10th. On December 15, Tebbutt found the comet "brighter than I expected." He also noted a slight condensation. Holetschek said the comet was 3′ across on the 18th, with a nucleus of magnitude 9.5, while Tebbutt noted the comet was fainter than on the 15th, but still "pretty well seen" in the 6-cm finder. He also noted a central condensation. Tebbutt said the comet was still visible in the finder on the 19th. On the 22nd, Tebbutt said the comet was not as distinct in the refractor as on the 19th, but was "still somewhat condensed." Engelhardt found the comet reasonably bright and about 1′ in diameter on the 26th. A nucleus was also visible. On the 27th, Tebbutt noted the comet was "much fainter and more difficult to observe." Holetschek found the coma 2.7′ across on the 28th, with a nucleus of magnitude 10.5. On December 30, Tebbutt said the comet was "very difficult to observe" and was "much diffused, but there was still a slight trace of a condensation." Thome described the comet as "very indistinct" on that same night.

On 1887 January 1, Thome noted the comet was "barely visible." W. Luther (Hamburg, Germany) saw the comet despite poor seeing on the 15th. He described it as diffuse and uncondensed. On the 16th and 25th, H. P. Hollis (Royal Observatory, Greenwich, England) observed with the 17-cm refractor and simply described the comet as very faint. Engelhardt described the comet as rather faint in the refractor on the 17th, although it was bright in the finder. The comet was 2′ across, and exhibited a very small, nearly stellar nucleus. That same night, Luther also described the comet as faint. Luther reported the comet was very faint on the 18th, faint and exhibiting a nucleus of magnitude 12 on the 23rd, and faint on the 25th. He said the nucleus was fainter than magnitude 12 on the 26th. On the 27th, Engelhardt found the comet "rather faint and diffuse," with a diameter of 2′. He added that a stellar point was seen north of the main nucleus. On January 28, Thome found the comet "very faint."

On February 9 and 10, Kammermann observed the comet and described it as very faint. On the 12th, Engelhardt found the comet faint and diffuse, with a granular appearance and several bright points, while Luther described the comet as a faint nebulosity with a nucleus of magnitude 13. On February 12 and 25, Plummer described the comet as "excessively faint." On the 26th, Boss observed with the 33-cm refractor and noted the "atmosphere appeared to me to be remarkably pure and quiet, yet the comet was so faint that I was not always able to see it continuously across the field of view." C. F. Pechüle (Copenhagen, Denmark) observed the comet on February 17 and then obtained one final observation on March 16. He was using a 27-cm refractor. The comet was last detected in the evening sky on April 12.9 by H. Struve (Pulkovo Observatory, Russia). He was using the 30-cm refractor and indicated a position of $\alpha = 5^h 38.8^m$, $\delta = +25° 40'$.

The first parabolic orbit was calculated by Boss using positions spanning the period of September 26 to October 1. He gave the perihelion date as 1886 November 21.49 and suggested this was a return of de Vico's comet of 1844 (see 54P/1844 Q1). Very similar orbits were calculated by J. Holetschek, H. Oppenheim, and H. C. F. Kreutz. Oppenheim said the orbit did not prove this comet was identical to de Vico's comet.

The first elliptical orbits were independently calculated by Holetschek and Boss. Holetschek's was published in the November 1 issue of the *Astronomische Nachrichten*. He used positions spanning the period of September 26 to October 19 and determined the perihelion date as November 23.56 and the period of 5.28 years. Boss' orbit was published in the November 2 issue of the *Astronomical Journal* and used positions from the period of September 26 to October 22. He determined the perihelion date as November 20.37 and the period as 4.32 years. Boss noted the discrepancies between the orbits of this comet and that of de Vico's comet, but commented, "these ... are exactly what we should expect to see, because the very fact, that calculated returns of the 1844 comet have been looked for in vain, indicates that the elements of that comet have undergone material perturbations from some cause not yet explained. That cause may perhaps be found among the small planets." Holetschek stated that the identity of Finlay's comet with that of de Vico is possible only if the comet suffered large perturbations. G. M. Searle and H. Oppenheim independently calculated elliptical orbits based on the incorrect assumption that this comet was identical to de Vico's and found a period of 5.28 years. Other elliptical orbits, based purely on the current positions, were calculated by C. N. A. Krueger, Boss, Holetschek, Finlay, and L. Schulhof (1893), and eventually revealed a period of about 6.7 years. Although Boss was the greatest proponent of a link between this comet and that of de Vico, he eventually noted that the strong perturbations required to adjust the orbit did not exist. Krueger also wrote that the possible identity of this comet with that of de Vico's "must be given up."

The most recent multiple apparition orbit was calculated by B. G. Marsden and Z. Sekanina (1972) using positions spanning the period of 1886–1906. Their orbit is given below. In a paper published by Marsden,

Sekanina, and D. K. Yeomans (1973) the nongravitational terms were determined as $A_1 = +0.53$ and $A_2 = +0.1266$.

T	ω	Ω (2000.0)	i	q	e
1886 Nov. 22.8897 (TT)	315.3372	53.8321	3.0362	0.997545	0.717833

ABSOLUTE MAGNITUDE: $H_{10} = 9.2$ (V1964)

FULL MOON: Sep. 13, Oct. 13, Nov. 11, Dec. 11, Jan. 9, Feb. 8, Mar. 9, Apr. 8, May 7

SOURCES: W. H. Finlay, *AN*, **115** (1886 Sep. 30), p. 223; E. Millosevich, H. J. A. Perrotin, T. Zona, J. Ritchie, Jr, and L. Boss, *AN*, **115** (1886 Oct. 7), pp. 237–40; E. Millosevich and J. Holetschek, *AN*, **115** (1886 Oct. 13), pp. 253–6; E. E. Barnard, E. Millosevich, H. Oppenheim, and J. Ritchie, Jr, *AN*, **115** (1886 Oct. 23), pp. 267–70; E. Millosevich, J. Holetschek, and H. C. F. Kreutz, *AN*, **115** (1886 Nov. 1), pp. 283–6; L. Boss, *AJ*, **7** (1886 Nov. 2), pp. 7–8; C. N. A. Krueger, *AN*, **115** (1886 Nov. 13), p. 319; G. M. Searle, *AJ*, **7** (1886 Nov. 24), pp. 15–16; H. C. Russell, *MNRAS*, **47** (1886 Dec.), pp. 68–9; E. Millosevich, *AN*, **116** (1886 Dec. 8), p. 27; L. Boss, *AJ*, **7** (1886 Dec. 9), pp. 21–4; B. von Engelhardt, E. Millosevich, H. Oppenheim, and J. Holetschek, *AN*, **116** (1886 Dec. 18), pp. 43–8; C. N. A. Krueger, *AN*, **116** (1887 Jan. 3), p. 77; E. A. Lamp, W. Luther, and B. von Engelhardt, *AN*, **116** (1887 Jan. 14), p. 111; L. Boss, *AJ*, **7** (1887 Jan. 19), pp. 43–7; C. N. A. Krueger, *AN*, **116** (1887 Jan. 22), p. 127; G. M. Searle; *AJ*, **7** (1887 Jan. 28), pp. 52–4; H. Kloock, *AN*, **116** (1887 Feb. 10), p. 173; A. Abetti, *AN*, **116** (1887 Feb. 28), p. 215; H. P. Hollis, *MNRAS*, **47** (1887 Mar.), pp. 275–6; W. H. Finlay, *MNRAS*, **47** (1887 Mar.), pp. 290–3; W. H. Finlay, *MNRAS*, **47** (1887 Mar.), p. 302; H. Pomerantzeff and B. von Engelhardt, *AN*, **116** (1887 Mar. 12), pp. 247–50; L. Boss, *AJ*, **7** (1887 Mar. 15), supplement; L. Boss, *AJ*, **7** (1887 Apr. 7), p. 84; C. N. A. Krueger, *AN*, **116** (1887 Apr. 15), p. 335; W. H. Finlay, *AN*, **116** (1887 Apr. 17), pp. 309–16; J. Palisa, H. Oppenheim, and J. Holetschek, *AN*, **116** (1887 Apr. 23), p. 347; A. Kammermann, *AN*, **117** (1887 May 25), p. 41; J. Tebbutt and C. W. Pritchett, *AN*, **117** (1887 Jun. 23), pp. 109–14; J. M. Thome, *AJ*, **7** (1887 Jul. 29), pp. 117–18; V. Knorre, *AN*, **117** (1887 Aug. 3), pp. 171, 197; J. M. Thome, *AN*, **117** (1887 Aug. 27), pp. 271–6; J. I. Plummer, *MNRAS*, **48** (1887 Nov.), pp. 55–6; C. F. Pechüle, *AN*, **118** (1887 Nov. 17), p. 73; H. C. F. Kreutz and H. Struve, *VJS*, **23** (1888), p. 13; L. Boss, AJ, **8** (1888 Oct. 17), pp. 113–14; W. Luther, *AN*, **120** (1888 Nov. 3), pp. 115, 131; O. C. Wendell, *AJ*, **9** (1890 Apr. 21), p. 191; V. Knorre, *AN*, **124** (1890 Jun. 17), pp. 345, 363–6; L. Schulhof, *AN*, **133** (1893 Jun. 29), pp. 51–6; W. H. Finlay, *Annals of the Cape Observatory*, **1** pt. 1 (1898), pp. 25–6; V1964, p. 62; B. G. Marsden and Z. Sekanina, *QJRAS*, **13** (1972), pp. 428–9; B. G. Marsden, Z. Sekanina, and D. K. Yeomans, *AJ*, **78** (1973 Mar.), pp. 213, 215; B. G. Marsden and Z. Sekanina, *CCO*, 12th ed. (1997), pp. 52–3.

C/1886 T1
(Barnard–
Hartwig)

1886 IX = 1886f

Discovered: 1886 October 5.45 ($\Delta = 2.27$ AU, $r = 1.53$ AU, Elong. $= 33°$)
Last seen: 1887 June 17.03 ($\Delta = 2.52$ AU, $r = 3.01$ AU, Elong. $= 109°$)
Closest to the Earth: 1886 December 5 (0.9653 AU)
Calculated path: SEX (Disc), LEO (Oct. 10), VIR (Oct. 27), BOO (Nov. 20), SER (Dec. 1), HER (Dec. 5), OPH (Dec. 16), HER-OPH (Dec. 19), AQL (Dec. 22), AQR (Jan. 17), CAP (Feb. 8), AQR (Feb. 27), PsA (Apr. 21), SCL (Apr. 23), PHE (Jun. 6)

E. E. Barnard (Vanderbilt University Observatory, Nashville, Tennessee, USA) discovered this comet in the morning sky with a 13-cm refractor on 1886 October 5.45. He described it as "bright, round and much condensed." On October 5.46, Barnard used a 15-cm refractor and estimated the position as $\alpha = 10^h 36.1^m$, $\delta = +0° 58'$. E. Hartwig (Bamberg, Germany) independently discovered the comet on October 6.17 and described it as bright and round. Hartwig said no movement was detected because of the approaching morning twilight. C. F. Pechüle (Copenhagen, Denmark) also independently discovered this comet in bright morning twilight on October 6.2, but no position could be measured. By the time he had confirmed the find on October 7.17, announcements had already been made by Barnard and Hartwig.

The comet was still about two months from its closest approach to both the sun and Earth and was widely observed during the remainder of October. The only actual estimate of its total magnitude came from L. Weinek (Prague, Czech Republic) on October 31, when he gave it as 5.6. Other estimates of the total brightness most likely referred to the nuclear condensation. Barnard did report that the comet appeared as an "ill-defined spot of light" to the naked eye on the 30th. Barnard also reported the comet seemed to fade on the 7th and 9th, but was again brightening on the 11th. A. Riccò (Palermo, Italy) was observing with a 25-cm refractor on the 8th and reported that 41 minutes before sunrise the comet disappeared in the twilight when 8th-magnitude stars were still visible, but not stars of the 9th magnitude. Once again, this probably referred to the nuclear condensation. Other magnitude estimates of the nuclear condensation were "brighter than 9" by Weinek and his colleague, G. Gruss, on the 10th, 9.5 by J. Palisa (Vienna, Austria) on the 18th, 8.5 by Barnard on the 20th, 8 by W. Luther (Hamburg, Germany) on the 28th and 29th, 8 or 9 by J. Lamp (Bothkamp, Germany) on the 30th, and 8 by Weinek on the 31st. The "nucleus" was frequently described as stellar, "very distinct", or nearly stellar. E. von Gothard (Astrophysical Observatory, Herény, Hungary) said the nucleus appeared yellow on the 30th. The coma was reported as round throughout the month and the diameter was given as 3' by Riccò on the 8th, 3.1' by Weinek and Gruss on the 10th, 3' by Palisa on the 18th, 5' by Riccò on the 26th, 2.5' by Gruss on the 28th, and 3.5' by Weinek on the 31st. The tail slowly developed during the month, with Barnard reporting most of the meager details for most of the month as he observed with his 15-cm refractor. He said the tail was faint and diffuse on the 6th, but saw no definite tail on the 7th. He reported "signs of brush-like tail preceding" on the 17th and described it as diffuse on the 19th. Barnard said a "strong brushy tail" extended toward the preceding side of the coma on the 20th, but noted only faint traces on the 23rd. On the 25th, Barnard described the tail as short, broad, "bushy", and not very distinct, while on the 30th he described it as large and "bushy" and suspected it was "bifurcated." A few other tail descriptions came from other astronomers. Riccò saw a trace of tail toward the west on the 8th and reported a faint tail extended 8' westward on the 26th. Gruss saw a trace of tail on the 28th. E. A. Lamp (Kiel, Germany) reported a short, broad tail on the 28th, 29th, and 30th. Weinek said the tail was about 6.5' long on October 31.

The comet was a naked-eye object throughout November, including the period of strong moonlight around mid-month. Barnard reported it was "as easy to naked eye as a sixth mag. star" on the 1st and looked like "a brightish spot of haze" to the naked eye on the 4th. Riccò was observing with a 25-cm refractor and said the comet disappeared about 35 minutes before sunrise on the 4th. On the 5th, Barnard said it was "brightish to eye, but no elongation seen," while on the 7th he wrote "brightish to naked eye, and faint extension away from sun as noticeable as a fifth mag. star." Barnard said it was "much easier to naked eye" with a faint tail on the 8th. T. W. Backhouse (West Hendon House Observatory, England) said the comet was "conspicuous with naked eye, as a star" on the 10th and estimated the magnitude as 5.3. Riccò said the comet remained visible in his refractor until 31 minutes before sunrise on the 10th. On the 14th, Backhouse found the comet "pretty plain to the naked eye, as a star" in moonlight. Moving his field-glasses out of focus, he concluded the total magnitude was 4.5. Backhouse said the comet appeared pale blue on the 16th. On the 19th, Barnard said the comet was as bright to the naked eye as a star of magnitude 4. On the 22nd, Backhouse said that when his field-glasses were "much out of focus" the comet was "rather brighter than υ Boötis" (magnitude 4.06 – SC2000.0). Barnard estimated the naked-eye brightness as at least 4th magnitude on the 24th. Backhouse gave the magnitude as 3.6 in slight twilight on the 25th. On the 26th, Riccò said the comet appeared bluish to the naked eye and was similar in appearance to a star of magnitude 5.3. Barnard said the comet was more conspicuous than a star of magnitude 4 on the 27th. On the 29th, Barnard said the comet was "fully as noticeable" as ζ Boötis (magnitude 3.78 – SC2000.0). On November 30, Barnard said the head was bright to the naked eye and was more noticeable than ε Boötis (magnitude 2.37 – SC2000.0), while Backhouse gave the magnitude as 3.3. The nucleus maintained a stellar appearance throughout the month, but estimates of its brightness were not as plentiful as in October. Its magnitude was estimated as 8 by Gothard on the 3rd and 5 by Weinek on the 28th. Barnard said the nucleus was slightly yellowish on the 7th, 27th, and 30th, and appeared diffuse at higher powers on the 29th. He added that it appeared "planetary, having a perceptible diameter" on the 30th.

The physical characteristics were also well observed during November. The coma diameter was given as 6' by Riccò on the 4th, 8' by Backhouse and 10' by Riccò on the 10th, 6' by Gothard and Weinek on the 28th, and 7–8' by Gothard on the 29th. Barnard said the head was better defined in the 15-cm refractor on the 5th than on previous nights. On the 7th, he noted the head was shaping up, but still did not exhibit a very definite outline. Backhouse said the head was "white, or bluish" on the 10th. Backhouse said the head was a pale blue on the 14th. The tail garnered the attention of most observers during the month, but it was Barnard who provided the best, most consistent series of observations with his 15-cm refractor. On the 1st, he said one tail extended about 2.5° toward PA 287°, while the other extended about 0.25° toward PA 134°. There was no nebulosity between the tails and Barnard noted "a slight bulging of the head of the comet" at PA 150°, and suspected it might "indicate the probable formation of a third tail." On the 4th, Barnard said the northernmost tail was traceable for 2° or 3°, while the elongation noted at PA 150° on the 1st could not be verified, although "there is suspicion of faint prolongation there." He said the northernmost tail was 2° long on the 5th. On the 7th, the northern tail was 2.5° long, while the southern was about 0.5° long. Also on the 7th, Barnard said the slight bulging of the head at PA 150° was again suspected. On the 8th, Barnard said the southern tail was longer and brighter than on previous nights. Barnard's next observation came on the 24th. He noticed a "slender train traceable for at least 7° or 8° " with the naked eye. Observing with the 15-cm refractor, he said there were now three tails, with the third tail having appeared between the other two and was about 0.25° long. The tails were distinct, "with dark sky between them." Barnard described the northern tail as peculiar. He noted that it was slender when it leaves the head, but then suddenly diffuses on its north side at a distance of about 1.5°. The south side of the northern tail is still sharply defined. On the 26th, Barnard said the tail was not as well seen with the naked eye as on previous nights. The refractor revealed the north and south tails were brighter and better defined than on previous days, and the middle tail "can now scarcely be separated" from the northern tail. On the 27th, Barnard said the north and south tails were "clear and distinct," with the northern still longest and brightest. The middle tail was no longer visible. The northern tail extended about 5° passed Arcturus, with the axis passing 0.6° north of that star. The axis of the northern tail also passed through the nucleus, while the southern tail's axis passed north of the nucleus. Barnard said, "The head, as noticed at former observations, is larger than at the junction of the tails, giving it the appearance of having a neck, from which the tails spring." On the 29th, Barnard said there was a faint, slender tail about 12° long to the naked eye, while the refractor showed the two tails as conspicuous objects. The southern tail could not be traced beyond about 1.25°, and there was still no longer any evidence of the middle tail. Barnard said the tail did not seem as bright to the naked eye on the 30th. The refractor showed a very bright head, while the two tails were once again brighter and better defined than on previous days. He noted "a narrow ray of light running along the axis" of the northern tail which emanated from the nucleus. This ray was "quite sharp and definite." Barnard added that the space between the two tails "is now filling with nebulosity." Another fine series of observations was provided by Backhouse. On November 10, he said two faint tails were visible in spite of the twilight. The north tail extended 1.5° in PA 300°, while the south one extended 30' in PA 250°. He added that the north tail was 25' wide, while the south tail was 9′ wide. He said the two tails extended toward PA 305° and PA 255° on the 14th, while the northern tail was 50′ long and the southern was 26′ long on the 16th. Moonlight and twilight virtually hid the southern tail on the 22nd, but field glasses revealed the northern tail extended 3° 50′ toward PA 305°. On November 25, Backhouse said field glasses revealed the north tail extending 5° in PA 320° and the south tail extending 40′ in PA 275°. The north tail was nearly 12′ wide. Backhouse also noted the south tail, though shorter than the north tail, was nevertheless the brighter of the two, and he added, "I suspect a tail in the opposite direction." On the 26th, Backhouse observed with binoculars and noted the north tail extended 9° toward PA 310°, while the south tail extended 1° 35′ toward PA 275°. On the 29th, Backhouse said the naked eye revealed the north tail extended 10°, while binoculars revealed it extended 12° toward PA 320°, with a width of 1° 14′. The south tail was visible in the 11-cm refractor (38×) and was at least 1° 13′ long, although he suspected it could be about 20′ longer. On the 30th, Backhouse said the north tail extended 21° 40′ toward PA 315° and was 3° 30′ wide.

The comet was at its best during December, as it passed closest to Earth early in the month and passed perihelion at mid-month. The comet reached its most northerly declination of +18° on the 4th. The comet continued to be a naked-eye object throughout the month, with Backhouse giving the longest series of brightness estimates. He gave magnitudes of 3.8 on the 1st, 3.3 on the 3rd, 4.0 on the 13th, 3.6 on the 16th, 3.3 on the 20th, 3.7 in the morning and 3.3 in the evening on the 23rd, and 3.6 on the 28th. Other magnitude estimates were 2.6 by Gruss and 3.7 by Riccò on the 8th and 4.0 by G. L. Tupman (Harrow, England) on the 20th. Weinek said the "nucleus" was between γ Herculis (magnitude 3.75 – SC2000.0) and β Herculis (magnitude 2.77 – SC2000.0) in brightness on the 6th, but this seems more in line with the total magnitudes given by others. Descriptions of the "nucleus" were not plentiful. Barnard said it was bright, but not stellar on the 2nd. Tupman said it was 15–20″ across on the 18th and gave the magnitude as 7 on the 19th. B. von Engelhardt (Dresden, Germany) said the nucleus looked like a bright bluish disk on the 26th. The coma was even more poorly observed than the nucleus. Although it was usually described as circular, the only estimates of the diameter were 10′ by Tupman on the 18th and 2′ by Engelhardt on the 26th. The tail caught the attention of most observers and was observed throughout the month, both with and without optical aid. On December 1, Backhouse reported that one tail extended 15° toward PA 330°, while the other extended 1° toward PA 280°. Barnard said a "slender, straight, and decidedly noticeable" tail about 10° long was seen with the naked eye on the 2nd, but the 15-cm refractor showed "the head is better formed, having lost that neck-like appearance, and the tails flowing symmetrically from the head." The bright ray situated within the northern tail (on November 30) had vanished and the space between the two tails was "decidedly filling with haze." That same night, Backhouse said the northern tail extended 11°

toward PA 320° and was 1° 50′ wide. A naked-eye observation by Backhouse on the 3rd revealed the north tail extended 19° toward PA 315°, while the south tail extended toward PA 275°. The north tail was 1° 50′ wide. On the 6th, Barnard said the comet was "very noticeable" to the naked eye. The tail was not bright but was "very straight and slender," and about 10° long. That same night, Backhouse said the naked eye revealed the north tail was extending 18° 20′ toward PA 345°, while the south tail extended 2° 5′ toward PA 290°. The north tail was 65′ wide, while the south tail was 16′ wide. On the 7th, Riccò said one tail extended about 2.3° toward PA 340°, while the other extended about 0.5° toward PA 280°. On the 8th, Riccò reported one tail extended about 3.5° toward PA 350°, while the other extended toward PA 290°. He noted that the long tail seemed slightly convex on the east side. Backhouse observed with spectacles on the 10th and noted the north tail extended 11° 30′ toward PA 355°, while the south tail extended 38′. The north tail was 1° 50′ wide, while the south tail was 7′ wide. Riccò said the tails seemed slightly convex on the east side on the 16th. Tupman said the northern tail was about 7° long on the 20th. On the 23rd, Backhouse observed with binoculars during the evening and noted the north tail extended 7° toward PA 10°, while the south tail extended 20′ toward PA 320°. On the 24th, Riccò noted one tail extended about 5.5° toward PA 15°, while the other extended about 20′ toward PA 320°. That same night, Backhouse observed with binoculars and noted the north tail extended 9° 55′ toward PA 350°, while the south tail extended 2.5° toward PA 310°. On the 25th, Backhouse noted the north tail extended 9° toward PA 10°, while the south tail extended 5° 45′ toward PA 350°. On the 28th, Backhouse said the north tail extended 6° 30′ toward PA 15°. On the 30th, he said the north tail extended 2° 50′ toward PA 15°, while the south tail extended 1° toward PA 340°.

As January began, the comet was becoming more challenging because of moonlight and its steady movement into twilight. On the 1st, Weinek said the nucleus was brighter than γ Aquilae (magnitude 2.72 – SC2000.0). On the 4th, Engelhardt said the comet was visible in a 13-cm finder in bright moonlight and bright twilight when no stars were visible. Engelhardt observed it with a 30-cm refractor on the 8th and said it was rather bright, but very diffuse, with a diameter of 0.5′. On January 10, J. I. Plummer (Orwell Park Observatory, England) simply described the comet as faint in a 25.4-cm refractor.

The comet was lost in the sun's glare for several weeks, during which time it passed less than 4° from the sun on February 11. The comet was picked up in the morning sky by W. H. Finlay (Royal Observatory, Cape of Good Hope, South Africa) on April 30.16 with the 18-cm refractor, while sweeping for comets. He observed it again the next morning and said the comet was small and faint, with slight central condensation. He remained the only observer of this comet from this moment on. He made another observation on the morning of May 31 and noted "occasional glimpses of a star-like nucleus." He estimated the magnitude of the nucleus as 11. The comet was last seen on June 17.03, when Finlay gave the position as $\alpha = 23^h\ 22.0^m$, $\delta = -44°\ 04'$. He said the comet was then faint, about 0.5' across, and with a slight central condensation.

Riccò said spectral observations on October 9 and 10 revealed the three bands of hydrocarbons. On November 2 and 10, he noted the yellow band was weaker than the blue. Gothard saw the three cometary bands at 5607 Å, 5156 Å, and 4706 Å on November 3. The middle band was sharp and moderately bright, while the other two were faint. Backhouse observed all three bands on several occasions during the period of November 16 to December 30. Pechüle saw all three bands on December 21.

The first parabolic orbit was calculated by H. Oppenheim. He used positions measured on October 7, 8, and 10, and determined the perihelion date as 1886 December 28.64. At about the same time, L. Boss and J. von Hepperger independently calculated orbits using observations from December 7–11. Boss' orbit had a perihelion date of December 11.90, while Hepperger's had a perihelion date of December 24.77. Additional orbits were determined by W. C. Winlock, H. Oppenheim, and C. N. A. Krueger before the end of the first month of observation. Krueger's was the first to represent the orbit accurately, as he used positions from October 7 to October 30 and determined the perihelion date as December 16.96. Additional parabolic orbits were calculated by S. C. Chandler, Jr, A. Svedstrup, W. H. Allen, and O. C. Wendell, before the comet faded from sight, but they only slightly improved upon the orbit of Krueger.

Nonparabolic orbits were calculated by J. Morrison (1887), C. G. T. Buschbaum (1889), and B. G. Marsden (1978). Morrison's orbit used positions spanning the period of October 8 to December 3 and was elliptical with a perihelion date of December 17.01 and a period of about 11 866 years. Buschbaum was the first person to use positions spanning the entire period of visibility. His orbit was hyperbolic and had a perihelion date of December 17.00 and an eccentricity of 1.0003824. Marsden used 62 positions spanning the entire period of visibility and revealed the hyperbolic orbit below.

T	ω	Ω (2000.0)	i	q	e
1886 Dec. 16.9968 (TT)	86.3484	138.9713	101.6154	0.663317	1.000408

ABSOLUTE MAGNITUDE: $H_{10} = 4.9$ (V1964)

FULL MOON: Sep. 13, Oct. 13, Nov. 11, Dec. 11, Jan. 9, Feb. 8, Mar. 9, Apr. 8, May 7, Jun. 5, Jul. 5

SOURCES: E. E. Barnard and E. Hartwig, *AN*, **115** (1886 Oct. 7), p. 239; C. F. Pechüle, S. Oppenheim, L. Weinek, and G. Gruss, *AN*, **115** (1886 Oct. 13), pp. 253–6; A. Riccò, H. Oppenheim, L. Boss, J. Hepperger, and H. C. F. Kreutz, *AN*, **115** (1886 Oct. 23), pp. 267–70; E. A. Lamp, W. Luther, J. Lamp, H. Oppenheim, and C. N. A. Krueger, *AN*, **115** (1886 Nov. 1), pp. 283–6; W. C. Winlock and E. Frisby, *AJ*, **7** (1886 Nov. 2), pp. 6, 8; H. Oppenheim, *AN*, **115** (1886 Nov. 13), p. 317; H. H. Turner and W. G. Thackeray, *MNRAS*, **47** (1886 Dec.), p. 65; A. Svedstrup, *AN*, **116** (1886 Dec. 1), p. 15; A. Riccò,

J. Franz, F. Porro, and E. A. Lamp, *AN*, **116** (1886 Dec. 8), pp. 27–30; S. C. Chandler, Jr, *AJ*, **7** (1886 Dec. 9), p. 23; F. Schwab, *AN*, **116** (1886 Dec. 18), pp. 43–6; E. Frisby, *AJ*, **7** (1886 Dec. 27), pp. 30–1; H. H. Turner, H. P. Hollis, and W. G. Browne, *MNRAS*, **47** (1887 Jan.), pp. 116–117; A. Svedstrup, *AN*, **116** (1887 Jan. 3), p. 77; E. E. Barnard, *AJ*, **7** (1887 Jan. 19), pp. 41–3; E. von Gothard, J. Lamp, and E. A. Lamp, *AN*, **116** (1887 Jan. 22), pp. 122–5; W. H. Allen, *AJ*, **7** (1887 Jan. 28), p. 55; L. Weinek and G. Gruss, *AN*, **116** (1887 Feb. 3), p. 155; B. von Engelhardt, *AN*, **116** (1887 Mar. 12), p. 247; E. Millosevich and A. Riccò, *AN*, **116** (1887 Mar. 25), p. 265; J. Palisa, S. Oppenheim, and J. Holetschek, *AN*, **116** (1887 Apr. 23), p. 347; J. Morrison, *MNRAS*, **47** (1887 May), p. 438; C. F. Pechüle, *AN*, **117** (1887 May 9), pp. 11–14; O. C. Wendell, *AN*, **117** (1887 Jun. 8), p. 59; W. H. Finlay, *AN*, **117** (1887 Sep. 20), p. 339; G. L. Tupman, *MNRAS*, **47** (1887, Supplementary Notice), p. 549; J. I. Plummer, *MNRAS*, **48** (1887 Nov.), pp. 57–8; W. H. Finlay, *MNRAS*, **48** (1888 Feb.), p. 181; W. Luther, *AN*, **120** (1888 Nov. 3), pp. 117, 133; C. G. T. Buschbaum, *Untersuchungen über die Bahn des Kometen 1886 IX (Barnard–Hartwig)*. Dissertation: Göttingen (1889), pp. 1–45; L. Weinek and G. Gruss, *AN*, **121** (1889 May 28), pp. 331–3; C. G. T. Buschbaum, *The Observatory*, **13** (1890 Jan.), pp. 61–2; W. H. Finlay, *Annals of the Cape Observatory*, **1** pt. 1 (1898), p. 27; T. W. Backhouse, *Publications of the West Hendon House Observatory*, no. 2 (1902), pp. 47–59; V1964, p. 63; B. G. Marsden, Z. Sekanina, and E. Everhart, *AJ*, **83** (1978 Jan.), p. 66; B. G. Marsden, *CCO*, 3rd ed. (1979), pp. 19, 47; SC2000.0, pp. 394, 397, 468, 493.

C/1887 B1 (Great Southern Comet)

1887 I = 1887a

Discovered: 1887 January 18.8 ($\Delta = 0.69$ AU, $r = 0.39$ AU, Elong. $= 18°$)

Last seen: 1887 January 30.50 ($\Delta = 0.61$ AU, $r = 0.77$ AU, Elong. $= 51°$)

Closest to the Earth: 1886 December 22 (0.5727 AU), 1887 January 27 (0.5986 AU)

Calculated path: MIC (Disc), GRU (Jan. 22), PHE (Jan. 28)

J. M. Thome (Cordoba, Argentina) discovered this comet in the evening sky on January 19.03. He noted, "it was then so faint and illusory in the twilight and denser atmosphere near the horizon, that I only suspected its nature." He said it was situated in Grus and looked similar to the bright comet of 1880. Although many sources have credited Thome as the first discoverer over the years, W. H. Finlay (Royal Observatory, Cape of Good Hope, South Africa) wrote in the March 1887 issue of the *Monthly Notices of the Royal Astronomical Society*, "The tail of this remarkable comet was first seen, so far as I am aware, by a farmer and a fisherman at Blauwberg (now known as Bloubergstrand, South Africa), on Tuesday night, January 18; the next evening it was seen at Grahamstown, Fraserburg, etc." The following rough dates can be estimated from Finlay's statement. The comet was seen at Blauwberg on January 18.8, Grahamstown on January 19.8, and Fraserburg on January 19.8. The comet had passed perihelion a week before discovery. Although it had passed closest to Earth about a month before discovery, the comet made a second slightly lesser approach about a week after the discovery.

The comet was widely observed by Southern Hemisphere observatories during the remainder of January. C. Todd (Adelaide, South Australia) said the head of the comet was hidden by the mists of the horizon on the 20th and indicated a tail length of about 25° on both that date and on the 21st. Also on the 21st, E. J. Molony (on board the ship *British Merchant*, in the southwestern portion of the Atlantic Ocean) described the comet as "a long slightly curved tail, originating in a point towards the Sun, and stretching for 18 1/2° towards the nearest (smallest) Magellan Cloud..." Several interesting discriptions were provided by Finlay, Todd, Molony, and Thome on the 22nd. Finlay described the comet as about 35° long. He said, "It presented the appearance of a pale narrow ribbon of light, quite straight, and of nearly uniform brightness throughout its length. There was no head or condensation of any kind visible near the end, the light simply fading away to nothing." Todd said, "The comet has a long narrow tail of about 30°, but no well-defined nucleus, resembling, in fact, very closely in appearance the comet of Feb. 1880...." Molony said the comet was "more indistinct" than on the 21st. Thome said the appearance of the comet was "precisely the same as the one observed here in 1880." He noted the comet was about 5° above the horizon, and exhibited a straight, narrow tail over 25° long. Although a careful search was made, no nucleus could be found. Thome found the comet virtually unchanged on the 23rd, except that it was "somewhat brighter." He said no nucleus or condensation could be found, but the tail was narrow and clearly outlined, with a length of "40° or more." Molony said the comet was faintly visible to the naked eye on the 25th, while Thome noted a tail 41° long, but "no distinctive" nucleus could be seen. Todd said the tail extended about 32° on the 26th, while Thome noted the comet was fading rapidly. On January 27, Todd indicated a tail length of about 29°. He added, "the head of the comet, a very diffused nebulous mass, appeared to be cut off by a wide break or rift, from the tail, and higher up there was another narrow break in the continuity of the tail." Although Todd said these breaks might be due to small clouds, he pointed out that faint stars were seen in one of the breaks. Tebbutt experienced cloudy weather until the evening of the 28th. As soon as it was dark he saw the tail "extending over many degrees." He noticed a 5th-magnitude star at the lower, bright extremity and assumed it was the nucleus, but upon turning the 11-cm refractor toward it he realized it was a star, which he later identified as σ Phoenicis. A careful search revealed no trace of the nucleus or a condensation. He described the tail as straight and indicated it extended about 20° toward PA 353°. Thome also saw the comet on the 28th and noted it was much fainter and "almost impossible to distinguish in the telescope." Finlay said the tail remained straight up to the 25th, and exhibited a "slight curvature" on the 27th, 28th, and 29th. The comet attained its most southerly declination of −49° on January 28.

The comet was last seen on January 30.50, when J. Tebbutt (Windsor, New South Wales, Australia) found it after the moon had set. At that time he noted the comet was excessively faint to the naked eye. Once again, his telescope revealed no nucleus or condensation, but he roughly measured the position as $\alpha = 0^h\ 26.1^m$, $\delta = -49°\ 26'$. Tebbutt said skies were cloudy on the 31st, but February 1 was clear. His search for the comet revealed nothing. Although L. Swift (Warner Observatory, Rochester, New York, USA) reported in the August 13 issue of the *Astronomische Nachrichten* that he possibly observed this comet with his 41-cm refractor on the evening of February 13, his position of $\alpha = 3^h\ 33^m 13^s$, $\delta = -37°\ 34'\ 36''$ was recognized as being several degrees away from the predicted position by H. C. F. Kreutz in the same issue. Swift was nevertheless convinced this object was a comet and noted that the "seeing was exquisite" and added, "It is remarkable that an object so large and bright should if a nebula have so long escaped detection." In the 1888 January 5 issue of the *Astronomische Nachrichten*, Swift indicated he still believed he had seen this comet.

The first orbit given for this comet was actually a hypothetical one based on the fact that this comet's appearance resembled the appearance of comets C/1843 D1, C/1880 C1, and C/1882 R1. H. C. F. Kreutz noted the similarity in the January 27 issue of the *Astronomische Nachrichten*. He subsequently provided the first ephemeris by taking the orbit of C/1880 C1 and changing the perihelion date to 1887 January 11.5.

Orbits based exclusively on positions measured for this comet did not immediately materialize for several reasons. First, the comet was only visible in the Southern Hemisphere and there were fewer observatories. Second, the astronomers who measured the positions frequently remarked that not only was a nucleus rarely, if ever, seen, but many were not even certain they were seeing the comet's head. Third, it took some time for all of the available positions to be collected. The first orbit was calculated by Finlay during the first half of February. In a letter he wrote on 1887 February 14 which was published in the 1887 March issue of the *Monthly Notices*, he said he used positions he had obtained on January 22, 25, and 28, and calculated a parabolic orbit with a perihelion date of January 11.74. Although he determined the perihelion distance as 0.015 AU, he realized the uncertainty of his orbit and said it "proved conclusively" that this comet belonged to the sungrazer family. Interestingly, Thome calculated an orbit from his positions measured on January 22, 23, and 24, but it was not sent to anyone until mid-March. It gave the perihelion date as January 9.33 and a much larger perihelion distance of 0.1966 AU. All available positions were published during the next couple of months and during May Oppenheim and S. C. Chandler, Jr, independently calculated orbits. Oppenheim's ultimately proved closer to the truth as he determined the perihelion date as January 11.91. Oppenheim (1889) revised the orbit using 22 positions obtained during the period of January 21–29. He determined the perihelion date as January 11.84 and reiterated Kreutz' earlier statement that Swift did not see this comet on 1887 February 13. Later orbits have been determined by Kreutz (1901) and Z. Sekanina (1978).

Sekanina's orbit is actually hypothetical and assumes "that the direction of perihelion is identical with that of members of the Kreutz sungrazing group and that the unobserved head of the comet was on a great circle through the Sun and inner part of the tail." Sekanina's orbit is given below.

T	ω	Ω (2000.0)	i	q	e
1887 Jan. 11.934 (UT)	83.513	4.585	144.383	0.00483	1.0

ABSOLUTE MAGNITUDE: $H_{10} = 6.3$ (V1964)

FULL MOON: Jan. 9, Feb. 8

SOURCES: J. M. Thome and H. C. F. Kreutz, *AN*, **116** (1887 Jan. 27), p. 143; H. C. F. Kreutz, *AN*, **116** (1887 Feb. 22), p. 205; J. Tebbutt, *MNRAS*, **47** (1887 Mar.), p. 293; W. H. Finlay and C. Todd, *MNRAS*, **47** (1887 Mar.), pp. 303–6; J. Tebbutt, *AN*, **116** (1887 Apr. 17), p. 319; E. J. Molony, *MNRAS*, **47** (1887 May), pp. 432–3; H. Oppenheim and W. H. Finlay, *AN*, **117** (1887 May 9), pp. 13–16; J. M. Thome and S. C. Chandler, Jr, *AJ*, **7** (1887 May 20), pp. 91–5; S. C. Chandler, Jr, *AJ*, **7** (1887 Jun. 6), p. 100; L. Swift and H. C. F. Kreutz, *AN*, **117** (1887 Aug. 13), pp. 217–22; J. M. Thome, *AN*, **117** (1887 Aug. 20), pp. 259–62; L. Swift, *AN*, **118** (1888 Jan. 5), p. 203; H. Oppenheim, *MNRAS*, **49** (1888 Dec.), pp. 68–9; H. Oppenheim and J. Tebbutt, *AN*, **121** (1889 May 31), pp. 337–42; W. H. Finlay, *Annals of the Cape Observatory*, **1** pt. 1 (1898), p. 27; H. C. F. Kreutz, *EAN*, **1** (1901), p. 55; V1964, p. 63; Z. Sekanina, *QJRAS*, **19** (1978), pp. 52–3.

C/1887 B2
(Brooks)

1887 II = 1887b

Discovered: 1887 January 23.00 ($\Delta = 1.41$ AU, $r = 1.78$ AU, Elong. $= 94°$)
Last seen: 1887 April 23.89 ($\Delta = 2.31$ AU, $r = 1.70$ AU, Elong. $= 42°$)
Closest to the Earth: 1887 February 11 (1.2879 AU)
Calculated path: DRA (Disc), CEP (Jan. 31), CAS (Feb. 10), CAM (Feb. 25), PER (Mar. 5), TAU-AUR (Apr. 2), TAU (Apr. 3)

W. R. Brooks (Phelps, New York, USA) discovered this comet on 1887 January 23.00. He gave the position as $\alpha = 18^h\ 00^m$, $\delta = +71°$, and said it was slowly moving eastward. Brooks simply described the comet as faint. O. C. Wendell (Harvard College Observatory, Cambridge, Massachusetts, USA) estimated the comet's magnitude as 9.5 on the 25th and 28th. G. Bigourdan (Paris, France) saw the comet on the 27th and said it was similar in brightness to a 12th-magnitude star. He added that it was round, about 1.5′ across, with a nearly stellar nucleus that was slightly off center. Upon the announcement of discovery, several astronomers noted the position was close to search ephemerides for P/Olbers (1815), but the direction of motion soon made it obvious this was not the expected comet.

The comet's perihelion distance was beyond the orbit of Mars and, as a result the comet's closest distance to Earth during the first half of February was a rather distant 1.29 AU. The comet attained its most northerly declination of $+80°$ on February 4. Jedrzejewicz (Plonsk) observed the comet with a 14.0-cm refractor on February 11 and described it as a faint, nebulous spot, with an intermittently visible nucleus of magnitude 10 or 11. W. Luther

(Hamburg, Germany) saw the comet with a 26-cm refractor on the 12th and said it looked like a star of magnitude 11 surrounded by a small and extraordinarily faint nebulosity. B. von Engelhardt (Dresden, Germany) saw the comet with a 30-cm refractor on the 15th and described it as bright, with a strong condensation and a star-like nucleus of magnitude 10. He added that the coma was 3′ across. Engelhardt also noted the comet was visible in the 13-cm finder. Luther said the comet looked like a nebulous star of magnitude 10 on the 16th. Engelhardt said the comet was quite bright on the 19th, with a condensed coma 1.5′ across, and a nucleus.

J. I. Plummer (Orwell Park Observatory, England) obtained several observations during March, but on each occasion the conditions made observation difficult. On the 2nd he said the comet appeared "very faint" because of moonlight and haze. On the 6th and 30th the comet was faint because of moonlight. Engelhardt described the comet as bright on the 11th, with a diameter of about 1′ and a good condensation and nucleus. On March 15, Luther said the 10th-magnitude nucleus stood out more sharply against the nebulosity than on previous nights. Engelhardt said the comet was bright on the 24th, with a diameter of about 1.5′ and with a well-defined nucleus.

Plummer found the comet "excessively faint" in the 25.4-cm refractor when the moon rose on April 10. Luther said the nucleus was magnitude 9 on the 11th. Luther said the comet was closer to the horizon when seen on the 12th and was distinctly fainter. He noted the nucleus was magnitude 10.5 and the coma was 2.5′ across.

The comet was last detected on April 23.89, when Plummer gave the position as $\alpha = 4^h 58.7^m$, $\delta = +20°\ 16′$. Observing at a low altitude with the 25.4-cm refractor, he said the comet was then faint.

The first orbit was calculated by H. A. Kobold. He used positions from January 25, 26, and 28, and determined the perihelion date as 1887 March 15.62. Orbits were calculated during the next few weeks by other astronomers like L. Boss, R. Spitaler, and H. Oppenheim, but the resulting perihelion dates ranged from March 8.45 to March 25.53. It was not until a full month of positions had become available that the perihelion date could be firmly established. The first to do this was Spitaler, who used positions from January 25 to February 25 and determined the perihelion date as March 17.51. Thereafter, Boss and Oppenheim included March positions in their calculations and obtained very similar orbits. C. Stechert (1888) was the first person to include the April positions in his calculations. Although he calculated a parabolic orbit similar to Spitaler's, the added positions indicated the orbit deviated from a parabola. Ultimately he found that an elliptical orbit with a perihelion date of March 17.87 and a period of about 1090 years fitted the positions best. Stechert (1904) refined his calculations by including the perturbations of Mercury to Saturn. The result was a perihelion date of March 17.89 and a period of about 1000 years. This orbit is given below.

T	ω	Ω (2000.0)	i	q	e
1887 Mar. 17.8904 (UT)	159.4229	281.5110	104.2737	1.630144	0.983692

ABSOLUTE MAGNITUDE: $H_{10} = 5.4$ (V1964)

FULL MOON: Jan. 9, Feb. 8, Mar. 9, Apr. 8, May 7

SOURCES: W. R. Brooks and H. A. Kobold, *AN*, **116** (1887 Jan. 27), pp. 143; W. R. Brooks, L. Boss, O. C. Wendell, and E. Frisby, *AJ*, **7** (1887 Jan. 28), pp. 55–6; G. Bigourdan, *CR*, **104** (1887 Jan. 31), pp. 276–7; E. A. Lamp, H. A. Kobold, G. A. P. Rayet, and L. Boss, *AN*, **116** (1887 Feb. 3), pp. 157–60; G. Bigourdan, G. Celoria, J. Palisa, S. H. Oppenheim, F. Bidschof, R. Spitaler, and H. Oppenheim, *AN*, **116** (1887 Feb. 10), p. 173; O. C. Wendell, *AN*, **116** (1887 Feb. 16), p. 191; L. Boss, E. Frisby, and E. E. Barnard, *AJ*, **7** (1887 Feb. 17), pp. 61–3; W. Luther, B von Engelhardt, and R. Spitaler, *AN*, **116** (1887 Feb. 22), pp. 203–6; H. Oppenheim, *AN*, **116** (1887 Feb. 28), p. 221; W. Schur and B. von Engelhardt, *AN*, **116** (1887 Mar. 12), p. 249; R. Spitaler, *AN*, **116** (1887 Mar. 12), p. 253; E. Frisby and C. L. Doolittle, *AJ*, **7** (1887 Mar. 15), pp. 78, 80; G. Agnello, B. von Engelhardt, and W. Schur, *AN*, **116** (1887 Mar. 25), pp. 265–8; H. H. Turner, A. M. W. Downing, W. G. Thackeray, and H. P. Hollis, *MNRAS*, **47** (1887 Apr.), pp. 392–3; L. Boss and E. Frisby, *AJ*, **7** (1887 Apr. 7), pp 85–7; B. von Engelhardt and H. Oppenheim, *AN*, **116** (1887 Apr. 17), p. 317; Jedrzejewicz, *AN*, **117** (1887 Sep. 10), p. 305; J. I. Plummer, *MNRAS*, **48** (1887 Nov.), pp. 58–60; C. Stechert, *AN*, **119** (1888 Aug. 13), pp. 331–4; W. Luther, *AN*, **120** (1888 Nov. 3), pp. 117, 133; C. Stechert, *AN*, **165** (1904 Jun. 13), pp. 321–8; V1964, p. 63.

C/1887 B3
(Barnard)

1886 VIII = 1887c

Discovered: 1887 January 24.44 ($\Delta = 2.16$ AU, $r = 1.68$ AU, Elong. $= 48°$)
Last seen: 1887 May 22.99 ($\Delta = 3.33$ AU, $r = 2.74$ AU, Elong. $= 47°$)
Closest to the Earth: 1886 July 28 (1.9433 AU), 1887 February 1 (2.1539 AU)
Calculated path: VUL (Disc), LYR (Jan. 24), VUL-CYG (Jan. 27), CEP (Mar. 15), CAS (Apr. 1)

E. E. Barnard (Nashville, Tennessee, USA) discovered this comet with a 13-cm refractor on 1887 January 24.44. He described the comet as magnitude 10, with a round coma and "very slight central condensation." Barnard confirmed the discovery with a 15-cm refractor on January 24.47, and gave the position as $\alpha = 19^h 07.7^m$, $\delta = +25° 22'$. The comet had passed perihelion nearly two months prior to discovery and had actually passed closest to Earth nearly six months earlier.

O. C. Wendell (Harvard College Observatory, Cambridge, Massachusetts, USA) saw the comet with the 38-cm refractor on January 25. He estimated the magnitude as 10, and noted a circular coma, less than 1' across, with some central condensation, but no tail. On the 27th, G. Bigourdan (Paris, France) said the comet resembled a nebula of class II, with a brightness similar to a star of magnitude 12. He added that the round coma was 1–1.5' across, with a diffuse nucleus 4–5" across.

The comet was discovered about a week before it made a second, lesser, approach to Earth as February began. H. Leitzmann (Göttingen, Germany) observed the comet with a 15-cm cometseeker on January 26 and said a nucleus was occasionally visible. A. H. P. Charlois (Nice, France) saw the comet on the 27th and described it as a faint, circular nebulosity, about 1' across, with a nucleus of magnitude 11. On February 17 and 18, W. Luther (Hamburg, Germany) observed with a 23-cm telescope and said the comet was very difficult to observe. J. I. Plummer (Orwell Park, England) described the comet as excessively faint in a 25.4-cm refractor on March 11. Luther noted the comet was faint, but distinct in his refractor on the 15th. Plummer said his refractor revealed a small and faint comet on the 16th. On April 15, J. Palisa (Vienna, Austria) said the coma was small, with a distinct nucleus. Charlois observed the comet with a 38-cm refractor on the 27th and noted a nucleus of magnitude 13 surrounded by faint nebulosity. On April 28, B. von Engelhardt (Dresden, Germany) observed the comet with a 30-cm refractor and described the comet as round, small, and faint, but added that it was still visible in a 13-cm seeker. The comet was last detected on May 22.99, when Palisa found it at a position of $\alpha = 3^h 02.3^m$, $\delta = +66° 39'$.

The first parabolic orbit was calculated by H. V. Egbert using positions obtained during the period spanning January 24–27. The resulting perihelion date was 1886 December 2.59. Additional orbits were calculated by E. Weiss, H. Oppenheim, Charlois, and Egbert, which eventually revealed a perihelion date of November 28.9. A definitive orbit was calculated by E. Fagerholm (1905). He began with 86 positions spanning the entire period of visibility and ultimately determined the orbit below. Although no details were given, Fagerholm noted that an elliptical orbit gave nearly the same residuals as the parabolic one.

T	ω	Ω (2000.0)	i	q	e
1886 Nov. 28.9056 (UT)	31.9115	259.7963	85.5867	1.480714	1.0

ABSOLUTE MAGNITUDE: $H_{10} = 4.8$ (V1964)

FULL MOON: Jan. 9, Feb. 8, Mar. 9, Apr. 8, May 7, Jun. 5

SOURCES: E. E. Barnard and O. C. Wendell, AN, **116** (1887 Jan. 27), p. 143; E. E. Barnard, O. C. Wendell, and H. V. Egbert, AJ, **7** (1887 Jan. 28), p. 56; G. Bigourdan, CR, **104** (1887 Jan. 31), pp. 276–7; A. H. P. Charlois, BA, **4** (1887 Feb.), p. 58; H. Leitzmann, G. Cacciatore, F. Courty, and E. Weiss, AN, **116** (1887 Feb. 3), pp. 157–60; H. Oppenheim, AN, **116** (1887 Feb. 10), p. 175; O. C. Wendell and E. Weiss, AN, **116** (1887 Feb. 16), p. 191; H. V. Egbert, L. Boss, and E. Frisby, AJ, **7** (1887 Feb. 17), pp. 61–4; H. Oppenheim, AN, **116** (1887 Feb. 22), p. 207; H. V. Egbert, AJ, **7** (1887 Feb. 28), p. 71; W. Luther and E. E. Barnard, AN, **116** (1887 Mar. 12), pp. 249–52; E. Frisby and E. E. Barnard, AJ, **7** (1887 Mar. 15), pp. 78–9; H. V. Egbert, AJ, **7** (1887 Apr. 7), pp. 83, 87; J. Palisa and H. C. F. Kreutz, AN, **116** (1887 Apr. 23), p. 367; J. Palisa, A. H. P. Charlois, B. von Engelhardt, and H. C. F. Kreutz, AN, **117** (1887 May 25), pp. 41–6; J. I. Plummer, MNRAS, **48** (1887 Nov.), p. 56; J. Palisa, AN, **119** (1888 May 13), p. 119; W. Luther, AN, **120** (1888 Nov. 3), pp. 117, 133; E. Fagerholm, AN, **169** (1905 Sep. 15), pp. 225–30; AMAF, **2** (1906 Apr. 26), pp. 1–22; V1964, p. 62.

C/1887 D1 *Discovered:* 1887 February 17.17 ($\Delta = 0.27$ AU, $r = 1.21$ AU, Elong. $= 139°$)
(Barnard) *Last seen:* 1887 April 10.88 ($\Delta = 1.65$ AU, $r = 1.03$ AU, Elong. $= 36°$)
Closest to the Earth: 1887 February 18 (0.2702 AU)
1887 III = 1887d *Calculated path:* PUP (Disc), MON (Feb. 18), ORI (Feb. 21), TAU (Feb. 23), ORI (Feb. 24), TAU (Feb. 25), PER (Mar. 10)

E. E. Barnard (Vanderbilt University Observatory, Nashville, Tennessee, USA) discovered this comet on 1887 February 17.17 and estimated the position as $\alpha = 8^h 04.1^m$, $\delta = -16° 10'$. He described it as "very faint, round, average size" and noted a rapid northwestward motion. The comet was confirmed by G. Bigourdan (Paris, France) on February 18.01 and by O. C. Wendell (Harvard College Observatory, Massachusetts, USA) on February 18.04. Wendell estimated the comet's magnitude as 11. The comet was then passing closest to Earth, but was still heading toward perihelion.

Barnard described the comet as very faint, round, and of average size on the February 19 and 23, but noted it was increasing in brightness. B. von Engelhardt (Dresden, Germany) said the comet was rather bright, but weakly condensed in the 30-cm refractor on the 24th, with a coma 1.5' across. He said the finder revealed a coma 4' across. Also on the 24th, E. Millosevich (Rome, Italy) noted the coma was slightly elliptical and uncondensed with an average diameter of 2', while A. Kammermann (Geneva, Switzerland) said the comet was faint, round, and slightly condensed at the center. On February 26, L. Boss (Dudley Observatory, New York, USA) said the comet "appeared to me to be at least three times as bright as the Finlay comet, which I had observed on the same evening." W. Luther (Hamburg, Germany) also saw the comet on the 26th and said the 26-cm refractor revealed a coma about 3' across and centrally condensed, but with no pronounced nucleus. J. I. Plummer (Orwell Park Observatory, England) noted the comet was excessively faint because of fog and moonlight on the 28th. He added that no central condensation was visible in the 25.4-cm refractor. On February 20 and 26, Wendell estimated the comet's magnitude as 11.

On March 1, Wendell estimated the comet's magnitude as 12. Engelhardt described the comet as faint on the 11th, with a slight condensation, but without a nucleus. Luther said the coma was about 2.5' across on the 13th, with no recognizable nucleus. On the 14th, Plummer said the 25.4-cm refractor revealed the comet as a "diffuse mass of light" in a hazy sky, while Luther said the 26-cm refractor revealed the comet was brighter than on the previous night and exhibited a nucleus of magnitude 12.5. Plummer said the comet was better seen on the 16th than on the 14th and he noted "a very faint nucleus was discernible towards the following side." On the 18th, Kammermann said the comet was very faint and only suspected when the threads of the micrometer were illuminated. Engelhardt said the comet was quite faint on the 24th, with a diameter of 2' and a faint condensation. He added that a stellar nucleus occasionally flashed. On March 28, Luther said the comet was very faint.

The comet was last detected on April 10.88, when Plummer saw it with a 25.4-cm refractor and gave the position as $\alpha = 2^h\,55.2^m$, $\delta = +36°\,40'$.

The first parabolic orbit was calculated by Boss using positions spanning the period of February 17–21. He determined a perihelion date of 1887 April 7.27. A few days later, Boss revised his calculations by adding a position on the 23rd and determined the perihelion date as March 28.97. This last calculation was generally confirmed during the next few weeks by the calculations of H. Oppenheim, J. Palisa, Wendell, and Barnard. P. A. Heinricius (1891) set out to calculate a definitive orbit. He determined both a parabolic and a hyperbolic orbit, with the latter having an eccentricity of 1.0004192. The parabolic orbit actually fitted the positions better and, because of the short period of visibility, this orbit is given below.

T	ω	Ω (2000.0)	i	q	e
1887 Mar. 28.9359 (UT)	36.5539	137.0419	139.7763	1.006455	1.0

ABSOLUTE MAGNITUDE: $H_{10} = 10.9$ (V1964)

FULL MOON: Feb. 8, Mar. 9, Apr. 8, May 7

SOURCES: E. E. Barnard, G. Bigourdan, and L. Boss, *AN*, **116** (1887 Feb. 22), p. 207; E. E. Barnard, E. F. Sawyer, O. C. Wendell, and L. Boss, *AJ*, **7** (1887 Feb. 28), p. 72; H. A. Kobold, B. von Engelhardt, J. Palisa, F. Hayn, W. Luther, and L. Boss, *AN*, **116** (1887 Feb. 28), pp. 221–4; E. E. Barnard, E. Millosevich, A. Kammermann, V. Knorre, J. Palisa, and H. Oppenheim, *AN*, **116** (1887 Mar. 12), pp. 251–6; E. Frisby, O. C. Wendell, E. E. Barnard, and L. Boss, *AJ*, **7** (1887 Mar. 15), pp. 78–9, supplement; O. C. Wendell, G. Agnello, B. von Engelhardt, H. A. Kobold, H. Oppenheim, *AN*, **116** (1887 Mar. 25), pp. 267, 271; L. Boss, *AJ*, **7** (1887 Apr. 7), p. 84; A. Kammermann, B. von Engelhardt, and O. C. Wendell, *AN*, **116** (1887 Apr. 17), pp. 315–18; E. E. Barnard, *AJ*, **7** (1887 May 20), p. 95; E. E. Barnard, *AN*, **117** (1887 Jun. 8), p. 59; J. I. Plummer, *MNRAS*, **48** (1887 Nov.), p. 61; W. Luther, *AN*, **120** (1888 Nov. 3), pp. 117, 133; P. A. Heinricius, *AN*, **128** (1891 Sep. 1), pp. 161–70; V1964, p. 63.

C/1887 J1   *Discovered:* 1887 May 13.19 ($\Delta = 0.48$ AU, $r = 1.48$ AU, Elong. = 168°)
(Barnard)   *Last seen:* 1887 August 12.14 ($\Delta = 0.78$ AU, $r = 1.61$ AU, Elong. = 127°)
   Closest to the Earth: 1887 June 8 (0.3992 AU)
1887 IV = 1887e   *Calculated path:* LUP (Disc), LIB (May 15), SCO (Jun. 6), OPH (Jun. 13)

E. E. Barnard (Nashville, Tennessee, USA) discovered this comet on 1887 May 13.19 and gave the position as $\alpha = 15^h\,10.8^m$, $\delta = -30°\,36'$ on May 13.21. He described it as faint, with a magnitude of about 11, and slowly moving northeastward.

The comet was well observed during the remainder of May as it approached both the sun and Earth. Several physical descriptions were provided on the 14th. O. C. Wendell (Harvard College Observatory, Cambridge, Massachusetts, USA) observed with the 38-cm refractor and gave the total magnitude as 10.2. L. Boss (Dudley Observatory, New York, USA) said the comet was extremely condensed, with a stellar nucleus of magnitude 11.5. Barnard described the comet as elongated, with a faint tail preceding. He added that he suspected a faint nucleus in an elongated condensation. E. A. Lamp (Kiel, Germany) described the comet as faint, round, with a nucleus. E. Millosevich (Rome, Italy) said the coma was small, faint, and irregular, but contained a nucleus of magnitude 11 or 12. On the 15th, G. Bigourdan (Paris, France) said the comet was magnitude 13, with a round coma about 1' across and a central condensation, while T. Zona (Palermo, Italy) said the comet was faint, with a nucleus. Lamp simply described the comet as faint on May 16, while C. Trépied (Alger, now al-Jazâ'ir, Algeria) said the nucleus was magnitude 12 in the 50-cm refractor, while the coma was elongated along the east–west line and was about 1.5' long. On the 19th, A. Kammermann (Geneva, Switzerland) said the comet was quite bright, with a magnitude of 9 or 10 in the 25.4-cm refractor. The coma was about 30" across, round, with a central condensation. That same night, B. von Engelhardt (Dresden, Germany) said the comet was quite bright, about 0.5' across, with a star-like nucleus. W. H. Finlay (Royal Observatory, Cape of Good Hope, South Africa) said the comet was small and round in the 18-cm refractor on the 20th, with a brightness equal to a star of magnitude 10. Engelhardt said the comet was rather bright, with a central condensation, and a small coma on the 22nd. Finlay estimated the nuclear magnitude as 11 on the 23rd, while Boss described the nucleus as stellar and magnitude 11.0 on the 24th. On the 27th, L. Weinek and G. Gruss (Prague, Czech Republic) said the comet was quite bright, with a distinct nucleus of magnitude 9.6. It was about 1' long. C. A. Young (Princeton College Observatory, New Jersey, USA) saw the comet with the 58-cm refractor on May 29 and said it was very diffuse and faint, with a diameter of about 1.5', but no central condensation. J. G. Lohse (Wiggleworth's Observatory, Scarborough, England) observed the comet with the 39-cm refractor on May 20, 22, and 30 and described it as "faint and small," with a nucleus.

The comet passed closest to both the sun and Earth during June. Young said the comet was faint and diffuse on June 6, with a diameter of about 4'. On the 8th, Finlay said the nucleus was about magnitude 11 in the 18-cm refractor. J. I. Plummer (Orwell Park Observatory, England) described the comet as "small, fairly bright, and with distinct nucleus" in the 25.4-cm refractor on the 9th. That same night, Wendell observed with the refractor and gave the magnitude as 9.8. On the 13th, Engelhardt said the comet was small, rather bright, and looked like a nebulous star of magnitude 9. Wendell saw the comet with the refractor on the 14th. He gave the magnitude as 9.7 and noted, "an incipient tail noticed, about 2' of arc in length." On the 15th, Clemens (Göttingen, Germany) said the comet was very faint. On the 16th, H. Battermann (Berlin, Germany) said the nucleus was between magnitude 10 and 11, while W. Luther (Hamburg, Germany) said the comet looked like a magnitude-10.5 star surrounded by a very tenuous nebulosity in the 26-cm refractor. Kammermann estimated the tail length as 2′ on the 17th, while Luther noted the coma was 20″ across and exhibited a nucleus of magnitude 10. On the 18th, E. Millosevich (Rome, Italy) said the tail was 5′ long and the nucleus was magnitude 10. That same night, Weinek said the comet had a distinct nucleus of magnitude 9.3 and a 1.4′ wide tail that extended 2.7–3.6′ toward PA 190–200°. Clemens described the comet as faint with an eccentric condensation on the 22nd. Luther said the comet was distinct with a nucleus of magnitude 10 on the 24th. Battermann said the comet was quite easy to see on the 26th and exhibited a faint condensation. That same night, Luther said the nucleus was less sharp than on previous nights and was surrounded by a small, faint nebula. On the 28th, Luther said the nucleus was stellar and magnitude 11. G. L. Tupman (Harrow, England) saw the comet on June 12, 15, 17, 19, and 22. He described it as "faint and small, condensed excentrically." The magnitude was given as 11 or 12.

The comet was moving away from both the sun and Earth during the remainder of its apparition. On July 8, Kammermann observed with a 25.4-cm refractor and said the comet still exhibited a short tail. On the 13th, Luther said the nucleus was magnitude 11 and was situated within a coma about 1′ across. Engelhardt saw the comet on the 16th and said it was quite bright, with a nucleus of magnitude 11, and a tail extending 1.5′ toward PA 239°. He added that the tail was 20″ wide. On the 17th, Luther said the comet appeared as a very faint nebulous star of magnitude 12, with very little coma. On July 20, Plummer said the comet was "faint and difficult to observe" in a 25.4-cm refractor. During the period of July 9–21, Barnard said a small, stellar nucleus was occasionally seen and was situated in the northern part of the coma. In addition, a faint fan-shaped tail was frequently noted. On the 27th, Plummer said the comet was faint, so that the position was uncertain, while Barnard said it was very faint. Plummer observed with a 25.4-cm refractor on the 28th and said, "The comet has become faint and difficult to observe, especially in declination, owing to its peculiar form. It is much elongated from N. to S., and narrow from W. to E. This peculiarity has been noticed on several occasions lately." The comet attained its most northerly declination of +9° on August 2.

The comet was last detected on August 11.14 and 12.14, when Barnard noted it was "excessively faint and difficult" in the 15-cm refractor. He indicated a position of $\alpha = 18^h\ 17.9^m$, $\delta = +8°\ 42′$. Barnard did not actually give a position for the comet on this date, but he did measure the distance of the comet from a star that the Author has determined as SAO 123381. The position above was determined by the Author after applying Barnard's measurements of the comet's distance from that star.

The first parabolic orbits were independently calculated by Boss and Lamp. Boss took positions from May 13, 15, and 16, and determined the perihelion date as 1887 June 27.18. Lamp took positions from May 13, 14, and 16, and determined the perihelion date as June 19.32. Additional orbits calculated during the next few days by A. Abetti, H. Oppenheim, and S. Oppenheim revealed the difficulty in establishing the perihelion date, as they gave values of June 15.83, June 21.61, and June 25.02, respectively. After further positions had been obtained, additional orbits by Wendell, S. Oppenheim, H. Oppenheim, and S. C. Chandler, Jr, indicated a perihelion date on June 17.

The first elliptical orbit was calculated by Chandler using positions obtained up to July 13. He, in fact, had calculated an improved parabolic orbit, but found rather large residuals in the middle places. Chandler then calculated an elliptical orbit with a perihelion date of June 17.16 and a period of about 5640 years.

Definitive orbits have been calculated by F. Muller (1888) and A. Abetti (1891). Both astronomers determined the perihelion date as June 17.16, but Muller gave the period as about 6725 years, while Abetti gave it as about 8298 years. Chandler (1891) expressed his opinion that Muller's orbit was the better of the two. Muller's orbit is given below.

T	ω	Ω (2000.0)	i	q	e
1887 Jun. 17.1634 (UT)	15.0876	246.8450	17.5479	1.393813	0.996088

ABSOLUTE MAGNITUDE: $H_{10} = 9.6$ (V1964)

FULL MOON: May 7, Jun. 5, Jul. 5, Aug. 3, Sep. 2

SOURCES: G. Bigourdan, *CR*, **104** (1887 May 14), p. 1360; E. E. Barnard, E. A. Lamp, T. Zona, and H. A. Kobold, *AN*, **117** (1887 May 18), p. 31; E. E. Barnard and L. Boss, *AJ*, **7** (1887 May 20), p. 96; G. Bigourdan, A. H. P. Charlois, J. Palisa, A. Abetti, E. Millosevich, E. A. Lamp, A. Kammermann, B. von Engelhardt, V. Knorre, S. Oppenheim, and H. Oppenheim, *AN*, **117** (1887 May 25), pp. 43–6; J. G. Lohse, *MNRAS*, **47** (1887 Jun.), p. 498; E. E. Barnard, C. A. Young, L. Boss, and S. C. Chandler, Jr, *AJ*, **7** (1887 Jun. 6), pp. 99, 102–4; E. E. Barnard, C. Trèpied, T. Zona, G. Agnello, B. von Engelhardt, L. Weinek, G. Gruss, H. Oppenheim, and S. Oppenheim, *AN*, **117** (1887 Jun. 8), pp. 57–62; A. Abetti, *AN*, **117** (1887 Jun. 16), pp. 101–4; O. C. Wendell and H. Oppenheim, *AN*, **117** (1887 Jun. 23), p. 119; Clemens, J. Lamp, W. Luther, and B. von Engelhardt, *AN*, **117** (1887 Jul. 1), p. 133; S. Oppenheim, *AN*, **117** (1887 Jul. 18), p. 165; O. C. Wendell, *AJ*, **7** (1887 Jul. 29), p. 119; B. von Engelhardt, *AN*, **117** (1887 Aug. 4), p. 215; O. C. Wendell, *AN*, **117** (1887 Aug. 13), p. 243; E. Millosevich and V. Cerulli, *AN*, **117** (1887 Aug. 27), pp. 265–72, 275; S. C. Chandler, Jr, *AJ*, **7** (1887 Sep. 3), pp. 121–2; W. H. Finlay, *AN*, **117** (1887 Sep. 20), p. 339; H. Battermann and E. E. Barnard, *AN*, **117** (1887 Oct. 11), pp. 385–8; G. L. Tupman, *MNRAS*, **47** (1887, Supplementary Notice), p. 550; J. I. Plummer, *MNRAS*, **48** (1887 Nov.), pp. 61–2; A. Kammermann, *AN*, **118** (1888 Jan. 18), p. 239; A. Kammermann, *AN*, **119** (1888 Apr. 16), pp. 39–42; F. Muller, *AJ*, **8** (1888 Jun. 27), pp. 44–56; W. Luther, *AN*, **120** (1888 Nov. 3), pp. 117, 133; L. Weinek, *AN*, **121** (1889 May 28), pp. 331–3; A. Abetti, *AN*, **126** (1891 Jan. 15), pp. 177–214; S. C. Chandler, Jr, *AJ*, **10** (1891 Feb. 21), pp. 166–7; V1964, p. 63.

13P/1887 Q1
(Olbers)

1887 V = 1887f

Recovered: 1887 August 25.37 ($\Delta = 2.14$ AU, $r = 1.37$ AU, Elong. $= 30°$)
Last seen: 1888 July 6.26 ($\Delta = 2.71$ AU, $r = 3.63$ AU, Elong. $= 151°$)
Closest to the Earth: 1887 October 9 (1.8782 AU)
Calculated path: CNC (Rec), LEO (Sep. 4), LMi (Sep. 12), LEO (Sep. 26), COM (Oct. 6), BOO (Oct. 28), SER (Nov. 24), OPH (Dec. 21), SER (Feb. 22), SCT (Mar. 17), SER (Apr. 18), SGR (May 23), OPH (May 30)

F. K. Ginzel (1882) redetermined the 1815 orbit of this comet and then applied perturbations from Jupiter, Saturn, Uranus, and Neptune, for the period of 1815 April 26 to 1887 January 19. The result was a prediction that the comet would next arrive at perihelion on 1886 December 17.40. Although he took great care in his calculations, he still indicated a potential uncertainty of ± 1.6 years in the period. The comet was accidentally discovered by W. R. Brooks (Phelps, New York, USA) on 1887 August 25.37. He estimated the position as $\alpha = 8^h 33^m$, $\delta = +29°$. J. Palisa (Vienna, Austria) observed the comet on August 28. The comet was just over one month from both perihelion and its closest approach to Earth.

A few physical descriptions were provided during the remainder of August. On the 28th, E. E. Barnard (Vanderbilt University, Nashville, Tennessee, USA) observed with the 15-cm refractor and noted the comet was bright, with a faint tail and a small, stellar nucleus. On the 29th, Barnard said the comet was bright and looked like "a large comet developing a head with small nucleus." There was also a faint tail. F. Porro (Turin, Italy) said the comet was small, but bright on the 30th. Although it was strongly condensed, no nucleus was visible. That same night, A. Kammermann (Geneva, Switzerland) said the comet was bright, at magnitude 7–8, with a nucleus and tail. G. Celoria (Milan, Italy) said a 20-cm refractor revealed a trace of tail on the 31st.

During the period of August 30 to September 3, C. Trépied, C. Rambaud, and F. Sy (Alger, now al-Jazâ'ir, Algeria) observed the comet with a 50-cm telescope. They noted a coma 2′ across and a nucleus of magnitude 10. They added that the nucleus seemed slightly south of the center of the coma. During the period of August 30 to September 2, L. J. Gruey (Besançon, France) said the circular coma was about 30″ across, with a small, central nucleus of magnitude 10–12.

Kammermann said the comet was visible in a 5-cm finder in moonlight on September 7, while a trace of tail was seen in a refractor. The comet attained its most northly declination of $+30°$ on the 9th. J. Holetschek (Vienna, Austria) estimated the nuclear magnitude as 8 on the 11th and 17th. G. Le Cadet (Lyon, France) observed with an 18-cm refractor on the 14th and described the comet as round and about 1′ across. A nucleus of magnitude 9.6 was located "a little behind the center" of the coma. On September 16, E. A. Lamp (Kiel, Germany) said the comet was faint and the nucleus was only occasionally seen, while G. Gruss (Prague, Czech Republic) said the comet was bright, 2′ across, and exhibited a distinct nucleus of magnitude 9.5. E. A.

Lamp said the comet was faint, with a sharp nucleus on the 19th, while on the 21st he noted the nucleus was larger and more diffuse. Also on the 21st, W. Luther (Hamburg, Germany) said the nucleus was about magnitude 10. J. Franz (Königsberg, now Kalingrad, Russia) said the tail was 10–15' long on the 22nd, while the comet's brightness was similar to a star of magnitude 9. That same night, Le Cadet said the comet was 5' across, with a condensation of magnitude 9.5. The coma seemed slightly spread out toward PA 305° and the condensation was slightly flattened in a direction perpendicular to this direction. J. Lamp (Bothkamp, Germany) said the nucleus was 13" in diameter on the 25th.

The comet passed about 35° from the sun on October 7, passed perihelion on the 8th, and then passed closest to Earth on the 9th. Franz said the comet was about as bright as an 8.8-magnitude star on the 9th and was noticeably fainter than an 8.3-magnitude star. On the 13th, Franz said the tail was 10' long. Holetschek gave the nuclear magnitude as 8.7 on the 14th. On October 18, Kammermann described the comet as very bright and noted the nucleus disappeared in the dawn light at the same time as stars of magnitude 9.5. Kammermann said the comet was fainter on the 28th than on the 25th.

The comet was moving away from both the sun and Earth from November onward. Although numerous positions were obtained, physical descriptions amounted to only a few a month. Luther said the nucleus was about magnitude 9 on November 16, but no tail was seen in the 25-cm refractor. The comet passed 33° from the sun on November 24. On December 12, Kammermann said the tail was about 5' long in a 25.4-cm telescope. A. H. P. Charlois (Nice, France) observed with a 38-cm refractor on December 24 and noted a bright nebulosity, with a nucleus of about magnitude 10. He added that the tail extended 20–25' toward PA 225°. According to Kammermann, the comet was very faint in moonlight on January 4. J. J. Plummer (Orwell Park Observatory, England) said the comet was "rather faint" on January 9 and faint because of fog and the brightening dawn on the 10th. On the 26th, B. von Engelhardt (Dresden, Germany) described the comet as quite bright, with a coma 1' across. He also noted a nucleus and condensation. On February 11, Plummer said the comet had "become sensibly fainter since the last observation" on January 10. On April 7, C. F. Pechüle (Copenhagen, Denmark) said the comet was extremely faint in a 27-cm refractor. E. E. Barnard (Lick Observatory, California, USA) observed the comet with a 30-cm refractor on June 16 and 18, as well as July 2, 3, and 5. He said the comet was very faint, "especially after the full moon in June, when it was observed only with the greatest difficulty."

The comet was last detected on July 6.26, when Barnard gave the position as $\alpha = 16^h\ 57.3^m$, $\delta = -20°\ 57'$. He was using the 30-cm refractor and said the comet "was then excessively faint and could be seen only once or twice as the feeblest glow." Barnard tried to observe the comet with the same instrument on July 7 and 8, "but no trace of it could be seen."

The first parabolic orbit was calculated by J. Franz using positions obtained during the period of August 27–29. He determined a perihelion date of 1887 October 14.13. Rambaud and Sy followed a few days later with an orbit using positions from August 30, September 1, and 3. The resulting perihelion date was October 14.44. These orbits were enough for astronomers to realize this was the return of Olbers' comet.

The first elliptical orbits were the result of a correction to Ginzel's prediction. Palisa said his position of August 28 indicated Ginzel's predicted perihelion date should be corrected to 1887 October 9.15. C. N. A. Krueger took Palisa's position and obtained a perihelion date of October 9.00 and a period of 72.61 years. O. Tetens took positions obtained up through September 21 and corrected Ginzel's perihelion date to October 8.96 and the period to 72.48 years.

The first elliptical orbit calculated exclusively from the 1887 positions was by H. V. Egbert. Using positions spanning the period of August 29 to September 24, he determined the perihelion date as October 9.05 and the period as 71.84 years. An additional orbit was calculated by G. M. Searle, which had a period of 72.26 years.

Later calculations used positions from the comet's 1815, 1887, and 1956 apparitions. The most recent investigations of this comet's orbit were conducted by H. Q. Rasmusen (1967), D. K. Yeomans (1978, 1986), and K. Kinoshita (2000). Kinoshita's orbit is given below.

T	ω	Ω (2000.0)	i	q	e
1887 Oct. 8.9741 (TT)	65.3558	86.0578	44.5717	1.199085	0.930966

ABSOLUTE MAGNITUDE: $H_{10} = 5.0$ (V1964), $H_0 = 5.0$ $n = 6$ (Kronk)

FULL MOON: Aug. 3, Sep. 2, Oct. 2, Oct. 31, Nov. 30, Dec. 30, Jan. 28, Feb. 27, Mar. 27, Apr. 26, May 25, Jun. 23, Jul. 23

SOURCES: F. K. Ginzel, *VJS*, **17** (1882), pp. 109–14; W. E. Plummer and F. K. Ginzel, *MNRAS*, **46** (1886 Feb.), pp. 230–1; W. R. Brooks, *AN*, **117** (1887 Aug. 27), p. 279; H. A. Kobold, E. Millosevich, F. Schwab, F. Porro, A. Kammermann, J. Palisa, and J. Franz, *AN*, **117** (1887 Sep. 3), pp. 293–6; C. Trèpied, C. Rambaud, F. Sy, and L. J. Gruey, *CR*, **105** (1887 Sep. 5), pp. 430–2; G. Celoria, A, Kammermann, and C. N. A. Krueger, *AN*, **117** (1887 Sep. 10), pp. 307–10; C. Trèpied, C. Rambaud, F. Sy, E. E. Barnard, F. Porro, Jedrzejewicz, and E. A. Lamp, *AN*, **117** (1887 Sep. 15), pp. 325–8; C. Rambaud and F. Sy, *CR*, **105** (1887 Sep. 19), p. 487; L. J. Gruey, J. Franz, H. A. Kobold, and E. A. Lamp, *AN*, **117** (1887 Sep. 20), p. 341; G. Le Cadet, *CR*, **105** (1887 Sep. 26), p. 512; E. A. Lamp, W. Luther, and O. Tetens, *AN*, **117** (1887 Sep. 28), pp. 355–8; J. Lamp, J. Franz, W. Luther, and F. K. Ginzel, *AN*, **117** (1887 Oct. 11), pp. 387–90; H. V. Egbert, *AJ*, **7** (1887 Oct. 18), pp. 135–6; J. Franz and J. Palisa, *AN*, **118** (1887 Nov. 9), pp. 41–4; A. Kammermann, *AN*, **118** (1887 Dec. 2), pp. 109–12; G. M. Searle, *AJ*, **7** (1887 Dec. 14), pp. 153–5; A. H. P. Charlois, *CR*, **106** (1888 Jan. 2), pp. 42–3; B. von Engelhardt, *AN*, **118** (1888 Feb. 3), p. 271; B. Matthiessen, *AN*, **118** (1888 Feb. 25), p. 319; K. Bohlin, A. Kammermann, and A. Abetti, *AN*, **118** (1888 Mar. 21), p. 379; A. Abetti, *AN*, **119** (1888 Apr. 16), p. 41; G. Gruss, *AN*, **119** (1888 Jul. 24), pp. 261–3; E. Millosevich, *AN*,

119 (1888 Aug. 29), p. 363; E. E. Barnard and F. Schwab, *AN*, **120** (1888 Oct. 15), pp. 65–8; W. Luther, *AN*, **120** (1888 Nov. 3), pp. 119, 134; J. J. Plummer, *MNRAS*, **49** (1888 Dec.), pp. 73–4; C. F. Pechüle, *AN*, **122** (1889 Jul. 25), pp. 137–40; J. Palisa and J. Holetschek, *AN*, **123** (1889 Nov. 2), p. 57; J. Holetschek, *DAWW*, **93** (1917), p. 238; H. Q. Rasmusen, *Publikationer og mindre Meddeler fra Kobenhavns Observatorium*, No. 147 (1948); V1964, p. 63; H. Q. Rasmusen, *The Definitive Orbit of Comet Olbers for the Periods 1815–1887–1956*. Copenhagen: Arnold Busck (1967), pp. 18–47, 56–64; D. K. Yeomans, *QJRAS*, **19** (1978), pp. 80–1, 88; D. K. Yeomans, *QJRAS*, **27** (1986), p. 604; personal correspondence from K. Kinoshita (2000).

C/1888 D1   *Discovered:* 1888 February 19.1 ($\Delta = 0.89$ AU, $r = 0.89$ AU, Elong. $= 56°$)
(Sawerthal)   *Last seen:* 1888 September 25.1 ($\Delta = 2.37$ AU, $r = 3.10$ AU, Elong. $= 128°$)
   Closest to the Earth: 1888 February 25 (0.8844 AU)
1888 I = 1888a   *Calculated path:* TEL (Disc), SGR (Feb. 26), MIC (Mar. 2), CAP (Mar. 9), AQR (Mar. 18), CAP (Mar. 19), AQR (Mar. 22), PEG (Mar. 31), AND (May 10), CAS (Jul. 2)

On the morning of 1888 February 19 H. Sawerthal (Royal Observatory, Cape of Good Hope, South Africa) was returning from the photographic observatory, where he had been working on the southern *Durchmusterung*, when he spotted something with the naked eye that "looked suspiciously like a comet." He at once confirmed his suspicion with the aid of an opera glass. He called his colleague, W. H. Finlay, who obtained a position of $\alpha = 19^h 11.5^m$, $\delta = -56° 04'$ on February 19.11. Finlay noted the comet was visible to the naked eye, and that a telescope revealed a stellar nucleus of magnitude 7 and a 2° long tail.

The comet passed closest to Earth a few days after its discovery, but was still a month from perihelion. It was exclusively a Southern Hemisphere object during the remainder of February. Finlay said the 18-cm refractor revealed a tail 3° long on the 21st. A. B. Biggs (Launceston, Tasmania) observed with a 20-cm reflector on the 24th and said the nucleus was "bright, well-defined, and surrounded with considerable coma." He indicated the nuclear magnitude was 8.4. On the 25th, L. A. Eddie (Graham's Town, Cape of Good Hope, South Africa) said the comet had a "very condensed nucleus, equal in magnitude to a 4th magnitude star." The curved tail could be traced to 8°, but it was especially bright up to 3° from the nucleus. In the 24-cm reflector, Eddie noted, "it appeared a magnificent object of a deep golden colour, with highly condensed nucleus possessing a nucleolus, or tiny point of brilliant light in the centre. The nucleus was more sharply defined on its posterior border, but no dark interval was traceable." Eddie said the comet was lost to the naked eye about 49 minutes before sunrise, but it continued to be followed with the telescope. As twilight increased he wrote, "At one time it appeared as a barbed spear, the barbed head resembling two wings; then these faded away and it resembled a ball of fire on the end of a luminous stick; lastly a fluffy ball of pale light alone remained, all vestige of tail and surrounding coma having faded away." On the 26th, J. Clarke (Captain of the SS *Olbers*, near the east coast of South America and about 400 miles north of Rio de Janeiro) said the tail was 4° long to the naked eye. Eddie said the comet was scarcely visible to the naked eye on the 28th because of the full moon. Although he suspected it had not changed in brightness since the 25th, he did note "a widening of the coma in advance of the nucleus, and a slight trace of dark intervals therein dividing it into cometic envelopes."

On March 4, Eddie said the comet was conspicuous to the naked eye with a 3° long tail in moonlight and he estimated its brightness as magnitude 3.5. In the 24-cm reflector he noted, "the tail appeared ruddy, and the nucleus a greenish yellow. The latter was very brilliant, and surrounded by a coma which seemed to proceed in a double jet of faintly luminous matter from the centre of the nucleus to a short distance in front of it, and was then bent back to the right and left into the tail, the interval between these jets being filled with cometic matter of still fainter luminosity." On the 7th, Biggs observed with a 20-cm reflector and said the nucleus was sharply defined, with less coma about the head. He estimated its magnitude as about 6.4. He added that the tail tapered toward the head. R. Belding (Captain of the barque *Atlantic*, sailing northward in the south Atlantic Ocean about 1000 miles off the coast of South America) first made sextant observations of the comet on the 10th. He noted the comet appeared brighter on the morning of the 11th and was still brighter and larger on the 12th. On the 13th, Eddie noted the tail was 6° long to the naked eye, but decreased to about 2.5° or 3° after Venus rose. On the 14th, Belding said the tail curved toward the southwest, while A. Riccò and T. Zona (Palermo, Italy) said the wide, divergent tail was directed toward the west-southwest. Riccò and Zona added that the nucleus was shining almost as bright as 27 Capricorni (magnitude 6.2 – SC2000.0). Biggs said the nucleus equaled 30 Capricorni (magnitude 5.43 – SC2000.0) on the 15th.

The comet passed perihelion on the 17th at which time Riccò and Zona saw it with the naked eye and noted the nucleus was only slightly less luminous than 18 Aquarii (magnitude 5.49 – SC2000.0). They saw a tail 2° long. Eddie last saw the comet on March 17. He said it still appeared equal in brightness to a 4th-magnitude star. The 24-cm reflector revealed the nucleus "to be slightly diffused on the preceding border." Belding also saw the comet on this morning. He said the nucleus appeared brighter than on previous mornings, but seemed to vary in brightness. The tail also looked "lighter and thinner." Finlay last detected the comet on March 18 and noted the tail extended toward PA 230°. On the 20th, Riccò and Zona said a telescope revealed "the nucleus seems lengthened and divided in two." The tail was bright to a length of 1° 28′, with a fainter portion visible to 2°. The south side was brighter than the north side, while the north side was more defined. That same morning, C. B. Hill (Chabot Observatory, Oakland,

California, USA) said the nucleus was magnitude 4.5 and the tail was about 1.5° long. On the 21st, Riccò and Zona said the tail was 1° 40' long and convex on the northern side, with the southern side the brighter. A. H. P. Charlois (Nice, France) said he observed with a 38-cm refractor on March 20 and 22, and noted the nucleus was elongated toward PA 240° and, at times, seemed double. On the 22nd, L. C. Dart (Captain of the ship *Alcester*, then sailing in the eastern Indian Ocean) noted the tail extended about 2° to the west-southwest of the nucleus. F. Porro (Turin, Italy) said the comet had a red color on the 23rd. It also exhibited a straight tail, which was directed toward the west. He added that the nucleus was slightly inferior in brightness to the comparison star (aka ξ Aquarii, magnitude 4.69 – SC2000.0). Belding last saw the comet on March 24, when his ship was about 500 miles south of the equator. Riccò and Zona saw the comet the same morning and said it was visible to the naked eye like a star of magnitude 5. The tail was easily visible to 1° 40', although a fainter portion was visible to 2°. Also on the 24th, Hill said the tail was less than 1° long and the nucleus was about 0.3 magnitude less than β Aquarii (magnitude 2.91 – SC2000.0). G. Bigourdan (Paris, France) said the nucleus was 20″ long and 15″ wide, with a diffuse border on the 26th. He added that the coma was 3' across, while the tail extended 20'. On the 26th, Riccò and Zona said the tail was 1° 40' long, while Hill said the tail was less than 0.5° long. Hill added that the nucleus was slightly less bright than on the previous morning. Hill said the comet was easily seen with the naked eye on the 27th and exhibited a nucleus of magnitude 4. On the 28th, L. Cruls (Rio de Janeiro, Brazil) revealed the nucleus was elongated, with three apparent condensations present, while H. A. Kobold (Strasbourg, France) said the comet was easy to see with the naked eye. On the 30th, F. Schwab (Kremsmünster, Austria) noted a bright nucleus and tail extending about 16' in moonlight. That same morning, Riccò and Zona said the nucleus was as bright as a star of magnitude 8, while the tail was convex toward the north. In addition, Cruls and H. Morize (Rio de Janeiro, Brazil) said the secondary nucleus was located 2.96″ from the primary in PA 243.1°.

The comet's distances from the sun and Earth were increasing during April. Riccò and Zona said the comet was still a naked-eye object on the 1st, while a telescope revealed a jet about 5' long. Hill said the tail was about 0.5° long in moonlight on the 2nd. On the 3rd, Cruls and Lacaille (Rio de Janeiro, Brazil) said the secondary nucleus was located 3.79″ from the primary in PA 243.6°. W. Luther (Hamburg, Germany) said the nucleus was magnitude 8, while the tail was directed westward on the 4th. That same night, E. A. Lamp (Kiel, Germany) noted the middle of the tail was wider than either end. On the 5th, Biggs said, "The whole figure appeared to have assumed quite a ruddy tint as compared with its former appearance – possibly atmospheric, but I think not." Hill reported two distinct nuclei were visible in his 21.6-cm refractor. Numerous descriptions were published for April 6. E. Millosevich (Rome, Italy) said the stellar nucleus was about magnitude

6, while the tail extended 50′ toward PA 260°. He added that the comet was visible to the naked eye. C. F. Pechüle (Copenhagen, Denmark) said the tail extended toward PA 260.5° ± 0.3°. He also noted that the second nucleus was situated 5″ from the main nucleus toward PA 247.5° ± 1.5°. Riccò and Zona said the nucleus was less distinguished than when previously seen, and the tail was 1° 38′ long and exhibited a well-defined northern border. A. Anguiano (Observatorio Astronómico Nacional Mexicano de Tacubaya, Mexico) said the comet was yellow-gold in color and exhibited a nucleus of magnitude 6–7 and a tail 2° long. He added that a brilliant line crossed the entire length of the tail. Another night of extensive observations was the 7th. W. Wickham (Radcliffe Observatory, Oxford, England) said the tail was distinctly visible to the naked eye. He added that the 25.4-cm refractor revealed the nucleus was "dead white, well defined, and elongated, thus at times I thought the nucleus double." Kobold estimated the comet's magnitude as 7. Hill said the comet was easily seen with the naked eye, with a tail 0.5° long. Pechüle said the tail extended toward PA 263° ± 2°, while Bigourdan said it extended toward PA 260.3°. Pechüle added that the secondary nucleus was situated toward PA 249° ± 0.1°. On the 8th, B. von Engelhardt (Dresden, Germany) estimated the magnitude of the comet's head was 7, while Schwab said the tail appeared 1° long in the finder. Riccò and Zona noted a red coloration to the nucleus. On the 9th, Riccò and Zona said the comet was visible to the naked eye and exhibited a tail 1° 38′ long, which was convex on the northern side. That same night, Kobold estimated the magnitude of the nucleus as 8 and noted the tail extended 20′ toward PA 262° 18′. Anguiano first noted the nucleus was divided on the 10th, while using a 38-cm refractor. He said the fainter component was located 4″ from the main nucleus in the direction of the tail, and noted a very weak condensation "in front of the previous ones." Hill said the double nucleus was again seen on the 11th in his 21.6-cm refractor, although not as distinctly as on the 5th. E. von Gothard (Herény, Hungary) obtained a 45-minute exposure of this comet on April 13. He said the plate showed a very bright, round nucleus, from which very narrow, straight lines extend to form the tail. The tail was slightly curved, with the convex side toward the north, and became more diffuse the further it was from the coma. That same night, Hill said two nuclei were seen, with the fainter component closer to the brighter than on the 5th. W. H. Robinson (Radcliffe Observatory, Oxford, England) said the nucleus was as bright as a 5th-magnitude star on the 16th, while the tail was about 5° long to the naked eye. Engelhardt and Pechüle provided additional observations that same night. Engelhardt said the tail extended about 50′ toward PA 260°, while Pechüle reported it extended 30′ toward PA 264° ± 0.5°. Engelhardt said the secondary nucleus was situated 6.3″ from the primary in PA 255° 09′, while Pechüle said it was located 5″ away in PA 258° ± 0.1°. On the 18th, N. de Konkoly (O'Gyalla Observatory, Hungary) noted the nucleus was as bright as a 5th-magnitude star, while the tail was over 1.5° long, and J. Franz (Königsberg, now Kaliningrad, Russia) said the tail was 1° long. On the 20th, Engelhardt said the 13-cm finder revealed a tail extending 75' toward PA 266° 58', which contained a bright linear ray than ran from the head to the middle of the tail. The 30-cm refractor again showed a double nucleus, with the companion being located in PA 254° 33'. Kobold said the nucleus was magnitude 9 on the 22nd, while Luther gave it as magnitude 9.5 on the 23rd. Also on the 23rd, Schwab said the tail appeared 16' long in moonlight. On the 26th, Luther reported the nucleus was slightly fainter than the magnitude 8.1 comparison star, while Hill said the comet was surprisingly bright, with only one nucleus. Pechüle noted the tail extended toward PA 270°. Schwab said the tail was 20' long on the 30th and curved with the convex side facing northward.

On May 2, Wickham said the comet was invisible to the naked eye. On the 4th, F. A. Bellamy (Radcliffe Observatory, Oxford, England) observed the central part of the elongated nucleus with the 25.4-cm refractor, while Millosevich said the comet was fainter than when previously seen, but still exhibited a tail 50' long. Wickham said the 25.4-cm refractor still revealed an elongated nucleus on the 5th. On the 6th, H. H. Turner (Royal Observatory, Greenwich, England) said the 17-cm refractor revealed a tail 20' long, while J. J. Plummer (Orwell Park Observatory, England) said the comet was "rather fainter than expected." That same night, E. Becker (Strasbourg, France) said the nucleus was brighter and more condensed than when previously seen, with a magnitude of 9. He added that the tail extended toward PA 265°. Anguiano said the tail was 2° long on the 8th, while the nucleus was magnitude 10–10.5. On the 9th, L. Schwarz and G. Gruss (Prague, Czech Republic) said the comet exhibited a nucleus of magnitude 9.0 and a tail which extended 20' toward PA 270°. L. Becker (Lord Crawford's Observatory, Dun Echt, Scotland) reported the comet was magnitude 9.5 on the 11th, while Pechüle said the tail extended toward PA 270°. Pechüle added that the second nucleus was situated 15" from the main nucleus toward PA 270°. On May 12, Turner said the comet was only visible to the naked eye by averted vision. On May 13, W. G. Thackeray (Royal Observatory, Greenwich, England) described the comet as "very faint" to the naked eye, while Franz said the comet was not seen with the naked eye. During the period of May 14–18, I. Y. Kortazzi (Nicolajew, now Mykolayiv, Ukraine) said the nucleus was elongated and was so faint it disappeared with the slightest illumination of the micrometer wire. A thin tail extended 60' toward PA 273° on the 16th. L. Weinek (Prague, Czech Republic) observed with a 9.1-cm refractor (54×) on the 14th and said the comet's head had a magnitude of 8.8, while the tail extended 31.3' toward 270°. Franz gave the tail length as 15–20' on the 16th. On the 18th, L. Becker said the comet looked "like a bright, pretty large nebula, hardly brighter in the middle." He added that a tail extended about a minute of arc toward the north-preceding side in "rather strong twilight." The tail length was given as 10' by J. Fènyi (Haynald Observatory, Kalocsa, Hungary), 50' by Gruss, and 1° 05' by Kobold. Kobold said the tail extended toward PA 268° 40'. Engelhardt noted an elliptical nucleus on the 19th and said the tail extended 7.5′ toward PA 271° 33′. On the 20th, F. Blumbach (Dorpat, now Tartu, Estonia) said the nucleus was magnitude 9–10, while Fènyi gave it as 9.3.

The comet experienced an outburst in brightness on May 21. Fènyi said it exhibited an easily visible, diffuse nucleus in the refractor, while the 5.4-cm finder revealed the comet's brightness was about equal to a star of magnitude 7.8. Kortazzi said the nucleus was condensed with the clarity of a star of magnitude 8 or even 7.8, while the tail appeared fainter than when previously seen. He noted the nucleus withstood a fairly well-lit field of view. Boss said the comet was "decidedly star-like" on the 22nd and about magnitude 7. Also, on the 22nd, L. Becker first reported a "remarkable change in the appearance of the comet" while observing with a 38-cm refractor. He said the nucleus appeared as "a well-defined object of a few seconds diameter, from which the matter of the comet appeared to flow out towards the north and south preceding sides, falling over in parabolic curves equally to the north and south." Becker also noted that the tail seen on May 18 extending toward the north-preceding side was gone and the magnitude of the nucleus was 9.1. Fènyi also reported the comet was remarkably bright, with the 5.4-cm finder revealing the total brightness as 6.8. Numerous observations were also made on the 23rd. Plummer said the sky was very clear and the comet was "brighter than expected." Schwarz said the comet had unexpectedly brightened since his last observation on the 21st. Blumbach said the nucleus was magnitude 7–8 and about 5″ in diameter. Franz said the comet had maintained its new brightness, but the nucleus was more diffuse and the tail was not visible. Fènyi said the 5.4-cm finder revealed the comet as magnitude 6.8. B. Matthiessen (Karlsruhe, Germany) noted the nucleus was remarkably bright and estimated the magnitude as 5–6. Another day of extensive observations was May 24. Thackeray said the comet had a well-defined nucleus of magnitude 4 or 5, which was "showing brightly in strong moonlight." He saw no tail. Robinson observed with the 25.4-cm refractor (100×) and said the comet was about 3′ wide. He wrote that it "resembles a pair of wings. The following half of the nucleus forms a well-defined and bright semicircular disc, the preceding part gradually fading down until lost in the nebulosity of the coma." Robinson added that the nucleus was magnitude 6.8 in the 8-cm finder. A faint tail was also visible that extended about 2′. L. Becker said the comet had not changed in appearance since the 22nd and the nucleus was still magnitude 9.1. Millosevich observed in moonlit skies, but noted a bright nucleus. Pechüle said the tail extended 15′ toward PA 270°. On the 25th, L. Becker commented, "The nucleus was larger than previously, and appeared decidedly brighter" than a star of magnitude 9.2. Kortazzi said the nucleus was less condensed than when seen on the 21st. E. Becker said the nucleus was magnitude 8–9. On the 26th, Kobold said the nucleus was magnitude 8.5, and the 46-cm refractor revealed that it exhibited a large fan-shaped emittance. Fènyi said the comet was very diffuse, large, and bright on the 28th. Pechüle described the nucleus as magnitude

9 on the 29th and described the tail as extending toward PA 275°. Kortazzi said the nucleus had become so faint on May 30 that it was very difficult to see when the field of view was illuminated.

During the first half of June, Plummer said his measurements of the comet's right ascension were uncertain "owing to the elongated form of the nucleus of the comet." On the 1st, Turner reported the comet was "about as bright as a 7th magnitude star," while Kortazzi said the nucleus was again elongated in the direction of the tail. Kortazzi added that the tail extended about 50' toward PA 272°. On the 2nd, E. Becker said the tail extended toward PA 271.7°, while Schwab noted the tail was 20' long. Turner reported the comet was "a little fainter than a 7.5 magnitude star" on the 4th, while Kobold said the nucleus was magnitude 9, with a tail extending toward PA 271° 45'. On the 6th, Turner said the comet was "about as bright as a 9th magnitude star," while Kortazzi described the tail as extending 24' toward PA 271°. E. E. Barnard (Lick Observatory, California, USA) scrutinized the comet on the 8th. He noted the head was "remarkably slender and pointed" and added that if the nucleus existed it could not be distinguished from the head itself. The tail was still traceable for several degrees. Barnard also noted, "The peculiar wing-like appendages of the head, though much less bright than the body, were very conspicuous and gave the comet a most ghostly and bird-like appearance. The shoulders of these wings were quite well defined, and were noticeably unsymmetrical with reference to the main body." Thackeray said the comet was "very faint" on the 8th and 11th. Schwab found the tail 15' long on the 9th. On the 11th, Barnard said the "wings were very striking and the shoulders singularly well defined." He added, "The body of the comet had undergone a decided change, being somewhat swollen and diffused into the wings." E. Becker said the nucleus was elongated and magnitude 9–10 on the 12th. On the 14th, Engelhardt said the tail extended 10' toward PA 270°, while the nucleus had a linear shape and was 1' long. E. Becker said the nucleus was magnitude 10–11 on the 27th and the tail was 7–8' long. On the 28th, A. M. W. Downing (Royal Observatory, Greenwich, England) noted the comet exhibited a stellar nucleus of magnitude 10, as well as a straight tail extending about 6'. On the 29th, Kortazzi said the nucleus was about magnitude 12.

On July 3, Wickham noted the "wing" appendages seen on May 24 were gone, while the tail was about 8' long. He added that the nucleus was only occasionally seen in the 25.4-cm refractor (100×). Bellamy estimated the nuclear magnitude as 11 on the 5th. On the 9th, Matthiessen said the comet was a pale nebulosity, without a recognizable nucleus, while Engelhardt said the comet was 10' long without a nucleus. Engelhardt added that the tail extended toward PA 262° 04'. On the 10th, Schwab noted the comet was only visible for brief instances as a narrow, 10' long nebulosity of almost even brightness. H. P. Hollis (Royal Observatory, Greenwich, England) reported the comet "was only a very faint patch of light" on the 11th. He added that there was "no perceptible nucleus." Robinson reported the comet was barely visible at times in the 25.4-cm refractor (100×) on the 12th, while the nucleus was "exceedingly faint, and only seen occasionally." E. Becker observed with a 46-cm refractor on the 15th and said the nucleus was magnitude 11–12. On the 16th, Engelhardt said the comet was quite bright and measured about 0.5′ wide and about 10′ long and that the 30.5-cm refractor may have shown a weak nucleus. That same night, Kortazzi said the nucleus was not visible, while the tail extended 6′ to 7′.

On August 2, Plummer reported he observed the comet, but noted a small star within the coma and decided he could not accurately measure the comet's position. E. Becker said the nucleus was magnitude 13, while a tail extended 18′ toward PA 260°. On the 8th, Plummer reported the comet was very faint and the coma had almost disappeared, but it exhibited a small, star-like nucleus. He only then realized that the "star" of the 2nd was actually the nucleus. During the period of August 4–8, A. Abetti (Padova, Italy) said the comet was like a star of magnitude 12–13 and was "lacking all characteristics of comets." Wickham noted the sky was "exceptionally fine" when he observed the comet with the 25.4-cm refractor on August 9. He said that when his eye became dark adapted he noted the field was "rich with small stars" and that the comet "could be distinguished as [a] very faint luminous haze." On the 10th, Kobold said the coma was an elongated nebulosity, with a long axis of 3′. He added that the coma was without condensation, except for a fine point of light. The comet attained its most northerly declination of +55° on August 26.

The final observations were made during September. Kobold saw the comet on the 3rd with a 46-cm refractor and described it as extremely faint, 5′ across, and without a condensation. No position could be measured. L. Swift (Warner Observatory, Rochester, New York, USA) turned his 41-cm refractor toward the predicted position of the comet on September 25 as soon as evening twilight had ended. The sky was "exceptionally fine" and Swift wrote, "After closely scanning the field, I at length noticed a large and excessively faint nebulous object" and gave the position as $\alpha = 23^h 53.7^m$, $\delta = +52° 19′$. Because of its size, Swift was uncertain as to whether this was the comet or a nebula; however, he examined the position on two other occasions during the next two months and did not detect anything nebulous. Swift did not provide the exact time the position was measured. He implied that the first observation occurred not long after the end of evening twilight, which would be around September 25.1.

J. Tebbutt (Windsor, New South Wales, Australia) and J. M. Thome (Cordoba, Argentina) gave brief summaries of their observations. Tebbutt observed the comet with the naked eye and with a 20-cm refractor nearly every morning from February 28 to March 5. He said moonlight prevented the comet from being conspicuous to the naked eye, and the tail was only visible for 2–3°. The telescope revealed a small, brilliant nucleus. He last saw the comet on April 3, with his 20-cm refractor. He noted the comet's nucleus was "considerably elongated." For the period of February

24 to April 10, Thome said the comet was visible to the naked eye. The maximum brightness was about 3.5, while the telescope revealed a dense and perfectly straight tail, which attained a maximum length of 5°. He noted, "The nucleus was a bright elliptical nebula of approximately 20″ by 13″, and continued visible for nearly half an hour after the micrometer threads were illuminated by morning light."

The spectrum of this comet was obtained by several observers during the period of mid-March until early May. Zona observed it on March 14 and 17, and noted traces of the carbon bands. E. W. Maunder (Royal Observatory, Greenwich, England) observed the comet on April 11, 20, and May 4 and said the spectrum was continuous on each morning, although the observations on the 20th revealed very feeble traces of the first and second carbon bands. The third carbon band was only suspected on that same morning. On April 18, de Konkoly obtained the spectrum and noted five bands. He measured these as 5614.7 Å, 5462.5 Å, 5158.8 Å, 5132.6 Å, and 4725.6 Å.

The first parabolic orbit was calculated by Finlay at the end of February. The resulting perihelion date was 1888 March 17.68. Additional orbits by E. Becker, E. C. Cooke, W. C. Winlock, and C. N. A. Krueger eventually established the perihelion date as March 17.50.

The first elliptical orbit was determined by A. Berberich using positions obtained on March 14, 25, and April 6. He determined the perihelion date as March 17.50 and the period as about 2639 years. About a week later, he revised the period to 2371 years. Additional elliptical orbits were calculated by Berberich (period = 2371 years), G. M. Searle (period = 1648 years), and L. Boss (period = 1615 years). Boss added, "I suspect that the true period will be decidedly greater than 2000 years."

A definitive orbit by J. F. Tennant (1889) began with positions spanning the period of February 19 to July 16. The final result was an elliptical orbit with a perihelion date of March 17.50 and a period of about 2182 years. Nearly 90 years later, B. G. Marsden (1978) became the first person to use positions obtained into September. Although the orbit was essentially unchanged, the period was slightly greater at about 2204 years. This orbit is given below.

Z. Sekanina (1979) obtained two possible solutions for the motions of the fragments. The first solution indicated the split happened on March 11 and utilized both precise and approximate observations of the nuclei spanning the period of March 30 to May 11. The second solution only utilized the precise positions from the period of March 30 to April 16 and indicated a split date of March 2. Sekanina said the second solution was the better fit. He also said there were not enough observations of the possible third nucleus to allow an analysis.

T	ω	Ω (2000.0)	i	q	e
1888 Mar. 17.5021 (TT)	359.9049	246.9622	42.2482	0.698772	0.995874

ABSOLUTE MAGNITUDE: $H_{10} = 4.7$ (V1964)

FULL MOON: Jan. 28, Feb. 27, Mar. 27, Apr. 26, May 25, Jun. 23, Jul. 23, Aug. 21, Sep. 20, Oct. 19

SOURCES: H. Sawerthal, *AN*, **118** (1888 Feb. 25), p. 319; H. Sawerthal, W. H. Finlay, and D. Gill, *MNRAS*, **48** (1888 Mar.), p. 295; W. H. Finlay and B. Matthiessen, *AN*, **118** (1888 Mar. 6), p. 351; A. Riccò and T. Zona, *AN*, **119** (1888 Mar. 26), p. 15; G. Bigourdan, *CR*, **106** (1888 Mar. 26), p. 919; H. H. Turner, L. A. Eddie, J. Clarke, and J. Tebbutt, *MNRAS*, **48** (1888 Apr.), pp. 309–12; A. H. P. Charlois, F. Porro, E. Millosevich, H. A. Kobold, E. A. Lamp, C. N. A. Krueger, and J. Möller, *AN*, **119** (1888 Apr. 3), pp. 27–30; A. H. P. Charlois, *CR*, **106** (1888 Apr. 3), pp. 1000–1; G. Bigourdan, *CR*, **106** (1888 Apr. 9), pp. 1060–1; J. Tebbutt, F. Schwab, L. Schwarz, W. Luther, E. Millosevich, B. von Engelhardt, E. A. Lamp, E. Becker, and A. Berberich, *AN*, **119** (1888 Apr. 16), pp. 41–6; W. C. Winlock, *AJ*, **8** (1888 Apr. 18), p. 16; B. von Engelhardt, *AN*, **119** (1888 Apr. 24), p. 79; J. Tebbutt, A. Riccò, T. Zona, W. Luther, E. A. Lamp, E. von Gothard, and A. Berberich, *AN*, **119** (1888 Apr. 30), pp. 89, 93–6; H. H. Turner, W. Wickham, W. H. Robinson, F. A. Bellamy, A. B. Biggs, and R. Belding, *MNRAS*, **48** (1888 May), pp. 341–3, 345–50; L. Boss and G. M. Searle, *AJ*, **8** (1888 May 4), pp. 22–4; W. Luther, W. H. Finlay, and O. Tetens, *AN*, **119** (1888 May 8), p. 107; N. de Konkoly, *AN*, **119** (1888 May 19), p. 141; L. Cruls, Lacaille, Morize, and L. Schwarz, *AN*, **119** (1888 May 26), pp. 155–9; H. H. Turner, W. G. Thackeray, W. H. Robinson, A. B. Biggs, J. Tebbutt, and L. Becker, *MNRAS*, **48** (1888 Jun.), pp. 373–81; L. Boss, *AJ*, **8** (1888 Jun. 4), pp. 36–7; A. Abetti, O. Tetens, F. Blumbach, J. Franz, and A. Kammermann, *AN*, **119** (1888 Jun. 4), pp. 167, 171; J. Fènyi and L. von Wutschichowski, *AN*, **119** (1888 Jun. 20), pp. 189–92; J. M. Thome, I. Y. Kortazzi, E. J. White, and P. Baracchi, *AN*, **119** (1888 Jun. 25), pp. 215–18; W. H. Finlay, L. Schwarz, Gruss, L. Weinek, H. Pomerantzeff, and A. Kammermann, *AN*, **119** (1888 Jul. 24), pp. 257–64; B. von Engelhardt, *AN*, **119** (1888 Jul. 31), pp. 301–4; B. Matthiessen and B. von Engelhardt, *AN*, **119** (1888 Aug. 22), p. 345; E. Millosevich, *AN*, **119** (1888 Aug. 29), p. 363; E. E. Barnard, *AN*, **120** (1888 Sep. 26), p. 43; H. A. Kobold, E. Becker, and I. Y. Kortazzi, *AN*, **120** (1888 Oct. 15), pp. 67–72; H. C. Wilson, *AN*, **120** (1888 Oct. 19), p. 85; C. Todd, E. C. Cooke, and L. C. Dart, *MNRAS*, **48** (1888, Supplementary Notice), pp. 397–8; W. Wickham, F. A. Bellamy, and W. H. Robinson, *MNRAS*, **49** (1888 Nov.), p. 33; J. J. Plummer, A. M. W. Downing, W. G. Thackeray, and H. P. Hollis, *MNRAS*, **49** (1888 Dec.), pp. 75–6, 79–80; L. Swift, *SM*, **7** (1888 Dec.), p. 454; F. Schwab and O. C. Wendell, *AN*, **120** (1888 Dec. 13), pp. 233–6; A. Abetti, *AN*, **120** (1889 Jan. 9), p. 285; H. Sawerthal, *MNRAS*, **49** (1889 Feb.), p. 184; C. B. Hill, *AN*, **120** (1889 Feb. 21), pp. 323–6; J. F. Tennant and E. W. Maunder, *MNRAS*, **49** (1889 Mar.), pp. 282–5, 307; E. Becker and H. A. Kobold, *AN*, **121** (1889 Apr. 25), pp. 201–6; C. F. Pechüle, *AN*, **122** (1889 Jul. 25), pp. 137–41; A. Anguiano, *AN*, **122** (1889 Aug. 17), pp. 225–8; J. Holetschek and J. Palisa, *AN*, **123** (1889 Nov. 2), p. 57; W. Wickham, W. H. Robinson, and F. A. Bellamy, *MNRAS*, **55** (1895 Jan.), pp. 161–3; W. H. Finlay, *Annals of the Cape Observatory*, **1** pt. 1 (1898), pp. 30–1; V1964, p. 63; B. G. Marsden, *AJ*, **83** (1978 Jan.), p. 66; B. G. Marsden, *CCO*, 3rd ed. (1979), pp. 19, 48; Z. Sekanina, *Icarus*, **38** (1979), pp. 303–5; SC2000.0, pp. 537, 541, 544, 548, 550.

2P/Encke

1888 II = 1888b

Recovered: 1888 July 8.35 ($\Delta = 0.98$ AU, $r = 0.43$ AU, Elong. $= 25°$)

Last seen: 1888 August 25.98 ($\Delta = 0.96$ AU, $r = 1.25$ AU, Elong. $= 78°$)

Closest to the Earth: 1888 July 31 (0.7129 AU)

Calculated path: CNC (Rec), LEO (Jul. 12), HYA (Jul. 13), SEX (Jul. 15), LEO-CRT (Jul. 24), CRV (Aug. 1), VIR-HYA (Aug. 8), CEN (Aug. 22), LUP (Aug. 25)

The 1888 June 4 issue of the *Astronomische Nachrichten* contained a prediction for this comet's return in 1888. O. Backlund and B. Seraphimoff determined an orbit for the 1885 apparition and then applied perturbations by Jupiter for the period of 1884 December 18 to 1888 March 7, and Venus, Earth, Mars, and Saturn for the period of 1884 December 18 to 1885 August 10. They predicted the comet would arrive at perihelion on 1888 June 28.49 and provided an ephemeris for the period of 1888 May 12 to August 28. J. Tebbutt (Windsor, New South Wales, Australia) recovered this comet with his 11-cm refractor on 1888 July 8.35 at a position of $\alpha = 8^h 47.9^m$, $\delta = +12° 49'$. He said it was found with the help of the ephemeris of Backlund and Seraphimoff, and was then located "In the band of twilight along the western horizon." Tebbutt described the comet as a "small bright well condensed nebula without a nucleus." He later revised his description to, "The comet presented the appearance of a bright round nebulous star about 1' in diameter, uniformly condensed and without coma." The sky location was very poor for Tebbutt because of the comet's low altitude and his observatory's buildings. He said he telegraphed the information to three other Australian observatories. The comet had already passed perihelion.

The comet was heading toward a moderately close approach to Earth at the end of July. Tebbutt remained the only observer to provide physical descriptions for most of July. With the use of his 11-cm refractor, he noted the comet was smaller and fainter on the 10th than on the 8th. He noted it was smaller still on the 11th. On the 15th and 16th, Tebbutt said the comet was excessively faint because of the moon. He described it as "of the last degree of faintness" on the 18th. Tebbutt observed the comet with his 20-cm refractor on the 25th and said it appeared excessively faint, 2' in diameter, without a condensation. Using the same telescope on July 31 and August 1, he said the comet was "of the last degree of faintness and observed with the greatest difficulty." Tebbutt again noticed no condensation. J. M. Thome (Cordoba, Argentina) saw the comet in dense haze on July 29.

Although Tebbutt had sent a formal notice to the *Astronomische Nachrichten* that was dated July 10, it was delayed on its way to that publication and was not published until the August 22 issue. In the meantime, the August 10 issue of the *Astronomische Nachrichten* contained the notice that W. H. Finlay (Royal Observatory, Cape of Good Hope, South Africa) had recovered the comet on August 3.76. On that night, as well as those of August 4, 5, 6, and 9, Finlay said the comet appeared as "a very faint patch of light without visible condensation, and about 2' in diameter." He added that it was "exceedingly faint and very difficult" to see in the 18-cm refractor.

The last observations were obtained by Thome. He described the comet as very faint on August 24 and saw it for the final time on August 25.98, when he gave the position as $\alpha = 15^h\ 01.4^m$, $\delta = -30°\ 37'$. He described the comet as "barely visible."

Refinements of this comet's orbit using positions from multiple apparitions, as well as planetary perturbations, have been published by several astronomers. Of most significance was the study of B. G. Marsden and Z. Sekanina (1974). They found a perihelion date of June 28.47 and a period of 3.31 years. Marsden and Sekanina determined the nongravitational terms as $A_1 = -0.14$ and $A_2 = -0.02263$. The orbit of Marsden and Sekanina is given below.

T	ω	Ω (2000.0)	i	q	e
1888 Jun. 28.4699 (TT)	183.9291	336.2255	12.9000	0.343093	0.845460

ABSOLUTE MAGNITUDE: $H_{10} = 9.7$ (V1964)

FULL MOON: Jun. 23, Jul. 23, Aug. 21, Sep. 20

SOURCES: O. Backlund and B. Seraphimoff, *BASP* (Series 4), **32** (1888), pp. 467–72; O. Backlund and B. Seraphimoff, *AN*, **119** (1888 Jun. 4), pp. 173–6; W. H. Finlay, *Nature*, **38** (1888 Aug. 9), p. 350; W. H. Finlay, *AN*, **119** (1888 Aug. 10), p. 319; J. Tebbutt, *AN*, **119** (1888 Aug. 22), p. 349; W. H. Finlay, *AN*, **120** (1888 Sep. 26), p. 43; J. Tebbutt, *AN*, **120** (1888 Dec. 11), p. 219; J. M. Thome, *AN*, **122** (1889 Sep. 6), p. 307; W. H. Finlay, *Annals of the Cape Observatory*, **1** pt. 1 (1898), p. 32; V1964, p. 63; B. G. Marsden and Z. Sekanina, *AJ*, **79** (1974 Mar.), pp. 415–16.

C/1888 P1
(Brooks)

1888 III = 1888c

Discovered: 1888 August 8.08 ($\Delta = 1.62$ AU, $r = 0.91$ AU, Elong. $= 31°$)
Last seen: 1888 October 30.74 ($\Delta = 2.42$ AU, $r = 1.77$ AU, Elong. $= 39°$)
Closest to the Earth: 1888 August 27 (1.5193 AU)
Calculated path: UMa (Disc), CVn (Aug. 22), BOO (Sep. 9), SER (Sep. 24), HER (Oct. 12), OPH (Oct. 20)

W. R. Brooks (Smith Observatory, Geneva, New York, USA) discovered this comet on 1888 August 8.08 and estimated the position as $\alpha = 10^h\ 05.0^m$, $\delta = +44°\ 30'$. He said it was moving easterly. The comet had just passed perihelion and would be closest to Earth at the end of August.

On August 9, H. Seeliger (Munich, Germany) simply described the comet as bright in a 27-cm refractor. The comet attained its most northerly declination of $+45°$ on the 10th, when H. A. Kobold (Strasbourg, France) said the comet was 0.5' across, with a sharp nucleus of magnitude 10. A tail-like extension protruded toward PA $= \pm250°$. That same night, Seeliger said the tail was about 1' long in the refractor. W. Luther (Hamburg, Germany) described the comet as small and round on the 14th, with a nucleus of magnitude 11. During the period of August 9–16, A. Abetti (Padova, Italy) said the comet remained near the horizon and in haze, but a trace of tail was visible.

On the 24th, C. N. A. Krueger (Kiel, Germany) said the comet was difficult to see in bright moonlight, while E. Millosevich (Rome, Italy) and Luther independently noted a distinct nucleus of magnitude 11. Millosevich also saw a tail extending 6–7′ toward the northwest. B. von Engelhardt (Dresden, Germany) said the comet was quite bright, round, and condensed in the 30.5-cm refractor on the 25th. He added that the coma was 0.5′ across and exhibited a stellar nucleus of magnitude 12. V. Knorre (Berlin, Germany) described the comet as faint in the 24.4-cm refractor (90×) on the 27th. Kobold observed with the 46-cm refractor on the 28th and noted a nucleus of magnitude 9.5, as well as a tail extending 5′ toward PA 308.2°. Engelhardt said the comet was quite bright on the 29th. He described the coma as 0.5′ across, with a granular condensation and a stellar nucleus. A faint tail extended 3′ toward PA 314.6°. On the 30th, O. C. Wendell (Harvard College Observatory, Cambridge, Massachusetts, USA) described the comet as rather low and faint in the 38-cm refractor. On August 31, A. Kammermann (Geneva, Switzerland) said the comet exhibited a rather bright, centrally condensed coma, with a short, faint tail (25.4-cm telescope). Luther reported the comet was distinctly brighter than nearby NGC 5033 (magnitude 10.1 – NGC2000.0). Krueger said the comet was difficult to see when the threads of the micrometer were illuminated.

During the period of August 28 to September 3, Abetti said the nucleus was magnitude 7–8, while the tail was 6′ long. Luther estimated the magnitude of the nucleus as 11 on September 1. On the 6th, Wendell said the comet was faint in the 38-cm refractor. Luther said the comet was distinctly faint on the 11th, with a nucleus of magnitude 12.5. Engelhardt found the comet small, faint, and round in the refractor on the 12th. He noted a granular condensation and a small, stellar nucleus. Kammermann observed with a 25.4-cm telescope on the 15th and said the comet was so faint in moonlight that no position could be measured. Engelhardt described the comet as small, round, and quite bright in the 30.5-cm refractor on the 23rd. He also noted a possible tail. On the 25th, Luther reported a distinct, stellar nucleus of magnitude 11.8. Millosevich noted the comet was faint in a 25-cm refractor on the 27th.

On October 2, Millosevich said the comet was faint, with a nucleus of magnitude 12. J. I. Plummer (Orwell Park Observatory, England) noted the comet was "rather faint" on the 5th, while Luther saw a nucleus of magnitude 13. On October 9 and 11, Wendell said the comet was very faint in the 38-cm refractor. The comet was last detected on October 30.74, when J. Palisa (Vienna, Austria) gave the position as $\alpha = 16^{\text{h}}\ 49.4^{\text{m}}$, $\delta = -0°\ 29′$. He was using a 68.0-cm refractor.

The first parabolic orbit was calculated by H. C. F. Kreutz using positions from August 9, 10, and 11. The resulting perihelion date was determined as 1888 July 16.66. Additional orbits were calculated by L. Stutz, L. Boss, Kreutz, W. C. Winlock, and H. V. Gummere.

Two definitive orbits were calculated in 1889, which indicated the comet was moving in an elliptical orbit. J. F. Tennant determined the perihelion date as July 31.60 and the period as about 8897 years. Millosevich determined the perihelion date as July 31.64 and the period as about 969 580 years.

T	ω	Ω (2000.0)	i	q	e
1888 Jul. 31.6364 (UT)	59.2168	103.0576	74.1904	0.902226	0.999908

ABSOLUTE MAGNITUDE: $H_{10} = 7.6$ (V1964)
FULL MOON: Jul. 23, Aug. 21, Sep. 20, Oct. 19, Nov. 18
SOURCES: W. R. Brooks and E. Weiss, *AN*, **119** (1888 Aug. 10), p. 319; F. Porro, H. A. Kobold, H. Seeliger, and H. C. F. Kreutz, *AN*, **119** (1888 Aug. 13), p. 335; F. Schwab, F. Porro, W. Luther, L. Stutz, L. Boss, and H. C. F. Kreutz, *AN*, **119** (1888 Aug. 22), pp. 347–50; L. Boss, *AJ*, **8** (1888 Aug. 23), p. 80; E. Millosevich, W. Luther, C. N. A. Krueger, and H. C. F. Kreutz, *AN*, **119** (1888 Aug. 29), pp. 365–8; H. A. Kobold, B. von Engelhardt, V. Knorre, E. Millosevich, W. Luther, C. N. A. Krueger, and L. J. Gruey, *AN*, **119** (1888 Sep. 6), p. 379; L. J. Gruey, C. N. A. Krueger, and I. Y. Kortazzi, *AN*, **120** (1888 Sep. 26), p. 45; G. A. P. Rayet, F. Courty, and J. Möller, *AN*, **120** (1888 Oct. 4), pp. 59, 63; W. C. Winlock, *AJ*, **8** (1888 Oct. 17), p. 118; B. von Engelhardt, *AN*, **120** (1888 Oct. 19), p. 89; E. Millosevich, *AN*, **120** (1888 Dec. 13), p. 235; O. C. Wendell and E. Stuyvaert, *AN*, **120** (1888 Dec. 19), pp. 251–4; F. Schwab, *AN*, **120** (1888 Dec. 28), p. 271; A. Abetti, *AN*, **120** (1889 Jan. 9), p. 285; J. I. Plummer, *MNRAS*, **49** (1889 Apr.), p. 370; H. A. Kobold, *AN*, **121** (1889 Apr. 25), pp. 201–5; H. Seeliger and V. Knorre, *AN*, **121** (1889 May 10), pp. 265–8; C. W. Pritchett, *AN*, **122** (1889 Jun. 25), p. 31; A. Kammermann, *AN*, **122** (1889 Jul. 19), p. 115; H. V. Gummere, *AJ*, **9** (1889 Oct. 31), p. 94; J. F. Tennant, *MNRAS*, **50** (1889 Nov.), pp. 43–5; J. Palisa, *AN*, **123** (1889 Nov. 2), p. 57; E. Millosevich, *AN*, **123** (1889 Nov. 19), p. 111; V. Knorre, *AN*, **124** (1890 Jun. 17), pp. 345, 362; W. Luther, *AN*, **127** (1891 Apr. 17), pp. 49–52, 71; V1964, p. 63; NGC2000.0, p. 157.

4P/Faye

1888 IV = 1888d

Recovered: 1888 August 10.12 ($\Delta = 1.98$ AU, $r = 1.75$ AU, Elong. $= 62°$)
Last seen: 1889 February 8.42 ($\Delta = 1.45$ AU, $r = 2.35$ AU, Elong. $= 149°$)
Closest to the Earth: 1888 December 26 (1.2265 AU)
Calculated path: TAU (Rec), ORI (Aug. 26), GEM (Sep. 7), ORI-GEM (Sep. 9), CMi (Oct. 9), CNC (Nov. 3), CMi (Nov. 6), HYA (Nov. 13), CMi (Dec. 23)

H. C. F. Kreutz (1888) took D. M. A. Möller's predicted orbit for the 1881 apparition, applied a correction to represent the positions of 1880–1, and simply advanced it to the 1888 apparition after noticing no significant perturbations by Jupiter. The result was a perihelion date of 1888 August 17.6. He published a sweeping ephemeris covering the period of 1888 July 15 to October 27. This ephemeris gave positions for the actual predicted perihelion date as well as for perihelion dates 8 days earlier and later. The comet was recovered by H. J. A. Perrotin (Nice, France) on 1888 August 10.12, using Kreutz' ephemeris and a 38-cm refractor. He said the comet was very faint, with a weak central condensation, and a circular coma about 1' across. The position was given as $\alpha = 5^h 00.5^m$, $\delta = +20° 01'$, which indicated the comet was 2.60 days later than the prediction. E. E. Barnard (Lick Observatory, California, USA) said the comet was easily found on the morning of August 11 with the 30-cm refractor. He noted, "I think it could have been observed here probably from one to two months earlier had its place been known."

The last observations made during August were by Perrotin on the 18th and Barnard on the 19th. Although the comet passed perihelion on August 20, it changed little in brightness during the next four months as it moved closer to Earth until near the end of December. It remained a difficult object to follow throughout this period. Barnard managed an observation on September 3, but the observatory at Cambridge, England reported their failure to find the comet on September 12 in a "perfectly clear" sky. They noted the Northumberland refractor was used to search along Kreutz' corrected ephemeris until daybreak. Barnard next saw the comet on October 8 and obtained at least three observations per month during the next few months. Meanwhile, other observatories continued to report problems. Searches for the comet in November were futile (except for Barnard's) because the comet was moving through a rich star field. E. Becker (Strasbourg, France) failed on the 10th, and B. von Engelhardt (Dresden, Germany) also searched in vain. R. Spitaler (Vienna, Austria) reported that he found the comet on November 12 with the 69-cm refractor, but then noted on the 14th that the object had not moved and was a hitherto undiscovered nebulosity (the object is now designated IC 498). Failure to find the comet was also reported at England's Orwell Park Observatory. They said, "Faye's Comet also was searched for on three occasions under favourable conditions as regards atmosphere, but it was not seen, and would appear to have been too faint for an aperture of [25-cm] at this apparition."

Aside from Barnard, the next astronomer to report a successful observation of the comet was Perrotin, who managed to find it with a 38-cm refractor on December 4 and noted it was about one-half degree from the position predicted by E. A. Lamp. E. J. M. Stephan (Marseille, France) saw the comet with an 80-cm refractor on the 6th and 7th. He said it was very faint, diffuse, and irregularly ovoid, with a diameter of about 0.50'. Stephan added that an eccentric point of magnitude 14 was also seen. On the 8th, Stephan said the comet was more noticeable than on previous nights, while the nucleus was more prominent. He added that a trace of tail was directed toward the northeast. Stephan said the comet was essentially unchanged between the 8th and 9th, although a nucleus of magnitude 11–12 was seen on the last date. Spitaler found the comet with the 69-cm refractor on December 15 and described it as very faint and round, with a coma about 40" across. There was a faint nucleus.

On 1889 January 2, E. Becker said the comet was about 0.3' across in his 46-cm refractor. Spitaler said the comet was very faint on the 4th and about

0.5′ across, with a hardly detectable nucleus. On the 29th, Spitaler said the comet was a very faint, round, diffuse nebulosity, without a noticeable nucleus. In the year-end report published in the February 1890 issue of the *Monthly Notices of the Royal Astronomical Society*, Radcliffe Observatory (Oxford, England) reported that the Barclay refractor was used "in a search for Faye's Comet on three nights (in 1889), which was, however, unsuccessful from the faintness of the comet."

Spitaler and Barnard were the last observers of this comet during February. On the 3rd, Spitaler said the comet was very faint and diffuse in the 69-cm refractor, with a diameter of about 0.5′. There was no nucleus. Barnard could not initially find the comet on the 5th, but close examination of a 9th-magnitude star revealed "a faint glow almost symmetrically surrounding it." About 15 minutes later, Barnard noticed the nebulosity was following the star by 2″ or 3″. The comet was last detected on February 8.42, when Barnard saw it with the 30-cm refractor and gave its position as $\alpha = 7^h$ 31.6^m, $\delta = +3° 05'$. He added, "Several attempts to refind the comet after the February moon proved futile, as no ephemeris was at hand, and later, when an ephemeris was received, our bad weather had set in and it was not seen again." Spitaler looked for the comet with the 69-cm refractor on February 23 and 24, but no trace was found.

Barnard summarized his observations of August 11 to February 8. He observed with a 30-cm refractor and wrote, "The smallness of the comet has been quite striking. It was at no time over 30″ in diameter – a little brighter in the middle – its light being more like that of one of the small indefinite nebulae than that of a comet. It was brightest and easiest to observe about the first part of December." Barnard saw it during every month.

Later calculations using multiple apparitions were published by B. G. Marsden and Z. Sekanina (1971). These revealed a perihelion date of August 20.18 and a period of 7.59 years. Marsden and Sekanina determined the nongravitational terms as $A_1 = +0.290$ and $A_2 = +0.01044$. This orbit is given below.

T	ω	Ω (2000.0)	i	q	e
1888 Aug. 20.1804 (TT)	201.1927	211.2906	11.2966	1.747990	0.547457

ABSOLUTE MAGNITUDE: $H_{10} = 7.4$ (V1964)

FULL MOON: Jul. 23, Aug. 21, Sep. 20, Oct. 19, Nov. 18, Dec. 18, Jan. 17, Feb. 15

SOURCES: D. M. A. Möller, *BAJ for 1882*, **107** (1880), p. 138; H. C. F. Kreutz, *AN*, **119** (1888 Jul. 24), p. 271; H. J. A. Perrotin, *AN*, **119** (1888 Aug. 10), p. 319; H. J. A. Perrotin, *CR*, **107** (1888 Aug. 20), p. 436; H. J. A. Perrotin, *CR*, **107** (1888 Aug. 27), pp. 456–7; H. C. F. Kreutz, *AN*, **119** (1888 Sep. 6), p. 381; H. J. A. Perrotin and A. H. P. Charlois, *AN*, **120** (1888 Sep. 26), p. 45; E. A. Lamp, *AN*, **120** (1888 Oct. 15), p. 77; E. A. Lamp, E. Becker, B. von Engelhardt, R. Spitaler, and H. J. A. Perrotin, *AN*, **120** (1888 Nov. 19), p. 171; E. J. M. Stephan, *CR*, **107** (1888 Dec. 10), p. 936; H. J. A. Perrotin, *AN*, **120**

(1888 Dec. 11), p. 219; R. Spitaler, *AN*, **120** (1888 Dec. 19), p. 253; *MNRAS*, **49** (1889 Feb.), pp. 192, 205; E. Becker and W. Valentiner, *AN*, **121** (1889 Apr. 25), pp. 203–6; E. E. Barnard, *AJ*, **9** (1889 Jun. 19), pp. 29–31; R. Spitaler, *AN*, **122** (1889 Jul. 29), p. 157; E. E. Barnard, *AN*, **122** (1889 Sep. 6), pp. 309–14; *MNRAS*, **50** (1890 Feb.), p. 203; V1964, p. 63; B. G. Marsden and Z. Sekanina, *AJ*, **76** (1971 Dec.), pp. 1136–7.

C/1888 R1   *Discovered:* 1888 September 3.52 ($\Delta = 2.96$ AU, $r = 2.59$ AU, Elong. $= 59°$)
(Barnard)   *Last seen:* 1890 September 8.18 ($\Delta = 6.35$ AU, $r = 6.37$ AU, Elong. $= 86°$)
                  Closest to the Earth: 1888 November 22 (1.0852 AU)
1889 I = 1888e *Calculated path:* MON (Disc), ORI (Oct. 17), ERI (Nov. 10), CET (Nov. 28), PSC (1889 Jan. 7), AQR (Jan. 24), PSC (Feb. 16), PEG (Jun. 7), AQR (Jun. 23), AQL (Jul. 21), SCT (Aug. 25), SER (Sep. 30), SCT (Nov. 6), AQL (1890 Feb. 12), SCT (Mar. 24), SER (May 20), OPH (Jun. 3)

E. E. Barnard (Lick Observatory, California, USA) had just finished observing periodic comet Faye on the morning of 1888 September 3, when he began comet seeking with a 10-cm "broken tube" cometseeker. He found a "faintish and suspicious object" on September 3.52, which he at once examined with a 30-cm refractor and estimated its position as $\alpha = 6^h 52.3^m$, $\delta = +10° 59'$. Since no nebula was located in that area of the sky, he assumed this was a comet. He described it as circular, with a diameter of 1', with a "tolerably well-defined nucleus." Barnard noted the magnitude was about 11 or fainter and there was no tail. An independent discovery was announced by W. R. Brooks (Smith Observatory, Geneva, New York, USA). He said the comet was found on the morning of September 4, just one day after Barnard, and that a telegram announcing Barnard's discovery did not reach him until September 5. He described the comet as faint, round, and with some central condensation.

The comet was discovered nearly 5 months prior to perihelion and nearly 3 months before its closest approach to Earth. Despite its faintness, the comet was well observed during the remainder of September. H. A. Kobold (Strasbourg, France) observed with a 46-cm refractor on the 5th and said the comet was round, about 2' across, with a central condensation of magnitude 11. On the 6th, G. Bigourdan (Paris, France) described the comet as a round nebulosity, between 1' and 1.5' in diameter, with a "rather stellar nucleus" of magnitude 11.5–12. The nucleus was not central, but shifted from the center of the coma toward PA 20°. B. von Engelhardt (Dresden, Germany) said the comet was bright and elongated on the 14th. It exhibited a stellar nucleus situated within a granular condensation. That same night, E. Millosevich (Rome, Italy) said the coma was round and about 1' across, with a nucleus of about magnitude 11, while L. Boss (Dudley Observatory, New York, USA) said the comet was indefinitely elongated toward PA 352°. On the 15th, W. Luther (Hamburg, Germany) said the coma and nucleus had both brightened since the 14th, with the nucleus at magnitude 10.5. On the 27th, Boss said the comet was faint with the moon 21° away.

On October 1, Millosevich noted an eccentric nucleus of magnitude 11 and a rudimentary tail extending toward the west-northwest. On the 2nd, Luther said the nucleus was magnitude 10, while Engelhardt said the comet's brightness had increased slightly and the coma was about 3' across. He added that a bright oval condensation appeared granular and gave the impression of multiple nuclei, while a stellar nucleus was about magnitude 10.5. E. A. Lamp (Kiel, Germany) said the nucleus was rather bright on the 4th and located within a round nebulosity. On the 6th, W. H. Robinson (Radcliffe Observatory, Oxford, England) noted the comet appeared as a circular nebulosity about 3.5' across and exhibited a distinct nucleus of magnitude 10. C. F. Pechüle (Copenhagen, Denmark) said the coma was about 2' across on the 13th, while the nucleus was about 5" in diameter in a 27-cm refractor. On the 13th and 14th, W. J. Crofton (Stony-hurst College Observatory, England) reported the comet exhibited a stellar nucleus of magnitude 10–11, which was "surrounded by a round nebulos-ity, but without tail." Boss said the comet was magnitude 9.0 on the 19th. E. Becker (Strasbourg, France) noted the sharp nucleus had a magnitude of 10–11 on the 27th and he suspected the beginning of a tail. R. Matthiessen (Karlsruhe, Germany) said the comet was quite bright on the 28th, with a coma 4–5' across. J. I. Plummer (Orwell Park Observatory, England) said the comet was faint because of fog and cloud on the 30th. On October 31, F. Schwab (Kremsmünster, Austria) said the coma was 4' across and con-tained a bright nucleus.

The comet was closest to Earth on November 22 (1.09 AU), but continued approaching the sun. On November 2, Engelhardt said the comet was bright and 6' in diameter, with a stellar nucleus of magnitude 9, while Boss said the comet was faint and large, and appeared more extended toward PA 0°. He added that there was a strong nuclear condensation. Pechüle reported the coma was 6' across on the 5th. On the 7th, E. Becker said the nucleus had a magnitude of 10. L. Weinek (Prague, Czech Republic) reported the comet was nearly round and about 4.5' across on the 8th, while the 16.3-cm refractor (74×) revealed a distinct nucleus of magnitude 9.2. On the 9th, Schwab said the longest diameter of the coma was 7', while Weinek noted the coma was about 5.5' across, with a nucleus that seemed more diffuse than on the previous morning. J. Bauschinger (Munich, Germany) said the nucleus was magnitude 9 on the 10th and the coma was about 4' across. On the 11th, Matthiessen described the comet as very bright, with a nucleus of magnitude 9–9.5, while E. Becker said the nucleus had a magnitude of 9. L. Schwarz and G. Gruss (Prague, Czech Republic) noted a distinct nucleus of magnitude 8 on the 12th, and a fan-shaped tail extending toward about PA 10°. On the 26th, T. Lewis (Royal Observatory, Greenwich, England) said the comet had a very bright nucleus, while F. A. Bellamy (Radcliffe Observatory, Oxford, England) said the nucleus was magnitude 9–10. On the 27th, the nuclear magnitude was given as 10 by W. Wickham (Radcliffe Observatory, Oxford, England), 8.5 by Bauschinger, and 8.5 by Schwarz and Gruss. Wickham described the nucleus as "planetary". Bauschinger said the coma was oval. Schwarz and Gruss said the coma was 5.5' across. Bellamy reported the nucleus was fainter on the 28th than on the 26th and was near magnitude 10.5. On November 29, Engelhardt said the comet was brighter than the 8.1-magnitude comparison star and exhibited a stellar nucleus of magnitude 7. He noted a short, fan-shaped tail.

On December 3, Engelhardt said the nucleus was about magnitude 9, while the tail was fan-shaped and extended about 10' toward PA 44°. E. Becker noted the comet had a fan-shaped tail extending over 30' toward PA 58°, while the nucleus was magnitude 10. Schwab said the tail was 24' long. On the 6th and 7th, Engelhardt reported the comet was bright and condensed, with a stellar nucleus of about magnitude 9. The tail was very wide and fan-shaped. On the 11th and 12th, H. P. Hollis and W. G. Thackeray (Royal Observatory, Greenwich, England) described the comet as very faint because of bright moonlight. On the 13th, Boss said the comet looked like a diffuse star, slightly brighter than magnitude 9. The comet attained its most southerly declination of −8° on the 19th. On December 22, W. H. Robinson (Radcliffe Observatory, Oxford, England) gave the magnitude of the nucleus as 10. Engelhardt said the comet was bright on the 27th, with a granular condensation and a stellar nucleus of about magnitude 9.3. There was also a wide, short tail. That same night, Schwarz said the comet was diffuse and about 1.5' across. He also noted a tail extending toward PA 310°. On the 28th, A. J. G. F. von Auwers (Berlin, Germany) said the comet was bright in clear skies. On December 31, Plummer said the comet was faint because of haze.

The comet passed perihelion on 1889 January 31. On January 2, Bauschinger said the coma was round and about 4' across, and exhibited a sharp nucleus of magnitude 9.2. J. Holetschek (Vienna, Austria) estimated the condensation magnitude as 8.5 and noted a diffuse nucleus. On the 4th, R. Spitaler (Vienna, Austria) said the comet was visible to the naked eye as a diffuse star and was also easily seen in the finder. The 69-cm refractor revealed a nearly stellar nucleus of magnitude 8 sitting within a coma at least 3' across. A fan-like tail extended northeastward for about 6'. Holetschek estimated the nuclear magnitude as 9 on the 5th. Auwers described the comet as extremely faint on the 7th, when only 4° from the moon. On the 26th, Holetschek described the nucleus as diffuse and magnitude 10. On January 28, E. Becker said the nucleus was about magnitude 10–11.

With the comet's low altitude in the evening sky, observations were quickly becoming scarce. Holetschek reported on February 3 that the 3-day-old moon made the comet difficult to see. He said the comet was easier to see at low altitude on the 5th. On the 8th, Plummer said the comet was faint because of moonlight. On the 9th, Robinson saw the comet at a low altitude and said it was very faint and "just discernible" in a refractor.

Robinson added that it looked like a "small circular nebula with central condensation." E. Becker was barely able to see the comet in a 46-cm refractor because of its small altitude on the 13th. The comet was followed until February 17, when Millosevich measured it low over the horizon.

The comet passed just 0.6° from the sun on March 14 and was then recovered in the morning sky by Barnard on April 29. The comet was well observed during the next several months. On June 3, Barnard reported that the comet had lost its normal tail and exhibited a sunward tail extending 1° toward PA 90.0°. This new tail was 2–3′ wide. On the 7th, E. Becker described it as a small, possibly oval, nebulosity, with a central condensation of magnitude 10. Spitaler observed the comet on the 25th with the 69-cm refractor. He gave the total magnitude as 8 and the nuclear magnitude as 9. Spitaler said the coma was oblong, 3′ across, and diffuse on the southeast side. He added that it was easily seen in his 10-cm finder. A. Abetti (Padova, Italy) estimated the magnitude as 9 on the 26th. Holetschek saw the comet in the 15-cm refractor on the 27th and said it was well defined on the side facing the sun. He reported a central condensation of magnitude 9.5 and said the comet remained visible in morning twilight for as long as stars of magnitude 10.5. J. Tebbutt (Windsor, New South Wales, Australia) saw the comet with his 20-cm refractor on June 23, 26, 29, and July 1. He noted it was "very small with a minute central condensation." Engelhardt described the comet as rather bright, small, round, and condensed on June 26 and July 22. On July 2, E. Becker described the comet as round, about 0.5′ across, with a condensation of magnitude 11. Abetti gave the magnitude as 10–11 on July 4 and 8. M. Brendel and V. Wellmann (Berlin, Germany) saw the comet on several occasions during the period spanning June 8 to July 25. They continually noted it was difficult to observe in a 24.4-cm refractor because of twilight. On July 6, L. Swift (Warner Observatory, Rochester, New York, USA) reported the discovery of a new comet, which turned out to be C/1888 R1. Holetschek estimated the magnitude of the central condensation as 10 on the 21st and 30th. Engelhardt said the comet was bright, small, and granular, with an oblong nucleus on July 24. On the 30th, Engelhardt said the comet was rather bright, with a strong condensation, and a coma 2′ across. During the period of July 4 to August 5, J. F. Schroeter (Strasbourg, France) continually described the comet as very faint in a 15-cm refractor. Holetschek gave the magnitude of the central condensation as 10.5 on August 3, 11 on the 17th, and 11.5 on the 24th. On August 5, E. Becker gave the nuclear magnitude as 10 and said it appeared diffuse, or possibly multiple. J. A. Pollock (Sydney, Australia) observed the comet with the 29-cm refractor on August 13 and 19. On the 17th, E. Becker gave the diameter as about 0.5′ and the nuclear magnitude as 11. On the 28th, he gave the diameter as about 15″ and the nuclear magnitude as 11. Engelhardt found the comet rather bright, small, and round on the 30th and 31st. During the period of August 26 to September 2, F. Renz (Pulkovo Observatory, Russia) observed with a 38-cm refractor. He noted a bright, stellar nucleus and a narrow, straight tail extending toward PA 150°. Holetschek was no longer certain he was seeing the comet in the 15-cm refractor on September 20. On September 23, Engelhardt said the comet was rather bright and oblong, with a tail extending 1′ toward PA 150°. He also noted a nucleus of magnitude 12. Renz observed on the 24th and found a sharply defined nucleus and a fainter, more fan-shaped tail.

On October 14 and 17, Renz described the comet as a faint, condensed nebulosity. On the 15th, E. Becker gave the diameter as about 15″ and on the 16th he simply described the comet as faint. Renz could not measure the comet's position after the 17th because of its nearness to the horizon. E. Becker described the comet as a small, dull, slightly condensed cloud. Spitaler observed with the 69-cm refractor on October 23 and described the comet as 1.5′ across, with a condensation of magnitude 12. That same evening, Abetti said it was at the limit of visibility in his refractor. The comet was detected by Barnard on November 15 and 16, and then became lost in the sun's glare for several months.

The comet passed about 11° from the sun on December 30. Spitaler recovered it as a very faint, but easy to see object in the 69-cm refractor on 1890 March 29. He noted it was small and round, with a central condensation. Spitaler said the comet was well seen on April 17. Barnard observed the comet on May 16 and said it was "still easily visible" in a 30-cm refractor. He continued to observe it with a 30-cm refractor during the next few months, but following his observation of August 12, he began using a 91-cm refractor. Meanwhile, Spitaler continued making observations with a 69-cm refractor. On May 26, he described the comet as a small, round nebulosity, with a magnitude of 14. He found the comet very faint and about 15″ across on June 26. It was about 0.5′ across on July 7, 9, and 19, with a magnitude of 14 on the 7th and exhibited a faint nucleus on the 9th. On August 18, Barnard described the comet as "round, and very gradually a little brighter in the middle, with no trace of a nucleus." He estimated the total magnitude as 13.5–14. Barnard saw the comet again on the next evening.

The comet was seen for the final time on September 8.18, when Barnard found it with the 91-cm refractor and gave the position as $\alpha = 16^h\ 49.5^m$, $\delta = -8°\ 47′$. Barnard said he carefully searched for this comet with this refractor during 1891, but did not find it.

There were two additional positions reported for this comet. The 1891 May 13 issue of the *Astronomische Nachrichten* published a letter from Spitaler which said he found the comet on 1891 May 2, very close to an ephemeris prepared by A. Berberich. Spitaler made the observation with a 69-cm refractor during a period of "extremely pure and transparent air." He described it as an extremely faint spot about 5″ across that was only perceptible "with the greatest effort of the eye." Movement was noticed in the course of an hour. Because of the comet's faintness, Spitaler could not it measure directly, but had to memorize its position when the micrometer eyepiece was used. The last of the two positions was obtained on May 2.03, and Spitaler gave it as $\alpha = 17^h\ 08.9^m$, $\delta = -8°\ 17'$. The Author notes that Spitaler's measurement of δ is about 6' south of the predicted position, which casts some doubt that this was C/1888 R1. Interestingly, G. van Biesbroeck (1940) said he examined the region with the 102-cm refractor at Yerkes Observatory "and found exactly in the place indicated by him a very diffuse round nebula some 30" in diameter which evidently misled the observer."

Abetti noted that observations made during the period of October 30 to December 14 revealed the nucleus brightened from magnitude 9 to 7, while the tail length reached 20'. It looked like an out-of-focus star of 8th magnitude.

The spectrum of this comet was analyzed during November and December by R. Copeland (Lord Crawford's Observatory, Dun Echt, Scotland) and E. W. Maunder (Royal Observatory, Greenwich, England). Both astronomers reported a continuous spectrum, Copeland on November 14 and Maunder on November 27, although Maunder did write that he barely detected the green carbon band. Copeland obtained additional observations on November 26, December 5, and December 8. He saw all three carbon bands and gave measurements of 5478 Å, 5125 Å, and 4746 Å.

The first parabolic orbits were independently calculated by C. N. A. Krueger and Berberich. Using positions from September 5, 9, and 14, Krueger determined the perihelion date as 1889 January 28.76, while Berberich determined it as January 29.56. During the next few weeks and months, J. Hackenberg, W. C. Winlock, Boss, E. Viennet, Berberich, C. W. Crockett, and O. C. Wendell published orbits with perihelion dates generally between January 31.6 and 31.8.

Although L. Becker had indicated a possible hyperbolic tendency during November of 1888 following his computations using positions spanning only the first two months of visibility, the first true hyperbolic orbit was calculated by Berberich. He took positions spanning over a year and determined the perihelion date as January 31.67 and the eccentricity as 1.0010863. Later orbits using longer periods of visibility were calculated. Van Biesbroeck (1940) determined the eccentricity as 1.001255. Z. Sekanina and B. G. Marsden (1978) determined the eccentricity as 1.001230 and their orbit is given below.

T	ω	Ω (2000.0)	i	q	e
1889 Jan. 31.6682 (TT)	340.4643	358.9750	166.3842	1.814916	1.001230

ABSOLUTE MAGNITUDE: $H_{10} = 3.6$ (V1964)

FULL MOON: Aug. 21, Sep. 20, Oct. 19, Nov. 18, Dec. 18, 1889 Jan. 17, Feb. 15, Mar. 17, Apr. 15, May 15, Jun. 13, Jul. 12, Aug. 11, Sep. 9, Oct. 9, Nov. 7, Dec. 7, 1890 Jan. 6, Feb. 5, Mar. 6, Apr. 5, May 4, Jun. 3, Jul. 2, Jul. 31, Aug. 30, Sep. 28

SOURCES: E. E. Barnard and H. A. Kobold, *AN*, **119** (1888 Sep. 6), p. 383; G. Bigourdan, *CR*, **107** (1888 Sep. 10), p. 495; F. Schwab, C. Wagner, G. Cacciatore, H. A. Kobold, E. Becker, W. Luther, B. von Engelhardt, C. N. A. Krueger, and A. Berberich, *AN*,

120 (1888 Sep. 19), p. 31; W. R. Brooks, J. Franz, B. von Engelhardt, W. Luther, and E. Millosevich, *AN*, **120** (1888 Sep. 26), pp. 37, 47; E. E. Barnard, W. C. Winlock, and L. Boss, *AJ*, **8** (1888 Oct. 2), pp. 108–11; F. Courty, B. von Engelhardt, and W. Luther, *AN*, **120** (1888 Oct. 4), pp. 59, 63; E. Millosevich, E. A. Lamp, J. Hackenberg, and A. Berberich, *AN*, **120** (1888 Oct. 15), pp. 77–80; H. V. Egbert, E. E. Barnard, E. Millosevich, and A. Berberich, *AN*, **120** (1888 Oct. 19), pp. 89–92, 95; E. Viennet, *CR*, **107** (1888 Oct. 22), pp. 646–7; E. Viennet, *BA*, **5** (1888 Nov.), pp. 497–8; W. J. Crofton, *MNRAS*, **49** (1888 Nov.), p. 34; B. von Engelhardt, *AN*, **120** (1888 Nov. 10), p. 157; L. J. Gruey, A. Hérique, and A. Berberich, *AN*, **120** (1888 Nov. 19), pp. 167–70; L. Boss, *AJ*, **8** (1888 Nov. 27), p. 132; R. Matthiessen and W. Luther, *AN*, **120** (1888 Nov. 28), p. 189; R. Copeland, T. Lewis, H. P. Hollis, W. G. Thackeray, and W. H. Robinson, *MNRAS*, **49** (1888 Dec.), pp. 70–2, 81–2, 84–5; B. Matthiessen, *AN*, **120** (1888 Dec. 13), p. 237; B. von Engelhardt, *AN*, **120** (1888 Dec. 19), p. 255; L. Boss, *AJ*, **8** (1889 Jan. 22), p. 151; E. W. Maunder, F. A. Bellamy, W. Wickham, and W. H. Robinson, *MNRAS*, **49** (1889 Mar.), pp. 307, 323–5; F. Schwab and A. Berberich, *AN*, **121** (1889 Mar. 12), pp. 39, 43–8; J. I. Plummer, *MNRAS*, **49** (1889 Apr.), pp. 371–3; A. J. G. F. von Auwers, *AN*, **121** (1889 Apr. 2), pp. 97–102; E. Millosevich, *AN*, **121** (1889 Apr. 11), p. 141; E. Becker and H. A. Kobold, *AN*, **121** (1889 Apr. 25), pp. 201–6; C. W. Crockett, *AJ*, **8** (1889 May 1), pp. 190–1; B. von Engelhardt, L. J. Gruey, A. Hérique, and O. C. Wendell, *AN*, **121** (1889 May 7), pp. 235–40; J. Bauschinger, A. Abetti, and E. A. Lamp, *AN*, **121** (1889 May 10), pp. 265–72; L. Weinek, G. Gruss, L. Schwarz, and R. Matthiessen, *AN*, **121** (1889 May 28), pp. 331–4; E. Millosevich, *AN*, **121** (1889 Jun. 5), p. 367; E. E. Barnard, *AN*, **122** (1889 Jun. 25), p. 27; L. Swift, E. Millosevich, and C. N. A. Krueger, *AN*, **122** (1889 Jul. 19), p. 117; C. F. Pechüle, *AN*, **122** (1889 Jul. 25), pp. 139–41; J. Palisa, E. Weiss, J. Holetschek, and R. Spitaler, *AN*, **123** (1889 Nov. 2), p. 59; R. Spitaler and J. Tebbutt, *AN*, **123** (1889 Nov. 14), pp. 87–90; B. von Engelhardt, *AN*, **123** (1889 Nov. 19), pp. 107–10; A. Berberich, *AN*, **123** (1890 Jan. 7), pp. 275–82; A. Abetti, *AN*, **123** (1890 Jan. 30), pp. 361–4; J. A. Pollock, *MNRAS*, **50** (1890 Mar.), p. 334; R. Spitaler, *AN*, **124** (1890 Apr. 5), p. 171; E. Becker, *AN*, **124** (1890 Apr. 16), pp. 203–7; M. Brendel, V. Wellmann, and F. Renz, *AN*, **124** (1890 Jun. 17), pp. 345, 362, 365–7; A. Berberich and E. E. Barnard, *AN*, **124** (1890 Jul. 1), pp. 407–10; R. Spitaler and J. Holetschek, *AN*, **125** (1890 Sep. 12), pp. 259–62, 267–9; R. Spitaler, *AN*, **125** (1890 Sep. 18), pp. 281–4; E. E. Barnard, *AJ*, **10** (1890 Sep. 19), p. 67; J. I. Plummer, *MNRAS*, **51** (1890 Nov.), p. 51; O. C. Wendell, *AJ*, **11** (1891 May 12), p. 4; R. Spitaler, *AN*, **127** (1891 May 13), p. 183; E. E. Barnard, *AJ*, **11** (1891 Oct. 10), pp. 51–4; G. van Biesbroeck, *Publications of the Yerkes Observatory*, **8** pt. IV (1940), pp. 10–26; V1964, p. 63; Z. Sekanina and B. G. Marsden, *AJ*, **83** (1978 Jan.), p. 66; B. G. Marsden and Z. Sekanina, *CCO*, 3rd ed. (1979), pp. 19, 48.

C/1888 U1	*Discovered:* 1888 October 31.54 ($\Delta = 1.77$ AU, $r = 1.67$ AU, Elong. $= 68°$)
(Barnard)	*Last seen:* 1889 May 23.26 ($\Delta = 2.36$ AU, $r = 2.89$ AU, Elong. $= 74°$)
	Closest to the Earth: 1889 January 15 (1.3878 AU)
1888 V = 1888f	*Calculated path:* HYA (Disc), SEX (Nov. 23), LEO (Jan. 9), LMi (Mar. 8)

E. E. Barnard (Lick Observatory, California, USA) discovered this comet on 1888 October 31.54 at a position of $\alpha = 9^h\ 43.4^m$, $\delta = -15°\ 19'$. He had just ended his observations with the 30-cm refractor, and had begun to sweep for comets with the 10-cm "broken-tube" cometseeker when a faint, suspicious object was found. Barnard went back to the refractor to examine the object further and realized it was a comet. He wrote, "the head was moderately well condensed, with an ill-defined nucleus, and a short faint tail preceding." He gave the diameter as 1' and estimated the magnitude as 11–12.

The comet had passed perihelion about a month and a half before it was discovered, but it was still decreasing its distance from Earth and faded slowly during the next couple of months. On November 2, J. Franz (Königsberg, now Kaliningrad, Russia) said the comet was as bright as a star of magnitude 10. On the 3rd, J. Palisa (Vienna, Austria) noted the comet was 2' across and exhibited a nucleus of magnitude 12. E. Millosevich (Rome, Italy) said the comet was bright on the 4th, with a nucleus of magnitude 11. There was also a tail 3' long. W. Luther (Hamburg, Germany) said the nucleus was magnitude 11.5 on the 7th. E. Becker (Strasbourg, France) said the coma was about 1' across and contained a stellar condensation of magnitude 9–10 on the 10th. On the 11th, E. A. Lamp (Kiel, Germany) reported the round coma was 1' across, with a faint nucleus, while Becker said the comet had a stellar nucleus of magnitude 10. On the 12th, B. von Engelhardt (Dresden, Germany) noted the comet was bright, 2' in diameter, with a condensation containing a stellar nucleus of magnitude 10.5. He added that the comet was immediately visible in a 13-cm cometseeker. J. I. Plummer (Orwell Park Observatory, England) said the comet was faint because of moonlight on the 14th. In moonlight and haze on the 28th, F. A. Bellamy (Radcliffe Observatory, Oxford, England) observed with the 25.4-cm refractor (100×) and described the comet as a "faint nebulous, almost circular, patch" about 1.3' in diameter. He added that a nucleus was occasionally seen and was similar to an 11.5-magnitude star. During the period spanning November 5 to December 14, A. Abetti (Padova, Italy) reported a coma 1' across, a nucleus of magnitude 10, and a trace of tail extending northward.

On December 4, Engelhardt described the comet as faint, small, and diffuse, with a nucleus of magnitude 13, while Becker said the comet was 1' across, with a stellar condensation of magnitude 11. Plummer said the comet was "rather faint," because of hazy skies on the 7th. On the 8th, Engelhardt noted the comet was moderately bright and round, with a coma 0.5' across and a nucleus of magnitude 12. Luther said the comet was faint. Engelhardt said the comet was noticeably fainter on the 14th than on the 8th. He added that it was round, about 1' across, with a condensation and a nucleus of magnitude 12. Plummer said the comet was faint on the 26th. Luther described the comet as very faint on the 28th. On December 31, Plummer said the comet was faint because of its proximity to a small star.

The comet made a moderately close approach to Earth on 1889 January 15 (1.39 AU). On the 1st, Becker said the nucleus was magnitude 11–12. On the 2nd, Plummer said the comet was fainter than on December 31, while R. Spitaler (Vienna, Austria) said the comet was 3' across, with a nucleus of magnitude 9. On the 4th, Spitaler reported the nucleus looked like a diffuse star of magnitude 10. Plummer said the comet was faint on the 8th. Engelhardt described the comet as small, faint, and condensed on the 11th, with a nucleus of magnitude 12.5. On the 22nd, W. J. Croften and W. Carlisle (Stonyhurst College Observatory, England) said the comet appeared "faint and diffuse, without any clearly marked nucleus." Becker said the comet was 1′ across on the 27th, with a quite sharp nucleus of magnitude 11. On the 29th, Bellamy reported the nucleus was stellar, although slightly elongated, and about magnitude 11.5–12. It was surrounded by a "moderately bright nebulosity" about 4′ across, but there was no tail. Spitaler said the comet was 4′ across, with a nucleus of magnitude 10. He added that the comet was still visible in the large refractor's finder. On January 30, Barnard observed with the 30-cm refractor and said the comet was "pretty bright with a star-like nucleus, and a tail 20′ long."

On February 2, W. Wickham (Radcliffe Observatory, Oxford, England) described the comet as "a hazy patch of about 1′ diameter." There was a stellar nucleus of magnitude 13. Millosevich said the nucleus was magnitude 12 on the 7th. On the 23rd, Spitaler reported the comet was still quite bright, with a diffuse nucleus. On the 24th, Spitaler said the coma was 2–3′ across, with a diffuse nucleus of magnitude 12–13. He noted the comet was approximately 2′ across on the 25th, with a diffuse nucleus of magnitude 12–13.

On March 3, Plummer noted, "The comet appears to maintain its lustre in a remarkable manner and is not sensibly fainter than it was at the end of January." On the 5th, Becker said the comet was about 0.5′ across, with a nucleus of magnitude 11–12. Millosevich said the nucleus was somewhat indistinct in a 25-cm refractor on the 6th. On the 7th, Spitaler wrote the coma was 1′ across in the 69-cm refractor and the total brightness of the comet was 12–13. During the period of March 4–8, J. Bauschinger (Munich, Germany) said the comet was an extraordinarily faint, diffuse nebulosity in a 27-cm refractor, with a coma 4′ across. On the 21st, Spitaler said the comet was 1–1.5′ across, with a stellar nucleus of magnitude 14. Millosevich said the comet was very faint on the 23rd, with a hardly perceptible nucleus. On the 24th, Becker noted a condensation of magnitude 12. On March 25, Plummer said the comet was "very faint and difficult to observe." He attributed this partly to the inability to block out all artificial light.

On April 4, Spitaler said the comet was approximately 2′ across, with a stellar nucleus of magnitude 13–14, while Barnard said the comet was very faint and difficult to see in the 30-cm refractor. Becker only suspected the comet in the 46-cm refractor on the 19th, under slightly hazy skies. The comet attained its most northerly declination of +38° on the 25th. On the 27th, Barnard said the comet was brighter than on the 4th and exhibited a stellar nucleus of magnitude 13. On April 29, Engelhardt said the comet was very faint and diffuse, with a diameter of 0.5′. There was also a stellar nucleus of magnitude 14 visible in a 30-cm refractor.

On May 1, Engelhardt said the comet was very faint and small, with a nucleus of magnitude 14, while Spitaler said the comet's total magnitude was about 13–14, with a centrally condensed coma about 1.5′ across. On the 2nd, Spitaler said the comet was faint, round, and about 1.5′ across. He reported the comet was faint and about 1–1.5′ across on the 20th. There was also a faint, stellar nucleus. On May 21, Spitaler said the comet was so faint that he had trouble measuring a position in the 69-cm refractor. He noted the comet appeared as a faint nebulous patch, with an occasionally visible stellar nucleus.

The comet was last detected on May 23.26, when Barnard gave the position as $\alpha = 9^{\mathrm{h}}\ 48.0^{\mathrm{m}}$, $\delta = +37°\ 03′$. In the 30-cm refractor, he noted the comet was very faint, with a diameter of about 10″, and exhibited a slight condensation.

The first parabolic orbit was calculated by J. M. Schaeberle using positions obtained by Barnard on November 1, 3, and 5. The result was a perihelion date of 1888 September 9.95. C. N. A. Krueger used positions from November 3 to November 6 and determined the perihelion date as September 8.97. Additional orbits calculated by W. C. Winlock, Spitaler, J. Hackenberg, J. Halm, and G. M. Searle eventually established the perihelion date as September 13.5.

The first elliptical orbit was calculated by Searle using positions obtained up to December 14. He determined the perihelion date as September 13.07 and the period as about 1231 years. Additional elliptical orbits were calculated by Searle (2613 and 2253 years).

Definitive orbits were calculated by A. Dinter (1903) and B. G. Marsden (1978). Dinter took 250 positions spanning the period of October 31 to May 23, applied perturbations from Earth, Jupiter, and Saturn, and determined the perihelion date as September 13.28 and the period as about 2367 years. Marsden used 54 positions spanning the period of October 31 to May 20, applied perturbations from all nine planets, and determined the perihelion date as September 13.28 and the period as about 2405 years. Marsden's orbit is given below.

T	ω	Ω (2000.0)	i	q	e
1888 Sep. 13.2824 (TT)	290.8086	139.0895	56.3425	1.527853	0.991488

ABSOLUTE MAGNITUDE: $H_{10} = 5.8$ (V1964)

FULL MOON: Oct. 19, Nov. 18, Dec. 18, Jan. 17, Feb. 15, Mar. 17, Apr. 15, May 15, Jun. 13

SOURCES: E. E. Barnard, *AN*, **120** (1888 Nov. 3), p. 143; J. Franz, J. Palisa, E. Millosevich, W. Luther, R. Spitaler, and C. N. A. Krueger, *AN*, **120** (1888 Nov. 10), pp. 157–60; E. Millosevich, F. Schwab, E. Becker, W. Luther, B. von Engelhardt, E. A. Lamp, and R. Spitaler, *AN*, **120** (1888 Nov. 19), pp. 167–70, 173–6; E. E. Barnard and W. C. Winlock, *AJ*, **8** (1888 Nov. 27), p. 134–6; E. Millosevich and J. Hackenberg, *AN*, **120** (1888 Nov. 28), p. 189; B. von Engelhardt and W. Luther, *AN*, **120** (1888 Dec. 11), p. 221;

E. E. Barnard and R. Spitaler, *AN*, **120** (1888 Dec. 13), pp. 237–40; G. M. Searle and J. M. Schaeberle, *AJ*, **8** (1888 Dec. 17), pp. 140, 144; W. Luther, *AN*, **120** (1889 Jan. 9), p. 287; G. M. Searle and W. C. Winlock, *AJ*, **8** (1889 Jan. 22), pp. 146–8; J. Halm and R. Spitaler, *AN*, **120** (1889 Jan. 24), pp. 301–4; G. M. Searle, *AJ*, **8** (1889 Apr. 15), pp. 181–2; F. A. Bellamy, W. Wickham, W. J. Crofton, and W. Carlisle, *MNRAS*, **49** (1889 Mar.), pp. 325–6, 328; F. Schwab, *AN*, **121** (1889 Mar. 21), p. 39; J. I. Plummer, *MNRAS*, **49** (1889 Apr.), pp. 373–4; E. Becker and H. A. Kobold, *AN*, **121** (1889 Apr. 25), pp. 203, 206; B. von Engelhardt, L. J. Gruey, and A. Hérique, *AN*, **121** (1889 May 7), pp. 235–8; A. Abetti, *AN*, **121** (1889 May 10), p. 269; A. Abetti, *AN*, **121** (1889 May 18), p. 299; C. F. Pechüle, *AN*, **122** (1889 Jul. 25), pp. 139–41; J. Bauschinger, *AN*, **122** (1889 Jul. 29), p. 157; E. E. Barnard, *AJ*, **9** (1889 Oct. 3), p. 87; E. Millosevich, *AN*, **122** (1889 Oct. 8), p. 405; J. Palisa, J. Holetschek, and R. Spitaler, *AN*, **123** (1889 Nov. 2), p. 59; R. Spitaler, *AN*, **123** (1889 Nov. 14), pp. 87–90; E. Becker, *AN*, **124** (1890 Apr. 16), pp. 203–6; J. I. Plummer, *MNRAS*, **51** (1890 Nov.), p. 51; A. Dinter, Inaugural-Dissertation der philosophischen Fakultät der Universität Breslau. Dissertation: Breslau (1903), 31 pp.; V1964, p. 63; B. G. Marsden, *AJ*, **83** (1978 Jan.), p. 66; B. G. Marsden, *CCO*, 3rd ed. (1979), pp. 19, 48.

X/1889 A1

1889a

W. R. Brooks (Smith Observatory, Geneva, New York, USA) was conducting a routine search for comets in the morning sky when he found this comet on 1889 January 15.47. He was using the 26-cm refractor (40×) and, with dawn approaching, he was only able to take circle readings to determine a rough position of $\alpha = 18^h 04^m$, $\delta = -21° 20'$. He described the comet as "nearly round, with slight central condensation" and added that it was "faintish", but was still "brighter than well-known nebulae in that region which were not seen." Brooks added, "the comet was in a well-marked field of telescopic stars, and so I was enabled to detect motion in a few minutes, which was rapid in a westerly course." Brooks was so sure this was a comet that he immediately telegraphed an announcement.

Cloudy weather prevented Brooks from confirming the comet during the next few days. Finally, on the morning of January 20th, the sky cleared and he was able to make careful sweeps in the direction of motion from the discovery position. Unfortunately, the moon was three days past full and this hampered the search. The object was not confirmed and another streak of bad weather followed. E. E. Barnard (Lick Observatory, California, USA) searched with the 30-cm refractor (80×) on several mornings during the period of January 19 to February 3. Not only was the comet not found, but no nebulae were seen in the immediate region. L. Swift was visiting Lick Observatory on the morning of January 28 and "searched thoroughly" with the 10-cm "broken-tube" cometseeker. Barnard said that Swift covered all of the southeastern sky and repeated the search on several later mornings.

SOURCES: W. R. Brooks, *AN*, **120** (1889 Jan. 24), p. 303; W. R. Brooks, *Nature*, **39** (1889 Jan. 24), p. 307; W. R. Brooks, *AN*, **120** (1889 Feb. 26), p. 415; E. E. Barnard and L. Swift,

AJ, **8** (1889 Mar. 4), p. 168; W. R. Brooks, E. E. Barnard, and L. Swift, *Nature*, **39** (1889 Mar. 7), p. 449; W. R. Brooks, *MNRAS*, **49** (1889 Mar.), p. 327.

C/1889 G1
(Barnard)

1889 II = 1889b

Discovered: 1889 April 1.17 ($\Delta = 2.57$ AU, $r = 2.40$ AU, Elong. $= 69°$)

Last seen: 1890 September 8.3 ($\Delta = 4.20$ AU, $r = 5.18$ AU, Elong. $= 165°$)

Closest to the Earth: 1889 January 31 (1.7307 AU), 1889 October 12 (1.7319 AU)

Calculated path: TAU (Disc), ORI (Apr. 4), TAU (Aug. 31), ERI (Sep. 15), CET (Oct. 13), AQR (Dec. 23), CET (1890 Feb. 25), AQR (Jul. 7)

E. E. Barnard (Lick Observatory, California, USA) discovered this comet on 1889 April 1.17 while using the 30-cm refractor. He described it as "a very small and extremely slender comet, the head not being over 10″ in diameter with a tail at least 15′ long." Barnard noted a stellar nucleus of magnitude 13, and said the comet's overall brightness was less than magnitude 12. On April 1.19, he measured the position as $\alpha = 5^\text{h}\ 20.9^\text{m}$, $\delta = +16°\ 07'$.

Although the comet had made its closest approach to Earth back in January of 1889, it was heading toward perihelion when discovered and would make a comparable approach to Earth later in the year. The comet was also dropping into evening twilight as April progressed. On the 2nd, Barnard said the tail was twice as broad as when first seen and was very diffuse, with a length of about 10′. Overall, the comet seemed brighter than when discovered the night before. On the 3rd, Barnard said the nucleus was magnitude 13, while the tail extended toward PA 80.0°. On the 4th, E. A. Lamp (Kiel, Germany) described the comet as faint, while C. F. Pechüle (Copenhagen, Denmark) said the comet was small and faint. That same night, R. Spitaler (Vienna, Austria) said the coma was about 1′ across and contained a nucleus of magnitude 11–12. Even though moonlight hampered the seeing conditions, he said the 69-cm refractor still revealed a faint tail extending 7–8′ toward PA 70°. Spitaler noted the comet was faint in the large refractor on the 5th. He said the comet displayed a stellar nucleus on the 8th. A good tail was also visible in spite of the moonlight. E. Becker (Strasbourg, France) said the comet was small, with a central condensation in the 46-cm refractor on the 9th. Spitaler described the comet as about 1′ across and very faint on the 11th, with a barely visible nucleus. Barnard said the comet was excessively faint because of moonlight and haze on the 15th. On the 17th, Barnard said the nucleus was perfectly stellar and magnitude 13, while the tail extended toward PA 75.0°. That same evening, Becker estimated the total magnitude as 11–12. On the 17th and 18th, A. Kammermann (Geneva, Switzerland) said the comet was faint and centrally condensed in the 25.4-cm refractor. He also noted the beginnings of a tail. Barnard said the tail extended toward PA 78.3° on the 18th. G. Bigourdan (Paris, France) saw the comet on April 18 and 19. He said it was very difficult to see, and noted a small, rather stellar nucleus was suspected. D. Klumpke (Paris, France) saw the comet on the

19th and described it as a very faint nebulosity with a central condensation. Becker also saw the comet on the 19th and said it was about magnitude of 12–13, without a distinct condensation. G. A. P. Rayet (Bordeaux, France) observed the comet with a 38-cm refractor on the 20th. He found it very faint and round, with a nucleus of magnitude 14. Barnard saw the comet on several occasions during the period spanning April 15–29. He said the tail was typically 10–15′ long, while a perfectly stellar nucleus was magnitude 13. Spitaler tried to see the comet in twilight with the 69-cm refractor on May 1 and 2, but nothing was detected.

The comet was lost in the sun's glare during the next couple of months and passed only 9° from the sun on June 7. It was recovered in the morning sky by W. W. Campbell (Ann Arbor, Michigan, USA) on July 26. Although many observations were made during the next few months, few physical descriptions were provided. On September 1, B. von Engelhardt (Dresden, Germany) observed with a 30-cm refractor and said the comet was faint and diffuse, with a stellar nucleus accompanied by nearby occasionally visible points. There was also a short, preceding tail. On September 24, Engelhardt said the comet was difficult to measure except for short intervals of improved seeing when a nucleus was revealed. On October 5, M. Brendel (Berlin, Germany) said the comet appeared very faint in the 24.4-cm refractor. On the 15th, Becker described the comet as small, with a nucleus of magnitude 12 on the 15th, while he said the nucleus was magnitude 10–11 and exhibited an emittance toward PA 357° on the 19th. Rayet said the comet was easily visible in the 36-cm refractor on October 23, while Spitaler observed the comet with a 69-cm refractor and said it was 1′ across, with a nucleus of magnitude 11 and a short tail extending toward the northeast. On October 31, J. I. Plummer (Orwell Park Observatory, England) said the comet was faint. On November 21, Spitaler described the comet as very faint, about 1.5′ across, with a diffuse nucleus of magnitude 14. The comet attained its most southerly declination of −17° on November 23.

The comet passed 9° from the sun on 1890 March 15. Following conjunction with the sun, Barnard unsuccessfully searched for the comet with a 30-cm refractor on several occasions during 1890. A. Berberich finally sent Barnard an improved ephemeris later in the year and the comet was subsequently found with a 91-cm refractor on 1890 August 24. Barnard described it as "very small and faint, and sensibly round, with no nucleus." He estimated the magnitude as 14.5 or 15. He saw the comet again on the next morning and observed it for the final time on September 8.3 at a probable position of $\alpha = 22^h 37.3^m$, $\delta = -18° 13'$.

The first orbits for this comet came from Lick Observatory. A. O. Leuschner took the first positions obtained at that observatory on April 1, 2, and 3, and calculated a parabolic orbit with a perihelion date of 1889 May 24.33. J. M. Schaeberle used similar positions and calculated a parabolic orbit with a perihelion date of 1889 May 26.94. Schaeberle noted the orbit was "extremely uncertain." The comet's large perihelion distance made it difficult to establish the perihelion date quickly. After J. von Hepperger took positions up to April 8, and determined the perihelion date as July 27.94, the next orbits were calculated using positions into the last half of April and beyond. The result was that C. N. A. Krueger, F. K. Ginzel and Berberich, Campbell, and E. Millosevich eventually established the perihelion date as June 11.

The first elliptical orbit came from Millosevich. Using positions from March 31 to November 21, he determined the perihelion date as July 11.27 and the period as about 323 000 years. Later revisions by F. K. Zweck (1926) and B. G. Marsden and Z. Sekanina (1973), using positions spanning the entire period of visibility, revealed a perihelion date of July 11.31. The period was given as about 580 000 years by Zweck and about 1.4 million years by Marsden and Sekanina. The last orbit is given below. Marsden and Sekanina indicated the period of the original orbit was probably near 35 000 years, while the period of the future orbit was near 22 000 years.

T	ω	Ω (2000.0)	i	q	e
1889 Jun. 11.3139 (TT)	236.0618	312.2212	163.8517	2.255596	0.999818

ABSOLUTE MAGNITUDE: $H_{10} = 5.3$ (V1964)

FULL MOON: Mar. 17, Apr. 15, May 15, Jun. 13, Jul. 12, Aug. 11, Sep. 9, Oct. 9, Nov. 7, Dec. 7, 1890 Jan. 6, Feb. 5, Mar. 6, Apr. 5, May 4, Jun. 3, Jul. 2, Jul. 31, Aug. 30, Sep. 28

SOURCES: E. E. Barnard, E. A. Lamp, E. Weiss, C. F. Pechüle, and J. M. Schaeberle, *AN*, **121** (1889 Apr. 11), p. 143; J. von Hepperger and J. Bauschinger, *AN*, **121** (1889 Apr. 13), p. 175; D. Klumpke, G. Bigourdan, and G. A. P. Rayet, *CR*, **108** (1889 Apr. 23), pp. 846–8; E. Becker, C. F. Pechüle, and C. N. A. Krueger, *AN*, **121** (1889 Apr. 25), p. 207; E. E. Barnard, F. K. Ginzel, and A. Berberich, *AN*, **121** (1889 Apr. 29), p. 223; J. Bauschinger, *AN*, **121** (1889 May 7), p. 239; E. E. Barnard and W. W. Campbell, *AJ*, **9** (1889 May 16), pp. 5–6; A. Abetti, E. E. Barnard, and A. Berberich, *AN*, **121** (1889 May 18), pp. 299–302; A. Kammermann, *AN*, **121** (1889 May 28), p. 333; J. M. Schaeberle, *AN*, **121** (1889 Jun. 5), p. 365; W. W. Campbell, *AJ*, **9**, (1889 Jul. 2), p. 37; A. O. Leuschner and E. Millosevich, *AN*, **122** (1889 Jul. 4), pp. 41–4; E. Millosevich, *AN*, **122** (1889 Sep. 6), p. 315; W. W. Campbell, *AJ*, **9**, (1889 Oct. 31), p. 94; R. Spitaler, *AN*, **123** (1889 Nov. 14), pp. 87–90; B. von Engelhardt and G. A. P. Rayet, *AN*, **123** (1889 Nov. 19), pp. 107–9; R. Spitaler, *AN*, **123** (1889 Dec. 3), p. 141; E. Millosevich, *AN*, **123** (1889 Dec. 12), p. 207; E. Becker, *AN*, **124** (1890 Apr. 16), pp. 203–7; M. Brendel, *AN*, **124** (1890 Jun. 17), pp. 347, 362; R. Spitaler, *AN*, **125** (1890 Sep. 12), pp. 261, 269; E. E. Barnard, *AJ*, **10** (1890 Sep. 19), p. 67; E. Millosevich, *AN*, **125** (1890 Oct. 7), p. 319; J. I. Plummer, *MNRAS*, **51** (1890 Nov.), p. 55; F. K. Zweck, *AN*, **227** (1926 Apr. 13), pp. 221–4; V1964, p. 63; B. G. Marsden and Z. Sekanina, *AJ*, **78** (1973 Dec.), pp. 1119–20.

D/1889 M1 *Discovered:* 1889 June 24.42 ($\Delta = 1.10$ AU, $r = 1.11$ AU, Elong. $= 63°$)
(Barnard 2) *Last seen:* 1889 August 7.47 ($\Delta = 1.39$ AU, $r = 1.33$ AU, Elong. $= 65°$)
Closest to the Earth: 1889 May 30 (1.0099 AU)
1889 III = 1889c *Calculated path:* AND (Disc), PER (Jul. 9), AUR (Aug. 11)

E. E. Barnard (Lick Observatory, California, USA) discovered this comet in Andromeda with a 17-cm refractor on 1889 June 24.42. He described the comet as faint, with no condensation or tail. He gave the position as $\alpha = 1^h\ 20.9^m$, $\delta = +38°\ 51'$ on June 24.45.

The comet had already passed the sun and Earth when discovered, but was well observed during the next few weeks in the morning sky. Unfortunately, despite extensive observations in June, the comet initially showed no signs of a condensation or nucleus and this probably lowered the precision of the early measured positions, especially for observers with smaller telescopes. Barnard switched to using a 30-cm refractor after the comet's discovery. He noted the comet was "probably somewhat more condensed" on the 25th and was "probably slightly condensed" on the 26th. Numerous other astronomers provided descriptions on the 26th as well. E. Becker (Strasbourg, France) described the comet as very faint in a 46-cm refractor. E. Millosevich (Rome, Italy) observed a round coma, but no distinct nucleus with his 15-cm refractor. J. Bauschinger (Munich, Germany) said the comet was "extraordinarily faint," with a diffuse coma 2' across. R. Spitaler (Vienna, Austria) observed with a 69-cm refractor and noted the comet was a "round, diffuse nebulosity of 2' diameter, without nucleus." The first mention of a nucleus came from F. P. Leavenworth (Haverford College Observatory, Massachusetts, USA) on June 28, when he described the comet as extremely faint, with a nucleus of magnitude 13.

The comet's increasing faintness made it a more difficult object as July progressed. Descriptive details are only available during the first half of the month, with the moon interfering during the second half. Barnard described the comet as smaller and more condensed in the 30-cm refractor on the 2nd. Although he noted it seemed somewhat brighter, he suggested this was because of the condensation. Becker saw the comet with a 46-cm refractor on the 4th and said it appeared faint and 1' across, without a distinct condensation. Spitaler simply described the comet as faint in the 69-cm refractor on the 5th. Leavenworth described the comet as extremely faint on the 6th and again noted a nucleus of magnitude 13. The comet attained its minimum observed solar elongation of 61° on July 11. That morning, Barnard reported he was only able to see the comet for a few minutes between moonset and dawn. He commented, "This is the last observation of the comet that can be made until the moon withdraws, if it can then be observed."

Spitaler and Barnard provided the only observations of the comet during August. Spitaler described the comet as a very difficult object to observe with the 69-cm refractor on the 1st, although he did estimate its diameter as 1'. Barnard obtained the final observation of this comet on August 7.47, using the 30-cm refractor. He measured the position as $\alpha = 4^h\ 28^m 49.2^s$, $\delta = +50°\ 01'\ 51''$ and wrote, "The comet was of the last degree of faintness, and was observed with the utmost difficulty." The comet reached it greatest northern declination of +50° on August 12.

A. O. Leuschner computed the first parabolic orbit. Using positions obtained by Barnard on June 24, 25, and 26, he determined the perihelion date as 1889 June 20.65. His computations were particularly remarkable because the perihelion date was only 14 hours earlier than what later orbits would reveal using positions spanning more than a month. What makes it even more remarkable is that the first orbits calculated by European astronomers using positions spanning the period of June 24–27 produced a wide range of perihelion dates, with H. C. F. Kreutz, R. Spitaler, and Bauschinger giving the perihelion date as July 3.35, July 31.86, and June 24.95, respectively. Later parabolic orbits by Spitaler and Campbell using positions acquired into early July determined the perihelion date within a few hours.

After the comet was no longer visible, A. Berberich (1889) investigated its orbit and soon realized it was elliptical. Using five positions from the period of June 25 to August 1, he determined the perihelion date as June 21.24 and the period as about 128 years. Several decades later, B. G. Marsden and Z. Sekanina (1972) redetermined the orbit using 19 positions spanning the period of June 24 to August 7. They found a perihelion date of June 21.25 and a period of 145±10 years. This orbit is given below.

T	ω	Ω (2000.0)	i	q	e
1889 Jun. 21.2488 (TT)	60.2321	272.6342	31.2468	1.104894	0.960033

ABSOLUTE MAGNITUDE: $H_{10} = 8.8 - 10$ (V1964)

FULL MOON: Jun. 13, Jul. 12, Aug. 11

SOURCES: E. E. Barnard, *AN*, **122** (1889 Jun. 25), p. 31; E. E. Barnard and A. O. Leuschner, *AJ*, **9** (1889 Jul. 2), p. 40; E. Becker, E. Millosevich, J. Bauschinger, H. C. F. Kreutz, and R. Spitaler, *AN*, **122** (1889 Jul. 4), pp. 43–6; A. O. Leuschner, J. Bauschinger, and R. Spitaler, *AN*, **122** (1889 Jul. 9), p. 103; W. W. Campbell, F. P. Leavenworth, and E. E. Barnard, *AJ*, **9** (1889 Jul. 17), pp. 46–7; E. E. Barnard, E. Becker, and J. Bauschinger, *AN*, **122** (1889 Jul. 19), pp. 115–18; E. E. Barnard, *AJ*, **9** (1889 Jul. 25), p. 55; A. O. Leuschner, *PASP*, **1** (1889 Jul. 27), p. 31; E. Millosevich and R. Spitaler, *AN*, **122** (1889 Aug. 13), p. 217; E. E. Barnard, *AJ*, **9** (1889 Oct. 3), p. 86; A. Berberich, *AN*, **123** (1889 Nov. 7), p. 77; E. Becker, *AN*, **124** (1890 Apr. 16), pp. 203–8; R. Spitaler and J. Palisa, *AN*, **125** (1890 Sep. 12), pp. 261, 269; V1964, p. 63; B. G. Marsden and Z. Sekanina, *QJRAS*, **13** (1972), pp. 428–9.

16P/1889 N1
(Brooks 2)

1889 V = 1889d

Discovered: 1889 July 7.23 ($\Delta = 1.47$ AU, $r = 2.08$ AU, Elong. $= 112°$)
Last seen: 1891 January 13.31 ($\Delta = 2.80$ AU, $r = 3.76$ AU, Elong. $= 165°$)
Closest to the Earth: 1889 September 20 (0.9498 AU)
Calculated path: AQR (Disc), CET (Jul. 13), PSC (Aug. 3), AQR (Sep. 26), PSC (Oct. 18), ARI (1890 Feb. 6), TAU (Apr. 2), GEM (Jun. 19), CNC (Aug. 31)

W. R. Brooks (Smith Observatory, Geneva, New York, USA) was sweeping the southeastern morning sky for comets with a 26-cm refractor when he found a nebulous object on 1889 July 7.23. He immediately suspected it was a comet and described it as "faintish with a short wide tail." The position was given as $\alpha = 23^h$ 45.0^m, $\delta = -9°$ 10'. E. E. Barnard (Lick Observatory, California, USA) confirmed the comet with the 30-cm refractor on July 9, 10, and 11, before moonlight blocked it from view. He described the comet as magnitude 11–12, about 1' across, with a nucleus of magnitude 12. He added that a tail extended 10' toward PA 245° on the 10th.

Observations ceased shortly after discovery because of moonlight, but they resumed after it had left the sky. As the comet was approaching both the sun and Earth, it brightened and observations became more widespread as July ended. On July 21, R. Spitaler (Vienna, Austria) estimated the comet's magnitude as 11 and noted it was 2' across, with a diffuse nucleus. On July 30, B. von Engelhardt (Dresden, Germany) said the comet was rather faint, but still visible in the 13-cm finder. He added that the comet was small and condensed, with a tail extending 2' toward the southeast. That same evening, Spitaler said the nucleus appeared oblong in the 69-cm refractor, while the coma was 1' across and a tail extended about 2.5' toward the southwest. On the 31st, Engelhardt said the comet was moderately bright and small, while J. Bauschinger (Munich, Germany) noted the comet was faint, with an occasionally visible nucleus. E. Becker (Strasbourg, France) said the comet had a possible fan shape and exhibited a distinct nucleus of magnitude 11–12.

On August 1, Bauschinger described the comet as quite faint. Becker observed with the 46-cm refractor, and simply noted the nucleus was magnitude 10 on the 3rd. Also on the 3rd, Barnard said the 30-cm refractor revealed a fan-shaped tail extending 15' toward PA 320°. He also noted a nucleus of magnitude 12 and a total magnitude of 11. On the 5th, W. Luther (Hamburg, Germany) said the comet was quite faint, while Becker reported the coma appeared granular. That same night, Spitaler said the comet disappeared in morning twilight at the same time as stars of magnitude 11. J. Palisa (Vienna, Austria) noted the comet was quite bright with a distinct tail on the 19th. On the 21st, Barnard said the 30-cm refractor revealed a tail 15–20' long. On the 24th, Palisa reported the nucleus was about magnitude 11, while the tail extended 5'. Becker noted a diffuse nucleus on the 28th and estimated its magnitude as 9.5. He added that the 46-cm refractor revealed a jet extending toward PA 240°. On the 30th, Barnard said the tail extended about 0.5° toward PA 251°.

A very important aspect of this apparition was the discovery of several fainter comets moving with the primary one. Barnard was the first to spot them. The 30-cm refractor needed repair as July progressed and Barnard had been continuing his observations with the 17-cm refractor during the remainder of that month. On the night of August 2, he was finally able to resume his observations with the 30-cm refractor and noted, "a very small and faint nebulosity close north-following the head of the comet. I also saw another a little larger some 4' distant north-following in the line with the first two bodies." He labelled these "A", "B", and "C", with "A" being the main comet. Barnard said "B" was located 64.14" from "A" toward PA 59.4° on that date. On the 3rd, Barnard noted that all three comets had advanced together. He said "B" was located 65.73" away toward PA 64.0°, while "C" was located 264.97" away toward PA 63.9°. Barnard added that four other nebulous objects were detected, which were not in any catalogs of nebulae, and he labelled these "D", "E", "F", and "G". Barnard continued making observations with the 91-cm refractor and Spitaler began his own set of measurements with the 69-cm refractor on the 5th. By the end of August, observations were also being made by A. H. P. Charlois (Nice, France) and G. Bigourdan (Paris, France), while other astronomers began making observations in September. But, according to Z. Sekanina (1977), it was Barnard who is considered to have made the best series of observations. Barnard said objects "F" and "G" were not seen after August 3, while the extremely faint object "E" was last seen on the August 5. The final observations of "D" were made on August 18, while "B" was last seen on September 6 and "C" was last seen on November 26. Interestingly, with a smaller telescope, Spitaler saw three companions until late October and the last companion during January of 1891. Sekanina's analysis of the motion of objects "A", "B", and "C" indicated that "A" and "C" apparently split around the time of a close approach to Jupiter during the summer of 1886, while "B" probably separated from "C" during early 1888. Not enough observations of the other objects were available to determine probable split dates.

The comet passed closest to both the sun and Earth during the second half of September. On September 17, Becker noted a sharp nucleus of magnitude 10. On the 25th, H. P. Hollis (Royal Observatory, Greenwich, England) said the comet was "exceedingly faint," with no nucleus and "no appearance of a part detached." Following his occasional observations of this comet from August 29 to September 23, J. I. Plummer (Orwell Park Observatory, England) saw the comet on September 25 and wrote, "Comet not quite so well seen as previously; the reason of this is not apparent. Sky clear."

The comet was moving away from both the sun and Earth as October began. On October 15, J. Holetschek (Vienna, Austria) observed with a 15-cm refractor and said the comet was 1' across, with a condensation of magnitude 11. Becker said the nucleus was magnitude 11 on the 15th and 10–11 on the 19th. On the 23rd, Holetschek gave the diameter as 1.5–2' across, and the total magnitude as 11.0, while Spitaler estimated the total magnitude as 10. Spitaler also estimated the magnitude as 10 on the 24th and gave the coma diameter as 3'. Barnard observed with the 91-cm refractor on the 29th and said the comet was somewhat round, condensed, and about magnitude 11.5, with a diffuse tail.

On November 15 and 25, Barnard said the magnitude was about 12.5. C. F. Pechüle (Copenhagen, Denmark) simply noted the comet was faint in the 27-cm refractor on the 16th. Spitaler estimated the total magnitude as 10 on the 20th. On November 21, V. Knorre (Berlin, Germany) described the comet as very faint in the 24.4-cm refractor. On December 20 and 27, Becker said the nucleus was magnitude 12.

Several astronomers were still making observations of the comet in the evening sky as the new year began. On January 12, 20, 22, and February 10, Plummer said the comet was faint. Spitaler reported the comet was magnitude 11.5 on the 16th, with an elongated coma 1.5–2′ across and a nucleus of magnitude 13. He reported the total magnitude as 11 on January 17 and the coma diameter as 2′. On the 22nd, Becker said the comet was about 0.5′ across, with a nucleus of magnitude 12. On January 23, F. P. Leavenworth (Haverford College Observatory, Pennsylvania, USA) described the comet as very faint. On February 10, Becker gave the nuclear magnitude as 12–13. Spitaler gave the total magnitude as 12 on the 11th and said the coma was about 1.5′ across. D. C. Miller (Halsted Observatory, New Jersey, USA) reported that the comet was easily seen with the 58-cm refractor on February 11, 13, 14, and 16. He noted that there were a nucleus and a tail. Spitaler said the comet was 2′ in diameter on the 15th. Becker reported the nucleus was magnitude 12–13 on February 16. On March 9, Miller said the comet was still easily seen in the 58-cm refractor, but the nucleus was gone. He noted the comet "was a mere diffuse patch of light." Spitaler said the coma was about 1.5′ across on the 11th, while the nucleus seemed speckled. He said the total magnitude was 14 on the 15th, while the coma was 1′ across. Spitaler said the comet was difficult to observe in the 69-cm refractor on March 17.

The comet passed 4° from the sun on June 20 and attained its most northerly declination of +28° on June 28. Early in October, A. Berberich sent an ephemeris to Barnard in the hope that the comet might still be visible in the 91-cm refractor. Barnard searched at every available opportunity during the next few weeks, but no trace could be found. The sky was exceptionally clear on the morning of November 22 and, after the moon had set, Barnard again began searching for the comet. After an hour he found "a most excessively faint and difficult object." He estimated the diameter as 10″ and said there was no condensation. The object was again found on the morning of the 23rd and the positions from these two days proved it was the sought for comet. Barnard added, "The comet is now the faintest and most difficult object that I have ever seen in the heavens, and from its present appearance no other telescope in existence can possibly show a trace of it." Barnard found the comet again on December 21, when the 91-cm refractor revealed it as an extremely difficult object about 6″ or 8″ across, with a magnitude between 16 and 17. He noted it was actually easier to observe than during the November observations.

The comet was last detected on 1891 January 13.31, when Barnard gave the position as $\alpha = 8^h 38.0^m$, $\delta = +26° 43′$. He was using the 91-cm refractor and further noted, "The comet was sought for several times after the January observation, but the seeing was always too poor to get even a trace of it."

Positions were hard to acquire shortly after discovery because of moonlight. Consequently, nearly two weeks passed before an orbit was determined. The first orbits were independently calculated by S. C. Chandler and K. Zelbr using positions from July 9, 11, and 21. The perihelion date was given as 1889 August 3.90 by Chandler and June 28.40 by Zelbr. Chandler added that the orbit was very uncertain, but that this was probably a short-period comet. O. Knopf took positions from July 9, 21, and 31, and determined the perihelion date as July 12.72.

The first elliptical orbit was calculated by Zelbr using positions from July 9, 21, and August 5. He determined a perihelion date of August 26.98 and a period of 12.33 years. A couple of weeks later, Zelbr and Knopf calculated orbits using slightly different sets of observations spanning the period of July 9 to August 20. Zelbr determined the perihelion date as September 19.76 and the period as 7.80 years. Knopf gave the perihelion date as September 27.46 and the period as 7.29 years. Additional orbits were calculated by H. C. F. Kreutz, Knopf, Chandler, and Bauschinger, which eventually established the perihelion date as September 30.85 and the period as 7.07 years. Zelbr briefly looked at the possibility that this comet was a return of D/1884 O1 (Barnard 1), but the orbit of Brooks 2 had too long a period and there was no way to explain the discrepancy.

Chandler (1889) and C. L. Poor (1891, 1894) investigated the probability that this comet was a return of Lexell's comet of 1770. Both astronomers investigated the close approach to Jupiter in 1886 and both astronomers concluded Brooks' comet was a probable return of Lexell's. Poor was very interested in absolutely confirming this identity and he set out to better model the 1886 approach to Jupiter, as well as a potential close approach to Saturn around 1845. He published the revised results in 1894 and found the pre-1886 orbit to have had a period of 31.38 years. He concluded that an identity with comet Lexell was unlikely.

The first definitive orbit utilizing positions spanning the entire apparition was published by Bauschinger (1898). He gave the perihelion date as September 30.83 and the period as 7.07 years. Multiple apparition orbits by A. D. Dubiago (1950), B. G. Marsden and Z. Sekanina (1972), I. Y. Evdokimov (1978), and Sekanina and D. K. Yeomans (1985) revealed the perihelion date as September 30.85 and the period as 7.07 years. Dubiago identified a secular acceleration in the motion. Marsden, Sekanina, and Yeomans (1973) determined nongravitational terms of $A_1 = +3.61$ and $A_2 = -0.3269$. V. V. Emel'yanenko (1985) calculated orbits for nuclei "B" and "C". The first orbit below is that of comet "A" by Sekanina and Yeomans, while the second and third are comets "B" and "C", respectively, by Emel'yanenko.

T	ω	Ω (2000.0)	i	q	e
1889 Sep. 30.8461 (TT)	343.6605	19.4634	6.0811	1.949832	0.470804
1889 Sep. 30.7978 (ET)	343.6524	19.4609	6.0820	1.949962	0.470819
1889 Sep. 30.7134 (ET)	343.6542	19.4612	6.0828	1.949548	0.470903

ABSOLUTE MAGNITUDE: $H_{10} = 7.2$ (V1964)

FULL MOON: Jun. 13, Jul. 12, Aug. 11, Sep. 9, Oct. 9, Nov. 7, Dec. 7, 1890 Jan. 6, Feb. 5, Mar. 6, Apr. 5, May 4, Jun. 3, Jul. 2, Jul. 31, Aug. 30, Sep. 28, Oct. 28, Nov. 26, Dec. 26, 1891 Jan. 24

SOURCES: W. R. Brooks and E. E. Barnard, *AJ*, **9** (1889 Jul. 17), p. 48; W. R. Brooks, E. E. Barnard, S. C. Chandler, Jr, and R. Spitaler, *AN*, **122** (1889 Jul. 19), pp. 117–20; E. E. Barnard, *AJ*, **9** (1889 Jul. 25), pp. 54–5; W. R. Brooks and E. Weiss, *AN*, **122** (1889 Jul. 25), p. 143; S. C. Chandler and K. Zelbr, *AN*, **122** (1889 Jul. 30), pp. 171–4; B. von Engelhardt, J. Bauschinger, and E. E. Barnard, *AN*, **122** (1889 Aug. 7), pp. 189–92; W. Luther, E. Becker, and O. Knopf, *AN*, **122** (1889 Aug. 13), pp. 217, 221; K. Zelbr, *AN*, **122** (1889 Aug. 24), p. 255; E. E. Barnard, *AN*, **122** (1889 Aug. 29), p. 267; O. Knopf, *AN*, **122** (1889 Sep. 4), p. 303; E. Weiss, R. Spitaler, H. C. F. Kreutz, A. H. P. Charlois, and G. Bigourdan, *AN*, **122** (1889 Sep. 6), pp. 313–16, 319; E. E. Barnard, *AJ*, **9** (1889 Sep. 16), pp. 77–8; K. Zelbr, *AN*, **122** (1889 Sep. 28), pp. 395–8; H. P. Hollis, *MNRAS*, **49** (1889, Supplementary Notice), p. 446; J. I. Plummer, *MNRAS*, **50** (1889 Nov.), p. 45; R. Spitaler, *AN*, **123** (1889 Nov. 7), p. 79; S. C. Chandler, *AJ*, **9** (1889 Nov. 25), pp. 100–3; O. Knopf, *AN*, **123** (1889 Nov. 26), pp. 123–6; O. Knopf, *AN*, **123** (1890 Feb. 10), p. 411; A. Abetti, *AN*, **124** (1890 Mar. 21), p. 111; E. Becker, *AN*, **124** (1890 Apr. 16), pp. 203–8; D. C. Miller, D. P. Hibberd, and F. P. Leavenworth, *AJ*, **9** (1890 Apr. 21), pp. 191–2; E. E. Barnard, *AJ*, **10** (1890 May 8), pp. 4–6; V. Knorre and F. Renz, *AN*, **124** (1890 Jun. 17), pp. 347, 362, 365–8; E. E. Barnard, *AN*, **125** (1890 Aug. 27), pp. 177–96; R. Spitaler, J. Palisa, and J. Holetschek, *AN*, **125** (1890 Sep. 12), pp. 261–4, 269; R. Spitaler, *AN*, **125** (1890 Sep. 18), pp. 281–5; C. F. Pechüle, *AN*, **126** (1890 Oct. 28), pp. 25–8; J. I. Plummer, *MNRAS*, **51** (1890 Nov.), pp. 53–4, 57; C. L. Poor, *AJ*, **10** (1890 Nov. 11), p. 91; E. E. Barnard and A. Berberich, *AJ*, **10** (1890 Dec. 8), p. 111; S. C. Chandler, *AJ*, **10** (1890 Dec, 15), pp. 118–21; E. E. Barnard, *AJ*, **10** (1891 Jan. 9), p. 136; J. Bauschinger, *AN*, **126** (1891 Jan. 15), p. 213; E. E. Barnard, *AJ*, **11** (1891 May 12), p. 5; C. L. Poor, *AJ*, **11** (1891 Jul. 24), pp. 30–1; C. L. Poor, *MNRAS*, **54** (1894 Feb.), p. 244; J. Bauschinger, *Neue Annalen der K. Sternwarte in München*, **3** (1898), pp. 3–40; A. D. Dubiago, *Trudy Kazanskaia Gorodkoj Astronomicheskoj Observatorii*, No. 31 (1950), p. 25; V1964, p. 63; B. G. Marsden and Z. Sekanina, *QJRAS*, **13** (1972), pp. 428–9; B. G. Marsden, Z. Sekanina, and D. K. Yeomans, *AJ*, **78** (1973 Mar.), p. 213; B. G. Marsden and Z. Sekanina, *CCO*, 2nd ed. (1975), pp. 19, 46; Z. Sekanina, *Icarus*, **30** (1977), pp. 578, 581–3; I. Y. Evdokimov, *QJRAS*, **19** (1978), pp. 80–1, 88; V. V. Emel'yanenko, *QJRAS*, **26** (1985), pp. 113–14; Z. Sekanina and D. K. Yeomans, *CCO*, 5th ed. (1986), pp. 16, 51; D. K. Yeomans, *QJRAS*, **27** (1986), p. 604.

C/1889 O1   *Discovered:* 1889 July 19.95 ($\Delta = 0.38$ AU, $r = 1.04$ AU, Elong. = 83°)
(Davidson)   *Last seen:* 1889 November 21.72 ($\Delta = 2.25$ AU, $r = 2.18$ AU, Elong. = 73°)
  Closest to the Earth: 1889 July 24 (0.3603 AU)
1889 IV = 1889e   *Calculated path:* CEN (Disc), HYA (Jul. 24), VIR (Jul. 27), LIB-VIR (Aug. 1), LIB (Aug. 4), VIR (Aug. 5), SER (Aug. 9), HER (Aug. 23), SER (Aug. 24), HER (Aug. 26), LYR (Oct. 29)

J. E. Davidson (Mackay, Queensland, Australia) discovered this comet on July 19.95. It was described as about 5′ in diameter, with a bright nucleus of magnitude 5. Although he noted that no tail was visible, he did mention an extension of nebulosity to the south-following side. He immediately sent a notice to Sydney Observatory and on July 22 he wired an announcement to Melbourne Observatory. R. L. J. Ellery (Melbourne, Australia) confirmed the discovery on July 22, when the comet was easily seen with the naked eye. He noted a sharp nucleus of magnitude 5–6, a coma about 5' across, and a tail extending about 30' toward the south-following direction.

The comet had passed perihelion on the day of discovery and was closest to Earth on July 24. On July 23 and 24, J. Tebbutt (Windsor, New South Wales, Australia) said the comet was conspicuous to the naked eye, with a magnitude of 4. His 20-cm refractor revealed a small nucleus surrounded by an extensive coma. On the 23rd and 25th, P. Baracchi (Melbourne, Australia) described the comet as a bright object, with a coma 4' or 5' across, and exhibiting a stellar nucleus of magnitude 5 or 6. The tail extended more than 30' toward the south-following side. He noted it was easily visible to the naked eye. W. H. Finlay (Royal Observatory, Cape of Good Hope, South Africa) saw the comet with the 18-cm refractor on the 25th and noted a bright, round nucleus of magnitude 8, which was "surrounded by faint nebulous matter, with a short tail." That same evening, A. Riccò (Palermo, Italy) observed with a 25-cm refractor and noted a stellar nucleus of magnitude 6.9 surrounded by a bright, diffuse coma that was slightly elongated. On the 26th, Riccò said the comet seemed fainter, while C. Trèpied (Alger, now al-Jazâ'ir, Algeria) noted the coma was elongated toward PA 120° and the nucleus was 13.7" across and about magnitude 8 in the 50-cm refractor. Also on the 26th, E. E. Barnard (Lick Observatory, California, USA) saw the comet in the 8.3-cm finder of the 30-cm refractor and described the comet as "large and bright, with a bushy tail." He also noted, "It was easily visible to the naked eye, appearing as a large oblong nebulous mass as bright as a 6th magnitude star." Finlay reported a fairly sharp nucleus and a faint tail on the 27th. On the 29th, A. Kammermann (Geneva, Switzerland) described the comet as having a very bright coma, with a sharp nucleus. He estimated the brightness as magnitude 6–7 and said it would probably be a naked-eye object in dark skies. That same evening, Riccò said the total brightness of the comet exceeded that of the magnitude 6.7 comparison star, while the nucleus was fainter than the star and diffuse. On the 30th, Baracchi said the comet was still visible to the naked eye, although it seemed a little fainter. With a refractor he noted the nucleus was diffused and no brighter than magnitude 6, while the tail extended 30' toward the south-following side. On the same evening, Riccò noted a faint, diffuse tail extending toward the east-southeast. He added that the comet was difficult to see with the naked eye, but an easy target in binoculars. During the period of July 26–30, E. Millosevich (Rome, Italy) observed with a 15-cm refractor and said the nucleus was about magnitude 5–6, while the tail extended 12–15' toward PA 135°. Several observers saw the comet on July 31. J. Bauschinger (Munich,

Germany) said the comet had a round coma and a nucleus of magnitude 8.5. B. von Engelhardt (Dresden, Germany) described the comet as bright, white, and condensed, with a faint coma 3' across. J. Palisa (Vienna, Austria) estimated the magnitude of the nucleus as 8.5. Riccò said the comet was visible to the naked eye, while a refractor revealed a tail extending 24' toward the east-southeast. Also on the 31st, Barnard obtained a 90-minute exposure with a 13-cm lens. He wrote, "The head of the comet shows as a neat round mass. The tail is fan-shaped, with its borders convex to the axis, and very narrow at the root. It can easily be traced 20' and it is evident for about 53'."

With the comet moving away from both the sun and Earth, it generally faded during the next several months. On August 1, Engelhardt said the comet was bright, small, and diffuse, with a faint coma about 3' across. He added that it was brighter than on the previous evening. J. Holetschek (Vienna, Austria) estimated the magnitude of the nucleus as 8 on the same evening. On the 2nd, Baracchi said the comet was still visible to the naked eye and was generally of a similar appearance to that of the 30th. On the 3rd, Riccò observed the comet despite the nearby moon and noted a double nucleus, with the fainter component east-southeast of the main nucleus. Holetschek then estimated the magnitude of the nucleus as 9. M. Brendel (Berlin, Germany) observed with a 23-cm refractor on the 4th and said the comet was magnitude 8, with a bright nucleus. The tail was barely visible in the evening twilight. That same evening, Riccò said the nucleus was fainter than the 6.5-magnitude comparison star and was still split, with the fainter component toward PA 96°. E. Becker (Strasbourg, France) gave the nuclear magnitude as 10 on the 5th. W. Wickham (Radcliffe Observatory, Oxford, England) said the nucleus was stellar and about magnitude 9 on the 6th. Riccò reported the nucleus was fainter than magnitude 8 on the 7th. He said the second nucleus was located toward PA 102° on that date and toward 111° on the 8th. On August 9, Riccò noted the nucleus was brighter than magnitude 9 and double, with the component to the east-southeast being rather large, while Holetschek estimated the magnitude of the nucleus as 9.0. Riccò said the nucleus was elongated, but the component was less distinguished than before. On the 11th, Riccò said the nuclei were magnitude 9 and 10. Finlay described the comet as noticeably fainter on the 12th, although the nucleus was still about magnitude 9.5. Several observers obtained physical details on the 17th. Engelhardt said the coma contained a disk-shaped, white nucleus and exhibited a broad, fan-shaped tail about 15' long. Kammermann estimated the comet's brightness as magnitude 7–8. He added that a second, fainter tail extended from the coma 65–70° further north than the main tail. Holetschek estimated the magnitude of the nucleus as 9 and gave the coma diameter as 4'. Palisa estimated the magnitude of the nucleus as 10 and gave the coma diameter as 5'. Becker gave the nuclear magnitude as 9–10 and said it was located just north of the coma's center. He added that the tail extended 17′ toward PA 125°. On the 18th, Baracchi said the comet was "much fainter" and the nucleus was "quite diffused." The tail extended about 20′ toward the south-following side. Wickham said a coma surrounded a nucleus of magnitude 10.5 on the 22nd. Becker said the 46-cm refractor revealed a nucleus of magnitude 9, as well as two jets: the short, bright one extending toward PA 22° and another at PA 62°. Holetschek gave the nuclear magnitude as 9 on the 24th and 31st. Engelhardt noted on the 30th that the comet was fainter than a star of magnitude 8.4 and was diffuse, with a short, broad tail. On the 31st, Engelhardt said the nucleus was bright and granular, while the tail was short and very broad.

On September 6, J. I. Plummer (Orwell Park Observatory, England) said the comet was faint in bright moonlight, while Wickham described it as a "small, faint patch," with a stellar nucleus that was only occasionally glimpsed in the 25.4-cm refractor (100×). Plummer said the comet was "too faint in the moonlight" for a satisfactory measurement of the comet's position on the 8th. Holetschek estimated the magnitude of the condensation as 9.7 on the 14th and 10 on the 15th. Engelhardt described the comet as rather bright and condensed on the 23rd. A. Abetti (Padova, Italy) described the nucleus as elongated and magnitude 9–10 on the 22nd and 24th. Bauschinger estimated the nuclear magnitude as 9 on the 24th. Holetschek estimated the magnitude of the condensation as 10.5 on the 26th. On September 30, Plummer noted the nucleus was elongated.

On October 1, Holetschek estimated the magnitude of the condensation as 11. Plummer said the comet was faint in moonlight on the 3rd and he noted the nucleus was elongated on the 12th. C. F. Pechüle (Copenhagen, Denmark) described the comet as 30″ in diameter and condensed on the 16th. On the 19th, Holetschek said the comet was at the limit of visibility in the 15-cm refractor. Abetti said the comet was difficult to measure on the 23rd, because of its faint, elongated appearance. On October 24, R. Spitaler (Vienna, Austria) saw the comet with the 69-cm refractor and said the coma was about 2′ across, with a nucleus of magnitude 11. He noted a trace of a tail toward the east-northeast which gave the coma the appearance of an ellipse. Pechüle said the comet was elongated toward PA 30° on the 26th.

On November 12, Plummer said the comet was faint and became invisible after moonrise. On the 13th, Plummer said the comet was faint and the measured position was doubtful. Pechüle said the comet was very faint in the 27-cm refractor on the 14th.

The comet was last seen on November 21.72, when Spitaler found it with the 69-cm refractor. He gave the position as $\alpha = 18^h 59.6^m$, $\delta = +37° 57′$ and described the comet as a round nebulosity, about 1.5′ across, with a faint nucleus. Even though he said the comet could presumably still be seen in December, he reported no further observations.

Riccò examined the spectrum and noted traces of the cometary bands near the nucleus on July 26, 29, and 31, as well as a continuous spectrum on the last date. On August 10, he noted a single band, and slight traces of others. J. E. Keeler saw the three carbon bands in the light of the coma and nucleus on August 1 and 2.

The first parabolic orbit was calculated by C. N. A. Krueger using positions from July 23, 26, and 27. He determined the perihelion date as 1889 July 19.71. A few days later, R. L. J. Ellery took Baracchi's positions from July 23, 26, and 29, and determined a perihelion date of July 19.79. Additional orbits were calculated by K. Zelbr, E. A. Lamp, T. H. Safford, L. Becker, and W. Bellamy.

The first elliptical orbit was calculated by W. W. Campbell using positions spanning the period of July 23 to September 28. He determined the perihelion date as July 19.77 and the period as about 2780 years. A. Berberich (1890), G. Horn (1904), and B. G. Marsden (1978) calculated orbits using positions spanning the entire period of visibility. Berberich determined a period of about 5127 years, Horn calculated it to be about 9740 years, and Marsden gave it as about 9079 years. Marsden's orbit is given below.

T	ω	Ω (2000.0)	i	q	e
1889 Jul. 19.7840 (TT)	345.8619	287.7114	65.9916	1.039721	0.997611

ABSOLUTE MAGNITUDE: $H_{10} = 6.5$ (V1964)

FULL MOON: Jun. 13, Jul. 12, Aug. 11, Sep. 9, Oct. 9, Nov. 7, Dec. 7

SOURCES: J. E. Davidson, *AN*, **122** (1889 Jul. 25), p. 143; J. E. Keeler, *PASP*, **1** (1889 Jul. 27), pp. 34–6; E. Millosevich, F. Porro, and C. N. A. Krueger, *AN*, **122** (1889 Jul. 30), pp. 171–4; E. Millosevich and K. Zelbr, *AN*, **122** (1889 Aug. 7), pp. 187–92; Miranda, A. Kammermann, J. Palisa, E. A. Lamp, J. Bauschinger, B. von Engelhardt, C. Wagner, M. Brendel, R. L. J. Ellery, and J. Möller, *AN*, **122** (1889 Aug. 13), pp. 219–24; A. Kammermann, *AN*, **122** (1889 Aug. 24), p. 253; C. Trèpied, F. Sy, and Renaux, *AN*, **122** (1889 Aug. 29), p. 269; L. Becker, *Nature*, **40** (1889 Aug. 29), p. 424; E. E. Barnard, *AJ*, **9** (1889 Aug. 31), pp. 66–7; T. H. Safford, *AJ*, **9** (1889 Sep. 16), p. 79; G. Celoria and J. Tebbutt, *AN*, **122** (1889 Sep. 16), p. 349; W. Bellamy, *AJ*, **9** (1889 Oct. 31), p. 95; Criswick and T. Lewis, *MNRAS*, **49** (1889, Supplementary Notice), p. 447; R. L. J. Ellery, P. Baracchi, H. C. Russell, and Pollock, *MNRAS*, **50** (1889 Nov.), pp. 46–9; J. E. Davidson and R. L. J. Ellery, *AN*, **123** (1889 Nov. 14), p. 91; B. von Engelhardt, *AN*, **123** (1889 Nov. 19), pp. 107–10; R. Spitaler, *AN*, **123** (1889 Dec. 3), p. 141; A. Riccò, *AN*, **123** (1889 Dec. 12), pp. 205–8; W. W. Campbell, *AJ*, **9** (1889 Dec. 18), p. 119; J. Bauschinger, *AN*, **123** (1890 Jan. 7), p. 281; A. Abetti, *AN*, **123** (1890 Jan. 30), pp. 361–4; A. Berberich, *AN*, **124** (1890 Mar. 29), pp. 147–50; E. Becker, *AN*, **124** (1890 Apr. 16), pp. 203, 207; J. E. Davidson, *The Observatory*, **13** (1890 Jul.), p. 247; R. Spitaler, *AN*, **125** (1890 Sep. 12), pp. 261, 269; J. I. Plummer, *MNRAS*, **51** (1890 Nov.), p. 52; E. E. Barnard, *AJ*, **11** (1891 Dec. 9), p. 75; W. Wickham, *MNRAS*, **55** (1895 Jan.), pp. 161–3; W. H. Finlay, *Annals of the Cape Observatory*, **1** pt. 1 (1898), pp. 32–3; G. Horn, *AN*, **165** (1904 Jun. 13), pp. 327–30; V1964, p. 63; B. G. Marsden, *AJ*, **83** (1978 Jan.), p. 66; B. G. Marsden, *CCO*, 3rd ed. (1979), pp. 19, 48.

64P/1889 W1
(Swift–Gehrels)

1889 VI = 1889f

Discovered: 1889 November 17.13 ($\Delta = 0.66$ AU, $r = 1.37$ AU, Elong. $= 110°$)
Last seen: 1890 January 22.24 ($\Delta = 1.02$ AU, $r = 1.50$ AU, Elong. $= 97°$)
Closest to the Earth: 1889 October 9 (0.6092 AU)
Calculated path: PEG (Disc), PSC-AND (Dec. 19), PSC (Dec. 30), TRI (Jan. 14)

L. Swift (Warner Observatory, Rochester, New York, USA) was engaged in the search for nebulae when he found an object of cometary appearance on 1889 November 17.13. It was located very close to ξ Pegasi and was described as "Pretty faint, large, little elongated." Swift gave the position as $\alpha = 22^h\ 40.6^m$, $\delta = +11°\ 35'$. No motion was detected during the next 30 minutes and he suspected it was one of two nebulae that W. Herschel discovered near this star late in the eighteenth century, although Swift's position did not agree. He reobserved the object on November 17.99 and noticed it had moved, whereupon he sent a discovery announcement.

The comet had passed closest to Earth over a month before discovery, but was heading toward a perihelion passage at the end of November. Several physical descriptions were provided during the remainder of that month. On November 20, J. Palisa (Vienna, Austria) observed with a 30-cm refractor and described the comet as a pale nebulous mass, about 3–4′ in diameter and with a faint condensation. That same night, T. Zona (Palermo, Italy) simply noted the comet was faint. E. E. Barnard (Lick Observatory, California, USA) saw the comet in a 30-cm refractor on the 21st and said it was faint and round, with a diameter of about 45″. It was very gradually brighter in the middle. On the 22nd, B. von Engelhardt (Dresden, Germany) observed with a 30-cm refractor and said the comet was very faint, about 0.5′ across, and without a condensation. He added that a star-like nucleus was occasionally visible. Barnard said the comet was about magnitude 13, with a diameter of about 45″. It was very slightly brighter in the middle. On the 23rd, Barnard said the comet was still faint, but better seen than on the previous nights, while Engelhardt said the comet appeared very faint, about 2′ in diameter, and without a condensation. It also exhibited a granular appearance in the 30-cm refractor. Engelhardt added that the comet was better seen in the 13-cm finder at low magnification. R. Spitaler (Vienna, Austria) observed with the 69-cm refractor and said the comet was about magnitude 13 and 2′ across.

The comet was moving away from both the sun and Earth as December began. Spitaler described it as 2.5′ across in the refractor on the 9th, and noted a granular center. On the 17th, he estimated the magnitude as 13 and said the coma was 1.5′ across. On December 11, 12, and 24, J. I. Plummer (Orwell Park Observatory, England) wrote, "Comet very faint, a diffused mass without central condensation. I failed to see this comet on December 20 and 25." C. A. Young (Halsted Observatory, New Jersey, USA) saw the comet on the 22nd, 23rd, and 24th, and described it as "nearly round, from 2–3′ in diameter, ill-defined, without any trace of a tail, and only slightly condensed at the center." E. Becker (Strasbourg, France) only suspected the comet in the 46-cm refractor on the 27th. Spitaler found the comet 2′ across on January 17.

The comet was last detected on 1890 January 22.24, when Barnard measured the position as $\alpha = 2^h\ 07.3^m$, $\delta = +28°\ 05'$. He described it as very faint in the 30-cm refractor.

The first parabolic orbit was calculated by K. Zelbr using positions from November 17, 20, and 22. He determined the perihelion date as 1889 December 11.03. Independent calculations were published by G. M. Searle and R. Schorr a few days later, using positions from November 19, 21, and 23. Searle gave the perihelion date as 1889 December 2.11, while Schorr gave it as November 28.98. Searle noted the small inclination and remarked that it "gives considerable probability of periodicity." A revision was published by Zelbr at the beginning of December.

The first elliptical orbit was calculated by Zelbr using positions spanning the period of November 19 to December 9. He determined the perihelion date as November 30.13 and the period as 6.91 years. A couple of weeks later, G. M. Searle took positions from November 19 to December 14 and calculated a perihelion date of November 30.32 and a period of 8.82 years. In the years following the comet's departure, J. R. Hind (1891) and J. Coniel (1896) calculated periods of 8.53 years and 8.92 years, respectively. Coniel commented that the uncertainty in the period was ±0.9 years, which would make it impracticable to provide ephemerides for the next return.

The comet was lost until T. Gehrels (Palomar Observatory, California, USA) accidentally photographed it on 1973 February 8. B. G. Marsden (1973) was the first person to identify this comet as a return of Swift's comet. After linking the two apparitions, Marsden gave the perihelion date as November 30.10 and the period as 9.15 years. This orbit was essentially confirmed by M. Bielicki and G. Sitarski (1991) and S. Nakano (1991). Bielicki and Sitarski gave nongravitational parameters of $A_1 = +0.5391$, $A_2 = +0.0092$, and $A_3 = +0.0380$, and their orbit is given below.

T	ω	Ω (2000.0)	i	q	e
1889 Nov. 30.1080 (TT)	69.8107	331.8754	10.3048	1.357683	0.689692

ABSOLUTE MAGNITUDE: $H_{10} = 11.5$ (V1964)

FULL MOON: Nov. 7, Dec. 7, Jan. 6, Feb. 5

SOURCES: L. Swift, *AN*, **123** (1889 Nov. 19), p. 111; L. Swift and E. E. Barnard, *AJ*, **9** (1889 Nov. 25), p. 104; T. Zona, J. Palisa, B. von Engelhardt, and K. Zelbr, *AN*, **123** (1889 Nov. 26), pp. 125–8; L. Swift and G. M. Searle, *SM*, **8** (1889 Dec.), pp. 463–4; G. M. Searle, *AJ*, **9** (1889 Dec. 3), p. 112; R. Spitaler, G. Bigourdan, O. Tetens, B. von Engelhardt, and R. Schorr, *AN*, **123** (1889 Dec. 3), pp. 141–4; K. Zelbr, *AN*, **123** (1889 Dec. 9), p. 191; E. E. Barnard, *AJ*, **9** (1889 Dec. 18), p. 119; G. M. Searle, *AJ*, **9** (1889 Dec. 26), p. 128; K. Zelbr, *AN*, **123** (1889 Dec. 30), p. 255; E. E. Barnard, *AJ*, **9** (1890 Jan. 10), p. 131; C. A. Young, *AJ*, **9** (1890 Feb. 4), pp. 141–2; O. C. Wendell, *AJ*, **9** (1890

Feb. 28), pp. 159–60; O. C. Wendell, *AN*, **124** (1890 Mar. 21), p. 113; E. Becker, *AN*, **124** (1890 Apr. 16), pp. 205–8; J. Palisa and R. Spitaler, *AN*, **125** (1890 Sep. 12), pp. 263, 270; J. I. Plummer, *MNRAS*, **51** (1890 Nov.), p. 55; J. R. Hind, *CR*, **113** (1891 Jul. 20), pp. 113–14; J. Coniel, *BA*, **13** (1896 Jul.), pp. 264–75; V1964, p. 63; T. Gehrels, *IAUC*, No. 2491 (1973 Feb. 12); B. G. Marsden, *IAUC*, No. 2500 (1973 Feb. 28); B. G. Marsden, *IAUC*, No. 2517 (1973 Apr. 4); M. Bielicki and G. Sitarski, *AA*, **41** (1991), pp. 315–16; S. Nakano, *Nakano Note* No. 546 (1991 Jan. 15).

C/1889 X1　*Discovered:* 1889 December 12.81 ($\Delta = 1.01$ AU, $r = 1.18$ AU, Elong. = 73°)
(Borrelly)　*Last seen:* 1890 January 20.27 ($\Delta = 0.82$ AU, $r = 0.35$ AU, Elong. = 20°)
　　　　　　　Closest to the Earth: 1890 January 14 (0.7991 AU)
1890 I = 1889g　*Calculated path:* HER (Disc), LYR (Dec. 16), HER (Jan. 1), OPH (Jan. 10), AQL-SER (Jan. 12), AQL (Jan. 14), SCT (Jan. 17), AQL (Jan. 18), SGR (Jan. 20)

This comet was discovered by A. L. N. Borrelly (Marseille, France) on 1889 December 12.81 at a position of $\alpha = 18^{\mathrm{h}}\ 07.0^{\mathrm{m}}$, $\delta = +48°\ 53'$. He described it as faint, diffuse, and about 2′ across.

The comet was heading toward both the sun and Earth when discovered and steadily brightened through the remainder of December. Although E. E. Barnard (Lick Observatory, California, USA) described the comet as "pretty bright" on December 16 and 18, B. von Engelhardt (Dresden, Germany) said it was "very faint", while J. Palisa (Vienna, Austria) estimated the magnitude as 9.5 on December 17. All three astronomers were using 30-cm refractor telescopes. The comet must have then been fairly diffuse, as Engelhardt said it was easily seen in the wider field of view of the 13-cm seeker. Barnard and Engelhardt said the coma was 2′ across, while Palisa estimated 3′. All three astronomers noted the comet was centrally condensed. On December 19, Engelhardt still noted the comet was quite faint in the 30-cm telescope. He added that the coma was still 2′ across, but seemed irregular in shape and unevenly bright. J. I. Plummer (Orwell Park Observatory, England) said the comet was fairly bright on December 25, with a central condensation. E. Becker (Strasbourg, France) described the comet as a round, centrally condensed mass, about 1′ across on the 27th.

The comet was continuing its trek toward both the sun and Earth as January began, but its solar elongation was also decreasing. J. Bauschinger (Munich, Germany) saw the comet on the 5th and said the comet appeared quite bright and was easy to find in a 27-cm refractor, despite moonlight and low altitude. On the 8th, he noted the comet's appearance was similar to a star of magnitude 7.5, although he added the coma was 3′ across. Bauschinger last saw the comet on January 9. He said the altitude was then less than 5° and noted that the 9th-magnitude comparison star was very faint and indefinite. T. W. Backhouse (Sunderland, England) saw the comet very low over the eastern horizon on the morning of the 16th and saw a tail extending 45′ in field-glasses (4×) and 40′ long in his 10.8-cm refractor (38×).

W. W. Campbell (Ann Arbor, Michigan, USA) measured the position as $\alpha = 18^h 47.7^m$, $\delta = -1° 37′$ on January 16.50 using a 15-cm refractor. Although this was the last time the position was determined, the final observation was made on January 20.27 by Backhouse. The comet was then very low over the eastern horizon, and the tail was 5′ long. The comet passed only 16° from the sun on January 26.

The first parabolic orbit calculated for this comet was by K. Zelbr and R. Froebe. They took positions from December 13, 15, and 17, and determined the perihelion date as 1890 January 28.21. A few days later, A. Berberich took positions from December 14 to December 23, determined four Normal positions, and calculated the perihelion date as January 26.98. This orbit was very accurate, as proven by later orbits by C. N. A. Krueger, G. A. Hill, Campbell, and H. S. Chase.

Definitive orbits were calculated by A. Seydler (1896) and F. G. Radelfinger (1896). The orbits were quite similar to one another, as well as those determined earlier by other astronomers. Radelfinger used more positions extending over a longer observation arc. Although the latter observations were made at low altitude, he said they were still satisfactorily represented by his orbit. R. J. Buckley (1979) selected the 40 most reliable positions obtained during this comet's apparition and calculated both a hyperbolic and a parabolic orbit. Each represented the orbit equally well, and since the comet was observed for barely over one month, he preferred the parabolic orbit, which is given below.

T	ω	Ω (2000.0)	i	q	e
1890 Jan. 26.9764 (TT)	199.8551	10.0181	56.7537	0.269796	1.0

ABSOLUTE MAGNITUDE: $H_{10} = 8.8$ (V1964)

FULL MOON: Dec. 7, Jan. 6, Feb. 5

SOURCES: A. L. N. Borrelly and E. E. Barnard, *AJ*, **9** (1889 Dec. 18), p. 120; A. L. N. Borrelly, E. A. B. Mouchez, and B. von Engelhardt, *AN*, **123** (1889 Dec. 19), p. 223; A. Abetti, G. Celoria, K. Zelbr, and R. Froebe, *AN*, **123** (1889 Dec. 23), p. 239; C. N. A. Krueger, *AJ*, **9** (1889 Dec. 26), p. 127; G. Celoria, *AN*, **123** (1889 Dec. 30), p. 255; A. L. N. Borrelly, H. J. A. Perrotin, A. H. P. Charlois, J. Palisa, R. Spitaler, B. von Engelhardt, A. Abetti, J. Bauschinger, and A. Berberich, *AN*, **123** (1890 Jan. 7), pp. 283–6; E. E. Barnard, E. Frisby, G. A. Hill, and W. W. Campbell, *AJ*, **9** (1890 Jan. 10), pp. 131–6; J. Bauschinger, *AN*, **123** (1890 Jan. 15), p. 319; W. Schur, *AN*, **123** (1890 Jan. 21), p. 335; T. W. Backhouse, *The Observatory*, **13** (1890 Feb.), p. 90; E. E. Barnard and W. W. Campbell, *AJ*, **9** (1890 Feb. 4), pp. 142–3; C. N. A. Krueger, *AN*, **124** (1890 Feb. 26), p. 47; O. C. Wendell, *AJ*, **9** (1890 Feb. 28), p. 160; A. Abetti and O. C. Wendell, *AN*, **124** (1890 Mar. 21), pp. 111–14; E. Becker, *AN*, **124** (1890 Apr. 16), pp. 205–8; H. S. Chase, *AJ*, **10** (1890 May 8), p. 7; J. Palisa and R, Spitaler, *AN*, **125** (1890 Sep. 12), p. 263; F. Schwab, *AN*, **125** (1890 Sep. 18), p. 277; J. I. Plummer, *MNRAS*, **51** (1890 Nov.), p. 56; F. G. Radelfinger and A. O. Leuschner, *AN*, **142** (1896 Nov. 30), pp. 65–76; V1964, p. 63; R. J. Buckley, *JBAA*, **89** (1979 Oct.), pp. 584–5.

C/1890 F1
(Brooks)

1890 II = 1890a

Discovered: 1890 March 20.38 ($\Delta = 2.73$ AU, $r = 2.11$ AU, Elong. $= 43°$)
Last seen: 1892 February 5.05 ($\Delta = 5.58$ AU, $r = 6.56$ AU, Elong. $= 176°$)
Closest to the Earth: 1890 June 4 (1.5650 AU)
Calculated path: EQU (Disc), PEG (Apr. 4), VUL (Apr. 17), CYG (Apr. 29), DRA (Jun. 1), UMa (Jul. 4), CVn (Jul. 19), COM (Sep. 16), CVn (1891 Jan. 18), COM (Jan. 22), CVn (Jan. 30), UMa (Feb. 1), LMi (Mar. 3), LEO (Apr. 6), CNC (1892 Feb. 10)

W. R. Brooks (Smith Observatory, Geneva, New York, USA) was engaged in a routine search for comets when he discovered this comet with the 26-cm refractor (40×) on 1890 March 20.38. The position was given as $\alpha = 21^h 09.0^m$, $\delta = +5° 35'$. Brooks said, "I at once felt confident that the object was a comet, for the region has been thoroughly searched many times, and I had no record of a nebula in that place." He confirmed the lack of a nebula in this position after consulting various catalogs. Brooks described the comet as "rather bright, telescopic, with stellar nucleus, and a short, broad tail." Morning twilight soon hid the comet, with Brooks only able to suspect a northward motion. He sent a notification of his find to Harvard, Lick, and Warner Observatories. His next three mornings were cloudy, but on March 24.38 he found the comet about 1.5° north of the previous position. In the meantime, O. C. Wendell (Harvard College Observatory, Massachusetts, USA) had confirmed the discovery on March 22.40. The comet was discovered just over two months prior to its closest approaches to the sun and Earth.

On March 24, E. E. Barnard (Lick Observatory, California, USA) estimated the comet's magnitude as 11. He said the 30-cm refractor revealed a 10th-magnitude nucleus was in the north portion of the coma and a faint, short, fan-shaped tail extended due south. J. Palisa (Vienna, Austria) estimated the magnitude as 10 on the same date. On the 28th and 29th, G. A. P. Rayet and L. Picart (Bordeaux, France) described the comet as round, with a nucleus. On the 29th, J. Holetschek (Vienna, Austria) said the comet disappeared in morning twilight at the same time as stars of magnitude 11, while G. Le Cadet (Lyon, France) said the condensation was magnitude 11.5. That same morning, G. Bigourdan (Paris, France) described the comet as a round nebulosity, about 40–50" across, with a pronounced, but not stellar nucleus. He added that the comet was about magnitude 10. On March 30, Holetschek estimated the magnitude of the condensation as 10.5, while E. A. Lamp (Kiel, Germany) described the comet as quite bright and round, with a nucleus of magnitude 9–10.

On April 4, D. Klumpke (Paris, France) said the comet was very faint because of moonlight. On April 12, W. Luther (Hamburg, Germany) described the comet as round, with a nucleus of magnitude 9.5. E. Millosevich (Rome, Italy) saw the comet on the 14th and said the nucleus was about magnitude 10.5, while the coma was not symmetrical and more developed to the south-southwest. On the 15th, R. Spitaler (Vienna, Austria) estimated the magnitude as 9 and said the nucleus was large and diffuse. He added that the 69-cm refractor revealed the coma was bulging toward the west and south as if forming a double tail. Spitaler and Holetschek gave the total magnitude as 9 and 9.5, respectively, on the 17th. Palisa gave the magnitude as 10 on the 19th, while Holetschek indicated it was between 9 and 9.5. Millosevich estimated the nuclear magnitude as 10 on the 23rd and noted the coma was not symmetrical and was most developed toward the south. On April 28, Holetschek said the comet disappeared in morning twilight at the same time as stars of magnitude 10.

On May 8, Millosevich estimated the nuclear magnitude as 10 and noted a rudimentary tail toward the south-southwest. On May 15, T. Lewis (Royal Observatory, Greenwich, England) simply described the comet as "very bright" in the 17-cm refractor, while Rayet noted a coma 4' across, a nucleus of magnitude 9 or 10, and a tail 10' long. Rayet also obtained a one-hour exposure, which showed a disk-shaped nucleus of sensible diameter, and a tail 2' long. B. von Engelhardt (Dresden, Germany) said the comet was bright in the 30-cm refractor on the 16th. He noted a granular, disk-shaped nucleus about 5–6" across, as well as a short, fan-shaped tail. Holetschek also made the first of his magnitude estimates with a 3.7-cm finder on the 16th and indicated a magnitude of 8.5. On the 17th, Holetschek said the comet was between magnitudes 8.0 and 8.7. That same night, the nuclear magnitude was given as 9.2 by Engelhardt and 10 by C. F. Pechüle (Copenhagen, Denmark). Pechüle added that the coma was 4' across and seemed to extend toward PA 180°. On the 18th, Engelhardt said the disk-shaped nucleus was about magnitude 9.5. On the 19th, W. G. Thackeray (Royal Observatory, Greenwich, England) simply described the comet as "very faint" in the 17-cm refractor, while Engelhardt said the nucleus was about magnitude 10 and granular. That same evening, Holetschek said the comet was near magnitude 8.1. On the 20th, Millosevich noted a nucleus of magnitude 10, as well as a tail extending 3' toward the south-southwest, while Holetschek said the comet was near magnitude 8.7. Holetschek gave the magnitude as 8.1 on the 21st. C. Trépied (Alger, now al-Jazâ'ir, Algeria) photographed the comet on May 22 and measured the comet's position with respect to two stars. Engelhardt estimated the nuclear magnitude as 9.8 in the 30-cm refractor on the 23rd, and noted the 13-cm finder revealed a short, fan-shaped tail. On the 26th, Engelhardt said the nucleus appeared diffuse, but brighter than on the 23rd, while Holetschek gave the magnitude as 8.6. On the 28th, A. D. Dubiago and Trotzki (Kazan, Russia) simply noted the comet was faint in the 24-cm refractor. On May 30, Thackeray said the comet was "rather faint" in the refractor.

The comet passed closest to the sun and Earth during the first days of June. Holetschek simply noted the comet's magnitude was between 7.3 and 8.6 on the 9th and between 7.8 and 8.7 on the 10th. On the 10th and 12th, J. Bauschinger (Munich, Germany) estimated the nuclear magnitude as 9. On the 12th, Holetschek indicated the comet was probably brighter than

8.0–8.2. The comet attained its most northerly declination of +66° on the 17th, at which time Holetschek gave the magnitude as 8.0. J. Kortazzi (Nicolajew, now Mykolayiv, Ukraine) said the tail extended toward PA 230°. Holetschek indicated the comet was about 8.5 on the 25th and around 8.3–8.4 on the 27th. Also on June 27, Millosevich estimated the nuclear magnitude as 8 and noted the tail extended 5–6' toward PA 98.7°. Interestingly, this position angle is about 90° short of the predicted value.

On July 4, Holetschek estimated the magnitude as 8.6. Holetschek gave the magnitude as 8 on the 7th and about 8.9 on the 9th. On the 13th, Engelhardt estimated the nuclear magnitude as 9 and noted a fan-shaped tail about 5' long. That evening, Kortazzi said the tail extended toward PA 270°. On the 14th, Engelhardt said the diffuse, disk-shaped nucleus was about magnitude 9.5, while the tail seemed fainter, but longer than on the 13th. That same evening, Holetschek said the magnitude was between 8.5 and 8.8. On the 15th, Engelhardt estimated the nuclear magnitude as 10 and noted the tail was 15' long, 12' wide, and exhibited bright strips. Holetschek said the comet was only slightly brighter than 8.9 on the 17th and about 8.6 on the 19th. On the 23rd, Holetschek said the magnitude was between 7.7 and 9.0.

On August 2, Krosnow (Kazan, Russia) said the comet appeared very faint in the 24-cm refractor. Trotzki found the comet faint on the 5th. On August 9, Holetschek said the magnitude was between 8.5 and 9.0. On the 10th, Holetschek said it was magnitude 8.3–8.4. Holetschek gave the total magnitude as 8.5 on August 13. On the 15th, Engelhardt gave the nuclear magnitude as 10.2 and noted a short, faint tail, while Holetschek indicated a total magnitude of 9.2. Pechüle said the 27-cm refractor revealed a tail extending 4' toward PA 80° on the 16th. Holetschek estimated the magnitude as 8.4 on the 16th and 8.9 on the 17th. Holetschek indicated a magnitude of 9 on the 19th.

On September 5, Engelhardt said the comet was near the horizon, but appeared quite bright and very small. Holetschek said the comet was not brighter than the 9.9-magnitude comparison star on the 9th. On the 14th, A. M. W. Downing (Royal Observatory, Greenwich, England) observed with the 17-cm refractor and described the comet as "a faint patch of light with distinct stellar nucleus, but difficult to observe." Holetschek estimated the magnitude as 9.7 on the 17th. Holetschek said the comet was fainter than the 10.3-magnitude comparison star on the 19th. Bauschinger said the comet was faint in the 27-cm refractor on the 27th and 29th because of moonlight. The comet passed 33° from the sun on September 29.

Few physical descriptions were published during the remainder of the year. Bauschinger said the comet was 2' across on October 4 and 6, and he saw traces of a faint tail on the 14th. Holetschek observed the comet in the morning sky with the 15-cm refractor and estimated the total magnitude as 10.5 on November 14 and 10–10.5 on the 16th. Holetschek's observations during December revealed a magnitude of 10 on the 10th and "not brighter than 10.5" on the 15th.

A few physical descriptions were also provided as the comet continued to be observed during 1891. Luther observed with the 25-cm refractor on January 16 and noted a faint, round nebulosity, with a nucleus of magnitude 12. Luther gave the nuclear magnitude as 12 on February 4, 11.6 on the 7th, and 12.5 on the 26th. Holetschek said the coma was 2′ across on February 9 and the total magnitude was 10.5 on the 14th. Engelhardt observed with the 30-cm refractor on February 28 and said the comet appeared quite bright and granular, with a nuclear magnitude of 12 and a fan-shaped tail about 1′ long. On March 5 and 7, Pechüle described the comet as 10″ in diameter, with a nucleus of magnitude 12. On March 6, Holetschek said the magnitude was 10.5. Luther gave the nuclear magnitude as 12 on March 12th. On March 14, O. Stone and N. M. Parrish (Leander McCormick Observatory, Virginia, USA) estimated the magnitude of the nucleus as 12 and noted a tail extending 1.5′ toward PA 230°. Luther gave the nuclear magnitude as 12.5 on April 3 and 26. Rayet and Picart also observed the comet on several occasions during the period of February 3 to April 29 and obtained precise positions. They spotted the comet one final time on May 30, but said it was too faint to measure in weak evening twilight. The comet passed about 4° from the sun on August 17.

S. Javelle (Nice, France) obtained several observations of this comet with the 76-cm refractor during the period of 1892 January 7 to 1892 February 5. He wrote in the May 1892 issue of the *Bulletin Astronomique*, "The comet was extremely weak, very badly defined, and one minute wide at most." His last observation on February 5 was also the final time the comet was seen anywhere. He gave the position as $\alpha = 9^{\mathrm{h}}\,21.0^{\mathrm{m}}$, $\delta = +11°\,37′$.

The first orbit was calculated by G. M. Searle using positions from the period spanning March 22–25. The result was a perihelion date of 1890 June 3.93. Additional orbits were calculated by W. W. Campbell, F. Bidschof, A. O. Leuschner, E. Viennet, and G. A. Hill, which eventually established the perihelion date as about June 3.

The first definitive orbit was calculated by Bidschof using positions encompassing the 1890 and 1891 positions. The result was a hyperbolic orbit with a perihelion date of June 2.04 and an eccentricity of 1.00037259. E. Strömgren (1896) took positions spanning the entire period of visibility and applied perturbations by Earth, Mars, Jupiter, and Saturn. The result was a perihelion date of June 2.04 and an eccentricity of 1.000266.

T	ω	Ω (2000.0)	i	q	e
1890 Jun. 2.0375 (UT)	68.9271	321.8779	120.5690	1.907582	1.000266

ABSOLUTE MAGNITUDE: $H_{10} = 3.3$ (V1964)

FULL MOON: 1890 Mar. 6, Apr. 5, May 4, Jun. 3, Jul. 2, Jul. 31, Aug. 30, Sep. 28, Oct. 28, Nov. 26, Dec. 26, 1891 Jan. 24, Feb. 23, Mar. 25, Apr. 24, May 23, Jun. 22, Jul. 21, Aug. 19, Sep. 18, Oct. 17, Nov. 16, Dec. 15, 1892 Jan. 14, Feb. 12

SOURCES: G. Bigourdan, G. A. P. Rayet, and L. Picart, *CR*, **110** (1890 Mar. 31), pp. 694–5; W. R. Brooks, *MNRAS*, **50** (1890 Apr.), pp. 375–6; O. C. Wendell, *The Observatory*, **13** (1890 Apr.), p. 160; E. A. Lamp and F. Bidschof, *AN*, **124** (1890 Apr. 5), pp. 171–5; W. R. Brooks, O. C. Wendell, E. E. Barnard, G. M. Searle, and W. W. Campbell, *AJ*, **9** (1890 Apr. 8), pp. 183–4; E. Viennet and D. Klumpke, *CR*, **110** (1890 Apr. 8), pp. 746–7; W. R. Brooks, *AN*, **124** (1890 Apr. 18), p. 221; E. E. Barnard and W. W. Campbell, *AJ*, **9** (1890 Apr. 21), pp. 189–90; G. Le Cadet, *CR*, **110** (1890 Apr. 21), p. 839; A. M. W. Downing, T. Lewis, and H. P. Hollis, *MNRAS*, **50** (1890 May), pp. 408–9; W. R. Brooks, *The Observatory*, **13** (1890 May), p. 188; E. Millosevich, A. O. Leuschner, and G. A. Hill, *AN*, **124** (1890 May 7), pp. 247–51; G. A. P. Rayet, *CR*, **110** (1890 May 19), pp. 1025–6; B. von Engelhardt, *AN*, **124** (1890 May 28), p. 285; T. Lewis, A. M. W. Downing, W. G. Thackeray, and H. P. Hollis, *MNRAS*, **50** (1890 Jun.), pp. 523–6; C. Trépied, *CR*, **110** (1890 Jun. 9), pp. 1182–4; B. von Engelhardt, *AN*, **125** (1890 Aug. 12), p. 121; E. Frisby, J. R. Eastman, and A. N. Skinner, *AJ*, **10** (1890 Aug. 30), pp. 62–3; A. D. Dubiago, Trotzki, and B. von Engelhardt, *AN*, **125** (1890 Sep. 27), pp. 297–300; Krosnow and Trotzki, *AN*, **125** (1890 Oct. 21), p. 429; W. G. Thackeray and H. P. Hollis, *MNRAS*, **50** (1890, Supplementary Notice), pp. 528–9; A. M. W. Downing, T. Lewis, and J. I. Plummer, *MNRAS*, **51** (1890 Nov.), pp. 49–50, 56; J. Kortazzi, *AN*, **126** (1890 Nov. 7), p. 61; W. Luther, *AN*, **127** (1891 Apr. 17), pp. 53–6, 73; J. Bauschinger, *AN*, **127** (1891 Apr. 28), pp. 107–9; O. C. Wendell, *AJ*, **11** (1891 May 12), p. 4; B. von Engelhardt, *AN*, **127** (1891 May 19), p. 197; O. Stone and N. M. Parrish, *AJ*, **11** (1891 Jun. 5), p. 11; G. A. P. Rayet and L. Picart, *CR*, **112** (1891 Jun. 8), pp. 1301–2; J. Palisa, J. Holetschek, and R. Spitaler, *AN*, **127** (1891 Jul. 2), pp. 329–32, 337–9; F. Bidschof, *AN*, **128** (1891 Sep. 16), pp. 201–4; E. E. Barnard, *AJ*, **11** (1891 Dec. 9), p. 79; S. Javelle, *AN*, **129** (1892 Jan. 15), p. 31; S. Javelle, *The Observatory*, **15** (1892 Feb.), p. 123; S. Javelle, *BA*, **9** (1892 May), pp. 222–3; J. Holetschek, *AN*, **130** (1892 Jul. 5), pp. 69–71; W. Luther, *AN*, **131** (1892 Nov. 7), pp. 81, 107; C. F. Pechüle, *AN*, **136** (1894 Oct. 20), pp. 305, 315; E. Strömgren, *Berechnung der Bahn des Komenten 1890 II*. Lund: E. Mälmström (1896), 106 pp.; V1964, p. 63.

C/1890 O1	*Discovered:* 1890 July 18.92 ($\Delta = 1.56$ AU, $r = 0.79$ AU, Elong. $= 26°$)
(Coggia)	*Last seen:* 1890 August 14.20 ($\Delta = 1.88$ AU, $r = 1.03$ AU, Elong. $= 23°$)
	Closest to the Earth: 1890 July 8 (1.5186 AU)
1890 III = 1890b	*Calculated path:* LYN (Disc), UMa (Jul. 21), LYN (Jul. 22), LMi (Jul. 25), LEO (Aug. 10)

J. E. Coggia (Marseille, France) discovered this comet on 1890 July 18.92 at a position of $\alpha = 8^h 48.2^m$, $\delta = +44° 43'$. An additional observation on July 19.89 established the motion as southeastward. Coggia described the comet as rather bright, with a slight central condensation. He added that the coma was about 1.5′ across.

The comet was moving away from both the sun and Earth when discovered. Although it was widely observed in the short time it was visible, moonlight and the comet's decreasing altitude affected the physical descriptions. For the period of July 20–25, A. L. N. Borrelly (Marseille, France) observed with the 26-cm refractor and said the comet was rather bright, round, and about 2' across. There was also a central condensation. He added that the brightness decreased during the same period. D. Klumpke (Paris, France) described the comet as a very brilliant, round nebulosity on the 21st. She noted a condensation of magnitude 10 or 11 and added that the comet was easily seen in the 10-cm finder. E. A. Lamp (Kiel, Germany) saw the comet in twilight on July 21 and 22 and noted a distinct nucleus and a faint coma. Klumpke also saw the comet at low altitude on the 22nd and said the nucleus was hardly detected. Also on the 22nd, A. H. P. Charlois (Nice, France) observed the comet with a 38-cm refractor and described it as a circular nebulosity about 1.5' across, with a rather brilliant central condensation. He added that the comet disappeared at the same time as stars of magnitude 10 when the field was illuminated. B. von Engelhardt (Dresden, Germany) saw the comet in bright moonlight and evening twilight on July 27. He described it as very faint, small, round, and condensed, as seen in the 30-cm refractor. R. Spitaler (Vienna, Austria) observed with the 69-cm refractor on the 27th and 28th and noted the comet appeared remarkably faint. Moonlight also caused L. Boss (Dudley Observatory, New York, USA) to describe the comet as extremely faint and difficult to see in a 33-cm refractor on the 28th. Spitaler saw the comet with the 69-cm refractor on the July 31 and said it was very faint and seen with difficulty under very clear skies and a full moon. Boss said the comet was barely visible through a light fog in the 33-cm refractor on August 5.

The last sightings of the comet were by E. E. Barnard (Lick Observatory, California, USA) on August 11.20, 13.20, and 14.20. He was using the 30-cm refractor and remarked, "At the last few observations, the comet was seen for only a few minutes. It was still bright, but set too early to permit accurate observations." On the final date, he was not able to measure α, but $\delta = +22° 45'$.

The first parabolic orbit was calculated by A. Berberich using positions from July 19, 21, and 22. The resulting perihelion date was 1890 July 8.44. The orbit was steadily refined during the next few weeks by A. O. Leuschner, Lubrano, P. Maitre, L. Fabry, W. Bellamy, L. Boss, and F. Bidschof. Interestingly, none of the August observations were applied to orbital calculations until 1893, when W. Ebert took positions measured during the period of July 19 to August 6 and calculated the perihelion date as July 9.04. J. Rheden (1904) took 66 positions from the period of July 21 to August 14 and determined six Normal positions. He then calculated a parabolic and a hyperbolic orbit. The parabolic one had a perihelion date of July 9.03. The hyperbolic one had a perihelion date of July 9.07 and an eccentricity of 1.005827. Although the hyperbolic orbit fitted the observations best, the parabolic one was preferred since the comet was observed for less than one month.

T	ω	Ω (2000.0)	i	q	e
1890 Jul. 9.0373 (UT)	85.6608	15.8347	63.3509	0.764087	1.0

ABSOLUTE MAGNITUDE: $H_{10} = 8.6$ (V1964)

FULL MOON: Jul. 2, Jul. 31, Aug. 30

SOURCES: A. L. N. Borrelly, J. E. Coggia, and A. H. P. Charlois, *BA*, **7** (1890), pp. 356–8, 379–83; J. E. Coggia, *CR*, **111** (1890 Jul. 21), pp. 152–3; J. E. Coggia and E. A. Lamp, *AN*, **125** (1890 Jul. 22), p. 63; J. E. Coggia, D. Klumpke, A. Riccò, E. A. Lamp, and A. Berberich, *AN*, **125** (1890 Jul. 26), pp. 77–80; J. E. Coggia, A. L. N. Borrelly, E. Stéphan, L. Fabry, L. Picart, F. Courty, and D. Klumpke, *CR*, **111** (1890 July 28), pp. 216–18, 223–4; D. Klumpke, J. E. Coggia, A. L. N. Borrelly, E. Stéphen, Lubrano, P. Maitre, L. Fabry, A. Abetti, E. Millosevich, R. Spitaler, A. Riccò, E. A. Lamp, B. von Engelhardt, and F. Bidschof, *AN*, **125** (1890 Aug. 4), pp. 90–6; L. Boss, W. W. Campbell, J. E. Coggia, and W. Bellamy, *AJ*, **10** (1890 Aug. 6), pp. 53–4, 56; L. Boss, *AJ*, **10** (1890 Aug. 30), pp. 61–2; A. O. Leuschner, *PASP*, **2** (1890 Sep.), p. 237; O. C. Wendell, *AJ*, **10** (1890 Sep. 19), p. 70; G. A. P. Rayet, L. Picart, and F. Courty, *CR*, **111** (1890 Sep. 29), p. 476; E. E. Barnard, *AJ*, **10** (1890 Dec. 1), p. 100; F. Schwab, *AN*, **127** (1891 May 5), p. 149; O. C. Wendell, *AJ*, **11** (1891 May 12), p. 4; R. Spitaler, *AN*, **127** (1891 Jul. 2), pp. 331, 339; W. Ebert, *AN*, **132** (1893 Feb. 9), pp. 97–106; W. Ebert, *The Observatory*, **16** (1893 Jun.), p. 241; V. Knorre, *AN*, **133** (1893 Jul. 8), p. 57; J. Rheden, *SAWW*, **113** Abt. IIa (1904), pp. 3–51; V1964, p. 63.

C/1890 O2 *Discovered:* 1890 July 23.99 ($\Delta = 1.53$ AU, $r = 1.58$ AU, Elong. = 74°)
(Denning) *Last seen:* 1890 November 8.01 ($\Delta = 2.16$ AU, $r = 1.43$ AU, Elong. = 32°)
  Closest to the Earth: 1890 August 26 (1.1925 AU)
1890 VI = 1890c *Calculated path:* UMi (Disc), DRA (Aug. 4), BOO (Aug. 14), CrB (Aug. 23), SER (Aug. 31), OPH (Sep. 22), SCO (Sep. 27), OPH (Oct. 13), SCO (Oct. 17)

W. F. Denning (Bristol, England) discovered this comet on 1890 July 23.99, while searching for comets. He estimated its position as $\alpha = 15^{\text{h}} 12^{\text{m}}$, $\delta = +78°$ and noted it was moving southwards at 55′ per day. Denning was using a 25.4-cm reflector and described the comet as a small, faint nebulosity, with a diameter of about 1′. The comet had attained its most northerly declination of +89° on July 9.

During the period of July 26–28, E. E. Barnard (Lick Observatory, California, USA) observed with the 30-cm refractor and noted a 13th-magnitude stellar nucleus and a faint diffuse tail. On August 7 and 10, R. Schorr (Hamburg, Germany) said the comet appeared very faint in the 25-cm refractor. L. Boss (Dudley Observatory, New York, USA) found the comet in the 33-cm refractor on the 8th and said it was extremely faint because of thin clouds. On the 9th, G. Agnello (Palermo, Italy) said the comet was at the limit of visibility in the 25-cm refractor, while it could not be found on the 10th. W. R. Brooks (Phelps, New York, USA) saw the comet on August 13 and 17 and described it as "faintish but not difficult." On the 14th and 16th, Schorr noted a faint condensation. On the 15th and 16th, Denning noted it was "decidedly brighter" than when discovered on July 23. A magnification of 97× revealed a nucleus of magnitude 12, "surrounded in a circular nebulosity" about 1′ across. He added, "The outlying parts of this were only visible by glimpses, and apparently flashed out now and then similar to the effect produced by faint nebulae which cannot be steadily held." B. von Engelhardt (Dresden, Germany) saw the comet with the 30-cm refractor on August 19 and described it as very faint, about 1' across, with an irregular border. He added that it appeared granular. On September 5, Engelhardt described the comet as rather faint, small, condensed, and noticeably granular. W. Luther (Hamburg, Germany) noted the comet was quite faint on the 8th and exhibited no marked condensation on the 13th. On the 17th, Engelhardt said the comet was very faint, rather small, and condensed, while Luther found it extraordinarily faint in the 23-cm refractor. Engelhardt said the comet was faint on the 18th, with a nucleus of magnitude 14. On September 19, Engelhardt said the comet was just as faint as on the 18th. On that same evening, C. F. Pechüle (Copenhagen, Denmark) reported the comet was faint and small in the 27-cm refractor, because of the low altitude. Barnard obtained the final Northern Hemisphere observation on October 7, when the comet's altitude was about 8°.

The first Southern Hemisphere observation was obtained on October 2 by J. M. Thome (Cordoba, Argentina), using a 28-cm refractor. He said the comet was barely visible in a dark field. On the 4th, glimpses of the comet were obtained by J. Tebbutt (Windsor, New South Wales, Australia) with a 20-cm refractor and it was described as excessively faint. Thome continued to observe the comet until it was last seen on November 8.01. He then gave the position as $\alpha = 16^h 45.8^m$, $\delta = -37° 11'$. He wrote, "Had it been situated nearer the zenith, I think it might have been followed a month longer."

The first orbits were independently calculated by A. Berberich and A. O. Leuschner. Berberich used positions from July 24 to July 28 and determined the perihelion date as 1890 September 25.22, while Leuschner took positions spanning the period of July 26–28 and determined the perihelion date as September 22.51. Further orbits were calculated during the next few weeks by Berberich, Leuschner, Boss, and C. N. A. Krueger, which revealed a perihelion date of September 25.0.

N. Bobrinskoy (1893) took 123 positions spanning the period of July 24 to November 8, determined six Normal positions, and calculated an elliptical orbit. The resulting perihelion date was September 25.01, while the period was over 57 000 years. This orbit is given below.

T	ω	Ω (2000.0)	i	q	e
1890 Sep. 25.0080 (UT)	163.0523	101.6583	98.9376	1.260225	0.999154

ABSOLUTE MAGNITUDE: $H_{10} = 8.1$ (V1964)

FULL MOON: Jul. 2, Jul. 31, Aug. 30, Sep. 28, Oct. 28, Nov. 26

SOURCES: W. F. Denning, *AN*, **125** (1890 Jul. 26), p. 79; W. F. Denning and A. H. P. Charlois, *The Observatory*, **13** (1890 Aug.), p. 287; A. Berberich, *AN*, **125** (1890 Aug. 4), p. 95; W. F. Denning, *AJ*, **10** (1890 Aug. 6), p. 56; A. Berberich, *AN*, **125** (1890 Aug. 12), pp. 123–6; R. Schorr, *AN*, **125** (1890 Aug. 16), p. 157; G. Agnello and R. Schorr,

AN, **125** (1890 Aug. 21), p. 175; A. O. Leuschner and L. Boss, *AJ*, **10** (1890 Aug. 30), pp. 61–2; W. F. Denning, *The Observatory*, **13** (1890 Sep.), p. 295; A. O. Leuschner, *PASP*, **2** (1890 Sep.), p. 237; B. von Engelhardt and C. N. A. Krueger, *AN*, **125** (1890 Sep. 2), p. 219; L. Boss, *AJ*, **10** (1890 Sep. 19), p. 68; B. von Engelhardt and W. Luther, *AN*, **125** (1890 Sep. 27), pp. 299–302; G. A. P. Rayet and L. Picart, *CR*, **111** (1890 Sep. 29), p. 476; C. N. A. Krueger, *AN*, **125** (1890 Oct. 7), p. 317; L. Boss, *AJ*, **10** (1890 Oct. 18), p. 86; E. E. Barnard, *AJ*, **10** (1890 Nov. 11), pp. 95–6; J. M. Thome, *AN*, **126** (1891 Jan. 19), pp. 229–32; W. F. Denning and W. R. Brooks, *The Observatory*, **14** (1891 Feb.), p. 89; J. Tebbutt, *AN*, **126** (1891 Feb. 28), p. 357; W. Luther, *AN*, **127** (1891 Apr. 17), pp. 55, 73; N. Bobrinskoy and W. F. Denning, *BASP* (Series 4), **36** (1893), pp. 227–46; N. Bobrinskoy, *AN*, **134** (1894 Jan. 17), pp. 243–6; V1964, p. 63.

6P/d'Arrest **1890 V = 1890d**	*Recovered:* 1890 October 7.23 ($\Delta = 0.85$ AU, $r = 1.34$ AU, Elong. $= 92°$) *Last seen:* 1890 December 14.05 ($\Delta = 1.49$ AU, $r = 1.65$ AU, Elong. $= 81°$) *Closest to the Earth:* 1890 August 6 (0.7193 AU) *Calculated path:* SGR (Rec), CAP (Oct. 18), SGR (Oct. 21), MIC (Oct. 23), CAP (Oct. 31), PsA (Nov. 7), AQR (Nov. 20)

G. Leveau (1883) look positions obtained during the apparitions of 1870 and 1877, applied perturbations by Mars, Jupiter, and Saturn, and predicted the comet would arrive at perihelion on 1884 January 14.08. Although the comet was not favorably placed for observation, some searches were conducted during 1883, when the comet's elongation from the sun could have allowed observations by larger telescopes. E. W. L. Tempel (Arcetri, Italy) unsuccessfully searched for the comet during May, while W. E. Plummer searched a little later on.

Leveau (1890) was still confident of the 1883 orbit, so he applied perturbations from Jupiter up to the comet's next apparition and determined a perihelion date of 1890 September 17.99. Searches were begun by E. E. Barnard (Lick Observatory, California, USA) on 1890 April 14, by R. Spitaler (Vienna, Austria) on April 16, and by F. Renz (Pulkovo Observatory, Russia) on April 17. These astronomers, as well as L. Swift (Warner Observatory, Rochester, New York, USA) continued to search through the summer months. Barnard appears to have made the final search on September 4. Since the comet should then have started fading, Barnard said further searches would be hopeless and "the comet dropped entirely from my memory." On October 7, Barnard had just wrapped up an observation of comet C/1890 O2 and begun one of his routine searches for new comets with the 30-cm refractor. During a sweep near the southern horizon he spotted a very faint diffuse nebulosity with a diameter of 2–3′. He gave the position as $\alpha = 19^h\ 13.5^m$, $\delta = -26°\ 08'$ on October 7.23. Barnard telegraphed news of his discovery and astronomers immediately realized the position and motion were similar to those Leveau had predicted for d'Arrest's periodic comet.

Although the comet was already past perihelion and its closest approach to Earth, numerous observations were made during the remainder of its apparition. On October 9, H. A. Kobold (Strasbourg, France) described the comet as a pale, oblong nebulosity, with a weak condensation. On the 10th, E. Millosevich (Rome, Italy) said it was faint, poorly defined, and without a nucleus. J. Bauschinger (Munich, Germany) saw the comet with a 27-cm refractor on the 11th and noted it was extraordinarily faint. That same evening, J. Palisa (Vienna, Austria) saw it with the 69-cm refractor and said it was very faint, with a central condensation. Spitaler said the 69-cm refractor revealed a diffuse nebulosity, about 2' across, with a bright center. On the 17th, Spitaler said the coma was 2' across and exhibited a nucleus of magnitude 13. The comet attained its most southerly declination of $-28°$ on the 26th. On the 31st, Spitaler described the comet as a diffuse nebulosity, with a granular center. He said the coma was 2' across and was oblong in the direction of the comet's movement. During the period of October 12 to November 6, J. Tebbutt (Windor, New South Wales, Australia) continuously noted the comet was excessively faint and diffuse in his 20-cm refractor. During early November, Barnard said the comet was occasionally seen in the 8.3-cm finder of the 30-cm refractor. He added that on several occasions he "swept it up, without thinking of its presence, with the 10-cm broken tube cometseeker." Spitaler saw the comet in a 69-cm refractor on the 9th and gave the total magnitude as 12. He said the coma was 2' across, although he occasionally suspected it was 4' across. The nucleus was about 0.5' across. Spitaler said the comet was 1–1.5' across on the 13th, while it was 1' across on the 14th and about magnitude 13. On December 10, O. Stone (Leander McCormick Observatory, Virginia, USA) noted a slight nucleus was seen with the 66-cm refractor.

The comet was last detected on December 14.05, when N. M. Parrish (Leander McCormick Observatory, Virginia, USA) observed it with the 66-cm refractor. He described it as faint and measured the position as $\alpha = 23^h\ 19.0^m$, $\delta = -19°\ 49'$.

Minor refinements of this comet's orbit using positions from multiple apparitions, as well as planetary perturbations, have been published by several astronomers. Of most significance were the studies of A. W. Recht (1939) and B. G. Marsden and Z. Sekanina (1972). These have typically revealed a perihelion date of September 18.10 and a period of 6.68 years. Marsden and Sekanina determined the nongravitational terms as $A_1 = +0.79$ and $A_2 = +0.0937$. Their orbit is given below.

T	ω	Ω (2000.0)	i	q	e
1890 Sep. 18.1035 (TT)	173.0497	147.7722	15.7174	1.321158	0.627486

ABSOLUTE MAGNITUDE: $H_{10} = 9.7$ (V1964)

FULL MOON: Sep. 28, Oct. 28, Nov. 26, Dec. 26

SOURCES: G. Leveau, *CR*, **96** (1883 Jan. 22), p. 229; G. Leveau, *AN*, **105** (1883 Mar. 14), pp. 19–22; E. W. L. Tempel, *AN*, **106** (1883 Jun. 23), p. 29; W. E. Plummer, *MNRAS*, **44** (1884 Feb.), p. 180; ARSI, part 1 (1886), p. 383; G. Leveau, *AN*, **124** (1890 Mar. 21), pp. 113–16; R. Spitaler and L. Swift, *AN*, **125** (1890 Sep. 18), p. 286; E. E. Barnard, *AN*, **125** (1890 Oct. 13), p. 367; H. A. Kobold, E. Millosevich, and J. Bauschinger, *AN*, **125** (1890 Oct. 17), pp. 381–4; E. E. Barnard, G. Leveau, and G. M. Searle, *AJ*, **10** (1890 Oct. 18), pp. 87–8; E. E. Barnard, *SM*, **9** (1890 Nov.), pp. 419–20; E. E. Barnard and G. Bigourdan, *The Observatory*, **13** (1890 Nov.), p. 368; F. Renz, *AN*, **126** (1890 Nov. 7), p. 59; E. E. Barnard, *AJ*, **10** (1890 Nov. 11), pp. 92–3; E. Frisby, *AJ*, **10** (1890 Dec. 1), p. 99; O. Stone and N. M. Parrish, *AJ*, **10** (1891 Jan. 9), pp. 135–6; J. Tebbutt, *AN*, **126** (1891 Feb. 28), p. 357; E. E. Barnard, *AJ*, **11** (1891 May 12), pp. 6–7; J. Palisa and R. Spitaler, *AN*, **127** (1891 Jul. 2), pp. 331, 339; A. W. Recht, *AJ*, **48** (1939 Jul. 17), pp. 65–78; V1964, p. 65; B. G. Marsden and Z. Sekanina, *QJRAS*, **13** (1972), pp. 428–9; B. G. Marsden, Z, Sekanina, and D. K. Yeomans, *AJ*, **78** (1973 Mar.), p. 213.

C/1890 V1 (Zona)   *Discovered:* 1890 November 15.90 ($\Delta = 1.48$ AU, $r = 2.37$ AU, Elong. $= 147°$)
Last seen: 1891 January 13.89 ($\Delta = 2.41$ AU, $r = 2.76$ AU, Elong. $= 100°$)
1890 IV = 1890e   *Closest to the Earth:* 1890 November 22 (1.4573 AU)
Calculated path: AUR (Disc), PER (Nov. 27), TRI (Dec. 18)

T. Zona (Palermo, Italy) discovered this comet on 1890 November 15.90 at a position of $\alpha = 5^h 35.9^m$, $\delta = +33° 23'$. He described it as "rather bright." Zona confirmed a motion toward the northwest on November 15.99.

The comet had passed perihelion over three months earlier, but was discovered a few days before it passed closest to Earth. On November 16, E. Millosevich (Rome, Italy) said the comet was round, about 1' across, and exhibited a nucleus of magnitude 11.5. On the 17th, R. Spitaler (Vienna, Austria) gave the total magnitude as 8, the nuclear magnitude as 9, and the coma diameter as 3'. B. von Engelhardt (Dresden, Germany) observed the comet with a 30-cm refractor on the 18th and said it appeared quite faint, 1' in diameter, and exhibited a nucleus. During the period of November 18–20, E. E. Barnard (Lick Observatory, California, USA) described the comet as round, with a magnitude of about 12 in the 30-cm refractor. It was slightly brighter toward the middle, but there was no definite nucleus. On the 19th, O. C. Wendell (Harvard College Observatory, Massachusetts, USA) said the comet was 1' across in the 38-cm refractor, with some central condensation and a "not quite stellar" nucleus. Wendell said the comet was well seen on the 20th and exhibited a rather sharply defined nucleus. On the 21st, Wendell said the nucleus was fairly well defined, while G. Bigourdan (Paris, France) said it was faint, with a nucleus of magnitude 12.5. Wendell said the comet was rather faint on the 22nd, but "pretty well defined." He reported the comet was faint in moonlight on the 23rd. The comet attained its most northerly declination of $+35°$ on the 29th. On November 30, J. Holetschek (Vienna, Austria) described the comet as a small faint nebulosity in the 15-cm refractor, which was about as easy to see as a star of magnitude 11. He added that at times it was almost unrecognizable.

On December 3 and 5, A. Abetti (Padova, Italy) indicated the condensation was about magnitude 12, while the diameter was 0.33' on the last date.

On the 4th, Holetschek said the comet was 1′ across and about magnitude 10.5. During the period of December 7–14, Millosevich said the comet was difficult to see in the 25-cm refractor and the nucleus was uncertain. On the 8th and 28th, G. A. P. Rayet (Bordeaux, France) noted the comet was very faint and disappeared when the moon rose on the last date. During the period of December 9–15, Abetti said the comet was small and faint. On the 10th, Barnard said the 30-cm refractor revealed the comet was about 0.5′ across, with an ill-defined nucleus. J. Bauschinger (Munich, Germany) observed the comet with a 27-cm refractor on the 11th and described it as 1′ in diameter with a faint central condensation. H. A. Kobold (Strasbourg, France) observed with a 46-cm refractor and said the comet was somewhat brighter in the middle on the 12th and 0.5′ across with a faint condensation on the 15th. Engelhardt observed with a 30-cm refractor and described the comet as very faint on the 15th and 29th, and extremely faint in haze and moonlight on the 17th. On December 31, Bauschinger said the comet was an extremely faint spot that was only occasionally visible in the refractor.

On 1891 January 2, Kobold said the comet was 0.5′ across, with only a slight central condensation. On the 10th, Barnard observed with a 30-cm refractor and described the comet as about magnitude 13.5, "with an indefinite brightening in the middle." The comet was last detected on January 13.89, when Kobold measured the position as $\alpha = 1^{\mathrm{h}}\,45.2^{\mathrm{m}}$, $\delta = +27°\,32'$. He said it appeared very faint in the 46-cm refractor.

The first parabolic orbits were independently calculated by G. M. Searle, F. Bidschof, and G. Agnello using positions spanning the period of November 15–19. Searle determined the perihelion date as 1890 August 9.73, Bidschof gave it as July 25.51, and Agnello determined it as July 28.18. All three orbits revealed a perihelion distance of about 2 AU, which accounted for the discrepancy in the perihelion dates since only a few observations were used in the calculation. J. R. Hind calculated an orbit using positions obtained up to November 21 and determined the perihelion date as August 8.94. Additional orbits by A. Berberich, Agnello, W. W. Campbell, and F. W. Ristenpart narrowed down the perihelion date to about August 7.6.

A definitive orbit was calculated by A. Venturi (1896). He began with about 120 positions and determined the elliptical orbit below, which has a period of about 11 040 years. J. Holetschek examined the motion and brightness of this comet and noted that it would have been as bright as it was at discovery back in March of 1890. Although it would also have been more favorably situated with respect to the sun, it was then at a declination that was favorable only to observers in the Southern Hemisphere.

T	ω	Ω (2000.0)	i	q	e
1890 Aug. 7.3827 (UT)	331.2842	86.9342	154.3070	2.046694	0.995872

ABSOLUTE MAGNITUDE: $H_{10} = 5.2$ (V1964)

FULL MOON: Oct. 28, Nov. 26, Dec. 26, Jan. 24

SOURCES: T. Zona and F. Bidschof, *AN*, **126** (1890 Nov. 24), p. 95; J. R. Hind, *MNRAS*, **51** (1890 Dec.), p. 98; T. Zona, E. E. Barnard, O. C. Wendell, G. M. Searle, and E. Frisby, *AJ*, **10** (1890 Dec. 1), pp. 101–4; E. Millosevich, B. von Engelhardt, G. Bigourdan, and F. Bidschof, *AN*, **126** (1890 Dec. 1), pp. 107, 111; E. E. Barnard, J. K. Rees, and H. Jacoby, *AJ*, **10** (1890 Dec. 8), p. 110; O. C. Wendell and A. Berberich, *AN*, **126** (1890 Dec. 12), pp. 121, 125; G. Agnello, *AN*, **126** (1890 Dec. 20), p. 143; W. W. Campbell, *AJ*, **10** (1891 Jan. 9), p. 134; F. W. Ristenpart, B. von Engelhardt, and G. A. P. Rayet, *AN*, **126** (1891 Jan. 12), pp. 167–70; J. Holetschek, *MNRAS*, **51** (1891 Feb.), p. 239; E. E. Barnard and M. Updegraff, *AJ*, **10** (1891 Feb. 4), p. 151; E. B. Frost, *AJ*, **10** (1891 Mar. 14), p. 175; E. Millosevich, A. Abetti, and J. Bauschinger, *AN*, **127** (1891 Apr. 28), pp. 103–10; F. L. Chase, *AJ*, **11** (1891 May 12), p. 3; H. A. Kobold, *AN*, **127** (1891 May 27), pp. 227–30; R. Spitaler and J. Holetschek, *AN*, **127** (1891 Jul. 2), pp. 331, 339; A. Venturi, *AN*, **140** (1896 May 6), p. 239; V1964, p. 63.

113P/1890 W1
(Spitaler)

1890 VII = 1890f

Discovered: 1890 November 17.14 ($\Delta = 0.90$ AU, $r = 1.82$ AU, Elong. $= 150°$)
Last seen: 1891 February 4.77 ($\Delta = 1.31$ AU, $r = 2.01$ AU, Elong. $= 121°$)
Closest to the Earth: 1890 December 1 (0.8791 AU)
Calculated path: AUR (Disc)

Having received a report that a new comet had been discovered by T. Zona (see C/1890 V1), R. Spitaler (Vienna, Austria) set out to make his first observations. He knew the discovery position as well as the direction and rate of motion, and, as the morning of the November 17 was clear, he began searching with a 69-cm refractor. It was not long before he found a faint nebulous object near the likely extrapolated position of the comet; however, he noted it seemed too faint to be Zona's comet. He began moving the telescope back and forth, and soon came upon the comet discovered by Zona. The position of Spitaler's faint comet was given as $\alpha = 5^h 27.3^m$, $\delta = +33° 37'$ on 1890 November 17.14.

Spitaler's comet was not immediately confirmed. Only the larger refractors could follow it, and where key European observatories were hampered by bad weather, observatories in the USA were hampered by a telegram that gave an erroneous discovery position. Conditions finally improved at Vienna on the morning of November 25. Spitaler immediately found the November 17 position was empty and began sweeping for the comet. He found a cometary object a few degrees northwest of the discovery position and estimated its position. Unfortunately, morning twilight quickly washed out the object. Moonlight and bad weather followed before Spitaler could again search for the comet. On December 4 he checked the position of the November 25 object and found it was still there. This object was not his November 17 comet, but a nebula. Spitaler began sweeping with the telescope and soon found an object that matched the appearance of his November 17 comet. He described it as about magnitude 13, with a round coma 0.5' across and a distinct nucleus. Observations during the next four hours confirmed the identity.

During the remainder of the comet's apparition it was moving away from both the sun and Earth. C. F. Pechüle (Copenhagen, Denmark) described the comet as very faint and small on December 6. B. von Engelhardt (Dresden, Germany) saw the comet with a 30-cm refractor on the 7th and wrote, "The comet is between small stars, is hardly visible and often completely disappears." E. E. Barnard (Lick Observatory, California, USA) observed the comet with the 30-cm refractor on the 7th and 8th. He described it as magnitude 13, with a coma "only a few seconds in diameter" and with "an indefinite brightening in the middle." On the 13th, H. A. Kobold (Strasbourg, France) observed with a 46-cm refractor and said the comet's central condensation was slightly fainter than a nearby star of magnitude 13. On the 29th, Spitaler noted the comet was round, about 0.5′ across, and exhibited a small bright nucleus. Kobold again saw the comet with the refractor on December 30 and described it as a hardly condensed, round, and faint nebula. He added that it was difficult to observe. On January 3, Kobold said the comet was distinctly fainter than when previously seen. Spitaler noted the comet was round, about 0.5′ across, with a central condensation on January 6. On the 27th, Spitaler said the comet was faint.

The comet was last detected on 1891 February 4.77, when Spitaler measured the position as $\alpha = 5^h 04.6^m$, $\delta = +39° 56′$. He found it with the 69-cm refractor and noted it was quite faint and diffuse. Although Spitaler also looked for the comet on February 7 and 9, and on March 6 and 7, when the sky conditions were very good, no trace was found.

By the time the comet had been recovered and enough positions had been acquired to determine an orbit, it immediately became obvious that it was moving in a short-period orbit. The first elliptical orbit was calculated by G. M. Searle using positions spanning the period of November 17 to December 8. He determined the perihelion date as 1890 September 30.89 and the period as 12.71 years. A few days later, G. Rosmanith used positions from November 17, December 4, and December 13 to calculate the orbit. The result was a perihelion date of 1890 October 26.97 and a period of 6.4 years. Spitaler noticed from this orbit that the comet would have passed very close to Jupiter in 1887. Additional orbits were calculated by Spitaler, Searle, J. F. Tennant, and J. R. Hind (1891). Tennant ruled out Spitaler's suggested close approach to Jupiter.

The comet's next expected apparition was in 1897 and Spitaler (1896) predicted it would pass perihelion on March 12.74. The comet was not recovered and, in fact, remained lost for the next century. The comet was accidentally rediscovered by J. V. Scotti on 1993 October 24. Scotti suggested this was Spitaler's comet and B. G. Marsden (1995) successfully linked the 1890 and 1994 apparitions. His orbit is given below.

T	ω	Ω (2000.0)	i	q	e
1890 Oct. 27.4872 (TT)	13.6050	46.5339	12.8150	1.815851	0.469616

ABSOLUTE MAGNITUDE: $H_{10} = 9.0$ (V1964)

FULL MOON: Oct. 27, Nov. 26, Dec. 26, Jan. 24, Feb. 23

SOURCES: R. Spitaler and E. Weiss, *AN*, **126** (1890 Nov. 24), p. 95; R. Spitaler, *AJ*, **10** (1890 Dec. 1), p. 104; R. Spitaler, *AN*, **126** (1890 Dec. 1), p. 111; R. Spitaler and C. F. Pechüle, *AN*, **126** (1890 Dec. 12), pp. 121–4; G. Rosmanith, *AN*, **126** (1890 Dec. 20), p. 143; O. C. Wendell, E. E. Barnard, and G. M. Searle, *AJ*, **10** (1890 Dec. 30), pp. 123, 125, 127; R. Spitaler, *The Observatory*, **14** (1891 Jan.), p. 71; R. Spitaler, *AN*, **126** (1891 Jan. 3), p. 157; O. C. Wendell, *AJ*, **10** (1891 Jan. 9), p. 135; B. von Engelhardt and R. Spitaler, *AN*, **126** (1891 Jan. 12), pp. 169–72; G. M. Searle, *AJ*, **10** (1891 Jan. 23), pp. 143–4; H. A. Kobold, *AN*, **127** (1891 May 27), pp. 227–30; R. Spitaler, *AN*, **127** (1891 Jul. 2), pp. 331, 339; J. R. Hind, *CR*, **113** (1891 Jul. 20), 113–14; J. F. Tennant, *MNRAS*, **52** (1891 Nov.), pp. 29–30; R. Spitaler, *AN*, **129** (1892 May 27), pp. 355–60; R. Spitaler, *AN*, **141** (1896 Aug. 11), pp. 173–6; V1964, p. 63; J. V. Scotti, *IAUC*, No. 5885 (1993 Oct. 28) ; B. G. Marsden, *CCO*, 10th ed. (1995), pp. 70–1.

C/1891 F1 (Barnard– Denning)

1891 I = 1891a

Discovered: 1891 March 30.17 ($\Delta = 1.23$ AU, $r = 0.82$ AU, Elong. $= 41°$)

Last seen: 1891 July 9.99 ($\Delta = 1.51$ AU, $r = 1.62$ AU, Elong. $= 76°$)

Closest to the Earth: 1891 March 4 (1.0781 AU), 1891 June 4 (0.9821 AU)

Calculated path: AND (Disc), TRI (Apr. 9), PSC (Apr. 14), ARI (Apr. 18), PSC (Apr. 29), CET (May 1), ERI (May 14), FOR (May 25), ERI (May 27), CAE (Jun. 3), COL (Jun. 5), PIC–PUP (Jun. 11), VEL (Jun. 24)

E. E. Barnard (Lick Observatory, California, USA) was sweeping the northwestern sky for comets with a 10-cm "broken-tube" cometseeker when he found this "small faint nebulous object" on 1891 March 30.17. Upon examination with the 30-cm refractor, he concluded it was a comet and gave the position as $\alpha = 1^h 00.1^m$, $\delta = +44° 48'$ on 1891 March 30.19. Barnard described it as about magnitude 10, with a diameter of less than 1' and a "tolerably well-defined nucleus." Barnard also noted a tail less than 30' long. An independent discovery was made by W. F. Denning (Bristol, England) on March 30.88, while sweeping for comets with his 25.4-cm reflector (40×). He described it as a bright, round, centrally condensed, nebulous object and, being familiar with the region, he concluded it was a comet. Denning next saw it on March 31.19 and said it was very obvious, in spite of the light from a gibbous moon. He then gave the daily motion as 70' toward the south-southeast.

Upon discovery, the comet was actually approaching both the sun and Earth. It was also quickly approaching evening twilight. On March 31, B. von Engelhardt (Dresden, Germany) observed the comet with a 30-cm refractor and described it as bright, round, about 1' across, with a condensation. He added that it was slightly visible in the 13-cm finder. On April 1, R. Spitaler (Vienna, Austria) saw the comet with a 69-cm refractor and described it as 4' across with a tail extending 5' toward the north-northeast. He estimated the total magnitude as 8 and the magnitude of the large diffuse nucleus as 10. That same evening, J. Holetschek (Vienna, Austria) saw the comet through the finder of a 15-cm refractor and said it had a nucleus of magnitude 8, while E. Millosevich (Rome, Italy) said the comet was 2' across with a nucleus of magnitude 10. On the 2nd, F. Renz (Pulkovo Observatory, Russia) observed with the 38-cm refractor and said the comet was very bright, with a round, granular coma exhibiting a central condensation. A faint tail several minutes long extended toward the north. W. Luther (Hamburg, Germany) saw the comet on the 3rd and noted it was bright and round, with a nucleus of magnitude 9.5. On the 4th, Denning said the comet was brighter than during previous observations and exhibited "a delicate tapering tail, so exceedingly diaphanous that it could only be caught by glimpses." He indicated the tail was at least 25' long. That same evening, Luther noted the comet was distinct, Renz said the comet appeared granular, with a faint tail 4' long, and C. F. Pechüle (Copenhagen, Denmark) said the comet was 1' in diameter, with a condensation of magnitude 9 or 10. Pechüle added that a tail extended 6–8' toward PA 12°. J. Bauschinger (Munich, Germany) observed with a 27-cm refractor on the 6th and described the comet as very bright and large, with a diffuse nucleus. H. A. Kobold (Strasbourg, France) saw the comet with a 46-cm refractor on the 9th and said it was 0.5' in diameter, with a central condensation of magnitude 10. No nucleus was seen. Luther last detected the comet in the morning sky on April 11 and said it was distinct in the bright dawn. The final Northern Hemisphere observation was made by Renz on April 13. Although he said the appearance of the comet was unchanged and, although he noted it had increased in brightness, he said it was difficult to measure because of the bright twilight.

The comet passed about 5° from the sun on April 25. J. Tebbutt (Windsor, New South Wales, Australia) first detected the comet on May 18, but was unable to secure an accurate position with his 11-cm refractor. W. H. Finlay (Royal Observatory, Cape of Good Hope, South Africa) saw the comet on June 9 and described it as a "faint diffused mass" about 1.5' across, with slight condensation. On the 14th and 15th, Finlay saw the comet in moonlight and said it was very faint and difficult. Cloudy weather prevented J. M. Thome (National Argentine Observatory, Cordoba, Argentina) from seeing the comet until June 17. On that evening he said the comet was faint in the strong moonlight. The comet attained its most southerly declination of $-48°$ on June 25. Tebbutt saw the comet with a 20-cm refractor on June 29, July 2, and 3, and described it as "a small patch of indistinct whiteness just visible by looking obliquely into the eyepiece." Tebbutt next saw the comet on July 6 with his 20-cm refractor, and he said, "it was little more than suspected, and was seen only by glimpses."

The comet was last detected on July 9.99 by Thome and his colleague R. H. Tucker. They noted, "its light had diminished so much that it was lost under the faintest illumination of the micrometer threads." Tucker measured the position as $\alpha = 9^h 52.0^m$, $\delta = -45° 34'$.

The first orbit was calculated by A. Berberich using positions spanning the period of March 31 to April 2. He determined the perihelion date as

1891 April 28.19. Similar orbits were calculated by J. M. Schaeberle, E. A. Lamp, and W. Bellamy. A definitive orbit was calculated by Lamp (1894) using positions spanning the entire period of visibility. His orbit is given below.

T	ω	Ω (2000.0)	i	q	e
1891 Apr. 28.0210 (UT)	178.7501	195.4535	120.5116	0.397896	1.0

ABSOLUTE MAGNITUDE: $H_{10} = 8.8$ (V1964)

FULL MOON: Mar. 25, Apr. 24, May 23, Jun. 22, Jul. 21

SOURCES: W. F. Denning, E. E. Barnard, and A. Berberich, *AN*, **127** (1891 Apr. 7), pp. 45–8; E. E. Barnard, W. F. Denning, and J. M. Schaeberle, *AJ*, **10** (1891 Apr. 8), pp. 183–4; B. von Engelhardt, E. Millosevich, H. A. Kobold, J. Bauschinger, and C. F. Pechüle, *AN*, **127** (1891 Apr. 17), p. 77; E. E. Barnard and W. Bellamy, *AJ*, **10** (1891 Apr. 21), pp. 190–1; E. A. Lamp, *AN*, **127** (1891 Apr. 23), pp. 93–6; W. F. Denning, *The Observatory*, **14** (1891 May), pp. 199–200; O. C. Wendell, *AJ*, **11** (1891 May 12), p. 4; W. H. Finlay, *AN*, **128** (1891 Sep. 1), p. 175; J. M. Thome and R. H. Tucker, *AN*, **128** (1891 Sep. 9), p. 189; T. Tebbutt, *AN*, **128** (1891 Sep. 16), p. 203; J. Tebbutt, *AN*, **128** (1891 Sep. 25), pp. 221–4; *The Observatory*, **14** (1891 Nov.), p. 389; R. Spitaler, *AN*, **129** (1892 May 27), pp. 357–60; J. Holetschek, *AN*, **130** (1892 Jul. 5), pp. 69–71; W. Luther, *AN*, **131** (1892 Nov. 7), pp. 81, 107; F. Renz, *AN*, **131** (1893 Dec. 16), pp. 273–7; E. A. Lamp, *Publication der Sternwarte in Kiel*, **9** no. 1 (1894), pp. 1–18; V1964, p. 63.

14P/1891 J1
(Wolf)

1891 II = 1891b

Recovered: 1891 May 2.06 ($\Delta = 2.31$ AU, $r = 2.01$ AU, Elong. $= 60°$)
Last seen: 1892 March 31.83 ($\Delta = 2.58$ AU, $r = 2.51$ AU, Elong. $= 74°$)
Closest to the Earth: 1891 October 27 (0.7981 AU)
Calculated path: PEG (Rec), AND (Jun. 13), PSC (Jun. 28), TRI (Jul. 20), ARI (Aug. 6), TAU (Aug. 28), ORI (Oct. 13), ERI (Nov. 2), ORI (Mar. 6)

With this comet still under observation during 1884, C. N. A. Krueger determined an elliptical orbit for that apparition based on an arc of only 48 days. He then determined the comet's previous perihelion dates back to 1864, as well as its next one in 1891. For the 1891 apparition he predicted a perihelion date of August 20.31. Although he realized this was not a definitive orbit, Krueger calculated it more to investigate how the comet's orbit could be changed by Jupiter. More reliable predictions were supplied by A. Berberich (1891), A. K. Thraen (1891), and G. W. L. Struve, with Berberich giving a perihelion date of September 4.33, Thraen giving it as September 3.23, and Struve giving it as September 4.37.

The comet was recovered by R. Spitaler (Vienna, Austria) on 1891 May 2.06, using a 69-cm refractor. He described it as a small, very faint nebulosity. Although the position was established as $\alpha = 22^h 28.1^m$, $\delta = +12° 38'$, no motion was detected, so the identity with that of the expected comet was not absolutely certain. Even more frustrating was the fact that a combination of cloudy weather and moonlight prevented his confirming his recovery of Wolf's comet until June 1.99, at which time the coma was about 20" across and exhibited a faint nucleus. E. E. Barnard independently recovered the comet on May 4.48. Barnard had actually been searching for the comet with the 30-cm refractor since Berberich had sent an ephemeris to him in 1890. When found, Barnard described the comet as extremely faint and small, with a magnitude of about 13.5. The diameter was given as 5–10". The recovery positions of both Spitaler and Barnard indicated Berberich's prediction required a correction of only −0.4 day. W. Luther (Hamburg, Germany) failed to find the comet during his searches with a 25-cm refractor during the period of May 5–11.

The comet was quite faint and not widely observed early in its apparition, but its decreasing distances from both the sun and Earth brought it into the range of the smaller telescopes during July. On July 8th, L. Fabry (Marseille, France) observed with a 26-cm refractor and said the comet was magnitude 12.5, with an ill-defined coma about 40" across. There was no nucleus. Luther reported the comet was faint on July 28. The comet attained its most northerly declination of +28° on August 1. On August 4, Luther said the nucleus was nonstellar and about magnitude 11. On the 6th, J. Holetschek (Vienna, Austria) observed with a 15-cm refractor and said the comet was 1.5' across, with a nuclear magnitude of about 10.5, while Luther gave the nuclear magnitude as 11.5. On the 10th, F. Renz (Pulkovo Observatory, Russia) observed with the 38-cm refractor and said the stellar nucleus was magnitude 11. On the 11th, Holetschek said the coma was 2' across, while L. Ambronn (Göttingen, Germany) said the coma was 3–4' across, with a clearly defined nucleus. On the 13th, Holetschek estimated the total magnitude as 10. On August 18, Ambronn said the comet was quite faint in the heliometer, and Fabry said the comet was magnitude 11, with a coma 40" across that exhibited a very diffuse border. That same night, G. Witt (Urania Observatory, Berlin, Germany) reported the nucleus was magnitude 12.5. Renz said the stellar nucleus was magnitude 11–12 on the 28th. On August 29, Luther noted the nucleus was stellar and magnitude 11. The coma was circular. E. Millosevich (Rome, Italy) estimated the nuclear magnitude as 11 in the 25-cm refractor on the 30th and noted a trace of tail.

At the beginning of September, F. Schwab (Kremsmünster, Austria) gave the magnitude as 10–11 and noted a round coma and a nucleus. On September 2, Luther gave the nuclear magnitude as 10.5. On September 3, this comet transitted 21 Tauri. S. W. Burnham (Lick Observatory, California, USA) and Barnard prepared for the event by carefully observing the declination differences between 21 and 22 Tauri, so they could detect any changes in declination that might be caused by refraction by the cometary material. Barnard used a 30-cm refractor, but noted no sensible change in declination for the star. Burnham observed with a 91-cm refractor and noted a slight change in the declination, which might indicate the comet did refract the star's light. The coma diameter was given as about 0.75–1' across and the nucleus came to within 10" of the star. On the 4th, Holetschek reported the comet was as easy to see as a star of magnitude 9.5. On the 5th, Holetschek reported the comet seemed fainter than on the previous morning. On the 8th, G. Le Cadet (Lyon, France) observed with a 36-cm refractor and said the magnitude was 12.5. He added that the coma was slightly elongated in the direction of the diurnal movement. On September 9 and 12, M. Updegraff (State University of Missouri, USA) said the comet was only faintly visible in the 19-cm refractor because of light clouds. On September 11, W. H. Robinson (Radcliffe Observatory, Oxford, England) noted that a 12th-magnitude nucleus was just within the coma on the following side (25.4-cm refractor, 100×). On September 12, W. Wickham (Radcliffe Observatory, Oxford, England) saw a nucleus surrounded by nebulosity (25.4-cm refractor, 100×). On the 13th, Holetschek estimated the total magnitude as 9.0, while Luther noted a distinct nucleus of magnitude 10.8. Around mid-September, Schwab gave the magnitude as 10 and the coma diameter as 1'. On the 23rd, Luther estimated the nuclear magnitude as 10. Luther said the comet was bright on the 30th, with a nucleus of magnitude 9.

Holetschek saw the comet on several occasions during the period of October 2 to November 9 and said it always appeared as distinct as a star of magnitude 8.5–9 in the finder of the 15-cm refractor. At the beginning of October, Schwab said the magnitude was brighter than 10. On October 4, Luther said the comet disappeared at the same time as the 9.4-magnitude comparison star when the field was illuminated. That same night, C. F. Pechüle (Copenhagen, Denmark) observed with a 27-cm refractor and said the comet was 1' across, with a tail 2' long and 1' wide that extended toward PA 270°. Renz noted a short, fan-shaped tail on the 9th and a nucleus of magnitude 11. Le Cadet indicated a magnitude of 10.5 on the 11th, while Luther said the comet looked like a 9.5-magnitude star surrounded by a weak nebulosity. W. Schur (Göttingen, Germany) measured the coma diameter as 62" on the 24th. On the 27th, Luther noted that when the field was illuminated, the comet disappeared at the same time as a star of magnitude 9.3. On October 28, Robinson estimated the magnitude of the nucleus as 12 (25.4-cm refractor, 100×), while Luther described the comet as a nebulous star of magnitude 10.

As November began, the comet was heading away from both the sun and Earth. On the 1st, Luther estimated the nuclear magnitude as 10, while Le Cadet gave it as 12. On the 2nd, Luther said the nucleus was magnitude 11.4. On the 10th, Luther said the nucleus was magnitude 10.5. Luther gave the nuclear magnitude as 11.8 on the 12th. On the 23rd, Luther noted the nuclear magnitude was 11.5. On November 25, the nuclear magnitude was given as 13 by Robinson.

On December 2, Holetschek said the comet was about magnitude 9.5 in the finder. On the 6th, Luther gave the nuclear magnitude as 11.5. On the 20th, Luther gave the nuclear magnitude as 12. On the 24th, Holetschek said the magnitude was between 10.5 and 11, while Spitaler noted the comet was

2' across, with a nearly stellar nucleus. On December 31, Spitaler said the coma was 1.5' across and occasionally seemed to extend up to 3' toward the north. The nuclear condensation was magnitude 10 and seemed brighter on the northern side than on the southern.

On 1892 January 1, Spitaler said the 69-cm refractor revealed the nuclear condensation was better defined on the southern side, while the coma seemed to extend up to 5' toward the north-northeast. On January 2, Schwab estimated the nuclear magnitude as 11. On January 4, Luther gave the nuclear magnitude as 12.2 and noted a very faint coma. On January 6, Robinson said the comet was "exceedingly faint" in moonlight (25.4-cm refractor, 100×). On January 19, Luther said the comet was extremely faint in the 25-cm refractor, partly because it was very close to a star of magnitude 11.5. On January 20, H. A. Kobold (Strasbourg, France) described the comet as 0.5' across with a central condensation of magnitude 11. That same night, Spitaler gave the total magnitude as 10. He noted the coma was 0.75–1' across, with the brightest portion about 0.5' across. Spitaler gave the diameter as 1' on the 21st and said the coma seemed elongated toward the northeast, with a length of 1.5'. On the 30th and 31st, Luther gave the nuclear magnitude as 12.5.

On February 2, Luther observed the comet with a 25-cm refractor while the moon was up and described it as extraordinarily faint, without a distinct nucleus. On the 15th, Luther said the comet was about magnitude 13. A. Kammermann (Geneva, Switzerland) saw the comet with a 25-cm refractor on February 18, 24, and 26. On the 23rd, Spitaler gave the total magnitude as 11 and the nuclear magnitude as 12.5. He added that the coma was about 0.5' across. Luther noted the comet was extraordinarily faint on the 24th, while Spitaler gave the total magnitude as 11 and the coma diameter as 0.5'. On February 25, Spitaler gave the total magnitude as 11.5 and the nuclear magnitude as 13.

On March 16, Spitaler said the comet was about magnitude 12.5, with a coma 0.5' across. On the 18th, he estimated the total magnitude as 12.5 and the nuclear magnitude as 14. The coma was 1' across. Kobold said the comet appeared as a faint, irregular nebulosity about 1' across on the 19th, with a nucleus of magnitude 14. Spitaler said the comet was 0.5–0.75' across on the 24th.

The comet was last seen on 1892 March 31.83, when Spitaler measured the position as $\alpha = 5^h 41^m 12^s$, $\delta = -1° 00'$. He said the comet was easy to see in the 69-cm refractor despite moonlight, and described it as round and about 15" across. Spitaler made another attempt to observe the comet on April 22, but nothing was seen as the comet would have been within evening twilight.

The orbit was revised by W. Bellamy using the recovery position of Barnard, with the resulting perihelion date being September 3.96. Thraen compared his ephemeris with positions obtained during the first month and a half of observations and concluded the perihelion date was September 3.97. Berberich (1892) reinvestigated the orbit of this comet. He once again took positions from the 1884–5 apparition, but applied perturbations by both Jupiter and Saturn. The result was a perihelion date of September 3.94.

The first multiple apparition orbits came from Thraen (1891, 1898). Using positions from the 1884–5 and 1891 apparitions, as well as planetary perturbations, he determined the perihelion date as September 3.93 and the period as 6.82 years. Very similar results were later obtained by M. Kamienski (1922, 1959) and D. K. Yeomans (1978). Yeomans (1975) determined the nongravitational terms as $A_1 = +0.139$ and $A_2 = -0.0081$ and his orbit is given below.

T	ω	Ω (2000.0)	i	q	e
1891 Sep. 3.9318 (TT)	172.7903	207.9089	25.2314	1.592792	0.557150

ABSOLUTE MAGNITUDE: $H_{10} = 7.8$ (V1964)

FULL MOON: Apr. 24, May 23, Jun. 22, Jul. 21, Aug. 19, Sep. 18, Oct. 17, Nov. 16, Dec. 15, Jan. 14, Feb. 12, Mar. 13, Apr. 12

SOURCES: C. N. A. Krueger, *AN*, **110** (1884 Nov. 14), p. 207; A. Berberich, *AJ*, **10** (1891 Mar. 14), p. 175; A. K. Thraen, *AN*, **127** (1891 Mar. 28), pp. 11–16; G. W. L. Struve, *AN*, **127** (1891 Apr. 7), p. 45; E. E. Barnard, *AJ*, **11** (1891 May 12), p. 7; A. Berberich and E. E. Barnard, *AJ*, **11** (1891 Jun. 5), pp. 12–13; R. Spitaler, *AN*, **127** (1891 Jun. 11), p. 303; A. Thraen, *AN*, **127** (1891 Jun. 20), pp. 315–20; W. Bellamy, *AJ*, **11** (1891 Jun. 30), pp. 17–18; A. K. Thraen, *AN*, **128** (1891 Jul. 28), p. 31; L. Ambronn, *AN*, **128** (1891 Aug. 26), p. 159; E. Frisby, *AJ*, **11** (1891 Aug. 29), p. 40; L. Ambronn, *AN*, **128** (1891 Sep. 1), p. 173; M. Updegraff, *AJ*, **11** (1891 Oct. 10), p. 55; L. Fabry, *BA*, **8** (1891 Nov.), pp. 503–4; W. G. Thackeray, T. Lewis, H. P. Hollis, and A. C. D. Crommelin, *MNRAS*, **52** (1891 Nov.), pp. 27–8; E. Frisby, *AJ*, **11** (1891 Nov. 7), p. 61; A. C. D. Crommelin, *MNRAS*, **52** (1891 Dec.), p. 118; E. Frisby, *AJ*, **11** (1891 Dec. 9), p. 80; A. Thraen, *AN*, **128** (1891 Dec. 22), pp. 421–4; E. Frisby, *AJ*, **11** (1891 Dec. 23), p. 88; E. Frisby and A. Berberich, *AJ*, **11** (1892 Jan. 12), pp. 94, 103–4; *The Observatory*, **15** (1892 Feb.), p. 128; F. P. Leavenworth and W. H. Collins, *AJ*, **11** (1892 Feb. 13), p. 120; M. Updegraff, *AJ*, **11** (1892 Mar. 3), p. 135; W. H. Collins, *AJ*, **11** (1892 Mar. 31), pp. 157–8; O. C. Wendell, *AJ*, **11** (1892 May 3), pp. 191–2; E. Millosevich, *AN*, **129** (1892 May 6), p. 291; G. Le Cadet, *BA*, **9** (1892 Jun.), pp. 249–50; A. Graham, *MNRAS*, **52** (1892 Jun.), pp. 570–1; F. Schwab, *AN*, **130** (1892 Jun. 18), pp. 9–14; J. Holetschek, *AN*, **130** (1892 Jul. 5), pp. 69–74; E. E. Barnard, *AJ*, **12** (1892 Oct. 4), pp. 87–8; W. Luther, *AN*, **131** (1892 Nov. 7), pp. 83–6, 107–9; T. Reed, *AJ*, **12** (1892 Dec. 5), pp. 134–5; R. Spitaler, *AN*, **131** (1893 Jan. 10), pp. 383–8; A. Kammermann, *AN*, **131** (1893 Jan. 12), p. 401; G. Witt, *AN*, **132** (1893 Jan. 17), pp. 7–12; W. Schur, *AN*, **132** (1893 Apr. 15), pp. 227–31; F. Renz, *AN*, **133** (1893 Aug. 24), pp. 245–51; H. A. Kobold, *AN*, **134** (1893 Dec. 21), pp. 169–73; C. F. Pechüle, *AN*, **136** (1894 Oct. 20), pp. 305, 315; W. H. Robinson and W. Wickham, *MNRAS*, **55** (1895 Jan.), pp. 161–3; A. K. Thraen, *AN*, **146** (1898 Mar. 10), pp. 11–14; M. Kamienski, *AOAT*, **5** No. 5 (1922), pp. 1–5; M. Kamienski, *AA*, **9** (1959), p. 58; V1964, p. 63; D. K. Yeomans, *PASP*, **87** (1975 Aug.), pp. 635–7; D. K. Yeomans, *QJRAS*, **19** (1978), pp. 52–3, 57.

2P/Encke

1891 III = 1891c

Recovered: 1891 August 2.50 ($\Delta = 1.58$ AU, $r = 1.51$ AU, Elong. $= 67°$)
Last seen: 1891 October 13.18 ($\Delta = 1.15$ AU, $r = 0.37$ AU, Elong. $= 18°$)
Closest to the Earth: 1891 September 20 (0.9170 AU)
Calculated path: TAU (Rec), PER (Aug. 5), AUR (Aug. 11), GEM (Sep. 1), CNC (Sep. 12), LEO (Sep. 20), VIR (Oct. 7)

The recovery of this comet began with O. Backlund (1891). He took the comet's orbit from the 1884–5 apparition and applied perturbations by Venus, Earth, Mars, Jupiter, and Saturn up to early 1888, and then perturbations by Jupiter up to early 1891. He predicted the comet would arrive at perihelion on 1891 October 18.48. E. E. Barnard (Lick Observatory, California, USA) had unsuccessfully searched for this comet on several occasions with the 30-cm refractor, but on the morning of 1891 August 2 he decided to examine the position predicted by Backlund with the 91-cm refractor. In a very short time, he found the comet and gave its position as $\alpha = 3^h 55.8^m$, $\delta = +30° 03'$ on August 2.50. Barnard described the comet as "extremely faint and diffused, and only very feebly brighter towards the middle." He estimated the magnitude as slightly brighter than 17 and gave the diameter as 45″. Barnard confirmed the recovery on August 4 and noted the comet appeared brighter.

The comet was heading toward both the sun and Earth when discovered. It passed closest to Earth during the second half of September. On August 13, W. H. Robinson (Radcliffe Observatory, Oxford, England) observed with the refractor and noted, "The Comet's light is of the feeblest, requiring averted vision to render it visible. No nucleus seen. Centre of patch observed." The comet attained its most northerly declination of $+35°$ on August 30. On September 2 and 3, F. Schwab (Kremsmünster, Austria) said the comet was diffuse, with a faint central condensation and a coma about 2′ across. F. Renz (Pulkovo Observatory, Russia) saw the comet in the 38-cm refractor on the 3rd and described it as a formless nebulosity, with a condensation in the preceding section. G. Witt (Berlin, Germany) saw the comet in a 30-cm refractor (125×) on the 4th and described it as a large, nebulous mass, without a nucleus. On September 6, W. Luther (Hamburg, Germany) described the comet as a bright, nebulous mass, with a distinct central condensation, but without a sharp, stellar nucleus. Luther noted the comet was moderately brighter on the 8th and was 2′ across with a nuclear condensation of magnitude 11. J. Holetschek (Vienna, Austria) estimated the magnitude as 8.0 with a 3.7-cm finder on the 10th. That same night, Schwab said the condensation was brighter and more distinct than when last seen. Robinson found the coma fan-shaped on the 11th and said it formed an obtuse angle on the western side. With a refractor, he saw a nucleus of magnitude 13.5 within the apex of this angle. Also on the 11th, Holetschek estimated the magnitude as 8.1–8.2 with the finder, while Witt saw a large condensation about 40″ across that exhibited a minor condensation toward PA 270°. Holetschek gave the magnitude as 8.1 on the 12th, while H. A. Kobold (Strasbourg, France) saw the nucleus was near the western edge of the coma with a 46-cm refractor. On the 13th, Holetschek said the comet was "somewhat brighter" than a star of magnitude 8.2 in the 3.7-cm finder, while Luther reported the comet was bright, with a nucleus of magnitude 11.5. Holetschek said the comet was about magnitude 8.0 on the 14th, while Renz noted a condensation was distinctly seen in the western portion of the coma. Renz described the comet as bright on the 15th, and noted the western part of the coma was condensed. Holetschek estimated the comet's brightness as 7.5 on the 26th. On September 30, Luther said the comet was very bright, with a nucleus of magnitude 8.

Although the comet was moving away from Earth, it was still heading toward perihelion and was well observed during the first two weeks of October. O. Knopf (Jena, Germany) saw the comet on September 30 and October 2, and noted it was as bright as M31. Robinson said the coma was still fan-shaped on October 1 and that the nucleus had brightened to magnitude 11. He noted, "strong twilight obliterated the fainter parts of the Comet." That same night, Holetschek said the comet was about magnitude 6.5, while Luther described it as very bright, with a central, well-defined stellar condensation of magnitude 8. W. Schur (Göttingen, Germany) saw the comet on October 1, 2, and 5, and noted a strong condensation, with a possible stellar nucleus. On the 3rd, the brightness was given as 6.0 by Holetschek and 6–7 by Renz. Renz added that the comet was a bright, circular nebulosity, with a condensation in the western part. Renz saw the comet in twilight with the 38-cm refractor on the 4th and described it as a bright, circular nebulosity, with a narrow, curved tail about 15' long. On October 5 and 6, Luther said the comet's low altitude made it appear fainter, with a nucleus of about magnitude 9. Renz again saw the comet at low altitude on the 6th and said the coma was about magnitude 6–7, while the tail was 8' long. A nucleus was occasionally seen. Holetschek found the nucleus slightly diffuse on the 11th. On the 12th, Robinson said the comet was "just visible in very strong twilight when stars below the 7th magnitude had disappeared from the field of view." That same morning, Holetschek watched the comet until it was lost in bright morning twilight.

The comet was last detected on October 13.18, when Holetschek viewed it with a 15-cm Fraunhofer refractor until 22 minutes after the beginning of morning twilight. He said it always remained indistinct and impossible to measure.

Refinements of this comet's orbit using positions from multiple apparitions, as well as planetary perturbations, have been published by a few astronomers. Of most significance were the studies of B. G. Marsden and Z. Sekanina (1974). This revealed a perihelion date of October 18.43 and a period of 3.30 years. Marsden and Sekanina determined the nongravitational terms as $A_1 = -0.14$ and $A_2 = -0.02263$. The orbit of Marsden and Sekanina is given below.

T	ω	Ω (2000.0)	i	q	e
1891 Oct. 18.4254 (TT)	183.9441	336.2250	12.9294	0.340450	0.846476

ABSOLUTE MAGNITUDE: $H_{10} = 9.1$ (V1964)

FULL MOON: Jul. 21, Aug. 19, Sep. 18, Oct. 17

SOURCES: O. Backlund, *AN*, **127** (1891 Jul. 8), pp. 427–32; E. E. Barnard, *AJ*, **11** (1891 Aug. 29), pp. 36–7; W. Schur, *AN*, **128** (1891 Oct. 20), p. 253; W. H. Robinson, *MNRAS*, **52** (1891 Dec.), pp. 119–20; J. Holetschek, *AN*, **129** (1892 Jan. 15), pp. 29–32; O. Knopf, *AN*, **130** (1892 Sep. 17), pp. 369–74; W. Luther, *AN*, **131** (1892 Nov. 7), pp. 81–114; F. Renz, *AN*, **131** (1892 Dec. 16), pp. 265–74; H. A. Kobold, *AN*, **131** (1893 Jan. 3), pp. 319–22; G. Witt, *AN*, **132** (1893 Jan. 17), pp. 7–12; V1964, p. 63; B. G. Marsden and Z. Sekanina, *AJ*, **79** (1974 Mar.), pp. 415–16.

11P/Tempel–
Swift–LINEAR

1891 V = 1891d

Recovered: 1891 September 28.20 ($\Delta = 0.37$ AU, $r = 1.27$ AU, Elong. $= 129°$)
Last seen: 1892 January 21.77 ($\Delta = 0.51$ AU, $r = 1.37$ AU, Elong. $= 131°$)
Closest to the Earth: 1891 December 2 (0.2386 AU)
Calculated path: AQR (Rec), EQU (Oct. 16), PEG (Oct. 30), AND (Dec. 5), PSC (Dec. 10), TRI (Dec. 18), ARI (Dec. 24), TAU (Jan. 3)

The story of the recovery of this comet begins with J. F. Bossert. Having been one of the first astronomers to determine an elliptical orbit for this comet in 1880, Bossert continued to investigate its motion. Early in 1886 he successfully linked the apparitions of 1869 and 1880. He then applied perturbations by Jupiter and Saturn and predicted the comet would return on 1886 May 9.93. He pointed out that this was a very unfavorable apparition. No observations were reported. Bossert (1891) published an update in the *Astronomische Nachrichten* on 1891 June 4 in which he integrated the orbit up to the next apparition and predicted the comet would arrive at perihelion on 1891 November 15.45. E. E. Barnard (Lick Observatory, California, USA) began searching for this comet with the 30-cm refractor in July, but was unsuccessful. On the evening of September 27/28 he was again unsuccessful and after a time began one of his regular comet hunting sessions. With the telescope already pointing to the predicted position of 11P, he moved it and began sweeping in the region just to the southwest. After a short time he found an exceedingly faint nebulosity about 3° from the predicted position of 11P. He gave the position as $\alpha = 20^h\ 54.2^m$, $\delta = -1°\ 32'$ on September 28.20. Barnard later remarked that he did not even consider that this was the comet he had been looking for because he believed the predicted position would be known quite accurately. He confirmed his find on September 29.20 and estimated the magnitude as slightly brighter than 14. After a further measurement on September 30, Barnard realized that this comet was moving at the same rate and in the same direction as were expected for 11P. An independent recovery was made by W. F. Denning (Bristol, England) on September 30.

C. F. Pechüle (Copenhagen, Denmark) saw the comet on October 1, 3, and 6, and described it as very faint, slightly condensed, and about 2′ across. R. Spitaler (Vienna, Austria) saw the comet with a 69-cm refractor on the 2nd and noted it was rather bright, about 2′ across, with a large diffuse nucleus. Spitaler also noted an outer coma that was between 5′ and 8′ across. While Spitaler described the comet as about 1.5′ across on the 6th, and noted a very diffuse central condensation, J. Holetschek (Vienna, Austria) could not find the comet in a 15-cm refractor. F. Renz (Pulkovo Observatory, Russia) said the comet appeared as a pale nebulous mass in the 38-cm refractor on the 7th, while it was brighter on the next evening. On the 9th, Spitaler said the comet was very diffuse and about 1.5′ across. H. A. Kobold (Strasbourg, France) observed the comet with a 46-cm refractor on the 22nd and described it as 1′ across, with a faint central condensation and a nucleus of magnitude 13. Spitaler found the comet to be 2′ in diameter on October 23 and 24. M. Chandrikoff (Kiev, Ukraine) described the comet as very faint, with a large, diffuse condensation on October 24. Spitaler said the comet was 3′ in diameter on the 25th. W. Luther (Hamburg, Germany) saw the comet with a 25-cm refractor and said it was a faint nebulosity with a nucleus of magnitude 12.5. B. von Engelhardt (Dresden, Germany) saw the comet with a 30-cm refractor on October 28 and described it as a very faint nebulosity, without a condensation. Luther also saw the comet on the 28th and said it was rather faint, about 2′ across, with a very faint nucleus.

The comet passed perihelion around mid-November. Spitaler described the comet as 2′ across in the 69-cm refractor on the 3rd and noted a diffuse nucleus 30″ across. Spitaler said the comet was 3′ across in the refractor on the 6th. Kobold noted the comet was 2′ across on the 7th and 28th. E. Millosevich (Rome, Italy) saw the comet in a 25-cm refractor on November 22 and described it as extremely faint, with little condensation. Luther said the comet was hardly visible in the 25-cm refractor on the 23rd. Engelhardt described the comet as very faint and about 1′ across on November 28.

The comet passed closest to Earth at the beginning of December. Kobold said the comet was 1′ across with a central condensation of magnitude 13. Spitaler said the comet was about 2.5′ across on the 24th, with a faint outer coma 6–8′ across, while Holetschek described the comet as a faint nebulous glow at the limit of the 15-cm refractor. The comet attained its most northerly declination of +28° on the 26th.

Spitaler said the centrally-condensed comet was 1.5′ across on 1892 January 1. The comet was last detected on January 21.77, when Spitaler measured the position as $\alpha = 4^{h}\ 40.5^{m}$, $\delta = +25°\ 08′$. He was using the 69-cm refractor and said the comet was very faint and diffuse, with a slight centrally condensed coma about 45″ across. After a long period of bad weather, Spitaler again looked for the comet on February 23, but no trace was found. Astronomers at Sydney Observatory searched for this comet on two mornings during 1891, but nothing was found.

The most recent analysis of this comet's orbit using multiple apparitions was published by B. G. Marsden (1971). He found a perihelion date
of November 17.87 and a period of 5.54 years. Marsden determined the
nongravitational terms as $A_1 = +0.04$ and $A_2 = -0.0459$. This orbit is given
below.

T	ω	Ω (2000.0)	i	q	e
1891 Nov. 17.8659 (TT)	106.6322	298.1458	5.3971	1.086420	0.652988

ABSOLUTE MAGNITUDE: $H_{10} = 13.8$ (V1964)

FULL MOON: Sep. 18, Oct. 17, Nov. 16, Dec. 15, Jan. 14, Feb. 12

SOURCES: J. F. Bossert, BA, **3** (1886 Feb.), pp. 65–78; J. F. Bossert, AN, **127** (1891 Jun. 4),
p. 271; E. E. Barnard and W. F. Denning, AN, **128** (1891 Oct. 6), p. 237; E. E. Barnard
and O. C. Wendell, AJ, **11** (1891 Oct. 10), p. 56; C. F. Pechüle, AN, **128** (1891 Oct. 20),
p. 253; M. Chandrikoff, AN, **128** (1891 Nov. 3), p. 305; O. C. Wendell and E. E. Barnard,
AJ, **11** (1891 Nov. 7), pp. 61–2; MNRAS, **52** (1892 Feb.), p. 266; B. von Engelhardt,
AN, **129** (1892 Feb. 10), pp. 57–60; O. C. Wendell, AJ, **11** (1892 May 3), pp. 191–2;
E. Millosevich, AN, **129** (1892 May 6), p. 291; R. Spitaler, AN, **129** (1892 May 27), pp.
357–60; J. Holetschek, AN, **130** (1892 Jul. 5), p. 73; W. Luther, AN, **131** (1892 Nov. 7),
pp. 87, 109; F. Renz, AN, **131** (1892 Dec. 16), pp. 273–8; H. A. Kobold, AN, **131** (1893
Jan. 3), pp. 319–22; V1964, p. 64; B. G. Marsden and Z. Sekanina, AJ, **76** (1971 Dec.),
pp. 1142–3.

C/1891 T1 *Discovered:* 1891 October 3.54 ($\Delta = 0.95$ AU, $r = 1.20$ AU, Elong. $= 76°$)
(Barnard) *Last seen:* 1891 December 7.32 ($\Delta = 1.65$ AU, $r = 1.05$ AU, Elong. $= 37°$)
  Closest to the Earth: 1891 October 7 (0.9441 AU)
1891 IV = 1891e *Calculated path:* PUP (Disc), VEL (Oct. 11), CEN (Oct. 30), CRU (Nov. 5), CEN
(Nov. 13), LUP (Nov. 28)

E. E. Barnard (Lick Observatory, California, USA) was conducting another
routine sweep for comets with his 10-cm "broken-tube" cometseeker when
he found this "nebulous object" on 1891 October 3.54. He examined the
object with the 30-cm refractor and determined it was a comet. The position
was given as $\alpha = 7^h 31.4^m$, $\delta = -27° 54'$. Barnard estimated the magnitude
as 12 and said the coma was 1' across.

Few physical descriptions were ever provided for this comet and Barnard
was the only person in the Northern Hemisphere to see the comet because
of its rapid southeastward motion. His final observation was made on the
morning of October 10. Thanks to the arrival of a telegram announcing the
discovery, R. P. Sellors (Sydney, New South Wales, Australia) was able to
view the comet with a 29-cm refractor on October 10 and 12. He described
it as "a faint, round body with slight central condensation." The comet
attained its most southerly declination of $-56°$ on November 8.

The comet was last detected on December 7.32, when C. W. Ljungstedt
(Cordoba, Argentina) observed with a 30-cm refractor and measured the

position as $\alpha = 14^h\ 44.3^m$, $\delta = -49°\ 39'$. A period of cloudy and rainy weather followed that lasted over three weeks. No trace of the comet was found thereafter.

The first parabolic orbit was calculated by W. W. Campbell. He took positions obtained at Lick Observatory on October 3, 4, and 5, and determined the perihelion date as 1891 November 9.25. Further calculations using the Lick positions through October 9 and 10 were published by E. B. Davis, A. Berberich, R. Froebe, and Campbell. These indicated a perihelion date November 13.

The first astronomer to utilize the Southern Hemisphere positions to improve the orbit was J. R. Hind (1894). He took the Cordoba positions from October 20, November 13, and December 4, and determined a perihelion date of November 14.05. H. A. Peck (1903) calculated a definitive orbit for this comet. He began with 41 positions from the Lick, Sydney, and Cordoba observatories, and ultimately determined a parabolic orbit with a perihelion date of November 14.04 and an elliptical orbit with a perihelion date of November 14.05 and a period of about 54 436 years. Although the residuals for the parabolic orbit indicated the comet was deviating from a parabola, the residuals for the elliptical were rather large, so the parabola, which is given below, was favored.

T	ω	Ω (2000.0)	i	q	e
1891 Nov. 14.0427 (UT)	269.5671	219.5276	77.9880	0.971089	1.0

ABSOLUTE MAGNITUDE: $H_{10} = 9.4$ (V1964)

FULL MOON: Sep. 18, Oct. 17, Nov. 16, Dec. 15

SOURCES: E. E. Barnard, *AN*, **128** (1891 Oct. 6), p. 239; E. E. Barnard and W. W. Campbell, *AJ*, **11** (1891 Oct. 10), p. 56; W. W. Campbell and H. C. F. Kreutz, *AN*, **128** (1891 Oct. 20), p. 255; E. E. Barnard, *The Observatory*, **14** (1891 Nov.), p. 389; E. E. Barnard, *AJ*, **11** (1891 Nov. 7), pp. 63–4; R. P. Sellors, *MNRAS*, **52** (1891 Dec.), p. 121; A. Berberich, *AN*, **128** (1891 Dec. 14), pp. 405–8; R. Froebe, *AN*, **128** (1891 Dec. 31), p. 439; W. W. Campbell, *AJ*, **11** (1892 Jan. 29), pp. 111–12; E. B. Davis, *AJ*, **11** (1892 Feb. 13), p. 119; J. M. Thome and C. W. Ljungstedt, *AN*, **129** (1892 May 6), pp. 285–8; J. R. Hind, *AN*, **135** (1894 Jun. 13), p. 383; H. A. Peck, *AJ*, **23** (1903 Aug.), pp. 163–6; V1964, p. 63.

C/1892 E1 (Swift) *Discovered:* 1892 March 7.42 ($\Delta = 1.13$ AU, $r = 1.15$ AU, Elong. $= 65°$)

Last seen: 1893 February 16.81 ($\Delta = 4.78$ AU, $r = 4.30$ AU, Elong. $= 55°$)

1892 I = 1892a *Closest to the Earth:* 1892 March 27 (1.0512 AU)

Calculated path: SGR (Disc), CAP (Mar. 20), AQR (Mar. 28), PEG (Apr. 10), AND (May 22), CAS (Jun. 29), AND-CAS (Jul. 8), AND (Sep. 28), PEG (Nov. 30), AND (Dec. 26), PSC (Feb. 9)

L. Swift (Warner Observatory, Rochester, New York, USA) was searching for comets with his 11-cm telescope, when he found this one on March 7.42. He described it as bright and gave the position as $\alpha = 18^h\ 59.0^m$, $\delta = -31°\ 20'$. Swift added that the comet was moving easterly.

The comet was well observed during the remainder of March as it approached the sun. It passed closest to Earth on March 27. E. E. Barnard (Lick Observatory, California, USA) said the comet was visible to the naked eye on March 8, as a large, hazy star of magnitude 5 or 6. He noted the head was very bright in the 8.3-cm finder, and a faint tail was visible. Through the 30-cm refractor the comet exhibited a large, round coma about 8' across. There was a central nucleus of magnitude 11, which was not quite stellar. J. Holetschek (Vienna, Austria) estimated the magnitude as 4 on the 10th. That same night, O. C. Wendell (Harvard College Observatory, Massachusetts, USA) observed the comet with the 38-cm refractor and said the coma was 3' across, with a pretty well-defined nucleus. W. H. Finlay (Royal Observatory, Cape of Good Hope, South Africa) said the comet was a naked-eye object and looked like a "hazy star of the fourth magnitude." There was also a sharp nucleus of about magnitude 9.5. Barnard said the comet was conspicuous to the naked eye and was more noticeable and bright than the Trifid nebula, but smaller. On the 11th, H. C. Russell and R. P. Sellors (Sydney, New South Wales, Australia) said a photographic exposure of $1^h\ 50^m$ revealed "five equidistant rays, the outside ones enclosing an angle of 25°." They added that the longest ray was 35' long and all were faint because of moonlight and the frequent passage of thin clouds. On the 12th, Wendell said the coma was 3' across, with a strong central condensation. Barnard said the comet was "quite easily seen with the naked eye" on the 16th, despite a nearly full moon. Wendell observed under a "rather bright sky" on the 21st and noted a "fairly well-defined" nucleus, while J. K. Rees and H. Jacoby (Columbia College Observatory, New York, USA) said the comet "showed a well-defined nucleus." That same night, H. C. Wilson (Goodsell Observatory, Minnesota, USA) noted the beginning of a "fan" on the sunward side. On March 22, Russell and Sellors said a visual observation with the 29-cm refractor "showed no tail except a slight hazy extension, and not a sign of any ray." However, they reported a photograph with an exposure of $2^h\ 23^m$ "shows no less than eight rays, two of which extend to the edge of the photograph and may have been even longer; the recorded length is . . .1° 10'." There were 28 minutes of cloud cover during the exposure, which, once again, caused the rays to be faint, but they did note that one of the rays seen on the photo of March 11 was "some distance from the head, and not visibly joined to it. The longest rays, as well as three new ones, were located "on the south side of the tail, and these seem to have a definite connection with a jet from the nucleus, which extends forwards and then bends round to these rays." That same night, Finlay said the nucleus was "well-defined" and similar to a star of magnitude 9.5. On the 25th, Holetschek estimated the magnitude as 4. Finlay said a "short faint tail" was seen in the finder on the 27th. On March 30, E. A. Lamp (Kiel, Germany) said the comet was easily seen with the naked eye, while the cometseeker revealed a tail about 2°

long. W. Luther (Hamburg, Germany) also estimated the nuclear magnitude as 8.5.

The comet passed perihelion on April 7 and, although well observed during April, the astronomers of the Southern Hemisphere had the best view. The Chinese text *Ch'ing Ch'ao Hsü Wên Hsien Thung Khao*, which was written by Liu Chin-Tsao around 1910, notes that Chinese astronomers discovered a "broom-star" comet sometime during the period of 1892 March 28 to April 26. E. B. Knobel (president of the Royal Astronomical Society) observed this comet during a trip into the Southern Hemisphere. He saw the comet while on board a ship on April 1 and noted the tail was 8° long. Also on April 1, Otto Knopf (Jena, Germany) said the comet was very bright in a 20-cm refractor, despite morning twilight, with a tail about 5′ long and G. Le Cadet (Lyon, France) described the comet as 10′ across, with a nuclear condensation 6″ across. Le Cadet added that the straight tail consisted of several rays and extended 1.5° toward PA 250°. Several observations were made on the 2nd. Knobel said the tail length had increased to 12°. Finlay noted a faint, straight tail about 4° long, as well as a sharp nucleus of magnitude 9. F. Porro (Torino, Italy) observed the comet with the naked eye and gave the magnitude as 4.62. In binoculars he noted a nucleus of magnitude 5 or 6 and said the tail extended westward and was better defined on the northern side than on the southern. Le Cadet said the nucleus disappeared in twilight at the same time as a star of magnitude 9. On April 2 and 8, J. M. Thome (Cordoba, Argentina) said the coma and nucleus were unusually large, with the coma estimated as 5′ across and the elliptical nucleus measuring 12″ by 8″. He noted the tail was 15° long and was faint and indistinct to the naked eye. Thome estimated the naked-eye brightness as "a little brighter" than magnitude 3. On the 3rd, Finlay said the nucleus was "not well defined." Porro said the tail was directed toward PA 250°. He also noted a jet extending from the nucleus in a direction opposite to the tail. Luther gave the nuclear magnitude as 7. On the 4th, Holetschek estimated the magnitude as 3–4. A. Kammermann (Geneva, Switzerland) said the comet was an easy naked-eye comet, with a nucleus of magnitude 4–5. Le Cadet said the comet was visible to the naked eye with averted vision as a diffuse star of magnitude 4. The tail extended 4° toward PA 245°. Knobel said the tail was 14° or 15° long. On April 6, Wilson reported the comet was conspicuous to the naked eye, with the head being equal to a 4th-magnitude star. The tail was faint and perfectly straight, with its length traced for about 2° 30′ with the naked eye. The 41-cm refractor showed a non-stellar nucleus, with a bright, indefinite fan. As daylight washed out the nebulosity, the fan remained and seemed as bright as the nucleus. Luther estimated the nuclear magnitude as 7. On April 7, E. E. Barnard (Lick Observatory, California, USA) obtained a 1-hour-5-minute exposure of the comet with a 15-cm Willard lens. He said the tail consisted of "two broad streams, the northern of which is very bright, and the southern faint. The two streams merge together near the head, and at this point there is a quick bend in its southern side. A great deal of detail is shown in the brighter component in the form of bright streaks and patches. Fine threads or short 'whisker tails' extend back from the head at considerable angles to the main tail. There are some indications present also of the remarkable disturbance which followed some twenty-four hours later." On the 8th, George F. Parson (Master of the ship *Earnock*, in the middle of the Atlantic Ocean) saw the small comet in the eastern sky. Barnard obtained a 50-minute exposure with the 15-cm Willard lens. This showed the tail separated into a system of thin strands, the largest and brightest of which contained a "remarkable development...which might be taken for a secondary comet with a system of tails of its own." He suggested, "The large mass or secondary comet was doubtless thrown off from the nucleus or head some time during the preceding twenty-four hours, and must have had a very considerable velocity." On the 9th, Luther said the coma was 6' across, while the nucleus was 15" across and magnitude 6.

On April 17, Porro said the nucleus was about magnitude 5 to the naked eye. Luther reported the comet was magnitude 4 on the 20th, with a tail 1° long. Knopf observed at the beginning of morning twilight on the 21st and said the coma was round, while the tail extended about 40' westward. C. Wagner (Kremsmünster, Austria) saw the comet with the naked eye on the 22nd. He noted that the coma was 4' across in the refractor and contained a bright nucleus, while the finder revealed a narrow tail about 1° long. That same night, Porro said the tail extended 1° 50' toward PA 280°. On the 23rd, Wilson said the comet was "very brilliant," and exhibited a tail about 5° long. The 41-cm refractor revealed the fan was very bright, while the nucleus was bright and well defined. On April 24, Holetschek estimated the magnitude as 4, while M. Chandrikoff (Kiev, Ukraine) said the tail extended about 3°. That same night, Le Cadet reported the comet was visible to the naked eye as a diffuse star of magnitude 4. The tail extended 4° toward PA 260°. On April 25, Barnard obtained a 2-hour-20-minute exposure with the 15-cm Willard lens. He noted, "The tail partially separates into a number of streams, and on the north side is very sharply defined by what appears to be a thin black rift; if this edge of the tail is continued to the comet, it will pass south of the centre of the head, and consequently does not appear due to a force at that moment seated in the nucleus. The south portion of the inner bright tail is irregular near the head." W. Schur (Göttingen, Germany) said the comet was very bright and centrally condensed. On the 27th, Barnard obtained a 2-hour-25-minute exposure with the 15-cm Willard lens. He said the tail "appears to be made up of a number of bright strands which centre in the head." Schur noted the tail extended about 1.5° toward PA 259° on the 28th. On April 30, T. W. Backhouse (Sunderland, England) observed the comet and noted "its narrow, straight tail, though faint, was visible to the naked eye to a length of about 11°; the sky there was slightly affected by twilight." Backhouse remarked that he "saw it best" on this date. A. C. D. Crommelin and T. Lewis (Royal Observatory, Greenwich, England) noticed a short tail while observing with the 17-cm refractor (55×).

As May began, the comet was moving away from both the sun and Earth. On May 1, Backhouse noted the comet's head was about 4th magnitude. Parson said the comet was very faint while making sextant observations on the 2nd. Wagner said the comet disappeared with stars of magnitude 9 in the morning twilight on the 3rd. On the 4th, Wagner saw a straight tail about 1.5° long, while Schur said the tail extended toward PA 265.3°. On May 8, Le Cadet said moonlight weakened the light of the comet so that a tail was not visible. On the other hand, he did note that the nucleus was elongated toward PA 261°. Knopf observed with a 20-cm refractor on the 10th and said the comet appeared as a bright, circular nebulosity, with a central condensation of about magnitude 9.3. On the 19th, Wendell noted the coma was "somewhat elongated." On the 24th, Crommelin said the comet was "very large and bright, with a bifurcation tail." On the 25th, Wendell noted the nucleus was rather faint, while the comet's total brightness was magnitude 5.5. He added that there was "no definite structural detail visible." Wilson said the comet was about magnitude 8, with a tail 2° 30′ long on the 26th, while F. Renz (Pulkovo Observatory, Russia) observed with the 38-cm refractor and said a small coma and nucleus of magnitude 10–10.5 were visible in the bright dawn. The nucleus was shifted toward the western portion of the coma. On the 27th, Schur saw the comet with the naked eye and said the tail was about 1.5° long and extended toward PA 264.5°. On May 28, Holetschek estimated the magnitude as 5.5. Wagner said the coma was 4′ across, while the tail extended 30′. Le Cadet noted the comet was visible to the naked eye as a small nebulosity.

On June 3, Holetschek estimated the magnitude as 6. On the 9th, Wilson said the comet exhibited a faint tail at least 30′ long and said it was still easily seen in the finder, despite moonlight. On June 10, Le Cadet said the comet was hardly visible to the naked eye, while the telescope revealed a 10th-magnitude nucleus. On the 16th, H. A. Kobold (Strasbourg, France) said the comet exhibited a nucleus of magnitude 10, while Wendell said the nucleus was fairly well defined and the coma was 4′ across. On the 21st, Holetschek gave the magnitude as 6.3. A. Abetti (Padua, Italy) was not able to see the comet with the naked eye on June 22. On June 27, Le Cadet described the comet as a round nebulosity, with a slightly eccentric nucleus of magnitude 10–11, while Abetti said the nucleus was magnitude 7. On June 28, Knopf said the comet exhibited a central condensation, and a faint tail extending about 4′ toward the west.

On July 5, Holetschek estimated the magnitude as 6.5. On the 6th, Kobold said the coma was 1′ across, with a central condensation of magnitude 11. On July 9, Le Cadet described the comet as small and very faint. Holetschek estimated the magnitude as 7 on the 14th and 19th. On July 21, Wendell said the coma was 4′ across and the nucleus was "a little eccentric." During the period of July 21–27, M. W. Whitney (Vassar College Observatory, New York, USA) simply described the comet as faint in the 30-cm refractor.

On August 5, Wagner said the coma was 2′ across and centrally condensed. On the 12th, Holetschek estimated the magnitude as 8. Renz described the comet as a bright coma with a nucleus on the 14th. Kammermann said the comet was visible in a 5-cm finder on the 18th, and appeared round and centrally condensed in the 25-cm telescope. The comet attained its most northerly declination of +53° on August 21. That same night, Renz estimated the nuclear magnitude as 11–12. On the 22nd, Knopf observed with the 20-cm refractor and said the comet resembled a nebula with a central condensation. No tail was seen. On August 30, Holetschek estimated the magnitude as 8.

On September 1 Kobold said the comet was 0.5′ across with a central condensation of magnitude 10.5. On the 3rd, Wendell said the nucleus was "fairly well defined." Holetschek estimated the magnitude as 8.5 on the 11th and 9 on the 19th. B. von Engelhardt (Dresden, Germany) saw the comet with a 30-cm refractor on the 18th and noted it was rather bright and 0.5′ in diameter. It was still well seen in his 13-cm finder. On the 20th, Knopf said the comet appeared rather faint, with the central condensation of magnitude 11.5, while Wilson said the comet was still easily visible in the finder, while the 41-cm refractor revealed a nucleus of magnitude 11. That same night, Renz observed a second point of light located 3–4″ northwest of the nucleus.

On October 1, Renz said he still suspected a second nucleus a short distance to the northwest of the main nucleus. Holetschek estimated the magnitude as 9.5 on the 14th. Renz said the nucleus was magnitude 13 on the 19th. Wilson reported the comet was still easily seen in the finder on the 20th, while the 41-cm refractor revealed a nucleus of magnitude 10.5 located within a coma 1′ across. The tail was 3′ long and spread to a width of 2′. On October 23, Crommelin noted the comet was "extremely faint and was only suspected." Holetschek estimated the magnitude as 10, while G. Gruss (Prague, Czech Republic) observed with a 20-cm refractor and noted a coma 4′ across and a very faint nucleus. On the 26th, Gruss said the comet appeared very faint with a very faint nucleus. On October 31, Backhouse noted the comet was still "a conspicuous object" in his 11-cm refractor. He said "it still had two faint tails, pointing in opposite directions, [preceding and following]; and the extension of the head towards [north] was decidedly less than towards [south]." He added that the total magnitude equaled that of an 11th-magnitude star.

Although positions continued to be obtained during the remainder of the comet's apparition, there were few physical descriptions. Holetschek gave the magnitude as 10.3 on November 8 and 11 and on December 11. Engelhardt gave the coma diameter as about 1′ on December 9 and about

30″ on the 11th. On December 16 Kobold said the comet was 1′ across, with a slight condensation. Holetschek estimated the magnitude as 11.5 on December 23. On 1893 January 11, Kobold said the comet was small and faint, with a slight condensation. On February 6, Kobold said the comet was very small and extremely faint, making it rather difficult to observe. The comet was seen for the final time on February 16.81, when Kobold measured its position as $\alpha = 0^h 51.1^m$, $\delta = +25° 23′$.

The comet's spectrum was observed by several astronomers. The three cometary bands of diatomic carbon (C_2) were detected early in April by N. de Konkoly (O'Gyalla Observatory, Hungary), W. W. Campbell (Lick Observatory, California, USA), W. H. Pickering (Arequipa, Peru), and E. von Gothard (Herény, Hungary), while Backhouse detected the same molecules near the end of the month. In addition, Pickering and Gothard noted spectral lines that are now attributed to cyanogen (CN) and triatomic carbon (C_3).

The first parabolic orbit was calculated by H. C. F. Kreutz using positions obtained on March 8, 9, and 10. The resulting perihelion date was 1892 March 27.32. G. M. Searle took positions from March 8, 12, and 13, and determined the perihelion date as April 27.49. Further calculations by E. A. Lamp and Searle continued to pin down the orbit, with Searle being the first to determine a perihelion date within an hour of the true date when he used positions spanning the period of March 11–29. Further computations by Searle, J. R. Hind, M. Updegraff, F. Bidschof, A. Berberich, and F. G. Wentworth brought only slight refinements.

Many of these parabolic orbits came out a month or more after the comet's discovery, but there were some surprises. First, during May, Searle examined why the parabolic orbits were no longer adequately representing the positions. By taking positions from March 11, 30, and April 23, he calculated a hyperbolic orbit with a perihelion date of April 7.13 and an eccentricity of 1.002986. During August, Berberich decided to conduct his own investigation of the increasing inability of the parabolic orbits to represent this comet's motion. Using positions spanning the period of March 9 to July 12, he calculated an elliptical orbit with a perihelion date of April 7.15 and a period of about 20 133 years.

Although the comet continued to be observed until early 1893, no further orbits were calculated until E. E. Kühne (1914) determined a definitive orbit. He took 1124 positions spanning the period of March 8 to February 16, grouped them into 29 Normal positions, applied the perturbations by Jupiter, and determined the perihelion date as April 7.15 and the period as about 24 479 years. A further investigation was made by B. G. Marsden (1978). He took 81 positions, applied the perturbations by all nine planets, and determined the perihelion date as April 7.15 and the period as about 23 017 years. Marsden's orbit is given below.

T	ω	Ω (2000.0)	i	q	e
1892 Apr. 7.1517 (TT)	24.4974	242.4297	38.7002	1.026832	0.998731

ABSOLUTE MAGNITUDE: $H_{10} = 3.2$ (V1964)

FULL MOON: Feb. 12, Mar. 13, Apr. 12, May 11, Jun. 10, Jul. 10, Aug. 8, Sep. 6, Oct. 6, Nov. 4, Dec. 4, 1893 Jan. 2, Feb. 1, Mar. 2

SOURCES: L. Swift, E. E. Barnard, and H. C. F. Kreutz, *AN*, **129** (1892 Mar. 11), p. 119; L. Swift, E. E. Barnard, and G. M. Searle, *AJ*, **11** (1892 Mar. 15), p. 144; E. A. Lamp, *AN*, **129** (1892 Mar. 26), p. 151; F. P. Leavenworth, O. C. Wendell, E. E. Barnard, M. Updegraff, J. K. Rees, H. Jacoby, and G. M. Searle, *AJ*, **11** (1892 Mar. 31), pp. 157–9; E. A. Lamp, A. H. P. Charlois, A. Kammermann, G. A. P. Rayet, and J. Holetschek, *AN*, **129** (1892 Apr. 2), pp. 159–62; G. M. Searle, H. C. Wilson, and W. W. Campbell, *AJ*, **11** (1892 Apr. 16), pp. 175–6; N. de Konkoly, W. Luther, and F. Bidschof, *AN*, **129** (1892 Apr. 25), pp. 259–62; H. C. Russell, A. C. D. Crommelin, and T. Lewis, *MNRAS*, **52** (1892 May), pp. 512–16; A. Berberich, *AN*, **129** (1892 May 2), pp. 277–80; O. C. Wendell, F. P. Leavenworth, G. L. Jones, and W. H. Collins, *AJ*, **11** (1892 May 3), pp. 191–2; G. M. Searle and J. R. Hind, *AJ*, **12** (1892 May 24), pp. 13–15; W. Schur, *AN*, **129** (1892 May 31), p. 391; A. C. D. Crommelin, W. H. Finlay, and G. F. Parson, *MNRAS*, **52** (1892 Jun.), pp. 566–9, 572; E. B. Knobel and T. W. Backhouse, *The Observatory*, **15** (1892 Jun.), pp. 234, 263; E. von Gothard, *AN*, **129** (1892 Jun. 8), pp. 405–8; C. Wagner, *AN*, **130** (1892 Jun. 18), pp. 11–14; F. P. Leavenworth, J. H. Dennis, E. H. Gifford, W. H. Collins, and J. B. Coit, *AJ*, **12** (1892 Jun. 22), pp. 39–40; J. M. Thome and W. H. Finlay, *AN*, **130** (1892 Jun. 30), pp. 53–8; G. Le Cadet, *BA*, **9** (1892 Jul.), pp. 294–6; M. W. Whitney, *AJ*, **12** (1892 Jul. 2), p. 48; M. Updegraff, *AJ*, **12** (1892 Jul. 22), pp. 50, 54; G. Le Cadet, *AN*, **130** (1892 Aug. 13), pp. 185–8; F. G. Wentworth, *AJ*, **12** (1892 Aug. 23), p. 72; A. Berberich, *AN*, **130** (1892 Aug. 26), p. 215; B. von Engelhardt, *AN*, **131** (1892 Oct. 11), pp. 45–8; M. W. Whitney and O. C. Wendell, *AJ*, **12** (1892 Oct. 22), pp. 93–5; A. C. D. Crommelin, R. Bryant, and G. M. Lourison, *MNRAS*, **52** (1892, Supplementary Notice), pp. 605–7; A. C. D. Crommelin, *MNRAS*, **53** (1892 Nov.), pp. 49–51; W. Luther, *AN*, **131** (1892 Nov. 7), pp. 87–90, 110; T. W. Backhouse, *The Observatory*, **15** (1892 Dec.), p. 452; F. Porro, *AN*, **131** (1892 Dec. 3), pp. 191–4; T. Reed, *AJ*, **12** (1892 Dec. 5), pp. 134–5; G. Gruss, *AN*, **131** (1892 Dec. 8), pp. 207–10; G. M. Searle, *AJ*, **12** (1892 Dec. 9), p. 140; G. Le Cadet, *AN*, **131** (1893 Jan. 3), pp. 321–4; A. Kammermann, *AN*, **131** (1893 Jan. 12), p. 401; B. von Engelhardt, *AN*, **132** (1893 Jan. 17), p. 13; A. Kammermann, *AN*, **132** (1893 Jan. 23), p. 27; H. C. Wilson, *AJ*, **12** (1893 Jan. 31), pp. 183–4; *MNRAS*, **53** (1893 Feb.), p. 245; F. Renz and C. Wagner, *AN*, **132** (1893 Feb. 13), pp. 119–24, 133, 137; W. Schur, *AN*, **132** (1893 Apr. 15), pp. 229–32; A. Abetti, *AN*, **132** (1893 May 23), pp. 347–9; J. Holetschek, *AN*, **133** (1893 Jul. 19), pp. 89–92, 95; H. A. Kobold and O. C. Wendell, *AN*, **134** (1893 Dec. 21), pp. 169, 173–5, 179; O. Knopf, *AN*, **134** (1894 Jan. 6), pp. 203, 207–10; M. Chandrikoff, *AN*, **136** (1894 Aug. 2), pp. 67–72; W. H. Pickering, *Annals of the Astronomical Observatory of Harvard College*, **32** (1895), pp. 267–95; E. E. Barnard, *MNRAS*, **59** (1899 Mar.), pp. 355–6; E. E. Kühne, *Definitive Bahnbestimmung des Kometen 1892 I (Swift) für die Oskulationsepoche 1892 März 21.0*. Königsberg: R. Leupold (1913), pp. 1–80; E. E. Kühne, *AN*, **197** (1914 Feb. 13), pp. 139–42; V1964, p. 64; HA1970, p. 88; B. G. Marsden, *AJ*, **83** (1978 Jan.), pp. 66, 68; B. G. Marsden, *CCO*, 3rd ed. (1979), pp. 15, 44.

C/1892 F1   *Discovered:* 1892 March 18.86 ($\Delta = 2.36$ AU, $r = 2.08$ AU, Elong. $= 61°$)
(Denning)   *Last seen:* 1893 January 12.83 ($\Delta = 2.89$ AU, $r = 3.45$ AU, Elong. $= 116°$)
               Closest to the Earth: 1892 March 12 (2.3557 AU), 1892 November 10 (2.1287
1892 II = 1892b   AU)

Calculated path: CEP (Disc), CAS (Mar. 21), CEP (Mar. 22), CAS (Mar. 24), PER (Apr. 22), CAM (May 13), PER (May 17), AUR (Jun. 13), GEM (Aug. 19), ORI (Oct. 1), ERI (Dec. 1), LEP (Dec. 2), ERI (Dec. 7)

W. F. Denning (Bristol, England) was sweeping for comets in the region of Cepheus on the evening of 1892 March 18. The sky was very clear and he was using his 25.4-cm reflector (40×), which had a field of view of 65'. On March 18.86, Denning found a "small, faint nebulosity situated 3° E.N.E. of δ Cephei, and at once suspected the cometary nature of the object, as I had never picked it up before." The position was estimated at $\alpha = 22^h 42^m$, $\delta = +59°$, and during the next 3 hours and 15 minutes Denning noted the comet moved about 50' toward the east. He reobserved the comet on March 19, 21, 22, 23, 24, and 28. In a letter written to *The Observatory* on March 29, he described the comet as a "very inconspicuous object" with "a nucleus of about the 13th magnitude." He noted that the comet was quite close to NGC 7510 on the nights of March 22 and 23.

On March 19, H. A. Kobold (Strasbourg, France) described the comet as rather bright, round, and 2' across, with a condensation of magnitude 11, while C. F. Pechüle (Copenhagen, Denmark) said the condensation was magnitude 12. During the period of March 19–21, O. Knopf (Jena, Germany) said the condensation was magnitude 12. On the 20th, O. C. Wendell (Harvard College Observatory, Cambridge, Massachusetts, USA) observed with the 38-cm refractor and said the coma was 1' across, while the nucleus was "fairly well defined." That same night, W. Luther (Hamburg, Germany) described the comet as a delicate, round nebulosity, with a central condensation of magnitude 12, while R. Spitaler (Vienna, Austria) observed with a 69-cm refractor and said the coma was 2' across with a nucleus of magnitude 10. On the 21st, Spitaler said the coma was about 1.5' across, with a fainter outer portion 3' across. The diffuse nucleus was about 15" across. On March 21 and 22, E. E. Barnard (Lick Observatory, California, USA) observed with the 30-cm refractor and estimated the magnitude as 12.5. He noted the coma was 1' across, with a slight brightening near the middle. On the 22nd, Wendell said the nucleus was fairly well defined, while the comet was magnitude 12, while Spitaler reported the coma was 1' across and the diffuse nucleus was about 30" across. On the 22nd and 25th, Luther said the nucleus was magnitude 11.5. Spitaler said the comet was 1' in diameter, with a magnitude of 10 on the 24th. He also noted a diffuse nucleus. During the period of March 25–29, F. Schwab (Kremsmünster, Austria) said the comet always appeared as a faint, roundish nebulosity, with a bright central condensation and a nucleus that occasionally flashed into view. On the 30th, B. von Engelhardt (Dresden, Germany) observed with a 30-cm refractor and said the comet was very small, round, and rather bright, with a stellar nucleus. That same night, Luther gave the nuclear magnitude as 12.3. On March 31, Kobold said the comet was round and 1' across, with a central condensation of magnitude 12, while Luther gave the nuclear magnitude as 11.8. Spitaler also saw the comet on this date, and reported it was about 30" across, with a magnitude of 11.5 in faint moonlight.

The comet attained its most northerly declination of +61° on April 3. On April 2, Luther noted a distinct nucleus of magnitude 11.5. Luther gave the nuclear magnitude as 11.8 on the 6th and 7th. In bright moonlight on the 9th, Spitaler said the comet was about 30" across and about magnitude 11.5. On the 15th, Luther gave the nuclear magnitude as 12.0. G. A. P. Rayet (Bordeaux, France) observed with a 38-cm refractor on the 16th and said the comet was round and noted a nucleus of magnitude 15. Luther said the comet was distinct, with a nucleus of magnitude 11.6 on the 17th. On the 19th, Luther described the comet as rather brighter, with a round coma and a nucleus of magnitude 11.4. On the 20th, O. Knopf (Jena, Germany) observed with a 20-cm refractor and said the comet exhibited a central condensation. Luther gave the nuclear magnitude as 11.4 on the 25th. On April 27, Luther gave the nuclear magnitude as 11.6.

On May 2, F. Renz (Pulkovo Observatory, Russia) noted a diffuse condensation and said the coma barely stood out against the bright field of view. On May 21, Kobold saw the comet despite haze and clouds and described it as very faint. Knopf found the comet difficult to observe with a 20-cm refractor on May 24 and 29, but still precisely measured its position.

The comet became a more difficult object during June, as its elongation from the sun steadily decreased to 23° by June 19. Positions were already being measured by Rayet on the mornings of June 27, 28, and 30. R. Schorr (Hamburg, Germany) began searching for the comet in the weak morning twilight during July. He finally succeeded on July 28 and noted it appeared extremely faint.

Although positions continued to be measured during the next few months, there were few physical descriptions. On August 21, C. Wagner (Kremsmünster, Austria) described the comet as a faint, diffuse nebulosity. On September 30, Schorr said the comet was very faint, with a nucleus of magnitude 12. On October 20, Kobold said the comet was rather small, round, and slightly condensed. On December 16, Kobold described the comet as 1' across nebulosity with a slight condensation. On December 19, Kobold said the comet seemed slightly smaller than on the 16th. The comet was last seen on 1893 January 12.83, when Kobold saw it with a 46-cm refractor and measured the position as $\alpha = 4^h\ 15.4^m$, $\delta = -17°\ 27'$. He described the comet as extremely faint and about 30" across, with a slight brightening in the middle.

The comet's rather large perihelion distance caused some initial problems in pinning down the perihelion date. The first orbit was calculated by F. Bidschof. It was a parabolic orbit that was telegraphed to various publications on March 24 and gave the perihelion date as 1892 May 12.75. A few days later, Schorr took positions from March 19, 22, and 25, and determined a perihelion date of May 6.60. G. Lorentzen followed in mid-April with a calculated perihelion date of May 10.48, and Schorr revised his calculations early in May, with a perihelion date of May 11.68.

The first definitive orbit was calculated by L. Steiner (1898). He took positions spanning the entire period of visibility, applied perturbations by Mercury, Venus, Earth, Jupiter, and Saturn, and calculated a hyperbolic orbit with a perihelion date of May 11.72 and an eccentricity of 1.000345. A similar orbit was calculated by Z. Sekanina (1978). He also used positions spanning the entire period of visibility, but applied perturbations by all nine planets. Sekanina's orbit is given below.

T	ω	Ω (2000.0)	i	q	e
1892 May 11.7275 (TT)	129.3189	254.9390	89.6959	1.970726	1.000371

ABSOLUTE MAGNITUDE: $H_{10} = 5.4$ (V1964)

FULL MOON: Mar. 13, Apr. 12, May 11, Jun. 10, Jul. 10, Aug. 8, Sep. 6, Oct. 6, Nov. 4, Dec. 4, Jan. 2, Feb. 1

SOURCES: W. F. Denning, R. Spitaler, H. A. Kobold, O. Knopf, and F. Bidschof, *AN*, **129** (1892 Mar. 26), pp. 149–52; W. F. Denning and E. E. Barnard, *AJ*, **11** (1892 Mar. 31), p. 160; W. F. Denning, *The Observatory*, **15** (1892 Apr.), p. 190; W. Luther, C. F. Pechüle, O. Knopf, and R. Schorr, *AN*, **129** (1892 Apr. 2), pp. 163–5; G. Lorentzen, *AN*, **129** (1892 Apr. 16), p. 243; O. C. Wendell, *AJ*, **11** (1892 May 3), pp. 191–2; R. Schorr, *AN*, **129** (1892 May 6), p. 295; B. von Engelhardt, *AN*, **129** (1892 May 7), p. 311; F. Schwab, *AN*, **130** (1892 Jun. 18), pp. 11–14; R. Schorr, *AN*, **130** (1892 Aug. 13), p. 191; G. A. P. Rayet, *CR*, **115** (1892 Sep. 5), pp. 377–8; G. A. P. Rayet and R. Schorr, *AN*, **131** (1892 Oct. 17), pp. 53–6, 63; W. Luther, *AN*, **131** (1892 Nov. 7), pp. 87–90, 111; R. Spitaler, *AN*, **131** (1893 Jan. 10), pp. 385–8; C. Wagner, *AN*, **132** (1893 Feb. 18), pp. 135–7; F. Renz, *AN*, **133** (1893 Aug. 24); H. A. Kobold, *AN*, **134** (1893 Dec. 21), pp. 169, 173; O. C. Wendell, *AN*, **134** (1893 Dec. 21), pp. 175, 179; O. Knopf, *AN*, **134** (1894 Jan. 6), pp. 203, 209; L. Steiner, *AN*, **145** (1898 Jan. 27), pp. 247–54; V1964, p. 64; B. G. Marsden, Z. Sekanina, and E. Everhart, *AJ*, **83** (1978 Jan.), p. 66; B. G. Marsden, *CCO*, 3rd ed. (1979), pp. 15, 44.

7P/Pons– Winnecke

1892 IV = 1892c

Recovered: 1892 March 18.90 ($\Delta = 0.77$ AU, $r = 1.70$ AU, Elong. $= 147°$)
Last seen: 1892 October 21.21 ($\Delta = 0.88$ AU, $r = 1.77$ AU, Elong. $= 141°$)
Closest to the Earth: 1892 July 9 (0.1235 AU)
Calculated path: COM (Rec), CVn (Mar. 22), UMa (Apr. 16), LMi (Jun. 18), LYN (Jun. 26), CNC (Jun. 29), GEM (Jul. 4), MON (Jul. 10), ORI (Jul. 11), ERI (Jul. 18), FOR (Aug. 7), SCL (Sep. 26)

E. F. von Haerdtl continued his long personal investigation of this comet's orbit and published two predictions during 1891. The first was published in the 1891 January 12 issue of the *Astronomische Nachrichten* and predicted a perihelion date of 1892 July 1.29. The second was published in the 1891 October 20 issue of the same journal and gave the perihelion date as July 1.39. The first searches were conducted by R. Spitaler (Vienna, Austria) using a 69-cm refractor during February of 1891, but were unsuccessful. Following the publication of Haerdtl's second paper, Spitaler was able to recover the comet on 1892 March 18.90 at a position of $\alpha = 12^h\ 43.5^m$, $\delta = +30°\ 36'$. He described it as extremely faint, about 5" in diameter, with a stellar nucleus of magnitude 16.

The comet was found nearly three and a half months prior to passing closest to both the sun and Earth. Spitaler obtained a few observations during the remainder of March. On the 20th, he described it as very faint, with a pale, round coma about 10" across and a weak central condensation. He still noted the comet was very faint on the 25th and estimated the coma diameter as 15". He said the comet was noticeably brighter on the 31st, with a diameter of about 15".

Observations during April were primarily made at observatories with large telescopes. On April 17, W. Luther (Hamburg, Germany) said the comet was very faint. He reported it was at the limit of visibility on the 19th, with a nucleus slightly brighter than magnitude 13. On the 20th, G. Witt (Urania Observatory, Berlin, Germany) described the comet as an uncommonly faint nebulosity in the 30-cm refractor. Luther found the comet substantially brighter on the 27th, with a nucleus of magnitude 12, while R. Schorr (Hamburg, Germany) noted a distinct nucleus of magnitude 12 and a coma about 1.5' across. On April 29, H. C. Wilson (Goodsell Observatory, Minnesota, USA) said the comet was visible in the 13-cm finder. The 41-cm refractor revealed a coma 2' across, which contained a well-defined nucleus of magnitude 11. That same evening, Luther noted a distinct nucleus of magnitude 12.5.

On May 12, the comet attained its most northerly declination of +45°. On the 14th, Spitaler gave the magnitude as about 9.5 and said the coma was 3' across with a diffuse nucleus. Spitaler said the comet was quite bright on the 18th, with a bright coma about 30" across and a faint outer coma about 5' across. The nucleus was sharp. Schorr gave the nuclear magnitude as 11.3 on the 21st and 11.6 on the 22nd. On the 22nd and 25th, J. Holetschek (Vienna, Austria) observed the comet with the 4-cm finder of a 15-cm refractor and gave the magnitude as 9.5. On May 24, H. A. Kobold (Strasbourg, France) observed with a 46-cm refractor and described the comet as a round nebulosity, about 2' across, with a bright center, but no nucleus. That same night, Spitaler said the round coma was 3–4' across, with a condensation about 30" across and about magnitude 10. On the 25th, Spitaler said the comet was 6' across and magnitude 9, with a condensation about 30" across. On the 25th and 27th, K. Oertel (Munich, Germany) observed with the 27-cm refractor and said the coma was elongated, measuring 1' by 1.5' on the last date. On May 26, Wilson said the comet was easily seen in the 13-cm finder, while the 41-cm refractor revealed a coma at least 3' across, which contained a well-defined nucleus of magnitude 11. Holetschek gave the magnitude as 9 on the 27th, while Spitaler said the comet was easily seen in the finder of the 69-cm refractor. He added that the refractor revealed a nucleus of magnitude 12 and a coma 6' across.

On June 2, Oertel described it as very faint. On the 6th, A. Kammermann (Geneva, Switzerland) observed with the 25-cm refractor and said the comet was a nebulous spot, without a central condensation. On June 12, Holetschek gave the magnitude as 8. On the 13th, O. Knopf (Jena, Germany) observed with a 20-cm refractor and said the comet appeared rather bright. Holetschek gave the magnitude as 7.0 on the 18th and 6.5–7 on the 21st. Schorr noted a distinct nucleus of magnitude 11. Oertel said the comet was rather bright on the 19th and very bright with a nucleus of magnitude 10 on the 24th. On June 27, A. C. D. Crommelin (Royal Observatory, Greenwich, England) observed with the 17-cm refractor and said the comet was very bright and "was readily visible in spite of the bright twilight." That same night, Oertel noted a tail toward PA 45°, despite moonlight. On the 28th, Kobold observed with the 46-cm refractor and said the comet was 1' across and round, with a central condensation. He also noted the beginning of a broad tail extending northward. During the last half of June, E. Millosevich (Rome, Italy) observed with a 25-cm refractor and continually noted a nucleus of magnitude 10 located in an asymmetrical coma. The last observation prior to conjunction with the sun was obtained by E. Frisby (US Naval Observatory, Washington, DC, USA) on June 30, while using the 23-cm refractor. Interestingly, Crommelin reported in the 1892 July issue of *The Observatory* that a telegram was received from Moss (Muravera) on June 15 that stated, "Bright comet near µ Ursae Majoris." Crommelin commented, "This was probably Winnecke's Comet, which was within a degree of this star on the night of June 14."

The comet passed 3° from the sun on July 7. Following conjunction, John Tebbutt (Windsor, New South Wales, Australia) found the comet with the 11-cm refractor on July 17.82. Tebbutt continued observing with a 20-cm telescope through September 27. On this last date, he said the comet was centrally located on a 10th-magnitude star, so that it was only visible as a faint, nebulous haze surrounding the star. The comet attained its most southerly declination of –32° on September 15.

The comet was last seen on October 21.21, when Wilson gave the position as $\alpha = 1^h \, 01.5^m$, $\delta = -26°\, 25'$. He said the 41-cm refractor revealed the comet as very faint and round, with a diameter of 1'. There was also a central condensation.

Later calculations using multiple apparitions and planetary perturbations were published by B. G. Marsden and Z. Sekanina (1972). These revealed a perihelion date of July 1.40 and a period of 5.82 years. Marsden, Sekanina, and D. K. Yeomans (1973) determined nongravitational terms of $A_1 = -0.01$ and $A_2 = -0.0021$. The Marsden and Sekanina (1972) orbit is given below.

T	ω	Ω (2000.0)	i	q	e
1892 Jul. 1.4039 (TT)	172.1654	105.5587	14.5201	0.886557	0.725997

ABSOLUTE MAGNITUDE: $H_{10} = 10.6$ (V1964)

FULL MOON: Mar. 13, Apr. 12, May 11, Jun. 10, Jul. 10, Aug. 8, Sep. 6, Oct. 6, Nov. 4

SOURCES: E. F. von Haerdtl, *AN*, **126** (1891 Jan. 12), pp. 171–4; E. F. von Haerdtl, *AN*, **128** (1891 Oct. 20), pp. 241–8; R. Spitaler, *AN*, **129** (1892 Mar. 26), p. 149; R. Spitaler, *AJ*, **11** (1892 Mar. 31), p. 160; E. F. von Haerdtl, *AN*, **129** (1892 Apr. 8), pp. 169–76; G. Witt, *AN*, **129** (1892 May 6), p. 289; E. Frisby, *AJ*, **12** (1892 Jun. 13), p. 30; A. C. D. Crommelin, *The Observatory*, **15** (1892 Jul.), p. 302; E. Frisby, *AJ*, **12** (1892 Jul. 22), p. 54; E. Millosevich, *AN*, **130** (1892 Sep. 3), p. 235; E. F. von Haerdtl, *AN*, **130** (1892 Sep. 8), pp. 241–6; A. C. D. Crommelin, *MNRAS*, **52** (1892, Supplementary Notice), pp. 605–6; W. Luther, *AN*, **131** (1892 Nov. 7), pp. 89, 111; J. Tebbutt, *MNRAS*, **53** (1892 Dec.), pp. 70–3; J. Holetschek, *AN*, **131** (1892 Dec. 16), pp. 277–80; R. Spitaler, *AN*, **131** (1893 Jan. 10), pp. 385–9; J. Tebbutt, *AN*, **131** (1893 Jan. 12), p. 405; A. Kammermann, *AN*, **132** (1893 Jan. 23), p. 27; K. Oertel, *AN*, **132** (1893 Jan. 25), pp. 57–60; H. C. Wilson, *AJ*, **12** (1893 Jan. 31), pp. 183–4; W. E. Plummer, *MNRAS*, **53** (1893 Feb.), p. 265; H. A. Kobold, *AN*, **134** (1893 Dec. 21), pp. 171–4; O. Knopf, *AN*, **134** (1894 Jan. 6), pp. 203, 209; R. Schorr, *AN*, **134** (1894 Mar. 1), pp. 373, 397; V1964, p. 64; B. G. Marsden and Z. Sekanina, *QJRAS*, **13** (1972), pp. 428–9; B. G. Marsden, Z. Sekanina, and D. K. Yeomans, *AJ*, **78** (1973 Mar.), pp. 214, 216.

C/1892 Q1   *Discovered:* 1892 August 28.80 ($\Delta = 2.34$ AU, $r = 2.15$ AU, Elong. $= 67°$)
(Brooks)   *Last seen:* 1893 July 19.2 ($\Delta = 2.27$ AU, $r = 3.11$ AU, Elong. $= 140°$)
  Closest to the Earth: 1892 December 12 (0.8753 AU)
1892 VI = 1892d   *Calculated path:* AUR (Disc), GEM (Sep. 11), CNC (Oct. 9), LEO (Nov. 1), SEX (Nov. 7), CRT (Nov. 22), CRV (Dec. 5), HYA (Dec. 10), CEN (Dec. 16), LUP (Jan. 7), NOR (Jan. 19), SCO (Jan. 27), CrA (Feb. 22), SGR (Mar. 14), OPH (Jun. 9)

W. R. Brooks (Geneva, New York, USA) discovered this comet on 1892 August 28.80 at a position of $\alpha = 5^h 59.0^m$, $\delta = +31° 52'$. He established the rate and direction of motion when he reobserved the comet on August 30.50. It was confirmed by G. Bigourdan (Paris, France) on August 31. He described the comet as a round nebulosity, about 30″ across, whose brightness equaled a star of magnitude 12.5–13.

The comet was found exactly four months prior to its passing perihelion and it was well observed during the days following its discovery. On September 1, H. A. Kobold (Strasbourg, France) observed the comet with a 46-cm refractor and described it as rather bright and 1′ across, with a condensation of magnitude 11, while E. A. Lamp (Kiel, Germany) observed with a refractor and said the comet could only be seen when the field was dark. That same night, G. Witt (Urania Observatory, Berlin, Germany) said the comet was reasonably bright in the 30-cm refractor and was comfortably seen in the finder. Several observations were made on the 2nd. O. C. Wendell (Harvard College Observatory, Cambridge, Massachusetts, USA) observed with the 38-cm refractor and said the comet was faint and about magnitude 12, with a diameter of about 0.5'. The nucleus was "pretty well defined" and there was no tail. H. C. Wilson (Goodsell Observatory, Minnesota, USA) said the comet was barely visible in the 13-cm finder, while it was easily seen in the 41-cm refractor and exhibited a coma 1' across, with a "pretty well defined" nucleus. A tail extended 3'. G. Le Cadet (Lyon, France) saw the comet with a 16.0-cm refractor and said the comet was very faint. F. W. Ristenpart (Karlsruhe, Germany) said the comet appeared as a faint nebulous mass about 15" across, without a nucleus. F. Renz (Pulkovo Observatory, Russia) saw the comet with the 38-cm refractor and described it as a rather faint nebulous mass, with an irregular border and a condensation. J. Holetschek (Vienna, Austria) gave the total magnitude as 10.5 on the 3rd, while Wendell said the nucleus was fairly well defined. On the 5th, Wilson said the comet appeared fainter in the refractor because of moonlight. The nucleus was about magnitude 11.

On November 15, F. Cohn (Königsberg, now Kaliningrad, Russia) observed in moonlight and said the comet was very faint, without a nucleus. On the 16th, Renz described the comet as a large, faint nebulous mass, with a condensation. On the 18th, B. von Engelhardt (Dresden, Germany) said the comet was rather bright, 1' in diameter, and exhibited an oval condensation, while Renz noted a diffuse nucleus. J. Möller (Bothkamp, Germany) saw a stellar nucleus of magnitude 11 on the 19th as well as a fan-shaped tail 3' long. Holetschek gave the magnitude as 9.5 on the 20th. Renz described the comet as a rather bright nebulous mass on the 24th and noted a central condensation and a nucleus of magnitude 12.0. On September 25, A. C. D. Crommelin (Royal Observatory, Greenwich, England) said the comet was very faint in the 17-cm refractor (55×). R. Schorr (Hamburg, Germany) saw the comet on the 26th and noted a diffuse nucleus of magnitude 10.0 situated within a coma about 45" across. He also saw the beginnings of a faint tail. That same night, Holetschek gave the magnitude as 9. G. Gruss (Prague, Czech Republic) observed with a 20-cm refractor and said the coma was elliptical and over 1' across on the 27th, with a faint tail. On the 29th, Wilson said the comet was brighter than when last observed. He added that there was a "strong condensation of nebulosity about the nucleus." On September 30, F. Schwab (Kremsmünster, Austria) gave the magnitude as 10, while Schorr gave the nuclear magnitude as 9.6 and noted a distinct tail.

On October 2, Holetschek gave the magnitude as 8.5, while Renz described the comet as a large bright nebulosity with a condensation. On the 5th, Cohn noted a rather distinct nucleus. Schorr said the comet was faint because of moonlight on the 8th, while Renz said the comet was very bright on the 19th. On the 21st, Wilson said the comet was quite bright in the 41-cm refractor, with a non-stellar nucleus and a tail 9' long, while Schwab gave the magnitude as 9. On the 24th, Crommelin described the comet as "large and round, with decided central condensation." Schorr gave the nuclear magnitude as 9.2 on the 26th, and reported a coma about 1.5' across. On the 27th, Ristenpart observed the comet with a 15-cm refractor and said the comet appeared diffuse, without a sharp nucleus. That same night, Holetschek gave the magnitude as 7.5. On October 28 Ristenpart said the comet appeared very faint in twilight, while Gruss said the coma was 3' across and exhibited a distinct nucleus of magnitude 10. Gruss added that the tail extended about 10' toward PA 90°.

On November 3, Renz said the comet was very bright, with a large coma, and a nucleus of magnitude 10.5. Crommelin said the comet was "very faint" because of moonlight on the 4th. On the 7th, R. Bryant (Royal Observatory, Greenwich) reported the comet was "fairly large" with a nucleus of magnitude 7.5 in the 17-cm refractor (55×). T. W. Backhouse (Sunderland, England) said the comet was plainly visible to the naked eye on the 17th and exhibited a tail 2° 35' long. Holetschek gave the magnitude as 6.3 on the 18th. On the 21st, Ristenpart said the comet was bright and about 1.5' across. There was also a nucleus of magnitude 7 which measured about 15" across. A cometseeker revealed a tail extending westward. Holetschek gave the magnitude as 5.5 on the 23rd. Gruss said the nucleus was magnitude 8 and the tail was about 20' long on the 24th. On the 25th, Cohn estimated the comet's magnitude as 6. Schorr saw the comet in twilight on the 26th and gave the nuclear magnitude as 7.5 on the 27th. Gruss gave the nuclear magnitude as 7–8 on the 26th and said the tail extended toward PA 60°. F. S. Archenhold (Halensee, Germany) obtained three 20-minute exposures on the 27th, which revealed a narrow tail 5° long. He said the tail was only 0.5° long in the cometseeker. Also on the 27th, Möller said the tail extended toward PA 280°, while Schorr gave the nuclear magnitude as 7.5. J. Tebbutt (Windsor, New South Wales, Australia) observed the comet with an 11-cm refractor on November 28 and 29.

The comet passed closest to Earth around mid-December and passed perihelion near the end of the month. On December 1 Ristenpart said the coma was 3' across. Schwab saw a stellar nucleus of magnitude 9 on the 3rd. Holetschek gave the magnitude as 5.7 on the 18th and obtained his final observation on December 20. Tebbutt saw the comet on several occasions during the period of December 8–29 and said it had a "fairly bright condensation," which grew fainter and smaller. A faint tail was occasionally seen in the 20-cm refractor.

Physical descriptions were not numerous during 1893. The comet attained a minimum elongation of 57° on January 14. W. H. Finlay (Royal Observatory, Cape of Good Hope, South Africa) said the comet was very diffuse on January 17 and 19. The comet attained its most southerly declination of −42° on January 24. On February 20, Finlay indicated the comet's brightness was similar to a 10th-magnitude star. Finlay said the comet was faint on April 15. Tebbutt said the comet was barely visible by averted vision in the 6-cm finder of his 20-cm refractor on April 17, while he described the comet as "a faint condensed point, with scarcely any surrounding coma" on the 20th. Finlay found the comet faint and difficult on April 24. Tebbutt saw the comet with the 20-cm refractor after the moon had left the sky on May 8, while he described the comet as the "faintest possible specimen" against one of the bright patches of the Milky Way on the 19th. L. Fabry (Marseille, France) observed with a 26-cm refractor on 1893 June 8, 13, 20, and 22. He described the comet as a very faint nebulosity, about 30" in diameter. F. Bidschof (Vienna, Austria) simply noted the comet was very faint on June 13. During the period of June 15–19, Tebbutt said the comet was still quite distinct and attributed this to its having moved away from the Milky Way.

The comet was measured for the final time on July 13.20, when E. O. Lovett (Leander McCormick Observatory, Virginia, USA) gave the position as $\alpha = 17^h 04.8^m$, $\delta = -20° 40'$. The comet was last detected on July 19.2, when H. C. Wilson (Goodsell Observatory, Minnesota, USA) found the comet was too faint and diffuse for measurement with the 41-cm refractor. Tebbutt tried to find the comet around the middle of July, but was unsuccessful.

The spectrum was observed by W. W. Campbell (Lick Observatory, California, USA) on November 10. Although he said a continuous spectrum was present, the three cometary bands were also seen.

The first orbit was calculated by A. Berberich using positions from September 1, 3, and 4. The result was a perihelion date of 1892 December 20.19. The next orbit was by Ristenpart and used positions from September 1, 6, and 14. He gave the perihelion date as December 28.75. Additional orbits were calculated by G. M. Searle, G. A. Hill, Ristenpart, and H. Oppenheim, which narrowed down the perihelion date to December 28.6.

G. Van Biesbroeck began work on the orbit of this comet and, following his death, C. D. Vesely and B. G. Marsden (1976) completed the orbit. Ultimately, 123 positions were used covering the period of 1892 September 1 to 1893 July 11. In addition, perturbations by all nine planets were considered. The result was a hyperbolic orbit with a perihelion date of 1892 December 28.59 and an eccentricity of 1.0004410. This orbit is given below. The original and future values of the semi-major axis were also determined, but, although they both proved to be hyperbolic, the authors wrote, "we do not consider this to be a proven case of an interstellar comet but suggest that it was rather probably a 'new' comet in the Oort sense."

T	ω	Ω (2000.0)	i	q	e
1892 Dec. 28.5864 (TT)	252.6530	266.0346	24.8012	0.975989	1.000441

ABSOLUTE MAGNITUDE: $H_{10} = 5.3$ (V1964)

FULL MOON: Aug. 8, Sep. 6, Oct. 6, Nov. 4, Dec. 4, 1893 Jan. 2, Feb. 1, Mar. 2, Apr. 1, Apr. 30, May 30, Jun. 29, Jul. 28

SOURCES: W. R. Brooks, E. A. Lamp, and G. Witt, AN, **130** (1892 Sep. 3), p. 239; G. Bigourdan and G. Le Cadet, CR, **115** (1892 Sep. 5), pp. 384–6; A. Berberich, AN, **130** (1892 Sep. 9), p. 343; W. R. Brooks, H. C. Wilson, O. C. Wendell, E. E. Barnard, and A. Berberich, AJ, **12** (1892 Sep. 17), p. 80; F. W. Ristenpart, AN, **130** (1892 Sep. 17),

p. 373; F. W. Ristenpart, *AN*, **130** (1892 Sep. 26), p. 421; G. M. Searle and H. C. Wilson, *AJ*, **12** (1892 Oct. 4), p. 86; R. Schorr, *AN*, **131** (1892 Oct. 6), p. 29; B. von Engelhardt, *AN*, **131** (1892 Oct. 11), pp. 45–8; R. Schorr, *AN*, **131** (1892 Oct. 17), p. 63; O. C. Wendell, *AJ*, **12** (1892 Oct. 22), pp. 94–5; A. C. D. Crommelin and R. Bryant, *MNRAS*, **53** (1892 Nov.), pp. 48–51; F. W. Ristenpart, *AN*, **131** (1892 Nov. 7), p. 115; G. A. Hill, *AJ*, **12** (1892 Nov. 14), supplement; G. A. Hill, *AJ*, **12** (1892 Nov. 18), p. 119; J. B. Coit, *AJ*, **12** (1892 Dec.), p. 142; T. W. Backhouse, *The Observatory*, **15** (1892 Dec.), p. 452; H. Oppenheim and F. W. Ristenpart, *AN*, **131** (1892 Dec. 1), pp. 175–80; R. Schorr, *AN*, **131** (1892 Dec. 3), p. 197; W. W. Campbell and F. S. Archenhold, *AN*, **131** (1892 Dec. 8), pp. 211, 215; G. M. Searle and J. B. Coit, *AJ*, **12** (1892 Dec. 9), p. 140, 142; G. Gruss, *AN*, **131** (1893 Jan. 10), p. 389; H. C. Wilson, *AJ*, **12** (1893 Jan. 31), pp. 183–4; J. Möller, *AN*, **132** (1893 Feb. 6), p. 91; F. Schwab, *AN*, **132** (1893 Feb. 18), pp. 135–7; W. E. Plummer, *MNRAS*, **53** (1893 Apr.), p. 381; W. H. Finlay, *MNRAS*, **53** (1893 Jun.), pp. 488–9; L. Fabry, *BA*, **10** (1893 Jul.), p. 281; F. Cohn, *AN*, **133** (1893 Jul. 8), pp. 65–7; J. Holetschek, *AN*, **133** (1893 Jul. 19), pp. 91–5; J. Tebbutt, *AN*, **133** (1893 Aug. 3), pp. 125–32; W. H. Finlay and J. Tebbutt, *AN*, **133** (1893 Aug. 15), pp. 193–8; F. Renz, *AN*, **133** (1893 Aug. 24), pp. 247, 252; J. Tebbutt, *AN*, **133** (1893 Sep. 19), p. 347; E. O. Lovett, *AJ*, **13** (1893 Sep. 20), pp. 133–4; H. C. Wilson, *AJ*, **13** (1893 Oct. 23), p. 149; F. W. Ristenpart, *AN*, **134** (1893 Oct. 27), pp. 17–18, 23; H. A. Kobold, *AN*, **134** (1893 Dec. 21), pp. 171–2, 174; O. C. Wendell, *AN*, **134** (1893 Dec. 21), pp. 175–6, 179; O. Knopf, *AN*, **134** (1894 Jan. 6), pp. 203–4, 210; R. Schorr, *AN*, **134** (1894 Mar. 1), pp. 373, 397; F. Bidschof, *AN*, **149** (1899 Jun. 15), p. 405; V1964, p. 64; G. Van Biesbroeck, C. D. Vesely, and B. G. Marsden, *AN*, **81** (1976 Feb.), pp. 125–6.

D/1892 T1 *Discovered:* 1892 October 13.15 ($\Delta = 1.11$ AU, $r = 1.57$ AU, Elong. $= 96°$)
(Barnard 3) *Last seen:* 1892 December 8.80 ($\Delta = 1.28$ AU, $r = 1.43$ AU, Elong. $= 76°$)
  Closest to the Earth: 1892 October 7 (1.1133 AU)
1892 V = 1892e *Calculated path:* AQL (Disc), DEL (Nov. 1), EQU (Nov. 12), AQR (Nov. 13)

This comet marked the beginning of a new era in cometary astronomy, as it was the first to be discovered by photography. Shortly after darkness fell on 1892 October 13, E. E. Barnard (Lick Observatory, California, USA) aimed the 15-cm Willard lens at the Milky Way west of α Aquilae and exposed a photographic plate for 4 hours and 20 minutes, which covered the period from October 13.15 to October 13.33. The lens had been guided to counter the effect of Earth's rotation, which resulted in the stars appearing as sharp dots on the plate. He immediately developed the plate and quickly noticed a "distinct hazy trail near the middle of it, and about a quarter of a degree long." The position was roughly measured as $\alpha = 19^h 32^m$, $\delta = +12° 50'$. Unfortunately, that area of the sky was then too near the horizon to make a search worthwhile. As soon as it got dark the next evening, Barnard looked for the object with the 30-cm refractor telescope and found a very faint, diffuse object about a degree southeast of the position on the photograph. Barnard estimated the magnitude as about 13, and noted the comet was round, with little central condensation.

The comet had passed closest to Earth about a week before discovery, but was still heading toward perihelion. J. Palisa (Vienna, Austria) described it as faint, with a nucleus on October 16. G. Bigourdan (Paris, France) said the diffuse, round comet was very faint on the 17th, with a brightness of 13.3–13.4. He noted that the coma was 40–50″ across, with a central condensation. R. Schorr (Hamburg, Germany) said the comet seemed to contain several nuclear points on the 17th and 18th. C. F. Pechüle (Copenhagen, Denmark) described the comet as small and faint on the 18th. F. Sy (Alger, now al-Jazâ'ir, Algeria) described the comet as very faint and round on the 19th, with a diameter of about 20″. He noted a slight condensation. F. Renz (Pulkovo Observatory, Russia) observed the comet with the 38-cm refractor and described it as a very faint nebulosity, with a suggestion of a condensation. The comet was well observed on October 20, with observations reported by H. A. Kobold (Strasbourg, France), O. C. Wendell (Harvard College Observatory, Cambridge, Massachusetts, USA), and Renz. Kobold said it was round, about 1′ across, with a faint condensation. Wendell observed with the 38-cm refractor and said the comet was faint, with a rather stellar nucleus. Renz described the comet as a small, faint nebulosity, about 30″ across, that seemed slightly brighter than on the previous night. H. C. Wilson (Goodsell Observatory, Minnesota, USA) saw the comet with a 41-cm refractor on the 21st and described it as faint and round, with a diameter of 1′. He added that it was centrally condensed and contained a nucleus of magnitude 12. Wilson added that the comet was barely visible in the 13-cm finder. Wilson was no longer able to find the comet thereafter. Palisa said the comet was very faint on the 27th.

Palisa said the comet was exceptionally faint on November 7, while Barnard noted the comet appeared excessively faint and difficult to see in the 30-cm refractor on the 8th. Barnard added, "It is scarcely probable this object can be followed very much longer." Palisa said the comet was exceptionally difficult to see on the 9th and 13th. On November 22, Barnard noted the comet was "excessively faint and diluted," as seen in the 30-cm refractor. He added, "Further observations will be impossible."

The comet was last detected on December 8.80, when S. Javelle (Nice, France) located it with the 76-cm refractor. He gave the position as $\alpha = 22^h\ 14.3^m$, $\delta = -3°\ 42'$.

The first parabolic orbit was calculated by W. W. Campbell. He used positions obtained on October 16, 17, and 18, and determined the perihelion date as 1892 August 26.64. He said the orbit was "naturally uncertain." During the next few weeks calculations were published by R. Schorr, G. E. Whitaker, and L. Schulhof which revealed perihelion dates in the range of November 21–28. Campbell then published a revised orbit using positions spanning the period of October 14–26, which gave the perihelion date as December 3.10. He suggested that this fact, "taken in connection with the direct motion, and the fairly small inclination, points strongly to an elliptic orbit."

The first elliptical orbit was calculated by C. N. A. Krueger. Using positions spanning the period of October 16–25, he determined the perihelion date as December 7.75 and the period as 10.38 years. A few days later, Schulhof calculated his first elliptical orbit and found a perihelion date of December 9.57 and a period of 6.14 years. He pointed out the similarity between the orbit of this comet and the orbit of periodic comet Wolf of 1891. He suggested that since comet Wolf had passed within Jupiter's sphere of influence in 1875, this comet may have split off from comet Wolf at that time. About a week later he added that it was also possible that the division could have occurred sometime prior to 1875. Additional elliptical orbits were calculated by Krueger, J. R. Hind (1895), J. G. Porter (1894, 1895), and J. Coniel (1895). Each astronomer found a perihelion date on December 11, but while Krueger and Porter found a period between 6.2 and 6.3 years, Hind gave it as 6.63 years, and Coniel gave orbits with periods of 6.23, 6.52, and 6.84 years.

The comet has never been recovered, despite predictions that it would return in 1899, 1905, and later. The latest investigation into the orbit was published by D. K. Yeomans (1975). He took 40 positions spanning the entire period of visibility and determined the orbit given below. He gave the period as 6.52 years and indicated this had a probable error of about 2 weeks. Interestingly, Yeomans was the first to investigate Schulhof's suggestion that this comet was related to periodic comet Wolf. He concluded that the two comets were not related.

T	ω	Ω (2000.0)	i	q	e
1892 Dec. 11.1746 (TT)	169.9945	208.0361	31.2608	1.432189	0.589631

ABSOLUTE MAGNITUDE: $H_{10} = 9.8$ (V1964)

FULL MOON: Oct. 6, Nov. 4, Dec. 4, Jan. 2

SOURCES: E. E. Barnard, *AN*, **131** (1892 Oct. 17), p. 63; E. E. Barnard and W. W. Campbell, *AJ*, **12** (1892 Oct. 22), p. 95; G. Bigourdan and L. Schulhof, *CR*, **115** (1892 Oct. 24), pp. 585–6; C. F. Pechüle and R. Schorr, *AN*, **131** (1892 Oct. 26), pp. 77–80; F. Sy and L. Schulhof, *CR*, **115** (1892 Oct. 31), pp. 643–6; E. E. Barnard, *MNRAS*, **53** (1892 Nov.), p. 36; E. E. Barnard, *AJ*, **12** (1892 Nov. 4), p. 102; L. Schulhof and C. N. A. Krueger, *AN*, **131** (1892 Nov. 7), pp. 115–18; W. W. Campbell, E. E. Barnard, G. E. Whitaker, and C. N. A. Krueger, *AJ*, **12** (1892 Nov. 14), pp. 111–12, supplement; F. Renz and L. Schulhof, *AN*, **131** (1892 Nov. 14), pp. 129–32; C. N. A. Krueger, E. E. Barnard, and G. E. Whitaker, *AJ*, **12** (1892 Nov. 18), pp. 118, 120; C. N. A. Krueger, *AN*, **131** (1892 Nov. 19), pp. 149–50; E. E. Barnard, *AJ*, **12** (1893 Jan. 2), pp. 157–8; H. C. Wilson, *AJ*, **12** (1893 Jan. 31), pp. 183–4; W. E. Plummer, *MNRAS*, **53** (1893 Apr.), p. 381; J. Palisa, *AN*, **133** (1893 Aug. 16), pp. 209–10, 214; F. Ristenpart, *AN*, **134** (1893 Oct. 27), p. 24; H. A. Kobold, *AN*, **134** (1893 Dec. 21), pp. 171–2, 174–5; O. C. Wendell, *AN*, **134** (1893 Dec. 21), pp. 175–6, 179; J. G. Porter, *AJ*, **13** (1894 Jan. 20), pp. 183–7; S. Javelle, *BA*, **12** (1895 Jan.), pp. 28–9; J. R. Hind, *AN*, **137** (1895 Jan. 15), pp. 109–10; J. Coniel, *BA*, **12** (1895 Jun.), pp. 245–61; J. G. Porter, *AJ*, **15** (1895 Jun. 19), p. 86; E. E. Barnard, *MNRAS*, **59** (1899 Mar.), p. 360; V1964, p. 64; D. K. Yeomans, *PASP*, **87** (1975 Aug.), pp. 635–7.

17P/1892 V1
(Holmes)

1892 III = 1892f

Discovered: 1892 November 6.98 (Δ = 1.51 AU, r = 2.39 AU, Elong. = 144°)
Last seen: 1893 April 6.9 (Δ = 3.58 AU, r = 2.96 AU, Elong. = 45°)
Closest to the Earth: 1892 October 7 (1.4081 AU)
Calculated path: AND (Disc), TRI (Jan. 19), PER (Mar. 5)

E. Holmes (London, England) was a regular observer of the Andromeda galaxy (M31), so he knew the region very well. On the evening of 1892 November 6, with skies that were not very favorable, he finished making a few observations of Jupiter and some double stars with his 32-cm reflector, and then decided to take a quick look at the faint companions of μ Andromedae and nearby M31 before quitting for the night. Upon turning the reflector toward that region, he saw what he thought was M31 enter the field of the finder, but when he looked through the eyepiece he saw something different. Holmes said he "called out involuntarily, 'What is the matter? There is something strange here.' My wife heard me and thought something had happened to the instrument and came to see." The object in the field of Holmes' telescope was a comet with a coma about 5′ across and with a bright nucleus. The date was then November 6.98. Holmes was able to determine a rough position of $\alpha = 0^h$ 46.8^m, δ = +38° 32′ on November 7.03, before clouds moved in. He immediately wrote to E. W. Maunder (Royal Observatory, Greenwich, England), W. H. Maw (England), and Kidd (Bramley, England). Kidd immediately expressed some skepticism about Holmes' find because of its nearness to M31; however, on November 7.75, Kidd and Bartlett (Bramley, England) spotted the comet with the naked eye. The comet was independently discovered by T. D. Anderson (Edinburgh, Scotland) on November 8.9 and by J. E. Davidson (Mackay, Queensland, Australia) on November 9.5.

The comet had passed perihelion nearly five months before discovery and had passed closest to Earth just a month before. As astronomers would later realize, it was discovered during an apparent outburst in brightness. Nearly every astronomer reported the comet was visible to the naked eye through the first half of November, but few made actual total magnitude estimates. E. E. Barnard (Lick Observatory, California, USA) said the comet "was easily visible to the naked eye, as a small hazy star, and almost exactly as bright as the brightest part" of the galaxy M31 on the 9th. On the 11th, Barnard indicated the comet was slightly fainter than magnitude 4.5. On the 11th and 12th, G. M. Searle (Catholic University, Washington, DC, USA) said the comet was plainly visible to the naked eye and was nearly as bright as M31. On the 13th, G. Bigourdan (Paris, France) said the comet was as brilliant as M31. On the 14th, J. B. Coit (Boston University, Massachusetts, USA) indicated the comet was about magnitude 5.4. Most observers reported a nucleus was visible, but the only magnitude estimates were 7.7 by O. C. Wendell (Harvard College Observatory, Massachusetts, USA) on the 13th and 7.5 by R. Schorr (Hamburg, Germany) on the 13th and 14th. Bigourdan reported the nucleus appeared diffuse and about 10″ across on the 9th, while F. W. Ristenpart (Karlsruhe, Germany) indicated it was 15" across on the 15th. On the 12th, J. K. Rees (Columbia College Observatory, New York, USA) observed with a 33-cm refractor and said the nucleus seemed "spread out toward northwest." He added, "There were indications of divisions in the nucleus." J. Möller (Bothkamp, Germany) said the nucleus appeared oblong on the 13th and Schorr noted it was elongated toward PA 122.5° on the 13th and 14th. L. Boss (Dudley Observatory, New York, USA) reported the nucleus was elongated toward PA 110° on the 14th. On the 15th, G. Le Cadet (Lyon, France) said the 32-cm refractor revealed the nucleus was elongated toward PA 120°. Some observers were reporting a fan-like jet coming from the nucleus. F. Hayn (Leipzig, Germany) said a 30-cm refractor revealed it extending toward PA 120° on the 9th. Searle noted two jets on the 11th and 12th. The diameter of the coma expanded during this period. On the 9th, Hayn, Barnard, Bigourdan, W. F. Denning (Bristol, England), and A. C. D. Crommelin and T. Lewis (Royal Observatory, Greenwich, England) all gave coma diameters between 4.5' and 5.7' across. Somewhat inconsistent estimates were given during the next few days, with Barnard giving the diameter as about 12' on the 10th, Wendell giving it as 7' and E. Millosevich (Rome, Italy) giving it as 6.4' on the 11th, Searle indicating a value of 7' on the 11th and 12th, Rees giving it as 8' and Hayn giving it as 7' on the 12th, Bigourdan and Möller giving values of 8' and 10', respectively, on the 13th, Schorr giving it as 9–10' on the 13th and 14th, Boss giving it as 9' and F. Schwab (Kremsmünster, Austria) giving it as 7' on the 14th, and Ristenpart and O. Knopf (Jena, Germany) giving it as 8' and 9', respectively, on the 15th. The appearance of the coma changed from round and symmetrical to slightly elliptical during this period. T. A. Skelton (Southampton, England) measured it as 9' by 10' on the 14th, while Le Cadet said the long axis was 10' long and directed toward PA 150° on the 15th. The edges of the coma also showed changes. Barnard said the coma was a "well-defined disc" on the 10th and "sharply defined like a planetary nebula" on the 11th. Searle said the edges were especially sharply defined on the edge opposite the jets on the 11th and 12th. On the 13th, Möller said the western edge was sharp, while the eastern edge was diffuse, "suggesting a possible tail." Le Cadet observed with a 32-cm refractor on the 15th and said the northern edge was sharply defined, while the southern edge was diffuse. Bigourdan reported an elliptical region inside the coma on the 9th. He described it as vague and said it measured 1.5–2' by 30", with the long axis directed toward PA 127.1°. On the 13th, Bigourdan said this region measured 2' by 30", with the long axis directed toward PA 116.8°. The only reports of a tail during this period came from Barnard. He noted a short, faint, hazy tail on the 10th. On the 11th, he obtained a photograph with the 15-cm Willard lens and said it revealed "an irregular nebulous appendage about a degree to the south-east of the comet and attached to it by a hazy connection."

The comet faded throughout the second half of November, but it was still a naked-eye object until the last days of this period. Although no actual total magnitude estimates were made, a couple of astronomers made observations that give clues to the comet's brightness. G. Gruss (Prague, Czech Republic) said it was very easy to see in the 5-cm finder on the 23rd and he said it seemed brighter on the 24th, while M. Updegraff (State University of Missouri, USA) said it was barely visible to the naked eye after the moon had set on the 26th. Also on the 26th, B. von Engelhardt (Dresden, Germany) said the galaxy M31 was brighter than the comet. The coma also continued to increase in size. Although there were numerous estimates of its size, Denning provided a good series of measurements for the period. He gave the diameter as 10′ 33″ on the 16th, 14′ 30″ on the 19th, 15′ on the 20th, and 20′ on the 26th. Although most other astronomers reported coma diameters similar to Denning's, F. Renz (Pulkovo Observatory, Russia) gave a much larger value of 30′ on the 23rd. During the latter days of the first half of November, several observers reported the coma as oval or elliptical. On the 18th, Skelton gave the dimensions as 12′ by 13′, while Gruss said the coma measured 10′ by 15′. Skelton said it measured 15′ by 18′ on the 20th. On the 26th, Skelton said it was 18′ by 20′ across, while Engelhardt said it was 15′ long and 13′ wide. The only magnitude estimates of the "nucleus" during this period were 12.5 by J. Palisa (Vienna, Austria) on the 17th, 10 by Wendell on the 19th, and 11.5 by Ristenpart on the 22nd, although this last probably referred to the condensation as seen in his 15-cm refractor. Other observers generally described it as absent or diffuse, depending on the telescope they were using. A few examples include Wendell describing it as "rather poorly defined" in the 38-cm refractor on the 18th, Rees noted it was "somewhat elongated in the east and west direction" in the 33-cm refractor on the 21st, and Schorr said it was a nebulous mass elongated toward PA 120° on the 25th. The condensation also faded and spread out, with A. Kammermann (Geneva, Switzerland) particularly noting it was barely visible in the 25-cm refractor on the 21st and 24th. A tail continued to be seen by some observers. Rees described it as faint on the 21st and said it extended eastward. On the 23rd, L. Boss (Dudley Observatory, New York, USA) noted, "a very faint extension of the nebulosity was noticed in position-angle 120°. This appendage or tail could be traced to about 12′ from the nucleus and was, perhaps, 8′ in breadth." That same night, Renz noted two tail-like extensions protruding southeastward. He said the tail extended toward the east-southeast on the 24th. Interestingly, the comet had a peculiar appearance on November 16 that was reported by both Denning and T. W. Backhouse (Sunderland, England). Denning observed with a 25.4-cm reflector and described the comet as "a slightly elongated mass with a bright streak, forming the nucleus." Backhouse observed with an 11.4-cm refractor and said the comet itself was a bright, round disk about 12′ in diameter; however, in the middle "was a strip of brighter light, whose brightest part, forming an indefinite nucleus, was considerably [north-preceding] the centre of the circular disk. I suspected other points of concentration of light in it. The strip formed the core, as it were, of a tail, the rest of which was very faint; both the core and

the faint part reached to 20′ from the nucleus at a position-angle of 140°."
A few interesting photographs were obtained. Isaac Roberts (Liverpool, England) obtained a photograph with a 51-cm reflector on the 18th and said the comet was nearly circular, with the north-preceding half well-defined and the south-following half undefined. He noted a dense condensation was projecting from the nucleus, which resembled a tail about 6′ long and 80″ wide. This tail was straight, but not sharply defined at its margins.

The comet was not visible because of moonlight on December 3, according to Boss. On the 6th, H. A. Kobold (Strasbourg, France) described the comet as a shapeless nebulosity, without a distinct condensation. Ristenpart said the comet could barely be seen by averted vision in the 15-cm refractor on the 7th. J. Tebbutt (Windsor, New South Wales, Australia) said the comet was "of the last degree of faintness" in the 11-cm refractor on the 7th and 8th. Engelhardt said the comet was bright and easily visible in the 13-cm finder on the 9th, while it appeared very faint in the 30-cm refractor. He added that the condensation had faded since November 26. Using a 20-cm refractor, Knopf was not able to see the comet during searches on December 9–13. Wendell said the nucleus was stellar on the 10th, while Schorr said it was only suspected as a pale glow, which vanished with the slightest illumination of the field. That same night, Gruss noted the comet was very faint, elongated, and 10–12′ across. On December 11, Barnard said a photograph with the 15-cm Willard lens revealed the coma was about one-half degree across and exhibited a central condensation, a nucleus, and a "diffusion of the head away from the Sun." That same night, Engelhardt said the condensation was very faint and diffuse. Wendell gave the nuclear magnitude as 12.1 on the 12th. Kobold said the comet was difficult to see and was without a condensation on the 13th, while it was a sharply bounded nebulosity about 5′ across with a condensation 30–40″ across in the northeastern portion of the coma on the 14th. Kammermann said the comet was very faint on the 13th, with a weak trace of condensation and a coma 3–4′ across. Also on the 14th, Le Cadet said the comet was extremely faint in the 32-cm refractor and added that the condensation was only visible by averted vision, while Schorr said it was a pale glow about 2′ across. On the 16th, Kobold said the comet exhibited a bright, fan-like extension with the middle directed toward PA 125°. The tail extended about 12–15′ and looked like a veil of mist. The head was round and about 30–40″ across, with a faint condensation. G. A. P. Rayet and L. Picart (Bordeaux, France) observed with the 38-cm refractor and said observations made on the 16th and 17th were very difficult because of the comet's faintness and diffuse nature. W. O. Lay (Dudley Observatory, New York, USA) said the comet looked like an extremely faint star, seen out of focus on the 21st. Wendell noted a central condensation on the 23rd.

On 1893 January 5, Barnard viewed the comet at low power in the 30-cm refractor and said the comet appeared very large and very faint, while Kobold said the 46-cm refractor revealed a faint spot of nebulosity about 2′

across which was only visible with great effort. Kobold noted the comet was next to a star of magnitude 10 on the 12th and was very difficult to see. The coma was about 30″ across.

The comet experienced another outburst in brightness around January 16. Kobold noted the comet was visible to the naked eye on January 16.81 and said a telescope revealed a nucleus of magnitude 8 and a coma 41″ across. At about the same time, Palisa also found the comet shining like a star of magnitude 8, and gave the coma diameter as 20″. Numerous observations were made on the 17th, with Cohn, Palisa, and Pechüle estimating the nuclear magnitude as 8, while it was given as 7.5 by Barnard, 7.2 by Schorr, 7.8 by Kammermann, and 8.2 by Kobold. The coma diameter was given as 20″ by Palisa, 30″ by Pechüle and E. A. Lamp, and 56.5″ by Kobold and the diameter of the nucleus was given as 5″ by both Schorr and Kobold. Although observers were still estimating the nuclear magnitude as about 8 on the 18th, the coma had increased in size. The coma diameter was given as 50″ by Kobold, 1′ by Gruss, 1.3′ by Palisa, and 1.5′ by Wendell and Schorr. Schorr said the nucleus was 2″ across, while Palisa measured it as 3.4″. A 30-minute exposure by Roberts revealed "a very dense circular nucleus surrounded by symmetrical nebulosity, which gave the comet the appearance of a nebulous star." He measured the coma as 39″ across and the nuclear condensation as 14″ across. On January 19, Ristenpart said the coma was about 45″ across and Kobold gave it as 78″. Schorr gave the nuclear magnitude as 7.7, while Wendell gave it as 8.2. On the 20th, Schwab said the nucleus appeared as a nebulous star of magnitude 8 situated within a coma 2′ across, while Palisa gave the coma diameter as 2.5′. Other observations on the 20th included Crommelin describing the comet as small and bright, W. Wickham (Radcliffe Observatory, Oxford, England) noting that the coma was fan-shaped, and Le Cadet describing it as a bright nebulous condensation of magnitude 9. Roberts obtained a 50-minute exposure on the 20th which revealed a "nebulous star" with a coma 145″ across and a nuclear condensation 53″ across. He added, "The nucleus and the nebulosity surrounding it are very dense and symmetrically circular, without sharply defined boundaries." On the 21st, Wendell said the coma was 1.8′ across, while Schorr said the nucleus was magnitude 8.5. Wendell said the coma was 2.5′ across on the 22nd, while Millosevich said it was 2′ across, with a nucleus of magnitude 9. On the 23rd, Schorr gave the nuclear magnitude as 9.3, while Cohn and Palisa gave the coma diameter as 1.5′ and 3′, respectively. That same night, Engelhardt said the disk-shaped condensation was about magnitude 10 and about 12″ across, while the fan-shaped tail was 1′ long. On the 27th, Roberts obtained a 15-minute exposure which revealed "a faint nucleus, a faint tail, and very faint nebulosity surrounding them. . . ." Kobold described the comet as a pale nebulosity, about 1.5′ across. He said no condensation was visible, although a nucleus of magnitude 13 occasionally appeared. K. Oertel (Munich, Germany) could hardly see the comet in moonlight on January 29.

On February 3, Palisa said the comet was oblong, while Kammermann said the tail was about 5′ long. On the 4th, Oertel said the nucleus was magnitude 9.8, while Schorr and Palisa gave the coma diameter as 3′ and Rayet estimated it as 4′. On the same night, Kobold said the 46-cm refractor revealed a faint tail extending at least 5′ toward PA 70° and a nucleus of magnitude 13. Schwab said the coma was 3′ across on the 4th and 5th. Ristenpart said the 15-cm refractor revealed a coma 5′ across with a nucleus of magnitude 11 on the 4th. Roberts photographed the comet with a 51-cm reflector on the 4th and said the 52-minute exposure revealed a nucleus, a faint coma, and a tail. On the 6th, Le Cadet said the comet was very faint and almost without condensation. He added that the nebulosity extended a little toward the northeast, with the length in this direction about 5′. That same night, A. Abetti (Padova, Italy) said the central condensation was about 2′ across. Abetti saw the comet again on the 9th and said the comet was elongated toward PA 53°. On the 12th, Gruss described the comet as very faint and 3–4′ across. Palisa said the coma was about 3′ across on the 13th. O. Stone (Leander McCormick Observatory, Virginia, USA) said the comet was faint and 2–3′ across on the 14th. Kobold described it on the 15th as 2′ in diameter and centrally condensed. On the 17th, Wendell gave the nuclear magnitude as 12.2 and the coma diameter of 3.5′, while Palisa said the coma was 2.5′ across. On February 18, Rayet said the coma was 3–4′ across.

On March 9, Wendell gave the nuclear magnitude as 13.0. Kobold said the comet was extremely faint and small on March 9, and he measured the position for the final time on the 10th. The comet's position was measured for the final time on March 13.79, when Palisa found it with the 69-cm refractor and described it as extremely faint. He gave the position as $\alpha = 2^h 49.8^m$, $\delta = +35° 19′$. The final observations of the comet were made by H. C. Wilson (Goodsell Observatory, Northfield, Minnesota, USA) on April 4 and Kobold on April 6.9. Wilson observed with a 41-cm refractor and described the comet as "exceedingly faint" with a coma about 2′ across and "very slight" central condensation. Kobold said it was "extremely faint" and added that it was impossible to measure the position "on this and on several following evenings." Wilson again looked for the comet in the 41-cm refractor on August 16 and September 14, but no trace was found. Using a 15-cm camera and an exposure time of 1 hour, he took a photograph of the region where the comet was expected to be on 1894 January 12. Although a slightly oval stain about 20′ across and with no condensation was seen at the correct spot, it was "so suspiciously like a dirty water stain that we hesitate to say anything about it without verification."

The spectrum was observed by W. W. Campbell (Lick Observatory, California, USA) on November 9 and 10. He said a continuous spectrum was present, although a slight trace of the green cometary band was detected. H. C. Vogel (Potsdam, Germany) saw a faint continous spectrum on the 13th, with a brightening in the yellowish-green region. Kammermann examined the comet on January 17 and noted only a continuous spectrum.

Interestingly, it should be mentioned that the first instance of a pre-discovery photographic image was also reported. Following the discovery announcement, W. Schooling (Hammersmith, England) re-examined a photograph he had taken of M31 on October 18 and found what looked like a comet. The plate had been exposed for an hour and a half on October 18.83–18.90. He sent the negative to Maunder and Russell and they were able to measure the position as $\alpha = 1^h\ 02^m10.7^s$, $\delta = +39°\ 55'\ 54''$ for October 18.86. Maunder added that upon receiving the negative he took one of the available ephemerides and extended it backwards and was impressed how close it came to Schooling's object. Both Maunder and Russell agreed that the object had a cometary appearance. E. Roberts took this position, as well as those obtained at the Royal Observatory (Greenwich, England) on November 9 and 26, and calculated an elliptical orbit with a perihelion date of 1892 April 15.06 and a period of 15.18 years. Although a great value was attributed to this image because it immediately extended the observed arc backwards nearly one month, later positions and subsequent orbital computations revealed Schooling's object could not be the same as Holmes' comet. This is not now considered a pre-discovery image, but is perhaps the first instance of a photographic emulsion flaw fooling astronomers.

S. J. Corrigan (1893) believed this comet came about as a result of a collision between two asteroids. He took all known asteroids and examined their orbits to see if he could identify one or both of the culprits, but found none. Nevertheless, he thought it reasonable to suppose that many more asteroids exist than were currently known.

Barnard discussed the possible circumstances leading to this comet's discovery in the March 1899 issue of the *Monthly Notices of the Royal Astronomical Society*. He said, "The failure to see the comet previous to its sudden apparition near the Andromeda Nebula, its uncometary appearance, its peculiar freaks, and final utter disappearance from the heavens, connected with the nebulous appendage shown in the photograph of November 10 [November 11 UT], would strongly suggest that the object was not a comet at all, but more probably a result of some celestial accident. I think there is no question but this 'comet' will never be seen again, and doubtless before now it has ceased to exist as an individual body."

The comet's perihelion distance of over 2 AU made the initial orbits somewhat discordant, especially in the perihelion date. The first orbit was calculated by H. C. F. Kreutz using positions from November 9, 10, and 11. The resulting perihelion date was 1892 August 16.24. During the next several days, the perihelion date was given as April 19.92 by E. Weiss and April 20.04 by A. Berberich. Using positions obtained up to November 15, L. Picart and F. Kromm determined the perihelion date as July 31.68, while L. Schulhof gave two perihelion dates of April 21.81 and June 25.54. Using positions obtained up to November 17, Kreutz demonstrated the difficulty in determining the orbit, as he gave four parabolic orbits with perihelion dates ranging from February 28.82 to June 7.30 that each fitted the available positions in a similar fashion.

Prominent astronomers on both sides of the Atlantic Ocean noted something particularly interesting about the orbit of this comet around mid-November. Berberich reported in a November 11 Circular issued by the *Astronomische Nachrichten* and Boss reported in the November 18 issue of the *Astronomical Journal* that this comet might be the lost comet Biela. The perihelion date was given as December 28 by Berberich and December 27 by Boss.

The first elliptical orbits were independently calculated by Kreutz and Searle using positions spanning the period of November 9–17. Kreutz gave the perihelion date as June 10.46 and the period as 7.09 years. Searle gave the perihelion date as October 12.48 and the period as 6.14 years. Additional orbits by Boss, Schulhof, Berberich, V. Cerulli, and J. R. Hind eventually established the perihelion date as June 13 and the period as 6.9 years. These orbits proved that this comet was not a return of comet Biela.

The first definitive orbit was calculated by H. J. Zwiers (1895) using positions spanning almost the entire period of visibility. He determined the perihelion date as June 13.82 and the period as 6.88 years. Zwiers (1897) later revised this orbit and determined the perihelion date as June 13.86 and the period as 6.90 years. E. Kohlschütter (1896) also calculated a definitive orbit with a perihelion date of June 13.90 and a period of 6.90 years.

The most significant studies using multiple apparitions and planetary perturbations were published by F. Koebcke (1948) and B. G. Marsden (1963). These revealed a perihelion date of June 13.89 and a period of 6.90 years. Koebcke's orbit is given below.

T	ω	Ω (2000.0)	i	q	e
1892 Jun. 13.8905 (UT)	14.2792	333.2083	20.8008	2.140739	0.409532

ABSOLUTE MAGNITUDE: $H_{10} = 0.0$–6.0 (V1964)

FULL MOON: Nov. 4, Dec. 4, Jan. 2, Feb. 1, Mar. 2, Apr. 1, Apr. 30

SOURCES: A. C. D. Crommelin and T. Lewis, *MNRAS*, **53** (1892 Nov.), p. 52; E. Holmes and A. Berberich, *AN* Circular (1892 Nov. 11); E. Holmes, E. E. Barnard, and G. M. Searle, *AJ*, **12** (1892 Nov. 14), p. 112; E. Holmes, F. Bidschof, A. Berberich, T. D. Anderson, and H. C. F. Kreutz, *AN*, **131** (1892 Nov. 14), pp. 133–6; G. Bigourdan, *CR*, **115** (1892 Nov. 14), pp. 782–3; O. C. Wendell, L. Boss, J. B. Coit, E. E. Barnard, G. M. Searle, *AJ*, **12** (1892 Nov. 18), p. 119, supplement; F. Hayn, F. W. Ristenpart, E. Millosevich, R. Schorr, J. Möller, E. Weiss, A. Berberich, and H. C. F. Kreutz, *AN*, **131** (1892 Nov. 19), pp. 145–52; L. Boss, J. B. Coit, E. E. Barnard, T. Reed, and G. M. Searle, *AJ*, **12** (1892 Nov. 26), pp. 126–8; G. Le Cadet, R. Schorr, F. Cohn, W. Schur, L. Picart, F. Kromm, L. Schulhof, and H. C. F. Kreutz, *AN*, **131** (1892 Nov. 26), pp. 163–8; I. Roberts, E. Roberts, W. Schooling, A. C. D. Crommelin, T. Lewis, R. Bryant, and C. Davidson, *MNRAS*, **53** (1892 Dec.), pp. 65–9; E. Holmes, Kidd, T. W. Backhouse, and W. Schooling, *The Observatory*, **15** (1892 Dec.), pp. 420, 441–3,

451, 455; L. Schulhof and A. Berberich, *AN*, **131** (1892 Dec. 1), pp. 179–82; L. Weinek, R. Spitaler, A. Kammermann, and R. Schorr, *AN*, **131** (1892 Dec. 3), p. 195; M. W. Whitney, J. K. Rees, H. Jacoby, and G. M. Searle, *AJ*, **12** (1892 Dec. 5), pp. 133, 135–6; G. Gruss, F. Renz, B. von Engelhardt, and W. W. Campbell, *AN*, **131** (1892 Dec. 8), pp. 207–12; O. C. Wendell, H. C. Wendell, L. Boss, and G. M. Searle, *AJ*, **12** (1892 Dec. 9), pp. 142–4; V. Cerulli, *AN*, **131** (1892 Dec. 20), pp. 293–6; G. M. Searle, M. Updegraff, E. E. Barnard, O. Stone, M. W. Whitney, L. Boss, and J. K. Rees, *AJ*, **12** (1892 Dec. 22), pp. 148–52; A. Abetti, B. von Engelhardt, R. Schorr, A. Kammermann, and G. Le Cadet, *AN*, **131** (1892 Dec. 27), pp. 305–8; W. Schooling, E. W. Maunder, Russell, and I. Roberts, *The Observatory*, **16** (1893 Jan.), pp. 34–6; W. O. Lay, *AJ*, **12** (1893 Jan. 2), pp. 158–9; L. Schulhof, *AN*, **131** (1893 Jan. 3), pp. 325–8; H. C. Vogel, *AN*, **131** (1893 Jan. 5), p. 373; J. E. Davidson, *AN*, **131** (1893 Jan. 10), p. 391; G. A. P. Rayet and L. Picart, *AN*, **131** (1893 Jan. 12), p. 403; H. A. Kobold and J. Palisa, *AN*, **132** (1893 Jan. 17), pp. 13–15; E. E. Barnard and J. Palisa, *AJ*, **12** (1893 Jan. 21), pp. 175–6; J. Tebbutt, F. Cohn, R. Schorr, E. A. Lamp. H. A. Kobold, J. Palisa, and J. Holetschek, *AN*, **132** (1893 Jan. 23), pp. 25–32; A. Kammermann and G. Gruss, *AN*, **132** (1893 Jan. 25), p. 61; T. D. Anderson and J. E. Davidson, *MNRAS*, **53** (1893 Feb.), pp. 238, 267; L. Schulhof, *The Observatory*, **16** (1893 Feb.), p. 120; E. Millosevich, B. von Engelhardt, F. Cohn, and R. Schorr, *AN*, **132** (1893 Feb. 1), p. 77; H. A. Kobold, *AN*, **132** (1893 Feb. 6), p. 93; W. Schooling, *AN*, **132** (1893 Feb. 9), p. 107; J. R. Hind, *AN*, **132** (1893 Feb. 13), p. 127; G. Le Cadet, *CR*, **116** (1893 Feb. 13), p. 304; F. Schwab and R. Schorr, *AN*, **132** (1893 Feb. 18), pp. 135–40; O. Stone, *AJ*, **13** (1893 Feb. 23), p. 8; I. Roberts, J. Palisa, A. C. D. Crommelin, T. Lewis, H. P. Hollis, R. Bryant, and T. Hudson, *MNRAS*, **53** (1893 Mar.), pp. 332–3, 344–5; W. F. Denning, T. A. Skelton, and S. J. Corrigan, *The Observatory*, **16** (1893 Mar.), pp. 142–4, 151; K. Oertel and H. A. Kobold, *AN*, **132** (1893 Mar. 8), pp. 167–72; H. A. Kobold, *AN*, **132** (1893 Mar. 14), p. 171; L. Boss, *AJ*, **13** (1893 Mar. 17), pp. 30–2; W. E. Plummer, *MNRAS*, **53** (1893 Apr.), pp. 377–82; I. Roberts, *The Observatory*, **16** (1893 Apr.), p. 154; A. Kammermann and V. Cerulli, *AN*, **132** (1893 Apr. 15), pp. 233–5; G. Le Cadet, *AN*, **132** (1893 May 1), p. 285; H. A. Kobold, *AN*, **132** (1893 May 8), pp. 303–4; H. C. Wilson, *AJ*, **13** (1893 May 11), pp. 61–2; A. Abetti, *AN*, **132** (1893 May 19), pp. 329–32; E. E. Barnard, *The Observatory*, **16** (1893 Jul.), pp. 250–1; G. Gruss, *AN*, **133** (1893 Jul. 8), p. 63; G. A. P. Rayet, *AN*, **133** (1893 Jul. 27), pp. 113–16; F. Hayn, *AN*, **133** (1893 Aug. 9), p. 183; J. Palisa, *AN*, **133** (1893 Aug. 16), pp. 207, 211–14; G. Witt, *AN*, **133** (1893 Oct. 6), pp. 385, 391; H. C. Wilson, *AJ*, **13** (1893 Oct. 23), p. 149; F. W. Ristenpart, *AN*, **134** (1893 Oct. 27), pp. 17, 23; F. Schwab, *AN*, **134** (1893 Dec. 11), pp. 149–52; H. A. Kobold and O. C. Wendell, *AN*, **134** (1893 Dec. 21), pp. 169–80; O. Knopf, *AN*, **134** (1894 Jan. 6), pp. 203, 209; *PA*, **1** (1894 Feb.), p. 285; R. Schorr, *AN*, **134** (1894 Mar. 1), pp. 371–4, 395–7; C. F. Pechüle, *AN*, **136** (1894 Oct. 20), pp. 311, 316; W. Wickham, *MNRAS*, **55** (1895 Jan.), pp. 162–3; H. J. Zwiers, *AN*, **138** (1895 Jun. 24), pp. 65–78; O. Knopf, *AN*, **138** (1895 Jul. 13), pp. 201, 210–11; H. J. Zwiers, *PA*, **3** (1895 Sep.), p. 39; H. J. Zwiers, *AN*, **138** (1895 Oct. 3), pp. 419–22; E. Kohlschütter, *AN*, **141** (1896 Aug. 27), pp. 241–50; H. J. Zwiers, *AN*, **142** (1897 Jan. 29), pp. 257–64; E. Kohlschütter, *AN*, **142** (1897 Mar. 5), pp. 401–6; E. E. Barnard, *MNRAS*, **59** (1899 Mar.), p. 357; O. C. Wendell, *AAOHC*, **33** pt. 8 (1900), pp. 152–7; L. Weinek, R. Spitaler, R. Lieblein, and C. Pin, *Astronomische Beobachtungen an der K. K. Sternwarte zu Prag* (1901), pp. 27–8; F. Koebcke, *Poznan Observatory Reprint*, No. 12 (1948); B. G. Marsden, *AJ*, **68** (1963 Dec.), pp. 795–801; V1964, p. 64; SC2000.0, pp. 18, 20.

C/1892 W1
(Brooks)

1893 I = 1892g

Discovered: 1892 November 20.4 ($\Delta = 1.76$ AU, $r = 1.41$ AU, Elong. $= 53°$)
Last seen: 1893 March 11.82 ($\Delta = 2.31$ AU, $r = 1.55$ AU, Elong. $= 31°$)
Closest to the Earth: 1893 January 4 (0.6993 AU)
Calculated path: VIR (Disc), COM (Nov. 24), BOO (Dec. 9), CVn (Dec. 12), BOO (Dec. 18), DRA (Dec. 27), CEP (Jan. 9), LAC (Jan. 15), AND (Jan. 19), PSC (Mar. 8)

W. R. Brooks (Geneva, New York, USA) discovered this "bright nebulous object" on November 20.4, at a position of $\alpha = 12^h 56^m$, $\delta = +12° 59'$. The comet was confirmed by O. C. Wendell (Harvard College Observatory, Massachusetts, USA) on November 22. He said the 38-cm refractor revealed the comet was round, about 1.5' across, and exhibited a "tolerably well-defined nucleus" of about magnitude 10.5.

The comet was found just over a month prior to passing closest to both the sun and Earth. On November 23, Wendell said the nucleus was rather stellar. O. Knopf (Jena, Germany) observed with a 20-cm refractor on the 24th and said the comet had a stellar central condensation. On the 25th, Wendell said the coma was 1' across, while C. F. Pechüle (Copenhagen, Denmark) said the comet was quite bright, round, and 1' across. F. Cohn (Königsberg, now Kaliningrad, Russia) saw the comet on the 26th and described it as quite faint and small, with a rather distinct nucleus. That same night, E. A. Lamp (Kiel, Germany) said the comet was quite faint, 1' across, with a central condensation, but no nucleus, while J. Holetschek (Vienna, Austria) estimated the magnitude as 10. On November 27, Knopf said the coma was about 1.5' across and exhibited a stellar central condensation of magnitude 10.3, while R. Schorr (Hamburg, Germany) said the coma was about 30" across, with a nucleus of magnitude 11.5.

On December 1, F. W. Ristenpart (Karlsruhe, Germany) observed with the 15-cm refractor and said the comet appeared as a faint nebulosity with a diameter of about 30", while Holetschek gave the magnitude as 9.5. H. A. Kobold (Strasbourg, France) described the comet as rather bright on the same date, with a round coma 2' across and a rather strong condensation. H. P. Hollis (Royal Observatory, Greenwich, England) observed with the 17-cm refractor (55×) on the 13th and did not see a nucleus. Hollis said the comet appeared fainter on the 16th and still exhibited no nucleus. A. C. D. Crommelin (Royal Observatory, Greenwich, England) observed with the 17-cm refractor (55×) and said the comet was pretty bright and small on the 16th, with a central condensation. On the 17th, Ristenpart said the coma was about 40" across, with a central condensation of magnitude 10, but with no nucleus. That same morning, Holetschek gave the magnitude as 9. L. Picart (Bordeaux, France) said the comet was round, with a nucleus on the 17th and 18th. On the 19th, Ristenpart said it appeared as a diffuse nebulous mass about 52" across, without a nucleus. On the 20th, Ristenpart described the comet as faint and diffuse, with a diameter of about 67", while Schorr said it was a round, faint nebulosity, about 1' across and without a nucleus. Also on the 20th, Holetschek gave the magnitude as 8.2. F. Schwab (Kremsmünster, Austria) noted a bright nucleus and a coma 2–3' across on the 21st. On the 23rd, Ristenpart said the comet appeared fainter and more diffuse than on the 20th. On the 24th, Ristenpart said the comet was fainter than on the 23rd, with a coma diameter of about 45", while Holetschek gave the magnitude as 7.8. That same night, Schorr said the stellar nucleus was magnitude 11.0. B. von Engelhardt (Dresden, Germany) said the comet was easily visible in the 13-cm finder on the 25th, while the 30-cm refractor revealed a round coma about 2' across, which exhibited a condensation and nucleus shifted slightly west of center. There was also a short, fan-shaped tail. Also on the 24th, Schorr noted a distinct stellar nucleus of magnitude 11.0. Schwab observed the comet on December 25, 27, and 28 and said the coma diameter was about 2', while the nucleus was distinct. He added that the nucleus was eccentrically situated in the following portion of the coma on the 28th. G. Le Cadet (Lyon, France) saw the comet on December 29 and 30. He described the comet as a bright, round, centrally condensed nebulosity, with a diameter of about 2'.

The comet passed closest to Earth on 1893 January 4 and then passed perihelion on the 6th. Also on the 6th, the comet attained its most northerly declination of +66°. Cohn said the comet was about magnitude 9 during the period of January 5–8. Ristenpart failed to find the comet in the 15-cm refractor on the 7th and attributed this to moonlight, while G. Gruss (Prague, Czech Republic) said the comet was faint in the 20-cm refractor and exhibited a coma 2–3' across, as well as a nucleus of magnitude 10. Kobold also detected the comet and reported a round coma 2' across and a condensation of magnitude 11. Schwab saw the comet on January 6, 8, and 13, and said the nucleus had become indistinct. Gruss said the comet was easily seen in the 6-cm finder on the 8th. Holetschek gave the magnitude as 7.5 on the 8th and 7.8 on the 10th. Schorr gave the nuclear magnitude as 10.2 on the 9th and 10.1 on the 10th. G. A. P. Rayet (Bordeaux, France) said the comet was rather brilliant, with a nucleus of magnitude 13 on the 11th, while Gruss said the coma was 4–5' across. Gruss noted the nucleus occasionally flashed on the 12th. Ristenpart said the coma was about 20" across without a nucleus on the 13th, while Holetschek gave the magnitude as 8.5. Cohn reported the comet was about magnitude 9.5 on the 14th and 15th. On the 15th, Ristenpart said the comet appeared about 15" across, while Gruss said it was 2–3' across with a distinct nucleus of magnitude 11–12. On the 16th, Ristenpart said the comet seemed brighter than on the 15th, Schwab noted the nucleus was no longer visible, and Holetschek gave the magnitude as 8.4. Holetschek estimated the magnitude as 9 on the 20th. On the 24th, J. G. Porter (Cincinnati, Ohio, USA) noted the comet was faint in moonlight. On January 25, Rayet said the nucleus was still visible.

On February 3 and 4, Cohn said the comet was fainter than magnitude 10. On the 4th, Kobold described the coma as round and 1' across, with a central condensation of magnitude 12, while Holetschek estimated the magnitude as 10. That same night, Schorr said the coma was about 1.5′ across with a nucleus of magnitude 11.7. Ristenpart said the comet was about 15″ across on the 5th, while Schwab said it looked like a nebula 1′ across. On the 6th, Ristenpart said the small comet was difficult to see because of twilight. On February 10, R. Bryant (Royal Observatory, Greenwich, England) observed with the 17-cm refractor (55×) and described the comet as "exceedingly faint and difficult to observe." That same night, Holetschek estimated the magnitude as 10. Holetschek gave the magnitude as 9.7 on the 13th and 10.5 on the 18th. Schorr said the comet was 1′ across on the 16th, with a nucleus of magnitude 12.0.

On March 8, Rayet said the coma was about 15″ across, while Holetschek estimated the magnitude as 11. This comet was last seen on 1893 March 11.82, when Rayet measured the position as $\alpha = 0^h\ 48.2^m$, $\delta = +20°\ 25′$.

The first orbit was calculated by A. Berberich using positions from November 22, 25, and 27. The result was a perihelion date of 1893 January 7.40. Additional orbits were calculated by H. C. F. Kreutz, P. Maitre, S. C. Chandler, J. G. Porter, Ristenpart, and P. Isham, which established the perihelion date as January 7.

J. Polak (1911) computed a definitive orbit using 276 positions obtained between 1892 November 22 and 1893 March 11. He computed a parabolic orbit and a hyperbolic orbit. He concluded that, although the parabolic solution fitted best, the hyperbolic orbit was the more likely. Perturbations by five planets were applied. The hyperbolic orbit is given below.

T	ω	Ω (2000.0)	i	q	e
1893 Jan. 6.9912 (UT)	85.2134	187.1312	143.8444	1.195186	1.001586

ABSOLUTE MAGNITUDE: $H_{10} = 7.8$ (V1964)

FULL MOON: Nov. 4, Dec. 4, Jan. 2, Feb. 1, Mar. 2, Apr. 1

SOURCES: W. R. Brooks and O. C. Wendell, *AJ*, **12** (1892 Nov. 26), p. 128; W. R. Brooks and C. F. Pechüle, *AN*, **131** (1892 Nov. 26), p. 167; W. R. Brooks, *The Observatory*, **15** (1892 Dec.), p. 455; A. Berberich, *AN*, **131** (1892 Dec. 1), p. 181; R. Schorr, C. F. Pechüle, F. Cohn, and H. C. F. Kreutz, *AN*, **131** (1892 Dec. 3), pp. 197–9; J. G. Porter and S. C. Chandler, *AJ*, **12** (1892 Dec. 5), p. 136; O. C. Wendell and J. G. Porter, *AJ*, **12** (1892 Dec. 9), pp. 142, 144; E. A. Lamp and W. R. Brooks, *AN*, **131** (1892 Dec. 12), pp. 245–8; P. Maitre, *AN*, **131** (1892 Dec. 20), p. 295; R. Schorr, *AN*, **131** (1892 Dec. 27), p. 309; A. C. D. Crommelin and H. P. Hollis, *MNRAS*, **53** (1893 Jan.), pp. 134–5; J. G. Porter, *AJ*, **12** (1893 Jan. 2), p. 159; G. Le Cadet, *CR*, **116** (1893 Jan. 2), p. 19; B. von Engelhardt, *AN*, **131** (1893 Jan. 3), p. 327; R. Schorr and F. W. Ristenpart, *AN*, **131** (1893 Jan. 5), pp. 371–6; J. G. Porter, *AJ*, **12** (1893 Jan. 21), pp. 174–5; G. Le Cadet, *AN*, **132** (1893 Jan. 23), pp. 27–9; J. G. Porter, *AJ*, **12** (1893 Feb. 8), p. 191; F. Schwab, *AN*, **132** (1893 Feb. 18), pp. 135–8; F. W. Ristenpart, *AN*, **132** (1893 Feb. 25), p. 155; H. P. Hollis, *MNRAS*, **53** (1893 Mar.), pp. 343–5; *MNRAS*, **53** (1893 Apr.), pp. 378–9; W. E. Plummer, *MNRAS*, **53** (1893 Apr.), p. 382; G. A. P. Rayet, *CR*, **116** (1893 May 1), pp. 940–2; J. G. Porter and P. Isham, *The Observatory*, **16** (1893 Jun.), p. 241; G. Gruss and F. Cohn, *AN*, **133** (1893

Jul. 8), pp. 63–8; J. Holetschek, *AN*, **133** (1893 Jul. 19), pp. 91–6; G. A. P. Rayet, *AN*, **133** (1893 Aug. 9), pp. 181–2; F. Ristenpart, *AN*, **134** (1893 Oct. 27), pp. 17–18, 23–4; F. Schwab, *AN*, **134** (1893 Dec. 11), pp. 149–52; H. A. Kobold, *AN*, **134** (1893 Dec. 21), pp. 171–2, 175; O. C. Wendell, *AN*, **134** (1893 Dec. 21), pp. 177–8, 180; O. Knopf, *AN*, **134** (1894 Jan. 6), pp. 203–4, 210; R. Schorr, *AN*, **134** (1894 Mar. 1), pp. 375, 397; J. Polak, *AN*, **189** (1911 Jul. 17), pp. 7–12; V1964, p. 64.

15P/1893 K1
(Finlay)

1893 III = 1893a

Recovered: 1893 May 18.2 ($\Delta = 1.31$ AU, $r = 1.26$ AU, Elong. $= 64°$)
Last seen: 1893 September 22.11 ($\Delta = 1.60$ AU, $r = 1.40$ AU, Elong. $= 60°$)
Closest to the Earth: 1893 June 17 (1.1835 AU)
Calculated path: AQR (Rec), PSC (May 21), CET (May 28), PSC (Jun. 3), CET (Jun. 18), ARI (Jun. 19), TAU (Jul. 5), GEM (Aug. 10), CNC (Sep. 17)

L. Schulhof (1893) took positions obtained during this comet's 1886 apparition and computed a perihelion date of 1893 July 12.68, with an uncertainty of ± 2 days in the perihelion date. He published an ephemeris on 1893 February 25, which covered the period of 1893 March 13 to 1893 June 1. W. H. Finlay (Royal Observatory, Cape of Good Hope, South Africa) searched for the comet at every available opportunity during March and April of 1893, but it remained too faint for his 18-cm refractor. He finally recovered it on 1893 May 18.2. Finlay reobserved it on May 19.13 and gave the position as $\alpha = 23^h\ 42.0^m$, $\delta = -5°\ 02'$. He described it as very diffuse, circular, and 1' across. He added that the comet appeared faint, with a magnitude of about 11, and exhibited no tail.

On June 24, Finlay said the comet was "much brighter than at any previous observation." He was using the 18-cm refractor and said it withstood the illumination of the wire as well as a star of magnitude 10.5. The coma was about 30" across, but showed no well-marked condensation. Finlay said the comet was faint on July 16 and very faint on the 25th. On July 25 and 26, J. Holetschek (Vienna, Austria) said the comet was at a low altitude and was as bright as a star of magnitude 10.5 in the 15-cm refractor. Its appearance was that of a small nebulosity. On August 10, G. Witt (Berlin, Germany) described the comet as very faint in the 30-cm refractor (125×). Witt noted on the 11th that it was "better than yesterday." On the 13th, J. G. Porter (Cincinnati, Ohio, USA) described the comet as exceedingly faint. H. C. Wilson (Goodsell Observatory, Minnesota, USA) saw the comet in the 41-cm refractor on August 17 and said it appeared faint, slightly condensed in the center, and about 1' across. The comet attained its most northerly declination of $+23°$ on the 19th. On August 24 and 26, Holetschek said the comet had increased its altitude since the July observations and was as bright as a star of magnitude 11.5. It still had the appearance of a small nebulosity in the 15-cm refractor. On September 10, Witt described the comet as very faint. Wilson looked for the comet with the 41-cm refractor on September 15, but found nothing.

The comet was last detected on September 22.11, when S. Javelle (Nice, France) saw it with the 76-cm refractor. He gave the position as $\alpha = 8^h\,05.6^m$, $\delta = +21°\,36'$. R. Schorr (Hamburg, Germany) looked for the comet on several occasions at the end of October and beginning of November, but was unable to find any trace.

The most recent multiple apparition orbit was calculated by B. G. Marsden and Z. Sekanina (1972) using positions spanning the period of 1886–1906. Their orbit is given below. In a paper published by Marsden, Sekanina, and D. K. Yeomans (1973) the nongravitational terms were determined as $A_1 = +0.53$ and $A_2 = +0.1266$.

T	ω	Ω (2000.0)	i	q	e
1893 Jul. 12.6131 (TT)	315.4562	53.7288	3.0416	0.989138	0.719471

ABSOLUTE MAGNITUDE: $H_{10} = 10.1$ (V1964)

FULL MOON: Apr. 30, May 30, Jun. 29, Jul. 28, Aug. 27, Sep. 25

SOURCES: L. Schulhof, *AN*, **132** (1893 Feb. 25), pp. 157–60; W. H. Finlay, *AJ*, **13** (1893 May 22), p. 72; W. H. Finlay, *AN*, **132** (1893 May 23), pp. 351–2; W. H. Finlay, *The Observatory*, **16** (1893 Jun.), p. 240; L. Schulhof, *AN*, **133** (1893 Jun. 29), pp. 51–6; J. G. Porter, *AJ*, **13** (1893 Aug. 25), p. 126; F. H. Finlay, *AN*, **133** (1893 Sep. 12), pp. 329–32; J. Holetschek, *AN*, **133** (1893 Sep. 19), pp. 347–50; H. C. Wilson, *AJ*, **13** (1893 Oct. 23), p. 149; R. Schorr, *AN*, **134** (1894 Mar. 1), p. 398; S. Javelle, *BA*, **12** (1895 Jan.), p. 29; G. Witt, *AN*, **137** (1895 Mar. 4), pp. 199–203; V1964, p. 64; B. G. Marsden and Z. Sekanina, *QJRAS*, **13** (1972), pp. 428–9; B. G. Marsden, Z. Sekanina, and D. K. Yeomans, *AJ*, **78** (1973 Mar.), pp. 213, 215; B. G. Marsden and Z. Sekanina, *CCO*, 12th ed. (1997), pp. 52–3.

C/1893 N1   *Discovered:* 1893 June 20.35 ($\Delta = 0.94$ AU, $r = 0.76$ AU, Elong. $= 46°$)

(Rordame–   *Last seen:* 1893 December 21.20 ($\Delta = 2.79$ AU, $r = 2.81$ AU, Elong. $= 81°$)

Quénisset)   *Closest to the Earth:* 1893 July 8 (0.4073 AU)

Calculated path: ARI (Disc), TAU (Jun. 28), PER (Jun. 30), AUR (Jul. 4), LYN

1893 II = 1893b   (Jul. 8), UMa (Jul. 10), LYN (Jul. 11), UMa (Jul. 12), LMi (Jul. 13), UMa (Jul. 18), LEO (Jul. 19), COM (Jul. 28), VIR (Aug. 5)

The actual discovery of this comet was a case of mistaken identity. W. E. Sperra (Randolph, Ohio, USA) found a comet in the morning sky on 1893 June 20.35. He thought he was observing comet 15P/Finlay, which had been recovered on May 18. When Sperra made this June observation both comet Finlay and C/1893 N1 were in Aries and were, coincidentally, at their closest angular distance from each other – about 11°. Sperra gave the position as $\alpha = 2^h\,43^m$, $\delta = +17°\,30'$ and described the comet as "about sixth magnitude or less, round, nebulous with condensation toward center, perhaps 3' in diameter." Not having an ephemeris for comet Finlay, he observed this comet for 11 more nights before word came of a new comet discovery and he realized his mistake.

The comet was approaching both the sun and Earth when found and, although it was brightening, its solar elongation was decreasing and reached a minimum of 26° on July 8. The comet attained its most northerly declination of +48° on the 9th. The moon was interfering at the end of June and during the first days in July. The first observers to report their discovery of this comet were A. Rordame (Salt Lake City, Utah, USA) and F. Quénisset (Juvisy, France). Rordame found the comet with the naked eye in Lynx on July 9.23. He said it was like a star of magnitude 3 and noted, "The tail became quite apparent to the naked eye after a few minutes' scrutiny." Quénisset found the comet on July 9.87 and described it as a beautiful naked-eye comet and made a sketch showing a bright nucleus surrounded by two concentric nebulous envelopes and a straight tail about 3° long, which tapered to a point.

Further independent discoveries were apparently made by M. Roso de Luna (Logrosan, Estremadura, Spain) on July 5.14, D. E. Hadden (Alta, Iowa, USA) on July 8.16, C. Johnson and J. Miller (Alta, Iowa, USA) on July 9.2, H. Filmer (Faversham, England), Waterhouse (Putney, England), and Mileham (Norwood, England) on July 9.96, L. Boss (Dudley Observatory, New York, USA) on July 10.09, and Merlin (Vólos, Greece) on July 10.85. Roso de Luna thought he had found a new star a few degrees south of α Aurigae. He estimated the magnitude as 4, but did not realize the importance of his find. He mailed a letter to Madrid Observatory on July 6, but a period of cloudy weather had settled in when the letter arrived on the 8th and several days passed before the Madrid astronomers were able to confirm the object and announce the discovery. Hadden said the comet looked like a large, hazy star, but he did not then suspect its true nature and no report was made until after Johnson and Miller had contacted him. Filmer said he found the comet with the naked eye about mid-way between Ursa Major and Gemini. He knew Finlay's comet was supposed to be in the region and assumed this must have been it. He examined the comet with an 8-cm refractor and noted it "had a very bright head and the tail was brighter than the afterglow upon which it was projected. The tail was about 15° in length." Filmer drew a sketch of the comet's head which indicated the presence of a fountain emanating from the nucleus. Boss described it as magnitude 3.5, "with a strong central condensation and a tail 4° long."

This comet was extremely well observed during the remainder of July. Magnitude estimates came from numerous observers, but it was E. F. Sawyer (Brighton, Massachusetts, USA) and J. Holetschek (Vienna, Austria) who provided the best series. Sawyer used an opera-glass (2.5×) and took the instrument slightly out of focus to compare the combined nucleus and coma with at least two stars each night. He obtained magnitudes of 3.30 on the 14th, 3.50 on the 15th, 3.67 on the 17th, 4.56 on the 20th, 4.48 on the 21st, 4.73 in moonlight on the 24th, 5.10 in moonlight on the 25th, and 5.76 on the 31st. Holetschek obtained magnitudes of 4 on the 16th, 5 on the 25th, 5.5 on the 27th, and 6 on the 31st. The comet was reported as visible to the naked eye through most of July and was last spotted in this fashion on July 25 by A. L. N. Borrelly (Marseille, France). Only a few observers estimated the brightness of the nucleus, but R. Schorr (Hamburg, Germany) produced the best series in July. Using a 26-cm refractor, he gave the magnitude as 6 on the 17th, 7.5 on the 23rd, 8.5 on the 25th, and 9 on the 31st. Schorr reported the nucleus as diffuse on every occasion. O. C. Wendell (Harvard College Observatory, Massachusetts, USA) observed with the 38-cm refractor and said the nucleus was "rather sharp" on the 11th, "quite stellar" on the 15th, and "fairly well defined" on the 20th. B. von Engelhardt (Dresden, Germany) observed with a 30-cm refractor and described the nucleus as disk-shaped on the 13th. A. C. D. Crommelin and A. Everett (Royal Observatory, Greenwich, England) observed with the 17-cm refractor (55×) on the 16th and said the nucleus was to the north-preceding side of the coma's center. L. Steiner (O'Gyalla Observatory, Hungary) noted the nucleus was diffuse in the 15-cm refractor on the 29th. Estimates of the coma diameter were somewhat discordant depending on the telescope that was used, but for the most part the diameter was around 4′ in mid-July and 2′ at the end of July. The head of the comet displayed color according to some observers. Engelhardt said it was bluish-white on the 13th, while F. S. Archenhold (Grunewald Observatory, Berlin, Germany) said it was blue on the 13th and 14th. W. J. Hussey (Lick Observatory, California, USA) obtained two photographs of the comet on the 14th and noted several condensations within the tail. He measured three condensations and their apparent velocities away from the nucleus. The velocity of the condensation at 1.87° was 45 miles per second, that of the condensation at 3.6° was 53 miles per second, and that of the condensation at 5.88° was 60 miles per second. The tail was generally described as fan-shaped throughout the month, but there was a great discordance in the estimates of the length until moonlight began interfering, with observers either describing it as only a couple of degrees long or more than 10° long. Hadden seems to have explained this as he noted on the 10th that the tail was "fully as bright as the coma" for the first 4° or 5°, with a faint extension out to 15°. Hadden produced the best series of tail estimates throughout the month, giving the length as 12° on the 11th, 17° on the 12th, 22° on the 13th, 13° long on the 14th, 5° on the 15th, 3° on the 18th, 1° on the 20th, and 0.5° on the 29th. Examples of other observers' tail lengths are 8–10° by Borrelly and 2.5° by Sawyer on the 14th, 2° by Wendell and Sawyer on the 15th, and 1.8° by Sawyer on the 17th.

On August 1, W. Laska (Prague, Czech Republic) said the comet was easy to see and noted a coma 1–3′ across. W. H. Finlay (Royal Observatory, Cape of Good Hope, South Africa) saw the comet with an 18-cm refractor on the 2nd and said it was faint. Boss saw the comet on the 3rd and determined the magnitude as 6.5 after defocusing the view and comparing it with a couple of nearby stars. On the 4th, Hadden said the nucleus was "decidedly stellar," while the tail was about 0.3° long. That same evening, Sawyer gave the magnitude as 5.93, while Holetschek gave it as 6.3. Laska said the coma appeared larger on the 5th than on the 1st. F. Schwab (Kremsmünster,

Austria) said the coma was round and 2′ across on the 7th. Holetschek gave the magnitude as 7.3 on the 8th. On the 10th, R. Bryant (Royal Observatory, Greenwich, England) said the comet was faint in the 17-cm refractor (55×). Holetschek gave the magnitude as 7.7 on the 12th. On August 14, Hadden said the comet was smaller, with no nucleus or tail present. The comet was becoming more difficult to observe shortly before mid-August. The final observations prior to conjunction with the sun were made by Finlay on August 16.73 and L. Picart (Bordeaux, France) on August 16.85, when the comet was situated low over the horizon in the evening sky.

The comet passed about 6° from the sun on September 30 and was recovered in the morning sky by V. Cerulli (Teramo, Italy) on November 4. He described it as very faint, but it was still rather sharp in his 39-cm refractor. Cerulli said the comet was easily visible in the refractor on the 20th and was well seen in spite of twilight on November 22. He saw the comet with the 39-cm Cooke refractor on December 17. Its apparent position was then between the galaxies NGC 4666 (magnitude 10.8 – NGC2000.0) and NGC 4668 (magnitude 13.1 – NGC2000.0). Cerulli said the comet was brighter than NGC 4668.

The comet was last seen on 1893 December 21.20, when Cerulli gave the position as $\alpha = 12^h\ 37.6^m$, $\delta = +0°\ 15′$. He was then using the 39-cm Cooke refractor.

W. W. Campbell (Lick Observatory, California, USA) said the carbon bands were easily detected visually with a spectroscope on the 91-cm refractor on July 12, 13, and 18, while photographs on July 14 and 17 revealed the bands of carbon and cyanogen.

The first orbits were independently calculated by Schorr and E. A. Lamp using three positions from July 11 and 12. Schorr gave the perihelion date as 1893 July 7.90 and Lamp gave it as July 7.78. Lamp's would prove to be a very accurate representation, as shown by later orbits by F. W. Ristenpart, J. G. Porter, G. M. Searle, F. L. Chase, Schorr, Lamp, Boss, W. E. Plummer, and Cerulli.

A definitive orbit was calculated by F. Kromm (1895) using positions spanning the entire period of visibility. Although a parabolic orbit was calculated, he found an elliptical orbit with a period of about 44 409 years fitted the positions best. This orbit is given below.

T	ω	$\Omega\ (2000.0)$	i	q	e
1893 Jul. 7.7715 (UT)	47.1177	338.8337	159.9804	0.674549	0.999462

ABSOLUTE MAGNITUDE: $H_{10} = 6.6$ (V1964), $H_0 = 6.42\ n = 2.24$ (Bobrovnikoff, 1942)
FULL MOON: May 30, Jun. 29, Jul. 28, Aug. 27, Sep. 25, Oct. 25, Nov. 23, Dec. 23
SOURCES: F. Quénisset, A. Rordame, R. Schorr, and E. A. Lamp, *AN*, **133** (1893 Jul. 13), p. 85–7; F. Quénisset, *CR*, **117** (1893 Jul. 17), p. 138; E. A. Lamp, R. Schorr, F. W. Ristenpart, W. Schur, B. von Engelhardt, and F. S. Archenhold, *AN*, **133** (1893 Jul. 19),

p. 101–3; F. S. Archenhold, *AN*, **133** (1893 Jul. 27), p. 117; F. Quénisset, H. Filmer, and R. Schorr, *The Observatory*, **16** (1893 Aug.), pp. 295–6, 303; M. Roso de Luna, *AN*, **133** (1893 Aug. 3), p. 135; W. W. Campbell, *AN*, **133** (1893 Aug. 9), pp. 149–52; J. G. Porter, G. M. Searle, F. L. Chase, and L. Boss, *AJ*, **13** (1893 Aug. 11), pp. 115–17; G. Gruss and R. Schorr, *AN*, **133** (1893 Aug. 16), pp. 213–16; D. E. Hadden, *PA*, **1** (1893 Sep.), p. 44; M. Roso de Luna, F. Quénisset, C. Johnson, J. Miller, Merlin, Waterhouse, and Mileham, *The Observatory*, **16** (1893 Sep.), pp. 330–2; A. Rordame, J. Miller, C. Johnson, D. E. Hadden, and M. Roso de Luna, *PASP*, **5** (1893 September 9), pp. 154–5; J. Holetschek, *AN*, **133** (1893 Sep. 19), p. 347; A. Abetti and F. Cohn, *AN*, **133** (1893 Sep. 28), pp. 377–82; L. Picart, *CR*, **117** (1893 Oct. 2), pp. 456–7; V. Cerulli, *AN*, **133** (1893 Oct. 6), p. 399; E. F. Sawyer, *AJ*, **13** (1893 Oct. 23), pp. 149–50; C. Davidson, A. C. D. Crommelin, A. Everett, and R. Bryant, *MNRAS*, **53** (1893, Supplementary Notice), pp. 502–4; W. E. Plummer, *MNRAS*, **54** (1893 Nov.), pp. 34–7; W. E. Sperra and V. Cerulli, *AN*, **134** (1893 Nov. 9), pp. 57, 63; O. C. Wendell, *AN*, **134** (1893 Dec. 21), pp. 177, 180; L. Steiner, *AN*, **134** (1894 Jan. 17), pp. 239–44; *MNRAS*, **54** (1894 Feb.), pp. 242–3; W. H. Finlay, *AN*, **134** (1894 Feb. 28), pp. 361–4; W. Laska, *AN*, **134** (1894 Feb. 28), pp. 363–6; R. Schorr, *AN*, **134** (1894 Mar. 1), pp. 375, 398; H. A. Kobold, *AN*, **137** (1894 Dec. 14), pp. 41–3; W. Wickham, *MNRAS*, **55** (1895 Jan.), pp. 162–3; F. Kromm, *BA*, **12** (1895 Feb.), pp. 76–8; W. J. Hussey, *PA*, **2** (1895 Jun.), pp. 445–9; N. T. Bobrovnikoff, *Contributions from the Perkins Observatory*, **16** (1942), p. 39; V1964, p. 64; NGC2000.0, p. 141.

C/1893 U1 | *Discovered:* 1893 October 17.37 ($\Delta = 1.72$ AU, $r = 0.63$ AU, Elong. $= 27°$)
(Brooks) | *Last seen:* 1894 January 27.1 ($\Delta = 2.06$ AU, $r = 2.27$ AU, Elong. $= 89°$)

Closest to the Earth: 1893 July 14 (1.1526 AU), 1893 December 6 (1.3079 AU)

1893 IV = 1893c | *Calculated path:* VIR (Disc), COM (Oct. 17), CVn (Nov. 10), UMa (Nov. 27), BOO (Dec. 2), UMa (Dec. 3), DRA (Dec. 6), UMi (Dec. 12), DRA (Dec. 15), UMi (Dec. 17), DRA (Dec. 20), CEP-DRA (Jan. 2), CEP (Jan. 4)

W. R. Brooks (Smith Observatory, Geneva, New York, USA) discovered this comet on 1893 October 17.37 at a position of $\alpha = 12^h\ 21^m$, $\delta = +12°\ 55'$. He said it was then about magnitude 7, with a distinct nucleus and a tail about 3° long. G. M. Searle (Catholic University, Washington, DC, USA) confirmed the comet on October 18.42. He said it was about magnitude 7 with a tail about 2° long.

Although the comet had passed perihelion one month earlier, it was still approaching Earth when discovered. On October 18, R. Schorr (Hamburg, Germany) described the comet as bright, about 1.5′ across, with a diffuse nucleus of about 9th magnitude. He added that there was a bright tail about 20′ long. That same morning, E. E. Barnard (Lick Observatory, California, USA) obtained a photograph with the 15-cm portrait lens that revealed a tail about 3.5° long. He wrote, "This tail irregularly divides into two slightly divergent branches." He said a northern ray originated from the nucleus and was about 30′ long, while a southern ray left the main tail 10–15′ from the coma and was also about 30′ long. F. Schwab (Kremsmünster, Austria) described the comet as bright on the 19th and noted a tail extending 7′ toward PA 290°. On the 20th, I. Benko von Boinik (Pula, Yugoslavia) observed with a 15-cm refractor and gave the magnitude as 7.0. He added that the tail extended about 2.5° toward PA 315°. On the 20th and 21st, J. Holetschek (Vienna, Austria) gave the magnitude as 7.7. Barnard obtained a 35-minute exposure with the 15-cm portrait lens on the 21st. The photographic plate showed a staight tail that gradually widened and became more diffuse as it extended northward. He added, "From the northern side of the head a short diffused tail stretches out for half a degree or more, at an angle of some thirty degrees to the main tail."

Barnard obtained another 35-minute exposure with the 15-cm portrait lens on November 22. He wrote, "There is an utter transformation of the comet in this picture. The tail is larger and brighter and very much distorted, as if it had encountered some resistance in its sweep through space. This disturbance seems to have disrupted the north-east edge of the tail." Barnard also obtained a 42-minute exposure with the 15-cm portrait lens on the 23rd. He wrote, "The tail appears a total wreck in this photograph, and is still more suggestive of a disaster. It is badly broken, and on the south-west side hangs in irregular cloud-like masses. Near the extremity a large gap exists in the tail, as if something had gone through it from the north-east, and a large mass is torn off beyond this break and seems to be drifting independent of the comet." Benko von Boinik said the comet was distinctly fainter on the 29th and exhibited a tail extending 5′ toward PA 300°, while A. Abetti (Padova, Italy) gave the nuclear magnitude as 9.5. Numerous observations were made on October 31. Schorr said the comet appeared fainter than on the 18th. The magnitude of the diffuse nucleus was determined as 9.4 and the coma diameter was 54″ across. He added that the tail extended 15′ toward PA 319.5°. Schorr also reported that a 12.2-cm cometseeker revealed a tail 2° long. G. Gruss (Prague, Czech Republic) observed with the 20-cm refractor and said the tail extended toward PA 317°. R. Bryant (Royal Observatory, Greenwich, England) observed with the 17-cm refractor (55×) and said, "The Comet was fairly bright, and had a diffuse nucleus. A distinct tail was seen, 15′ in length, pointing in north preceding direction."

On November 1, Schorr said the nucleus was not as diffuse as on the previous morning and noted the tail was about 8′ long. Barnard obtained a 75-minute exposure with the 15-cm portrait lens on the 3rd. He said, "The tail looks as if it were beating against a resisting force, and it seems to be encountered... on the advancing side of the tail." On November 4 E. Millosevich (Rome, Italy) observed with a 25-cm refractor and estimated the nuclear magnitude as 10. G. Le Cadet (Lyon, France) observed the comet with a 32-cm refractor (75×) on the 5th and said the coma was round, without a distinct central condensation. A tail extended about 35′ toward the northwest. That same night, Abetti indicated the tail extended toward

PA 318.5°. On the 7th, H. A. Kobold (Strasbourg, France) said a pale tail extended about 30′ toward PA 300°. A. C. D. Crommelin (Royal Observatory, Greenwich, England) observed with the 17-cm refractor (55×) on the 10th and noted a "distinct straight tail in south preceding direction." That same night, Gruss gave the total magnitude as 9, the nuclear magnitude as 11, and said the tail extended toward PA 318°. He also noted a faint secondary tail. Barnard obtained a 97-minute exposure with the 15-cm portrait lens on the 12th. He noted the tail appeared straight and added, "It consists, at some distance from the head, essentially of two branches. The western branch is sinuous, as if matter were streaming irregularly back from the head, while the northern is very straight. At the end of the tail is a condensation which is nearly separated from the main tail." On the 14th, Gruss said it had faded since the 10th. He described it as elliptical, with a short tail, and noted it was not visible in the finder. That same morning, Holetschek gave the magnitude as 8.4. Millosevich said the tail extended 8–10′ toward the northwest on the 15th. He also noted several shining points in place of the nucleus. Bryant said the comet was extremely faint and very diffuse in a bright sky on the 27th. On November 28, Le Cadet said the 19-day-old moon diminished the apparent brightness of the comet. A magnification of 100× revealed a nebulosity 25″ in diameter, without a notable condensation. There was a perceptible elongation extending about 3′ toward PA 327°.

On December 2, Kobold said the comet was difficult to observe and appeared as a shapeless nebulosity with a slight condensation. On the 3rd, Holetschek gave the magnitude as 9.0, while Schorr said the comet appeared a rather faint, elongated, and uncondensed nebulosity. That same night, Gruss said the condensation was very faint and Bryant said the comet was extremely faint. Bryant said the comet was very faint, with hardly any condensation on the 4th. Holetschek gave the magnitude as 9.7 on the 7th. On the 10th, Schorr said the comet exhibited a faint nebulosity about 45″ across, with a faint, eccentrically situated condensation. On the 12th and 14th, Schorr said the comet was extremely faint, without condensation. Holetschek gave the magnitude as 10.0 on the 15th and 16th. Bryant said the comet was brighter and more condensed on the 16th than on the 7th. Holetschek gave the magnitude as 10.0 on the 25th. On the 27th and 29th, Le Cadet described the comet as "extremely faint, vague, and diffuse." The comet attained its most northerly declination of +76° on the 29th. Holetschek gave the magnitude as 11.5 on December 30.

On 1894 January 7, H. C. Wilson (Goodsell Observatory, Minnesota, USA) said the comet was very faint in the 41-cm refractor and invisible in the 13-cm finder. He noted a small central condensation within a coma about 2′ across. The comet's position was determined for the final time on January 8.84, when Kobold gave it as $\alpha = 21^{\rm h} 38.8^{\rm m}$, $\delta = +72°44′$. He said the comet was rather faint and only seen with difficulty in the 46-cm refractor. The comet was last detected on 1894 January 27.1, when Wilson found a very faint glow near the predicted position, but it was too faint to measure in the 41-cm refractor.

The spectrum was visually observed by W. W. Campbell (Lick Observatory) on October 18, 19, 26, and November 3. On every occasion, five bands were detected overlaying the continuous spectrum. Although three of the bands were the usual carbon bands found in all comets, the two remaining bands had seldom been seen previously. Campbell measured the position of one of these rare bands as 4862 Å and attributed it to hydrogen. The other band was measured as 4557 Å and was attributed to cyanogen.

The first orbits were independently calculated by F. Bidschof, J. G. Porter, and H. C. F. Kreutz. Bidschof used positions from October 18 to October 24 and determined a perihelion date of 1893 September 20.16. Porter used positions from October 19 to October 24 and determined a perihelion date of September 21.01. Kreutz used positions from October 18 to October 25 and determined a perihelion date of September 19.67. Additional orbits were calculated by C. N. A. Krueger, Porter, P. Isham, Searle, and L. Schulhof, which eventually established a perihelion date of September 19.7.

A definitive orbit was calculated by D. Peyra (1895). He began with 153 positions spanning the entire period of visibility and ultimately determined both parabolic and elliptical orbits. The elliptical one was preferred and gave a period of about 3516 years. This orbit is given below.

T	ω	Ω (2000.0)	I	q	e
1893 Sep. 19.7223 (UT)	347.4515	176.4141	129.8233	0.811991	0.996489

ABSOLUTE MAGNITUDE: $H_{10} = 6.6$ (V1964)

FULL MOON: Sep. 25, Oct. 25, Nov. 23, Dec. 23, Jan. 21, Feb. 20

SOURCES: W. R. Brooks and R. Schorr, *AN*, **134** (1893 Oct. 18), p. 15; W. R. Brooks and G. M. Searle, *AJ*, **14** (1893 Oct. 23), p. 150; R. Schorr, F. Schwab, F. Bidschof, and H. C. F. Kreutz, *AN*, **134** (1893 Oct. 27), pp. 29–32; E. E. Barnard, *AJ*, **13** (1893 Oct. 31), p. 157; R. Bryant, *MNRAS*, **54** (1893 Nov.), pp. 38–9; W. R. Brooks and J. G. Porter, *PA*, **1** (1893 Nov.), pp. 143–4; G. M. Searle, *AJ*, **13** (1893 Nov. 14), p. 165; R. Schorr, *AN*, **134** (1893 Nov. 9), pp. 59–64; L. Schulhof, *CR*, **117** (1893 Nov. 13), p. 659; H. A. Kobold and E. Millosevich, *AN*, **134** (1893 Nov. 23), p. 85; W. W. Campbell, *PASP*, **5** (1893 Nov. 25), pp. 208–10; C. N. A. Krueger, *AN*, **134** (1893 Nov. 30), p. 103; A. C. D. Crommelin and R. Bryant, *MNRAS*, **54** (1893 Dec.), pp. 125–6; J. G. Porter, P. Isham, and W. R. Brooks, *PA*, **1** (1893 Dec.), pp. 186–8; R. Schorr, *AN*, **134** (1893 Dec. 11), p. 151; G. Le Cadet, H. A. Kobold, and R. Schorr, *AN*, **134** (1893 Dec. 13), p. 165; R. Bryant, *MNRAS*, **54** (1894 Jan.), pp. 143–4; C. N. A. Krueger, *AN*, **134** (1894, Jan. 6), p. 213; R. Schorr and G. Le Cadet, *AN*, **134** (1894 Jan. 17), pp. 237–40; G. Gruss, *AN*, **134** (1894 Feb. 28), pp. 363–6; R. Schorr, *AN*, **134** (1894 Mar. 1), pp. 375–8, 398–9; H. C. Wilson, *AJ*, **14** (1894 May 1), pp. 29–30; A. Abetti, *AN*, **136** (1894 Sep. 12), pp. 147; J. Holetschek, *AN*, **136** (1894 Oct. 18), pp. 299–301; H. A. Kobold, *AN*, **137** (1894 Dec. 14), p. 43; D. Peyra, *AN*, **137** (1895, Apr. 9), pp. 273–90; I. Benko von Boinik, *AN*, **141** (1896 Jul. 25), pp. 105, 111; F. W. Ristenpart, *AN*, **142** (1897 Feb. 27), pp. 379, 385; E. E. Barnard, *MNRAS*, **59** (1899 Mar.), pp. 358–60; V1964, p. 64.

D/1894 F1 *Discovered:* 1894 March 26.90 ($\Delta = 0.40$ AU, $r = 1.29$ AU, Elong. $= 131°$)
(Denning) *Last seen:* 1894 June 5.92 ($\Delta = 1.24$ AU, $r = 1.84$ AU, Elong. $= 109°$)
Closest to the Earth: 1894 February 12 (0.2657 AU)
1894 I = 1894a *Calculated path:* LMi (Disc), LEO (Apr. 2), LMi (Apr. 6), LEO (Apr. 12), VIR (May 13)

W. F. Denning (Bristol, England) was conducting a routine sweep for comets when he found this faint object on 1894 March 26.90. He roughly measured the position as $\alpha = 9^h\ 55^m$, $\delta = +32°\ 15'$.

This comet was discovered over a month and a half after having passed closest to both the sun and Earth. It was widely observed on March 27, 28, and 29, before its decreasing light began taking it out of view of the smaller telescopes. On March 27, F. Bidschof (Vienna, Austria) and F. Ristenpart (Göttingen, Germany) indicated a total magnitude near 10.5, while R. Schorr (Hamburg, Germany) and F. Renz (Pulkovo Observatory, Russia) gave the nuclear magnitude as 12. J. Bauschinger (Munich, Germany) said the coma was 2′ across, while Ristenpart said the condensation was 15″ across. Several observers described the comet as fan-shaped, while others said it exhibited a tail. The tail length was given as 3′ by Bidschof, while O. Knopf (Jena, Germany) said it extended toward PA 135°. Bauschinger said the coma was granular in appearance.

The best night of observations came on March 28. There was a wide variation in the magnitude estimates of both the comet and nucleus – so much so that there was overlap. All but one of the brightness estimates fell within the range of magnitude 11–13, with B. von Engelhardt's (Dresden, Germany) estimate that the stellar nucleus was of magnitude 13–14 being most discordant. Engelhardt noted that the comet was not visible in the 13-cm finder. The coma was again frequently described as fan-shaped, with the nucleus being in the northern portion. It was also generally estimated as 1–2′ across. The tail was said to extend toward PA 165° by J. Hartmann (Leipzig, Germany) and toward PA 150° by L. Boss (Dudley Observatory, New York, USA). Boss gave the tail length as 5′, while E. Millosevich (Rome, Italy) gave it as 3′. K. Oertel (Munich, Germany) said the coma appeared granular in the 27-cm refractor.

A wide variation in the estimates of the comet's total and nuclear brightness was again reported on March 29. In general, it would seem that no estimates of the total brightness were actually made, but a magnitude of 11 for the condensation is likely. Both Engelhardt and G. Le Cadet (Lyon, France) specifically noted a stellar nucleus of magnitude 12.5. Engelhardt and H. C. Wilson (Goodsell Observatory, Minnesota, USA) independently noted that the comet was perceptible in their 13-cm finders. The coma was again estimated as 1–2′ across. While most observers described the coma as fan-shaped, A. Kammermann (Geneva, Switzerland) said it was shaped like a parabola. Wilson said the tail was 2′ long. Oertel noted the nucleus appeared granular.

On March 30, Ristenpart gave the magnitude as 9 or 10, with a coma diameter of about 45″, while Engelhardt described the comet as faint, elongated, and diffuse, with a condensation and a stellar nucleus. Also on the 30th, G. Gruss (Prague, Czech Republic) said the comet was brighter than on the 29th and exhibited a distinct nucleus, while Wilson said the tail was 1′ wide and extended 3′ toward PA 164.0°. On March 31, L. Ambronn (Göttingen, Germany) gave the magnitude as 11–12, while C. F. Pechüle (Copenhagen, Denmark) noted a condensed nebulosity extending toward about PA 100°.

On April 1, Engelhardt described the comet as very faint, diffuse, elongated, and condensed. On the 2nd, I. Benko von Boinik (Pula, Yugoslavia) gave the magnitude as 11.0 and estimated the coma as 5′ across and the condensation as 1′ across. That same evening, Wilson said the tail extended toward PA 170.2°. On the 7th, J. Guillaume (Lyon, France) said the magnitude was 11 and the stellar nucleus was magnitude 12–13, while Ristenpart said the comet was so faint that it sometimes disappeared. That same night, Oertel said the comet appeared extremely faint. Knopf said the comet was at the limit of visibility on the 8th. Gruss described the comet as very faint on the 9th, while Schorr gave the nuclear magnitude as 12. On the 21st, Renz said the comet was very faint and appeared as a small, pale nebula, with recognizable condensation. On the 22nd, Renz described the comet as a faint, oblong nebulosity, whose northern section was condensed. Ristenpart was unable to see the comet with the 15-cm refractor on the 23rd, while H. A. Kobold (Strasbourg, France) observed with the 46-cm refractor and described the comet as rather small, with a nucleus of magnitude 12. He said the coma extended like a tail. Schorr could not see the comet with the 26-cm refractor on the 25th.

Although a few observatories were still measuring the comet's position during May and into June, the only physical description came from Kobold on May 5, when he said the comet was very small and faint. The comet was last detected on June 5.92, when S. Javelle (Nice, France) found it with a 76-cm refractor. He gave the position as $\alpha = 12^{\mathrm{h}}\ 26.5^{\mathrm{m}}$, $\delta = +3°\ 51′$.

The first orbit was calculated by L. Schulhof using positions from March 27, 29, and 31. He determined the perihelion date as 1894 February 13.70. Boss took positions spanning the period of March 28 to April 2 and determined the perihelion date as February 17.77. He wrote, "The computations indicate possible eccentricity." Additional parabolic orbits were published by Schulhof, Boss, and C. N. A. Krueger during the next few weeks that seemed to isolate the perihelion date as February 14. Schulhof began suspecting a deviation from parabolic motion in the April 16 issue of the *Astronomische Nachrichten*.

The first elliptical orbit was calculated by Schulhof using positions spanning the period of March 27 to April 25. It indicated a perihelion date of February 9.58 and a period of 6.74 years. Later calculations would reveal the general correctness of this early elliptical orbit, but there continued to be some discrepancy in the period. During the next few weeks the period was given as 7.94 years by Boss, 6.79 years by Schulhof, 6.76 years by J. R. Hind, and 7.70 years by Hind. Several months after the comet was last seen, Schulhof (1895) gave a couple of likely orbits, but still considered the uncertainty in each as ±30 days. The resulting periods were 7.42 years and 7.55 years. As orbits were computed for this comet, Hind and Lamp suggested it might be the lost comet Brorsen. Schulhof concluded that the comets could not be identical, but said it was probable they came from the same source.

R. J. Buckley (1979) used 73 positions obtained between March 27 and June 5, applied perturbations by Venus to Neptune, and determined a perihelion date of February 9.94 and a period of 7.39 years. He noted the comet had been well placed for observation in the northern sky for three months prior to discovery. The consequently short observation arc brings an uncertainty of about 2 weeks in the period.

T	ω	Ω (2000.0)	I	q	e
1894 Feb. 9.9349 (TT)	46.3504	85.7329	5.5274	1.147000	0.697887

ABSOLUTE MAGNITUDE: $H_{10} = 10.4$ (V1964)

FULL MOON: Mar. 21, Apr. 20, May 19, Jun. 18

SOURCES: W. F. Denning, *AN*, **135** (1894 Mar. 28), p. 103; A. C. D. Crommelin, *MNRAS*, **54** (1894 Apr.), pp. 384–5; W. E. Plummer, *MNRAS*, **54** (1894 Apr.), pp. 386–7; L. Boss, *AJ*, **14** (1894 Apr. 4), pp. 15–16; C. W. L. M. Ebell, R. Schorr, E. A. Lamp, C. F. Pechüle, J. Bauschinger, F. Renz, E. Millosevich, and L. Schulhof, *AN*, **135** (1894 Apr. 4), pp. 115–19; I. Benko von Boinik, F. W. Ristenpart, C. F. Pechüle, J. Hartmann, G. Le Cadet, and C. N. A. Krueger, *AN*, **135** (1894 Apr. 11), p. 131–5; B. von Engelhardt and L. Schulhof, *AN*, **135** (1894 Apr. 16), pp. 149–51; P. Chofardet and H. Petit, *AN*, **135** (1894 Apr. 19), p. 165; W. F. Denning, *PA*, **1** (1894 May), p. 421; W. R. Brooks, *PA*, **1** (1894 May), p. 423; H. C. Wilson and L. Boss, *AJ*, **14** (1894 May 1), pp. 29–31; L. Ambronn and L. Schulhof, *AN*, **135** (1894 May 4), pp. 197–200; L. Boss, *AJ*, **14** (1894 May 12), p. 39; F. Ristenpart and H. A. Kobold, *AN*, **135** (1894 May 22), pp. 243–6; L. Schulhof, *AN*, **135** (1894 May 25), p. 261; J. R. Hind, *AN*, **135** (1894 Jun. 13), p. 383; J. R. Hind, *AN*, **136** (1894 Aug. 15), p. 93; G. Gruss, *AN*, **136** (1894 Aug. 23), p. 123; K. Oertel, *AN*, **136** (1894 Aug. 29), pp. 139–42; S. Javelle, *BA*, **12** (1895 Jan.), p. 30; G. Gruss and A. Kammermann, *AN*, **137** (1895 Feb. 11), pp. 171–4; L. Schulhof, *AN*, **137** (1895 Feb. 21), p. 191; E. Lamp, *AN*, **137** (1895 Mar. 15), pp. 209–20; J. R. Hind, E. Lamp, and L. Schulhof, *PA*, **2** (1895 Apr.), p. 384; F. Renz, *AN*, **138** (1895 Jul. 6), pp. 181–4; I. Benko von Boinik, *AN*, **141** (1896 Jul. 25), pp. 105, 111; R. Schorr, *AN*, **143** (1897 Jun. 2), pp. 265, 281; F. Bidschof, *AN*, **149** (1899 Jun. 15), p. 407; V1964, p. 64; R. J. Buckley, *JBAA*, **89** (1979 Apr.), pp. 261, 263.

C/1894 G1 (Gale)   *Discovered:* 1894 April 1.4 ($\Delta = 0.91$ AU, $r = 1.01$ AU, Elong. $= 63°$)

Last seen: 1894 August 21.82 ($\Delta = 2.65$ AU, $r = 2.24$ AU, Elong. $= 55°$)

1894 II = 1894b   *Closest to the Earth:* 1894 May 1 (0.3367 AU)

Calculated path: HOR (Disc), RET (Apr. 11), DOR (Apr. 14), PIC (Apr. 17), COL (Apr. 23), PUP (Apr. 26), CMa (Apr. 27), PUP (Apr. 29), HYA (May 2), CNC (May 7), LEO (May 10), LMi (May 17), UMa (May 27), CVn (Jun. 27)

This comet was discovered on 1894 April 1.4 by W. F. Gale (Sydney, New South Wales, Australia), while he was observing with an 8-cm telescope. It was then in the evening sky. Gale sent a telegram to J. Tebbutt (Windsor, New South Wales, Australia) on the 2nd. Clouds blocked Tebutt's attempt to find the comet that evening, but clear skies were present the next night and he found it with the 11-cm telescope on April 3.44. The position was determined as $\alpha = 2^h 30.8^m$, $\delta = -55° 35'$. Tebbutt said the comet was a round, bright nebula, exhibiting a "pretty well condensed" coma and a faint tail. Tebbutt then sent notification to Melbourne Observatory on the 4th.

The comet remained an easy object for observers in the Southern Hemisphere throughout most of April. Although it passed perihelion around mid-month, the comet continued approaching Earth as April ended. On April 5, H. C. Russell (Sydney, New South Wales, Australia) obtained a 1-hour exposure of the comet and noted a coma 3' across, with a "well-marked central condensation." Two faint rays were also detected extending about 10'. Gale saw the comet with the naked eye on the 7th, then on the 12th he said the coma was 12' in diameter, while the tail was 2° long and 1' wide. Also on the 12th, W. H. Finlay (Royal Observatory, Cape of Good Hope, South Africa) observed with the 18-cm refractor and said the comet was bright and large in moonlight, with no well-defined condensation. J. M. Thome (Cordoba, Argentina) observed with a 30-cm refractor and wrote, "About perihelion, a short, straight tail, from 7–10' in length, was faintly visible in dark field...." Gale said the comet was circular, with no trace of a tail on the 15th, but noted a "diffuse extension" on the 20th, which subsequently appeared as an ill-defined tail 40' long. Russell photographed the comet on the 23rd and noted a coma 6' across and two rays forming the outer borders of the tail that extended about 20'. Russell obtained a 55-minute exposure on the 24th, which revealed a coma about 7.5' across with a central condensation. There was also a narrow tail extending about 1.25° to the edge of the photographic plate and measuring about 2' across. Russell added, "Upon careful examination it appeared that at [45'] from the comet the tail forked, but this feature is so faint that I cannot be sure of it." Russell said a visual examination of the comet with the large refractor revealed a tail 2° long. That same night, Finlay said the comet exhibited a short, faint tail. On the 25th, Gale reported the comet resembled a star of magnitude 4, while Russell made a 1-hour-44-minute exposure that revealed a coma 9' across and a tail about 0.75° long. A visual examination through the 29-cm refractor revealed a very distinct coma with a very marked central condensation and an exceedingly faint tail. On the 26th, Russell obtained a 1-hour-16-minute exposure which showed a coma 10' across and a "broad, diffused, and fan-shaped" tail. Russell obtained a 3-hour exposure on the 27th, which was a particularly clear night. He noted the coma was 25' across, while the tail was 3' wide and extended about 1.25° to the edge of the plate. A visual examination through the 28-cm refractor revealed the tail was 2° long. Russell obtained a

2-hour exposure on the 29th and noted a coma 15' across. He added, "there is no defined tail, only a very faint and diffused something." No tail was seen in the 29-cm refractor. The comet was first seen in the Northern Hemisphere on April 26, when A. E. Douglass (Lowell Observatory, Flagstaff, Arizona, USA) said it was about magnitude 5, with a coma about 4' across. There was also a central nucleus and a narrow tail 8' long. On April 30, W. E. Sperra (Randolph, Ohio, USA) saw the comet with an 8-cm telescope near the horizon. He said it was round, about 15' across, with a strong central condensation. There was no tail.

The comet passed closest to Earth on May 1. Sperra saw it with the naked eye on the 2nd and estimated its magnitude as 3. On the 4th, E. Millosevich (Rome, Italy) observed with a 25-cm refractor (200×) and described the comet as round and 5–6' across. He added that the nucleus was magnitude 9 and somewhat eccentric. The comet was extensively observed during the next 2 days. On the 5th, J. Holetschek (Vienna, Austria) observed with opera-glasses and gave the magnitude as 4.6, while J. Guillaume and G. Le Cadet (Lyon, France) estimated the magnitude as 4 or 5 with the naked eye. Although W. Schur (Göttingen, Germany) and G. Gruss (Prague, Czech Republic) gave the coma diameter as 6–8' across, Guillaume and Le Cadet said it was about 15' across. Le Cadet said the 32-cm refractor (75×) revealed a distinct elliptical nucleus about 20" across, which was about magnitude 9.5. Le Cadet also reported a thread-like tail extending about 1° toward PA 120°. F. W. Ristenpart (Karlsruhe, Germany) observed with a 15-cm refractor and said the nucleus was magnitude 7.5 and 15" across. On May 6, everyone reported the comet was visible to the naked eye, while the coma diameter was given as 9' by E. Hartwig (Bamberg, Germany) and 10' by B. von Engelhardt (Dresden, Germany). R. Spitaler (KK Observatory, Prague, Czech Republic) said a bright, diffuse nucleus was about 1' across in the 10-cm refractor (48×), while R. Schorr (Hamburg, Germany) gave the nuclear magnitude as 8. Photographs were obtained by E. E. Barnard (Lick Observatory, California, USA) and M. F. J. C. Wolf (Heidelberg, Germany). Barnard obtained a 150-minute exposure and wrote, "In this picture the tail is thread-like for some distance from the head. Further away it broadens out slightly, and separates into two or more parts. The northern edge of the tail appears to have a double curvature." Wolf obtained a 46-minute exposure and said it revealed a generally faint and vague tail extending toward the southeast, from which "several faint and three brighter branches show up." The brighter branches were the longest.

On May 7, R. Bryant (Royal Observatory, Greenwich, England) observed with the 17-cm refractor (55×) and said the comet had a stellar nucleus situated south of the coma's center and there was a slight elongation of the coma toward the north. On the 8th, Holetschek gave the magnitude as 4.8, while Guillaume gave it as 5. Guillaume said the stellar nucleus was magnitude 9.5, while Gruss gave it as 10. The coma diameter was given as 6' by Gruss and 2' by C. W. L. M. Ebell (Berlin, Germany). On the 9th,

Spitaler said the comet was visible to the naked eye in moonlight, while Holetschek gave the magnitude as 4.7. H. A. Kobold (Strasbourg, France) noticed a stellar nucleus of magnitude 10. The coma diameter was given as 4' by F. Schwab (Kremsmünster, Austria), 5' by Spitaler, 6–7' by W. Winkler (Jena, Germany), and 8' by Schorr. Winkler said the coma was 5' across on the 12th and 4' across on the 14th. Schorr gave the nuclear magnitude as 9 on the 15th, while Kobold gave it as 11. Winkler said the coma was 3' across on the 19th. On May 21, Holetschek observed with a 4-cm finder and gave the magnitude as 6. H. P. Hollis (Royal Observatory, Greenwich, England) observed with the 17-cm refractor (55×) and described the comet as "brilliant, approximately circular, elongated on the preceding side." He also noted a nucleus. Winkler said the coma was 3–4' across on the 28th and 4' across on the 29th. Le Cadet described the comet as round and 3' across on the 29th. On the 30th, Holetschek gave the magnitude as 6.9, while Guillaume said the coma was 3' across, with a nucleus of magnitude 11.5 and a faint tail toward the northwest. Ristenpart gave the total magnitude as 7.15 on May 31 and said the coma was 3' across.

Holetschek gave the magnitude as 7.3 on June 1 and 7.7 on the 8th. Le Cadet observed the comet on the 15th and said it had rapidly diminished in brightness and was completely hidden by the light of a 12-day old moon. On the 21st, Kobold described the comet as a round nebulosity about 2' across, with little condensation. Guillaume said the comet was 2.5' in diameter with a diffuse border on the 22nd, while Schwab said it was 2' across with a diffuse nucleus. On June 24, Hollis said the comet was very faint and diffuse. Holetschek gave the magnitude as 8.8 on the 27th, while Kobold noted a distinct nucleus and said the comet was seen without difficulty in the finder. On June 28, Hollis said the comet was exceedingly faint. Holetschek gave the magnitude as 9.0 on the 29th and 30th.

On July 2, Hollis said the comet was excessively faint and very hard to see. On the 2nd and 4th, Guillaume said the comet was only visible using averted vision in the 16-cm refractor. A. C. D. Crommelin (Royal Observatory, Greenwich, England) said the comet was very faint and ill defined in the 17-cm refractor (55×) on the 3rd. On the 6th, Holetschek began estimating the brightness using a 15-cm refractor and gave the magnitude as 10. Crommelin said the comet was excessively faint on the 7th. The comet attained its most northerly declination of +44° on July 10. On the 11th, E. W. Maunder (Royal Observatory, Greenwich, England) said the comet was exceedingly diffuse and faint. He added that it took partly averted vision to see the comet in the 71-cm refractor and that no shape or apparent nucleus could be seen. Kobold could no longer find the comet in the 46-cm refractor on July 24. Holetschek gave the magnitude as 11.5 on the 24th and 27th. During the period of July 21–28, Le Cadet said the comet was extremely faint in the 32-cm refractor and could hardly be seen.

On August 5 and 6, Holetschek gave the magnitude as about 12. Although observations abruptly ended shortly thereafter because of moonlight, the comet was seen one last time on August 21.82, when S. Javelle (Nice, France) gave the position as $\alpha = 13^h\ 39.3^m$, $\delta = +42°\ 24'$.

The comet's spectrum was examined by W. W. Campbell (Lick Observatory, California, USA). He reported 25 bright lines and said the spectrum was "identical in every respect with that of Comet [C/1893 N1]." He added, "Its principal constituents were therefore carbon and cyanogen." A. Fowler (London, England) observed the spectrum on May 7 and noted the three carbon bands, as well as a fairly bright continuous spectrum.

The earliest parabolic orbits came from Australia. T. Roseby took Tebbutt's positions for April 3, 6, and 12, and calculated a perihelion date of 1894 April 14.26, while R. L. J. Ellery and R. T. A. Innes independently took positions spanning the period of April 3–11 and determined perihelion dates of April 14.25 and April 13.92, respectively. Additional orbits by H. C. F. Kreutz, Tebbutt, Roseby, E. Kohlschütter, and Graham eventually established the perihelion date as April 14.0. The first elliptical orbit was calculated by Roseby using positions spanning a 78-day arc. The result was a perihelion date of April 13.89 and a period of 1001 years. H. A. Peck (1901) calculated a perihelion date of April 13.91 and a period of about 1143 years. During 1907, Peck revised his orbit by giving the perihelion date as April 13.89 and the period as about 960 years. This last orbit is given below.

T	ω	Ω (2000.0)	i	q	e
1894 Apr. 13.5390 (UT)	324.1843	207.8850	86.9665	0.983032	0.989889

ABSOLUTE MAGNITUDE: $H_{10} = 6.3$ (V1964)

FULL MOON: Mar. 21, Apr. 20, May 19, Jun. 18, Jul. 17, Aug. 16, Sep. 15

SOURCES: W. F. Gale, *AN*, **135** (1894 Apr. 11), pp. 135–6; W. F. Gale, *AJ*, **14** (1894 Apr. 14), p. 24; H. C. F. Kreutz, *AN*, **135** (1894 Apr. 16), pp. 149–52; R. L. J. Ellery, *AN*, **135** (1894 Apr. 25), pp. 183–4; R. Bryant, *MNRAS*, **54** (1894 May), p. 450; W. F. Gale, *PA*, **1** (1894 May), p. 421; H. C. F. Kreutz and A. E. Douglass, *AJ*, **14** (1894 May 1), p. 32; A. E. Douglass, A. H. P. Charlois, and H. C. F. Kreutz, *AN*, **135** (1894 May 4), pp. 197–200; A. Fowler, *Nature*, **50** (1894 May 10), pp. 36–7; E. Millosevich, W. Schur, and R. Schorr, *AN*, **135** (1894 May 11), p. 211; M. F. J. C. Wolf, W. F. Gale, J. Tebbutt, E. Kohlschütter, and R. T. A. Innes, *AN*, **135** (1894 May 25), pp. 257–64; H. C. Russell, R. P. Sellors, A. C. D. Crommelin, R. Bryant, H. P. Hollis, C. Davidson, W. F. Gale, T. Roseby, R. S. Ball, and J. Tebbutt, *MNRAS*, **54** (1894 Jun.), pp. 547–56; A. E. Douglass, *PA*, **1** (1894 Jun.), p. 470; E. Hartwig, B. von Engelhardt, R. Schorr, G. Le Cadet, and J. Guillaume, *AN*, **135** (1894 Jun. 2), pp. 275–80; H. C. Russell, R. P. Sellors, and R. L. J. Ellery, *AN*, **135** (1894 Jun. 8), pp. 343–6; W. Winkler, *AN*, **136** (1894 Jul. 14), p. 29; J. Tebbutt, *AN*, **136** (1894 Aug. 15), pp. 95–6; W. H. Finlay, *AN*, **136** (1894 Aug. 23), p. 123; W. E. Sperra, *PA*, **2** (1894 Sep.), p. 44; C. W. L. M. Ebell, *AN*, **136** (1894 Sep. 12), p. 155; J. Guillaume and G. Le Cadet, *AN*, **136** (1894 Sep. 27), pp. 195–200; J. M. Thome, *AN*, **136** (1894 Oct. 9), pp. 265–8; J. Holetschek, *AN*, **136** (1894 Oct. 18), p. 302; H. P. Hollis, A. C. D. Crommelin, E. W. Maunder, R. S. Ball Graham, and W. H. Finlay, *MNRAS*, **54** (1894, Supplementary Notice), pp. 581–6; W. E. Plummer, *MNRAS*, **55**

(1894 Nov.), pp. 45–9; H. A. Kobold, *AN*, **137** (1894 Dec. 14), pp. 43–5; W. W. Campbell, *PASP*, **7** (1895 Jan. 1), p. 63; W. E. Plummer and W. F. Gale, *MNRAS*, **55** (1895 Feb.), p. 234; G. Gruss, *AN*, **137** (1895 Feb. 11), pp. 171–4; F. Schwab, *AN*, **139** (1896 Feb. 8), p. 335; T. Roseby, *MNRAS*, **56** (1896 Mar.), pp. 329–30; F. W. Ristenpart, *AN*, **142** (1897 Feb. 27), pp. 379, 385; E. E. Barnard, *MNRAS*, **59** (1899 Mar.), pp. 360–1; ABP (1901), p. 28; H. A. Peck and S. Javelle, *AJ*, **21** (1901 Apr. 29), pp. 121–34; H. A. Peck, *AJ*, **25** (1907 Sep.), pp. 181–2; V1964, p. 64.

10P/Tempel 2

1894 III = 1894c

Recovered: 1894 May 9.16 ($\Delta = 1.69$ AU, $r = 1.36$ AU, Elong. = 54°)
Last seen: 1894 August 9.07 ($\Delta = 1.52$ AU, $r = 1.76$ AU, Elong. = 85°)
Closest to the Earth: 1894 October 14 (1.3283 AU)
Calculated path: AQR (Rec), PSC (May 10), CET (May 20), PSC (Jun. 20), CET (Jun. 25), TAU (Jul. 31)

This comet was predicted to pass perihelion on 1883 November 20 and 1889 February 2, but both apparitions were not considered favorable. The 1883 apparition was considered slightly better than that of 1889, but searches by E. W. L. Tempel and E. Hartwig were unsuccessful. As the 1894 apparition approached, L. Schulhof took the orbit for the 1878 return and advanced it to 1894. He found the comet would next arrive at perihelion on 1894 April 24.13, with a likely uncertainty of ±2 days. The comet was recovered by W. H. Finlay (Royal Observatory, Cape of Good Hope, South Africa) on 1894 May 9.16, at a position of $\alpha = 23^h 45.3^m$, $\delta = -4° 51'$. Finlay described it as circular, less than 1' across, and about magnitude 11 or fainter. There was some central condensation, but no tail. Although Finlay's position indicated Schulhof's prediction required a correction of only −0.4 day, the May 16 issue of the *Astronomische Nachrichten* included a revision by Schulhof that had been mailed on April 30. This new prediction gave a perihelion date of April 23.81, which was just 1.6 hours later than the comet's actual perihelion date.

The comet was already past perihelion when discovered and its closest approach to Earth would not occur until it had faded from view. Observations were not plentiful and physical descriptions were even more scarce. Finlay reobserved the comet on May 10 and 11, and found it unchanged in appearance since the 9th. J. Tebbutt (Windsor, New South Wales, Australia) saw this comet with a 20-cm refractor on May 12 and 13. He described it as excessively faint. Finlay also reported it as excessively faint on June 3 and 15.

The comet was last detected on August 5.10 and 9.07, when S. Javelle (Nice, France) found it at the limit of visibility in a 76-cm refractor. For the last date he gave the position as $\alpha = 3^h 30.9^m$, $\delta = +4° 25'$.

Minor refinements of this comet's orbit using positions from multiple apparitions, as well as planetary perturbations, have been published over the years. The most recent were by B. G. Marsden and Z. Sekanina (1971) and Sekanina (1985). These have typically revealed a perihelion date of

April 23.74 and a period of 5.22 years. Sekanina determined the nongravitational terms as $A_1 = +0.08$ and $A_2 = +0.0021$ and his orbit is given below.

T	ω	Ω (2000.0)	i	q	e
1894 Apr. 23.7442 (TT)	185.1388	122.5930	12.7331	1.350670	0.551105

ABSOLUTE MAGNITUDE: $H_{10} = 9.5$ (V1964)

FULL MOON: Apr. 20, May 19, Jun. 18, Jul. 17, Aug. 16

SOURCES: W. E. Plummer, *MNRAS*, **44** (1884 Feb.), p. 180; *ARSI*, part 1 (1886), p. 383; L. Schulhof, *AN*, **120** (1888 Nov. 19), p. 173; L. Schulhof, *AN*, **135** (1894 Mar. 22), p. 45; L. Schulhof, *PA*, **1** (1894 May), p. 422; W. H. Finlay, *AN*, **135** (1894 May 11), p. 215; L. Schulhof, *AN*, **135** (1894 May 16), p. 229; L. Schulhof, *BA*, **11** (1894 Jun.), pp. 254–6; W. H. Finlay, *PA*, **1** (1894 Jun.), p. 471; W. H. Finlay, *AN*, **135** (1894 Jun. 13), p. 383; L. Schulhof, *AN*, **136** (1894 Aug. 15), p. 91; W. H. Finlay, *AN*, **136** (1894 Aug. 23), p. 125; J. Tebbutt, *AN*, **136** (1894 Oct. 5), pp. 241–6; S. Javelle, *BA*, **13** (1896 Jan.), p. 18; V1964, p. 64; B. G. Marsden and Z. Sekanina, *AJ*, **76** (1971 Dec.), pp. 1137–8; Z. Sekanina, *QJRAS*, **26** (1985), p. 114; B. G. Marsden and Z. Sekanina, *CCO*, 12th ed. (1972), pp. 50–1.

2P/Encke

1895 I = 1894d

Recovered: 1894 October 31.77 ($\Delta = 0.91$ AU, $r = 1.76$ AU, Elong. $= 134°$)

Last seen: 1895 January 25.74 ($\Delta = 0.63$ AU, $r = 0.44$ AU, Elong. $= 19°$)

Closest to the Earth: 1894 November 13 (0.8982 AU), 1895 January 29 (0.6177 AU)

Calculated path: PEG (Rec), AQR (Jan. 7)

A. Berberich sent an ephemeris to a few observatories in late October of 1894 that was based on his prediction that the comet would return to perihelion on 1895 February 5.24. This comet was recovered by M. F. J. C. Wolf (Heidelberg, Germany) on 1894 October 31.77 through the use of photography. The position was estimated as $\alpha = 23^h 16.6^m$, $\delta = +14° 32'$. Wolf described it as very faint, with a total magnitude near 13, but without a recognizable nucleus. It was independently recovered by H. J. A. Perrotin (Nice, France) on October 31.90 and V. Cerulli (Teramo, Italy) on November 1.8. Perrotin described it as a broad spot and said it was at the extreme limit of the 76-cm refractor. Cerulli said the comet was very faint and diffuse.

C. Rambaud and F. Sy (Alger, now al-Jazâ'ir, Algeria) said the comet was difficult to measure on November 19 and 20, because of its faintness. H. A. Kobold (Strasbourg, France) saw the comet with a 46-cm refractor on the 24th and said it was very faint. On November 28, E. D. Swift (Lowe Observatory, Echo Mountain, California, USA) was observing with a 41-cm refractor and located a "large, bright comet which, for a few minutes, he thought was a new one." L. Swift added that after a few minutes, he realized his mistake as he did not know comet Encke had become so bright. L. Swift said the comet was also visible in the 9-cm finder. J. Holetschek (Vienna, Austria) made a series of magnitude estimates during the month, using a

15-cm Fraunhofer refractor. It was about 12 on November 18 and 20, and 11.7 on the 23rd.

G. Le Cadet (Lyon, France) saw the comet with a 32-cm refractor (180×) on December 1. He described it as extremely faint and hardly distinct, with a diameter of about 5′. He added that averted vision seemed to reveal an eccentric condensation near the edge of the coma. W. E. Plummer (Liverpool Observatory, Liverpool, England) said the comet was still very feeble, without any noticeable condensation on the 19th. On December 28, W. Winkler (Jena, Germany) observed with a 15-cm refractor (60×) and said the comet was bright, about 3′ across, but without a central condensation. Plummer said the comet was well seen on the 30th. R. Bryant (Royal Observatory, Greenwich, England) saw the comet on the same date with the 17-cm refractor and noted it was bright and distinct, with a stellar condensation. Winkler said the comet appeared less bright than on the 28th. Holetschek continued to estimate the comet's total magnitude during the month. He found it at 10.5 on December 2, 8.7 on the 17th, 8.5 on the 22nd, 8.1 on the 26th, and 7.8 on the 31st.

W. E. Sperra (Randolph, Ohio, USA) was initially uncertain whether he was seeing this comet or not. Using his 8-cm refractor, he first searched for the comet on December 14. He wrote, "I succeeded in picking up what appeared to be an 8th magnitude star with diffuse nebulosity...." Although he was unable to identify the object with any known nebula, he looked at the object on the 15th and noticed no change of position. Although he then believed it was a nebula, his next opportunity to observe it came on the 18th, at which time he noted it had moved -1^m in α and $-25'$ in δ. Having no ephemeris at hand to confirm the comet's motion, Sperra wrote to J. E. Keeler (Allegheney, Pennsylvania, USA) who checked on the matter and responded that Sperra was indeed seeing Encke's comet.

As the new year began the comet was again approaching both the sun and Earth. C. W. L. M. Ebell (Berlin, Germany), Bryant, and W. H. Robinson (Radcliffe Observatory, Oxford, England) all reported the comet was faint because of moonlight, during the first 8 days of January. Ebell said the comet was about 1′ across on January 1, with a magnitude of 12. After the moon had left the sky, Bryant reported the comet was very bright on the 13th, and Ebell estimated the comet's magnitude as 8.5 on the 14th and said the coma was about 2′ across. F. Schwab (Kremsmünster, Austria) saw the comet on January 15 and 18, and said it was round, 2′ across, with an eccentrically situated nucleus. Sperra said the comet was near naked-eye visibility on the 17th. Although he thought he could occasionally see a whitish object at the comet's position, he could not be certain. With a telescope, Sperra said the comet was round, with a stellar nucleus. He added that the tail was at least a degree long, slightly curved southward, and slightly wider at the end. Most notable was that the tail was much brighter for the first half degree and contained several "bright patches or condensations." He said these condensations were located at distances of 15′, 25′, and 40′ from the head, with the last being the faintest and somewhat doubtful. H. C. Wilson (Northfield, Minnesota, USA) said a faint tail could be traced for about 1.5° on the 16th. He added that the comet was visible to the naked eye.

Twilight began being reported as interfering with observations on January 18. On that date, Winkler said the comet was very bright at low altitude, with a central condensation. W. Wickham (Radcliffe Observatory, Oxford, England) said the comet was faint in the 25.4-cm refractor, with no tail and only a slight brightening following its center. He noted it was brighter on the 21st, with an occasionally visible nucleus. A. C. D. Crommelin (Royal Observatory, Greenwich, England) saw the comet in fairly bright twilight on January 23, while using the 17-cm refractor. On the same date, Robinson said the nucleus was a "diffused disc" of magnitude 9 or 10. Holetschek estimated the total magnitude as 6.6 on January 15, 6.3 on the 17th, and about 5.2 on the 24th and 25th.

The comet was last detected on January 25.74, when Robinson gave the position as $\alpha = 21^h 31.0^m$, $\delta = -6° 57'$. He said the comet was visible in strong twilight as a diffused mass in the 25-cm refractor (100×). Plummer said he also detected the comet on the 25th, but he was unable to measure its position because of its low altitude and a lack of a nearby comparison star. Although no time was given, his observations on previous days would indicate a likely time of January 25.73.

An orbit was calculated by A. Iwanow (1898) using only those positions obtained during this apparition. He determined the perihelion date as 1895 February 5.24 and a period of 3.30 years. Refinements of this comet's orbit using positions from multiple apparitions, as well as planetary perturbations, have been published by several astronomers. Of most significance were the studies of O. Backlund (1910) and B. G. Marsden and Z. Sekanina (1974). Both revealed a perihelion date of February 5.25 and a period of 3.30 years. Marsden and Sekanina determined the nongravitational terms as $A_1 = -0.18$ and $A_2 = -0.02109$. The orbit of Marsden and Sekanina is given below.

T	ω	Ω (2000.0)	i	q	e
1895 Feb. 5.2496 (TT)	183.9420	336.2230	12.9186	0.341076	0.846242

ABSOLUTE MAGNITUDE: $H_{10} = 9.3$ (V1964)

FULL MOON: Oct. 14, Nov. 13, Dec. 12, Jan. 11, Feb. 9

SOURCES: M. F. J. C. Wolf and V. Cerulli, AN, **136** (1894 Nov. 3), p. 351; V. Cerulli, AJ, **14** (1894 Nov. 9), p. 136; M. F. J. C. Wolf, V. Cerulli, and H. J. A. Perrotin, AN, **136** (1894 Nov. 9), p. 367; A. Berberich, AN, **136** (1894 Nov. 12), p. 379; M. F. J. C. Wolf, Nature, **51** (1894 Nov. 15), p. 64; M. Wolf and W. T. Lynn, Nature, **51** (1894 Nov. 29), pp. 108–9; V. Cerulli, PA, **2** (1894 Dec.), p. 189; G. Le Cadet, C. Rambaud, and F. Sy, AN, **137** (1894 Dec. 31), p. 77; R. Bryant and W. H. Robinson, MNRAS, **55** (1895 Jan.), pp. 159–63; E. D. Swift and L. Swift, PA, **2** (1895 Jan.), p. 236; E. Frisby and J. G. Porter, AJ, **14** (1895 Jan. 31), pp. 191–2; H. C. Wilson and W. E. Sperra, PA, **2** (1895 Feb.), pp. 279–80;

G. Le Cadet, *AN*, **137** (1895 Feb. 1), p. 143; F. Valle, *AN*, **139** (1896 Feb. 3), p. 297; G. Le Cadet and F. Schwab, *AN*, **139** (1896 Feb. 8), pp. 329, 333, 335; W. Winkler, *AN*, **137** (1895 Feb. 11), pp. 173–6; H. C. Wilson, *AJ*, **15** (1895 Feb. 28), p. 16; R. Bryant, H. Furner, H. P. Hollis, and A. C. D. Crommelin, *MNRAS*, **55** (1895 Mar.), pp. 320–1; W. E. Sperra, *PA*, **2** (1895 Mar.), p. 333; W. S. Eichelberger, *AJ*, **15** (1895 Mar. 12), p. 23; J. Holetschek, *AN*, **137** (1895 Mar. 21), pp. 237–40; P. Isham and J. G. Porter, *AJ*, **15** (1895 Mar. 27), pp. 35, 37; C. W. L. M. Ebell, *AN*, **137** (1895 May 6), pp. 363–6; E. Frisby, *AJ*, **15** (1895 May 8), p. 63; M. E. Byrd, *AJ*, **15** (1895 Jul. 16), p. 103; F. Valle, *MNRAS*, **56** (1895 Nov.), pp. 41–2; W. Wickham and W. H. Robinson, *MNRAS*, **56** (1895 Dec.), pp. 81–3; H. J. A. Perrotin, *BA*, **13** (1896 Jan.), p. 18; F. Schwab, *AN*, **139** (1896 Feb. 8), p. 335; H. A. Kobold, *AN*, **140** (1896 May 30), p. 315; W. E. Plummer, *MNRAS*, **56** (1896 Jun.), pp. 504–5; W. Laska, *AN*, **142** (1897 Mar. 5), p. 399; A. Iwanow, *AN*, **146** (1898 Apr. 25), p. 159; O. Backlund, *MNRAS*, **70** (1910 Mar.), pp. 429–42; V1964, p. 64; B. G. Marsden and Z. Sekanina, *AJ*, **79** (1974 Mar.), pp. 415–16.

54P/1894 W1 (de Vico–Swift–NEAT) 1894 IV = 1894e

Discovered: 1894 November 21.18 ($\Delta = 1.03$ AU, $r = 1.46$ AU, Elong. $= 92°$)
Last seen: 1895 January 30.18 ($\Delta = 1.84$ AU, $r = 1.81$ AU, Elong. $= 72°$)
Closest to the Earth: 1894 August 7 (0.6513 AU)
Calculated path: AQR (Disc), PSC (Dec. 18), CET (Jan. 4), PSC (Jan. 5)

Following this comet's 1844 apparition, predictions were made for its return. The first of 1850 was considered unfavorable and apparently no searches were made. The next was expected in 1855, but J. R. Hind suggested that persistently cloudy skies over Europe probably prevented the comet from being found. F. F. E. Brünnow (1859) investigated the orbit of this comet and provided predictions for the next few returns. Searches were made at Harvard College Observatory (Massachusetts, USA) during the expected 1860 return, but nothing was found. Hind distributed search ephemerides for the 1866 return, but nothing was reported. Further searches at the expected 1871 return were also futile, and Hind (1871) remarked that, after so many missed returns, the predicted perihelion date for that apparition was probably several months in error. No further predictions were made.

E. D. Swift (Lowe Observatory, Echo Mountain, California, USA) accidentally discovered this comet on 1894 November 21.18, and estimated the position as $\alpha = 22^h 18.4^m$, $\delta = -13° 07'$. He described it as faint and elongated toward the sun. The comet had already passed perihelion and its closest approach to Earth.

Only a couple of physical descriptions were provided before the end of November. The first confirmation came from E. E. Barnard (Lick Observatory, California, USA) on November 22.13. Using a 30-cm refractor, he described it as small and fainter than magnitude 13. Barnard added that there was an indefinite brightening toward the middle and a faint trace of a short tail. G. Bigourdan (Paris, France) said the comet appeared very faint on the 25th and estimated the brightness as 13.4. He described it as nearly stellar, with weak nebulosity.

Only a few descriptions were provided during the remainder of the comet's apparition. H. A. Kobold (Strasbourg, France) described the comet as very small and faint on December 1, with a stellar nucleus of magnitude 12. H. C. Wilson (Goodsell Observatory, Northfield, Minnesota, USA) observed the comet with a 41-cm refractor on the 19th and described it as 30″ across, with a slight condensation, and "an exceedingly faint tail 3′ long." V. Cerulli (Teramo, Italy) saw the comet with a 39-cm Cooke telescope on December 27 and said the comet was round, with a diameter of 1′. H. A. Howe (Chamberlin Observatory, Colorado, USA) described the comet as an extremely faint patch of nebulosity on January 20. He added that it was devoid of condensation. After an extended period of bad weather, Barnard tried to find the comet with the 30-cm telescope on January 25, but was unsuccessful. He turned the 91-cm refractor toward the predicted position of the comet on the 26th and found it with the "greatest difficulty." He described it as excessively faint and about 10–15″ across. On this and the following nights, Barnard said the brightness could not have exceeded that of a 17th-magnitude star. The comet was last detected on January 30.18, when Barnard measured its position as $\alpha = 1^h\ 22.4^m$, $\delta = +8°\ 57′$. He was using the 91-cm refractor.

An orbit had not yet been calculated for this comet when A. Berberich made an interesting announcement on November 23. He suggested this comet was the short-period comet found by de Vico in 1844 and added, "The delay of approximately 13 months for this return is probably due to disturbances by Jupiter." He suggested Brünnow's orbit for the 1844 apparition was probably close to correct and said the comet had passed close to Jupiter in 1863, 1874, and 1886. Finally, he added that the perihelion date would be found close to 1894 October 11.

The first parabolic orbit calculated for this comet came from A. O. Leuschner. He took positions obtained by Barnard on November 22, 23, and 24, and determined a perihelion date of October 5.40. L. Schulhof followed with an orbit using positions spanning the period of November 22–29 and determined a perihelion date of October 19.99. Schulhof also expressed his opinion that this comet was identical to de Vico's comet of 1844. He subsequently gave a hypothetical elliptical orbit with a perihelion date of October 12.66 and a period of 5.8 years. F. H. Seares (1899) calculated a definitive orbit for this comet, but he did not attempt to link it to the 1844 apparition. He began with 75 positions obtained at the apparition of 1894–5 and ultimately obtained a perihelion date of October 12.70 and a period of 5.86 years. Interestingly, the fact that de Vico's comet of 1844 and Swift's comet of 1894 were identical was not absolutely proven until 1963, when B. G. Marsden successfully linked the two apparitions. For the 1894 apparition, he determined the perihelion date as October 12.7 and the period as 5.86 years. Marsden's prediction enabled the comet to finally be recovered in 1965. Using positions from multiple apparitions, S. D. Shaporev (1978) and S. Nakano (1992, 1999) indicated a perihelion date of October 12.63 and a period of 5.86 years. Nakano's last orbit is given below.

T	ω	Ω (2000.0)	i	q	e
1894 Oct. 12.6265 (UT)	296.7799	49.9681	2.9750	1.391598	0.571922

ABSOLUTE MAGNITUDE: $H_{10} = 10.2$ (V1964)

FULL MOON: Nov. 13, Dec. 12, Jan. 11, Feb. 9

SOURCES: F. F. E. Brünnow, *Astronomical Notices*, No. 3 (1859), p. 4; J. R. Hind, *MNRAS*, **31** (1871 May), pp. 216–17; J. R. Hind, *AN*, **78** (1871 Oct. 9), p. 223; E. Swift and E. E. Barnard, *AN*, **137** (1894 Nov. 24), p. 15; G. Bigourdan, *CR*, **119** (1894 Nov. 26), p. 894; E. Swift, *PA*, **2** (1894 Dec.), p 189; E. E. Barnard, A. O. Leuschner, A. Berberich, F. F. Tisserand, and L. Schulhof, *AN*, **137** (1894 Dec. 4), p. 31; E. E. Barnard, *AJ*, **14** (1894 Dec. 10), p. 151; L. Schulhof, H. A. Kobold, and J. Palisa, *AN*, **137** (1894 Dec. 14), pp. 37–42; E. Swift and L. Swift, *PA*, **2** (1895 Jan.), pp. 234, 236; V. Cerulli, *AN*, **137** (1895 Jan. 10), p. 95; E. E. Barnard, *AJ*, **15** (1895 Feb. 12), supplement; E. E. Barnard, *AJ*, **15** (1895 Feb. 28), pp. 9–10; H. A. Howe, *AJ*, **15** (1895 Mar. 12), p. 23; H. A. Kobold, *AN*, **140** (1896 May 30), p. 315; F. H. Seares and E. E. Barnard, *AN*, **151** (1899 Dec. 9), pp. 81–102; V1964, p. 64; S. D. Shaporev, *QJRAS*, **19** (1978), pp. 52–3, 57; S. Nakano, *Nakano Note* No. 569 (1992 Apr. 22); S. Nakano, *Nakano Note* No. 569R (1999 Oct. 31).

95P/Chiron *Discovered:* 1895 April 24.15 ($\Delta = 7.52$ AU, $r = 8.41$ AU, Elong. $= 150°$)
Closest to the Earth: 1895 March 27 (7.4076 AU)
Calculated path: VIR (Disc)

The discovery of this comet did not occur until 1977, and even at that time, this object was considered a minor planet until a tail was photographed in 1989.

C. T. Kowal (Palomar Observatory, California, USA) discovered this slow-moving object on photographs made in 1977. After the period had been reasonably determined and it was realized that this object could reach a magnitude of 14.5 at perihelion, the search was on for pre-discovery images. Searches concentrated on 1895 and 1945, with images being found on photographs obtained in both of these years, as well as several other years.

The 1895 image was found by W. Liller and L. J. Chaisson (Center for Astrophysics, Cambridge, Massachusetts, USA) on a 60-minute photograph obtained at Harvard College Observatory (Cambridge, Massachusetts, USA) with the 61-cm Bruce astrograph on April 24.15. The position was given as $\alpha = 12^h$ 12.2^m, $\delta = -4° 30'$. The object was described as "extremely weak".

Although this object was originally designated minor planet 2060, 12 years after the discovery K. J. Meech and M. J. S. Belton reported they had obtained images showing a coma, so that this "minor planet" was really a comet. The 1895 orbit was calculated by K. Ziolkowski (1995) and is given below.

T	ω	Ω (2000.0)	i	q	e
1895 Mar. 16.8215 (TT)	336.2382	210.5085	6.9854	8.404520	0.376808

ABSOLUTE MAGNITUDE: $H_{10} = 3.0$ (Kronk, 2001)
FULL MOON: Apr. 9, May 8
SOURCES: C. T. Kowal, *IAUC*, No. 3129 (1977 Nov. 4); B. G. Marsden, *IAUC*, No. 3145 (1977 Nov. 30); W. Liller and L. J. Chaisson, *IAUC*, No. 3151 (1977 Dec. 13); K. J. Meech and M. J. S. Belton, *IAUC*, No. 4770 (1989 Apr. 11); B. G. Marsden and K. Ziolkowski, *CCO*, 10th ed. (1995), pp. 68–9.

X/1895 M1

L. Swift observed a nebulous object on 1895 June 30.50, at a position of $\alpha = 1^h\ 20^m$, $\delta = +2°\ 55'$. Bad weather prevented a follow-up search until July 5, at which time nothing was at that position and the object was not seen again. Although Swift suggested this was possibly a return of Barnard's comet of 1884, it was concluded that the position was too far from that expected for the 1884 comet.

SOURCES: L. Swift, *AN*, **138** (1895 Jul. 6), p. 183; L. Swift, *PA*, **3** (1895 Sep.), p. 38.

D/1895 Q1
(Swift)

1895 II = 1895a

Discovered: 1895 August 21.42 ($\Delta = 0.35$ AU, $r = 1.30$ AU, Elong. $= 139°$)
Last seen: 1896 February 6.19 ($\Delta = 2.04$ AU, $r = 2.26$ AU, Elong. $= 90°$)
Closest to the Earth: 1895 August 31 (0.3451 AU)
Calculated path: PSC (Disc), CET (Dec. 21), ARI (Jan. 8)

L. Swift (Lowe Observatory, California, USA) discovered this comet on 1895 August 21.42 at a position of $\alpha = 0^h\ 27.7^m$, $\delta = +5°\ 30'$. He described it as faint and round, with a bright elliptical central condensation. E. E. Barnard (Lick Observatory, California, USA) confirmed the discovery on August 22.31. Interestingly, Swift found the comet while trying to secure a more accurate position for the last nebula he had discovered at Warner Observatory (Rochester, New York, USA). Upon setting the 41-cm refractor on the position of the nebula, Swift looked through and immediately noticed "a beautiful comet instead of the expected nebula."

The comet was discovered on the day of perihelion and passed closest to Earth on the last day of August. Therefore, not long after astronomers began observing this comet, it started to fade. Observers, whose telescopes ranged from 25-cm to 32-cm refractors, typically described the comet as either faint or extremely faint during the remainder of August. G. Le Cadet (Lyon, France) described the comet as very faint and very diffuse on the 23rd and noted a central condensation was only visible by averted vision. W. H. Robinson (Radcliffe Observatory, Oxford, England) said the comet was about 2' across on the 25th, with a condensation. On the same night, W. E. Plummer (Liverpool, England) said the comet appeared ill defined, while

R. Schorr (Hamburg, Germany) indicated the condensation was 20" across. Le Cadet estimated the coma diameter as 5' on the 26th and noted a central condensation about 30" across. H. A. Kobold (Strasbourg, France) said the comet was 1' across on the 28th, with a slight condensation. J. Holetschek (Vienna, Austria) attempted to estimated the total brightness of the comet during this period. He said it was about magnitude 11 on August 24 and 27, but was closer to 12 on the 30th.

Moonlight had become a factor as September began and, although Le Cadet described the comet as extremely faint on the 1st, most observers did not provided descriptions during the first half of the month, except for H. A. Howe (Chamberlin Observatory, Colorado, USA). Howe was observing with a 51-cm refractor and said the comet was quite bright on the 2nd, despite moonlight and dawn. By the 12th, however, the comet's steadily increasing distance from the sun and Earth caused it to appear very faint in moonlight, according to Howe. Le Cadet noted the comet was hardly visible in his 32-cm refractor on the 13th, and added that there was little condensation. Kobold noted the comet was extremely faint in a 46-cm refractor on the 16th, while Robinson said it was feeble and about 3' across in the 25.4-cm refractor on the 18th. Both astronomers noted a faint nucleus or condensation. The comet must have been a very diffuse object, not well suited for large refractors, because where Le Cadet could only see it with averted vision in a 32-cm refractor on the 19th and Robinson said it was extremely feeble in a 25.4-cm refractor on the 20th, V. Cerulli (Teramo, Italy) noted it was visible in a 6.4-cm seeker during the period of September 18–20. Cerulli added that the 39-cm refractor (250×) revealed a round nucleus 20–30" across. Interestingly, Robinson said the comet seemed slightly brighter on the 25th than when seen on the 20th. By the 26th the comet was again being affected by moonlight. Howe said it was just visible in the 51-cm refractor on the 26th, with a first-quarter moon in the sky. The comet became fainter as moonlight increased. The nucleus was reported as more distinct during the last days of September. P. Chofardet (Besançon, France) said it was magnitude 13 during the period of September 24–29. Interestingly, a tail was suspected by Howe on the 30th and he estimated it was about 5' long in the 51-cm refractor.

Although Howe continually said the comet was easily seen in the 51-cm refractor through October and November, W. A. Villiger (Munich, Germany) said the comet was hardly visible in a 27-cm refractor on October 19 and Robinson said it was a "small faint haze" on the 24th with a 25-cm refractor. The comet maintained a round shape during this time, but the nucleus became increasingly difficult to see in smaller telescopes near the end of October. Howe made two interesting observations with the 51-cm refractor during November which indicated how faint the coma had become. On the 9th he said the comet appeared as an excessively faint star, with an occasional hazy appearance. On the 15th he noted a faint, tenous coma occasionally visible that was nearly 9' across. Kobold saw the comet with a

46-cm refractor on the 21st and noted the coma was only 0.5′ across, but this probably referred to the inner portion of the coma, which some observers were reporting as a condensation.

Howe provided the only descriptions during December and indicated that even the 51-cm refractor was beginning to be challenged by this comet. He said the moon nearly blotted out the comet on the 5th and it was very hard to see on the 10th without moonlight. Howe described the comet as "not excessively difficult" on the 20th and stated it would probably be observed in January. Unfortunately, bad weather or poor sky conditions prevented him from continuing observations after December.

The final observations of this comet were made at Lick Observatory on February 6.14, by W. J. Hussey, and February 6.19, by W. W. Campbell. Campbell gave the position as $\alpha = 3^h\,01.3^m$, $\delta = +13°\,42'$.

The first parabolic orbit was calculated by Hussey using positions obtained on August 22–24. The resulting perihelion date was 1895 October 5.70. The orbit was refined as the observation arc increased, with A. Berberich and L. Boss independently using three positions obtained between August 22 and 25, and determining perihelion dates of September 3.83 and August 25.04, respectively.

The first elliptical orbit was calculated by Berberich using positions obtained during the period of August 22–28. He determined the perihelion date as August 25.96 and the period as 3.22 years. Berberich said the elements were still somewhat uncertain and that the likely period would probably be 5 or 6 years. Berberich revised his calculations nearly 10 days later. On this occasion he took positions spanning the period of August 22 to September 16, and determined the perihelion date as August 21.35 and the period as 7.06 years. Additional orbits were calculated by L. Schulhof (1895) and H. R. Morgan (1898) that slightly revised the perihelion date to August 21.3 and the period to 7.2 years. Despite several later predictions, the comet has never been recovered. The latest redetermination of the 1895 orbit was calculated by N. A. Belyaev and O. J. Stal'bovskij (1972). This orbit is given below.

T	ω	Ω (2000.0)	i	q	e
1895 Aug. 21.3156 (TT)	167.7817	171.7537	2.9923	1.297763	0.652013

ABSOLUTE MAGNITUDE: $H_{10} = 11.4$ (V1964)

FULL MOON: Aug. 5, Sep. 4, Oct. 3, Nov. 2, Dec. 2

SOURCES: L. Swift, E. E. Barnard, E. A. Lamp, V. Cerulli, S. Javelle, G. Witt, R. Schorr, *AN*, **138** (1895 Aug. 26), p. 319; G. Le Cadet, *CR*, **121** (1895 Aug. 26), p. 371; L. Swift, *AJ*, **15** (1895 Aug. 29), p. 136; A. Berberich, Hussey, L. Boss, and H. A. Kobold, *AN*, **138** (1895 Aug. 31), pp. 333–6; J. Hartmann, S. Javelle, and A. Berberich, *AN*, **138** (1895 Sep. 11), p. 351; A. Berberich, *AN*, **138** (1895 Sep. 21), p. 367; V. Cerulli, *AN*, **138** (1895 Sep. 30), p. 383; L. Swift and E. E. Barnard, *PA*, **3** (1895 Oct.), pp. 95–6; V. Cerulli, *AN*, **139** (1895 Oct. 10), pp. 13–16; P. Chofardet, *AN*, **139** (1895 Oct. 22), p. 27; W. A. Villiger and H. A. Kobold, *AN*, **139** (1895 Oct. 29), pp. 43–6; L. Boss and A. Berberich, *PA*, **3** (1895 Nov.), pp. 151–2; L. Schulhof, *CR*, **121** (1895 Nov. 4), pp. 628–9; G. Le Cadet, *AN*, **139** (1895 Nov. 12), pp. 77–80; L. Schulhof, *AN*, **139** (1895 Nov. 20), p. 92; W. H. Robinson, *MNRAS*, **56** (1895 Dec.), pp. 81–3; V. Cerulli, *AN*, **139** (1895 Dec. 14), p. 175; H. C. Wilson, *AJ*, **16** (1895 Dec. 20), p. 6; G. Le Cadet, *BA*, **13** (1896 Jan.), pp. 23–4; W. A. Villiger, *AN*, **139** (1896 Jan. 14), p. 253; J. Halm, *MNRAS*, **56** (1896 Mar.), pp. 323–5; W. W. Campbell, *AJ*, **16** (1896 Mar. 21), pp. 70–1; O. Stone, H. A. Howe, and J. Dickerman, *AJ*, **16** (1896 Mar. 27), pp. 79–80; W. W. Campbell and W. J. Hussey, *AJ*, **16** (1896 Apr. 16), p. 96; H. A. Kobold, *AN*, **140** (1896 May 30), pp. 315–18; W. E. Plummer, *MNRAS*, **56** (1896 Jun.), p. 506; H. A. Howe, *AN*, **141** (1896 Aug. 18), pp. 215–22; C. F. Pechüle, *AN*, **141** (1896 Sep. 26), pp. 305, 309; H. A. Howe, *AN*, **142** (1897 Jan. 11), p. 187; R. Schorr, *AN*, **143** (1897 Jun. 2), pp. 267, 282; H. R. Morgan, *AJ*, **19** (1898 Dec. 12), p. 155; J. Holetschek, *AN*, **149** (1899 Apr. 10), p. 53; V1964, p. 64; N. A. Belyaev and O. J. Stal'bovskij, *QJRAS*, **13** (1972), pp. 428–9, 434.

4P/Faye

1896 II = 1895b

Recovered: 1895 September 27.00 ($\Delta = 1.55$ AU, $r = 2.36$ AU, Elong. $= 133°$)
Last seen: 1896 January 15.99 ($\Delta = 2.19$ AU, $r = 1.84$ AU, Elong. $= 56°$)
Closest to the Earth: 1895 September 11 (1.5261 AU)
Calculated path: AQR (Rec), PSC (Dec. 27)

S. Javelle (Nice, France) recovered this comet with a 76-cm refractor on 1895 September 27.00 at a position of $\alpha = 21^h\ 08.2^m$, $\delta = -1°\ 54'$. He described it as rather faint and round, with a diameter of 20–25″. He confirmed the recovery on September 28.04.

O. Stone (Leander McCormick Observatory, Virginia, USA) saw the comet with the 66-cm refractor on several occasions during the period of October 9–22 and wrote, "The Comet was at all times extremely faint and small." Javelle said the comet had not changed in appearance when seen with the 76-cm refractor on October 11, 12, 19, and 20. H. A. Kobold (Strasbourg, France) saw the comet with the 46-cm refractor on October 15 and said it was at the extreme limit of visibility in slightly hazy skies. Kobold estimated the diameter as 0.5′ on the 17th and noted a central condensation. He said the comet was only barely seen in the 46-cm refractor on the October 21. During the period of October 17 to November 9, W. W. Campbell (Lick Observatory, California, USA) said the comet was at the limit of visibility in the 30-cm refractor, while it was an easy object in the 91-cm refractor and was described as "ill-defined, about 20″ in diameter, with an inconspicuous (not stellar) nucleus." Stone saw the comet on several occasions during the period of November 13–19 and wrote, "The comet was always very faint and small."

The comet was last seen on 1896 January 15.99, when S. J. Brown (US Naval Observatory, Washington, DC, USA) gave the position as $\alpha = 23^h\ 27.6^m$, $\delta = -1°\ 27'$. Observing with the 66-cm telescope (175×), he said the comet was very distinct. With a magnification of 400× he noted a "decided condensation at the center of its faint nebulous mass 30″ diameter."

Later calculations using multiple apparitions were published by B. G. Marsden and Z. Sekanina (1971). They revealed a perihelion date of March 19.52 and a period of 7.56 years. Marsden and Sekanina determined the nongravitational terms as $A_1 = +0.290$ and $A_2 = +0.01044$. This orbit is given below.

T	ω	Ω (2000.0)	i	q	e
1896 Mar. 19.5194 (TT)	201.2488	211.2414	11.3160	1.736934	0.548973

ABSOLUTE MAGNITUDE: $H_{10} = 7.8$ (V1964)
FULL MOON: Sep. 4, Oct. 3, Nov. 2, Dec. 2, Dec. 31, Jan. 30
SOURCES: S. Javelle, *AN*, **138** (1895 Sep. 30), p. 383; S. Javelle, *AN*, **139** (1895 Nov. 5), p. 63; O. Stone, *AJ*, **15** (1895 Nov. 13), p. 181; O. Stone, *AJ*, **16** (1896 Jan. 23), p. 31; S. J. Brown, *AJ*, **16** (1896 Feb. 17), p. 48; W. W. Campbell, *AJ*, **16** (1896 Mar. 21), p. 69; H. A. Kobold, *AN*, **140** (1896 May 30), pp. 315–18; V1964, p. 64; B. G. Marsden and Z. Sekanina, *AJ*, **76** (1971 Dec.), pp. 1136–7.

C/1895 W1
(Perrine)

1895 IV = 1895c

Discovered: 1895 November 17.56 ($\Delta = 1.59$ AU, $r = 0.94$ AU, Elong. $= 33°$)
Last seen: 1896 August 10.26 ($\Delta = 3.71$ AU, $r = 4.01$ AU, Elong. $= 100°$)
Closest to the Earth: 1895 December 14 (0.7936 AU)
Calculated path: VIR (Disc), LIB (Dec. 1), SCO (Dec. 10), OPH (Dec. 14), SCO (Dec. 15), SGR (Dec. 17), AQL (Jan. 14), SGE (Apr. 20), HER (May 5)

C. D. Perrine (Lick Observatory, California, USA) discovered this comet on 1895 November 17.56 at a position of $\alpha = 13^h 44^m$, $\delta = +1°20'$. He described it as bright, with a tail. Perrine confirmed his find on November 18.56.

The comet was approaching both the sun and Earth and was well observed on the 19th, when about a dozen astronomers provided physical descriptions. E. A. Lamp (Kiel, Germany) said the comet was bright, but not visible to the naked eye. Estimates of the brightness of the nucleus were quite discordant. R. Schorr (Hamburg, Germany) said the nucleus was oblong and magnitude 5, J. Halm (Royal Observatory, Edinburgh, Scotland) observed with the 38-cm refractor and said the "nucleus" was about magnitude 6, and E. Esmiol (Marseille, France) observed with a 26-cm refractor and said it was magnitude 9 or 10. Both I. Benko von Boinik (Pula, Yugoslavia) and E. Millosevich (Rome, Italy) gave the nuclear magnitude as 7. Although no estimates of the coma diameter were made, most observers provided information on the tail. Schorr said the tail extended 10′ toward PA 320°, while Millosevich said it extended 12–15′ toward PA 330°. G. Gruss and W. Laska (Prague, Czech Republic) described the tail as broad.

During the period of November 19–21, A. Pannekoek (Leiden, Netherlands) said the nucleus was 10″ across. On the 20th, A. A. Nijland (Utrecht, Netherlands) observed with a 26-cm refractor and said the nucleus was magnitude 8, while the tail extended 21' toward PA 329°. The tail was also 130" wide. That same night, Benko von Boinik said the comet was about magnitude 7, but was not visible to the naked eye. He added that the tail extended 3' toward PA 340° and fanned across an angle of 20°. On the 21st, F. Bidschof (Vienna, Austria) described the comet as 3' across, with a nucleus of magnitude 6.5, while Schorr gave the nuclear magnitude as 7 and B. von Engelhardt (Dresden, Germany) said the tail was 20' long. J. Holetschek (Vienna, Austria) observed with a 15-cm refractor on the 21st, 24th, and 27th, and gave the total magnitude as 6. On the 21st and 22nd, G. A. P. Rayet and L. Picart (Bordeaux, France) said the coma was 1' across, with a nucleus of magnitude 7 or 8. There was also a tail extending 15'. Those same two nights, P. Chofardet (Besançon, France) described the comet as bright, with a condensation of magnitude 7. Schorr gave the nuclear magnitude as 7.5 on the 22nd and said the tail was 8' long and 1.5' wide. On the 23rd, Josef and Jan Fric (Prague, Czech Republic) obtained a 30-minute exposure that revealed a bright, stellar nucleus, which was surrounded by a faint coma. The coma was fan-shaped "to the rear" from which two tails extended. The first tail was curved and extended 30' toward PA 335°, while the second extended 3° 30' toward PA 305°. F. Schwab (Kremsmünster, Austria) said the tail was 7' long on the 24th. During the period of November 20–26, E. Hartwig (Göttingen, Germany) said the tail was fan-like and about 1° long. G. Le Cadet (Lyon, France) observed with the 32-cm refractor on the 27th and gave the nuclear magnitude as 9. That same night, Nijland said the tail extended 40' toward PA 315° and was 150" wide. On the 29th, both Schorr and Engelhardt reported a disk-shaped nucleus, with Schorr giving the nuclear magnitude as 5.5. On November 30, Josef and Jan Fric obtained a 20-minute exposure which revealed the northern tail extended about 50' toward PA 330–350°, while the southern tail extended 7° toward PA 304°.

The comet's altitude rapidly declined during the first half of December as the comet approached its closest distances to both the sun and Earth. On December 2 and 3, W. E. Plummer (Liverpool, England) said the comet was bright and easily seen, despite its low altitude. On the 3rd, Le Cadet gave the nuclear magnitude as 8.2. He added that the tail extended 12' toward the northeast and was convex toward the south. A. C. D. Crommelin (Royal Observatory, Greenwich, England) said the comet was very bright on the 7th, with a distinct tail pointing westward from the nucleus. On December 13, an astronomer at the Royal Observatory, Cape of Good Hope (South Africa) saw the comet near the eastern horizon in twilight in the same field as Antares. He watched the comet fade as the twilight brightened.

The comet was closest to Earth on December 13 and passed 8° from the sun on the 16th. It attained its most southerly declination of −31° on December 17 and passed perihelion late on the 18th. With the solar elongation slowly increasing and the comet's brightness at its peak, the south declination favored observers in the Southern Hemisphere. A. E. Parker (Hungerford, Queensland, Australia) said that he and several other people spotted this object "shortly after sunset" on December 21. It was located in the west-southwest, about 15° above the horizon when first seen, and it remained visible for about an hour. Parker wrote, "The tail was apparently 5 feet long by about 18 inches to 6 inches wide. Nucleus small, fairly bright. It was quite distinct." He added, "Several people on Caiwarro Station, forty miles distant N.E., saw it" on the 22nd. Parker continued his own observations on December 22, 23, and 24, and said the comet became brighter and brighter. On the final two dates it became visible very near the horizon not long after sunset. It was not seen thereafter. J. Tebbutt (Windsor, New South Wales, Australia) wrote, "A brilliant comet was reported to be visible in the south-west at several places in Victoria and South Australia, but no chance was afforded here for getting a view of the stranger. It probably passed quickly into the northern hemisphere." H. C. Russell (Sydney, New South Wales, Australia) said the "comet" was apparently seen in South Australia and New South Wales. Clouds blocked his view of the sky on the evenings the object was seen by Parker. A diligent search thereafter revealed nothing. Russell commented, "In December our sky is so bright near sunset that you could not see a first-magnitude star in the position the comet is said to have occupied. . . . " An attempt was made to see the comet near perihelion in daylight at the Royal Observatory (Cape of Good Hope, South Africa), but nothing was found. H. C. F. Kreutz wrote that there was no doubt that Parker's object was Perrine's comet. The comet moved out to an elongation of 13° by December 23, and then began moving back toward the sun. It again passed 8° away on January 8.

The comet was finally recovered in the morning sky on February 14, when it was detected by C. F. Pechüle (Copenhagen, Denmark) and Lamp. Pechüle observed with the 36-cm refractor and described the comet as moderately bright and fan-shaped. Lamp described it as rather large, with a possible tail, and a stellar nucleus of magnitude 11–12. Josef and Jan Fric obtained a 40-minute exposure on the morning of the 16th that revealed a stellar coma of magnitude 10 and a tail extending 1° toward PA 120°. Engelhardt said the comet was extremely faint in the 30-cm refractor on the 16th and 21st because of morning twilight. During the period of February 18–24, Schwab described the comet as small, with little condensation. J. Hartmann (Leipzig, Germany) observed with a 30-cm refractor on the 20th and said the comet was a diffuse nebulosity about 3' across, with a condensation in the eastern half. Holetschek gave the magnitude as 9 on the 21st. Also on the 21st, Josef and Jan Fric obtained a 50-minute exposure which revealed a fan-shaped tail extending about 1.5° toward PA 120°. On the morning of the 22nd, they obtained a 45-minute exposure which revealed a 2° long tail, while A. A. Nijland (Utrecht, Netherlands) gave the nuclear magnitude as 11. W. H. Robinson (Radcliffe Observatory, Oxford, England) saw the comet with the 25-cm refractor (100×) on the 24th. He said it was 1' in diameter and exhibited a nucleus of magnitude 11 or 12.

Physical descriptions became fewer during the following months. On March 4 and 10, Halm observed with the 38-cm refractor and said the comet was very faint. Holetschek gave the magnitude as 10.5 on the 18th and 11 on the 23rd. Hartmann said the comet was seen with difficulty on March 23. During the period of May 2–13, Pechüle described the comet as small, very faint, with some condensation. On May 11, H. A. Kobold (Strasbourg, France) said the comet was extremely faint and was seen with great trouble in the 46-cm refractor. The comet was last detected on August 10.26, when Perrine gave the position as $\alpha = 16^h\ 26.2^m$, $\delta = +17° 13'$. He saw it with the 91-cm refractor.

The first parabolic orbits came from astronomers working at Lick Observatory and used positions obtained on November 17, 18, and 19. A. O. Leuschner gave the perihelion date as 1895 December 18.77, while W. W. Campbell gave the perihelion date as December 18.91. Additional orbits were calculated by F. Kromm, Lamp, A. O. Leuschner, G. M. Searle, and G. Lorentzen, which eventually established the perihelion date as December 18.8. Definitive orbits were calculated by G. van Biesbroeck (1973) and Z. Sekanina (1978). Van Biesbroeck indicated the orbit was elliptical with a period of about 1.4 million years, while Sekanina indicated it was slightly hyperbolic. Sekanina's orbit is given below.

T	ω	Ω (2000.0)	i	q	e
1895 Dec. 18.8292 (TT)	272.6665	321.9553	141.6263	0.191978	1.000051

ABSOLUTE MAGNITUDE: $H_{10} = 5.2$ (V1964)

FULL MOON: Nov. 2, Dec. 2, Dec. 31, Jan. 30, Feb. 28, Mar. 29, Apr. 27, May 26, Jun. 25, Jul. 24, Aug. 23

SOURCES: C. D. Perrine, E. A. Lamp, and I. Benko von Boinik, *AN*, **139** (1895 Nov. 20), p. 95; C. D. Perrine, A. O. Leuschner, and W. W. Campbell, *AJ*, **15** (1895 Nov. 23), p. 192; E. Esmiol, *CR*, **121** (1895 Nov. 25), p. 760; A. C. D. Crommelin, *MNRAS*, **56** (1895 Dec.), pp. 79–80; R. Schorr, E. Millosevich, J. Franz, A. Pannekoek, B. von Engelhardt, E. Hartwig, P. Chofardet, and E. A. Lamp, *AN*, **139** (1895 Dec. 2), pp. 120–4; G. A. P. Rayet, L. Picart, and F. Kromm, *CR*, **121** (1895 Dec. 2), p. 802; G. M. Searle and A. O. Leuschner, *AJ*, **15** (1895 Dec. 5), p. 198; F. Schwab, R. Schorr, B. von Engelhardt, and I. Benko von Boinik, *AN*, **139** (1895 Dec. 7), pp. 139–42; G. Lorentzen and E. A. Lamp, *AN*, **139** (1895 Dec. 10), p. 157; E. Lamp. *AN*, **139** (1896 Jan. 9), p. 239; A. A. Nijland, *AN*, **139** (1896 Jan. 21), p. 267; G. Le Cadet, Josef and Jan Fric, *AN*, **139** (1896 Jan. 27), p. 287; J. Tebbutt, *MNRAS*, **56** (1896 Feb.), pp. 226–7, 247; E. A. Lamp, *AN*, **139** (1896 Feb. 15), p. 351; C. F. Pechüle, *AN*, **139** (1896 Feb. 22), p. 381; J. Halm, A. E. Parker, and H. C. Russell, *MNRAS*, **56** (1896 Mar.), pp. 323–7, 337; B. von Engelhardt, *AN*, **140** (1896 Mar. 6), p. 45; Josef and Jan Fric, *AN*, **140** (1896 Mar. 11), p. 63; H. C. F. Kreutz, *MNRAS*, **56** (1896 Apr.), p. 391; J. Halm, *AN*, **140** (1896 Apr. 8), p. 139; A. A. Nijland, *AN*, **140** (1896 May 5), pp. 221–4; F. Schwab, *AN*, **140** (1896 May 18), p. 267; W. E. Plummer, *MNRAS*, **56** (1896 Jun.), p. 507; G. Gruss and W. Laska, *AN*, **140** (1896

Jun. 15), pp. 381–4; J. Hartmann, *AN*, **141** (1896 Aug. 27), p. 251; C. F. Pechüle, *AN*, **141** (1896 Sep. 26), pp. 305, 309; W. H. Robinson, *MNRAS*, **57** (1896 Nov.), p. 19; J. Holetschek, *AN*, **142** (1896 Dec. 22), p. 141; C. D. Perrine, *AJ*, **17** (1897 Jan. 5), pp. 60–2; H. A. Kobold, *AN*, **143** (1897 Mar. 10), pp. 13–15; R. Schorr, *AN*, **143** (1897 Jun. 2), pp. 267, 282; F. Bidschof, *AN*, **149** (1899 Jun. 15), p. 407; V1964, p. 64; G. van Biesbroeck, *QJRAS*, **14** (1973), pp. 404–6; B. G. Marsden, Z. Sekanina, and E. Everhart, *AJ*, **83** (1978 Jan.), p. 66; B. G. Marsden and Z. Sekanina, *CCO*, 3rd ed. (1979), pp. 20, 48.

C/1895 W2   *Discovered:* 1895 November 22.30 ($\Delta = 0.41$ AU, $r = 1.03$ AU, Elong. $= 83°$)
(Brooks)   *Last seen:* 1895 December 20.86 ($\Delta = 0.48$ AU, $r = 1.36$ AU, Elong. $= 133°$)
   Closest to the Earth: 1895 December 4 (0.2772 AU)
1895 III = 1895d   *Calculated path:* HYA (Disc), SEX (Nov. 24), HYA (Nov. 26), LEO (Nov. 30), CNC (Dec. 1), LYN (Dec. 6), CAM (Dec. 14)

William R. Brooks (Smith Observatory, Geneva, New York, USA) was sweeping for comets with his 25.4-cm refractor on the morning of 1895 November 22, when he found a moderately bright, "large nebulous mass" on November 22.30. He measured the position as $\alpha = 9^h 51.8^m$, $\delta = -17° 40'$. Although the sky clouded up after only a few minutes of observations, Brooks was able to determine that the comet was rapidly moving northward. He kept the telescope pointed on that spot, by use of the clock drive, and the sky cleared about an hour later, which allowed him to confirm the direction of motion. He sent a telegraph announcing the discovery that morning.

The comet was found about a month after it passed perihelion, but it was about two weeks from making a close approach to Earth. I. Benko von Boinik (Pula, Yugoslavia) described the comet as a large, diffuse nebulosity, about 10' across, without a detectable condensation on November 24 and 26. On the last date he noted a faint condensation toward PA 345°. On the 27th, J. Holetschek (Vienna, Austria) estimated the magnitude as 8–8.5 and gave the coma diameter as 5'. E. Millosevich (Rome, Italy) saw the comet on November 27 and 28, and said it looked like a cloudy heap without a true nucleus. Brooks said the comet was larger and brighter on the 28th than it had been on the 22nd, and he even noted it had brightened still more by the 29th. B. von Engelhardt (Dresden, Germany) also saw the comet on the 29th and said it was bright and diffuse in a 12.7-cm cometseeker. His 30-cm refractor showed the comet as a faint, irregular nebulosity, without condensation. The last comment was also used by W. E. Plummer (Liverpool, England) on the same date. On the 30th, Engelhardt said the comet was 2.5' across and granular in appearance, as seen in the 30-cm refractor.

Having passed closest to Earth within the first days of December, the comet faded rapidly. Holetschek estimated the magnitude as 9 on the 7th. J. Halm (Royal Observatory, Edinburgh, Scotland) observed with the 38-cm refractor during the period of December 7–13, and described the comet as "exceedingly faint and of irregular outline." C. F. Pechüle (Copenhagen, Denmark) simply described it as faint on December 8. Interestingly, on the 11th, H. A. Kobold (Strasbourg, France) noted the comet was easily seen and exhibited a faint condensation. W. A. Villiger (Munich, Germany) said the comet was very faint on December 12 and 13, and Brooks said the comet was then considerably fainter than when seen on November 29. Brooks noticed a slight central condensation. J. Hartmann (Leipzig, Germany) estimated the coma diameter as 4' on the 13th, as seen with a 30-cm refractor. On December 15, A. C. D. Crommelin (Royal Observatory, Greenwich, England) said the comet was a large, faint, and diffused nebulosity, with no nucleus (17-cm refractor, 55×). That same night, G. Bigourdan (Paris, France) said the comet appeared very faint and very diffuse, with a coma 2–3' across, and a central condensation about 25" across. He added that the condensation had a granular appearance. Brooks said the comet was large, round, and faint on the 16th. Kobold saw the comet with a 46-cm refractor on the same night and described it as a faint, diffuse nebulosity, with only a slight condensation.

The comet was seen for the final time on December 20.86, when Halm measured the position as $\alpha = 4^h 49.8^m$, $\delta = +68° 29'$ with the 38-cm refractor. He described it as "exceedingly faint and of irregular outline."

The first parabolic orbit was calculated by H. C. F. Kreutz using positions from November 25, 26, and 27. The resulting perihelion date was 1895 October 21.35. A few days later, A. Berberich calculated an orbit using positions spanning the period of November 25–30, and determined the perihelion date as October 21.50. These early orbits would prove to be quite close to later computations using all available positions. Additional calculations were made by Berberich and A. O. Leuschner. A definitive orbit was calculated by A. Wassilief (1897). He took 78 positions, determined four Normal positions, and found a perihelion date of October 21.55. This orbit is given below. He noted that the result was satisfactory and that there was no reason to assume that the orbit deviated from a parabola.

F. Deichmüller (1895) pointed out the remarkable similarity between the orbit of this comet and that of C/1652 Y1. Wassilief thought the idea was worthy of mentioning in his investigation of the orbit of Brooks' comet, but drew no conclusions of his own.

T	ω	Ω (2000.0)	i	q	e
1895 Oct. 21.5527 (UT)	298.7831	84.5465	76.2461	0.843034	1.0

ABSOLUTE MAGNITUDE: $H_{10} = 10.5$ (V1964)

FULL MOON: Nov. 2, Dec. 2, Dec. 31

SOURCES: W. R. Brooks, *AJ*, **15** (1895 Nov. 23), p. 192; W. R. Brooks, I. Benko von Boinik, S. Javelle, and C. F. Pechüle, *AN*, **139** (1895 Nov. 27), p. 111; S. Javelle, B. von Engelhardt, and H. C. F. Kreutz, *AN*, **139** (1895 Dec. 2), p. 127; I. Benko von Boinik, E. Millosevich, B. von Engelhardt, L. Ambronn, and A. Berberich, *AN*, **139** (1895 Dec. 7), pp. 141–4; F. Deichmüller, A. O. Leuschner, and H. C. F. Kreutz, *AN*, **139** (1895

Dec. 10), p. 159; C. F. Pechüle, *AN*, **139** (1895 Dec. 14), p. 175; G. Bigourdan, *CR*, **121** (1895 Dec. 16), pp. 929–30; A. Berberich, *AN*, **139** (1895 Dec. 21), p. 205; A. C. D. Crommelin, *MNRAS*, **56** (1896 Jan.), p. 135; V. Cerulli, J. Halm, R. Copeland, H. A. Kobold, W. A. Villiger, and W. H. M. Christie, *AN*, **139** (1896 Jan. 6), pp. 215–18; C. Rambaud and F. Sy, *AN*, **139** (1896 Jan. 14), p. 251; J. Hartmann, *AN*, **139** (1896 Feb. 22), p. 379; J. Halm, *MNRAS*, **56** (1896 Mar.), pp. 323–5; W. R. Brooks, *MNRAS*, **56** (1896 Mar.), pp. 328–9; H. A. Kobold, *AN*, **140** (1896 May 30), pp. 315–18; W. E. Plummer, *MNRAS*, **56** (1896 Jun.), p. 507; J. O. Backlund and A. Wassilief, *BASP*, **6** (1897), pp. 505–40; A. Wassilief, *AN*, **143** (1897 May 18), pp. 229–32; J. Holetschek, *AN*, **149** (1899 Apr. 10), p. 53; V1964, p. 64.

C/1896 C1   *Discovered:* 1896 February 15.59 ($\Delta = 0.53$ AU, $r = 0.67$ AU, Elong. $= 39°$)
(Perrine–Lamp)   *Last seen:* 1896 April 16.89 ($\Delta = 2.05$ AU, $r = 1.59$ AU, Elong. $= 49°$)
   Closest to the Earth: 1896 February 24 (0.3754 AU)
1896 I = 1896a   *Calculated path:* AQL (Disc), DEL (Feb. 20), VUL (Feb. 21), CYG (Feb. 23), LAC (Feb. 25), AND (Feb. 27), CAS (Feb. 29), PER (Mar. 4), AND (Mar. 8), PER (Mar. 9)

The discovery of this comet is among the strangest ever. E. A. Lamp (Kiel, Germany) observed comet C/1895 W1 on February 14 and sent a telegram to Boston, Massachusetts to notify observers. By the time a telegram had reached Lick Observatory (California), the position had accidentally been altered. Lick astronomers had already found C/1895 W1 at the end of January and C. D. Perrine noted that the transmitted position was not on the predicted path of that comet. Perrine decided to examine that area of the sky and on February 15.59 he found a comet at a position of $\alpha = 19^h\ 21.9^m$, $\delta = -2°\ 49'$. He subsequently sent a telegram back to Kiel to inform Lamp that he had not seen C/1895 W1 on February 14, but had found a new comet. Lamp was puzzled by this statement, as his observation on the 14th had revealed that C/1895 W1 was exactly in the predicted position. He reobserved C/1895 W1 on February 16 and again noted it was in the predicted position. Lamp decided to sweep around the region and found a new comet on February 16.21. Several days passed before the matter was sorted out. Comets C/1895 W1 and C/1896 C1 passed about 4° from one another on February 16.

    The comet was well observed during the remainder of February. It had actually passed perihelion on the 1st, but passed closest to Earth on the 24th. On February 17, J. Holetschek (Vienna, Austria) observed with the 15-cm refractor and gave the total magnitude as 7.6. B. von Engelhardt (Dresden, Germany) observed with the 30-cm refractor and said the coma was 2' across. He added that the comet was easily seen in the 13-cm finder. On the 17th and 18th, F. Schwab (Kremsmünster, Austria) said the coma was round and 2' across. He noted that the comet disappeared in morning twilight at the same time as stars of magnitude 9. On the 17th and 19th, J. Hartmann

(Leipzig, Germany) said the comet was 2′ in diameter with a central condensation. On the 18th, W. Villiger (Munich, Germany) noted an elliptical coma
about 1.5–2.0′ across and estimated the total magnitude as 7. L. Ambronn
(Göttingen, Germany) and A. A. Nijland (Utrecht, Netherlands) both gave
the coma diameter as 3′. G. Le Cadet (Lyon, France) observed with the
32-cm refractor and said the nucleus was 4″ across and magnitude 9.8. On
the 19th, R. Schorr (Hamburg, Germany) noted that the comet exhibited a
coma 2′ across and a stellar nucleus of magnitude 9.5. C. F. Pechüle (Copenhagen, Denmark) said the coma was 4′ across on the 20th. On the 21st,
Holetschek gave the magnitude as 7.5, while Engelhardt gave the nuclear
magnitude as 10. The coma diameter was given as 2′ by Engelhardt and 3′ by
Hartmann. A. Stupar (Pula, Yugoslavia) observed with the 15-cm refractor
and said the condensation was about 20″ across. During February 21–24,
Schwab said the coma was 3′ across, with a moderate condensation. On the
22nd, E. Hartwig (Bamberg, Germany) said the coma was 6′ across, while G.
Gruss (Prague, Czech Republic) said it was 5′ across. Nijland gave the nuclear magnitude as 8. Josef and Jan Fric (Prague, Czech Republic) obtained
a 40-minute exposure and said the fan-shaped tail extended 30′ long toward
PA 275–325°. They also noted the condensation was about magnitude 7–8.
W. E. Plummer (Liverpool, England) said the comet was "tolerably bright"
on the 23rd, while the Fric brothers said the tail extended about 2.5° toward
PA 320°. Holetschek gave the total magnitude as 7.0 on the 24th, while the
coma diameter was given as 5′ by Gruss. That same night, the Fric brothers said the tail extended toward PA 325°, while W. H. Robinson (Radcliffe
Observatory, Oxford, England) noted a diffuse nucleus of magnitude 9 or
10. A. C. D. Crommelin (Royal Observatory, Greenwich, England) observed
with the 17-cm refractor (55×) on the 25th and described the comet as faint
and difficult to see because of moonlight and slight haze. On February 26,
W. Seraphimoff (Pulkovo Observatory, Russia) observed with the 38-cm
refractor and said the coma was 2′ across, with a distinct nucleus.

With the comet moving away from both the sun and Earth, it faded very
rapidly during March. The comet attained its most northerly declination of
+52° on March 4. Holetschek provided the best series of magnitude estimates, as he gave values of greater than 6 on the 4th, 6.7 on the 8th, 7.0 on
the 11th, 8.6 on the 14th, 9.0 on the 16th, and 8.7 on the 18th. The nucleus also
rapidly faded, although the estimates are somewhat discordant because of
the differing telescopes being used. The nuclear magnitude was given as
9.5 by Schorr on the 4th, 10.5 by I. Benko von Boinik (Pula, Yugoslavia)
on the 6th, 10.5 by Benko von Boinik on the 9th, 9.5 by Nijland and 11.0
by Benko von Boinik on the 10th, 11 by Nijland on the 13th, 10 by Robinson on the 14th, 12 by Nijland and 11 by Robinson on the 16th, and 13 by
W. Wickham (Radcliffe Observatory, Oxford, England) on the 19th. Robinson said the nucleus was 15″ across on the 3rd, while Benko von Boinik said
the nucleus was elongated on the 6th, with the major axis extending from the
east-southeast to the west-northwest. Estimates of the coma diameter were

quite discordant, with A. Abetti (Arcetri, Italy), Hartmann, and Schwab frequently noting it as 1', while Seraphimoff found it 4' across on the 9th and 5' across on the 11th. Several observers reported the comet was 2–3' across during the early days of March. H. A. Kobold (Strasbourg, France) observed with the 46-cm refractor on the 4th and reported a narrow, straight tail extending toward PA 37°. During the last days of March, moonlight interfered with observations. R. Bryant (Royal Observatory, Greenwich, England) observed with the 17-cm refractor (55×) on the 22nd and said the comet was extremely faint, while J. Halm (Royal Observatory, Edinburgh, Scotland) noted it was very faint in the 38-cm refractor. On March 31, Seraphimoff described the comet as faint and said a faint nucleus was occasionally seen.

On April 1, Crommelin said the comet was "still easily visible," but rather ill defined. Holetschek said the magnitude was less than 12 on the 7th. Robinson said the coma was 1' across, slightly condensed, and with no nucleus on the 9th, while it was "large, but feeble," with a condensation on the 10th. Bryant said the comet was "exceedingly faint" in the 17-cm refractor on the 12th and 13th.

The comet was last detected on April 16.89, when Kobold gave the position as $\alpha = 4^h 39.8^m$, $\delta = +40° 24'$. He observed with the 46-cm refractor and described the comet as a faint, diffuse nebulosity, about 1' across, without a condensation.

The first orbit was calculated by A. O. Leuschner and F. H. Seares using positions from February 15, 16, and 17. The resulting perihelion date was 1896 February 1.29. This was almost a perfect representation as proved by later orbits by E. A. Lamp, E. Weiss, G. Ravené, S. C. Chandler, and H. Buchholz. L. Schulhof took positions spanning the period of February 16 to March 15 and calculated a hyperbolic orbit with an eccentricity of 1.003579. A definitive orbit was calculated by W. K. Hristoff (1926). Although a hyperbolic orbit was given, he preferred the parabolic one because of the comet's relatively short period of observation. This orbit is given below.

T	ω	Ω (2000.0)	i	q	e
1896 Feb. 1.2766 (UT)	358.3153	210.2788	155.7381	0.587289	1.0

ABSOLUTE MAGNITUDE: $H_{10} = 8.8$ (V1964)

FULL MOON: Jan. 30, Feb. 28, Mar. 29, Apr. 27

SOURCES: E. A. Lamp, *AN*, **139** (1896 Feb. 21), p. 365–7; B. von Engelhardt, J. Hartmann, L. Ambronn, W. Villiger, F. Schwab, R. Schorr, and C. F. Pechüle, *AN*, **139** (1896 Feb. 22), pp. 381–4; A. O. Leuschner, F. H. Seares, C. D. Perrine, and S. C. Chandler, *AJ*, **16** (1896 Feb. 29), pp. 55–6; I. Benko von Boinik, A. Stupar, G. Le Cadet, B. von Engelhardt, E. Hartwig, E. Weiss, G. Ravené, E. A. Lamp, and C. D. Perrine, *AN*, **140** (1896 Feb. 29), pp. 25–31; A. C. D. Crommelin, H. P. Hollis, and J. Halm, *MNRAS*, **56** (1896 Mar.), pp. 321–2, 326–7; Josef Fric and Jan Fric, *AN*, **140** (1896 Mar. 11), p. 63; E. S. Holden, *AN*, **140** (1896 Mar. 16), p. 79; L. Ambronn, *AN*, **140** (1896 Mar. 21), p. 95; B. von Engelhardt, *AN*, **140** (1896 Mar. 24), p. 111; H. A. Kobold, *AN*, **140** (1896 Mar. 30), p. 125; H. P. Hollis, R. Bryant, and A. C. D. Crommelin, *MNRAS*, **56** (1896

Apr.), pp. 392–3; H. A. Kobold, *AN*, **140** (1896 Apr. 30), p. 207; R. Bryant, *MNRAS*, **56** (1896 May), pp. 432–4; A. A. Nijland, *AN*, **140** (1896 May 5), p. 221; L. Schulhof, *AN*, **140** (1896 May 6), p. 239; F. Schwab, *AN*, **140** (1896 May 18), p. 269; A. Abetti, *AN*, **140** (1896 May 30), pp. 311–14; H. Buchholz, *AN*, **140** (1896 Jun. 2), pp. 333–6; G. Gruss, *AN*, **140** (1896 Jun. 15), p. 383; A. A. Nijland, *AN*, **141** (1896 Jun. 30), p. 37; I. Benko von Boinik, *AN*, **141** (1896 Jul. 25), pp. 105, 111; W. Seraphimoff, *AN*, **141** (1896 Aug. 14), p. 205; J. Hartmann, *AN*, **141** (1896 Aug. 27), p. 251; J. Halm, *MNRAS*, **57** (1896 Nov.), pp. 20, 25; J. Holetschek, *AN*, **142** (1896 Dec. 22), p. 141; W. H. Robinson and W. Wickham, *MNRAS*, **57** (1897 Jan.), p. 181; H. A. Kobold, *AN*, **143** (1897 Mar. 10), pp. 13–16; W. E. Plummer, *MNRAS*, **57** (1897 May), pp. 551–2, 554; R. Schorr, *AN*, **143** (1897 Jun. 2), pp. 267, 282; F. Schwab, *AN*, **145** (1898 Feb. 24), p. 331; W. K. Hristoff, *Untersuchungen über die Bahn des Kometen 1896 I*. Dissertation: Sofia (1926), 64 pp.; V1964, p. 64.

C/1896 G1
(Swift)

1896 III = 1896b

Discovered: 1896 April 14.18 ($\Delta = 0.62$ AU, $r = 0.57$ AU, Elong. $= 31°$)
Last seen: 1896 June 21.44 ($\Delta = 1.19$ AU, $r = 1.42$ AU, Elong. $= 79°$)
Closest to the Earth: 1896 April 22 (0.5700 AU)
Calculated path: TAU (Disc), PER (Apr. 21), CAS (May 4), CEP (May 24)

L. Swift (Lowe Observatory, California, USA) discovered this comet on 1896 April 14.18 at a very low altitude over the western horizon near the end of evening twilight. He described it as bright, with a short tail and a slow westerly motion. He only kept the comet in view for 1 or 2 minutes and, although a position could not be obtained, he did note its location with respect to the Pleiades and Aldebaran. The next night was cloudy, but Swift was able to locate the comet with the 11-cm cometseeker near the end of evening twilight on April 16. He then switched to the 41-cm refractor and gave the position as $\alpha = 3^h 39^m$, $\delta = +15° 40'$ on April 16.2. As the sky continued to darken, he noticed a very faint tail.

During the remainder of April the comet was widely observed as it passed closest to both the sun and Earth. During the period of April 18–20, C. Rambaud and F. Sy (Alger, now al-Jazâ'ir, Algeria) observed with the 32-cm refractor and said the comet was round, with a central condensation of magnitude 10. On the 19th, R. Schorr (Hamburg, Germany) said the comet was 2' across, with an irregular border and a nucleus of magnitude 10.5, while E. A. Lamp (Kiel, Germany) noted the nucleus was not central. On the 20th, R. Bryant and A. C. D. Crommelin (Royal Observatory, Greenwich, England) observed with the 17-cm refractor (55×) and said the comet was bright, with a condensation, and visible in bright twilight. That same night, W. H. Robinson (Radcliffe Observatory, Oxford, England) observed with the 25-cm refractor (100×) and said the comet was nearly round and about 2' across, with a condensation. On the 21st, W. Villiger (Munich, Germany) said the comet was magnitude 7, while H. A. Kobold (Strasbourg, France) observed with the 46-cm refractor and said the comet was 1' across, with a nucleus of magnitude 8.5. W. E. Plummer (Liverpool, England) said the comet was "tolerably bright, especially towards the south-east." On the 24th, the coma diameter was given as about 1.5′ by L. Ambronn (Göttingen, Germany) and 1.7′ by A. A. Nijland (Utrecht, Netherlands). J. Hartmann (Leipzig, Germany) observed with the 30-cm refractor and said the condensation was 10″ across and magnitude 9, while Nijland observed with the 26-cm refractor and said the nucleus was magnitude 9.5. Schorr said the tail extended about 4′ toward the southeast. On the 25th, A. Stupar (Pula, Yugoslavia) observed with a 15-cm refractor and estimated the magnitude as 7.5. On the 27th, 28th, and 29th, A. Abetti (Arcetri, Italy) observed with a 28-cm refractor and noted a round coma 1′ across of about magnitude 8. On the 28th, J. Holetschek (Vienna, Austria) observed with a 15-cm refractor and gave the total magnitude as 6.5, while Robinson said the coma was 3′ across and the diffuse nucleus was 10–15″ across. That same night, Schorr said the diffuse nucleus was magnitude 9.7, while Plummer noted the comet was well seen, in spite of bright moonlight. On the 29th, J. Halm (Royal Observatory, Edinburgh, Scotland) gave the nuclear magnitude as about magnitude 10, while Nijland gave it as 9. Nijland also noted a tail extending 8′ toward PA 21°, while Hartmann said the coma was 3′ across. On April 30, C. F. Pechüle (Copenhagen, Denmark) said the coma was 2′ across and exhibited a tail extending toward PA 20°.

On May 1, H. P. Hollis (Royal Observatory, Greenwich, England) observed with the 17-cm refractor (55×) and said the comet exhibited a distinct condensation, while W. Wickham (Radcliffe Observatory, Oxford, England) observed with a transit circle (80×) and described the comet as a large, diffuse mass about 0.8′ across, with no definite condensation. That same night, Schorr noted a tail extending 3′ toward the north, as well as a diffuse nucleus of magnitude 9.5. Nijland noted a nucleus of magnitude 9 on the 3rd and 4th, and said the coma diameter changed from 2.6 to 2.2′. On the 5th, Holetschek gave the magnitude as 6.7, while Nijland observed with the 7.5-cm finder and gave it as 7.5. Nijland added that the coma was 2.4′ across and the nucleus was magnitude 10. I. Benko von Boinik (Pula, Yugoslavia) observed with the 15-cm refractor on the 6th and noted a distinct nucleus of magnitude 11.0. On the 8th, Schorr said the coma was 3′ across and exhibited a diffuse nucleus of magnitude 10.5, while B. von Engelhardt (Dresden, Germany) observed with the 30-cm refractor and said the coma was 2′ across, with a condensation about 30″ across. On the 9th, Robinson said the coma was 3′ across, while Schorr said the nucleus appeared granular. On May 10, F. Bidschof (Vienna, Austria) described the comet as rather bright and about 3′ in diameter. On the 10th and 12th, Benko von Boinik noted a nucleus of magnitude 12.0. On the 11th, Holetschek gave the magnitude as 7.3, while Robinson said the coma was 3′ across and exhibited a generally diffuse condensation. Plummer found the comet exceedingly faint on the 13th, while Abetti indicated a magnitude of 10. Wickham described the comet as a large, diffuse mass on the 14th. On the 15th, Holetschek gave the magnitude as 8.3, while Wickham obtained a momentary glimpse of the nucleus in the

25-cm refractor (100×). On May 16, Plummer said the comet was still barely visible, while Hartmann said it was very diffuse and faint, with a diameter of 2'. Bidschof noted the total magnitude was about 10 on the 17th, with a diameter of 4'. He noted a weak central condensation, which contained a nucleus. On the 18th, Kobold said the comet was about 1.5' across, with a hardly discernible nucleus. Holetschek gave the magnitude as 9 on the 19th, while Schorr said the coma was 1' across. Crommelin reported the comet was very faint and ill-defined because of moonlight on the 20th. On the 26th, Halm noted the comet was exceedingly faint and scarcely visible in the 38-cm refractor. Holetschek gave the magnitude as 10.5 on the 28th. On the 29th, Wickham described the comet as a "hazy, faint patch" which exhibited a tail extending toward PA 300°. Bryant said the comet was extremely faint on the 30th, with very slight central condensation. On May 31, Halm noted the comet was exceedingly faint and scarcely visible.

On June 1, Crommelin said the comet was very faint. Holetschek gave the magnitude as 10.3 on the 2nd, while Plummer said the comet was only occasionally glimpsed and J. Palisa (Vienna, Austria) described the comet as rather faint and centrally condensed. Crommelin and Bryant said the comet was exceedingly faint and ill-defined in the 17-cm refractor (55×) on the 5th and 6th. The comet attained its most northerly declination of +73° on the 7th. Holetschek gave the magnitude as 11 on the 8th. On June 15, Holetschek gave the magnitude as 12, while Robinson said the comet was "only very occasionally discernible" in the 25-cm refractor (100×).

The comet was last detected on June 21.44, when W. J. Hussey (Lick Observatory, California, USA) gave the position as $\alpha = 21^h\ 20.8^m$, $\delta = +71°\ 29'$. He was using the 91-cm refractor. Schorr reported he could not find the comet on the 26th with the 26-cm refractor.

The first parabolic orbit was calculated by Schorr using positions from April 16, 19, and 20. He determined the perihelion date as 1896 April 17.97. Later orbits by Schorr and Bidschof established the perihelion date as April 18.14.

A definitive orbit was calculated by R. G. Aitken (1899). He began with 433 positions and reduced them to seven Normal positions. No planetary perturbations were applied. The result was the hyperbolic orbit given below.

T	ω	Ω (2000.0)	i	q	e
1896 Apr. 18.1473 (UT)	1.7387	179.7008	55.5600	0.566286	1.000476

ABSOLUTE MAGNITUDE: $H_{10} = 10.3$ (V1964)

FULL MOON: Mar. 29, Apr. 27, May 26, Jun. 25

SOURCES: L. Swift, R. Schorr, and E. A. Lamp, *AN*, **140** (1896 Apr. 20), p. 191; L. Swift, *AJ*, **16** (1896 Apr. 30), p. 104; C. Rambaud, F. Sy, R. Schorr, W. Villiger, and L. Ambronn, *AN*, **140** (1896 Apr. 30), pp. 203–8; R. Bryant, A. C. D. Crommelin and H. P. Hollis, *MNRAS*, **56** (1896 May), pp. 432–4; L. Swift, *PA*, **3** (1896 May), pp. 482–3; A. Abetti,

R. Schorr, and B. von Engelhardt, *AN*, **140** (1896 May 18), pp. 269–72; F. Bidschof, *AN*, **140** (1896 May 30), p. 317; A. C. D. Crommelin and R. Bryant, *MNRAS*, **56** (1896 Jun.), pp. 501–3; I. Benko von Boinik, *AN*, **140** (1896 Jun. 15), p. 381; J. Hartmann, *AN*, **141** (1896 Aug. 27), p. 251; A. A. Nijland, *AN*, **141** (1896 Sep. 2), p. 271; C. F. Pechüle, *AN*, **141** (1896 Sep. 26), pp. 307, 310; J. Halm, *MNRAS*, **57** (1896 Nov.), pp. 20, 25; J. Holetschek, *AN*, **142** (1896 Dec. 22), pp. 141–4; A. Abetti, *AN*, **142** (1897 Jan. 29), p. 269; W. Wickham and W. H. Robinson, *MNRAS*, **57** (1897 Mar.), pp. 425–6; H. A. Kobold, *AN*, **143** (1897 Mar. 10), p. 15; J. Palisa, *AN*, **143** (1897 Mar. 24), pp. 49, 53; W. H. Robinson, E. E. McClellan, and W. Wickham, *MNRAS*, **57** (1897 Apr.), pp. 498–500; W. E. Plummer, *MNRAS*, **57** (1897 May), pp. 552–4; W. J. Hussey, *AJ*, **17** (1897 May 20), p. 150; A. A. Rambaut, *MNRAS*, **57** (1897 Jun.), pp. 605–8; R. Schorr, *AN*, **143** (1897 Jun. 2), pp. 267, 282–3; R. G. Aitken, *AN*, **148** (1899 Mar. 2), pp. 337–74; F. Bidschof, *AN*, **149** (1899 Jun. 15), p. 409; V1964, p. 64.

16P/1896 M1	*Recovered:* 1896 June 21.08 ($\Delta = 1.59$ AU, $r = 2.26$ AU, Elong. $= 118°$)
(Brooks 2)	*Last seen:* 1897 February 26.17 ($\Delta = 2.68$ AU, $r = 2.17$ AU, Elong. $= 49°$)
	Closest to the Earth: 1896 September 3 (1.0346 AU)
1896 VI = 1896c	*Calculated path:* AQR (Rec), PSC (Dec. 15)

C. L. Poor (1894) took J. Bauschinger's orbit for the 1889 apparition, applied complete perturbations by Earth, Jupiter, and Saturn, as well as partial perturbations by Mars, and predicted the comet would next arrive at perihelion on 1896 November 4.44. S. Javelle (Nice, France) recovered the comet on 1896 June 21.08, at a position of $\alpha = 22^h 25.6^m$, $\delta = -18° 34'$.

L. Cruls and H. Morize (Rio de Janeiro, Brazil) noted an extremely faint nebulosity in a 23-cm refractor on July 29. On the 31st, Cruls and Morize said it was very faint, about 1′ across, with a nucleus of magnitude 12. On August 5, Cruls and Morize noted the comet seemed to have faded. H. A. Kobold (Strasbourg, France) described the comet as faint and 0.5′ across on August 11, and noted a central condensation. On August 8, 16, and 18, J. Holetschek (Vienna, Austria) estimated the total magnitude as 11.5. On the 13th, Cruls and Morize said the comet seemed slightly brighter than on the 5th, with a nucleus of magnitude 11. J. Palisa (Vienna, Austria) described the comet as a small, centrally condensed nebula on the 18th, with a total magnitude of 13. During the period spanning August 30 to September 12, J. M. Thome (Cordoba, Argentina) continually found the comet very faint. On September 8 and 12, Holetschek estimated the total magnitude as 10.5. Holetschek gave the total magnitude as 11 on October 2, while he gave it as 11.5 on the 4th, 6th, and 8th. On October 9, W. H. Robinson (Radcliffe Observatory, Oxford, England) said the comet was about 1′ across, with a faint, but distinct nucleus (25.4-cm refractor, 100×). That same evening, F. Bidschof (Vienna, Austria) estimated the total magnitude as 11. On November 1, Holetschek gave the total magnitude as about 11.8. On November 25, Kobold said the comet was 1′ across, with a condensation of magnitude 12. On December 2, Palisa said the comet was quite faint, but still displayed a good condensation. Kobold found the comet quite faint in the 46-cm refractor on the 5th, and noted a nucleus of magnitude 13. On January 26, Kobold described the comet as a diffuse spot about 1′ across.

The comet was last detected on 1897 February 26.17, when W. J. Hussey (Lick Observatory, California, USA) measured the position as $\alpha = 1^h 38.2^m$, $\delta = +12° 50′$. He was using the 91-cm refractor.

Hussey wrote in the 1897 July 2 issue of the *Astronomical Journal*, "In July, August, September, and October, I searched carefully, at every opportunity, for the companions which accompanied this comet in 1889, without success."

The spectrum was visually observed by W. W. Campbell on August 15 and October 6. The continuous spectrum was seen on both occasions and Campbell noted "I was pretty certain that a trace of the green carbon band was visible, but not absolutely certain."

Multiple apparition orbits by A. D. Dubiago (1950), B. G. Marsden and Z. Sekanina (1972), I. Y. Evdokimov (1978), and Sekanina and D. K. Yeomans (1985) revealed the perihelion date as November 4.63 and the period as 7.10 years. Dubiago identified a secular acceleration in the motion. Marsden, Sekanina, and Yeomans (1973) determined nongravitational terms of $A_1 = +3.61$ and $A_2 = -0.3269$. The orbit of Sekanina and Yeomans is given below.

T	ω	Ω (2000.0)	i	q	e
1896 Nov. 4.6314 (TT)	343.8427	19.4169	6.0713	1.959218	0.469433

ABSOLUTE MAGNITUDE: $H_{10} = 8.0$ (V1964)

FULL MOON: May 26, Jun. 25, Jul. 24, Aug. 23, Sep. 21, Oct. 21, Nov. 20, Dec. 20, Jan. 18, Feb. 17, Mar. 18

SOURCES: C. L. Poor, *AJ*, **14** (1894 Jun. 12), p. 63; C. L. Poor, *PA*, **2** (1895 May), p. 424; S. Javelle, *AN*, **141** (1896 Jun. 22), p. 15; H. A. Kobold, *AN*, **141** (1896 Aug. 14), p. 205; W. H. Robinson, *MNRAS*, **57** (1896 Dec.), p. 83; L. Cruls and H. Morize, *AN*, **142** (1897 Feb. 10), p. 341; J. M. Thome, *AN*, **142** (1897 Feb. 15), p. 355; W. W. Campbell, *APJ*, **5** (1897 Apr.), p. 238; J. Palisa, *AN*, **143** (1897 Mar. 24), pp. 49, 53; H. A. Kobold, *AN*, **143** (1897 May 18), pp. 245–7; W. J. Hussey, *AJ*, **17** (1897 Jul. 2), pp. 182–3; J. Holetschek, *AN*, **149** (1899 Apr. 10), p. 53; F. Bidschof, *AN*, **149** (1899 Jun. 15), p. 409; A. D. Dubiago, *Trudy Kazanskaia Gorodkoj Astronomicheskoj Observatorii*, No. 31 (1950), p. 25; V1964, p. 64; B. G. Marsden and Z. Sekanina, *QJRAS*, **13** (1972), p. 428; B. G. Marsden, Z. Sekanina, and D. K. Yeomans, *AJ*, **78** (1973 Mar.), p. 213; B. G. Marsden and Z. Sekanina, *CCO*, 2nd ed. (1975), pp. 20, 47; I. Y. Evdokimov, *QJRAS*, **19** (1978), pp. 80–1, 88; Z. Sekanina and D. K. Yeomans, *CCO*, 5th ed. (1986), pp. 17, 52; D. K. Yeomans, *QJRAS*, **27** (1986), p. 604.

C/1896 R1   *Discovered:* 1896 September 1.10 ($\Delta = 1.70$ AU, $r = 1.40$ AU, Elong. $= 55°$)
(Sperra)   *Last seen:* 1896 October 6.82 ($\Delta = 1.77$ AU, $r = 1.76$ AU, Elong. $= 73°$)
                  Closest to the Earth: 1896 April 9 (1.5483 AU), 1896 September 14 (1.6768 AU)
1896 IV = 1896e *Calculated path:* UMa (Disc), BOO (Sep. 8), HER (Sep. 24)

W. E. Sperra (Randolph, Ohio, USA) discovered this comet near ζ Ursae Majoris on 1896 September 1.10, and gave the position as $\alpha = 13^h 08.4^m$, $\delta = +55° 40'$. He described it as a nebulous object. Sperra sent a letter to J. E. Keeler (Allegheny Observatory, Pennsylvania, USA) and Keeler relayed the announcement to Harvard College Observatory (Massachusetts, USA). Another letter to Keeler gave a position for September 2. W. R. Brooks received a notice of the discovery on September 4 and began looking for the comet that evening. He quickly found it.

The comet was nearly two months past perihelion and although it had been closest to Earth nearly five months earlier (1.5 AU), it was found about two weeks prior to a second "close" approach (1.7 AU). It was observed at several observatories on September 7 and 8. I. Benko von Boinik (Pula, Yugoslavia) described the comet as very faint on the 7th. W. Villiger (Munich, Germany) saw the comet on the 7th and 8th, and described it as about magnitude 12, with a diameter of 0.5'. Several other observers obtained their first observations on the 8th. J. Holetschek (Vienna, Austria) estimated the magnitude as about 11.5. A. Abetti (Arcetri, Italy) described the comet as a faint, ill-defined nebulosity, roughly 1' across. R. Schorr (Hamburg, Germany) said the comet was a faint, pale nebulous mass, about 1.5' across, but without a condensation. E. A. Lamp (Kiel, Germany) said the comet was fainter, but larger than the nebula near M101. The Author believes Lamp was comparing the comet to NGC 5474 (magnitude 10.9, size 4.5' – NGC2000.0). W. J. Hussey (Lick Observatory, California, USA) saw the comet with a 30-cm refractor on the 7th and 9th, and described it as faint and diffuse. On September 9, W. Wickham (Radcliffe Observatory, Oxford, England) described the comet as a diffuse, nebulous mass in a 25.4-cm refractor (100×), while Holetschek estimated the magnitude as 12 and Villiger simply noted the comet was very diffuse. On the 12th, Holetschek estimated the magnitude as about 11.3. H. A. Kobold (Strasbourg, France) saw the comet on the 13th and said it was diffuse and 2' across, with a slight condensation. J. Halm (Royal Observatory, Edinburgh, Scotland) observed the comet with the 38-cm refractor on the 14th and said it was exceedingly faint.. On the 15th, W. H. Robinson (Radcliffe Observatory, Oxford, England) observed with the 25.4-cm refractor (100×) and described the comet as a faint haze about 2' across, with a feeble, occasionally invisible, condensation. That same night, W. Seraphimoff (Pulkovo Observatory, Russia) said the comet appeared extremely faint in a 38-cm refractor, and was only slightly condensed. On September 28, H. C. Wilson (Goodsell Observatory, Northfield, Minnesota, USA) observed with the 41-cm refractor and noted the comet was "exceedingly faint" in moonlight. It was 1–2' in diameter and exhibited very slight condensation.

The comet was seen for the final times on October 4.77 and October 6.82, when J. Palisa (Vienna, Austria) found it with the 69-cm refractor. He described it as an extremely faint, "nebulous wisp." On the last date, Palisa gave the position as $\alpha = 16^h 52.6^m$, $\delta = +44° 07'$.

The first orbit was calculated by Lamp using positions spanning the period of September 7–10. The perihelion date was given as 1896 July 17.10. A couple of days later, independent orbits were calculated by Hussey, as well as F. H. Seares and R. T. Crawford, using positions obtained at Lick Observatory on September 7, 9, and 12. The perihelion date was determined as July 10.91 by Hussey and July 11.14 by Seares and Crawford. The only other orbit published in 1896 was a revision by Lamp around mid-October. The next orbit was calculated by H. A. Peck (1901) and was the first to utilize all available positions. This orbit is given below.

T	ω	Ω (2000.0)	i	q	e
1896 Jul. 11.4428 (UT)	41.0582	152.4398	88.4215	1.142895	1.0

ABSOLUTE MAGNITUDE: $H_{10} = 8.4$ (V1964)

FULL MOON: Aug. 23, Sep. 21, Oct. 21

SOURCES: W. R. Brooks, *AN*, **141** (1896 Sep. 8), p. 287; E. A. Lamp, *AN*, **141** (1896 Sep. 15), p. 303; W. E. Sperra, W. R. Brooks, F. H. Seares, and R. T. Crawford, *AJ*, **16** (1896 Sep. 17), p. 212; W. Villiger, A. Abetti, R. Schorr, E. A. Lamp, H. A. Kobold, I. Benko von Boinik, W. R. Brooks, and W. E. Sperra, *AN*, **141** (1896 Sep. 26), pp. 313–18; E. A. Lamp, *AN*, **141** (1896 Oct. 5), p. 357; W. J. Hussey, F. H. Seares, and R. T. Crawford, *AJ*, **17** (1896 Oct. 6), p. 7; H. C. Wilson, *AJ*, **17** (1896 Oct. 16), p. 16; W. Seraphimoff, *AN*, **142** (1897 Jan. 21), p. 239; J. Halm, *MNRAS*, **57** (1897 Mar.), pp. 420, 425; J. Palisa, *AN*, **143** (1897 Mar. 24), pp. 49, 53; W. Wickham and W. H. Robinson, *MNRAS*, **57** (1897 Apr.), pp. 499–500; R. Schorr, *AN*, **143** (1897 Jun. 2), pp. 267, 283; W. Villiger, *AN*, **144** (1897 Aug. 26), pp. 145, 171; J. Holetschek, *AN*, **149** (1899 Apr. 10), p. 53; H. A. Peck, *AJ*, **22** (1901 Oct. 15), pp. 35–7; V1964, p. 64; NGC2000.0 (1988), p. 170.

D/1896 R2
(Giacobini)

1896 V = 1896 d

Discovered: 1896 September 4.84 ($\Delta = 1.12$ AU, $r = 1.57$ AU, Elong. = 94)

Last seen: 1897 January 5.18 ($\Delta = 1.85$ AU, $r = 1.63$ AU, Elong. = 61°)

Closest to the Earth: 1896 July 7 (1.0048 AU)

Calculated path: OPH (Disc), SER (Sep. 20), SCT (Oct. 3), SGR (Oct. 16), CAP (Nov. 6), AQR (Nov. 16), CAP (Nov. 30), AQR (Dec. 10)

M. Giacobini (Nice, France) discovered this comet on 1896 September 4.84 and estimated the position as $\alpha = 17^h\ 10.5^m$, $\delta = -7°\ 29'$. The comet was confirmed by F. Sy (Alger, now al-Jazâ'ir, Algeria) on September 5.82. He described it as faint, diffuse, and about 30″ across, with an occasionally seen central condensation. An independent confirmation was made by W. A. Villiger (Munich, Germany) on September 5.83, while using a 27-cm refractor. He said the comet appeared as a round nebulosity, 1′ across, without a distinct nucleus. He determined the magnitude as 11.3. The comet had passed closest to Earth two months earlier, but was still approaching perihelion.

The comet was well observed during September, although it was not an easy object to see. H. J. A. Perrotin (Nice, France) simply described it as faint on the 6th. On September 7, A. Abetti (Arcetri, Italy) said the comet was small and faint, while H. A. Kobold (Strasbourg, France) observed with a 46-cm refractor and said the comet appeared "small, round, and rather faint, with a small condensation." He added that the nucleus was magnitude 12. G. Le Cadet (Lyon, France) described the comet as very faint and nearly round in the 32-cm refractor on the 10th and 11th. On September 8, Abetti said the comet was small and faint, while R. Schorr (Bergedorf Observatory, Hamburg, Germany) commented that the comet was small and extraordinarily faint. Villiger described the comet as very faint on the 9th, while Sy said the comet was faint at low altitude in the 32-cm refractor on the 10th. Sy added that the comet seemed elongated toward PA 160°. Perrotin said the coma was slightly elongated toward PA 90° on the 11th and indicated this extension was about 1′ long. He also noted a nucleus of magnitude 13. Perrotin noticed something unusual on September 26, 27, and 28. On each night he suspected the presence of an extremely faint, badly-defined companion very near the nucleus toward PA 225°.

For the remainder of the comet's apparition, the number of observers quickly declined as the comet continued to grow fainter. Perrotin reported that his observations on October 5 and 8 revealed the comet was so faint it was at the limit of his telescope. The comet attained its most southerly declination of −14° on November 3. Kobold observed with the 46-cm refractor on November 7 and said the comet appeared as a very faint, diffuse nebulosity, with a central condensation. The comet was last observed on 1897 January 5.18, when W. J. Hussey (Lick Observatory, California, USA) saw it with the 91-cm refractor and measured the position as $\alpha = 23^h 12.7^m$, $\delta = -6° 06′$.

The first parabolic orbit was calculated by H. C. F. Kreutz using positions spanning the period of September 5–7. The perihelion date was 1896 October 8.47. Additional calculations by Giacobini, Kreutz, and Perrotin during the next few weeks eventually advanced the perihelion date up to October 17.55.

The first elliptical orbit was calculated by Perrotin and Giacobini using positions spanning the period of September 4–27. They determined the perihelion date as October 28.80 and the period as 6.55 years. Additional calculations by Hussey, Perrotin, and C. W. L. M. Ebell (1903) eventually established the perihelion date as October 28.54 and the period as 6.65 years.

This comet has not been recovered since this apparition. The most recent investigation into its orbit was published by N. A. Belyaev, V. V. Emel'yanenko, and N. Y. Goryajnova (1974). They took 60 positions obtained during this apparition, applied perturbations by Venus to Saturn, and determined the perihelion date as October 28.53 and the period as 6.65 years. This orbit is given below.

T	ω	Ω (2000.0)	i	q	e
1896 Oct. 28.5342 (TT)	140.5091	194.8976	11.3507	1.454747	0.588483

ABSOLUTE MAGNITUDE: $H_{10} = 9.9$ (V1964)

FULL MOON: Aug. 23, Sep. 21, Oct. 21, Nov. 20, Dec. 20, Jan. 18

SOURCES: H. J. A. Perrotin and W. A. Villiger, *AN*, **141** (1896 Sep. 8), p. 287; H. C. F. Kreutz, *AN*, **141** (1896 Sep. 9), supplement; H. C. F. Kreutz, *AN*, **141** (1896 Sep. 15), pp. 301–4; H. J. A. Perrotin and M. Giacobini, *CR*, **123** (1896 Sep. 21), pp. 473–4; W. A. Villiger, A. Abetti, R. Schorr, F. Sy, and G. Le Cadet, *AN*, **141** (1896 Sep. 26), pp. 311–14; H. J. A. Perrotin, G. Le Cadet, and F. Sy, *CR*, **123** (1896 Sep. 28), pp. 473–4, 481–2; H. C. F. Kreutz, *AN*, **141** (1896 Oct. 5), p. 359; H. J. A. Perrotin and M. Giacobini, *CR*, **123** (1996 Oct. 12), pp. 555–6; H. C. F. Kreutz, *AN*, **141** (1896 Oct. 23), p. 423; H. A. Kobold, *AN*, **142** (1896 Nov. 30), p. 61; H. J. A. Perrotin, *CR*, **123** (1896 Nov. 30), pp. 925–8; W. J. Hussey, *AN*, **142** (1896 Dec. 14), p. 109; R. Schorr, *AN*, **143** (1897 Jun. 2), pp. 269, 283; W. A. Villiger, *AN*, **144** (1897 Aug. 26), pp. 145, 171; W. J. Hussey, *AJ*, **19** (1898 Jun. 14), p. 46; C. W. L. M. Ebell, *AN*, **161** (1903 Feb. 25), p. 139; V1964, p. 64; N. A. Belyaev, V. V. Emel'yanenko, and N. Y. Goryajnova, *QJRAS*, **15** (1974), pp. 450–1, 459.

X/1896 S1 This object was discovered by L. Swift (Lowe Observatory, California, USA) on 1896 September 21.08, when he looked out the window as the sun was setting behind "a spur of the Sierra Madre range." About one-quarter of the sun had disappeared, when he noticed "a luminous object about one degree above the sun" that equaled Venus in brightness. He at first believed the object was just an imperfection in the glass, but he moved his head and noticed the object remained stationary with respect to the sun. He wrote, "Going out on the veranda the object was seen more distinctly. At first it occurred to me that it might be a small fire on the mountain, but this idea was quickly dispelled, as one-half of the sun's disk was still above the mountain, and the object still higher." Swift then observed the object with an opera-glass and saw a much fainter companion some 30′ north of the brighter object. After 4 minutes of observing, the sun and the objects had set behind the mountain. Swift wrote, "Both objects were seen by some fifteen people." Swift searched for the objects the next morning with the 11-cm cometseeker, but failed "as the sun is not visible from the Observatory until 15° above the horizon." That evening, Swift began searching for the objects with the cometseeker when the sun was still 10° above the horizon, but it was not until one-half of the sun had set that he found one object on September 22.08. He wrote, "its faintness surprised me, for it was less bright through the telescope than with the naked eye the previous evening, though it is possible, and, perhaps probable, that it was the companion that was seen." Swift estimated that he observed the object for about 5 seconds as he rushed into the observatory to look at it with the much larger telescope. Unfortunately, with the telescope set to view the horizon, the eyepiece was out of reach and by the time he got back to the cometseeker, the object had set. Swift believed the object had "disappeared simultaneously with the sun's upper limb."

Searches were conducted at Lick Observatory (California, USA) by W. J. Hussey and C. D. Perrine. Hussey received a telegram on September 21 from Swift and that afternoon he searched the region several degrees around the sun with the 30-cm refractor at low power but found nothing. As the sun set, he searched the region north, south, and east of the sun, but without success, even though Swift later reported that he saw one object at that time. Hussey searched the region north, south, and west of the sun the following morning. Perrine searched "many degrees about the sun" during several of the following evenings and mornings with the cometseeker, but found nothing as well.

SOURCES: L. Swift, *AN*, **141** (1896 Sep. 26), p. 317; L. Swift, *AJ*, **17** (1896 Oct. 6), p. 8; L. Swift, *MNRAS*, **57** (1897 Feb.), p. 276; L. Swift, *The Observatory*, **20** (1897 Mar.), p. 140; W. J. Hussey and C. D. Perrine, *AJ*, **17** (1897 Mar. 5), p. 103.

C/1896 V1   *Discovered:* 1896 November 3.30 ($\Delta = 1.52$ AU, $r = 1.86$ AU, Elong. $= 93°$)
(Perrine)   *Last seen:* 1897 May 6.03 ($\Delta = 1.14$ AU, $r = 1.73$ AU, Elong. $= 107°$)
Closest to the Earth: 1897 April 12 (0.7853 AU)
1897 I = 1896f   *Calculated path:* VUL (Disc), SGE (Nov. 7), AQL (Nov. 15), SGR (Jan. 20), TEL (Mar. 20), PAV (Mar. 28), APS (Apr. 5), CHA (Apr. 12), MUS (Apr. 15), CAR (Apr. 16), VEL (Apr. 25)

C. D. Perrine (Lick Observatory, California, USA) discovered this comet on 1896 November 3.30, with the 30-cm refractor, and gave the position as $\alpha = 20^h 21.6^m$, $\delta = +25° 07'$. He described the comet as round, about $2'$ across, with a well-marked condensation. Perrine added that the magnitude was about 10–11, and no tail was visible.

The comet was found about three months prior to perihelion and about five months prior to its closest approach to Earth. It was well observed on November 4 and 5. On the 4th, R. Schorr (Hamburg, Germany) said the coma was $2'$ across, with a diffuse nucleus of magnitude 12. W. Villiger (Munich, Germany) described the comet as a round, diffuse nebula about $15''$ across. C. Davidson (Royal Observatory, Greenwich, England) observed with the 17-cm refractor ($55\times$) and said the comet was extremely faint and difficult to observe. W. H. Robinson (Radcliffe Observatory, Oxford, England) observed with the 25-cm refractor ($100\times$) and said the coma was $1' 40''$ across, while the condensation was distinct, but not stellar, and located in the north-preceding part of the coma. F. Bidschof (Vienna, Austria) gave the magnitude as 12 and the diameter as $45''$. On the 5th, J. Holetschek (Vienna, Austria) gave the magnitude as about 11.3. W. Wickham (Radcliffe Observatory, Oxford, England) observed with the 25-cm refractor ($100\times$) and said the comet was very faint and exhibited a stellar nucleus of magnitude 12. J. Halm (Royal Observatory, Edinburgh, Scotland) observed with the 38-cm refractor and said the comet was exceedingly faint and scarcely visible. W. E. Plummer (Liverpool, England) said the comet was very faint and difficult to see. H. A. Kobold (Strasbourg, France) observed with the 46-cm refractor and said the comet was $1'$ across, with a nucleus of magnitude 12.

On November 6, Wickham said the comet was extremely faint. On the 12th, A. Abetti (Arcetri, Italy) reported the comet was small and hardly visible in the light of the first quarter moon. On the 13th, Halm said the comet resembled a faint nebula with a distinct nucleus of about magnitude 13, while Abetti said the comet could not be seen. Halm noted the comet was very faint because of fog and moonlight on the 16th and was well seen, but faint on the 18th. On the 25th and 26th, Holetschek gave the magnitude as about 10.5, while the coma was 1.5' across on the former date. Numerous observations were provided on November 26. Plummer said the comet was very faint, with a central condensation only suspected. Robinson said the comet was fairly bright, with a 12th-magnitude nucleus. B. von Engelhardt (Dresden, Germany) observed with a 30-cm refractor and said the comet was as bright as a Herschel nebula of class II. He noted it was round, about 0.5' across, and exhibited a nucleus of magnitude 13. Schorr noted the comet was very faint and about 1.5' across. He added that the faint condensation seemed granular. W. Winkler (Jena, Germany) observed with a 15-cm refractor and said the comet was 0.5' across and about magnitude 11. On November 28, Schorr said the comet had brightened and exhibited a diffuse nucleus of magnitude 11. On the 29th, R. Bryant (Royal Observatory, Greenwich, England) observed with the 17-cm refractor (55×) and said the comet exhibited a fairly defined nucleus, while Holetschek gave the magnitude as about 9.8 and the coma diameter as 2.5'. That same night, Engelhardt said the comet was rather bright and granular, with several bright points visible in the coma instead of a nucleus. He added that it was easily seen in the 13-cm finder. On November 30, Davidson said the comet had a well-defined nucleus, while Wickham said it was brighter than when last seen on the 9th. He noted a possible faint, widely fan-shaped tail extending northward.

The comet became a challenging object as December progressed. First, moonlight interfered around mid-month, and then the comet moved into evening twilight at the end of the month. On December 2, Holetschek gave the magnitude as about 9.8 and the coma diameter as 2.2'. On the 3rd, Winkler said the comet was 1' across on the 3rd and estimated the magnitude as 11.5 on the 4th. Robinson said the coma was about 1.5' across on the 7th and exhibited a nucleus of magnitude 12. On December 8, Winkler said the comet was 2' across and about magnitude 11, while Holetschek gave the coma diameter as 2.7' and the magnitude as about 9.5. Halm said the comet was very faint because of a hazy sky and moonlight on the 13th and described it as a very distinct, round nebulosity on the 23rd. On the 29th, Holetschek gave the magnitude as less than 9.3. E. Frisby (US Naval Observatory, Washington, DC, USA) obtained the last Northern Hemisphere observation on December 31, when he found the comet low over the horizon with the 30-cm refractor.

The comet passed 10° from the sun on 1897 January 19 and continued heading southward. It was finally recovered in the morning sky on February 23, when J. Tebbutt (Windsor, New South Wales, Australia) found it at low altitude and said it was extremely faint because of the haze near the horizon.

On March 4, L. Cruls (Rio de Janeiro, Brazil) observed with a 23-cm refractor and said the comet was very faint and centrally condensed, with a coma diameter of about 3'. That same morning, W. H. Finlay (Royal Observatory, Cape of Good Hope, South Africa) observed with the 18-cm refractor and said the comet was round, 1' across, and centrally condensed. He noted that it faded under bright illumination at the same time as stars of magnitude 10.5. Cruls said the comet was 3–4' across on the 13th, with a central condensation. On March 16, he noted the comet was about 4' across.

The comet passed closest to Earth during April, but it was not a particularly close approach and the comet's increasing distance from the sun was taking a toll on its brightness. On April 7, Finlay said the comet was faint in the 18-cm refractor. On the 11th, the comet attained its most southerly declination of $-79°$. Cruls said the comet was diffuse, with a nucleus of about magnitude 10. J. M. Thome (Cordoba, Argentina) saw the comet on several occasions during the period spanning April 12–30. He wrote that the comet "was barely visible in dark field, and it faded away under the faintest illumination." Tebbutt said the comet was "of the last degree of faintness" in the 20-cm refractor on the 20th. He experienced exceptionally clear skies on the 26th and 27th and observed the comet again with the 20-cm refractor. He wrote, "By much eye-straining and looking obliquely into the eyepiece I could just see it as a faint stain on the blue of the heavens."

The comet was last observed on May 6.03, when Cruls gave the position as $\alpha = 9^h 26.7^m$, $\delta = -41° 10'$. He was observing with a 23-cm refractor and described the comet as very faint and at the limit of visibility.

The first parabolic orbits was calculated by Perrine, W. J. Hussey, and O. Knopf. Perrine and Hussey used positions obtained only by Perrine on November 3, 4, and 5, and determined a perihelion date of 1897 January 19.11. Knopf took positions from the same days, but from three different observatories, and determined the perihelion date as February 5.11. Knopf's orbit would prove much closer to the truth, although the perihelion date was still over three days too early. More refined orbits were later published by H. C. F. Kreutz, Knopf, and C. J. Merfield.

Definitive orbits were calculated by J. Möller (1902) and Z. Sekanina (1978). Both astronomers determined hyperbolic orbits and Sekanina's is given below.

T	ω	Ω (2000.0)	i	q	e
1897 Feb. 8.6040 (TT)	172.3394	87.9313	146.1365	1.062791	1.000983

ABSOLUTE MAGNITUDE: $H_{10} = 7.0 - -8.3$ (V1964)
FULL MOON: Oct. 21, Nov. 20, Dec. 20, Jan. 18, Feb. 17, Mar. 18, Apr. 17, May 16
SOURCES: C. Davidson, W. H. Robinson, and W. Wickham, *MNRAS*, **57** (1896 Nov.), pp. 17–18; C. D. Perrine, R. Schorr, W. Viliger, W. J. Hussey, H. C. F. Kreutz, and

O. Knopf, *AN*, **142** (1896 Nov. 12), pp. 29–32; C. D. Perrine and W. J. Hussey, *AJ*, **17** (1896 Nov. 13), p. 31; O. Knopf, *AN*, **142** (1896 Nov. 22), p. 47; H. A. Kobold and A. Abetti, *AN*, **142** (1896 Nov. 30), p. 61; R. Bryant and C. Davidson, *MNRAS*, **57** (1896 Dec.), p. 82; B. von Engelhardt and R. Schorr, *AN*, **142** (1896 Dec. 7), p. 91; B. von Engelhardt, *AN*, **142** (1896 Dec. 22), p. 143; O. Knopf, *AN*, **142** (1897 Jan. 4), p. 157; W. Winkler, *AN*, **142** (1897 Feb. 4), pp. 281–4; J. Halm, *MNRAS*, **57** (1897 Mar.), pp. 420, 425; W. E. Plummer, W. Wickham, and W. H. Robinson, *MNRAS*, **57** (1897 May), pp. 554, 558–9; C. J. Merfield, *MNRAS*, **57** (1897 Jun.), pp. 608–9; R. Schorr, *AN*, **143** (1897 Jun. 2), pp. 269, 283; J. Tebbutt, *AN*, **143** (1897 Jun. 10), pp. 303–6; J. Tebbutt, *AN*, **144** (1897 Jul. 29), p. 79; C. J. Merfield, *AN*, **144** (1897 Aug. 2), p. 95; J. M. Thome, *AN*, **144** (1897 Aug. 26), p. 175; E. Frisby, *AJ*, **18** (1897 Oct. 15), p. 52; L. Cruls, *CR*, **125** (1897 Nov. 2), pp. 637–8; L. Cruls, *AN*, **145** (1898 Jan. 3), pp. 201–4; W. H. Finlay, *AN*, **146** (1898 May 23), p. 203; J. Holetschek, *AN*, **149** (1899 Apr. 10), p. 54; F. Bidschof, *AN*, **149** (1899 Jun. 15), p. 409; J. Möller, *AJB*, **3** (1902), pp. 170, 190–1; V1964, p. 65; B. G. Marsden, Z. Sekanina, and E. Everhart, *AJ*, **83** (1978 Jan.), p. 66; B. G. Marsden and Z. Sekanina, *CCO*, 3rd ed. (1979), pp. 20, 48.

18D/1896 X1
(Perrine–Mrkos)

1896 VII = 1896g

Discovered: 1896 December 9.35 ($\Delta = 0.26$ AU, $r = 1.13$ AU, Elong. $= 117°$)
Last seen: 1897 March 4.14 ($\Delta = 1.04$ AU, $r = 1.67$ AU, Elong. $= 110°$)
Closest to the Earth: 1896 December 4 (0.2582 AU)
Calculated path: PSC (Disc), CET (Dec. 19), TAU (Jan. 1), ERI (Jan. 5), ORI (Jan. 22), MON (Mar. 1)

C. D. Perrine (Lick Observatory, California, USA) discovered this comet on 1896 December 9.35 at a position of $\alpha = 0^h\ 52.4^m$, $\delta = +6°\ 25'$. He estimated the magnitude as 8 and said there was a well-defined nucleus, as well as a fan-shaped tail less than 30' long. The comet was again seen by Perrine on December 10.15. The comet had already passed perihelion in late November and had passed closest to Earth five days earlier.

Word of the discovery quickly reached the European observatories, and several astronomers measured the comet's position late on December 10. Among them were E. Hartwig (Bamberg, Germany), A. Abetti (Arcetri, Italy), and C. Rambaud and F. Sy (Alger, now al-Jazâ'ir, Algeria). Hartwig noted the coma was diffuse and 5' across, while the nucleus seemed to flash. Abetti said the comet was small, like a star of magnitude 11, with a faint nebulosity 0.5' across. Rambaud and Sy described the comet as round in the 31.8-cm refractor, with a diameter of about 2' and a very apparent central condensation. W. Wickham (Radcliffe Observatory, Oxford, England) and W. E. Plummer (Liverpool, England) independently observed the comet on the 11th. Wickham used the 25.4-cm refractor (100×) and said the comet was about magnitude 9.5–10 in the moonlight. Plummer said the comet was easily seen in the moonlight. W. H. Robinson (Radcliffe Observatory, Oxford, England) and H. C. Wilson (Goodsell Observatory, Northfield, Minnesota, USA) independently saw the comet on the 12th. Robinson observed with the 25.4-cm refractor (100×) and said the comet exhibited a very faint nebulous condensation. Wilson said the comet was rather faint in moonlight, but was visible in the 13-cm finder. He estimated the coma diameter as 5' and said the 41-cm refractor revealed a strong central condensation, a stellar nucleus of about magnitude 12, but no tail. Perrine described the comet as faint in moonlight on the 18th and 21st.

Several observations were made during the final days of December as the moon had finally left the sky. J. Halm (Royal Observatory, Edinburgh, Scotland) saw the comet with the 38-cm Dunecht refractor on the 22nd and said it was faint, while on the 23rd he noted it was oval with a nucleus of magnitude 11. B. von Engelhardt (Dresden, Germany) saw the comet on the 26th and noted it was quite bright and extended in the finder, while the 30-cm telescope showed it small and faint. He added that the coma was condensed, but instead of a nucleus, he noted bright flashing points within the condensed region. A. A. Nijland (Utrecht, Netherlands) said the comet was round and rather bright in a 26-cm refractor on the 27th. He added that the coma was 3.5' across and contained a nucleus of magnitude 10.5. Wilson also saw the comet on that evening and described it as bright, with a distinct nucleus of about magnitude 12. He added that the coma was 5–6' across, with a tail on the north following side. J. Holetschek (University Observatory, Vienna, Austria) determined the magnitude as 8.4 on the 28th and estimated the coma diameter as 4.5'. F. Schwab (Kremsmünster, Austria) saw the comet on the same night and said it had a bright condensation and was 1' across. F. Bidschof (Vienna, Austria) said the comet was 2–3' across on the 29th, with a nucleus of magnitude 11.

As the new year began the comet continued to be followed from several observatories. Holetschek gave the total magnitude as 8.5 on January 2 and said the coma was 3.3' across. H. A. Kobold (Strasbourg, France) saw the comet on the same evening and estimated the magnitude as 10.5. Engelhardt saw the comet on the 5th and said it was small and rather bright in the 30-cm refractor, and even brighter in the finder. He said a 13th-magnitude nucleus was in the front of the coma. Nijland and Schwab also saw the comet that same night. Nijland observed with a 26-cm refractor and said the nucleus was very diffuse. Schwab simply noted the comet was small with a bright middle. Nijland and P. Chofardet (Besançon, France) independently saw the comet on the 6th. Nijland said the coma was 5–6' across, while Chofardet said the comet was round with a central condensation as bright as a star of magnitude 12. He added that the coma was 1' 30" across. Wilson said the comet was bright and easily observed on the 9th. He also noticed a faint, broad tail about 10' long. The comet reached its most southerly declination of −1° on January 12.

Following the January full moon, the comet was showing obvious signs of fading, although Plummer's observation of the 20th noted the comet was "brighter than expected." Holetschek estimated the total magnitude as 10.5 on January 20. Abetti observed with a magnification of 124× on the 25th and said the coma was 0.5' across. Kobold said the comet was still rather bright in a 46-cm refractor on the 26th and he noted a condensation of magnitude 12. On January 27, W. A. Villiger (Munich, Germany) said the comet appeared very faint in a 27-cm refractor, while Holetschek estimated the magnitude as 10.75. Holetschek added that the coma was 2' across. Wilson saw the comet the same evening and said it was visible in a 13-cm finder and was very diffuse. He added that a 41-cm refractor revealed a small condensation. Holetschek said the comet had faded to 11 by the 28th. Robinson saw the comet on the 23rd and 26th. He noted it was a faint nebulosity about 2' across, with a condensation.

The comet was becoming more challenging during February and fewer precise positions were reported for the month. R. Schorr (Hamburg, Germany) saw the comet on the 3rd and said it appeared small and faint, with a distinct, star-like condensation of magnitude 12.5. Wilson saw the comet on the 23rd and described it as exceedingly faint in the 41-cm refractor. Kobold said the comet was barely seen in a 46-cm refractor on the 24th. He described it as "a pale wisp" about 1' across.

The final observations were made at the beginning of March. Kobold said the comet was still barely visible in the 46-cm refractor on the 1st and Wilson obtained the last observation on March 4.14, while using the 41-cm refractor. He gave the position as $\alpha = 6^h \, 20.9^m$, $\delta = +3° \, 27'$ and noted the comet was "most exceedingly faint, but certainly seen."

The earliest parabolic orbits were published on December 14. The first was computed by W. J. Hussey and Perrine from observations obtained on December 9, 10, and 11. It indicated a perihelion date of 1896 November 26.17. The second was calculated by F. W. Ristenpart and indicated a perihelion date of December 1.21. By mid-January, Ristenpart, as well as Hussey and Perrine, had recognised the comet's short-period orbit, as well as a resemblance to the orbit of the lost periodic comet Biela. The orbit of Perrine's comet was indeed similar to that of Biela except for ω, which differed by 60°. Ristenpart computed a revised elliptical orbit which was published on February 4. The perihelion date was November 25.12 and the period was 6.44 years. He examined several hypotheses as to a possible relationship between this comet and the lost periodic comet Biela. Even with the gravitational effects of Jupiter being considered the orbits of the two comets could never be made to match up at any given time between the disappearance of Biela and the discovery of Perrine's comet. A definitive orbit was calculated by H. Osten (1898). Beginning with 24 positions obtained throughout the comet's period of visibility, he applied perturbations by Earth, Mars, Jupiter, and Saturn, and determined a perihelion date of November 25.13 and a period of 6.44 years. The most recent orbital analysis was published by B. G. Marsden (1978). He took 125 positions obtained during the apparitions of 1896, 1909, and 1955, applied perturbations by all nine planets, and found a perihelion date of November 25.12 and a period of 6.42 years. He also determined the nongravitational terms as $A_1 = -0.08$ and $A_2 = -0.0597$. This orbit is given below.

T	ω	$\Omega\,(2000.0)$	i	q	e
1896 Nov. 25.120 (UT)	163.848	248.068	13.665	1.11022	0.67866

ABSOLUTE MAGNITUDE: $H_{10} = 9.9$ (V1964)

FULL MOON: Nov. 20, Dec. 20, Jan. 18, Feb. 17, Mar. 18

SOURCES: C. D. Perrine, E. Hartwig, H. C. F. Kreutz, and F. Ristenpart, *AN*, **142** (1896 Dec. 14), pp. 109–12; C. D. Perrine, *AJ*, **17** (1896 Dec. 23), p. 55; C. Rambaud and F. Sy, *AN*, **124** (1897 Jan. 4), p. 22; B. von Engelhardt, H. A. Kobold, and F. Ristenpart, *AN*, **142** (1897 Jan. 11), pp. 189–92; W. J. Hussey and C. D. Perrine, *AN*, **142** (1897 Jan. 21), p. 239; A. Abetti and B. von Engelhardt, *AN*, **142** (1897 Jan. 29), pp. 269–72; F. Ristenpart, *AN*, **142** (1897 Feb. 4), pp. 283–6; A. A. Nijland, A. Abetti, and R. Schorr, *AN*, **142** (1897 Feb. 10), p. 339; C. D. Perrine, *AJ*, **17** (1897 Feb. 19), p. 91; J. Halm, *MNRAS*, **57** (1897 Mar.), pp. 420, 425; P. Chofardet and H. Petit, *AN*, **142** (1897 Mar. 5), pp. 397–400; W. Wickham and W. H. Robinson, *MNRAS*, **57** (1897 May), pp. 558–9; H. A. Kobold, *AN*, **143** (1897 May 18), pp. 245–8; H. C. Wilson, *AJ*, **18** (1897 Jul. 30), pp. 4–5; W. A. Villiger, *AN*, **144** (1897 Aug. 26), pp. 145, 171; F. Schwab, *AN*, **145** (1898 Feb. 24), p. 333; H. Osten, *AN*, **145** (1898 Mar. 4), pp. 349–52; P. Chofardet, *AN*, **146** (1898 Jun. 2), p. 299; W. E. Plummer, *MNRAS*, **59** (1898 Dec.), pp. 100, 103; J. Holetschek, *AN*, **149** (1899 Apr. 10), p. 53; R. Schorr, *AN*, **149** (1899 Apr. 19), pp. 145, 155; F. Bidschof, *AN*, **149** (1899 Jun. 15), p. 409; F. W. Ristenpart, *AN*, **161** (1903 Jan. 23), p. 11; V1964, p. 65; B. G. Marsden, *QJRAS*, **19** (1978), pp. 80–1, 88.

6P/d'Arrest

1897 II = 1897a

Recovered: 1897 June 29.48 ($\Delta = 1.42$ AU, $r = 1.39$ AU, Elong. $= 68°$)

Last seen: 1897 October 4.39 ($\Delta = 1.26$ AU, $r = 1.98$ AU, Elong. $= 122°$)

Closest to the Earth: 1897 October 12 (1.2559 AU)

Calculated path: CET (Rec), TAU (Jul. 29), ERI (Sep. 18)

G. Leveau (1897) took the orbit from this comet's 1890 apparition and advanced it to 1897 without consideration of planetary perturbations. The resulting perihelion date was 1897 May 23. He published an ephemeris for the period of March 6 to August 5. C. D. Perrine (Lick Observatory, California, USA) recovered this comet with the 30-cm refractor on 1897 June 29.48, at a position of $\alpha = 2^h\,01.4^m$, $\delta = +6°\,14'$. He said the observation was made in morning twilight and added that the comet was very faint. Perrine confirmed this was comet d'Arrest on the 30th and described the comet as 2′ across, with a poorly defined central condensation. He noted it "seems to be slightly extended north-preceding" and added that it was barely detectable in the 8.3-cm finder.

The comet was a month past perihelion when recovered and few magnitude estimates were made during the remainder of the apparition. During the early part of July, Perrine examined the comet on a few occasions with the 91-cm refractor (270×) and described it as 2′ in diameter with a bright central condensation 20–30″ across. He added that a nucleus was occasionally seen, which was about 3–4″ across. The nucleus was not central, "but a little south, preceding the center." L. Cruls (Rio de Janeiro, Brazil) saw the comet on July 2 with a 25-cm refractor and described it as a circular neb-ulosity, about 3′ across, without a well-defined nucleus. C. Rambaud and F. Sy (Alger, now al-Jazâ'ir, Algeria) saw the comet with a 31.8-cm refrac-tor on the morning of the 4th and 5th. They noted the comet appeared very faint in weak twilight, but was 1′ across and exhibited a condensation. Cruls noted a central nucleus of magnitude 11 on July 9. S. J. Brown (US Naval Observatory, Washington, DC, USA) said the comet was visible in the 13-cm finder of the 66-cm refractor on July 10. Perrine noted a well-marked central condensation about 10″ across in the 91-cm refractor on August 9. V. Cerulli (Teramo, Italy) saw the comet on August 26 and 28, with the 39-cm Cooke telescope, and simply described it as faint. Perrine found the comet "much fainter" on August 28 than it had been on the 9th. He estimated the diam-eter as 20–30″ and said it exhibited little condensation as viewed through the 91-cm refractor. Perrine gave the magnitude as 14–15. Perrine noted the comet was still fainter on September 1. On September 26, Perrine said the comet was faint, with a magnitude between 15 and 16 in the 91-cm refractor.

The comet was last detected on October 4.39, when Perrine measured the position as $\alpha = 4^h 32.2^m$, $\delta = -2° 38′$. He was using the 91-cm refractor and said the comet was difficult to see and measure. Perrine could not find the comet on October 5 and 25.

The first comprehensive investigation of this comet's motion was con-ducted by A. W. Recht (1939). He took 32 Normal positions covering eight apparitions spanning 1851–1923 and applied perturbations by Mercury to Neptune. For this apparition he determined a perihelion date of May 23.59. Although he recognized a large secular deceleration, he did not apply this to his computations. Nongravitational effects were first determined for the orbit of this apparition by B. G. Marsden and Z. Sekanina (1972, 1973). The perihelion date was May 23.84 and the period was 6.69 years, but the values for the nongravitational terms of A_1 and A_2 were determined as +0.79 and +0.0937, respectively.

T	ω	Ω (2000.0)	i	q	e
1897 May 23.8350 (TT)	173.0235	147.7967	15.6986	1.325977	0.626601

ABSOLUTE MAGNITUDE: $H_{10} = 8.0$–11 (V1964)

FULL MOON: Jun. 14, Jul. 14, Aug. 12, Sep. 11, Oct. 10

SOURCES: G. Leveau, *BA*, **14** (1897 Jan.), p. 30; G. Leveau, *AN*, **142** (1897 Feb. 15), p. 357; C. D. Perrine, *AJ*, **17** (1897 Jul. 2), p. 184; C. D. Perrine, *AN*, **143** (1897 Jul. 5), p. 415; C. Rambaud and F. Sy, *CR*, **125** (1897 Jul. 12), p. 83; C. D. Perrine, *AJ*, **17** (1897 Jul. 19), p. 192; R. G. Aitken, *AJ*, **18** (1897 Aug. 24), p. 24; C. D. Perrine, *AJ*, **18** (1897 Sep. 14), p. 29; S. J. Brown, *AJ*, **18** (1897 Nov. 15), p. 79; C. D. Perrine, *AJ*, **18** (1897 Dec. 14), p. 104; V. Cerulli, *AN*, **145** (1898 Feb. 24), p. 329; L. Cruls, *AN*, **146** (1898 Apr. 4), p. 91; P. Chofardet and H. Petit, *AN*, **146** (1898 Jun. 2), p. 299; W. E. Plummer, *MNRAS*, **59** (1898 Dec.), pp. 101, 103; A. W. Recht, *AJ*, **48** (1939 Jul. 17), pp. 65–78; V1964, p. 65; B. G. Marsden and Z. Sekanina, *QJRAS*, **13** (1972), pp. 428–9; B. G. Marsden, Z. Sekanina, and D. K. Yeomans, *AJ*, **78** (1973 Mar.), p. 213.

C/1897 U1
(Perrine)

1897 III = 1897b

Discovered: 1897 October 17.24 ($\Delta = 0.81$ AU, $r = 1.56$ AU, Elong. $= 118°$)
Last seen: 1897 November 28.02 ($\Delta = 1.08$ AU, $r = 1.37$ AU, Elong. $= 83°$)
Closest to the Earth: 1897 October 23 (0.7993 AU)
Calculated path: CAM (Disc), CAS (Oct. 18), CEP (Oct. 24), DRA-CEP (Nov. 2), DRA (Nov. 4)

C. D. Perrine (Lick Observatory, California, USA) discovered this comet on 1897 October 17.24 at a position of $\alpha = 3^h 36.1^m$, $\delta = +66°$ 47'. He said it was about 2' across, with a total magnitude of about 8. There was also a well-defined nucleus and a tail extending less than 30'. Perrine confirmed the comet on the 18th and noted the tail was extending toward PA 209.3°.

This comet was widely observed during the remainder of October and passed moderately close to Earth about a week after its discovery. It attained its most northerly declination of $+82°$ on the 30th. Although it was heading toward an early December perihelion and should have changed little in brightness for several weeks, the comet definitely faded by the end of October. The best series of magnitude estimates came from J. Holetschek (Vienna, Austria). He gave the magnitude as 8.7 on the 19th, 9.0 on the 24th, 9.2 on the 25th, 9.3 on the 26th, 9.5 on the 27th, 9.0 on the 28th, 9.5 on the 29th, and 10 on the 30th. The best estimates of the nuclear magnitude were made by V. Cerulli (Teramo, Italy). He gave values of 12 on the 25th, 12.5 or 13 on the 27th, and 13.5 on the 30th. C. Rambaud and F. Sy (Alger, now al-Jazâ'ir, Algeria) observed with the 32-cm refractor and said the nucleus considerably diminished in brightness between observations made on the 21st and 28th. Many observers reported the nucleus was stellar early in this period, while most reported the nucleus was absent or very diffuse at the end of the month. A wide range of coma diameters were given during this period, although Holetschek provided the best series. He gave the diameter as 2' on the 19th, 2.5' on the 26th, and 3' on the 28th. G. Bigourdan (Paris, France) provided coma diameters of 12" on the 18th and 19th, and 30" on the 21st, but these probably referred to the central condensation. Other coma diameters included 5' by K. Mysz (Pula, Yugoslavia) on the 18th, 2' by F. Bidschof (Vienna, Austria) and 5' by F. S. Archenhold (Treptow Observatory, Germany) on the 19th, 1.3' by W. Schur (Göttingen, Germany) on the 24th, and 2' by F. Schwab (Kremsmünster, Austria) and Cerulli on the 25th. On the 29th, R. Schorr (Hamburg, Germany) said the coma was elliptical and measured 2' long and 15" wide. Other observers began noticing unusual characteristics of the coma and nucleus as October progressed. On the 20th, J. Möller (Kiel, Germany) said the nucleus appeared oblong and diffuse in the 20-cm refractor, while L. Picart (Bordeaux, France) said the comet had an elliptical shape. Schwab noted the comet was somewhat elongated and weakly condensed on the 22nd. On the 23rd, Bigourdan reported the comet was more diffuse than on previous nights. On the 24th, Bigourdan said the comet was much less clear than on the previous night and he noted two stellar points within the coma. On the 25th, he said the coma was granular and presented a stellar nucleus, with one or two other nuclei suspected. The tail was extensively observed at several observatories. The greatest estimates of the length came from W. J. Hussey (Lick Observatory, California, USA) on the 18th and W. Villiger (Munich, Germany) on the 23rd. Both astronomers said it was 10′ long, with Hussey's estimate coming from a photograph, while Villiger's was visual. A. A. Nijland (Utrecht, Netherlands) obtained the best data on the tail size and direction. He said it extended 1.8′ toward PA 206.5° on the 19th, 1.95′ toward PA 201.9°on the 21st, 3.2′ toward PA 170.1° on the 26th, and 6.3′ toward PA 158.2° on the 27th. He also noted it extended toward PA 180.0° on the 25th and PA 139.6° on the 29th.

Several observations were obtained on November 1. W. E. Plummer (Liverpool, England) described the comet as "very feeble," while Perrine saw neither a stellar nucleus nor a condensation. Perrine wrote that the comet presented "the appearance of a streak of nebulosity of almost uniform brightness about 3′ long, followed by a fainter streak of 2′ length." Cerulli said the tail extended toward PA 107°. On the 2nd, H. C. Wilson (Goodsell Observatory, Minnesota, USA) observed with the 41-cm refractor and noted a stellar nucleus of magnitude 12, a bright narrow streak of tail 4′ or 5′ long, and a broader, exceedingly faint tail extending about 10′. There was a slight condensation. That same night, Kobold observed in moonlight and described the comet as an exceptionally diffuse, nebulous stroke, with a diffuse condensation. On the 22nd, Kobold said the comet was an exceptionally faint, round nebulosity, about 1′ across, that was elongated toward PA 45°. Kobold last saw the comet on November 25 and 26, but did not measure a position because it was difficult to ascertain the comet's center. The comet was last detected on November 28.02 by H. R. Morgan (Leander McCormick Observatory, Virginia, USA).

The first parabolic orbit was calculated by Hussey and R. G. Aitken using positions from October 17, 18, and 19. The resulting perihelion date was 1897 December 9.73. Additional orbits by H. C. F. Kreutz, Hussey, Aitken, Perrine, Möller, Wilson, and R. T. Crawford, eventually established the perihelion date as December 9.1. A definitive orbit was calculated by E. Wessell (1900). He began with 168 positions, reduced them down to five Normal positions, and determined the parabolic orbit given below.

T	ω	Ω (2000.0)	i	q	e
1897 Dec. 9.1426 (UT)	65.9082	33.4877	69.6101	1.356679	1.0

ABSOLUTE MAGNITUDE: $H_{10} = 8.0$ (V1964)

FULL MOON: Oct. 10, Nov. 9, Dec. 9

SOURCES: C. D. Perrine, H. A. Kobold, I. Benko von Boinik, F. Bidschof, W. J. Hussey, R. G. Aitken, A. Schwassmann, and J. Möller, *AN*, **144** (1897 Oct. 20), pp. 335–6; G. Bigourdan, *CR*, **125** (1897 Oct. 26), pp. 592–4; C. D. Perrine, W. J. Hussey, R. G. Aitken, J. G. Porter, and T. Reed, *AJ*, **18** (1897 Oct. 27), pp. 63–4; I. Benko von Boinik, K. Mysz, J. Möller, R. Schorr, L. Picart, and H. C. F. Kreutz, *AN*, **144** (1897 Oct. 28), pp. 349–52; W. J. Hussey and R. G. Aitken, *AJ*, **18** (1897 Nov. 5), p. 72; F. S. Archenhold, E. Millosevich, and W. Schur, *AN*, **144** (1897 Nov. 5), p. 381; C. Rambaud and F. Sy, *CR*, **125** (1897 Nov. 8), pp. 690–1; J. Möller, *AN*, **145** (1897 Nov. 20), pp. 45–8; C. D. Perrine, *AJ*, **18** (1897 Nov. 24), p. 88; H. C. Wilson, *AJ*, **18** (1897 Dec. 4), p. 93; W. J. Hussey and R. T. Crawford, *AJ*, **18** (1897 Dec. 24), pp. 108–10; V. Cerulli and F. Schwab, *AN*, **145** (1898 Feb. 24), pp. 329–34; A. A. Nijland, *AN*, **146** (1898 May 23), pp. 201–4; C. D. Perrine, *AJ*, **19** (1898 Jun. 2), p. 38; H. A. Kobold, *AN*, **147** (1898 Aug. 15), pp. 121–4; W. Villiger, *AN*, **147** (1898 Sep. 30), pp. 275, 279; W. E. Plummer, *MNRAS*, **59** (1898 Dec.), pp. 101, 103; J. Holetschek, *AN*, **149** (1899 Apr. 10), p. 54; R. Schorr, *AN*, **149** (1899 Apr. 19), pp. 145, 155; F. Bidschof, *AN*, **149** (1899 Jun. 15), p. 409; E. Wessell, C. D. Perrine, and H. R. Morgan, *AN*, **151** (1900 Jan. 19), pp. 209–18; V1964, p. 65.

7P/Pons–
Winnecke

1898 II = 1898a

Recovered: 1898 January 2.55 ($\Delta = 1.75$ AU, $r = 1.43$ AU, Elong. $= 55°$)
Last seen: 1898 March 1.57 ($\Delta = 1.35$ AU, $r = 0.97$ AU, Elong. $= 45°$)
Closest to the Earth: 1898 February 26 (1.3508 AU)
Calculated path: LIB (Rec), OPH (Jan. 12), SER (Jan. 30), OPH (Feb. 4), SER (Feb. 5), SCT (Feb. 12), SGR (Feb. 19)

C. Hillebrand (1897) began with E. F. von Haerdtl's work for the 1892 apparition, made the appropriate correction by using positions from that apparition, and then applied von Haerdtl's perturbations by Jupiter from 1892 to June of 1898. The result was a perihelion date of 1898 March 21.00. The comet was recovered by C. D. Perrine (Lick Observatory, California, USA) on 1898 January 2.55, while using the 91.4-cm refractor. He gave the position as $\alpha = 15^h 18.9^m$, $\delta = -3° 59'$. Additional positions were obtained on January 3 and 4. Perrine described the comet as very faint, slightly condensed, and about 10–15" across.

Perrine remained the only observer of this comet, even though most of his observations were made with the 30-cm refractor. Using the 91-cm refractor, he saw the comet on January 21 and said it was very distinct, about 30" across, and exhibited a central condensation. A nucleus was occasionally suspected. He said the comet was as bright as a star of magnitude 14. The 30-cm refractor was used on January 26, and Perrine described the comet as faint and difficult to measure. He was again using the larger refractor on January 29 and said the comet was 30–40" across, with a nucleus that "sometimes appears stellar." He estimated the nuclear magnitude as 16. Still using the larger refractor on January 30, Perrine estimated the comet's magnitude as 13 or 14. Although he noted a "decided central condensation," no stellar point was seen. Perrine used the smaller refractor on February 19. He noted a coma about 1' across, with a central condensation of magnitude about 12.5 or 13. On February 22, the smaller refractor revealed the comet as about magnitude 12, with a centrally condensed coma about 1' across.

The comet was last detected on March 1.57, when Perrine measured the position as $\alpha = 19^h 45.7^m$, $\delta = -14° 26'$. He noted that the measurements were made in morning twilight and were "somewhat uncertain." The comet was not observed after perihelion.

Perrine's initial positions were used to correct Hillebrand's predicted perihelion date. The result was March 20.89. Later astronomers established the perihelion date as March 20.87 and the period as 5.83 years. The comet's nongravitational forces were first examined for this apparition by B. G. Marsden, Z. Sekanina, and D. K. Yeomans (1973). They obtained values of $A_1 = +0.01$ and $A_2 = +0.0008$. K. Kinoshita (2000) has conducted one of the more recent investigations of this comet's past motion and determined the comet's nongravitational terms as $A_1 = -0.43$ and $A_2 = +0.0037$. His orbit is given below.

T	ω	Ω (2000.0)	i	q	e
1898 Mar. 20.8669 (TT)	173.4086	102.2358	16.9909	0.923820	0.714827

ABSOLUTE MAGNITUDE: $H_{10} = 9.6$ (V1964)
FULL MOON: Dec. 9, Jan. 8, Feb. 6, Mar. 8
SOURCES: C. Hillebrand, *AN*, **144** (1897 Sep. 22), pp. 233–6; C. D. Perrine and E. F. von Haerdtl, *AJ*, **18** (1898 Jan. 14), p. 127; C. Hillebrand and C. D. Perrine, *AN*, **145** (1898 Jan. 14), p. 235; C. D. Perrine, *AJ*, **19** (1898 Jul. 8), pp. 50–1; V1964, p. 65; B. G. Marsden, Z. Sekanina, and D. K. Yeomans, *AJ*, **78** (1973 Mar.), pp. 214, 216; personal correspondence from K. Kinoshita (2000).

C/1898 F1 *Discovered:* 1898 March 20.54 ($\Delta = 1.59$ AU, $r = 1.10$ AU, Elong. $= 43°$)
(Perrine) *Last seen:* 1898 November 16.33 ($\Delta = 2.66$ AU, $r = 3.48$ AU, Elong. $= 140°$)
Closest to the Earth: 1898 March 26 (1.5785 AU)
1898 I = 1898b *Calculated path:* PEG (Disc), LAC (Apr. 8), AND (Apr. 12), CAS (Apr. 26), PER (May 10), CAS (May 12), PER (May 14), CAM (Jun. 1), AUR (Jul. 4), LYN (Sep. 7), AUR (Sep. 30)

C. D. Perrine (Lick Observatory, California, USA) discovered this comet on March 20.54 and gave the position as $\alpha = 21^h 18.6^m$, $\delta = +16° 43'$. He described it as about equal in brightness to a star of magnitude 6–7. The coma was 2' across and exhibited a condensation 10" across. Perrine added that a broad, fan-shaped tail extended 1° toward PA 281.2° in the 10-cm comet-seeker. The comet was confirmed by Perrine's colleague, W. J. Hussey, on March 22.05, and Perrine noted it was distinctly visible to the naked eye.

The comet was well observed during the remainder of March. Although it had passed perihelion before the discovery, it passed closest to Earth late in the month. Among the many estimates of the total magnitude, the most complete were made by J. Holetschek (Vienna, Austria). He gave values of 6.7 on March 22, 24, and 27, 6.1 on the 28th, and 5.7 on the 29th. Interestingly, R. Schorr (Hamburg, Germany), F. Bidschof (Vienna, Austria), and E. Hartwig (Bamberg, Germany) gave the magnitude as 6.5 on the 22nd.

Combined with Holetschek's estimate for the same date, this marked the first time there was such consistency in magnitude estimates. Although the "nucleus" was usually described as stellar, there was a wide variation in brightness estimates because of the variety of telescopes used. It is likely that smaller telescopes only saw the condensation, while larger instruments saw the nucleus. Magnitude estimates of the condensation included 8 by W. E. Plummer (Liverpool, England) on the 22nd, 7.5 by K. Mysz (Pula, Yugoslavia) and 7 or 8 by G. Ciscato and A. Antoniazzi (Padova, Italy) on the 23rd, 7 by H. A. Kobold (Strasbourg, France) on the 27th, and 7.5 by M. F. J. C. Wolf, and A. Schwassmann (Heidelberg, Germany) and 8 by Schwab on the 29th. Magnitude estimates of the nucleus included 9.5 by Bidschof on the 22nd, 8.9 by Hartwig on the 23rd, "not brighter than 9" by F. Hayn (Leipzig, Germany) on the 29th, and 9 by Kobold on the 31st. The coma diameter was given as about 1.5' by Schorr and 2' by Hartwig on the 22nd, 2' by Hartwig on the 23rd, and 1' by Hayn on the 29th. The coma was frequently described as "parabolic" on "fan-shaped" by astronomers, because of the tail. Although most estimates of the tail length were less than 15', Hartwig and Holetschek reported it was much longer. On the 22nd, both astronomers gave the length as 1°. During the remainder of the month, it was given as 1° by Hartwig on the 23rd, 1° by Holetschek on the 28th, and 30' by Holetschek on the 30th. The direction of the tail was given as PA 300° by Mysz on the 23rd, PA 290° by Mysz on the 24th, PA 290° by Kobold on the 27th, and PA 290° by W. Villiger (Hamburg, Germany) on the 28th. Schwab noted the tail was straight on the 22nd. A. Abetti (Arcetri, Italy) said the coma and tail formed a parabolic shape on the 23rd, while Ciscato and Antoniazzi noted the tail was fan-shaped.

The comet continued to be well observed during April, but was heading away from both the sun and Earth. Once again, Holetschek supplied the best series of total magnitude determinations. He gave values of 5.6 on the 6th, 6.4 on the 9th and 12th, 6.3 on the 15th, 6.4 on the 17th, 6.8 on the 26th and 27th, and 7.3 on the 30th. The magnitude of the condensation was given as 7 by A. A. Nijland (Utrecht, Netherlands) and 7.5 by Schorr on the 1st, 7 by Nijland on the 2nd, 8 by Schorr on the 3rd, 7.5 by E. Millosevich (Rome, Italy) on the 4th, 7 by Ciscato and Antoniazzi on the 5th, 9 by Kobold and 7 by F. Ristenpart (Kiel, Germany) on the 6th, 7.0 by Nijland on the 7th, 8 by Ciscato and Antoniazzi on the 8th, 7.1 by Nijland on the 14th, "brighter than 7" by Ristenpart on the 15th, 6.9 by Nijland on the 21st, 7.4 by Nijland on the 24th, 7.5 by Nijland on the 25th, and 8.0 by Schorr on the 26th. The magnitude of the nucleus was given as 9 by Nijland on the 1st, 8.5 by Nijland on the 2nd, 10 by Kobold on the 6th, 9 by Schwab on the 6th, 7th, and 8th, 8.5 by Nijland on the 7th, 9 by Kobold and 9.0 by Nijland on the 14th, 8.5 by Schorr on the 15th, 9 by Nijland on the 21st, 9.0 by Nijland on the 24th, and 9.5 by A. Sokolow (Pulkovo Observatory, Russia) on the 25th. Coma diameter estimates were 3' by Plummer on the 1st, 3' by Schorr on the 2nd, 1' by Kobold and 30" by Ristenpart in moonlight on the 6th, 2' by Kobold on the 14th, and 2′ by Schorr on the 26th. Although nearly every observer reported measurements of the tail, Nijland gave the most consistent details throughout the month. He said it extended 32′ toward PA 275° on the 1st, 24′ toward PA 275° on the 2nd, 31′ toward PA 286° on the 14th, 27′ toward PA 287° on the 21st, 28′ toward PA 291° on the 22nd, 21′ toward PA 289° on the 24th, 20′ toward PA 294° on the 25th, and 21′ toward PA 298° on the 28th. E. E. Barnard (Yerkes Observatory, Wisconsin, USA) visually observed the comet with the 102-cm refractor on April 6 and described it as "bright with a rather definite outline to the head. It looked like a great comet. There was a small strong condensation, but no real nucleus."

Although observations began to curtail during May, Holetschek continued to provide total magnitude estimates throughout the month. He gave values of 7.4 on the 1st, 7.8 on the 14th, 8.0 on the 15th, 7.9 on the 16th and 18th, 8.4 on the 21st, 8.6 on the 22nd, 8.5 on the 23rd, and 8.6 on the 25th. Holetschek and others provided the following additional details of this comet during May. On the 1st, Holetschek said the tail was 5′ long, while Schwab gave the nuclear magnitude as 10. Nijland estimated the magnitude as 8.3 on the 2nd and gave the nuclear magnitude as 9.3. On the 3rd, Plummer described the comet as a faint nebulosity about 2′ across, while Abetti said the comet was still well seen in morning twilight, with a parabolic coma. On the 10th, Abetti estimated the magnitude as 8.5. On the 11th, Nijland said the comet was magnitude 7.8, with a nuclear magnitude of 9.5 and a tail extending 14′ toward PA 313.5°. Kobold said the bright nebulosity had a parabolic form on the 13th, as well as a nucleus of magnitude 10. On the 15th, Abetti said the comet was an easily visible round spot, with a trace of tail. On the 15th and 21st, Schwab said the comet was under magnitude 10 and the coma was 1′ across. There was no distinct nucleus. On the 20th, Kobold said the condensation was magnitude 11. On the 21st, Schorr said the comet was magnitude 11.0. On the 22nd, Abetti gave the magnitude as 9.5. On May 23, Villiger said the coma was 4′ across. Nijland said the magnitude was 9.3, with a nucleus of about magnitude 12.

The comet attained its most northerly declination of +56° on June 3. On June 8, Holetschek gave the total magnitude as 9.5. On June 13, Abetti said the comet appeared like a slightly diffuse star of magnitude 11. On June 20, Kobold noted the comet was "still rather bright, but with little condensation." On June 26, Abetti said the magnitude was 12.5. On July 12, Kobold reported the comet was a rather weak nebulosity about 1′ across, with a somewhat bright center. No nucleus was visible. On July 15, Kobold said the sky was better than on the 12th and the comet appeared as a small, rather bright, circular nebulosity, with a central condensation. There was also a nucleus of magnitude 12–13. On July 19, Villiger described the comet as a round, diffuse nebulosity about 30″ across.

Although Perrine had been observing the comet since its discovery, he began using the 91-cm refractor to follow it after July and provided the only physical descriptions for the remainder of the comet's apparition. He said it was magnitude 13–14 on August 13, with a diameter of about 30", and about magnitude 14 on the 14th. Perrine noted a coma 45–60" across on the 15th and 21st and a nucleus of about magnitude 16.5 on the first date and gave the total magnitude as 15 on the last. On August 25, he gave the magnitude as 14 and noted the "faintest possible nucleus." On September 10 and 18, Perrine said the comet was magnitude 16, while the coma was about 10" across on the last date in moonlight. The total magnitude was given as 15.5–16 on the 19th. Perrine said the comet was magnitude 16.5 on October 15 and was about 10–15" across on the 20th. On November 7, Perrine indicated a magnitude of 16.75. He added that the coma was 10" across, with slight central condensation. Perrine said the magnitude was 16.5–17 on November 9. The comet was last seen on November 16.33, when Perrine measured the position as $\alpha = 5^h\ 33.2^m$, $\delta = +48°\ 41'$. He said the comet was around magnitude 16.5–17 in the 91-cm refractor.

The first orbits were calculated by Hussey and Perrine, R. T. Crawford and H. K. Palmer, and Ristenpart using positions from March 20, 22, and 23. The perihelion date was given as 1898 March 19.17 by Hussey and Perrine, April 9.11 by Crawford and Palmer, and March 18.96 by Ristenpart. Additional orbits were calculated by I. Lagarde, Hussey, H. C. F. Kreutz, B. Cohn, and S. K. Winther, which established the perihelion date as March 17.9.

The first elliptical orbit was calculated by Perrine using positions spanning the period of March 20 to April 29. He gave the perihelion date as March 17.57. H. D. Curtis (1899) took 26 positions spanning the period of March 22 to July 19, formed four Normal positions, and revealed a perihelion date of March 17.62 and a period of about 403 years. B. G. Marsden (1978) took 171 positions spanning the period of March 20 to November 13, applied perturbations by all nine planets, and determined an elliptical orbit with a perihelion date of March 17.63 and a period of about 419 years.

T	ω	Ω (2000.0)	i	q	e
1898 Mar. 17.6320 (TT)	47.3073	263.8681	72.5292	1.095261	0.980454

ABSOLUTE MAGNITUDE: $H_{10} = 5.7$ (V1964)

FULL MOON: Mar. 8, Apr. 6, May 6, Jun. 4, Jul. 3, Aug. 2, Aug. 31, Sep. 29, Oct. 29, Nov. 28

SOURCES: C. D. Perrine, R. Schorr, F. Bidschof, E. Hartwig, and K. Mysz, *AN*, **146** (1898 Mar. 23), p. 47; C. D. Perrine, W. J. Hussey, R. T. Crawford, and H. K. Palmer, *AJ*, **18** (1898 Mar. 28), p. 220; R. Schorr, F. Schwab, E. Hartwig, F. Bidschof, C. D. Perrine, and W. J. Hussey, *AN*, **146** (1898 Mar. 28), pp. 77–80; I. Lagarde, *CR*, **126** (1898 Mar. 28), p. 945; H. A. Kobold, H. C. F. Kreutz, and B. Cohn, *AN*, **146** (1898 Apr. 4), pp. 93–6; K. Mysz, A. Abetti, R. Schorr, M. F. J. C. Wolf, and A. Schwassmann, *AN*, **146** (1898 Apr. 11), p. 125; C. D. Perrine, W. J. Hussey, R. T. Crawford, and H. K. Palmer, *AJ*, **19** (1898 Apr. 14). pp. 6–8; E. Millosevich, *AN*, **146** (1898 Apr. 20), p. 143; R. Bryant, *MNRAS*, **58** (1898 May), pp. 411–12; W. J. Hussey and C. D. Perrine, *AJ*, **19** (1898 May 11), pp. 23–4; A. C. D. Crommelin, *MNRAS*, **58** (1898 Jun.), p. 463; R. Schorr, *AN*, **146** (1898 Jun. 2), p. 307; S. K. Winther, *AN*, **146** (1898 Jun. 16), pp. 339–42; A. Sokolow, *AN*, **147** (1898

Sep. 1), p. 189; W. Villiger, *AN*, **147** (1898 Sep. 30), pp. 275, 280; J. Holetschek, *AN*, **147** (1898 Oct. 19), pp. 329–32; H. A. Kobold and F. Schwab, *AN*, **148** (1898 Nov. 21), pp. 7–10; W. E. Plummer, *MNRAS*, **59** (1898 Dec.), pp. 101–3; G. Ciscato, A. Antoniazzi, and A. Abetti, *AN*, **148** (1899 Jan. 7), pp. 113, 117, 121–4; F. W. Ristenpart, *AN*, **148** (1899 Jan. 12), pp. 155–8; F. Hayn, *AN*, **148** (1899 Feb. 15), pp. 299, 303; H. D. Curtis, *AJ*, **19** (1899 Feb. 20), p. 195; C. D. Perrine, *AJ*, **20** (1899 Apr. 8), pp. 17–22; R. Schorr, *AN*, **149** (1899 Apr. 19), pp. 145, 156; A. A. Nijland, *AN*, **149** (1899 Apr. 24), pp. 173–6; W. Villiger, *AN*, **150** (1899 Aug. 22), pp. 225, 233; E. E. Barnard, *AJ*, **41** (1931 Dec. 15), pp. 145, 149; V1964, p. 65; B. G. Marsden, Z. Sekanina, and E. Everhart, *AJ*, **83** (1978 Jan.), p. 66; B. G. Marsden, *CCO*, 3rd ed. (1979), pp. 20, 48.

C/1898 L1	*Discovered:* 1898 June 10.23 ($\Delta = 1.10$ AU, $r = 2.11$ AU, Elong. = 170°)
(Coddington–	*Last seen:* 1899 December 7.37 ($\Delta = 4.34$ AU, $r = 5.27$ AU, Elong. = 160°)
Pauly)	*Closest to the Earth:* 1898 June 19 (1.0787 AU)
	Calculated path: OPH (Disc), SCO (Jun. 11), LUP (Jun. 19), CEN (Jul. 7), LUP
1898 VII = 1898c	(Jul. 14), CEN (Jul. 25), CIR (Sep. 26), APS (Oct. 12), OCT (Nov. 9), HYI (Dec. 6), TUC (Dec. 9), HYI (1899 Jan. 2), ERI (Jan. 3), PHE (Jan. 10), ERI (Jan. 11), PHE (Jan. 14), FOR (Feb. 2), ERI (Mar. 7), ORI (Jun. 20), TAU (Nov. 29)

E. F. Coddington (Lick Observatory, California, USA) obtained a 2-hour exposure with the Crocker photographic telescope of the nebulous region north of Antares beginning on 1898 June 10.23. Because of changes being made in the darkroom, it was not until June 12 that the photographic plate was developed. Coddington then found a strong nebulous trail 2° or 3° northeast of Antares. Although the rate of motion was easy to determine, it was uncertain if the comet was heading southwestward or northeastward. Coddington visually observed the comet with the 30-cm telescope on June 12 and described it as a nebulous mass less than 1′ across with a nucleus of about magnitude 8. His colleague, W. J. Hussey, gave the position as $\alpha = 16^h\ 24.9^m$, $\delta = -25°\ 14'$ on June 12.22. An independent discovery was made by W. Pauly (Bucharest, Romania) on June 14.86. He was using a 7.5-cm telescope (28×) to observe the globular cluster M4, when he noticed a small nebula toward the southwest. Not possessing a catalog, he sketched the region. The sky was cloudy on the 15th, but clear on the 16th. A search in the region around M4 revealed the nebula had vanished, but he found it on June 16.91, southwest of its previous position.

The comet was near opposition when discovered and would pass closest to Earth later in June. It was still three months from perihelion. It was well observed during the remainder of June. On June 13, P. Harzer (Kiel, Germany) said the comet was hardly visible. On June 14, R. Schorr (Bergedorf Observatory, Hamburg, Germany) observed with the 26-cm refractor and said the diffuse nucleus was magnitude 9.5, while the coma was about 20″ across. That same night, G. Bigourdan (Paris, France) observed with a 30-cm refractor and saw the comet at low altitude and said it looked like a nebula of Herschel's class I. He added that it was round, with a diameter of about 50″. There was a granulous condensation about 15″ across at the center. Also on the 14th, E. Millosevich (Rome, Italy) observed with a 25-cm refractor and said the nucleus was magnitude 8, while a tail extended toward PA 20°. On the 14th and 15th, E. E. Barnard (Yerkes Observatory, Wisconsin, USA) observed with the 102-cm refractor and said the comet was magnitude 8 or 9, with a strong, nearly stellar condensation, and a "short brush of tail to the north." On the 15th, Bigourdan said an 11-cm refractor showed the comet near magnitude 10, while L. Cruls (Rio de Janeiro, Brazil) observed with a 24-cm refractor and said the comet was 4–5′ across, with a nucleus of magnitude 8. L. Picart (Bordeaux, France) saw the comet on June 15, 16, and 17, and described it as faint with a nucleus. The comet was photographed at Harvard College Observatory (Massachusetts, USA) on the 15th and 16th, with O. C. Wendell estimating the magnitude of the nucleus as 7.7. On the 16th, L. Ambronn (Göttingen, Germany) observed with a 15-cm refractor and said the comet was magnitude 9, while the coma was 1–1.5′ across. On June 17, Bigourdan said the 30-cm refractor revealed the comet was about magnitude 9, while the 11-cm refractor revealed a round, centrally condensed nebulosity about 45″ in diameter. That same night, E. Hartwig (Bamberg, Germany) said the comet was magnitude 8.5 and about 3′ in diameter. On the 18th, A. Abetti (Arcetri, Italy) observed with a 28-cm refractor and said the comet was magnitude 9.5, with a trace of tail extending northward. Abetti gave the nuclear magnitude as 10 on the 19th and 20th, noting only a trace of coma each night. On the 23rd, V. Cerulli (Teramo, Italy) said the comet was very bright and fan-shaped, with a length of about 1.5′. He said the "fan" opened up toward PA 30°. On the 24th, Abetti said the comet was like a nebulous star of magnitude 9, while Cruls said the comet exhibited a slightly eccentric nucleus. On June 26, Cruls noted a bright nucleus and commented that the coma diameter had increased.

Although the comet was initially visible in both the Northern and Southern Hemispheres, it was moving southward and was followed into July by only H. A. Howe (Chamberlin Observatory, Colorado, USA) and Hussey. Howe last saw the comet on July 9 with the 51-cm refractor, while Hussey last saw it on July 20 with the 30-cm refractor. On the last date, it was very low over the southern horizon. Observations during the remainder of the year and into the early days of February 1899 were exclusively made by observers in the Southern Hemisphere and the comet attained its most southerly declination of −84° on November 17. J. Tebbutt (Windsor, New South Wales, Australia) observed the comet on every available occasion from June 12 to October 18 with his 20-cm refractor. He simply noted, "The comet was small throughout, with a condensation in its centre. . . . " He added that the comet was so faint after the October 18 observation that he doubted he would recover it after the full moon. Nevertheless, he recovered the comet on October 31 and continued to observe it on every clear night up to 1899 March 3.

This final observation was made with a 20-cm refractor and the comet was "of the last degree of faintness."

The only Northern Hemisphere observer during the remainder of the comet's apparition was Coddington. He recovered the comet low over the horizon on 1899 February 7 and observed it on several nights up to the 16th. He noted that it was "still readily observable" in the 30-cm refractor. Following the February full moon, he again picked up the comet in the 30-cm refractor on February 28. Coddington next saw the comet on August 11 and 12, and on September 2 and 8. On the last date he estimated the magnitude as 15. The comet was last detected on 1899 December 7.37, when Coddington measured the position as $\alpha = 4^h 32.0^m$, $\delta = +3° 19'$.

The spectrum was observed by W. W. Campbell and W. H. Wright (Lick Observatory, California, USA) on June 12 and 17. On both occasions the three carbon bands were very faint compared to the continuous spectrum. Wright commented that the bands seemed stronger in the coma than in the nucleus on the 12th, but the nucleus was stronger on the 17th.

The first orbits were calculated by astronomers at Lick Observatory. First, Hussey and Coddington took positions obtained on June 12, 14, and 16, and determined the perihelion date as 1898 September 10.81. Second, W. W. Crawford took positions obtained on June 12, 13, and 14, and determined the perihelion date as September 8.86. Additional orbits by A. Berberich, Millosevich, and C. J. Merfield eventually established the perihelion date as September 14.

The first hyperbolic orbit was calculated by Merfield (1899) using positions spanning the period of June 18 to October 4. The resulting perihelion date was September 14.55, while the eccentricity was given as 1.0007539. A definitive orbit was calculated by Merfield (1901) which used positions spanning the entire period of visibility. It revealed a perihelion date of September 14.54 and an eccentricity of 1.0010336. Another hyperbolic orbit was calculated by Z. Sekanina (1978). It took 69 positions spanning the entire period of visibility and revealed the orbit below.

T	ω	Ω (2000.0)	i	q	e
1898 Sep. 14.5453 (TT)	233.2693	75.4087	69.9358	1.701595	1.000934

ABSOLUTE MAGNITUDE: $H_{10} = 5.0$ (V1964)

FULL MOON: Jun. 4, Jul. 3, Aug. 2, Aug. 31, Sep. 29, Oct. 29, Nov. 28, Dec. 27, 1899 Jan. 26, Feb. 25, Mar. 27, Apr. 25, May 25, Jun. 23, Jul. 22, Aug. 21, Sep. 19, Oct. 18, Nov. 17, Dec. 17

SOURCES: E. F. Coddington and W. J. Hussey, *AJ*, **19** (1898 Jun. 14), p. 48; E. F. Coddington, R. Schorr, and P. Harzer, *AN*, **146** (1898 Jun. 16), p. 341; E. F. Coddington and J. Tebbutt, *Nature*, **58** (1898 Jun. 16), p. 160; G. Bigourdan and L. Picart, *CR*, **126** (1898 Jun. 20), pp. 1768–72; W. Pauly, E. Millosevich, L. Ambronn, E. Hartwig, E. F. Coddington, W. J. Hussey, and A. Berberich, *AN*, **146** (1898 Jun. 23), pp. 355–60; E. Millosevich, *AN*, **146** (1898 Jun. 29), p. 375; L. Picart, *CR*, **127** (1898 Jul. 4), pp. 39–40;

W. J. Hussey, E. F. Coddington, W. W. Crawford, A. Berberich, and E. E. Barnard, *AJ*, **19** (1898 Jul. 8), pp. 52, 55; V. Cerulli and O. C. Wendell, *AN*, **147** (1898 Jul. 13), p. 11; E. F. Coddington, *PASP*, **10** (1898 Aug.), pp. 146–8; H. A. Howe, *AJ*, **19** (1898 Aug. 22), pp. 78–9; C. J. Merfield, *AN*, **147** (1898 Oct. 19), p. 333; H. A. Lenehan and R. P. Sellors, *MNRAS*, **58** (1898, Supplementary Notice), pp. 526–7; E. Millosevich, *AN*, **147** (1898 Nov. 14), p. 397; J. Tebbutt, *MNRAS*, **59** (1898 Dec.), pp. 93–9; L. Cruls, *AN*, **148** (1899 Jan. 21), p. 203; C. J. Merfield, *AN*, **148** (1899 Feb. 10), p. 287; A. Abetti, *AN*, **149** (1899 Mar. 23), pp. 17–21; J. Tebbutt, *MNRAS*, **59** (1899 Apr.), pp. 388–92; E. F. Coddington, *AJ*, **20** (1899 Apr. 8), pp. 23–4; V. Cerulli, *AN*, **149** (1899 Apr. 10), pp. 51–3; R. Schorr, *AN*, **149** (1899 Apr. 19), pp. 145, 157; E. F. Coddington, *AJ*, **20** (1899 Apr. 26), p. 32; J. Tebbutt, *AN*, **150** (1899 Jun. 30), p. 45; W. H. Wright and W. W. Campbell, *APJ*, **10** (1899 Oct.), p. 173; E. F. Coddington, *PASP*, **11** (1899 Oct. 1), p. 203; E. F. Coddington, *AN*, **152** (1900 Apr. 7), p. 157; C. J. Merfield, *AN*, **154** (1901 Feb. 1), pp. 229–68; V1964, p. 65; B. G. Marsden, Z. Sekanina, and E. Everhart, *AJ*, **83** (1978 Jan.), pp. 66, 68; B. G. Marsden, *CCO*, 3rd ed. (1979), pp. 20, 48.

2P/Encke

1898 III = 1898d

Recovered: 1898 June 7.28 ($\Delta = 0.73$ AU, $r = 0.44$ AU, Elong. $= 23°$)

Last seen: 1898 July 10.40 ($\Delta = 0.28$ AU, $r = 1.03$ AU, Elong. $= 84°$)

Closest to the Earth: 1898 July 7 (0.2745 AU)

Calculated path: GEM (Rec), MON (Jun. 11), GEM–MON (Jun. 13), CMi (Jun. 14), MON (Jun. 21), HYA (Jun. 26), PUP (Jun. 27), HYA (Jun. 28), PYX (Jun. 29), HYA–PYX (Jun. 30), ANT (Jul. 4), VEL (Jul. 8), CEN (Jul. 10)

A. Iwanow (1898) used the orbit determined from this comet's apparition of 1895, and applied perturbations from Jupiter for the period between 1894 and 1898. The resulting predicted date of perihelion for the upcoming return was 1898 May 27.30. He provided an ephemeris for the period of 1898 June 1 to 1898 July 31. D. Smart (1898) also provided an ephemeris for the period of May 1 to June 14, based on an assumed perihelion date of May 24. This comet was recovered by John Grigg (Thames, New Zealand) on 1898 June 7.28. He computed his own ephemeris and found the comet about 3° above the western horizon. J. Tebbutt (Windsor, New South Wales, Australia) made an independent recovery using Iwanow's ephemeris on June 11.35. The observation was obtained with an 11-cm refractor and he described the comet as "a small round nebula condensed a little towards the centre, and . . . just on the confines of the band of bright sky along the western horizon." On June 12.34, Tebbutt determined the position as α = 6^h 53.5^m, δ = +11° 34'. After observing the comet in an 11-cm telescope, he wrote, "It appeared as a round well condensed nebula about 30" in diameter on the confines of the bright band of twilight along the horizon."

Tebbutt said the comet was still immersed in twilight when he saw it on June 15, and he decided to wait a few days until it was higher in the sky. His next attempt to find the comet came on June 25, but despite "a beautifully clear sky" he could not locate the comet. The 26th was another very clear evening and Tebbutt again attempted to locate the comet. He finally succeeded "with great difficulty" in detecting "an extremely faint patch of light" near the predicted position. He estimated the comet's diameter as 2–3'. The comet was again invisible on June 27.

The comet was last detected on July 10.40, when Tebbutt, using a 20-cm telescope and averted vision in a low-power eyepiece, detected "a faint whitness about 5' or 6' in diameter." From his observations, Tebbutt concluded that the comet "becomes rapidly expanded and diffused as it recedes from perihelion...." He estimated the position as roughly $\alpha = 11^h\ 08.0^m$, $\delta = -43°\ 34'$.

B. G. Marsden and Z. Sekanina (1974) used 86 positions obtained between 1891 and 1905, perturbations by all nine planets, and nongravitational acceleration terms to determine the perihelion date as May 27.37. The nongravitational terms were $A_1 = -0.18$ and $A_2 = -0.02109$. This orbit is given below.

T	ω	Ω (2000.0)	i	q	e
1898 May 27.3690 (TT)	183.9690	336.2096	12.9207	0.340757	0.846377

ABSOLUTE MAGNITUDE: $H_{10} = 10.7$ (V1964)

FULL MOON: Jun. 4, Jul. 3

SOURCES: A. Iwanow, *AN*, **146** (1898 Apr. 25), p. 159; D. Smart, *AJ*, **19** (1898 Apr. 25), p. 16; A. Iwanow, *AJ*, **19** (1898 May 21), p. 32; J. Tebbutt, *AJ*, **19** (1898 Jun. 14), p. 48; J. Tebbutt, *AN*, **146** (1898 Jun. 16), p. 341; J. Tebbutt and J. Grigg, *The Observatory*, **21** (1898 Aug.), pp. 315–16, 319; J. Tebbutt and J. Grigg, *AN*, **147** (1898 Oct. 8), pp. 313–18; V1964, p. 65; B. G. Marsden and Z. Sekanina, *AJ*, **79** (1974 Mar.), pp. 415–16.

C/1898 L2	*Discovered:* 1898 June 15.47 ($\Delta = 1.95$ AU, $r = 1.37$ AU, Elong. $= 42°$)
(Perrine)	*Last seen:* 1898 August 11.52 ($\Delta = 1.46$ AU, $r = 0.64$ AU, Elong. $= 21°$)
	Closest to the Earth: 1898 August 25 (1.4238 AU)
1898 VI = 1898e	*Calculated path:* CAM (Disc), AUR (Jul. 1), GEM (Jul. 25), CNC (Aug. 9)

C. D. Perrine (Lick Observatory, California, USA) discovered this comet with a 30-cm refractor on 1898 June 15.47, at a position of $\alpha = 3^h\ 29.0^m$, $\delta = +58°\ 35'$. The position was measured in morning twilight and was not considered as precise as normal. Perrine confirmed the find on June 16.43. He said the centrally condensed comet was fainter than magnitude 10.5 and measured 1.5' across.

Although the comet was approaching both the sun and Earth, it remained a small, faint object throughout its apparition. G. Bigourdan (Paris, France) saw the comet with the 30.5-cm refractor on June 18 and described it as a very diffuse, round, nebulous spot measuring 1–1.5' across. Its brightness was comparable to a Herschel nebula of class II. The central portion was diffuse and rather granular. E. Hartwig (Bamberg, Germany) saw the comet in a heliometer that same evening and described it as magnitude 9, with a round coma 2′ across, but no nucleus. A. Abetti (Arcetri, Italy) observed with a 28.4-cm refractor on June 18 and 20, and saw no marked nucleus. Although Perrine also noted no well-defined nucleus on the 20th, Bigourdan did report that the nuclear condensation was about magnitude 13.2 in the 30.5-cm refractor. Perrine did report a fairly sharp nucleus on the 21st, while Abetti still noted the comet was extremely faint in the 28.4-cm refractor. R. Schorr (Hamburg, Germany) described the comet as a small nebulous mass on June 23. He added that it exhibited a condensation of magnitude 11.5. Perrine reported the nucleus was magnitude 13 on the 24th. On June 25, Perrine estimated the comet's magnitude as 9.5, while Abetti said it looked like an out-of-focus star of magnitude 9. Abetti described the comet as a round spot, about 1′ across, without a trace of tail. Perrine observed with the 91-cm refractor on the 26th and said the comet gradually brightened to a central condensation, but exhibited no nucleus. Abetti reported the comet was faint because of moonlight on the 28th, as seen in the 28.4-cm refractor. Perrine observed with the 30-cm telescope on the 29th and said the nucleus was much sharper than when previously seen, with a magnitude of 10.5 or 11. G. Ciscato and A. Antoniazzi (Padova, Italy) said the comet was very bright with a distinct nucleus on June 29 and 30.

The comet continued approaching both the sun and Earth during July. Perrine noted on the 9th that the nucleus seemed sharp in the 30-cm refractor, while the comet's total magnitude was 9.0. He estimated the nuclear magnitude as 11–12 on the 12th. H. A. Kobold (Strasbourg, France) observed with a 46-cm refractor on the 13th and described the comet as a bright, round nebulosity, about 1′ across, with a strong central condensation. That same night, Perrine gave the total magnitude as 8.5 and the nuclear magnitude as 10.5–11. Ciscato and Antoniazzi said the comet was very bright with a distinct nucleus on July 15 and 16. On the 17th, Perrine and Abetti independently gave total magnitudes between 8.5 and 9. Perrine gave the total magnitude as 8.5 and the nuclear magnitude as 10 on the 18th. He also noted a short tail extending toward PA 305°. Perrine gave the total magnitude as 8 on the 20th. Perrine and Abetti estimated magnitudes between 8 and 9 on the 22nd, with Perrine adding that the sharp nucleus was distinctly visible. Perrine indicated that the total magnitude steadily increased to 7.5 by the 30th, while the nuclear magnitude reached 9.5. He noted a short, bushy tail on the 25th and said the tail extended 3–4′ on the 30th.

The comet was still approaching the sun and Earth as August began, but many astronomers had lost it in twilight. Perrine continued making observations. He estimated the nuclear magnitude as 9.0 on the 5th and 9.1 on the 8th. The comet was last detected on August 11.51 and 11.52, by Perrine, while using the 30-cm refractor. He measured the position as $\alpha = 7^h\ 56.7^m$, $\delta = +16°\ 07'$ on the first date, but was only able to partially measure the position on the second because of interference from clouds. Perrine looked for the comet in the morning sky of August 17, "but its place was too low in the dawn."

The first parabolic orbit was calculated by Perrine and R. G. Aitken. Using positions from June 15, 16, and 17, they determined the perihelion date as 1898 August 17.90. Other early orbits came from A. Berberich and G. Fabry. Berberich took positions from June 15, 16, and 17, and determined the perihelion date as August 5.21. Fabry used positions obtained on June 17, 18, and 19, and determined the perihelion date as August 3.30. The perihelion date derived by Perrine and Aitken was just over a day after the actual perihelion date, which later orbits by Berberich and Perrine eventually established as August 16.7.

A definitive orbit was calculated by H. D. Curtis and L. S. Richardson (1909). They took 122 positions spanning the entire period of visibility, deduced seven Normal places, and determined an elliptical orbit with a perihelion date of August 16.71 and a period of about 73 000 years. Noticing that the size of the probable error of the eccentricity made the orbit essentially parabolic, they adopted a parabolic orbit with a perihelion date of August 16.71 as best representing the positions. This orbit is given below.

T	ω	Ω (2000.0)	i	q	e
1898 Aug. 16.7056 (UT)	205.6135	260.5279	70.0300	0.626438	1.0

ABSOLUTE MAGNITUDE: $H_{10} = 7.6$ (V1964)

FULL MOON: Jun. 4, Jul. 3, Aug. 2, Aug. 31

SOURCES: G. Bigourdan and G. Fabry, *CR*, **126** (1898 Jun. 20), pp. 1767–8, 1770–1; C. D. Perrine and E. Hartwig, *AN*, **146** (1898 Jun. 23), p. 359; C. D. Perrine, R. G. Aitken, and A. Berberich, *AN*, **146** (1898 Jun. 29), pp. 371–3; A. Berberich, *AN*, **146** (1898 Jun. 29), supplement; C. D. Perrine, *AJ*, **19** (1898 Jul. 8), p. 54; R. Schorr, *AN*, **147** (1898 Jul. 13), p. 13; E. Frisby, *AJ*, **19** (1898 Jul. 21), p. 63; W. E. Plummer, *MNRAS*, **59** (1898 Dec.), p. 102; G. Ciscato and A. Antoniazzi, *AN*, **148** (1899 Jan. 7), pp. 115–18; H. A. Kobold, *AN*, **148** (1899 Mar. 8), pp. 385–8; A. Abetti, *AN*, **149** (1899 Mar. 23), pp. 17–22; C. D. Perrine, *AJ*, **20** (1899 Sep. 28), pp. 99–102; H. D. Curtis and L. S. Richardson, *AN*, **182** (1909 Oct. 29), pp. 337–58; V1964, p. 65.

14P/Wolf
1898 IV = 1898f

Recovered: 1898 June 17.47 ($\Delta = 2.10$ AU, $r = 1.61$ AU, Elong. $= 48°$)

Last seen: 1899 March 11.24 ($\Delta = 2.33$ AU, $r = 2.75$ AU, Elong. $= 103°$)

Closest to the Earth: 1898 November 29 (1.3960 AU)

Calculated path: ARI (Rec), TAU (Jul. 9), ORI (Aug. 15), TAU (Aug. 29), ORI (Aug. 31), MON (Sep. 17), CMa (Nov. 17), MON (Feb. 14)

Two predictions were made for this comet's 1898 apparition. A. Berberich (1892) worked out the orbit of this comet for the apparition of 1891 and predicted the next return to perihelion as 1898 June 30. A. K. Thraen (1898) waded through 681 positions measured for this comet during the apparitions of 1884 and 1891–2, and determined 16 Normal positions. He then calculated an orbit for the 1891 appearance, applied perturbations by Earth, Mars, Jupiter, and Saturn, and predicted the next perihelion date would be 1898 July 5.06. W. J. Hussey (Lick Observatory, California, USA) recovered this comet with a 91-cm refractor on 1898 June 17.47 at a position of $\alpha = 2^h\ 16.3^m$, $\delta = +19°\ 43'$. The position indicated the correction to Thraen's ephemeris was "practically insignificant."

On July 15, A. Abetti (Arcetri, Italy) said the comet was quite faint in his 28-cm refractor, because of moonlight, weak morning twilight, and haze near the horizon. He added that nearby 10th-magnitude stars were as difficult to see as the comet. On the 16th, Abetti said conditions had improved since the previous morning, but the comet was about as bright as a 12th-magnitude star. On the 17th, H. A. Kobold (Strasbourg, France) observed with a 46-cm refractor and described the comet as a small, round, nebulosity, with a central condensation of magnitude 11. On July 17 and 18, Abetti said the sky was "most splendid," but the comet was faint and difficult to measure. On August 13, W. Villiger (Munich, Germany) observed with a 27-cm refractor and said the nucleus was distinctly visible. Abetti said the comet was exceptionally faint during the period of August 18–21. On August 22 and 23, R. Schorr (Hamburg, Germany) observed with a 26-cm refractor and described the comet as very faint, about 1' across, with a distinct condensation of magnitude 12. On September 10, Villiger said the comet was difficult to see in moonlight. C. F. Pechüle (Copenhagen, Denmark) estimated the magnitude as 11 on the 14th, while Kobold said the coma was 2' across and exhibited a condensation of magnitude 12. On the 15th, F. Schwab (Kremsmünster, Austria) described the comet as a small, roundish nebulosity. During the period of September 17–21, Abetti estimated the comet's magnitude as about 12. On October 14, F. Cohn (Königsberg, now Kaliningrad, Russia) observed with a 32-cm refractor and described the comet as very faint. Villiger said the sky was very transparent on the morning of the 21st, but the comet was fainter than in September. On November 20, Kobold noted the comet was rather faint, about 1.5' across, with slight condensation. Cohn said the comet was reasonably well seen on the 21st. On November 23, Schorr described the comet as very faint and difficult to observe. On December 8, Cohn described the comet as quite faint. The comet attained its most southerly declination of $-17°$ on December 28.

On 1899 January 9, Kobold saw the comet in the evening sky and said it was magnitude 12, about 1' in diameter, with a slight central condensation. The comet was last detected on 1899 March 11.24, when Hussey gave the position as $\alpha = 6^h\ 23.0^m$, $\delta = -6°\ 45'$.

The most elaborate investigations of this comet's orbit using multiple apparitions have come from M. Kamienski (1922, 1959) and D. K. Yeomans (1978). They determined similar orbits with a perihelion date of July 5.07 and a period of 6.85 years. Yeomans (1975) determined the nongravitational terms as $A_1 = +0.139$ and $A_2 = -0.0081$ and his orbit is given below.

T	ω	Ω (2000.0)	i	q	e
1898 Jul. 5.0670 (TT)	172.8615	207.8958	25.1934	1.603058	0.555336

ABSOLUTE MAGNITUDE: $H_{10} = 7.8$ (V1964)

FULL MOON: Jun. 4, Jul. 3, Aug. 2, Aug. 31, Sep. 29, Oct. 29, Nov. 28, Dec. 27, Jan. 26, Feb. 25, Mar. 27

SOURCES: A. Berberich, *AJ*, **11** (1892 Jan. 12), p. 104; A. K. Thraen, *AN*, **146** (1898 Mar. 10), pp. 11–14; W. J. Hussey, *AN*, **146** (1898 Jun. 23), p. 359; W. J. Hussey and A. K. Thraen, *AJ*, **19** (1898 Jul. 8), pp. 53–6; R. Schorr, *AN*, **147** (1898 Sep. 1), p. 207; C. F. Pechüle, *AN*, **147** (1898 Sep. 15), p. 255; F. Schwab, *AN*, **148** (1898 Nov. 21), p. 9; H. A. Kobold, *AN*, **148** (1899 Mar. 8), pp. 385–8; A. Abetti, *AN*, **149** (1899 Mar. 23), pp. 17, 21; R. Schorr, *AN*, **149** (1899 Apr. 19), pp. 145, 156; H. A. Kobold, *AN*, **149** (1899 Jun. 2), pp. 343–7; F. Cohn, *AN*, **150** (1899 Aug. 14), pp. 217, 222; W. Villiger, *AN*, **150** (1899 Aug. 22), pp. 225, 233; W. J. Hussey, *AN*, **151** (1899 Dec. 28), p. 171; M. Kamienski, *AOAT*, **5** No. 5 (1922), pp. 1–5; M. Kamienski, *AA*, **9** (1959), p. 58; V1964, p. 65; D. K. Yeomans, *PASP*, **87** (1975 Aug.), pp. 635–7; D. K. Yeomans, *QJRAS*, **19** (1978), pp. 52–3, 57.

C/1898 M1 (Giacobini)

1898 V = 1898g

Discovered: 1898 June 19.02 ($\Delta = 0.66$ AU, $r = 1.59$ AU, Elong. $= 142°$)

Last seen: 1898 August 16.85 ($\Delta = 1.74$ AU, $r = 1.53$ AU, Elong. $= 61°$)

Closest to the Earth: 1898 June 29 (0.5319 AU)

Calculated path: CAP (Disc), SGR (Jun. 22), OPH (Jul. 1), SCO (Jul. 6), LIB (Jul. 8), VIR (Jul. 20)

M. Giacobini (Nice, France) discovered this comet on 1898 June 19.02 at a position of $\alpha = 20^h 36.5^m$, $\delta = -21° 14'$. It was described as rather faint, with an elongated nucleus. The comet was confirmed on June 19.96 by T. Zona (Palermo, Italy) and on June 19.99 by L. Picart (Bordeaux, France).

The comet was widely observed following its discovery as it headed for perihelion, as well as a close approach to Earth near the end of June. On June 20, E. J. M. Stephan (Marseille, France) observed with a 26-cm refractor and described the comet as rather faint, with a slightly grainy appearance. That same night, G. Bigourdan (Paris, France) observed with a 30-cm refractor and described the comet as very faint, with a magnitude near 13.2 and a diameter of 30″. The central condensation was "granulous" and about 15″ across. A. Abetti (Arcetri, Italy) observed with a 28-cm refractor (124×) and said the comet appeared faint on both the 20th and 21st and compared it to a magnitude-12 star, with slight nebulosity. Stephan said the comet appeared fainter and more condensed on the 21st than on the 20th. E. Hartwig (Bamberg, Germany) observed with a 25-cm refractor and estimated the comet's total magnitude as near 10 on June 22. He described it as round, 2′ across, with a noncentral condensation. On the 24th, Stephan said the comet was brighter than on the 20th and was strongly condensed around a nucleus of magnitude 10. V. Cerulli (Teramo, Italy) said the comet was visible in the 6-cm finder of his refractor. He described it as round. A. Antoniazzi

(Padova, Italy) observed with a 19-cm refractor and saw the comet during the period spanning June 27–30 and noted that it remained faint because of moonlight. He added that the nucleus was never very distinct. Abetti saw the comet in strong moonlight on the 28th and noted the comet was as visible as stars of magnitude 10 or 11. On June 29, Stephan said the comet was difficult to see in moonlight.

Moonlight prevented observations during the period of July 1–4. P. Chofardet (Besançon, France) said the comet was "excessively faint" on July 9. H. A. Kobold (Strasbourg, France) observed with a 46-cm refractor and described the comet as an "extremely faint nebulous mass" on July 16 and noted a 15th-magnitude nucleus. S. Javelle (Nice, France) was the only observer to follow this comet after July 18. Moonlight interfered during the last days of July and first days of August, but he resumed his observations on August 6. The comet was last detected on August 16.85, when Javelle measured the position as $\alpha = 13^{\text{h}}\ 42.0^{\text{m}}$, $\delta = -2°\ 52'$. He said the comet was at the extreme limit of visibility in a 76-cm refractor.

The first parabolic orbit was calculated by H. C. F. Kreutz using positions spanning the period of June 20–22. The resulting perihelion date was 1898 July 6.70. As June progressed, and additional positions became available, I. Lagarde determined the perihelion date as July 26.65 and Javelle determined it as July 25.39. During early August, A. Stichtenoth took positions from June 20 to July 18 and determined the perihelion date as July 26.01. This orbit was found to very accurately represent the comet's motion. Over two years later A. Hnatek (1901) examined this orbit. He used 84 positions obtained between June 19 and August 16, and computed a parabolic orbit with a perihelion date of July 26.00. An elliptical orbit was also calculated with a perihelion date of July 26.00 and a period of 42 043 years. Although both orbits very closely represented the available positions, Hnatek favored the parabolic orbit because of the short period of observation. This orbit is given below.

T	ω	Ω (2000.0)	i	q	e
1898 Jul. 25.9966 (UT)	22.3475	279.6618	166.8542	1.501318	1.0

ABSOLUTE MAGNITUDE: $H_{10} = 9.6$ (V1964)

FULL MOON: Jun. 4, Jul. 3, Aug. 2, Aug. 31

SOURCES: P. Chofardet, E. J. M. Stephan, and J. E. Coggia, *BA*, **15** (1898), pp. 322, 422, 467; M. Giacobini and G. Bigourdan, *CR*, **126** (1898 Jun. 20), pp. 1767–70; I. Lagarde, *CR*, **126** (1898 Jun. 27), pp. 1851–2; M. Giacobini, S. Javelle, H. A. Kobold, E. Hartwig, and H. C. F. Kreutz, *AN*, **146** (1898 Jun. 29), pp. 373–6; L. Picart, *CR*, **127** (1898 Jul. 4), pp. 39–40; M. Giacobini, *AJ*, **19** (1898 Jul. 8), p. 53–4; T. Zona, H. A. Kobold, and S. Javelle, *AN*, **147** (1898 Jul. 13), pp. 13–16; A. Stichtenoth, *AN*, **147** (1898 Aug. 15), p. 123; H. A. Kobold, *AN*, **147** (1898 Nov. 14), p. 395; G. Ciscato and A. Antoniazzi, *AN*, **148** (1899 Jan. 7), pp. 115–18; L. Cruls, *AN*, **148** (1899 Jan. 21), pp. 203–6; A. Abetti, *AN*, **149** (1899 Mar. 23), pp. 19–22; V. Cerulli, *AN*, **149** (1899 Apr. 10), pp. 51–3; A. Hnatek and S. Javelle, *SAWW*, **110** Abt. IIa (1901), pp. 231–88; A. Hnatek, *AN*, **158** (1902 Feb. 17), pp. 23–8; V1964, p. 65.

C/1898 R1
(Perrine–
Chofardet)
1898 IX = 1898h

Discovered: 1898 September 13.54 ($\Delta = 1.56$ AU, $r = 0.98$ AU, Elong. $= 38°$)
Last seen: 1898 October 10.55 ($\Delta = 1.37$ AU, $r = 0.50$ AU, Elong. $= 16°$)
Closest to the Earth: 1898 October 7 (1.3729 AU)
Calculated path: LEO (Disc), VIR (Oct. 5)

C. D. Perrine (Lick Observatory, California, USA) discovered this comet with the 30-cm refractor on 1898 September 13.54 at a position of $\alpha = 9^h 35.8^m$, $\delta = +31° 05'$. He estimated the magnitude as 8–9 and noted a short tail toward the north-following side. He confirmed the discovery on September 14.51 and noted the coma was round and 5' across, with a sharp, but not stellar, central condensation. The tail extended toward PA 306.8°. P. Chofardet (Besançon, France) independently discovered the comet on September 15.16. He described it as round, about 3' across, with a strong central condensation.

Physical descriptions were in abundance during the period of September 16–19 as the major observatories of the Northern Hemisphere turned their telescopes toward the comet. The total magnitude was estimated as 6–7 by A. Antoniazzi (Padova, Italy) and 7–8 by F. W. Ristenpart (Kiel, Germany), J. Holetschek (Vienna, Austria), A. A. Nijland (Utrecht, Netherlands), A. Abetti (Arcetri, Italy), and Perrine. The nuclear condensation was estimated as magnitude 9.5–10 by H. A. Kobold (Strasbourg, France), R. Schorr (Hamburg, Germany), and Ristenpart. During the same period the coma diameter was estimated as 1' by Abetti, 1.5' by Schorr, and 2' by Kobold, Holetschek, and E. Hartwig (Bamberg, Germany). The tail was estimated as 5' long by Holetschek, 10' long by Nijland, and 30' long by 2' wide by Hartwig, while V. Cerulli (Teramo, Italy), Kobold, and Perrine independently said it extended toward PA 300–310°.

During the remainder of September, only a few astronomers provided physical descriptions. Perrine estimated the total magnitude as 7 on the 20th and 21st, while the nuclear magnitude was between 9.5 and 10. Holetschek determined the magnitude as 6.8 on the 22nd and estimated a coma diameter of about 2.5' and a tail length of 10'. Perrine reported the comet was "distinctly visible to the naked eye" on the 23rd, while the sharp nucleus was about magnitude 9.0. On the 25th, Abetti estimated the total magnitude as 6–7, while Cerulli said the tail extended about 10' toward PA 310°. Nijland saw the comet on the 26th and determined the total magnitude as 7.3 and the nuclear magnitude as 9.0. He also said the tail extended 9.6' toward PA 318.4°. W. Villiger (Munich, Germany) saw the comet on the same morning and noted a tail extending 10' toward PA 311°. On September 27, Perrine said the comet was very bright, with a nucleus of magnitude 8.8. The tail was also brighter than on the 23rd. Villiger saw the comet on the same morning and said the tail extended 15' toward PA 322°.

Although the comet passed Earth a week into October, it was becoming a difficult object to view as it dropped into morning twilight. On October 3, Nijland gave the total magnitude as 6.5 and noted the tail extended toward

PA 324.3°. Perrine noted the nucleus was not as sharp and about magnitude 8.0 on the 4th. He said the comet was low in the dawn on the 6th and exhibited a faint tail, but no nucleus. The comet was last detected on October 10.55, when Perrine measured the position as $\alpha = 12^h\ 28.6^m$, $\delta = +6°\ 56'$ with the 30-cm refractor. The comet was continually seen through clouds and was very near the horizon. Perrine looked for the comet on the 11th and 12th, but found nothing in the dawn and low altitude. The comet passed 0.7° from the sun on October 23.

The first parabolic orbit was calculated by Perrine and R. G. Aitken. Using positions obtained by Perrine on September 13, 14, and 15, the perihelion date was determined as 1898 October 20.52. Further calculations by A. Berberich and H. A. Peck (1902) narrowed the perihelion date down to October 21.0. Peck noted that a general solution revealed a hyperbolic orbit with an eccentricity of 1.00012, but he placed little weight on this because of the short period of visibility and stayed with the parabolic orbit. Peck's orbit is given below.

T	ω	Ω (2000.0)	i	q	e
1898 Oct. 21.0396 (UT)	162.3762	36.2952	28.8596	0.420373	1.0

ABSOLUTE MAGNITUDE: $H_{10} = 7.0$ (V1964)

FULL MOON: Aug. 31, Sep. 29, Oct. 29

SOURCES: C. D. Perrine and P. Chofardet, *AN*, **147** (1898 Sep. 15), p. 255; C. D. Perrine and R. G. Aitken, *AJ*, **19** (1898 Sep. 20), pp. 95–6; R. Schorr, F. W. Ristenpart, E. Hartwig, C. D. Perrine, R. G. Aitken, and A. Berberich, *AN*, **147** (1989 Sep. 20), pp. 269–72; P. Chofardet, H. A. Kobold, and E. Hartwig, *AN*, **147** (1898 Sep. 30), pp. 283–6; A. Berberich, *AN*, **147** (1898 Oct. 19), pp. 333–6; A. C. D. Crommelin, *MNRAS*, **58** (1898, Supplementary Notice), p. 525; W. E. Plummer, *MNRAS*, **59** (1898 Dec.), p. 103; A. Antoniazzi, *AN*, **148** (1899 Jan. 7), pp. 115–18; A. Abetti, *AN*, **149** (1899 Mar. 23), pp. 19–22; V. Cerulli and J. Holetschek, *AN*, **149** (1899 Apr. 10), pp. 51–4; R. Schorr, *AN*, **149** (1899 Apr. 19), pp. 147, 157; A. A. Nijland, *AN*, **149** (1899 May 6), pp. 237–40; W. Villiger, *AN*, **150** (1899 Aug. 22), pp. 225, 234; C. D. Perrine, *AJ*, **20** (1899 Sep. 28), pp. 99–102; H. A. Peck, *AJ*, **22** (1902 Oct. 15), pp. 169–73; V1964, p. 65.

C/1898 U1   *Discovered:* 1898 October 21.01 ($\Delta = 0.62$ AU, $r = 1.00$ AU, Elong. $= 71°$)
(Brooks)   *Last seen:* 1898 November 26.75 ($\Delta = 1.38$ AU, $r = 0.76$ AU, Elong. $= 32°$)
    Closest to the Earth: 1898 October 23 (0.6134 AU)
1898 X = 1898i   *Calculated path:* DRA (Disc), BOO-DRA-HER (Oct. 24), OPH (Nov. 8), SER (Nov. 18)

W. R. Brooks (Smith Observatory, Geneva, New York, USA) was sweeping for comets in the northern sky with a 25.4-cm refractor when he found this comet on 1898 October 21.01. He measured the position as $\alpha = 14^h\ 35.2^m$ (or 32.2^m), $\delta = +60°\ 26'$. He said, "The comet was quite large, round, with bright central condensation, and at times a minute stellar nucleus was noted." A rapid, southeasterly motion was noted in only a few minutes. Since the comet was circumpolar, Brooks was able to determine another position on October 21.42.

The comet passed closest to Earth just two days after discovery, but was still a month from perihelion. Numerous observations were made during the remainder of October. On October 22, the total magnitude was given as 7 by F. W. Ristenpart (Kiel, Germany) and 7.5 by A. A. Nijland (Utrecht, Netherlands). Ristenpart also gave the coma diameter as 2'. Brooks found the comet was brighter on the 23rd than when discovered. He noted that, in bright moonlight, the comet was conspicuous in the 25-cm refractor and an easy object in the 8-cm finder. The comet was also seen by J. Holetschek (Vienna, Austria) and E. Hartwig (Bamberg, Germany) on the 23rd. Holetschek gave the magnitude as 7.8 and the coma diameter as 2.5', while Hartwig said the coma was 7–8' across and a little oblong in moonlight. He also noted a stellar nucleus of about magnitude 8.5. On the 24th, W. E. Plummer (Liverpool, England) said the comet was a "fairly bright object, with well-defined centre." Brooks next saw the comet on October 25, but moonlight prevented further observations during the next week or so. On the 27th, Holetschek gave the magnitude as 8.0, while F. Bidschof and J. Palisa (Vienna, Austria) estimated it as 9.0. That same night, H. A. Kobold (Strasbourg, France) observed with a 46-cm refractor and said the comet was a bright, round nebulosity in very bright moonlight. It was 2' in diameter, with a strong central condensation, but with no actual nucleus. On the 28th and 30th, Holetschek gave the magnitude as 7.9 and noted the coma diameter was 1.9' on the last date. On the 30th, R. Schorr (Hamburg, Germany) observed in bright moonlight and said the coma was about 45" across and exhibited a diffuse nucleus of magnitude 10.5. On October 31, Nijland said the comet was magnitude 8.3 in moonlight.

On November 1, Nijland gave the magnitude as 8.2. On the 2nd, Holetschek gave the magnitude as 7.6, while Brooks said the comet appeared brighter than when seen on October 25 and he glimpsed a "broad short tail." That same night, W. Schur (Göttingen, Germany) described the comet as a bright mass, with a concentric condensation, and a diameter of 70". On the 3rd, Holetschek gave the magnitude as 7.7 and the coma diameter as 2', while F. Schwab (Kremsmünster, Austria) said the coma was 3' across and centrally condensed. On the 3rd and 5th, V. Cerulli (Teramo, Italy) observed the comet in a 6-cm finder and said it looked like a diffuse star of magnitude 6. On the 4th, Schur described it as bright, but without a tail, while Schorr noted a distinct nucleus of magnitude 10.5 and a coma about 1.5' across. Kobold said the condensation was magnitude 11 on the 6th. On November 7, Holetschek gave the magnitude as 8.1, while Nijland gave it as 8.4. That same night, Plummer said the comet was bright, but "very ill-defined," while Kobold gave the total magnitude as 10. Also on the 7th, L. Ambronn (Göttingen, Germany) observed with a 15-cm cometseeker and reported a centrally condensed nebulous mass 2–3' across. On November 8,

Schur said the comet was fainter, while Schwab estimated the magnitude as 9 and said the coma was about 1.5′ across with a stellar nucleus. On the 8th, 9th, and 10th, A. Abetti (Arcetri, Italy) said the comet was about 1′ across, with a trace of tail. On the 9th, Holetschek gave the magnitude as 7.8 and the coma diameter as 2′, while Ambronn said the comet was 1–2′ across. On the 10th, Holetschek gave the magnitude as 8.2, while Nijland gave it as 8.6. On the 12th, Brooks said, "two tails were plainly seen nearly at right angles to each other. The more prominent one was pointed away from the Sun, the second tail to the northward." On the 13th, W. Winkler (Jena, Germany) observed with a 15-cm refractor (60×) and said the comet was round and 4–5′ across. Winkler said the comet was diffuse and 3–4′ across on the 14th.

On November 15 and 16, Abetti described the comet as a "beautiful round spot" of about magnitude 8. On November 16, Brooks said only one tail was visible. On the 17th, Kobold said the comet was about 1′ in diameter, with a strong condensation. During the period spanning November 17–20, Schwab estimated the magnitude as 10 and noted a coma about 30″ across. On November 18, G. Bischlager (Royal Observatory, Greenwich, England) observed with the 17-cm refractor (55×) and said the comet was very faint. On the 18th and 19th, Winkler said the comet was quite faint and 2–3′ across. On the 19th, Holetschek gave the magnitude as 8.3, while Nijland gave it as 8.6. That same night, Ambronn said the comet was noticeably fainter, while Kobold said it was 1′ across, with a condensation of magnitude 12. Holetschek gave the magnitude as 8.4 on the 20th. On the 26th, Holetschek gave the magnitude as 8.9, while Winkler said the comet was well seen despite moonlight, while the nucleus seemed to flash. Brooks noted in an article published in the December 1898 issue of the *Monthly Notices of the Royal Astronomical Society*, "The comet at its brightest was just visible to the naked eye, and readily picked up with a good opera or field glass."

The comet was last seen on November 26.75, when C. Rambaud and F. Sy (Alger, now al-Jazâ'ir, Algeria) measured the position as $\alpha = 18^{\mathrm{h}}\ 12.4^{\mathrm{m}}$, $\delta = -7° 50′$. They said the comet was then very near the horizon, and violent winds were shaking the telescope. The comet passed less than 1° from the sun on December 26.

The spectrum was observed by W. W. Campbell and W. H. Wright (Lick Observatory, California, USA) on November 4. They said the three carbon bands were easily visible, with the green band being "four to six times as bright as the others." The continuous spectrum was very faint and only visible in the nucleus.

The first orbit was calculated by W. Ristenpart and J. Möller using positions from October 22 and 23. They gave the perihelion date as 1898 November 23.77. Additional orbits by W. J. Hussey, H. C. Wilson, and Ristenpart established the perihelion date as November 23.6. A definitive orbit was calculated by S. Scharbe (1904). He began with 266 positions spanning the entire period of visibility and ultimately determined both parabolic and

779

elliptical orbits. The parabolic one had a perihelion date of November 23.65. The elliptical orbit had a perihelion date of November 23.66 and a period of about 158 700 years. Although Scharbe accepted the elliptical orbit as the definitive solution because it marginally fitted the observations better, the Author includes only the parabolic solution below, because of the comet's short period of observation.

T	ω	Ω (2000.0)	i	q	e
1898 Nov. 23.6524 (UT)	123.5520	97.7443	140.3448	0.756011	1.0

ABSOLUTE MAGNITUDE: $H_{10} = 9.2$ (V1964)

FULL MOON: Sep. 29, Oct. 29, Nov. 28

SOURCES: W. R. Brooks, *AN*, **147** (1898 Oct. 21), p. 351; W. R. Brooks, W. J. Hussey, and H. C. Wilson, *AJ*, **19** (1898 Oct. 28), p. 120; F. W. Ristenpart, E. Hartwig, A. A. Nijland, and W. J. Hussey, *AN*, **147** (1898 Oct. 29), p. 365; T. Lewis, A. C. D. Crommelin, C. Davidson, H. Furner, and G. Bischlager, *MNRAS*, **59** (1898 Nov.), pp. 17–25; F. W. Ristenpart, J. Möller, R. Schorr, F. Bidschof, and J. Palisa, *AN*, **147** (1898 Nov. 5), pp. 379–82; W. Schur, V. Cerulli and F. W. Ristenpart, *AN*, **147** (1898 Nov. 14), pp. 395–9; W. J. Hussey, *AJ*, **19** (1898 Nov. 22), p. 148; G. Bischlager, H. Furner, and C. Davidson, *MNRAS*, **59** (1898 Dec.), pp. 90–1; W. R. Brooks, *MNRAS*, **59** (1898 Dec.), pp. 92–3; H. A. Kobold, *AN*, **148** (1898 Dec. 28), p. 109; A. Abetti and L. Ambronn, *AN*, **148** (1899 Jan. 26), pp. 215, 221; H. A. Kobold, *AN*, **148** (1899 Mar. 8), pp. 385–8; J. Holetschek and W. Winkler, *AN*, **149** (1899 Apr. 10), pp. 53–8; R. Schorr, *AN*, **149** (1899 Apr. 19), pp. 147, 157; A. A. Nijland, *AN*, **149** (1899 May 6), pp. 237–9; W. H. Wright and W. W. Campbell, *APJ*, **10** (1899 Oct.), p. 173; F. Schwab, *AN*, **151** (1899 Dec. 28), p. 169; C. Rambaud and F. Sy, *AN*, **151** (1900 Feb. 8), pp. 349–52; W. E. Plummer, *MNRAS*, **60** (1900 Apr.), p. 520; S. Scharbe, *AN*, **164** (1904 Mar. 10), p. 377; V1964, p. 65.

C/1898 V1
(Chase)

1898 VIII = 1898j

Discovered: 1898 November 15.24 ($\Delta = 2.20$ AU, $r = 2.37$ AU, Elong. $= 87°$)

Last seen: 1899 June 27.21 ($\Delta = 3.91$ AU, $r = 3.73$ AU, Elong. $= 72°$)

Closest to the Earth: 1899 January 23 (1.8467 AU)

Calculated path: LEO (Disc), LMi (Dec. 1), LEO (Dec. 29), UMa (Jan. 3), LMi (Mar. 18), UMa (Apr. 30), LMi (May 7), UMa (May 24), LEO (Jun. 9)

Astronomers at Yale Observatory (Connecticut, USA) were intent on photographing the Leonid meteor shower in 1898. They set up two photographic stations, one at the observatory in New Haven and the other two miles away near Hamden, which included a total of ten cameras with lens apertures ranging from 5 to 8 inches. Although bad weather hampered some of the planned nights, they managed to obtain 60 photographic plates during the period of November 13–16. A few plates were quickly scanned for meteors, but it was not until November 21 that F. L. Chase sat down to carefully examine each plate. As the day progressed, Chase suddenly found a hazy object, elongated in the direction of the parallel, on a plate exposed on November 15.24. The position was given as $\alpha = 10^h 04.6^m$, $\delta = +23° 08'$. Additional images were found on plates exposed on November 15.37, November 16.24, and November 16.37. Even though the comet showed a consistent movement between exposures, the fact that all four plates were obtained with a Voightlander lens of 17-cm aperture made the Yale astronomers consider the possibility that the object was a ghost image. Since the night of November 21/22 was clear they set up two cameras and exposed two plates with each. Meanwhile, Chase swept over that area of the sky with a 20-cm refractor, but saw nothing with certainty. When the four plates were developed the next morning, a diffuse object of magnitude 11 was located very near the position extrapolated from the earlier plates and an announcement was sent to Harvard. E. C. Pickering (Harvard College Observatory, Massachusetts, USA) examined plates obtained around the time of the Leonids and found the comet on two plates exposed on November 15.40 and November 15.44.

The comet was already two months past perihelion and its fading was only slowed by the forthcoming rather distant close approach to Earth in late January. Physical descriptions were not plentiful and were generally only obtained from observatories with large telescopes. On November 24, 25, and 26, E. F. Coddington (Lick Observatory, California, USA) estimated the magnitude as 11. On December 8, E. E. Barnard (Yerkes Observatory, Wisconsin, USA) estimated the magnitude as 11–12 (102-cm refractor). The coma was 20″ in diameter and possessed a faint nucleus, while there was probably a faint tail preceding. C. F. Pechüle (Copenhagen, Denmark) described the comet as very faint on the 9th and 14th. During the period of January 4–6, A. Abetti (Arcetri, Italy) gave the magnitude of the condensation as 12–13 and noted it was surrounded by a very faint nebulosity. H. A. Kobold (Strasbourg, France) saw the comet in a 46-cm refractor on the 9th and described it as round, faint, and 1′ across. On the 10th, R. Schorr (Hamburg, Germany) described the comet as a very faint nebulosity in a 24-cm refractor. On February 4, Kobold described the comet as a rather faint nebulosity, about 1.5′ across, and with slight condensation. He gave the total magnitude as 11. On March 2, Kobold said the comet was a rather faint, round nebulosity of magnitude 12. He noted the coma was 1′ across and granular in appearance. The comet attained its most northerly declination of +38° on March 15. On April 4, Kobold said the comet was round and very faint, with a coma 1′ across and a nucleus of magnitude 14.

The comet was last detected on 1899 June 27.21, when H. A. Howe (Chamberlin Observatory, Colorado, USA) gave the position as $\alpha = 11^h 29.6^m$, $\delta = +26° 05′$. He saw the comet in the 51-cm telescope and noted the comet was a difficult object to see and measure.

The first orbit was calculated by Coddington and H. K. Palmer, using positions from November 24, 25, and 26. They determined the perihelion date as 1899 April 11.15. J. Möller and H. C. F. Kreutz followed a short time later with an orbit using positions from November 15, 24, and 26. It gave the perihelion date as 1898 September 8.65. The somewhat discordant orbits were a result of the comet's large perihelion distance of 2.3 AU. As it turned out, the longer arc used by Möller and Kreutz produced an orbit closer to the truth. Later orbits by Möller, B. Cohn, Coddington, and S. C. Chandler eventually established the perihelion date as September 20. Möller noted that observations from an arc of November 15 to December 5 indicated a hyperbolic orbit with an eccentricity of 1.025.

The first elliptical orbit was calculated by G. J. Fayet using positions spanning the period of November 24 to December 14. The result was a perihelion date of September 21.28 and a period of about 183 000 years. Additional elliptical orbits were calculated by M. Wasnetzoff (1914) and B. G. Marsden and Z. Sekanina (1973). Wasnetzoff gave the period as 211 000 years, while Marsden and Sekanina gave it as 315 000 years. Marsden and Sekanina indicated the original orbit was probably hyperbolic, with an eccentricity of 1.00016, while the future orbit was elliptical with a period near 65 000 years. Their orbit is given below.

T	ω	Ω (2000.0)	i	q	e
1898 Sep. 20.6177 (TT)	4.6417	97.2443	22.5046	2.284591	0.999507

ABSOLUTE MAGNITUDE: $H_{10} = 5.6$ (V1964)

FULL MOON: Oct. 29, Nov. 28, Dec. 27, Jan. 26, Feb. 25, Mar. 27, Apr. 25, May 25, Jun. 23, Jul. 22

SOURCES: F. L. Chase, E. F. Coddington, H. K. Palmer, F. Möller, and H. C. F. Kreutz, *AN*, **148** (1898 Dec. 5), pp. 29–32; F. L. Chase, E. F. Coddington, H. K. Palmer, E. C. Pickering, and S. C. Chandler, *AJ*, **19** (1898 Dec. 12), pp. 151–4; F. Möller and B. Cohn, *AN*, **148** (1898 Dec. 13), p. 47; C. F. Pechüle, *AN*, **148** (1898 Dec. 14), p. 63; G. J. Fayet and C. F. Pechüle, *AN*, **148** (1898 Dec. 28), p. 111; E. F. Coddington, *AJ*, **19** (1899 Jan. 16), p. 172; A. Abetti and R. Schorr, *AN*, **148** (1899 Jan. 30), p. 239; W. E. Plummer and F. L. Chase, *MNRAS*, **59** (1899 Feb.), p. 278; H. A. Kobold, *AN*, **149** (1899 Jun. 2), pp. 345–7; M. Wasnetzoff, *AN*, **197** (1914 Feb. 4), pp. 121–30; E. E. Barnard, *AJ*, **41** (1931 Dec. 15), pp. 146, 150; V1964, p. 65; B. G. Marsden and Z. Sekanina, AJ, **78** (1973 Dec.), pp. 1119–20.

C/1899 E1 (Swift)

1899 I = 1899a

Discovered: 1899 March 4.16 ($\Delta = 0.78$ AU, $r = 1.06$ AU, Elong. $= 72°$)

Last seen: 1899 August 12.86 ($\Delta = 2.54$ AU, $r = 2.41$ AU, Elong. $= 71°$)

Closest to the Earth: 1899 February 21 (0.7108 AU), 1899 May 29 (0.5493 AU)

Calculated path: ERI (Disc), FOR (Mar. 5), ERI (Mar. 7), CET (Mar. 21), PSC (Apr. 3), AND (Apr. 30), PEG (May 2), AND (May 13), LAC (May 16), CYG (May 21), CEP-CYG (May 26), DRA (May 29), HER (Jun. 5), BOO (Jun. 10), VIR (Jul. 31)

L. Swift (Lowe Observatory, California, USA) discovered this comet on 1899 March 4.16. It was then low over the southwest horizon in the evening sky and Swift estimated the position as $\alpha = 3^h 45^m$, $\delta = -29°$. He described it as bright, large, and visible to the naked eye. There was also a short tail.

Although the comet was approaching perihelion, it had passed moderately close to Earth nearly two weeks prior to discovery and was moving away from our planet; however, this did not prevent the comet from being widely observed during most of the remainder of March. Numerous magnitude estimates were made during the month, while other astronomers indicated the comet was flirting with naked-eye visibility. E. Hartwig (Bamberg, Germany) estimated the magnitude as 6.0 on the 5th, while A. A. Nijland (Utrecht, Netherlands) said it was 6 at low altitude and in bright twilight. Hartwig gave the magnitude as 7.0 on the 6th, while A. Abetti (Arcetri, Italy) said it was 5 or 6, P. Chofardet (Besançon, France) estimated it as 7, and M. Wolf (Heidelberg, Germany) said it was easy to see with the naked eye. W. E. Plummer (Liverpool, England) said the comet was visible on the 7th, "long before any star of comparison could be seen in the light sky of twilight owing to the low altitude." Abetti estimated it as magnitude 6 on the 16th. L. A. Eddie (Grahamstown, South Africa) could not detect the comet with the naked eye on the 17th. Abetti estimated the magnitude as 5 in moonlight on the 17th and 18th, with the help of the telescope's finder. The coma diameter was estimated as 7' by Hartwig and 1.5' by Abetti on the 6th, 3' by W. Winkler (Jena, Germany) and 3–4' by L. Ambronn (Göttingen) on the 13th, 2' by F. Schwab (Kremsmünster, Austria) on the 14th, and 1' by Abetti on the 16th. All observers agreed that the comet was round and centrally condensed during the month, with many noticing a distinct nucleus. Although Wolf noticed a trace of a tail in a 15-cm telescope on the 6th, there were no other reports of the tail until the 13th, when Eddie described it as faint, long, and straight in a 23-cm reflector. He said the tail was 30' long on the 17th. Eddie provided some of the more descriptive details of the comet during the month. For the 13th, he wrote that the comet was "fairly large and bright, of a very undefined outline, but considerably condensed in the centre, though showing no stellar nucleus or defined cometic envelopes. It was very fluffy and extremely ragged, with woolly protuberances on its northern edge. It shone with a bluish-white light." Eddie said the nucleus was more condensed on the 17th than on the 13th, and he found the nucleus "decidedly stellar" on the 22nd. Also on the 22nd, Eddie said the tail appeared broader, with a streaky, hair-like structure. Eddie found the comet visible in bright twilight on the 24th. The last observation before the comet was lost in the sun's glare was acquired by J. Tebbutt (Windsor, New South Wales, Australia) on March 31.36.

The comet remained lost in the sun's glare throughout most of April, during which time it passed 0.3° from the sun on the 12th and passed perihelion on the 13th. It then began heading toward its closest approach to Earth. One of the earliest recoveries was made by V. Laska (Lemberg, now L'viv, Ukraine) on April 27, when he described it as very bright and about magnitude 4–5. Although magnitude estimates were quite plentiful, J. Holetschek (Vienna, Austria) provided the best and probably most accurate estimates. Using binoculars, he estimated the comet's magnitude as 3.0 on the

8th, 4.0 on the 11th, 4.3 on the 14th, 4.6 on the 15th, and 5.5 on the 31st. The only other useful total magnitude estimates were 5 by Schwab on the 10th (naked eye), 5.5 by E. E. Barnard (Yerkes Observatory, Wisconsin, USA) on the 19th (naked eye), and 6–6.3 by Ambronn on the 30th (binoculars). The nucleus remained well defined throughout the month and its magnitude was estimated as magnitude 7.5 by R. Schorr (Bergedorf Observatory, Hamburg, Germany) on the 30th and 8.5 by A. Scheller (Bergedorf Observatory, Hamburg, Germany) on the 31st. Hartwig described it as disk-shaped and strongly eccentric within the coma on the 17th. The comet maintained a circular coma during the month, with the diameter being given as 5′ by W. M. Witchell (Royal Observatory, Greenwich, England) on the 7th, 3′ by Hartwig on the 17th, 6′ by Schorr and 6–8′ by Ambronn on the 30th, and 4′ by Scheller and 7′ by Holetschek on the 31st. The tail was well observed. Nijland saw it on several occasions with his refractor's 7.4-cm finder. He said it extended 90′ toward PA 269° on the 6th, 60′ toward PA 267° on the 14th, 70′ toward PA 264° on the 15th, and 55′ toward PA 262° on the 16th. Holetschek described it as faint in binoculars on the 15th. Hartwig said it extended 20′ toward PA 262° on the 17th. I. Benko von Boinik (Pula, Yugoslavia) estimated it as 1° long on the 17th. Barnard said no decided tail was visible to the naked eye on the 19th, but the finder of the refractor revealed a faint tail of 1–1.5°, while a photograph revealed a length of 6° or 8°. Holetschek said it was 20–30′ long on the 31st. Barnard watched the comet pass over a 10th-magnitude star on the 31st. Using the 102-cm refractor he commented, "The star shone through the comet apparently unchanged. It must have been almost a perfectly central transit. The nucleus must have passed within a fraction of a second north of the star." The comet attained its most northerly declination of +57° on May 30.

Most notable was the appearance of a second nucleus during May. This nucleus was independently observed by C. D. Perrine (Lick Observatory, California, USA), using the 91-cm refractor, and Barnard, using the 102-cm refractor. Perrine had seen the nucleus on the 8th and had described it as sharp and about magnitude 7 or brighter. He added that to the north-following side of the nucleus was an appendage about 10″ long and about 5″ wide that was brighter than the rest of the coma. On the 12th he noted a secondary nucleus was distinctly visible and was located 12.5″ from the primary toward PA 263.5°. He estimated the primary nucleus as magnitude 8.0 and the secondary one as 9.5. He made measurements of the separation between the two nuclei almost nightly between the 12th and 21st. In particular, he noted the secondary was located 18.2″ away toward PA 256.6° on the 15th and 29.4″ toward PA 241.0° on the 21st. On the last date he said the primary was magnitude 9.5, while the secondary was magnitude 11.5. Barnard saw the secondary on May 21, 22, 23, and 24. On the first date he said it was separated 28.84″ toward PA 241.9°, while on the last he said it was separated 38.16″ toward PA 228.0°. Perrine was uncertain of the secondray nucleus on May 27 and could not see it on June 4. It was again visible on June 5, 7, 8, 9, and 10. On the last date, he said the secondary was very faint and located 16.9″ away toward PA 149.0°.

The comet was moving away from both the sun and Earth as June began. Holetschek continued to provide the best total magnitude estimates and even observed an outburst that peaked on the 5th. He gave the magnitude as 5.4 on the 1st, 5.7 on the 2nd, 6.0 on the 3rd, 5.3 on the 4th, 4.4 on the 5th, 4.6–4.8 on the 6th, 4.6 on the 7th, 6 on the 9th, 5.7 on the 10th, 6.2 on the 11th, 6.5 on the 13th, 7.0 on the 16th, 7.3 on the 17th, 8 on the 20th, and 8.9 on the 27th and 28th. Nijland also provided several estimates using binoculars. He gave the magnitude as 5.8 on the 7th, 6.8 on the 12th, 6.7 on the 13th, 6.8 on the 13th, and 6.7 on the 15th. When the comet attained its maximum outburst magnitude on the 5th, its total magnitude was independently estimated as 5 by both Schorr and Hartwig, 4.5 by Ambronn, and 4–5 by W. Valentiner (Königstuhl Observatory, Heidelberg, Germany). Magnitude estimates of the nuclear condensation were also plentiful, with Holetschek again producing the best series. Using a 15-cm Fraunhofer refractor he gave the magnitude as 9–10 on the 1st and 2nd, 7–8 on the 4th, 7 on the 5th, 7–8 on the 6th, 8–9 on the 7th, 9–10 on the 9th, 9 on the 10th, 10 on the 11th and 13th, 9 on the 16th, 10.5 on the 17th, and 10 on the 20th. H. A. Kobold (Strasbourg, France) observed with a 46-cm refractor and frequently saw a stellar condensation. He estimated its magnitude as 10 on the 5th, 11 on the 7th, 10 on the 8th, and 11 on the 12th, 16th, and 27th. Interestingly, he could not see the nucleus on the 19th. Estimates of the coma diameter were quite numerous and also reflected the outburst. Holetschek gave the diameter as 6′ on the 1st, 2nd, and 3rd, 4′ on the 4th, 7′ on the 5th, 10–12′ on the 7th, 7′ on the 9th and 10th, 5′ on the 11th, 5–6′ on the 13th, 4–5′ on the 16th, 4′ on the 17th, 3′ on the 20th, about 2.8′ on the 27th, and about 2.5′ on the 28th. On the night when the comet had reached peak outburst brightness, the coma diameter was given as 9′ by Schorr and 12′ by Hartwig. The tail had faded considerably as June began. Holetschek saw a trace of a wide tail on the 1st and noted a distinct tail on the 4th. He then reported that no tail was visible on the 5th, 6th, 7th, 9th, and 10th.

Observations were declining as July began. Holetschek still reported an admirable number of total magnitudes. He estimated the brightness as 9.2 on the 1st, 9.5 on the 3rd, 10.5 on the 8th, 10.5–10.8 on the 10th and 11th, 11 on the 12th, and 11–11.5 on the 13th. Nijland estimated the magnitude as 10 on the 2nd, J. Palisa (Vienna, Austria) gave it as 12 on the 10th, and Perrine said it was 13 or fainter on the 31st. Holetschek also estimated the coma diameter as 2′ on the 1st and 3rd, and 1.5′ on the 10th and 11th. Palisa estimated the diameter as 1′ on the 10th. Kobold estimated the magnitude of the stellar nucleus as 12 on the 8th. V. Cerulli (Teramo, Italy) said the tail extended toward PA 90° on the 27th.

Observations were not plentiful during August. Perrine estimated the total magnitude as 12.5 and the nuclear magnitude as 15 on the 1st. Cerulli said the comet was in the vicinity of NGC 5551 on the 7th. The comet was last detected on August 12.86, when Cerulli gave the position as $\alpha = 14^h\ 16.4^m$, $\delta = +4°\ 47'$.

The first parabolic orbit was calculated by W. J. Hussey, who took positions from March 5, 6, and 7, and determined the perihelion date as 1899 April 13.76. A few days later, H. C. F. Kreutz refined the orbit using positions from March 5, 6, and 9, and gave the perihelion date as April 13.49. Additional orbits by C. J. Merfield, Kreutz, A. Stichtenoth, and others only slightly improved upon Kreutz' initial orbit.

The first nonparabolic orbit was calculated by Merfield. Using positions spanning the period of March 5 to July 13, he calculated a hyperbolic orbit with a perihelion date of April 13.48 and an eccentricity of 1.00039453. Similar hyperbolic orbits using positions spanning the entire period of visibility were later published by Merfield (1901), A. Wedemeyer (1903), and B. G. Marsden, Z. Sekanina, and E. Everhart (1978). The last orbit is given below.

T	ω	Ω (2000.0)	i	q	e
1899 Apr. 13.4781 (TT)	8.7101	26.4080	146.2690	0.326576	1.000357

ABSOLUTE MAGNITUDE: $H_{10} = 5.4$ (V1964)

FULL MOON: Feb. 25, Mar. 27, Apr. 25, May 25, Jun. 23, Jul. 22, Aug. 21

SOURCES: L. Swift, *AJ*, **20** (1899 Mar. 8), p. 8; L. Swift and E. Hartwig, *AN*, **148** (1899 Mar. 8), p. 387; W. J. Hussey and H. C. F. Kreutz, *AN*, **149** (1899 Mar. 16), p. 13; A. A. Nijland, E. Hartwig, A. Abetti, M. Wolf, P. Chofardet, W. Winkler, and L. Ambronn, *AN*, **149** (1899 Mar. 23), pp. 27–30; E. Hartwig, A. Abetti, H. C. F. Kreutz, *AN*, **149** (1899 Apr. 10), pp. 59–62; L. A. Eddie, *MNRAS*, **59** (1899 May), pp. 503–5; A. Stichtenoth and A. A. Nijland, *AN*, **149** (1899 May 17), p. 271; W. M. Witchell, *MNRAS*, **59** (1899 Jun.), p. 542; J. Holetschek and E. Hartwig, *AN*, **149** (1899 Jun. 2), p. 351; E. Hartwig, R. Schorr, and J. Holetschek, *AN*, **149** (1899 Jun. 9), p. 383; C. J. Merfield and J. Holetschek, *AN*, **149** (1899 Jun. 13), pp. 397–400; E. E. Barnard and C. D. Perrine, *AJ*, **20** (1899 Jun. 14), pp. 60–1; L. Ambronn, *AN*, **151** (1899 Nov. 14), pp. 9–12; C. J. Merfield, *AN*, **151** (1899 Nov. 16), pp. 23–8; F. Schwab, *AN*, **151** (1899 Dec. 28), p. 169; A. Abetti and H. A. Kobold, *AN*, **151** (1900 Jan. 27), pp. 283, 289–92; J. Holetschek, *AN*, **151** (1900 Feb. 2), p. 299; V. Laska, *AN*, **151** (1900 Feb. 10), pp. 369–72; I. Benko von Boinik, *AN*, **151** (1900 Feb. 14), p. 383; A. A. Nijland, *AN*, **151** (1900 Feb. 20), pp. 391–4; C. D. Perrine, *AN*, **20** (1900 Mar. 31), p. 186; W. E. Plummer, *MNRAS*, **60** (1900 Apr.), pp. 521–2; J. Palisa, *AN*, **152** (1900 Apr. 7), p. 151; W. Valentiner, *AN*, **152** (1900 Jun. 2), pp. 283–6; O. Knopf, *AN*, **154** (1901 Feb. 28), pp. 365–72; R. Schorr, *AN*, **156** (1901 Jun. 22), pp. 51–8; V. Cerulli, *AN*, **157** (1901 Nov. 9), pp. 85–94; C. J. Merfield, *AN*, **157** (1901 Nov. 11), pp. 37–76; A. Wedemeyer, *AJB*, **4** (1903), pp. 173, 188–9; E. E. Barnard, *AJ*, **41** (1931 Dec. 15), pp. 146, 150; V1964, p. 65; B. G. Marsden, Z. Sekanina, and E. Everhart, *AJ*, **83** (1978 Jan.), p. 66.

8P/Tuttle

1899 III = 1899b

Recovered: 1899 March 5.82 ($\Delta = 1.78$ AU, $r = 1.35$ AU, Elong. $= 49°$)

Last seen: 1899 July 10.73 ($\Delta = 1.82$ AU, $r = 1.42$ AU, Elong. $= 51°$)

Closest to the Earth: 1899 May 28 (1.6994 AU)

Calculated path: PSC (Rec), TRI (Mar. 8), ARI (Mar. 22), TAU (Apr. 8), ORI (Apr. 29), MON (May 28), CMa (Jun. 14), PUP (Jun. 16), PYX (Jul. 3)

J. Rahts (1894) took positions obtained for this comet during the apparitions of 1871 and 1885, and calculated orbits for each of those years. He then took the 1885 orbit, applied perturbations by Jupiter and Saturn for the period of 1885 September 9 to 1899 June 2, and predicted the comet would next arrive at perihelion on 1899 June 3.14. Rahts revised his prediction early in 1899, giving the perihelion date as May 14.51 and providing an ephemeris for the period of March 5 to June 3.

M. F. J. C. Wolf (Königstuhl Observatory, Heidelberg, Germany) recovered this comet on 1899 March 5.82. He had obtained a photographic plate of the field and, with the help of A. Schwassmann, he found the comet several degrees from the expected place and measured its position as $\alpha = 1^h 16.0^m$, $\delta = +31° 38'$. He estimated the magnitude as 11–12.

Despite the fact that the comet was approaching both the sun and Earth, this was not one of its most favorable returns. Consequently, the comet brightened very slowly. C. D. Perrine (Lick Observatory, California, USA) estimated the magnitude as 11.5–12 on March 10 and 11, with the 91-cm refractor revealing a sharp nucleus of magnitude 16 on the 11th. The comet was reported as round and about 1' across by W. Seraphimoff (Pulkovo Observatory, Russia) on March 10 and 13, as well as by H. A. Kobold (Strasbourg, France) on the 14th. Moonlight interfered with observations thereafter.

The comet continued to slowly brighten during April. Perrine estimated the magnitude as 10.5–11 on the 5th, and 10 through the remainder of the month. Although some astronomers reported neither a condensation nor a nucleus, the 91-cm refractor at Lick Observatory allowed Perrine to see a "sharp condensation" and a nucleus on several occasions throughout the month. On the 15th, Perrine estimated the nucleus was about magnitude 12–13. Perrine reported a coma diameter of 1' on the 5th, 1.5–2' on the 6th, and then continually estimated it as 2' across during the rest of the month. With smaller telescopes, W. E. Plummer (Liverpool Observatory, Liverpool, England) estimated the coma diameter as 1' on the 4th, while A. Abetti (Arcetri, Italy) estimated it as 1' on the 5th. Early in the month Abetti remarked on the irregular shape of the coma.

The comet was steadily moving southward and its solar elongation was steadily decreasing. The comet passed 32° from the sun on May 1 and a short time later it was lost from view. The comet was found by R. T. A. Innes (Royal Observatory, Cape of Good Hope, South Africa) on June 26 with a 17.8-cm refractor. He noted on the 29th that the comet was very faint when the reticle was illuminated. On the 30th, Innes said the comet was barely seen.

Innes was again barely able to see the comet in the 17.8-cm refractor on July 1. He saw it for the last time on July 10.73 and noted it was very faint in the refractor. The position was then $\alpha = 8^h 51.7^m$, $\delta = -23° 24'$.

Innes indicated that the comet was not visible on the last date when a 6th-magnitude star entered the field of view.

The most recent investigation using multiple apparitions was by B. G. Marsden and Z. Sekanina (1972). They determined the perihelion date as May 4.55 and the period as 13.62 years, with Marsden and Sekanina giving nongravitational terms of $A_1 = +0.32$ and $A_2 = +0.0131$.

T	ω	Ω (2000.0)	i	q	e
1899 May 4.5531 (TT)	206.6276	271.2447	54.4891	1.013808	0.822255

ABSOLUTE MAGNITUDE: $H_{10} = 8.5$ (V1964)

FULL MOON: Feb. 25, Mar. 27, Apr. 25, May 25, Jun. 23, Jul. 22

SOURCES: J. Rahts, *AN*, **136** (1894 Aug. 2), pp. 65–8; J. Rahts and M. F. J. C. Wolf, *AN*, **148** (1899 Mar. 8), p. 391; M. F. J. C. Wolf, A. Schwassmann, and C. D. Perrine, *AN*, **149** (1899 Mar. 23), p. 31; *MNRAS*, **59** (1899 Apr.), p. 392; A. Abetti, *AN*, **149** (1899 Apr. 29), p. 205; H. A. Kobold, *AN*, **149** (1899 Jun. 2), pp. 345–8; W. Seraphimoff, *AN*, **150** (1899 Jul. 18), p. 79; R. T. A. Innes, *AN*, **151** (1899 Dec. 9), pp. 109–12; C. D. Perrine, *AN*, **151** (1899 Dec. 28), pp. 171–4; W. E. Plummer, *MNRAS*, **60** (1900 Apr.), p. 523; V1964, p. 65; B. G. Marsden and Z. Sekanina, *QJRAS*, **13** (1972), pp. 427–9; B. G. Marsden, *CCO*, 12th ed. (1997), pp. 50–1.

10P/Tempel 2　　*Recovered:* 1899 May 7.41 ($\Delta = 0.87$ AU, $r = 1.64$ AU, Elong. $= 121°$)

Last seen: 1899 December 2.15 ($\Delta = 1.45$ AU, $r = 1.89$ AU, Elong. $= 99°$)

1899 IV = 1899c　　*Closest to the Earth:* 1899 July 28 (0.3746 AU)

Calculated path: SCT (Rec), AQL (May 7), CAP (Jun. 30), MIC (Aug. 8), PsA (Aug. 24), SCL (Oct. 31), AQR (Nov. 9)

L. Schulhof (1899) published an approximate prediction for the comet's approaching 1899 return. The resulting perihelion date was 1899 July 29.00. He said a minor approach to Jupiter during the period of 1894–1899 had delayed the perihelion date by more than 15 days. He provided an ephemeris covering the period of April 2 to June 25. The comet was recovered by C. D. Perrine (Lick Observatory, California, USA) on 1899 May 7.41 at a position of $\alpha = 18^h 53.0^m$, $\delta = -4° 32'$. He was using the 91-cm refractor and described the comet as a centrally condensed, round object, no more than 10″ across and about magnitude 15.5–16. He also suspected a very faint stellar point near the center. Perrine confirmed the comet on the 8th.

The comet was moving toward both the sun and Earth when recovered. Perrine remained the only observer of this comet until nearly the end of May. He estimated the magnitude as 14.5–15 on the 12th and noted a decided central condensation. On the 13th he found the magnitude near 13 and said the coma was 15″ across, with a bright central condensation. He found the comet almost stellar on the 20th, with a magnitude of 14 or fainter and a diameter of about 5″. On the 21st, Perrine said the comet was condensed and 5" or 6" across, with a magnitude of 14.5. H. A. Kobold (Strasbourg, France) saw the comet with the 46-cm refractor on May 30 and described it as small, round, and rather faint.

The comet continued its trek towards the sun and Earth during June, but it remained out of reach for many astronomers. Perrine continued to estimate the comet's magnitude at every opportunity. He gave values of 13 on the 4th, 12 on the 5th, 11 on the 10th and 11th, 10.5 on the 15th, and 10 on the 20th and 22nd. Kobold gave the magnitude as 11 on the 10th, but attributed this to the central condensation. Perrine remarked on the sharp, nearly stellar condensation during virtually every observation. He estimated the magnitude of this "nucleus" as 12 on the 11th, 16th, and 17th, and 11.5 or 12 on the 20th. Perrine also estimated the coma diameter and gave values of 15" on the 4th, 20–30" on the 10th, 30" on the 11th, 1' on the 15th, 2' on the 17th, and 1.5–2' on the 20th. To these observations, Perrine added that he saw a fan-shaped extension about 30" long on the 16th and noted the coma was slightly brighter on the north-following side on the 20th.

A considerable number of observations were made during July, as the comet passed closest to the sun and Earth near the end of the month. Estimates of the total magnitude were primarily provided by J. Holetschek (Vienna, Austria) and Perrine. Holetschek gave magnitudes of 9 on the 10th, 8.8 on the 11th, 9 on the 13th, and 8.5 on the 20th. Perrine gave values of 9 on the 1st, 9.5 on the 5th and 6th, and 9 on the 10th, 12th, 13th, and 14th. Although Holetschek estimated the magnitude of the "nucleus" as 10 on the 10th and 9.5 on the 20th with his 15-cm refractor, Perrine obtained fainter values with his much larger refractor. He estimated the magnitude of the "nucleus" as 10 on the 1st, 10.5 on the 5th, 11 on the 6th, 10.5 on the 7th, 11 on the 8th, 10.5 or 11 on the 10th, 11 on the 12th, 10.5 on the 13th, 14th, and 15th, and 10 on the 30th. Perrine examined the nucleus at magnifications of 270× and 520× on the 1st and said it was "fully as sharp" as a nearby star. The coma diameter was estimated by several astronomers. Holetschek gave it as over 1.5' on the 10th, 2' on the 11th, and 2–3' on the 20th. Perrine estimated it as 2' on the 1st and 5' on the 8th. A tail was not widely reported, but some descriptions were obtained. Perrine described it as a fan-shaped feature on the north-preceding side on the 2nd, and simply noted it was fan-shaped during his remaining July observations. He gave the length as 1' on the 5th, 30" on the 7th, and 1' on the 10th and 30th. Kobold saw the comet on the 17th and said the short, fan-shaped tail extended toward PA 340°. W. E. Plummer (Liverpool, England) saw the comet on the 26th and said it appeared "faint and ill-defined."

The comet was moving away from the sun and Earth as August began. V. Cerulli (Teramo, Italy) said the tail extended toward PA 350° on the 1st. Although most Northern Hemisphere observatories ceased to follow the comet around mid-month as it moved into a more southerly region of the sky, several Italian astronomers continued to follow it. A. Abetti (Arcetri,

Italy) obtained several magnitude estimates, with values of 9 on August 2 and 4, 10 on the 6th, 9 on the 11th, 10 on the 12th, and 12 on the 28th, 29th, and 31st. Although observations from the Southern Hemisphere had begun early in July, physical descriptions were a rarity until L. A. Eddie (Grahamstown, South Africa) published his observations for the last half of August. For August 15, Eddie wrote, "Head sharply defined on preceding edge, narrow in anterior portion, spreading out immediately at a wide angle, and very much diffused posteriorly. Condensation in central portion of head." He found the comet centrally condensed, but faint in moonlight on the 18th. On August 24, Eddie said the comet was well seen in the finder scope, while the 23-cm reflector revealed the nucleus as globular in form and well condensed, with a diameter of about 2'. The nucleus was also surrounded by a "very rare coma." He added, "A broad, though extremely faint, tail could be traced to about 16'." For the 25th, Eddie wrote, "Cometic envelopes, though faint, well defined, with two dark intervals in preceding portion, fairly conspicuous condensed nucleus, with dark rift behind, and broad, faint tail about 16' in length."

The comet attained its most southerly declination of $-36°$ on September 12, and Northern Hemisphere observations picked up later in the month. Perrine estimated the magnitude as 12 on the 24th and noted a coma was 30" across with a faint nucleus. He gave the magnitude as 10 on the 25th and said the coma was 1–2' across. He also saw a nucleus of magnitude 13.

Only a handful of observers followed the comet after September. Cerulli noted it passed NGC 7267 on October 2 and estimated the nuclear magnitude as 14. Perrine said the nucleus was still visible on the 3rd. Kobold saw the comet on November 6 and said it was extraordinarily faint and visible with great difficulty in the 46-cm refractor. He added that the coma was 0.5' across and exhibited only slight condensation. On November 23, Perrine estimated the magnitude as 14 on the 23rd, and added that the coma was about 30" across.

The comet was last detected on December 2.15, when Perrine gave the position as $\alpha = 23^h\ 50.9^m$, $\delta = -18°\ 54'$. He described it as very faint, with a magnitude near 15. The comet was very diffuse, about 1' across, with almost no condensation.

Minor refinements of this comet's orbit using positions from multiple apparitions, as well as planetary perturbations, have been published over the years. The most recent were by B. G. Marsden and Z. Sekanina (1971) and Sekanina (1985). These have typically revealed a perihelion date of July 29.04 and a period of 5.28 years. Sekanina determined the nongravitational terms as $A_1 = +0.08$ and $A_2 = +0.0021$ and his orbit is given below.

T	ω	Ω (2000.0)	i	q	e
1899 Jul. 29.0363 (TT)	185.6557	122.3254	12.6403	1.388620	0.542082

ABSOLUTE MAGNITUDE: $H_{10} = 9.4$ (V1964)

FULL MOON: Apr. 25, May 25, Jun. 23, Jul. 22, Aug. 21, Sep. 19, Oct. 18, Nov. 17, Dec. 17

SOURCES: L. Schulhof, *AN*, **149** (1899 Mar. 23), pp. 23–6; C. D. Perrine, *AN*, **149** (1899 May 10), p. 255; C. D. Perrine, *AJ*, **20** (1899 May 15), p. 40; C. D. Perrine, *AJ*, **20** (1899 May 22), p. 45; V. Cerulli, *AN*, **150** (1899 Oct. 17), p. 375; L. A. Eddie, *MNRAS*, **59** (1899 Supplementary Notice), pp. 570–1; A. Abetti, *AN*, **151** (1900 Jan. 27), pp. 285–90; J. Holetschek, *AN*, **151** (1900 Feb. 2), p. 299; G. A. P. Rayet, *CR*, **130** (1900 Feb. 5), pp. 302–3; C. D. Perrine, *AJ*, **20** (1900 Feb. 12), pp. 167–8; H. A. Kobold, *AN*, **152** (1900 Mar. 6), pp. 59–62; W. E. Plummer, *MNRAS*, **60** (1900 Apr.), p. 523; J. Palisa, *AN*, **152** (1900 Apr. 7), p. 151; V. Cerulli, *AN*, **157** (1901 Nov. 9), pp. 85–94; V1964, p. 65; B. G. Marsden and Z. Sekanina, *AJ*, **76** (1971 Dec.), pp. 1137–8; Z. Sekanina, *QJRAS*, **26** (1985), p. 114; B. G. Marsden and Z. Sekanina, *CCO*, 12th ed. (1972), pp. 50–1.

17P/1899 L1	*Recovered:* 1899 June 11.46 ($\Delta = 2.54$ AU, $r = 2.15$ AU, Elong. $= 56°$)
(Holmes)	*Last seen:* 1900 January 21.26 ($\Delta = 2.42$ AU, $r = 2.84$ AU, Elong. $= 105°$)
	Closest to the Earth: 1899 October 26 (1.6422 AU)
1899 II = 1899d	*Calculated path:* PSC (Rec), ARI (Jun. 27), TRI (Jul. 15), PER (Aug. 8), AND (Nov. 5)

Predictions for this return came from E. Kohlschütter (1896) and H. J. Zwiers (1895, 1897, 1899). Both astronomers made careful investigations of the comet's discovery apparition of 1892–3, but where Kohlschütter did not apply perturbations between that apparition and 1899, Zwiers carefully determined the effects Jupiter and Saturn would have on the comet's motion. The ultimate result was that Kohlschütter predicted a perihelion date of 1899 May 8.51, while Zwiers predicted it as April 28.17. Zwiers (1899) wrote that the comet was favorably situated for recovery during the Spring of 1898 for observers in the Southern Hemisphere though it could have been just out of range of their telescopes. The failure to find the comet made a recovery during the autumn of 1899 very important. The comet was recovered by C. D. Perrine (Lick Observatory, California, USA) with a 91-cm refractor (270×) on 1899 June 11.46. He gave the position as $\alpha = 1^{\text{h}}\ 15.5^{\text{m}}$, $\delta = +17°\ 30'$, indicating Zwiers' perihelion date required a correction of only +0.43 day. Perrine said the comet was not brighter than magnitude 16. He described it as a "round nebulous mass about 30″ in diameter, with only a slight brightening at the center." The comet had passed about 4° from the sun on February 6.

The comet was only observed at Lick Observatory and Yerkes Observatory (Wisconsin, USA) using their large 91-cm and 102-cm refractors, respectively. Although the magnitude estimates seem to jump around a bit, their sparse distribution and lack of confirming observations makes it difficult to determine if the variations were because the comet was still experiencing outbursts or because of other factors.

Perrine noted on June 16 that the comet had not appreciably brightened since discovery, while on the 17th and 18th he simply said it was very faint. Perrine said the comet was slightly brighter by July 7 and estimated the magnitude as 15. He gave the magnitude as 14 on July 10 and 15.5 on the 16th. He added that a faint nucleus was visible on the 10th, while a condensed coma 20–30″ across was noted on the 16th.

As the comet continued to move away from the sun, E. E. Barnard (Yerkes Observatory) estimated the magnitude as 13 on August 16 and 13.5 on the 17th. On the last date he said the coma was round and diffuse, with a feeble nucleus, and was estimated as 0.5–0.75′ in diameter. Barnard estimated the total magnitude as 15 on September 11, while the nuclear magnitude was given as 16.5.

Both Perrine and Barnard obtained observations during October and November. Perrine estimated the magnitude as 14.5 on October 1 and noted the centrally condensed coma was 15″ across. Perrine estimated the magnitude as 14 on the 29th and said the centrally condensed coma was 20″ across. The comet attained its most northerly declination of +49° on October 31. On that date, Barnard estimated the magnitude as 15–16. He added that the coma seemed "a little brighter in the center" and estimated the total diameter as 8–10″. Barnard estimated the magnitude as 15.5 on November 5. He again noted a slight brightening in the center and gave the diameter as 15″. Perrine estimated the magnitude as 15 on November 7.

Barnard did not see the comet after November, but Perrine continued measuring its position. On December 25, he estimated the magnitude as 16 and said the coma was 10–15″ across.

The comet was last detected on 1900 January 21.26, when Perrine gave the position as $\alpha = 2^h\ 23.8^m$, $\delta = +40°\ 14'$. The observation was made with the 91-cm refractor and Perrine described the comet as very faint, with a magnitude of 16.

In preparation for the comet's 1906 return, Zwiers (1906) took 21 positions from this apparition and corrected his earlier prediction. He determined the perihelion date as April 28.60 and the period as 6.87 years. F. Koebcke (1948) linked the apparitions of 1892, 1899, and 1906. For this apparition, he revealed a perihelion date of April 28.60 and a period of 6.87 years. Koebcke's orbit is given below.

T	ω	Ω (2000.0)	i	q	e
1899 Apr. 28.5985 (UT)	14.0821	333.1418	20.8147	2.128154	0.411336

ABSOLUTE MAGNITUDE: $H_{10} = 9.5$ (V1964)

FULL MOON: May 25, Jun. 23, Jul. 22, Aug. 21, Sep. 19, Oct. 18, Nov. 17, Dec. 17, Jan. 15

SOURCES: H. J. Zwiers, *AN*, **138** (1895 Oct. 3), pp. 419–22; E. Kohlschütter, *AN*, **141** (1896 Aug. 27), pp. 241–50; H. J. Zwiers, *AN*, **142** (1897 Jan. 29), pp. 257–64; H. J. Zwiers, *AN*, **149** (1899 Mar. 16), pp. 9–12; C. D. Perrine, C. N. A. Krueger, and H. J. Zwiers, *AN*, **149** (1899 Jun. 13), p. 399; C. D. Perrine, *AJ*, **20** (1899 Jun. 14), p. 64; C. D. Perrine, *AJ*, **20** (1899 Jun. 30), p. 72; C. D. Perrine, *AJ*, **20** (1900 Mar. 31), pp. 187–8;

H. J. Zwiers, *AN*, **171** (1906 Apr. 26), pp. 65–72; E. E. Barnard, *AJ*, **41** (1931 Dec. 15), pp. 146, 150; F. Koebcke, *Poznan Observatory Reprint*, No. 12 (1948); V1964, p. 65.

C/1899 S1   *Discovered:* 1899 September 29.81 ($\Delta = 2.05$ AU, $r = 1.80$ AU, Elong. $= 61°$)
(Giacobini)   *Last seen:* 1899 December 24.10 ($\Delta = 2.79$ AU, $r = 2.19$ AU, Elong. $= 43°$)
    Closest to the Earth: 1899 July 14 (1.2468 AU)
1899 V = 1899e   *Calculated path:* OPH (Disc), HER (Dec. 7)

Michel Giacobini (Nice, France) discovered this comet on 1899 September 29.81 at a position of $\alpha = 16^h 26.5^m$, $\delta = -5° \ 10'$. He simply described the comet as faint.

Even though the comet was faint at discovery, and was steadily moving away from both the sun and Earth, its rather large perihelion distance caused the fading to be slow during October. Total magnitude estimates were made by E. Marchetti (Pula, Yugoslavia) and F. Cohn (Königsberg, now Kaliningrad, Russia) from the 1st to the 4th and they indicated it was between 10 and 11. Magnitude estimates of the condensation for the same period were made by S. Javelle (Nice, France) and P. Chofardet (Besançon, France) and they indicated it was between 12 and 13. Cohn and Chofardet both estimated the coma diameter as about 1' across, and everyone indicated the coma was centrally condensed, but exhibited neither nucleus nor tail. The comet changed little during the next two weeks. F. W. Ristenpart (Kiel, Germany) estimated the total magnitude as 11 on the 8th and C. D. Perrine (Lick Observatory) indicated the magnitude of the condensation was 12–13 on the 16th. Various observers reported the comet was diffuse, with ill-defined or diffuse edges. Large telescopes continued to see the condensation, while observers with smaller telescopes began having problems seeing the comet. As an example, Marchetti noted the comet was extremely faint in his 15.2–cm refractor on the 8th and A. Abetti (Arceti, Italy) said the comet was at the limit of his 28.4-cm refractor on the 8th and 9th. Cohn did note on the 9th that his 33.0-cm refractor revealed the condensation was located in the northern portion of the coma. Moonlight interfered for a few days after the 16th. Cohn described the comet as very faint when he next saw it as the moon was rising on the 21st. E. Hartwig (Bamberg, Germany) estimated the total magnitude as 11 on the 23rd and H. A. Kobold (Strasbourg, France) said the condensation was about magnitude 14 on the 27th. Although most observers were still estimating the coma diameter as 1', Kobold and W. Seraphimoff (Pulkovo Observatory, Russia) were using larger telescopes and consistently reported a round coma about 2' across during the remainder of October. With his 46-cm refractor, Kobold said he occasionally saw a nucleus of magnitude 15 on the 26th, while Seraphimoff reported that his 38-cm refractor revealed a possible nucleus on the same date. Marchetti saw the comet for the final time in his 15-cm refractor on the 27th.

Most observers who saw the comet during November reported how faint it was. J. Palisa (Vienna, Austria) turned a 68.6-cm refractor toward the comet on the 1st and said it was very faint and centrally condensed. Cohn, observing with a 33.0-cm refractor, said the comet was very faint on the 2nd and faint with an occasionally visible condensation on the 3rd. Seraphimoff said the coma was still about 2′ across on the 1st and 3rd. He added that he still occasionally suspected a small nucleus. Perrine said the coma was about 1′ across on the 24th. Kobold's only description of the comet in November came on the 4th, when he saw the comet at an altitude of 9.5°. At that time he reported the comet was very difficult to see in the 46-cm refractor. The only brightness estimates obtained during the month were by Perrine. He said the comet was about magnitude 12 on the 2nd and 7th, and about magnitude 13 on the 21st and 24th.

Only observers with the larger refractors were still observing the comet during December. Perrine saw the comet on December 2 and reported it was 1′ across. He added that the nebulosity was more pronounced on the southern side of the nucleus. Perrine said the nucleus was about magnitude 14. On December 8, Kobold was observing in moonlight and hazy skies and saw something that might have been the comet with the 46-cm refractor. He admitted though that the object could have been a star near the ephemeris position. Moonlight and unfavorable weather prevented Kobold from seeing the comet again or from confirming the December 8 observation. On December 10, Perrine said the comet was about magnitude 13, with a round coma about 0.75′ to 1′ across.

The comet was last detected on December 24.10, when Perrine gave the position as $\alpha = 18^{\mathrm{h}}\ 44.5^{\mathrm{m}}$, $\delta = +19°\ 06′$. He described the comet as magnitude 13, about 20″ across, with a central condensation.

The first astronomers to calculate parabolic orbits were at a disadvantage because of the comet's slow motion and rather large perihelion distance. Consequently, the first computations varied considerably. Although Giacobini was the first to produce an orbit, it was later found that he had incorporated an erroneous position which meant that his perihelion date of 1899 October 25.70 was completely unreliable. The first orbit therefore came from J. Möller, who took positions from October 1–4, and determined the perihelion date as August 27.17. A few days later, A. Hobe, Y. Kuno, and S. C. Phipps took positions from October 3, 4, and 5, and determined the perihelion date as July 25.81, while H. J. A. Perrotin took positions from September 30 to October 6 and calculated a perihelion date of September 12.13. Perrotin's orbit represented the comet's motion quite well, although later calculations by S. K. Winther and Perrine would ultimately adjust the perihelion date to September 15.

K. Lous (1913) calculated a definitive orbit. He began with 107 positions and determined seven Normal positions. After the perturbations of Venus to Jupiter were calculated, he produced a parabolic orbit with a perihelion date of September 15.40. This orbit is given below.

T	ω	Ω (2000.0)	i	q	e
1899 Sep. 15.4003 (UT)	10.7733	273.6234	76.9442	1.785838	1.0

ABSOLUTE MAGNITUDE: $H_{10} = 6.5$ (V1964)

FULL MOON: Sep. 19, Oct. 18, Nov. 17, Dec. 17

SOURCES: M. Giacobini, E. Marchetti, S. Javelle, and J. Möller, *AN*, **150** (1899 Oct. 7), p. 359; A. Hobe, Y. Kuno, S. C. Phipps, H. J. A. Perrotin, and P. Chofardet, *AN*, **150** (1899 Oct. 17), p. 375; H. A. Kobold, E. Hartwig, F. W. Ristenpart, H. Thiele, and S. K. Winther, *AN*, **150** (1899 Oct. 21), pp. 387–90; S. K. Winther, C. D. Perrine, and H. A. Kobold, *AN*, **150** (1899 Nov. 7), pp. 429–32; E. Hartwig, *AN*, **151** (1899 Nov. 14), p. 15; V. Cerulli, *AN*, **151** (1899 Nov. 29), p. 63; A. Abetti, *AN*, **151** (1900 Jan. 27), p. 285; F. Cohn, *AN*, **151** (1900 Feb. 10), pp. 367–70; E. Marchetti, *AN*, **151** (1900 Feb. 14), p. 385; T. J. J. See, *AN*, **152** (1900 Feb. 26), pp. 27–30; H. A. Kobold, *AN*, **152** (1900 Mar. 6), p. 61; C. D. Perrine, *AJ*, **20** (1900 Mar. 21), pp. 179–80; W. E. Plummer, *MNRAS*, **60** (1900 Apr.), p. 524; J. Palisa, *AN*, **152** (1900 Apr. 7), p. 151; W. Seraphimoff, *AN*, **152** (1900 Jun. 18), p. 359; F. Cohn, *AN*, **153** (1900 Aug. 6), pp. 111, 115; K. Lous, *AN*, **193** (1913 Feb. 15), pp. 411–20; V1964, p. 65.

Appendix 1
Uncertain Objects and Mistaken Identities

1803 Reissig (Kassel, Germany) was observing the region near the double star 36 Ophiuchi with his cometseeker of 30-inches focal length when he found a "star" on 1803 February 2.20, which had not been seen 5 days earlier with his 7-foot focal length telescope. The "star" was magnitude 5 or 6 and he measured a rough position of $\alpha = 16^h\ 55.2^m$, $\delta = -26°\ 19'$ with his 3-foot focal length reflector. He wrote to J. E. Bode and said a magnification of $400\times$ revealed no sensible nebulosity, although the object was larger than a star. Reissig said the comet had moved westward when seen on February 4.16. Although the object was again seen on the morning of February 7, no position could be measured as it was faint because of the light from the full moon. A rough position was again measured on the 8th. Reissig last saw the object on February 9.20 and gave the position as $\alpha = 16^h\ 35.4^m$, $\delta = -25°\ 11'$. Bode commented that the observations did not lie on a regular curve, but attributed this to their roughness. He added that the object was moving in a retrograde orbit and was probably "a very distant comet."

No further discussion of this object arose until an article appeared in the 1876 August 10 issue of *Nature*. The article was probably written by J. R. Hind, who frequently published orbits of comets in *Nature*. Hind acknowledged the roughness of the positions, but still produced the parabolic orbit given below. He said the orbit indicated a very close approach to Earth at the end of January and that perturbations by Earth "probably would" have affected the orbit.

T	ω	Ω (2000.0)	i	q	e
1803 Feb. 10.664 (UT)	197.39	311.61	0.94	0.96015	1.0

SOURCES: Reissig and J. E. Bode, *BAJ for 1806*, **31** (1803), p. 266; J. R. Hind, *Nature*, **14** (1876 Aug. 10), p. 311.

1808 J. C. Thulis announced that J. L. Pons had discovered a comet within the neck of the Giraffe during September of 1808. No other details are to be found. L. Schulhof (1885) was examining the motion of the comet 1873 V1 and its apparent link to comet 1818 D1. Among the periods he suggested was one of 9.3 years, which he said could indicate Pons' 1808 comet was an early appearance. As it turned out, the period of the comet was nearly 28 years and an identity with Pons' 1808 object was impossible.

SOURCES: L. Schulhof, *AN*, **113** (1885 Dec. 15), p. 143.

1808   F. W. Bessel (Lilienthal, Germany) was searching for comet C/1807 R1 early in November of 1808. During his search on the 7th he spotted two nebulae and thought one of them might be the comet. Interestingly, although both nebulae were still present on the next evening, one of them had "lost a lot of its nebulosity and looked similar to a faint star." Another search was conducted on the 9th and on November 9.78 Bessel found a "delicate nebulous spot." He was able to observe it for 25 minutes before clouds moved in, and gave the position as $\alpha = 4^h$ 20.3^m, $\delta = +58°$ $50'$. The sky was hazy on the 10th, but very clear skies on the 11th allowed him to resume searching. He first noted that the "nebulous spot" seen on the 9th was gone and a short time later he saw something nebulous at a position of $\alpha = 4^h$ 17.4^m, $\delta = +58°$ $42'$. A further search on the 12th revealed only a star at the position noted on the 11th, and nothing else of a nebulous nature in the vicinity. His search on the 13th was also unsuccessful. Bessel said he was certain of what he saw on the 9th and noted the object was too far from the expected position of comet C/1807 R1 to have been that comet. He suggested that "the nebulous appearance of the little stars of the 7th and 11th could be caused by an occultation" of a comet.

SOURCES: F. W. Bessel, *BAJ for 1812*, **37** (1809), pp. 126–7.

1820   J. Reeves (Macao, now Aomen, China) discovered this comet in Centaurus on 1820 May 5. J. F. Davis (1857) mentioned the comet and said, "its position being such as to be cut by two straight lines, one of them drawn through α and β, or the foot and easternmost arm of the Cross, and produced N.E., the other through ε and β, or the western foot of Centaur. After the first observation it became more visible by degrees, and then slowly disappeared towards the north-east."

SOURCES: Sir John Francis Davis, *China: A General Description of that Empire and its Inhabitants*. London: W. Clowes and Sons (1857), volume 2, pp. 249–50.

1823   H. W. M. Olbers received a report from Horner that a "very obvious comet" was seen in the evening sky by Swiss hunters on 1823 December 1.8. It was in the west-northwest. Olbers said much of Europe was experiencing cloudy skies on that date. He concluded it could not have been an early observation of C/1823 Y1, which would then have been south of the sun and invisible to European astronomers.

SOURCES: Horner, *CA*, **10** (1824), pp. 83, 185; H. W. M. Olbers, *BAJ for 1827*, **52** (1824), p. 186.

1826   J. F. A. Gambart (Royal Observatory, Marseilles, France) independently discovered comet 3D/1826 D1 (Biela) on 1826 March 9, and in a letter dated
1826 III   March 22 Gambart informed H. Flaugergues (Viviers, France) of the discovery. He said the comet was found in Cetus and had moved into Taurus.
Flaugergues first looked for the comet on the evening of March 29, and in a
letter to F. Baily wrote, "in traversing this part of the heavens, [I] perceived
under the left arm of *Orion*, a round, white, scarce visible, nebulosity. As my
instruments were not then well disposed, and thinking it to be the comet of
M. Gambart, I contented myself with making a configuration of its position
with the neighbouring stars." Flaugergues was observing with a 6.6-cm refractor and added that equally rough observations were obtained on March
30 and 31, which revealed considerable motion toward the northwest and a
considerable decrease in brightness. Flaugergues did not observe the comet
during the next two days as he was busy adapting a circular micrometer
to his refractor. On the evening of April 3, he was able to make "tolerable"
measurements of the comet's position, while on the evenings of April 4, 5,
and 6, Flaugergues said he made "very good ones." He pointed out that the
comet had grown fainter during the days it was visible. He described it as
"scarcely visible" on April 6, while he could no longer find it on April 7.

Flaugergues did not reduce his observations, as he conceded that other
astronomers with better telescopes had probably obtained more precise positions; however, a few months later, he was reading the details of comet
3D/1826 D1 when he noticed that the motion was different than the one
he had observed. In his letter to Baily he wrote that the motion of comet
3D/1826 D1 "was direct and nearly in the parallel of declination 10° N., and
it remained visible till the beginning of May. The comet, which I observed
was apparently retrograde; it passed in a few days over 10° of declination
towards the north, and remained visible only till the 7th of April."

During September 1826, Flaugergues determined the comet's positions
for April 4–6, and computed a parabolic orbit, with a perihelion date of
1826 April 27.45, a perihelion distance of 0.646 AU, and an inclination of
9.5°. Clüver (1835) examined the positions and determined an orbit with
a perihelion date of April 29.54, a perihelion distance of 0.188 AU, and an
inclination of 174.7°.

No further discussion of this comet came until 1914, when W. Hassenstein
published an interesting paper in the *Astronomische Nachrichten*. His investigation revealed, "Flaugergues erred with the identification of the comparison stars in a nearly unbelievable way." After studying Flaugergues'
original observation log, Hassenstein found that Flaugergues consistently
misidentified the stars he was using to measure the comet's position, and
even made mistakes in the actual measurements. Therefore, Flaugergues

was actually observing 3D/1826 D1. Hassenstein suggested the decrease in brightness was simply due to the comet's decreasing altitude.

ABSOLUTE MAGNITUDE: $H_{10} = 9.5$ (V1964)
SOURCES: H. Flaugergues, *MAS*, **3** (1829), pp. 95–7; H. Flaugergues and Clüver, *AN*, **12** (1835 Jul. 1), pp. 281–4; W. Hassenstein, *AN*, **198** (1914 Jul. 22), pp. 449–58; V1964, p. 55.

1832 F. W. Bessel (Königsberg, now Kaliningrad, Russia) recorded a nebula at a position of $\alpha = 2^h\ 42.1^m$, $\delta = +36°\ 47'$ on 1832 November 8.92. The magnitude was given as 9. Two decades later, H. L. d'Arrest (Leipzig, Germany) was trying to reobserve previously reported nebulosities. On two occasions during January 1856 he observed the position of Bessel's object. On both occasions he found a 9th-magnitude star, but no trace of nebulosity. An anonymous author writing in the 1879 April 17 issue of *Nature* suggested Bessel's object might have been a comet.

SOURCES: F. W. Bessel and H. L. d'Arrest, *Nature*, **19** (1879 Apr. 17), p. 555.

1835 The Chinese text *Ch'ing Ch'ao Hsü Wên Hsien Thung Khao*, which was written by Liu Chin-Tsao around 1910, noted that Chinese astronomers discovered a "broom-star" comet on 1835 August 5.

SOURCES: HA1970, p. 86.

1841 C. K. L. Rümker (Hamburg, Germany) discovered a "nebula" on 1841 May 27.87 at a position of $\alpha = 13^h\ 52.6^m$, $\delta = +45°\ 36'$. Only a 9th-magnitude star has been detected in that position by other observers.

SOURCES: *Nature*, **30** (1884 Jun. 26), p. 201.

1849 The first report of this object came to light when the Reverend J. Curley (Georgetown College, Washington, DC, USA) reported that his colleague the Reverend T. M. Jenkins saw a naked-eye comet while sailing from Baltimore to Rio de Janeiro aboard the *Maryland*. The observation was made on 1849 November 28, at about 7:15 p.m. The comet was in the western sky "nearly in the track of the sun, about 14° above the horizon. . . ." It was described as exhibiting a very distinct nucleus, about as large as Mars, while a tail extended about 1° southward. The tail was longer when viewed with a telescope. The "comet" remained visible for 20 minutes and was observed by the entire crew, until covered by clouds. It was not observed again. Curley suggested this comet might have been the expected comet of 1264 and 1556.

J. R. Hind (London, England) immediately studied the information and said the description indicated $\alpha = 18^h\ 30^m$. He pointed out that if the comet of 1556 had been at perihelion on 1849 November 13, it would have been at the indicated α on November 28, "but on this supposition the comet's declination would be rather less than that of the sun, which seems to be contradicted by the direction of the tail."

Attempts were made to gather additional information. During 1850 May, M. F. Maury (Washington, DC, USA) forwarded a new letter from Curley, who, in turn, was forwarding a letter from Captain R. Horner of the ship *Maryland*, to the staff of the *Astronomical Journal*. Horner said Jenkins had died in Rio de Janeiro and he stated that mistakes were made in the original report. According to Horner, the observation was made on November 15, at 7:30 p.m. (November 15.91 UT). The comet's head was then $48°$ above the western horizon and it remained visible for 1 hour. Hind responded that the comet's position was now probably at $\alpha = 20^h\ 36.6^m$, $\delta = +4°\ 18'$ and that it was not possible for this to have been the expected comet.

SOURCES: *MNRAS*, **10** (1850 Mar. 8), pp. 122–3; *AN*, **30** (1850 Apr. 29), pp. 275–6; Jenkins, *AJS* (Series 2), **9** (1850 May), p. 442; M. F. Maury, J. Curley, R. Horner, and Jenkins, *AJ*, **1** (1850 Jun. 10), p. 79; *MNRAS*, **10** (1850 Jun. 14), pp. 192–3.

1851   I. Calandrelli (Rome, Italy) found a "comet...in the morning twilight" on 1851 November 30.20, at a position of $\alpha = 14^h\ 21.6^m$, $\delta = +1°\ 47'$. He believed this was an observation of periodic comet 5D/Brorsen and said the position was not far from that predicted by P. van Galen who said the comet would return to perihelion on 1851 November 10. It was later found that van Galen's prediction was a month and a half late, so that Calandrelli's "comet" was not this periodic comet.

SOURCES: *Nature*, **15** (1877 Jan. 25), pp. 281–2.

1859   The Chinese text *Ch'ing Ch'ao Hsü Wên Hsien Thung Khao*, which was written by Liu Chin-Tsao around 1910, noted that Chinese astronomers discovered a "broom-star" sometime during 1859. A further note in the text said the Astronomer Royal "reported to the Emperor that the light of the comet [probably its tail – suggested Ho Peng-Yoke and Ang Tian-Se] swept the *Guànsuò* [α, β, γ, δ, ε, θ, ι, π, and ρ Coronae Borealis] asterism in Corona Borealis." The Author notes that comet Tempel of 1859 was in the opposite part of the sky from Corona Borealis when at its brightest.

SOURCES: HA1970, p. 87.

1860   B. Valz was attempting to compute an orbit for C/1860 U1, when he received a letter from H. P. Tuttle (Cambridge, Massachusetts, USA) announcing that he had found a very faint comet on 1860 November 14 near Polaris. Since Valz had determined that C/1860 U1 would pass through this region, he assumed Tuttle's object was a new observation, even though C/1860 U1 should have passed through this region several days later. No further observations were reported.

SOURCES: E. W. L. Tempel, H. P. Tuttle, and B. Valz, *AN*, **55** (1861 Mar. 30), pp. 79–80.

1865   E. J. Lowe (England) sent the *Monthly Notices of the Royal Astronomical Society* a few details of two comets he saw on 1865 August 27.85. The first comet was seen at a position of $\alpha = 15^h\ 15^m$, $\delta = -3°\ 50'$, while the second was at $\alpha = 15^h\ 00^m$, $\delta = -7°\ 30'$. He claimed to have made 17 observations "and the range of error is very small, not 20' of space." Lowe added, "From an account I see in the newspapers of a Comet seen at $3^h\ 45^m$ a.m. in the East, three days ago, I have little doubt this is one of the Comets I saw in August. The weather has been overcast in the mornings for more than a week, so that I have been prevented seeing the Comet myself. The description as to large size and shape (no tail) agrees with the first Comet I myself saw." The only comet he could have been referring to was 4P/Faye.

SOURCES: E. J. Lowe, *MNRAS*, **25** (Supplemental Notice, 1865), p. 278.

1865   The 1865 return of comet Biela was expected to be a favorable one. J. R. Hind published an ephemeris and numerous searches were made. Although most were unsuccessful, there were a few reports of nebulous objects. The following may or may not be associated with one another and their relationship with comet Biela is considered doubtful.

C. G. Talmage (Leyton, England) was sweeping for Biela's comet in mostly cloudy skies on 1865 November 4. On November 4.89 he "got a glimpse of a cometic-looking object" through a break in the clouds. He described it as "exceedingly faint" and said the area was covered by clouds before he could compare the position with nearby stars; however, the instrumental position indicated $\alpha = 22^h\ 55^m\ 04^s$, $\delta = +13°\ 25'\ 55''$. Talmage added in a letter to Hind on November 9 that every night thereafter had been cloudy. In the meantime, Hind wrote a letter to Talmage that was dated November 8 and stated that on that evening he "had a close search here, and certainly suspected a nebulosity not far from the place with [a perihelion date of] Jan. 27, but could not decide about it; only clear for twenty minutes." Talmage finally had clear skies again on November 16. Although the object was not seen, he did take the time to determine the instrumental positions of α Pegasi and ζ Pegasi in order to compare them with the published values. A correction was adopted and applied to his previous position of the November 4 object, which resulted in a refined position of $\alpha = 22^h\ 54^m\ 45.47^s$, $\delta = +13°\ 26'\ 21''$.

Hind again wrote to Talmage on November 20 and said, "Mr. Barber caught a glimpse on the 18th of a cometary object nearly on the declination of my ephemeris for Jan. 27.66, and following something less than a minute. I expect this is about the most likely position for it to turn up." Following the receipt of this letter, Talmage compared his revised position to the January 27.66 perihelion date and determined errors of $\Delta\alpha = -26.07^s$, $\Delta\delta = -1'02''$.

J. Buckingham (Walworth Common, London, England) began using Hind's ephemeris during early November to search for comet Biela. On November 9.90, he was sweeping near the ephemeris position when he "found two round vapory bodies near each other. . . . " He referred to them as "A" and "B". Object B was fairly close to a faint star of magnitude 12 or 14 (according to Buckingham) and he determined its position as $\alpha = 23^h\ 19^m\ 12.75^s$, $\delta = +12°\ 35'$. Object A's position was incomplete and was given as $\alpha = 23^h\ 19^m\ 03^s$, $\delta =$ "some minutes north of B". In talking to Hind on the morning of November 10, Buckingham said Hind commented "the bodies observed could not be Biela, in consequence of the places not agreeing with the calculated ones, and so close together, but that they were nebulae."

W. H. S. Monck (1892) suggested the two nebulous objects seen by Buckingham on 1865 November 9 might be identical to N. R. Pogson's object of 1872 and suggested a possible return in 1893, which did not happen.

SOURCES: C. G. Talmage and J. R. Hind, *MNRAS*, **26** (1866 Apr.), pp. 241–3; J. Buckingham, *MNRAS*, **26** (1866 May), pp. 271–2; W. H. S. Monck, N. R. Pogson, J. Buckingham, and C. G. Talmage, *PASP*, **4** (1892 Nov. 26), pp. 253–4.

1871 A. C. Ranyard reported a comet was visible during the solar eclipse of 1871 December 12. E. W. Maunder (1899) compared it to the eclipse comets reported in 1882 and 1893, and said Ranyard's was "less convincing."

SOURCES: E. E. Barnard, *The Observatory*, **16** (1893 Feb.), pp. 92–5; E. W. Maunder, *The Indian Eclipse 1898*. London: Hazell, Watson, & Viney (1899).

1871 During a search for 2P/Encke during September of 1871, J. R. Hind (Twickenham, England) spotted "an extremely faint nebulous object" on September 22.96. He determined the position as $\alpha = 1^h\ 58^m\ 47.0^s$, $\delta = +31°\ 21'\ 38''$. Hind said he presumed this was 2P/Encke "though the difference of position from the ephemeris of Mr. Glasenapp is larger than we have been accustomed to look for in predictions relating to this body." W. E. Plummer (Twickenham, England) measured an additional position of $\alpha = 1^h\ 58^m\ 44.9^s$, $\delta = +31°\ 22'\ 24''$ on September 23.01. Bad weather prevented another search for 2P/Encke until October 8, at which time the comet was found quite close to the ephemeris position. Hind concluded, "Whatever the object may have been which we observed here on Sept. 22 (and I shall examine the matter further on the first really good night) it appears certain that it was not Encke's comet, as I had first supposed."

SOURCES: J. R. Hind and W. E. Plummer, *AN*, **78** (1871 Oct. 9), pp. 223–4; J. R. Hind, *AN*, **78** (1871 Oct. 24), pp. 253–4.

1876 The *Wanganui Herald*, a New Zealand newspaper, reported in the 1876 January 20 edition that a "small comet was visible in the south, in the constellation of Argo Navis, for about two hours last night, the rising moon rendering it invisible." The newspaper added that the object "appeared to be rapidly moving towards the east." There were no confirming reports or additional details.

SOURCES: *Nature*, **13** (1876 Apr. 6), p. 449.

1881 The Chinese text *Ch'ing Ch'ao Hsü Wên Hsien Thung Khao*, which was written by Liu Chin-Tsao around 1910, noted that Chinese astronomers discovered a "broom-star" sometime during the lunar month of 1881 February 28 to March 29.

SOURCES: HA1970, p. 87.

1881 In a letter written to the *Astronomische Nachrichten* that was dated 1881 June 16, B. A. Gould (Cordoba, Argentina) told of an unusual observation he made while observing C/1881 K1. He observed the comet in the evening sky on 1881 June 11.97. He said it was found with little difficulty because of its diffuse appearance, although it was tailless in the bright twilight and low altitude. Just after obtaining a rough position using the setting circles of the refractor, Gould noted a star at the edge of the field, which he assumed was "one of the many bright stars of Orion...." He noted it "appeared only a little fainter than the comet itself, and not very dissimilar in aspect; since although its apparent diameter was much less than the comet's, it was greatly blurred by the exceptionally thick haze and the mists of the horizon...." Gould added, "I do not think it would have been below the third magnitude, and could rather believe it to have been as bright as the second." For the date given above, Gould concluded the position was $\alpha = 5^h$ $10^m 16^s$, $\delta = -9° 30'$.

Four comparisons of the comet's position with respect to this star were made before the field dropped below the horizon. Interestingly, when Gould checked his catalogs to find the precise position of this star, he found it was not listed. Upon searching the region the next evening, Gould did not find the object, although he did note that Rigel "was much brighter than the missing object." He then looked at the comet, which had moved about 3° northward, and saw "no visible object in its vicinity."

Gould said he withheld the details of this observation for a few days because he feared "some gross eror in reading the circles." He finally felt confident there was no error and said later, precise orbits would confirm his rough position. Indeed, the Author finds that Gould's rough setting circle position was no more than 1' off in both α and δ.

C. N. A. Krueger published details of his brief investigation into the matter in the 1881 July 26 issue of the *Astronomische Nachrichten*. He used an orbit determined by Hind, based on an arc of just over a week, and confirmed the closeness of Gould's rough position to the computed position of the comet. He also derived a position for the "star" and confirmed that nothing was present in that position in the *Berliner Academischen Charten* (*BAC*). Krueger did note one strange circumstance for this observation: the comet did not move as far northward as it should have with respect to the "star". Krueger questioned whether the object might have been a mirage caused by the atmospheric conditions near the horizon, or, possibly, a second comet.

W. Bone (Castlemaine, Victoria, Australia) may also have seen this same object about a day and a half earlier. He presented the details in a letter to the *Monthly Notices* that was dated 1881 November 6. He also included a telegram he transmitted to Melbourne Observatory on 1881 June 10 at 8:10 p.m. local time, which was about an hour and a half after his observation. Bone began observing C/1881 K1 on 1881 June 10.344. After a few minutes of measuring the comet's position he "noticed a peculiar discordance in each succeeding measure, and at length found that the star (?) from which I was measuring was a rapidly-moving body." He carefully inspected the object and noted "it was somewhat discoid, but its light, although bright, was diffused and hazy." Bone also said it was visible to the naked eye and estimated the brightness as 2.5. Eventually he determined the object was moving at a rate of $+24^s$ in α and $-6'$ in δ every 34^m 34^s. On June 10.381 he measured the position as $\alpha = 5^h$ 18^m 30^s, $\delta = -14° 24'$. C/1881 K1 and this object should have set within the next 15 minutes.

Since C/1881 K1 was then visible in the morning and evening sky in the Southern Hemisphere, Bone attempted to find the object the next morning, but was met with bad weather. On the evening of the 11th he attempted to find the object, but was unsuccessful.

J. Tebbutt (Windsor, New South Wales, Australia) wrote a letter to the *Astronomische Nachrichten* that was dated 1881 October 10. Published in the 1882 January 16 issue, it gave details of his observations of C/1881 K1 during the period of 1881 June 11.847–11.858, in which he had noted there were no comparison stars of 7th magnitude or brighter and that he had to make comparisons with Rigel. Tebbutt suggested Gould might have seen the blurred images of two stars designated BAC 1592 and 1597, the last also known as λ Eridani.

Gould responded in a letter written to the *Astronomische Nachrichten* on 1882 February 23. Published in the 1882 June 6 issue, Gould wrote that his own experiments with stars of various brightnesses and different observing conditions at low altitudes indicated he had not seen diffuse stars. He also pointed out that the object could not have been a companion to C/1881 K1 because of the difference in motion. Tebbutt responded to Gould in this journal in the November 8 issue and basically stood his ground. He also brought up the object seen by Bone and said that, in light of his own failure to see such an object and the motion indicated by Gould, Bone's object must be different than that of Gould.

SOURCE: B. A. Gould, *AN*, **100** (1881 Jul. 26), pp. 113–16; C. N. A. Krueger, *AN*, **100** (1881 Jul. 26), p. 115; *Nature*, **24** (1881 Aug. 11), p. 342; W. Bone, *MNRAS*, **42** (1882 Jan.), pp. 105–6; J. Tebbutt, *AN*, **101** (1882 Jan. 16), pp. 171–4; B. A. Gould, *AN*, **102** (1882 Jun. 6), pp. 145–8; J. Tebbutt, *AN*, **103** (1882 Nov. 8), p. 311.

1882 According to the 1882 January 28 edition of the *New York Times*, a "brilliant comet" was seen by several people in San Francisco at 5 a.m. (local time) on January 13. It was described as exhibiting a "head or nucleus" that appeared as "larger and brighter than Jupiter." The comet was situated "a little east of south, not over about 20° above the horizon." The tail was described as "short, somewhat bushy, and slightly arched and pointed easterly." The comet remained visible for one hour. It was seen again the following morning, but only for just over half an hour. There were no additional details or confirming observations.

SOURCES: *New York Times* (1882 Jan. 28), p. 5, col. 1.

1882 The 1882 December 22 issue of the *Dundee Advertiser* reported that an object was seen by several people at Broughty Ferry (near Dundee, Scotland) between 10 and 11 a.m. on December 21. The correspondent wrote, "The sun at the same time was shining brightly, being about due south, when a star was seen in close proximity to it. The star was a little above the sun's path, and the peculiar phenomenon was seen by various persons, who had their attention directed to it. Being daytime, the star did not have the brilliant luminous radiance stars exhibit at night, but was of a milky white appearance, and seemed, when seen through a glass, to be of a crescent shape." A letter in the December 25 issue of the *Dundee Advertiser* said the object was also seen at Dundee and that "The description which he [the previous correspondent] gives of it makes it quite certain that it was the planet Venus, and no other. It is at present a little to the west of the sun, and above it." The December 22 article was also published in the December 29 issue of *Knowledge*. J. E. Gore wrote in the 1883 January 5 issue of *Knowledge*, "The object could not have been the planet Venus, which was situated about 23° west of the sun on the day in question."

SOURCES: *Dundee Advertiser* (1882 Dec. 22), p. 5, col. 5; *Dundee Advertiser*, (1882 Dec. 25), p. 7, col. 4; *Knowledge*, **2** (1882 Dec. 29), p. 489; *Knowledge*, **3** (1883 January 5), p. 13; personal correspondence from Martin Taylor (1994).

1883 The 1883 February 3 issue of the *Scientific American* reported a comet was seen by W. L. Burton (second officer of the steamship *City of Savannah*) at 2:00 a.m. (local time) on the morning of January 12. The position was simply described as "southeast of Orion." It was then faintly visible to the naked eye and was seen again at 9:00 local time that evening. The 1883 May 12 issue of the *Scientific American* reported Burton's comet was actually "the shadowy form of the great comet of 1882."

SOURCE: W. L. Burton, *SA*, **48** (1883 Feb. 3), p. 65; *SA*, **48** (1883 May 12), p. 288.

1883 The 1883 February 1 issue of *Nature* reported a telegram, dated January 23, had been received describing a comet found at Puebla, Mexico. It was supposedly located near Jupiter. No further word came out of Puebla, but searches of the region did reveal that Jupiter was then near the supernova remnant M1 (the Crab Nebula).

SOURCE: *Nature*, **27** (1883 Feb. 1), p. 324.

1883 The *New York Times*, and, later, *Nature*, reported that E. Hartwig (Strasbourg, France) had managed to recover periodic comet d'Arrest with a 51-cm refractor on April 4.11. The position was given as $\alpha = 13^h\ 55.4^m$, $\delta = +8°$ 16′ and the comet was described as very faint. No motion was given in Hartwig's announcement, but when C. N. A. Krueger (Kiel, Germany) received the telegram, he issued a new telegram announcing the recovery and giving the daily motion comet d'Arrest was expected to have at that time. The *New York Times* announced on April 7 that Hartwig had not seen comet d'Arrest, but had discovered a new nebula.

SOURCE: *New York Times* (1883 Apr. 5), p. 2, col. 2; *New York Times* (1883 Apr. 7), p. 5, col. 3; *Nature*, **27** (1883 Apr. 12), p. 567; *Nature*, **27** (1883 Apr. 19), p. 589; *Nature*, **27** (1883 Apr. 26), p. 618; *Nature*, **30** (1884 May 1), p. 21; *ARSI*, part 1 (1886), p. 383.

1883 L. Swift (Warner Observatory, Rochester, New York, USA) reported to Harvard College Observatory that he had discovered a new comet on 1883 September 11.50 at a position of $\alpha = 18^h\ 41.9^m$, $\delta = +73°\ 09′$. Swift apparently confirmed motion on September 14.00 and estimated the position as $\alpha = 18^h$ 26.0^m, $\delta = +73°$ 08′. The Harvard announcement, when it appeared in the 1883 September 21 issue of the *Astronomische Nachrichten*, added, "Has not been seen by any one else."

The 1883 September 28 issue of the *Astronomische Nachrichten* contained letters by E. Hartwig (Strasbourg, France), J. Palisa (Vienna, Austria), and E. A. Lamp (Kiel, Germany) giving details of their searches. Hartwig was searching for Swift's object on September 19, when he found a nebulosity at $\alpha = 17^h$ 51^m 47.86^s, $\delta = +72°$ 03′ 19.8″ on September 19.82. He added that there was no movement after one hour and suggested it was probably a nebula. Palisa said he finally was able to search for Swift's object on September 20, with the 30-cm refractor. He located a nebula at $\alpha = 18^h$ 25.1^m, $\delta = +73°$ 08′. He described it as a star of magnitude 12–13, surrounded by a nebulosity about 1′ across. Palisa suggested this was the same object seen by Swift on the 14th, and that the almost exactly 4° deviation between the 11th and 14th was probably an error. Lamp found the same nebulosity as Palisa when he searched for Swift's object on September 23, but more precisely measured the position as $\alpha = 18^h$ 26^m 42.26^s, $\delta = +73°$ 06′ 59.3″.

SOURCE: L. Swift, *AN*, **106** (1883 Sep. 21), p. 335; E. Hartwig, J. Palisa, and E. A. Lamp, *AN*, **106** (1883 Sep. 28), p. 381.

1883 The Chinese text *Ch'ing Ch'ao Hsü Wên Hsien Thung Khao* (written by Liu Chin-Tsao) noted that Chinese astronomers discovered a "broom-star" comet sometime during the period of 1883 September 1 to 30.

SOURCES: HA1970, p. 88.

1884 L. Schulhof examined the eight positions available for Tuttle's comet of 1858 (later known as 41P/Tuttle–Giacobini–Kresak) and said he suspected it moved in a short-period orbit with a period near 6.5 years. Noting it would apparently reappear around 1884, he published sweeping ephemerides.

R. Spitaler (Vienna, Austria) was searching for the comet with the 69-cm refractor along the path predicted by Schulhof, when, on May 26, he found three small, faint nebulae that did not appear in any catalog. Clouds were coming in and Spitaler obtained an approximate place for each nebulosity. Three weeks of bad weather followed and it was not until June 17 that Spitaler could recheck the same field. The nebulous object at a position of $\alpha = 17^h$ 40^m 50^s, $\delta = +35°$ 33′ was missing and he confirmed its absence on June 18 and 20. He described this "nebula" as faint and round, with a central condensation.

Although Spitaler's nebulosity was near the positions predicted by Schulhof, Schulhof said he thought it was improbable that this was the missing comet Tuttle, as, if it was that comet, it should have been considerably brighter. Another astronomer noted that if it was Tuttle's comet it would have faded quickly, so that by June 17 another observation would have been hopeless.

SOURCES: L. Schulhof, R. Spitaler, and E. Weiss, *AN*, **109** (1884 Jun. 21), p. 111; H. C. F. Kreutz, *AN*, **109** (1884 Jun. 25), p. 127; E. Weiss, *AN*, **109** (1884 Jun. 30), p. 143; *Nature*, **30** (1884 Jul. 10), p. 253; W. E. Plummer, *MNRAS*, **45** (1885 Feb.), p. 243.

1887 The Chinese text *Ch'ing Ch'ao Hsü Wên Hsien Thung Khao*, which was written by Liu Chin-Tsao around 1910, noted that Chinese astronomers discovered a "broom-star" sometime during the lunar month of 1887 September 17 to 1887 October 16.

SOURCES: HA1970, p. 88.

1889 L. Swift (Warner Observatory, Rochester, New York, USA) found this ob-
1889a ject during his program of routine comet searches on 1889 July 6.33. He estimated the position as $\alpha = 21^h\ 24.3^m$, $\delta = +0°\ 49'$ and sent a telegram to Massachusetts announcing his find. A telegram was subsequently sent from Boston to the *Astronomische Nachrichten*, where the announcement was published. Unfortunately, an error in the telegram indicated a position of $\alpha = 22^h\ 52.5^m$, $\delta = +0°\ 49'$ and this is what was printed. Interestingly, E. Millosevich (Rome, Italy) reported on July 8 that the comet was not found and suggested it was C/1888 R1. C. N. A. Krueger noted that C/1888 R1 was 21° away in α at that time, which would seem to indicate a link was impossible. He requested a confirmation from Boston and the response was that α was in error and should have read $21^h\ 24.3^m$. Therefore, Millosevich had been correct and Swift's comet was none other than C/1888 R1.

SOURCES: L. Swift and E. Millosevich, *AN*, **122** (1889 Jul. 9), supplement; L. Swift, E. Millosevich, and C. N. A. Krueger, *AN*, **122** (1889 Jul. 19), p. 117.

1889 L. Swift (Warner Observatory, Rochester, New York, USA) found this object on 1889 December 23 at a position of $\alpha = 3^h\ 36.1^m$, $\delta = -5°\ 02'$. He believed it was a comet and said he could not locate it during subsequent searches and noted it was not found at Harvard College Observatory. Swift said the object was in line with NGC 1417 and NGC 1418. It seems likely this was the galaxy IC344.

SOURCES: L. Swift, *AN*, **126** (1890 Nov. 7), p. 49; L. Swift, *AN*, **126** (1891 Jan. 19), p. 225.

1890   L. A. Eddie (Cape Town, South Africa) reported that an interesting "naked-eye comet" was discovered on 1890 October 27.77. The object was first detected almost due west at 7:45 p.m. local time, with a tail about 30° long which was tilted to the south by about 45°. A southerly movement was almost immediatedly detected with the naked eye. The object moved around the western and southern horizons at an altitude of 20–25°. The tail steadily grew in length and at maximum it measured 90° long and only 1° wide, running parallel to the southern horizon. The comet was last detected at 8:32 p.m. local time, when it had faded in the southeastern part of the sky. There was never any independent confirmation, and it was suggested late in 1891 that this object might have been auroral in character.

SOURCES: L. A. Eddie, *Nature*, **43** (1890 Nov. 27), pp. 89–90; W. E. Wilson, *Nature*, **44** (1891 Sep. 24), p. 494; H. Rix and J. L. E. Dreyer, *Nature*, **44** (1891 Oct. 8), p. 541.

1892   M. F. J. C. Wolf (Heidelberg, Germany) exposed three photographic plates with a 15-cm portrait lens on 1892 March 19.86, March 19.95, and March 20.87. Around noon on March 22, he sat down to investigate them and immediately noticed an object on the earliest plate at a position of $\alpha = 11^h$ 31.8^m, $\delta = +10° 30'$ (1855.0). He described it as a narrow, cigar-shaped object, about 15' long, which was centrally condensed. The object was also found on the other two plates and the position was found to have changed in a consistent manner between each one. On the night of March 22, Wolf obtained another exposure of the expected position of the object and noted a possible weak trace near the expected position. Wolf considered the possibility that the plates of the 19th and 20th had captured a reflection of Saturn, which was about 8° away, but an exposure of the appropriate region on the 22nd did not reveal any unexpected objects. Subsequently, Wolf sent the position and motion details to C. N. A. Krueger (Kiel, Germany).

Krueger reported that searches on March 22 and 23 were unsuccessful. He further reported that messages from Harvard College Observatory (Cambridge, Massachusetts, USA) and Lick Observatory (California, USA) also reported no success. R. Spitaler (Vienna, Austria) wrote that he searched for this object with a 69-cm refractor on 1892 March 24, 25, and 31. Although the skies were very clear, he was unsuccessful. Spitaler added that he did discover several new nebulae.

SOURCES: M. F. J. C. Wolf, *AN*, **129** (1892 Apr. 8), p. 181; R. Spitaler, *AN*, **131** (1893 Jan. 10), p. 389.

1892   Freeman (Brighton, England) reported that he had discovered a faint comet on 1892 November 24.89 at a position of $\alpha = 0^h$ 29.0^m, $\delta = +30° 09'$. The daily motion was given as 3° 12' due south. E. Weiss (Vienna, Austria) announced on November 27 that he was unable to find the comet, but curiously suggested on the 28th that comet Freeman was a "probable" fragment of the lost periodic comet Biela, with a perihelion date of December 28. W. H. M. Christie (Greenwich, England) announced on November 28 that Freeman's object was probably a nebula. The Author has checked the various catalogs of deep sky objects and has found only one "nebula" close to the position of Freeman's object. It is NGC 140, which is a 14th-magnitude galaxy about 11' from Freeman's position. In the 1892 December 3 issue of the *Astronomische Nachrichten*, C. N. A. Krueger summarized the details of Freeman's "comet" and included the note that he believed the discovery was in error.

SOURCES: Freeman, *The New York Times* (1892 Nov. 27), p. 2, col. 5; C. N. A. Krueger, Freeman, E. Weiss, and W. H. M. Christie, *AN*, **131** (1892 Dec. 3), pp. 199–200; Freeman, *Nature*, **47** (1892 Dec. 8), p. 133.

1893   The total solar eclipse of 1893 April 16 was extensively photographed by Northern Hemisphere astronomers who traveled to South America and Africa. Results of the expeditions were announced during the next few months. One particularly interesting result was announced by J. M. Schaeberle (Lick Observatory, California, USA) before the Astronomical Congress at Chicago in August of 1893. He said images obtained by the Lick Observatory eclipse expedition to Chile showed a "comet-like structure visible on all the... negatives of the outer corona." The images spanned less than three minutes of time and no motion was detected. Schaeberle requested prints of the negatives obtained by British astronomers in Brazil and Africa. Although these did not arrive until May of 1894, Schaeberle announced that they also revealed the "comet" and said it exhibited a daily motion of about 3.25° toward PA 200°. E. S. Holden (Lick Observatory) wrote in the 1894 September 27 issue of the *Astronomische Nachrichten* that the "comet" was also present on three plates obtained during the Harvard College Observatory eclipse expedition. Schaeberle became the seventeenth recipient of the Donohoe Comet Medal from the Astronomical Society of the Pacific for the discovery of this object.

E. W. Cliver (1989) suggested this "comet" was actually a disconnected coronal mass ejection. He compared the drawings of the object with such a coronal mass ejection detected on 1980 March 15 and 16. He also compared the object's appearance with the comets X/1882 K1 and C/1948 V1, which had been detected during solar eclipses, as well as six comets observed by the SOLWIND solar satellite. He said the object was far larger than the comets observed near the sun, but was comparable in size to the 1980 event. Further coronal mass ejection events were looked at by D. F. Webb and E. W. Cliver (1995) and a photograph of an event observed on 1980 April 5 bore an uncanny resemblance to the 1893 "comet."

SOURCES: J. M. Schaeberle, *AJ*, **14** (1894 May 23), p. 46; J. M. Schaeberle and E. S. Holden, *PASP*, **6** (1894 Aug. 15), pp. 228, 237–8; E. S. Holden, *AN*, **136** (1894 Sep. 27), pp. 203–4; J. M. Schaeberle and W. H. Wesley, *Nature*, **51** (1894 Nov. 8), p. 40; E. W. Cliver, *Solar Physics*, **122** (1989), pp. 319–33; D. F. Webb and E. W. Cliver, *Journal of Geophysical Research*, **100** (1995 Apr. 1), pp. 5853–70.

1893 The Chinese text *Ch'ing Ch'ao Hsü Wên Hsien Thung Khao*, which was written by Liu Chin-Tsao around 1910, noted that Chinese astronomers discovered a "broom-star" sometime during the lunar month of 1893 October 10 to November 7.

SOURCES: HA1970, p. 88.

1894 E. Holmes (London, England) discovered a "bright comet's tail" on 1894 April 9.9, and measured a position of $\alpha = 17^h\ 58^m$, $\delta = +71° 30'$ on April 10.0. No direction or rate of motion was given. Searches were made in Europe and the USA, but no trace of a comet was found. It was announced on April 16 that Holmes had not found a comet, but had observed NGC 6503.

SOURCES: E. Holmes, *AJ*, **14** (1894 Apr. 14), p. 24; E. Holmes, *AN*, **135** (1894 Apr. 16), pp. 151–2; E. Holmes and W. R. Brooks, *PA*, **1** (1894 May), pp. 421, 423; E. Holmes, *AJ*, **14** (1894 May 1), p. 32.

1895 A. M. du Celliée Muller (Nymegen, Netherlands) reported that he had discovered a comet close to Venus on 1895 December 8.27 at a position of $\alpha = 13^h\ 54.7^m$, $\delta = -8° 47'$. He was using a 6.1-cm refractor (65×) and was accompanied by G. J. van Dyk. The "comet" was described as bright, with a tail 3–4' long inclined 35° to the horizon. There was a strong nucleus, which Dyk described as elliptical, but Muller found to be round. Muller was confident it was not a false image of Venus. Searches were made at several observatories, but nothing was found. C. N. A. Krueger noted that the angle of the tail also did not fit with respect to the direction of the sun.

SOURCES: C. N. A. Krueger, A. M. du Celliée Muller, and G. J. van Dyk, *AN*, **139** (1896 Feb. 5), pp. 319–20; W. E. Plummer, *MNRAS*, **57** (1897 Feb.), p. 276.

1899 The 1899 October 29 edition of the *New York Times* reported that observers in Santiago, Chile had recovered Biela's comet. It was apparently visible to the naked eye. Two days later, the same newspaper reported the "comet" was actually a cluster of stars.

SOURCES: *New York Times* (1899 Oct. 29), p. 7, col. 3; *New York Times* (1899 Oct. 31), p. 6, col. 7.

Appendix 2
Periodical Abbreviations

To help keep the "Sources" listings from overwhelming the text, the following abbreviations have been used for the primary periodicals consulted while researching *Cometography*.

AA	*Acta Astronomica*
AAOHC	*Annals of the Astronomical Observatory of Harvard College*
AAP	*Astronomy and Astrophysics*
ARSI	*Annual Report of the Board of Regents of the Smithsonian Institution*
AMAF	*Arkiv För Matematik, Astronomi Och Fysik*
ABSB	*Astronomische Beobachtungen auf der Königlichen Sternwarte zu Berlin*
AJ	*Astronomical Journal*
AJB	*Astronomischer Jahresbericht*
AJS	*American Journal of Science*
AN	*Astronomische Nachrichten*
AOAT	*Annales de l'Observatoire Astronomique de Tokyo*
AR	*Astronomical Register*
APJ	*Astrophysical Journal*
BA	*Bulletin Astronomique*
BAC	*Bulletin of the Astronomical Institutes of Czechoslovakia*
BAJ	*Berliner Astronomisches Jahrbuch*
BASP	*Bulletin de l'Académie Impériale des Sciences de St. Pétersbourg*
BIOP	*Bulletin International de l'Observatoire de Paris*
CA	*Correspondance Astronomique, Geographique, Hydrographique et Statistique du Baron de Zach*
CCO	*Catalog of Cometary Orbits* (Cambridge: Smithsonian Astrophysical Observatory)
CDT	*Connaissance des Temps*
CR	*Comptes Rendus Hebdomadaires des Séances de l'Académie des Sciences*
DAWW	*Denkschriften der Mathematisch-Naturwissenschaftlichen Classe der Kaiserlichen Akademie der Wissenschaften Wien*
EAN	*Ergänzungshefte zu den Astronomischen Nachrichten*
IAJ	*Irish Astronomical Journal*
IAUC	*International Astronomical Union Circulars*

JRASC	*Journal of the Royal Astronomical Society of Canada*
MASP	*Mémoires de l'Académie Impériale des Sciences de St. Pétersbourg*
MAS	*Memoirs of the Astronomical Society of London* (later became the *Memoirs of the Royal Astronomical Society*)
MINS	*Memoires de L'Institut National des Sciences et Arts*
MSPG	*Mémoires de la Société de Physique et d'Histoire Naturelle de Genève*
MNRAS	*Monthly Notices of the Royal Astronomical Society*
MRAS	*Memoirs of the Royal Astronomical Society*
MC	*Monatliche Correspondenz*
OAP	*Observations Astronomiques Faites à l'Observatoire Royal de Paris*
PA	*Popular Astronomy*
PASJ	*Publications of the Astronomical Society of Japan*
PMJ	*The Philosophical Magazine and Journal*
PRSL	*Proceedings of the Royal Society of London*
PTRSL	*Philosophical Transactions of the Royal Society of London*
QJRAS	*Quarterly Journal of the Royal Astronomical Society*
SA	*Scientific American*
SAWW	*Sitzungsberichte der Kaiserlichen Akademie der Wissenschaften in Wien*
SM	*The Sidereal Messenger*
ST	*Sky & Telescope*
TAPS	*Transactions of the American Philosophical Society*
VA	*Vistas in Astronomy*
VJS	*Vierteljahrsschrift Astronomische Gesellschaft*
ZA	*Zeitschrift für Astronomie*

Appendix 3
Source Abbreviations

What follows are the publication details of the primary sources used while researching *Cometography*. For each entry, the text appearing before the dash was used in the Introduction and each comet's "Sources" listing.

G1894 – Galle, J. G., *Cometenbahnen*. Leipzig (1894), pp. 1–315.

HA1970 – Ho Peng-Yoke and Ang Tian-Se, *Oriens Extremus*, **17** (1970 Dec.), pp. 63–99.

NGC2000.0 – *NGC 2000.0*, edited by Roger W. Sinnott, Cambridge: Sky Publishing Corporation (1988).

O1899 – *Neue Reduktion der von Wilhelm Olbers Im Zeitraum von 1795 bis 1831 auf seiner Sternwarte in Bremen Angestellten Beobachtungen von Kometen und Kleinen Planeten*, edited by Wilhelm Schur and Albert Stichtenoth, Berlin: Julius Springer (1899).

SC2000.0 – *Sky Catalogue 2000.0*, Volume 1, edited by Alan Hirshfeld and Roger W. Sinnott, Cambridge: Sky Publishing Corporation (1982).

V1964 – *Physical Characteristics of Comets*, by Sergey Konstantinovich Vsekhsvyatskij, Jerusalem: Israel Program for Scientific Translations (1964).

Person Index

Comet Designation Index

This is an index containing all comets with formal designations given in this volume. The well-known periodic comets are listed first, since their designation does not include a year, and the remaining comets are listed chronologically thereafter. There are four prefixes that were adopted by the International Astronomical Union in 1994 as part of the new designation system. These are as follows:

"P/" represents the predictable comets and these have well-established orbits. When a number precedes this letter it indicates astronomers have either seen it at more than one apparition or are confident the comet can be followed throughout its orbit. In general, these comets have been referred to as short-period comets and have had orbits with periods of less than 200 years.

"C/" indicates comets observed well enough for an orbit to be determined. Sometimes this orbit is only parabolic, indicating the observations were either too few or too imprecise to allow the computation of an elliptical or hyperbolic orbit. Elliptical orbits indicate the comet will eventually return, while hyperbolic ones indicate the comet will likely never return.

"D/" indicates comets observed well enough for a short-period orbit to be determined, but which have either ceased to exist or may suffer from a large enough uncertainty in the orbit to seriously hamper future recovery. Some of these comets will have a number in front of them because they were once returning and being observed at regular intervals. In this volume, comet 3D/Biela is an example. It broke up and vanished in the middle of the nineteenth century.

"X/" indicates comets which were very likely real, but whose positions are too few or so poor that a reliable orbit cannot be determined. For the purpose of this volume, the Author has chosen to give comets this designation if corroborating observations are available from at least two different cultures. This at least confirms the reality and nature of the object.